Mathematics for CBSE Class X

Mathematics for CBSE Class X

Tarun Gulati
BSc Mathematics, MSc Mathematics
Director, Educare Academy, New Delhi

Universities Press

MATHEMATICS FOR CBSE CLASS X

UNIVERSITIES PRESS (INDIA) PRIVATE LIMITED

Registered office
3-6-747/1/A & 3-6-754/1, Himayatnagar, Hyderabad 500 029, Telangana, India
info@universitiespress.com; www.universitiespress.com

Distributed by
Orient Blackswan Private Limited

Registered office
3-6-752 Himayatnagar, Hyderabad 500 029, Telangana, India

Other offices
Bengaluru / Chennai / Guwahati / Hyderabad / Kolkata
Mumbai / New Delhi / Noida / Patna / Visakhapatnam

ISBN: 978-93-89211-69-6

501788

Typeset in Minion Pro 10/12 *by*
SRS Technologies, Puducherry

Printed in India by
Navya Printers, Hyderabad 500 013

Published by
Universities Press (India) Private Limited
3-6-747/1/A & 3-6-754/1, Himayatnagar, Hyderabad 500 029, Telangana, India

Preface

It is my great pleasure to present this book to my students.

This book has primarily been designed for students preparing for the Class X exams. All the exercises have been developed according to the new syllabus prescribed by the Central Board of Secondary Education (CBSE).

This book encompasses the full spectrum of knowledge for every topic: from the fundamentals to advanced concepts, so as to help students achieve top-level results in the Class X Board exam. It covers both levels of the CBSE Board: Basic and Standard, as per the new pattern.

This book has been presented in a simple manner. The approach is that the student must first understand the topic and then try to solve the questions.

Exercises include NCERT questions, NCERT Exemplar problems, various school question papers, previous years' Board question papers, and so on. Different types of questions are provided: solved, objective, fill in the blank, reason and assertion, and subjective.

It is an absolutely new concept for teaching and studying Maths and focuses on comprehension, execution and practice. It is advised that students work out all the steps involved in every question thoroughly.

Sincere hard work is the key to success, as has been said in Shri-Mad Bhagwad Gita.

"Yogastha kuru karmani" – Do your work with full devotion and dedication.

Despite all my efforts, some errors may have crept into the content. I shall be grateful to readers if they can send me their feedback. I also know that there is always scope for further improvement, and hence, all such suggestions are welcome.

Tarun Gulati

Acknowledgements

Special thanks are due to all the students who motivated me to distill my 25 years of experience to write this book.

I am grateful to my family and friends for their unflagging encouragement: Manju Gulati, my wife; Rakesh Kumar Ahuja, PGT Physics; Vimal Gulati, PGT Maths; Vinod Aggarwal, PGT Maths; Rajeev Kumar Chauhan, PGT Social Science; Naresh Kumar Ahuja, TGT Maths and Science; Kiran Kapoor Chopra, PGT Maths; Giriraj Soni, PGT Maths; and Anil Kumar, typist.

Tarun Gulati

Contents

To
My uncle, the late Sh. Madan Lal Gulati,
and all my family members

CHAPTER

1 Real Numbers

Key Concepts

- HCF and LCM by prime factorisation method
- Fundamental Theorem of Arithmetic
- Euclid's division lemma to find HCF
- Prove that a number is irrational
- Decimal expansion of rational numbers

Concept: Real Numbers

Real numbers represent the collection of all rational and irrational numbers.

Concept: Composite Numbers

A natural number greater than 1 and not prime is called a composite number. For example, 4, 6, 8, 9, 10, 12, 14, … are composite numbers.

Concept: Fundamental Theorem of Arithmetic

Every composite number can be expressed (factorised) as a product of primes, and this factorisation is unique, apart from the order in which the prime factors occur.

Concept: Computing the HCF and LCM of Two Positive Integers (Using Fundamental Theorem)

(i) HCF of two numbers is the product of the smallest power of each common prime factor in the given numbers.

(ii) LCM of two or more numbers is the product of the greatest power of each prime factor involved in the numbers.

(iii) For any two positive integers a and b, HCF $(a, b) \times$ LCM $(a, b) = a \times b$.

Example 1 Express 180 as the product of its prime factors.

Solution:

$$180 = 2 \times 2 \times 3 \times 3 \times 5$$

$$= 2^2 \times 3^2 \times 5$$

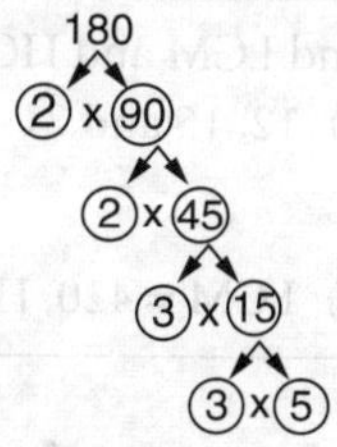

1 Try to Solve

1. Express the following as a product of its prime factors:
   (a) 240
   (b) 156
   (c) 140
   (d) 420
   (e) 3825
   (f) 7429 [NCERT]

Answers

1. (a) $24 \times 3 \times 5$ (b) $22 \times 3 \times 13$ (c) $22 \times 5 \times 7$ (d) $22 \times 3 \times 5 \times 7$ (e) $32 \times 52 \times 17$ (f) $17 \times 19 \times 23$

Example 2 Find HCF using the prime factorisation method of 70 and 30.

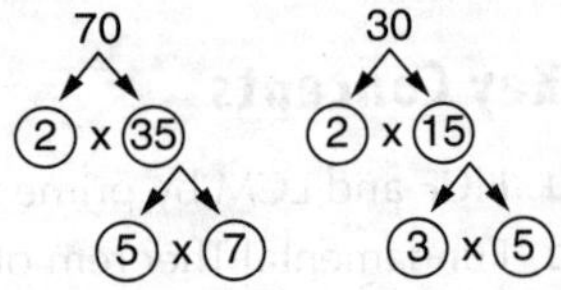

Solution: $70 = 2 \times 5 \times 7,\ 30 = 2 \times 3 \times 5$

$$\text{HCF} = 2 \times 5 = 10$$

2 Try to Solve

1. Find HCF using the prime factorisation method:
(a) 32 and 54 (b) 18 and 24 (c) 336 and 54 (d) 510 and 92

Answers

1. (a) 2 (b) 6 (c) 6 (d) 2

Example 3 Find LCM and HCF applying the prime factorisation method of 40, 36 and 136.

Solution:

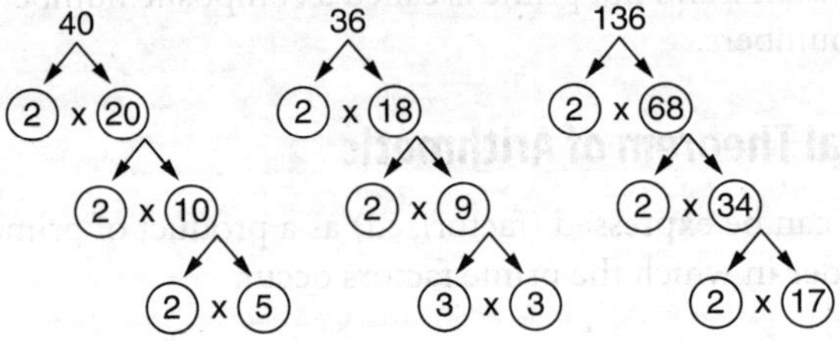

$$40 = 2 \times 2 \times 2 \times 5 = 2^3 \times 5$$

$$36 = 2 \times 2 \times 3 \times 3 = 2^2 \times 3^2$$

$$136 = 2 \times 2 \times 2 \times 17 = 2^3 \times 17$$

$$\text{HCF} = 2^2 = 4$$

$$\text{LCM} = 2^3 \times 3^2 \times 5 \times 17 = 6120$$

3 Try to Solve

1. Find LCM and HCF applying the prime factorisation method:
(a) 12, 15 and 21 (b) 8, 9 and 25 (c) 24, 15 and 36 [NCERT]

Answers

1. (a) LCM = 420, HCF = 3 (b) LCM = 1800, HCF = 1 (c) LCM = 360, HCF = 3

Concept: Formulae for Finding the HCF and LCM of Two Numbers

$$\text{LCM} = \frac{1^{\text{st}}\text{ no.} \times 2^{\text{nd}}\text{ no.}}{\text{HCF}},\quad \text{HCF} = \frac{1^{\text{st}}\text{ no.} \times 2^{\text{nd}}\text{ no.}}{\text{LCM}},\quad \text{LCM} \times \text{HCF} = 1^{\text{st}}\text{ no.} \times 2^{\text{nd}}\text{ no.}$$

$$1^{st}\text{ no.} = \frac{LCM \times HCF}{2^{nd}\text{ no.}},\ 2^{nd}\text{ no.} = \frac{LCM \times HCF}{1^{st}\text{ no.}}$$

Example 4 Given that HCF (306, 657) = 9, find LCM. **[NCERT]**

Solution:

$$LCM = \frac{1^{st}\text{ no.} \times 2^{nd}\text{ no.}}{HCF} = \frac{306 \times \cancel{657}}{\cancel{9}}$$

$$= 306 \times 73 = 22338.$$

4 Try to Solve

1. Given the HCF (81, 54) = 27, find LCM.
2. Find the LCM and HCF of 312 and 27, and verify LCM × HCF = Product of the numbers. **[CBSE (CCE) 2012]**
3. Find the LCM and HCF of the following pairs of integers and verify that LCM × HCF = Product of the two numbers. **[NCERT]**
   (a) 26 and 91 (b) 510 and 92
   (c) 336 and 54 (d) 404 and 96 **[CBSE 2018]**

Answers

1. 162 **2.** 2808, 3 **3.** (a) 182, 13 (b) 23460, 2 (c) 3024, 6 (d) 9696, 4

Example 5 The HCF of two numbers is 23 and their LCM is 1449. If one of the numbers is 161, find the other number.

Solution: Let the two numbers be a and b. Here, $a = 161$ (given), HCF = 23 and LCM = 1449.

We know that $a \times b = \text{HCF}(a, b) \times \text{LCM}(a, b)$

$\Rightarrow$ $161 \times b = 23 \times 1449$

$\Rightarrow$ $b = \dfrac{23 \times 1449}{161}$ Remember [HCF × LCM = $a \times b$]

$\therefore$ $b = 207$

Hence, the other number is 207.

5 Try to Solve

1. The LCM and HCF of two numbers is 180 and 6, respectively. If one of the numbers is 30, find the other number.
2. The HCF of two numbers is 16 and their product is 3072. Find their LCM.
3. The HCF of two numbers is 145 and their LCM is 2175. If one number is 725, find the other number.
4. What is the HCF of the smallest prime number and the smallest composite number? **[CBSE 2018]**

Answers

1. 36 **2.** 192 **3.** 435 **4.** 2

Concept: Euclid's Division Lemma

Given two positive integers a and b, there exist unique whole numbers q and r satisfying: $a = bq + r$, $0 \le r < b$.

Computing the HCF of Two Positive Integers

Euclid's division lemma provides a convenient method to compute the HCF of two positive integers. When we use it for this purpose, the systematic procedure adopted is called Euclid's division algorithm, which is stated as follows:

To obtain the HCF of two positive integers, say c and d, with $c > d$:

(1) Apply Euclid's division lemma to c and d, and find whole numbers q and r such that $c = dq + r$, $0 \le r < d$.

(2) If $r = 0$, d is the HCF of c and d. If $r \neq 0$, apply the division lemma to d and r.

(3) Continue the process till the remainder is zero. The divisor at this stage will be the required HCF.

Example 6 Use Euclid's division algorithm to find the HCF of 135 and 225. **[NCERT]**

Solution: Apply Euclid's division lemma to 225 and 135 (since 225 > 135).

$$225 = 135 \times 1 + 90$$

Now, the remainder $90 \neq 0$, so apply the division lemma to 135 and 90.

$$135 = 90 \times 1 + 45$$

Again, the remainder $45 \neq 0$, and thus, apply the division lemma to 90 and 45.

$$90 = 45 \times 2 + 0$$

Since the remainder becomes 0, the divisor at this stage, that is, 45, is the HCF of 135 and 225.

6 Try to Solve

1. Use Euclid's division algorithm to find the HCF of:

(a) 55 and 210
(b) 135 and 225
(c) 1260 and 7344
(d) 196 and 38220 **[NCERT]**
(e) 336 and 1260
(f) 867 and 255 **[NCERT, CBSE 2018 (C)]**
(g) 4052 and 12576 **[NCERT]**

Answers

1. (a) 5 (b) 45 (c) 36 (d) 196 (e) 84 (f) 51 (g) 4

Concept: Irrational Numbers

A number is said to be irrational if it cannot be written in the form of $\frac{p}{q}$, where p and q are integers and $q \neq 0$. For example: $\sqrt{2}$, $\sqrt{3}$, $\sqrt{5}$, π, etc.

Example 7 Prove that $\sqrt{2}$ is irrational. **[NCERT, (CBSE 2008, 09, 19)]**

Solution: Let us assume that $\sqrt{2}$ is rational.

So, $\sqrt{2} = \frac{a}{b}$, where a and b are coprime integers and $b \neq 0$.

$\Rightarrow$ $$b\sqrt{2} = a$$

Squaring both sides, we get $2b^2 = a^2$ ...(i)

$\Rightarrow$ 2 divides a^2

$\Rightarrow$ 2 divides a also [$\because$ 2 is a prime number and divides a^2]

Let $a = 2c$, where c is some integer. Inserting $a = 2c$ in (i), we get

$$2b^2 = 4c^2$$

$\Rightarrow$ $b^2 = 2c^2$

$\Rightarrow$ 2 divides b^2

$\Rightarrow$ 2 divides b also [$\because$ 2 is a prime number and divides b^2]

Thus, a and b have a common factor 2.

This contradicts the fact that a and b have no common factor other than 1 (as a and b are coprime). This contradiction has arisen because of our incorrect assumption that $\sqrt{2}$ is rational. Hence, $\sqrt{2}$ is an irrational number.

7 Try to Solve

1. Prove that $\sqrt{3}$ is irrational.
2. Prove that $\sqrt{5}$ is irrational. [NCERT]
3. Prove that $\sqrt{7}$ is irrational.

Example 8 Prove that $\frac{1}{\sqrt{2}}$ is irrational. [NCERT]

Solution: Let us assume that $\frac{1}{\sqrt{2}}$ is rational. So $\frac{1}{\sqrt{2}} = \frac{b}{a}$, where a and b are coprime integers.

$\Rightarrow$ $a = \sqrt{2}\, b$

$\Rightarrow$ $a^2 = 2b^2$ [On squaring both sides] ...(i)

$\Rightarrow$ 2 divides a^2

$\Rightarrow$ 2 divides a also [$\because$ 2 is a prime number and divides a^2]

Let $a = 2c$, where c is some integer. Inserting $a = 2c$ in (i), we get

$$4c^2 = 2b^2$$

$$2c^2 = b^2$$

$\Rightarrow$ 2 divides b^2

$\Rightarrow$ 2 divides b also

Thus, a and b have a common factor 2.

This contradicts the fact that a and b have no common factor other than 1 (as a and b are coprime). This contradiction has arisen because of our incorrect assumption that $\frac{1}{\sqrt{2}}$ is rational. Hence, $\frac{1}{\sqrt{2}}$ is irrational.

8 Try to Solve

1. Prove that $\frac{1}{\sqrt{3}}$ is irrational.
2. Prove that $\frac{1}{\sqrt{5}}$ is irrational.
3. Prove that $\frac{1}{\sqrt{7}}$ is irrational.

Example 9 Prove that $3\sqrt{7}$ is irrational.

Solution: Let us assume that $3\sqrt{7}$ is rational. So, $3\sqrt{7} = \frac{a}{b}$, where a and b are coprime integers.

$$\Rightarrow \quad \sqrt{7} = \frac{a}{3b}$$

Since a, b and 3 are integers, $\frac{a}{3b}$ is rational, and so, $\sqrt{7}$ is rational. But this contradicts the fact $\sqrt{7}$ is irrational.

This contradiction has arisen because of our incorrect assumption that $3\sqrt{7}$ is rational. Hence, $3\sqrt{7}$ is irrational.

9 Try to Solve

1. Prove that $2\sqrt{7}$ is irrational.
2. Prove that $5\sqrt{3}$ is irrational.
3. Prove that $6\sqrt{2}$ is irrational. [CBSE (CCE) 2009]

Example 10 Prove that $\frac{1}{2-\sqrt{5}}$ is an irrational number. [NCERT Exemplar]

Solution: Let us assume that $\frac{1}{2-\sqrt{5}}$ is rational. Let $\frac{1}{2-\sqrt{5}} = \frac{1}{a}$, where a is rational.

$$\Rightarrow \quad 2-\sqrt{5} = a$$

$$\Rightarrow \quad 2 = a + \sqrt{5}$$

$$\Rightarrow \quad \sqrt{5} = 2 - a$$

Since a is rational, $(2 - a)$ is rational, and so, $\sqrt{5}$ is rational.

But this contradicts the fact that $\sqrt{5}$ is irrational. This contradiction has arisen because of our incorrect assumption that $\frac{1}{2-\sqrt{5}}$ is rational. Hence, $\frac{1}{2-\sqrt{5}}$ is irrational.

10 Try to Solve

1. Prove that $\frac{1}{5-\sqrt{3}}$ is irrational.
2. Prove that $\frac{1}{3-\sqrt{2}}$ is irrational.
3. Prove that $\frac{3}{\sqrt{3}+2}$ is irrational.

Example 11 Prove that $\sqrt{3} + \sqrt{5}$ is irrational. [CBSE, (CCE) 2012]

Solution: Let us assume that $\sqrt{3} + \sqrt{5}$ is rational. Let $\sqrt{3} + \sqrt{5} = a$, where a is a rational number.

$$\Rightarrow \quad \sqrt{3} = a - \sqrt{5}$$

On squaring both sides, we get $\quad (\sqrt{3})^2 = (a - \sqrt{5})^2$

$\Rightarrow \quad 3 = a^2 + 5 - 2\sqrt{5}\,a$ [Using $(a - b)^2 = a^2 + b^2 - 2ab$]

$\Rightarrow \quad 2\sqrt{5}\,a = a^2 + 5 - 3$

$\Rightarrow \quad 2\sqrt{5}\,a = a^2 + 2$

$\therefore \quad \sqrt{5} = \dfrac{a^2+2}{2a} = \text{Rational}$ [$\because$ a is rational, so $\dfrac{a^2+2}{2a}$ is rational]

But $\sqrt{5}$ is irrational. This contradiction has arisen because of our incorrect assumption that $\sqrt{3} + \sqrt{5}$ is rational. Hence, $\sqrt{3} + \sqrt{5}$ is an irrational number.

11 Try to Solve

1. Prove that $\sqrt{2} + \sqrt{3}$ is irrational.
2. Prove that $\sqrt{7} + \sqrt{11}$ is irrational.
3. Prove that $\sqrt{5} - \sqrt{2}$ is irrational.

Example 12 Prove that $(2 - 3\sqrt{5})$ is irrational. **[CBSE 2010]**

Solution: Let us assume that $2 - 3\sqrt{5}$ is rational. Let $2 - 3\sqrt{5} = a$, where a is rational.

$\Rightarrow \quad 2 = a + 3\sqrt{5}$ ...(i)

$\Rightarrow \quad 3\sqrt{5} = 2 - a$

$\Rightarrow \quad \sqrt{5} = \dfrac{2-a}{3} = \text{Rational}$ [$\because$ a is rational]

But, we know $\sqrt{5}$ is an irrational number.

This contradiction has arisen because of our incorrect assumption that $2 - 3\sqrt{5}$ is rational. Hence, $2 - 3\sqrt{5}$ is an irrational number.

Example 13 Given that $\sqrt{2}$ is irrational, prove that $(5 + 3\sqrt{2})$ is an irrational number. **[CBSE 2018]**

Solution: Let $5 + 3\sqrt{2} = \dfrac{p}{q}$ be a rational number, where p and q have no common factor other than 1 and $q \neq 0$.

$\Rightarrow \quad 3\sqrt{2} = \dfrac{p}{q} - 5$

$\Rightarrow \quad \sqrt{2} = \dfrac{p-5q}{3q}$

For any values of p and $q(q \neq 0)$, RHS $= \dfrac{p-5q}{3q}$ is rational, but LHS $= \sqrt{2}$ is irrational.

This contradicts the fact. So, our assumption is wrong.

$\therefore \quad 5 + 3\sqrt{2}$ is an irrational number.

12 Try to Solve

1. Prove that $5 - \sqrt{3}$ is irrational. [NCERT]
2. Prove that is $3 + 2\sqrt{5}$ irrational.
3. Prove that $4 - 5\sqrt{3}$ is irrational.
4. Given that $\sqrt{3}$ is an irrational number, prove that $(2 + \sqrt{3})$ is an irrational number. **[CBSE 2018 (C)]**
5. Prove that $2 + 5\sqrt{3}$ is an irrational number, given that $\sqrt{3}$ is an irrational number. **[CBSE Allahabad 2019]**

Concept: Decimal Expansion of Rational Numbers

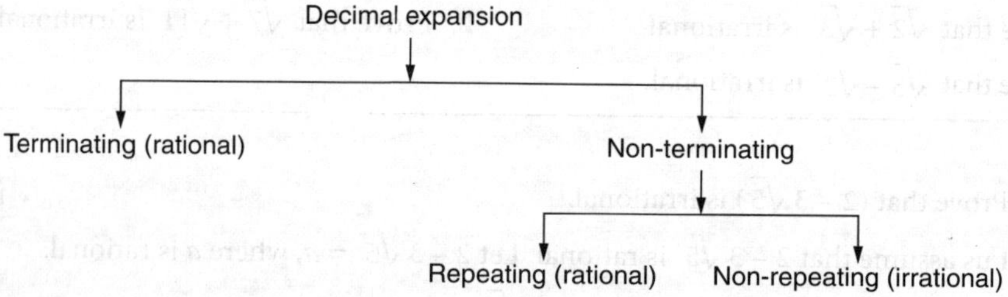

(i) The decimal expansion of a rational number is either terminating or non-terminating recurring. The converse of this result is also true, that is, a number whose decimal expansion is terminating or non-terminating recurring is rational.

Stated below are some important results that help you to find out exactly when the decimal expansion of a rational number is terminating and when it is non-terminating recurring:

(a) Let $x = \frac{p}{q}$ be a rational number, such that the prime factorisation of q is of the form 2^n5^m, where n, m are non-negative integers. Then, x has a decimal expansion which terminates.

(b) Let $x = \frac{p}{q}$ be a rational number, such that the prime factorisation of q is not in the form 2^n5^m, where n, m are non-negative integers. Then, x has a decimal expansion which is non-terminating and recurring.

(c) Let x be a rational number whose decimal expansion terminates. Then, x can be expressed in the form $\frac{p}{q}$, where p and q are coprime, and the prime factorisation of q is in the form 2^n5^m, where n, m are non-negative integers.

(d) Let x be a rational number whose decimal expansion is non-terminating recurring. Then, x can be expressed in the form $\frac{p}{q}$, where p and q are coprime, and the prime factorisation of q is not in the form 2^n5^m; it has a factor other than 2 or 5.

Example 14 Without actual division, state whether the following rational numbers will have a terminating decimal expansion or a non-terminating repeating decimal expansion:

(i) $\frac{11}{2^3 \times 3}$ (ii) $\frac{129}{2^2 \times 5^7 \times 7^{17}}$ (iii) $\frac{23}{2^3 \times 5^2}$ **[NCERT]**

(iv) $\frac{77}{210}$ [NCERT] (v) $\frac{171}{800}$ (vi) $\frac{13}{125}$

(vii) $\frac{19}{3125}$ (viii) $\frac{33}{300}$

Solution:

(i) We have $\frac{11}{2^3 \times 3}$: The denominator $2^3 \times 3$ is not of the form $2^m \times 5^n$. So the decimal expansion is non-terminating and repeating.

(ii) $\frac{129}{2^2 \times 5^7 \times 7^{17}}$: The denominator $2^3 \times 5^2$ is not of the form $2^m \times 5^n$. So the decimal expansion is non-terminating and repeating.

(iii) $\frac{23}{2^3 \times 5^2}$: The denominator $2^3 \times 5^2$ is of the form $2^m \times 5^n$. So the decimal expansion is terminating.

(iv) We have $\frac{77}{210} = \frac{11}{30} = \frac{11}{5 \times 2 \times 3}$: The denominator $(5 \times 2 \times 3)$ is not of the form $2^m \times 5^n$. So the decimal expansion is non-terminating and repeating.

(v) We have $\frac{171}{800} = \frac{171}{2^5 \times 5^2}$: The denominator 800 is of the form $2^m \times 5^n$. So the decimal expansion is terminating.

(vi) $\frac{13}{125} = \frac{13 \times 2^3}{5^3 \times 2^3}$: The denominator 125 can be expressed in the form $2^m \times 5^n$. So the decimal expansion is terminating.

(vii) $\frac{19}{3125} = \frac{19 \times 2^5}{5^5 \times 2^5}$: The denominator 3125 can be expressed in the form $2^m \times 5^n$. So the decimal expansion is terminating.

(viii) $\frac{33}{300} = \frac{11}{100} = \frac{11}{2^2 \times 5^2}$: The denominator 100 is of the form $2^m \times 5^n$. So, the decimal expansion is terminating.

13 Try to Solve

1. Without actually performing long division, state whether the following rational numbers will have a terminating decimal expansion or a non-terminating repeating decimal expansion:

(a) $\frac{13}{3125}$ (b) $\frac{17}{8}$ (c) $\frac{64}{455}$ (d) $\frac{15}{1600}$ (e) $\frac{29}{343}$ (f) $\frac{23}{2^3 5^2}$

(g) $\frac{129}{2^3 5^7 7^5}$ (h) $\frac{6}{15}$ (i) $\frac{35}{50}$ (j) $\frac{77}{210}$ [NCERT]

Answers

1. (a) Terminating (b) Terminating (c) Non-terminating (d) Terminating
(e) Non-terminating (f) Terminating (g) Non-terminating
(h) Non-terminating (i) Terminating (j) Non-terminating

Example 15 What type of decimal expansion will $\frac{69}{60}$ represent? After how many places will the decimal expansion terminate?

Solution: We have

$$\frac{69}{60} = \frac{23}{20} = \frac{23}{2^2 \times 5} = \frac{23 \times 5}{2^2 \times 5^2}$$

$$\frac{115}{(2 \times 5)^2} = \frac{115}{(10)^2} = \frac{115}{100}$$

The denominator 20 is of the form $2^m \times 5^n$. So the decimal expansion is terminating.

Here $m = 2$, $n = 1$. As the largest exponent of the denominator is 2, the decimal expansion will terminate after two places.

Example 16 Write the denominator of $\frac{91}{1250}$ in the form $2^m \times 5^n$, where m and n are non-negative integers. Also write its decimal expansion without performing actual division. **[CBSE 2019]**

Solution: We have $\frac{91}{1250} = \frac{91}{2 \times 5^4} = \frac{91 \times 2^3}{2^4 \times 5^4}$ [On multiplying numerator and denominator by 2^3]

$$= \frac{728}{10000} = 0.0728$$

Example 17 Write whether $\frac{2\sqrt{45} + 3\sqrt{20}}{2\sqrt{5}}$ on simplification gives an irrational or a rational number. **[CBSE 2018 (C)]**

Solution:

$$\frac{2\sqrt{45} + 3\sqrt{20}}{2\sqrt{5}} = \frac{2 \times 3\sqrt{5} + 3 \times 2\sqrt{5}}{2\sqrt{5}} = \frac{6\sqrt{5} + 6\sqrt{5}}{2\sqrt{5}} = \frac{(6+6)}{2} \times \frac{\sqrt{5}}{\sqrt{5}} = 6 \text{ (Rational Number)}$$

Example 18 If $\frac{p}{q}$ is a rational number $(q \neq 0)$, what is the condition of q so that the decimal representation of $\frac{p}{q}$ is terminating?

Solution: For any rational number $\frac{p}{q}$ with terminating decimal representation, the prime factorisation of q is of the form $2^n \cdot 5^m$, where n and m are non-negative integers.

Example 19 Find a rational number between $\sqrt{2}$ and $\sqrt{3}$. **[Delhi 2019]**

Solution: Let p be rational numbers $\sqrt{2}$ and $\sqrt{3}$.

$$\therefore \quad \sqrt{2} < p < \sqrt{3}$$

On squaring throughout, we have $2 < p^2 < 3$. The perfect squares that lie between 2 and 3 are 2.25, 2.56, 2.89.

So $$2 < 2.25 < 2.56 < 2.89 < 3.$$

Taking square root throughout, $\sqrt{2} < 1.5 < 1.6 < 1.7 < \sqrt{3}$.

Rational numbers between $\sqrt{2}$ and $\sqrt{3}$ are 1.5, 1.6, 1.7 and more.

14 Try to Solve

1. After how many decimal expansions will the following expansion terminate? Also, write its decimal expansion without performing actual division.

   (a) $\frac{15}{1600}$ (b) $\frac{13}{125}$ (c) $\frac{43}{2^4 \times 5^3}$ (d) $\frac{107}{1250}$

2. Find a rational number between $\sqrt{3}$ and $\sqrt{5}$.

Answers

1. (a) 6, 0.009375 (b) 3, 0.104 (c) 4, 0.0215 (d) 4, 0.0856
2. 1.8, 1.9, 2.0 and many more

FOR MORE PRACTICE AND KNOWLEDGE

Example 20 State the Fundamental Theorem of Arithmetic. **[CBSE 2012]**

Solution: Every composite number can be expressed (factorised) as a product of primes, and this factorisation is unique, apart from the order in which the prime factors occur.

Example 21 Express the number $0.3\overline{178}$ in the form of rational number $\frac{a}{b}$.

Solution: Let $x = 0.3\overline{178}$ ...(i)

Multiplying both sides of equation (i) by 10

$$10x = 3.\overline{178} \quad \text{...(ii)}$$

$$10x = 3.178\overline{178} \quad \text{...(iii)}$$

Multiplying both sides of equation (iii) by 1000

$$10000x = 3178.178 \quad \text{...(iv)}$$

Subtracting equation (ii) from equation (iv), we get

$$9990x = 3175$$

$$x = \frac{3175}{9990} = \frac{635}{1998}$$

Example 22 Express the HCF of 234 and 111 as $234x + 111y$, where x and y are integers. Also find x and y.

Solution: Since 234 > 111. Dividing 234 by 111, we get

```
111 ) 234 ( 2
      222
   12 ) 111 ( 9
        108
     03 ) 12 ( 4
          12
           X
```

$$234 = 111 \times 2 + 12 \quad \text{...(A)}$$

Since $12 \neq 0$

$$111 = 12 \times 9 + 3 \quad \text{...(B)}$$

Since $3 \neq 0$

$$12 = 3 \times 4 + 0$$

Since $r = 0$

$$\text{HCF} = 3$$

According to the question $3 = 234x + 111y$ ...(i)

From the above, we get $3 = 111 - 12 \times 9$ **[From (B)]**

$3 = 111 - (234 - 111 \times 2) \times 9$ **[From (A)]**

$$3 = 111 - 234 \times 9 + 111 \times 18$$

$$3 = 111 \times (19) + 234\,(-9) \qquad \ldots\text{(ii)}$$

Comparing (i) and (ii), we get $x = -9$, and $y = 19$

Example 23 Two alarm clocks ring their alarms at regular intervals of 50 seconds and 48 seconds. If they first beep together at 12 noon, at what time will they beep together again?

Solution: Factorise 50 and 48.

$$50 = 2 \times 5 \times 5,\ 48 = 2 \times 2 \times 2 \times 2 \times 3$$

$\therefore$ LCM of 50 and 48 $= 2 \times 2 \times 2 \times 2 \times 3 \times 5 \times 5 = 1200$

$\therefore$ 1200 sec = 20 min

Hence, they will beep together again at 12.20 pm.

Example 24 Three sets of physics, chemistry and mathematics books have to be stacked in such a way that all the books are stored topic-wise and the number of books in each stack is the same. The number of physics books is 192, the number of chemistry books is 240 and the number of mathematics books is 168. Determine the number of stacks of physics, chemistry and mathematics books.

Solution: Here, we have to find the HCF of 192, 240 and 168 because the HCF will be the largest number which divides 192, 240 and 168 exactly.

$$192 = 2^6 \times 3;\ 240 = 2^4 \times 3 \times 5$$

$$168 = 2^3 \times 3 \times 7$$

Now, the HCF of 192, 240 and 168 is $= 2^3 \times 3 = 24$.

There must be 24 books in each stack.

$\therefore$ Number of stacks of physics books $= \dfrac{192}{24} = 8$

Number of stacks of chemistry books $= \dfrac{240}{24} = 10$

Number of stacks of mathematics books $= \dfrac{168}{24} = 7$

Example 25 The length, breadth and height of a room are 8 m 25 cm, 6 m 75 cm and 4 m 50 cm, respectively. Determine the length of the longest rod that can measure the three dimensions of the room exactly.

Solution: Length of the longest rod = HCF of 825 cm, 675 cm and 450 cm.

$$825 = 3 \times 5 \times 5 \times 11$$

$$675 = 3 \times 3 \times 3 \times 5 \times 5$$

$$450 = 2 \times 3 \times 3 \times 5 \times 5$$

$\Rightarrow$ HCF = 75

$\therefore$ Length of the longest rod = 75 cm

Example 26 State Euclid's division lemma. **[CBSE 2013]**

Solution: Given positive integers a and b, there exist unique whole numbers q and r 'satisfying' $a = bq + r,\ 0 \le r < b$.

Example 27 Find HCF (865, 255) using Euclid's division lemma. [CBSE 2013]

Solution: The given numbers are 255 and 865.

On applying Euclid's division algorithm, we have

$$865 = 255 \times 3 + 100$$

$$255 = 100 \times 2 + 55$$

$$100 = 55 \times 1 + 45$$

$$55 = 45 \times 1 + 10$$

$$45 = 10 \times 4 + 5$$

$$10 = 5 \times 2 + 0$$

```
255 )865( 3
     765
     100 )255( 2
          200
          55)100( 1
              55
              45) 55 ( 1
                  45
                  10) 45 ( 4
                      40
                       5) 10 ( 2
                          10
                           0
```

$\therefore$ At the last stage, the divisor is 5.

$\therefore$ The HCF of 255 and 865 is 5.

Example 28 Show that the square of any positive odd integer is of the form $8m + 1$, for some integer m.

Solution: Let $a = 2q + 1$ be any positive odd integer.

Now $a^2 = (2q + 1)^2 = 4q^2 + 4q + 1$

$= 4q(q + 1) + 1 = 4(2m) + 1$

[$\because q$ and $(q + 1)$ are consecutive numbers; so one of them must be even and of the form $2m$]

$= 8m + 1$

$\therefore$ Square of any positive odd integer is of the form $8m + 1$, for some integer m.

Example 29 Find the value of: $(-1)^n + (-1)^{2n} + (-1)^{2n+1} + (-1)^{4n+2}$, where n is any positive odd integer.

Solution: To find $(-1)^n + (-1)^{2n} + (-1)^{2n+1} + (-1)^{4n+2}$ ...(i)

As 'n' is any positive odd integer, $2n$ and $4n + 2$ are even positive integers.

Now $(-1)^n = -1$

$(-1)^{2n} = +1$

$(-1)^{2n+1} = (-1)^{2n}(-1)^1 = 1(-1) = -1\ (-1)$

$(-1)^{4n+2} = (-1)^{4n}(-1)^2 = 1(1) = 1$

Inserting all the values in the above expression (i), we get

$$-1 + 1 - 1 + 1 = 0.$$

Example 30 Use Euclid's division lemma to show that the square of any positive integer is of the form $3m$ or $3m + 1$, for some integer m.

Solution: Let N be any positive number. According to Euclid's division lemma

$N = 3q$ or $3q + 1$ or $3q + 2$, where q is some integer

When $N = 3q$

$\Rightarrow$ $N^2 = (3q)^2 = 9q^2 = 3(3q^2) = 3m$, where $m = 3q^2$

When $N = 3q + 1$

$\Rightarrow \quad N^2 = (3q+1)^2 = 9q^2 + 6q + 1 = 3(3q^2 + 2q) + 1$

$= 3m + 1$, where $m = 3q^2 + 2q$

When $\quad N = (3q+2) \Rightarrow N^2 = 9q^2 + 12q + 4 = 9q^2 + 12q + 3 + 1$

$\Rightarrow \quad N^2 = 3(3q^2 + 4q + 1) + 1$

$= 3m + 1$, where $m = 3q^2 + 4q + 1$

Hence, the square of any positive integer is of two types, $3m$ or $3m + 1$, for some integer m.

Example 31 Use Euclid's division lemma to show that the cube of any positive integer is of the form $9m$, $9m + 1$ or $9m + 8$.

Solution: Let N be a positive integer. Using Euclid's lemma

$N = 3q + r$, where q is an integer.

And $\quad r = 0, 1$ or 2

$\therefore \quad N = 3q, 3q + 1$ or $3q + 2$

(i) When $\quad N = 3q$

$N^3 = (3q)^3 = 27q^3 = 9(3q^3) = 9m$, where $m = 3q^3$

(ii) When $\quad N = 3q + 1$

$\Rightarrow \quad (N)^3 = (3q+1)^3 \Rightarrow N^3 = 27q^3 + 27q^2 + 9q + 1$

$\Rightarrow \quad N^3 = 9(3q^3 + 3q^2 + q) + 1 = 9m + 1$, where $m = 3q^3 + 3q^2 + q$

(iii) When $\quad N = 3q + 2$

$N^3 = (3q+2)^3 = 27q^3 + 54q^2 + 36q + 8$

$= 9(3q^3 + 6q^2 + 4q) + 8$

$= 9m + 8$, where $m = 3q^3 + 6q^2 + 4q$

Hence, the cube of any positive integer is of the form $9m$, $9m + 1$ or $9m + 8$ for some integer m.

Example 32 Show that the reciprocal of $3 + 2\sqrt{2}$ is an irrational number.

Solution: We have to prove that $\dfrac{1}{3+2\sqrt{2}}$ is an irrational number.

First rationalise it, $\dfrac{1}{3+2\sqrt{2}} \times \dfrac{3-2\sqrt{2}}{3-2\sqrt{2}} = \dfrac{3-2\sqrt{2}}{(3)^2-(2\sqrt{2})^2}$

$= \dfrac{3-2\sqrt{2}}{9-8} = \dfrac{3-2\sqrt{2}}{1} = 3 - 2\sqrt{2}$

Let us assume that $3 - 2\sqrt{2}$ is rational.

$\therefore$ There exist coprime integers a and b $(b \neq 0)$ such that

$3 - 2\sqrt{2} = \dfrac{a}{b} \Rightarrow 2\sqrt{2} = 3 - \dfrac{a}{b}$

$\Rightarrow \quad 2\sqrt{2} = \dfrac{3b-a}{b}$

$\Rightarrow \quad \sqrt{2} = \dfrac{3b-a}{2b} = \dfrac{3}{2} - \dfrac{a}{2b}$

Since a and b are integers, $\frac{3}{2} - \frac{a}{2b}$ is rational and so, $\sqrt{2}$ is rational.

But this contradicts the fact that $\sqrt{2}$ is irrational. This contradiction has arisen because of our incorrect assumption that $3 - 2\sqrt{2}$ is rational. Hence, we conclude that $3 - 2\sqrt{2}$ is irrational.

Example 33 Express $\left(\frac{17}{8} + \frac{7}{80}\right)$ as a decimal representation.

Solution: $$\frac{17}{8} + \frac{7}{80} = \frac{170+7}{80} = \frac{177}{80} = 2.2125.$$

Example 34 If $\frac{278}{2^3 m}$ has a terminating decimal expansion and m is a positive integer such that $2 < m < 9$, find the value of m.

Solution: $\because$ $\frac{278}{2^3 \cdot m}$ has a terminating decimal expansion.

$\therefore$ $2^3 \times m$ should be of the type $2^p \cdot 5^q$, where p and q are whole numbers.

$\Rightarrow$ $2^3 \times m = 2^p \times 5^q$

$\Rightarrow$ $m = 5^q$

$\because$ $2 < m < 9$

$\therefore$ $m = 5$ and $q = 1$

$\Rightarrow$ $m = 5$

Multiple Choice Questions

Choose the correct answer for each of the following questions:

1. Which of the following is a pair of coprimes?
   (a) (14, 35) (b) (18, 25) (c) (31, 93) (d) (32, 62)
2. If $a = (2^2 \times 3^3 \times 5^4)$ and $b = (2^3 \times 3^2 \times 5)$, then HCF (a, b) = ?
   (a) 90 (b) 180 (c) 360 (d) 540
3. LCM of $(2^3 \times 3 \times 5)$ and $(2^4 \times 5 \times 7)$ is
   (a) 40 (b) 560 (c) 1120 (d) 1680
4. What is the largest number that divides 1152 and 1664 exactly?
   (a) 32 (b) 64 (c) 128 (d) 256
5. What is the largest number that divides 70 and 125, leaving remainders 5 and 8, respectively?
   (a) 13 (b) 9 (c) 3 (d) 585
6. What is the largest number that divides 245 and 1029, leaving remainder 5 in each case?
   (a) 15 (b) 16 (c) 9 (d) 5
7. A number when divided by 143 leaves 31 as remainder. What will be the remainder when the same number is divided by 13?
   (a) 0 (b) 1 (c) 3 (d) 5
8. Which of the following is an irrational number?
   (a) $\frac{22}{7}$ (b) 3.1416 (c) $3.\overline{1416}$ (d) 3.141141114 …
9. π is
   (a) An integer (b) A rational number (c) An irrational number (d) None of these

10. a and b are two positive integers such that the least prime factor of a is 3 and the least prime factor of b is 5. Then, the least prime factor of $(a + b)$ is

(a) 2 (b) 3 (c) 5 (d) 8

11. What is the least number that is divisible by all the natural numbers from 1 to 10 (both inclusive)?

(a) 100 (b) 1260 (c) 2520 (d) 5040

12. $0.\overline{68} + 0.\overline{73} = ?$

(a) $1.\overline{41}$ (b) $1.\overline{42}$ (c) $0.\overline{141}$ (d) None of these

13. If HCF of 65 and 117 can be expressed in the form $65m - 117$, then the value of m is

(a) 1 (b) 2 (c) 3 (d) 4

14. If two positive integers a and b are written as $a = x^5y^2$ and $b = x^2y^3$ (x and y are prime numbers), then HCF (a, b) is

(a) xy (b) x^5y^3 (c) x^3y^3 (d) x^2y^2

15. When Euclid's lemma is used for $a < b$ as $a = bq + r$, which of the following is true?

(a) $b = 0$ (b) $r = 0$ (c) $q = 0$ (d) $a = 0$

16. If 'a' and 'b' are coprime, then a^2 and b^2 are

(a) Coprime (b) Not coprime (c) Odd numbers (d) Even numbers

17. The number $(\sqrt{2} + \sqrt{3})^{10} (\sqrt{2} - \sqrt{3})^{10}$ is a

(a) Rational number (b) Irrational number (c) Prime number (d) Negative number

18. The number 3.24636363 … is

(a) An integer (b) A rational number (c) An irrational number (d) None of these

19. Arrange the following in ascending order $\frac{17}{18}, \frac{43}{45}, \frac{59}{60}$ and $\frac{31}{36}$:

(a) $\frac{17}{18} < \frac{59}{60} < \frac{43}{45} < \frac{31}{36}$ (b) $\frac{31}{36} < \frac{17}{18} < \frac{43}{45} < \frac{59}{60}$

(c) $\frac{43}{45} < \frac{59}{60} < \frac{31}{36} < \frac{17}{18}$ (d) $\frac{59}{60} < \frac{43}{45} < \frac{31}{36} < \frac{17}{18}$

20. Three numbers are in the ratio 3:4:5 and their LCM is 1200. Then, their HCF is

(a) 40 (b) 60 (c) 20 (d) 120

21. If n is a natural number, then $25^{2n} - 9^{2n}$ is always divisible by

(a) 16 (b) 34 (c) Both 16 and 34 (d) None of these

22. If two positive integers a and b are written as $a = x^3y^2$ and $b = xy^3$; x, y being prime numbers, then HCF (a, b) is

(a) xy (b) xy^2 (c) x^3y^3 (d) x^2y^2

23. If two positive integers p and q can be expressed as $p = ab^2$ and $q = - a^3b$; a, b being prime numbers, then LCM (p, q) is

(a) ab (b) a^2b^2 (c) a^3b^2 (d) a^3b^3

24. For some integer q, every odd integer is of the form

(a) q (b) $q + 1$ (c) $2q$ (d) $2q + 1$

25. The decimal expansion of the rational number $\frac{33}{2^2.5}$ will terminate after

(a) One decimal place (b) Two decimal places

(c) Three decimal places (d) More than three decimal places

26. The decimal expansion of the rational number $\frac{14857}{1250}$ will terminate after

(a) One decimal place (b) Two decimal places
(c) Three decimal places (d) Four decimal places

Fill in the Blanks

Complete the following statements with appropriate answers(s) in the blank space(s):

27. Given positive integers a and b, there exist unique integers q and r satisfying $a = bq + r$, $0 \leq r < b$. This statement is known as ________.
28. The least number that is divisible by all the numbers from 1 to 10 (both inclusive) is ________.
29. The HCF of prime numbers is ________.
30. "The product of three consecutive positive integers is divisible by 6" can be written in form of n as ________. **[NCERT Exemplar]**

Assertion–Reason

The following questions consist of two statements: Assertion (A) and Reason (R). Answer these questions by selecting the appropriate option given below.

(A) Both A and R are true and R is the correct explanation for A.
(B) Both A and R are true and R is not the correct explanation for A.
(C) A is true but R is false.
(D) A is false but R is true.

31. **Assertion (*A*)**: 6^n ends with the digit zero, where n is a natural number.
Reason (*R*): Any number ends with the digit zero if its prime factor is of the form $2^m \times 5^n$, where m, n are natural numbers.
32. **Assertion (*A*)**: When a positive integer a is divided by 3, the values of the remainder can be 0, 1 or 2.
Reason: (*R*): According to Euclid's division lemma, $a = bq + r$, where $0 \leq r < b$ and r is an integer.
33. **Assertion (*A*)**: The rational number $\frac{7}{75}$ can be expressed as a terminating decimal expansion.
Reason (*R*): The rational number with denominator $2^m \times 5^n$ can be expressed as a terminating decimal expansion.
34. **Assertion (*A*)**: $\frac{29}{990}$ will have a non-terminating repeating decimal expansion.
Reason (*R*): Let $a = \frac{p}{q}$ be a rational number such that p and q are coprime and the prime factorisation of q is of the form $2^n \times 5^m$, where n and m are non-negative integers (whole numbers). Then, a has a decimal expansion, which is non-terminating repeating.

Subjective Questions

35. Find the HCF of 81445 and 687897.
36. If the HCF of 152 and 272 is expressible in the form $272 \times 8 + 152x$, find x.
37. Complete the missing entries in the following factor tree:

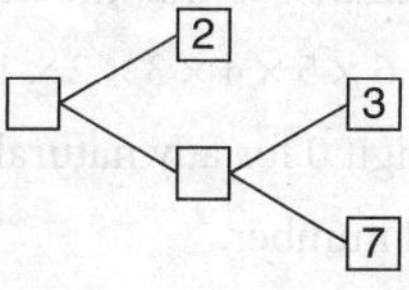

38. Let d be the HCF of 24 and 36. Find two number a and b, such that $d = 24a + 36b$. **[CBSE (CCE) 2014]**

39. A forester wants to plant 66 apple trees, 88 banana trees and 110 mango trees in equal rows (in terms of the number of trees). He also wants to make distinct rows of trees (only one type of tree in one row). Find the minimum number of rows required.

40. In a seminar, the number of participants in Mathematics, Physics and Biology are 336, 240 and 96, respectively. Find the minimum number of rooms required if in each room the same number of participants are to be seated and all of them belong to the same subject.

41. The HCF of 2472, 1284 and a third number N is 12. If their LCM is $2^3 \times 3^2 \times 5 \times 103 \times 107$, find the number N.

42. On a morning walk, three persons step together and their steps measure 40 cm, 42 cm and 45 cm, respectively. What is the minimum distance each should walk so that each can cover the same distance in complete steps?

43. Determine the values of p and q so that the prime factorisation of 2520 is expressible as $2^3 \times 3^p \times q \times 7$.

44. Fifty people work in a cooperative society. They all use their own conveyance: 20 people use scooters, 12 go by car, 16 go by public transport and 2 use bicycles.
(a) Find the HCF of 20, 12, 16 and 2.
(b) One day, they all decide to go by public transport. What is the value shown by them?

45. Priya and Ravish planned to participate in a cycle race to be organised for national integration. They decide to practice on a circular path around a sports field. Priya takes 18 minutes to complete one round, while Ravish takes 12 minutes for the same. Suppose they both start at the same time and travel in the same direction:
(a) After how many minutes will they meet again at the starting point?
(b) What is the value shown by Priya and Ravish?

46. A charitable trust donates 28 different books on maths, 16 different books on science and 12 different books on social science to poor students. Each student is given a maximum number of only one subject of their interest and all the students received an equal number of books.
(a) Find the number of books each student received.
(b) Find the total number of students who received books.
(c) How does it help your society?

47. What is the greatest possible speed at which a man can walk 52 km and 91 km in an exact number of minutes?

48. If $A = 2n + 13$ and $B = n + 7$, where n is a natural number, then find the HCF of A and B.

49. Find the greatest number that will divide 445, 572 and 699, leaving remainders 4, 5 and 6, respectively.

50. On MG Road, three consecutive traffic lights change after 36, 42 and 72 seconds. If the lights are first switched on at 9.00 am, at what time will they change simultaneously?

51. Two tankers contain 850 litres and 680 litres of petrol, respectively. Find the maximum capacity of a container which can measure the petrol of either tanker in exact number of times.

52. Find the HCF and LCM of 6, 72 and 120 using the prime factorisation method. **[NCERT]**

53. A street shopkeeper prepares 396 gulab jamuns and 342 rasgullas. He packs them in combination, each container consisting of either gulab jamuns or rasgullas but having an equal number of pieces. Find the number of pieces he should put in each box so that the number of boxes is least. **[CBSE 2016]**

54. Explain why $7 \times 11 \times 13 + 13$ and $7 \times 6 \times 5 \times 4 \times 3 \times 2 \times 1 + 5$ are composite numbers. **[NCERT]**

55. Check whether 6^n can end with the digit 0 for any natural number n. **[NCERT Exemplar]**

56. Prove that $15 + 17\sqrt{3}$ is an irrational number. **[CBSE (CCE) 2011]**

57. Show that $(\sqrt{3}+\sqrt{5})^2$ is an irrational number. **[CBSE (CCE) 2015]**

58. Prove that $2\sqrt{3}-1$ is an irrational number. **[CBSE (CCE) 2010]**

59. Show that $5-\sqrt{3}$ is irrational. **[NCERT]**

60. Prove that $3+2\sqrt{5}$ is irrational. **[NCERT]**

61. Prove that $6+\sqrt{2}$ is irrational. **[NCERT]**

62. Without actually performing long division, find if $\frac{987}{10500}$ will have terminating or non-terminating (repeating) decimal expansion. Give reasons for your answer. **[NCERT Exemplar]**

63. Find two numbers which on multiplication with $\sqrt{180}$ give a rational number. Are these numbers rational or irrational? **[CBSE (CCE) 2015]**

64. Find whether the decimal expansion of $\frac{13}{64}$ is a terminating or non-terminating decimal. If it terminates, find the number of decimal places its decimal expansion has. **[NCERT]**

65. The following real numbers have decimal expansions as given below. In each case, decide whether they are rational or not. If they are rational and of the form $\frac{p}{q}$, what can you say about the prime factors of q? **[NCERT Exemplar]**

(i) 43.123456789 (ii) 0.120120012000120000 … (iii) $43.\overline{123456789}$

66. Express the number 0.3178 in the form of a rational number $\frac{a}{b}$. **[NCERT Exemplar]**

67. A rational number in its decimal expansion is 327.7081. What can you say about the prime factors of q, when this number is expressed in the form $\frac{p}{q}$? Give reasons. **[NCERT Exemplar]**

68. Show that only one of the numbers n, $n+2$ and $n+4$ is divisible by 3. **[NCERT Exemplar]**

69. Show that p^2 will leave a remainder 1 when divided by 8, if p is an odd positive integer.

70. Prove that the product of any three consecutive positive integers is divisible by 6. **[CBSE 2016]**

71. If the HCF of 144 and 180 is expressed in the form $13m-3$, find the value of m. **[CBSE 2014]**

72. Show that any positive odd integer is of the form $4m+1$ or $4m+3$, where m is some integer. **[CBSE 2013]**

73. Find the HCF of 65 and 117 and find a pair of integral values of m and n such that HCF $= 65m+117n$. **[NCERT Exemplar]**

74. An army contingent of 616 members is to march behind an army band of 32 members in a parade. The two groups are to march in the same number of columns. What is the maximum number of columns in which they can march? **[NCERT]**

75. Use Euclid's division lemma to show that the square of any positive integer is of the form $3m$ or $3m+1$ for some integer m. **[NCERT]**

76. Use Euclid's division lemma to show that the cube of any positive integer is of the form $9m$, $9m+1$ or $9m+8$. **[NCERT]**

77. Show that every positive even integer is of the form $2q$, and that every positive odd integer is of the form $2q+1$, where q is some integer. **[NCERT]**

78. Show that any positive odd integer is of the form $4q+1$ or $4q+3$, where q is some integer. **[NCERT]**

79. A sweets seller has 420 kajubarfis and 130 badambarfis. She wants to stack them in such a way that each stack has the same number, and they take up the least area of the tray. What is the number of barfis that can be placed in each stack for this purpose? **[NCERT]**

80. Write whether every positive integer can be of the form $4q + 2$, where q is an integer. Justify your answer. **[NCERT Exemplar]**
81. Show that the square of any positive integer cannot be of the form $5q + 2$ or $5q + 3$, for any integer q. **[NCERT Exemplar]**
82. Show that the square of any odd integer is of the form $4q + 1$, for some integer q. **[NCERT Exemplar]**
83. If n is an odd integer, then show that $n^2 - 1$ is divisible by 8. **[NCERT Exemplar]**
84. Prove that if x and y are both odd positive integers, then $x^2 + y^2$ is even but not divisible by 4. **[NCERT Exemplar]**
85. Use Euclid's division algorithm to find the HCF of 441, 567, 693. **[NCERT Exemplar]**
86. Prove that one and only one out of n, $n + 2$ and $n + 4$ is divisible by 3, where n is any positive integer. **[NCERT Exemplar]**
87. For any positive integer n, prove that $n^3 - n$ is divisible by 6. **[NCERT Exemplar]**
88. Show that one and only one out of n, $n + 4$, $n + 8$, $n + 12$ and $n + 16$ is divisible by 5, where n is any positive integer. **[NCERT Exemplar]**

Answers

Multiple Choice Questions

1. (b) **2.** (b) **3.** (d) **4.** (c) **5.** (a) **6.** (b) **7.** (d) **8.** (d) **9.** (c) **10.** (a)
11. (c) **12.** (b) **13.** (b) **14.** (d) **15.** (b) **16.** (a) **17.** (a) **18.** (b) **19.** (b) **20.** (c)
21. (c) **22.** (b) **23.** (c) **24.** (d) **25.** (b) **26.** (d)

Fill in the Blanks

27. Euclid's division lemma **28.** 2520 **29.** 1 **30.** $n(n+1)(n+2)$

Assertion–Reason

31. (D) **32.** (A) **33.** (D) **34.** (C)

Subjective Questions

35. 1253 **36.** $\frac{-271}{19}$ **37.** 42, 21 **38.** 5, –3 **39.** 12 **40.** 14 **41.** 180 **42.** 2520 **43.** $p = 2$ and $q = 5$
44. 2, environment awareness **45.** 36, love for their country **46.** 4, 14, charity for poor students
47. 13 km/min **48.** 1 **49.** 63 **50.** 9 : 08 : 24 **51.** 170 litres **52.** 6, 360
53. 18 **62.** Yes, terminating decimals
63. Numbers are $\sqrt{5}$ and 1. One is irrational and other is rational.
64. Terminating, 6 places of decimals
65. (i) Rational prime, factors of q will be either 2 or 5 or both only.
(ii) Irrational
(iii) Rational, prime factors of q will also have a factor other than 2 or 5.
66. 635/1998
67. Since 327.7081 is a terminating decimal number, q must be of the form $2^m \cdot 5^n$; m, n are natural numbers.
71. 3 **73.** 2, –1 **74.** 8 columns **79.** 10
80. No, because an integer can be written in the form $4q$, $4q + 1$, $4q + 2$, $4q + 3$ **85.** 63

Real Numbers (Basic)

SELF-ASSESSMENT TEST

Time allowed: 90 min. Max. Marks: 40

General Instructions

1. *All questions are compulsory. There are 20 questions in all.*
2. *This question paper has four sections: Section A, Section B, Section C and Section D.*
3. *Section A contains 10 questions of one mark each, Section B contains three questions of two marks each, Section C contains four questions of three marks each, and Section D contains three questions of four marks each.*

Section A [1 Mark Each]

Direction: (Q. Nos. 1–5) These are Multiple Choice Questions. Select the most appropriate answer for the given questions.

1. Which of the following is an irrational number?

(a) $\frac{22}{7}$ (b) 3.1416 (c) $3.\overline{1416}$ (d) 3.141141114 ...

2. π is

(a) an integer (b) a rational number (c) an irrational number (d) none of these

3. LCM of $(2^3 \times 3 \times 5)$ and $(2^4 \times 5 \times 7)$ is

(a) 40 (b) 560 (c) 1120 (d) 1680

4. If $a = (2^2 \times 3^3 \times 5^4)$ and $b = (2^3 \times 3^2 \times 5)$, then HCF (a, b) = ?

(a) 90 (b) 180 (c) 360 (d) 540

5. The decimal expansion of the rational number $\frac{33}{2^2 . 5}$ will terminate after

(a) one decimal place (b) two decimal places

(c) three decimal places (d) more than three decimal places

6. The HCF of prime numbers is ___________.

7. The least number that is divisible by all the numbers from 1 to 10 (both inclusive) is ___________.

8. Express $\left(\frac{17}{8} + \frac{7}{80}\right)$ as a decimal representation.

9. Given that HCF (306, 657) = 9, find LCM.

10. Complete the missing entries in the following factor tree:

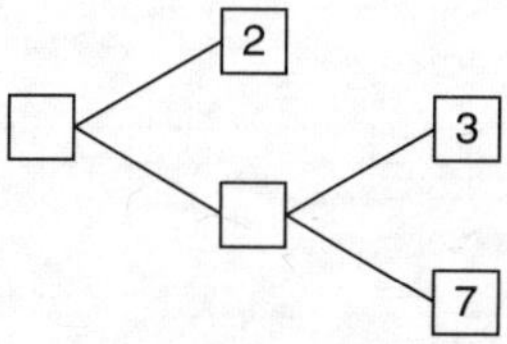

Section B [2 Marks Each]

11. Prove that $\sqrt{2}$ is irrational.
12. The HCF of two numbers is 23 and their LCM is 1449. If one of the numbers is 161, find the other.
13. Explain why $7 \times 11 \times 13 + 13$ and $7 \times 6 \times 5 \times 4 \times 3 \times 2 \times 1 + 5$ are composite numbers.

Section C [3 Marks Each]

14. If the HCF of 152 and 272 is expressible in the form $272 \times 8 + 152x$, then find x.
15. Prove that $3\sqrt{7}$ is irrational.
16. Show that the square of any positive odd integer is of the form $8m + 1$, for some integer m.
17. Find the HCF (865, 255) using Euclid's division lemma.

Section D [4 Marks Each]

18. Prove that $\sqrt{3} + \sqrt{5}$ is irrational.
19. Given that $\sqrt{2}$ is irrational, prove that $(5 + 3\sqrt{2})$ is an irrational number.
20. A street shopkeeper prepares 396 gulab jamuns and 342 rasgullas. He packs them in combination, each container consisting of either gulab jamuns or rasgullas but having an equal number of pieces. Find the number of pieces he should put in each box so that the number of boxes is least.

Answers

1. (d) **2.** (c) **3.** (d) **4.** (b) **5.** (b) **6.** 1 **7.** 2520 **8.** 2.2125 **9.** 22338

10. 42, 21 **12.** 207 **14.** $-\frac{271}{19}$ **17.** 5 **20.** 18

Real Numbers (Standard)

SELF-ASSESSMENT TEST

Time allowed: 90 min. Max. Marks: 40

General Instructions

1. *All questions are compulsory. There are 20 questions in all.*
2. *This question paper has four sections: Section A, Section B, Section C and Section D.*
3. *Section A contains 10 questions of one mark each, Section B contains three questions of two marks each, Section C contains four questions of three marks each, and Section D contains three questions of four marks each.*

Section A [1 Mark Each]

Direction: (Q. Nos. 1–5) These are Multiple Choice Questions. Select the most appropriate answer for the given questions.

1. If two positive integers a and b are written as $a = x^3y^2$ and $b = xy^3$, where x, y are prime numbers, then HCF (a, b) is

(a) xy (b) xy^2 (c) x^3y^3 (d) x^2y^2

2. What is the least number that is divisible by all the natural numbers from 1 to 10 (both inclusive)?

(a) 100 (b) 1260 (c) 2520 (d) 5040

3. If the HCF of 65 and 117 is expressible in the form 65 m – 117, then the value of m is

(a) 1 (b) 2 (c) 3 (d) 4

4. The number $(\sqrt{2}+\sqrt{3})^{10}\ (\sqrt{2}-\sqrt{3})^{10}$ is a

(a) rational number (b) irrational number (c) prime number (d) negative number

5. $0.\overline{68} + 0.\overline{73} = ?$

(a) $1.\overline{41}$ (b) $1.\overline{42}$ (c) $0.\overline{141}$ (d) none of these

6. Given positive integers a and b, there exist unique integers q and r satisfying $a = bq + r$, $0 \le r < b$. This statement is known as ____________.

7. "The product of three consecutive positive integers is divisible by 6" can be written in form of n as ____________.

8. Check whether 6^n can end with the digit 0 for any natural number n.

9. State the Fundamental Theorem of Arithmetic.

10. Find the value of $(-1)^n + (-1)^{2n} + (-1)^{2n+1} + (-1)^{4n+2}$, where n is any positive odd integer.

Section B [2 Marks Each]

11. Use Euclid's division lemma to show that the cube of any positive integer is of the form $9m$, $9m + 1$ or $9m + 8$.

12. Show that the reciprocal of $3 + 2\sqrt{2}$ is an irrational number.

13. If $A = 2n + 13$ and $B = n + 7$, where n is a natural number, then find the HCF of A and B.

Section C [3 Marks Each]

14. Let d be the HCF of 24 and 36. Find two numbers a and b, such that $d = 24a + 36b$.

15. Prove that $6+\sqrt{2}$ is irrational.

16. Show that one and only one out of n, $n + 4$, $n + 8$, $n + 12$ and $n + 16$ is divisible by 5, where n is any positive integer.

17. Prove that one and only one out of n, $n + 2$ and $n + 4$ is divisible by 3, where n is any positive integer.

Section D [4 Marks Each]

18. Three sets of physics, chemistry and mathematics books have to be stacked in such a way that all the books are stored topic-wise and the number of books in each stack is the same. The number of physics books is 192, the number of chemistry books is 240 and the number of mathematics books is 168. Determine the number of stacks of physics, chemistry and mathematics books.

19. Express the HCF of 234 and 111 as $234x + 111y$, where x and y are integers.

20. Priya and Ravish planned to participate in a cycle race to be organised for national integration. They decided to practice on a circular path around a sports field. Priya takes 18 minutes to complete one round, while Ravish takes 12 minutes for the same. Suppose they both start at the same time and move in the same direction:

(a) After how many minutes will they meet again at the starting point?

(b) What is the value depicted by Priya and Ravish?

Answers

1. (b) **2.** (c) **3.** (b) **4.** (a) **5.** (b) **6.** Euclid's division lemma

7. $n(n + 1)(n + 2)$ will always be divisible by 6, at least one of the factors will be divisible by 2 and at least one of the factors will be divisible by 3. **8.** No

9. Given positive integers a and b, there exist unique whole numbers q and r 'satisfying' $a = bq + r$, $0 < r < b$.

10. 0 **13.** 1 **14.** 5, – 3 **18.** 8 (Physics), 10 (Chemistry), 7 (Mathematics) **19.** $x = -9$ and $y = 19$

20. 36, Love for their country

CHAPTER

2 Polynomials

Key Concepts

- Definition of polynomial
- Zeros of polynomials
- Relationship between zeros and coefficients of a polynomial
- Formation of quadratic polynomial
- Division in polynomials

Concept: Polynomials

(i) A function $p(x)$ of the form $p(x)=a_0x^n+a_1x^{n-1}+a_2x^{n-2}+...+a_n$ is called a polynomial in x over reals, if $a_0, a_1, a_2, a_3...a_n$ are real numbers and n is a non-negative integer.

(ii) If $p(x)$ is a polynomial in x, the highest power of x in $p(x)$ is called the degree of the polynomial $p(x)$.

(iii) Polynomials of degrees 1, 2 and 3 are called linear, quadratic and cubic polynomials, respectively.

(iv) A quadratic polynomial in x with real coefficients is of the form ax^2+bx+c, where a, b, c are real numbers with $a \neq 0$.

(v) If $p(x)$ is a polynomial in x, and if k is any real number, then the value obtained by replacing x with k in $p(x)$ is called the value of $p(x)$ at $x=k$, and is denoted by $p(k)$.

Concept: Zeros of a Polynomial

(i) A real number k is said to be a zero of a polynomial $p(x)$, if $p(k)=0$. In other words, the zeros of a polynomial $p(x)$ are the solutions of the polynomial equation $p(x)=0$.

(ii) Geometrically, the zeros of a polynomial $p(x)$ are the x coordinates of the points where the graph of $y=p(x)$ intersects the x axis.

(iii) A quadratic polynomial can have at most two zeros and a cubic polynomial can have at most three zeros.

(iv) A polynomial of degree n can have at most n zeros.

(v) Zero may be considered as a zero of a polynomial.

(vi) A non-zero constant polynomial has no zeros.

(vii) Every real number can be considered as a zero of the polynomial.

Example 1 Given below are the graphs of some polynomials. For each of the graphs, find the number of zeros of the corresponding polynomial.

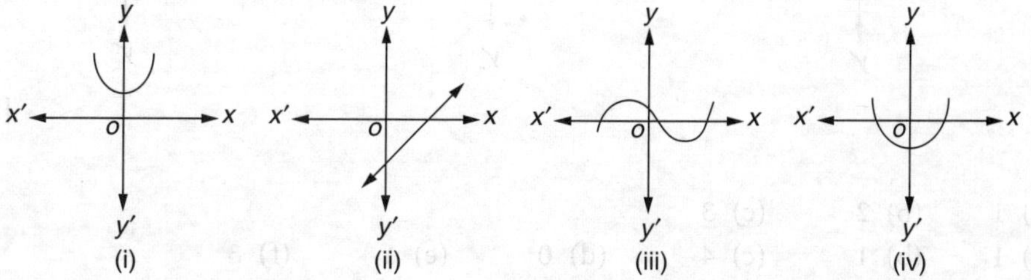

[NCERT]

Solution:

(i) Since the graph does not intersect the *x* axis, the polynomial has no zeros.

(ii) Number of zeros is 1, as the graph intersects the *x* axis at one point only.

(iii) As the graph intersects the *x* axis at three points, the polynomial has three zeros.

(iv) Number of zeros is 2, as the graph intersects the *x* axis at two points.

1 Try to Solve

1. Look at the graphs given below. Each is a graph of $y = p(x)$, where $p(x)$ is a polynomial. For each of the graphs, find the number of zeros of $p(x)$.

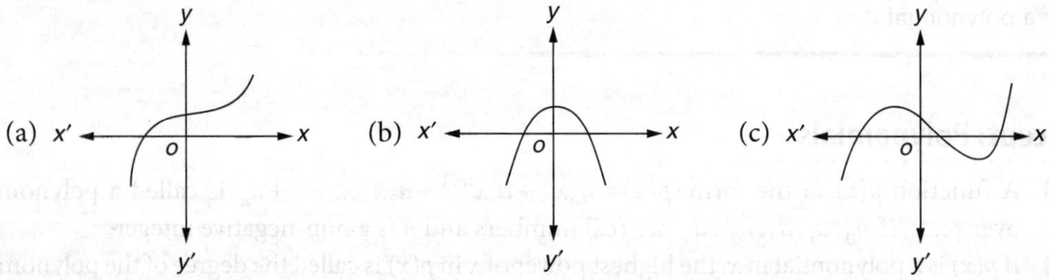

[NCERT]

2. The graphs of $y = p(x)$ are given below for some polynomial $p(x)$. Find the number of zeros of $p(x)$ in each case.

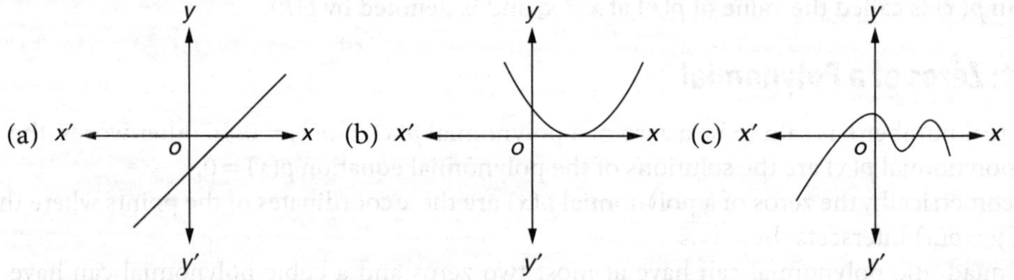

[NCERT]

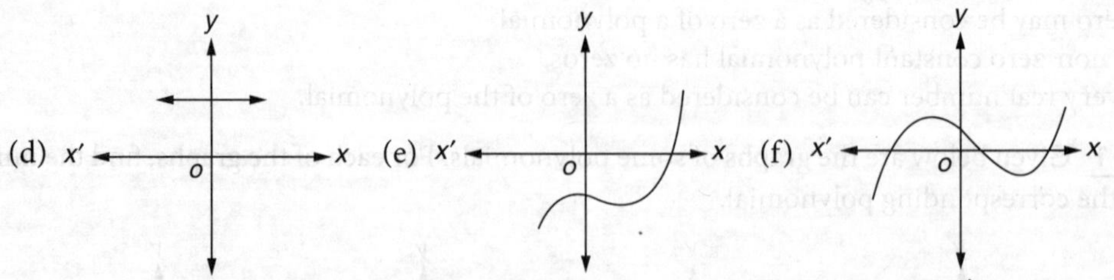

[NCERT]

Answers

1. (a) 1 (b) 2 (c) 3

2. (a) 1 (b) 1 (c) 4 (d) 0 (e) 1 (f) 3

Example 2 Find the zeros (root) of the polynomial in each of the following cases:

(i) $f(x) = x - 5$ (ii) $2x + 5$

Solution: For zeros of the polynomial $f(x) = 0$

(i) $f(x) = x - 5$, so $x - 5 = 0$, $x = 5$

$x = 5$ is the root of $f(x)$ as well as the zero of $f(x)$

(ii) $f(x) = 2x + 5$, and $2x + 5 = 0$, $2x = -5$

$x = -\frac{5}{2}$ is a root of $f(x) = 2x + 5$

2 Try to Solve

1. Find the zeros of the polynomials:

(a) $2x + 2$ (b) $-x + 5$ (c) $5x - 6$

Answers

1. (a) $x = -1$ (b) $x = 5$ (c) $x = \frac{6}{5}$

Example 3 Find the zeros of the quadratic polynomials:

(i) $x^2 + 5x + 6$ (ii) $x^2 - 2x - 8$

Solution: (i) Let

$$f(x) = x^2 + 5x + 6$$
$$= x^2 + (2 + 3)x + 6 = x^2 + 2x + 3x + 6$$
$$= x(x + 2) + 3(x + 2) = (x + 3)(x + 2)$$

Here $x + 3 = 0$ and $x + 2 = 0$

$$x = -3 \text{ and } x = -2$$

Hence, $f(x)$ has two zeros, $x = -3$ and $x = -2$.

(ii) Let

$$f(x) = x^2 - 2x - 8$$
$$= x^2 - (4 - 2)x - 8 = x^2 - 4x + 2x - 8$$
$$= x(x - 4) + 2(x - 4) = (x + 2)(x - 4)$$

Here $x + 2 = 0$ and $x - 4 = 0$.

$$x = -2 \text{ and } x = 4$$

Hence, $f(x)$ has two zeros, $x = -2$ and $x = 4$.

3 Try to Solve

1. Find the zeros of the quadratic polynomials:

(a) $4x^2 - 4x + 1$ (b) $4x^2 + 8x$ (c) $6x^2 - 3 - 7x$

(d) $x^2 + 7x + 10$ (e) $x^2 - 3$

Answers

1. (a) $x = \frac{1}{2}, \frac{1}{2}$ (b) $x = 0, -2$ (c) $x = -\frac{1}{3}, \frac{3}{2}$

(d) $-2, -5$ (e) $\sqrt{3}, -\sqrt{3}$

Concept: Relationship between Zeros and Coefficients of a Quadratic Polynomial

If α and β are the two zeros of the quadratic polynomial $p(x) = ax^2 + bx + c, a \neq 0$,

then $\alpha + \beta = -\frac{b}{a}$ and $\alpha\beta = \frac{c}{a}$ OR

$$\text{Sum of zeros} = \alpha + \beta = -\frac{b}{a} = \frac{-(\text{Coefficient of } x)}{\text{Coefficient of } x^2}$$

$$\text{Product of zeros} = \alpha\beta = \frac{c}{a} = \frac{\text{Constant term}}{\text{Coefficient of } x^2}$$

Example 4 Find the zeros of the following quadratic polynomials and verify the relationship between the zeros and the coefficients:

(i) $x^2 - 2x - 8$ **(ii)** $4s^2 - 4s + 1$ **(iii)** $4u^2 + 8u$ **[NCERT]**

Solution:

(i) Let $f(x) = x^2 - 2x - 8 = x^2 - 4x + 2x - 8$

$= x(x - 4) + 2(x - 4) = (x - 4)(x + 2)$

Now $f(x) = 0 \Rightarrow (x - 4)(x + 2) = 0.$

That is $x = 4$ or $x = -2$

$\therefore$ The zeros of $f(x) = 4$ and -2

$$\text{Sum of zeros} = 4 + (-2) = 2 = \frac{-(-2)}{1} = \frac{-(\text{Coefficient of } x)}{\text{Coefficient of } x^2}$$

$$\text{Product of zeros} = 4 \times (-2) = -8 = \frac{-8}{1} = \frac{\text{Constant term}}{\text{Coefficient of } x^2}$$

(ii) Let $f(s) = 4s^2 - 4s + 1 = 4s^2 - 2s - 2s + 1$

$= 2s(2s - 1) - 1(2s - 1)$

Now $f(s) = 0 \Rightarrow (2s - 1)(2s - 1) = 0.$

That is $s = 1/2, 1/2$

The zeros of $f(s)$ are $1/2, 1/2$.

$$\text{Sum of zeros} = \frac{1}{2} + \frac{1}{2} = 1 = \frac{-(-4)}{4} = \frac{-(\text{Coefficient of } s)}{\text{Coefficient of } s^2}$$

$$\text{Product of zeros} = \frac{1}{2} \times \frac{1}{2} = \frac{1}{4} = \frac{\text{Constant term}}{\text{Coefficient of } s^2} = \frac{-(\text{Coefficient of } s)}{\text{Coefficient of } s^2}$$

(iii) Let $f(u) = 4u^2 + 8u = 4u(u + 2)$

Now $f(u) = 0 \Rightarrow 4u(u + 2) = 0.$

That is $u = 0$ or $u = -2$

$\therefore$ The zeros of $f(u)$ are 0 and -2.

$$\text{Sum of zeros} = 0 + (-2) = -2 = \frac{-(8)}{4} = \frac{-(\text{Coefficient of } u)}{\text{Coefficient of } u^2}$$

$$\text{Product of zeros} = 0 \times (-2) = 0 = \frac{0}{4} = \frac{\text{Constant term}}{\text{Coefficient of } u^2}$$

4 Try to Solve

1. Find the zeros of the following quadratic polynomials and verify the relationship between the zeros and the coefficients:
   (a) $6x^2 - 3 - 7x$ (b) $t^2 - 15$ (c) $3x^2 - x - 4$ [NCERT]

Answers

1. (a) $-\frac{1}{3}, \frac{3}{2}$ (b) $\pm\sqrt{15}$ (c) $-1, \frac{4}{3}$

Concept: Formation of Quadratic Polynomial

(i) Formula: $k(x^2 - Sx + P)$
where S = sum of roots = $\alpha + \beta$ and P = product of roots = $\alpha \times \beta$ (where α and β are the zeros of the polynomial or values of 'x').

Example 5 Form a quadratic polynomial whose roots are –2 and 3.

Solution: We know that the quadratic polynomial is $k(x^2 - Sx + P)$.
Here, $\alpha = -2$, and $\beta = 3$.

$$S = \alpha + \beta = -2 + (3) = -2 + 3 = 1$$
$$p = \alpha \times \beta = (-2)(3) = -6$$

So, $k(x^2 - 1x - 6)$ is the required quadratic polynomial.

5 Try to Solve

1. Form a quadratic polynomial whose roots are 5 and 6.
2. Form a quadratic polynomial whose roots are – 4 and 5.
3. Form a quadratic polynomial whose roots are $\frac{1}{4}$ and 1.

Answers

1. $k(x^2 - 11x + 30)$ 2. $k(x^2 - 1x - 20)$ 3. $k(4x^2 - 5x + 1)$

Example 6 Form a quadratic polynomial with $\frac{1}{4}, -1$ as the sum and product of its zeros, respectively. [NCERT, CBSE (CCE) 2015]

Solution: Let $ax^2 + bx + c$ be the polynomial and α and β be its zeros.

$$\therefore \quad \alpha + \beta = \frac{1}{4} \text{ and } \alpha\beta = -1$$
$$P(x) = k(x^2 - Sx + P)$$
$$P(x) = k\left[x^2 - \frac{1}{4}x - 1\right] = k\left[\frac{4x^2 - x - 4}{4}\right]$$
$$= \frac{k}{4}(4x^2 - x - 4)$$

$\therefore$ The required polynomial is $k(4x^2 - x - 4)$.

6 Try to Solve

1. Find a quadratic polynomial each, with the given numbers as the sum and product of its zeros, respectively:

(a) $\sqrt{2}, \frac{1}{3}$ (b) $0, \sqrt{5}$ (c) $1, 1$ (d) $-\frac{1}{4}, \frac{1}{4}$ (e) $4, 1$ **[NCERT]**

Answers

1. (a) $3x^2 - 3\sqrt{2}\,x + 1$ (b) $x^2 + \sqrt{5}$ (c) $x^2 - x + 1$
(d) $4x^2 + x + 1$ (e) $x^2 - 4x + 1$

Concept: Division in Polynomials

Example 7 Divide $2x^2 + 3x + 1$ by $(x + 2)$. **[NCERT]**

Solution:

$$
\begin{array}{r|l}
x+2 & 2x^2 + 3x + 1 \;(\,2x - 1 \\
 & \cancel{2x^2} + 4x \\
 & \;\;-\quad\;\; - \\
\hline
 & -\cancel{x} + 1 \\
 & -\cancel{x} - 2 \\
 & \;\;+\quad\;\; + \\
\hline
 & \qquad 3
\end{array}
$$

Here, remainder = 3 and quotient = $2x - 1$.

Concept: Division Algorithm

Dividend = Divisor × Quotient + Remainder

For example, $\frac{10}{3} = 10 \div 3$.

$$
\begin{array}{rr|l}
 & & 3 \longrightarrow \text{Quotient} \\
\text{Divisor} \longleftarrow & 3 & 10 \longrightarrow \text{Dividend} \\
 & & \underline{-9} \\
 & & 1 \longrightarrow \text{Remainder}
\end{array}
$$

Example 8 Divide $x^4 - 3x^2 + 2x + 1$ by $x - 1$.

Solution:

$$
\begin{array}{r|l}
 & x^3 + x^2 - 2x \\
x-1 & \cancel{x^4} - 0x^3 - 3x^2 + 2x + 1 \;(\\
 & \cancel{x^4} - x^3 \\
 & -\quad + \\
\hline
 & \cancel{x^3} - 3x^2 + 2x + 1 \\
 & \cancel{x^3} - x^2 \\
 & -\quad + \\
\hline
 & -2\cancel{x^2} + 2\cancel{x} + 1 \\
 & -2\cancel{x^2} + 2\cancel{x} \\
 & +\qquad - \\
\hline
 & \qquad\qquad 1
\end{array}
$$

Hence, remainder = 1; quotient = $x^3 + x^2 - 2x$

7 Try to Solve

1. Divide the polynomial $p(x)$ by the polynomial $g(x)$ and find the quotient and remainder in each of the following:
   (a) $p(x) = x^3 - 3x^2 + 5x - 3,\ g(x) = x^2 - 2$ **[NCERT]**
   (b) $p(x) = x^4 - 3x^2 + 4x + 5,\ g(x) = x^2 + 1 - x$
   (c) $p(x) = x^4 - 5x + 6,\ g(x) = 2 - x^2$
2. Divide $3x^3 + x^2 + 2x + 5$ by $1 + 2x + x^2$. **[NCERT]**

Answers

1. (a) Quotient $= x - 3$ and remainder $= 7x - 9$
   (b) Quotient $= x^2 + x - 3$ and remainder $= 8$
   (c) Quotient $= -x^2 - 2$, remainder $= -5x + 10$
2. Quotient $= 3x - 5$, remainder $= 9x + 10$

Example 9 Find all the zeros of the polynomial $P(x) = 2x^3 + 3x^2 - 11x - 6$, if one of its zeros is -3. **[CBSE (CCE) 2011]**

Solution: We have

$P(x) = 2x^3 + 3x^2 - 11x - 6$ and one zero is -3.

As -3 is a zero of $P(x)$, $(x + 3)$ is a factor of $P(x)$.

Now, we divide $P(x)$ by $(x + 3)$ to find the other zeros of $P(x)$.

$$\begin{array}{r|l} & 2x^2 - 3x - 2 \\ \hline x+3 & 2x^3 + 3x^2 - 11x - 6 \\ & 2x^3 + 6x^2 \\ & - \quad - \\ \hline & -3x^2 - 11x - 6 \\ & -3x^2 - 9x \\ & + \quad + \\ \hline & -2x - 6 \\ & -2x - 6 \\ & + \quad + \\ \hline & \times \end{array}$$

$$\frac{2x^3}{x} = 2x^2, \quad \frac{-3x^2}{x} = -3x, \quad \frac{-2x}{x} = -2$$

By division algorithm, we have

$$\begin{aligned} 2x^3 + 3x^2 - 11x - 6 &= (x + 3)(2x^2 - 3x - 2) \\ &= (x + 3)(2x^2 - 4x + x - 2) \\ &= (x + 3)\{2x(x - 2) + 1(x - 2)\} \\ &= (x + 3)(x - 2)(2x + 1) \end{aligned}$$

Hence, the zeros of $P(x)$ are -3, 2 and $-1/2$.

8 Try to Solve

1. Obtain all the zeros of $f(x) = x^3 + 13x^2 + 32x + 20$, if one of its zeros is -2.
2. Find all the zeros of the polynomial $3x^3 + 10x^2 - 9x - 4$, if one of its zeros is 1. **[Delhi 2019]**

Answers

1. $-10, -1, -2$
2. -4 and $-\dfrac{1}{3}$

Example 10 If -1 and 2 are two zeros of the polynomial $2x^3 - x^2 - 5x - 2$, find its third zero.

[CBSE (CCE 2013)]

Solution: -1 and 2 are zeros of $p(x) = 2x^3 - x^2 - 5x - 2s$

$\therefore \{x - (-1)\}(x - 2) = (x + 1)(x - 2) = x^2 - 2x + x - 2 = x^2 - x - 2$ is a factor of $p(x)$

For the other zeros $\quad 2x + 1 = 0 \Rightarrow x = \dfrac{-1}{2}$

$\therefore \quad$ Other zero $= \dfrac{-1}{2}$.

$$\begin{array}{r|l} & 2x + 1 \\ x^2 - x - 2 & 2x^3 - x^2 - 5x - 2 \\ & \underline{\overset{-}{2x^3} \overset{+}{-\,2x^2} \overset{+}{-\,4x}} \\ & x^2 - x - 2 \\ & \underline{\overset{-}{x^2} \overset{+}{-\,x} \overset{+}{-\,2}} \\ & \text{X} \end{array}$$

Example 11 Find all the zeros of the polynomial $2x^3 + x^2 - 6x - 3$, if two of its zeros are $-\sqrt{3}$ and $\sqrt{3}$.

Solution: $-\sqrt{3}$ and $\sqrt{3}$ are zeros of $p(x) = 2x^3 + x^2 - 6x - 3$

$\therefore \quad \{x - (-\sqrt{3})\}(x - \sqrt{3}) = (x + \sqrt{3})(x - \sqrt{3}) = x^2 - 3$ is a factor of $p(x)$

For the other factor:

For the other zeros, $2x + 1 = 0 \Rightarrow x = \dfrac{-1}{2}$.

The zeros of the polynomial are $-\sqrt{3}$, $\sqrt{3}$ and $\dfrac{-1}{2}$.

$$\begin{array}{r|l} & 2x^2 + 1 \\ x^2 - 3 & 2x^3 + x^2 + 6x - 3 \\ & \underline{\overset{-}{2x^3} \quad \overset{-}{-\,6x}} \\ & x^2 - 3 \\ & \underline{x^2 - 3} \\ & 0 \end{array}$$

9 Try to Solve

1. Find all the zeros of the polynomial $f(x) = x^4 + x^3 - 34x^2 - 4x + 120$, if two of its zeros are 2 and -2.
2. Given that $\sqrt{2}$ is the zero of a cubic polynomial $6x^3 + \sqrt{2}\,x^2 - 10\,x - 4\sqrt{2}$, find its other two zeros. **[NCERT Exemplar]**

Answers

1. $-6, 5$
2. $\sqrt{2}, -\dfrac{1}{\sqrt{2}}$ and $\dfrac{-4}{3\sqrt{2}}$

Example 12 Obtain all the zeros of the polynomial $p(x) = x^4 - 3x^3 - x^2 + 9x - 6$, if two zeros are $-\sqrt{3}$ and $\sqrt{3}$. **[CBSE (CCE) 2011]**

Solution: As $-\sqrt{3}$ and $\sqrt{3}$ are the zeros of $P(x)$, $(x+\sqrt{3})$ and $(x-\sqrt{3})$ are the factors of $P(x)$.

$\therefore$ $$g(x) = (x+\sqrt{3})(x-\sqrt{3}) = x^2 - 3 \quad [\text{Using } a^2 - b^2 = (a-b)(a+b)]$$

On dividing $P(x)$ by $g(x)$, we get

$$\begin{array}{r|l} x^2-3 & \overline{x^4 - 3x^3 - x^2 + 9x - 6}\ \ (x^2 - 3x + 2 \\ & \underline{\overset{-}{x^4} \quad\quad \overset{+}{-3x^2}} \\ & -3x^3 + 2x^2 + 9x - 6 \\ & \underline{\overset{+}{-3x^3} \quad\quad \overset{-}{+9x}} \\ & 2x^2 - 6 \\ & \underline{\overset{-}{2x^2} \overset{+}{-6}} \\ & \times \end{array}$$

$$\frac{x^4}{x^2} = x^2, \quad \frac{-3x^3}{x^2} = -3x, \quad \frac{2x^2}{x^2} = 2$$

By division algorithm, we have

$$x^4 - 3x^3 - x^2 - 9x - 6 = (x^2-3)(x^2 - 3x + 2)$$
$$= (x-\sqrt{3})(x+\sqrt{3})(x^2 - 2x - x + 2)$$
$$= (x-\sqrt{3})(x+\sqrt{3})\{x(x-2) - 1(x-2)\}$$
$$= (x-\sqrt{3})(x+\sqrt{3})(x-1)(x-2)$$

$\therefore$ The zeros of $P(x)$ are: $\sqrt{3}, -\sqrt{3}$, 1 and 2.

10 Try to Solve

1. Obtain all the zeros of the polynomial $f(x) = 2x^4 + x^3 - 14x^2 - 19x - 6$, if two of its zeros are -2 and -1.
2. Find all the zeros of $2x^4 - 3x^3 - 3x^2 + 6x - 2$, if you know that two of its zeros are $\sqrt{2}$ and $-\sqrt{2}$. **[NCERT, CBSE (CCE) 2012]**

Answers

1. $-\frac{1}{2}, 3, -2, -1$ **2.** $1, \frac{1}{2}$

Example 13 Obtain all the zeros of $3x^4 + 6x^3 - 2x^2 - 10x - 5$, if two of its zeros are $\sqrt{\frac{5}{3}}$ and $-\sqrt{\frac{5}{3}}$. **[NCERT]**

Solution: Since $\sqrt{\frac{5}{3}}$ and $-\sqrt{\frac{5}{3}}$ are two zeros of $p(x)$.

$$\therefore \quad \left(x - \sqrt{\frac{5}{3}}\right)\left(x + \sqrt{\frac{5}{3}}\right) = \left(x^2 - \frac{5}{3}\right) = \frac{1}{3}(3x^2 - 5)$$

$\therefore$ $(3x^2 - 5)$ is a factor of $3x^4 + 6x^3 - 2x^2 - 10x - 5$.

Now, by division algorithm:

For the other zeros $\quad x^2 + 2x + 1 = 0$

$\Rightarrow \quad (x + 1)(x + 1) = 0$

$\Rightarrow \quad x = -1, x = -1$

$\therefore$ The zeros are $\sqrt{\frac{5}{3}}, -\sqrt{\frac{5}{3}}, -1$ and -1.

$$
\begin{array}{r}
x^2 + 2x + 3 \\
3x^2 - 5 \;\big)\; 3x^4 + 6x^3 - 2x^2 - 10x - 5 \\
\underset{-}{3x^4} \qquad \underset{+}{-5x^2} \qquad\qquad \\
\hline
6x^3 + 3x^2 - 10x - 5 \\
\underset{-}{6x^3} \qquad \underset{+}{-10x} \qquad \\
\hline
3x^2 - 5 \\
\underset{-}{3x^2} \underset{+}{- 5} \\
\hline
\text{X}
\end{array}
$$

11 Try to Solve

1. Find all the zeros of the polynomial $f(x) = 2x^4 - 2x^3 - 7x^2 + 3x + 6$, if two of its zeros are $-\sqrt{\frac{3}{2}}$ and $\sqrt{\frac{3}{2}}$. **[CBSE (CCE) 2010]**
2. Obtain all the other zeros of the polynomial $x^4 - 3\sqrt{2}\,x^3 + 3x^2 + 3\sqrt{2}\,x - 4$, if two of its zeros are $\sqrt{2}$ and $2\sqrt{2}$.

Answers

1. $2, -1, \sqrt{\frac{3}{2}}, -\sqrt{\frac{3}{2}}$ **2.** $1, -1$

Concept: Proof of Relationship between the Zeros and Coefficients of Quadratic and Cubic Polynomials

(i) Relationship between the zeros and coefficients of a quadratic polynomial.

Let α and β be the zeros of a quadratic polynomial.

$$p(x) = ax^2 + bx + c, \text{ where } a \neq 0$$

Say, $(x - \alpha)$ and $(x - \beta)$ are the factors of $p(x)$.

$$\therefore \quad ax^2 + bx + c = k(x - \alpha)(x - \beta), \text{ where } k \text{ is some constant}$$

$$ax^2 + bx + c = k[x^2 - (\alpha + \beta)x + \alpha\beta]$$

$$ax^2 + bx + c = kx^2 - k(\alpha + \beta)x + k(\alpha\beta)$$

On comparing both sides, we get

$$k = a, -k(\alpha + \beta) = b \quad \text{and} \quad k(\alpha\beta) = c$$

$$\Rightarrow \quad -a(\alpha + \beta) = b \quad \text{and} \quad a(\alpha\beta) = c$$

$$\Rightarrow \quad \alpha + \beta = -\frac{b}{a} \quad \text{and} \quad \alpha\beta = \frac{c}{a}$$

$$\therefore \quad \text{Sum of zeros} = \frac{-(\text{Coefficient of } x)}{(\text{Coefficient of } x^2)}$$

$$\text{Product of zeros} = \frac{(\text{Constant term})}{(\text{Coefficient of } x^2)}$$

(ii) Relationship between the zeros and coefficient of a cubic polynomial.

Let $p(x) = ax^3 + bx^2 + cx + d$, where $a \neq 0$ be a cubic polynomial whose zeros are α, β and γ. It means $(x - \alpha)$, $(x - \beta)$ and $(x - \gamma)$ are the factors of $p(x)$.

$\therefore \quad (ax^3 + bx^2 + cx + d) = k(x - \alpha)(x - \beta)(x - \gamma)$ for some constant k.

$$= k[x^3 - (\alpha + \beta + \gamma)x^2 + (\alpha\beta + \beta\gamma + \gamma\alpha)x - (\alpha\beta\gamma)]$$

$$ax^3 + bx^2 + cx + d = kx^3 - k(\alpha + \beta + \gamma)x^2 + k(\alpha\beta + \beta\gamma + \gamma\alpha)x - k(\alpha\beta\gamma)$$

On comparing both sides, we get

$$k = a, -k(\alpha + \beta + \gamma) = b, k(\alpha\beta + \beta\gamma + \gamma\alpha) = c, -k(\alpha\beta\gamma) = d$$

$$\Rightarrow \quad -a(\alpha + \beta + \gamma) = b, a(\alpha\beta + \beta\gamma + \gamma\alpha) = \frac{c}{a}, -a(\alpha\beta\gamma) = d$$

$$\Rightarrow \quad \alpha + \beta + \gamma = -\frac{b}{a}, \alpha\beta + \beta\gamma + \gamma\alpha = \frac{c}{a}, \alpha\beta\gamma = -\frac{d}{a}$$

FOR MORE PRACTICE AND KNOWLEDGE

Example 14 The graph of $x = f(y)$ is given below. Find the zeros and number of zeros of $f(y)$.

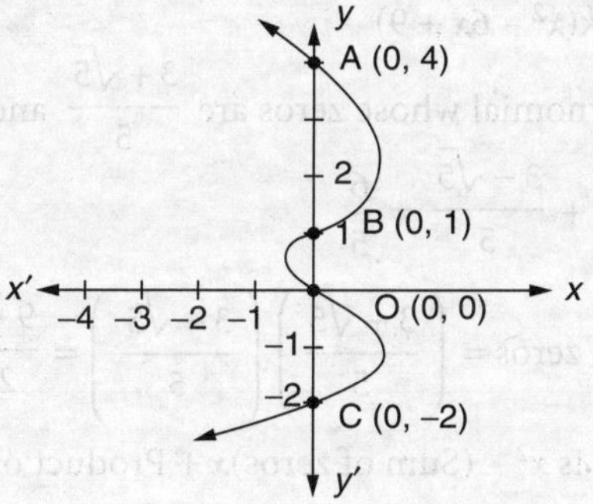

Solution: The graph of $x = f(y)$ intersects the y axis at $A(0, 4)$, $B(0, 1)$, $O(0, 0)$ and $C(0, -2)$.

$\therefore$ The zeros are $y = 4, 1, 0$ and -2 and the number of zeros is four.

Example 15 k is a zero of the polynomial $p(x) = x^2 - 11x + 24$. If k is a prime number, then find the value of k.

Solution: $\because$ k is a zero of $p(x) = x^2 - 11x + 24$

$$\therefore \quad k^2 - 11k + 24 = 0$$

$$k^2 - 8k - 3k + 24 = 0$$

$$\Rightarrow \quad (k - 8)(k - 3) = 0$$

$$\Rightarrow \quad k = 3, 8$$

$\because$ k is a prime number.

$$\therefore \quad k = 3$$

Example 16 Form a quadratic polynomial, one of whose zeros is $\sqrt{5}$ and where the product of the zeros is $-2\sqrt{5}$.

Solution: One zero (α) = product of zeros, $P = \alpha\beta = -2\sqrt{5}$.

$$\therefore \text{ Other zero } (\beta) = \frac{-2\sqrt{5}}{\alpha} = \frac{-2\sqrt{5}}{\sqrt{5}} = -2$$

∴ Sum of zeros $(\alpha+\beta)$ $S = -2+\sqrt{5}$

Quadratic polynomial $p(x) = x^2 - (S)x + P$

$$= k[x^2 - (-2+\sqrt{5})x + (-2\sqrt{5})] = k[x^2 + (2-\sqrt{5})x - 2\sqrt{5}]$$

Example 17 If α and β are zeros of a polynomial $x^2 + 6x + 9$, form a polynomial whose zeros are $-\alpha$ and $-\beta$. **[CBSE 2014]**

Solution: Let $p(x) = x^2 + 6x + 9$. ∵ α, β are the zeros of $p(x)$.

So, the sum of zeros, $\alpha+\beta = -6$ and the product of zeros, $\alpha\beta = 9$.

Now, $-\alpha$ and $-\beta$ are the zeros of the required quadratic polynomial.

So, the sum of zeros of the required polynomial $= -\alpha-\beta = -(\alpha+\beta) = 6$.

The product of zeros of the required polynomial $= (-\alpha)(-\beta) = \alpha\beta = 9$.

∴ Required quadratic polynomial is given by

$$q(x) = k[x^2 - (\text{Sum of zeros})x + \text{Product of zeros}]$$
$$= k(x^2 - 6x + 9)$$

Example 18 Find a quadratic polynomial whose zeros are $\frac{3+\sqrt{5}}{5}$ and $\frac{3-\sqrt{5}}{5}$. **[CBSE 2013]**

Solution: Sum of zeros $= \frac{3+\sqrt{5}}{5} + \frac{3-\sqrt{5}}{5} = \frac{6}{5}$

$$\text{Product of zeros} = \left(\frac{3+\sqrt{5}}{5}\right)\left(\frac{3-\sqrt{5}}{5}\right) = \frac{9-5}{25} = \frac{4}{25}$$

∴ Required quadratic polynomial is x^2 – (Sum of zeros)x + Product of zeros

$$k\left(x^2 - \frac{6}{5}x + \frac{4}{25}\right) \text{ or } k(25x^2 - 30x + 4)$$

Example 19 If the sum of the zeros of the quadratic polynomial $ky^2 + 2y - 3k$ is equal to twice their product, find the value of k.

Solution: $p(y) = ky^2 + 2y - 3k$.

$$a = k,\ b = 2,\ c = -3k$$

According to the question Sum of zeros = 2 × Product of zeros

⇒ $\frac{-b}{a} = 2 \times \frac{c}{a}$

⇒ $\frac{-2}{k} = 2 \times \frac{-3k}{k}$

$$-2 = -6k$$

$$k = \frac{1}{3}$$

Example 20 If one zero of the polynomial $(k+1)x^2 - 5x + 5$ is the multiplicative inverse of the other, then find the zeros of $kx^2 - 3kx + 9$, where k is constant.

Solution: Here $f(x) = (k+1)x^2 - 5x + 5$

$$a = k+1,\ b = -5,\ c = 5$$

Let one zero = α. $\therefore$ According to the question, other zero = $\frac{1}{\alpha}$. Now, the product of zeros = $\frac{c}{a}$.

$$\Rightarrow \qquad \alpha \times \frac{1}{\alpha} = \frac{5}{k+1}$$

$$\Rightarrow \qquad 1 = \frac{5}{k+1} \Rightarrow k+1 = 5 \Rightarrow k = 4$$

Now, inserting $k = 4$ in the polynomial $p(x) = kx^2 - 3kx + 9$, we get

$$p(x) = 4x^2 - 12x + 9$$

For zeros of $p(x)$ $\qquad 4x^2 - 12x + 9 = 0$

$$\Rightarrow \qquad (2x-3)(2x-3) = 0$$

$$\Rightarrow \qquad x = \frac{3}{2}, x = \frac{3}{2}$$

Hence, the zeros are $\frac{3}{2}, \frac{3}{2}$.

Example 21 If the product of the zeros of the polynomial $kx^2 + 41x + 42$ is 7, find the zeros of the polynomial $(k-4)x^2 + (k+1)x + 5$.

Solution: Here $\qquad f(x) = kx^2 + 41x + 42$

$$a = k, b = 41, c = 42$$

According to the question $\qquad$ Product of zeros = 7

$$\Rightarrow \qquad \frac{c}{a} = 7$$

$$\Rightarrow \qquad \frac{42}{k} = 7 \Rightarrow 42 = 7k \Rightarrow k = 6$$

Inserting $k = 6$ in the polynomial $p(x) = (k-4)x^2 + (k+1)x + 5$, we get

$$p(x) = (6-4)x^2 + (6+1)x + 5$$

$\Rightarrow$ For zeros of $f(x)$ $\qquad 2x^2 + 7x + 5 = 0$

$$2x^2 + 5x + 2x + 5 = 0$$

$$\Rightarrow \qquad x(2x+5) + 1(2x+5) = 0$$

$$\Rightarrow \qquad (x+1)(2x+5) = 0$$

$$x = -1, x = -\frac{5}{2}$$

$\therefore$ The zeros are $-1, -\frac{5}{2}$.

Example 22 If α, β are zeros of the polynomial $p(x) = 5x^2 + 5x + 1$, find the value of (i) $\alpha^2 + \beta^2$ and (ii) $\alpha^{-1} + \beta^{-1}$.

Solution: The given polynomial is $p(x) = 5x^2 + 5x + 1$. Hence, $a = 5, b = 5$ $c = 1$.

(i) Now, $\alpha^2 + \beta^2 = (\alpha + \beta)^2 - 2\alpha\beta$ $\qquad [\because (\alpha+\beta)^2 = a^2 + b^2 + 2a\beta \Rightarrow (\alpha+\beta)^2 - 2a\beta = a^2 + b^2]$

$\Rightarrow \alpha^2 + \beta^2 = \left(\frac{-b}{a}\right)^2 - 2\frac{c}{a}$ $\qquad [\because$ Sum of zeros, that is, $\alpha + \beta = \frac{-b}{a}$, and the product of zeros, that is, $\alpha\beta = \frac{c}{a}$]

$$\Rightarrow a^2 + b^2 = \left[\frac{-5}{5}\right]^2 - 2 \times \frac{1}{5} = 1 - \frac{2}{5} = \frac{3}{5}$$

(ii) $\alpha^{-1} + \beta^{-1} = \frac{1}{\alpha} + \frac{1}{\beta} = \frac{\beta + \alpha}{\alpha\beta} = \frac{-b/a}{c/a} = \frac{-5/5}{1/5} = -5.$

Example 23 α, β, γ are the zeros of the cubic polynomial $kx^3 - 5x + 9$. If $\alpha^3 + \beta^3 + \gamma^3 = 27$, find the value of k.

Solution: Here, $a = k, b = 0, c = -5, d = 9$.

$$\alpha + \beta + \gamma = -\frac{b}{a} = 0 \text{ and } \alpha\beta\gamma = -\frac{d}{a} = -\frac{9}{k}$$

$\because$ $$\alpha + \beta + \gamma = 0$$

$\therefore$ $$\alpha^3 + \beta^3 + \gamma^3 = 3\alpha\beta\gamma$$

$\Rightarrow$ $$27 = 3 \times \left(\frac{-9}{k}\right) \Rightarrow k = -1$$

Example 24 Give examples of polynomials $p(x)$, $g(x)$, $q(x)$ and $r(x)$, which satisfy the division algorithm and

(i) $\deg p(x) = \deg q(x)$ (ii) $\deg q(x) = \deg r(x)$ (iii) $\deg r(x) = 0$ [NCERT]

Solution: (i) $p(x), g(x), q(x), r(x)$

$$\deg p(x) = \deg q(x)$$

$\therefore$ Both $g(x)$ and $r(x)$ are constant terms.

$$p(x) = 2x^2 - 2x + 14$$
$$g(x) = 2$$
$$q(x) = x^2 - x + 7$$
$$r(x) = 0$$

(ii) $\deg q(x) = \deg r(x)$

$\therefore$ This is possible when the degree of both $q(x)$ and $r(x)$ is less than $p(x)$ and $g(x)$.

$$p(x) = x^3 + x^2 + x + 1$$
$$g(x) = x^2 - 1$$
$$q(x) = x + 1, r(x) = 2x + 2$$

(iii) $\deg r(x)$ is 0.

This is possible when the product of $q(x)$ and $g(x)$ forms a polynomial whose degree is equal to the degree of $p(x)$ and the constant term.

Example 25 If the polynomial $6x^4 + 8x^3 + 17x^2 + 21x + 7$ is divided by another polynomial $3x^2 + 4x + 1$, the remainder is $(ax + b)$. Find a and b.

Solution:

$$\begin{array}{r|l} & 2x^2 + 5 \\ \hline 3x^2 + 4x + 1 & 6x^4 + 8x^3 + 17x^2 + 21x + 7 \\ & 6x^4 + 8x^3 + 2x^2 \\ & - \quad - \quad - \\ \hline & 15x^2 + 21x + 7 \\ & 15x^2 + 20x + 5 \\ & - \quad - \quad - \\ \hline & x + 2 \end{array}$$

But, the given remainder is $ax + b$. Therefore, $ax + b = x + 2 \Rightarrow a = 1, b = 2$.

Example 26 If 1 and – 1 are zeros of the polynomial $Lx^4 + Mx^3 + Nx^2 + Rx + P = 0$, show that $L + N + P = M + R = 0$.

Solution: $p(1) = 0$

$\Rightarrow$ $L(1)^4 + M(1)^3 + N(1)^2 + R(1) + P = 0$

$\Rightarrow$ $L + M + N + R + P = 0$ ...(i)

Also $p(-1) = 0$

$\Rightarrow$ $L - M + N - R + P = 0$

$\Rightarrow$ $L + N + P = M + R$ ...(ii)

From (i) and (ii) $2(M + R) = 0 \Rightarrow M + R = 0$

From (ii) $L + N + P = 0$

Hence $L + N + P = M + R = 0$.

Example 27 If the polynomial $x^4 - 6x^3 + 16x^2 - 25x + 10$ is divided by $(x^2 - 2x + k)$, the remainder is $x + a$. Find k and a.

Solution: $P(x) = x^4 - 6x^3 + 16x^2 - 25x + 10, g(x) = x^2 - 2x + k$

$r(x) = x + a$. let $q(x) = px^2 + qx + r$

$\because$ $p(x) = g(x)\, q(x) + r(x)$

$\therefore$ $x^4 - 6x^3 + 16x^2 - 25x + 10 = (x^2 - 2x + k)\,(px^2 + qx + r) + (x + a)$

$\Rightarrow$ $p(x) = px^4 + qx^3 + rx^2 - 2px^3 - 2qx^2 - 2rx + kpx^2 + kqx + kr + x + a.$

$\Rightarrow$ $P(x) = px^4 + (q - 2p)x^3 + (r - 2q + kp)x^2 + (kq - 2r + 1)x + kr + a$

$\Rightarrow$ $x^4 - 6x^3 + 16x^2 - 25x + 10 = px^4 + (q - 2p)x^3 + (r - 2q + kp)\, x^2 + (kq - 2r + 1)x + (kr + a)$

Equating the coefficients, we get $p = 1, q - 2p = -6 \Rightarrow q - 2 \times 1 = -6 \Rightarrow q = -4$

Also $kq - 2r + 1 = -25$

$\Rightarrow$ $k \times (-4) - 2$

$r = -26 \Rightarrow -4k - 2r = -26$

$\Rightarrow$ $2k + r = 13$ (i)

Using $r - 2q + kp = 16$

$\Rightarrow$ $r - 2 \times (-4) + k \times 1 = 16 \Rightarrow r + k = 8$ (ii)

From (i) and (ii), we get $r = 3, k = 5$

$\Rightarrow$ $kr + a = 10 \Rightarrow 5 \times 3 + a = 10 \Rightarrow a = -5$

Multiple Choice Questions

Choose the correct answer for each of the following questions:

1. Given that two of the zeros of the cubic polynomial $ax^3 + bx^2 + cx + d$ are 0, the third zero is

(a) $\frac{-b}{a}$ (b) $\frac{b}{a}$ (c) $\frac{c}{a}$ (d) $-\frac{d}{a}$ [NCERT Exemplar]

2. If one of the zeros of the quadratic polynomial $(k - 1)x^2 + kx + 1$ is -3, then the value of k is

(a) $\frac{4}{3}$ (b) $\frac{-4}{3}$ (c) $\frac{2}{3}$ (d) $\frac{-2}{3}$ **[NCERT Exemplar]**

3. If the zeros of the quadratic polynomial $x^2 + (a + 1)x + b$ are 2 and -3, then

(a) $a = -7, b = -1$ (b) $a = 5, b = -1$ (c) $a = 2, b = -6$ (d) $a = 0, b = -6$

[NCERT Exemplar]

4. Which of the following is not a graph of a quadratic polynomial?

(a) 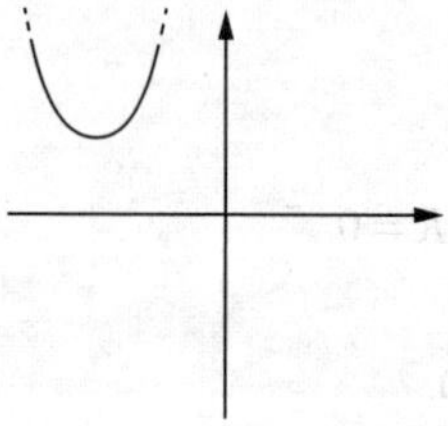

(b) 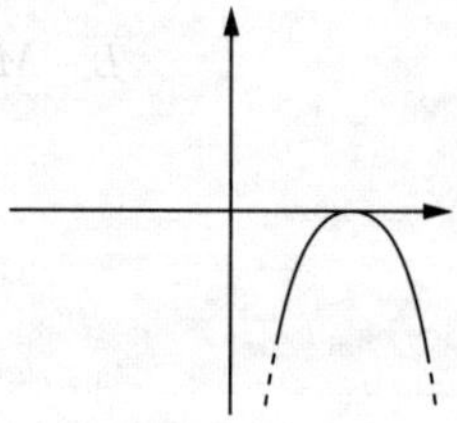

(c) 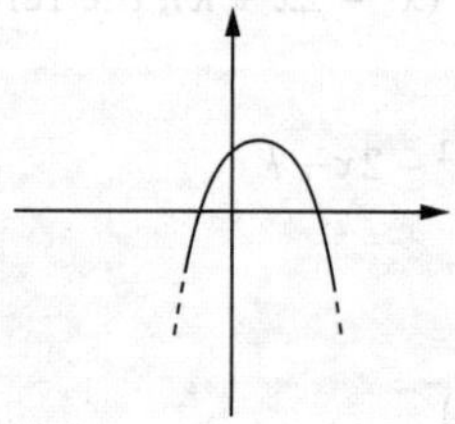

(d) 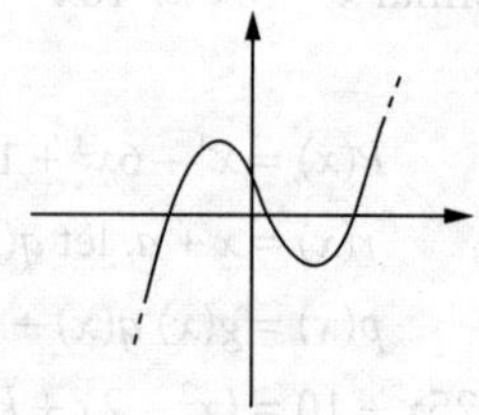

[NCERT Exemplar]

5. The zeros of the quadratic polynomial $x^2 + 99x + 127$ are

(a) Both positive (b) Both negative

(c) One positive and one negative (d) Both equal **[NCERT Exemplar]**

6. If one of the zeros of the cubic polynomial $x^3 + ax^2 + bx + c$ is -1, then the product of the other two zeros is

(a) $b - a + 1$ (b) $b - a - 1$ (c) $a - b + 1$ (d) $a - b - 1$

[NCERT Exemplar]

7. Which of the following is a polynomial?

(a) $x^2 - 5x + 4\sqrt{x} + 3$ (b) $x^{3/2} - x + x^{1/2} + 1$ (c) $\sqrt{x} + \frac{1}{\sqrt{x}}$ (d) $\sqrt{2}x^2 - 3\sqrt{3}x + \sqrt{6}$

8. The zeros of the polynomial $x^2 + \frac{1}{6}x - 2$ are

(a) $-3, 4$ (b) $\frac{-3}{2}, \frac{4}{3}$ (c) $\frac{-4}{3}, \frac{3}{2}$ (d) None of these

9. The zeros of the quadratic polynomial $x^2 + 88x + 125$ are

(a) Both positive (b) Both negative

(c) One positive and one negative (d) Both equal

10. If -2 and 3 are the zeros of the quadratic polynomial $x^2 + (a + 1)x + b$, then

(a) $a = -2, b = 6$ (b) $a = 2, b = -6$ (c) $a = -2, b = -6$ (d) $a = 2, b = 6$

11. If the sum of the zeros of the quadratic polynomial $kx^2 + 2x + 3k$ is equal to the product of its zeros, then $k = ?$

(a) $\frac{1}{3}$ (b) $\frac{-1}{3}$ (c) $\frac{2}{3}$ (d) $\frac{-2}{3}$

12. If α, β, γ are the zeros of the polynomial $x^3 - 6x^2 - x + 30$, then $(a\beta + a\gamma + g\alpha) = ?$

(a) -1 (b) 1 (c) -5 (d) 30

13. If α, β, γ are the zeros of the polynomial $2x^3 + x^2 - 13x + 6$, then $ab\gamma = ?$

(a) -3 (b) 3 (c) $\frac{-1}{2}$ (d) $\frac{-13}{2}$

14. If α, β, γ are the zeros of the polynomial $p(x)$ such that $(\alpha + \beta + \gamma) = 3$, $(a\beta + b\gamma + g\alpha) = -10$ and $ab\gamma = -24$, then $p(x) = ?$

(a) $x^3 + 3x^2 - 10x + 24$ (b) $x^3 + 3x^2 + 10x - 24$

(c) $x^3 - 3x^2 - 10x + 24$ (d) None of these

15. If α, β are the zeros of the polynomial $2x^2 + 5x + k$ such that $a^2 + b^2 + a\beta = \frac{21}{4}$, then $k = ?$

(a) 3 (b) -3 (c) -2 (d) 2

16. Which of the following is a true statement?

(a) $x^2 + 5x - 3$ is a linear polynomial (b) $x^2 + 4x - 1$ is a binomial

(c) $x + 1$ is a monomial (d) $5x^3$ is a monomial

17. In the given figure, the graph of a polynomial $p(x)$ is shown. The number of zeros of $p(x)$ is

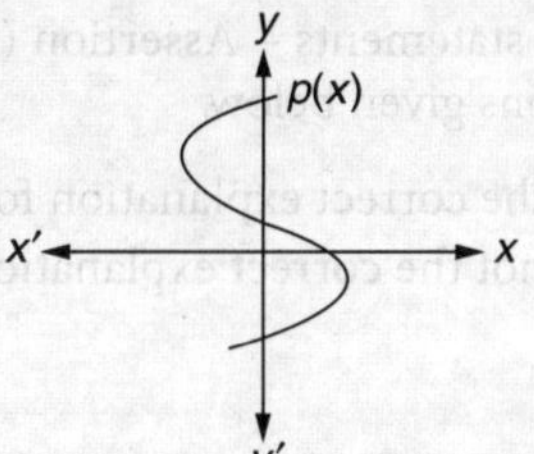

(a) 1 (b) 2 (c) 3 (d) 4

18. If $(x + 1)$ is a factor of $x^2 - 3ax + 3a - 7$, the value of a is

(a) 1 (b) − 1 (c) 0 (d) 2

19. If one of the zeros of the quadratic polynomial $(k - 1)x^2 + kx + 1$ is (-3), then k is equal to

(a) $\frac{4}{3}$ (b) $-\frac{4}{3}$ (c) $\frac{2}{3}$ (d) $-\frac{2}{3}$

20. The quadratic polynomial whose sum of zeros is 3 and whose product of zeros is − 2 is

(a) $x^2 + 3x - 2$ (b) $x^2 - 2x + 3$ (c) $x^2 - 3x + 2$ (d) $x^2 - 3x - 2$

21. The graph of a quadratic polynomial is a/an

(a) Straight line (b) Circle (c) Ellipse (d) Parabola

22. If α and β are the zeros of the polynomial $2 - 3x - x^2$, then $\alpha + \beta =$

(a) 2 (b) 3 (c) 1 (d) None of these

23. If α and β are the zeros of $p(x) = x^2 - 5x + 6$, then the value of $\alpha + \beta - 3\alpha\beta$ is

(a) -23 (b) -13 (c) 13 (d) 23

24. The zeros of the polynomial $4\sqrt{3}x^2 + 5x - 2\sqrt{3}$ are

(a) $\frac{\sqrt{3}}{4}, \frac{2}{\sqrt{3}}$ (b) $\frac{\sqrt{3}}{4}, \frac{-2}{\sqrt{3}}$ (c) $\frac{-\sqrt{3}}{4}, \frac{-2}{\sqrt{3}}$ (d) $\frac{-\sqrt{3}}{4}, \frac{-2}{\sqrt{3}}$

25. The number of zeros of a cubic polynomial is

(a) More than 3 (b) At most 3 (c) Only 3 (d) None

26. The sum and product of the zeros, respectively, of the polynomial $9x^2 - 5$ are

(a) $0, \frac{-9}{5}$ (b) $0, \frac{9}{5}$ (c) $0, \frac{5}{9}$ (d) $0, \frac{-5}{9}$

Fill in the Blanks

Complete the following statements with appropriate answers(s) in the blank space(s):

27. The sum of the zeros of $2x^2 - 8x + 6$ is ________.

28. A quadratic polynomial can have at most ________ and a cubic polynomial can have at most ________.

29. The graph of a linear polynomial is always a ________.

30. In polynomial $p(x)$, if the curve intersects the x axis at three points, the number of zeros is ________.

31. If the graph of a quadratic polynomial does not intersect the x axis, the number of zeros is ________.

Assertion–Reason

The following questions consist of two statements – Assertion (A) and Reason (R). Answer these questions by selecting the appropriate options given below.

(A) Both A and R are true and R is the correct explanation for A.

(B) Both A and R are true and R is not the correct explanation for A.

(C) A is true but R is false.

(D) A is false but R is true.

32. **Assertion (*A*):** If 1 is a zero of the polynomial $ax^2 - 3(a - 1)x - 1$, then $a = 1$.
Reason (*R*): k is said to be a zero of the polynomial $p(x)$, if $p(k) = 0$.

33. **Assertion (*A*):** If both zeros of a quadratic polynomial $x^2 - 2kx + 2$ are equal in magnitude but opposite in sign, then the value of k is $\frac{1}{2}$.
Reason (*R*): Sum of zeros of a quadratic polynomial $ax^2 + bc + c$ is $-\frac{b}{a}$.

34. **Assertion (*A*):** If one zero of the polynomial $(a^2 + 25)x^2 + 36x + 10a$ is the reciprocal of the other, then the value of a is 5.
Reason (*R*): Sum of zeros $= \alpha + \beta = -\frac{b}{a}$.

Subjective Questions

35. Find the zeros of the polynomial $(x - 2)^2 + 4$.

36. Manish engages a labourer to perform some repair work. The charges to be paid for this work are zeros of the polynomial $x^2 - 300x + 22500$.

(a) Find the zeros of this polynomial.

(b) The labourer claims ₹125 for the whole work. Manish paid the actual amount. What value is depicted by Manish?

37. If 2 and –3 are the zeros of the quadratic polynomial $x^2 + (a + 1)x + b$, what are the values of a and b? **[CBSE 2011]**

38. For what value of k is 3 a zero of the polynomial $2x^2 + x + k$? **[CBSE 2010]**

39. For what value of k is (– 4) a zero of the polynomial $x^2 - x - (2k + 2)$? **[CBSE 2009]**

40. For what value of p is (– 4) a zero of the polynomial $x^2 - 2x - (7p + 3)$? **[CBSE 2009]**

41. If 1 is a zero of the polynomial $p(x) = ax^2 - 3(a - 1)x - 1$, then find the value of a. **[CBSE 2009]**

42. Find a quadratic polynomial, the sum and product of whose zeros is –3 and 2, respectively. **[NCERT]**

43. Verify that $3, -1, -\frac{1}{3}$ are the zeros of the cubic polynomial $p(x) = 3x^3 - 5x^2 - 11x - 3$, and then verify the relationship between the zeros and the coefficients. **[NCERT]**

44. Find a cubic polynomial with sum, sum of the product of its zeros taken two at a time, and product of its zeros as 2, –7, –14, respectively. **[NCERT]**

45. For each of the following, find a quadratic polynomial whose sum and product, respectively, of the zeros are as follows. Also find the zeros of these polynomials by factorisation.

(a) $\frac{-8}{3}, \frac{4}{3}$ (b) $\frac{21}{8}, \frac{5}{16}$ (c) $-2\sqrt{3}, -9$ (d) $\frac{-3}{2\sqrt{5}}, -\frac{1}{2}$

[NCERT Exemplar]

46. A group consists of 12 honest people and 8 dishonest people. Write a quadratic polynomial whose roots are equal to the number of honest people and number of dishonest people. Which value do you prefer?

47. Form a quadratic polynomial whose zeros are $\frac{3-\sqrt{3}}{5}$ and $\frac{3+\sqrt{3}}{5}$.

48. Quadratic polynomial $2x^2 - 3x + 1$ has zeros as α and β. Now form a quadratic polynomial whose zeros are 3α and 3β. **[CBSE 2016]**

49. If α and β are zeros of a polynomial $x^2 + 6x + 9$, then form a polynomial whose zeros are $-\alpha$ and $-\beta$. **[CBSE 2014]**

50. Find a quadratic polynomial whose zeros are $3 + \sqrt{2}$ and $3 - \sqrt{2}$. **[CBSE 2013]**

Find the zeros of the following polynomials by factorisation method and verify the relations between the zeros and the coefficients of the polynomial for Q53 to Q57.

51. $2x^2 + \frac{7}{2}x + \frac{3}{4}$ **[NCERT Exemplar]**

52. $4x^2 + 5\sqrt{2}x - 3$ **[NCERT Exemplar]**

53. $2s^2 - (1 + 2\sqrt{2})s + \sqrt{2}$ **[NCERT Exemplar]**

54. $y^2 + \frac{3}{2}\sqrt{5}y - 5$ **[NCERT Exemplar]**

55. $7y^2 - \frac{11}{3}y - \frac{2}{3}$ **[NCERT Exemplar]**

56. Find a quadratic polynomial, the sum and product of whose zeros is $\sqrt{2}$ and $-\frac{3}{2}$, respectively. Also find its zeros.

57. If the product of the zeros of the polynomial $ax^2 - 6x - 6$ is 4, then find the value of a. Also find the sum of zeros of the polynomial. **[CBSE (CCE) 2014]**

58. If the sum of the zeros of the polynomial $5x^2 - (3+k)x + 7$ is zero, then find the zeros of the polynomial $2x^2 - 2(k+11)x + 30$.

59. Form a quadratic polynomial, one of whose zeros is $\sqrt{5}$ and whose product of zeros is $-2\sqrt{5}$.

60. If α, β are zeros of the polynomials $x^2 - 6x + k$, find the value of k such that $(\alpha+\beta)^2 - 2\alpha\beta = 40$.

61. Find a cubic polynomial with the sum, sum of the product of its zeros taken two at a time, and product of its zeros as 5, – 6 and –20, respectively.

62. If α, β are the zeros of the polynomial $p(x) = x^2 - p(x+1) - c$ such that $(\alpha+1)(\beta+1) = 0$, what is the value of c?

63. If α, β are the zeros of the polynomial $x^2 - (k+6)x + 2(2k-1)$, find the value of k if $\alpha + \beta = \frac{1}{2}\alpha\beta$.

64. If m, n are zeros of $ax^2 - 5x + c$, find the values of a and c if $m + n = m \cdot n = 10$.

65. If α, β are zeros of $x^2 + 7x + 7$, find the value of $\frac{1}{\alpha} + \frac{1}{\beta} - 2\alpha\beta$.

66. If one zero of polynomial $3x^2 = 8x + 2k + 1$ is seven times the other, find the zeros and value of k.

67. Find the value of k such that the polynomial $x^2 - (k+6)x + 2(2k-1)$ has its sum of its zeros equal to half of their product. **[Delhi 2019]**

68. α, β, γ are zeros of the cubic polynomial $x^3 - 12x^2 + 44x + c$. If $2\beta = \alpha + \gamma$, find the value of c.

69. α, β, γ are zeros of cubic polynomial $x^3 - 2x^2 + qx - r$. If $\alpha + \beta = 0$, then show that $2q = r$.

70. α, β and γ are zeros of the polynomial $x^3 + px^2 + qx + 2$ such that $\alpha \times \beta + 1 = 0$. Find the value of $2p + q + 5$.

71. Find the zeros of the quadratic polynomial $3x^2 - 2$ and verify the relationship between the zeros and the coefficients.

72. If one zero of the quadratic polynomial $f(x) = 4x^2 - 8kx + 8x - 9$ is the negative of the other, then find the zeros of $kx^2 + 3kx + 2$.

73. If the zeros of the polynomial $x^2 + px + q$ are double in value to the zeros of $2x^2 - 5x - 3$, find the values of p and q.

74. Divide $3x^2 - x^3 - 3x + 5$ by $x - 1 - x^2$ and verify the division algorithm. **[NCERT]**

75. On dividing $x^3 - 3x^2 + x + 2$ by a polynomial $g(x)$, the quotient and remainder were $x - 2$ and $-2x + 4$, respectively. Find $g(x)$. **[NCERT]**

76. If two zeros of the polynomial $x^4 - 6x^3 - 26x^2 + 138x - 35$ are $2 \pm \sqrt{3}$, find other zeros. **[NCERT]**

77. Find all the zeros of the polynomial $(2x^4 - 9x^3 + 5x^2 + 3x - 1)$ if two of its zeros are $(2+\sqrt{3})$ and $(2-\sqrt{3})$. **[CBSE 2018]**

78. If the polynomials $x^4 - 6x^3 + 16x^2 - 25x + 10$ is divided by another polynomial $x^2 - 2x + k$, the remainder is $x + a$. Find k and a. **[NCERT]**

79. If the zeros of the polynomial $x^3 - 3x^2 + x + 1$ are $a - b, a, a + b$, find a and b. **[NCERT]**

80. If the remainder on division of $x^3 + 2x^2 + kx + 3$ by $x - 3$ is 21, find the quotient and the value of k. Hence, find the zeros of the cubic polynomial $x^3 + 2x^2 + kx - 18$. **[NCERT Exemplar]**

81. Find k so that $x^2 + 2x + k$ is a factor of $2x^4 + x^3 - 14x^2 + 5x + 6$. Also find all the zeros of the two polynomials. **[NCERT Exemplar]**

82. Given that $x - \sqrt{5}$ is a factor of the cubic polynomial $x^3 - 3\sqrt{5}x^2 + 13x - 3\sqrt{5}$, find all the zeros of the polynomial. **[NCERT Exemplar]**

83. For which values of a and b are the zeros of $q(x) = x^3 + 2x^2 + a$ also the zeros of the polynomial $p(x) = x^5 - x^4 - 4x^3 + 3x^2 + 3x + b$? Which zeros of $p(x)$ are not the zeros of $q(x)$? **[NCERT Exemplar]**

84. Divide $ax^2 + (b + ac)x + bc$ by $x + c$. **[NCERT Exemplar]**

85. $p(x)$ is a polynomial of degree more than 2. When $p(x)$ is divided by $x - 2$, it leaves remainder 1 and when it is divided by $x - 3$, it leaves remainder 3. Find the remainder when $p(x)$ is divided by $(x - 2)(x - 3)$.

86. If the polynomial $x^4 - 6x^3 + 16x^3 - 25x + 10$ is divided by $(x^2 - 2x + m)$, the remainder is $x + n$. Find m and n.

87. Find the polynomial of least degree which should be subtracted from the polynomial $x^4 + 2x^3 - 4x^2 + 6x - 3$ so that it is exactly divisible by $x^2 - x + 1$.

88. Given that the zeros of the cubic polynomial $x^3 - 6x^2 + 3x + 10$ are of the form a, $a + b$, $a + 2b$ for some real numbers a and b, find the values of a and b as well as the zeros of the given polynomial.

89. Find the values of a and b so that $x^4 + x^3 + 8x^2 + ax - b$ is divisible by $x^2 + 1$. **[CBSE (CCE) 2015]**

90. If a polynomial $x^4 - 3x^3 - 6x^2 + kx - 16$ is exactly divisible by $x^2 - 3x + 2$, then find the value of k. **[CBSE (CCE) 2014]**

91. Obtain all other zeros of the polynomial $x^4 - 3\sqrt{2}x^3 + 3x^2 + 3\sqrt{2}x - 4$, if two of its zeros are $\sqrt{2}$ and $2\sqrt{2}$.

92. Divide the polynomial $x^4 - 17x^2 + 34x - 12$ by $x - 2$ and find the quotient and the remainder. Also verify the division algorithm.

93. An NGO decided to distribute books and pencils to the students of a school run by some other NGO. For this, they collected some amount from different people. The total amount collected is represented by $4x^4 + 2x^3 - 8x^2 + 3x - 7$. From this fund, each student received an equal amount. The number of students who received the amount is represented by $x - 2 + 2x^2$. After distribution, $5x - 11$ amount is left with the NGO, which they donated to the school for their infrastructure. Find the amount received by each student from the NGO. What value has been depicted here?

94. Obtain all other zeros of the polynomial $x^4 - 17x^2 - 36x - 20$, if two of its zeros are $+5$ and -2. **[CBSE 2016]**

95. When a polynomial $6x^4 + 8x^3 + 29x^2 + 21x + 7$ is divided by another polynomial $3x^2 + 4x + 1$, the remainder is in the form $ax + b$. Find a and b. **[CBSE 2014]**

96. If a polynomial $x^4 - 3x^3 - 6x^2 + kx - 16$ is exactly divisible by $x^2 - 3x + 2$, find the value of k. **[CBSE 2014]**

97. Obtain all the other zeros of the polynomial $x^4 - 3\sqrt{2}x^3 + 3x2 + 3\sqrt{2} - 4$, if two of its zeros are $\sqrt{2}$ and $2\sqrt{2}$. **[CBSE 2014]**

98. On dividing $f(x) = 3x^3 - 2x^2 + 5x - 5$ by polynomial $p(x)$, the quotient and remainder are $x^2 - x + 2$ and -7, respectively. Find $p(x)$. **[CBSE (CCE) 2015]**

99. Find the value of b for which $(2x + 3)$ is a factor of $2x^3 + 9x^2 - x - b$.

100. What must be subtracted or added to $p(x) = 8x^4 + 14x^3 - 2x^2 + 8x - 12$ so that $4x^2 + 3x - 2$ is a factor of $p(x)$? **[CBSE 2012]**

101. Divide $x^4 - 3x^2 + 4x + 5$ by $x^2 - x + 1$, and find the quotient and remainder. **[CBSE 2011]**

102. On dividing the polynomial $4x^4 - 5x^3 - 39x^2 - 46x - 2$ by the polynomial $g(x)$, the quotient and remainder were $x^2 - 3x - 5$ and $-5x + 8$, respectively. Find $g(x)$. **[CBSE 2011]**

103. Divide $2x^4 - 9x^3 + 5x^2 + 3x - 8$ by $x^2 - 4x + 1$ and verify the division algorithm. **[CBSE 2011]**

104. If the polynomial $6x^4 + 8x^3 + 17x^2 + 21x + 7$ is divided by another polynomial $3x^2 + 4x + 1$, the remainder is $(ax + b)$. Find a and b. **[CBSE 2009]**

105. If the polynomial $x^4 + 2x^3 + 8x^2 + 12x + 18$ is divided by another polynomial $x^2 + 5$, the remainder is $px + q$. Find the values of p and q. **[CBSE 2009]**

106. Verify that the numbers given alongside the cubic polynomials below are their zeros. Also verify the relationship between the zeros and the coefficient in each case:

(i) $2x^3 + x^2 - 5x + 2$; $\frac{1}{2}, 1, -2$ (ii) $x^3 - 4x^2 + 5x - 2$; 2, 1, 1 **[NCERT]**

Answers

Multiple Choice Questions

1. (a) **2.** (a) **3.** (d) **4.** (d) **5.** (b) **6.** (a) **7.** (d) **8.** (b) **9.** (b) **10.** (c)
11. (d) **12.** (a) **13.** (a) **14.** (c) **15.** (d) **16.** (d) **17.** (a) **18.** (a) **19.** (a) **20.** (d)
21. (d) **22.** (b) **23.** (b) **24.** (b) **25.** (b) **26.** (d)

Fill in the Blanks

27. 4 **28.** 2 zeros and 3 zeros **29.** Straight line **30.** 3 **31.** 0

Assertion–Reason

32. (A) **33. (D)** **34. (B)**

Subjective Questions

35. No real zero **36.** (a) 150 (b) Labour demanded ₹125, but Manish paid ₹150; shows his honesty

37. $a = 0, b = -6$ **38.** $k = -21$ **39.** $k = 9$ **40.** $p = 3$ **41.** $a = 1$ **42.** $k(x^2 + 3x + 2)$

44. $x^3 - 2x^2 - 7x + 14$

45. (a) $-2, \frac{2}{3}$ (b) $\frac{5}{2}, \frac{1}{8}$ (c) $-3\sqrt{3}, \sqrt{3}$ (d) $\frac{\sqrt{5}}{5}, \frac{-\sqrt{5}}{2}$ **46.** $k(x^2 - 20x + 96)$, Honesty

47. $k(25x^2 - 30x + 6)$ **48.** $\frac{k}{2}(2k^2 - 9x + 9)$ **49.** $k(x^2 - 6x + 9)$ **50.** $k(x^2 - 6x + 7)$ **51.** $\frac{3}{2}, -\frac{1}{4}$

52. $\frac{\sqrt{2}}{4}, \frac{-3\sqrt{2}}{2}$ **53.** $\frac{1}{2}, \sqrt{2}$ **54.** $-2\sqrt{5}, \frac{\sqrt{5}}{2}$ **55.** $\frac{2}{3}, -\frac{1}{7}$ **56.** $-\frac{1}{\sqrt{2}}, \frac{3}{\sqrt{2}}$

57. $a = -\frac{3}{2}$ and sum of zeros $= -4$. **58.** 3, 5 **59.** $x^2 + (2 - \sqrt{5})x - 2\sqrt{5}$ **60.** $k = -2$

61. $x^3 - 5x^2 - 6x + 20$ **62.** $c = 1$ **63.** $k = 7$ **64.** $a = \frac{1}{2}, c = 5$ **65.** −15

66. $\frac{1}{3}, \frac{7}{3}, -\frac{5}{3}$ **67.** $k = 7$ **68.** −48 **69.** 0 **70.** 0

71. $+\sqrt{\frac{2}{3}}$ **72.** −1, −2 **73.** $p = -5, q = -6.$ **75.** $g(x) = x^2 - x + 1$

76. – 5, 7 **77.** 1, $-\frac{1}{2}$ **78.** $k = 5$ and $a = -5$ **79.** $a = 1, b = +\sqrt{2}$ **80.** 3, –2, –3

81. $k = -3$, Zeros of $2x^4 + x^3 - 14x^2 + 5x + 6$ are 1, –3, 2, $-\frac{1}{2}$, Zeros of $x^2 + 2x - 3$ are 1, –3.

82. $\sqrt{5}, \sqrt{5}+\sqrt{2}, \sqrt{5}-\sqrt{2}$ **83.** $a = -1, b = -2$, 1 and 2 are the zeros of $q(x)$ which are not the zeros of $P(x)$.

84. Quotient = $ax + b$, remainder = 0 **85.** $r(x) = 2x - 3$. **86.** $m = 5$ and $n = -5$ **87.** $x - 1$

88. $a = -1$ or $a = 5, b = +3$, Zeros are: –1, 2, 5 or 5, 2, –1. **89.** $a = 1, b = -7$ **90.** $k = 24$

91. 1, –1 **92.** $q(x) = x^3 + 2x^2 - 13x + 8, r(x) = 4$

93. Amount received by each student = $2x^2 - 2$; Value: Humanity and socialism.

94. –2 and –1 **95.** $a = -15$ and $b = -2$ **96.** 24. **97.** –1 and 1 **98.** $p(x) = 3x + 1$

99. $b = 15$ **100.** If we subtract $15x - 14$ or add $-15x + 14$ to obtain remainder 0.

101. Quotient = $x^2 + x - 3$, Remainder = 8. **102.** $4x^2 + 7x + 2$ **103.** Quotient = $2x^2 - x - 1$, Remainder = – 7

104. $a = 1, b = 2$ **105.** $p = 2$ and $q = 3$.

Polynomials (Basic)

SELF-ASSESSMENT TEST

Time allowed: 90 min. Max. Marks: 40

General Instructions

1. *All questions are compulsory. There are 20 questions in all.*
2. *This question paper has four sections: Section A, Section B, Section C and Section D.*
3. *Section A contains 10 questions of one mark each, Section B contains three questions of two marks each, Section C contains four questions of three marks each, and Section D contains three questions of four marks each.*

Section A [1 Mark Each]

Direction: (Q. Nos. 1–5) These are Multiple Choice Questions. Select the most appropriate answer for the given questions.

1. Which of the following is a polynomial?

   (a) $x^2 - 5x + 4\sqrt{x} + 3$ (b) $x^{3/2} - x + x^{1/2} + 1$

   (c) $\sqrt{x} + \frac{1}{\sqrt{x}}$ (d) $\sqrt{2}x^2 - 3\sqrt{3}x + \sqrt{6}$

2. The zeros of the quadratic polynomial $x^2 + 88x + 125$ are

   (a) both positive (b) both negative

   (c) one positive and one negative (d) both equal

3. If α and β are the zeros of $p(x) = x^2 - 5x + 6$, then the value of $\alpha + \beta - 3\alpha\beta$ is

   (a) –23 (b) – 13 (c) 13 (d) 23

4. The number of zeros of a cubic polynomial is

   (a) more than 3 (b) at most 3 (c) only 3 (d) none

5. The sum and product, respectively, of the zeros of the polynomial $9x^2 - 5$ are

   (a) $0, \frac{-9}{5}$ (b) $0, \frac{9}{5}$ (c) $0, \frac{5}{9}$ (d) $0, \frac{-5}{9}$

6. The sum of the zeros of $2x^2 - 8x + 6$ is ____________

7. The graph of a linear polynomial is always a ____________.

8. Find the zeros (root) of the polynomial $f(x) = x - 5$.

9. Find the zeros of the quadratic polynomial $x^2 + 5x + 6$.

10. Form a quadratic polynomial whose roots are –2 and 3.

Section B [2 Marks Each]

11. Divide $2x^2 + 3x + 1$ by $(x + 2)$.

12. Find the zeros of the polynomial $(x - 2)^2 + 4$.

13. Find the zeros of the quadratic polynomial and verify the relationship between the zeros and the coefficients of $x^2 - 2x - 8$.

Section C [3 Marks Each]

14. Find all the zeros of the polynomial $P(x) = 2x^3 + 3x^2 - 11x - 6$, if one of its zeros is -3.

15. If α and β are zeros of a polynomial $x^2 + 6x + 9$, then form a polynomial whose zeros are $-\alpha$ and $-\beta$.

16. Divide $x^4 - 3x^2 + 4x + 5$ by $x^2 - x + 1$; find the quotient and remainder.

17. The graphs of $y = p(x)$ are given in the figure for some polynomial $p(x)$. Find the number of zeros of $p(x)$ in each case.

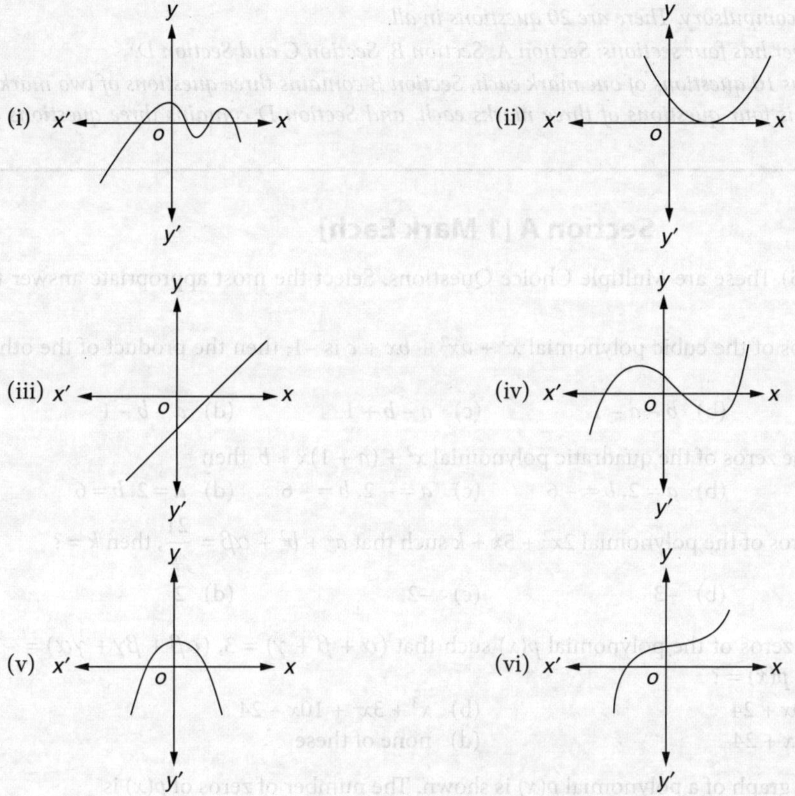

Section D [4 Marks Each]

18. Find all the zeros of the polynomial $2x^3 + x^2 - 6x - 3$, if two of its zeros are $-\sqrt{3}$ and $\sqrt{3}$.

19. Find the value of k such that the polynomial $x^2 - (k + 6)x + 2(2k - 1)$ has the sum of its zeros equal to half of their product.

20. Find all the zeros of $2x^4 - 3x^3 - 3x^2 + 6x - 2$, if you know that two of its zeros are $\sqrt{2}$ and $-\sqrt{2}$.

Answers

1. (d) **2.** (b) **3.** (b) **4.** (b) **5.** (d) **6.** 4 **7.** Straight line **8.** 5

9. $-2, -3$ **10.** $k(x^2 - x - 6)$ **11.** Quotient = $2x - 1$, Remainder = 3 **12.** No real zero

13. 4, -2 **14.** $-3, 2, -\frac{1}{2}$ **15.** $k(x^2 - 6x + 9)$ **16.** Quotient = $x^2 + x - 3$, Remainder = 8

17. 4, 1, 1, 3, 2, 1 **18.** $\sqrt{3}, \sqrt{3}, -\frac{1}{2}$ **19.** $k = 7$ **20.** $1, \frac{1}{2}$

Polynomials (Standard)

SELF-ASSESSMENT TEST

Time allowed: 90 min. Max. Marks: 40

General Instructions

1. *All questions are compulsory. There are 20 questions in all.*
2. *This question paper has four sections: Section A, Section B, Section C and Section D.*
3. *Section A contains 10 questions of one mark each, Section B contains three questions of two marks each, Section C contains four questions of three marks each, and Section D contains three questions of four marks each.*

Section A [1 Mark Each]

Direction: (Q. Nos. 1–5) These are Multiple Choice Questions. Select the most appropriate answer for the given questions.

1. If one of the zeros of the cubic polynomial $x^3 + ax^2 + bx + c$ is -1, then the product of the other two zeros is

(a) $b - a + 1$ (b) $b - a - 1$ (c) $a - b + 1$ (d) $a - b - 1$

2. If -2 and 3 are the zeros of the quadratic polynomial $x^2 + (a + 1)x + b$, then

(a) $a = -2, b = 6$ (b) $a = 2, b = -6$ (c) $a = -2, b = -6$ (d) $a = 2, b = 6$

3. If α, β are the zeros of the polynomial $2x^2 + 5x + k$ such that $a^2 + b^2 + \alpha\beta = \frac{21}{4}$, then $k = ?$

(a) 3 (b) -3 (c) -2 (d) 2

4. If α, β, γ be the zeros of the polynomial $p(x)$ such that $(\alpha + \beta + \gamma) = 3$, $(\alpha\beta + \beta\gamma + \gamma\alpha) = -10$ and $\alpha\beta\gamma = -24$, then $p(x) = ?$

(a) $x^3 + 3x^2 - 10x + 24$ (b) $x^3 + 3x^2 + 10x - 24$

(c) $x^3 - 3x^2 - 10x + 24$ (d) none of these

5. In the figure, the graph of a polynomial $p(x)$ is shown. The number of zeros of $p(x)$ is

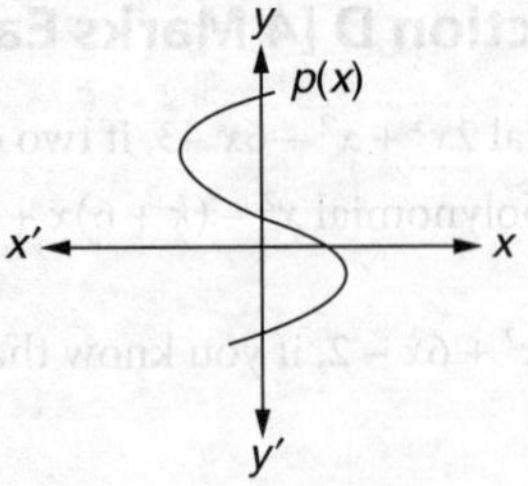

(a) 1 (b) 2 (c) 3 (d) 4

6. In polynomial $p(x)$, if the curve intersects the x axis at three points, then the number of zeros is __________.

7. A quadratic polynomial can have at most __________ and a cubic polynomial can have at most __________.

8. For what value of p is (-4) a zero of the polynomial $x^2 - 2x - (7p + 3)$?

9. The graph of $x = f(y)$ is given below. Find the zeros and number of zeros of $f(y)$.

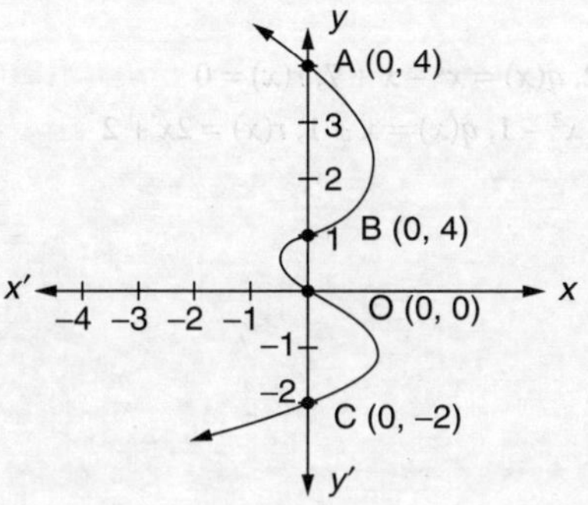

10. If α and β are the zeros of a polynomial $x^2 + 6x + 9$, then form a polynomial whose zeros are $-\alpha$ and $-\beta$.

Section B [2 Marks Each]

11. Verify that $3, -1, -\frac{1}{3}$ are the zeros of the cubic polynomials $p(x) = 3x^3 - 5x^2 - 11x - 3$, and then verify the relationship between the zeros and the coefficients.

12. If α, β are the zeros of polynomials $x^2 - 6x + k$, find the value of k such that $(\alpha + \beta)^2 - 2\alpha\beta = 40$.

13. If one zero of the quadratic polynomial $f(x) = 4x^2 - 8kx + 8x - 9$ is the negative of the other, then find the zeros of $kx^2 + 3kx + 2$.

Section C [3 Marks Each]

14. If two zeros of the polynomial $x^4 - 6x^3 - 26x^2 + 138x - 35$ are $2 \pm \sqrt{3}$, find the other zeros.

15. What must be subtracted or added to $p(x) = 8x^4 + 14x^3 - 2x^2 + 8x - 12$ so that $4x^2 + 3x - 2$ is a factor of $p(x)$?

16. When a polynomial $6x^4 + 8x^3 + 29x^2 + 21x + 7$ is divided by another polynomial $3x^2 + 4x + 1$, the remainder is in the form $ax + b$. Find a and b.

17. If the zeros of the polynomial $x^3 - 3x^2 + x + 1$ are $a - b, a, a + b$, find a and b.

Section D [4 Marks Each]

18. If 1 and – 1 are the zeros of polynomial $Lx^4 + Mx^3 + Nx^2 + Rx + P = 0$, show that $L + N + P = M + R = 0$.

19. Give examples of polynomials $p(x)$, $g(x)$, $q(x)$ and $r(x)$, which satisfy the division algorithm and

(i) $\deg p(x) = \deg q(x)$ (ii) $\deg q(x) = \deg r(x)$ (iii) $\deg r(x) = 0$

20. If the polynomial $x^4 - 6x^3 + 16x^2 - 25x + 10$ is divided by $(x^2 - 2x + k)$, the remainder is $x + a$. Find k and a.

Answers

1. (a) 2. (c) 3. (d) 4. (c) 5. (a) 6. 3 7. 2 zeroes and 3 zeroes
8. $p = 3$ 9. 4 10. $k\,(x^2 - 6x + 9)$ 12. $k = -2$ 13. $-1, -2$ 14. $-5, 7$
15. We need to subtract $15x - 14$ or add $-15x + 14$ to obtain remainder 0 16. $a = -16$ and $b = -2$
17. $a = 1, b = +\sqrt{2}$
19. (i) $p(x) = 2x^2 - 2x + 14, g(x) = 2, q(x) = x^2 - x + 7, r(x) = 0$
(ii) $p(x) = x^3 + x^2 + x + 1, g(x) = x^2 - 1, q(x) = x + 1, r(x) = 2x + 2$
(iii) Try to solve it by yourself
20. $k = 5$ and $a = -5$

CHAPTER

3 Linear Equations in Two Variables

Key Concepts

- Definition of linear equation
- Solving linear equations
  - Using the elimination method
  - Using the substitution method
  - Using the cross-multiplication method
- Graphical solutions of linear equations
- Solution of linear equations
  - Unique solution
  - No solution
  - Infinitely many solutions
- Word problems based on linear equations in two variables

Important Terms and Definitions

Linear Equation

(i) An equation of the form $ax + b = 0$, where a and b are real numbers, $a \neq 0$ and x is a variable is known as a linear equation in one variable.

(ii) An equation of the form $ax + by + c = 0$, where a, b and c are real numbers, $a \neq 0$, $b \neq 0$ and x and y are variables is known as a linear equation in two variables. Any pair of x and y which satisfies the equation $ax + by + c = 0$ is called its solution.

(iii) A pair of linear equations in two variables is said to form a system of simultaneous linear equations.

(iv) The general form of a system of two simultaneous linear equations in two variables x and y is given by:

$$a_1x + b_1y + c_1 = 0 \text{ and } a_2x + b_2y + c_2 = 0$$

A pair of values of x and y variables satisfying each one of the given equations is called a solution of the system.

(v) A system of simultaneous linear equations is said to be consistent if it has at least one solution.

(vi) A system of simultaneous linear equations is said to be inconsistent if it has no solution.

Concept: Solving Linear Equations

$$\begin{array}{l} \boxed{a_1}\,x + \boxed{b_1}\,y = \boxed{c_1} \\ \boxed{a_2}\,x + \boxed{b_2}\,y = \boxed{c_2} \end{array}$$

a_1, a_2 ↓ Coefficient of x

b_1, b_2 ↓ Coefficient of y

c_1, c_2 → Constant terms

System of linear equations

1. **Elimination Method:** Eliminate the coefficient of x or y, making it either the same sign or opposite sign by subtracting or adding.

Example 1 Solve: $x + y = 5$

$x - y = 1$

Solution:

Case 1: $1x + y = 5$ ...(i)

$1x - y = 1$ ...(ii)

Subtract (ii) from (i)

$$\begin{array}{r} \cancel{x} + y = 5 \\ \cancel{x} - y = 1 \\ - \quad + \quad - \\ \hline 2y = 4 \end{array}$$

$$y = \frac{\cancel{4}}{\cancel{2}} = 2$$

Insert the value of y in (i)

$$x + 2 = 5$$

$$x = 5 - 2$$

$$x = 3$$

[**Ans.** $x = 3, y = 2$]

Case 2: $x + y = 5$ ...(i)

$x - y = 1$ ...(ii)

Add (i) and (ii)

$$\begin{array}{r} x + \cancel{y} = 5 \\ x - \cancel{y} = 1 \\ \hline 2x \quad = 6 \end{array}$$

$$x = \frac{\cancel{6}}{\cancel{2}} = 3$$

Insert the value of x in (i)

$$3 + y = 5$$

$$y = 5 - 3$$

$$y = 2$$

[**Ans.** $x = 3$ and $y = 2$]

1 Try to Solve

1. Solve using the elimination method:
   (a) $x + y = 7, x - y = 1$ (b) $x + y = 15, x - y = 5$ (c) $x + y = 12, x - y = 2$

Answers

1. (a) $x = 4, y = 3$ (b) $x = 10, y = 5$ (c) $x = 7, y = 5$

Example 2 Solve: $2x + 3y = 11$ and $x - 4y = -22$ and hence find the value of 'm' for which $y = mx + 3$. [NCERT]

Solution: $2x + 3y = 11$ ...(i)

$x - 4y = -22$ ...(ii)

Multiplying equation (ii) by 2 and (i) by 1, we get

$2x + 3y = 11$ ...(iii)

$2x - 8y = -44$ ...(iv)

Subtracting (iv) from (iii), we get

$$\begin{aligned} 2x + 3y &= 11 \\ 2x - 8y &= -44 \\ - \quad + \quad &\quad + \\ \hline 11y &= 55 \\ y = \frac{55}{11} &= 5 \end{aligned}$$

Inserting the value of y in (ii), we get

$$x - 4 \times 5 = -22$$

$$x - 20 = -22$$

$$x = -22 + 20 = -2$$

Hence, $x = -2$ and $y = 5$. Inserting the value of x and y in $y = mx + 3$

$$5 = m(-2) + 3 \qquad 5 = -2m + 3$$

$$2m = 3 - 5 = -2 \qquad m = -\frac{2}{2} = -1$$

[Ans. $x = -2, y = 5$ and $m = -1$]

2 Try to Solve

1. Solve using the elimination method:
   (a) $x + y = 7, 12x + 5y = 7$
   (b) $3x + 4y = 10, 2x - 2y = 2$ [NCERT]
   (c) $3x - 5y - 4 = 0, 9x = 2y + 7$ [NCERT]
   (d) $0.2x + 0.3y = 1.3$, $0.4x + 0.5y = 2.3$ [NCERT]

Answers

1. (a) $x = -4, y = 11$ (b) $x = 2, y = 1$ (c) $x = \frac{9}{13}, y = \frac{-5}{15}$ (d) $x = 2$ and $y = 3$

Example 3 Solve: $\frac{x}{2} + \frac{2y}{3} = -1$ and $x - \frac{y}{3} = 3$.

Solution:

$$\frac{x}{2} + \frac{2y}{3} = -1 \qquad x - \frac{y}{3} = 3$$

$$\frac{3x + 4y}{6} = -1 \qquad \frac{3x - y}{3} = 3$$

$$3x + 4y = -6 \quad \text{...(i)} \qquad 3x - y = 9 \quad \text{...(ii)}$$

Subtracting (ii) from (i)

$$\begin{aligned} 3x + 4y &= -6 \\ 3x - \; y &= 9 \\ - \quad + \quad &\quad - \\ \hline 5y &= -15 \\ y &= -\frac{15}{5} \\ y &= -3 \end{aligned}$$

Inserting the value of y in (i) $3x + 4(-3) = -6$

$$3x - 12 = -6$$

$$3x = -6 + 12, \quad \Rightarrow 3x = 6$$

$$x = \frac{6}{3} = 2$$

[Ans. $x = 2$ and $y = -3$]

3 Try to Solve

1. Solve using the elimination method:
   (a) If $2x + y = 35$ and $3x + 4y = 65$, find $\frac{x}{y}$. (b) $x + \frac{y}{2} = 4$; $\frac{x}{3} + 2y = 5$
   (c) $\frac{x+1}{2} + \frac{y-1}{3} = 8, \frac{x-1}{3} + \frac{y+1}{3} = 9$ (d) $\frac{3}{2}x - \frac{5}{3}y = -2, \frac{x}{3} + \frac{y}{2} = \frac{13}{6}$

Answers

1. (a) $x = 15, y = 5, \frac{x}{y} = 3$ (b) $x = 3, y = 2$ (c) 7, 13 (d) $x = 2$ and $y = 3$

Example 4 Solve: $\frac{2}{x} + \frac{2}{3y} = \frac{1}{6}$; $\frac{3}{x} + \frac{2}{y} = 0$ and hence find the value of 'a' for which $y = ax + 4$. **[CBSE (CCE) 2012]**

Solution: $\frac{2}{x} + \frac{2}{3y} = \frac{1}{6}$ and $\frac{3}{x} + \frac{2}{y} = 0$

$$2 \times \frac{1}{x} + \frac{2}{3} \times \frac{1}{y} = \frac{1}{6}, \quad 3 \times \frac{1}{x} + 2 \times \frac{1}{y} = 0$$

Insert $\frac{1}{x} = u$ and $\frac{1}{y} = v$

$$\frac{2u}{1} + \frac{2v}{3} = \frac{1}{6}$$

$$\frac{6u + 2v}{3} = \frac{1}{6}$$

$$6u + 2v = \frac{3}{6} = \frac{1}{2}$$

$$12u + 4v = 1 \quad \text{...(i)}$$

Now

$$3u + 2v = 0 \quad \text{...(ii)}$$

Multiply equation (ii) by 4 and (i) by 1

$$12u + 4v = 1 \quad \text{...(iii)}$$

$$12u + 8v = 0 \quad \text{...(iv)}$$

Subtract (iv) from (iii)

$$\begin{array}{r} 12u + 4v = 1 \\ 12u + 8v = 0 \\ - \quad - \quad + \\ \hline -4v = 1 \end{array}$$

$$v = -\frac{1}{4}, \text{ insert the value of } v \text{ in (ii)}$$

$$3u + 2\left(-\frac{1}{4}\right) = 0$$

$$3u - \frac{1}{2} = 0$$

$$3u = \frac{1}{2}$$

$$u = \frac{1}{2\times 3} = \frac{1}{6}$$

But $$u = \frac{1}{x} = \frac{1}{6}$$

$$x = 6$$

$$v = \frac{1}{y} = -\frac{1}{4}$$

$$y = -4.$$

Insert the value of x and y in $y = ax + 4$

$$-4 = a(6) + 4$$

$$-4 - 4 = 6a$$

$$-8 = 6a$$

$$-\frac{8}{6} = a$$

$$-\frac{4}{3} = a$$

$$a = -\frac{4}{3}$$

4 Try to Solve

1. Solve using the elimination method:

(a) $\frac{2}{x} + \frac{3}{y} - 2 = 0;\ \frac{1}{x} - \frac{4}{y} - 6 = 0;\ (x \neq 0, y \neq 0)$ (b) $\frac{1}{2x} - \frac{1}{y} = -1;\ \frac{1}{x} + \frac{1}{2y} = 8, x \neq 0, y \neq 0$

(c) $\frac{7x - 2y}{xy} = 5;\ \frac{8x + 7y}{xy} = 15$[CBSE (CCE) 2014] (d) $\frac{2xy}{x+y} = \frac{3}{2};\ \frac{xy}{2x - y} = \frac{-3}{10},\ x + y \neq 0,$

$2x - y \neq 0$

Answers

1. (a) $\frac{11}{26}, \frac{-11}{10}$ (b) $x = \frac{1}{6}, y = \frac{1}{4}$ (c) $x = 1, y = 1$ (d) $x = \frac{1}{2}$ and $y = -\frac{3}{2}$

2. Substitution Method

(i) Find out the value of any one of the two variables in terms of the other, from any one of the two given variables.
(ii) Substitute this value in the other equation to make it a linear equation in one variable.
(iii) Solve the above equation and obtain the value of the variable.
(iv) Insert this value in any one of the given equations and obtain the value of the other variable.

Example 5 Solve: $x + y = 7$ and $12x + 5y = 7$.

Solution:

$$x + y = 7 \quad \text{...(i)}$$

$$12x + 5y = 7 \quad \text{...(ii)}$$

From (i) $$y = 7 - x \quad \text{...(iii)}$$

Substitute the value of 'y' in (ii) $$12x + 5(7 - x) = 7$$

$$12x + 35 - 5x = 7$$

$$7x = 7 - 35$$

$$7x = -28$$

$$x = -\frac{28}{7}, x = -4$$

Insert the value of x in (iii) $$y = 7 - (-4) = 7 + 4 = 11$$ **[Ans.** $x = -4$ and $y = 11$]

5 Try to Solve

1. Solve using the substitution method:
   (a) $x + y = 7, x - y = 1$ (b) $x + y = 15, x - y = 5$ (c) $x + y = 12, x - y = 2$

Answers

1. (a) $x = 4, y = 3$ (b) $x = 10, y = 5$ (c) $x = 7, y = 5$

Example 6 Solve: $\frac{x}{2} + \frac{2y}{3} = -1$ and $x - \frac{y}{3} = 3$.

Solution:

$$\frac{x}{2} + \frac{2y}{3} = -1$$

$$\frac{3x + 4y}{6} = -1$$

$$3x + 4y = -6 \quad \text{...(i)}$$

$$\frac{x}{1} - \frac{y}{3} = 3$$

$$\frac{3x - y}{3} = 3$$

$$3x - y = 9 \quad \text{...(ii)}$$

From (ii) $$3x - 9 = y$$

Substitute the value of 'y' in (i) $$3x + 4(3x - 9) = -6$$

$$3x + 12x - 36 = -6$$

$$15x = 36 - 6 \Rightarrow x = \frac{30}{15} = 2$$

Insert the value of x in (i) $$3(2) + 4y = -6$$

$$6 + 4y = -6$$

$$4y = -6 - 6$$

$$4y = -12$$

$$y = -\frac{12}{4} = -3$$ **[Ans.** $x = 2$ and $y = -3$]

6 Try to Solve

1. Solve using the substitution method:

(a) $\frac{2}{x}+\frac{3}{y}-2=0; \frac{1}{x}-\frac{4}{y}-6=0; (x \neq 0, y \neq 0)$ (b) $\frac{1}{2x}-\frac{1}{y}=-1; \frac{1}{x}+\frac{1}{2y}=8, x \neq 0, y \neq 0$

(c) $\frac{7x-2y}{xy}=5; \frac{8x-7y}{xy}=15$ (d) $\frac{2xy}{x+y}=\frac{3}{2}; \frac{xy}{2x-y}=\frac{-3}{10}, x+y \neq 0, 2x-y \neq 0$

Answers

1. (a) $\frac{11}{26}, \frac{-11}{10}$ (b) $x=\frac{1}{6}, y=\frac{1}{4}$ (c) $x=1, y=1$ (d) $x=\frac{1}{2}$ and $y=-\frac{3}{2}$

3. Cross-Multiplication Method

$$a_1x+b_1y=c_1$$
$$a_2x+b_2y=c_2$$

Use the formulae to solve a linear equation by the cross-multiplication method.

$$\frac{x}{b_1c_2-b_2c_1}=\frac{y}{c_1a_2-c_2a_1}=\frac{-1}{a_1b_2-a_2b_1}$$

$$x=\frac{b_2c_1-b_1c_2}{a_1b_2-a_2b_1},\ y=\frac{c_2a_1-c_1a_2}{a_1b_2-a_2b_1}$$

Example 7 Solve: $x+y=5$ and $x-y=1$.

Solution:

$$x+y=5 \quad \ldots\text{(i)}$$
$$x-y=1 \quad \ldots\text{(ii)}$$

Here

$$a_1=1, b_1=1, c_1=5$$
$$a_2=1, b_2=-1, c_2=1$$

So, using

$$\frac{x}{b_1c_2-b_2c_1}=\frac{y}{c_1a_2-c_2a_1}=\frac{-1}{a_1b_2-a_2b_1}.$$

$$x=\frac{b_2c_1-b_1c_2}{a_1b_2-a_2b_1} \qquad y=\frac{c_2a_1-c_1a_2}{a_1b_2-a_2b_1}$$

$$x=\frac{(-1)(5)-(1)(1)}{(1)(-1)-(1)(1)} \qquad y=\frac{(-1)(1)-(5)(1)}{(1)(-1)-(1)(1)}$$

$$=\frac{-5-1}{-1-1}=\frac{-4}{-2} \qquad =\frac{-5-1}{-1-1}=\frac{-4}{-2}$$

$$x=\frac{+6}{+2} \qquad y=\frac{-4}{-2}$$

$$x=3 \qquad y=2$$

[**Ans.** $x=3$ and $y=2$]

7 Try to Solve

1. Solve using the cross-multiplication method:
   (a) $x+y=7, x-y=1$ (b) $x+y=15, x-y=5$ (c) $x+y=12, x-y=2$

Answers

1. (a) $x=4, y=3$ (b) $x=10, y=5$ (c) $x=7, y=5$

Example 8 Solve: $\frac{x}{2}+\frac{2y}{3}=-1$ and $x-\frac{y}{3}=3$ using the cross-multiplication method. **[NCERT]**

Solution:

$$\frac{x}{2}+\frac{2y}{3}=-1 \Rightarrow \frac{3y+4y}{6}=-1$$

$$\Rightarrow 3x+4y=-6 \quad \text{...(i)}$$

$$\frac{x}{1}-\frac{y}{3}=3$$

$$\frac{3x-y}{3}=3$$

$$3x-y=9 \quad \text{...(ii)}$$

Here $a_1=3, b_1=4, c_1=-6$

$a_2=3, b_2=-1, c_2=9$

So, using

$$\frac{x}{b_1c_2-b_2c_1}=\frac{y}{c_1a_2-c_2a_1}=\frac{-1}{a_1b_2-a_2b_1}$$

$$x=\frac{b_2c_1-b_1c_2}{a_1b_2-a_2b_1}, \quad y=\frac{c_2a_1-c_1a_2}{a_1b_2-a_2b_1}$$

$$x=\frac{(-1)(-6)-(4)(9)}{(3)(-1)-(3)(4)}; y=\frac{(9)(3)-(-6)(3)}{(3)(-1)-(3)(4)}$$

$$x=\frac{6-36}{-3-12}; \quad y=\frac{27+18}{-3-12}$$

$$x=\frac{-30}{-15}; \quad y=\frac{\cancel{45}}{-\cancel{15}}=-3$$

$$x=\frac{\cancel{30}}{\cancel{15}}$$

$$x=2$$

[Ans. $x=2$ and $y=-3$]

8 Try to Solve

1. Solve using the cross-multiplication method:

   (a) $\frac{2}{x}+\frac{3}{y}-2=0; \frac{1}{x}-\frac{4}{y}-6=0; (x\neq 0, y\neq 0)$ **[CBSE (CCE) 2011]**

(b) $\dfrac{1}{2x} - \dfrac{1}{y} = -1;\ \dfrac{1}{x} + \dfrac{1}{2y} = 8,\ x \neq 0, y \neq 0$

(c) $\dfrac{7x-2y}{xy} = 5;\ \dfrac{8x+7y}{xy} = 15$ [NCERT]

(d) $\dfrac{2xy}{x+y} = \dfrac{3}{2};\ \dfrac{xy}{2x-y} = \dfrac{-3}{10},\ x+y \neq 0,\ 2x-y \neq 0$ [CBSE (CCE) 2013]

Answers

(a) $\dfrac{11}{26}, \dfrac{-11}{10}$ (b) $x = \dfrac{1}{6}, y = \dfrac{1}{4}$ (c) $x = 1, y = 1$ (d) $x = \dfrac{1}{2}$ and $y = -\dfrac{3}{2}$

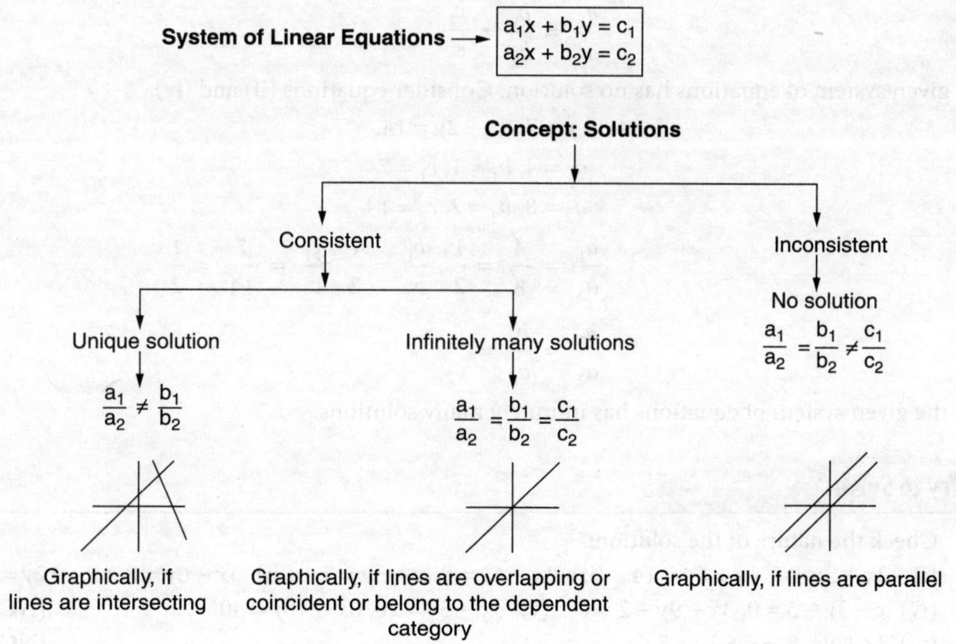

Example 9 Check the pair of equations having a unique solution, no solution or infinitely many solutions: $2x + 3y = 8$, $4x + y = 7$, $4x + 6y = 7$ and $8x + 2y = 14$.

Solution: There are four equations

$$2x + 3y = 8 \quad \text{...(i)}$$
$$4x + y = 7 \quad \text{...(ii)}$$
$$4x + 6y = 7 \quad \text{...(iii)}$$
$$8x + 2y = 14 \quad \text{...(iv)}$$

From (i) and (ii) $2x + 3y = 8,\ 4x + y = 7$

Here $a_1 = 2, a_2 = 4$

$b_1 = 3, b_2 = 1$

$c_1 = 8, c_2 = 7$

Here $\frac{a_1}{a_2} = \frac{\cancel{4}}{\cancel{2}} = \frac{1}{2}; \frac{b_1}{b_2} = \frac{3}{1}$

Hence $\frac{a_1}{a_2} \neq \frac{b_1}{b_2}$

So, the given equations have a unique solution.

Consider equations (i) and (iii) $2x + 3y = 8, 4x + 6y = 7$

Here $a_1 = 2, b_1 = 3, c_1 = 8$

$a_2 = 4, b_2 = 6, c_2 = 7$

$$\frac{a_1}{a_2} = \frac{\cancel{2}}{\cancel{4}} = \frac{1}{2}, \frac{b_1}{b_2} = \frac{\cancel{3}}{\cancel{6}} = \frac{1}{2}, \frac{c_1}{c_2} = \frac{8}{7}$$

Here $\frac{a_1}{a_2} = \frac{b_1}{b_2} \neq \frac{c_1}{c_2}$

So, the given system of equations has no solution. Consider equations (ii) and (iv)

$4x + y = 7, 8x + 2y = 14.$

Here $a_1 = 4, b_1 = 1, c_1 = 7$

$a_2 = 8, b_2 = 2, c_2 = 14$

$$\frac{a_1}{a_2} = \frac{\cancel{4}}{\cancel{8}} = \frac{1}{2}, \frac{b_1}{b_2} = \frac{1}{2}, \frac{c_1}{c_2} = \frac{\cancel{7}}{\cancel{14}} = \frac{1}{2}$$

Here $\frac{a_1}{a_2} = \frac{b_1}{b_2} = \frac{c_1}{c_2}$

So, the given system of equations has infinitely many solutions.

9 Try to Solve

1. Check the nature of the solution:
   (i) $2x + 3y = 7, 4x + 6y = 14$ (ii) $4x - 6y = 8, x + 3y = 7$ (iii) $3x - 6y = 8, 6x - 12y = 7$
   (iv) $x - 3y - 3 = 0, 3x - 9y - 2 = 0$ (v) $3x - 5y = 20, 6x - 10y = 40$ [NCERT]
   (vi) $2x + y = 5, 3x + 2y = 8$ [NCERT]

Answers

1. (i) Infinitely many solutions (ii) Unique solution (iii) No solution (iv) No solution
   (v) Infinitely many solutions (vi) Unique solution

Example 10 Find the value of k for which the system of equations: $x - 2y = 3, 3x - ky = 1$ has a unique solution.

Solution: $x - 2y = 3, 3x - ky = 1$

Here $a_1 = 1, b_1 = -2, c_1 = 3$

$$a_2 = 3,\ b_2 = -k,\ c_2 = 1$$

For a unique solution
$$\frac{a_1}{a_2} \neq \frac{b_1}{b_2}$$
$$\frac{1}{3} \neq \frac{-2}{-k}$$
$$\frac{1}{3} \neq \frac{\not{-}2}{\not{-}k}$$
$$\frac{1}{3} \neq \frac{2}{k}$$
$$k \neq 6$$

For a unique solution, $k \neq 6$.

10 Try to Solve

1. Find the value of k for which the system of equations has a unique solution:
   (a) $2x + ky = 1;\ 2x - 3y = 1$ (b) $2x + 5y = 7;\ 3x - ky = 5$
   (c) $x + 2y = 5;\ 3x + ky + 15 = 0$

Answers

1. (a) $k \neq -3$ (b) $k \neq -15/2$ (c) $k \neq 6$

Example 11 For what value of k will the following system of linear equations have no solution?

$$3x + y = 1,\ (2k - 1)x + (k - 1)y = 2k + 1$$ **[NCERT, CBSE (CCE) 2011]**

Solution:
$$3x + y = 1,\ (2k - 1)x + (k - 1)y = 2k + 1$$

Here
$$a_1 = 3,\ b_1 = 1,\ c_1 = 1$$
$$a_2 = 2k - 1,\ b_2 = k - 1,\ c_2 = 2k + 1$$

For no solution
$$\frac{a_1}{a_2} = \frac{b_1}{b_2} \neq \frac{c_1}{c_2}$$
$$\frac{3}{2k-1} = \frac{1}{k-1} \neq \frac{1}{2k+1}$$

$\frac{3}{2k-1} = \frac{1}{k-1}$	$\frac{1}{k-1} \neq \frac{1}{2k+1}$
$3(k-1) = 2k-1$	$2k+1 \neq k-1$
$3k-3 = 2k-1$	$2k-k \neq -1-1$
$3k-2k = -1+3$	$k \neq -2$
$k = 2$	

For no solution, $k = 2$.

11 Try to Solve

1. For what value of k will the following system of linear equations have no solution or be inconsistent?
   (a) $2x + 4y = -9;\ kx + 3y - 8 = 0$ (b) $kx + 3y = 3,\ 12x + ky = 6$
   (c) $kx + 3y = (k - 3);\ 12x + ky = k$

Answers

1. (a) $k = \frac{3}{2}$ (b) $k = -6$ (c) $k = -6$

Example 12 For what value of k will the following system of equations have an infinite number of solutions?

$$kx + 3y = k - 3;\ 12x + ky = k$$

Solution:

$$kx + 3y = k - 3 \quad \text{...(i)}$$

$$12x + ky = k \quad \text{...(ii)}$$

Here

$$a_1 = k,\ b_1 = 3,\ c_1 = k - 3$$

$$a_2 = 12,\ b_2 = k,\ c_2 = k$$

For IMS

$$\frac{a_1}{a_2} = \frac{b_1}{b_2} = \frac{c_1}{c_2}$$

$$\frac{k}{12} = \frac{3}{k} = \frac{k-3}{k}$$

$$\frac{k}{12} = \frac{3}{k}$$

$$k^2 = 36$$

$$12 = \pm\sqrt{36}$$

$$k = \pm 6$$

$$\frac{3}{\cancel{k}} = \frac{k-3}{\cancel{k}}$$

$$3 = k - 3$$

$$3 + 3 = k$$

$$k + 6$$

$$\frac{k-3}{k} = \frac{k}{12}$$

$$12(k-3) = k \times k$$

$$12k - 36 = k^2$$

$$k^2 - 12k + 36 = 0$$

$$k^2 - 6k - 6k + 36 = 0$$

$$k(k-6) - 6(k-6) = 0$$

$$(k-6)(k-6) = 0$$

$$k - 6 = 0$$

$$k = 6$$

For IMS, $k = 6$.

12 Try to Solve

1. For what value of k will the following pairs of linear equations have infinitely many solutions or coincident lines?
   (a) $2x - 3y + 10 = 0$ and $3x + ky + 15 = 0$ (b) $3x + 4y + 2 = 0$ and $9x + 12y + k = 0$
   (c) $2x + 3y - 5 = 0$ and $6x + ky - 15 = 0$ (d) $kx + 3y - (k - 3) = 0$ and $12x + ky - k = 0$

[NCERT, CBSE (CCE) 2015]

Answers

1. (a) $k = -9/2$ (b) $k = 6$ (c) $k = 9$ (d) $k = 6$

Concept: Solve the System of Equations Graphically

Graphical Representation of a Pair of Linear Equations in Two Variables

(i) A pair of linear equations in two variables will be graphically represented by a pair of straight lines. Only one of the following three possibilities can occur:

(a) The two lines intersect at one point.
(b) The two lines do not intersect, that is, they are parallel.
(c) The two lines are coincident or overlapping.

Graphical Solution of a Pair of Linear Equations in Two Variables

In order to find a graphical solution of a pair of linear equations in two variables,

1. Draw the graph of each one of the two given equations. Let the lines l_1 and l_2 represent these graphs.
2. If the lines l_1 and l_2 intersect at a unique point, the coordinates of that point, say α and β, represent the solution of the system, given by: $x = \alpha$, $y = \beta$.
3. If the lines l_1 and l_2 are coincident, every point on this line is a solution, that is, the system has infinite solutions.
4. If the lines l_1 and l_2 are parallel, that is, no point is common to them, the system has no solution.

Example 13 Solve the following pair of equations graphically:

$$x + y = 3$$
$$2x + 5y = 12$$

Solution: We have $x + y = 3$ ...(i)

If $x = 1$ $y = 2$

If $x = 2$ $y = 1$

The two solutions are

x	1	2
y	2	1

And $2x + 5y = 12$

If $x = 1$ $2 \times 1 + 5y = 12$

$\Rightarrow$ $5y = 10$

$\therefore$ $y = 2$

If $x = 6$ $2 \times 6 + 5y = 12$

$\Rightarrow$ $5y = 0$

$y = 0$

$\therefore$ The two solutions are

x	1	6
y	2	0

The figure given below shows the graph of the equations. The two lines intersect at the point (1, 2).

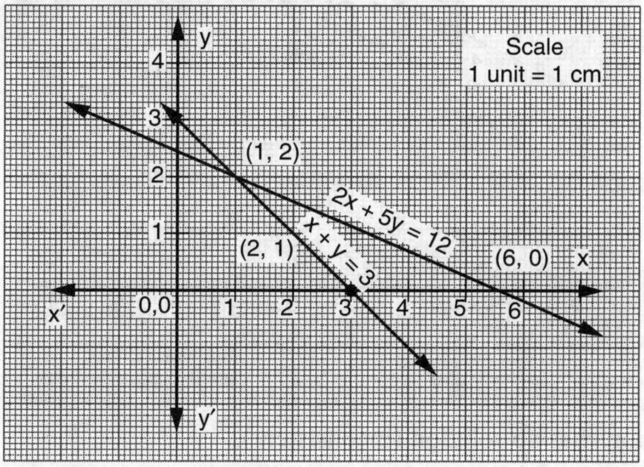

Hence, the solution of the given pair of equations is $x = 1$, $y = 2$.

13 Try to Solve

1. Solve the system of equations graphically:
   (a) $x + y = 5, 2x + 2y = 10$ [NCERT]
   (b) $2x + y - 6 = 0, 4x - 2y - 4 = 0$ [NCERT]
   (c) $2x - 2y - 2 = 0, 4x - 4y - 5 = 0$ [NCERT]

Answers

1. (a) IMS (b) Consistent at $x = 2$ and $y = 2$ (c) No solution

Example 14 Show graphically that the following system of equations is inconsistent:

$$2y - x = 9\ ;\ 6y - 3x = 21$$

Solution: We have $2y - x = 9$ ...(i)

If $x = 1$ $y = 5$

If $x = 3$ $y = 6$

The two solutions are

x	1	3
y	5	6

And $6y - 3x = 21$

If $x = 3$ $6y = 30$

$\therefore$ $y = 5$

If $x = 5$ $6y = 36$

$\therefore$ $y = 6$

The two solutions are

x	3	5
y	5	6

The figure shows a graph of the given equations. Both the lines are parallel to each other.

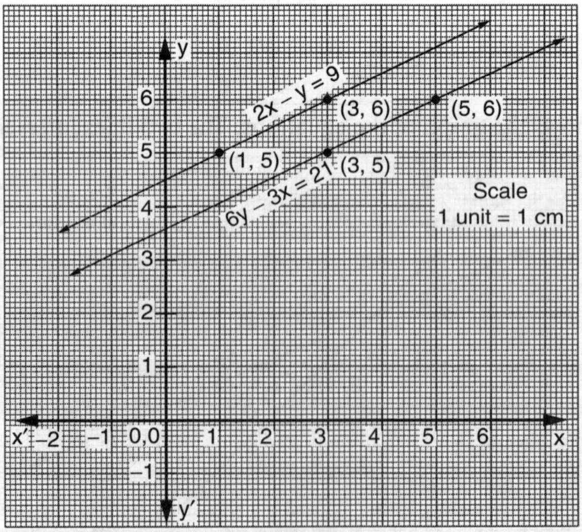

Hence, the pair of equations has no solution (inconsistent).

14 Try to Solve

1. Check graphically whether the pair of equations $x + 3y = 6$, $2x - 3y = 12$ is consistent. If so, solve them graphically.
2. Show graphically that each one of the following pair of equations has no solution (inconsistent):

   $x - 2y = 6$ $\quad\quad$ $3x - 6y = 0$
3. Show graphically that each one of the following pair of equations has infinitely many solutions:

   $3x + y = 8$ $\quad\quad$ $6x + 2y = 16$

Answer

1. Unique solution at (6, 0) consistent.

Example 15 Draw graphs of the equations $4x - y = 4$ and $4x + y = 12$. Determine the vertices of the triangle formed by the line representing these equations and the x axis. Shade the triangular region so formed.

Solution:

$$4x - y = 4 \quad \text{...(i)}$$
$$4x + y = 12 \quad \text{...(ii)}$$

From (i) $\quad y = 4x - 4$

From (ii) $\quad y = 12 - 4x$

$4x - y = 4$

x	1	2
y	0	4

$4x + y = 12$

x	3	2
y	0	4

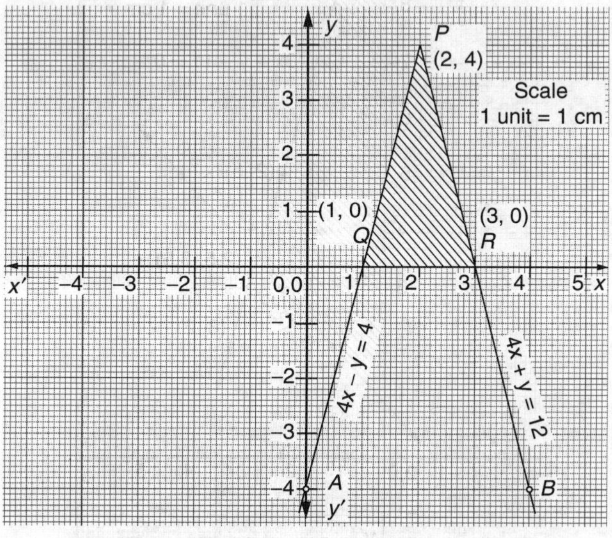

From the graph, we observe that $P(2, 4)$, $Q(1, 0)$ and $R(3, 0)$ are the three vertices of the triangle formed by the given lines and the x axis.

Example 16 Solve the following system of linear equations graphically:

$$3x + y - 11 = 0; x - y + 1 = 0.$$

Shade the region bounded by these lines and the y axis. Also, find the area of the region bounded by the lines and the x axis. **[CBSE (CCE) 2010]**

Solution: We have $3x + y - 11 = 0$

$\Rightarrow$ $3x + y = 11$

$\therefore$ $y = 11 - 3x$

The two solutions are

x	2	3
y	5	2

And $x - y + 1 = 0$

$\therefore$ $x = y - 1$

The two solutions are

x	2	1
y	3	2

The figure shows a graph of the equations. The two lines intersect at point (2.5, 3.5). Hence, the two equations have a unique solution, which is $x = 2.5, y = 3.5$.

Area of the region bounded by the lines and the y axis

$$= \frac{1}{2} \times b \times h = \frac{1}{2} \times 10 \times 2.5 = 12.5$$ **[Ans.** = 12.5 sq. units**]**

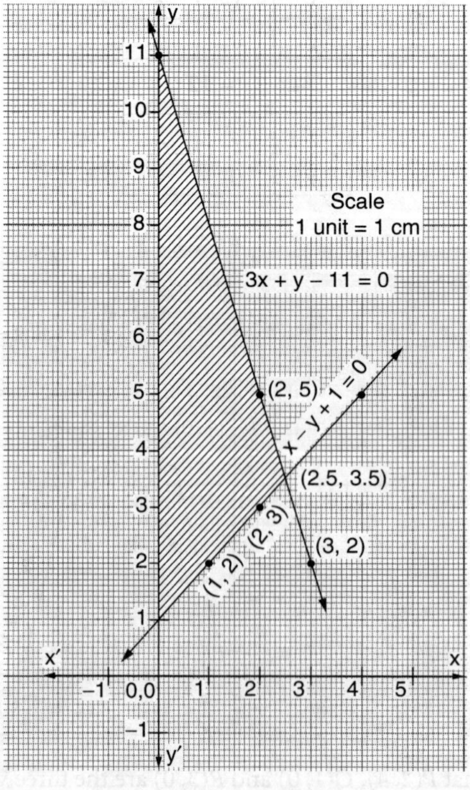

15 Try to Solve

1. Draw a graph of $x - y + 1 = 0$ and $2x + y - 10 = 0$. Calculate the area bounded by these lines and the x axis. **[NCERT, CBSE (CCE) 2012]**
2. Draw the graphs of the following equations on the same graph paper: $3x - 2y = 6$; $3x + y = 15$. Find the coordinates of the vertices of the triangle formed by these two straight lines and the x axis. **[CBSE (CCE) 2011]**
3. Solve graphically the system of linear equations: $4x - 3y + 4 = 0$; $4x + 3y - 20 = 0$. Find the area of the region bounded by these lines and the x axis. **[CBSE (CCE) 2009]**
4. Draw the graph of the equations $x - y + 1 = 0$ and $3x + 2y - 12 = 0$. Determine the coordinates of the vertices of the triangle formed by these lines and the x axis and shade the triangular region. **[NCERT, CBSE (CCE) 2011]**

Answers

1. 12 sq. units **2.** (4, 3); (2, 0); (5, 0) **3.** (2, 4); 12 cm^2 **4.** (−1, 0); (4, 0); (2, 3)

Concept: Word Problems

The concept of word problems based on linear equations can be understood in the following ways:

- Problems based on simple day-to-day life problems
- Problems based on fractions
- Problems based on two-digit numbers
- Problems based on age
- Problems based on speed, distance and time
- Problems based on area
- Problems based on angles
- Miscellaneous word problems

Concept: Problems Based on Simple Day-to-Day Life Problems

Example 17 The total cost of 8 buckets and 5 mugs is ₹92 and the total cost of 5 buckets and 8 mugs is ₹ 77. Find the cost of 2 mugs and 3 buckets.

Solution: Let the cost of one bucket = ₹x

The cost of one mug = ₹y

According to the question $8x + 5y = 92$ ...(i)

$5x + 8y = 77$ ...(ii)

Multiplying equation (i) by 5 and (ii) by 8, we get and subtract

$$40x + 25y = 460$$
$$40x + 64y = 616$$
$$- \quad - \quad -$$
$$-39y = -156$$
$$-39y = -156$$
$$y = \frac{156}{39} = 4$$

Insert the value of y in (i) $$8x + 5(4) = 92$$

$$8x = 92 - 20$$

$$8x = 72$$

$$x = \frac{72}{8} = 9$$

Hence, the cost of one bucket = ₹9 and the cost of one mug = ₹4

But, we require the cost of 2 mugs + 3 buckets = $2 \times 4 + 3 \times 9 = 8 + 27 =$ ₹35

16 Try to Solve

1. In a parking lot, the parking charges for 15 cars and 7 scooters is ₹67 and the parking charges for 10 cars and 3 scooters is ₹43. Find the parking charges for 1 car and 1 scooter.
2. Two audio cassettes and 3 video cassettes cost ₹425 and 3 audio cassettes and 2 video cassettes cost ₹350. What is the price of an audio cassette and of a video cassette?
3. Five pencils and 7 pens together cost ₹50, whereas 7 pencils and 5 pens together cost ₹46. Find the cost of one pencil and one pen. **[NCERT]**

Answers

1. ₹4, ₹1 2. ₹40, ₹115 3. ₹3 and ₹5

Concept: Problems Based on Fractions

$$\text{Fraction} = \boxed{\frac{x}{y}} \begin{matrix} \rightarrow \text{Numerator} \\ \rightarrow \text{Denominator} \end{matrix}$$

Example 18 A fraction becomes $\frac{1}{3}$ when 1 is subtracted from the numerator and it becomes $\frac{1}{4}$ when 8 is added to its denominator. Find the fraction. **[NCERT]**

Solution: Let the numerator of the fraction be x and denominator of the fraction be y.

$$\therefore \quad \text{Fraction} = \frac{x}{y}$$

According to question, $\frac{x-1}{y} = \frac{1}{3}$.

$$3(x-1) = y$$

$$3x - 3 = y$$

$$3x - y - 3 = 0 \quad \text{...(i)}$$

And $$\frac{x}{y+8} = \frac{1}{4}$$

$$4x = y + 8$$

$$4x - y - 8 = 0 \quad \text{...(ii)}$$

Subtracting (i) and (ii), we get

$$4x - \not{y} - 8 = 0$$
$$3x - \not{y} - 3 = 0$$
$$- \quad + \quad +$$
$$x \quad - 5 = 0$$

$\therefore$ $x = 5$

Inserting $x = 5$ in equation (ii)

$$4 \times 5 - y = 8$$
$$20 - y = 8$$
$$-y = 8 - 20$$
$$-y = 8 - 20$$
$$-y = -12$$

$\therefore$ $y = 12$

$$\text{Fraction} = \frac{x}{y} = \frac{5}{12}$$

17 Try to Solve

1. A fraction becomes $\frac{9}{11}$ if 2 is added to both the numerator and the denominator. If 3 is added to both the numerator and the denominator, it becomes $\frac{5}{6}$. Find the fraction. **[NCERT(CCE) 2012]**
2. A fraction is such that if the numerator is multiplied by 3 and the denominator is reduced by 3, we get 18/11; but if the numerator is increased by 8 and the denominator is doubled, we get 2/5. Find the fraction. **[NCERT(CCE) 2011]**
3. The sum of the numerator and denominator of a fraction is 12. If the denominator is increased by 3, the fraction becomes 1/2. Find the fraction. **[NCERT(CCE) 2011]**
4. If we add 1 to the numerator and subtract 1 from the denominator, a fraction reduces to 1. It becomes $\frac{1}{2}$ if we add 1 to the denominator. What is the fraction? **[NCERT]**

Answers

1. $\frac{7}{9}$ **2.** 12/25 **3.** $\frac{5}{7}$ **4.** $\frac{3}{5}$

Concept: Problems Based on Two-Digit Numbers

Numbers from 10 to 99 are known as two-digit numbers.

(63)	=	(6)×10	+	(3)×1
↓		↓		↓
Two-digit no.		Digit at the ten's place		Digit at the unit's place or one's place

Example 19 In a two-digit number, the unit's digit is twice the value of the ten's digit. If 27 is added to the number, the digits interchange their places. Find the number. [CBSE (CCE) 2014]

Solution: Let the digit at the ten's place = x and the digit at the unit's place = y.

Two-digit number = $10x + y$

Reversing the two-digit number = $10y + x$

According to the question
$$y = 2x$$
$$\Rightarrow \quad 2x - y = 0 \quad \text{...(i)}$$
$$27 + 10x + y = 10y + x$$
$$9x - 9y = -27$$
$$9(x - y) = -27$$
$$x - y = -\frac{27}{9}$$
$$x - y = -3 \quad \text{...(ii)}$$

On solving equations (i) and (ii)
$$2x - y = 0 \quad \text{...(i)}$$
$$x - y = -3 \quad \text{...(ii)}$$

Subtract (ii) from (i)
$$2x - y = 0$$
$$x - y = -3$$
$$- \quad + \quad +$$
$$x = 3$$

Insert the value of x in (i)
$$2x - y = 0$$
$$2(3) - y = 0$$
$$6 - y = 0$$
$$6 = y$$

Hence, the two-digit number $= 10x + y \times 1 = 10 \times 3 + 3 \times 1$
$$= 30 + 6 = 36$$

18 Try to Solve

1. A number consists of two digits whose sum is 10. If 72 is subtracted from the number, the digits are reversed. Find the number.
2. The sum of a number of two digits and the number obtained on interchanging the digits is 99. If 5 is added to the number then it is 4 less than 6 times the sum of the digits. Find the number. [CBSE (CCE) 2015]
3. The sum of a two-digit number and the number obtained by reversing the digits is 66. If the digits of the number differ by 2, find the number. How many such numbers are there? [NCERT]

Answers

1. 91 **2.** 45 **3.** 42 and 24

Concept: Problems Based on Age

Example 20 I am three times as old as my son. Five years later, I shall be two and a half times as old as my son. How old am I and how old is my son?

Solution: Suppose my age is x years and my son's age is y years. Then

$$x = 3y \quad \text{...(i)}$$

Five years later, my age will be $(x + 5)$ years and my son's age will be $(y + 5)$ years.

$$\therefore \quad x + 5 = \frac{5}{2}(y + 5) \quad \text{[Given]}$$

$$2(x + 5) = 5(y + 5)$$

$$2x + 10 = 5y + 25$$

$$\Rightarrow \quad 2x - 5y - 15 = 0 \quad \text{...(ii)}$$

Inserting $x = 3y$ in equation (ii), we get

$$2(3y) - 5y - 15 = 0$$

$$6y - 5y - 15 = 0 \Rightarrow y = 15$$

Inserting $y = 15$ in equation (i), we get $x = 45$

Hence, my present age is 45 years and my son's present age is 15 years.

19 Try to Solve

1. A father is three times as old as his son. After 12 years, his age will be twice that of his son. Find their present ages.
2. Ten years later, A will be twice as old as B and five years ago, A was three times as old as B. What are the present ages of A and B?
3. Five years ago, Nuri was thrice as old as Sonu. Ten years later, Nuri will be twice as old as Sonu. How old are Nuri and Sonu? **[NCERT, CBSE (CCE) 2012]**

Answers

1. Father's age = 36 years, Son's age = 12 years
2. A's present age = 50 years, B's present age = 20 years
3. 50 years and 20 years

Concept: Problems Based on Speed, Distance and Time

Example 21 Points A and B are 90 km apart on a highway. A car starts from A and another from B at the same time. If they move in the same direction, they meet in 9 hours, and if they move in opposite directions, they meet in 9/7 hours. Find their speeds. **[CBSE (CCE) 2013]**

Solution: Let X and Y be two cars starting from points A and B, respectively. Let the speed of car X be x km/h and that of car Y be y km/h.

Case I: When two cars move in the same direction:

Suppose two cars meet at point Q. Then

Distance travelled by car $X = AQ$.

Distance travelled by car $Y = BQ$.
It is given that two cars meet in 9 hours.
Distance travelled by car X in 9 hours $= 9x$ km.

$\Rightarrow$ $$AQ = 9x$$

Distance travelled by car y in 9 hours $= 9y$ km.

$\Rightarrow$ $$BQ = 9y$$

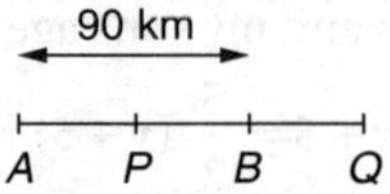

Clearly $AQ - BQ = AB$

$\Rightarrow$ $$\cancel{9}\,(x - y) = \cancel{90} \qquad [\because AB = 90 \text{ km}]$$

$\Rightarrow$ $$x - y = 10 \qquad \text{...(i)}$$

Case II: When two cars move in opposite directions:
Suppose two cars meet at point P. Then
Distance travelled by car $X = AP$.
Distance travelled by car $Y = BP$.
In this case, two cars meet in 9/7 hours.

$\therefore$ Distance travelled by car X in $\frac{9}{7}$ hours $= \frac{9}{7}x$ km

$\Rightarrow$ $$BP = \frac{9}{7}y$$

Clearly $$AP + BP = AB$$

$$\frac{9}{7}x + \frac{9}{7}y = 90$$

$$\frac{9}{7}(x + y) = 90,\; x + y = \frac{\cancel{90} \times 7}{\cancel{9}}$$

$$x + y = 70 \qquad \text{...(ii)}$$

Solving equations (i) and (ii), we get $x = 40$ and $y = 30$. [**Ans.** 40 km/h and 30 km/h]

20 Try to Solve

1. Points A and B are 70 km apart on a highway. A car starts from A and another car starts from B simultaneously. If they travel in the same direction, they meet in 7 hours, but if they travel towards each other, they meet in one hour. Find the speed of the two cars.
2. Places A and B are 80 km apart on a highway. A car starts from A and another from B at the same time. If they move in the same direction, they meet in 8 hours and if they move in opposite directions, they meet in 1 hour and 20 minutes. Find the speeds of the cars.
3. Places A and B are 100 km apart on a highway. One car starts from A and the other from B at the same time. If they move in the same direction at different speeds, they meet in 5 hours. If they travel towards each other, they meet in 1 hour. What are the speeds of the two cars? **[NCERT, CBSE (CCE) 2011]**

Answers

1. 40 km/h, 30 km/h **2.** 35 km/h and 25 km/h **3.** 60 km/m, 40 km/h

Example 22 Ved travels 600 km to his home partly by train and partly by car. He takes 8 hours if he travels 120 km by train and the rest by car. He takes 20 minutes longer if he travels 200 km by train and the rest by car. Find the speed of the train and the car. **[CBSE (CCE) 2016]**

Solution: Let the speed of the train be x km/h and the speed of the car be y km/h.

Case I: When he travels 120 km by train and the rest by car.

If Ved travels 120 km by train:

Distance covered by car is (600 – 120) km = 480 km.

Now, time taken to cover 120 km by train = $\frac{120}{y}$ h.

Time taken to cover 480 km by car = $\frac{120}{x}$ h. $\left[\because \text{Time} = \frac{\text{Distance}}{\text{Speed}}\right]$

It is given that the total time of the journey is 8 hours.

$$\therefore \quad \frac{120}{x} + \frac{480}{y} = 8$$

$$\Rightarrow \quad 8\left(\frac{15}{x} + \frac{60}{y}\right) = 8$$

$$\Rightarrow \quad \frac{15}{x} + \frac{60}{y} = 1$$

$$\Rightarrow \quad \frac{15}{x} + \frac{60}{y} - 1 = 0 \quad \text{...(i)}$$

Case II: When he travels 200 km by train and the rest by car.

If Ved travels 200 km by train: Distance travelled by car is (600 – 200) km = 400 km.

Now, time taken to cover 200 km by train = $\frac{200}{x}$ h.

Time taken to cover 400 km by car = $\frac{400}{y}$ h.

In this case, the total time of journey is 8 hours 20 minutes

$$\therefore \quad \frac{200}{x} + \frac{400}{y} = 8 \text{ hours } 20 \text{ minutes}$$

$$\Rightarrow \quad \frac{200}{x} + \frac{400}{y} = 8\frac{1}{3} \qquad \left[\because 8 \text{ hours } 20 \text{ minutes} = 8\frac{20}{60}\text{ h} = 8\frac{1}{3}\text{ h.}\right]$$

$$\Rightarrow \quad \frac{200}{x} + \frac{400}{y} = \frac{25}{3}$$

$$\Rightarrow \quad 25\left(\frac{8}{x} + \frac{16}{y}\right) = \frac{25}{3}$$

$$\Rightarrow \quad \frac{8}{x} + \frac{16}{y} = \frac{1}{3}$$

$$\Rightarrow \quad \frac{24}{x} + \frac{48}{y} = 1$$

$$\Rightarrow \quad \frac{24}{x} + \frac{48}{y} - 1 = 0 \quad \text{...(ii)}$$

Inserting $\frac{1}{x} = u$ and $\frac{1}{y} = v$ in equations (i) and (ii), we get

$$15u + 60v - 1 = 0 \quad \text{...(iii)}$$

$$24u + 48v - 1 = 0 \quad \text{...(iv)}$$

On solving (iii) and (iv), we get

$$\Rightarrow \quad u = \frac{1}{60} \text{ and } v = \frac{1}{80}$$

Now $\quad u = \frac{1}{x} \Rightarrow \frac{1}{60} = \frac{1}{x} \Rightarrow x = 60$

And $\quad v = \frac{1}{y} \Rightarrow \frac{1}{80} = \frac{1}{y} \Rightarrow y = 80$

Hence, the speed of the train is 60 km/h and the speed of the car is 80 km/h.

21 Try to Solve

1. Roohi travels 300 km to her home partly by train and partly by bus. She takes 4 hours, if she travels 60 km by train and the remaining distance by bus. If she travels 100 km by train and the remaining distance by bus, she takes 10 minutes longer. Find the speed of the train and the bus.
2. While covering a distance of 30 km, Ajeet takes 2 hours more than Amit. If Ajeet doubles his speed, he would take 1 hour less than Amit. Find their speed of walking.
3. Ramesh travels 760 km to his home partly by train and partly by car. He takes 8 hours if he travels 160 km by train and the rest by car. He takes 12 minutes more if the travels 240 km by train and the rest by car. Find the speed of the train and the car, respectively. **[CBSE (CCE) 2012]**
4. A takes 3 hours more than B to walk a distance of 30 km. But, if A doubles his pace (speed), he is ahead of B by $1\frac{1}{2}$ hours. Find their speeds of walking.

Answers

1. 60 km/h, 80 km/h
2. 5 km/h, 7.5 km/h
3. 80 km/h and 100 km/h
4. $\frac{10}{3}$ km/h, 5 km/h

Example 23 A boat travels 32 km upstream and 36 km downstream in 7 hours. Also, it travels 40 km upstream and 48 km downstream in 9 hours. Find the speed of the boat in still water and that of the stream. **[CBSE (CCE) 2014]**

Solution: Let the speed of the boat in still water be x km/h and the speed of the stream be y km/h.
Speed upstream = $(x - y)$ km/h

Speed downstream = $(x + y)$ km/h

Now, time taken to cover 32 km upstream = $\frac{32}{x - y}$ h

Time taken to cover 36 km downstream $= \frac{36}{x+y}$ h

But, total time of the journey is 7 hours.

$$\therefore \quad \frac{32}{x-y} + \frac{36}{x+y} = 7 \quad \text{...(i)}$$

Time taken to cover 40 km upstream $= \frac{40}{x-y}$

Time taken to cover 48 km downstream $= \frac{48}{x+y}$

In this case, total time of the journey is given to be 9 hours.

$$\therefore \quad \frac{40}{x-y} + \frac{48}{x+y} = 9 \quad \text{...(ii)}$$

Inserting $\frac{1}{x-y} = u$ and $\frac{1}{x+y} = v$ in equations (i) and (ii), we get

$$32u + 36v = 7 \quad \text{...(iii)}$$

$$40u + 48v = 9 \quad \text{...(iv)}$$

On solving these equations, we get $u = \frac{1}{8}$ and $v = \frac{1}{12}$.

Now $\quad u = \frac{1}{8} \Rightarrow \frac{1}{x-y} = \frac{1}{8} \Rightarrow x - y = 8 \quad$ ...(v)

And $\quad v = \frac{1}{12} \Rightarrow \frac{1}{x+y} = \frac{1}{12} \Rightarrow x + y = 12 \quad$ ...(vi)

Solving equations (v) and (vi), we get $x = 10$ and $y = 2$.

Hence, speed of the boat in still water = 10 km/h.

Speed of the stream = 2 km/h.

22 Try to Solve

1. A boat moves 45 km upstream and 65 km downstream in 14 hours. In 11 hours, it can go 35 km upstream and 52 km downstream. Find the speed of the stream and that of the boat in still water.
2. A sailor travels 8 km downstream in 40 minutes and returns in 1 hour. Determine the speed of the sailor in still water and the speed of the current.
3. A boat travels 30 km upstream and 44 km downstream in 10 hours. In 13 hours, it can go 40 km upstream and 55 km downstream. Determine the speed of stream and that of the boat in still water. **[NCERT, CBSE (CCE) 2015]**

Answers

1. 4 km/h, 9 km/h **2.** 10 km/h, 2 km/h **3.** 3 km/h, 8 km/h

Example 24 A train covered a certain distance at a uniform speed. If the train had been 6 km/h faster, it would have taken 4 hours less than the scheduled time. And, if the train had been slower by 6 km/h, it would have taken 6 hours more than the scheduled time. Find the length of the journey.

[CBSE (CCE) 2013]

Solution: Let the actual speed of the train be x km/h and the actual time taken be y hours.

Then, distance covered = (xy) km ...(i) [Distance = Speed × Time]

If the speed is increased by 6 km/h, the time of journey is reduced by 4 hours; that is, when speed is $(x + 6)$ km/h, time of journey is $(y - 4)$ hours.

$\therefore$ Distance covered = $(x + 6)(y - 4)$

$\Rightarrow$ $xy = (x + 6)(y - 4)$ [Using (i)]

$\Rightarrow$ $\not{xy} = \not{xy} - 4x + 6y - 24$

$\Rightarrow$ $-4x + 6y = 24$

$\Rightarrow$ $-2(2x - 3y) = 24$

$\Rightarrow$ $2x - 3y = -12$ ...(ii)

When the speed is reduced by 6 km/h, the time of journey is increased by 6 hours; that is, when speed is $(x - 6)$ km/h, time of journey is $(y + 6)$ hours.

$\therefore$ Distance covered = $(x - 6)(y + 6)$

$\Rightarrow$ $xy = (x - 6)(y + 6)$ **[using (i)]**

$\Rightarrow$ $\not{xy} = \not{xy} + 6x - 6y - 36$

$\Rightarrow$ $6x - 6y = 36$

$\Rightarrow$ $x - y = 6$ ...(iii)

On solving equations (ii) and (iii), we get

$\Rightarrow$ $x = 30$ and $y = 24$

Inserting the values of x and y in equation (i), we obtain

Distance = (30×24) km = 720 km.

Hence, the length of the journey is 720 km.

23 Try to Solve

1. A train covered a certain distance at a uniform speed. If the train had been 10 km/h faster, it would have taken 2 hours less than the scheduled time. And, if the train had been slower by 10 km/h, it would have taken 3 hours more than the scheduled time. Find the distance covered by the train.
2. A bus covered a certain distance at a uniform speed. If the bus had been 6 km/h faster, it would have taken 4 hours less than the scheduled time. And if the bus had been slower by 6 km/h, it would have taken 6 hours more than the scheduled time. Find the length of the journey.

Answers

1. 600 km 2. 720 km

Concept: Problems Based on Area

Example 25 The area of a rectangle is reduced by 9 square units if its length is reduced by 5 units and the breadth is increased by 3 units. If we increase the length by 3 units and breadth by 2 units, the area is increased by 67 square units. Find the length and breadth of the rectangle.

[NCERT, CBSE (CCE) 2015, 16]

Solution: Let the length and breadth of the rectangle be x and y units, respectively. Then

$$\text{Area} = xy \text{ sq. units.}$$

If length is reduced by 5 units and breadth is increased by 3 units, area is reduced by 9 square units.

$\therefore$ $$xy - 9 = (x - 5)(y + 3)$$

$\Rightarrow$ $$\cancel{xy} - 9 = \cancel{xy} + 3x - 5y - 15$$

$\Rightarrow$ $$3x - 5y = 6 \quad \text{...(i)}$$

When length is increased by 3 units and breadth by 2 units, area is increased by 67 sq. units.

$\therefore$ $$xy + 67 = (x + 3)(y + 2)$$

$\Rightarrow$ $$\cancel{xy} + 67 = \cancel{xy} + 2x + 3y + 6$$

$\Rightarrow$ $$2x + 3y = 61 \quad \text{...(ii)}$$

Thus, we obtain the following system of linear equations:

$$3x - 5y = 6 \quad \text{...(i)}$$

$$2x + 3y = 61 \quad \text{...(ii)}$$

On solving (i) and (ii), we get $x = 17$ and $y = 9$.

Hence, the length and breadth of the rectangle are 17 units and 9 units, respectively.

24 Try to Solve

1. The perimeter of a rectangular plot is 32 m. If the length is increased by 2 m and the breadth is decreased by 1 m, the area of the plot remains the same. Find the dimensions of the plot.
2. The area of a rectangle increases by 200 sq. m if the length is increased by 8 m and the breadth by 3 m. The area increases by 255 sq. m if the length is increased by 3 m and the breadth by 8 m. Find the length and breadth of the rectangle.

Answers

1. 10 m, 8 m **2.** 24 m, 13 m

Concept: Problems Based on Angles

Example 26 In a ΔABC, $\angle C = 3\angle B = 2(\angle A + \angle B)$. Find the three angles. **[NCERT]**

Solution: Let $\angle A = x°$, $\angle B = y°$. Then

$$\angle C = 3\angle B \Rightarrow \angle C = 3y°$$

$\therefore$ $$3\angle B = 2(\angle A + \angle B)$$

$$3y = 2x + 2y$$
$$3y - 2y = 2x$$
$$y = 2x \Rightarrow 2x - y = 0 \quad \text{...(i)}$$

Since $\angle A, \angle B$ and $\angle C$ are angles of a triangle

$$\angle A + \angle B + \angle C = 180° \quad [\because \text{ Angle sum property of } \Delta]$$
$$x + y + 3y = 180° \Rightarrow x + 4y = 180° \quad \text{...(ii)}$$

Inserting $y = 2x$ in equation (ii), we get

$$x + 8x = 180 \Rightarrow 9x = 180 \Rightarrow x = 20°$$

Inserting the value of x in equation (i), we get $y = 40°$.

Hence, $\angle A = 20°, \angle B = 40°$ and $\angle C = 3y° = (3 \times 40°) = 120°$.

25 Try to Solve

1. In a ΔABC, $\angle A = x°$, $\angle B = (3x - 2)°$ and $\angle C = y°$. Also $\angle C - \angle B = 9°$. Find the three angles.
2. In a ΔABC, $\angle A = x°$, $\angle B = 3x°$ and $\angle C = y°$. If $3y - 5x = 30$, prove that the triangle is right-angled. **[CBSE (CCE) 2012]**

Answers

1. 25°, 73°, 82° 2. 30°, 90°, 60°

Example 27 Find the four angles of a cyclic quadrilateral $ABCD$ in which $\angle A = (2x - 1)°$, $\angle B = (y + 5)°$ $\angle C = (2y + 15)°$ and $\angle D = (4x - 7)°$. **[NCERT]**

Solution: We know that the sum of the opposite angles of a cyclic quadrilateral is 180°.
In the cyclic quadrilateral $ABCD$, angles A and C and angles B and D form pairs of opposite angles.

$$\therefore \quad \angle A + \angle C = 180° \quad \text{and} \quad \angle B + \angle D = 180°$$
$$\Rightarrow \quad 2x - 1 + 2y + 15 = 180 \quad \text{and} \quad y + 5 + 4x - 7 = 180$$
$$\Rightarrow \quad 2x + 2y = 166 \quad\quad 4x + y = 182$$
$$\Rightarrow \quad x + y = 83 \quad \text{...(i)} \quad\quad 4x + y = 182 \quad \text{...(ii)}$$

Subtracting equation (i) from equation (ii), we get

$$3x = 99 \Rightarrow x = 33$$

Substituting $x = 33$ in equation (i), we get $y = 50$.

Hence $\angle A = (2 \times 33 - 1)° = 65°, \angle B = (y + 5)° = (50 + 5)° = 55°$

$$\angle C = (2y + 15)° = (2 \times 50 + 15)° = 115° \text{ and } \angle D = (4 \times 33 - 7)° = 125°$$

26 Try to Solve

1. Find the four angles of a cyclic quadrilateral $ABCD$ in which $\angle A = (2x + 4)°$, $\angle B = (y + 3)°$ $\angle C = (2y + 10)°$, $\angle D = (4x - 5)°$.
2. The angles of a cyclic quadrilateral $ABCD$ are $\angle A = (6x + 10)°$, $\angle B = 5x°$, $\angle C = (x + y)°$ and $\angle D = (3y - 10)°$. Find x and y, and hence the values of the four angles. **[CBSE (CCE) 2015]**

3. $ABCD$ is a cyclic quadrilateral in which $\angle A = 4y + 20$, $\angle B = 3y - 5$, $\angle C = -4x$, $\angle D = -7x + 5°$. Find the angles of the cyclic quadrilateral.

Answers

1. 70°, 53°, 110°, 127° 2. 130°, 100°, 50°, 80°, $x = 20°$ and $y = 30°$ 3. 120°, 70°, 60°, 110°

FOR MORE PRACTICE AND KNOWLEDGE

Example 28 Solve for x and y using the elimination method:

$$4x + \frac{6}{y} = 15; x - \frac{4}{y} = 7, y \neq 0.$$

Hence find the value of p if $2y = 3px + 7$.

Solution:

$$4x + \frac{6}{y} = 15 \quad ...(i)$$

$$x - \frac{4}{y} = 7 \quad ...(ii) \qquad (y = 0)$$

Inserting $\frac{1}{y} = a$, equations (i) and (ii) become

$$4x + 6a = 15 \quad ...(iii)$$

$$x - 4a = 7 \quad ...(iv)$$

Now, multiplying equation (iv) by 4, we get

$$4x - 16a = 28 \quad ...(v)$$

Subtracting (v) from (iii), we get

$$4x + 06a = 15$$
$$4x - 16a = 28$$
$$- \quad + \quad -$$
$$22a = -13$$

$$a = -\frac{13}{22}$$

When $a = \frac{-13}{22}$, equation (iii) becomes

$$4x + 6\left(\frac{-13}{22}\right) = 15$$

$$4x - \frac{39}{11} = 15 \Rightarrow 4x = 15 + \frac{39}{11}$$

$$4x = \frac{165 + 39}{11} \Rightarrow 4x = \frac{204}{11}$$

$$\Rightarrow \quad x = \frac{51}{11}$$

So $$a = \frac{1}{y} = \frac{-13}{22}$$

$$\therefore \quad y = \frac{1}{a}$$

$$y = \frac{-22}{13}$$

$$\therefore \quad x = \frac{51}{11},\ y = \frac{-22}{13}$$

Now $$2y = 3px + 7$$

$$2 \times \left(-\frac{22}{13}\right) = 3p \times \frac{51}{11} + 7$$

$$-\frac{44}{13} = \frac{153p}{11} + 7$$

$$\Rightarrow \quad -\frac{44}{13} - 7 = \frac{153p}{11}$$

$$\Rightarrow \quad \frac{-44-91}{13} = \frac{153p}{11}$$

$$\Rightarrow \quad \frac{-\cancel{135}}{13} \times \frac{11}{\cancel{153}} = p$$

$$\Rightarrow \quad \frac{-15 \times 11}{13 \times 17} = p$$

$$\therefore \quad \boxed{p = \frac{-165}{221}}$$

Example 29 Solve for x and y using the elimination method:

$$\frac{3}{3x-y} + \frac{2}{3x+y} = 2,\ \frac{10}{3x-y} + \frac{2}{3x+y} = \frac{11}{2} (3x - y \neq 0, 3x + y \neq 0)$$

Solution:

$$\frac{3}{3x-y} + \frac{2}{3x+y} = 2 \quad \text{...(i)}$$

$$\frac{10}{3x-y} + \frac{2}{3x+y} = \frac{11}{2} \quad \text{... (ii)}$$

Inserting $\frac{1}{3x-y} = A$ and $\frac{1}{3x+y} = B$ in equations (i) and (ii), we get

$$3A + 2B = 2 \quad \text{...(iii)}$$

$$10A + 2B = \frac{11}{2} \quad \text{...(iv)}$$

Subtracting (iv) from (iii), we get

$$3A + 2B = 2$$
$$10A + 2B = \frac{11}{2}$$
$$- \quad - \quad -$$
$$-7A = 2 - \frac{11}{2}$$
$$-7A = \frac{4-11}{2} \Rightarrow -7A = -\frac{7}{2}$$

$\Rightarrow$ $$A = \frac{1}{2}.$$

When $A = \frac{1}{2}$, equation (iii) becomes

$$3 \times \frac{1}{2} + 2B = 2 \Rightarrow \frac{3}{2} + 2B = 2$$
$$2B = 2 - \frac{3}{2} \Rightarrow 2B = \frac{1}{2}$$

$\Rightarrow$ $$B = \frac{1}{4}$$

When $$A = \frac{1}{2} \Rightarrow \frac{1}{3x - y} = \frac{1}{2}$$

$\Rightarrow$ $$3x - y = 2 \quad \text{...(v)}$$

When $$B = \frac{1}{4} \Rightarrow \frac{1}{3x + y} = \frac{1}{4}$$

$\Rightarrow$ $$3x + y = 4 \quad \text{...(vi)}$$

Adding (v) and (vi), we get

$$3x - y = 2$$
$$3x + y = 4$$
$$6x = 6$$
$$x = 1$$

Inserting 'x' in equation (vi), we get

$$3 \times 1 + y = 4$$

$\Rightarrow$ $$y = 4 - 3$$
$$y = 1$$

$\therefore$ $$x = 1, y = 1$$

Example 30 Solve for x and y using the cross-multiplication method:

$$ax + by = 1; \; bx + ay = \frac{(a+b)^2}{a^2 + b^2} - 1$$

Solution:

$$ax + by = 1$$
$$ax + by - 1 = 0 \quad \text{...(i)}$$

And $$bx + ay = \frac{(a-b)^2 - (a^2 + b^2)}{a^2 + b^2}$$

$$bx + ay = \frac{\cancel{a^2} + \cancel{b^2} + 2ab - \cancel{a^2} - \cancel{b^2}}{a^2 + b^2}$$

$$bx + ay = \frac{2ab}{a^2 + b^2}$$

$\Rightarrow$ $(a^2 + b^2)bx + (a^2 + b^2)ay - 2ab = 0$ ...(ii)

Here $a_1 = a,\ b_1 = b,\ c_1 = -1$

$a_2 = (a^2 + b^2)b,\ b_2 = (a^2 + b^2)a,\ c_2 = -2ab$

By the cross-multiplication method, we have

$$\Rightarrow \frac{x}{b_1c_2 - b_2c_1} = \frac{y}{c_1a_2 - c_2a_1} = \frac{1}{a_1b_2 - a_2b_1}$$

$$\Rightarrow \frac{x}{b\times(-2ab) - (a^2 + b^2)a\times(-1)} = \frac{y}{(-1)\times(a^2 + b^2)b - (-2ab)\times a} = \frac{1}{a(a^2 + b^2)a - (a^2 + b^2)b\times b}$$

$$\Rightarrow \frac{x}{-2ab + a^3 + b^2a} = \frac{y}{a^2b - b^3} = \frac{1}{(a^2 + b^2)(a^2 - b^2)}$$

$$\Rightarrow \frac{x}{a^3 - ab^2} = \frac{y}{a^2b - b^3} = \frac{1}{(a^2 + b^2)(a^2 - b^2)}$$

$$\Rightarrow \frac{x}{a(a^2 - b^2)} = \frac{y}{b(a^2 - b^2)} = \frac{1}{(a^2 + b^2)(a^2 - b^2)}$$

$$\Rightarrow \frac{x}{a} = \frac{y}{b} = \frac{1}{(a^2 + b^2)}$$

$$\Rightarrow \frac{x}{a} = \frac{1}{a^2 + b^2} \text{ and } \frac{y}{b} = \frac{1}{a^2 + b^2}$$

$$\Rightarrow x = \frac{a}{a^2 + b^2} \text{ and } y = \frac{b}{a^2 + b^2}$$

Example 31 Solve the following pair of linear equations for x and y:

$$2(ax - by) + (a + 4b) = 0;\ 2(bx + ay) + (b - 4a) = 0.$$ [CBSE 2010]

Solution: $2(ax - by) + (a + 4b) = 0$

$\Rightarrow$ $2ax - 2by = -a - 4b$ ...(i)

$2(bx + ay) + (b - 4a) = 0$

$2bx + 2ay = 4a - b$ ...(ii)

Multiplying (i) by 'a' and (ii) by 'b', we get

$2a^2x - 2aby = -a^2 - 4ab$ ...(iii)

$2b^2x + 2aby = 4ab - b^2$ ...(iv)

Adding (iii) and (iv), we get $2a^2x + 2b^2x = -a^2 - \cancel{4ab} + \cancel{4ab} - b^2$

$2(a^2 + b^2)x = -a^2 - b^2$

$$2(a^2 + b^2)x = -\,(a^2 + b^2)$$

$$2x = -\;-\frac{(a^2 + b^2)}{a^2 + b^2}$$

$$2x = -\,1$$

$\Rightarrow$ $$x = -\,\frac{1}{2}$$

Substituting in (i), we get $$-\,a - 2by = -\,a - 4b$$

$$-2\,\cancel{b}\;y = -\,4\,\cancel{b}$$

$$y = 2$$

$\therefore$ So, the solution is $x = -\,\dfrac{1}{2}$ and $y = 2$.

Example 32 Solve for x and y using the substitution method:

$$a^2x - b^2y = a^2 - 2b^2;$$

$$b^2x + a^2y = b^2 + 2a^2$$

Solution:

$$a^2x - b^2y = a^2 - 2b^2 \qquad \text{...(i)}$$

$$b^2x + a^2y = b^2 + 2a^2 \qquad \text{...(ii)}$$

From equation (i), we get $$a^2x - b^2y = a^2 - 2b^2$$

$\Rightarrow$ $$a^2x = a^2 - 2b^2 + b^2y$$

$\Rightarrow$ $$x = \frac{a^2 - 2b^2 + b^2y}{a^2} \qquad \text{...(iii)}$$

Substituting '$x = \dfrac{a^2 - 2b^2 + b^2y}{a^2}$' in equation (ii), we get

$$b^2\left(\frac{a^2 - 2b^2 + b^2y}{a^2}\right) + a^2y = b^2 + 2a^2$$

$\Rightarrow$ $$\frac{b^2 - 2b^2 + b^4y}{a^2} + a^2y = b^2 + 2a^2$$

$\Rightarrow$ $$\frac{b^2a^2 - 2b^4 + b^4y + a^4y}{a^2} = b^2 + 2a^2$$

$\Rightarrow$ $$b^2a^2 - 2b^4 + (b^4 + a^4)y = a^2b^2 + 2a^4$$

$\Rightarrow$ $$(b^4 + a^4)y = \cancel{a^2b^2} + 2a^4 - \cancel{a^2b^2} + 2b^4$$

$\Rightarrow$ $$(b^4 + a^4)y = 2\,(a^4 + b^4)$$

$\Rightarrow$ $$y = \frac{2\,\cancel{(a^2 + b^2)}}{\cancel{(b^4 + a^4)}} \Rightarrow y = 2.$$

So, when $y = 2$, equation (iii) becomes $x = \frac{a^2 - 2b^2 + b^2 \times 2}{a^2}$

$$x = \frac{a^2 - 2b^2 + 2b^2}{a^2}$$

$$x = \frac{a^2}{a^2}$$

$\Rightarrow$ $x = 1$

Example 33 Solve for x and y:

$$99x + 101y = 499$$

$$101x + 99y = 501$$

[CBSE (CCE) 2011]

Solution:

$$99x + 101y = 499 \quad \text{...(i)}$$

$$101x + 99y = 501 \quad \text{...(ii)}$$

Adding (i) and (ii), we get

$$99x + 101y = 499$$

$$101x + 99y = 501$$

$$200x + 200y = 1000$$

$$200(x + y) = 1000$$

$$x + y = \frac{1000}{200}$$

$$x + y = 5 \quad \text{...(iii)}$$

Now, subtracting (ii) from (i), we get

$$99x + 101y = 499$$

$$101x + 99y = 501$$

$$- \quad - \quad -$$

$$-2x + 2y = -2$$

$\Rightarrow$ $x - y = 1$ ...(iv)

Now, adding (iii) and (iv)

$$x + y = 5$$

$$x - y = 1$$

$\Rightarrow$ $2x = 6$

$\Rightarrow$ $x = \frac{6}{2} = 3 \Rightarrow x = 3$

So, when $x = 3$, equation (iii) becomes $3 + y = 5$

$$y = 5 - 3$$

$$y = 2$$

[Ans. $x = 3, y = 2$**]**

Example 34 For what value of k will the following system of equations $2x + ky = 1$, $3x - 5y = 7$ have (i) a unique solution, (ii) no solution.

Solution: (i) For a unique solution $\frac{a_1}{a_2} \neq \frac{b_1}{b_2}$

$$\Rightarrow \quad \frac{2}{3} \neq \frac{k}{-5}$$

$$\Rightarrow \quad \frac{-10}{3} \neq k$$

(ii) For no solution $\frac{a_1}{a_2} = \frac{b_1}{b_2} \neq \frac{c_1}{c_2}$

$$\Rightarrow \quad \frac{2}{3} = \frac{k}{-5} \neq \frac{1}{7}$$

$$\Rightarrow \quad k = \frac{-10}{3}$$

Example 35 For what value of k will the equations $x + 2y + 7 = 0$, $2x + ky + 14 = 0$ represent coincident lines?

Solution: For coincident lines, we have infinitely many solutions.

$$\therefore \quad \frac{a_1}{a_2} = \frac{b_1}{b_2} = \frac{c_1}{c_2}$$

Here

$$x + 2y + 7 = 0$$
$$2x + ky + 14 = 0$$

$$\therefore \quad a_1 = 1, b_1 = 2, c_1 = 7$$

$$\therefore \quad a_2 = 2, b_2 = k, c_2 = 14$$

$$\frac{1}{2} = \frac{2}{k} = \frac{7}{14}$$

So

$$\frac{1}{2} = \frac{2}{k} \qquad \frac{2}{k} = \frac{7}{14}$$
$$k = 4 \qquad 7k = 28$$
$$k = \frac{28}{7} \Rightarrow 4$$

$\therefore$ For $k = 4$, the lines will coincide.

Example 36 Find the values of a and b for which the following system of linear equations has an infinite number of solutions.

$$2x - 3y = 7, (a + b)x - (a + b - 3)y = 4a + b$$

Solution:

$$2x - 3y = 7$$
$$(a + b)x - (a + b - 3)y = 4a + b$$

Here $a_1 = 2,\ b_1 = -3,\ c_1 = -7$

$$a_2 = (a + b),\ b_2 = -(a + b - 3),\ c_2 = -(4a + b)$$

For infinitely many solutions $\dfrac{a_1}{a_2} = \dfrac{b_1}{b_2} = \dfrac{c_1}{c_2}$

$$\Rightarrow \quad \frac{2}{(a+b)} = \frac{-3}{-(a+b-3)} = \frac{-7}{-4(a+b)}$$

Taking $\dfrac{2}{(a+b)} = \dfrac{3}{(a+b-3)}$

$$\Rightarrow \quad \frac{2}{a+b} = \frac{3}{a+b-3}$$

$$\Rightarrow \quad 2a + 2b - 6 = 3a + 3b$$

$$\Rightarrow \quad a + b + 6 = 0 \quad \text{...(i)}$$

$$\frac{3}{a+b-3} \quad \frac{7}{4a+b}$$

Now $\dfrac{3}{+(a+b-3)} = \dfrac{7}{(4a+b)}$

$$\Rightarrow \quad \frac{3}{a+b-3} = \frac{7}{4a+b}$$

$$\Rightarrow \quad 12a + 3b = 7a + 7b - 21$$

$$\Rightarrow \quad 5a - 4b + 21 = 0 \quad \text{...(ii)}$$

Now, we have $a + b = -6$

$$5a - 4b = -21$$

Multiplying (i) by 5 and (ii) by 1, we get

$$5a + 5b = -30 \quad \text{...(iii)}$$

$$5a - 4b = -21 \quad \text{...(iv)}$$

Subtracting (iv) from (iii), we get

$$\begin{array}{r} 5a + 5b = -30 \\ 5a - 4b = -21 \\ - \quad + \qquad + \\ \hline 9b = -9 \end{array}$$

$$b = -\frac{\not{9}}{\not{9}}$$

$$b = -1$$

Inserting 'b' in (i), we get $a+(-1)=-6$

$\Rightarrow$ $a-1=-6$

$\Rightarrow$ $a=-6+1$

$\Rightarrow$ $a=-5$ **[Ans.** $a=-5, b=-1$**]**

Example 37 For what value of k will the following pair of linear equation have no solution?

$$2x+ky=1$$
$$6x+4y=3$$

Solution:

$$2x+ky=1$$
$$6x+4y=3$$

For no solution $\dfrac{a_1}{a_2}=\dfrac{b_1}{b_2}\neq\dfrac{c_1}{c_2}$

$\Rightarrow$ $\dfrac{2}{6}=\dfrac{k}{4}\neq\dfrac{1}{3}$

$\Rightarrow$ $\dfrac{2}{6}=\dfrac{k}{4}$ and $\dfrac{k}{4}\neq\dfrac{1}{3}$

$\Rightarrow$ $\dfrac{1}{3}=\dfrac{k}{4}$ and $\dfrac{k}{4}\neq\dfrac{1}{3}$

$\Rightarrow$ $k=\dfrac{4}{3}$ and $k\neq\dfrac{4}{3}$

$\therefore$ No value of k has no solution.

Example 38 Draw the graphs of the equations $x-y+1=0$ and $3x+2y-12=0$. Determine the coordinates of the vertices of the triangle formed by these lines and the x axis, and shade the triangular region.

Solution: First equation is $x-y+1=0$

$$x=y-1$$

x	1	–2	2
y	2	–1	3

Second equation is $3x+2y=12$

$\Rightarrow$ $3x=12-2y \Rightarrow x=\dfrac{12-2y}{3}$

x	4	2	0
y	0	3	6

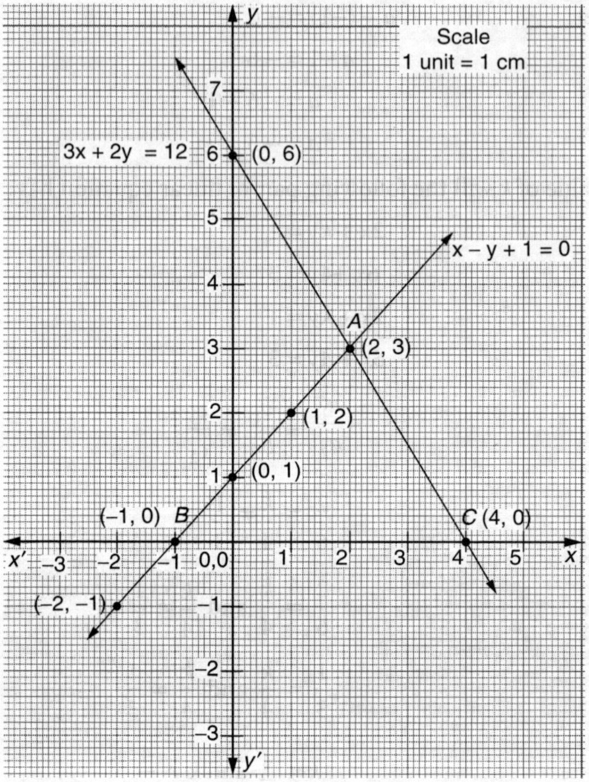

The required triangle is *ABC*. The coordinates of its vertices are $A(2, 3)$, $B(-1, 0)$, $C(4, 0)$.

Example 39 Two numbers are in the ratio 4:5. If 30 is subtracted from each number, the ratio becomes 1:2. Form the pair of linear equations for the above situation and represent them graphically.

Solution: Let the numbers be x and y.

$\therefore$ According to the question $\quad \dfrac{x}{y} = \dfrac{4}{5} \Rightarrow 5x = 4y$

$\Rightarrow \quad 5x - 4y = 0$

When 30 is subtracted from both, the ratio becomes 1:2.

$\Rightarrow \quad \dfrac{x-30}{y-30} = \dfrac{1}{2}$

$\Rightarrow \quad 2x - 60 = y - 30$

$\Rightarrow \quad 2x - y - 30 = 0$

The pair of linear equations is $\quad 5x - 4y = 0 \quad$ ...(i)

$2x - y - 30 = 0 \quad$ ...(ii)

$5x - 4y = 0$ is

x	0	4	8
y	0	5	10

$2x - y - 30 = 0$ is

x	15	12	10
y	0	−6	−10

Graphical representation

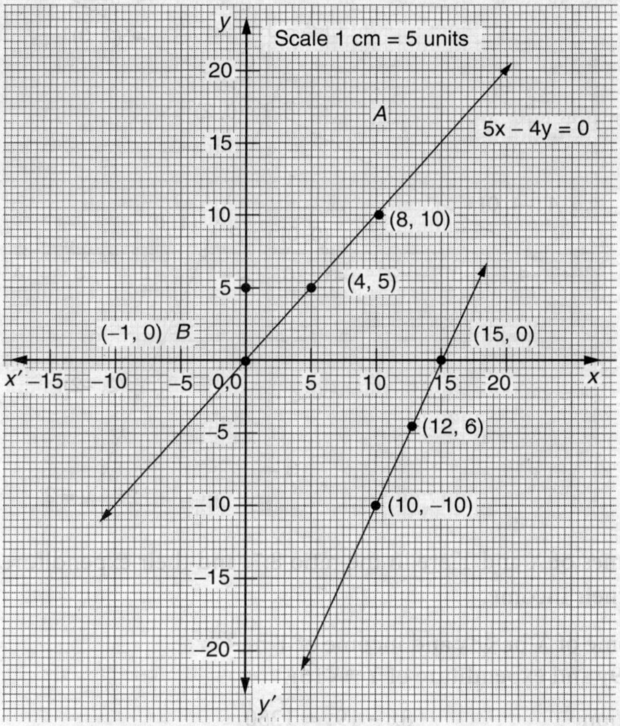

Example 40 Solve the equations graphically: $2x + y = 2$; $2y - x = 4$. What is the area of the triangle formed by the two lines and the line $y = 0$? **[CBSE (CCE) 2011]**

Solution:

$$2x + y = 2 \quad \text{...(i)}$$

$$2y - x = 4 \quad \text{...(ii)}$$

$$2x + y = 2$$

x	1	0	2
y	0	2	−2

$$2y - x = 4$$

x	0	− 4	2
y	2	0	3

Area of $\Delta = \frac{1}{2} \times AB \times CO = \frac{1}{2} \times 5 \times 2 = 5$ square units.

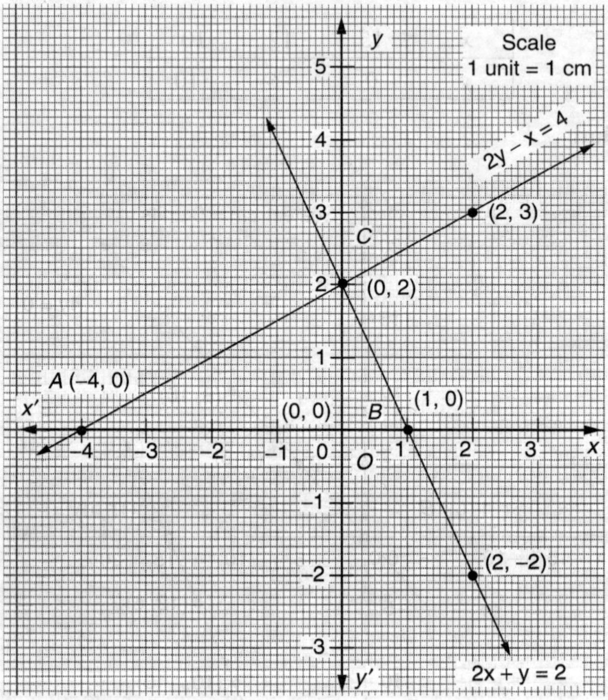

Example 41 The incomes of X and Y are in the ratio 8:7, and their expenditures are in the ratio 19:16. If X saves ₹1250, find their income.

Solution: Let the income of X be ₹$8x$ and income of Y be ₹$7x$.
Expenditure of X = ₹$19y$ and expenditure of Y = ₹$16y$.

According to the question $\quad 8x - 19y = 1250 \quad$ ...(i)

$$7x - 16y = 1250 \quad \text{...(ii)}$$

Multiplying equation (i) by 7 and equation (ii) by 8, we get

$$56x - 133y = 8750 \quad \text{...(iii)}$$

$$56x - 128y = 10000 \quad \text{...(iv)}$$

Subtracting (iv) from (iii), we get

$$\begin{array}{r} 56x - 133y = 8750 \\ 56x - 128y = 10000 \\ - \quad + \quad - \\ \hline +5y = +1250 \end{array}$$

$$y = \frac{1250}{5} \Rightarrow 250$$

Inserting $y = 250$ in equation (i), we get

$$8x - 19 \times 250 = 1250$$

$$8x - 4750 = 1250$$

$$8x = 1250 + 4750$$

$$8x = 6000$$

$$x = \frac{6000}{8} \Rightarrow x = 750$$

$\therefore$ Income of $X = 8 \times 750 = ₹6000$. Income of $Y = 7 \times 750 = ₹5250$.

Example 42 A man travels 600 km partly by train and partly by car. If he covers 400 km by train and the rest by car, it takes him 6 hours and 30 minutes. However, if he travels 200 km by train and the rest by car, he takes half an hour longer. Find the speed of the train and car.

Solution: Total distance = 600 km

Let the speed of the train be x km/h and the speed of the car be y km/h.

Case I: When 400 km is covered by train and the rest by car.

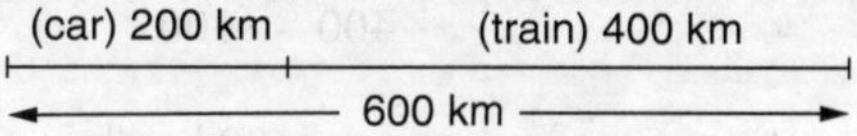

$\Rightarrow$ $$\frac{400}{x} + \frac{200}{y} = \frac{13}{2} \qquad \text{...(i)}$$

Case II: When 200 km is covered by train and the rest by car.

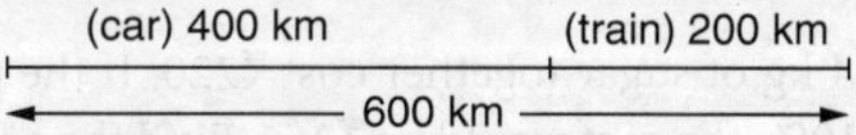

$\Rightarrow$ $$\frac{200}{x} + \frac{400}{y} = 7 \qquad \text{...(ii)}$$

Inserting $\frac{1}{x} = p$ and $\frac{1}{y} = q$ in (i) and (ii), we get

$$400p + 200q = \frac{13}{2} \qquad \text{...(iii)}$$

$$200p + 400q = 7 \qquad \text{...(iv)}$$

Subtracting (iv) from (iii), we get

$$\begin{aligned} 400p + 200q &= \frac{13}{2} \\ 200p + 400q &= 7 \\ - \quad - \quad & \quad - \\ \hline 200p - 200q &= -\frac{1}{2} \\ \hline 200(p - q) &= -\frac{1}{2} \end{aligned}$$

$\Rightarrow$ $$p = q - \frac{1}{400} \qquad \text{...(v)}$$

Inserting $p = q - \frac{1}{400}$ in (iii), we get

$$400\left(q - \frac{1}{400}\right) + 200q = \frac{13}{2}$$

$$\Rightarrow \quad 400q - 1 + 200q = \frac{13}{2}$$

$$\Rightarrow \quad 600q = \frac{13}{2} + 1$$

$$\Rightarrow \quad 600q = \frac{15}{2}$$

$$\Rightarrow \quad q = \frac{15}{2} \times \frac{1}{600} = \frac{1}{80}$$

Inserting $q = \frac{1}{80}$ in (v), we get $p = \frac{1}{80} - \frac{1}{400} = \frac{5-1}{400}$

$$p = \frac{4}{400} = \frac{1}{100}$$

$$\therefore \quad x = \frac{1}{p} = 100 \text{ km/h}$$

And $y = \frac{1}{q} = 80$ km/h

$\therefore$ The speed of the train = 100 km/h and the speed of the car = 80 km/h.

Example 43 One kg of tea and 4 kg of sugar together cost ₹220. If the price of sugar increases by 50% and the price of tea increases by 10%, the cost would be ₹266. Find the original cost per kilogram of each.

Solution: Let the price of sugar = ₹x/kg and the price of tea = ₹y/kg.

$$\therefore \quad 4x + y = 220 \quad \text{...(i)}$$

New price of sugar $= x + \frac{50}{100}x = ₹\frac{3}{2}x$/kg

New price of tea $= y + \frac{10}{100}y = ₹\frac{11}{10}y$/kg

According to the question $4 \times \frac{3}{2}x + \frac{11}{10}y = 266$

$$\Rightarrow \quad 6x + \frac{11}{10}y = 266 \quad \text{...(ii)}$$

Multiplying equation (i) by 3 and equation (ii) by 2, we get

$$12x + 3y = 660 \quad \text{...(iii)}$$

$$12x + \frac{11}{5}y = 532 \quad \text{...(iv)}$$

Subtracting (iv) from (iii), we get

$$\cancel{12x} + 3y = 660$$

$$\cancel{12x} + \frac{11}{5}y = 532$$

$$- \quad - \quad -$$

$$\frac{4}{5}y = 128$$

$$\Rightarrow \quad \frac{4}{5}y = 128$$

$$\Rightarrow \quad y = \frac{128 \times 5}{4}$$

$$\Rightarrow \quad y = 32 \times 5 \Rightarrow y = 160$$

Inserting $y = 160$ in equation (i), we get $4x + 160 = 220$

$$\Rightarrow \quad 4x = 220 - 160$$

$$\Rightarrow \quad 4x = 60$$

$$\Rightarrow \quad x = \frac{60}{4} = 15$$

∴ Price of sugar = ₹15/kg and price of tea = ₹160/kg.

Example 44 In a deer park, the number of heads and the number of legs of deer and human visitors were counted and it was found that there were 39 heads and 132 legs. Find the number of deer and human visitors in the park.

Solution: Let the number of humans be x and deer be y.

Then $\quad x + y = 39 \quad$ ...(i)

And $\quad 2x + 4y = 132$

$$x + 2y = 66 \quad \text{...(ii)}$$

Subtracting (ii) from (i), we get

$$x + y = 39$$
$$x + 2y = 66$$
$$- \quad - \quad -$$
$$-y = -27$$
$$y = 27$$

Inserting 'y' in (i), we get $\quad x + 27 = 39$

$$x = 39 - 27$$
$$x = 12$$

∴ The number of humans in the park is 12 and the number of deer is 27.

Example 45 A number consists of three digits whose sum is 17. The middle one exceeds the sum of the other two by 1. If the digits are reversed, the number is diminished by 396. Find the number.

Solution: Let the digit at the unit's place = x and the digit at the hundred's place = y.

∴ Middle digit = $x + y + 1$

∴ Number = $100y + 10(x + y + 1) + x$

Number obtained by reversing the digits = $100x + 10(x + y + 1) + y$

According to the question $\quad x + y + (x + y + 1) = 17$

$$\Rightarrow \quad 2x + 2y = 16 \; x + y = 8 \quad \text{... (i)}$$

And $\quad 100x + 10(x + y + 1) + y = 100y + 10(x + y + 1) + x - 396$

$$\Rightarrow \quad 99x - 99y = -396$$

$$\Rightarrow \quad x - y = \frac{396}{99}$$

$$\Rightarrow \quad x - y = -4 \quad \text{...(ii)}$$

Adding (i) and (ii), we get $2x = 4$

$$x = 2$$

∴ Inserting 'x' in (i), we get $2 + y = 8$

$$y = 6$$

∴ $x = 2$, $y = 6$ and numbers = 692.

Example 46 A railway half-ticket costs half the full fare and the reservation charge is the same as on the full ticket. One reserved first-class ticket from Mumbai to Ahmedabad costs ₹216 and one full and one half reserved first-class ticket costs ₹327. What is the basic first-class full fare and what is the reservation charge?

Solution: Let the full fare = ₹x and reservation charges = ₹y.

According to the question $x + y = 216$

$$\Rightarrow \quad y = 216 - x \quad \text{...(i)}$$

And

$$(x + y) + \left(\frac{1}{2}x + y\right) = 327$$

$$\Rightarrow \quad \frac{3}{2}x + 2y = 327$$

$$\Rightarrow \quad 3x + 4y = 654 \quad \text{...(ii)}$$

Using (i), we get $3x + 4(216 - x) = 654$

$$-x = 654 - 864$$

$$+x = +210$$

$$x = 210.$$

When $x = 210$, equation (i) becomes $y = 216 - 210 = 6$.

∴ Full fare = ₹210 and reservation charge = ₹6.

Example 47 Four men and 6 boys can finish a piece of work in 5 days while 3 men and 4 boys can finish it in 7 days. Find the time taken by 1 man alone or that by a boy alone. **[CBSE (CCE) 2011]**

Solution: One man can finish the work in x days and one boy can finish the work in y days.

∴ One man's one day's work = $\frac{1}{x}$ and one boy's one day's work = $\frac{1}{y}$.

According to the question

$$\frac{4}{x} + \frac{6}{y} = \frac{1}{5} \quad \text{...(i)}$$

And

$$\frac{3}{x} + \frac{4}{y} = \frac{1}{7} \quad \text{...(ii)}$$

Inserting $\frac{1}{x} = A$ and $\frac{1}{y} = B$, equations (i) and (ii) become

$$4A + 6B = \frac{1}{5} \quad \text{...(iii)}$$

$$3A + 4B = \frac{1}{7} \quad \text{...(iv)}$$

From equation (iii), we get $4A = \frac{1}{5} - 6B$

$$4A = \frac{1-30B}{5}$$

$$\Rightarrow \quad A = \frac{1-30B}{20} \quad \text{...(v)}$$

Substituting $A = \frac{1-30B}{20}$ in equation (iv), we get

$$\frac{3\times(1-30B)}{20} + 4B = \frac{1}{7} \Rightarrow \frac{3-90B+80B}{20} = \frac{1}{7}$$

$$\Rightarrow \quad 3 - 10B = \frac{20}{7} \Rightarrow -10B = \frac{20}{7} - 3$$

$$\Rightarrow \quad 10B = \frac{1}{7} \Rightarrow B = \frac{1}{70}$$

When $B = \frac{1}{70}$, equation (v) becomes $A = \frac{1-30\times\frac{1}{70}}{20}$

$$\Rightarrow \quad A = \frac{1-\frac{3}{7}}{20}$$

$$\Rightarrow \quad A = \frac{4}{7\times 20} \Rightarrow A = \frac{1}{35}$$

When $A = \frac{1}{35} \Rightarrow \frac{1}{x} = \frac{1}{35} \Rightarrow x = 35$

When $B = \frac{1}{70} \Rightarrow \frac{1}{y} = \frac{1}{70} \Rightarrow y = 70$

$\therefore$ Time taken by one man = 35 days. Time taken by one boy = 70 days.

Multiple Choice Questions

Choose the correct answer for each of the following questions:

1. If $2^{x+y} = 2^{x-y} = \sqrt{8}$, then the value of y is

(a) $\frac{1}{2}$ (b) $\frac{3}{2}$ (c) 0 (d) None of these

2. If the lines given by $3x + 2ky = 2$ and $2x + 5y + 1 = 0$ are parallel, then the value of k is

(a) $\frac{-5}{4}$ (b) $\frac{2}{5}$ (c) $\frac{3}{2}$ (d) $\frac{15}{4}$

3. If a pair of linear equations is consistent, their graph lines will be
(a) Parallel (b) Always coincident
(c) Always intersecting (d) Intersecting or coincident

4. If a pair of linear equations is inconsistent, their graph lines will be
(a) Parallel (b) Always coincident
(c) Always intersecting (d) Intersecting or coincident

5. In a ΔABC, $\angle C = 3\angle B = 2(\angle A + \angle B)$, then $\angle B = ?$
(a) 20° (b) 40° (c) 100° (d) 110°

6. In a cyclic quadrilateral *ABCD*, it is given that $\angle A = (x + y + 10)°$, $\angle B = (y + 20)°$, $\angle C = (x + y - 30)°$ and $\angle D = (x + y)°$, then $\angle B = ?$
(a) 70° (b) 80° (c) 100° (d) 110°

7. The sum of the digits of a two-digit number is 15. The number obtained by interchanging the digits exceeds the given number by 9. The number is
(a) 96 (b) 69 (c) 87 (d) 78

8. In a given fraction, if 1 is subtracted from the numerator and 2 is added to the denominator, it becomes $\frac{1}{2}$. If 7 is subtracted from the numerator and 2 is subtracted from the denominator, it becomes $\frac{1}{3}$.
The fraction is
(a) $\frac{13}{24}$ (b) $\frac{15}{26}$ (c) $\frac{16}{27}$ (d) $\frac{16}{21}$

9. Five years hence, the age of a man shall be 3 times the age of his son, while five years earlier, the age of the man was 7 times the age of his son. The current age of the man is
(a) 45 years (b) 50 years (c) 47 years (d) 40 years

10. The graphs of the equations $5x - 15y = 8$ and $3x - 9y = \frac{24}{5}$ are two lines that are
(a) Coincident (b) Parallel
(c) Intersecting exactly at one point (d) Perpendicular to each other

11. If $x = -y$ and $y > 0$, which of the following is incorrect?
(a) $x^2 y > 0$ (b) $x + y = 0$ (c) $xy < 0$ (d) $\frac{1}{x} - \frac{1}{y} = 0$

12. A man has some hens and cows. If the number of heads is 48 and the number of feet is 140, how many cows are there?
(a) 20 (b) 22 (c) 24 (d) 21

13. If $\frac{2x}{3} - \frac{y}{2} + \frac{1}{6} = 0$ and $\frac{x}{2} + \frac{2y}{3} = 3$, then
(a) $x = 2, y = 3$ (b) $x = -2, y = 3$ (c) $x = 2, y = -3$ (d) $x = -2, y = -3$

14. If $\frac{2x + y + 2}{5} = \frac{3x - y + 1}{3} = \frac{3x + 2y + 1}{6}$, then
(a) $x = 1, y = 1$ (b) $x = -1, y = = -1$
(c) $x = 1, y = 2$ (d) $x = 2, y = 1$

15. If $4x + 6y = 3xy$ and $8x + 9y = 5xy$, then
(a) $x = 2, y = 3$ (b) $x = 1, y = 2$ (c) $x = 3, y = 4$ (d) $x = 1, y = 1$

16. Find the value of 'a' so that the point (3, a), lies on the line represented by $2x - 3y = 5$.

(a) 0 (b) $\frac{1}{3}$ (c) 3 (d) $\frac{2}{3}$

17. Find the co-ordinate where the line $x - y = 8$ will intersect the y axis.

(a) (0, –8) (b) (8, 0) (c) (–8, 0) (d) (0, 8)

18. The graphical representation of the equations $x + 2y = 3$ and $2x + 4y + 7 = 0$ gives a pair of ________ lines.

(a) Intersecting (b) Parallel (c) Coincident (d) Overlapping

19. The pair of equations $y = 2$ and $y + 7 = 0$ has

(a) One solution (b) Two solutions
(c) Infinitely many solutions (d) No solution

20. If $x = a$, $y = b$ is the solution of the pair of equations $x - y = 2$ and $x + y = 4$, then the respective values of a and b are

(a) 3, 5 (b) 5, 3 (c) 3, 1 (d) –1, –3 **[CBSE 2018 (C)]**

21. The sum of the digits of a two-digit number is 9. If 27 is added to it, the digits of the number are reversed. The number is

(a) 25 (b) 72 (c) 63 (d) 36 **[NCERT Exemplar]**

22. The pair of equations $x = a$ and $y = b$ graphically represents lines which are

(a) Parallel (b) Intersecting at (b, a) (c) Coincident (d) Intersecting at (a, b)

[NCERT Exemplar]

23. For what value of k do the equations $3x - y + 8 = 0$ and $6x - ky = -16$ represent coincident lines?

(a) $\frac{1}{2}$ (b) $-\frac{1}{2}$ (c) 2 (d) –2 **[NCERT Exemplar]**

24. One equation of a pair of dependent linear equations is $-5x + 7y = 2$. The second equation can be

(a) $10x + 14y + 4 = 0$ (b) $-10x - 14y + 4 = 0$ (c) $-10x + 14y + 4 = 0$ (d) $10x - 14y = -4$

[NCERT Exemplar]

25. A fraction becomes $\frac{4}{5}$ when 1 is added to the numerator and denominator. However, if we subtract 5 from them, it becomes $\frac{1}{2}$. Then, the numerator of the fraction is

(a) 6 (b) 7 (b) 8 (d) 9

26. The sum of the digits of a two-digit number is 9. If 27 is added to it, the digits of the number are reversed. The number is

(a) 25 (b) 72 (c) 63 (d) 36

27. The father's age is six times his son's age. Four years hence, the age of the father will be four times his son's age. The current ages, in years, of the son and the father are, respectively,

(a) 4 and 24 (b) 5 and 30 (c) 6 and 36 (d) 3 and 24

[NCERT Exemplar]

28. If a pair of linear equations is consistent, then the lines will be

(a) Always intersecting (b) Parallel
(c) Always coincident (d) Intersecting or coincident

29. The pair of equations $ax + 2y = 7$ and $3x + by = 16$ represents parallel lines if

(a) $a = b$ (b) $3a = 2b$ (c) $2a = 3b$ (d) $ab = 6$

30. The lines represented by the equations $a_1x + b_1y + c_1 = 0$ and $a_2x + b_2y + c_2 = 0$ are coincident if

(a) $\frac{a_1}{a_2} \neq \frac{b_1}{b_2}$ (b) $\frac{a_1}{a_2} = \frac{b_1}{b_2} = \frac{c_1}{c_2}$ (c) $\frac{a_1}{a_2} = \frac{b_1}{b_2} \neq \frac{c_1}{c_2}$ (d) $\frac{a_1}{a_2} \neq \frac{b_1}{b_2} \neq \frac{c_1}{c_2}$

31. The value of c for which the pair of equations $cx - y = 2$ and $6x - 2y = 4$ will have infinitely many solutions is

(a) 3 (b) – 3 (c) – 12 (d) No value

32. The pair of equations $x = 2$ and $y = 3$ graphically represents lines which are

(a) Coincident (b) Parallel
(c) Intersecting at (3, 2) (d) Intersecting at (2, 3)

33. The pair of linear equations $x - 2y = 5$ and $2x - 4y = 1$ has

(a) Many solutions (b) No solution (c) One solution (d) Two solutions

34. The pair of linear equations $x - 2y = 0$ and $3x + 4y = 20$ has

(a) One solution (b) Two solutions (c) Many solutions (d) No solutions

35. The solution of the system of equations $\sqrt{2}\,x - \sqrt{5}\,y = 0$ and $\sqrt{5}\,x + \sqrt{2}\,y = 0$ is

(a) $x = \sqrt{2}, y = \sqrt{5}$ (b) $x = \sqrt{5}, y = \sqrt{2}$ (c) $x = 0, y = 0$ (d) $x = \sqrt{2}, y = 0$

36. The value of k for which the pair of linear equations $4x + 6y - 1 = 0$ and $2x + ky - 7 = 0$ represents parallel lines is

(a) $k = 3$ (b) $k = 2$ (c) $k = 4$ (d) $k = -2$

37. If $x = a$, $y = b$ is the solution of the equations $x - y = 2$ and $x + y = 4$, then the values of a and b are, respectively,

(a) 3 and 1 (b) 5 and 3 (c) 3 and 5 (d) – 1 and – 3

38. The pair of linear equations $3x - 5y = 7$ and $6x - 10y = 7$ has

(a) One solution (b) Two solutions
(c) Infinitely many solutions (d) No solution

Fill in the Blanks

Complete the following statements with appropriate answers(s) in the blank space(s):

39. On solving the following equations: $\frac{5}{x} + 6y = 3$; $\frac{-5}{x} + 3y = 6$, the values of x and y are ________ and ________, respectively.

40. On solving the following pair of linear equations: $2x - y = 2$; $5x + 2y = 14$ by substitution method, the values of x and y are ____________ and ____________, respectively.

41. The pair of linear equations $2ax + by = a$, $6ax + 3by = 3a$; $a, b \neq 0$ is ______________.

42. The graphic representation of the equations $x + 2y = 3$ and $2x + 4y + 7 = 0$ gives a pair of ____________ lines.

43. If $x = a$ and $y = b$ is the solution of pair of equations $x - y = 2$ and $x + y = 4$, then $a =$ ____________ and $b =$ ____________.

44. There are two positive numbers such that sum of twice the first and thrice the second is 39, while the sum of thrice the first and twice the second is 36. The larger of the two is ______________.

True or False

45. For the pair of equations: $\lambda x + 3y = -7$, $2x + 6y = 14$ to have infinitely many solutions, the value of λ should be 1. Is the statement true? Given reasons. **[NCERT Exemplar]**

46. For all real values of c, the pair of equations $x - 2y = 8$; $5x - 10y = c$ has a unique solution. Justify whether it is true or false. **[NCERT Exemplar]**

Assertion–Reason

The following questions consist of two statements – Assertion (A) and Reason (R). Answer these questions by selecting the appropriate options given below.

(A) Both A and R are true and R is the correct explanation for A.
(B) Both A and R are true and R is not the correct explanation for A.
(C) A is true but R is false.
(D) A is false but R is true.

47. **Assertion (*A*):** If in a cyclic quadrilateral, one angle is 40° then its opposite angle is 140°.
Reason (*R*): Sum of opposite angles in a cyclic quadrilateral is equal to 360°.

48. **Assertion (*A*):** $4x + 3y = 18$ is a line which is parallel to the x axis.
Reason (*R*): The graph of linear equation $ax = b$ where $a \neq 0$ is parallel to the y axis.

49. **Assertion (*A*):** The graphs of the equations $2x + 3y - 12 = 0$ and $x - 9y - 8 = 0$ are two lines which intersect at one point.
Reason (*R*): If the pair of linear equations has unique solutions, then the lines on the graph are parallel.

50. **Assertion (*A*):** For $k = 6$, the system of linear equations $x + 2y + 3 = 0$ and $3x + ky + 6 = 0$ is inconsistent.
Reason (*R*): The system of linear equations $a_1x + b_1y + c_1 = 0$ and $a_2x + b_2y + c_2 = 0$ is inconsistent if $\frac{a_1}{a_2} = \frac{b_1}{b_2} = \frac{c_1}{c_2}$.

Subjective Questions

51. If $2x + y = 23$ and $4x - y = 19$, find the values of $5y - 2x$ and $\frac{y}{x} - 2$. **[NCERT Exemplar]**

52. Find the solution of the pair of equations $\frac{x}{10} + \frac{y}{5} - 1 = 0$ and $\frac{x}{8} + \frac{y}{6} = 15$. Hence, find λ if $y = \lambda x + 5$. **[NCERT Exemplar]**

53. Solve the pair of equations:

$$43x + 67y = -24$$
$$67x + 43y = 24$$

[NCERT Exemplar]

54. Solve the pair of equations: $\frac{x}{a} + \frac{y}{b} = a + b$; $\frac{x}{a^2} + \frac{y}{b^2} = 2$; $a, b \neq 0$ **[NCERT Exemplar]**

55. Solve the following pair of equations:

$$21x + 47y = 110$$
$$47x + 21y = 162$$

[NCERT Exemplar]

Solve the following pairs of linear equations by the substitution method:

56. $0.2x + 0.3y = 1.3$
$0.4x + 0.5y = 2.3$ **[NCERT]**

57. $\sqrt{2}\,x+\sqrt{3}\,y=0,\ \sqrt{3}\,x-\sqrt{8}\,y=0$ **[NCERT]**

58. $\frac{3x}{2}-\frac{5y}{2}=-2,\ \frac{x}{3}+\frac{y}{2}=\frac{13}{6}$ **[NCERT]**

59. $\sqrt{7}x+\sqrt{13}y=0,\ \sqrt{5}x+\sqrt{17}y=0$

60. $\frac{ax}{b}-\frac{by}{a}=a+b\ \ ax-by=2ab$

61. Solve for x and y: $a^2x+b^2y=c^2; b^2x+a^2y=d^2$.

62. Solve for x and y: $x+20y-1000=0$

$x+26y-1180=0$

63. Solve for x and y: $\frac{x}{a}-\frac{y}{b}=0,\ ax+by=a^2+b^2$.

Solve the following pairs of equations by reducing them to a pair of linear equations:

64. $\frac{2}{\sqrt{x}}+\frac{3}{\sqrt{y}}=2,\ \frac{4}{\sqrt{x}}-\frac{9}{\sqrt{y}}=-1$ **[NCERT]**

65. $\frac{5}{x-1}+\frac{1}{y-2}=2,\ \frac{6}{x-1}-\frac{3}{y-2}=1$ **[NCERT]**

66. $6x+3y=6xy,\ 2x+4y=5xy$ **[NCERT]**

67. $\frac{1}{3x+y}+\frac{1}{3x-y}=\frac{3}{4};\ \frac{1}{2(3x+y)}-\frac{1}{2(3x-y)}=\frac{1}{8}$ **[NCERT]**

68. Solve the following equations for x and y.

$\frac{a^2}{x}-\frac{b^2}{y}=0;\ \frac{a^2b}{x}+\frac{b^2a}{y}=a+b;\ x, y\neq 0$

69. Solve for x and y: $\frac{6}{x+y}=\frac{7}{x-y}+3;\ \frac{1}{2(x+y)}=\frac{1}{3(x-y)}$, where $x+y\neq 0$ and $x-y\neq 0$.

70. Solve for u and v: $2(3u-v)=5uv;\ 2(u+3v)=5uv$.

71. Solve for x and y: $\frac{2}{y}+\frac{3}{x}=\frac{7}{xy};\ \frac{6}{y}+\frac{9}{x}=\frac{11}{xy},\ x\neq 0, y\neq 0$.

72. Solve for x and y: $4x+\frac{6}{y}=15;\ x-\frac{4}{y}=7,\ y\neq 0$. Hence find the value of p if $2y=3px+7$.

73. Solve for x and y: $47x+31y=63;\ 31x+47y=15$.

74. Solve the system of equations for x and y: $\frac{b^2x}{a}-\frac{a^2y}{b}=ab\,(a+b)$ and $b^2x-a^2y=2a^2b^2$.

75. Solve the following equations for x and y: $mx-ny=m^2+n^2,\ x+y=2m$.

76. Solve the following systems of linear equations:

$a(x+y)+b(x-y)=a^2-ab+b^2;\ a(x+y)-b(x-y)=a^2+ab+b^2$

77. Solve for x and y: $bx + ay = a + b$; $ax\left(\frac{1}{a-b} - \frac{1}{a+b}\right) + by\left(\frac{1}{b-a} - \frac{1}{b+a}\right) = \frac{2a}{a+b}$.

78. Solve for x and y: $\frac{10}{x+1} + \frac{2}{y-1} = \frac{5}{2}, \frac{5}{x+1} - \frac{2}{y-1} = \frac{1}{2}; x \neq -1\, y \neq 1.$

79. Solve for x and y: $3x - \frac{y+7}{11} + 2 = 10, 2y + \frac{x+11}{7} = 10.$

80. Solve for x and y. $631x + 279y = 910; 279x + 631y = 910.$

81. Solve for x and y: $254x + 309y = -55; 309x + 254y = -55.$

82. Solve for x and y: $3x + 2y = 2x + y + 3 = 4x + 3y - 3.$

83. Solve for x and y: $6x + 3y = 8x + 9y - 5 = 10x + 12y - 8.$

84. Solve for x and y: $\frac{1}{x+1} + \frac{1}{y+1} = 10; \frac{1}{x+1} - \frac{1}{y+1} = 4.$ **[CBSE 2015]**

85. Solve for x and y: $\frac{57}{x+y} + \frac{6}{x-y} = 5; \frac{38}{x+y} + \frac{21}{x-y} = 9.$

86. Determine the value of k for which the following system for equations has no solution:

$$kx + 2y - 1 = 0, 5x - 3y + 2 = 0$$

87. For what value of k does the following pair of linear equations have a unique solution?

$$kx + 3y = 3, 12x + ky = 6$$

88. Find whether the following pair of equations has no solution, unique solution or infinitely many solutions.

$$5x - 8y + 1 = 0; 3x - \frac{24}{5}y + \frac{3}{5} = 0$$

89. For what value of 'a' does the following pair of linear equations have infinitely many solutions?

$$ax - 3y = 1, -12x + ay = 2$$

90. For what value of p does the pair of equations given below have a unique solution?

$$4x + py + 8 = 0 \text{ and } 2x + 2y + 2 = 0$$

91. Find the value of 'm' for which the pair of linear equations $2x + 3y - 7 = 0$ and $(m-1)x + (m+1)y = (3m-1)$ has infinitely many solutions.

92. For what value of p will the following pair of linear equations have infinitely many solutions?

$$(p-3)x + 3y = p; px + py = 12$$ **[Foreign 2010]**

93. Find the value of α for which the following pair of equation has infinitely many solutions.

$$x - 4y - 9 = 0, 2x - \alpha y - 27 = 0.$$

94. Are the following pairs of linear equations consistent? Justify your answer.

(i) $2ax + by = a$;
$4ax + 2by - 2a = 0$; $a, b \neq 0$

(ii) $\frac{3}{5}x - y = \frac{1}{2}$;
$\frac{1}{5}x - 3y = \frac{1}{6}$

(iii) $x + 3y = 11; 2(2x + 6y) = 22$ **[NCERT Exemplar]**

95. Solve the following pair of equations:

$$x + y = 3.3, \quad \frac{0.6}{3x - 2y} = -1, \; 3x - 2y \neq 0$$ **[NCERT Exemplar]**

96. Solve the following pair of equations:

$$\frac{2xy}{x + y} = \frac{3}{2}$$

$$\frac{xy}{2x - y} = -\frac{3}{10}, \; x + y \neq 0, \; 2x - y \neq 0$$ **[NCERT Exemplar]**

97. Do the equations $4x + 3y - 1 = 5$ and $12x + 9y = 15$ represent a pair of coincident lines? Justify your answer. **[NCERT Exemplar]**

98. Is the pair of equations $x + 2y - 3 = 0$ and $6y + 3x - 9 = 0$ consistent? Justify your answer. **[NCERT Exemplar]**

99. For what values of p and q will the following pair of linear equations have infinitely many solutions?

$$4x + 5y = 2;$$

$$(2p + 7q)x + (p + 8q)y = 2q - p + 1$$ **[NCERT Exemplar]**

100. (a) For what values of λ does the pair of linear equations. $\lambda x + y = \lambda^2$ and $x + \lambda y = 1$ have

(i) No solution? (ii) Infinitely many solutions? (iii) A unique solution?

(b) Find the values of p and q in the following pair of linear equations:

$$2x + 3y = 7 \text{ and } 2px + py = 28 - qy$$

if the pair of equations has infinitely many solutions.

101. (a) For what values of k will the pair of equations $kx + 3y = k - 3$ and $12x + ky = k$ have no solution? **[NCERT Exemplar]**

(b) Two straight paths are represented by the equation $x - 3y = 2$ and $-2x + 6y = 5$. Check whether the paths cross each other or not.

102. (a) For which value of 'a' and 'b' will the following pairs of linear equations have infinitely many solutions?

$$x + 2y = 1$$

$$(a - b)x + (a + b)y = a + b - 2$$ **[NCERT Exemplar]**

(b) Write a pair of linear equations which has the unique solution $x = -1, y = 3$. How many such pairs can you write?

103. Given the linear equations $3x + 4y - 8 = 0$, write another linear equation in two variables such that the geometrical representation of the pair so formed is parallel lines.

104. For what value of m and n the does the following system of linear equations have infinitely many solutions? **[CBSE 2018 (C)]**

$$3x + 4y = 12, \; (m + n)x + 2(m - n)y = 5m - 1.$$

105. Find the value(s) of k so that the pair of equations $x + 2y = 5$ and $3x + ky + 15 = 0$ has a unique solution. **[CBSE Allahabad 2019]**

106. Find c if the system of equations $cx + 3y + (3 - c) = 0$; $12x + cy - c = 0$ has infinitely many solutions. **[Delhi 2019]**

107. Draw the graphs of the following equations:

$$x + y = 5, \; x - y = 5$$

(i) Find the solution of the equations from the graph.
(ii) Shade the triangular region formed by the line and the y axis. **[CBSE (CCE) 2011]**

108. Determine, by drawing graphs, whether the following system of linear equations has a unique solution or not: $2x - 3y = 6,\ x + y = 1$

109. Given the linear equation $2x + 3y - 8 = 0$, write another linear equation in two variables, such that the geometrical representation of the pair so formed is:
(i) Intersecting lines (ii) Parallel lines (iii) Coincident lines

110. Form the pair of linear equations in the following problems, and find their solutions graphically.
(i) 10 students of class X took part in a Mathematics quiz. If the number of girls is 4 more than the number of boys, find the number of boys and girls who took part in the quiz.
(ii) 5 pencils and 7 pens together cost ₹50, whereas 7 pencils and 5 pens together cost ₹46. Find the cost of one pencil and one pen.

111. Aftab tells his daughter, "Seven years ago, I was seven times as old as you were then. Also, three years from now, I shall be three times as old as you will be." Represent this situation algebraically and graphically.

112. The coach of a cricket team buys 3 bats and 6 balls for ₹3,900. Later, she buys another bat and 3 more balls of the same kind for ₹1,300. Represent this situation algebraically and graphically.

113. The cost of 2 kg of apples and 1 kg of grapes on a day was found to be ₹160. After a month, the cost of 4 kg of apples and 2 kg of grapes was ₹300. Represent the situation algebraically and graphically.

114. Determine by drawing graphs whether the following pair of linear equations has a unique solution or not: $2x - 5 = 0;\ y + 4 = 0$

115. Determine by drawing graphs whether the following pair of linear equations has an infinite numbers of solution or not: $y = 5;\ y + 3 = 0$

116. From a stationary shop, Archana bought two pencils and three pens for ₹40 and Indu bought one pencil and two pens for ₹25. Find the price of one pencil and one pen graphically.

117. The perimeter of a rectangle is 52 cm, where the length is 6 cm more than the width of the rectangle. Form the pair of linear equations for the above situation and find the dimensions of the rectangle graphically.

118. Draw the graphs of the equations: $4x - 3y - 6 = 0;\ x + 3y - 9 = 0$
Determine the co-ordinates of the vertices of the triangle formed by the lines and the y axis.

119. Use a single graph paper and draw the graph of the following equations:
$$2y - x = 8;\ 5y - x = 14;\ y - 2x = 1$$

120. Solve the following system of linear equations graphically: $2x + y = 10,\ 4x - y = 8$.
Does the point (1, –4) lie on any of the lines? Write its equation.

121. Draw the graph $x - y + 1 = 0$ and $2x + y - 10 = 0$. Shade the region bounded by these lines and the x axis. Find the area of the shaded region.

122. The line represented by $x = 7$ is parallel to the x axis. Justify whether the statement is true or not. **[NCERT Exemplar]**

123. Draw the graphs of the pair of linear equations $x - y + 2 = 0$ and $4x - y - 4 = 0$. Calculate the area of the triangle formed by the lines so drawn and the x axis. **[NCERT Exemplar]**

124. Draw the graph of the pair of equations $2x + y = 4$ and $2x - y = 4$. Write the vertices of the triangle formed by these lines and the y axis. Also find the area of this triangle. **[NCERT Exemplar]**

125. Write an equation of a line passing through the point representing the solution of the pair of linear equations $x + y = 2$ and $2x - y = 1$. How many such lines can we find? **[NCERT Exemplar]**

126. Draw the graphs of the lines $x = -2$ and $y = 3$. Write the vertices of the figure formed by these lines, the x axis and the y axis. Also, find the area of the figure. **[NCERT Exemplar]**

127. Graphically solve the following pair of equation:

$$2x + y = 6;\ 2x - y + 2 = 0$$

Find the ratio of the areas of the two triangles formed by the lines representing these equations with the x axis and the lines with the y axis. **[NCERT Exemplar]**

128. Draw the graphs of the equations $x = 3$, $x = 5$ and $2x - y - 4 = 0$. Also find the area of the quadrilateral formed by the lines and the x axis. **[NCERT Exemplar]**

129. Find whether the lines representing the following pair of linear equations intersecting at a point are parallel or coincident.

$$2x - 3y + 6 = 0;\ 4x - 5y + 2 = 0$$ **[CBSE 2015]**

130. Given a linear equations $3x - 5y = 1$, form another linear equation in these variables such that the geometric representation of pair so formed is:

(i) Intersecting lines (ii) Coincident lines (iii) Parallel lines **[CBSE 2015]**

131. A man travels 600 km partly by train and partly by car. It takes 8 hours and 40 minutes if he travels 320 km by train and the rest by car. It would take 30 minutes more if he travels 200 km by train and the rest by car. Find the speed of the train and the car separately. **[CBSE (CCE) 2011]**

132. After covering a distance of 30 km with uniform speed, there occurs some defect in a train engine and thereafter its speed is reduced to 4/5 of its original speed. Consequently, the train reaches its destination late by 45 minutes. Had it happened after covering 18 km more, the train would have reached 36 minutes late. Find the speed of the train and the distance of the journey.

133. The sum total of the ages of father and son is 55 years. If the father were to live till his son's age equals his present age, the total of their ages would be 93 years. Find their present ages.

134. A lab assistant has a solution of 50% acid and another which has 25% acid. How much of each should be mixed to make 10 litres of a 40% acid solution?

135. Vijay had some bananas and he divided them into two lots, A and B. He sold the first lot at the rate of ₹2 for 3 bananas and the second lot at the rate of ₹1 per banana, and received a total of rupees ₹400. If he had sold the first lot at the rate of ₹1 per banana, and the second lot at the rate of ₹4 for 5 bananas, his total collection would have been ₹460. Find the total number of bananas he had. **[NCERT Exemplar]**

136. The taxi charge in a city comprises a fixed charge and a charge for the distance covered. For a journey of 10 km, the charge paid is ₹75 and for a journey of 15 km, the charge is ₹110. What will a person have to pay for travelling a distance of 25 km?

137. There are two examination rooms A and B. If 10 candidates are sent from A to B, the numbers of students in each room is the same. If 20 candidates are sent from B to A, the number of students in A is double the number of students in B. Find the number of students in each room.

138. The area of a rectangle is reduced by 80 sq. units if its length is reduced by 5 units and the breadth is increased by 2 units. If we increase the length by 10 units and decrease the breadth by 5 units, the area is increased by 50 sq. units. Find the length and breadth of the rectangle.

139. Ten years ago, the sum of the ages of two sons was one-third of their father's age. One son is two years older than the other and the sum of their current ages is 14 years less than the father's current age. Find the current ages of all.

140. A car goes uphill at the rate of 30 km an hour and downhill at the rate of 50 km an hour. After 15 hours, it has covered 650 km. How long does it go downhill and uphill, respectively?

141. A man bought 4 horses and 9 cows for ₹1,340. He sells the horses at a profit of 10% and the cows at a profit of 20% and his whole gain is ₹188. What price did he paid for the horse?

142. If three times the larger of the two numbers is divided by the smaller one, we get 4 as quotient and 3 as reminder. Also, if seven times the smaller number is divided by the larger one, we get 5 as quotient and 1 as remainder. Find the numbers.

143. On selling a tea set at 5% loss and a lemon set at 15% gain, a crockery seller gains ₹7. If he sells the tea set at 5% gain and lemon set at 10% gain, he gains ₹13. Find the actual price of the tea set.

144. In the figure below, *ABCDE* is a pentagon with $BE \| CD$ and $BC \| DE$. *BC* is perpendicular to *CD*. If the perimeter of *ABCDE* is 21 cm, find the value of *x* and *y*. **[CBSE (CCE) 2011]**

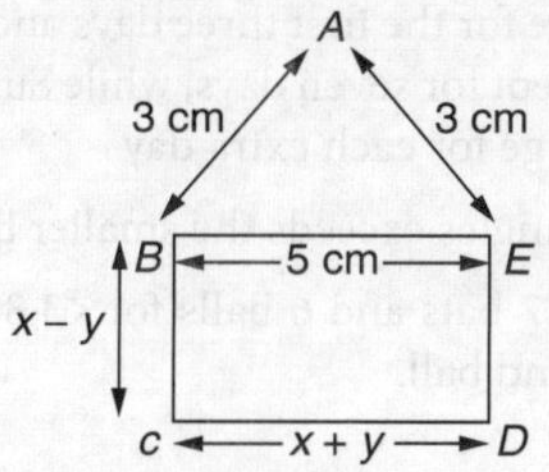

145. Find the four angles of cyclic quadrilateral *ABCD* in which $\angle A = (2x-1)°$, $\angle B = (y+5)°$, $\angle C = (2y+15)°$ and $\angle D = (4x-7)°$.

146. A test consists of 'true' or 'false' questions. One mark is awarded for every correct answer while $\frac{1}{4}$ mark is deducted for every wrong answer. A student knows answers to some of the questions. The rest of the questions he attempted by cheating. He answered 120 questions and obtained 90 marks. If the answers to all the questions he attempted by cheating were wrong, then how many questions did he answer correctly? How will the habit of cheating affect his character?

147. A man wished to donate some money to a group of poor people. He decided to give ₹120 to each person and found that he fell short by ₹60. He, therefore, distributed ₹90 to each person and found that ₹90 was left over. How much money did he have and how many people were there? What value is depicted by the man?

148. A trust has ₹30,000. It is invested in two different types of bonds. The first bond pays 5% interest per annum which will be given to an orphanage and the second bond pays 7% interest per annum which will be given to an NGO for cancer aid. The trust obtains an annual interest of ₹1,800.
(a) How much amount is invested in each type of bond? (b) Which value is depicted by the trust?

149. Ritu can row downstream 20 km in 2 hours and upstream 4 km in 2 hours. Find her speed of rowing in still water and the speed of the current. **[NCERT]**

150. Two women and 5 men can together finish an embroidery work in 4 days, while 3 women and 6 men can finish it in 3 days. Find the time taken by one woman alone to finish the work, and also that taken by 1 man alone. **[NCERT]**

151. A fraction becomes $\frac{1}{3}$ when 1 is subtracted from the numerator, and it becomes $\frac{1}{4}$ when 8 is added to its denominator. Find the fraction. **[NCERT]**

152. Yash scored 40 marks in a test, getting 3 marks for each right answer and losing one mark for each wrong answer. Had 4 marks been awarded for each correct answer and 2 marks been deducted for each incorrect answer, Yash would have scored 50 marks. How many questions are there in the test? **[NCERT]**

153. The sum of a two-digit number and the number formed by reversing the order of digits is 154. If the two digits differ by 4, find the number.

154. The students of a class are made to stand in rows. If there are 4 additional students in a row, there would be two rows less. If there are 4 fewer students in a row, there would be four more rows. Find the numbers of students in the class. **[NCERT]**

155. The sum of the digits of a two-digit number is 9. Also, nine times this number is twice the number obtained by reversing the order of the digits. Find the number. **[NCERT]**

156. Meena went to a bank to withdraw ₹2000. She asked the cashier to give her ₹50 and ₹100 notes only. Meena got 25 notes in all. Find out how many notes of ₹50 and ₹100 she received. **[NCERT]**

157. A lending library has a fixed charge for the first three days and an additional charge for each day thereafter. Saritha paid ₹27 for a book kept for seven days, while Susy paid ₹21 for a book she kept for 5 days. Find the fixed charge and the charge for each extra day. **[NCERT]**

158. The larger of two supplementary angles exceeds the smaller by 18 degrees. Find them. **[NCERT]**

159. The coach of a cricket team buys 7 bats and 6 balls for ₹3,800. Later, she buys 3 bats and 5 balls for ₹1,750. Find the cost of each bat and ball. **[NCERT]**

160. A fraction becomes $\frac{9}{11}$ if 2 is added to both the numerator and the denominator. If 3 is added to both the numerator and denominator, it becomes $\frac{5}{6}$. Find the fraction. **[NCERT]**

161. Five years hence, the age of Jacob will be three times that of his son. Five years ago, Jacob's age was 7 times that of his son. What are their current ages?

162. The larger of two supplementary angles exceeds the smaller by 20 degrees. Find the angles.

163. The angles of a triangle are x, y and 40°. The difference between the two angles x and y is 30°. Find x and y. **[NCERT Exemplar]**

164. The age of his father is twice the sum of the ages of his two children. After 20 years, his age will be equal to the sum of the ages of the children. Find the age of the father. **[NCERT Exemplar]**

165. Two numbers are in the ratio 5:6. If 8 is subtracted from each of the numbers, the ratio becomes 4:5. Find the numbers. **[NCERT Exemplar]**

166. A man bought 4 horses and 9 cows for ₹1,340. He sells the horses at a profit of 10% and cows at a profit of 20% and his whole gain is ₹188. What price did he pay for the horse?

167. A shopkeeper gives books on rent for reading. She takes a fixed charge for the first two days, and an additional charge for each day thereafter. Latika paid ₹22 for a book kept for 6 days, while Anand paid ₹16 for a book kept for 4 days. Find the fixed charges and the charge for each extra day. **[NCERT Exemplar]**

168. In a competitive examination, one mark is awarded for each correct answer while $\frac{1}{2}$ mark is deducted for every wrong answer. Jayanti answered 120 questions and got 90 marks. How many questions did she answer correctly? **[NCERT Exemplar]**

169. Jamila sold a table and chair for ₹1,050, thereby making a profit of 10% on the table and 25% on the chair. If she had taken a profit of 25% on the table and 10% on the chair, she would have got ₹1,065. Find the cost price of each. **[NCERT Exemplar]**

170. It can take 12 hours to fill a swimming pool using two pipes. If the pipe of larger diameter is used for 4 hours and the pipe of smaller diameter for 9 hours, only half the pool will be filled. How long would it take for each pipe to fill the pool separately? **[NCERT Exemplar]**

171. The cost of 4 pens and 4 pencil boxes is ₹100. Three times the cost of a pen is ₹15 more than the cost of the pencil box. Form the pair of linear equations for the above situation. Find the cost of a pen and a pencil box. **[NCERT Exemplar]**

172. Ankita travels 14 km to her home partly by rickshaw and partly by bus. She takes half an hour if she travels 2 km by rickshaw, and the remaining distance by bus. On the other hand, if she travels 4 km by rickshaw and the remaining distance by bus, she takes 9 minutes longer. Find the speed of the rickshaw and of the bus. **[NCERT Exemplar]**

173. A person rowing at the rate of 5 km/h in still water takes thrice as much time in moving 40 km upstream as in going 40 km downstream. Find the speed of the stream. **[NCERT Exemplar]**

174. A motor boat can travel 30 km upstream and 28 km downstream in 7 hours. It can travel 21 km upstream and return in 5 hours. Find the speed of the boat in still water and the speed of the stream. **[NCERT Exemplar]**

175. A two-digit number is obtained by either multiplying the sum of the digits by 8 and then subtracting 5 or by multiplying the difference of the digits by 16 and then adding 3. Find the number. **[NCERT Exemplar]**

176. A railway half-ticket costs half the full fare, but the reservation charges are the same as on a full ticket. One reserved first-class ticket from station *A* to station *B* costs ₹2,530. Also, one reserved first-class ticket and one reserved first-class half-ticket from *A* to *B* costs ₹3,810. Find the full first-class fare from station *A* to station *B*, and also the reservation charges for a ticket. **[NCERT Exemplar]**

177. Suman invested a certain amount of money in two schemes *A* and *B*, which offer interest at the rate of 8% per annum and 9% per annum, respectively. She received ₹1,860 as annual interest. However, had she interchanged the amount of investments in the wo schemes, she would have received ₹20 more as annual interest. How much money did she invest in each scheme? **[NCERT Exemplar]**

178. A shopkeeper sells a saree at 8% profit and a sweater at 10% discount, thereby obtaining the sum of ₹1,008. If she had sold the saree at 10% profit and the sweater at 8% discount, she would have got ₹1,028. Find the cost price of the saree and the list price (price before discount) of the sweater. **[NCERT Exemplar]**

179. The figure *ABCD* is a rectangle. Find the values of *x* and *y*. **[CBSE 2018]**

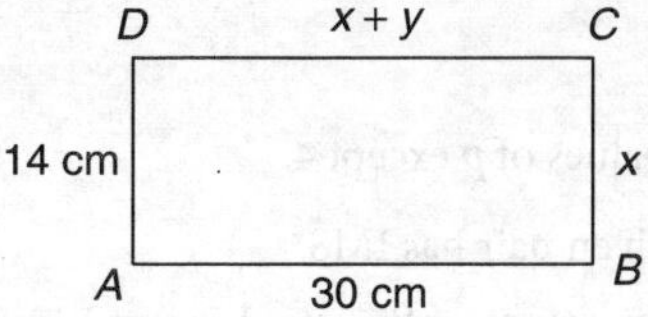

Answers

Multiple Choice Questions

1. (c)	**2.** (d)	**3.** (d)	**4.** (a)	**5.** (b)	**6.** (b)	**7.** (d)	**8.** (b)	**9.** (d)	**10.** (a)
11. (d)	**12.** (b)	**13.** (a)	**14.** (a)	**15.** (c)	**16.** (b)	**17.** (a)	**18.** (b)	**19.** (d)	**20.** (c)

21. (d) 22. (d) 23. (c) 24. (d) 25. (b) 26. (d) 27. (c) 28. (d) 29. (d) 30. (b)
31. (a) 32. (d) 33. (b) 34. (a) 35. (c) 36. (a) 37. (a) 38. (d)

Fill in the Blanks

39. $\frac{-5}{3}, 1$ 40. 2, 2 41. Consistent. 42. Parallel 43. $a = 3$ and $b = 1$

True/False

44. Second number 45. The given statement is not true 46. False

Assertion–Reason

47. (C) 48. (D) 49. (C) 50. (C)

Subjective Questions

51. $x = 7, y = 9$, and $5y - 2x = 31, \frac{y}{x} - 2 = \frac{-5}{7}$ 52. $x = 340, y = -165$, and $\lambda = -\frac{1}{2}$ 53. $x = 1, y = -1$
54. $x = a^2, y = b^2$ 55. $x = 3, y = 1$ 56. $x = 2, y = 3$ 57. $x = 0, y = 0$
58. $x = 2, y = 3$ 59. $x = 0, y = 0$ 60. $x = b, y = -a$ 61. $x = \frac{c^2a^2 - b^2d^2}{a^4 - b^4}, y = \frac{a^2d^2 - b^2c^2}{a^4 - b^4}$
62. $x = 400, y = 30$ 63. $x = a, y = b$ 64. $x = 4, y = 9$ 65. $x = 4, y = 5$ 66. $x = 1, y = 2$
67. $x = 1, y = 1$ 68$x = a^2, y = b^2$ 69. $x = \frac{-5}{4}$ and $y = \frac{-1}{4}$ 70. $u = 2, v = 1$
71. It has no solution as $\frac{a_1}{a_2} = \frac{b_1}{b_2} \neq \frac{c_1}{c_2}$ 72. $x = \frac{51}{11}, y = -\frac{22}{13}$ and $p = \frac{-165}{221}$ 73. $x = 2, y = -1$
74. $x = a^2, y = -b^2$ 75. $x = m + n, y = m - n$ 76. $x = \frac{1}{6}, y = \frac{3}{6}$ 77. $x = \frac{a^2 + b^2}{b(a + b)}, y = \frac{2b}{a + b}$
78. $x = 4, y = 5$ 79. $x = 3, y = 4$ 80. $x = 1, y = 1$ 81. $x = 1, y = -1$
82. It has infinitely many solutions 83. $x = \frac{1}{2}, y = \frac{2}{3}$ 84. $a = \frac{-6}{7}, b = -\frac{2}{3}$ 85. $x = 11, y = 8$
86. $k = -\frac{10}{3}$ 87. $k = -6$ 88. The pair equation has IMS
89. $a = -6$ 90. $p \neq 4$, ∴ All values of p except 4. 91. $m = 5$ 92. $p = 6$
93. There is no value of α for which the given pair has IMS.
94. (i) Consistent (ii) Consistent (iii) Inconsistent 95. $x = 1.2, y = 2.1$ 96. $x = \frac{1}{2}, y = -\frac{3}{2}$
97. No, they do not represent a pair of coincident lines 98. Yes 99. $p = -1, q = 2$
100. (a) (i) $\lambda = +1$ (ii) $\lambda = +1$ (iii) All real values except $\lambda = 1, -1$ (b) $p = 4, q = 8$
101. (a) $k = -6$(b) The two straight paths do not cross each other
102. (a) $a = 3, b = 1$(b) Infinite number of pairs of a system of equations are possible
103. One linear equation can be $6x + 8y + k = 0$, where k is a constant not equal to -16.
104. $m = 5, n = 1$ 105. $k \neq 6$ 106. $c = 6$ 107. (i) Point is (5, 0)

109. (i) Equations are $2x + 3y = 8$ and $3x + 2y = 4$. (ii) Equations are $2x + 3y = 8$ and $2x + 3y = 4$
(iii) Equations are $2x + 3y = 8$ and $2x + 3y = 8$

110. (i) 7 girls and 3 boys. (ii) Cost of one pencil = ₹3, Cost of one pen = ₹5

111. Algebraic equations are $x - 7y + 42 = 0$ and $x - 3y - 6 = 0$

112. Algebraic equation $3x + 6y = 3900$ and $x + 3y = 1300$

113. Algebraic equations are $2x + y = 160$ and $4x + 2y = 300$

114. It has a unique solution

115. No, this pair does not have an infinite number of solutions

116. Price of one pencil = ₹5, Price of one pen = ₹10

117. Length = 16 cm, Breadth = 10 cm

118. Vertices of Δ (0, 3), (3, 2) and (0, –2)

119. Vertices of Δ (1, 3), (– 4, 2) and (2, 5).

120. $x = 3$ and $y = 4$. Yes, (1, – 4) lies on the line $4x - y = 8$

121. Shaded area is 12 sq. units

122. False

123. Area of Δ is 6 sq. units

124. Vertices of Δ are (0, 4), (0, – 4) and (2, 0) Area of Δ = 8 sq. units

125. $x = 1, y = 1$ and we can make infinite lines

126. Vertices of rectangle are (0, 0), (0, 3), (–2, 3), (–2, 0); Area of rectangle = 6 sq. units

127. Ratio of areas is 4 : 1

128. Area of quadrilateral is 8 sq. units

129. Intersecting lines

130. (i) Intersecting line $4x + 6y = 9$ (ii) Coincident line $6x - 10y = 2$ (iii) Parallel lines $6x - 10y = 5$

131. Speed of train = 80 km/h, speed of car = 60 km/h.

132. Speed of training 30 km/h. Distance = 120 km

133. Father's present age = 37 years, Son's present age = 18 years

134. 50% acid should be 6 litres and 25% acid should be 4 litres

135. Total numbers of bananas is 500.

136. The amount which a person has to pay for travelling a distance of 25 km will be ₹180.

137. No. of students in room $A = 100$, No. of students in room $B = 80$

138. Length = 40, Breadth = 30 units

139. Father's age = 46 years, elder son's age = 17 years, younger son's age = 15 years

140. Downhill journey = 10 hours, Uphill journey = 5 hours.

141. Cost price of house is ₹200.

142. Numbers are 25 and 18

143. Cost of tea set = ₹100.

144. $x = 5, y = 0$

145. $\angle A = 65°$, $\angle B = 55°$, $\angle C = 115°$, $\angle D = 125°$

146. He correctly answered 96 questions.

147. Money he has = ₹540 and number of people = 5, Value: Concern for others

148. (a) ₹15,000 is invested in each type of bond. (b) Value: Helping poor people

149. Ritu's speed in still water = 6 km/h; speed of current = 4 km/h.

150. Time taken by 1 woman = 18 days, Time taken by 1 man = 36 days

151. Fraction = $\frac{5}{12}$

152. Total question = 20

153. Numbers are 59 and 95

154. Total students = 96

155. Number is 18

156. Number of notes of ₹50 = 10, Number of notes of ₹100 = 15

157. Fixed charges is ₹15 and charge for each extra day is ₹3

→ Form the pair of linear equations for the following problems and find their solution by substitution method:

158. One angle is 81° and other angle is 99°.

159. Cost of 1 bat is ₹500, Cost of 1 ball is ₹50

160. Fraction is $\frac{7}{9}$

161. Jacob's present age is 40 years, and his son is 10 years old.

162. Larger angle is 100° and smaller angle is 80° **163.** $x = 85°$, $y = 55°$.

164. Age of father is 40 years **165.** Numbers are 40 and 48 **166.** ₹200

167. Fixed charges for first two days = ₹10, Additional charges per day after 2 days = ₹3 per day

168. Jayanti answered 100 questions correctly.

169. Cost price of table = ₹500, Cost price of chair = ₹400

170. Pipe with larger diameter will take 20 hours. Pipe with smaller diameter will take 30 hours.

171. Cost of a pen is ₹10, and cost of a pencil box is ₹15.

172. Speed of rickshaw = 10 km/h and speed of bus = 40 km/h.

173. Speed of stream is 2.5 km/h.

174. The speed of motor boat in still water = 10 km/h and speed of stream is 4 km/h.

175. Numbers is 83.

176. Full fare = ₹2,500 and Reservation charges of ticket station *A* to *B* = ₹30.

177. The amount invested in scheme *A* is ₹12,000 and amount invested in scheme *B* is ₹10,000.

178. CP of the saree is ₹600 and LP of the sweater is ₹400. **179.** $x = 22$ cm, $y = 8$ cm

Linear Equations in Two Variables (Basic)

SELF-ASSESSMENT TEST

Time allowed: 90 min. Max. Marks: 40

General Instructions

1. *All questions are compulsory. There are 20 questions in all.*
2. *This question paper has four sections: Section A, Section B, Section C and Section D.*
3. *Section A contains 10 questions of one mark each, Section B contains three questions of two marks each, Section C contains four questions of three marks each, and Section D contains three questions of four marks each.*

Section A [1 Mark Each]

Direction: (Q. Nos. 1–5) These are Multiple Choice Questions. Select the most appropriate answer for the given questions.

1. The lines represented by the equations $a_1x + b_1y + c_1 = 0$ and $a_2x + b_2y + c_2 = 0$ are coincident if

   (a) $\frac{a_1}{a_2} \neq \frac{b_1}{b_2}$ (b) $\frac{a_1}{a_2} = \frac{b_1}{b_2} = \frac{c_1}{c_2}$ (c) $\frac{a_1}{a_2} = \frac{b_1}{b_2} \neq \frac{c_1}{c_2}$ (d) $\frac{a_1}{a_2} \neq \frac{b_1}{b_2} \neq \frac{c_1}{c_2}$

2. The value of c for which the pair of equations $cx - y = 2$ and $6x - 2y = 4$ will have infinitely many solutions is

   (a) 3 (b) –3 (c) – 12 (d) no value

3. The pair of linear equations $x - 2y = 5$ and $2x - 4y = 1$ has

   (a) many solutions (b) no solution (c) one solution (d) two solutions

4. Find the value of 'a' so that the point $(3, a)$, lies on the line represented by $2x - 3y = 5$.

   (a) 0 (b) $\frac{1}{3}$ (c) 3 (d) $\frac{2}{3}$

5. The graphic representation of the equations $x + 2y = 3$ and $2x + 4y + 7 = 0$ gives a pair of ________ lines.

   (a) intersecting (b) parallel (c) coincident (d) overlapping

6. The graphic representation of the equations $x + 2y = 3$ and $2x + 4y + 7 = 0$ gives a pair of ________ lines.
7. For all real values of c, the pair of equations $x - 2y = 8$; $5x - 10y = c$ has a unique solution. Justify whether it is true or false.
8. Solve: $x + y = 7$ and $12x + 5y = 7$.
9. Find the value of k for which the system of equations: $x - 2y = 3$, $3x - ky = 1$ has a unique solution.
10. Five pencils and 7 pens together cost ₹50, whereas 7 pencils and 5 pens together cost ₹46. Find the cost of one pencil and of one pen.

Section B [2 Marks Each]

11. Solve using the elimination method: $2x + y = 35$ and $3x + 4y = 65$. Find $\frac{x}{y}$.

12. Solve $x + y = 5$ and $x - y = 1$ using the cross-multiplication method.

13. For what value of k will the following system of linear equation have no solution?

$$3x + y = 1,\ (2k - 1)x + (k - 1)y = 2k + 1$$

Section C [3 Marks Each]

14. Solve: $\frac{x}{2} + \frac{2y}{3} = -1$ and $x - \frac{y}{3} = 3$.

15. Solve the following pair of equations graphically:

$$x + y = 3$$

$$2x + 5y = 12$$

16. Draw the graphs of the equations $4x - y = 4$ and $4x + y = 12$.

Determine the vertices of the triangle formed by the line representing these equations and the x axis. Shade the triangular region so formed.

17. Solve the following system of linear equations graphically:

$$3x + y - 11 = 0;\ x - y + 1 = 0$$

Shade the region bounded by these lines and the y axis. Also find the area of the region bounded by the lines and the x axis.

Section D [4 Marks Each]

18. In a two-digit number, the unit's digit is twice the ten's digit. If 27 is added to the number, the digits interchange their places. Find the number.

19. Points A and B are 90 km from each other on a highway. A car starts from A and another from B at the same time. If they travel in the same direction, they meet in 9 hours, and if they travel in opposite directions, they meet in 9/7 hours. Find their speed.

20. A boat covers 32 km upstream and 36 km downstream in 7 hours. Also, it covers 40 km upstream and 48 km downstream in 9 hours. Find the speed of the boat in still water and that of the stream.

Answers

1. (b) **2.** (a) **3.** (b) **4.** (b) **5.** (b) **6.** Parallel **7.** No, because here, $\frac{a_1}{a_2} = \frac{b_1}{b_2}$

8. $x = -4$ and $y = 11$ **9.** $k \neq 6$ **10.** ₹ 3 and ₹ 5 **11.** $x = 15, y = 5, \frac{x}{y} = 3$

12. $x = 3, y = 2$ **13.** $k = 2$ **14.** $x = 2$ and $y = -3$ **15.** Unique solution at (1, 2)

16. (2, 4), (1, 0), (3, 0) **17.** (2.5, 3.5), 12.5 sq. units **18.** 36 **19.** 40 km/h; 30 km/h

20. 10 km/h; 2 km/h

Linear Equations in Two Variables (Standard)

SELF-ASSESSMENT TEST

Time allowed: 90 min. Max. Marks: 40

General Instructions

1. *All questions are compulsory. There are 20 questions in all.*
2. *This question paper has four sections: Section A, Section B, Section C and Section D.*
3. *Section A contains 10 questions of one mark each, Section B contains three questions of two marks each, Section C contains four questions of three marks each, and Section D contains three questions of four marks each.*

Section A [1 Mark Each]

Direction: (Q. Nos. 1–5) These are Multiple Choice Questions. Select the most appropriate answer for the given questions.

1. The father's age is six times his son's age. Four years hence, the age of the father will be four times his son's age. The present ages, in years, of the son and the father are, respectively,

(a) 4 and 24 (b) 5 and 30 (c) 6 and 36 (d) 3 and 24

2. Find the coordinate where the line $x - y = 8$ will intersect the y axis.

(a) (0, –8) (b) (8, 0) (c) (–8, 0) (d) (0, 8)

3. If $2^{x+y} = 2^{x-y} = \sqrt{8}$, then the value of y is

(a) $\frac{1}{2}$ (b) $\frac{3}{2}$ (c) 0 (d) none of these

4. If $\frac{2x+y+2}{5} = \frac{3x-y+1}{3} = \frac{3x+2y+1}{6}$, then

(a) $x = 1, y = 1$ (b) $x = -1, y = = -1$ (c) $x = 1, y = 2$ (d) $x = 2, y = 1$

5. In a cyclic quadrilateral $ABCD$, it is given that $\angle A = (x + y + 10)°$, $\angle B = (y + 20)°$, $\angle C = (x + y - 30)°$ and $\angle D = (x + y)°$. Then, $\angle B = ?$

(a) 70° (b) 80° (c) 100° (d) 110°

6. On solving the following equations: $\frac{5}{x} + 6y = 3$; $\frac{-5}{x} + 3y = 6$, the values of x and y are ____________ and ____________, respectively.

7. For the pair of equations: $\lambda x + 3y = -7$ and $2x + 6y = 14$ to have infinitely many solutions, the value of λ should be 1. Is the statement true? Given reasons.

8. Solve for x and y. $631x + 279y = 910$; $279x + 631y = 910$.

9. For what value of p will the following pair of linear equations have infinitely many solutions?

$$(p - 3)x + 3y = p; px + py = 12$$

10. For which values of λ does the pair of linear equations $\lambda x + y = \lambda^2$ and $x + \lambda y = 1$ have

(i) no solution? (ii) infinitely many solutions? (iii) a unique solution?

Section B [2 Marks Each]

11. For which values of p and q will the following pair of linear equations have infinitely many solutions?

$$4x + 5y = 2$$
$$(2p + 7q)x + (p + 8q)y = 2q - p + 1.$$

12. For which value of 'a' and 'b' will the following pair of linear equations have infinitely many solutions?

$$x + 2y = 1$$
$$(a - b)x + (a + b)y = a + b - 2$$

13. Yash scored 40 marks in a test, getting 3 marks for each correct answer and losing one mark for each wrong answer. Had 4 marks been awarded for each correct answer and 2 marks been deducted for each incorrect answer, Yash would have scored 50 marks. How many questions are there in the test?

Section C [3 Marks Each]

14. Draw the graphs of the equations $x = 3$, $x = 5$ and $2x - y - 4 = 0$. Also find the area of the quadrilateral formed by the lines and the x axis.

15. A railway half-ticket costs half the full fare, but the reservation charge is the same on the half-ticket as on the full ticket. One reserved first-class ticket from Mumbai to Ahmedabad costs ₹ 216 and one full and one half reserved first-class ticket costs ₹ 327. What is the basic first-class full fare and what is the reservation charge?

16. Solve the equations graphically:

$$2x + y = 2;\ 2y - x = 4$$

What is the area of the triangle formed by the two lines and the line $y = 0$?

17. Two numbers are in the ratio 4:5. If 30 is subtracted from each number, the ratio becomes 1:2. Form the pair of linear equations for the above situation and represent them graphically.

Section D [4 Marks Each]

18. Four men and 6 boys can finish a piece of work in 5 days, while 3 men and 4 boys can finish it in 7 days. Find the time taken by 1 man alone or that by 1 boy alone.

19. A man travels 600 km partly by train and partly by car. If he covers 400 km by train and the rest by car, it takes him 6 hours and 30 minutes. But if he travels 200 km by train and the rest by car, he takes half an hour longer. Find the speed of the train and that of the car.

20. On selling a tea set at 5% loss and a lemon set at 15% gain, a crockery seller gains ₹7. If he sells the tea set at 5% gain and the lemon set at 10% gain, he gains ₹13. Find the actual price of the tea set.

Answers

1. (c) **2.** (a) **3.** (c) **4.** (a) **5.** (b) **6.** $-\frac{5}{3}, 1$ **7.** False **8.** $x = 1$ and $y = 1$

9. $p = 6$ **10.** $k = -6$ **11.** $p = -1$ $q = 2$ **12.** $a = 3$, and $b = 1$ **13.** 20 **14.** 8 sq. units

15. ₹ 210 and ₹ 6 **16.** 5 sq. units **18.** 35 days, 70 days **19.** 100 km/h; 80 km/h

20. ₹ 100

CHAPTER

4 Quadratic Equations

Key Concepts

- Meaning of quadratic equation
- Solution of quadratic equation
- Solution by:
  - Splitting the middle term
  - Using quadratic formulae
  - Completing the square method
- Word problems

Concept: Meaning of Quadratic Equations

A polynomial that is quadratic (degree 2) in nature and with $p(x) = 0$, where $p(x) = ax^2 + bx + c$ and $ax^2 + bx + c = 0$, if and only if $a \neq 0$, where a, b and c are real constants and x is a real variable.

Example 1 Which of the following are quadratic equations?

(i) $x^2 - 6x + 4 = 0$ **(ii)** $x^2 = 0$ **(iii)** $x + \frac{1}{x} = x^2$

(iv) $x + \frac{1}{x} = 3$ **(v)** $\left(\sqrt{x} - \frac{2}{\sqrt{x}}\right)^2 = 1$

Solution:

(i) Yes, it is a quadratic polynomial, and hence, it is a quadratic equation.

(ii) Yes, it is a quadratic polynomial, and hence, it is a quadratic equation.

(iii) $x + \frac{1}{x} = x^2$

$$\Rightarrow \frac{x^2 + 1}{x} = x^2$$

$$\Rightarrow x^2 + 1 = x^3$$

$$\Rightarrow x^3 - x^2 - 1 = 0$$

No, it is a cubic polynomial and not quadratic in nature. Hence, it is not a quadratic equation.

(iv) $x + \frac{1}{x} = 3$

$$\Rightarrow \frac{x^2 + 1}{x} = 3$$

$$\Rightarrow x^2 + 1 = 3x$$

$$\Rightarrow x^2 - 3x + 1 = 0$$

Yes, it is a quadratic polynomial, and hence, it is a quadratic equation.

(v) $\left(\sqrt{x} - \frac{2}{\sqrt{x}}\right)^2 = 1$

Using $(a - b)^2 = a^2 - 2ab + b^2$

$$= (\sqrt{x})^2 - 2 \times \sqrt{x} \times \frac{2}{\sqrt{x}} + \left(\frac{2}{\sqrt{x}}\right)^2 = 1$$

$$= x - 4 + \frac{4}{x} = 1$$

$$= \frac{x^2 - 4x + 4}{x} = 1$$

$$= x^2 - 4x + 4 = x$$

$$= x^2 - 4x - x + 4 = 0$$

$$= x^2 - 5x + 4 = 0$$

Yes, it is a quadratic polynomial, and hence, it is a quadratic equation.

1 Try to Solve

1. Which of the following are quadratic equations?
   (a) $x^2 + 2x + 3 = 0$ (b) $x^4 + x^3 + 5x^2 + 3x + 1 = 0$
   (c) $x^2 + 3x + 1 = (x - 2)^2$ (d) $x^3 + 5x^2 - 3x = x(x^2 + x + 1)$
   (e) $(x - 2)(x + 1) = (x - 1)(x + 3)$ [NCERT]
   (f) $(x + 1)^2 = 2(x - 3)$ [NCERT]

Answers

1. (a) Yes, (b) No (c) No (d) Yes (e) No (f) Yes

Concept: Solution of Quadratic Equations

Example 2 Determine which of the following are solutions of the given quadratic equations:

(i) $x^2 - 2x + 1 = 0, x = 1$ **(ii)** $x^2 - 9 = 0, x = 3$ **(iii)** $x^2 - 3x + 2 = 0, x = 2, x = -1$

Solution:

(i) $x^2 - 2x + 1 = 0$

Insert $x = 1$ LHS $= (1)^2 - 2(1) + 1 = 1 - 2 + 1 = 2 - 2 = 0$

Hence LHS = RHS = 0

So, $x = 1$ is the solution of the given quadratic equation.

(ii) $x^2 - 9 = 0$

Insert $x = 3$ LHS $= (3)^2 - 9 = 9 - 9 = 0$

Hence LHS = RHS = 0

So, $x = 3$ is the solution of the given quadratic equation.

(iii) $x^2 - 3x + 2 = 0$

Insert $x = 2$ LHS $= (2)^2 - 3(2) + 2 = 4 - 6 + 2 = 6 - 6 = 0$

Hence LHS = RHS = 0

So, $x = 2$ is the solution of the quadratic equation.

Insert $x = -1$ LHS $= (-1)^2 - 3(-1) + 2 = 1 + 3 + 2 = 6 \neq 0.$

Hence LHS ≠ RHS.

So, $x = -1$ is not the solution of the quadratic equation.

2 Try to Solve

1. Determine which of the following are solutions of the given quadratic equation:

(a) $6x^2 - x - 2 = 0,\ x = \frac{-1}{2},\ x = \frac{2}{3}$ (b) $9x^2 - 3x - 2 = 0,\ x = -\frac{1}{3},\ x = 2$

(c) $2x^2 - 5x + 2 = 0,\ x = \frac{1}{2},\ x = 2$

Answers

1. (a) $x = -\frac{1}{2}$ is a solution, $x = \frac{2}{3}$ is also a solution.

(b) $x = -\frac{1}{3}$ is not a solution, $x = 2$ is also not a solution.

(c) $x = \frac{1}{2}$ and $x = 2$ are the solutions.

Example 3 Determine the value of k for $kx^2 + 2x - 3 = 0$, where $x = 2$ is the solution.

Solution: Since $x = 2$ is the solution of

$$kx^2 + 2x - 3 = 0$$
$$k(2)^2 + 2(2) - 3 = 0$$
$$4k + 4 - 3 = 0$$
$$4k + 1 = 0$$
$$4k = -1$$
$$k = -\frac{1}{4}$$

3 Try to Solve

1. In each of the following, find k if the solution is given:

(a) $7x^2 + kx - 3 = 0,\ x = \frac{2}{3}$ (b) $3x^2 + 2kx - 3 = 0,\ x = -\frac{1}{2}$ (c) $x^2 + 2kx + 1 = 0,\ x = 1$

Answers

1. (a) $k = -\frac{1}{6}$ (b) $k = -9/4$ (c) $k = -1$

Concept: Solving a Quadratic Equation by Factorisation

(i) A quadratic equation ($ax^2 + bx + c = 0$) can be solved by factorising the corresponding quadratic polynomial ($ax^2 + bx + c$) into two linear factors.
If $(px + q)$ and $(rx + s)$ are the two linear factors of the quadratic polynomial $ax^2 + bx + c$, then for the corresponding quadratic equation $ax^2 + bx + c = 0$, we have:

$$(px + q)(rx + s) = 0$$

$\Rightarrow$ $px + q = 0$ or $rx + s = 0$

That is $x = -q/p$ or $x = -s/r$

$\therefore$ $x = -\dfrac{q}{p}$ and $x = -\dfrac{s}{r}$ are the two roots or the solutions of the quadratic equation

$$ax^2 + bx + c = 0$$

In order to solve a quadratic equation by factorisation:

1. Clear all fractions and square roots (if any).
2. Write the equation in the form $p(x) = 0$.
3. Factorise the left-hand-side into a product of two linear factors.
4. Use the zero-product rule to solve the resulting equation.

Zero-Product Rule

(i) If a and b are two numbers or expressions and if $ab = 0$, then either $a = 0$ or $b = 0$ or both $a = 0$ and $b = 0$.

Example 4 Find the roots of the following quadratic equations by factorisation:

(i) $x^2 - 3x - 10 = 0$ (ii) $2x^2 + x - 6 = 0$ (iii) $\sqrt{2}\,x^2 + 7x + 5\sqrt{2} = 0$ [NCERT]

Solution:

(i) $x^2 - 3x - 10 = 0$

$\Rightarrow$ $x^2 - 5x + 2x - 10 = 0$

$\Rightarrow$ $x(x - 5) + 2(x - 5) = 0$

$\Rightarrow$ $(x - 5)(x + 2) = 0$

$\Rightarrow$ $x = 5$ or $x = -2$

$\therefore$ $x = 5, -2$ are the required solutions.

(ii) $2x^2 + x - 6 = 0$

$\Rightarrow$ $2x^2 + 4x - 3x - 6 = 0$

$\Rightarrow$ $2x(x + 2) - 3(x + 2) = 0$

$\Rightarrow$ $(x + 2)(2x - 3) = 0$

$\Rightarrow$ $x = -2$ or $x = \dfrac{3}{2}$

$\therefore$ $x = -2, \dfrac{3}{2}$ are the required solutions.

(iii) $\sqrt{2}x^2 + 7x + 5\sqrt{2} = 0$

$\Rightarrow$ $\sqrt{2}x^2 + 2x + 5x + 5\sqrt{2} = 0$

$\Rightarrow$ $\sqrt{2}x(x + \sqrt{2}) + 5(x + \sqrt{2}) = 0$

$\Rightarrow$ $(x + \sqrt{2})(\sqrt{2}x + 5) = 0$

$\Rightarrow \quad x = -\sqrt{2}$

or $\quad x = -\frac{5}{\sqrt{2}} = -\frac{5}{\sqrt{2}} \times \frac{\sqrt{2}}{\sqrt{2}} = -\frac{5\sqrt{2}}{2}$

$\therefore \quad x = -\sqrt{2}, -\frac{5\sqrt{2}}{2}$ are the required solutions.

4 Try to Solve

1. Find the roots of the following quadratic equations by factorisation:

(a) $x^2 + 6x + 8 = 0$

(b) $x^2 + 30x + 216 = 0$

(c) $100x^2 - 20x + 1 = 0$ **[NCERT]**

(d) $\sqrt{5}x^2 - 6x - 8\sqrt{5} = 0$

(e) $\sqrt{2}x^2 + 7x + 5\sqrt{2} = 0$ **[NCERT]**

(f) $2x^2 - x + \frac{1}{8} = 0$ **[NCERT]**

Answers

1. (a) $x = -2, x = -4$ (b) $x = -18, -12$ (c) $\frac{1}{10}, \frac{1}{10}$

(d) $-\frac{4}{\sqrt{5}}, 2\sqrt{5}$ (e) $-\sqrt{2}, -\frac{5}{\sqrt{2}}$ (f) $\frac{1}{4}, \frac{1}{4}$

Example 5 Solve: $\frac{1}{x-2} + \frac{2}{x-1} = \frac{6}{x}$. **[CBSE (CCE) 2013]**

Solution:

$$\frac{1}{x-2} + \frac{2}{x-1} = \frac{6}{x}$$

$$\frac{x-1+2(x-2)}{(x-2)(x-1)} = \frac{6}{x}$$

$$\frac{x-1+2x-4}{x^2-x-2x+2} = \frac{6}{x}$$

$$\frac{3x-5}{x^2-3x+2} = \frac{6}{x}$$

$$x(3x-5) = 6(x^2-3x+2)$$

$$3x^2 - 5x = 6x^2 - 18x + 12$$

$$6x^2 - 18x + 12 - 3x^2 + 5x = 0$$

$$3x^2 - 13x + 12 = 0$$

$$3x^2 - 9x - 4x + 12 = 0$$

$$3x(x-3) - 4(x-3) = 0$$

$$(3x-4)(x-3) = 0$$

Either $3x - 4 = 0$, $3x = 4$, $x = \frac{4}{3}$

Or $x - 3 = 0$, $x = 3$

$\left[\textbf{Ans. } x = 3, \frac{4}{3}\right]$

5 Try to Solve

1. Find the roots of the following quadratic equations by factorisation:

   (a) Solve: $\frac{4}{x} - 3 = \frac{5}{2x+3}, x \neq 0, \frac{-3}{2}$ [CBSE (CCE) 2013]

   (b) Solve: $\frac{x}{x+1} + \frac{x+1}{x} = \frac{34}{15}, x \neq 0, -1$ [CBSE (CCE) 2012]

   (c) Solve: $\frac{x+3}{x-2} - \frac{1-x}{x} = \frac{17}{4}, x \neq 2, 0$ [CBSE (CCE) 2013]

Answers

1. (a) $x = -2$ or $x = 1$ (b) $x = \frac{3}{2}$ or $x = -\frac{5}{2}$ (c) $x = 4$ or $x = -\frac{2}{9}$

Example 6 $\frac{1}{p+q+x} = \frac{1}{p} + \frac{1}{q} + \frac{1}{x}$. [CBSE 2005]

Solution: $\frac{1}{p+q+x} = \frac{1}{p} + \frac{1}{q} + \frac{1}{x}$

$$\frac{1}{p+q+x} - \frac{1}{x} = \frac{1}{p} + \frac{1}{q}$$

$$\frac{x-(p+q+x)}{(p+q+x)(x)} = \frac{q+p}{pq}$$

$$\frac{\cancel{x} - p - q - \cancel{x}}{px + qx + x^2} = \frac{p+q}{pq}$$

$$\frac{-(p+q)}{px+qx+x^2} = \frac{(p+q)}{pq}$$

$$\frac{-\cancel{(p+q)}}{px+qx+x^2} = \frac{\cancel{(p+q)}}{pq}$$

$$-pq = px + qx + x^2$$

$$x^2 + px + qx + pq = 0$$

$$x(x+p) + q(x+p) = 0$$

$$(x+q)(x+p) = 0$$

Either $x + q = 0$ Or $x + p = 0$

$x = -q$ $x = -p$ [**Ans.** $x = -q$ and $x = -p$]

6 Try to Solve

1. Find the roots of the following quadratic equations by factorisation:

(a) Solve: $\dfrac{x+3}{x+2}=\dfrac{3x-7}{2x-3}$

(b) Solve: $\dfrac{63}{x}+\dfrac{72}{x+6}=3$

(c) Solve: $\dfrac{1}{x}+\dfrac{1}{8-x}=\dfrac{8}{15}, x\neq 8$

Answers

1. (a) $x=-1, 5$ (b) 42, –3 (c) 3, 5

Example 7 Solve: $\dfrac{x-1}{2x+1}+\dfrac{2x+1}{x-1}=\dfrac{5}{2}$. **[CBSE (CCE) 2012]**

Solution: Insert $\dfrac{x-1}{2x+1}=t$

So
$$\frac{2x+1}{x-1}=\frac{1}{t}$$

$$t+\frac{1}{t}=\frac{5}{2},\quad \frac{t^2+1}{t}=\frac{5}{2}$$

$$2t^2+2=5t$$

$$2t^2-5t+2=0$$

$$2t^2-4t-t+2=0$$

$$2t(t-2)-1(t-2)=0$$

$$(2t-1)(t-2)=0$$

Either

$$2t-1=0$$
$$2t=1$$
$$t=\frac{1}{2}$$
$$t=\frac{x-1}{2x+1}$$

So, $\dfrac{x-1}{2x+1}=\dfrac{1}{2}$

$$2(x-1)=2x+1$$
$$\not{2}x-2=\not{2}x+1$$

Hence, x has no value.

Or

$$t-2=0$$
$$t=2$$
$$t=\frac{x-1}{2x+1}$$

So, $\dfrac{x-1}{2x+1}=\dfrac{2}{1}$

$$x-1=2(2x+1)$$
$$x-1=4x+2$$
$$4x+2-x+1=0$$
$$3x+3=0$$
$$3x=-3$$
$$x=-1$$

[Ans. $x=-1$**]**

7 Try to Solve

1. Solve: $\frac{x+1}{x-1} - \frac{x-1}{x+1} = \frac{5}{6}, x \neq 1, -1$ **[CBSE (CCE) 2014]**

2. Solve: $\frac{x-3}{x+3} - \frac{x+3}{x-3} = \frac{48}{7}, x \neq 3, x \neq -3$ **[CBSE (CCE) 2014]**

3. Solve: $\left(\frac{2x-3}{x-1}\right) - 4\left(\frac{x-1}{2x-3}\right) = 3, x \neq 1, x \neq \frac{3}{2}$ **[CBSE (CCE) 2014]**

Answers

1. $5, -\frac{1}{5}$
2. $-4, \frac{9}{4}$
3. $\frac{1}{2}, \frac{4}{3}$

Concept: Solving Quadratic Equations by the Method of Completing the Square

1. Write the given quadratic equation in the standard form.
2. Make the coefficient of x^2 unity, if it is not.
3. Shift the constant term to the RHS.
4. Add the square of half of the coefficient of x on both sides.
5. Express the LHS as the perfect square of the suitable binomial expression and simplify the RHS.
6. Take the square root of both the sides.
7. Obtain the values of x by shifting the constant term from the LHS to the RHS.

Example 8 Find the roots of the equation $x^2 - 2x - 8 = 0$ using the completing the square method.

Solution: $x^2 - 2x - 8 = 0$

$$\underbrace{x^2 - 2(1)(x) + (1)^2} - (1)^2 - 8 = 0$$

Using $(a - b)^2 = a^2 - 2ab + b^2$

So
$$(x-1)^2 - 1 - 8 = 0$$
$$(x-1)^2 = 9$$
$$(x-1)^2 = (3)^2$$
$$(x-1) = \pm 3$$

Either
$$x - 1 = 3$$
$$x = 3 + 1$$
$$x = 4$$

Or
$$x - 1 = -3$$
$$x = -3 + 1$$
$$x = -2$$

[**Ans.** $x = 4$ and -2]

8 Try to Solve

1. Find the roots using the completing the square method:
   (a) $x^2 - 5x + 4 = 0$ (b) $x^2 + 7x + 6 = 0$ (c) $x^2 - 4x + 4 = 0$

Answers

1. (a) $x = 1$ or $x = 4$ (b) $x = -1$, or $x = -6$ (c) $x = 2$

Example 9 Find the roots of the equation $2x^2 - 7x + 3 = 0$ using the method of completing the square. **[NCERT]**

Solution: $2x^2 - 7x + 3 = 0$.

Dividing both sides by 2, we get $x^2 - \frac{7}{2}x + \frac{3}{2} = 0$

$$\Rightarrow \quad x^2 - \frac{7}{2}x = -\frac{3}{2}$$

$$\Rightarrow \quad x^2 - 2\times\frac{7}{4}x + \left(\frac{7}{4}\right)^2 = \left(\frac{7}{4}\right)^2 - \frac{3}{2} \qquad \left[\text{Adding } \left(\frac{1}{2}\text{coefficient of } x\right)^2 \text{ on both sides}\right]$$

$$\Rightarrow \quad \left(x - \frac{7}{4}\right)^2 = \frac{49}{16} - \frac{3}{2} = \frac{49-24}{16} = \frac{25}{16} \qquad [\because (a-b)^2 = a^2 - 2ab + b^2]$$

$$\therefore \quad x - \frac{7}{4} = \pm\frac{5}{4}$$

$$\Rightarrow \quad x - \frac{7}{4} = \frac{5}{4} \qquad x - \frac{7}{4} = -\frac{5}{4}$$

$$x - \frac{7}{4} - \frac{5}{4} = 0 \qquad x - \frac{7}{4} + \frac{5}{4} = 0$$

$$x - \frac{12}{4} = 0 \qquad x - \frac{2}{4} = 0$$

$$x = 3 \qquad x = \frac{1}{2}$$

$\therefore$ $x = 3, \frac{1}{2}$ are the required roots.

9 Try to Solve

1. Find the roots of the following equations using the method of completing the square:
   (a) $5x^2 - 3x - 2 = 0$
   (b) $8x^2 - 22x - 21 = 0$
   (c) $9x^2 - 3x - 2 = 0$
   (d) $5x^2 - 6x - 2 = 0$ **[NCERT]**
   (e) $2x^2 - 5x + 3 = 0$
   (f) $2x^2 - 7x + 3 = 0$ **[NCERT]**

Answers

1. (a) $1, -\frac{2}{5}$ (b) $\frac{7}{2}, -\frac{3}{4}$ (c) $\frac{2}{3}, -\frac{1}{3}$
   (d) $x = \frac{3 \pm \sqrt{19}}{5}$ (e) $x = \frac{3}{2}, 1$ (f) $x = \frac{1}{2}, 3$

Concept: Solving a Quadratic Equation by Quadratic Formula

The expression $x = \dfrac{-b \pm \sqrt{b^2 - 4ac}}{2a}$ is known as the quadratic formula for solving the quadratic equation $ax^2 + bx + c = 0$.

Here, '$b^2 - 4ac$' is usually called the *discriminant* of the quadratic equation $ax^2 + bx + c = 0$ and is denoted by D. Depending upon the value of D, that is, $b^2 - 4ac$, three cases are possible.

Case (i): When $b^2 - 4ac = 0$

We have $$x = \frac{-b \pm \sqrt{0}}{2a}$$

That is $$x = \frac{-b}{2a}, \frac{-b}{2a}$$

∴ The quadratic equation $ax^2 + bx + c = 0$ has two real and equal roots, given by:

$$\alpha = -b/2a \text{ and } \beta = -b/2a$$

Case (ii): When $b^2 - 4ac > 0$

We have $$x = \frac{-b + \sqrt{b^2 - 4ac}}{2a} \text{ and } x = \frac{-b - \sqrt{b^2 - 4ac}}{2a}$$

The quadratic equation $ax^2 + bx + c = 0$ has two distinct real roots, given by:

$$\alpha = \frac{-b + \sqrt{b^2 - 4ac}}{2a} \text{ and } \beta = \frac{-b - \sqrt{b^2 - 4ac}}{2a}$$

Case (iii): When $b^2 - 4ac < 0$, that is, negative. The quadratic equation $ax^2 + bx + c = 0$ does not have any real roots. So, a quadratic equation $ax^2 + bx + c = 0$ has

(i) Two equal real roots if $b^2 - 4ac = 0$ $(D = 0)$

(ii) Two distinct real roots if $b^2 - 4ac > 0$ $(D > 0)$

(iii) No real roots if $b^2 - 4ac < 0$ $(D < 0)$

To solve the given quadratic equation using the quadratic formula:

1. Clear all the fractions and square roots (if any).
2. Write the equation in the form $ax^2 + bx + c = 0$.
3. Compare the equation obtained above with the equation $ax^2 + bx + c = 0$, to obtain the values of a, b and c.

(iv) Use the formula $x = \dfrac{-b \pm \sqrt{b^2 - 4ac}}{2a}$, to obtain the required solution.

Example 10 Write the discriminant of

(i) $3x^2 + 2x - 1 = 0$ **(ii)** $x^2 - 5x + 6 = 0$ **(iii)** $\sqrt{3}x^2 + 2\sqrt{2}x - 2\sqrt{3} = 0$

Solution: We know that $D = b^2 - 4ac$

(i) $3x^2 + 2x - 1 = 0$

$$a = 3, b = 2 \text{ and } c = -1$$
$$D = b^2 - 4ac = (2)^2 - 4 \times 3 \times (-1)$$
$$= 4 + 12 = 16$$

(ii) $x^2 - 5x + 6 = 0$

$$a = 1, b = -5, c = 6$$
$$D = b^2 - 4ac = (-5)^2 - 4 \times 1 \times 6$$
$$= 25 - 24 = 1$$

(iii) $\sqrt{3}x^2 + 2\sqrt{2}x - 2\sqrt{3} = 0$

$$a = \sqrt{3},\ b = 2\sqrt{2},\ c = -2\sqrt{3}$$

$$D = b^2 - 4ac = (2\sqrt{2})^2 - 4 \times \sqrt{3} \times (-2\sqrt{3})$$

$$= 8 + 4\sqrt{3} \times 2\sqrt{3} = 8 + 8 \times 3$$

$$= 8 + 24 = 32.$$

10 Try to Solve

1. Find the discriminant of the following:

   (a) $2x^2 + x + 1 = 0$ (b) $x^2 + 4x + 5 = 0$ (c) $\sqrt{3}\,x^2 + 2\sqrt{2}\,x - \sqrt{3} = 0$

Answers

1. (a) –7 (b) – 4 (c) 20

Example 11 Without finding the roots, comment upon the nature of the roots of:

(i) $x^2 - 5x + 6 = 0$ **(ii)** $x^2 - 4x + 4 = 0$ **(iii)** $2x^2 + x + 1 = 0$

Solution: We know that $D = b^2 - 4ac$

(i) $x^2 - 5x + 6 = 0$

$$a = 1,\ b = -5,\ c = 6$$

$$D = (-5)^2 - 4 \times 1 \times 6$$

$$= 25 - 24 = 1 > 0$$

Since $D > 0$, it has real and distinct roots.

(ii) $x^2 - 4x + 4 = 0$

$$a = 1,\ b = -4,\ c = 4$$

$$D = b^2 - 4ac$$

$$= (-4)^2 - 4 \times 1 \times 4 = 16 - 16 = 0$$

$$D = 0$$

So, the given equation has real and equal roots.

(iii) $2x^2 + x + 1$

$$a = 2,\ b = 1 \text{ and } c = 1$$

$$D = b^2 - 4ac$$

$$= (1)^2 - 4 \times 2 \times 1 = 1 - 8$$

$$D = -7$$

Since $D < 0$, it has no real roots.

11 Try to Solve

1. Without finding the roots, comment upon the nature of the roots:

   (a) $x^2 - 3x + 2 = 0$ (b) $x^2 + x + 1 = 0$

   (c) $x^2 - 2x + 1 = 0$ (d) $2x^2 - 6x + 3 = 0$ [NCERT]

   (e) $3x^2 - 4\sqrt{3}\,x + 4 = 0$ [NCERT]

Answers

1. (a) Real and distinct roots (b) No real roots
   (c) Real and equal roots (d) Real and distinct roots
   (e) Real and equal roots

Example 12 Find k if $2x^2 - kx + 1 = 0$ has real and equal roots.

Solution: $2x^2 - kx + 1 = 0$

$$a = 2,\ b = -k,\ c = 1$$
$$D = b^2 - 4ac$$
$$= (-k)^2 - 4 \times 2 \times 1 = k^2 - 8$$

For real and equal roots, $D = 0$

$$k^2 - 8 = 0$$
$$k^2 = 8$$
$$k = \pm\sqrt{8}$$
$$k = \pm 2\sqrt{2}$$

12 Try to Solve

1. Find 'k' if the following equations have real and equal roots:
   (a) $kx^2 + 4x + 1 = 0$ (b) $3x^2 + kx + 3 = 0$
   (c) $12x^2 + 4kx + 3 = 0$ (d) $kx\,(x - 2) + 6 = 0$ **[NCERT]**

Answers

1. (a) $k = 4$ (b) $k = \pm 6$ (c) ± 3 (d) $k = 6$

Example 13 Find k if $(k - 12)\,x^2 + 2\,(k - 12)\,x + 2 = 0$ has real and equal roots. **[CBSE (CCE) 2015]**

Solution: $(k - 12)x^2 + 2\,(k - 12)x + 2 = 0$

Here, $a = k - 12$, $b = 2\,(k - 12)$, $c = 2$

$$D = b^2 - 4ac$$
$$= [2(k - 12)]^2 - 4(k - 12)^2 = 4(k - 12)^2 - 8(k - 12)$$

For real and equal roots, $D = 0$

$$\Rightarrow \quad 4(k - 12)^2 - 8(k - 12) = 0$$
$$\Rightarrow \quad 4(k - 12)\,[k - 12 - 2] = 0$$
$$\Rightarrow \quad 4(k - 12)\,[k - 14] = 0$$
$$(k - 12)\,(k - 14) = 0$$

Either $k - 12 = 0$, $k = 12$ Or $k - 14 = 0$, $k = 14$

$k \neq 12$ because it does not obey the definition $ax^2 + bx + c = 0$, $a \neq 0$. So, $k = 14$.

13 Try to Solve

1. Find k if the given equations have equal roots:
   (a) $x^2 - 2x(1 + 3k) + 7(3 + 2k) = 0$ (b) $4x^2 - 2(k+1)x + (k+4) = 0$
   (c) $x^2 - 2xk - k = 0$

Answers

1. (a) $2, -\frac{10}{9}$ (b) 5, –3 (c) 0, –1

Example 14 Find the roots, if they exist, of the following equations using the quadratic formula:

(i) $x - \frac{1}{x} = 3; (x \neq 0)$ (ii) $\frac{1}{x+4} - \frac{1}{x-7} = \frac{11}{30}; (x \neq -4, 7)$ [NCERT]

Solution:

(i) $x - \frac{1}{x} = 3$

$\Rightarrow \quad x^2 - 1 = 3x$

$\Rightarrow \quad x^2 - 3x - 1 = 0$

This is a quadratic equation for which

$$D = (-3)^2 - 4(1)(-1)$$
$$= 9 + 4 = 13 \ (> 0)$$

$\therefore$ The given equation has two roots:

$$\alpha = \frac{-b+\sqrt{D}}{2a} \qquad \beta = \frac{-b-\sqrt{D}}{2a}$$

$$\alpha = \frac{-(-3)+\sqrt{13}}{2(1)} \qquad \beta = \frac{-(-3)-\sqrt{13}}{2(1)}$$

$$\alpha = \frac{3+\sqrt{13}}{2} \qquad \beta = \frac{3-\sqrt{13}}{2}$$

(ii) $\frac{1}{x+4} - \frac{1}{x-7} = \frac{11}{30}$

$$\Rightarrow \quad \frac{(x-7)-(x+4)}{(x+4)(x-7)} = \frac{11}{30}$$

$$\Rightarrow \quad \frac{\cancel{x} - 7 - \cancel{x} - 4}{x^2 - 7x + 4x - 28} = \frac{11}{30}$$

$$\Rightarrow \quad \frac{-11}{x^2 - 3x - 28} = \frac{11}{30}$$

$$\Rightarrow \quad x^2 - 3x - 28 = -30$$

$$\Rightarrow \quad x^2 - 3x + 2 = 0$$

That is the quadratic equation for which

$$D = (-3)^2 - 4(1)(2)$$
$$= 9 - 8 = 1 \ (> 0)$$

∴ The given equation has two roots:

$$\alpha = \frac{-b+\sqrt{D}}{2a} \qquad \beta = \frac{-b-\sqrt{D}}{2a}$$

$$\alpha = \frac{-(-3)+\sqrt{1}}{2.1} \qquad \beta = \frac{-(-3)-\sqrt{1}}{2.1}$$

$$\alpha = \frac{3+1}{2} \qquad \beta = \frac{3-1}{2}$$

Or $\alpha = 2 \qquad \beta = 1$

14 Try to Solve

1. Find the roots, if they exist, of the following equations using the quadratic formula:

(a) $2x^2 + 14x + 9 = 0$ (b) $x + \frac{4}{x} = -4, x \neq 0$ (c) $\frac{60}{15+x} + \frac{60}{15-x} = 9, x \neq 15, -15$

[CBSE (CCE) 2014]

Answers

1. (a) $\frac{-7 \pm \sqrt{31}}{2}$ (b) -2 (c) $5, -5$

Concept: Find the Roots, If they Exist, of the Following Equations using the Quadratic Formula

Example 15 Find the roots, if they exist, of the following equations using the quadratic formula:

(i) $2x^2 - 5x + 3 = 0$ **(ii)** $x^2 + 4x + 5 = 0$

(iii) $3x^2 + 2\sqrt{5}\,x - 5 = 0$ **(vi)** $2x^2 - 2\sqrt{2}\,x + 1 = 0$

Solution:

(i) We have $2x^2 - 5x + 3 = 0$

Here $a = 2, b = -5, c = 3$

Discriminant $D = b^2 - 4ac$

$$= (-5)^2 - 4 \times 2 \times 3$$
$$= 25 - 24 = 1 > 0.$$

∴ The given equation has real roots that are given by

$$x = \frac{-b + \sqrt{b^2 - 4ac}}{2a}$$

$$= \frac{-(-5)+1}{2\times 2} = \frac{5+1}{4} = \frac{3}{2}$$

And $$x = \frac{-b-\sqrt{b^2-4ac}}{2a}$$

$$= \frac{-(-5)-1}{2\times 2} = \frac{5-1}{4} = \frac{4}{4} = 1$$

Hence, the roots of the given equation are $\frac{3}{2}$ and 1.

(ii) We have $$x^2 + 4x + 5 = 0$$

Here $$a = 1, b = 4, c = 5$$

Discriminant $$D = b^2 - 4ac$$

$$= (4)^2 - 4\times 1\times 5$$

$$= 16 - 20 = -4 < 0.$$

∴ The given equation has no real roots.

(iii) We have $$3x^2 + 2\sqrt{5}\,x - 5 = 0$$

Here $$a = 3, b = 2\sqrt{5}, c = -5$$

Discriminant $$D = b^2 - 4ac$$

$$= (2\sqrt{5})^2 - 4\times 3\times(-5)$$

$$= 20 + 60 = 80 > 0.$$

∴ The given equation has real roots that are given by

$$x = \frac{-b+\sqrt{b^2-4ac}}{2a}$$

$$= \frac{-2\sqrt{5}+\sqrt{80}}{2\times 3} = \frac{-2\sqrt{5}+4\sqrt{5}}{6}$$

$$= \frac{2\sqrt{5}}{6} = \frac{\sqrt{5}}{3}$$

And $$x = \frac{-b-\sqrt{b^2-4ac}}{2a}$$

$$= \frac{-2\sqrt{5}-\sqrt{80}}{2\times 3} = \frac{-2\sqrt{5}-4\sqrt{5}}{6} = \frac{-6\sqrt{5}}{6} = -\sqrt{5}$$

Hence, the roots of the given equation are $\frac{\sqrt{5}}{3}$ and $-\sqrt{5}$.

(iv) We have $$2x^2 - 2\sqrt{2}\,x + 1 = 0$$

Here $$a = 2, b = -2\sqrt{2}, c = 1$$

Discriminant $$D = b^2 - 4ac$$

$$= (-2\sqrt{2})^2 - 4\times 2\times 1$$

$$= 8 - 8 = 0$$

∴ The given equation has real roots that are given by

$$x = \frac{-b+\sqrt{b^2-4ac}}{2a} = \frac{-(-2\sqrt{2})+\sqrt{0}}{2\times 2}$$

$$= \frac{2\sqrt{2}+0}{2\times 2} = \frac{2\sqrt{2}}{4} = \frac{\sqrt{2}}{2}$$

And $$x = \frac{-b-\sqrt{b^2-4ac}}{2a} = \frac{2\sqrt{2}-0}{2\times 2} = \frac{\sqrt{2}}{2}$$

Hence, the roots of the given equation are $\frac{\sqrt{2}}{2}$ and $\frac{\sqrt{2}}{2}$.

15 Try to Solve

1. Find the roots, if they exist, of the following equations using the quadratic formula:

(a) $6x^2 + 7x - 10 = 0$ (b) $2x^2 + 5\sqrt{3}\,x + 6 = 0$

(c) $3x^2 + 2\sqrt{5}\,x - 5 = 0$ (d) $\frac{1}{x+1} + \frac{2}{x+2} = \frac{4}{x+4}, x \neq -1, -2, -4$

Answers

1. (a) $-2, \frac{5}{6}$ (b) $\frac{-\sqrt{3}}{2}, -2\sqrt{3}$ (c) $\frac{\sqrt{5}}{3}, -\sqrt{5}$ (d) $2 \pm 2\sqrt{3}$

Concept: Word Problems

The concept of word problems based on quadratic equations can be understood in the following ways:

- Problems based on numbers
- Problems based on speed, distance and time
- Problems based on simple day-to-day life problems
- Problems based on age
- Problems based on area

Concept: Problems Based on Numbers

Example 16 The product of two consecutive positive integers is 240. Find the integers.

Solution: Let two consecutive positive integers be x and $(x + 1)$.

$$(x)\,(x+1) = 240$$

$$x^2 + x = 240$$

$$x^2 + x - 240 = 0$$

$$x^2 + x - 240 = 0$$

$$x^2 + 16\,x - 15x - 240 = 0$$

$$x(x + 16) - 15(x + 16) = 0$$
$$(x - 15)(x + 16) = 0$$

Either	Or
$x - 15 = 0$	$x + 16 = 0$
$x = 15$	$x = -16$

$x = -16$ is not possible because we require positive integers. So, $x = 15$ is possible.

Hence, the numbers are x and $x + 1$, that is, 15 and 16.

16 Try to Solve

1. The product of two consecutive (+)ve integers is 182. Find the (+)ve integers.
2. The product of two consecutive (+)ve integers is 420. Find the (+)ve integers.
3. The product of two consecutive positive integers is 306. We need to find the integer. [NCERT]
4. Johan and Jivanti together have 45 marbles. Both of them lost 5 marbles each, and the product of the number of marbles they now have is 124. We would like to find out how many marbles they had to start with. Write the equation. [NCERT]

Answers

1. 13 and 14 2. 20 and 21 3. 17 and 18 4. $x^2 - 45x + 324 = 0$

Example 17 Divide 29 into two parts so that the sum of the squares of the parts is 425. Find the numbers.

Solution:

Let the first number be = x. Let the second number be = $(29 - x)$.

29 ↙ ↘ x $29 - x$

According to the question

$$x^2 + (29 - x)^2 = 425$$
$$x^2 + (29)^2 + (x)^2 - 2 \times 29 \times x = 425 \quad [\because \text{Using } (a - b)^2 = a^2 - 2ab + b^2]$$
$$x^2 + 841 + x^2 - 58x = 425$$
$$2x^2 - 58x + 841 - 425 = 0$$
$$2x^2 - 58x + 416 = 0$$
$$2x^2 - (32 + 26)x + 416 = 0$$
$$2x^2 - 32x - 26x + 416 = 0$$
$$2x(x - 16) - 26(x - 16) = 0$$
$$(2x - 26)(x - 16) = 0$$

Either	Or
$2x \;\; 26 = 0$	$x \;\; 16 = 0$
$2x = 26$	$x = 16$
$x = \dfrac{26}{2}$	
$x = 13$	
First no. = 13	First no. = 16
Second no. = 29 − 13	Second no. = 29 − 16
= 16	= 13

17 Try to Solve

1. Divide 15 into two parts so that the sum of the squares is 117. Find the numbers.
2. Divide 39 into two parts so that the sum of the squares is 765. Find the numbers.
3. Find two numbers whose sum is 27 and product is 182.

Answers

1. 6 and 9 2. 18 and 21 3. 13 and 14

Example 18 Find two consecutive odd positive integers, the sum of whose squares is 514.

Solution: Let the two consecutive odd positive integers be $2x - 1$ and $2x + 1$.

According to the question $(2x - 1)^2 + (2x + 1)^2 = 514$

$\Rightarrow \quad 4x^2 - 4x + 1 + 4x^2 + 4x + 1 = 514$

$\Rightarrow \quad 8x^2 + 2 = 514$

$\Rightarrow \quad 4x^2 + 1 = 257$ [On dividing both sides by 2]

$\Rightarrow \quad 4x^2 = 256$

$\Rightarrow \quad x^2 = \frac{256}{4} = 64$

$\Rightarrow \quad x^2 = \pm 8$

If $x = 8$, $2x - 1 = 16 - 1 = 15$ and $2x + 1 = 2 \times 8 + 1 = 17$

If $x = -8$, $2x - 1 = -16 - 1 = -17$ and $2x + 1 = -16 + 1 = -15$

$\therefore$ -17 and -15 are not possible, being –ve.

Hence, 15 and 17 are two consecutive odd positive integers.

18 Try to Solve

1. Find the consecutive odd positive integers, the sum of whose squares is 34.
2. Find two consecutive even positive integers, the sum of whose squares is 52.
3. Find the consecutive positive integers, the sum of whose squares is 365. [NCERT]

Answers

1. 3 and 5. 2. 4 and 6 [Hint: (+)ve even integers are $2x$ and $2x + 2$]
3. 13 and 14

Example 19 A two-digit number is such that the product of its digits is 18. When 63 is subtracted from the number, the digits interchange their places. Find the number. **[CBSE (CCE) 2010]**

Solution: Let the ten's digit be x. Then, the unit's digit $= \frac{18}{x}$.

$\therefore$ Number $= 10x + \frac{18}{x}$, and number obtained by interchanging the digits $= 10 \times \frac{18}{x} + x$

$$\therefore \quad \left(10x + \frac{18}{x}\right) - \left(10 \times \frac{18}{x} + x\right) = 63$$

$$\Rightarrow \quad 10x + \frac{18}{x} - \frac{180}{x} - x = 63$$

$$\Rightarrow \quad 9x - \frac{162}{x} - 63 = 0$$

$$\Rightarrow \quad 9x^2 - 63x - 162 = 0$$

$$\Rightarrow \quad 9(x^2 - 7x - 18) = 0$$

$$\Rightarrow \quad x^2 - 7x - 18 = 0$$

$$\Rightarrow \quad x^2 - 9x + 2x - 18 = 0$$

$$\Rightarrow \quad x(x - 9) + 2(x - 9) = 0$$

$$\Rightarrow \quad (x - 9)(x + 2) = 0$$

$$\Rightarrow \quad x = 9 \text{ or } x = -2$$

But, a digit can never be negative. So, $x = 9$. Hence, the required number $= 10 \times 9 + \frac{18}{9} = 92$.

19 Try to Solve

1. A two-digit number is such that the product of its digits is 12. When 9 is added to the number, the digits interchange their places. Determine the number.
2. A two-digit number is such that the product of the digits is 14. When 45 is added to the number, the digits are reversed. Find the number.

Answers

1. 34 2. 27

Example 20 The sum of two numbers is 8. Determine the numbers if the sum of their reciprocals is $\frac{8}{15}$.

[CBSE (CCE) 2012]

Solution: Let one of the numbers be x. Then, the other number $= (8 - x)$.

According to the question $\frac{1}{x} + \frac{1}{8 - x} = \frac{8}{15}$

$$\Rightarrow \quad \frac{8 - x + x}{x(8 - x)} = \frac{8}{15}$$

$$\Rightarrow \quad \frac{8}{8x - x^2} = \frac{8}{15}$$

$$\Rightarrow \quad \frac{1}{8x - x^2} = \frac{1}{15}$$

$$\Rightarrow \quad 15 = 8x - x^2$$

$$\Rightarrow \quad x^2 - 8x + 15 = 0$$

$$\Rightarrow \quad x^2 - 5x - 3x + 15 = 0$$

$$\Rightarrow \quad x(x - 5) - 3(x - 5) = 0$$

$$\Rightarrow \quad (x - 3)(x - 5) = 0$$

$$\Rightarrow \quad x - 3 = 0 \text{ or } x - 5 = 0$$

$$\therefore \quad x = 3 \text{ or } x = 5$$

First number = 3 or 5. Second number = 8 – 3 = 5 or 8 – 5 = 3.

Hence, the two numbers are 3 and 5.

20 Try to Solve

1. The sum of two numbers is 16. The sum of their reciprocals is 1/3. Find the numbers.
2. The sum of a number and its reciprocal is $\frac{17}{4}$. Find the number.

Answers

1. 4, 12
2. 4 or $\frac{1}{4}$

Concept: Problems Based on Speed, Distance and Time

Example 21 A train travelling at a uniform speed for 360 km would have taken 48 minutes less to travel the same distance if its speed were 5 km/h more. Find the original speed of the train.

Solution: Let the original speed of the train be x km/h.

Time taken to travel 360 km at original speed = $\frac{360}{x}$ h.

When the speed is increased by 5 km/h

$$\text{Speed of the train} = (x + 5) \text{ km/h}$$

Time taken to travel 360 km at $(x + 5)$ km/h = $\frac{360}{x+5}$ h.

According to the question $\quad \frac{360}{x+5} = \frac{360}{x} - \frac{48}{60} \qquad \left[\because 48 \text{ mins} = \frac{48}{60} \text{ h}\right]$

$$\Rightarrow \quad \frac{360}{x+5} - \frac{360}{x} = \frac{-4}{5}$$

$$\Rightarrow \quad 360\left(\frac{1}{x+5} - \frac{1}{x}\right) = \frac{-4}{5}$$

$$\Rightarrow \quad 360\frac{(x - x - 5)}{(x+5)x} = \frac{-4}{5}$$

$$\Rightarrow \quad \frac{360(-5)}{x^2 + 5x} = \frac{-4}{5}$$

$$\Rightarrow \quad \not{-}\frac{1800}{x^2 + 5x} = \not{-}\frac{4}{5}$$

$$\Rightarrow \quad 4x^2 + 20x = 9000$$

$$\Rightarrow \quad 4x^2 + 20x - 9000 = 0$$

$\Rightarrow \quad x^2 + 5x - 2250 = 0$ [On dividing both sides by 4]

$\Rightarrow \quad x^2 + 50x - 45x - 2250 = 0$

$\Rightarrow \quad x(x + 50) - 45(x + 50) = 0$

$\Rightarrow \quad (x + 50)(x - 45) = 0$

$x + 50 = 0$ or $x - 45 = 0$

$\therefore \quad x = -50$ (Not possible) or $x = 45$

Hence, the original speed of the train is 45 km/h.

21 Try to Solve

1. A passenger train takes one hour less for a journey of 150 km if its speed is increased by 5 km/h from its usual speed. Find the usual speed of the train.
2. A train covers a distance of 90 km at a uniform speed. Had the speed been 15 km/h more, it would have taken 30 minutes less for the journey. Find the original speed of the train. **[CBSE (CCE) 2014]**

Answers

1. 25 km/h 2. 45 km/h

Example 22 A fast train takes 3 hours less than a slow train for a journey of 600 km. If the speed of the slow train is 10 km/h less than that of the fast train, find the speed of the two trains.

[CBSE (CCE) 2015]

Solution: Let the speed of the slow train be x km/h. Then, the speed of the fast train is $(x + 10)$ km/h.

$$T_1 = \text{Time taken by the slow train to cover 600 km} = \frac{600}{x} \text{ hours}$$

$$T_2 = \text{Time taken by the fast train to cover 600 km} = \frac{600}{x+10} \text{ hours}$$

According to the question, $T_1 - T_2 = 3$.

$$\therefore \quad \frac{600}{x} - \frac{600}{x+10} = 3$$

$$\Rightarrow \quad \frac{600(x+10) - 600x}{x(x+10)} = 3, \quad \frac{\cancel{600x} + 6000 - \cancel{600x}}{x^2 + 10x} = 3$$

$$\Rightarrow \quad \frac{6000}{x^2 + 10x} = 3$$

$$\Rightarrow \quad \cancel{3}(x^2 + 10x) = \cancel{6000}$$

$$\Rightarrow \quad x^2 + 10x = 2000$$

$$\Rightarrow \quad x^2 + 10x - 2000 = 0$$

$$\Rightarrow \quad x^2 + 50x - 40x - 2000 = 0$$

$$\Rightarrow \quad x(x + 50) - 40(x + 50) = 0$$

$$\Rightarrow \quad (x + 50)(x - 40) = 0$$

$\Rightarrow \quad x = -50$ or $x = 40 \Rightarrow x = 40$ [$\because$ x cannot be negative]

Hence, the speed of the two trains is 40 km/h and 50 km/h.

22 Try to Solve

1. A fast train takes one hour less than a slow train for a journey of 200 km. If the speed of the slow train is 10 km/h less than that of the fast train, find the speed of the two trains.
2. A passenger train takes 3 hours less for a journey of 360 km, if its speed is increased by 10 km/h from its usual speed. What is the usual speed?
3. An express train takes 1 hour less than a passenger train to travel 132 km between Mysore and Bengaluru (without taking into consideration the time they stop at intermediate stations). If the average speed of the express train is 11 km/h more than that of the passenger train, find the average speed of the two trains. **[CBSE (CCE) 2013]**

Answers

1. 40 km/h, 50 km/h   2. 30 km/h   3. 33 km/h, 44 km/h

Example 23 In a flight of 600 km, an aircraft was slowed down due to bad weather. Its average speed for the trip was reduced by 200 km/h and the time of the flight increased by 30 minutes. Find the duration of the flight. **[CBSE (CCE) 2016]**

Solution: Let the original speed of the aircraft be x km/h. Then

New speed = $(x - 200)$ km/h

T_1 = Duration of flight at original speed = $\left(\frac{600}{x}\right)$ hour

T_2 = Duration of flight at reduced speed = $\left(\frac{600}{x-200}\right)$ hour

According to the question $\quad T_2 - T_1 = \frac{30}{60} = \frac{1}{2}$

$$\therefore \quad \frac{600}{x-200} - \frac{600}{x} = \frac{1}{2}$$

$$\Rightarrow \quad \frac{600x - 600(x-200)}{x(x-200)} = \frac{1}{2}$$

$$\Rightarrow \quad \frac{\cancel{600x} - \cancel{600x} + 120000}{x^2 - 200x} = \frac{1}{2}$$

$$\Rightarrow \quad \frac{120000}{x^2 - 200x} = \frac{1}{2}$$

$$\Rightarrow \quad x^2 - 200x - 240000 = 0$$

$$\Rightarrow \quad x^2 - 600x + 400x - 240000 = 0$$

$$\Rightarrow \quad x(x - 600) + 400(x - 600) = 0$$

$$\Rightarrow \quad (x - 600)(x + 400) = 0$$

$$\Rightarrow \quad x - 600 = 0 \text{ or, } x + 400 = 0$$

$$\Rightarrow \quad x = 600 \text{ or, } x = -400 \Rightarrow x = 600 \quad [\because x \text{ cannot be negative}]$$

So, the original speed of the aircraft was 600 km/h.

Hence, duration of flight = $\left(\frac{600}{x}\right)$ h = $\left(\frac{600}{600}\right)$ h = 1 h.

23 Try to Solve

1. In a flight of 6000 km, an aircraft was slowed down due to bad weather. Its average speed for the trip was reduced by 400 km/h and the time increased by 30 minutes. Find the original duration of the flight.
2. A train travels 360 km at a uniform speed. If the speed had been 5 km/h more, it would have taken 1 hour less for the same journey. Find the speed of the train. [CBSE (CCE) 2015]

Answers

1. 2.5 h   2. 40 km/h

Example 24 A motor boat whose speed is 20 km/h in still water takes 1 hour more to travel 21 km upstream than to return downstream to the same spot. Find the speed of the stream.

Solution: Let the speed of the stream be x km/h. The speed of the boat downstream = $(20 + x)$ km/h

Speed of boat upstream = $(20 - x)$km/h

Time taken by boat to travel downstream = $\frac{21}{20+x}$ h

Time taken by boat to travel upstream = $\frac{21}{20-x}$ h

According to the question $\frac{21}{20-x} - \frac{21}{20+x} = 1$

$$\Rightarrow \quad 21\left[\frac{(20+x-20+x)}{(20+x)(20-x)}\right] = 1$$

$$\Rightarrow \quad \frac{21\times 2x}{400-x^2} = 1$$

$$\Rightarrow \quad 42x = 400 - x^2$$

$$\Rightarrow \quad x^2 + 42x - 400 = 0$$

$$\Rightarrow \quad x^2 + 50x - 8x - 400 = 0$$

$$\Rightarrow \quad x(x+50) - 8(x+50) = 0$$

$$\Rightarrow \quad (x+50)(x-8) = 0$$

$$\Rightarrow \quad x + 50 = 0 \text{ or } x - 8 = 0$$

$$\therefore \quad x = -50 \text{ (not possible) or } x = 8$$

Hence, the speed of the stream = 8 km/h.

24 Try to Solve

1. A motorboat whose speed in still water is 15 km/h travels 30 km downstream and returns to the starting point in a total time of 4 hours and 30 minutes. Find the speed of the stream.
2. The speed of a boat in still water is 11 km/h. It can travel 12 km upstream and return downstream to the original point in 2 hours 45 minutes. Find the speed of the stream. [CBSE (CCE) 2012]
3. A motor boat whose speed is 18 km/h in still water takes 1 hour more to travel 24 km upstream than to return downstream to the same spot. Find the speed of the stream. [NCERT]

Answers

1. 5 km/h   2. 5 km/h   3. 6 km/h

Concept: Problems Based on Simple Day-to-Day Life Problems

Example 25 If the list price of a toy is reduced by ₹2, a person can buy 2 toys more for ₹360. Find the original price of the toy. **[CBSE 2006 C]**

Solution: Let the original price of the toy be ₹x.

$\therefore$ Number of toys purchased for ₹360 = $\dfrac{360}{x}$

Reduced list price of toy = ₹$(x - 2)$

$\therefore$ Number of toys purchased for ₹360 = $\dfrac{360}{x-2}$

According to the question

$$\frac{360}{x-2} - \frac{360}{x} = 2$$

$$\Rightarrow \quad 360\left(\frac{1}{x-2} - \frac{1}{x}\right) = 2$$

$$\Rightarrow \quad \frac{1}{x-2} - \frac{1}{x} = \frac{2}{360}$$

$$\Rightarrow \quad \frac{x-x+2}{x(x-2)} = \frac{1}{180}$$

$$\Rightarrow \quad \frac{2}{x^2-2x} = \frac{1}{180}$$

$$\Rightarrow \quad x^2 - 2x = 360$$

$$\Rightarrow \quad x^2 - 20x + 18x - 360 = 0$$

$$\Rightarrow \quad x(x - 20) + 18(x - 20) = 0$$

$$\Rightarrow \quad (x- 20)\,(x + 18) = 0$$

$$\Rightarrow \quad x- 20 = 0 \text{ or } x + 18 = 0$$

$$\therefore \quad x = 20 \text{ or } x = -18 \text{ (Not possible)}$$

Hence, the original price of the toy is ₹20.

25 Try to Solve

1. A person on tour has ₹ 360 for his daily expenses. If he extends his tour for 4 days, he has to cut down his daily expenses by ₹ 3. Find the original duration of the tour. **[CBSE (CCE) 2011]**
2. A shopkeeper buys a number of books for ₹ 80. If he had bought 4 more for the same amount, each book would have cost ₹ 1 less. How many books did he buy?

Answers

1. 20 days
2. 16

Concept: Problems Based on Age

Example 26 The sum of the ages of a father and his son is 50 years. Five years ago, the product of their ages was 175. Find their present ages.

Solution: Let the age of the father be x years and the age of his son = $(50 - x)$ years

Five years ago Age of father = $(x - 5)$ years

Age of his son = $\{(50 - x) - 5\}$ years

$= (45 - x)$ years

According to the question $(x - 5)(45 - x) = 175$

$$45x - 225 - x^2 + 5x = 175$$

$$-x^2 + 50x - 225 - 175 = 0$$

$$-(x^2 - 50x + 400) = 0$$

$$x^2 - 50x + 400 = 0$$

$$x^2 - 40x - 10x + 400 = 0$$

$$x(x - 40) - 10(x - 40) = 0$$

$$(x - 10)(x - 40) = 0$$

$\therefore$ $x = 40$ or $x = 10$

Age of father cannot be 10 years. So, age of father = 40 years.

Age of his son = 50 – 40 = 10 years.

26 Try to Solve

1. The sum of the ages of Puneet and his father is 45 years and the product of their ages is 126. Find the ages of the father and the son.
2. The product of Meena's age five years ago and her age 8 years later is 30, her age at both times being given in years. Find her present age.
3. Rohan's mother is 26 years older than him. The product of their ages (in years) 3 years from now will be 360. We would like to find out Rohan's present age expressed in an equation. **[NCERT]**

Answers

1. 3 years, 42 years **2.** 7 years **3.** $x^2 + 32x - 273 = 0$

Concept: Problems Based on Area

Example 27 The area of a right-angled triangle is 600 cm^2. If the base of the triangle exceeds the altitude by 10 cm, find the dimensions of the triangle.

Solution: Let the altitude BC of the right-angled triangle ABC be x cm. Then

$$\text{Base} = BC = (x + 10) \text{ cm}$$

$$\therefore \quad \text{Area} = \frac{1}{2} (\text{Base} \times \text{Height})$$

$$\Rightarrow \quad \text{Area} = \frac{1}{2} (x + 10)\, x \text{ cm}^2$$

$$\Rightarrow \quad \frac{1}{2} x(x + 10) = 600 \qquad [\because \text{Area} = 600 \text{ cm}^2]$$

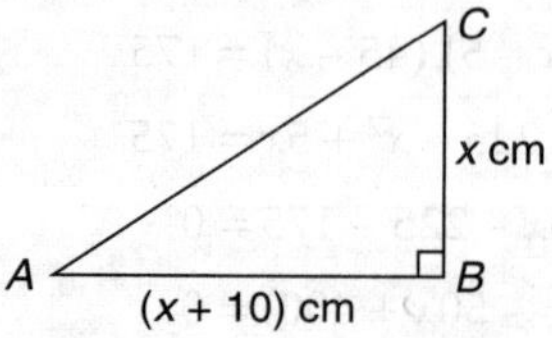

$$\Rightarrow \quad x^2 + 10x = 1200$$

$$\Rightarrow \quad x^2 + 10x - 1200 = 0$$

$$\Rightarrow \quad x^2 + 40x - 30x - 1200 = 0$$

$$\Rightarrow \quad x(x + 40) - 30(x + 40) = 0$$

$$\Rightarrow \quad (x + 40)(x - 30) = 0$$

$$\Rightarrow \quad x = 30, -40$$

$$\Rightarrow \quad x = 30 \qquad [\because x > 0]$$

Hence, Base = (30 + 10) cm = 40 cm and Altitude = 30 cm.

Example 28 The sum of the areas of two squares is 468 m². If the difference of their perimeters is 24 m, find the sides of the two squares. **[CBSE (CCE) 2014]**

Solution: Sum of areas of the two squares = 468 m²

Now, let the side of the smaller square be x m.

∴ Perimeter $\quad P_1 = 4 \times x = 4x$

It is given that $\quad P_2 - P_1 = 24$

Perimeter of the bigger square $\quad P_2 = 24 + P_1 = 24 + 4x$

Therefore, the side of the bigger square $= \dfrac{24 + 4x}{4} = \dfrac{4(6 + x)}{4} = (6 + x)$

Area of the smaller square, $A_1 = x^2$

Area of the bigger square, $A_2 = (6 + x)^2$

According to the question $\quad A_1 + A_2 = 468 \text{ m}^2$

$$\Rightarrow \quad x^2 + (6 + x)^2 = 468 \text{ m}^2$$

$$\Rightarrow \quad x^2 + 36 + x^2 + 12x = 468$$

$$\Rightarrow \quad 2x^2 + 12x = 468 - 36$$

$$\Rightarrow \quad 2x^2 + 12x - 432 = 0$$

$\Rightarrow \quad x^2 + 6x - 216 = 0$

$\Rightarrow \quad x^2 + 18x - 12x - 216 = 0$

$\Rightarrow \quad x(x + 18) - 12(x + 18) = 0$

$\Rightarrow \quad (x - 12)(x + 18) = 0$

$\therefore \quad x = 12 \text{ or } -18$

$x = -18$, not possible

Side of the smaller square = 12 m. Side of the bigger square = 12 + 6 = 18 m.

27 Try to Solve

1. The hypotenuse of a right-angled Δ is 6 m more than twice the shortest side. If the third side is 2 m less than the hypotenuse, find the sides of the Δ.
2. The altitude of a right-angled triangle is 7 cm less than its base. If the hypotenuse is 13 cm, find the other two sides.
3. The side of a square exceeds the side of another square by 4 cm and the sum of the areas of the two squares is 400 cm^2. Find the sides of the two squares. **[CBSE (CCE) 2012]**

Answers

1. 10 m, 24 m and 26 m **2.** 5 cm and 12 cm **3.** 12 cm and 16 cm

FOR MORE PRACTICE AND KNOWLEDGE

Example 29 Check whether the following are quadratic equations:

(i) $(x - 2)^2 + 1 = 2x - 3$ **(ii)** $x(x + 1) + 8 = (x + 2)(x - 2)$

(iii) $x(2x + 3) = x^2 + 1$ **(iv)** $(x + 2)^3 = x^3 - 4$ **[NCERT Exemplar]**

Solution: (i) LHS $= (x - 2)^2 + 1 = x^2 - 4x + 4 + 1 = x^2 - 4x + 5$

Therefore, $(x - 2)^2 + 1 = 2x - 3$ can be rewritten as

$$x^2 - 4x + 5 = 2x - 3$$

That is $\quad x^2 - 6x + 8 = 0$

It is of the form $\quad ax^2 + bx + c = 0.$

Therefore, the given equation is a quadratic equation.

(ii) Since $\quad x(x + 1) + 8 = x^2 + x + 8$ and $(x + 2)(x - 2) = x^2 - 4$

Therefore $\quad x^2 + x + 8 = x^2 - 4$

That is $\quad x + 12 = 0$

It is not of the form $\quad ax^2 + bx + c = 0.$

Therefore, the given equation is not a quadratic equation.

(iii) Here $\quad$ LHS $= x(2x + 3) = 2x^2 + 3x$

So $\quad x(2x + 3) = x^2 + 1$ can be rewritten as

$$2x^2 + 3x = x^2 + 1$$

Therefore, we get $\quad x^2 + 3x - 1 = 0$

It is of the form $\quad ax^2 + bx + c = 0.$

So, the given equation is a quadratic equation.

(iv) Here $\text{LHS} = (x+2)^3 = x^3 + 6x^2 + 12x + 8$

Therefore $(x+2)^3 = x^3 - 4$ can be rewritten as

$$x^3 + 6x^2 + 12x + 8 = x^3 - 4$$

That is $6x^2 + 12x + 12 = 0$ or, $x^2 + 2x + 2 = 0$

It is of the form $ax^2 + bx + c = 0.$

So, the given equation is a quadratic equation.

Example 30 Find the roots of the quadratic equation $3x^2 - 2\sqrt{6}\,x + 2 = 0.$ **[NCERT]**

Solution:

$$3x^2 - 2\sqrt{6}\,x + 2 = 3x^2 - \sqrt{6}\,x - \sqrt{6}\,x + 2$$
$$= \sqrt{3}\,x(\sqrt{3}\,x - \sqrt{2}) - \sqrt{2}\,(\sqrt{3}\,x - \sqrt{2})$$
$$= (\sqrt{3}\,x - \sqrt{2})(\sqrt{3}\,x - \sqrt{2})$$

So, the roots of the equation are the values of x for which

$$(\sqrt{3}x - \sqrt{2})(\sqrt{3}x - \sqrt{2}) = 0$$

Now $\sqrt{3}\,x - \sqrt{2} = 0$ for $x = \sqrt{\dfrac{2}{3}}$

So, this root is repeated twice, one for each repeated factor $\sqrt{3}\,x - \sqrt{2}$.

Therefore, the roots of $3x^2 - 2\sqrt{6}\,x + 2 = 0$ are $\sqrt{\dfrac{2}{3}}, \sqrt{\dfrac{2}{3}}$.

Example 31 If the roots of the quadratic equation $x^2 + 2px + mn = 0$ are real and equal, show that the roots of the quadratic equation $x^2 - 2(m+n)x + (m^2 + n^2 + 2p^2) = 0$ are also equal. **[Foreign 2016]**

Solution: It is given that the quadratic equation $x^2 + 2px + mn = 0$ has real and equal roots.

$\therefore$ $(2p)^2 - 4(mn)(1) = 0$ [$\because$ For real and equal roots $b^2 - 4ac = 0$]

$\Rightarrow$ $4p^2 - 4mn = 0 \Rightarrow p^2 - mn = 0$

$p^2 = mn$ ...(i)

Now, for the quadratic equation $x^2 - 2(m+n)x + (m^2 + n^2 + 2p^2) = 0$

$$D = b^2 - 4ac$$
$$D = [-2(m+n)]^2 - 4(1)(m^2 + n^2 + 2p^2)$$
$$D = 4(m+n)^2 - 4(m^2 + n^2 + 2mn) \quad \text{[From (i)]}$$
$$D = 4(m+n)^2 - 4(m+n)^2$$
$$D = 0$$

$\therefore$ The quadratic equation $x^2 - 2(m+n)x + (m^2 + n^2 + 2p^2) = 0$ also has equal roots.

Example 32 If the roots of the quadratic equation $(a-b)x^2 + (b-c)x + (c-a) = 0$ are equal, prove that $2a = b + c$. **[AI 2016]**

Solution: $(a-b)x^2 + (b-c)x + (c-a) = 0$

For equal roots, discriminant $D = 0$

$\Rightarrow$ $(b-c)^2 - 4(a-b)(c-a) = 0$

$\Rightarrow$ $b^2 - 2bc + c^2 - 4(ac - a^2 - bc + ab) = 0$

$\Rightarrow$ $b^2 - 2bc + c^2 - 4ac + 4a^2 + 4bc - 4ab = 0$

$$\Rightarrow \quad 4a^2 + b^2 + c^2 - 4ab + 2bc - 2ac = 0$$

$$\Rightarrow \quad (2a - b - c)^2 = 0$$

$$\Rightarrow \quad 2a - b - c = 0$$

$$\Rightarrow \quad 2a = b + c$$

Example 33 Solve the following quadratic equation for x: $9x^2 - 6b^2x - (a^4 - b^4) = 0$. [CBSE (CCE) 2015]

Solution: $9x^2 - 6b^2x - (a^4 - b^4) = 0$

$$\Rightarrow \quad x = \frac{6b^2 \pm \sqrt{36b^4 + 4 \times 9 \times (a^4 - b^4)}}{2 \times 9} \qquad \left[\because x = \frac{-b \pm \sqrt{b^2 - 4ac}}{2a}\right]$$

$$\Rightarrow \quad x = \frac{6b^2 \pm \sqrt{36b^4 + 36a^4 - 36b^4}}{2 \times 9}$$

$$\Rightarrow \quad x = \frac{6b^2 \pm \sqrt{36a^4}}{2 \times 9}$$

$$\Rightarrow \quad x = \frac{6b^2 \pm 6a^2}{2 \times 3 \times 3}$$

$$\Rightarrow \quad x = \frac{b^2 \pm a^2}{3}$$

$$\Rightarrow \quad x = \frac{b^2 + a^2}{3}, \frac{b^2 - a^2}{3}$$

Example 34 Solve the following quadratic equation for x: $x^2 - 2ax - (4b^2 - a^2) = 0$.

Solution: $x^2 - 2ax - (4b^2 - a^2) = 0$

$$D = (-2a)^2 - 4 \times [-(4b^2 - a^2)] = 4a^2 + 16b^2 - 4a^2 = 16b^2$$

$$\therefore \quad x = \frac{-(-2a) \pm \sqrt{16b^2}}{2 \times 1} = \frac{2a \pm 4b}{2} = a \pm 2b \qquad \left[\because x = \frac{-b \pm \sqrt{D}}{2a}\right]$$

$$\Rightarrow \quad x = a + 2b,\ a - 2b$$

Example 35 Find the roots of $4x^2 + 3x + 5 = 0$ by the method of completing the square. [NCERT]

Solution: $4x^2 + 3x + 5 = 0$

Dividing both sides by 4, we get $\quad x^2 + \frac{3}{4}x + \frac{5}{4} = 0$

$$x^2 + 2 \times \frac{3}{8}x + \left(\frac{3}{8}\right)^2 - \left(\frac{3}{8}\right)^2 + \frac{5}{4} = 0$$

$$\Rightarrow \quad \left(x + \frac{3}{8}\right)^2 - \frac{9}{64} + \frac{5}{4} = 0$$

$$\Rightarrow \qquad \left(x+\frac{3}{8}\right)^2 = \frac{-71}{64}$$

That is $$\left(x+\frac{3}{8}\right)^2 = \frac{-71}{64} < 0$$

But $\left(x+\frac{3}{8}\right)^2$ cannot be negative for any real value of x. So, there is no real value of x that satisfies the given equation. Therefore, the given equation has no real roots.

Example 36 Two water taps together can fill a tank in 6 hours. The tap of larger diameter takes 9 hours less than the smaller one to fill the tank separately. Find the time in which each tap can separately fill the tank. **[CBSE (CCE) 2010]**

Solution: Let x hours be the time taken by the smaller water tap.

$\therefore (x-9)$ hours is the time taken by the larger water tap. According to the question

$$\frac{1}{x}+\frac{1}{x-9}=\frac{1}{6}$$

$$\frac{x-9+x}{x(x-9)}=\frac{1}{6}$$

$$\Rightarrow \qquad 6(2x-9) = x(x-9)$$

$$\Rightarrow \qquad 12x - 54 = x^2 - 9x$$

$$\Rightarrow \qquad x^2 - 21x + 54 = 0$$

$$\Rightarrow \qquad x^2 - 18x - 3x + 54 = 0$$

$$\Rightarrow \qquad x(x-18) - 3(x-18) = 0$$

$$\Rightarrow \qquad (x-3)(x-18) = 0$$

$$\Rightarrow \qquad x = 3 \text{ or } x = 18$$

$\therefore$ Tap of smaller diameter takes 18 hours and that of larger diameter takes $(18 - 9) = 9$ hours.

Example 37 A pole has to be erected at a point on the boundary of a circular park of diameter 17 m in such a way that the difference of its distances from two diametrically opposite fixed gates A and B on the boundary is 7 m. Find the distances from the two gates where the pole is to be erected. **[CBSE (CCE) 2016]**

Solution: Let C be the position of the pole.

$$\angle ACB = 90^\circ \text{ (angle in a semi-circle in a right angle)}$$

By Pythagoras' theorem $$AB^2 = AC^2 + CB^2$$

$$17^2 = AC^2 + CB^2 \qquad \text{...(i)}$$

$$AC - CB = 7$$

$$\Rightarrow \qquad AC = 7 + CB \qquad \text{...(ii)}$$

Squaring both sides, we get $$(AC - CB)^2 = 49$$

$$\Rightarrow \qquad AC^2 + CB^2 - 2AC \cdot CB = 49 \qquad \text{...(iii)}$$

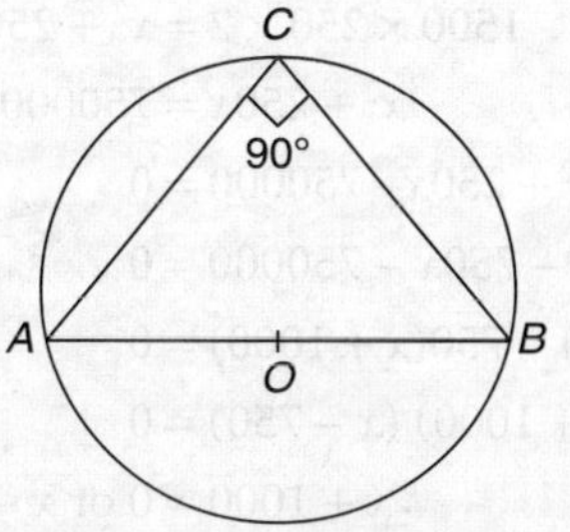

From (i) and (iii), we have $2AC \cdot CB = 289 - 49$

$\Rightarrow \quad AC \cdot CB = \frac{240}{2}$

$\Rightarrow \quad 120 = AC \cdot CB$ ...(iv)

From (ii) and (iv), we have $120 = (7 + CB) \cdot CB$

Let $CB = x$

$120 = (7 + x)\, x$

$\Rightarrow \quad x^2 + 7x - 120 = 0$

$\Rightarrow \quad x^2 + 15x - 8x - 120 = 0$

$\Rightarrow \quad x(x + 15) - 8(x + 15) = 0$

$\Rightarrow \quad (x - 8)(x + 15) = 0$

$x = 8$ or $x = -15$ (Rejected)

$CB = x = 8$ m

$AC = 7 + CB = 7 + 8 = 15$ m

Example 38 A passenger, while boarding a plane, slipped on the stairs and got injured. The pilot took the passenger to the emergency clinic at the airport for treatment. Due to this, the plane was delayed by half an hour. To reach the destination 1500 km away on time so that passengers could catch their connecting flights, the speed of the plane was increased by 250 km/h (greater than the usual speed). Find the usual speed of the plane. What value is depicted in this question? **[CBSE (CCE) 2016]**

Solution: Let the usual speed of the plane be x km/h.

Time taken to cover 1500 km at usual speed $(t_1) = \frac{1500}{x}$ hours.

When the speed of the plane is increased, new speed = $(x + 250)$ km/h.

Time taken to cover 1500 km at the new speed $(x + 250)$ km/h$(t_2) = \frac{1500}{x + 250}$ hours.

According to the question $t_1 - t_2 = \frac{1}{2}$

$\Rightarrow \quad \frac{1500}{x} - \frac{1500}{x + 250} = \frac{1}{2}$

$\Rightarrow \quad \frac{1500[\cancel{x} + 250 - \cancel{x}]}{(x)(x + 250)} = \frac{1}{2}$

$$1500 \times 250 \times 2 = x^2 + 250x$$

$$x^2 + 250x = 750000$$

$$x^2 + 250x - 750000 = 0$$

$\Rightarrow$ $x^2 + 1000x - 750x - 750000 = 0$

$\Rightarrow$ $x(x + 1000) - 750(x + 1000) = 0$

$\Rightarrow$ $(x + 1000)(x - 750) = 0$

$\Rightarrow$ $x + 1000 = 0$ or $x - 750 = 0$

$\Rightarrow$ $x = -1000$ or $x = 750$

$\Rightarrow$ $x = 750$ [$\because$ Speed cannot be negative]

Hence, the usual speed of the plane is 750 km/h.

In this question, the pilot's caring behaviour towards the passenger is shown; he is also cautious and alert in his duty to reach the destination at the scheduled time.

Multiple Choice Questions

Choose the correct answer for each of the following questions:

1. Which of the following is a quadratic equation?

(a) $x^2 + 2x + 1 = (4 - x)^2 + 3$ (b) $-2x^2 = (5 - x)\left(2x - \frac{2}{5}\right)$

(c) $(k + 1)x^2 + \frac{3}{2}x = 7$, where $k = -1$ (d) $x^3 - x^2 = (x - 1)^3$ **[NCERT Exemplar]**

2. Which of the following is not a quadratic equation?

(a) $2(x - 1)^2 = 4x^2 - 2x + 1$ (b) $2x - x^2 = x^2 + 5$

(c) $(\sqrt{2}x + \sqrt{3})^2 + x^2 = 3x^2 - 5x$ (d) $(x^2 + 2x)^2 = x^4 + 3 + 4x^3$ **[NCERT Exemplar]**

3. Which constant must be added and subtracted to solve the quadratic equation $9x^2 + \frac{3}{4}x - \sqrt{2} = 0$ by the method of completing the square?

(a) $\frac{1}{8}$ (b) $\frac{1}{64}$ (c) $\frac{1}{4}$ (d) $\frac{9}{64}$ **[NCERT Exemplar]**

4. Which of the following is not a quadratic equation?

(a) $3x - x^2 = x^2 + 5$ (b) $(x + 2)^2 = 2(x^2 - 5)$

(c) $(\sqrt{2}x + 3)^2 = 2x^2 + 6$ (d) $(x - 1)^2 = 3x^2 + x - 2$

5. The ratio of the sum and product of the roots of the equation $7x^2 - 12x + 18 = 0$ is

(a) 7:12 (b) 7:18 (c) 3:2 (d) 2:3

6. If α and β are the roots of the equation $3x^2 + 8x + 2 = 0$, then $\left(\frac{1}{\alpha} + \frac{1}{\beta}\right) = ?$

(a) $\frac{-3}{8}$ (b) $\frac{2}{3}$ (c) -4 (d) 4

7. The roots of the equation $ax^2 + bx + c = 0$ will be the reciprocal of each other if

(a) $a = b$ (b) $b = c$ (c) $c = a$ (d) None of these

8. If the equation $9x^2 + 6kx + 4 = 0$ has equal roots, then $k = ?$

(a) 2 or 0 (b) -2 or 0 (c) 2 or -2 (d) 0 only

9. If the equation $4x^2 - 3kx + 1 = 0$ has equal roots, then $k = ?$

(a) $\pm\frac{2}{3}$ (b) $\pm\frac{1}{3}$ (c) $\pm\frac{3}{4}$ (d) $\pm\frac{4}{3}$

10. The roots of the equation $2x^2 - 6x + 3 = 0$ are

(a) Real, unequal and rational (b) Real, unequal and irrational

(c) Real and equal (d) Imaginary

11. If the equation $x^2 - kx + 1 = 0$ has no real roots, then

(a) $k < -2$ (b) $k > 2$ (c) $-2 < k < 2$ (d) None of these

12. For what values of k will the equation $kx^2 - 6x - 2 = 0$ have real roots?

(a) $k < \frac{-9}{2}$ (b) $k > \frac{-9}{2}$ (c) $k < 2$ (d) None of these

13. The length of a rectangular field exceeds its breadth by 8 m and the area of the field is 240 m^2. The breadth of the field is

(a) 20 m (b) 30 m (c) 12 m (d) 16 m **[CBSE 2014]**

14. If one root of the equation $px^2 - 14x + 8 = 0$ is six times the other, then p is equal to

(a) 2 (b) 3 (c) 1 (d) None of these

15. The roots of $x^2 - 2x - (r^2 - 1) = 0$ are

(a) $1 - r, r - 1$ (b) $1 - r, r + 1$ (c) $1, r$ (d) $1 - r, r$

16. If the sum and product of the roots of the quadratic equation $ax^2 - 5x + c = 0$ are each equal to 10, then the values of a and c are

(a) $\frac{1}{2}$ and -5 (b) $\frac{1}{2}$ and 5 (c) 5 and $\frac{3}{2}$ (d) $\frac{3}{2}$ and 5

17. If discriminant = 0, then the root of a quadratic equation is

(a) Real and equal (b) No real roots (c) Real and distinct (d) None of these

18. If the roots of the equation $3x^2 - kx + 3 = 0$ are real and equal, than the value of k must be:

(a) ± 6 (b) Between -6 and 6

(c) Greater than 6 (d) Greater than 6 or less than -6

19. The values of k for which the quadratic equation $2x^2 - kx + k = 0$ has equal roots are

(a) 0 only (b) 4 (c) 8 only (d) 0, 8

20. The quadratic equation whose roots are $(2 + \sqrt{3})$ and $(2 - \sqrt{3})$ is

(a) $2x^2 - 4x + 1 = 0$ (b) $x^2 - 4x + 1 = 0$

(c) $x^2 - 6x + 4 = 0$ (d) None of these

21. One of the roots of the quadratic equation $6x^2 - x - 2 = 0$ is

(a) $\frac{1}{2}$ (b) $-\frac{1}{2}$ (c) $-\frac{2}{3}$ (d) -1

22. If -4 is a root of the equation $x^2 + px - 4 = 0$ and the quadratic equation $x^2 + px + k = 0$ has equal roots, then the value of k is

(a) $\frac{9}{4}$ (b) $\frac{4}{9}$ (c) $-\frac{9}{4}$ (d) $-\frac{4}{9}$

23. If the equation $x^2 + 4x + k = 0$ has real and distinct roots, then

(a) $k < 4$ (b) $k > 4$ (c) $k > 4$ (d) $k < 4$

24. The roots of the equation $x^2 - x - p(p + 1) = 0$, where p is a constant, are

(a) $p, p + 1$ (b) $-p, p + 1$ (c) $p, -p(p + 1)$ (d) $-p, -(p + 1)$

25. If $x^2 - 2x + 1 = 0$, then the value of $x + \frac{1}{x}$ is

(a) -2 (b) 2 (c) $-\frac{2}{3}$ (d) $\frac{2}{3}$

26. If the roots of $ax^2 + bx + c = 0$ are equal in magnitude but opposite in sign, then

(a) $a = c$ (b) $b = 0$ (c) $c = 0$ (d) None of these

27. The quadratic equation $2x^2 - \sqrt{5}\,x + 1 = 0$ has

(a) Two distinct real roots (b) Two equal real roots
(c) No real roots (d) More than two real roots **[NCERT]**

28. Which of the following equations has two distinct real roots?

(a) $2x^2 - 3\sqrt{2}\,x + \frac{9}{4} = 0$ (b) $x^2 + x - 5 = 0$ (c) $x^2 + 3x + 2\sqrt{2} = 0$ (d) $5x^2 - 3x + 1 = 0$ **[NCERT]**

29. Which of the following equations has no real roots?

(a) $x^2 - 4x + 3\sqrt{2} = 0$ (b) $x^2 + 4x - 3\sqrt{2} = 0$ (c) $x^2 - 4x - 3\sqrt{2} = 0$ (d) $3x^2 + 4\sqrt{3}\,x + 4 = 0$ **[NCERT]**

30. $(x^2 + 1)^2 - x^2 = 0$ has

(a) Four real roots (b) Two real roots (c) No real roots (d) One real root **[NCERT]**

Fill in the Blanks

Complete the following statements with appropriate answers(s) in the blank space(s):

31. If one root of the equation $(k - 1)x^2 - 10x + 3 = 0$ is the reciprocal of the other, then the value of k is __________.

32. The value of k for which the roots of the equation $kx(3x - 4) + 4$ are equal is __________.

33. If the discriminant of a quadratic polynomial is equal to zero, then the graph (parabola) represented by it touches __________.

34. The sum of the zeros of $2x^2 - 8x + 6$ is __________.

35. The positive value of k for which the equations $x^2 + kx + 64 = 0$ and $x^2 - 8x + k = 0$ will have real roots is __________.

True/False

36. Write whether the following statements are true or false. Justify your answers.

(i) Every quadratic equation has exactly one root.
(ii) Every quadratic equation has at least one real root.
(iii) Every quadratic equation has at least two roots.
(iv) Every quadratic equation has at most two roots.
(v) If the coefficient of x^2 and the constant term of a quadratic equation have opposite signs, then the quadratic equation has real roots.
(vi) If the coefficient of x^2 and the constant term have the same sign and if the coefficient of the x term is zero, then the quadratic equation has no real roots. **[NCERT Exemplar]**

Assertion–Reason

The following questions consist of two statements – Assertion (A) and Reason (R). Answer these questions by selecting the appropriate option given below.

(A) Both A and R are true and R is the correct explanation for A.

(B) Both A and R are true and R is not the correct explanation for A.

(C) A is true but R is false.
(D) A is false but R is true.

37. **Assertion (A):** The equation (m being real) $mx^2 + 2x + m = 0$ has two distinct roots, if $m \neq 0$.
Reason (R): The value equation $ax^2 + bx + c = 0$ has two distinct roots if $b^2 - 4ac > 0$.

38. **Assertion (A):** The value of $k = 2$ if one root of the quadratic equation $6x^2 - x - k = 0$ is $\frac{2}{3}$.
Reason (R): The quadratic equation $ax^2 + bx + c = 0$, $a \neq 0$ has two roots.

39. **Assertion (A):** The discriminant of the quadratic equation $(x + 5)^2 = 2(5x - 3)$ is -124.
Reason (R): $D = b^2 - 4ac$.

40. **Assertion (A):** The equation $2x^2 + \frac{1}{x} - 7 = 0$ is a quadratic equation.
Reason (R): Degree of quadratic equation is 2.

41. **Assertion (A):** 3 is the one and only one root of the quadratic equation $x^2 - x - 6 = 0$.
Reason (R): A quadratic equation has two roots.

42. **Assertion (A):** $5x^2 + 14x + 10 = 0$ has no real roots.
Reason (R): $ax^2 + bx + c = 0$ has no real roots if $b^2 < 4ac$.

43. **Assertion (A):** The sum of the roots of the quadratic equation $3x^2 + (2k + 1)x - (k + 4) = 0$ is equal to the product of the roots for $k = 3$.
Reason (R): Sum of roots $= -\frac{b}{a}$, Product of roots $= \frac{c}{a}$.

44. **Assertion (A):** The nature of the roots of the quadratic equation $x(3x - 4) + 5 = 0$ can be determined by finding its discriminant.
Reason (R): Discriminant $(D) = b^2 - 4ac$.

45. **Assertion (A):** The roots of the quadratic equation $\frac{2}{5}x^2 - x - \frac{3}{5} = 0$ are 3 and $-\frac{1}{2}$.
Reason (R): $ax^2 + bx + c = 0$, $a \neq 0$ is called the standard form of a quadratic equation.

Subjective Questions

46. Check whether the following are quadratic equations:
(i) $(x + 2)^3 = 2x(x^2 - 1)$ (ii) $x^3 - 4x^2 - x + 1 = (x - 2)^3$ **[NCERT]**

47. If $x = \frac{2}{3}$ and $x = -3$ are roots of the quadratic equation $ax^2 + 7x + b = 0$, find the values of a and b. **[CBSE 2016]**

48. Find the roots of the following quadratic equations by the factorisation method:
(i) $3\sqrt{2}\,x^2 - 5x - \sqrt{2} = 0$ (ii) $3x^2 + 5\sqrt{5}\,x + 10 = 0$ (iii) $21x^2 - 2x + \frac{1}{21} = 0$
[NCERT Exemplar]

49. Solve for x: $\sqrt{6x + 7} - (2x - 7) = 0$. **[CBSE (CCE) 2016]**

50. Solve for x: $\sqrt{3}x^2 - 2\sqrt{2}x - 2\sqrt{3} = 0$ **[CBSE (CCE) 2016]**

51. Solve for x: $\frac{2x}{x-3} + \frac{1}{2x+3} + \frac{3x+9}{(x-3)(2x+3)} = 0$, $x \neq 3, -\frac{3}{2}$. **[CBSE (CCE) 2016]**

52. Solve for x: $\frac{x+1}{x-1} + \frac{x-2}{x+2} = 4 - \frac{2x+3}{x-2}$; $x \neq -2, 1, 2$. **[CBSE (CCE) 2016]**

53. Solve the following quadratic equation for x:

$$x^2 + \left(\frac{a}{a+b} + \frac{a+b}{a}\right)x + 1 = 0$$ **[CBSE (CCE) 2016]**

54. Solve the given quadratic equation for x: $9x^2 - 9(a+b)x + (2a^2 + 5ab + 2b^2) = 0$. **[CBSE (CCE) 2016]**

55. Solve for x (in terms of a and b): $\frac{a}{x-b} + \frac{b}{x-a} = 2, x \neq a, b.$ **[CBSE (CCE) 2016]**

56. Find x in terms of a, b and c: $\frac{a}{x-a} + \frac{b}{x-b} = \frac{2c}{x-c}$ $x \neq a, b, c.$ **[CBSE (CCE) 2016]**

57. Solve the following quadratic equation for x: $4x^2 - 4a^2x + (a^4 - b^4) = 0.$ **[CBSE (CCE) 2015]**

58. Solve for x: $x^2 - (2b-1)x + (b^2 - b - 20) = 0.$ **[CBSE (CCE) 2015]**

59. Solve for x: $\frac{x-2}{x-3} + \frac{x-4}{x-5} = \frac{10}{3}; x \neq 3, 5.$ **[CBSE (CCE) 2014]**

60. Solve for x: $\left(\frac{2x-1}{x+3}\right) - 3\left(\frac{x+3}{2x-1}\right) = 5; x \neq -3, \frac{1}{2}.$ **[CBSE (CCE) 2014]**

61. Solve for x: $\frac{x+3}{x+2} = \frac{3x-7}{2x-3}, x \neq -2, \frac{3}{2}.$ **[CBSE 2017 (C)]**

62. Solve for x: $\frac{1}{x+1} + \frac{3}{5x+1} = \frac{5}{x+4}, x \neq -4, -1, -\frac{1}{5}.$ **[AI 2017]**

63. Solve for x: $3\left(\frac{3x-1}{2x+3}\right) - 2\left(\frac{2x+3}{3x-1}\right) = 5; x \neq \frac{1}{3}, -\frac{-3}{2}.$ **[CBSE (CCE) 2014]**

64. Solve for x: $\frac{1}{2a+b+2x} = \frac{1}{2a} + \frac{1}{b} + \frac{1}{2x}.$ **[CBSE (CCE) 2013]**

65. Solve the following quadratic equation in variable x: $abx^2 = (a+b)^2(x-1).$

66. Find the discriminant of the equation $3x^2 - 2x + \frac{1}{3} = 0$ and hence find the nature of its roots. Find the roots, if they are real. **[NCERT]**

67. Find the values of k for each of the following quadratic equations so that they have two equal roots:
(i) $2x^2 + kx + 3 = 0$ (ii) $kx(x-2) + 6 = 0$ **[NCERT]**

68. State whether the following quadratic equations have two distinct real roots. Justify your answer.
(i) $(x - \sqrt{2})^2 - 2(x+1) = 0$ (ii) $\sqrt{2}x^2 - \frac{3}{\sqrt{2}}x + \frac{1}{\sqrt{2}} = 0$
(iii) $x(1-x) - 2 = 0$ (iv) $(x-1)(x+2) + 2 = 0$
(v) $(x+1)(x-2) + x = 0$ **[NCERT Exemplar]**

69. Does there exist a quadratic equation whose coefficients are rational but both of its roots are irrational? Justify your answer. **[NCERT Exemplar]**

70. Does there exist a quadratic equation whose coefficients are all distinct irrational but both the roots are rational? Why? **[NCERT Exemplar]**

71. Is 0.2 a root of the equation $x^2 - 0.4 = 0$? Justify. **[NCERT Exemplar]**

72. Find out whether the following equations have real roots. If real roots exist, find them.

(i) $8x^2 + 2x - 3 = 0$ (ii) $-2x^2 + 3x + 2 = 0$

(iii) $5x^2 - 2x - 10 = 0$ (iv) $\frac{1}{2x-3} + \frac{1}{x-5} = 1, x \neq \frac{3}{2}, 5$

(v) $x^2 + 5\sqrt{5}\,x - 70 = 0$ **[NCERT Exemplar]**

73. Find the value of p for which the quadratic equation $(p+1)x^2 - 6(p+1)x + 3(p+9) = 0, p \neq -1$ has equal roots. Hence, find the roots of the equation. **[Delhi 2015]**

74. Find the non-zero value of k for which the quadratic equation $kx^2 + 1 - 2(k-1)x + x^2 = 0$ has equal roots. Hence, find the roots of the equation. **[Delhi 2015]**

75. If $x = -2$ is a root of the equation $3x^2 + 7x + p = 0$, find the values of k so that the roots of the equation $x^2 + k(4x + k - 1) + p = 0$ are equal. **[Foreign 2015]**

76. If $x = 3$ is the root of the equation $x^2 - x + k = 0$, find the value of p so that the roots of the equation $x^2 + k(2x + k + 2) + p = 0$ are equal. **[Foreign 2015]**

77. If 2 is a root of the quadratic equation $3x^2 + px - 8 = 0$ and the quadratic equation $4x^2 - 2px + k = 0$ has equal roots, find the value of k. **[Foreign 2015]**

78. If 1 is a root of the quadratic equation $3x^2 + ax - 2 = 0$ and the quadratic equation $a(x^2 + 6x) - b = 0$ has equal roots, find the value of b. **[Foreign 2014]**

79. Find the value of p for which the quadratic equation $(2p + 1)x^2 - (7p + 2)x + (7p - 3) = 0$ has equal roots. Also find these roots. **[Delhi 2014]**

80. For what values of k are the roots of the quadratic equation $x^2 - (3k - 1)x + 2k^2 + 2k - 11 = 0$ equal? **[Foreign 2013]**

81. For what values of m are the roots of the quadratic equation $x^2 - 2x(1 + 3m) + 7(3 + 2m) = 0$ equal? **[Foreign 2012]**

82. Find the value of p for which the roots of the quadratic equation $(p+3)x^2 + 2(p+3)x + 4 = 0$ are equal. **[Foreign 2012]**

83. For what value of k are the roots of the equations $x^2 + 4x + k = 0$ real? **[Delhi 2019]**

84. Find the value of k for which the equation $x^2 + k(2x + k - 1) = 0$ has real and equal roots. **[Delhi 2017]**

85. If the equation $(1 + m^2)x^2 + 2mcx + c^2 - a^2 = 0$ has equal roots, then show that $c^2 = a^2(1 + m^2)$. **[Delhi 2017]**

86. If $ad \neq bc$, prove that the equation $(a^2 + b^2)x^2 + 2(ac + bd)x + (c^2 + d^2) = 0$ has no real roots. **[AI 2017]**

87. Show that if the roots of the following quadratic equation are equal, then $ad = bc$.

$$x^2(a^2 + b^2) + 2(ac + bd)x + (c^2 + d^2) = 0$$ **[Delhi 2017 (C)]**

88. Find the positive value of k for which the equations $x^2 + kx + 64 = 0$ and $x^2 - 8x + k = 0$ will both have real roots. **[Foreign 2016]**

89. Write all the values of p for which the quadratic equation $x^2 + px + 16 = 0$ has equal roots. Find the roots of the equation so obtained. **[CBSE Allahabad 2019]**

90. Find the roots, if they exist, of the quadratic equation using the quadratic formula:

$$2x^2 - 2\sqrt{2}\,x + 1 = 0$$ **[NCERT]**

91. Find the roots of the quadratic equations by using the quadratic formula in each of the following:

(i) $-x^2 + 7x - 10 = 0$ (ii) $x^2 + 2\sqrt{2}\,x - 6 = 0$

(iii) $x^2 - 3\sqrt{5}\,x + 10 = 0$ (iv) $\frac{1}{2}x^2 - \sqrt{11}\,x + 1 = 0$ **[NCERT Exemplar]**

92. Find the roots of the equation $5x^2 - 6x - 2 = 0$ by the method of completing the square. **[NCERT]**

93. Check whether the equation $6x^2 - 7x + 2 = 0$ has real roots, and if it has, find them by the method of completing the squares. **[NCERT Exemplar]**

94. Represent the following situations mathematically:
A cottage industry produces a certain number of toys in a day. The cost of production of each toy (in rupees) was found to be ₹55 minus the number of toys produced in a day. On a particular day, the total cost of production was ₹750. We would like to find out the number of toys produced on that day. **[NCERT]**

95. A cottage industry produces a certain number of pottery articles in a day. It was observed on a particular day that the cost of production of each article (in rupees) was 3 more than twice the number of articles produced on that day. If the total cost of production on that day was ₹90, find the number of articles produced and the cost of each article. **[NCERT]**

96. In a class test, the sum of Shefali's marks in Mathematics and English is 30. Had she got 2 marks more in Mathematics and 3 marks less in English, the product of their marks would have been 210. Find her marks in the two subjects. **[NCERT]**

97. Two water taps together can fill a tank in $9\frac{3}{8}$ hours. The tap of larger diameter takes 10 hours less than the smaller one to fill the tank separately. Find the time in which each tap can separately fill the tank. **[NCERT]**

98. The sum of the areas of two squares is 468 m^2. If the difference of their perimeters is 24 m, find the sides of the two squares. **[NCERT]**

99. Had Ajita scored 10 more marks in her mathematics test out of 30 marks, 9 times these marks would have been the square of her actual marks. How many marks did she get in the test? **[NCERT Exemplar]**

100. A train travels at a certain average speed for a distance of 63 km and then travels a distance of 72 km at an average speed of 6 km/h more than its original speed. If it takes 3 hours to complete the total journey, what is its original average speed? **[CBSE 2018, NCERT Exemplar]**

101. Find the natural number whose square diminished by 84 is equal to thrice of 8 more than the given number. **[NCERT Exemplar]**

102. If Zeba were younger by 5 years than what she really is, then the square of her age (in years) would have been 11 more than five times her actual age. What is her age now? **[NCERT Exemplar]**

103. At present, Asha's age (in years) is 2 more than the square of her daughter Nisha's age. When Nisha grows to her mother's present age, Asha's age would be one year less than 10 times the present age of Nisha. Find the present ages of both Asha and Nisha. **[NCERT Exemplar]**

104. In the centre of a rectangular lawn of dimensions 50×40 m, a rectangular pond has to be constructed so that the area of the grass surrounding the pond would be 1184 m^2 (see figure). Find the length and breadth of the pond. **[NCERT Exemplar]**

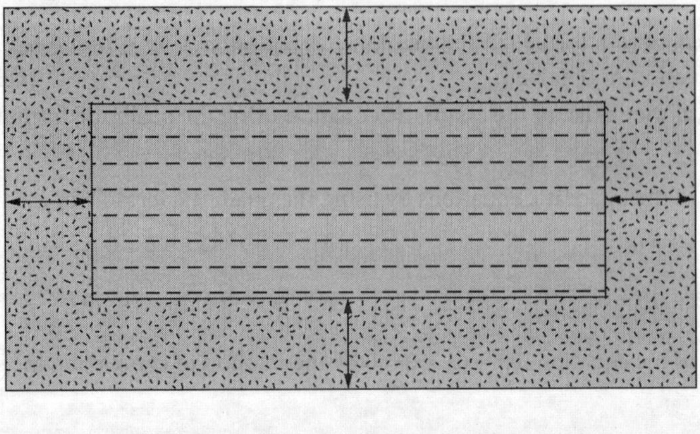

105. At t minutes past 2 pm, the time needed by the minute hand of a clock to show 3 pm was found to be 3 minutes less than $\frac{t^2}{4}$ minutes. Find t. **[NCERT Exemplar]**

106. Two water taps together can fill a tank in 9 hours 36 minutes. The tap of larger diameter takes 8 hours less than the smaller one to fill the tank separately. Find the time in which each tap can separately fill the tank. **[CBSE (CCE) 2016]**

107. The time taken by a person to cover 150 km was $2\frac{1}{2}$ hours more than the time taken on the return journey. If he returned at a speed of 10 km/h more than the speed while travelling onward, find the speed per hour in each direction. **[CBSE (CCE) 2016]**

108. A rectangular park is to be designed whose breadth is 3 m less than its length. Its area is to be 4 square metres more than the area of a park that has already been made in the shape of an isosceles triangle with its base as the breadth of the rectangular park and of altitude 12 m. Find the length and breadth of the rectangular park. **[CBSE (CCE) 2016]**

109. The denominator of a fraction is one more than twice its numerator. If the sum of the fraction and its reciprocal is $2\frac{16}{21}$, find the fraction. **[CBSE (CCE) 2016]**

110. A shopkeeper buys some books for ₹80. If he had bought 4 more books for the same amount, each book would have cost ₹1 less. Find the number of books he bought. **[CBSE (CCE) 2012]**

111. A two-digit number is such that the product of its digits is 14. When 45 is added to the number, the digits interchange their places. Find the number. **[CBSE (CCE) 2012]**

112. A motor boat whose speed is 20 km/h in still water takes 1 hour more to go 48 km upstream than to return downstream to the same spot. Find the speed of the stream. **[Delhi 2011]**

113. Some students planned a picnic. The total budget for food was ₹2,000. But 5 students failed to attend the picnic and thus the cost of food for each member increased by ₹20. How many students attended the picnic and how much did each student pay for the food? **[Foreign 2010]**

114. Divide 27 into two parts such that the sum of their reciprocals is $\frac{3}{20}$. Find the two parts. **[CBSE 2018 (C)]**

115. A plane left 30 minutes later than its scheduled time and in order to reach the destination 1500 km away on time, it had to increases its speed by 100 km/h from the usual speed. Find its usual speed. **[CBSE 2018]**

116. The speed of a boat in still water is 15 km/h. It moves 30 km upstream and returns to the same point in 4 hours 30 minutes. Find the speed of the stream. **[Delhi 2017]**

117. Find two consecutive odd natural numbers, the sum of whose squares is 394. **[Delhi 2017 (C)]**

118. A motor boat whose speed is 18 km/h in still water takes 1 hour more to move 24 km upstream than to return downstream to the same spot. Find the speed of the stream. **[AI 2014, CBSE 2018]**

119. The cost price of an article is ₹x and is sold at a profit of $(x + 10)$ %. Find the cost price of the article, if its selling price is ₹$(2x - 20)$.

120. Two taps running together can fill a tank in $3\frac{1}{13}$ hours. If one tap takes 3 hours more than the other to fill the tank, then how much time will each tap take to fill the tank? **[AI 2017]**

Answers

Multiple Choice Questions

1. (d)	2. (c)	3. (b)	4. (c)	5. (d)	6. (c)	7. (c)	8. (c)	9. (d)	10. (b)
11. (c)	12. (b)	13. (c)	14. (b)	15. (b)	16. (b)	17. (a)	18. (a)	19. (d)	20. (b)
21. (b)	22. (a)	23. (a)	24. (b)	25. (b)	26. (b)	27. (c)	28. (b)	29. (a)	30. (c)

Fill in the Blanks

31. 4 32. 3 33. *X* axis 34. 4 35. 16

True/False

36. (i) False, for example: $x^2 = 1$ is a quadratic equation with two roots.
 (ii) False, for example $x^2 + 1 = 0$ has no real root.
 (iii) False, because every quadratic equation has at most two zeros.
 (iv) True, because every quadratic polynomial has at most two zeros.
 (v) True, because if in $ax^2 + bx + c = 0$, a and c have opposite signs, then $ac < 0$ and so $b^2 - 4ac > 0$.
 (vi) True, because if in $ax^2 + bx + c = 0$, a and c have the same sign and $b = 0$, then $b^2 - 4ac = -4ac < 0$.

Assertion–Reason

37. (D) 38. (B) 39. (A) 40. (D) 41. (D) 42. (A) 43. (A) 44. (A) 45. (B)

Subjective Questions

46. (i) No (ii) Yes
47. $a = 3, b = -6$.
48. (i) $\sqrt{2}, -\frac{\sqrt{2}}{6}$ (ii) $-\sqrt{5}, -\frac{2\sqrt{5}}{3}$ (iii) $\frac{1}{21}, \frac{1}{21}$
49. $x = 7$.
50. $-\frac{\sqrt{6}}{3}, \sqrt{6}$
51. $x = -1$
52. $x = -5$ and $x = \frac{6}{5}$
53. $x = -\left(\frac{a}{a+b}\right)$ or $x = -\left(\frac{a+b}{a}\right)$
54. $x = \frac{2a+b}{3}$ and $x = \frac{a+2b}{3}$
55. $x = a + b$ or $x = \frac{a+b}{2}$
56. $x = 0$ or $x = \frac{2ab - ac - bc}{a + b - 2c}$
57. $x = \frac{a^2+b^2}{2}, \frac{a^2-b^2}{2}$
58. $x = b - 5, b + 4$
59. $x = \frac{7}{2}, 6$
60. $x = -\frac{1}{5}, -10$
61. 5, – 1
62. $\frac{-11}{17}, 1$
63. $x = 0, -7$
64. $x = -a$ or $x = -\frac{b}{2}$
65. $x = \frac{a+b}{a}, \frac{a+b}{b}$
66. $\frac{1}{3}, \frac{1}{3}$
67. (i) $\pm 2\sqrt{6}$, (ii) 6
68. (i) Yes, because discriminant = $(2 + 2\sqrt{2})^2 > 0$. (ii) Yes, because discriminant = 1 > 0.
 (iii) No, because discriminant = – 7 < 0. (iv) Yes, because discriminant = 1 > 0.
 (v) Yes, because discriminant = 8 > 0.
69. $x^2 - 6x + 7 = 0$, which has roots $3 + \sqrt{2}, 3 - \sqrt{2}$
70. Yes. $\sqrt{3}x^2 - 7\sqrt{3}x + 12\sqrt{3} = 0$, which has roots 3, 4.
71. No, because if the roots are rational, the product of the roots will be rational; if the roots are irrational, then the product can be both rational and irrational.

72. (i) Real roots exist; roots are $\frac{1}{2}, \frac{-3}{4}$ (ii) Real roots exist; roots are $2, -\frac{1}{2}$

(iii) Real roots exist; roots are $\frac{1}{5}+\frac{\sqrt{51}}{5}, \frac{1}{5}-\frac{\sqrt{51}}{5}$ (iv) Real roots exist; roots are $4+\frac{3\sqrt{2}}{2}, 4-\frac{3\sqrt{2}}{2}$

(v) Real roots exist; roots are $-7\sqrt{5}, 2\sqrt{5}$

73. $p = 3$ and roots are 3, 3 **74.** $k = 3$; roots are $\frac{1}{2}, \frac{1}{2}$ **75.** $k = -1, k = 2/3$ **76.** $k = -6, p = 12$

77. $k = 1$ **78.** $b = 9$ **79.** $p = 4$ and $p = -\frac{4}{7}$ roots are 7, 7 or $\frac{5}{3}, \frac{5}{3}$. **80.** $k = 9, k = 5$

81. $m = 2, m = \frac{-10}{9}$ **82.** $p = 1$ **83.** $k < 4$ **84.** $k = 0$ **88.** $k = 16$

89. If $p = 8, x = -4, -4$. If $p = -8, x = 4, 4$. **90.** $\frac{1}{\sqrt{2}}, \frac{1}{\sqrt{2}}$

91. (i) 5, 2(ii) $-3\sqrt{2}$, $\sqrt{2}$ (iii) $\sqrt{5}$, $2\sqrt{5}$ (iv) $\sqrt{11}+3$, $\sqrt{11}-3$ **92.** $\frac{3 \pm \sqrt{19}}{5}$

93. Yes, it has real roots, $\frac{3}{2}, \frac{1}{2}$ **94.** $x^2 - 55x + 750 = 0$

95. Number of articles = 6 cost of each article = ₹15

96. Marks in mathematics = 12, Marks in English = 18, or Marks in Mathematics = 13, Marks in English = 17 marks.

97. 15 hours, 25 hours **98.** 18 m, 12 m **99.** 15 **100.** 42 km/h **101.** 12

102. 14 years **103.** Nisha's age is 5 years and Asha's age is 27 years.

104. Length = 34 m and breath = 24 m **105.** 14 **106.** 16 hours and 24 hours

107. Velocity of going 20 m/h and velocity of returning 30 km/h **108.** Length = 7 m and Breath = 4 m

109. $\frac{3}{7}$ **110.** 16 **111.** 27 **112.** 4 km/h **113.** No. of student = 25 **114.** 15, 12

115. 500 km/h **116.** 5 km/h **117.** 13, 15 **118.** 6 km/h. **119.** ₹50 or ₹40

120. 8 hours, 5 hours.

Quadratic Equations (Basic)

SELF-ASSESSMENT TEST

Time allowed: 90 min. Max. Marks: 40

General Instructions

1. *All questions are compulsory. There are 20 questions in all.*
2. *This question paper has four sections: Section A, Section B, Section C and Section D.*
3. *Section A contains 10 questions of one mark each, Section B contains three questions of two marks each, Section C contains four questions of three marks each, and Section D contains three questions of four marks each.*

Section A [1 Mark Each]

Direction: (Q. Nos. 1–5) These are Multiple Choice Questions. Select the most appropriate answer for the given questions.

1. If α and β are the roots of the equation $3x^2 + 8x + 2 = 0$, then $\left(\frac{1}{\alpha}+\frac{1}{\beta}\right) = ?$

(a) $\frac{-3}{8}$ (b) $\frac{2}{3}$ (c) -4 (d) 4

2. One of the roots of the quadratic equation $6x^2 - x - 2 = 0$ is

(a) $\frac{1}{2}$ (b) $-\frac{1}{2}$ (c) $-\frac{2}{3}$ (d) -1

3. If discriminant = 0, then the roots of the quadratic equation are

(a) real and equal (b) no real roots (c) real and distinct (d) none

4. The ratio of the sum and product of the roots of the equation $7x^2 - 12x + 18 = 0$ is

(a) 7:12 (b) 7:18 (c) 3:2 (d) 2:3

5. The quadratic equation $2x^2 - \sqrt{5}\,x + 1 = 0$ has

(a) two distinct real roots (b) two equal real roots
(c) no real roots (d) more than two real roots

6. The sum of the zeros of $2x^2 - 8x + 6$ is __________.
7. The value of k for which the roots of the equation $kx(3x - 4) + 4$ are equal is __________.
8. Determine the value of k for $kx^2 + 2x - 3 = 0$, where $x = 2$ is the solution.
9. Find k if $2x^2 - kx + 1 = 0$ has real and equal roots.
10. Without finding the roots, comment on the nature of the roots of $2x^2 + x + 1 = 0$.

Section B [2 Marks Each]

11. Solve: $\frac{4}{x} - 3 = \frac{5}{2x+3}, x \neq 0, \frac{-3}{2}$.
12. Find the roots of the equation $x^2 - 2x - 8 = 0$ using the completing the square method.
13. Find the roots of the equation $6x^2 + 7x - 10 = 0$ using the quadratic formula.

Section C [3 Marks Each]

14. Find k if $(k - 12) x^2 + 2 (k - 12) x + 2 = 0$ has real and equal roots.

15. Divide 29 into two parts so that the sum of the squares of the parts is 425. Find the numbers.

16. Write all the values of p for which the quadratic equation $x^2 + px + 16 = 0$ has equal roots. Find the roots of the equation so obtained.

17. A shopkeeper buys some books for ₹80. If he had bought 4 more books for the same amount, each book would have cost ₹1 less. Find the number of books he bought.

Section D [4 Marks Each]

18. Solve: $\dfrac{1}{p+q+x} = \dfrac{1}{p} + \dfrac{1}{q} + \dfrac{1}{x}$.

19. The sum of the area of two squares is 468 m². If the difference of their perimeters is 24 m, find the sides of the two squares.

20. A motor boat whose speed is 20 km/h in still water takes 1 hour more to travel 48 km upstream than to return downstream to the same spot. Find the speed of the stream.

Answers

1. (c) **2.** (b) **3.** (a) **4.** (d) **5.** (c) **6.** 4 **7.** 3 **8.** $k = -\dfrac{1}{4}$

9. $k = +2\sqrt{2}$ **10.** No real roots **11.** –2, 1 **12.** 4, – 2 **13.** –2, $\dfrac{5}{6}$ **14.** $k = 12$ or 20

15. 16, 13 **16.** If $p = 8$, $x = -4, -4$ and if $p = -8$, $x = 4, 4$ **17.** 16 **18.** $-q, -p$ **19.** 12 m and 18 m

20. 4 km/h

Quadratic Equations (Standard)

SELF-ASSESSMENT TEST

Time allowed: 90 min. Max. Marks: 40

General Instructions

1. *All questions are compulsory. There are 20 questions in all.*
2. *This question paper has four sections: Section A, Section B, Section C and Section D.*
3. *Section A contains 10 questions of one mark each, Section B contains three questions of two marks each, Section C contains four questions of three marks each, and Section D contains three questions of four marks each.*

Section A [1 Marks Each]

Direction: (Q. Nos. 1–5) These are Multiple Choice Questions. Select the most appropriate answer for the given questions.

1. Which of the following is not a quadratic equation?
   (a) $3x - x^2 = x^2 + 5$ (b) $(x + 2)^2 = 2(x^2 - 5)$
   (c) $(\sqrt{2}\,x + 3)^2 = 2x^2 + 6$ (d) $(x - 1)^2 = 3x^2 + x - 2$
2. If α and β are the roots of the equation $3x^2 + 8x + 2 = 0$, then $\left(\frac{1}{\alpha} + \frac{1}{\beta}\right) = ?$
   (a) $\frac{-3}{8}$ (b) $\frac{2}{3}$ (c) -4 (d) 4
3. The roots of $x^2 - 2x - (r^2 - 1) = 0$ are
   (a) $1 - r, r - 1$ (b) $1 - r, r + 1$ (c) $1, r$ (d) $1 - r, r$
4. If the sum and product of the roots of the quadratic equation $ax^2 - 5x + c = 0$ are each equal to 10, then the values of a and c are
   (a) $\frac{1}{2}$ and -5 (b) $\frac{1}{2}$ and 5 (c) 5 and $\frac{3}{2}$ (d) $\frac{3}{2}$ and 5
5. If $x^2 - 2x + 1 = 0$, then the value of $x + \frac{1}{x}$ is
   (a) -2 (b) 2 (c) $-\frac{2}{3}$ (d) $\frac{2}{3}$
6. The positive value of k for which the equations $x^2 + kx + 64 = 0$ and $x^2 - 8x + k = 0$ will have real roots is __________.
7. If the discriminant of a quadratic polynomial is equal to zero, then the graph (parabola) represented by it touches __________ .
8. If $x = \frac{2}{3}$ and $x = -3$ are the roots of the quadratic equation $ax^2 + 7x + b = 0$, find the values of a and b.
9. Solve for x: $\sqrt{6x + 7} - (2x - 7) = 0$.
10. Find the positive value of k for which the equations $x^2 + kx + 64 = 0$ and $x^2 - 8x + k = 0$ will both have real roots.

Section B [2 Marks Each]

11. Solve for x: $\frac{1}{2a+b+2x} = \frac{1}{2a} + \frac{1}{b} + \frac{1}{2x}$.

12. Solve the following quadratic equation in variable x: $abx^2 = (a + b)^2 (x - 1)$.

13. If 2 is a root of the quadratic equation $3x^2 + px - 8 = 0$ and the quadratic equation $4x - 2px + k = 0$ has equal roots, find the value of k.

Section C [3 Marks Each]

14. Show that if the roots of the following quadratic equation are equal, then $ad = bc$.

$$x^2 (a^2 + b^2) + 2(ac + bd)x + (c^2 + d^2) = 0$$

15. A shopkeeper buys some books for ₹80. If he had bought 4 more books for the same amount, each book would have cost ₹1 less. Find the number of books he bought.

16. If the roots of the quadratic equation $(a - b)x^2 + (b - c)x + (c - a) = 0$ are equal, prove that $2a = b + c$.

17. Solve the following quadratic equation for x: $9x^2 - 6b^2x - (a^4 - b^4) = 0$.

Section D [4 Marks Each]

18. The cost price of an article is ₹x and it is sold at a profit of $(x + 10)$ %. Find the cost price of the article if its selling price is ₹$(2x - 20)$.

19. A plane left 30 minutes later than its scheduled time and, in order to reach the destination 1500 km away on time, it had to increase its speed by 100 km/h from the usual speed. Find its usual speed.

20. A train travelling at a uniform speed for 360 km would have taken 48 minutes less to travel the same distance if its speed were 5 km/h more. Find the original speed of the train.

Answers

1. (c) **2.** (c) **3.** (b) **4.** (b) **5.** (b) **6.** 16 **7.** x axis **8.** $a = 3, b = -6$

9. $x = 7$ **10.** $k = 16$ **11.** $-a, -\frac{b}{2}$ **12.** $\frac{a+b}{2}, \frac{a+b}{2}$ **13.** $k = 1$ **15.** 16 **17.** $\frac{b^2 \pm a^2}{3}$

18. ₹50 or ₹40 **19.** 500 km/h **20.** 45 km/h

CHAPTER

5 Arithmetic Progression

Key Concepts

- Meaning of sequence
- Meaning of progression
- Meaning of arithmetic progression
- General terms of an arithmetic progression
- The nth term from the end of an arithmetic progression
- Relationship between the terms of an arithmetic progression
- Sum of n-terms of an arithmetic progression

Concept: Sequence

(i) **Sequence:** A set of numbers arranged in definite order, according to a definite rule. For example: 2, 4, 6, 8, 10, …

(ii) **Term:** Each number in a sequence. For example, 2 is the first term in the above sequence.

(iii) **Finite sequence:** If a sequence ends at a particular term. For example, 2, 4, 6, 8, …, 24.

(iv) **Infinite sequence:** If a sequence does not end at a particular term. For example, 2, 4, 6, 8, 10, …. Here, the last term is not known.

Example 1 Write the nth term of the following sequences:

(i) 3, 5, 7, 9, 11 **(ii)** 2, 5, 10, 17

Solution:

(i) We have 3, 5, 7, 9, 11 …

$3 = 2 \times 1 + 1$ 1st term

$5 = 2 \times 2 + 1$ 2nd term

$7 = 2 \times 3 + 1$ 3rd term

$9 = 2 \times 4 + 1$ 4th term

$11 = 2 \times 5 + 1$ 5th term

$\therefore$ The nth term $= 2n + 1$

(ii) We have 2, 5, 10, 17, …

$2 = 1^2 + 1$ 1st term

$5 = 2^2 + 1$ 2nd term

$10 = 3^2 + 1$ 3rd term

$17 = 4^2 + 1$ 4th term

The nth term $= n^2 + 1$

Example 2 Write the first four terms of the following sequences:

(i) $a_n = 3n - 1$ (ii) $a_n = 2n^2 + 1$

Solution:

(i) We have $a_n = 3n - 1$

Inserting $n = 1, 2, 3$ and 4, we get

$$a_1 = 3 \times 1 - 1 = 2$$
$$a_2 = 3 \times 2 - 1 = 5$$
$$a_3 = 3 \times 3 - 1 = 8$$
$$a_4 = 3 \times 4 - 1 = 11$$

Hence, 2, 5, 8 and 11 are the first four terms of the sequence.

(ii) We have $a_n = 2n^2 + 1$

Inserting $n = 1, 2, 3$ and 4, we get

$$a_1 = 2 \times 1^2 + 1 = 3$$
$$a_2 = 2 \times 2^2 + 1 = 9$$
$$a_3 = 2 \times 3^2 + 1 = 19$$
$$a_4 = 2 \times 4^2 + 1 = 33$$

Hence, 3, 9, 19 and 33 are the first four terms of the sequence.

Example 3 Find the 4th and 9th terms of the sequence whose n^{th} term is given by $a_n = 7n - 9$.

[CBSE (CCE) 2013]

Solution: $a_n = 7n - 9$ [Given]

For $n = 4, a_4 = 7 \times 4 - 9 = 19$

$n = 9, a_9 = 7 \times 9 - 9 = 54$

Hence, 4th term = 19 and 9th term = 54

1 Try to Solve

1. Write the first five terms of each of the following sequences whose n^{th} terms are:

   (i) $a_n = 3n + 4$ (ii) $a_n = \frac{n-4}{3}$ (iii) $a_n = n^2 + n$

2. Find the indicated terms in each of the following sequences whose n^{th} terms are:

   (i) $a_n = 3n - 4$; a_5 and a_9 (ii) $a_n = \frac{3n+2}{2n-1}$; a_7 and a_{10} (iii) $a_n = n(n+1)$; a_{12} and a_{20}

Answers

1. (i) 7, 10, 13, 16, 19 (ii) $-1, \frac{-2}{3}, \frac{-1}{3}, 0 \frac{1}{3}$ (iii) 2, 6, 12, 20, 30

2. (i) $a_5 = 11, a_9 = 23$ (ii) $a_7 = \frac{23}{13}, a_{10} = \frac{32}{19}$ (iii) $a_{12} = 156, a_{20} = 420$

Concept: Progression

An arrangement of numbers (sequence) where the terms increase (or decrease) numerically. For example, 2, 4, 6, 8, … and 40, 36, 32, 28, 24, … are progressions.

Concept: Arithmetic Progression

(i) A progression where the difference of each term, except the first, from its preceding term is always the same. This constant difference is called common difference, denoted by 'd' and the first term is denoted by 'a'.

(ii) The nth term (general term) of an AP: Consider an AP with first term 'a' and common difference 'd'. The successive terms of this AP are

$$a_1 = a = a + (1-1)d$$

$$a_2 = a_1 + d = a + d = a + (2-1)d$$

$$a_3 = a_2 + d = (a+d) + d = a + 2d = a + (3-1)d$$

$$a_n = a + (n-1)d$$

Hence, the above formula represents the nth term (general term) of an AP whose first term is 'a' and whose common difference is d.

Example 4 For the AP, $\frac{3}{2}, -\frac{1}{2}, -\frac{1}{2}, -\frac{3}{2}, \ldots$, write the first term a and the common difference d. **[NCERT]**

Solution: Here, $a_1 = a = \frac{3}{2}$, $d = \frac{1}{2} - \frac{3}{2} = -1$.

2 Try to Solve

1. For the following APs, write the first term and the common difference:

(a) 3, 1, –1, –3, …

(b) –5, –1, 3, 7, …

(c) $\frac{1}{3}, \frac{5}{3}, \frac{9}{3}, \frac{13}{3}, \ldots$

(d) 0.6, 1.7, 2.8, 3.9, … **[NCERT]**

Answers

1. (a) $a = 3, d = -2$

(b) $a = -5, d = 4$

(c) $a = \frac{1}{3}, d = \frac{4}{3}$

(d) $a = 0.6, d = 1.1$

Example 5 Which of the following lists of numbers form an AP? If they form an AP, write the next two terms:

(i) 4, 10, 16, 22, …

(ii) 1, –1, –3, –5, …

(iii) –2, 2, –2, 2, –2, …

(iv) 1, 1, 1, 2, 2, 2, 3, 3, 3, … **[NCERT]**

Solution:

(i) We have $a_2 - a_1 = 10 - 4 = 6$

$a_3 - a_2 = 16 - 10 = 6$

$a_4 - a_3 = 22 - 16 = 6$

That is, $a_{k+1} - a_k$ is the same every time.

So, the given list of numbers forms an AP with common difference $d = 6$.

The next two terms are: $22 + 6 = 28$ and $28 + 6 = 34$.

(ii) $a_2 - a_1 = -1 - 1 = -2$

$a_3 - a_2 = -3 - (-1) = -3 + 1 = -2$

$a_4 - a_3 = -5 - (-3) = -5 + 3 = -2$

That is, $a_{k+1} - a_k$ is the same every time.

So, the given list of numbers forms an AP with common difference $d = -2$.

The next two terms are: $-5 + (-2) = -7$ and $-7 + (-2) = -9$

(iii) $a_2 - a_1 = 2 - (-2) = 2 + 2 = 4$

$a_3 - a_2 = -2 - 2 = -4$

As $a_2 - a_1 \neq a_3 - a_2$, the given list of numbers does not form an AP.

(iv) $a_2 - a_1 = 1 - 1 = 0$

$a_3 - a_2 = 1 - 1 = 0$

$a_4 - a_3 = 2 - 1 = 1$

Here $a_2 - a_1 = a_3 - a_2 \neq a_4 - a_3$.

So, the given list of numbers does not form an AP.

3 Try to Solve

1. Which of the following are APs? If they form an AP, find the common difference d and write three more terms.

(a) 2, 4, 8, 16, …

(b) $-10, -6, -2, 2, \ldots$

(c) 0.2, 0.22, 0.222, 0.2222, …

(d) $-\frac{1}{2}, -\frac{1}{2}, -\frac{1}{2}, -\frac{1}{2}, \ldots$ [NCERT]

Answers

1. (a) No

(b) Yes, $d = 4$; 6, 10, 14.

(c) No

(d) Yes, $d = 0$; $-\frac{1}{2}, -\frac{1}{2}, -\frac{1}{2}$

Example 6 Write the first four terms of an AP, when term $(a) = 10$ and common difference $(d) = 10$. [NCERT]

Solution: Since

$$a_1 = 10$$
$$a_2 = a_1 + d = 10 + 10 = 20$$
$$a_3 = a_2 + d = 20 + 10 = 30$$
$$a_4 = a_3 + d = 30 + 10 = 40$$

Hence, the first four terms are 10, 20, 30 and 40.

4 Try to Solve

1. Write the first four terms of an AP, when
   (a) $a = -2, d = 0$ (b) $a = 4, d = -3$ (c) $a = -1, d = \frac{1}{2}$

Answers

1. (a) −2, −2, −2, −2 (b) 4, 1, −2, −5 (c) $-1, -\frac{1}{2}, 0, \frac{1}{2}$ [NCERT]

Concept: General Term of an AP

The general or n^{th} term of an AP:

$$a_n = a_1 + (n - 1)d$$

where a_1 represents the first term, n represents the number of terms and d is the common difference.

For example, if we want to find the 12th term

$$a_{12} = a_1 + (12 - 1)d = a_1 + 11d$$
$$a_{50} = a_1 + (50 - 1)d = a_1 + 49d$$

Example 7 Find the 4th and 9th terms of the sequence whose n^{th} term is given by $a_n = 5n - 9$.

Solution: $a_n = 5n - 9$ [Given]

For $n = 4, a_4 = 5 \times 4 - 9 = 11$

$n = 9, a_9 = 5 \times 9 - 9 = 36$

Hence, 4th term = 11 and 9th term = 36

5 Try to Solve

1. Write the first five terms of the following sequences:
   (a) $a_n = 3n + 4$ (b) $a_n = 2^n$.
2. Write the a_5 and a_9 of $a_n = 3n - 4$.

Answers

1. (a) 7, 10, 13, 16, 19 (b) 2, 4, 8, 16, 32
2. 11, 23

Example 8 Find the 10th and 18th terms of the AP: 1, 4, 7, 10, …

Solution: We have 1, 4, 7, 10 … an AP.

Here, first term $a = 1$.

Common difference $d = 4 - 1 = 3$

The nth term of an AP is given by $a_n = a + (n - 1)d$

10th term $a_{10} = 1 + (10 - 1) \times 3$ $[\because n = 10]$

$= 1 + 27 = 28$

18th term $a_{18} = 1 + (18 - 1) \times 3$ $[\because n = 18]$

$= 1 + 51 = 52$

Hence, 10th term = 28 and 18th term = 52

Example 9 How many terms are there in the AP: 6, 10, 14, 18, …, 178?

Solution: We have 6, 10, 14, 18, … 178, an AP.

Here, first term $a = 6$, common difference $d = 10 - 6 = 4$.

Let n be the number of terms in the AP

$\therefore$ $a_n = 178$ [Given]

$\therefore$ $a + (n - 1)d = 178$

$\Rightarrow$ $6 + (n - 1) \times 4 = 178$

$\Rightarrow$ $(n - 1) \times 4 = 178 - 6 = 172$

$\Rightarrow$ $(n - 1) = \dfrac{172}{4} = 43$

$\therefore$ $n = 43 + 1 = 44$

6 Try to Solve

1. Choose the correct option in the following and justify:
   (i) 30th term of the AP: 10, 7, 4, …, is
   (a) 97 (b) 77 (c) −77 (d) −87
   (ii) 11th term of the AP: $-3, -\frac{1}{2}, 2, \ldots,$ is **[NCERT]**
   (a) 28 (b) 22 (c) −38 (d) $-48\frac{1}{2}$
2. Find the number of terms in each of the following APs: **[NCERT]**
   (a) 7, 13, 19, …, 205 (b) $18, 15\frac{1}{2}, 13, \ldots, -47$
3. Which term of the AP: 3, 8, 13, 18, …, is 78? **[NCERT]**

Answers

1. (i) c (ii) b 2. (a) 34 (b) 27
3. 16th term

Example 10 Is –150 a term of the AP 11, 8, 5, 2, … ?

Solution: Let the nth term be 150. The nth term of an AP is given by

$$a_n = a + (n - 1)d \qquad \text{...(i)}$$

Here $a = 11, d = 8 - 11 = -3, a_n = -150$

Inserting these values in (i), we get

$$-150 = 11 + (n - 1) \times -3$$

$$-150 - 11 = -3(n - 1)$$

$$\frac{-161}{-3} = n - 1$$

$$n - 1 = \frac{161}{3}$$

$$\therefore \quad n = \frac{161}{3} + 1 = \frac{164}{3}$$

Since n is not a natural number, –150 is not a term of the given AP.

Example 11 Which term of the AP 134, 129, 124, …, is its first negative term? **[CBSE (CCE) 2013]**

Solution: We have 134, 129, 125, … an AP.

Here $a = 134, d = 129 - 134 = -5$

Let the nth term be the first negative term.

That is $a_n < 0$

$$\Rightarrow \quad a + (n - 1)d < 0$$

$$\Rightarrow \quad 135 + (n - 1)(-5) < 0$$

$$\Rightarrow \quad 134 - 5n + 5 < 0 \qquad [\because d = -5]$$

$$\Rightarrow \quad 139 - 5n < 0$$

$$\Rightarrow \quad 139 < 5n$$

$$\Rightarrow \quad n > \frac{139}{5}$$

$$\Rightarrow \quad n > 27\frac{4}{5}$$

$$\Rightarrow \quad n > 28$$

Hence, the 28th term of the AP is the first negative term.

7 Try to Solve

1. Is 51 a term of the AP, 5, 8, 11, 14, …?
2. Which term of the AP 121, 117, 113, … is its first negative term?
3. Which term of the AP 20, $19\frac{1}{4}$, $18\frac{1}{2}$, $17\frac{3}{4}$, … is its first negative term? [CBSE 2017]

Answers

1. No 2. 32nd term 3. 28th term

Example 12 Find the 11th term from the end of the AP: 10, 7, 4, … – 62. [CBSE (CCE) 2013]

Solution: Rewriting the AP in reverse order, we get

$$-62, \ldots 4, 7, 10$$

Here $a = -62, d = 7 - 4 = 3, n = 11$

We know that $a_n = a + (n - 1)d$

$$a_{11} = -62 + (11 - 1) \times 3$$
$$= -62 + 30$$

$\therefore$ $a_{11} = -32$

Hence, the 11th term from the last term of the AP is – 32.

8 Try to Solve

1. Find the 8th term from the end of the AP 7, 10, 13, …, 184. [CBSE 2005]
2. Find the 6th term from the end of the AP 17, 14, 11, …, (–40). [CBSE 2005]

Answers

1. 163 2. –25

Example 13 If the 6th term of an AP is –10 and its 10th term is –26, find the 15th term of a AP. [CBSE (CCE) 2013]

Solution: Given:

6th term $a_6 = -10$

10th term $a_{10} = -26$

Let a be the first term and d be the common difference of the AP.

$$a_6 = a + (6 - 1)d$$

$\Rightarrow$ $-10 = a + 5d$ …(i)

And $a_{10} = a + (10 - 1)d$

$\Rightarrow$ $-26 = a + 9d$ …(ii)

Subtracting (i) from (ii), we get

$$
\begin{aligned}
-26 &= a + 9d \\
-10 &= a + 5d \\
+ \quad & - \quad - \\
\hline
-16 &= 4d
\end{aligned}
$$

$\therefore$ $d = -4$

Inserting $d = -4$ in (i), we get $-10 = a + 5 \times (-4)$

$-10 = a - 20$

$\therefore$ $a = 10$

15th term $a_{15} = a + (15 - 1)d$

$= 10 + 14 \times (-4) = -46$

Hence, the 15th term of the AP is – 46.

Example 14 Determine an AP whose 3rd term is 16, and when the 5th term is subtracted from the 7th term, we get 12. **[CBSE (CCE) 2013]**

Solution: Let a be the first term and d be the common difference of the AP.

Now, the 3rd term $a_3 = 16$

$\Rightarrow$ $a + (3 - 1)d = 16$ $[\because n = 3]$

$\Rightarrow$ $a + 2d = 16$ ...(i)

And $a_7 - a_5 = 12$ [Given]

$\Rightarrow$ $[a + (7 - 1)d] - [a + (5 - 1)d] = 12$

$\Rightarrow$ $a + 6d - a - 4d = 12$

$\Rightarrow$ $2d = 12$

$\therefore$ $d = 6$

Inserting $d = 6$ in (i), we get $a + 2 \times 6 = 16$

$a = 4$

$a_2 = a + d = 4 + 6 = 10$

$a_3 = a + 2d = 4 + 2 \times 6 = 16$

$a_4 = a + 3d = 4 + 3 \times 6 = 22$

Hence, the AP is 4, 10, 16, 22, ...

9 Try to Solve

1. Find the 31st term of an AP whose 11th term is 38 and 16th term is 73. **[NCERT]**
2. An AP consists of 50 terms of which the 3rd term is 12 and the last term is 106. Find the 29th term. **[NCERT]**
3. If the 3rd and 9th terms of an AP are 4 and –8, respectively, which term of this AP is zero? **[NCERT]**

Answers

1. 178 2. 64 3. 5th term

Example 15 Which term of the AP 3, 15, 27, 39, … will be 132 more than its 54th term? **[NCERT]**

Solution: 3, 15, 27, 39, … are in AP.

$$a = 3, d = 12$$

Let the nth term be 132 more than the 54th term.

$$a_{54} = a + 53d$$
$$= 3 + 53 \times 12$$
$$= 3 + 636$$
$$= 639$$

$\therefore$ $$a_n = a_{54} + 132 \text{ [Given]}$$
$$= 639 + 132$$
$$= 771$$

Again $$a_n = a + (n - 1)d$$
$$771 = 3 + (n - 1)12$$
$$771 = 3 + 12n - 12$$
$$771 + 9 = 12n$$
$$12n = 780$$

$\therefore$ $$n = 65$$

Hence, the 65th term is 132 more than the 54th term.

Example 16 In the following APs, find the missing terms:

(i) 2, □, 26 (ii) □, 13, □, 3 (iii) 5, □, □, $9\frac{1}{2}$

(iv) –4, □, □, □, □, 6 (v) □, 38, □, □, □, –22 **[NCERT]**

Solution:

(i) 2, __, 26

$$a = 2$$
$$a_3 = 26$$
$$a_3 = a + 2d$$
$$26 = 2 + 2d$$
$$\frac{24}{2} = d$$

$\therefore$ $$d = 12$$
$$a_2 = a + d = 2 + 12 = \boxed{14}$$

(ii) __, 13, __, 3

$$a_2 = a + d$$

$\Rightarrow$ $$a + d = 13 \quad \text{...(i)}$$

$$a_4 = a + 3d$$

$\Rightarrow$ $$a + 3d = 3 \quad \text{...(ii)}$$

Subtracting (i) and (ii), we get

$$2d = -10$$

$$\therefore \quad d = -5$$

Inserting the value of 'd' in (i), we get

$$a + d = 13$$

$$a + (-5) = 13$$

$$a = 13 + 5 = \boxed{18}$$

$$a_3 = a + 2d$$

$$= 18 + 2 \times (-5) = \boxed{8}$$

(iii) $5, \_\_, \_\_, 9\frac{1}{2}$

$$a_1 = 5$$

$$a_4 = \frac{19}{2} = a + 3d = 5 + 3d$$

$$3d = \frac{19}{2} - 5 = \frac{9}{2}$$

$$\therefore \quad d = \frac{3}{2}$$

$$a_2 = a + d$$

$$= 5 + \frac{3}{2}$$

$$= \frac{10 + 3}{2} = \boxed{\frac{13}{2}}$$

$$a_3 = a + 2d$$

$$= 5 + 2 \times \frac{3}{2} = \boxed{8}$$

(iv) $-4, \_\_, \_\_, \_\_, \_\_, 6$

$$a = -4$$

$$a_6 = a + 5d = 6$$

$$\Rightarrow \quad -4 + 5d = 6$$

$$d = \frac{10}{5}$$

$$d = 2$$

$$a_2 = a + d = -4 + 2$$

$$= \boxed{-2}$$

$$a_3 = a + 2d = -4 + 2 \times 2$$
$$= \boxed{0}$$
$$a_4 = a + 3d = -4 + 3 \times 2$$
$$= -4 + 6 = \boxed{2}$$
$$a_5 = a + 4d = -4 + 4 \times 2$$
$$= -4 + 8 = \boxed{4}$$

(v) __, 38, __, __, __, −22

$$a_2 = a + d = 38 \quad \ldots\text{(i)}$$
$$a_6 = a + 5d = -22 \quad \ldots\text{(ii)}$$

Subtracting (i) and (ii), we get

$$-4d = 60$$

∴ $d = -15$

Now $a + d = 38$

⇒ $a = 38 + 15$

∴ $a = 53$

$$a_1 = \boxed{53}$$
$$a_3 = 53 + 2 \times (-15) = \boxed{23}$$
$$a_4 = 53 + 3 \times (-15) = \boxed{8}$$
$$a_5 = 53 + 4 \times (-15) = \boxed{-7}$$

10 Try to Solve

1. Which term of the AP 8, 14, 20, 26, … will be 72 more than its 41st term? **[CBSE 2017]**
2. Which term of the AP 5, 15, 25, … will be 130 more than its 31st term? **[CBSE 2006 C]**

Answers

1. 53rd term 2. 44th term

Example 17 For what value of n are the nth terms of the following two APs the same: 13, 19, 25, … and 69, 68, 67, …? Also, find this term. **[CBSE 2006 C, 2011]**

Solution: Let the nth terms of the given progressions be t_n and T_n, respectively.

The first AP is 13, 19, 25, …

Let its first term be a and common difference be d. Then

$$a = 13 \text{ and } d = (19 - 13) = 6.$$

So, its nth term is given by

$$t_n = a + (n - 1)d$$

⇒ $t_n = 13 + (n - 1) \times 6$

⇒ $t_n = 6n + 7 \quad \ldots\text{(i)}$

The second AP is 69, 68, 67, …

Let its first term be A and common difference be D. Then

$$A = 69 \text{ and } D = (68 - 69) = -1$$

So, it n^{th} term is given by $T_n = A + (n-1) \times D$

$\Rightarrow$ $T_n = 69 + (n-1) \times (-1)$

$\Rightarrow$ $T_n = 70 - n$

Now $t_n = T_n \Rightarrow 6n + 7 = 70 - n$ …(ii)

$\Rightarrow$ $7n = 63 \Rightarrow n = 9.$

Hence, the 9th term of each AP is the same.

This term = 70 – 9 = 61 $[\because T_n = (70 - n)]$

11 Try to Solve

1. For what value of n are the nth terms of two APs 63, 65, 67, … and 3, 10, 17, … equal? **[NCERT]**
2. For what value of n are the nth terms of two APs 15, 12, 9, … and –15, –13, –11, … equal? **[CBSE (CCE) 2014]**

Answers

1. $n = 13$ 2. $n = 7$

Example 18 Find the middle term of the AP: 213, 205, 197, …, 37. **[CBSE 2015]**

Solution: Here, $a = 213$; $d = (205 - 213) = -8$.

Let the given AP contain n terms. Then

$$T_n = 37$$

$\Rightarrow$ $a + (n-1)d = 37$

$\Rightarrow$ $213 + (n-1) \times (-8) = 37$

$\Rightarrow$ $213 - 8n + 8 = 37$

$\Rightarrow$ $221 - 8n = 37$

$\Rightarrow$ $8n = 221 - 37 = 184$

$\Rightarrow$ $n = 23.$

Thus, the given AP contains 23 terms.

$\therefore$ Its middle term = $\frac{1}{2}(23+1)^{th}$ term = 12th term.

So Middle term = $T_{12} = (a + 11d)$

$= 213 + 11 \times (-8) = 213 - 88 = 125.$

Hence, the middle term is 125.

12 Try to Solve

1. Find the middle term of the AP: 6, 13, 20, …, 216. [CBSE 2015]
2. Find the middle term of the AP: 10, 7, 4, …, (−62). [CBSE 2009C]

Answers

1. 111 2. −26

Example 19 How many three-digit numbers are divisible by 7? [NCERT, CBSE 2013]

Solution: All three-digit numbers divisible by 7 are

$$105, 112, 119, \ldots, 994.$$

Clearly, these numbers form an AP with

$$a = 105,\ d = (112 - 105) = 7 \text{ and } l = 994.$$

Let it contain n terms. Then

$$T_n = 994$$

$$\Rightarrow \quad a + (n-1) \times d = 994$$

$$\Rightarrow \quad 105 + (n-1) \times 7 = 994$$

$$\Rightarrow \quad 105 + 7n - 7 = 994$$

$$\Rightarrow \quad 7n + 98 = 994$$

$$\Rightarrow \quad 7n = 994 - 98$$

$$\Rightarrow \quad 7n = 896$$

$$\Rightarrow \quad n = \frac{896}{7}$$

$$\Rightarrow \quad n = 128$$

Hence, there are 128 three-digit numbers divisible by 7.

Example 20 How many multiples of 4 lie between 10 and 250? [NCERT, CBSE 2011]

Solution: All multiples of 4 lying between 10 and 250 are

$$12, 16, 20, 24, \ldots, 248.$$

Clearly, these numbers form an AP with

$$a = 12,\ d = (16 - 12) = 4 \text{ and } l = 248.$$

Let it contain n terms. Then

$$T_n = 248$$

$$\Rightarrow \quad a + (n-1)d = 248$$

$$\Rightarrow \quad 12 + (n-1) \times 4 = 248$$

$$\Rightarrow \quad 4n = 240$$

$$\Rightarrow \quad n = 60.$$

Hence, there are 60 multiples of 4 lying between 10 and 250.

13 Try to Solve

1. How many two-digit numbers are divisible by 6? [CBSE 2011]
2. How many two-digit numbers are divisible by 3? [CBSE 2012]
3. How many three-digit numbers are divisible by 9? [CBSE 2013]

Answers

1. 15 2. 30 3. 100

Concept: Sum of '*n*' Terms of an AP

Let's say an AP is as follows:

$$a, a + d, a + 2d, \ldots, a_n(l).$$

$$S_n = \frac{n}{2}[2a + (n - 1)d]$$

Or
$$S_n = \frac{n}{2}[a + a_n]$$

Or
$$S_n = \frac{n}{2}[a + l]$$

where a = first term, n = number of terms, d = common difference and $a_n = l$ = last term of an AP.

Example 21 Find the sum of the first 22 terms of the AP: 8, 3, –2, … [NCERT]

Solution: Here, $a = 8$, $d = 3 - 8 = -5$, $n = 22$.

We know that
$$S_n = \frac{n}{2}[2a + (n - 1)d]$$

Therefore
$$S_{22} = \frac{22}{2}[16 + 21(-5)] = 11(16 - 105)$$

$$= 11(-89) = -979$$

So, the sum of the first 22 terms of the AP is –979.

14 Try to Solve

1. Find the sum of the following APs:
   (a) 2, 7, 12, …, to 10 terms. (b) –37, –33, –29, …, to 12 terms.
   (c) 0.6, 1.7, 2.8, …, to 100 terms. [NCERT] (d) $\frac{1}{15}, \frac{1}{12}, \ldots, \frac{1}{10}$, to 11 terms.

Answers

1. (a) 245 (b) –180 (c) 5505 (d) $\frac{33}{20}$

Example 22 Find the sum of 25 + 28 + 31 + … + 100. **[CBSE 2006 C]**

Solution: Here, $a = 25$, $d = (28 - 25) = 3$ and $l = 100$.

Let the total number of terms be n. Then

$$T_n = 100$$

$$\Rightarrow \quad a + (n-1)d = 100$$

$$\Rightarrow \quad 25 + (n-1) \times 3 = 100$$

$$\Rightarrow \quad 3n + 22 = 100$$

$$\Rightarrow \quad 3n = 78$$

$$\Rightarrow \quad n = 26.$$

$$\text{Required sum} = \frac{n}{2} \cdot (a + l)$$

$$= \frac{26}{2} \cdot (25 + 100) = 13 \times 125 = 1625.$$

Hence, the required sum is 1625.

15 Try to Solve

1. Find the sums given below:

(a) $7 + 10\frac{1}{2} + 14 + \ldots + 84$ (b) $34 + 32 + 30 + \ldots + 10$

(c) $-5 + (-8) + (-11) + \ldots + (-230)$ **[NCERT]**

Answers

1. (a) $1046\frac{1}{2}$ (b) 286 (c) -8930

Example 23 How many terms of the AP 3, 5, 7, 9, … must be added to obtain the sum 120?

Solution: Here, $a = 3$ and $d = (5 - 3) = 2$.

Let the required number of terms be n. Then

$$S_n = 120$$

$$\Rightarrow \quad \frac{n}{2} \cdot \{2a + (n-1)d\} = 120$$

$$\Rightarrow \quad \frac{n}{2} \cdot \{2 \times 3 + (n-1) \times 2\} = 120$$

$$\Rightarrow \quad \frac{n}{2} \cdot (2n + 4) = 120$$

$$\Rightarrow \quad n^2 + 2n - 120 = 0$$

$$\Rightarrow \quad n^2 + 12n - 10n - 120 = 0$$

$$\Rightarrow \quad n(n + 12) - 10(n + 12) = 0$$

$$\Rightarrow \quad (n + 12)(n - 10) = 0$$

$$\Rightarrow \quad n + 12 = 0 \text{ or } n - 10 = 0$$

$$\Rightarrow \quad n = -12 \text{ or } n = 10$$

$$\Rightarrow \quad n = 10 \quad [\because \text{Number of terms cannot be negative}]$$

Hence, the required number of terms is 10.

16 Try to Solve

1. How many terms of the AP 21, 18, 15, … must be added to obtain the sum 0?
2. How many terms of the AP 9, 17, 25, … must to taken so that their sum is 636? **[CBSE 2017]**
3. How many terms of the AP 63, 60, 57, 54, … must be taken so that their sum is 693? Explain the double answer.

Answers

1. 15 2. 12 3. $n = 21$ or $n = 22$, 22nd term is 0.

Example 24 The sum of the first 14 terms of an AP is 1505 and its first term is 10. Find its 25th term. **[CBSE 2012]**

Solution: Here, $a = 10$, and let d be the common difference. Then

$$S_{14} = 1505$$

$$\Rightarrow \quad \frac{n}{2}[2a + (n - 1)d] = 1505, \text{ where } n = 14 \text{ and } a = 10$$

$$\Rightarrow \quad \frac{14}{2} \cdot (20 + 13d) = 1505$$

$$\Rightarrow \quad (20 + 13d) = \frac{1505}{7} = 215$$

$$\Rightarrow \quad 13d = 215 - 20$$

$$\Rightarrow \quad 13d = 195$$

$$\Rightarrow \quad d = 15.$$

Thus, $a = 10$ and $d = 15$

$$\therefore \quad T_{25} = (a + 24d) = (10 + 24 \times 15) = 370.$$

Hence, the 25th term is 370.

Example 25 In an AP of 50 terms, the sum of the first 10 terms is 210 and the sum of its last 15 terms is 2565. Find the AP. **[CBSE 2014]**

Solution: Let a be the first term and d be the common difference of the given AP. Then, the sum of the first n terms is given by

$$S_n = \frac{n}{2} \cdot [2a + (n - 1)d]$$

$$\therefore \quad S_{10} = \frac{10}{2} \cdot (2a + 9d)$$

$\Rightarrow \quad 5(2a + 9d) = 210$

$\Rightarrow \quad 2a + 9d = 42 \quad$ ...(i)

Sum of the last 15 terms = $(S_{50} - S_{35})$.

$\therefore \quad (S_{50} - S_{35}) = 2565$

$\Rightarrow \quad \frac{50}{2}(2a + 49d) - \frac{35}{2}(2a + 34d) = 2565$

$\Rightarrow \quad 25(2a + 49d) - 35(a + 17d) = 2565$

$\Rightarrow \quad 50a + 1225d - 35a - 595d = 2565$

$\Rightarrow \quad (50a - 35a) + (1225d - 595d) = 2565$

$\Rightarrow \quad 15a + 630d = 2565$

$\Rightarrow \quad a + 42d = 171 \quad$ ...(ii)

On solving (i) and (ii), we get $a = 3$ and $d = 4$.

Hence, the required AP is 3, 7, 11, 15, 19, …

17 Try to Solve

1. The 14th term of an AP is twice its 8th term. If its 6th term is –8, find the sum of its first 20 terms. **[CBSE 2015]**
2. The sum of the 4th and 8th terms of an AP is 24 and the sum of its 6th and 10th terms is 44. Find the sum of the first 10 terms of the AP. **[CBSE 2012]**
3. Find the sum of the first 51 terms of an AP whose second and third terms are 14 and 18, respectively? **[NCERT]**

Answers

1. –340 **2.** 95 **3.** 5610

Example 26 Find the sum of all multiples of 7 lying between 500 and 900. **[CBSE 2012]**

Solution: All multiples of 7 lying between 500 and 900 are

504, 511, 518, …, 896

This is an AP in which $a = 504$, $d = 7$ and $l = 896$.

Let the given AP contain n terms. Then

$T_n = 896$

$\Rightarrow \quad a + (n - 1)d = 896$

$\Rightarrow \quad 504 + (n - 1) \times 7 = 896$

$\Rightarrow \quad 497 + 7n = 896$

$\Rightarrow \quad 7n = 399$

$\Rightarrow \quad n = 57.$

$$\therefore \quad \text{Required sum} = \frac{n}{2}(a+l)$$

$$= \frac{57}{2} \cdot (504 + 896) = \left(\frac{57}{2} \times 1400\right) = 39900.$$

Hence, the required sum is 39900.

Example 27 Find the sum of all three-digit natural numbers that are multiples of 11. **[CBSE 2009, 2012]**

Solution: All three-digit natural number that are multiples of 11 are given as 110, 121, 132, ..., 990.

This is an AP in which $a = 110$, $d = 11$ and $l = 990$.

Let the given AP contain n terms. Then

$$T_n = 990$$

$$\Rightarrow \quad a + (n-1)d = 990$$

$$\Rightarrow \quad 110 + (n-1) \times 11 = 990$$

$$\Rightarrow \quad 99 + 11n = 990$$

$$\Rightarrow \quad 11n = 891$$

$$\Rightarrow \quad n = 81$$

$$\therefore \quad \text{Required sum} = \frac{n}{2}(a+l) = \frac{81}{2} \times (110 + 990)$$

$$= \frac{81}{2} \times 1100 = 44550.$$

Hence, the required sum is 44550.

18 Try to Solve

1. Find the sum of the first 40 positive integers divisible by 6. **[NCERT]**
2. Find the sum of the first 15 multiples of 8. **[NCERT]**
3. Find the sum of the odd numbers between 0 and 50. **[NCERT]**

Answers

1. 4920 **2.** 960 **3.** 625

Example 28 Find the sum of first n terms of an AP whose n^{th} term is $(5n - 1)$. Hence, find the sum of the first 20 terms. **[CBSE 2011]**

Solution: The nth term of the AP is given by $T_n = (5n - 1)$.

Let a be the first term and d be the common difference of this AP. Then

$$a = T_1 = (5 \times 1 - 1) = 4, \; T_2 = (5 \times 2 - 1) = 9.$$

$$\therefore \quad d = (T_2 - T_1) = (9 - 4) = 5$$

$$\therefore \quad \text{Sum of the first } n \text{ terms} = \frac{n}{2} \cdot \{2a + (n-1)d\}$$

$$= \frac{n}{2} \times \{2 \times 4 + (n-1) \times 5\} = \frac{1}{2}\, n(5n+3)$$

$\therefore$ Sum of the first 20 terms $= \frac{1}{2} \times 20 \times (5 \times 20 + 3) = 1030$

Hence $S_n = \frac{1}{2} n(5n + 3)$ and $S_{20} = 1030$.

19 Try to Solve

1. Show that $a_1, a_2, \ldots$, an forms an AP, where a_n is defined as given below: **[NCERT]**
   (a) $a_n = 3 + 4n$ (b) $a_n = 9 - 5n$
2. Find the sum of the first 24 terms of a list of numbers whose n^{th} term is given by

$$a_n = 3 + 2n$$

Answers

1. (a) 525 (b) – 465 2. 672

Example 29 Find the number of natural numbers between 101 and 999 that are divisible by both 2 and 5. **[CBSE (CCE) 2014]**

Solution: Numbers between 101 and 999 that are divisible by both 2 and 5 (that is, by 10) are 110, 120, 130, ... 990.

An AP is formed with $a = 110$, $d = 10$ and $a_n = 990$.

Now, we know that $a_n = a + (n - 1)d$

$\Rightarrow$ $990 = 110 + (n - 1)10$

$\Rightarrow$ $880 = (n - 1)10$

$\Rightarrow$ $88 = n - 1$

$\Rightarrow$ $n = 89$

$\therefore$ Natural numbers that are divisible by both 2 and 5 are 89.

20 Try to Solve

1. Find the sum of all the two-digit numbers that are multiples of 2 or 3. **[CBSE (CCE) 2015]**

Answer

1. 3285

FOR MORE PRACTICE AND KNOWLEDGE

Sum of *n*-Terms of an AP

The sum of *n*-terms of an AP with the first term *a* and common difference *d* is given by

$$S_n = \frac{n}{2}[2a + (n - 1)d]$$

Proof: Let the AP be $a, a + d, a + 2d, \ldots a + (n - 2) d, a + (n - 1) d$

Sum of the first n-terms

$$S_n = a + (a + d) + (a + 2d) + \ldots + [a + (n-2)d] + [a + (n-1)d] \quad \ldots\text{(i)}$$

Writing (i) in reverse order, we get

$$S_n = [a + (n-1)d] + [a + (n-2)d] + \ldots + (a + d) + a \quad \ldots\text{(ii)}$$

Adding (i) and (ii), we get

$$\Rightarrow \quad 2S_n = [2a + (n-1)d] + [2a + d + (n-2)d] + [2a + 2d + (n-3)d] + \ldots + [2a + (n-1)d]$$

$$\Rightarrow \quad 2S_n = [2a + (n-1)d] + [2a + (n-1)d] + \ldots + [2a + (n-1)d]$$

$$\Rightarrow \quad 2S_n = n \times [2a + (n-1)d] \qquad [\because \text{There are } n\text{-terms of } (2a + (n-1)d)]$$

$$\therefore \quad S_n = \frac{n}{2}[2a + a(n-1)d]$$

Again $\quad S_n = \frac{n}{2}[a + a + (n-1)d]$

$$\therefore \quad S_n = \frac{n}{2}(a + a_n)$$

where $\quad a_n = \text{Last term} = a + (n-1)d$

An important result: If S_n and S_{n-1} are the sums of the first n-terms and $(n-1)$ terms of an AP, the nth term a_n of the AP can be determined using the formula

$$a_n = S_n - S_{n-1}$$

Example 30 For what value of k will the consecutive terms $2k + 1$, $3k + 3$ and $5k - 1$ form an AP? **[CBSE 2016]**

Solution: Given that $2k + 1$, $3k + 3$ and $5k - 1$ are in AP.

$$(3k + 3) - (2k + 1) = (5k - 1) - (3k + 3)$$

$$\Rightarrow \quad k + 2 = 2k - 4$$

$$\Rightarrow \quad 2k - k = 2 + 4$$

$$\Rightarrow \quad k = 6$$

Example 31 The nth term of an AP cannot be $n^2 + 1$. Justify your answer. **[NCERT Exemplar]**

Solution: Here $\quad a_n = n^2 + 1$

So $\quad a_1 = 1^2 + 1 = 2$

$$a_2 = 2^2 + 1 = 5$$

$$a_3 = 3^2 + 1 = 10$$

List of numbers becomes 2, 5, 10, ….

Here, $5 - 2 \neq 10 - 5$, so it does not form an AP.

Here, $a^n = n^2 + 1$ is not a linear polynomial in n. So, it cannot be the nth term of an AP.

Example 32 The pth, q^{th} and r^{th} terms of an AP are a, b and c, respectively. Show that $a(q - r) + b(r - p) + c(p - q) = 0$. **[Foreign]**

Solution: Let A and d be the first term and common difference of the given AP. Then

$$a_p = A + (p-1)d = a \quad \ldots\text{(i)}$$

$$a_q = A + (q-1)d = b \quad \ldots\text{(ii)}$$

$$a_r = A + (r-1)d = c \quad \ldots\text{(iii)}$$

Multiply both sides of equation (i) by $(q - r)$

$$A(q - r) + (p - 1)(q - r)d = a(q - r) \quad \text{...(iv)}$$

Multiply both sides of equation (ii) by $(r - p)$

$$A(r - p) + (q - 1)(r - p)d = b(r - p) \quad \text{...(v)}$$

Multiply both sides of equation (iii) by $(p - q)$

$$A(p - q) + (r - 1)(p - q)d = c(p - q) \quad \text{...(vi)}$$

Adding (iv), (v) and (vi), we get

$$A[q - r + r - p + p - q] + [pq - pr - q + r + qr - pq - r + p + pr - qr - p + q]\, d$$
$$= a(q - r) + b(r - p) + c(p - q)$$
$$A(0) + (0)d = a(q - r) + b(r - p) + c(p - q)$$
$$0 = a(q - r) + b(r - p) + c(p - q)$$

Example 33 How many three-digit numbers are divisible by 12? **[AI 2012]**

Solution: The three-digit numbers divisible by 12 are 108, 120, 132, ..., 996.

Here $a = 108, d = 12, a_n = 996 = n^{\text{th}}$ term

Now $a_n = a + (n - 1)d$

$\Rightarrow \quad 996 = 108 + (n - 1)12$

$\Rightarrow \quad 996 - 108 = (n - 1)12$

$\Rightarrow \quad 888 = (n - 1)12$

$\Rightarrow \quad 74 = n - 1$

$\Rightarrow \quad n = 75$

$\therefore$ There are 75 three-digit numbers divisible by 12.

Example 34 If the ratio of the sum of the first m and n terms of an AP is $m^2 : n^2$, show that the ratio of its mth and n^{th} terms is $(2m-1) : (2n-1)$. **[Foreign 2016]**

Solution: Let S_m and S_n be the sum of the first m and n terms of the AP. Let the first term and common difference of an AP be a and d, respectively. Then

$$\frac{S_m}{S_n} = \frac{\frac{m}{2}[2a + (m-1)d]}{\frac{n}{2}[2a + (n-1)d]} = \frac{m^2}{n^2}$$

$\Rightarrow \quad \dfrac{2a + (m-1)d}{2a + (n-1)d} = \dfrac{m}{n}$

$\Rightarrow \quad 2an + (mn - n)d = 2am + (mn - m)d$

$\Rightarrow \quad 2a(n - m) = (mn - m - mn + n)d$

$\Rightarrow \quad 2a(n - m) = (n - m)d$

$\Rightarrow \quad d = 2a$

Consider $$\frac{a_m}{a_n}=\frac{a+(m-1)d}{a+(n-1)d}=\frac{a+(m-1)(2a)}{a+(n-1)(2a)}=\frac{a+2am-2a}{a+2an-2a}$$

$$=\frac{2am-a}{2an-a}=\frac{2m-1}{2n-1}$$

Hence, the ratio of the mth and n^{th} terms is $2m - 1 : 2n - 1$.

Example 35 A thief, after committing a theft, runs at a uniform speed of 50 m/minute. After 2 minutes, a policeman runs to catch him. He travels 60 m in the first minute and increases his speed by 5 m/minute every succeeding minute. After how many minutes will the policeman catch the thief? **[Delhi 2016]**

Solution: Suppose the policeman catches the thief after t minutes.

Uniform speed of thief = 50 m/min.

As the thief ran for 2 minutes before the policeman start running

$\therefore$ Distance covered by thief in $(t + 2)$ minutes

$$= 50 \text{ m/min} \times (t + 2) \text{ min} = 50(t + 2)\text{m}$$

An AP is formed in the case of the policeman, that is, 60, 65, 70, ….

$\therefore$ Distance covered by policeman in t minutes

$$= \frac{t}{2}[2 \times 60 + (t - 1) \times 5] = 60t + \frac{5t}{2}(t - 1)$$

Now, when the policeman catches the thief, we have

$$\Rightarrow \quad 60t + \frac{5t^2}{2} - \frac{5t}{2} = 50t + 100$$

$$\Rightarrow \quad t^2 + 3t - 40 = 0$$

$$\Rightarrow \quad (t + 8)(t - 5) = 0$$

$$\Rightarrow \quad t + 8 = 0 \text{ or } t - 5 = 0$$

$$\Rightarrow \quad t = -8 \text{ or } t = 5$$

$$\Rightarrow \quad t = 5, \text{ since } t \text{ cannot be negative.}$$

Thus, the policeman catches the thief after 5 minutes.

Example 36 The sum of the first n terms of an AP is $4n^2 + 2n$. Find the nth term of this AP.

[Foreign 2014]

Solution: Given $S_n = 4n^2 + 2n$

$$S_{n-1} = 4(n - 1)^2 + 2(n - 1) = 4(n^2 - 2n + 1) + 2n - 2$$

$$= 4n^2 - 8n + 4 + 2n - 2 = 4n^2 - 6n + 2$$

$$\therefore \quad a_n = S_n - S_{n-1} = n^{th} \text{ term} = (4n^2 + 2n) - (4n^2 - 6n + 2)$$

$$= 4n^2 + 2n - 4n^2 + 6n - 2 = 8n - 2.$$

Multiple Choice Questions

Choose the correct answer for each of the following questions:

1. If the first term of an AP is –5 and the common difference is 2, then the sum of the first 6 terms is
(a) 0 (b) 5 (c) 6 (d) 15 **[NCERT Exemplar]**

2. In an AP, if $a = 1$, $a_n = 20$ and $S_n = 399$, then n is
(a) 19 (b) 21 (c) 38 (d) 42 **[NCERT Exemplar]**

3. The sum of the first five multiples of 3 is
(a) 45 (b) 55 (c) 65 (d) 75 **[NCERT Exemplar]**

4. In an AP, if $a = 3.5$, $d = 0$, $n = 101$, then a_n will be
(a) 0 (b) 3.5 (c) 103.5 (d) 104.5 **[NCERT Exemplar]**

5. The 21st term of the AP whose first two terms are –3 and 4 is
(a) 17 (b) 137 (c) 143 (d) –143 **[NCERT Exemplar]**

6. The 4th term from the end of the AP –11, –8, –5, …, 49 is
(a) 37 (b) 40 (c) 43 (d) 58 **[NCERT Exemplar]**

7. The famous mathematician associated with finding the sum of the first 100 natural numbers is
(a) Pythagoras (b) Newton (c) Gauss (d) Euclid **[NCERT Exemplar]**

8. The common difference of the AP $\frac{1}{p}, \frac{1-p}{p}, \frac{1-2p}{p}, \ldots$ is
(a) p (b) $-p$ (c) -1 (d) 1 **[CBSE 2013]**

9. The common difference of the AP $\frac{1}{3}, \frac{1-3b}{3}, \frac{1-6b}{3}, \ldots$ is
(a) $\frac{1}{3}$ (b) $\frac{-1}{3}$ (c) b (d) $-b$ **[CBSE 2013]**

10. The next term of the AP $\sqrt{7}, \sqrt{28}, \sqrt{63}, \ldots$ is
(a) $\sqrt{70}$ (b) $\sqrt{84}$ (c) $\sqrt{98}$ (d) $\sqrt{112}$ **[CBSE 2014]**

11. If the nth term of an AP is $(2n + 1)$, then the sum of its first three terms is
(a) $6n + 3$ (b) 15 (c) 12 (d) 21 **[CBSE 2012]**

12. The sum of the first n terms of an AP is $(3n^2 + 6n)$. The common difference of the AP is
(a) 6 (b) 9 (c) 15 (d) –3 **[CBSE 2014]**

13. The 7th term of an AP is –1 and its 16th term is 17. The nth term of the AP is **[CBSE 2014]**
(a) $(3n + 8)$ (b) $(4n - 7)$ (c) $(15 - 2n)$ (d) $(2n - 15)$

14. The 13th term of an AP is 4 times its 3rd term. If its 5th term is 16, then the sum of its first 10 terms is
(a) 150 (b) 175 (c) 160 (d) 135 **[CBSE 2015]**

15. An AP 5, 12, 19, … has 50 terms. Its last term is
(a) 343 (b) 35 (c) 348 (d) 362 **[CBSE 2015]**

16. The sum of the first 20 odd natural numbers is
(a) 100 (b) 210 (c) 400 (d) 420 **[CBSE 2012]**

17. How many three-digit numbers are divisible by 9?
(a) 86 (b) 90 (c) 96 (d) 100 **[CBSE 2013]**

18. What is the 20th term from the end of the AP 3, 8, 13, …, 253?

(a) 163 (b) 158 (c) 153 (d) 148

19. (5 + 13 + 21 + … + 181) = ?

(a) 2476 (b) 2337 (c) 2219 (d) 2139

20. The 17th term of an AP exceeds its 10th term by 21. The common difference of the AP is

(a) 3 (b) 2 (c) –3 (d) –2

21. Which term of the AP 5, 2, – 1, …… is – 49?

(a) 19th (b) 15th (c) 16th (d) 20th

22. If the nth term of an AP is $(2n + 1)$, then the sum of its first five terms is

(a) $6n + 3$ (b) 35 (c) 12 (d) 21

23. What is the sum of all the natural numbers from 1 to 100?

(a) 4780 (b) 5050 (c) 5150 (d) 4550

24. The sum of the first n terms of the series $\sqrt{2} + \sqrt{8} + \sqrt{18} + \ldots$ is:

(a) $\frac{n(n+1)}{2}$ (b) $\sqrt{2n}$ (c) $\frac{n(n+1)}{\sqrt{2}}$ (d) 1

25. If the 4th and 8th terms of an AP are 11 and 23, respectively, the common difference is

(a) 5 (b) 4 (c) 6 (d) 3

26. The first term of an AP is – 2 and the 10th term is 28. What is the value of the common difference?

(a) 14 (b) $\frac{10}{3}$ (c) $\frac{3}{10}$ (d) $\frac{20}{3}$

27. $(8K + 4), (6K - 2), (2K - 7)$ are three consecutive terms of an AP. The value of K is

(a) $\frac{3}{2}$ (b) 2 (c) $\frac{1}{2}$ (d) $-\frac{1}{2}$

28. In an AP, if $a = -7.2$, $d = 3.6$, $a_n = 7.2$, then n is

(a) 1 (b) 4 (c) 4 (d) 5

29. What is the common difference of an AP whose $a_{18} - a_{14} = 32$?

(a) 4 (b) – 4 (c) 8 (d) – 8

30. In an AP, if $a = 1$, $a_n = 41$ and $S_n = 399$, then n is

(a) 19 (b) 21 (c) 38 (d) 42

31. If S_n denotes the sum of the first n terms of an AP and $S_{2n} = 3S_n$, then $S_{3n} : S_n$ is equal to

(a) 10 (b) 8 (c) 6 (d) 4

32. The sum of n terms of the series $\sqrt{3} + \sqrt{12} + \sqrt{27} + \sqrt{48} + \ldots$ is

(a) $\frac{n(n+1)}{\sqrt{3}}$ (b) $\frac{\sqrt{3}}{2}n(n+1)$ (c) $\frac{n(n+1)}{\sqrt{2}}$ (d) 1

33. If the sum of the first n even natural numbers is equal to K times the sum of the first n odd natural numbers, then K =

(a) $\frac{n+1}{n}$ (b) $\frac{n-1}{n}$ (c) $\frac{n+1}{2n}$ (d) $\frac{1}{n}$

34. Which term of the AP 20, 17, 14, ….is the first negative term?

(a) 6th (b) 7th (c) 8th (d) 9th

35. In an AP, if $a = 3.5$, $d = 0$, $n = 101$, then a_n will be
(a) 0 (b) 3.5 (c) 103.5 (d) 104.5 **[NCERT Exemplar]**

36. The 21st term of the AP whose first two terms are – 4 and 4 is
(a) 17 (b) 156 (c) 143 (d) –143

37. The 4th term from the end of the AP –11, –8, –5, …, 49 is
(a) 37 (b) 37 (c) 43 (d) 58

38. The common difference of an AP in which $a_{25} - a_{12} = -52$ is
(a) – 6 (b) –14 (c) – 4 (d) 5

True/False

State whether the following statements are True/False and also justify:

39. The 22th term of AP exceeds its 10th term by 36. The common difference of the AP is 3.

40. The 4th term from the end of the AP –11, –8, –5, ……, 49 is 37.

Assertion–Reason

The following questions consist of two statements – Assertion (A) and Reason (R). Answer these questions by selecting the appropriate options given below.

(A) Both A and R are true and R is the correct explanation for A.
(B) Both A and R are true and R is not the correct explanation for A.
(C) A is true but R is false.
(D) A is false but R is true.

41. **Assertion (*A*):** The 10th term from the end of the AP 6, 13, 20, …, 216 is 150.
Reason (*R*): The nth term from the end of an AP is $l - (n - 1)d$.

42. **Assertion (*A*):** The 11th term from the last term of the AP 27, 23, 19, …, –65 is –25.
Reason (*R*): $a_n = a + (n - 1)d$

43. **Assertion (*A*):** The 8th term from the end of the AP 18, 14, 10, 6,…, –30 is –2.
Reason (*R*): The nth term from the end of an AP with last term l and common difference d is $l - (n - 1)d$.

44. **Assertion (*A*):** The 11th term from the end of the AP 10, 7, 4, …, – 62 is – 32.
Reason (*R*): The nth term from the end $= l - (n - 1)d$.

45. **Assertion (*A*):** The 16th term from the beginning of the AP 5, 8, 11, 14, …, is 51.
Reason (*R*): The nth term from the beginning of an AP with first term a and common difference d is $a + (n - 1)d$

Subjective Questions

46. In which of the following situations does the list of numbers involved make an arithmetic progression, and why?
(i) The taxi fare after each km when the fare is ₹15 for the first km and ₹8 for each additional km.
(ii) The amount of air present in a cylinder when a vacuum pump removes $\frac{1}{4}$ of the air remaining in the cylinder at a time.

(iii) The cost of digging a well after every metre of digging, when it costs ₹150 for the first metre and rises by ₹50 for each subsequent metre.

(iv) The amount of money in the account every year, when ₹10,000 is deposited at compound interest of 8% per annum. **[NCERT]**

47. Which of the following are APs? If they form an AP, find the common difference d and write three more terms.

(i) $\sqrt{2}, \sqrt{8}, \sqrt{18}, \sqrt{32}, \ldots$ (ii) $\sqrt{3}, \sqrt{6}, \sqrt{9}, \sqrt{12}, \ldots$

(iii) $1^2, 3^2, 5^2, 7^2, \ldots$ (iv) $1^2, 5^2, 7^2, 73, \ldots$ **[NCERT]**

48. Which of the following form an AP? Justify your answer.

(i) $-1, -1, -1, -1, \ldots$ (ii) $0, 2, 0, 2, \ldots$ (iii) $1, 1, 2, 2, 3, 3, \ldots$ (iv) $11, 22, 33, \ldots$

(v) $\frac{1}{2}, \frac{1}{3}, \frac{1}{4}, \ldots$ (vi) $2, 2^2, 2^3, 2^4, \ldots$ (vii) $\sqrt{3}, \sqrt{12}, \sqrt{27}, \sqrt{48}, \ldots$ **[NCERT Exemplar]**

49. Two APs have the same common difference. The first term of one AP is 2 and that of the other is 7. The difference between their 10th terms is the same as the difference between their 21st terms, which is the same as the difference between any two corresponding terms. Why? **[NCERT Exemplar]**

50. Is 0 a term of the AP 31, 28, 25, ... ? Justify your answer. **[NCERT Exemplar]**

51. The taxi fare after each km, when the fare is ₹15 for the first km and ₹8 for each additional km, does not form an AP as the total fare (in ₹) after each km is 15, 8, 8, 8, ...
Is the statement true? Give reasons. **[NCERT Exemplar]**

52. In which of the following situations do the lists of numbers involved form an AP? Give reasons for your answers.

(i) The fee charged from a student every month by a school for the whole session, when the monthly fee is ₹400.

(ii) The fee charged every month by a school from Classes I to XII, when the monthly fee for Class I is ₹250 and it increases by ₹50 for the next higher class.

(iii) The amount of money in the account of Varun at the end of every year when ₹1,000 is deposited at simple interest of 10% per annum.

(iv) The number of bacteria in a certain food item after each second, when they double every second. **[NCERT Exemplar]**

53. Justify whether it is true to say that the following are the nth terms of an AP:

(i) $2n - 3$ (ii) $3n^2 + 5$ (iii) $1 + n + n^2$ **[NCERT Exemplar]**

54. If the numbers $n - 2$, $4n - 1$ and $5n + 2$ are in AP, find the value of n. **[NCERT Exemplar]**

55. Find the value of the middle-most term(s) of the AP:

$$-11, -7, -3, \ldots, 49$$ **[NCERT Exemplar]**

56. The sum of the first three terms of an AP is 33. If the product of the first and the third terms exceeds the second term by 29, find the AP. **[NCERT Exemplar]**

57. Verify that each of the following is an AP, and then write its next three terms.

(i) $0, \frac{1}{4}, \frac{1}{2}, \frac{3}{4}, \ldots$ (ii) $5, \frac{14}{3}, \frac{13}{3}, 4, \ldots$

(iii) $\sqrt{3}, 2\sqrt{3}, 3\sqrt{3}, \ldots$ (iv) $a + b, (a + 1) + b, (a + 1) + (b + 1),$

(v) $a, 2a + 1, 3a + 2, 4a + 3, \ldots$ **[NCERT Exemplar]**

58. Write the first three terms of the APs when a and d are as given below:

(i) $a = \frac{1}{2}, d = -\frac{1}{6}$ (ii) $a = -5, d = -3$ (iii) $a = \sqrt{2}, d = \frac{1}{\sqrt{2}}$ **[NCERT Exemplar]**

59. For what value of k will $k + 9$, $2k - 1$ and $2k + 7$ be the consecutive terms of an AP? **[AI 2016]**

60. Find the 11th term from the last term (towards the first term) of the AP 10, 7,4, … – 62. **[NCERT]**

61. A sum of ₹1000 is invested at 8% simple interest per year. Calculate the interest at the end of each year. Do these interests form an AP? If so, find the interest at the end of 30 years, making use of this fact. **[NCERT]**

62. In a flower bed, there are 23 rose plants in the first row, 21 in the second, 19 in the third, and so on. There are 5 rose plants in the last row. How many rows are there in the flower bed? **[NCERT]**

63. How many multiples of 4 lie between 10 and 250? **[NCERT]**

64. For what value of n are the nth terms of two APs 63, 65, 67, … and 3, 10, 17, … equal? **[NCERT]**

65. Ramkali saved ₹5 in the first week of a year and then increased her weekly savings by ₹1.75. If in the nth week, her weekly savings become ₹20.75, find n. **[NCERT]**

66. Which term of the AP 121, 117, 113, …, is its first negative term? [**Hint:** Find n for $a_n < 0$] **[NCERT]**

67. The 26th, 11th and last terms of an AP are 0, 3 and $-\frac{1}{5}$, respectively. Find the common difference and the number of terms. **[NCERT Exemplar]**

68. Find the 20th term of the AP whose 7th term is 24 less than the 11th term, the first term being 12. **[NCERT Exemplar]**

69. Find whether 55 is a term of the AP 7, 10, 13…, or not. If yes, find out which term it is. **[NCERT Exemplar]**

70. Determine k so that $k^2 + 4k + 8$, $2k^2 + 3k + 6$, $3k^2 + 4k + 4$ are three consecutive terms of an AP. **[NCERT Exemplar]**

71. The angles of a triangle are in AP. The greatest angle is twice the least. Find all the angles of the triangle. **[NCERT Exemplar]**

72. If the nth terms of the two APs 9, 7, 5, … and 24, 21, 18, … are the same, find the value of n. Also find that term. **[NCERT Exemplar]**

73. Find the 12th term from the end of the AP– 2, – 4, – 6, …, –100. **[NCERT Exemplar]**

74. Which term of the AP 53, 48, 43,… is the first negative term? **[NCERT Exemplar]**

75. Find the number of natural numbers between 100 and 999 that are divisible by both 2 and 5. **[AI 2012]**

76. Find the common difference of an AP whose first term is 5 and the sum of its first four terms is half the sum of the next four terms. **[Foreign 2012]**

77. A sum of ₹1,600 is to be used to give 10 cash prizes to students of a school for their overall academic performance. If each prize is ₹20 less than its preceding prize, find the value of each prize. **[Foreign 2012]**

78. A manufacturer of TV sets produced 600 sets in the third year and 700 sets in the seventh year. Assuming that production increases uniformly by a fixed number every year, find:

(i) The production in the 1st year (ii) The production in the 10th year

(iii) The total production in the first 7 years **[NCERT]**

79. If the sum of the first n terms of an AP is $4n - n^2$, what is the first term (that is, S_1)? What is the sum of the first two terms? What is the second term? Similarly, find the 3rd, 10th and n^{th} terms. **[NCERT]**

80. A contract on a construction job specifies a penalty for delay of completion beyond a certain date as follows: ₹200 for the first day, ₹250 for the second day, ₹300 for the third day, etc., the penalty for each succeeding day being ₹50 more than for the preceding day. How much money will the contractor have to pay as penalty, if he has delayed the work by 30 days? **[NCERT]**

81. A sum of ₹700 is to be used to give seven cash prizes to students of a school for their overall academic performance. If each prize is ₹20 less than its preceding prize, find the value of each of the prizes. **[NCERT]**

82. A spiral is made up of successive semicircles, with centres alternately at A and B, starting with centre at A, of radii 0.5 cm, 1.0 cm, 1.5 cm, 2.0 cm, … as shown in the figure. What is the total length of such a spiral made up of 13 consecutive semicircles? [Take $\pi = \frac{22}{7}$]

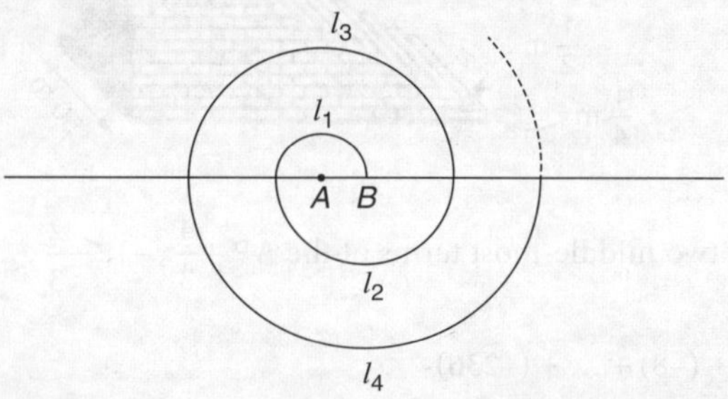

[NCERT]

83. Two hundred logs are stacked in the following manner: 20 logs in the bottom row, 19 in the next row, 18 in the row next to it and so on (see figure). In how many rows are the 200 logs placed and how many logs are in the top row?

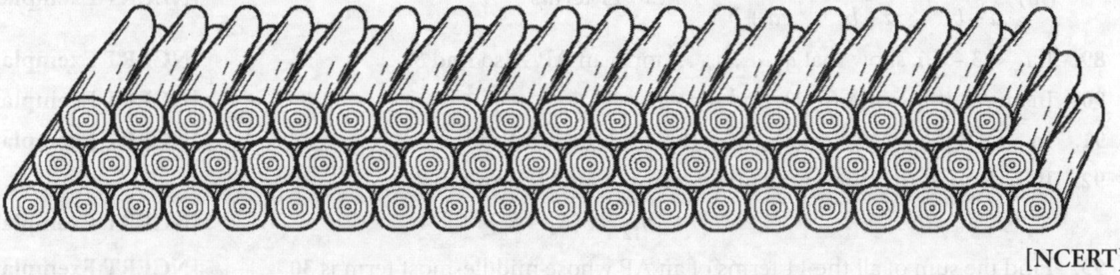

[NCERT]

84. In a potato race, a bucket is placed at the starting point, which is 5 m from the first potato, and the other potatoes are placed 3 m apart in a straight line. There are 10 potatoes in the line (see figure).

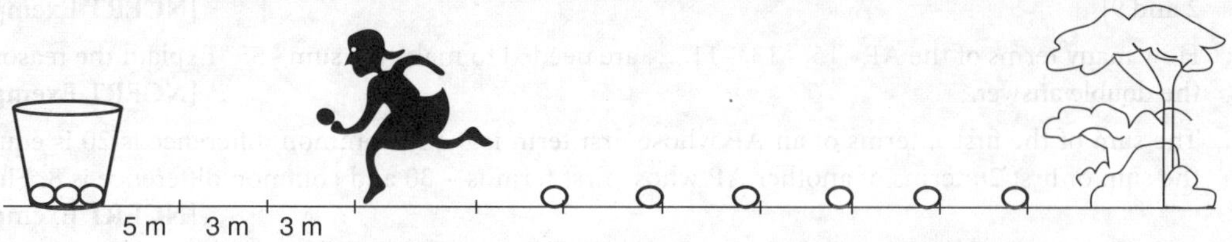

A competitor starts from the bucket, picks up the nearest potato, runs back with it, drops it in the bucket, runs back to pick up the next potato, runs to the bucket to drop it in, and continues in the same way until all the potatoes are in the bucket. What is the total distance the competitor has to run? **[NCERT]**

85. The sum of the 3rd and 7th terms of an AP is 6 and their product is 8. Find the sum of the first 16 terms of the AP. **[NCERT]**

86. A small terrace at a football ground comprises 15 steps, each of which is 50 m long and built of solid concrete. Each step has a rise of $\frac{1}{4}$ m and a tread of m (see figure). Calculate the total volume of concrete required to build the terrace.

[**Hint:** Volume of concrete required to build the first step = $\frac{1}{4} \times \frac{1}{2} \times 50$ m^3]

[NCERT]

87. Find the sum of the two middle-most terms of the AP $-\frac{4}{3}, -1, -\frac{2}{3}, \ldots, 4\frac{1}{3}$. **[NCERT Exemplar]**

88. Find the sum:

(i) $1 + (-2) + (-5) + (-8) + \ldots + (-236)$

(ii) $\left(4-\frac{1}{n}\right)+\left(4-\frac{2}{n}\right)+\left(4-\frac{3}{n}\right)+\ldots$, up to n terms

(iii) $\frac{a-b}{a+b} + \frac{3a-2b}{a+b} + \frac{5a-3b}{a+b} + \ldots$ to 11 terms **[NECRT Exemplar]**

89. If $a_n = 3 - 4n$, show that $a_1, a_2, a_3, \ldots$ form an AP. Also find S_{20}. **[NCERT Exemplar]**

90. In an AP, if $S_n = n(4n + 1)$, find the AP. **[NCERT Exemplar]**

91. In an AP, if $S_n = 3n + 5n$ and $a_k = 164$, find the value of k. **[NCERT Exemplar]**

92. If S_n denotes the sum of the first n terms of an AP, prove that

$$S_{12} = 3(S_8 - S_4)$$

[NCERT Exemplar]

93. Find the sum of all the 11 terms of an AP whose middle-most term is 30. **[NCERT Exemplar]**

94. Find the sum of last 10 terms of the AP 8, 10, 12, … , 126. **[NCERT Exemplar]**

95. Find the sum of first seven numbers which are multiples of 2 as well as of 9. [**Hint:** Take the LCM of 2 and 9] **[NCERT Exemplar]**

96. How many terms of the AP –15, –13, –11, … are needed to make the sum –55? Explain the reason for the double answer. **[NCERT Exemplar]**

97. The sum of the first n terms of an AP whose first term is 8 and common difference is 20 is equal to the sum of first $2n$ terms of another AP whose first term is – 30 and common difference is 8. Find n. **[NCERT Exemplar]**

98. Yasmeen saves ₹32 during the first month, ₹36 in the second month and ₹40 in the third month. If she continues to save in this manner, in how many months will she save ₹2,000? **[NCERT Exemplar]**

99. The sum of four consecutive numbers in an AP is 32 and the ratio of the product of the first and last terms to the product of the two middle terms is 7:15. Find the numbers.

[CBSE 2018, NCERT Exemplar]

100. Solve the equation: $1 + 4 + 7 + 10 + \ldots + x = 287$ **[NCERT Exemplar]**

101. Find the

(i) Sum of those integers between 1 and 500 which are multiples of 2 as well as of 5.

(ii) Sum of those integers from 1 to 500 which are multiples of 2 as well as of 5 .

(iii) Sum of those integers from 1 to 500 which are multiples of 2 or 5.

[**Hint** (iii): These numbers will be multiples of 2 + multiples of 5 – multiples of 2 as well as of 5]

[NCERT Exemplar]

102. An AP consists of 37 terms. The sum of the three middle-most terms is 225 and the sum of the last three is 429. Find the AP. **[NCERT Exemplar]**

103. The ratio of the 11th term to the 18th term of an AP is 2:3. Find the ratio of the 5th term to the 21st term, and also the ratio of the sum of the first five terms to the sum of the first 21 terms.

[NCERT Exemplar]

104. Show that the sum of an AP whose first term a, second term b and last term c is equal to

$$\frac{(a+c)(b+c-2a)}{2(b-a)}.$$

[NCERT Exemplar]

105. Solve the equation $-4 + (-1) + 2 + \ldots + x = 437$ **[NCERT Exemplar]**

106. Jaspal Singh repays his total loan of ₹1,18,000 by paying every month, starting with the first installment of ₹1,000. If he increases the installment by ₹100 every month, what amount will be paid by him in the 30th installment? What amount of loan does he still have to pay after the 30th installment?

[NCERT Exemplar]

107. The students of a school decided to beautify the school on the Annual Day by fixing colourful flags on the straight passage of the school. They have 27 flags to be fixed at intervals of 2 m. The flags are stored at the position of the middle-most flag. Ruchi was given the responsibility of placing the flags. Ruchi kept her books where the flags were stored. She could carry only one flag at a time. How much distance did she cover in completing this job and returning to collect her books? What is the maximum distance she travelled carrying a flag? **[NCERT Exemplar]**

108. If the ratio of the sum of first n terms of two APs is $(7n + 1) : (4n + 27)$, find the ratio of their m^{th} terms.

[AI 2016]

109. The sums of first n terms of three arithmetic progressions are S_1, S_2 and S_3, respectively. The first term of each AP is 1 and their common differences are 1, 2 and 3, respectively. Prove that $S_1 + S_3 = 2S_2$.

[AI 2016]

110. Divide 56 in four parts in an AP such that the ratio of the product of their extremes (1st and 4th) to the product of means (2nd and 3rd) is 5:6. **[Foreign 2016]**

111. The sums of the first n terms of three APs are S_1, S_2 and S_3. The first term of each is 5 and their common differences are 2, 4 and 6, respectively. Prove that $S_1 + S_3 = 2S_2$. **[Foreign 2016]**

112. A thief runs at a uniform speed of 100 m/minute. After one minute, a policeman runs after the thief to catch him. He moves at a speed of 100 m/minute in the first minute and increases his speed by 10 m/minute every succeeding minute. After how many minutes will the policeman catch the thief?

[Delhi 2016]

113. The sum of three numbers in an AP is 12 and the sum of their cubes is 288. Find the numbers.

[Delhi 2016]

114. Reshma wanted to save at least ₹6,500 to send her daughter to school the next year (after 12 months). She saved ₹450 in the first month and raised her savings by ₹20 every month. How much will she be able to save in the next 12 months? Will she be able to send her daughter to school the next year? What value is reflected in this question? **[Foreign 2016]**

115. In an AP, if $S_5 + S_7 = 167$ and $S_{10} = 235$, find the AP, where S_n denotes the sum of its first n terms. **[AI 2015]**

116. If the sum of the first n-terms of an AP is $(3n^2 + 7n)$, then find its nth term. Hence write its 20th term. **[Delhi 2015]**

117. If S_n denotes the sum of the first n-terms of an AP, prove that $S_{30} = 3\,[S_{20} - S_{10}]$. **[Delhi 2015]**

118. The 16th term of an AP is five times its third term. If its 10th term is 41, then find the sum of its first 15 terms. **[AI 2015]**

119. Ramkali required ₹2,500 after 12 weeks to send her daughter to school. She saved ₹100 in the first week and increased her weekly saving by ₹20 every week. Find out whether she will be able to send her daughter to school after 12 weeks. What value is generated in the above situation? **[Delhi 2015]**

120. Find the middle term of the sequence formed by all three-digit-numbers that leave a remainder 3 when divided by 4. Also find the sum of all numbers on both sides of the middle term, separately. **[Foreign 2015]**

121. Find the middle term of the sequence formed by all numbers between 9 and 95, which leave a remainder 1 when divided by 3. Also find the sum of the numbers on both sides of the middle term, separately. **[Foreign 2015]**

122. Find the middle term of the sequence formed by all three-digit numbers which leave a remainder 5 when divided by 7. Also find the sum of all numbers on both sides of the middle term, separately. **[Foreign 2015]**

123. The sum of the first seven terms of an AP is 182. If its 4th and the 17th terms are in the ratio 1:5, find the AP. **[AI 2014]**

124. Find the number of terms of the AP $-12, -9, -6, \ldots 21$. If 1 is added to each term of this AP, find the sum of all terms of the AP thus obtained. **[CBSE 2013]**

125. The sum of the first q terms of an AP is $63q - 3q^2$. If its pth term is -60, find the value of p. Also find the 11th term of this AP. **[AI 2011]**

126. Find the sum of all three-digit natural numbers, which are multiples of 9. **[Delhi 2012]**

127. In an AP, the first term is 12 and the common difference is 6. If the last term of the AP is 252, find its middle term. **[NCERT]**

128. In an AP, the first term is 5 and the common difference is 2. If the last term of the AP is 53, find its middle term. **[Foreign 2012]**

129. Find the sum of all multiples of 8 lying between 201 and 950. **[AI 2012]**

130. The sum of the first five terms of an AP is 25 and the sum of its next five terms is -75. Find the 10th term of the AP. **[Foreign 2012]**

131. The sum of the 3rd and 7th terms of an AP is 40 and the sum of its 6th and 14th terms is 70. Find the sum of first 10 terms of the AP. **[Foreign 2012]**

132. Find the next term of the sequence $\dfrac{1}{1+\sqrt{x}}, \dfrac{1}{1-x}, \dfrac{1}{1-\sqrt{x}}, \ldots$

133. If the ratio of the sum of the first n terms of two APs is $(7n + 1) : (4n + 27)$, find the ratio of their 9th terms. **[AI 2017]**

134. Find the sum of the first 24 terms of an AP whose n^{th} term is given by $a_n = 3 + 2n$. **[AI 2017 (C)]**

135. If the ratio of the 11th term of an AP to its 18th term is 2:3, find the ratio of the sum of the first five terms to the sum of its first 10 terms. **[Delhi 2017]**

136. If S_n the sum of the first n terms of an AP is given by $S_n = 3n^2 - 4n$, find the nth term. **[Delhi 2019]**

137. Which terms of the AP –7, –12, –17, –22, … will be –82. Is –100 any term of the AP? Given reasons for your answer. **[CBSE, Allahabad 2019]**

138. In an AP, if the common difference $(d) = -4$ and the 7th term (a_7) is 4, find the first term. **[CBSE, 2018]**

139. What is the common difference of an AP in which $a_{21} - a_7 = 84$? **[AI 2017]**

140. Which term of the progression $20, 19\frac{1}{4}, 18\frac{1}{2}, 17\frac{3}{4}, \ldots$ is the first negative term? **[AI 2017]**

Answers

Multiple Choice Questions

1. (a)	**2.** (c)	**3.** (a)	**4.** (b)	**5.** (b)	**6.** (b)	**7.** (c)	**8.** (c)	**9.** (d)	**10.** (d)
11. (b)	**12.** (a)	**13.** (d)	**14.** (b)	**15.** (c)	**16.** (c)	**17.** (d)	**18.** (b)	**19.** (d)	**20.** (a)
21. (a)	**22.** (b)	**23.** (b)	**24.** (b)	**25.** (d)	**26.** (b)	**27.** (c)	**28.** (d)	**29.** (c)	**30.** (c)
31. (c)	**32.** (b)	**33.** (a)	**34.** (c)	**35.** (b)	**36.** (b)	**37.** (b)	**38.** (c)		

True/False

39. True **40.** False

Assertion–Reason

41. (D) **42.** (B) **43.** (A) **44.** (A) **45.** (D)

Subjective Questions

46. (i) Yes, 15, 23, 31, … (ii) No, $v, \frac{3v}{4}, \left(\frac{3}{4}\right)^2 v, \ldots$ (iii) Yes, 150, 200, 250, …

(iv) No, amounts are $10000\left(1+\frac{8}{100}\right), 10000\left(1+\frac{8}{100}\right)^2, 10000\left(1+\frac{8}{100}\right)^3, \ldots$

47. (i) Yes, $d = \sqrt{2}$; $\sqrt{50}, \sqrt{72}, \sqrt{98}$ (ii) No (iii) No

(iv) Yes, $d = 24$; 97, 121, 145

48. (i) Yes, (ii) No, (iii) No (iv) Yes (vi) No (vi) No, (vii) Yes

49. The difference between any two corresponding terms of such APs is the same as the difference between their first terms.

50. No. **51.** False, as the total fare (in ₹) after each km is 15, 23, 31, 39,… **52.** (i), (ii) and (iii)

53. (i) Yes (ii) No (iii) No **54.** $n = 1$ **55.** 17 and 21 **56.** 2, 11, 20, … and 20, 11, 2, …

57. (i) $1, \frac{5}{4}, \frac{3}{2}$ (ii) $\frac{11}{3}, \frac{10}{3}, 3$ (iii) $4\sqrt{3}, 5\sqrt{3}, 6\sqrt{3}$

(iv) $(a+2)+(b+1), (a+2)+(b+2), (a+3)+(b+2)$ (v) $5a+4, 6a+5, 7a+6$

58. (i) $\frac{1}{2}, \frac{1}{3}, \frac{1}{6}$ (ii) –5, –8, –11 (iii) $\sqrt{2}, \frac{3}{\sqrt{2}}, \frac{4}{\sqrt{2}}$

59. 18 **60.** –32 **61.** ₹2,400 **62.** 10 rows **63.** 60 **64.** 13 **65.** 10

66. 32nd term **67.** $d = -\frac{1}{5}$, $n = 27$ **68.** 126 **69.** Yes, 17th term **70.** $k = 0$

71. 40°, 60°, 80° **72.** 16th term; –21 **73.** –78 **74.** 12th term **75.** 89 **76.** $d = 2$

77. ₹250, 230, 210, 190, 170, 150, 130, 110, 90, 70.

78. (i) 550 (ii) 775 (iii) 4375

79. $S_1 = 3, S_2 = 4; a_2 = S_2 - S_1 = 1; S_3 = 3, a_3 = S_3 - S_2 = -1, a_{10} = S_{10} - S_9 = -15; a_n = S_n - S_{n-1} = 5 - 2n.$

80. ₹ 27,750 **81.** Values of the prizes (in ₹) are 160, 140, 120, 100, 80, 60, 40. **82.** 143 cm

83. 16 rows, 5 logs placed in the top row. By inserting $S = 200$, $a = 20$, $d = -1$ in the formula $S = \frac{n}{2}[2a + (n-1)d]$, we get $41n - n^2 = 400$. On solving, $n = 16, 25$. Therefore, the number of rows is either 16 or 25. $a_{25} = a + 24d = -4$

That is, the number of logs in the 25th row is – 4, which is not possible. Therefore $n = 25$ is not possible. For $n = 16$, $a_{16} = 5$. Therefore, there are 16 rows and 5 logs are placed in the top row.

84. 370 m **85.** $S_{16} = 20, 76.$ **86.** 750 m^3 **87.** 3

88. (i) –9400 (ii) $\frac{7n-1}{2}$ (iii) $\frac{11(11a-6b)}{a+b}$

89. –780 **90.** 5, 13, 21…, **91.** $k = 27$ **93.** 330 **94.** 1170 **95.** 504

96. $n = 5, 11$ **97.** 11 **98.** 25 months **99.** $a = 8$, $d = 2$, numbers are 2, 6, 10, 14.

100. $n = 14, x = 40$ **101.** (i) 12250 (ii) 12750 (iii) 75250 **102.** 3, 7, 11, 15, … **103.** 1 : 3; 5 : 49

105. 50 **106.** ₹3,900; ₹44,500 **107.** 728 m; 26 m **108.** . $14m - 6 : 8m + 23$

110. 8, 12, 16, 20 **112.** 5 minutes **113.** 2, 4, 6

114. ₹6,720. Yes, she will be able to send her daughter to school because she saved more than the required sum. Value: Shows dedication towards her daughter's education.

115. 1, 6, 11, … **116.** 62 **117.** No **118.** 495

119. ₹2,520. Yes, she will be able to send her daughter to school because she saved more than the required sum. Value: Shows dedication towards her daughter's education.

120. 551, 87024 **121.** 52, 1043 **122.** 551, 49824. **123.** 2, 10, 18, … **124.** $n = 12, S_n = 66$

125. $p = 21, a_{11} = 0$ **126.** 55350 **127.** 132 **128.** 29 **129.** 53568

130. –23 **131.** 215 **132.** $\frac{1+2\sqrt{x}}{1-x}$ **133.** 24 : 19 **134.** 672

135. $\frac{6}{17}$ **136.** $6n - 7$

137. $n = 19\frac{3}{5}$, which is not an integral number. Hence, –100 is not a term of the AP; 16th term is – 82.

138. 28 **139.** 6 **140.** 28th term

Arithmetic Progression (Basic)

SELF-ASSESSMENT TEST

Time allowed: 90 min. Max. Marks: 40

General Instructions

1. *All questions are compulsory. There are 20 questions in all.*
2. *This question paper has four sections: Section A, Section B, Section C and Section D.*
3. *Section A contains 10 questions of one mark each, Section B contains three questions of two marks each, Section C contains four questions of three marks each, and Section D contains three questions of four marks each.*

Section A [1 Mark Each]

Direction: (Q. Nos. 1–5) These are Multiple Choice Questions. Select the most appropriate answer for the given questions.

1. If the first term of an AP is –5 and the common difference is 2, then the sum of the first 6 terms is
   (a) 0 (b) 5 (c) 6 (d) 15
2. The sum of the first 20 odd natural numbers is
   (a) 100 (b) 210 (c) 400 (d) 420
3. $(5 + 13 + 21 + ... + 181) = ?$
   (a) 2476 (b) 2337 (c) 2219 (d) 2139
4. The 17th term of an AP exceeds its 10th term by 21. The common difference of the AP is
   (a) 3 (b) 2 (c) –3 (d) –2
5. Which term of the AP 5, 2, – 1, …… is – 49?
   (a) 19th (b) 15th (c) 16th (d) 20th
6. Find the 4th and 9th terms of the sequence whose n^{th} term is given by $an = 7n - 9$.
7. For the AP: $\frac{3}{2}, \frac{1}{2}, -\frac{1}{2}, -\frac{3}{2}, ...,$ write the first term a and the common difference d.
8. How many terms are there in the AP 6, 10, 14, 18,..., 178?
9. Is –150 a term of the AP 11, 8, 5, 2,...?
10. Find the sum of the first 15 multiples of 8.

Section B [2 Marks Each]

11. Find the sum 25 + 28 + 31 + ... + 100.
12. Find the sum of all multiples of 7 lying between 500 and 900.
13. Find the middle term of the AP 6, 13, 20,..., 216.

Section C [3 Marks Each]

14. The sum of the first 14 terms of an AP is 1505 and its first term is 10. Find its 25th term.

15. Which term of the AP 134, 129, 124, ..., is its first negative term?

16. Determine an AP whose 3rd term is 16; when the 5th term is subtracted from the 7th term, we get 12.

17. Which term of the AP 3, 15, 27, 39, ... will be 132 more than its 54th term?

Section D [4 Marks Each]

18. In the following APs, find the missing terms in the boxes:

(i) 2, ☐, 26 (ii) ☐, 13, ☐, 3

(iii) 5, ☐, ☐, $9\frac{1}{2}$ (iv) –4, ☐, ☐, ☐, ☐, 6

(v) ☐, 38, ☐, ☐, ☐, –22

19. Find the sum of the first 24 terms of an AP whose n^{th} term is given by $a_n = 3 + 2n$.

20. Find the sum of all three-digit natural numbers that are multiples of 9.

Answers

1. (a) **2.** (c) **3.** (d) **4.** (a) **5.** (a) **6.** 19, 54 **7.** $a = \frac{3}{2}, d = -1$ **8.** 44

9. No **10.** 960 **11.** 1625 **12.** 39900 **13.** 111 **14.** 370 **15.** 28th term

16. 4, 10, 16, 22,... **17.** 65th term **18.** (i) 14 (ii) 18, 8 (iii) $\frac{13}{2}$, 8 (iv) –2, 0, 2, 4 (v) 53, 23, 8, –7

19. 672 **20.** 55350

Arithmetic Progression (Standard)

SELF-ASSESSMENT TEST

Time allowed: 90 min. Max. Marks: 40

General Instructions

1. *All questions are compulsory. There are 20 questions in all.*
2. *This question paper has four sections: Section A, Section B, Section C and Section D.*
3. *Section A contains 10 questions of one mark each, Section B contains three questions of two marks each, Section C contains four questions of three marks each, and Section D contains three questions of four marks each.*

Section A [1 Mark Each]

Direction: (Q. Nos. 1–5) These are Multiple Choice Questions. Select the most appropriate answer for the given questions.

1. The common difference of the AP $\frac{1}{p}, \frac{1-p}{p}, \frac{1-2p}{p}, \ldots$ is
   (a) p (b) $-p$ (c) -1 (d) 1
2. The famous mathematician associated with finding the sum of the first 100 natural numbers is
   (a) Pythagoras (b) Newton (c) Gauss (d) Euclid
3. The 7th term of an AP is -1 and its 16th term is 17. The nth term of the AP is
   (a) $(3n + 8)$ (b) $(4n - 7)$ (c) $(15 - 2n)$ (d) $(2n - 15)$
4. If S_n denotes the sum of the first n terms of an AP and $S_{2n} = 3S_n$, then $S_{3n}:S_n$ is equal to
   (a) 10 (b) 8 (c) 6 (d) 4
5. The sum of n terms of the series $\sqrt{3} + \sqrt{12} + \sqrt{27} + \sqrt{48} + \ldots$ is
   (a) $\frac{n(n+1)}{\sqrt{3}}$ (b) $\frac{\sqrt{3}}{2} n(n + 1)$ (c) $\frac{n(n+1)}{\sqrt{2}}$ (d) 1
6. Is 0 a term of the AP 31 ,28, 25, ... ? Justify your answer.
7. If the numbers $n - 2$, $4n - 1$ and $5n + 2$ are in AP, find the value of n.
8. How many multiples of 4 lie between 10 and 250?
9. Find the 20th term of the AP whose 7th term is 24 less than the 11th term, the first term being 12.
10. Find the 12th term from the end of the AP: $-2, -4, -6, \ldots, -100$.

Section B [2 Marks Each]

11. Find the middle term of the AP 213, 205, 197, ..., 37.
12. Find the common difference of an AP whose first term is 5 and the sum of it first four terms is half the sum of the next four terms.
13. Find the sum: $\frac{a-b}{a+b} + \frac{3a-2b}{a+b} + \frac{5a-3b}{a+b} + \ldots$ to 11 terms.

Section C [3 Marks Each]

14. The ratio of the 11th term to the 18th term of an AP is 2:3. Find the ratio of the 5th term to the 21st term, and also the ratio of the sum of the first five terms to the sum of the first 21 terms.

15. Show that the sum of an AP whose first term is a, second term is b and last term is c is equal to

$$\frac{(a+c)(b+c-2a)}{2(b-a)}$$

16. The sum of the first q terms of an AP is $63q - 3q^2$. If its pth term is -60, find the value of p. Also find the 11th term of this AP.

17. The sum of three numbers in AP is 12 and the sum of their cubes is 288. Find the numbers.

Section D [4 Marks Each]

18. A thief, after committing a theft, runs at a uniform speed of 50 m/minute. After 2 minutes, a policeman runs to catch him. He travels 60 m in the first minute and increases his speed by 5 m/minute every succeeding minute. After how many minutes will the policeman catch the thief?

19. If the ratio of the sum of the first m and n terms of an AP is $m^2: n^2$, show that the ratio of its mth and n^{th} terms is $(2m-1): (2n-1)$.

20. The pth, q^{th} and r^{th} terms of an AP are a, b and c, respectively. Show that $q(q-r) + b(r-p) + c(p-q) = 0$.

Answers

1. (c) **2.** (c) **3.** (d) **4.** (c) **5.** (b) **6.** No **7.** $n = 1$ **8.** 60 **9.** 126 **10.** -78

11. $n = 23$, middle term = 125 **12.** $d = 2$ **13.** $\frac{11(11a-6b)}{a+b}$ **14.** 1:3; 5:49 **16.** $p = 21, a_{11} = 0$

17. 2, 4, 6 **18.** 5 minutes

CHAPTER

6 Triangles

Key Concepts

- Definition of triangles
- Congruence of triangles
- Similarity of triangles
- Basic Proportionality Theorem (BPT) or Thales' Theorem
- Converse of Basic Proportionality Theorem
- Area Theorem
- Pythagoras' Theorem
- Converse of Pythagoras' Theorem

Concept: Triangle

A three-sided closed figure, consisting of three sides and three angles.

1. **Congruency of triangle (≡):** Two triangles are said to be congruent iff (if and only if) their corresponding sides are equal and their corresponding angles are equal.

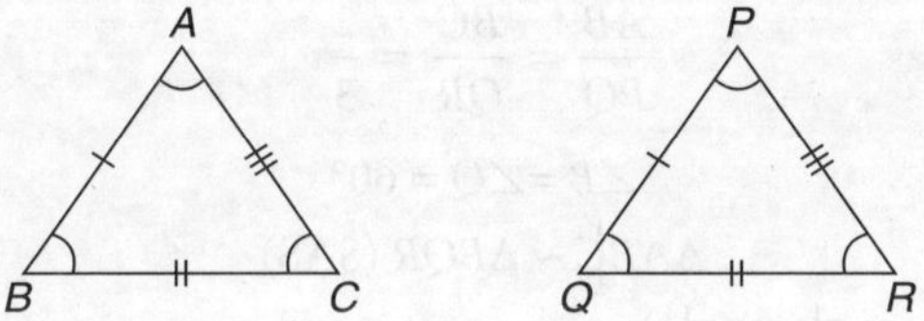

Corresponding sides of $AB = PQ$
Corresponding sides of $BC = QR$
Corresponding sides of $CA = RP$
Corresponding angle of $\angle A = \angle P$
Corresponding angle of $\angle B = \angle Q$
Corresponding angle of $\angle C = \angle R$

Rules of congruency of triangles

(a) SAS (side-angle-side) (b) SSS (side-side-side) (c) ASA(angle-side-angle)
(d) AAS (angle-angle-side) (e) RHS (right-angle-hypotenuse-side)

2. **Similarity of two figures**

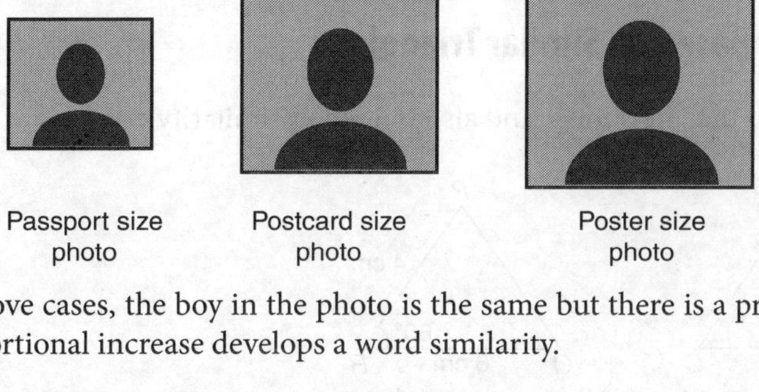

Passport size photo Postcard size photo Poster size photo

In all of the above cases, the boy in the photo is the same but there is a proportional increase in size. This proportional increase develops a word similarity.

3. **Similar triangles (~):** Two Δs are said to be similar iff (if and only if) their corresponding sides are in proportion and their corresponding angles are equal.
   (a) SSS (side-side-side)

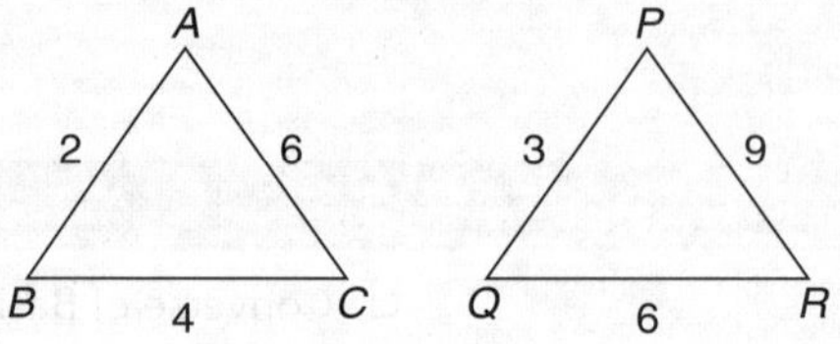

Here $$\frac{AB}{PQ} = \frac{BC}{QR} = \frac{AC}{PR} = \frac{2}{3}$$

Hence $$\Delta ABC \sim \Delta PQR \text{ (SSS)}$$

   (b) SAS (side-angle-side)

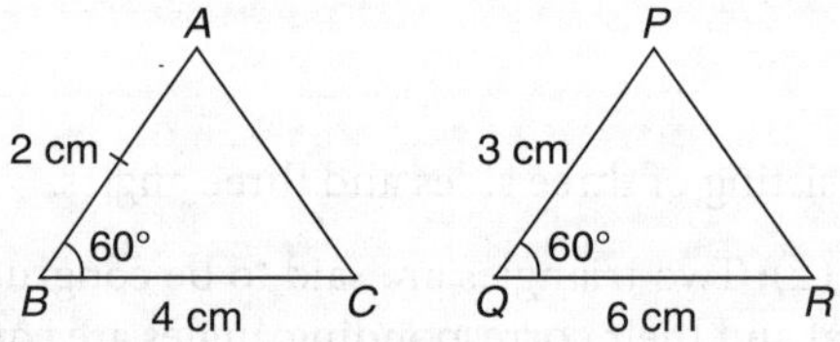

$$\frac{AB}{PQ} = \frac{BC}{QR} = \frac{2}{3}$$

$$\angle B = \angle Q = 60°$$

$$\Delta ABC \sim \Delta PQR \text{ (SAS)}$$

   (c) AA or AAA (angle-angle-angle)

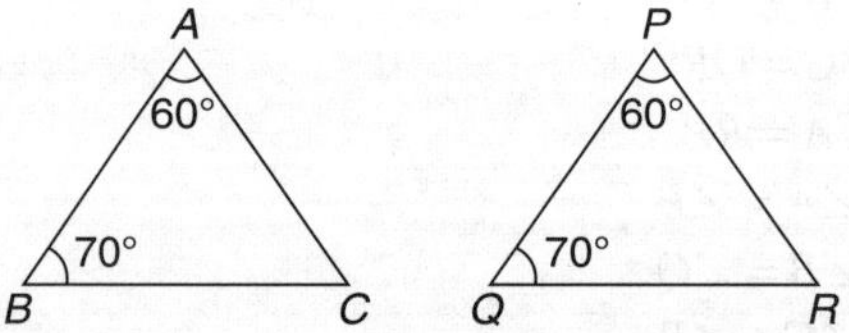

$$\angle A = \angle P = 60°$$

$$\angle B = \angle Q = 70°$$

$$\angle C = \angle R = 50° \quad [\because \text{ Angle sum property of } \Delta s]$$

$$\Delta ABC \sim \Delta PQR \quad \text{[AA or AAA]}$$

Concept: Problems Based on Similar Triangles

Example 1 State the Δs that are similar and also state their similarity criteria.

(a)

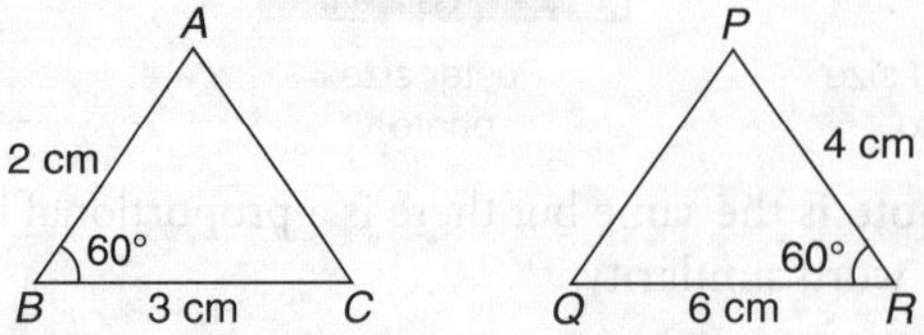

(b) 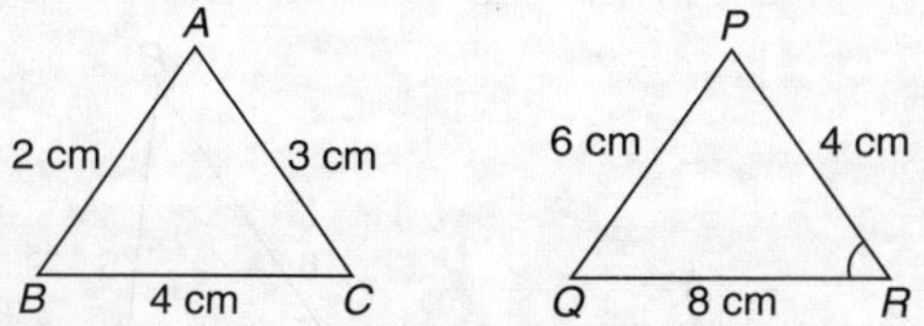

(c) 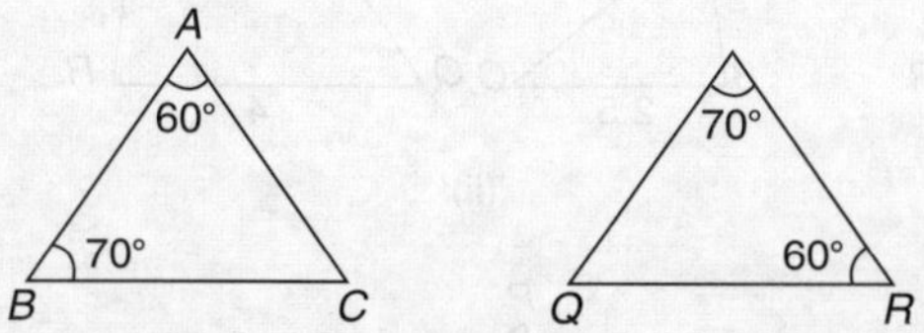

(d) 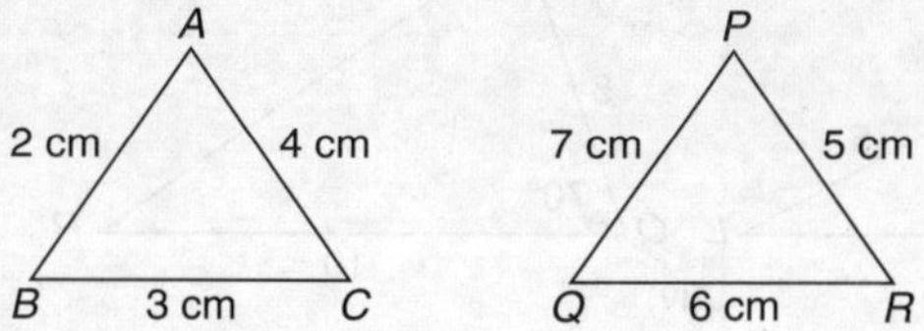

Solution:

(a) $\dfrac{AB}{PR} = \dfrac{\cancel{2}}{\cancel{4}} = \dfrac{1}{2}$ ...(i)

$$\dfrac{BC}{QR} = \dfrac{\cancel{3}}{\cancel{6}} = \dfrac{1}{2} \quad \text{...(ii)}$$

$$\angle B = \angle R = 60° \quad \text{...(iii)}$$

$$\Delta ABC \sim \Delta PRQ \quad [\because \text{SAS}]$$

(b) $\dfrac{AB}{PR} = \dfrac{\cancel{2}}{\cancel{4}} = \dfrac{1}{2}$ ...(i)

$$\dfrac{AC}{PQ} = \dfrac{\cancel{3}}{\cancel{6}} = \dfrac{1}{2} \quad \text{...(ii)}$$

$$\dfrac{BC}{QR} = \dfrac{4}{8} = \dfrac{1}{2} \quad \text{...(iii)}$$

$$\Delta ABC \sim \Delta PRQ \quad [\because \text{SSS}]$$

(c) $\angle A = \angle R = 60°$

$$\angle B = \angle P = 70°$$

$$\Delta ABC \sim \Delta RPQ \quad [\because \text{AA}]$$

(d) No sides are in proportion. Hence no similarity occurs.

1 Try to Solve

1. State which pairs of triangles in the figure are similar. Write the similarity criterion used by you for answering the question and also write the pairs of similar triangles in symbolic form. **[NCERT]**

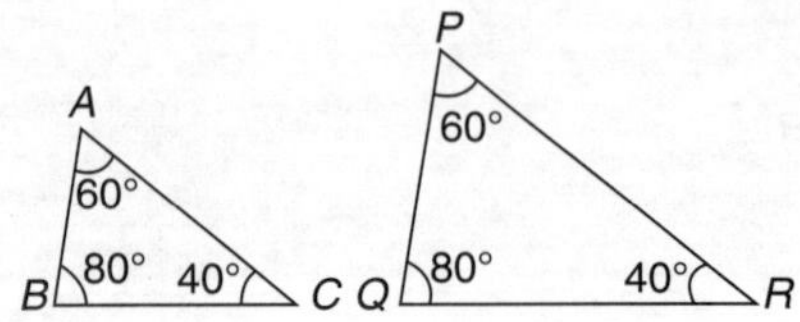

(i)

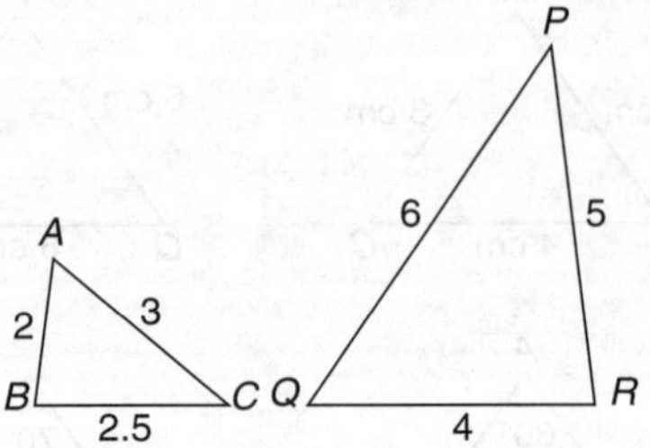

(ii)

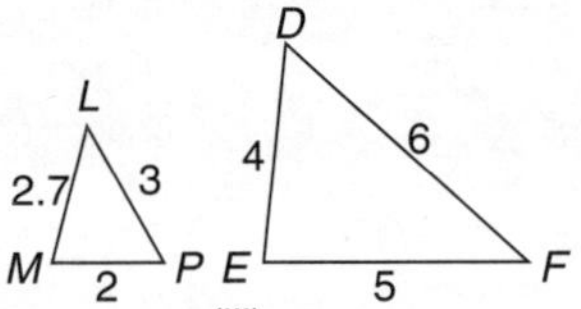

(iii)

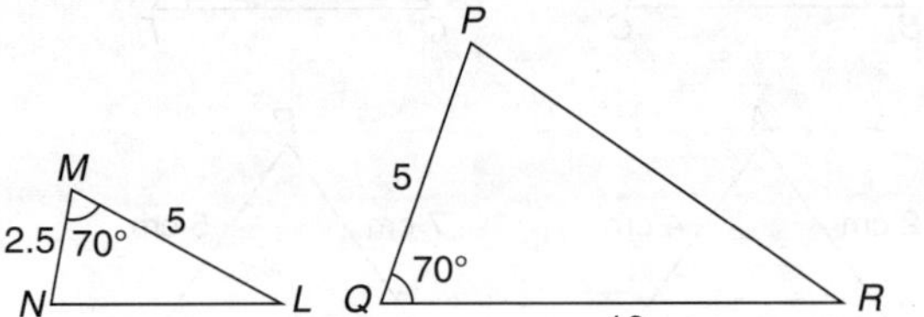

(iv)

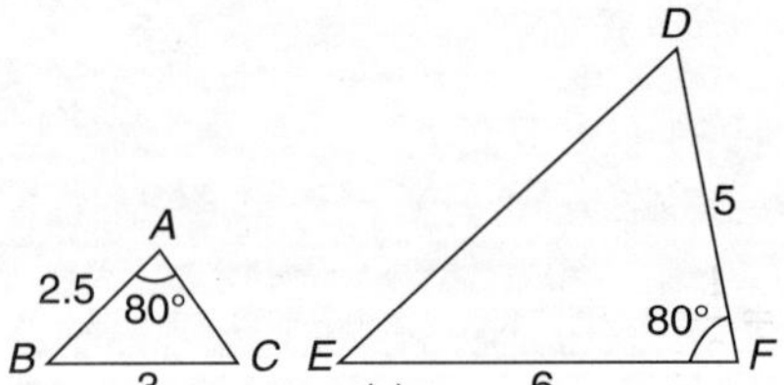

(v)

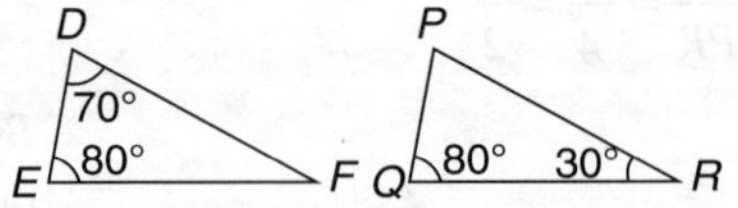

(vi)

Answers

1. (a) Yes, AAA, $\Delta ABC \sim \Delta PQR$ (b) Yes, SSS, $\Delta ABC \sim \Delta QRP$ (c) No
(d) Yes, SAS, $\Delta MNL \sim \Delta QPR$ (e) No (f) Yes, AA, $\Delta DEF \sim \Delta PQR$

Example 2 If $PQ \parallel RS$, prove that $\Delta POQ \sim \Delta SOR$. [NCERT]

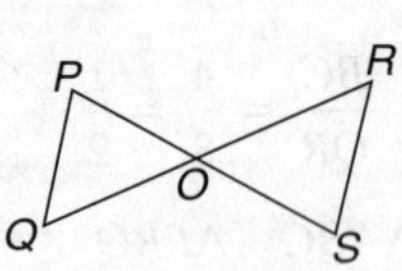

Solution: Since $PQ \parallel RS$, PS is a transversal

$\angle P = \angle S$ [∵ Alternate interior angles] ...**(i)**

$PQ \parallel RS$, QR is a transversal

$\angle Q = \angle R$ [∵ Alternate interior angles] ...**(ii)**

Consider ΔPOQ and ΔSOR

$\angle P = \angle S$ [From (i)]

$\angle Q = \angle R$ [From (ii)]

$\Delta POQ \sim \Delta SOR$ [AA]

2 Try to Solve

1. In the given figure, if $AB \parallel CD$, prove that $\Delta AOB \sim \Delta DOC$.

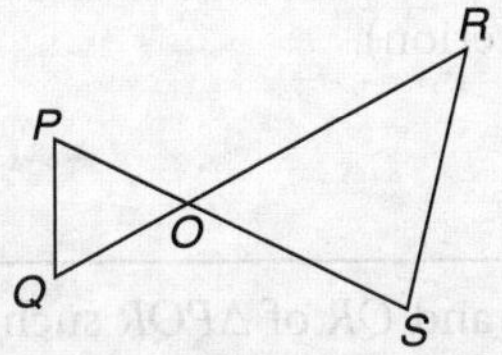

2. In the given figure, $\Delta ODC \sim \Delta OBA$, $\angle BOC = 125°$ and $\angle CDO = 70°$. Find $\angle DOC$, $\angle DCO$ and $\angle OAB$. **[NCERT]**

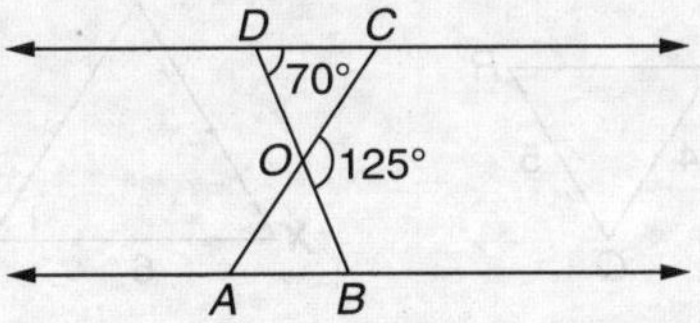

3. Diagonals AC and BD of a trapezium $ABCD$ with $AB \parallel DC$ intersect each other at the point O. Using a similarity criterion for two triangles, show that $\frac{AO}{OC} = \frac{OB}{OD}$. **[NCERT]**

Answers

2. $\angle DOC = 55°$, $\angle DCO = 55°$ and $\angle AOB = 55°$

Example 3 In the given figure, $\frac{QR}{QS} = \frac{QT}{PR}$ and $\angle 1 = \angle 2$. Show that $\Delta PQS \sim \Delta TQR$.

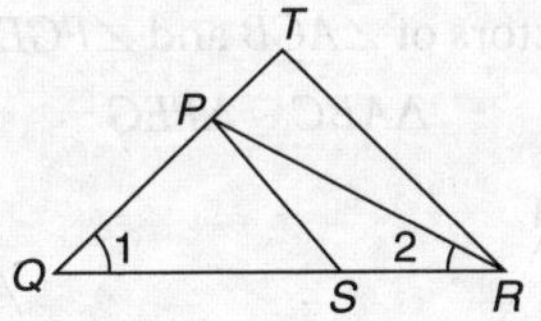

[NCERT, CBSE (CCE) 2016]

Solution: *Given*: $\frac{QR}{QS} = \frac{QT}{PR}$ and $\angle 1 = \angle 2$

To prove: $\Delta PQS \sim \Delta TQR$

Proof: $\frac{QR}{QS} = \frac{QT}{PR}$ [Given]

In ΔPQR $\angle 1 = \angle 2$ [Given]

$\therefore$ $PQ = PR$ ($\because$ In triangles, the sides opposite to equal angles are equal) ...(i)

$\Rightarrow$ $\frac{QR}{QS} = \frac{QT}{PQ}$ [From (i)] ...(ii)

In ΔTQR and ΔPQS $\quad \angle 1 = \angle 1$ [Common]

$$\frac{QR}{QS} = \frac{QT}{PQ} \qquad \text{[From (ii)]}$$

$\therefore$ $\Delta TQR \sim \Delta PQS$ (by SAS similarity criterion).

3 Try to Solve

1. S and T are points on sides PR and QR of ΔPQR such that $\angle P = \angle RTS$. Show that $\Delta RPQ \sim \Delta RTS$. **[NCERT]**
2. In the figure, $\Delta PQR \sim \Delta XYZ$. If $PQ = 4$ cm, $QR = 5$ cm and $XY = 6$ cm, find YZ. **[CBSE 2015]**

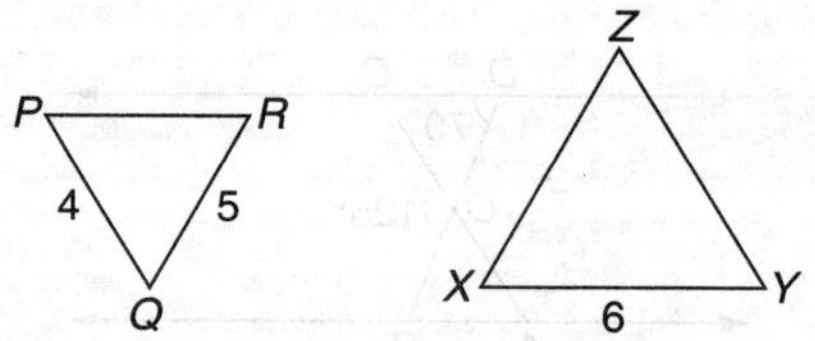

3. If the areas of two similar triangles are equal, prove that they are congruent. **[CBSE 2018]**

Answers

2. 7.5 cm

Example 4 CD and GH are, respectively, the bisectors of $\angle ACB$ and $\angle EGF$ such that D and H lie on sides AB and FE of ΔABC and ΔEFG, respectively. If $\Delta ABC \sim \Delta FEG$, show that

(i) $\Delta DCA \sim \Delta HGF$ (ii) $\frac{CD}{GH} = \frac{AC}{FG}$ (iii) $\Delta DCB \sim \Delta HGE$ **[NCERT]**

Solution: *Given*: CD and GH are bisectors of $\angle ACB$ and $\angle FGE$, respectively.

$$\Delta ABC \sim \Delta FEG$$

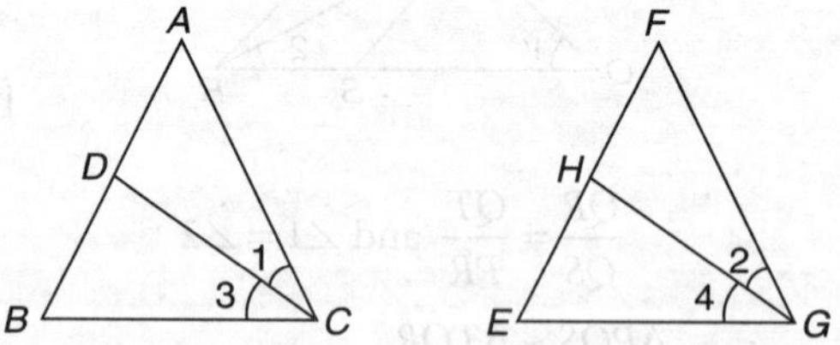

To prove:

(i) $\Delta DCA \sim \Delta HGF$ (ii) $\frac{CD}{GH} = \frac{AC}{FG}$ (iii) $\Delta DCB \sim \Delta HGE$

Proof: $\Delta ABC \sim \Delta FEG$ [Given]

$\therefore$
$$\angle A = \angle F$$
$$\angle B = \angle E$$

$\angle C = \angle G$ [∵ If two triangles are similar, their corresponding angles are equal]

$\Rightarrow \quad \frac{1}{2}\angle C = \frac{1}{2}\angle G$

$\Rightarrow \quad \angle 1 = \angle 2$ or $\angle 3 = \angle 4$

(i) In ΔDCA and ΔHGF $\quad \angle A = \angle F$ and $\angle 1 = \angle 2$ [Proved above]

$\therefore \Delta DCA \sim \Delta HGF$ [By AA similarity criterion]

(ii) $\frac{CD}{GH} = \frac{AC}{FG}$ [∵ If two triangles are similar, their corresponding sides are in proportion]

(iii) In ΔDCB and ΔHGE $\quad \angle B = \angle E$ [Proved above]

And $\quad \angle 3 = \angle 4$ [Proved above]

$\therefore \Delta DCB \sim \Delta HGE$ [By AA similarity criterion]

Example 5 *D* is a point on the side *BC* of a triangle *ABC* such that $\angle ADC = \angle BAC$. Show that $CA^2 = CB.CD$. **[NCERT, CBSE (CCE) 2013]**

Solution: *Given*: $\angle ADC = \angle BAC$

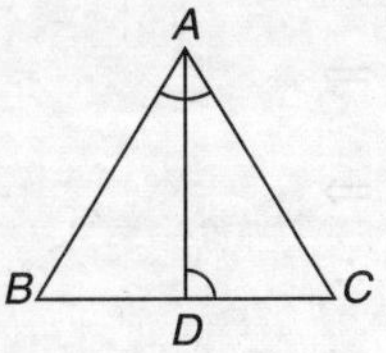

To prove: $CA^2 = CB \times CD$

Proof: In ΔADC and ΔBAC $\quad \angle C = \angle C$ [Common]

$\angle ADC = \angle BAC$ [Given]

$\therefore \quad \Delta ADC \sim \Delta BAC$ [By AA similarity criterion]

$\Rightarrow \quad \frac{DC}{AC} = \frac{AC}{BC}$ [∵ If two triangles are similar, their corresponding sides are in proportion]

$\Rightarrow \quad CA^2 = BC \times DC$

$\therefore \quad CA^2 = CB \times CD$

4 Try to Solve

1. In the given figure, if $\frac{AB}{BD} = \frac{BC}{EF}$ and $\angle 1 = \angle 2$, prove that $\Delta ABC \sim \Delta DEC$.

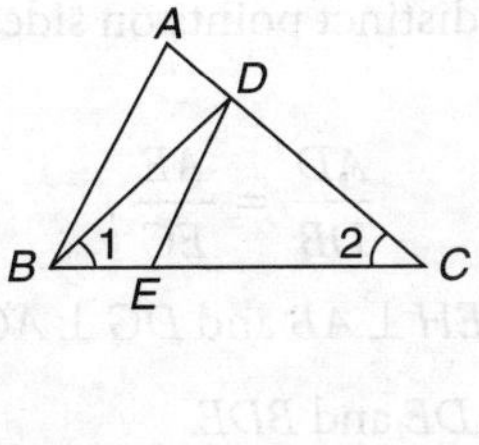

2. D is a point on the side BC of ΔABC such that $\angle ADB = \angle BAC$. Prove that $BA^2 = BD \times BC$.
3. In the figure, PQR and QST are two right triangles, right-angled at R and T, respectively.
   Prove that $QR \times QS = QP \times QT$. **[CBSE 2015]**

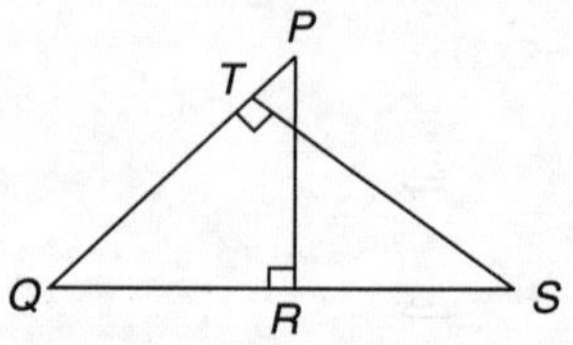

Example 6 A vertical pole of length 6 m casts a shadow 4 m long on the ground and at the same time a tower casts a shadow 28 m long. Find the height of the tower. **[NCERT, CBSE (CCE) 2016]**

Solution: *To find*: Height of tower, h

In ΔABC and ΔDEF

$\angle A = \angle D$ [Angle made by the sun at the same time]

$\angle B = \angle E$ [Each 90°]

$\therefore$ $\Delta ABC \sim \Delta DEF$ [By AA similarity criterion]

$\Rightarrow$ $\frac{AB}{DE} = \frac{BC}{EF}$ [$\because$ If two triangles are similar, their corresponding sides are in proportion]

$\Rightarrow$ $\frac{h}{6} = \frac{28}{4} = 7$

$\Rightarrow$ $h = 6 \times 7$

$\therefore$ $h = 42$ m

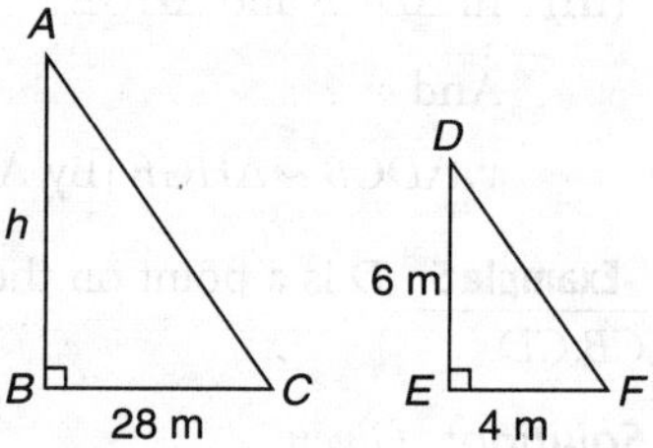

5 Try to Solve

1. A boy of height 150 cm is walking away from the base of a lamp post at a speed of 1.6 m/s. If the lamp is 4.5 m above the ground, find the length of his shadow after 5 seconds.

Answer

1. 4 m

Concept: Basic Proportionality Theorem (BPT)

State and prove the Basic Proportionality Theorem. **[CBSE Frequently Asked]**

Statement: If a line is drawn parallel to one side of a triangle to intersect the other two sides at distinct points, the other two sides are divided in the same ratio. Prove.

Given: ΔABC, in which D and E are two distinct points on sides AB and AC, respectively, and $DE \parallel BC$.

To prove: $$\frac{AD}{DB} = \frac{AE}{EC}$$

Construction: Join BE and CD and drop $EH \perp AB$ and $DG \perp AC$.

Proof: Consider the ratio of areas of Δs ADE and BDE.

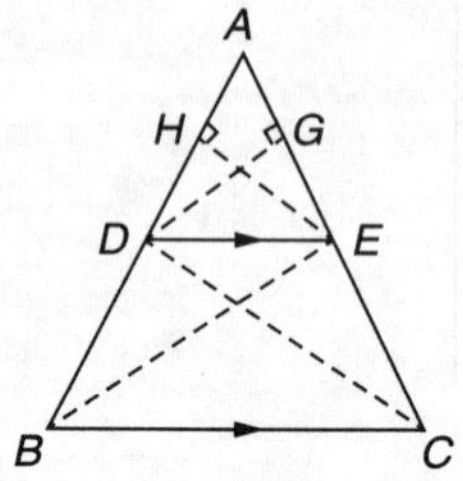

$$\frac{ar\Delta\ ADE}{ar\Delta\ BDE} = \frac{\frac{1}{2} \times AD \times EH}{\frac{1}{2} \times DB \times EH} = \frac{AD}{DB} \quad \ldots\text{(i)}$$

Similarly
$$\frac{ar\Delta\ ADE}{ar\Delta\ DEC} = \frac{\frac{1}{2} \times AE \times DG}{\frac{1}{2} \times EC \times DG} = \frac{AE}{EC} \quad \ldots\text{(ii)}$$

Since ΔBDE and ΔDEC are on the same base DE and between $DE \parallel BC$

$$ar\Delta BDE = ar\Delta DEC$$ [∵ Δs on the same base and between the same || lines are equal in area] ...(iii)

$$ar\Delta ADE = ar\Delta ADE \quad \ldots\text{(iv)}$$

Divide (iv) by (iii)
$$\frac{ar\Delta\ ADE}{ar\Delta\ BDE} = \frac{ar\Delta\ ADE}{ar\Delta\ DEC} \quad \ldots\text{(v)}$$

From (i), (ii) and (v), we get
$$\frac{AD}{DB} = \frac{AE}{EC}$$

Hence proved.

Results of BPT

(i) $\frac{BD}{DA} = \frac{CE}{EA}$ (ii) $\frac{AD}{AB} = \frac{AE}{AC}$ (iii) $\frac{AB}{AD} = \frac{AC}{AE}$ (iv) $\frac{DB}{BA} = \frac{EC}{CA}$ (v) $\frac{AB}{DB} = \frac{AC}{EC}$

Example 7 In ΔABC, D and E are points on the sides AB and AC, respectively, such that $DE \parallel BC$. If $AD = 6$ cm, $DB = 9$ cm and $AE = 8$ cm, find AC. **[CBSE 2006C]**

Solution: *Given*: $DE \parallel BC$, $AD = 6$ cm, $DB = 9$ cm, $AE = 8$ cm

To find: AC

Proof: In ΔABC, $DE \parallel BC$

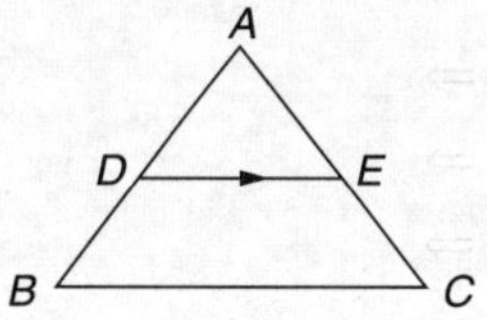

∴ $$\frac{AD}{DB} = \frac{AE}{EC} \quad \text{[By BPT]}$$

⇒ $$\frac{6}{9} = \frac{8}{EC}$$

⇒ $$EC = \frac{8 \times 9}{6}$$

∴ $$EC = 12 \text{ cm}$$

Now $$AC = AE + EC = 8 + 12 = 20 \text{ cm}$$

Example 8 D and E are points on the sides AB and AC, respectively, of a ΔABC such that $DE \parallel BC$. If $\frac{AD}{DB} = \frac{4}{7}$ and $AC = 6.6$ cm, find AE. **[CBSE 2008C]**

Solution: *Given:* In ΔABC, $DE \parallel BC$ and $\frac{AD}{DB} = \frac{4}{7}$, $AC = 6.6$ cm.

To find: AE

Let $AE = x$ cm

$EC = AC - AE = (6.6 - x)$cm

In ΔABC, $DE \parallel BC$ $\quad \frac{AD}{DB} = \frac{AE}{EC}$ [By BPT]

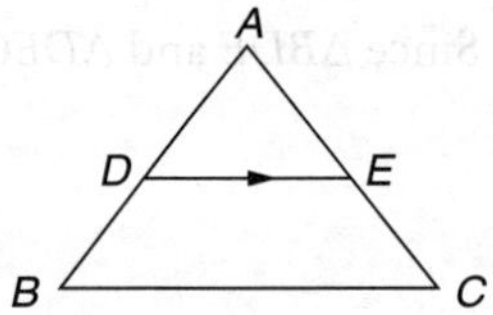

$\Rightarrow \quad \frac{4}{7} = \frac{x}{6.6 - x}$

$\Rightarrow \quad 4(6.6 - x) = 7x$

$\Rightarrow \quad 26.4 - 4x = 7x$

$\Rightarrow \quad 26.4 = 7x + 4x$

$\Rightarrow \quad 11x = 26.4$

$\therefore \quad x = \frac{26.4}{11} = 2.4$ cm

Hence $AE = 2.4$ cm

Example 9 In the given figure, $MN \parallel PR$. If $QM = 4x - 3$, $QN = 8x - 7$, $PM = 3x - 1$ and $RN = 5x - 3$, find the value of x. **[CBSE 2002]**

Solution: *Given:* $MN \parallel PR$

To find: x

In ΔPQR, $MN \parallel PR$ $\quad \frac{QM}{PM} = \frac{QN}{RN}$ [By BPT]

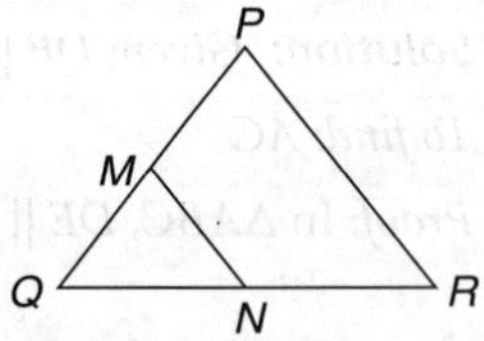

$\Rightarrow \quad \frac{4x - 3}{3x - 1} = \frac{8x - 7}{5x - 3}$

$\Rightarrow \quad (4x - 3)(5x - 3) = (3x - 1)(8x - 7)$

$\Rightarrow \quad 20x^2 - 15x - 12x + 9 = 24x^2 - 8x - 21x + 7$

$\Rightarrow \quad 20x^2 - 27x + 9 = 24x^2 - 29x + 7$

$\Rightarrow \quad 4x^2 - 2x - 2 = 0$

$\Rightarrow \quad 2x^2 - x - 1 = 0$ [On dividing both sides by 2]

$\Rightarrow \quad 2x^2 - 2x + x - 1 = 0$

$\Rightarrow \quad 2x(x - 1) + 1(x - 1) = 0$

$\Rightarrow \quad (2x + 1)(x - 1) = 0$

Either $\quad x - 1 = 0$ or $2x + 1 = 0$

$\therefore \quad x = 1 \text{ or } x = -\frac{1}{2}$ [Not possible, being –ve]

Hence, the value of $x = 1$.

6 Try to Solve

1. In the given figure, $DE \parallel BC$. If $AD = x$ cm, $DB = x - 3$, $AE = x + 3$ and $EC = x - 2$, find the value of x. [NCERT]

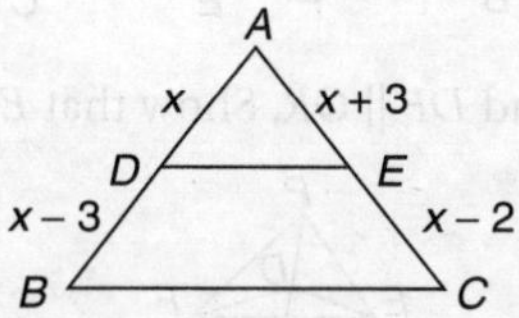

2. In the figure, $DE \parallel BC$. If $AD = 1.7$ cm, $AB = 6.8$ cm and $AC = 9$ cm, find AE.

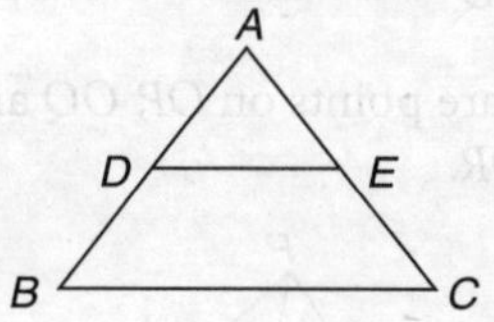

Answers

1. $x = 4.5$ cm 2. 2.25 cm

Example 10 In the given figure, $PM \parallel AB$ and $PN \parallel AD$. Prove that $\frac{BM}{CM} = \frac{DN}{CN}$.

Solution: *Given:* $PM \parallel AB$ and $PN \parallel AD$

To prove: $\frac{BM}{CM} = \frac{DN}{CN}$

Proof: In ΔABC, $PM \parallel AB$ $\quad \frac{CM}{BM} = \frac{CP}{AP}$ [By BPT] ...(i)

In ΔACD, $PN \parallel AD$ $\quad \frac{CN}{DN} = \frac{CP}{AP}$ [By BPT] ...(ii)

From (i) and (ii), we get $\quad \frac{CM}{BM} = \frac{CN}{DN}$

$\therefore \quad \frac{BM}{CM} = \frac{DN}{CN}$

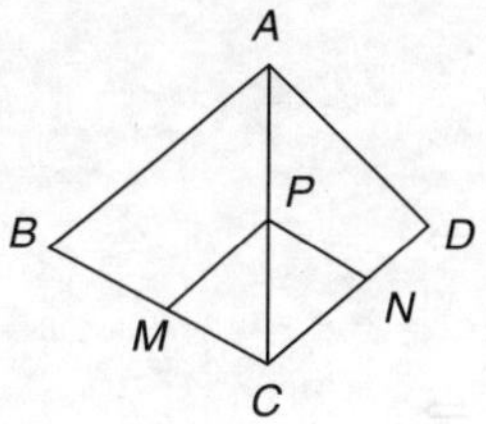

7 Try to Solve

1. In the given figure, $DE \parallel AC$ and $DF \parallel AE$. Prove that $\frac{BF}{FE} = \frac{BE}{EC}$. [NCERT]

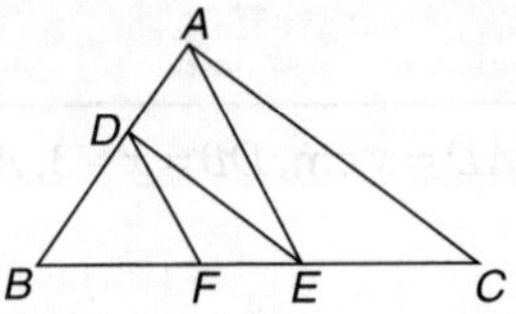

2. In the given figure, $DE \parallel OQ$ and $DF \parallel OR$. Show that $EF \parallel QR$. [NCERT, CBSE (CCE) 2012]

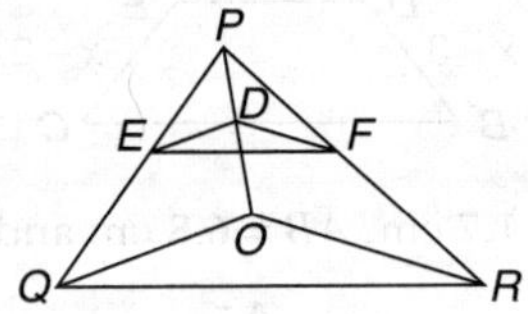

3. In the given figure, A, B and C are points on OP, OQ and OR, respectively, such that $AB \parallel PQ$ and $AC \parallel PR$. Show that $BC \parallel QR$. [NCERT, CBSE (CCE) 2012]

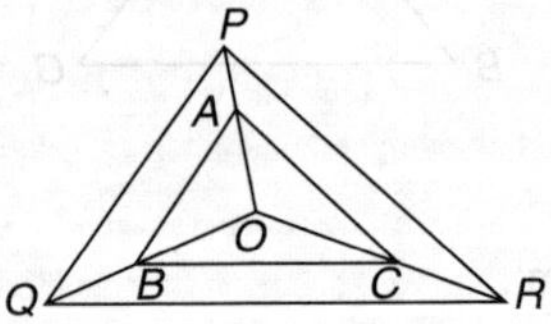

Example 11 Using BPT, prove that a line drawn through the mid-point of one side of a triangle parallel to another side bisects the third side. [NCERT]

Solution: *Given:* In ΔABC, $AD = DB$, $DE \parallel BC$.

To prove: E is the mid-point of AC, that is $AE = EC$

Proof: In ΔABC, $DE \parallel BC$

$$\frac{AD}{BD} = \frac{AE}{CE} \quad \text{[By BPT]}$$

$$\Rightarrow \quad \frac{AD}{AD} = \frac{AE}{CE} \quad [\because AD = DB \text{ given}]$$

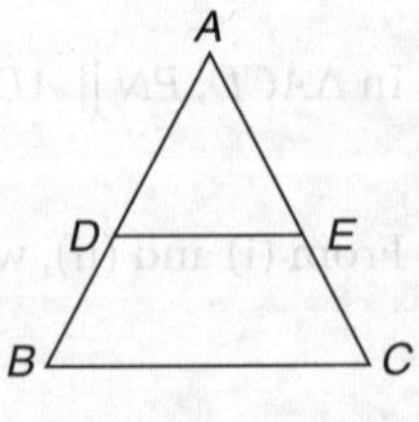

$$\Rightarrow \quad \frac{\cancel{AD}}{\cancel{AD}} = \frac{AE}{CE}$$

$$\Rightarrow \quad 1 = \frac{AE}{EC}$$

$$\Rightarrow \quad EC = AE$$

So, E is the mid-point of AC.

Example 12 *ABCD* is a trapezium in which $AB \parallel DC$ and its diagonals intersect each other at the point O.

Show that $\frac{AO}{BO} = \frac{CO}{DO}$. **[NCERT, CBSE (CCE) 2011]**

Solution: *Given: ABCD* is a trapezium in which $AB \parallel DC$.

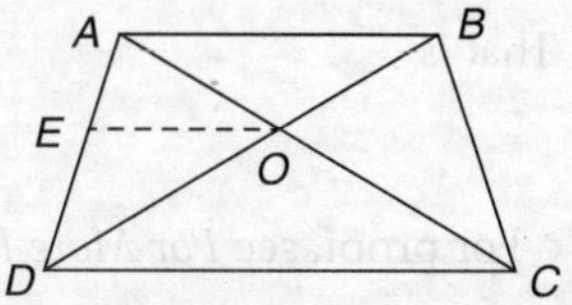

To prove: $\frac{AO}{BO} = \frac{CO}{DO}$

Construction: Draw a line $EO \parallel DC$.

Proof: In ΔADC, $EO \parallel DC$

$$\frac{AE}{DE} = \frac{AO}{CO} \quad \text{[By BPT]} \qquad \text{(i)}$$

In ΔADB, $EO \parallel AB$ ($\because EO \parallel DC$ and $AB \parallel DC$)

$$\therefore \quad \frac{DE}{AE} = \frac{DO}{BO} \qquad \text{[By BPT]}$$

$$\Rightarrow \quad \frac{AE}{DE} = \frac{BO}{DO} \qquad \text{(ii)}$$

From (i) and (ii), we get $\frac{AO}{CO} = \frac{BO}{DO}$

$$\Rightarrow \quad \frac{AO}{BO} = \frac{CO}{DO} \qquad [\because \text{Using Alternando}]$$

8 Try to Solve

1. In the given figure, $AB \parallel DE$ and $BD \parallel EF$. Prove that $DC^2 = CF \times AC$.

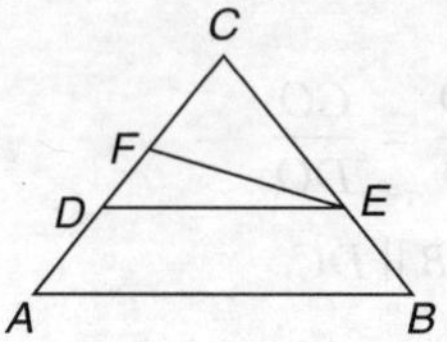

2. If $\angle P = \angle RTS$, show that $\angle PQR = \angle RST$.

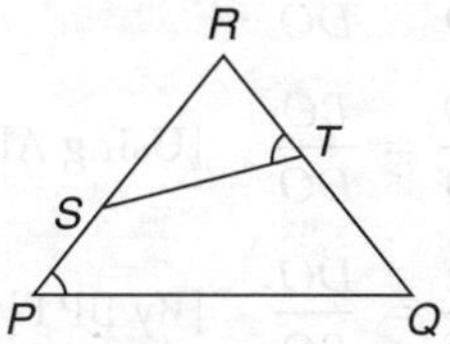

Concept: Converse of Basic Proportionality Theorem

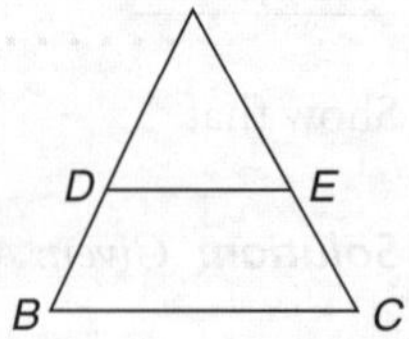

Statement: If a line divides any two sides of a triangle in the same ratio, then the line is parallel to the third side.

That is $\frac{AD}{DB} = \frac{AE}{EC}$ [Given]

$DE \parallel BC$ [Result]

* For proof, see *For More Practice and Knowledge*

Example 13 In ΔABC, D and E are points on the sides AB and AC, respectively. Show that $DE \parallel BC$, if $AB = 5.6$ cm, $AD = 1.4$ cm, $AC = 7.2$ cm and $AE = 1.8$ cm.

Solution: *Given:* $AB = 5.6$ cm, $AD = 1.4$ cm, $AC = 7.2$ cm and $AE = 1.8$ cm.

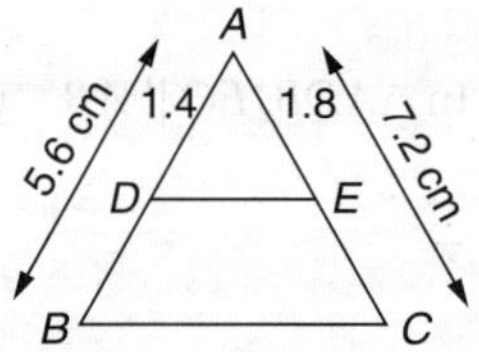

To prove: $DE \parallel BC$

Proof:
$$DB = AB - AD$$
$$= 5.6 - 1.4 = 4.2 \text{ cm}$$
$$EC = AC - AE$$
$$= 7.2 - 1.8 = 5.4 \text{ cm}$$

Now $\frac{AD}{DB} = \frac{1.4}{4.2} = \frac{1}{3}$

And $\frac{AE}{EC} = \frac{1.8}{5.4} = \frac{1}{3}$

As $\frac{AD}{DB} = \frac{AE}{EC}$

$\therefore$ $DE \parallel BC$ [By converse of BPT]

Example 14 The diagonals of a quadrilateral $ABCD$ intersect each other at the point O such that $\frac{AO}{BO} = \frac{CO}{DO}$. Show that $ABCD$ is a trapezium. **[NCERT CBSE (CCE) 2013]**

Solution: *Given:* $\frac{AO}{BO} = \frac{CO}{DO}$

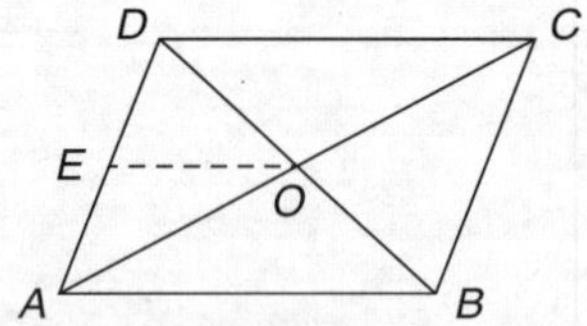

To prove: $ABCD$ is a trapezium, that is, $AB \parallel DC$.

Construction: Draw a line $EO \parallel AB$.

Proof: $\frac{AO}{BO} = \frac{CO}{DO}$ [Given]

$\Rightarrow$ $\frac{AO}{CO} = \frac{BO}{DO}$ [Using Alternando] ...(i)

In ΔADB, $EO \parallel AB$ $\frac{DE}{AE} = \frac{DO}{BO}$ [By BPT]

$$\Rightarrow \qquad \frac{AE}{DE} = \frac{BO}{DO} \qquad \ldots\textbf{(ii)}$$

From (i) and (ii), we get $\quad \dfrac{AO}{CO} = \dfrac{AE}{DE}$

In ΔADC $\quad \dfrac{AO}{CO} = \dfrac{AE}{DE}$

$\therefore \quad EO \parallel DC$ [By converse of BPT] ...**(iii)**

But $\quad EO \parallel AB$ [By construction] ...**(iv)**

From (iii) and (iv), we get $AB \parallel DC$.

Since in a quadrilateral *ABCD*, one pair of opposite sides is parallel, it is a trapezium.

9 Try to Solve

1. *R* and *S* are points on the sides *DE* and *EF*, respectively, of a ΔDEF such that $ER = 5$ cm, $RD = 2.5$ cm, $SE = 1.5$ cm and $FS = 3.5$ cm. Find whether $RS \parallel DF$ or not.

2. Prove that if a line is drawn parallel to one side of a triangle to intersect the other two sides in distinct points, the other two sides are divided in the same ratio.

Using the above result, perform the following: In the given figure, $DE \parallel BC$ and $BD = CE$. Prove that ΔABC is an isosceles triangle. **[CBSE Frequently Asked]**

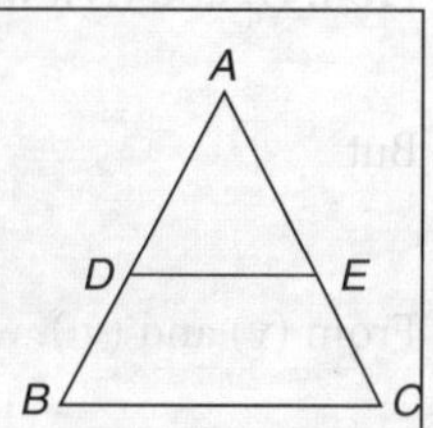

Concept: Area Theorem

State and prove the Area Theorem. **[CBSC Frequently Asked]**

Statement: The ratio of the areas of two similar triangles is equal to the square of the ratio of their corresponding sides.

Given: $\quad \Delta ABC \sim \Delta DEF$

That is $\quad \angle A = \angle D, \angle B = \angle E$, and $\angle C = \angle F$

$$\frac{AB}{DE} = \frac{BC}{EF} = \frac{AC}{DF}$$

To prove: $\quad \dfrac{ar\Delta\, ABC}{ar\Delta\, DEF} = \dfrac{AB^2}{DE^2} = \dfrac{BC^2}{EF^2} = \dfrac{AC^2}{DF^2}$

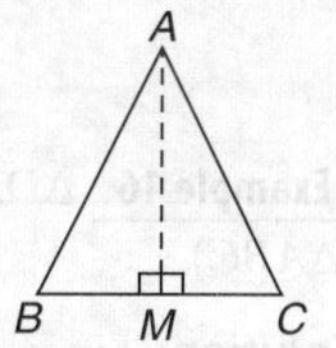

Construction: Drop $AM \perp BC$ and $DN \perp EF$.

Proof: Consider the ratio of areas of Δs.

$$\frac{ar\Delta\, ABC}{ar\Delta\, DEF} = \frac{\frac{1}{2} \times BC \times AM}{\frac{1}{2} \times EF \times DN} = \frac{BC \times AM}{EF \times DN} \qquad \ldots\textbf{(i)}$$

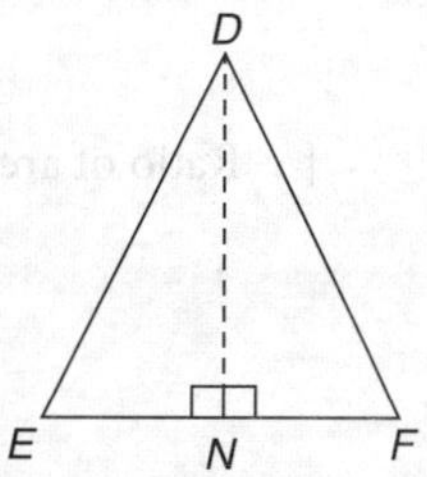

To prove the above result, we need to prove ΔAMB and ΔDNE are similar.

Consider ΔAMB and ΔDNE:

$$\angle AMB = \angle DNE \text{ [Each } 90^\circ]$$

$$\angle B = \angle E \text{ [Given]}$$

$$\angle AMB \sim \angle DNE \text{ [AA]}$$

$$\frac{AM}{DN} = \frac{AB}{DE} \quad [\because \text{ If two } \Delta\text{s are similar, their corresponding sides are in proportion}] \quad \ldots\text{(ii)}$$

But $$\frac{AB}{DE} = \frac{BC}{EF} \text{ [Given]} \quad \ldots\text{(iii)}$$

From (ii) and (iii) $$\frac{AM}{DN} = \frac{BC}{EF} \quad \ldots\text{(iv)}$$

From (i) and (iv), we get $$\frac{ar\Delta ABC}{ar\Delta DEF} = \frac{BC}{EF} \times \frac{BC}{EF} = \frac{BC^2}{EF^2} \quad \ldots\text{(v)}$$

But $$\frac{BC}{EF} = \frac{AB}{DE} = \frac{AC}{DF} \text{ [Given]} \quad \ldots\text{(vi)}$$

From (v) and (vi), we get $$\frac{ar\Delta ABC}{ar\Delta DEF} = \frac{AB^2}{DE^2} = \frac{BC^2}{EF^2} = \frac{AC^2}{DF^2}$$

Example 15 If it is given that $\Delta ABC \sim \Delta PQR$ with $\frac{BC}{QR} = \frac{1}{3}$, then find $\frac{ar\,(\Delta PQR)}{ar\,(\Delta ABC)}$.

Solution:

$$\frac{BC}{QR} = \frac{1}{3} \quad \text{[Given]}$$

$$\frac{ar\,(\Delta PQR)}{ar\,(\Delta ABC)} = \frac{(QR)^2}{(BC)^2} \quad [\because \text{ Ratio of area of similar triangles is equal to the ratio of the square of their corresponding sides}]$$

$$\left(\frac{QR}{BC}\right)^2 = \left(\frac{3}{1}\right)^2 = \frac{9}{1} = 9:1$$

Example 16 $\Delta DEF \sim \Delta ABC$, if $DE : AB = 2:3$ and ar(ΔDEF) is equal to 44 square units. Find the area (ΔABC). **[CBSE (CCE) 2010]**

Solution: Since

$$\Delta DEF \sim \Delta ABC$$

$$\frac{ar\,(\Delta DEF)}{ar\,(\Delta ABC)} = \frac{(DE)^2}{(AB)^2}$$

[∵ Ratio of area of similar triangles is equal to the ratio of the square of their corresponding sides]

$$\frac{44}{ar(\Delta ABC)} = \left(\frac{2}{3}\right)^2$$

$$\Rightarrow \quad ar(\Delta ABC) = \frac{44 \times 9}{4}$$

So $\quad ar(\Delta ABC) = 99 \text{ cm}^2$

Example 17 Let $\Delta ABC \sim \Delta DEF$ and their areas be, respectively, 64 cm^2 and 121 cm^2. If $EF = 15.4$ cm, find BC. **[NCERT]**

Solution: *Given:*

$$ar(\Delta ABC) = 64 \text{ cm}^2$$
$$ar(\Delta DEF) = 121 \text{ cm}^2$$
$$EF = 15.4 \text{ cm}^2$$

To find: BC

Proof: $\quad \Delta ABC \sim \Delta DEF \quad$ [Given]

$$\therefore \quad \frac{ar\,(ABC)}{ar\,(DEF)} = \left(\frac{BC}{EF}\right)^2$$

[∵ Ratio of area of similar triangles is equal to the ratio of the square of their corresponding sides]

$$\frac{64}{121} = \left(\frac{BC}{15.4}\right)^2$$

$$\left(\frac{8}{11}\right)^2 = \left(\frac{BC}{15.4}\right)^2$$

$$\Rightarrow \quad \frac{8}{11} = \frac{BC}{15.4}$$

$$\Rightarrow \quad BC = \frac{8 \times 15.4}{11}$$

$$\therefore \quad BC = 11.2 \text{ cm}$$

10 Try to Solve

1. If the areas of two similar triangles ABC and DEF are in the ratio 36:64 and $DE = 6.2$ cm, find AB.
2. In the given figure, $\Delta ACB \sim \Delta APQ$. If $BC = 10$ cm, $PQ = 5$ cm, $BA = 6.5$ cm and $AP = 2.8$ cm, find CA and AQ. Also find the area (ΔACB): area(ΔAPQ).

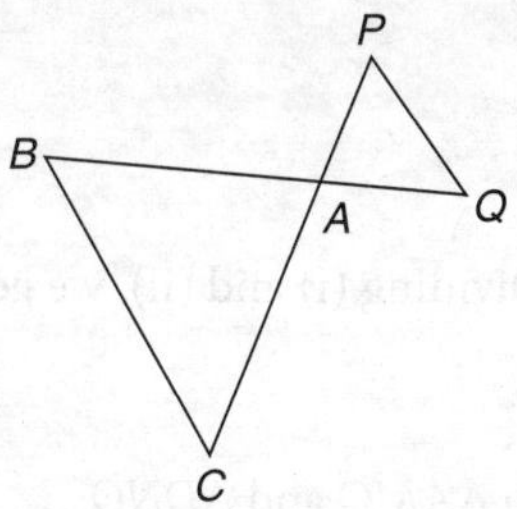

3. If the areas of two similar triangles are in the ratio 25:64, write the ratio of their corresponding sides. **[CBSE 2009]**
4. Given $\Delta ABC \sim \Delta PQR$, if $\frac{AB}{PQ} = \frac{1}{3}$, find $\frac{ar\Delta ABC}{ar\Delta PQR}$. **[CBSE 2018]**

Answers

1. $AB = 4.65$ cm 2. $CA = 5.6$ cm, $AQ = 3.25$ cm and 4:1 3. 5:8 4. 1:9

Example 18 The diagonals of a trapezium *ABCD* with *AB* || *DC* intersect each other at the point *O*. If *AB* = 2 *CD*, find the ratio of the areas of triangles *AOB* and *COD*. **[NCERT, CBSE (CCE) 2009]**

Solution: *Given*: $AB \parallel DC$

$$AB = 2\,CD$$

To find: $\dfrac{ar(\Delta AOB)}{ar(\Delta COD)} = ?$

Proof: *ABCD* is a trapezium in which *AB* || *CD*.

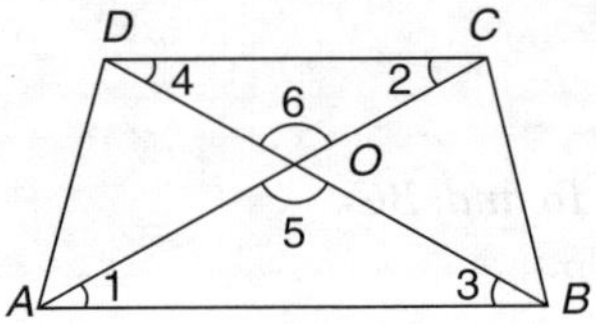

Now, in Δ*AOB* and ΔCOD, *AB* || *CD*

$\angle 1 = \angle 2$ [Alternate interior angles]

$\angle 3 = \angle 4$ [Alternate interior angles]

$\angle 5 = \angle 6$ [Vertically opposite angles]

$\therefore$ $\Delta AOB \sim \Delta COD$ [By AAA similarity criterion]

$$\frac{ar(\Delta AOB)}{ar(\Delta COD)} = \left(\frac{AB}{CD}\right)^2$$ [∵ The ratio of areas of two similar Δs is equal to the square of the ratios of their corresponding sides]

$$\therefore \quad \frac{ar(\Delta AOB)}{ar(\Delta COD)} = \left(\frac{2CD}{CD}\right)^2 = \frac{4}{1}$$

Example 19 In the given figure, *ABC* and *DBC* are two triangles on the same base *BC*. If *AD* intersects *BC* at *O*, show that $\dfrac{ar(\Delta ABC)}{ar(\Delta DBC)} = \dfrac{AO}{DO}$. **[NCERT, CBSE (CCE) 2009]**

Solution: *Given:* Δ*ABC* and Δ*DBC* are on the same base *BC*.

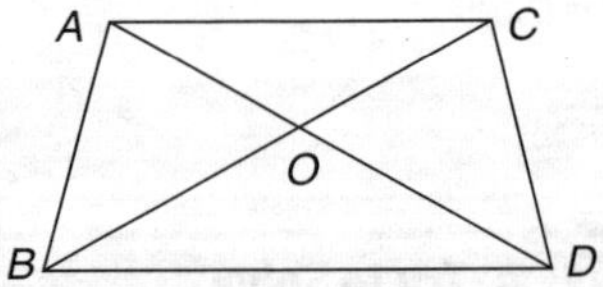

To prove: $\dfrac{ar(\Delta ABC)}{ar(\Delta DBC)} = \dfrac{AO}{DO}$

Construction: Draw *AM* and *DN* perpendicular from *A* and *D* on *BC*.

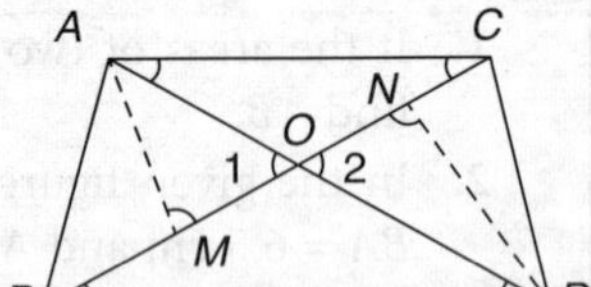

Proof: $ar(\Delta ABC) = \frac{1}{2} \times BC \times AM$ ...(i)

$ar(\Delta DBC) = \frac{1}{2} \times BC \times DN$ ...(ii)

Dividing (i) and (ii), we get $$\frac{ar(\Delta ABC)}{ar(\Delta DBC)} = \frac{\cancel{\frac{1}{2}} \times \cancel{BC} \times AM}{\cancel{\frac{1}{2}} \times \cancel{BC} \times DN} = \frac{AM}{DN}$$ ...(iii)

In Δ*AMO* and Δ*DNO* $\angle 1 = \angle 2$ (VOAs)

$\angle M = \angle N$ [Each 90°]

$\therefore \Delta AMO \sim \Delta DNO$ [AA similarity criterion]

$\Rightarrow \quad \frac{AM}{DN} = \frac{AO}{DO}$ [∵ If two Δs are similar, their corresponding sides are in proportion] ...(iv)

From (iii) and (iv), we get $\frac{ar(\Delta ABC)}{ar(\Delta DBC)} = \frac{AO}{DO}$

11 Try to Solve

1. $\Delta ABC \sim \Delta DEF$ such that the area of ΔABC is 9 cm^2 and the area of ΔDEF is 16 cm^2 and $BC = 2.1$ cm. Find the length of EF.
2. In the given figure, PB and QA are perpendicular to segment AB. If $PO = 5$ cm, $QO = 7$ cm and area $\Delta POB = 150$ cm^2, find the area of ΔQOA.

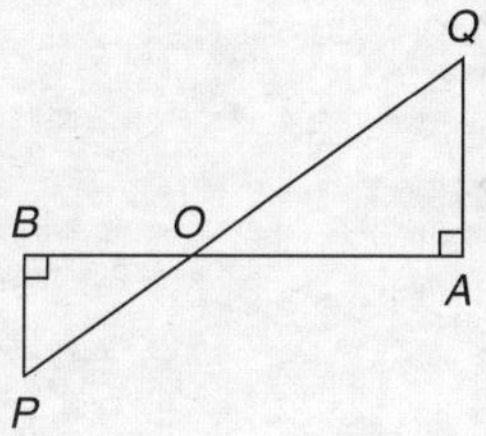

Answers

1. 2.8 cm
2. 294 cm^2

Example 20 Prove that the area of an equilateral triangle described on one side of a square is equal to half the area of the equilateral triangle described on one of its diagonals. **[NCERT, CBSE (CCE) 2012, 2018]**

Solution: *Given*: ΔDEC is an equilateral Δ described on side DC and ACF is the equilateral Δ described on the diagonal of square $ABCD$.

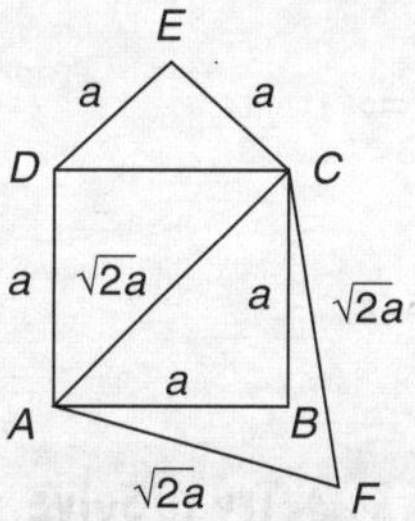

To prove: $ar(\Delta DEC) = \frac{1}{2} ar(\Delta AFC)$

Proof: Let the side of the square be 'a'.

∴ Diagonal $AC = \sqrt{a^2 + b^2} = \sqrt{2}\, a$

In ΔDEC and ΔACF $\angle E = \angle F$ [Each 60°]

$\angle D = \angle A$ [Each 60°]

∴ $\Delta DEC \sim \Delta AFC$ [By AA similarity criterion]

∴ $\frac{ar(\Delta DEC)}{ar(\Delta AFC)} = \frac{DC^2}{AC^2}$ [∵ The ratio of areas of two similar Δs is equal to the square of the ratios of their corresponding sides]

$= \frac{a^2}{(\sqrt{2a})^2} = \frac{a^2}{2a^2} = \frac{1}{2}$

∴ $ar\Delta DEC = \frac{1}{2} ar\, \Delta AFC$

Example 21 In the given figure, the line segment PQ is parallel to BC of ΔABC and it divides the triangle into two parts of equal areas. Find the ratio $\frac{PB}{AB}$.

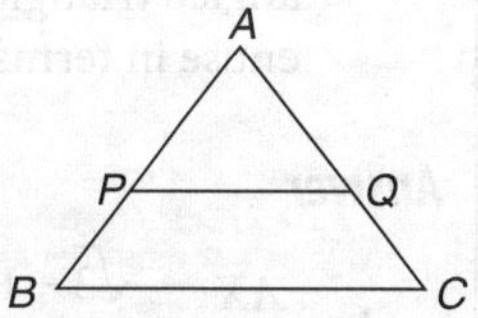

Solution: *Given:* $PQ \parallel BC$

$$ar\Delta ABC = 2\,ar\Delta APQ$$

To find: $PQ \parallel BC$ [Given]

$\therefore$ $\angle APQ = \angle ABC$ [Corresponding angles]

And $\angle AQP = \angle ACB$ [Corresponding angles]

$\therefore$ $\Delta APQ \sim \Delta ABC$ [By AA similarity criterion]

$$\Rightarrow \quad \frac{ar(\Delta APQ)}{ar(\Delta ABC)} = \frac{AP^2}{AB^2}$$

$$\Rightarrow \quad \frac{ar(\Delta APQ)}{2ar(\Delta ABC)} = \frac{AP^2}{AB^2}$$

$$\Rightarrow \quad \frac{AP^2}{AB^2} = \frac{1}{2}$$

$$\Rightarrow \quad \frac{AP}{AB} = \frac{1}{\sqrt{2}}$$

$$\Rightarrow \quad 1 - \frac{AP}{AB} = 1 - \frac{1}{\sqrt{2}}$$

$$\Rightarrow \quad \frac{AB - AP}{AB} = \frac{\sqrt{2}-1}{\sqrt{2}}$$

$$\Rightarrow \quad \frac{PB}{AB} = \frac{\sqrt{2}-1}{\sqrt{2}} \times \frac{\sqrt{2}}{\sqrt{2}}$$

$$\therefore \quad \frac{PB}{AB} = \frac{2-\sqrt{2}}{2}$$

12 Try to Solve

1. In the given figure, the line segment XY is parallel to side AC of ΔABC and it divides the triangle into two parts of equal areas. Find the ratio $\frac{AX}{AB}$. **[NCERT]**

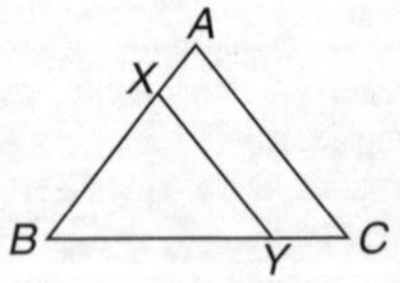

2. Prove that the equilateral triangles described on two sides of a right-angled triangle are together equal to the equilateral triangle on the hypotenuse in terms of their area. **[CBSE 2008]**

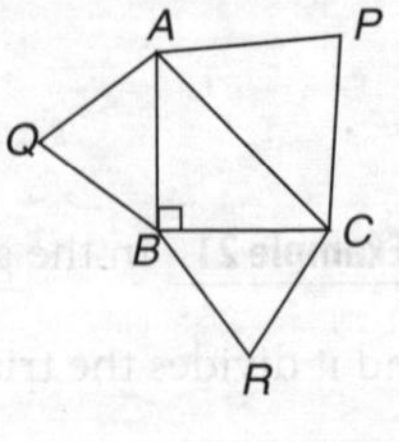

Answer

1. $\frac{AX}{AB} = \frac{\sqrt{2}-1}{\sqrt{2}}$

Concept: Pythagoras' Theorem

State and prove Pythagoras' Theorem. **[CBSE Frequently Asked]**

Statement: In a right triangle, the square of the hypotenuse is equal to the sum of the squares of the other two sides.

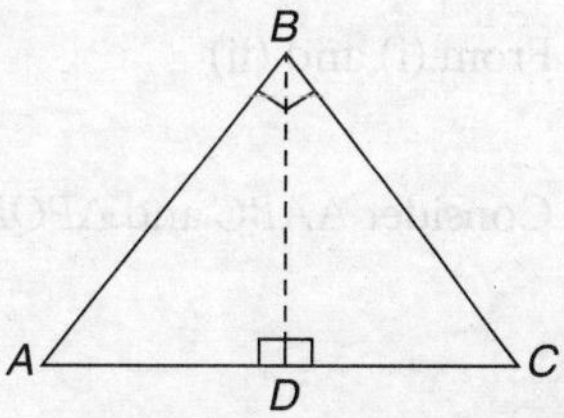

Given: Right ΔABC, right-angled at '*B*'.

To prove: $AC^2 = AB^2 + BC^2$

Construction: Drop $BD \perp AC$.

Proof: Consider ΔADB and ΔABC

$$\angle ADB = \angle ABC \quad [\because \text{ Each } 90°]$$

$$\angle A = \angle A \quad [\text{Common}]$$

$$\Delta ADB \sim \Delta ABC \quad [AA]$$

Hence $\dfrac{AD}{AB} = \dfrac{AB}{AC}$ [∵ If two Δs are ~ then their corresponding sides are in proportion]

$$AB^2 = AD \times AC \quad \ldots\text{(i)}$$

Similarly $\Delta DCB \sim \Delta BCA$ [AA]

$\dfrac{DC}{BC} = \dfrac{CB}{CA}$ [∵ If two Δs are ~ then their corresponding sides are in proportion]

$$BC^2 = CD \times CA \quad \ldots\text{(ii)}$$

Adding (i) and (ii), we get $AB^2 + BC^2 = AD \times AC + CD \times CA = AC\,(AD + CD) = AC \times AC$

$$AB^2 + BC^2 = AC^2$$

Hence proved.

Concept: Converse of Pythagoras' Theorem

State and prove the Converse of Pythagoras' Theorem. **[CBSE Frequently Asked]**

Statement: In a triangle, if the square of one side is equal to the sum of the squares of the other two sides, then the angle opposite the first side is a right angle.

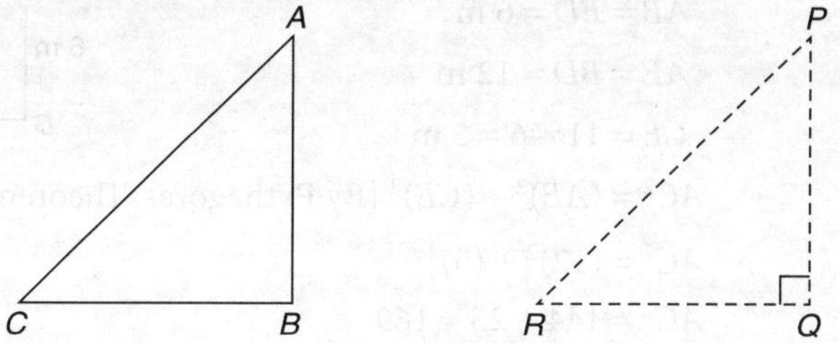

Given: ΔABC in which $AC^2 = AB^2 + BC^2$

To prove: $\angle ABC = 90°$

Construction: Construct a right ΔPQR in which the right angle is at '*Q*' and $PQ = AB$ and $BC = QR$.

Proof: In right ΔABC $AC^2 = AB^2 + BC^2$ [Given] (i)

In right ΔPQR $PR^2 = PQ^2 + QR^2$ [∵ Pythagoras' Theorem]

$PR^2 = AB^2 + BC^2$ [∵ By construction, $PQ = AB$ and $BC = QR$] (ii)

From (i) and (ii) $AC^2 = PR^2$

$AC = PR$ (iii)

Consider ΔABC and ΔPQR $AC = PR$ [From (iii)]

$AB = PQ$ [∵ By construction]

$BC = QR$ [∵ By construction]

$\Delta ABC \cong \Delta PQR$ [SSS]

$\angle\theta = 90°$ [By construction]

$\angle B = \angle Q = 90°$ [CPCT]

Hence, ΔABC is a right Δ with the right angle at B.

Example 22 A ladder is placed against a wall such that its foot is at a distance of 2.5 m from the wall and its top reaches a window 6 m above the ground. Find the length of the ladder.

Solution: *Given:* $AB = 6$ m, $BC = 2.5$ m

To find: Length of ladder, AC

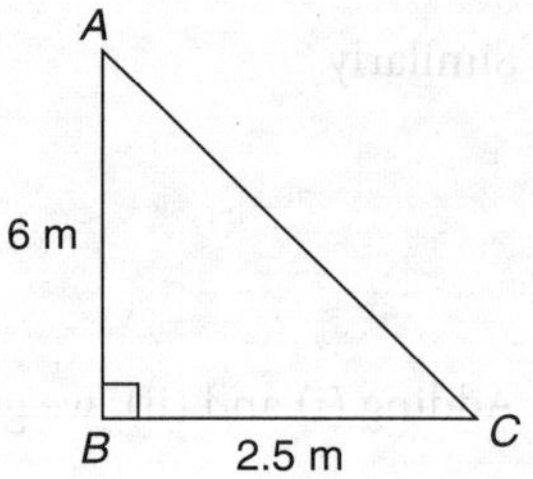

In right ΔABC $AC^2 = AB^2 + BC^2$ [By Pythagoras' Theorem]

$\Rightarrow$ $AC^2 = 6^2 + (2.5)^2 = 36 + 6.25$

$\Rightarrow$ $AC^2 = 42.25$

$AC = \sqrt{42.25} = 6.5$ m

The length of the ladder = 6.5 m.

Example 23 Two poles of height 6 m and 11 m stand on a plane ground. If the distance between their feet is 12 m, find the distance between their tops. **[NCERT, CBSE (CCE) 2009]**

Solution: Given: $AB = 6$ m, $CD = 11$ m, $BD = 12$ m

To find: AC

Construction: Draw $AE \parallel BD$.

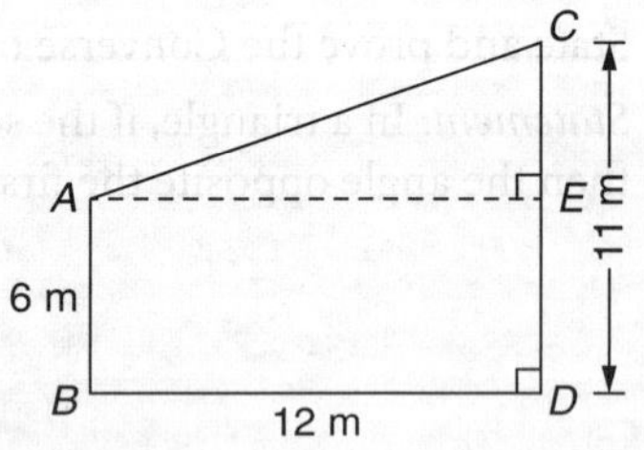

Solution: $AB = ED = 6$ m

$AE = BD = 12$ m

$CE = 11 - 6 = 5$ m

In right ΔAEC $AC^2 = (AE)^2 + (CE)^2$ [By Pythagoras' Theorem]

$\Rightarrow$ $AC^2 = (12)^2 + (5)^2$

$\Rightarrow$ $AC^2 = 144 + 25 = 169$

$\therefore$ $AC = 13$ m

Distance between the tops of the two poles = 13 m.

Example 24 From an airport, two aeroplanes start at the same time. If the speed of the first aeroplane due north is 500 km/h and that of the other due east is 650 km/h, find the distance between two aeroplanes after 2 hours. **[CBSE (CCE) 2016]**

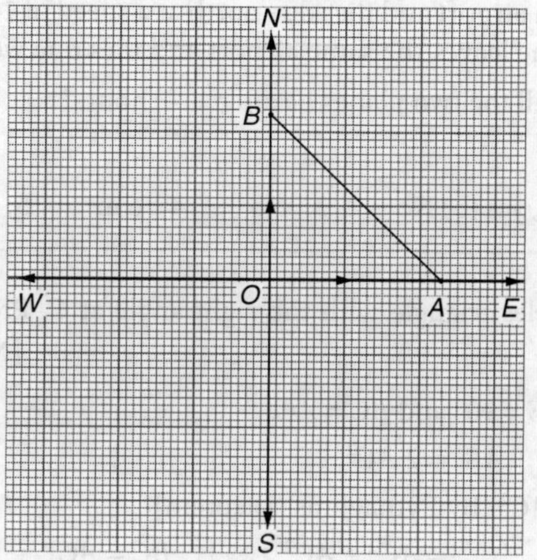

Solution: Speed of aeroplane along north = 500 km/h

Speed of aeroplane along east = 650 km/h

Distance travelled by aeroplane in 2 hours in the north direction

$= OB = 500 \times 2 = 1000$ km

Distance travelled by aeroplane in 2 hours in the east direction

$= OA = 650 \times 2 = 1300$ km

Distance between both aeroplanes after 2 hours = AB

$\therefore \quad AB^2 = OB^2 + OA^2$ [By Pythagoras' Theorem in ΔAOB]

$= (1000)^2 + (1300)^2 = 1000000 + 1690000 = 2690000$

$AB = 100\sqrt{269}$ km

13 Try to Solve

1. A ladder is placed against a wall such that its foot is at a distance of 5 m from the wall and its top reaches a window $5\sqrt{3}$ m above the ground. Find the length of the ladder. **[CBSE (CCE) 2015]**
2. The sides of a triangle are given below. Determine which of them are right triangles. In the case of a right triangle, write the length of its hypotenuse.
   (a) 7 cm, 24 cm and 25 cm (b) 3 cm, 8 cm and 6 cm **[NCERT]**
3. Is a triangle with sides 12 cm, 16 cm and 18 cm a right triangle? Give reasons.
4. A ladder 10 m long reaches a window 8 m above the ground. Find the distance of the foot of the ladder from the base of the wall. **[NCERT]**

5. A guy wire attached to a vertical pole of height 18 m is 24 m long and has a stake attached to the other end. How far from the base of the pole should the stake be driven so that the wire will be taut? **[NCERT]**

6. An aeroplane leaves an airport and flies due north at a speed of 1000 km per hour. At the same time, another aeroplane leaves the same airport and flies due west at a speed of 1200 km per hour. How far apart will the two planes be after $1\frac{1}{2}$ hours? **[NCERT]**

Answers

1. 10 m **2.** (a) 25 cm, yes (b) No **3.** No **4.** 6 m **5.** $6\sqrt{7}$ m **6.** $300\sqrt{61}$ km

Example 25 *ABC* is an isosceles triangle with the right angle at *C*. Prove that $AB^2 = 2AC^2$. **[NCERT]**

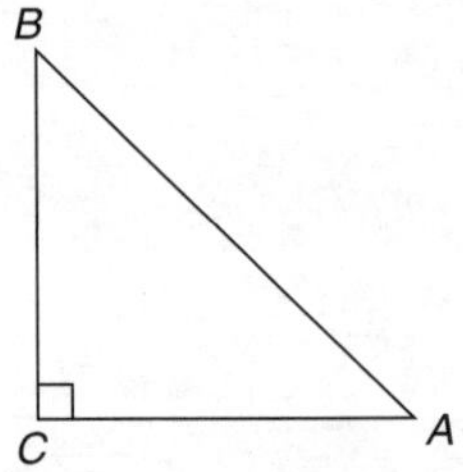

Solution: *Given*: In ΔBCA, $\angle C = 90°$ and $BC = AC$.

To prove: $AB^2 = 2\,AC^2$

Proof: In ΔBCA, $AB^2 = BC^2 + AC^2$ [By Pythagoras' Theorem]

$= AC^2 + AC^2$ [$\therefore BC = AC$]

$= 2AC^2$

Example 26 *ABC* is an isosceles triangle with $AC = BC$. If $AB^2 = 2AC^2$, prove that *ABC* is a right triangle.

Solution: *Given:* In ΔABC, $AC = BC$ **[NCERT]**

$AB^2 = 2AC^2$

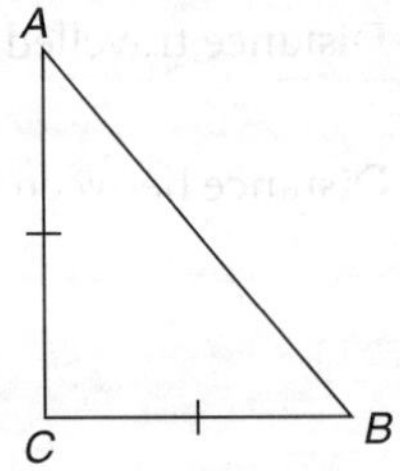

To prove: ABC is a right Δ.

Proof: $AB^2 = 2AC^2$ [Given]

$(AB)^2 = AC^2 + AC^2$

$(AB)^2 = (AC)^2 + (BC)^2$ [$\because AC = BC$]

$\therefore$ $\angle ACB = 90°$ [By Converse of Pythagoras' Theorem]

Example 27 Prove that the sum of the squares of the sides of a rhombus is equal to the sum of the squares of its diagonals. **[NCERT, CBSE (CCE) 2012]**

Solution: *Given: ABCD* is a rhombus.

To prove: $AB^2 + BC^2 + CD^2 + AD^2 = AC^2 + BD^2$

Proof: We know that the diagonals of a rhombus bisect each other at right angles.

$\therefore$ $OA = OC$ and $OB = OD$

$\angle AOB = \angle BOC = \angle COD = \angle AOD = 90°$

In ΔAOB, $AB^2 = AO^2 + BO^2$ [$\because$ Pythagoras' Theorem]

$\Rightarrow$ $AB^2 = \left(\frac{1}{2}AC\right)^2 + \left(\frac{1}{2}BD\right)^2$

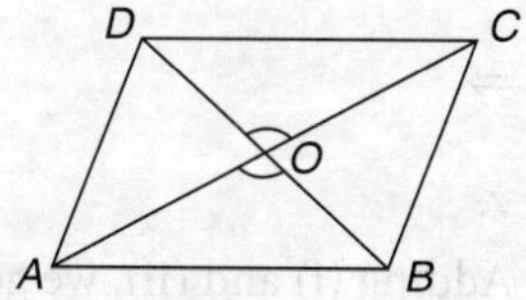

$\Rightarrow \quad AB^2 = \frac{1}{4} AC^2 + \frac{1}{4} BD^2$

$\Rightarrow \quad 4AB^2 = AC^2 + BD^2$

$\Rightarrow \quad AB^2 + AB^2 + AB^2 + AB^2 = AC^2 + BD^2$

$\therefore \quad AB^2 + BC^2 + CD^2 + DA^2 = AC^2 + BD^2$ (Since all sides of a rhombus are equal)

14 Try to Solve

1. The lengths of the diagonals of a rhombus are 24 cm and 10 cm. Find each side of the rhombus.
2. In the given figure, if $AD \perp BC$, prove that $AB^2 + CD^2 = BD^2 + AC^2$. **[CBSE 2005, 09]**

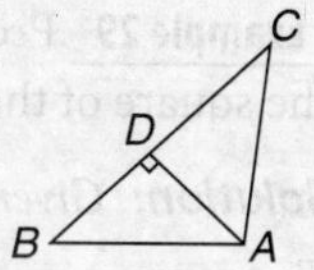

Answer

1. $AB = BC = CD = DA = 13$ cm

Example 28 BL and CM are medians of a ΔABC with right angles at A. Prove that $4(BL^2 + CM^2) = 5\,BC^2$.

[NCERT, CBSE (CCE) 2015]

Solution: *Given:* In ΔABC, BL and CM are medians.

To prove: $4(BL^2 + CM^2) = 5\,BC^2$

Proof: In right ΔBAC

$BC^2 = AB^2 + AC^2$ [By Pythagoras' Theorem]

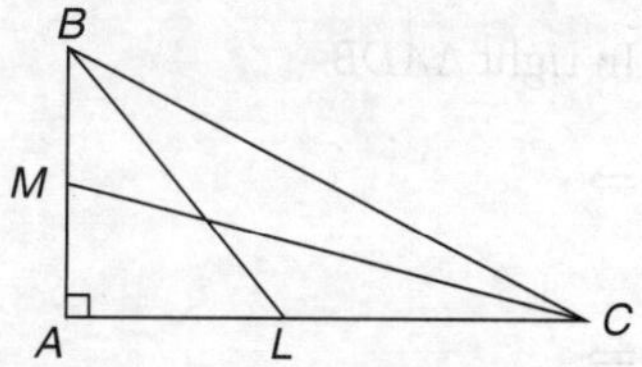

In right ΔBAL

$BL^2 = AB^2 + AL^2$ [By Pythagoras' Theorem]

$\Rightarrow \quad BL^2 = AB^2 + \left(\frac{AC}{2}\right)^2$ [$\because$ L is the mid-point of AC]

$\Rightarrow \quad BL^2 = AB^2 + \frac{AC^2}{4}$

$\Rightarrow \quad BL^2 = \frac{4AB^2 + AC^2}{4}$

$\therefore \quad 4BL^2 = 4AB^2 + AC^2 \quad \ldots$(i)

In right ΔCAM $\quad CM^2 = AC^2 + AM^2$ [By Pythagoras' Theorem]

$\Rightarrow \quad CM^2 = AC^2 + \left(\frac{AB}{2}\right)^2$ [$\because$ M is the mid-point of AB]

$\Rightarrow \quad CM^2 = AC^2 + \frac{AB^2}{4}$

$\Rightarrow$ $$CM^2 = \frac{4AC^2 + AB^2}{4}$$

$\therefore$ $$4CM^2 = 4AC^2 + AB^2 \quad \ldots\text{(ii)}$$

Adding (i) and (ii), we get $$4\,BL^2 + 4CM^2 = 4AB^2 + AC^2 + 4AC^2 + AB^2$$

$\Rightarrow$ $$4\,(BL^2 + CM^2) = 5\,AB^2 + 5\,AC^2$$

$\Rightarrow$ $$4\,(BL^2 + CM^2) = 5\,(AB^2 + AC^2)$$

$\Rightarrow$ $$4\,(BL^2 + CM^2) = 5\,BC^2 \quad [\because AB^2 + AC^2 = BC^2]$$

Example 29 Prove that three times the square of any side of an equilateral triangle is equal to four times the square of the altitude. **[NCERT, CBSE 2008]**

Solution: *Given*: ABC is an equilateral Δ. $AD \perp BC$.

To prove: $3AB^2 = 4AD^2$

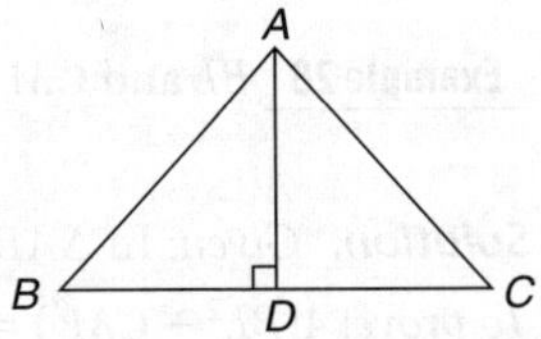

Proof: In ΔADB and ΔADC

$AB = AC$ [$\because$ ΔABC is equilateral]

$AD = AD$ [Common]

$\angle ADB = \angle ADC$ [Each 90°]

$\therefore$ $\Delta ADB \cong \Delta ADC$ [By RHS]

So $BD = CD$ [By CPCT]

$\Rightarrow$ $$BD = \frac{1}{2}\,BC = \frac{1}{2}\,AB \quad [\because BC = AB] \quad \ldots\text{(i)}$$

In right ΔADB $$AB^2 = AD^2 + BD^2$$

$\Rightarrow$ $$AB^2 = AD^2 + \left(\frac{1}{2}AB\right)^2 \quad \text{[From (i)]}$$

$\Rightarrow$ $$AB^2 = AD^2 + \frac{AB^2}{4}$$

$\Rightarrow$ $$AB^2 - \frac{AB^2}{4} = AD^2$$

$$\frac{4AB^2 - AB^2}{4} = AD^2$$

$$3AB^2 = 4AD^2$$

Example 30 The perpendicular from A on side BC of a ΔABC intersects BC at D such that $DB = 3\,CD$ (see figure). Prove that $2AB^2 = 2AC^2 + BC^2$. **[NCERT, CBSE (CCE) 2013, Chennai 2019]**

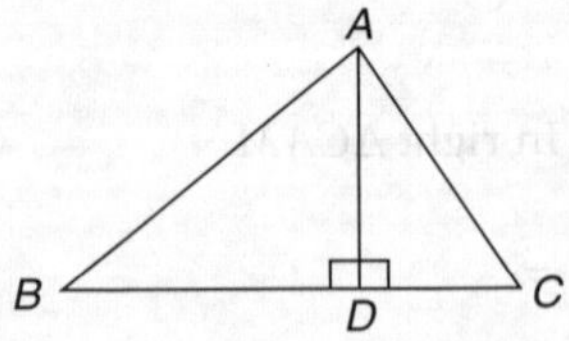

Solution: Since $DB = 3\,CD$, $4CD = BC$ and $CD = \frac{1}{4}\,BC$

In ΔADB $$AB^2 = AD^2 + BD^2 = AD^2 + (BC - DC)^2$$
$$= AD^2 + BC^2 + DC^2 - 2BC \cdot DC$$

$$= (AD^2 + DC^2) + BC^2 - 2BC \cdot DC$$

$$= AC^2 + BC^2 - 2BC \cdot \frac{1}{4} BC \quad [\text{Since in right } \Delta ADC,\ AC^2 = AD^2 + DC^2]$$

$$AB^2 = AC^2 + BC^2 - \frac{1}{2} BC^2$$

$$\Rightarrow \quad 2AB^2 = 2AC^2 + 2BC^2 - BC^2 = 2AC^2 + BC^2$$

Hence $2AB^2 = 2AC^2 + BC^2$

Example 31 In an equilateral triangle *ABC*, *D* is a point on side *BC* such that $BD = \frac{1}{3} BC$. Prove that $9\ AD^2 = 7\ AB^2$. **[NCERT, CBSE Frequently Asked]**

Solution: *Given*: In ΔABC, $AB = BC = AC$

$$BD = \frac{1}{3} BC$$

To prove: $9AD^2 = 7AB^2$

Construction: Draw $AP \perp BC$.

Proof: ΔABC is an equilateral Δ, $AP \perp BC$.

[∵ ⊥ drawn from the vertex of equilateral Δ bisects the base]

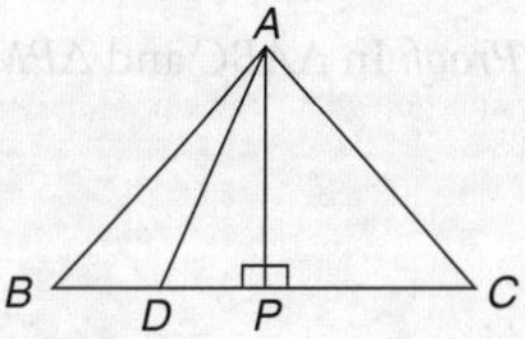

$$BP = \frac{1}{2} BC, \quad PC = \frac{1}{2} BC$$

$$BD = \frac{1}{3} BC, \quad DC = \frac{2}{3} = BC$$

$$\therefore \quad DP = BP - BD = \frac{1}{2} BC - \frac{1}{3} BC = \frac{1}{6} BC$$

In right ΔAPB $\quad AB^2 = BP^2 + AP^2 = BP^2 + AD^2 - DP^2$ [∵ Pythogoras' Theorem]

$$= \left(\frac{1}{2} BC\right)^2 + AD^2 - \left(\frac{1}{6} BC\right)^2$$

$$AB^2 = \frac{1}{4} BC^2 + AD^2 - \frac{1}{36} BC^2 \quad [\because AB = BC = AC]$$

$$AB^2 = \frac{9AB^2 + 36AD^2 - AB^2}{36}$$

$$36AB^2 = 9AB^2 + 36\ AD^2 - AB^2$$

$$36AB^2 = 8AB^2 + 36\ AD^2$$

$$\Rightarrow \quad (36 - 8)\ AB^2 = 36\ AD^2$$

$$\Rightarrow \quad 28\ AB^2 = 36\ AD^2$$

Dividing both sides by 4, we get

$$7AB^2 = 9\ AD^2$$

FOR MORE PRACTICE AND KNOWLEDGE

Theorem 1 (AAA Similarity) If in two triangles, the corresponding angles are equal, then the triangles are similar.

Solution: *Given:* ABC and PQR are two triangles in which $\angle A = P$, $\angle B = Q$ and $\angle C = R$.

To prove: $\Delta ABC \sim \Delta PQR$

Construction: In ΔPQR, cut $PM = AB$, $PN = AC$ and join MN.

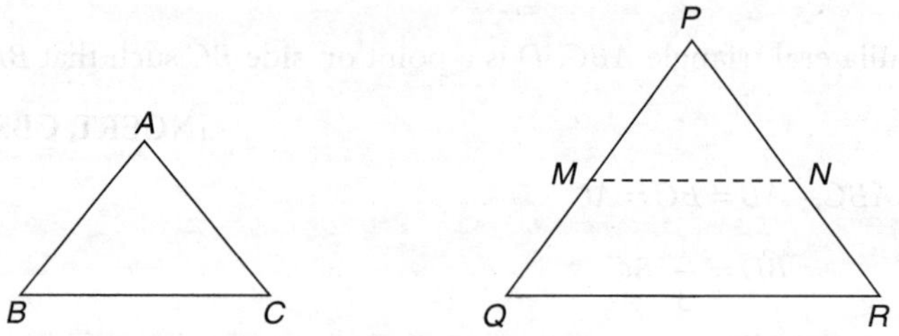

Proof: In ΔABC and ΔPMN $AB = PM$ [By construction]

$\angle A = P$ [Given]

$AC = PN$ [By construction]

$\therefore$ $\Delta ABC \cong \Delta PMN$ [By SAS]

So $\angle B = \angle M$ [By CPCT] ...(i)

$\angle B = Q$ [Given] ...(ii)

$\therefore$ $\angle M = Q$ [From (i) and (ii)]

$\angle M$ and Q are corresponding angles and are equal.

$\therefore$ $MN \parallel QR$

In ΔPQR $MN \parallel QR$

$$\frac{PM}{PQ} = \frac{PN}{PR} \quad \text{[By BPT]}$$

$\therefore$ $\dfrac{AB}{PQ} = \dfrac{AC}{PR}$ [$\because$ $PM = AB$ and $PN = AC$] [By construction] ...(iii)

Similarly, we can prove that $\dfrac{AB}{PQ} = \dfrac{BC}{QR}$ ...(iv)

From (iii) and (iv), we get $\dfrac{AB}{PQ} = \dfrac{BC}{QR} = \dfrac{AC}{PR}$

Thus, in ΔABC and ΔPQR $\angle A = P$, $\angle B = Q$, $\angle C = R$ and $\dfrac{AB}{PQ} = \dfrac{BC}{QR} = \dfrac{AC}{PR}$

$\Delta ABC \sim \Delta PQR$

Corollary (AA Similarity): If two angles of one triangle are, respectively, equal to two angles of another triangle, then the two triangles are similar.

Proof: In ΔABC and ΔPQR, $\angle A = P$, $\angle B = Q$. Then, by angle sum property, $\angle C = R$.

Thus, ΔABC and ΔPQR are equiangular, and hence, similar.

Theorem 2 (SSS Similarity) If the corresponding sides of two triangles are proportional, then they are similar.

Solution: *Given:* In ΔABC and ΔPQR $\dfrac{AB}{PQ} = \dfrac{BC}{QR} = \dfrac{AC}{PR}$

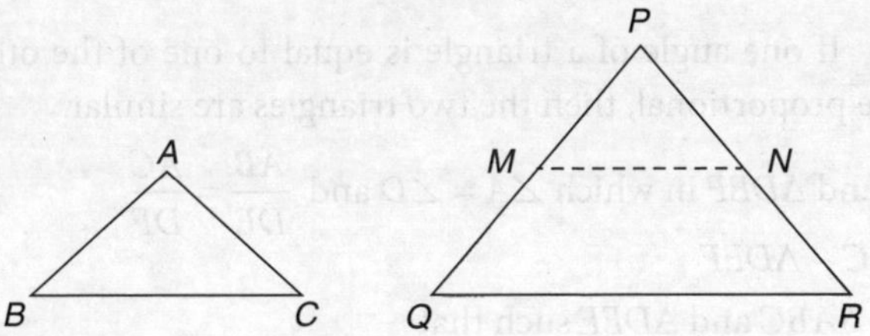

To prove: $\Delta ABC \sim \Delta PQR$

Construction: Mark points M and N on PQ and PR, respectively, such that $PM = AB$ and $PN = AC$. Join MN.

Proof: In ΔABC and ΔPQR $\dfrac{AB}{PQ} = \dfrac{AC}{PR}$ [Given]

$\Rightarrow$ $\dfrac{PM}{PQ} = \dfrac{PN}{PR}$ [$\because AB = PM$ and $AC = PN$]

$\therefore$ $MN \,||\, QR$ [By converse of BPT]

$\Rightarrow$ $PMN = Q$ and $PNM = R$ [Corresponding angles]

In ΔPMN and ΔPQR $PMN = Q$ and $PNM = R$

$\therefore$ $\Delta PMN \sim \Delta PQR$ [By AA Similarity]

$\Rightarrow$ $\dfrac{PM}{PQ} = \dfrac{MN}{QR}$

$\Rightarrow$ $\dfrac{AB}{PQ} = \dfrac{MN}{QR}$ [$\because AB = PM$ by construction] ...(i)

But $\dfrac{AB}{PQ} = \dfrac{BC}{QR}$ ...(ii)

From (i) and (ii), we get $\dfrac{MN}{QR} = \dfrac{BC}{QR}$

$\Rightarrow$ $MN = BC$

Thus, in ΔABC and ΔPMN $AB = PM, AC = PN$ [By construction]

And $BC = MN$ [Proved above]

$\Delta ABC \cong \Delta PMN$ [By SSS]

So $\angle A = P, \angle B = \angle M, \angle C = N$ [By CPCT]

$\angle M = Q$ and $N = R$ [As $MN \parallel QR$]

$\therefore$ $\angle A = P, \angle B = Q$ and $\angle C = R$

So, ΔABC and ΔPQR are equiangular, and hence, they are similar.

Theorem 3 (SAS Similarity) If one angle of a triangle is equal to one of the other triangle and the sides including these angles are proportional, then the two triangles are similar.

Solution: *Given:* ΔABC and ΔDEF in which $\angle A = \angle D$ and $\dfrac{AB}{DE} = \dfrac{AC}{DF}$

To prove: $\Delta ABC \sim \Delta DEF$

Construction: Let us take ΔABC and ΔDEF such that

$$\frac{AB}{DE} = \frac{AC}{DF} (< 1) \text{ and } \angle A = \angle D.$$

Cut $DP = AB$ and $DQ = AC$. Join PQ.

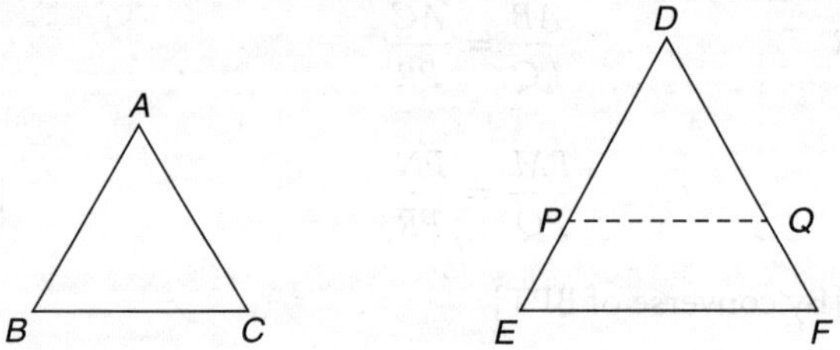

Proof: In ΔABC and ΔDPQ, we have

$AB = DP$ [By construction]

$\angle A = \angle D$ [Given]

$AC = DQ$ [By contraction]

$\therefore \Delta ABC \cong \Delta DPQ$ [By SAS Congruence]

$\therefore$ $\angle A = \angle D, \angle B = \angle P$ and $\angle C = \angle Q$

Now $\dfrac{AB}{DE} = \dfrac{AC}{DF}$ [Given]

$\Rightarrow$ $\dfrac{DP}{DE} = \dfrac{DQ}{DF}$ [$AB = DP$ and $AC = DQ$ by construction]

$\Rightarrow$ $PQ \parallel EF$ [By the converse of Thales' Theorem]

$\Rightarrow$ $\angle P = \angle E$ and $\angle Q = \angle F$ [Corresponding $\angle$s]

$\therefore$ $\angle A = \angle D, \angle B = \angle P = \angle E$ and $\angle C = \angle Q = \angle F$

Thus $\angle A = \angle D, \angle B = \angle E$ and $\angle C = \angle F.$

So, the given triangles are equiangular, and hence, similar.

Theorem 4 (Converse of Basic Proportionality Theorem) If a line divides any two sides of a triangle in the same ratio, then the line is parallel to the third side.

Solution: *Given:* In ΔABC, line DE intersects AB and AC at D and E, respectively, such that $\frac{AD}{DB} = \frac{AE}{EC}$.

To prove: $DE \parallel BC$

Construction: If possible, let DE not be parallel to BC but $DE' \parallel BC$.

Proof: In ΔABC $DE' \parallel BC$

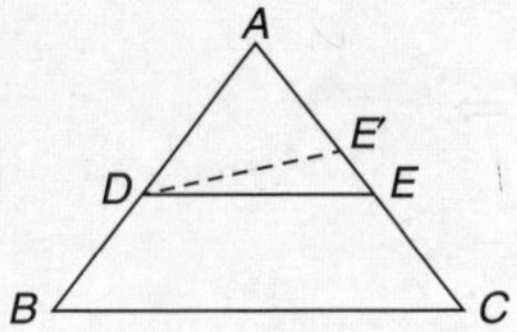

$\therefore$ $\frac{AD}{DB} = \frac{AE'}{E'C}$ [By BPT] ...(i)

But $\frac{AD}{DB} = \frac{AE}{EC}$ [Given] ...(ii)

From (i) and (ii), we get $\frac{AE'}{E'C} = \frac{AE}{EC}$

$\Rightarrow$ $\frac{AE'}{E'C} + 1 = \frac{AE}{EC} + 1$ [Adding 1 on both sides]

$\Rightarrow$ $\frac{AE' + E'C}{E'C} = \frac{AE + EC}{EC}$

$\Rightarrow$ $\frac{AC}{E'C} = \frac{AC}{EC}$

$\Rightarrow$ $\frac{1}{E'C} = \frac{1}{EC}$

$\Rightarrow$ $E'C = EC$

This is possible only if E' and E coincide.

Hence $DE \parallel BC$

Theorem 5 (Internal and External Bisectors of an Angle of a Triangle) The internal bisectors of an angle of a triangle divide the opposite side internally in the ratio of the sides containing the angle.

Solution: *Given:* ABC is a Δ in which AD is the bisector of $\angle A$.

That is $1 = 2.$

To prove: $\frac{BD}{DC} = \frac{AB}{AC}$

Construction: Draw $CE \parallel DA$ which meets BA produced at E.

Proof: $DA \parallel CE$ and AC is the transversal.

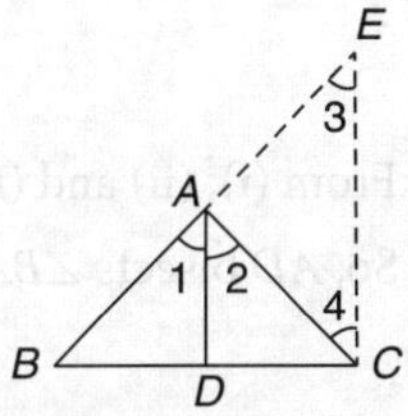

$2 = \angle 4$ [Alternate interior angles] ...(i)

$DA \parallel CE$ and BE is the transversal.

$\therefore$ $1 = 3$ [Corresponding angles] ...(ii)

$1 = 2$ [Given] ...(iii)

From (i), (ii) and (iii), we get $3 = \angle 4$

In ΔAEC $3 = \angle 4$ [From (iv)]

$AC = AE$ [$\because$ Sides opposite to equal angles are equal]

Now, in ΔBCE, $DA \parallel CE$ $\frac{BD}{DC} = \frac{BA}{AE}$ [By BPT]

$\Rightarrow$ $\frac{BD}{DC} = \frac{BA}{AC}$ [$\because AE = AC$]

$\therefore$ $\frac{BD}{DC} = \frac{AB}{AC}$

Theorem 6 If a line through one vertex of a triangle divides the opposite side in the ratio of the other two sides, then the line bisects the angle at the vertex.

Solution: *Given*: ABC is a Δ in which D is a point on BC such that $\frac{BD}{DC} = \frac{AB}{AC}$.

To prove: AD is the bisector of $\angle BAC$, that is, $1 = 2$.

Construction: Extend BA to E such that $AC = AE$. Join EC.

Proof: In ΔAEC, $AE = AC$ (By construction)

$3 = \angle 4$ [$\because$ In a Δ, the angles opposite the equal sides are equal]

$\frac{BD}{DC} = \frac{AB}{AC}$ [Given]

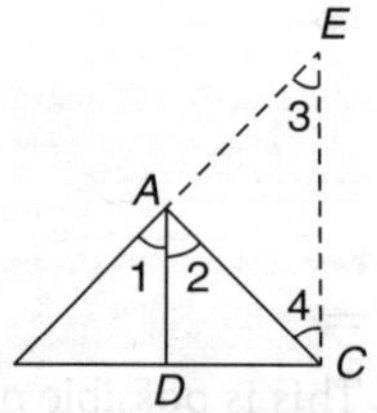

$\Rightarrow$ $\frac{BD}{DC} = \frac{AB}{AE}$ [$\because AC = AE$]

In ΔBCE $\frac{BD}{DC} = \frac{BA}{AE}$

$\therefore$ $DA \parallel CE$ [By converse of BPT]

Now, $DA \parallel CE$ and BE is the transversal.

$\therefore$ $1 = 3$ [Corresponding angles] ...(i)

$DA \parallel CE$ and AC is the transversal.

$\therefore$ $2 = \angle 4$ [Alternate interior angles] ...(ii)

$3 = \angle 4$ [Proved above] ...(iii)

From (i), (ii) and (iii), we get $1 = 2$

So, AD bisects $\angle BAC$.

Theorem 7 The external bisector of an angle of a triangle divides the opposite side externally in the ratio of the sides containing the angle.

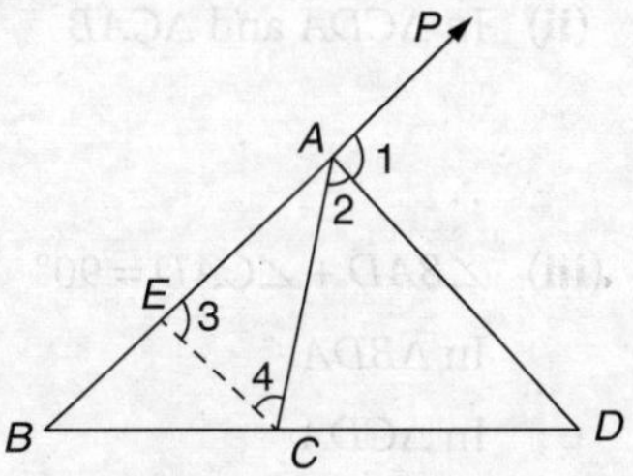

Solution: *Given*: ABC is a Δ in which AP is the bisector of exterior PAC, that is, $1 = 2$.

To prove: $\frac{BD}{CD} = \frac{AB}{AC}$

Construction: Draw $CE \parallel DA$ which meets AB at E.

Proof: $CE \parallel DA$ and BP is the transversal.

$1 = 3$ [Corresponding angles] ...(i)

$CE \parallel DA$ and AC is the transversal.

$2 = \angle 4$ [Alternate interior angles] ...(ii)

$1 = 2$ [Given] ...(iii)

From (i), (ii) and (iii) we get $3 = \angle 4$ ...(iv)

In ΔAEC $3 = \angle 4$ [From (iv)]

$AE = AC$ [$\because$ Sides opposite equal angles are equal] ...(v)

In ΔABD $CE \parallel DA$

$\therefore$ $\frac{BD}{CD} = \frac{BA}{EA}$ [By BPT]

$\Rightarrow$ $\frac{BD}{CD} = \frac{BA}{AC}$ [From (v)]

$\therefore$ $\frac{BD}{CD} = \frac{AB}{AC}$

Theorem 8 If a perpendicular is drawn from the vertex of the right angle of a right triangle to the hypotenuse, then the triangles on both sides of the perpendicular are similar to the whole triangle and also to each other.

Solution: *Given*: ABC is a Δ in which $\angle A = 90°$ and $AD \perp BC$.

To prove:

(i) $\Delta BDA \sim \Delta BAC$

(ii) $\Delta CDA \sim \Delta CAB$

(iii) $\Delta BDA \sim \Delta ADC$

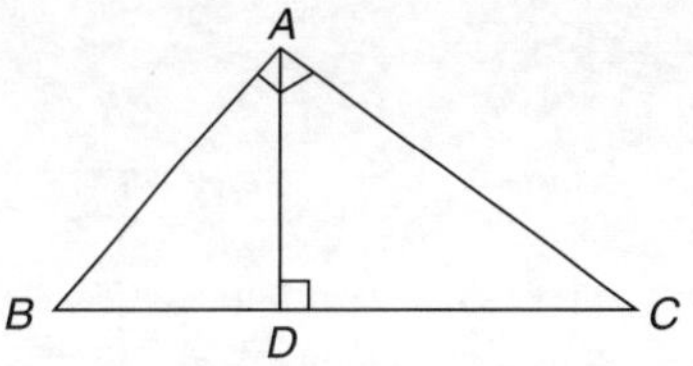

Proof:

(i) In ΔBDA and ΔBAC $\angle BDA = \angle BAC$ [Each 90°]

$\angle ABD = \angle CBA$ [Common]

$\therefore$ $\Delta BDA \sim \Delta BAC$ [By AA Similarity]

(ii) In ΔCDA and ΔCAB $\quad \angle CDA = \angle CAB$ [Each 90°]

$\angle ACD = \angle BCA$ [Common]

$\therefore \quad \Delta CDA \sim \Delta CAB$ [By AA Similarity]

(iii) $\angle BAD + \angle CAD = 90°$ ...(i)

In ΔBDA $\quad \angle B + \angle BAD = 90°$ $\quad [\because \angle BDA = 90°]$ ...(ii)

In ΔCDA $\quad \angle C + \angle CAD = 90°$ $\quad [\because \angle CDA = 90°]$ ...(iii)

From (i) and (ii), we get $\angle B + \angle BAD = \angle BAD + \angle CAD$

$\angle B = \angle CAD$

From (i) and (iii), we get $\angle C + \angle CAD = \angle BAD + \angle CAD$

$\angle C = \angle BAD$

In ΔBDA and ΔADC $\quad \angle B = \angle CAD$ [Proved above]

$\angle BAD = \angle C$ [Proved above]

$\therefore \quad \Delta BDA \sim \Delta ADC$ [By AA Similarity]

Example 32 Two poles of height 'a' and 'b' ($b > a$) are 'c' metres apart. Prove that the height 'h' in metres of the point of intersection of the lines joining the top of each pole to the foot of the opposite pole is $\dfrac{ab}{a+b}$.

Solution: *Given:* Two poles $AB = a$ and $CD = b$, ($b > a$). $BD = c$, AD and BC, intersect at E (say).

To find: EF

Consider Δs ABD and EFD $\quad \angle ABD = \angle EFD$ [Each 90°]

And $\quad \angle ADB = \angle EDF$ [Common]

$\therefore \quad \Delta ABD \sim \Delta EFD$ [AA Similarity]

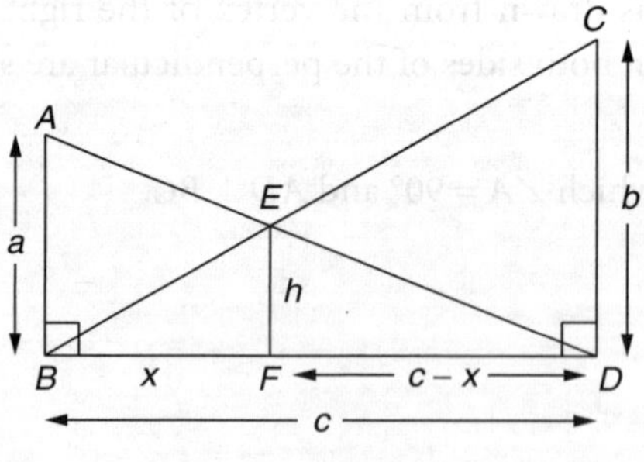

$$\therefore \quad \frac{EF}{AB} = \frac{FD}{BD}$$

$$\Rightarrow \quad \frac{h}{a} = \frac{c-x}{c}. \quad \text{...(i)}$$

Similarly, $\Delta CDB \sim \Delta EFB$ $\quad \dfrac{EF}{CD} = \dfrac{BF}{BD}$ [AA Similarity]

$\Rightarrow \quad \dfrac{h}{b} = \dfrac{x}{c}$

$\Rightarrow \quad x = \dfrac{hc}{b}$

Substituting in (i), we get $\quad \dfrac{h}{a} = \dfrac{c - \dfrac{hc}{b}}{c}$

$\Rightarrow \quad \dfrac{h}{a} = \dfrac{\dfrac{bc - hc}{b}}{c}$

$\Rightarrow \quad \dfrac{h}{a} = \dfrac{b-h}{b}$

$\Rightarrow \quad bh = ab - ah$

$\Rightarrow \quad (a+b)h = ab$

$\Rightarrow \quad h = \dfrac{ab}{a+b}.$

Example 33 In the figure, $AB \parallel PQ \parallel CD$, $AB = x$ units, $CD = y$ units and $PQ = z$ units. Prove that $\dfrac{1}{x} + \dfrac{1}{y} = \dfrac{1}{z}$. **[CBSE (CCE) 2011]**

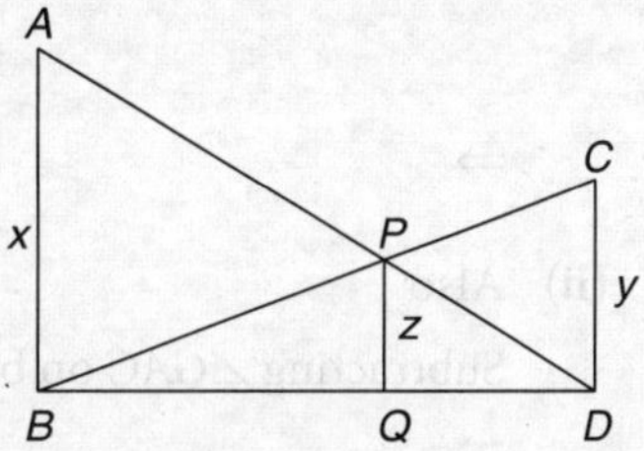

Solution: In ΔPQD and ΔABD, $PQ \parallel AB$.

$\angle 6 = \angle 6$ [Common]

$\angle 1 = \angle 2$ [$\because$ Corresponding angles]

$\Delta PQD \sim \Delta ABD$ [AA]

$\dfrac{PQ}{AB} = \dfrac{QD}{BD}$ [$\therefore$ If two Δs are similar, then their corresponding sides are in proportion]

$\dfrac{z}{x} = \dfrac{b}{a+b}$ ...(i)

Similarly $\quad \Delta PQB \sim \Delta CDB$

So $\quad \dfrac{PQ}{CD} = \dfrac{QB}{DB}$ [$\therefore$ If two Δs are similar, then their corresponding sides are in proportion]

$$\frac{z}{y} = \frac{a}{a+b} \quad \ldots\textbf{(ii)}$$

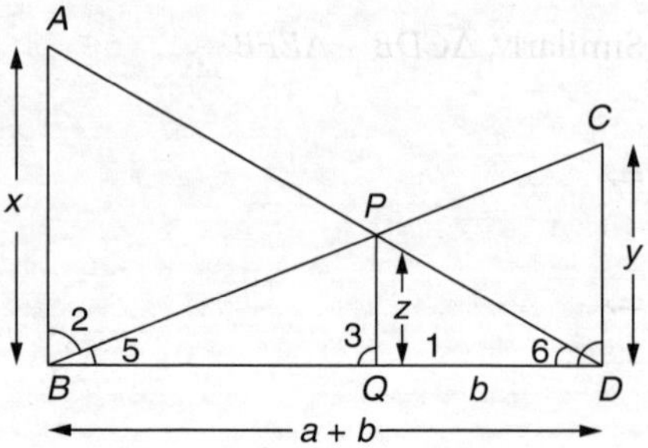

Adding (i) and (ii), we get $\frac{z}{x} + \frac{z}{y} = \frac{b}{a+b} + \frac{a}{a+b}$

$$z\left[\frac{1}{x} + \frac{1}{y}\right] = \frac{b+a}{a+b} = 1$$

$$\frac{1}{x} + \frac{1}{y} = \frac{1}{z}$$

Example 34 In the given figure, *ABCD* and *AEFG* are squares. Prove that:

(i) $\frac{AF}{AG} = \frac{AC}{AD}$ **(ii)** $\Delta ACF \sim \Delta ADG$

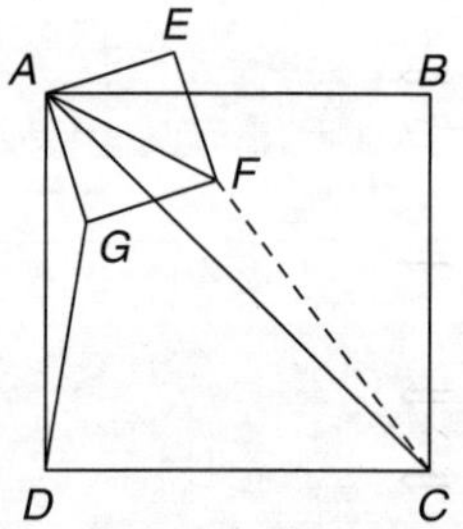

Solution:

(i) In ΔAGF and ΔADC $\angle AGF = \angle ADC$ [Each 90°]

$\angle GAF = \angle DAC$ [Each 45°]

$\therefore$ $\Delta AGF \sim \Delta ADC$ [AA Similarity]

$\Rightarrow$ $\frac{AG}{AD} = \frac{AF}{AC}$

[$\because$ If two Δs are similar, then their corresponding sides are in proportion] ...**(i)**

$\Rightarrow$ $\frac{AC}{AD} = \frac{AF}{AG}$ [Using Alternando] ...**(ii)**

(ii) Also $\angle DAC = \angle GAF$ [Each 45°]

Subtracting $\angle GAC$ on both sides, we get

$\Rightarrow$ $\angle DAC - \angle GAC = \angle GAF - \angle GAC$

$\Rightarrow$ $\angle DAG = \angle CAF$ ...**(iii)**

In ΔACF and ΔADG $\frac{AG}{AD} = \frac{AF}{AC}$ [From (i)]

$\angle DAG = \angle CAF$ [From (iii)]

$\therefore$ $\Delta ACF \sim \Delta ADG$ [SAS]

Example 35 In the figure, $\Delta FEC \cong \Delta GBD$ and $\angle 1 = \angle 2$. Prove that $\Delta ADE \sim \Delta ABC$.

Solution: *Given:* $\Delta FEC \cong \Delta G$ and $1 = 2$

To prove: $\Delta ADE \sim \Delta ABC$

Proof: In ΔADE $\angle 1 = \angle 2$

$\Rightarrow$ $AE = AD$ [Sides opposite equal angles are equal] ...**(i)**

$\Delta FEC \cong \Delta GBD$ [Given]

$$EC = BD \quad \text{[CPCT]} \quad \dots\text{(ii)}$$

Adding (i) and (ii), we get $AE + EC = AD + BD$

$$AC = AB \quad \dots\text{(iii)}$$

Dividing (i) by (iii) $\quad \dfrac{AD}{AB} = \dfrac{AE}{AC}$

And $\quad \angle A = \angle A \quad$ [Common]

$\therefore \quad \Delta ADE \sim \Delta ABC \quad$ [SAS Similarity]

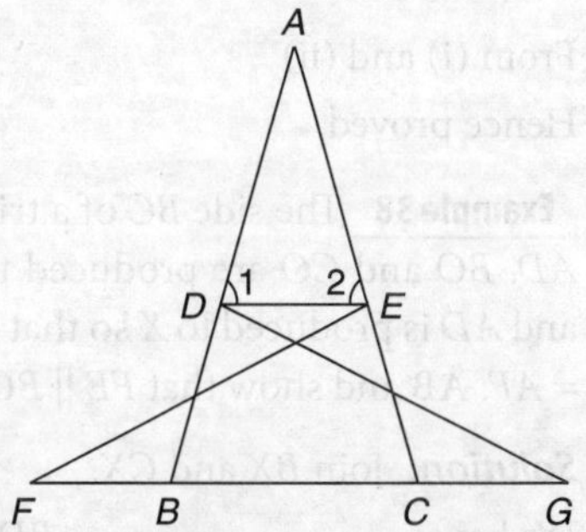

Example 36 In the figure, if $AB \parallel DC$ and AC and PQ intersect each other at the point O, prove that $OA \,.\, CQ = OC \,.\, AP$.

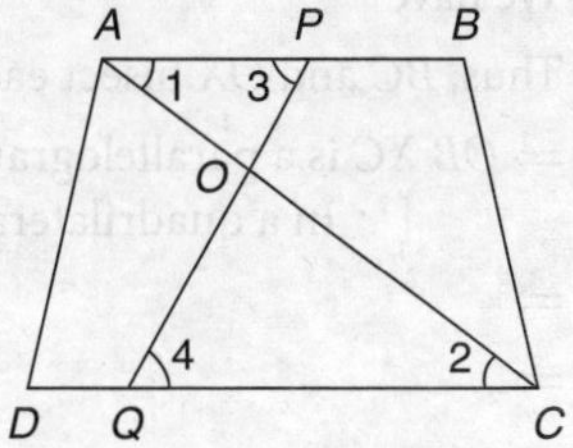

Solution: $\quad AB \parallel DC$

In ΔAPO and $\Delta CQO \quad \angle 1 = \angle 2 \quad$ [Alternate interior angles]

$\angle 3 = \angle 4 \quad$ [Alternate interior angles]

$\Delta APO \sim \angle CQO \quad$ [AA]

$\dfrac{OA}{OC} = \dfrac{AP}{CQ} \quad$ [If two Δs are similar, then their corresponding sides are in proportion]

$\Rightarrow \quad OA.CQ = OC.AP$

Example 37 P is the mid-point of BC of ΔABC and Q is the mid-point of AP. BQ when produced meets AC at L. Prove that $AL = \dfrac{1}{3} AC$. **[CBSE 2006C]**

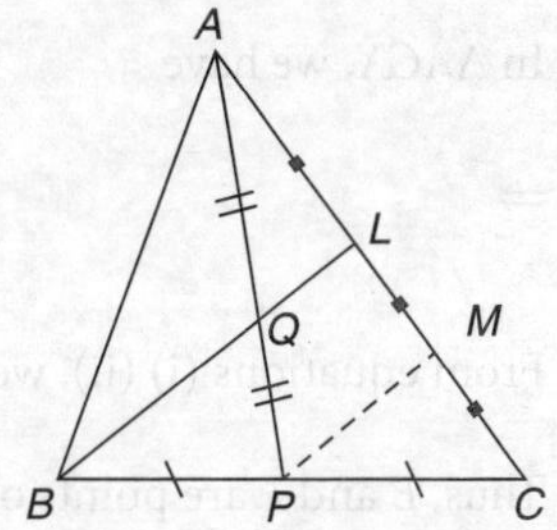

Solution: *Given:* In ΔABC, P is the mid-point of BC and Q the mid-point of AP. BQ when produced meets AC at L.

To prove: $\quad AL = \dfrac{1}{3} AC$

Construction: Draw $PM \parallel BL$ meeting AC at M.

Proof: In ΔBLC, $PM \parallel BL$

$\Rightarrow \quad \dfrac{BP}{PC} = \dfrac{LM}{MC} \quad$ [Basic Proportionality Theorem]

But $\quad BP = PC \quad$ [Given]

$\Rightarrow \quad \dfrac{LM}{MC} = 1 \Rightarrow LM = MC \quad \dots\text{(i)}$

Consider ΔAPM, Q is the mid-point of AP, and $QL \parallel PM$

$\therefore \quad \dfrac{AQ}{QP} = \dfrac{AL}{LM}$ [Basic Proportionality Theorem]

But $\quad AQ = QP \quad$ [Given]

$\Rightarrow \quad \dfrac{AL}{LM} = 1 \Rightarrow AL = LM \quad \dots\text{(ii)}$

From (i) and (ii) $AL = LM = MC \Rightarrow AL = \frac{1}{3}AC.$

Hence proved.

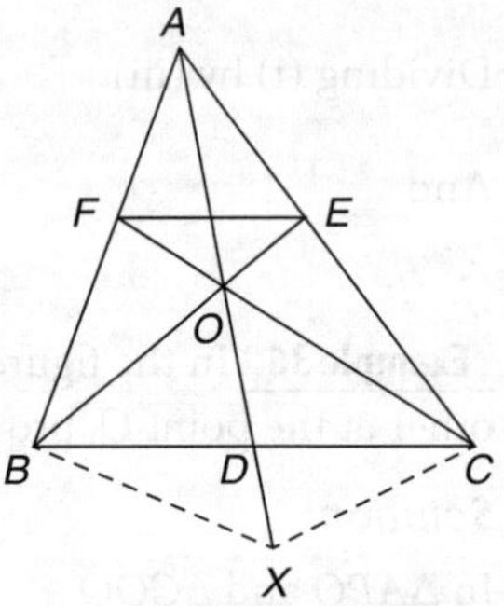

Example 38 The side *BC* of a triangle *ABC* is bisected at *D*. *O* is any point in *AD*. *BO* and *CO* are produced to meet *AC* and *AB* in *E* and *F*, respectively, and *AD* is produced to *X* so that *D* is the mid-point of OX. Prove that *AO*: *AX* = *AF*: AB and show that *FE* || *BC*.

Solution: Join *BX* and *CX*.

We have $BD = CD$ and $OD = DX$. [Given]

Thus, *BC* and *OX* bisect each other.

⇒ *OB XC* is a parallelogram.

[∵ In a quadrilateral, if diagonals bisect each other, then it is a || gm]

$\Rightarrow BX \parallel CO$ and $CX \parallel BO$

$\Rightarrow BX \parallel CF$ and $CX \parallel BE$

$\Rightarrow BX \parallel OF$ and $CX \parallel OE$

In Δ*ABX*, we have $BX \parallel OF$

$$\Rightarrow \frac{AO}{AX} = \frac{AF}{AB} \quad \text{...(i)}$$

In Δ*ACX*, we have $CX \parallel OE$

$$\Rightarrow \frac{AO}{AX} = \frac{AE}{AC} \quad \text{...(ii)}$$

From equations (i) (ii), we get $\frac{AF}{AB} = \frac{AE}{AC}$

Thus, *E* and *F* are points on *AB* and *AC* such that they divide *AB* and *AC*, respectively, in the same ratio. Therefore, by the converse of BPT, *FE* || *BC*.

Example 39 Two triangles *ABC* and *DBC* lie on the same side of the base *BC*. From a point *P* on *BC*, *PQ* || *AB* and *PR* || *BD* are drawn. They meet *AC* in *Q* and *DC* in *R*, respectively. Prove that *QR* || *AD*.

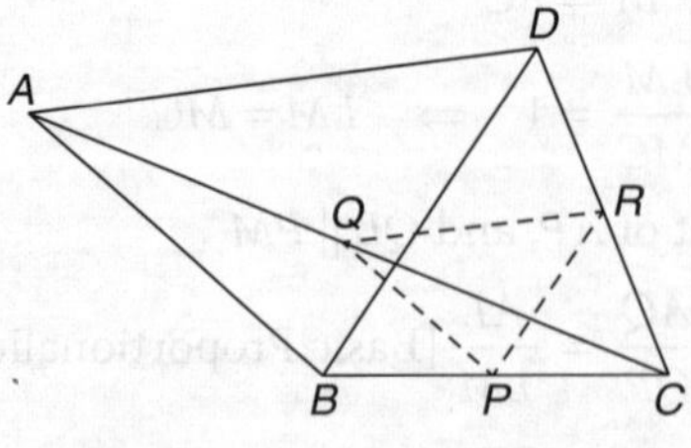

Given: Two triangles *ABC* and *DBC* lie on the same side of the base *BC*. Points *P*, *Q* and *R* are points on *BC*, *AC* and *CD*, respectively, such that *PR* || *BD* and *PQ* || *AB*.

To prove: $QR \parallel AD$

Proof: In Δ*ABC*, we have $PQ \parallel AB$

$$\therefore \quad \frac{CP}{PB} = \frac{CQ}{QA} \quad \ldots\text{(i)} \quad \text{[By Basic Proportionality Theorem]}$$

In ΔBCD, we have $PR \parallel BD$

$$\therefore \quad \frac{CP}{PB} = \frac{CR}{RD} \quad \ldots\text{(ii)} \quad \text{[By BPT]}$$

From (i) and (ii), we have $\dfrac{CQ}{QA} = \dfrac{CR}{RD}$

Thus, in ΔACD, Q and R are points on AC and CD, respectively, such that

$$\frac{CQ}{QA} = \frac{CR}{RD}$$

$$\Rightarrow \quad QR \parallel AD \quad \text{[By the converse of BPT]}$$

Example 40 ABC is an isosceles triangle right-angled at B. Two equilateral triangles are constructed with sides BC and AC, as shown in the figure. Prove that area of $\Delta BCD = \dfrac{1}{2}$ area of ΔACE.

Solution: Let $AB = x$, $BC = x$.

Area of equilateral triangle of side $a = \dfrac{\sqrt{3}}{4} a^2$

$$\text{Area of } \Delta BCD = \frac{\sqrt{3}}{4} x^2 \quad \ldots\text{(i)}$$

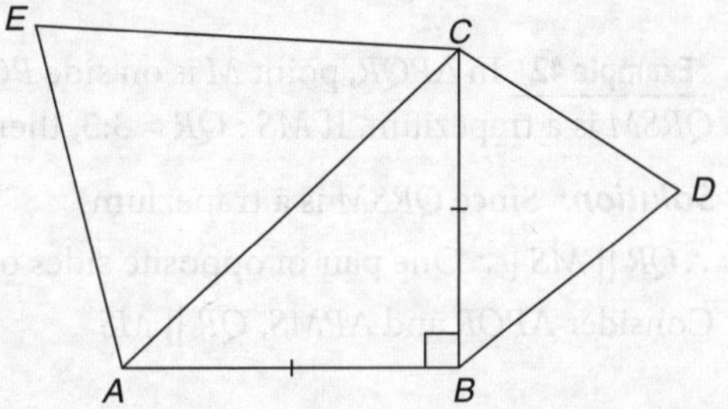

Now, side $AC = \sqrt{x^2 + x^2} = \sqrt{2x^2} = \sqrt{2}x$

[By Pythagoras' Theorem]

$$\text{Area of } \Delta ACE = \frac{\sqrt{3}}{4}(\sqrt{2}x)^2 = 2.\frac{\sqrt{3}}{4}x^2$$

$$= 2 \,.\, \text{Area of } \Delta BCD \quad \text{[From (i)]}$$

$$\therefore \quad ar\,\Delta BCD = \frac{1}{2} \text{ Area of } \Delta ACE. \text{ Hence proved.}$$

Example 41 Two isosceles triangles have equal angles and their areas are in the ratio 81:25. Find the ratio of their corresponding heights.

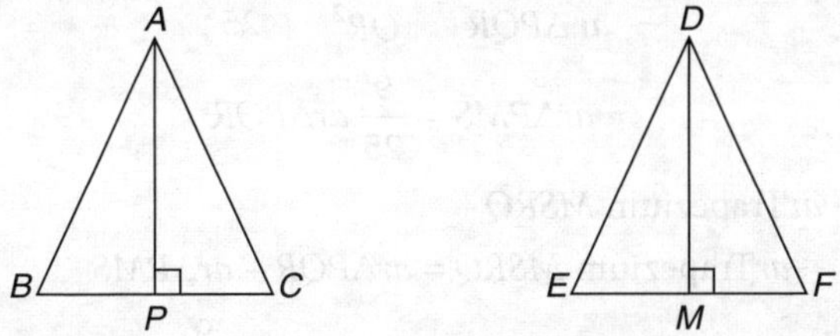

Solution: Consider ΔABC and ΔDEF

$$\frac{AB}{AC} = \frac{DE}{DF} = 1 \quad [\because \Delta\text{s are isosceles}]$$

$$\frac{AB}{DE} = \frac{AC}{DF}$$

As $\angle A = \angle D$ [Given]

$\therefore$ $\Delta ABC \sim \Delta DEF$ [SAS]

$\Rightarrow$ $$\frac{ar(\Delta ABC)}{ar(\Delta DEF)} = \frac{BC^2}{EF^2}$$ [$\because$ Area Theorem]

$\Rightarrow$ $$\frac{81}{25} = \frac{BC^2}{EF^2}$$

$\Rightarrow$ $$\frac{BC}{EF} = \frac{9}{5}$$ ...(i)

Also $$\frac{ar(\Delta ABC)}{ar(\Delta DEF)} = \frac{\frac{1}{2} BC \times AP}{\frac{1}{2} EF \times DM}$$

$\Rightarrow$ $$\frac{81}{25} = \frac{9}{5} \times \frac{AP}{DM}$$ [From (i)]

$\Rightarrow$ $$\frac{AP}{DM} = 9:5.$$

Example 42 In ΔPQR, point M is on side PQ and point S is on side PR such that $QRSM$ is a trapezium. If $MS : QR = 3:5$, then find area (ΔPMS) : area ($QRSM$).

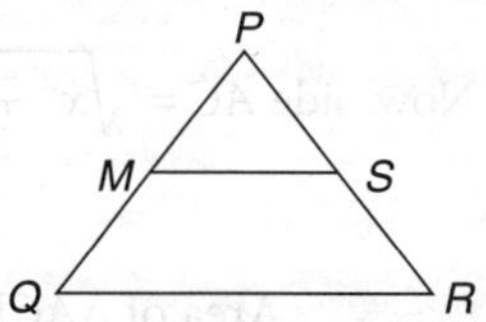

Solution: Since $QRSM$ is a trapezium

$\therefore QR \parallel MS$ [$\because$ One pair of opposite sides of a trapezium is ||]

Consider ΔPQR and ΔPMS, $QR \parallel MS$

$$\frac{PM}{PQ} = \frac{PS}{PR}$$ [$\because$ BPT]

$\angle P = \angle P$ [Common]

$\Delta PMS \sim \Delta PQR$ [SAS]

$$\frac{ar\Delta PMS}{ar\Delta PQR} = \frac{MS^2}{QR^2} = \frac{9}{25}$$ [$\because$ Area Theorem]

$$ar\Delta PMS = \frac{9}{25}\, ar\Delta PQR$$ ...(i)

Since $ar\Delta PQR = ar\Delta PMS + ar\text{Trapezium } MSRQ$

$$ar\text{Trapezium } MSRQ = ar\Delta PQR - ar\Delta PMS$$

$$= ar\Delta PQR - \frac{9}{25}\, ar\Delta PQR$$ [From (i)]

$$ar\text{Trapezium } MSRQ = \frac{16}{25}\, ar\Delta PQR$$ ...(ii)

So $$\frac{\text{ar}\Delta PMS}{\text{ar Trapezium } MSRQ} = \frac{\frac{9}{25}\text{ar}\Delta PQR}{\frac{16}{25}\text{ar}\Delta PQR}$$ [From (i) and (ii)]

$$\frac{\text{ar}\Delta PMS}{\text{ar Trapezium } MSRQ} = \frac{9}{16}$$

Example 43 A right triangle has a hypotenuse of length p and one side of length q cm. If $p - q = 1$, find the length of the third side of the triangle. **[CBSE (CCE) 2012]**

Solution: Let the third side be x cm. Then, by Pythagoras' Theorem, we have

$$p^2 = q^2 + x^2$$

$\Rightarrow$ $$x^2 = p^2 - q^2 = (p - q)(p + q) = p + q$$ $[\because p - q = 1]$

$\Rightarrow$ $$x = \sqrt{q+1+q} = \sqrt{2q+1}$$ $[\because p - q = 1 \therefore p = q + 1]$

Hence, the length of the third side is $\sqrt{2q+1}$ cm.

Example 44 ABC is a right triangle right-angled at C. Let $BC = a$, $CA = b$, $AB = c$ and let p be the length of the perpendicular from C on AB. Prove that **[CBSE 2002]**

(i) $cp = ab$ **(ii)** $\frac{1}{p^2} = \frac{1}{a^2} + \frac{1}{b^2}$

Solution: **(i)** Let $CD \perp AB$. Then, $CD = p$.

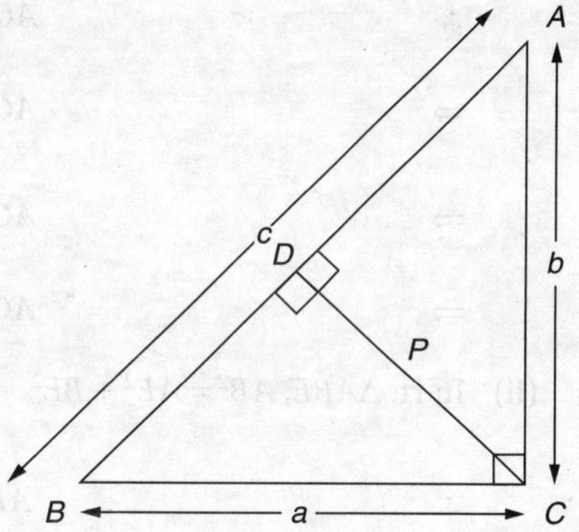

$\therefore$ Area of $\Delta ABC = \frac{1}{2}$ (Base × Height)

$\Rightarrow$ Area of $\Delta ABC = \frac{1}{2}(AB \times CD) = \frac{1}{2}cp$ ...**(i)**

Also Area of $\Delta ABC = \frac{1}{2}(BC \times AC) = \frac{1}{2}ab$ ...**(ii)**

From (i) and (ii), we get

$\therefore$ $$\frac{1}{2}cp = \frac{1}{2}ab$$

$\Rightarrow$ $$cp = ab$$

(ii) Since ΔABC is a right triangle right-angled at C

$\therefore$ $$AB^2 = BC^2 + AC^2$$ [$\because$ Pythagoras' Theorem]

$\Rightarrow$ $$c^2 = a^2 + b^2$$

$\Rightarrow$ $$\left(\frac{ab}{p}\right)^2 = a^2 + b^2$$ $\left[\because cp = ab \therefore c = \frac{ab}{p}\right]$

$\Rightarrow$ $$\frac{a^2b^2}{p^2} = a^2 + b^2$$

$\Rightarrow$ $$\frac{1}{p^2} = \frac{a^2 + b^2}{a^2b^2}$$

$$\Rightarrow \quad \frac{1}{p^2} = \frac{1}{b^2} + \frac{1}{a^2}$$

$$\Rightarrow \quad \frac{1}{p^2} = \frac{1}{a^2} + \frac{1}{b^2}$$

Example 45 In a triangle ABC, $AC > AB$, D is the mid-point of BC and $AE \perp BC$. Prove that:

(i) $AC^2 = AD^2 + BC \cdot DE + \frac{1}{4} BC^2$

(ii) $AB^2 = AD^2 - BC \cdot DE + \frac{1}{4} BC^2$

(iii) $AB^2 + AC^2 = 2AD^2 + \frac{1}{2} BC^2$

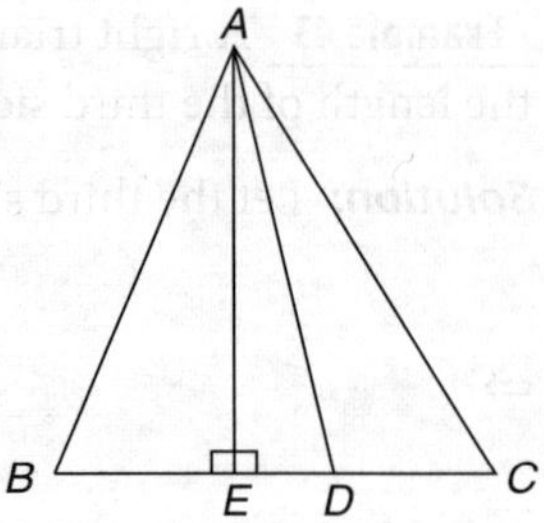

Solution: We have $\angle AED = 90°$

$\therefore \quad \angle ADE < 90°$ and $\angle ADC > 90°$.

That is, $\angle ADE$ is acute and $\angle ADC$ is obtuse and $BD = DC$.

(i) In ΔABC $\quad BD = \frac{1}{2} BC$ and $DC = \frac{1}{2} BC$

In rt. ΔAEC $\quad AC^2 = AE^2 + EC^2$ [$\because$ Pythagoras' Theorem]

$$= AD^2 - (DE)^2 + (ED + DC)^2$$

$$\therefore \quad AC^2 = AD^2 + DC^2 + 2\, DC \cdot DE$$

$$\Rightarrow \quad AC^2 = AD^2 + \left(\frac{1}{2} BC\right)^2 + 2 \cdot \frac{1}{2} BC \cdot DE$$

$$\Rightarrow \quad AC^2 = AD^2 + \frac{1}{4} BC^2 + BC \times DE$$

$$\Rightarrow \quad AC^2 = AD^2 + BC \cdot DE + \frac{1}{4} BC^2 \qquad \text{(i)}$$

(ii) In rt. ΔABE, $AB^2 = AE^2 + BE^2$

$$= AD^2 - DE^2 + (BD - DE)^2$$

$$\therefore \quad AB^2 = AD^2 + BD^2 - 2BD \cdot DE \qquad \text{[Same as proved above]}$$

$$\Rightarrow \quad AB^2 = AD^2 + \left(\frac{1}{2} BC\right)^2 - 2 \cdot \frac{1}{2} BC \cdot DE$$

$$AB^2 = AD^2 + \frac{1}{4} BC^2 - BC \times DE$$

$$\Rightarrow \quad AB^2 = AD^2 - BC \cdot DE + \frac{1}{4} BC^2 \qquad \text{(ii)}$$

(iii) Adding (i) and (ii), we get

$$\Rightarrow \quad AB^2 + AC^2 = 2AD^2 + \frac{1}{2} BC^2$$

Example 46 In the given figure, D and E trisect BC. Prove that $8AE^2 = 3AC^2 + 5AD^2$. **[CBSE2006C]**

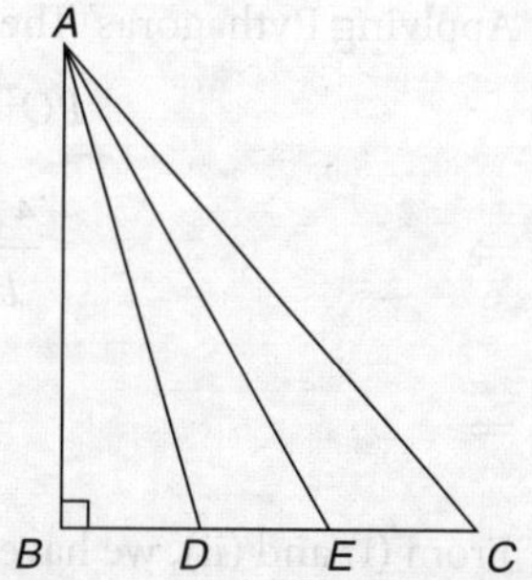

Solution: Since D and E are the points of trisection of BC

$$BD = DE = CE.$$

Let $BD = DE = CE = x.$

Then $BE = 2x$ and $BC = 3x.$

In right-triangles ABD, ABE and ABC, we have

$\Rightarrow$ $AD^2 = AB^2 + BD^2$

$\Rightarrow$ $AD^2 = AB^2 + x^2$ ...(i)

$AE^2 = AB^2 + BE^2$

$AE^2 = AB^2 + 4x^2$ ...(ii)

And $AC^2 = AB^2 + BC^2$

$\Rightarrow$ $AC^2 = AB^2 + 9x^2$ (iii)

$\text{LHS} = 8AE^2 = 8\,(AB^2 + 4x^2)$ [From (ii)]

$= 8AB^2 + 32x^2$ ...(iv)

$\text{RHS} = 3AC^2 + 5AD^2 = 3\,(AB^2 + 9x^2) + 5(AB^2 + x^2)$ [From (i) and (iii)]

$= 3AB^2 + 27x^2 + 5AB^2 + 5x^2 = 8AB^2 + 32x^2$ ...(v)

From (iv) and (v), LHS = RHS.

Example 47 If A is the area of a right triangle and b one of the sides containing the right angle, prove that the length of the altitude on the hypotenuse is $\dfrac{2Ab}{\sqrt{b^4 + 4A^2}}$.

Solution: Let PQR be a right triangle right-angled at Q such that $QR = b$ and A = Area of ΔPQR.

Draw QN perpendicular to PR.

We have $A = \text{Area of } \Delta PQR$

$\Rightarrow$ $A = \dfrac{1}{2}(QR \times PQ)$

$\Rightarrow$ $A = \dfrac{1}{2}(b \times PQ)$

$\Rightarrow$ $PQ = \dfrac{2A}{b}$ ...(i)

Now, in Δs PNQ and PQR, we have $PNQ = PQR$ [Each equal to 90°]

And $QPN = QPR$

So, by AA Similarity, we have $\Delta PNQ \sim \Delta PQR$

$\Rightarrow$ $\dfrac{PQ}{PR} = \dfrac{NQ}{QR}$

[∵ If two Δs are similar, then their corresponding sides are in proportion] ...(ii)

Applying Pythagoras' Theorem in ΔPQR, we have

$$PQ^2 + QR^2 = PR^2$$

$$\Rightarrow \quad \frac{4A^2}{b^2} + b^2 = PR^2$$

$$\Rightarrow \quad PR = \sqrt{\frac{4A^2 + b^4}{b^2}} = \frac{\sqrt{4A^2 + b^4}}{b} \quad \ldots\text{(iii)}$$

From (i) and (ii), we have

$$\frac{2A}{b \times PR} = \frac{NQ}{b}$$

$$\Rightarrow \quad NQ = \frac{2A}{PR}$$

$$\Rightarrow \quad NQ = \frac{2Ab}{\sqrt{4A^2 + b^4}} \quad \left[\because PR = \frac{\sqrt{4A^2 + b^4}}{b}\right] \quad \text{[From (iii)]}$$

Multiple Choice Questions

Choose the correct answer for each of the following questions:

1. The corresponding sides of two similar triangles are in the ratio 5:7. The areas of these triangles are in the ratio
   (a) $\sqrt{5}:\sqrt{7}$ (b) 7:5 (c) 25:49 (d) 49:25
2. In the given figure, if $\angle ADE = \angle ABC$, then CE is equal to

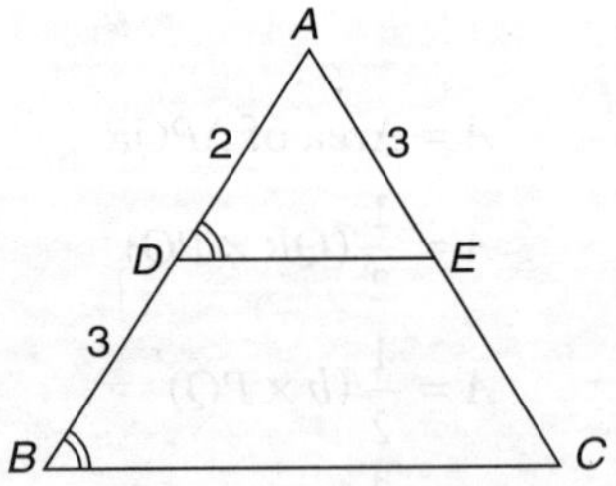

   (a) 2 (b) 5 (c) 4.5 (d) 3
3. The lengths of the diagonals of a rhombus are 24 cm and 32 cm. The perimeter of the rhombus is
   (a) 9 cm (b) 128 cm (c) 80 cm (d) 56 cm
4. Which of the following are not the sides of a right triangle?
   (a) 9 cm, 15 cm, 12 cm (b) 2 cm, 1 cm, $\sqrt{5}$ cm
   (c) 400 mm, 300 mm, 500 mm (d) 7 cm, 5 cm, 7 cm
5. In ΔXYZ and ΔPQR, $\frac{XZ}{PQ} = \frac{YX}{RQ} = \frac{YZ}{PR}$
   (a) $\Delta YXZ \sim \Delta PQR$ (b) $\Delta XYZ \sim \Delta RQP$ (c) $\Delta YXZ \sim \Delta RQP$ (d) $\Delta ZYX \sim \Delta PQR$

6. In the given figure, *ABCD* is a trapezium in which $AB \parallel EF \parallel DC$. The length of *AE* is

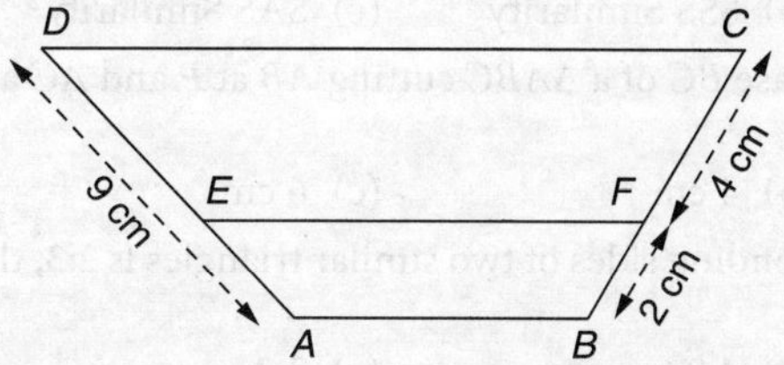

(a) 2 cm (b) 3 cm (c) 4 cm (d) 7 cm

7. If $\Delta PQR \sim \Delta XYZ$, $\angle Q = 50°$ and $\angle R = 70°$, then $\angle X$ is equal to

(a) 70° (b) 50° (c) 120° (d) 60°

8. In the given figure, if $DE \parallel BC$, then *x* equals

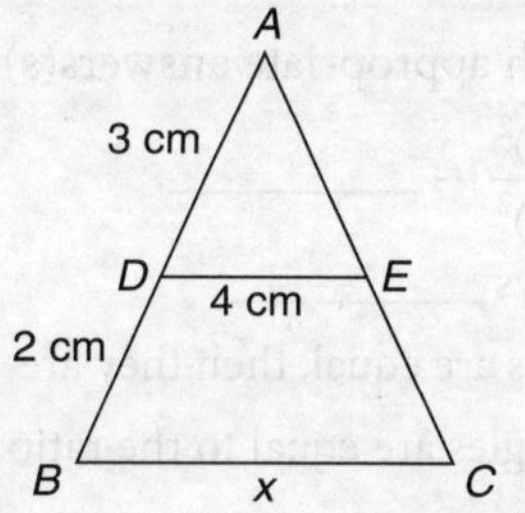

(a) 3 cm (b) 2 cm (c) 4 cm (d) 6.7 cm

9. In triangles *ABC* and *DEF*, $\angle B = \angle E$, $\angle C = \angle F$ and $AB = 3DE$. Then, the two triangles are

(a) Congruent but not similar (b) Neither congruent nor similar
(c) Similar but not congruent (d) Congruent as well as similar

10. $\Delta ABC \sim \Delta PQR$. If $ar(\Delta ABC) = 2.25\ \text{m}^2$, $ar(\Delta PQR) = 6.25\ \text{m}^2$, $PQ = 0.5$ m, then the length of *AB* is

(a) 30 cm (b) 0.5 m (c) 50 cm (d) 3 m

11. In the given figure, $DE \parallel AC$; which of the following is true?

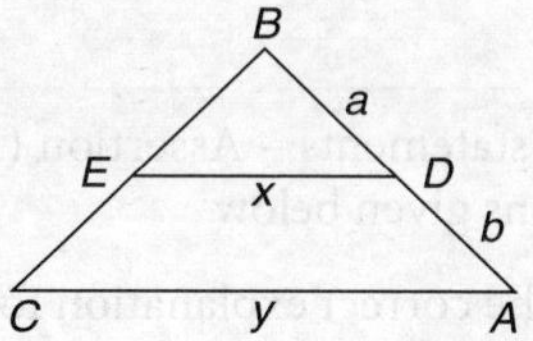

(a) $x = \dfrac{a+b}{xy}$ (b) $y = \dfrac{ax}{a+b}$ (c) $x = \dfrac{ay}{a+b}$ (d) $\dfrac{x}{y} = \dfrac{a}{b}$

12. In the given figure, $\Delta ABC \sim \Delta DEF$. If $AB = 3DE$, $ar(ABC) = 144\ \text{cm}^2$, then ar (*DEF*) is

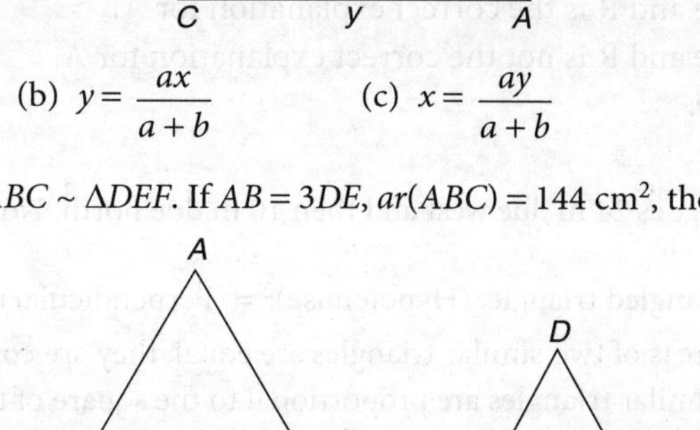

(a) 46 cm^2 (b) 16 cm^2 (c) 12 cm^2 (d) 432 cm^2

13. Two equilateral triangles are similar by
(a) AAA Similarity (b) SSS Similarity (c) SAS Similarity (d) All

14. PQ is drawn parallel to base BC of a ΔABC cutting AB at P and AC at Q. If $AB = 4BP$ and $CQ = 2$ cm, then AQ is equal to
(a) 2 cm (b) 4 cm (c) 6 cm (d) 8 cm

15. If the ratio of the corresponding sides of two similar triangles is 2:3, then the ratio of their corresponding altitudes is
(a) 3:2 (b) 4:9 (c) 2:3 (d) 16:81

16. Tick the correct answer and justify: In ΔABC, $AB = 6\sqrt{3}$ cm, $AC = 12$ cm and $BC = 6$ cm. The angle B is
(a) 120° (b) 60° (c) 90° (d) 45°

Fill in the Blanks

Complete the following statements with appropriate answers(s) in the blank space(s):

17. In similar ΔABC and ΔPQR, $\frac{(AB)^2}{(PQ)^2} =$ __________.

18. Two congruent triangles are always __________.

19. If the areas of two similar triangles are equal, then they are __________.

20. The ratio of the areas of two triangles are equal to the ratio of __________ of any two corresponding sides.

21. All equilateral triangles are __________.

22. The statement of the theorem is "If a line is drawn parallel to one side of a triangle to intersect the other two sides in distinct points, then the other two sides are divided in the same ratio." This statement is known as __________.

23. The ratio of the perimeter of two similar triangles is the same as the ratio of their __________.

24. If $\Delta ABC \sim \Delta PQR$, then $\angle B$ is equal to __________.

Assertion–Reason

The following questions consist of two statements – Assertion (A) and Reason (R). Answer these questions by selecting the appropriate options given below.

(A) Both A and R are true and R is the correct explanation for A.
(B) Both A and R are true and R is not the correct explanation for A.
(C) A is true but R is false.
(D) A is false but R is true.

25. **Assertion (*A*):** A man goes 24 m due west and then 10 m due north. Now he is 26 m from his starting point.
Reason (*R*): In a right-angled triangle, $(\text{Hypotenuse})^2 = (\text{Perpendicular})^2 + (\text{Base})^2$

26. **Assertion (*A*):** If the areas of two similar triangles are equal, they are congruent.
Reason (*R*): Areas of similar triangles are proportional to the square of their corresponding sides.

27. **Assertion (*A*):** The areas of two similar triangles ABC and PQR are in the ratio 9:16. If $BC = 4.5$ cm, then $QR = 6$ cm.
Reason (*R*): The ratio of the areas of two similar triangles is equal to the ratio of the squares of the corresponding sides.

28. **Assertion (*A*):** In a ΔABC, if $DE \parallel BC$ intersects AB in D and AC in E, then $\frac{AD}{AB} = \frac{AE}{AC}$.
Reason (*R*): If a line is drawn parallel to one side of a triangle to intersect the other two sides in distinct points, then these sides are divided in the same ratio.

29. **Assertion (*A*):** ΔABC is an isosceles triangle right-angled at C, then $AB^2 = 2AC^2$.
Reason (*R*): In right ΔABC, right-angled at B, $AC^2 = AB^2 + BC^2$.

Subjective Questions

30. In the figure, CM and RN are, respectively, the medians of ΔABC and ΔPQR. If $\Delta ABC \sim \Delta PQR$, prove that

(i) $\Delta AMC \sim \Delta PNR$ (ii) $\frac{CM}{RN} = \frac{AB}{PQ}$

(iii) $\Delta CMB \sim \Delta RNQ$

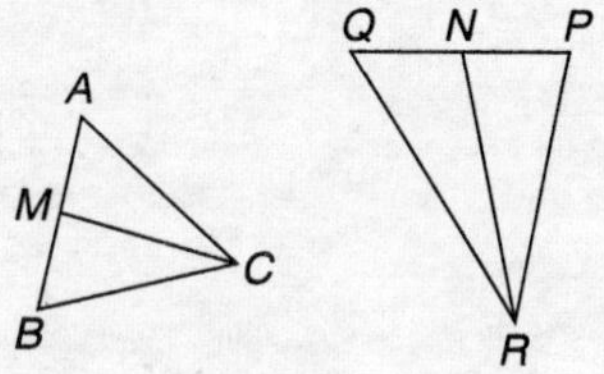

31. The perimeter of two similar triangles ABC and LMN are 60 cm and 48 cm, respectively. If $LM = 8$ cm, then what is the length of AB? **[Delhi 2008C]**

32. In the figure, $\angle M = \angle N = 46°$, express x in terms of a, b and c, where a, b and c are the lengths of LM, MN and NK, respectively. **[Delhi 2009]**

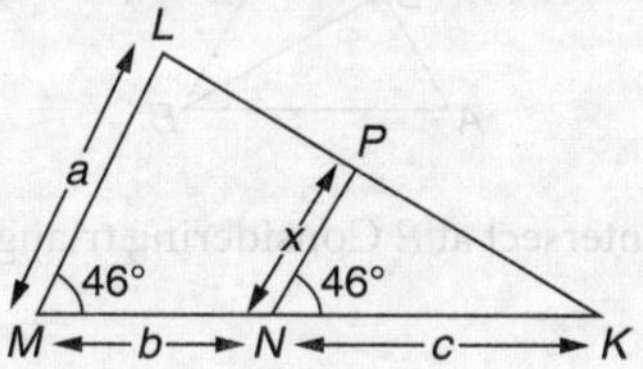

33. In the figure, $DE \parallel BC$ in ΔABC such that $BC = 8$ cm, $AB = 6$ cm and $DA = 1.5$ cm. Find DE. **[AI 2010]**

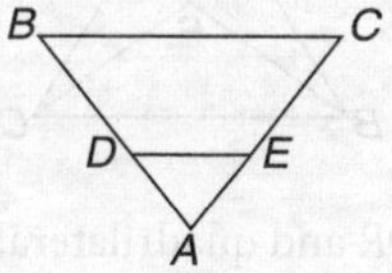

34. State whether the following quadrilaterals are similar or not:

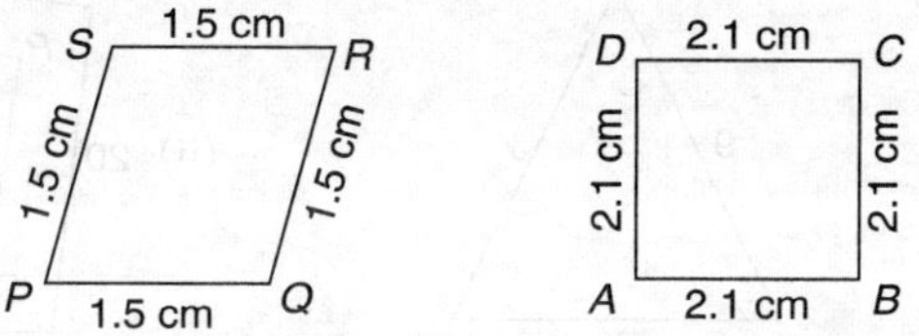

35. If ΔABC and ΔDEF are similar such that $\angle A = 57°$ and $\angle E = 83°$, find $\angle C$.

36. In ΔABC, D and E are points on the sides AB and AC, respectively, such that $DE \parallel BC$. If $AD = x$, $DB = x - 2$, $AE = x + 2$ and $EC = x - 1$, find the value of x.

37. In the figure, $\angle 1 = \angle 3$, $\angle 2 = \angle 4$, $DE = 4$, $CE - x + 1$, $AE = 2x + 4$; $BE = 4x - 2$. Find x.

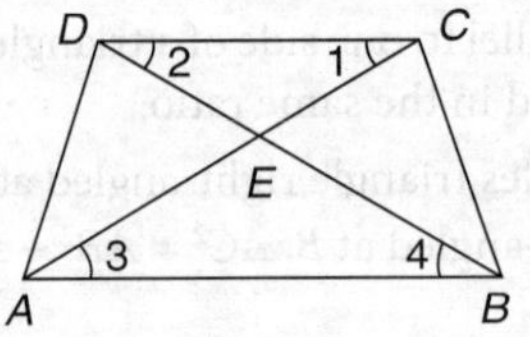

38. In the given figure, if $AB \parallel DC$, find the value of x.

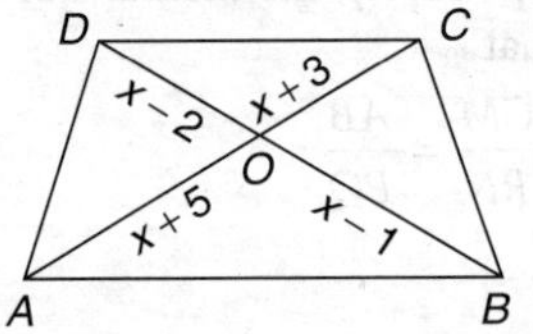

39. In the figure, $AB \parallel DE$ and $BD \parallel EF$. Prove that $DC^2 = CF \times AC$. **[Delhi 2007]**

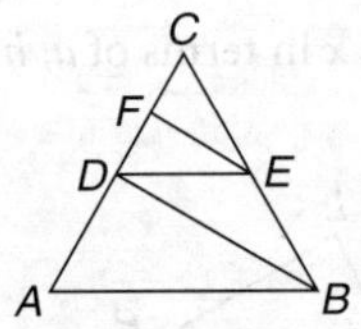

40. In the given figure, BD and CE intersect at P. Considering triangles BEP and CPD, prove that $BP \times PD = EP \times PC$.

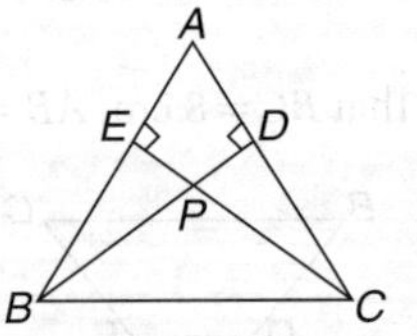

41. In the figure, given that $\Delta ABC \sim \Delta PQR$ and quadrilateral $ABCD \sim$ quadrilateral $PQRS$. Determine the values of x, y, z in each case.

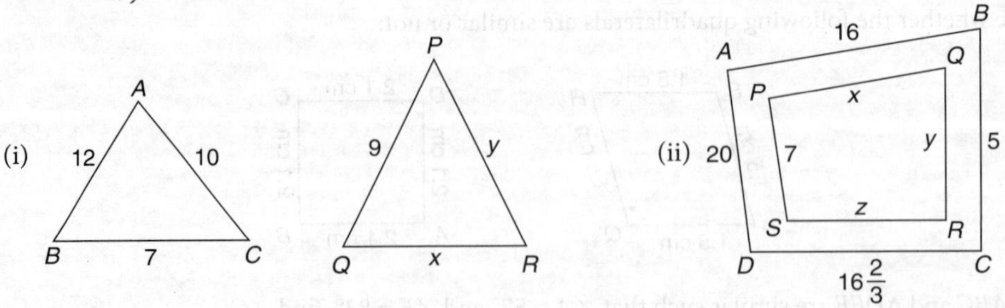

42. In an isosceles ΔABC, the base AB is produced both ways to P and Q such that $AP \times BQ = (AC)^2$. Prove that $\Delta ACP \sim \Delta BCQ$.

43. In the given figure, $\Delta AMB \sim \Delta CMD$. Determine MD in terms of x, y and z.

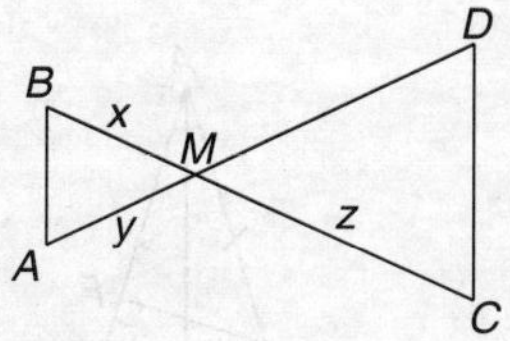

44. In a ΔABC, P and Q are points on sides AB and AC, respectively, such that $PQ \parallel BC$. If $AP = 4$ cm, $PB = 6$ cm and $PQ = 3$ cm, determine BC.

45. In the figure, $\angle 1 = \angle 2$ and $\angle 3 = \angle 4$. Show that $PT \times QR = PR \times ST$.

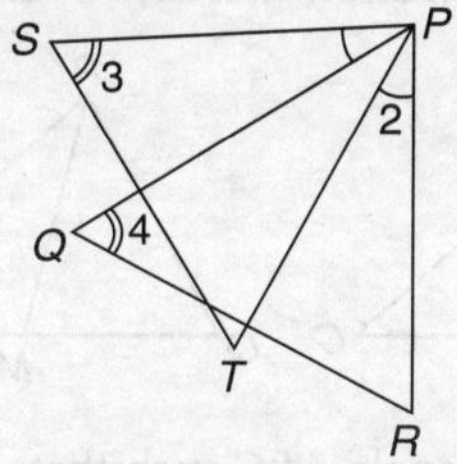

46. S and T are points on sides PR and QR of ΔPQR such that $\angle P = \angle RTS$. Show that $\Delta RPQ \sim \Delta RTS$. **[NCERT]**

47. In this figure, if $\Delta ABE \cong \Delta ACD$, show that $\Delta ADE \sim \Delta ABC$. **[NCERT]**

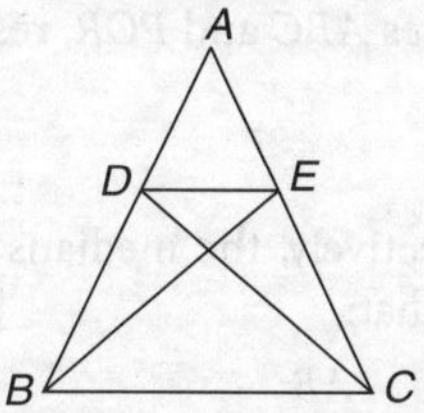

48. In this figure, altitudes AD and CE of ΔABC intersect each other at the point P. Show that: **[NCERT]**

(i) $\Delta AEP \sim \Delta CDP$ (ii) $\Delta ABD \sim \Delta CBE$

(iii) $\Delta AEP \sim \Delta ADB$ (iv) $\Delta PDC \sim \Delta BEC$

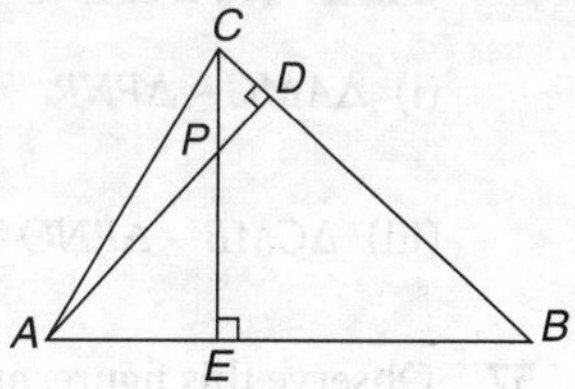

49. E is a point on the side AD produced of a parallelogram $ABCD$ and BE intersects CD at F. Show that $\Delta ABE \sim \Delta CFB$. **[NCERT]**

50. In this figure, ABC and AMP are two right triangles, right-angled at B and M, respectively. Prove that: **[NCERT]**

(i) $\Delta ABC \sim \Delta AMP$ (ii) $\dfrac{CA}{PA} = \dfrac{BC}{MP}$

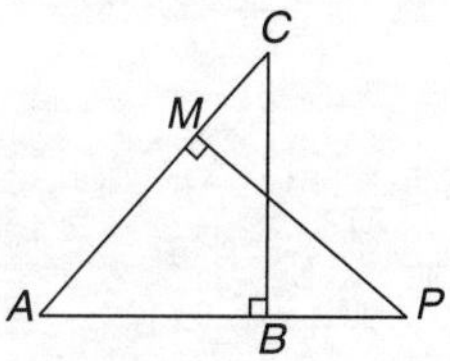

51. In this figure, E is a point on side CB produced of an isosceles triangle ABC with $AB = AC$. If $AD \perp BC$ and $EF \perp AC$. Prove that $\Delta ABD \sim \Delta ECF$. **[NCERT]**

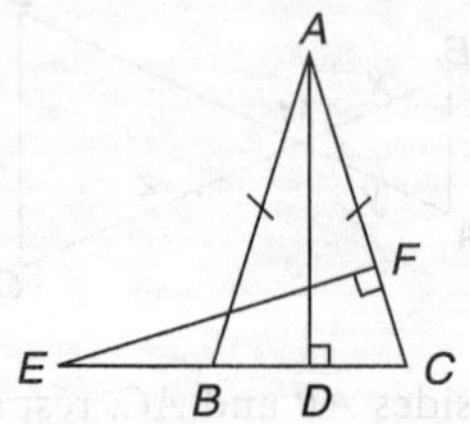

52. Sides AB and BC and median AD of a triangle ABC are, respectively, proportional to sides PQ and QR and median PM of ΔPQR (see figure). Show that $\Delta ABC \sim \Delta PQR$. **[NCERT]**

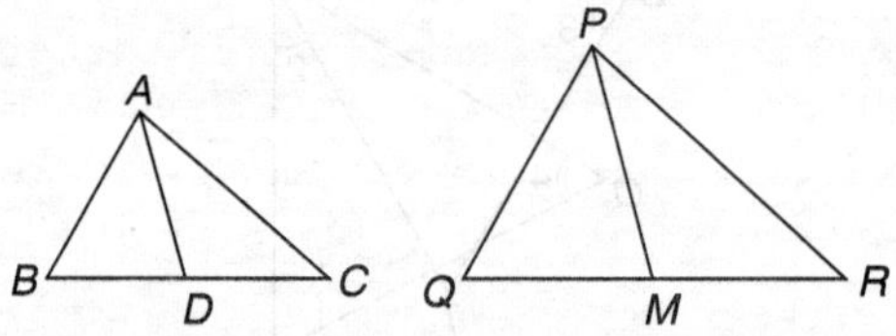

53. D is a point on the side BC of a triangle ABC such that $\angle ADC = \angle BAC$. Show that $CA^2 = CB.CD$. **[NCERT]**

54. Sides AB and AC and median AD of a triangle ABC are, respectively, proportional to sides PQ and PR and median PM of another triangle PQR. Show that $\Delta ABC \sim \Delta PQR$. **[NCERT]**

55. If AD and PM are medians of triangles ABC and PQR, respectively, where $\Delta ABC \sim \Delta PQR$, prove that $\dfrac{AB}{PQ} = \dfrac{AD}{PM}$.

56. In the figure, CM and RN are, respectively, the medians of ΔABC and ΔPQR. If $\Delta ABC \sim \Delta PQR$, prove that: **[NCERT]**

(i) $\Delta AMC \sim \Delta PNR$ (ii) $\dfrac{CM}{RN} = \dfrac{AB}{PQ}$

(iii) $\Delta CMB \sim \Delta RNQ$

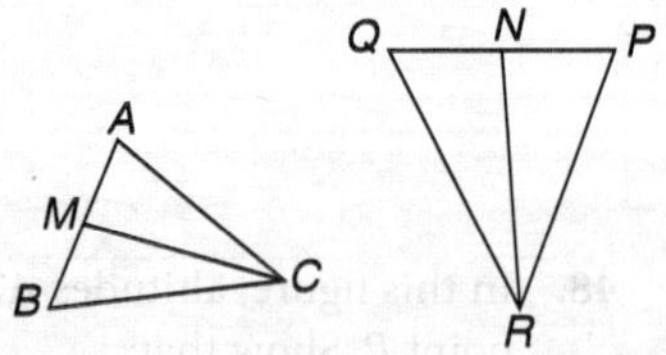

57. Observe this figure, and then find $\angle P$. **[NCERT]**

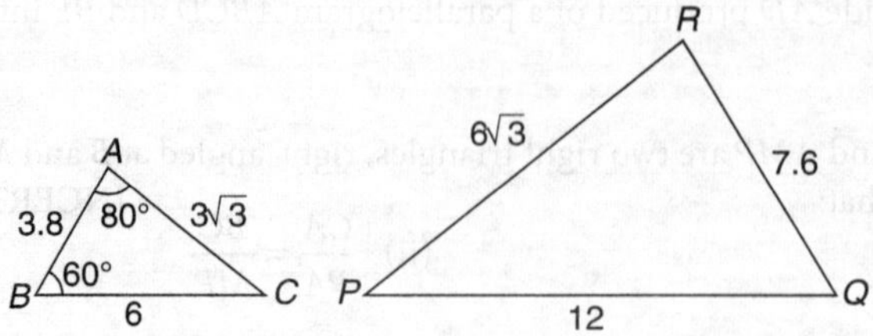

58. In the figure, $OA \times OB = OC \times OD$. Show that $\angle A = \angle C$ and $\angle B = \angle D$.

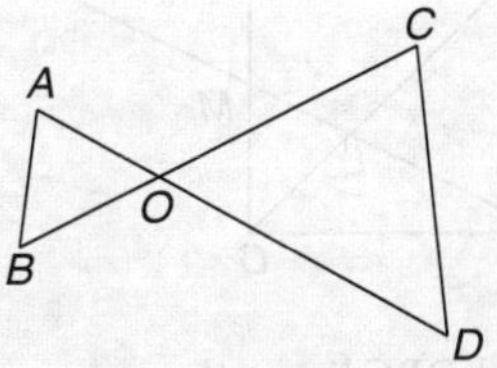

59. A girl of height 90 cm is walking away from the base of a lamp post at a speed of 1.2 m/s. If the lamp is 3.6 m above the ground, find the length of her shadow after 4 seconds. **[NCERT]**

60. In the figure, $LM \parallel NQ$ and $LN \parallel PQ$. If $MP = \frac{1}{3} MN$, find the ratio of the areas of ΔLMN and ΔQNP.

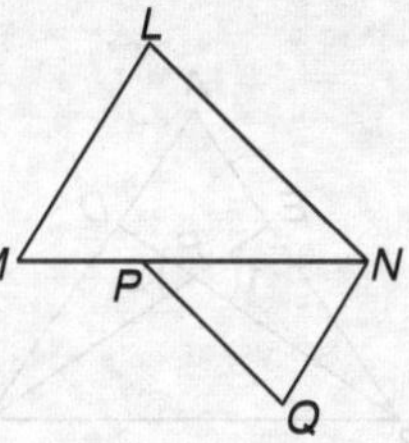

61. In the figure, $DB \perp BC$, $DE \perp AB$ and $AC \perp BC$. Prove that $\frac{BE}{DE} = \frac{AC}{BC}$. **[AI 2008]**

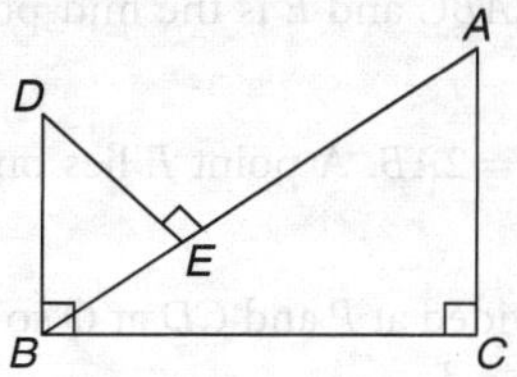

62. In the figure, M is the mid-point of side CD of a parallelogram $ABCD$. The line BM is drawn intersecting AC at L and AD produced at E. Prove that $EL = 2BL$. **[AI 2009]**

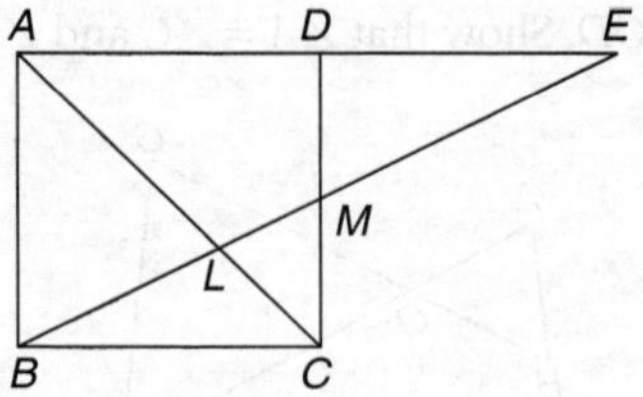

63. In the figure, two triangles ABC and DBC lie on the same side of base BC. P is a point on BC such that $PQ \parallel BA$ and $PR \parallel BD$. Prove that $QR \parallel AD$. **[Foreign 2009, CBSE (CCE) 2012]**

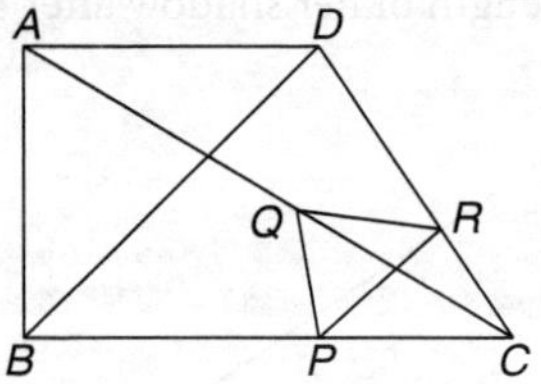

64. In the figure, ABC is an isosceles triangle in which $AB = AC$. E is a point on the side CB produced, such that $FE \perp AC$. If $AD \perp CB$, prove that $AB \times EF = AD \times EC$.

65. If the angles of one triangle are, respectively, equal to another triangle, prove that the ratio of the corresponding sides is the same as the ratio of the corresponding:
(i) Medians (ii) Bisectors (iii) Altitudes

66. In ΔABC, $BD \perp AC$, $CE \perp AB$. Prove that $BP \times PD = EP \times PC$.

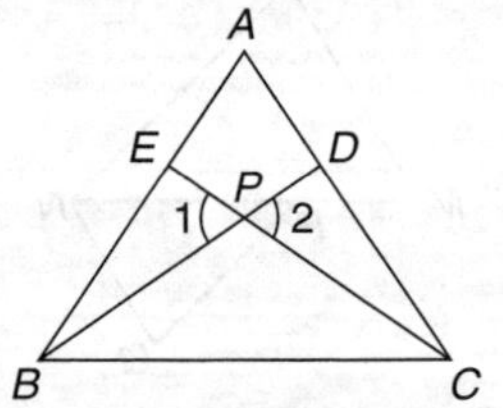

67. If one diagonal of a trapezium divides the other diagonal in the ratio 1:2, prove that one of the parallel sides is double the other.

68. D is the mid-point of side BC of ΔABC and E is the mid-point of AD. BE produced meets AC at the point M. Prove that $BE = 3\,EM$.

69. $ABCD$ is a rectangle in which $BC = 2AB$. A point E lies on ray CD such that $CE = 2BC$. Prove that $BE \perp AC$.

70. $ABCD$ is a parallelogram. AB is divided at P and CD at Q so that $AP{:}PB = 3{:}2$ and $CQ{:}QD = 4\ 1$. If PQ meets AC at R, then prove that $AR = \frac{3}{7}AC$.

71. It is given that $\Delta DEF \sim \Delta RPQ$. Is it true to say that $\angle D = \angle R$ and $\angle F = \angle P$?

72. In triangles PQR and MST, $\angle P = 55°$, $\angle Q = 25°$, $\angle M = 100°$ and $\angle S = 25°$. Is $\Delta QPR \sim \angle TSM$? Why?

73. In the figure, *PA, QB, RC* and *SD* are all perpendiculars to a line *l*, $AB = 6$ cm, $BC = 9$ cm, $CD = 12$ cm and $SP = 36$ cm. Find *PQ, QR* and *RS*.

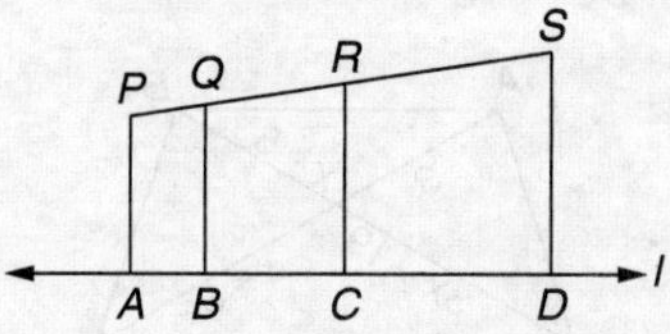

74. *P* is the mid-point of side *BC* of ΔABC. *Q* is the mid-point of *AP*. *BQ* when produced meets *AC* at *L*. Prove that $AL = \frac{1}{3} AC$.

75. The side *BC* of a triangle *ABC* is bisected at *D*. *O* is any point in *AD*. *BO* and *CO* produced meet *AC* and *AB* in *E* and *F*, respectively. *AD* is produced to *X* so that *D* is the mid-point of *OX*. Prove that $AO:AX = AF:AB$ and show that $FE \parallel BC$.

76. Two triangles *ABC* and *DBC* lie on the same side of the base *BC*. Point *R* on *BC*, $PQ \parallel AB$ and $PR \parallel BD$ are drawn. They meet *AC* in *Q* and *DC*, respectively. Prove that $QR \parallel AD$.

77. If a line intersects sides *AB* and *AC* of a ΔABC at *D* and *E*, respectively, and is parallel to *BC*, prove that $\frac{AD}{AB} = \frac{AE}{AC}$ (see figure). **[NCERT]**

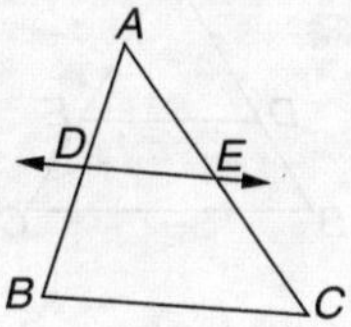

78. In this figure, $\frac{PS}{SQ} = \frac{PT}{TR}$ and $\angle PST = \angle PRQ$. Prove that *PQR* is an isosceles triangle. **[NCERT]**

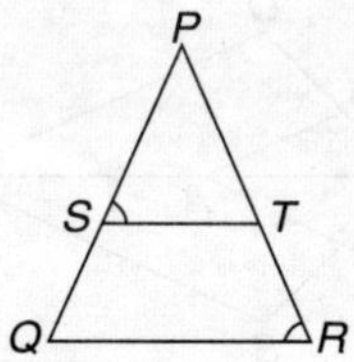

79. In the figure, $DE \parallel BC$. If $BD = x - 3$. $AB = 2x$. $CE = x - 2$ and $AC = 2x + 3$, find *x*.

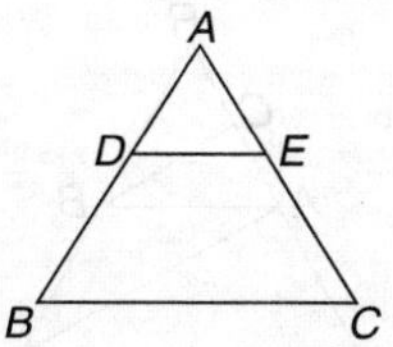

80. If a line is drawn parallel to one side of a triangle, prove that the other two sides are divided in the same ratio. In the following figure, $AB \parallel CD$. Using the above result, solve the following:

(i) Find the value of x from the given figure **[Foreign 2006]**

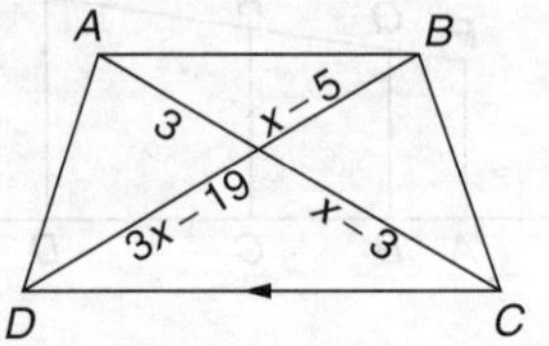

(ii) In the figure, $AB \parallel DE$ and $BC \parallel EF$. Prove that $AC \parallel DF$. **[Foreign 2008]**

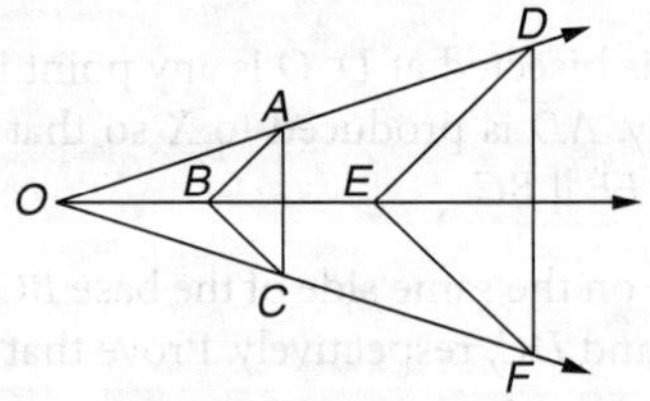

(iii) In the figure, $DE \parallel BC$ and $BD - CE$. Prove that ΔABC is an isosceles triangle. **[Delhi 2009]**

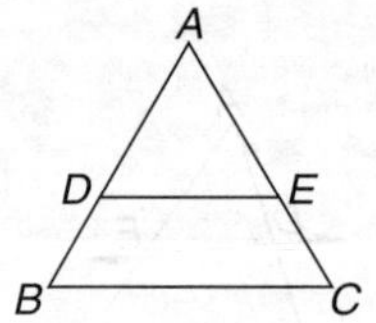

(iv) In the figure, $BA \parallel QR$, and $CA \parallel SR$. Prove that $\frac{QB}{BP} = \frac{SC}{CP}$. **[AI 2009]**

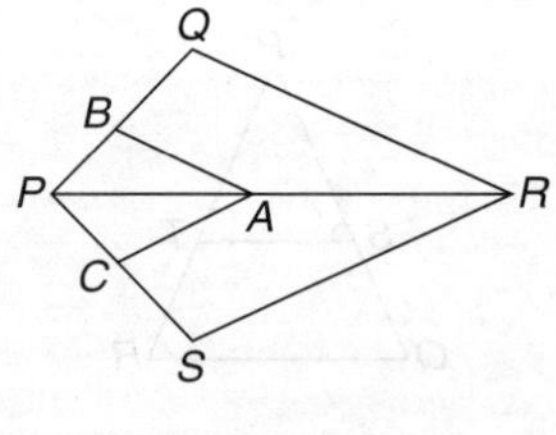

(v) In the figure, $PQ \parallel AB$ and $AQ \parallel CB$. Prove that $AR^2 = PR.CR$. **[Foreign 2010]**

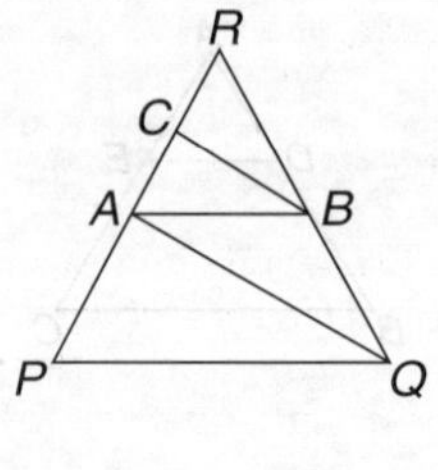

81. In figures (i) and (ii), $DE \parallel BC$. Find EC in (i) and AD in (ii). **[NCERT]**

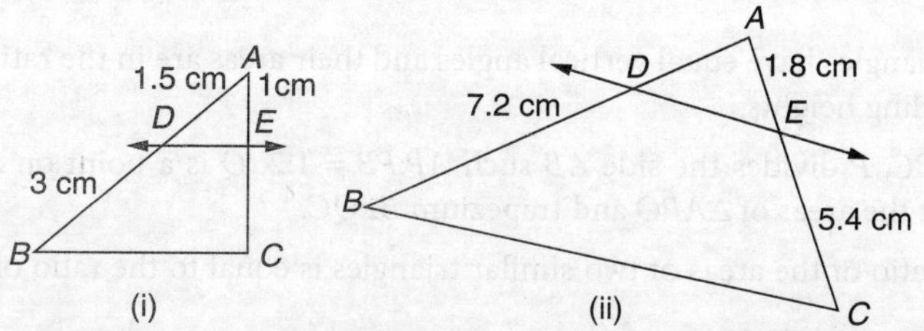

82. E and F are points on the sides PQ and PR, respectively, of a ΔPQR. For each of the following cases, state whether $EF \parallel QR$:

(i) $PE = 3.9$ cm, $EQ = 3$cm, $PF = 3.6$ cm and $FR = 2.4$ cm

(ii) $PE = 4$ cm, $QE = 4.5$ cm, $PF = 8$ cm and $RF = 9$ cm

(iii) $PQ = 1.28$ cm, $PR = 2.56$ cm, $PE = 0.18$ cm and $PF = 0.36$ cm **[NCERT]**

83. In this figure, if $LM \parallel CB$ and $LN \parallel CD$. Prove that $\frac{AM}{AB} = \frac{AN}{AD}$. **[NCERT]**

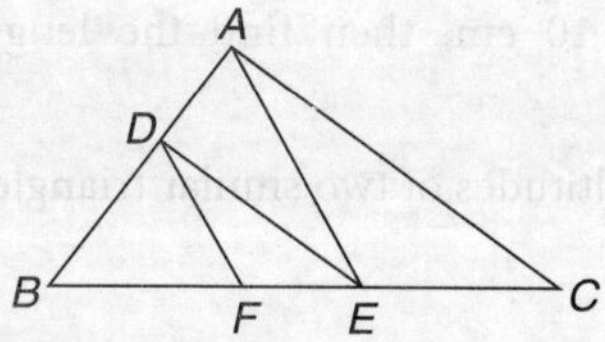

84. Using BPT, prove that a line drawn through the mid-point of one side of a triangle parallel to another side bisects the third side. (Recall that you have proved it in Class IX). **[NCERT]**

85. Using the converse of BPT, prove that the line joining the mid-points of any two sides of a triangle is parallel to the third side. (Recall that you have proved it in Class IX). **[NCERT]**

86. $ABCD$ is a trapezium with $AB \parallel DC$. E and F are points on non-parallel sides AD and BC, respectively, such that EF is parallel to AB (see figure). Show that $\frac{AE}{ED} = \frac{BF}{FC}$. **[NCERT]**

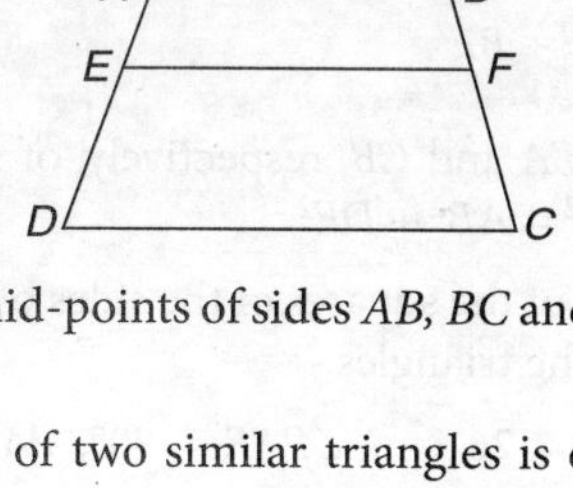

87. D, E and F are, respectively, the mid-points of sides AB, BC and CA of ΔABC. Find the ratio of the areas of ΔDEF and ΔABC.

88. Prove that the ratio of the areas of two similar triangles is equal to the square of the ratio of their corresponding medians.

89. In the figure, $PQ \parallel BC$ and $AP : PB = 1:2$. Find $\frac{ar\,(\Delta APQ)}{ar\,(\Delta ABC)}$. **[Foreign 2008]**

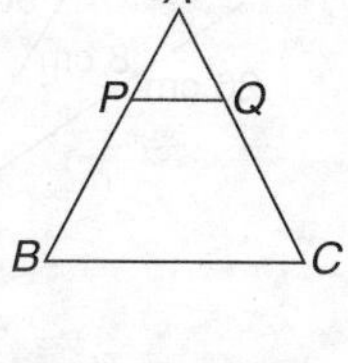

90. If the areas of two similar triangles are in ratio 25:64, write the ratio of their corresponding sides. **[AI 2009]**

91. Two isosceles triangles have equal vertical angles and their areas are in the ratio 9:16. Find the ratio of their corresponding heights.

92. In a triangle *ABC*, *P* divides the side *AB* such $AP:PB = 1:2$, *Q* is a point on *AC* such that $PQ \parallel BC$. Find the ratio of the areas of ΔAPQ and trapezium *BPQC*.

93. Prove that the ratio of the areas of two similar triangles is equal to the ratio of squares of their corresponding sides.

(i) Using the above, perform the following:
If *D* is a point on *AB* of ΔABC such $AD:DB = 3:2$ and *E* is a point on *BC* such that $DE \parallel AC$, find the ratio of the areas of ΔABC and ΔBDE. **[Delhi 2008 C]**

(ii) Using the above, perform the following:
In a trapezium *ABCD*, *AC* and *BD* intersect at *O*, $AB \parallel DC$ and $AB = 2CD$. If the area of $\Delta AOB = 84\text{ cm}^2$, find the area of ΔCOD. **[Delhi 2009]**

(iii) *AD* is an altitude of an equilateral triangle *ABC*. With *AD* as base, another equilateral triangle *ADE* is constructed. Prove that area (ΔADE): area (ΔABC) = 3:4. **[AI 2010]**

(iv) The areas of two similar triangles are 49 cm² and 64 cm², respectively. If the difference of the corresponding altitudes is 10 cm, then find the lengths of the altitudes (in centimetres). **[CBSE (CCE) 2011]**

94. The ratio of the corresponding altitudes of two similar triangles is $\frac{3}{5}$. Is it correct to say that the ratio of their areas is $\frac{6}{5}$? Why? **[NCERT Exemplar]**

95. To go to city *B* from city *A*, there is a route via city *C* such that $AC \perp CB$, $AC = 2x$ km and $CB = 2(x+7)$ km. It is proposed that a 26-km highway be constructed which directly connects the two cities *A* and *B*. Find out how much distance will be saved in reaching city *B* from city *A* after the construction of the highway. **[NCERT Exemplar]**

96. The diagonals of a trapezium *PQRS* intersect each other at the point *O*, $PQ \parallel RS$ and $PQ = 3RS$. Find the ratio of the areas of ΔPOQ and ΔROS.

97. In this figure, *O* is a point in the interior of a triangle *ABC*, $OD \perp BC$. $OE \perp AC$ and $OF \perp AB$. Show that **[NCERT]**

(i) $OA^2 + OB^2 + OC^2 - OD^2 - OE^2 - OF^2 = AF^2 + BD^2 + CE^2$

(ii) $AF^2 + BD^2 + CE^2 = AE^2 + CD^2 + BF^2$.

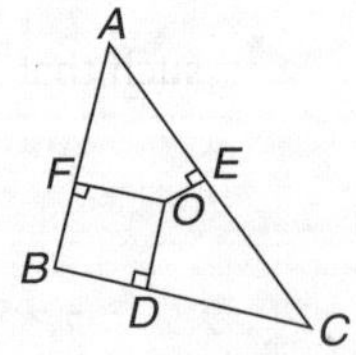

98. *D* and *E* are points on the sides *CA* and *CB*, respectively, of a triangle *ABC* right-angled at *C*. Prove that $AE^2 + BD^2 = AB^2 + DE^2$. **[NCERT]**

99. Prove that three times of the sum of the squares of the sides of a triangle is equal to four times the sum of the squares of the medians of the triangles.

100. In this figure, $PQ = 24$ cm, $QR = 26$ cm, $\angle PAR = 90°$, $PA = 6$ cm and $AR = 8$ cm. Find $\angle QPR$. **[Delhi 2008]**

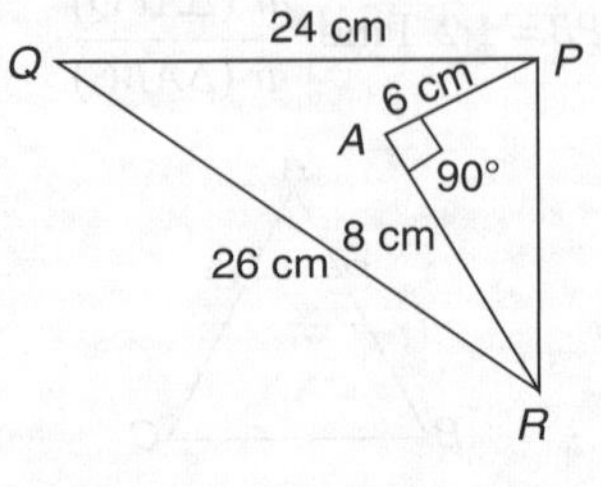

101. The lengths of the diagonals of a rhombus are 30 cm and 40 cm. Find the side of the rhombus. **[Foreign 2008]**

102. The perimeters of two similar triangles *ABC* and *LMN* are 60 cm and 48 cm, respectively. If $LM = 8$ cm, then what is the length of *AB*? **[Delhi 2008C]**

103. Triangle *ABC* is right-angled at *B*, and *D* is mid-point of *BC*. Prove that $AC^2 = 4AD^2 - 3AB^2$. **[AI 2008 C, Delhi 2008]**

104. A right triangle has hypotenuse of length *q* cm and one side of length *p* cm. If $(q - p) = 2$, express the length of the third side of the right triangle in terms of *q*.

105. Two poles of height 7 m and 12 m stand on a plane ground. If the distance between their feet is 12 m, find the distance between their tops.

106. In a triangle, *AD* is drawn perpendicular to BC. Prove that $(AB)^2 - (BD)^2 = (AC)^2 - (CD)^2$.

107. *L* and *M* are the mid-points of *AB* and *BC*, respectively, of ΔABC, right-angled at *B*. Prove that $4\,LC^2 = AB^2 + 4BC^2$.

108. In the figure, $AD \perp BC$ and $BD = \frac{1}{3}\,CD$. Prove that $2CA^2 = 2\,AB^2 + BC^2$. **[AI 2009]**

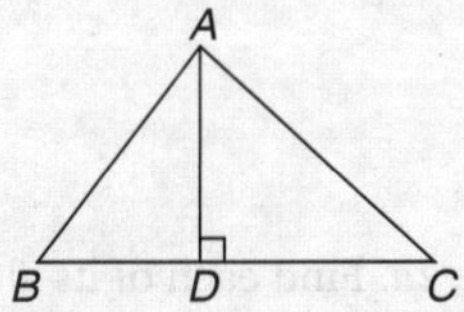

109. In the figure given below, $AD \perp BC$. Prove that $AB^2 + CD^2 = BD^2 + AC^2$: **[CBSE (CCE) 2011]**

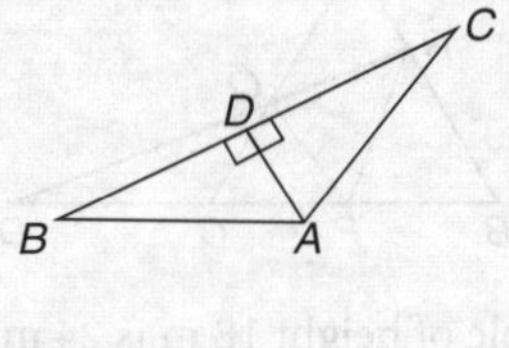

110. In a quadrilateral *ABCD*, $\angle B = 90°$, $AD^2 = AB^2 + CD^2 + BC^2$. Prove that $\angle ACD = 90°$.

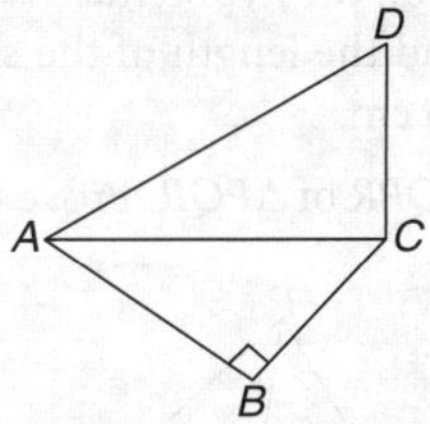

111. In this figure, $\angle ACB = 90°$ and $CD \perp AB$. Prove that $\dfrac{BC^2}{AC^2} = \dfrac{BD}{AD}$. [NCERT]

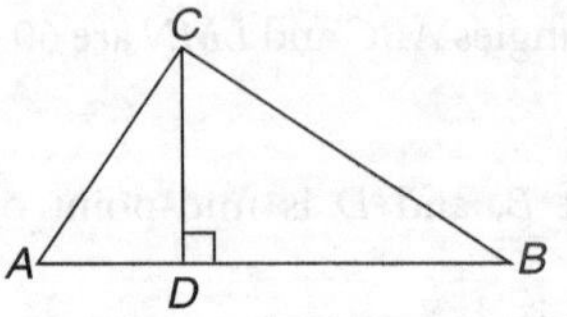

112. A ladder is placed against a wall such that its foot is at a distance of 2.5 m from the wall and its top reaches a window 6 m above the ground. Find the length of the ladder. [NCERT]

113. O is any point inside a rectangle $ABCD$ (see figure). Prove that $OB^2 + OD^2 = OA^2 + OC^2$. [NCERT]

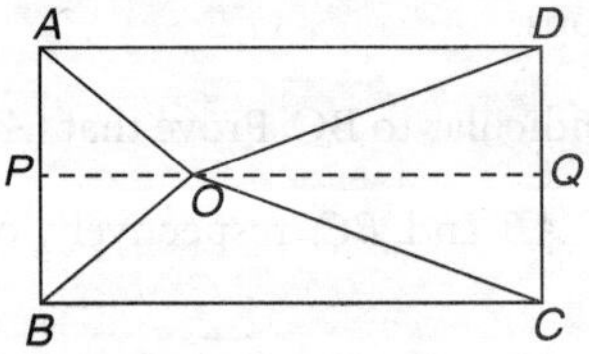

114. PQR is a triangle right-angled at P and M is a point on QR such that $PM \perp QR$. Show that $PM^2 = QM.MR$.

115. In this figure, ABD is a triangle right-angled at A and $AC \perp BD$. Show that

(i) $AB^2 = BC \cdot BD$

(ii) $AC^2 = BC \cdot DC$

(iii) $AD^2 = BD \cdot CD$

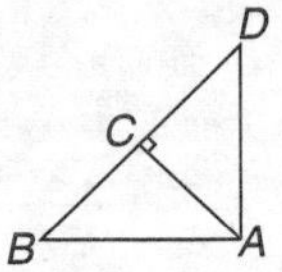

116. ABC is an equilateral triangle of side $2a$. Find each of its altitudes.

117. In the given figure, $PQ \parallel BA$; $PR \parallel CA$. If $PD = 12$ cm. Find $BD \times CD$. [CBSE (CCE) 2012]

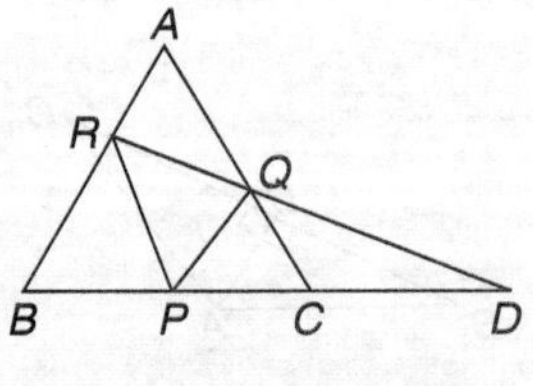

118. A guy wire attached to a vertical pole of height 18 m is 24 m long and has a stake attached to the other end. How far from the base of the pole should the stake be driven so that the wire will be taut?

119. In a right-angled triangle, the square on the hypotenuse is equal to the sum of the squares on the other two sides. Using the above result, find the length of the second diagonal of a rhombus whose side is 5 cm and one of whose diagonals is 6 cm.

120. In this figure, PS is the bisector of $\angle QPR$ of ΔPQR. Prove that $\dfrac{QS}{SR} = \dfrac{PQ}{PR}$. [NCERT]

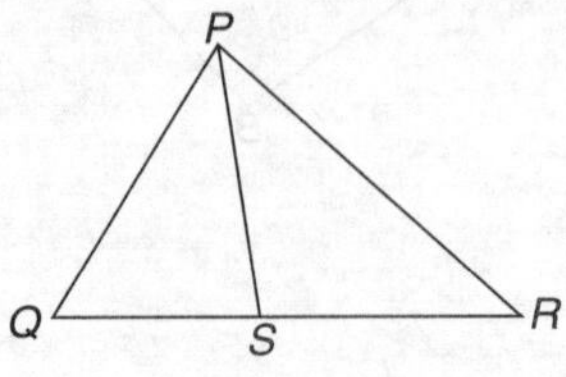

121. In this figure, D is a point on hypotenuse AC of ΔABC, such that $BD \perp AC$, $DM \perp BC$ and $DN \perp AB$. Prove that: **[NCERT]**

(i) $DM^2 = DN.MC$

(ii) $DN^2 = DM.AN$

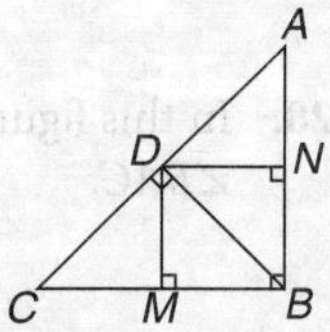

122. In this figure, ABC is a triangle in which $\angle ABC > 90°$ and $AD \perp CB$ produced. Prove that $AC^2 = AB^2 + BC^2 + 2BC \cdot BD.$ **[NCERT]**

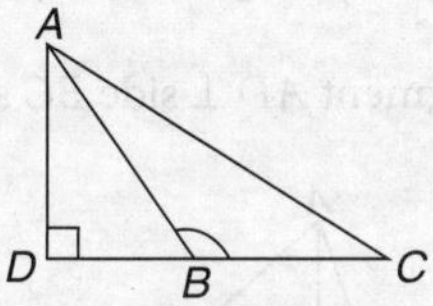

123. In this figure, ABC is a triangle in which $\angle ABC < 90°$ and $AD \perp BC$. Prove that $AC^2 = AB^2 + BC^2 - 2BC.BD.$ **[NCERT]**

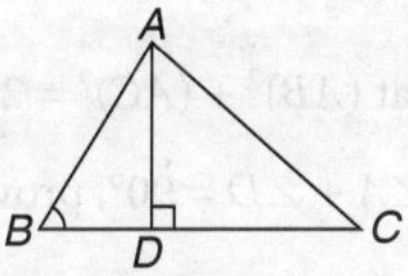

124. In this figure, AD is a median of a triangle ABC and $AM \perp BC$. Prove that: **[NCERT]**

(i) $AC^2 = AD^2 + BC \cdot DM + \left(\dfrac{BC}{2}\right)^2$ (ii) $AB^2 = AD^2 - BC \cdot DM + \left(\dfrac{BC}{2}\right)^2$

(iii) $AC^2 + AB^2 = 2\,AD^2 + \dfrac{1}{2}\,BC^2$

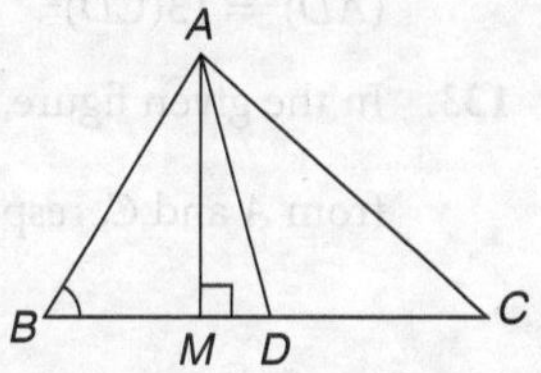

125. Prove that the sum of the squares of the diagonals of parallelogram is equal to the sum of the squares of its sides. **[NCERT]**

126. In this figure, two chords AB and CD intersect each other at the point P. Prove that: **[NCERT]**

(i) $\Delta APC \sim \Delta DPB$

(ii) $AP \cdot PB = CP \cdot DP$

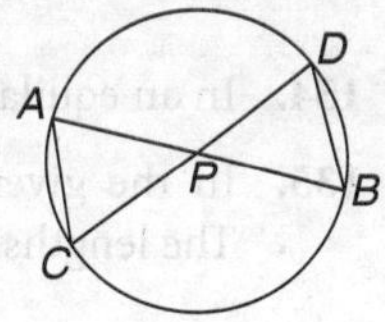

127. In this figure, two chords AB and CD of a circle intersect each other at the point P (when produced) outside the circle. Prove that: **[NCERT]**

(i) $\Delta PAC \sim \Delta PDB$

(ii) $PA \cdot PB = PC \cdot PD$

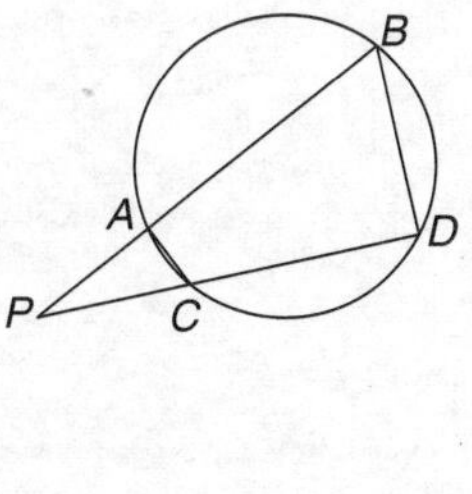

128. In this figure, D is a point on side BC of ΔABC such that $\frac{BD}{CD} = \frac{AB}{AC}$. Prove that AD is the bisector of $\angle BAC$. **[NCERT]**

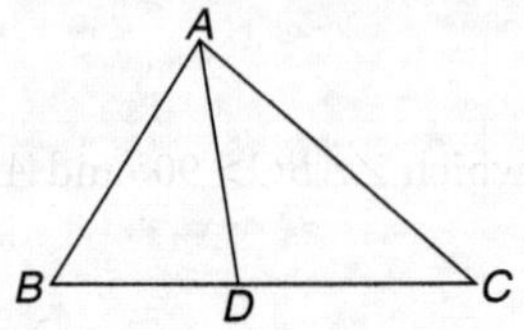

129. In the given figure, $\angle B < 90°$ and segment $AD \perp$ side BC show that $b^2 = h^2 + a^2 + x^2 - 2ax$.

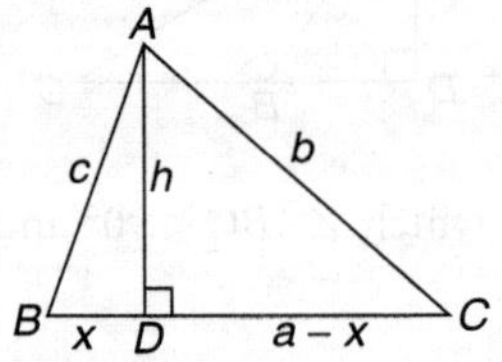

130. In a ΔABC, AD is a median. Prove that $(AB)^2 + (AC)^2 = 2(AD)^2 + 2(DC)^2$.

131. In a quadrilateral $ABCD$, given that $\angle A + \angle D = 90°$, prove that $(AC)^2 + (BD)^2 = (AD)^2 + (BC)^2$.

132. A point D is on the side BC of an equilateral triangle ABC such that $DC = \frac{1}{4}BC$. Prove that $(AD)^2 = 13(CD)^2$.

133. In the given figure, ABC is a right triangle, right-angled at B. AD and CE are the two medians drawn from A and C, respectively. If $AC = 5$ cm and $AD = \frac{3\sqrt{5}}{2}$ cm, find the length of CE.

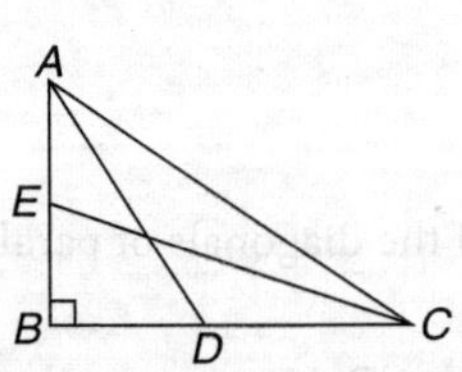

134. In an equilateral triangle PQR, the side QR is trisected at S. Prove that $9PS^2 = 7PQ^2$.

135. In the given figure, altitude BD is drawn to the hypotenuse AC of a right-angled triangle ABC. The lengths of different line segments are marked. Find x and y.

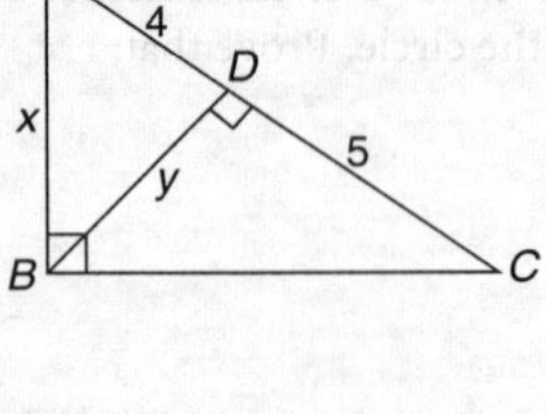

136. In the figure, P is the mid-point of BC and Q is the mid-point of AP. If BQ when produced meets AC at R, prove that $AR = \frac{1}{3} AC$.

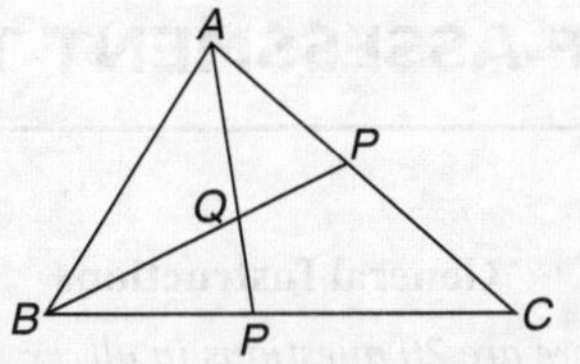

137. In a right triangle ABC, right-angled at C, P and Q are points on the sides CA and CB, respectively, which divide these sides in the ratio 1:2. Prove that

(i) $9AQ^2 = 9AC^2 + 4BC^2$ (ii) $9BP^2 = 9BC^2 + 4AC^2$ (iii) $9(AQ^2 + BP^2) = 13AB^2$

138. In ΔPQR, $PD \perp QR$ such that D lies on QR. If $PQ = a$, $PR = b$, $QD = c$ and $DR = d$, prove that $(a + b)(a - b) = (c + d)(c - d)$.

Answers

Multiple Choice Questions

1. (c) **2.** (c) **3.** (c) **4.** (d) **5.** (c) **6.** (b) **7.** (d) **8.** (d) **9.** (c) **10.** (a)
11. (c) **12.** (b) **13.** (a) **14.** (c) **15.** (c) **16.** (c)

Fill in the Blanks

17. $\frac{ar.\Delta ABC}{ar.\Delta PQR}$ **18.** Similar **19.** Congruent **20.** Square **21.** Similar
22. Basic Proportionality Theorem (BPT) **23.** Corresponding sides **24.** $\angle Q$

Assertion– Reason

25. (A) **26.** (A) **27.** (A) **28.** (A) **29.** (A)

Subjective Questions

31. 10 cm **32.** $x = \frac{ac}{b+c}$ **33.** = 2 cm **35.** 40° **36.** $x = 4$ **37.** = 3 **38.** = 7

41. (i) $x = \frac{21}{4}, y = \frac{15}{2}$ (ii) $x = \frac{28}{5}, y = \frac{7}{4}, z = \frac{35}{6}$ **42.** 7.5 cm **43.** $\frac{xz}{y}$ **44.** 7.5 cm **57.** 40°

59. = 1.6 m **60.** 9 : 4 **72.** Yes **73.** $PQ = 8$ cm, $QR = 12$ cm, $RS = 16$ cm. **79.** 9

81. (i) 2 cm (ii) 2.4 cm **82.** (i) No, (ii) Yes (iii) Yes **87.** 1 : 4 **89.** $\frac{1}{9}$ **90.** 5 : 8 **91.** 3 : 4

92. 1 : 8. **93.** (i) 25 : 4 (ii) 21 cm² (iii) 70 cm, 80 cm **94.** No, ratio will be $\frac{9}{25}$. **95.** 8 km

96. 9 : 1 **100.** 90° **101.** 25 cm **102.** 10 cm **104.** $2\sqrt{q-1}$ **105.** 13 cm

112. 6.5 m **116.** $\sqrt{3}\, a$ unit **117.** 144 cm² **118.** $6\sqrt{7}$ m **119.** 8 cm

133. $2\sqrt{5}$ cm **135.** $x = 6, y = 2\sqrt{5}$

Triangles (Basic)

SELF-ASSESSMENT TEST

Time allowed: 90 min. Max. Marks: 40

General Instructions

1. *All questions are compulsory. There are 20 questions in all.*
2. *This question paper has four sections: Section A, Section B, Section C and Section D.*
3. *Section A contains 10 questions of one mark each, Section B contains three questions of two marks each, Section C contains four questions of three marks each, and Section D contains three questions of four marks each.*

Section A [1 Mark Each]

Direction: (Q. Nos. 1–5) These are Multiple Choice Questions. Select the most appropriate answer for the given questions.

1. The corresponding sides of two similar triangles are in the ratio 5:7. The area of these triangles is in the ratio
   (a) $\sqrt{5}:\sqrt{7}$ (b) 7:5 (c) 25:49 (d) 49:25
2. Which of the following are not the sides of a right triangle?
   (a) 9 cm, 15 cm, 12 cm (b) 2 cm, 1 cm, $\sqrt{5}$ cm
   (c) 400 mm, 300 mm, 500 mm (d) 7 cm, 5 cm, 7 cm
3. In the figure, if $DE \parallel BC$, then x equals

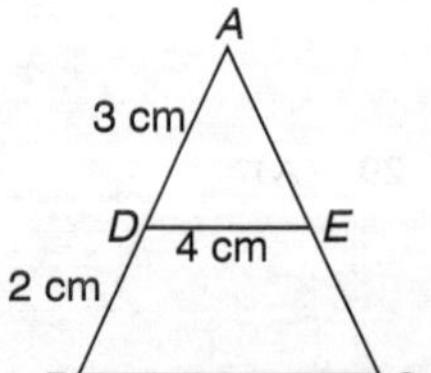

   (a) 3 cm (b) 2 cm (c) 4 cm (d) 6.7 cm
4. $\Delta ABC \sim \Delta PQR$. If $ar(\Delta ABC) = 2.25\ \text{m}^2$, $ar(\Delta PQR) = 6.25\ \text{m}^2$, $PQ = 0.5$ m, then the length of AB is
   (a) 30 cm (b) 0.5 m (c) 50 cm (d) 3 m
5. Two equilateral triangles are similar by
   (a) AAA similarity (b) SSS similarity (c) SAS similarity (d) all
6. In similar Δs ABC and PQR, $\frac{(AB)^2}{(PQ)^2} =$ ________.
7. If $\Delta ABC \sim \Delta PQR$, then $\angle B$ is equal to ________.
8. In ΔABC, D and E are points on the sides AB and AC, respectively, such that $DE \parallel BC$. If $AD = 6$ cm, $DB = 9$ cm and $AE = 8$ cm, find AC.
9. Let $\Delta ABC \sim \Delta DEF$ and their areas be, respectively, 64 cm² and 121 cm². If $EF = 15.4$ cm, find BC.
10. Two poles of height 6 m and 11 m stand on a plane ground. If the distance between their feet is 12 m, find the distance between their tops.

Section B [2 Marks Each]

11. ABC is an isosceles triangle with $AC = BC$. If $AB^2 = 2AC^2$, prove that ABC is a right triangle.

12. The diagonals of a trapezium $ABCD$ with $AB \parallel DC$ intersect each other at the point O. If $AB = 2\,CD$, find the ratio of the area of triangles AOB and COD.

13. $ABCD$ is a trapezium in which $AB \parallel DC$ and its diagonals intersect each other at the point O. Show that $\dfrac{AO}{BO} = \dfrac{CO}{DO}$.

Section C [3 Marks Each]

14. The perpendicular from A on side BC of a ΔABC intersects BC at D such that $DB = 3\,CD$ (see figure). Prove that $2AB^2 = 2AC^2 + BC^2$.

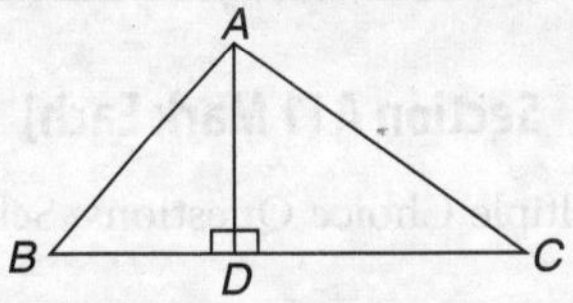

15. In the figure, ABC and DBC are two triangles on the same base BC. If AD intersects BC at O, show that $\dfrac{ar(ABC)}{ar(DBC)} = \dfrac{AO}{DO}$.

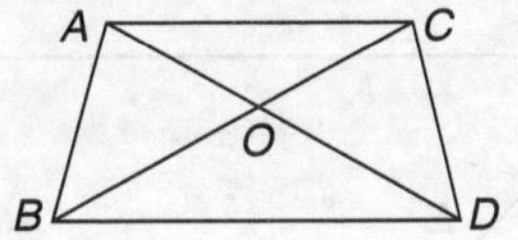

16. In the figure, $\dfrac{QR}{QS} = \dfrac{QT}{PR}$ and $\angle 1 = \angle 2$. Show that $\Delta PQS \sim \Delta TQR$.

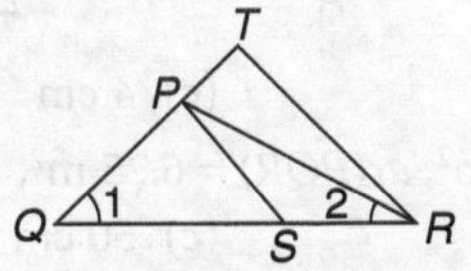

17. Prove that the area of an equilateral triangle described on one side of a square is equal to half the area of the equilateral triangle described on one of its diagonals.

Section D [4 Marks Each]

18. In an equilateral triangle ABC, D is a point on side BC such that $BD = \dfrac{1}{3}\,BC$. Prove that $9\,AD^2 = 7\,AB^2$.

19. State and prove Pythagoras' Theorem.

20. State and prove the Area Theorem.

Answers

1. (c) **2.** (d) **3.** (d) **4.** (a) **5.** (a) **6.** $\dfrac{ar\Delta ABC}{ar\Delta PQR}$ **7.** $\angle Q$ **8.** 20 cm

9. 11.2 cm **10.** 13 m **12.** $\dfrac{4}{1}$

Triangles (Standard)

SELF-ASSESSMENT TEST

Time allowed: 90 min. Max. Marks: 40

General Instructions

1. *All questions are compulsory. There are 20 questions in all.*
2. *This question paper has four sections: Section A, Section B, Section C and Section D.*
3. *Section A contains 10 questions of one mark each, Section B contains three questions of two marks each, Section C contains four questions of three marks each, and Section D contains three questions of four marks each.*

Section A [1 Mark Each]

Direction: (Q. Nos. 1–5) These are Multiple Choice Questions. Select the most appropriate answer for the given questions.

1. The lengths of the diagonals of a rhombus are 24 cm and 32 cm. The perimeter of the rhombus is
   (a) 9 cm (b) 128 cm (c) 80 cm (d) 56 cm
2. In the figure, $ABCD$ is a trapezium in which $AB \parallel EF \parallel DC$. The length of AE is

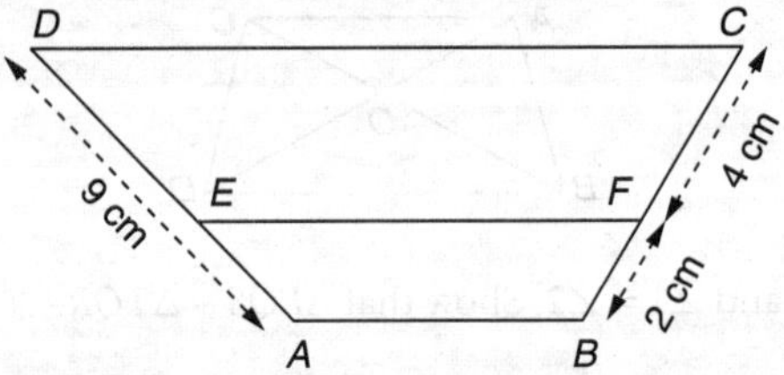

   (a) 2 cm (b) 3 cm (c) 4 cm (d) 7 cm
3. $\Delta ABC \sim \Delta PQR$. If $ar(ABC) = 2.25$ m^2, $ar(PQR) = 6.25$ m^2, $PQ = 0.5$ m, then the length of AB is
   (a) 30 cm (b) 0.5 m (c) 50 cm (d) 3 m
4. If the ratio of the corresponding sides of two similar triangles is 2:3, then the ratio of their corresponding altitudes is
   (a) 3:2 (b) 4:9 (c) 2:3 (d) 16:81
5. In the figure, $\Delta ABC \sim \Delta DEF$. If $AB = 3DE$, $ar(ABC) = 144$ cm^2, then $ar(DEF)$ is

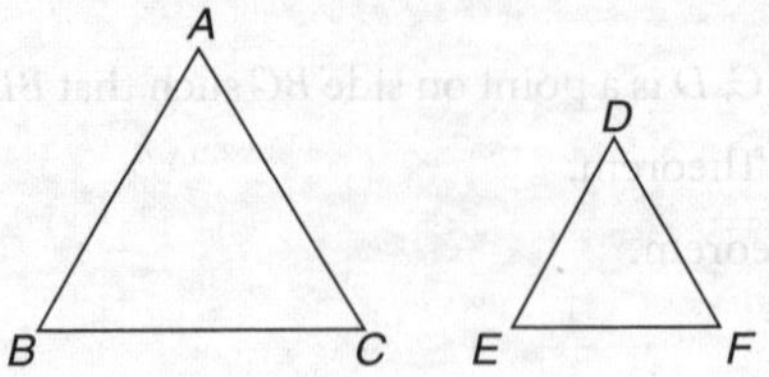

   (a) 46 cm^2 (b) 16 cm^2 (c) 12 cm^2 (d) 432 cm^2
6. If the areas of two similar triangles are equal, then they are __________.
7. The perimeter of two similar triangles ABC and LMN are 60 cm and 48 cm, respectively. If $LM = 8$ cm, then what is the length of AB?

8. In the figure, $\angle M = \angle N = 46°$. Express x in terms of a, b and c, where a, b and c are the lengths of LM, MN and NK, respectively.

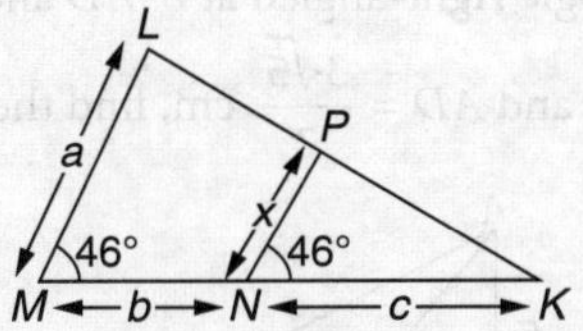

9. In the figure, if $AB \parallel DC$, find the value of x.

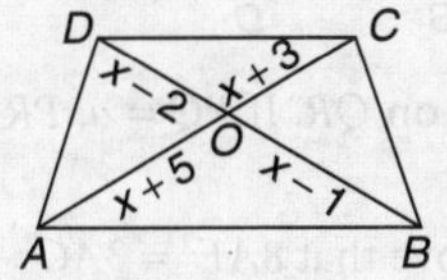

10. In the figure, $PQ = 24$ cm, $QR = 26$ cm, $\angle PAR = 90°$, $PA = 6$ cm and $AR = 8$ cm. Find $\angle QPR$.

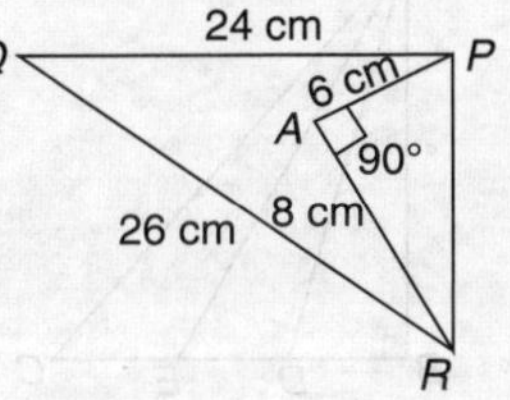

Section B [2 Marks Each]

11. In the figure, $\angle B < 90°$ and segment $AD \perp$ side BC. Show that $b^2 = h^2 + a^2 + x^2 - 2ax$.

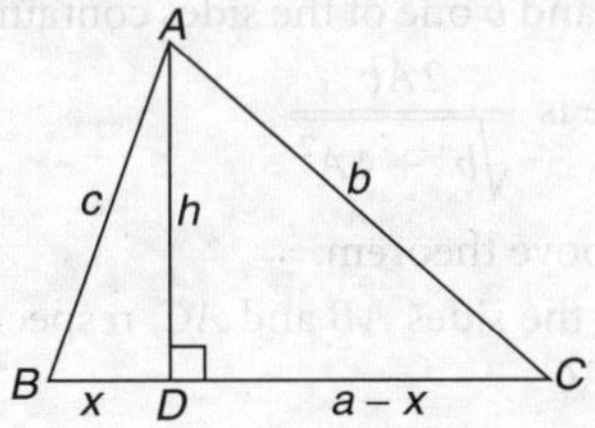

12. In the figure, $AB \parallel PQ \parallel CD$, $AB = x$ units, $CD = y$ units and $PQ = z$ units. Prove that $\dfrac{1}{x} + \dfrac{1}{y} = \dfrac{1}{z}$.

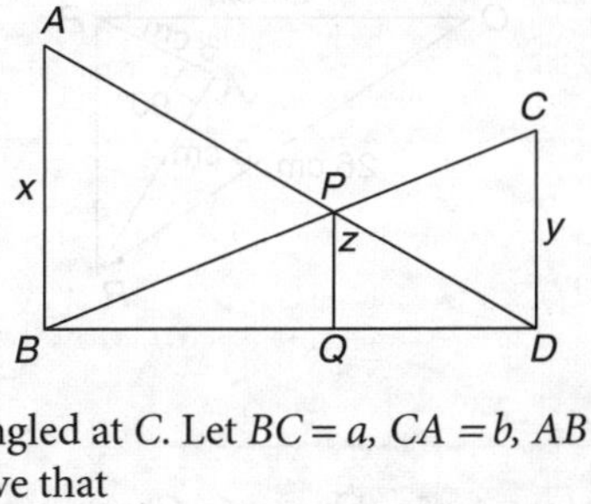

13. ABC is a right triangle right-angled at C. Let $BC = a$, $CA = b$, $AB = c$ and let p be the length of the perpendicular from C on AB. Prove that

(i) $cp = ab$ (ii) $\dfrac{1}{p^2} = \dfrac{1}{a^2} + \dfrac{1}{b^2}$

Section C [3 Marks Each]

14. In the figure, ABC is a right triangle right-angled at B. AD and CE are the two medians drawn from A and C, respectively. If $AC = 5$ cm and $AD = \frac{3\sqrt{5}}{2}$ cm, find the length of CE.

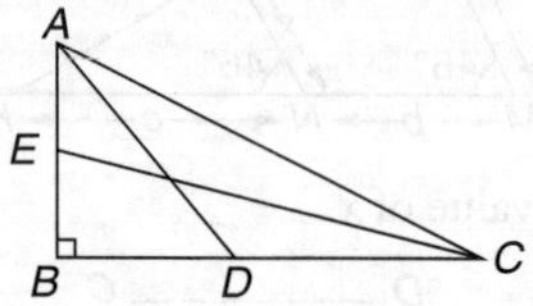

15. In ΔPQR, $PD \perp QR$ such that D lies on QR. If $PQ = a$, $PR = b$, $QD = c$ and $DR = d$, prove that $(a + b)(a - b) = (c + d)(c - d)$.

16. In the figure, D and E trisect BC. Prove that $8AE^2 = 3AC^2 + 5AD^2$.

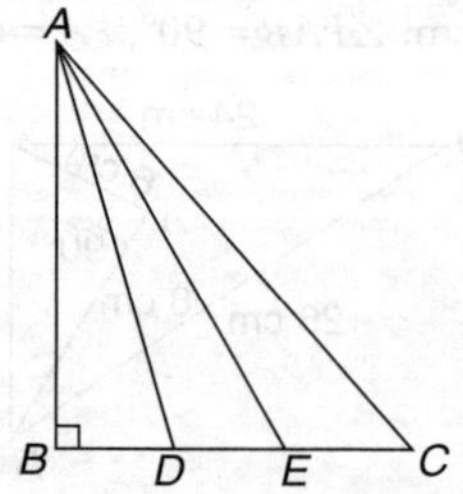

17. In an equilateral triangle ABC, D is a point on side BC such that $BD = \frac{1}{3}BC$. Prove that $9AD^2 = 7AB^2$.

Section D [4 Marks Each]

18. If A is the area of a right triangle and b one of the sides containing the right angle, prove that the length of the altitude on the hypotenuse is $\frac{2Ab}{\sqrt{b^4 + 4A^2}}$.

19. State and prove BPT using the above theorem.
In ΔABC, D and E are points on the sides AB and AC, respectively, such that $DE \parallel BC$. If $AD = 6$ cm, $DB = 9$ cm and $AE = 8$ cm, find AC.

20. State and prove the converse of Pythagoras' Theorem.
In the figure, $PQ = 24$ cm, $QR = 26$ cm, $\angle PAR = 90°$, $PA = 6$ cm and $AR = 8$ cm. Find $\angle QPR$.

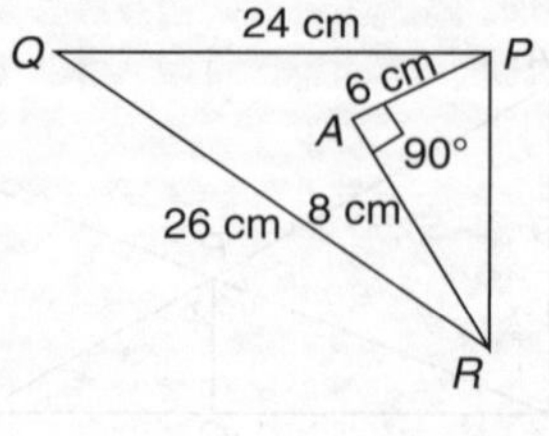

Answers

1. (c) **2.** (b) **3.** (a) **4.** (c) **5.** (b) **6.** Congruent **7.** 10 cm **8.** $x = \frac{ac}{b+c}$
9. 7 **10.** 90° **14.** $2\sqrt{5}$ cm **19.** 20 cm

CHAPTER

7 Coordinate Geometry

Key Concepts

- Distance formula
- Section formula
- Area of a triangle

Concept: Distance Formula

Before discussing the distance formula, we need to keep in mind that the distance of a point from the *y* axis is called its *x* coordinate or abscissa and the distance of a point from the *x* axis is called its *y* coordinate, or ordinate. The coordinates of a point on the *x* axis are of the form $(x, 0)$, and of a point on the *y* axis are of the form $(0, y)$.

The distance between two points $A(x_1, y_1)$ and $B(x_2, y_2)$ is given by the formula

$$AB = \sqrt{(x_2 - x_1)^2 + (y_2 - y_1)^2}$$

Questions that arise from the distance formula:

(i) To prove that the given three points are collinear: By collinear, we mean that all the three points lie on the same line. For example, points *A*, *B* and *C* (see figure).
If the sum of any two of these distances is equal to the third distance, then the given points are collinear. For example:

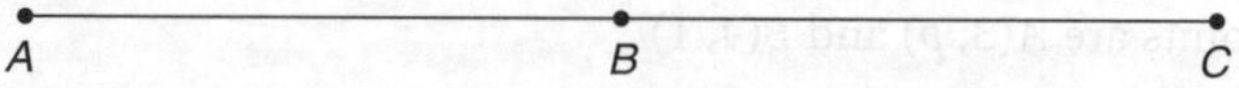

Here, $AB + BC = AC$ means that all the three points are collinear.

(ii) To prove that a triangle is isosceles or equilateral: If the lengths of two sides are equal, then it is an isosceles triangle; and if all three sides are equal, then it is said to be an equilateral triangle.

(iii) To prove whether three points form a triangle, or not: If the sum of any two sides is greater than the third side, or not.

(iv) For miscellaneous questions, we must keep in mind the following properties of various types of quadrilaterals.

Concept: Properties of Types of Quadrilaterals

A quadrilateral is a:

(i) Parallelogram: If its opposite sides are equal and parallel.
(ii) Rhombus: If all of its sides are equal and the diagonals are not equal.
(iii) Rectangle: If its opposite sides are equal and the diagonals are equal.

(iv) **Square:** If all of its sides are equal and the diagonals are equal.
(To prove that a given quadrilateral is a parallelogram, rhombus, rectangle or square, we need to prove the above properties accordingly.)

(v) **Parallelogram but not a rectangle:** If its opposite sides are equal and the diagonals are not equal.

(vi) **Rhombus but not a square:** If all of its sides are equal and the diagonals are not equal.

Example 1 Find the distance between the points A (2, 4) and B (–1, 5).

Solution: The given points are $A(2, 4)$ and $B(-1, 5)$

Here $x_1 = 2, x_2 = -1$

$y_1 = 4, y_2 = 5$

$\therefore \quad AB = \sqrt{(x_2 - x_1)^2 + (y_2 - y_1)^2} = \sqrt{(-1-2)^2 + (5-4)^2}$

$= \sqrt{9+1} = \sqrt{10}$ units.

1 Try to Solve

1. Find the distance between the following pairs of points:
(a) (2, 3), (4, 1) (b) (–5, 7), (–1, 3) (c) $(a, b), (-a, -b)$ [NCERT]

Answers

1. (a) $2\sqrt{2}$ (b) $4\sqrt{2}$ (c) $2\sqrt{a^2+b^2}$

Example 2 Find the value of p when the distance between the points $A(3, p)$ and $B(4, 1)$ is $\sqrt{10}$.

Solution: The given points are $A(3, p)$ and $B(4, 1)$

$AB = \sqrt{10}$

Here $x_1 = 3, y_1 = p$

$x_2 = 4, y_2 = 1$

$\therefore \quad AB = \sqrt{(x_2 - x_1)^2 + (y_2 - y_1)^2}$

$\Rightarrow \quad \sqrt{10} = \sqrt{(4-3)^2 + (1-p)^2}$

$\Rightarrow \quad \sqrt{10} = \sqrt{1+1+p^2-2p}$ [Using $(a - b)^2 = a^2 + b^2 - 2ab$]

$\Rightarrow \quad \sqrt{10} = \sqrt{p^2-2p+2}$

$\Rightarrow \quad 10 = p^2 - 2p + 2$ [On squaring both the sides]

$\Rightarrow \quad p^2 - 2p - 8 = 0$

$$\Rightarrow \quad p^2 - 4p + 2p - 8 = 0$$
$$\Rightarrow \quad p(p-4) + 2(p-4) = 0$$
$$\Rightarrow \quad (p-4)(p+2) = 0$$
$$\Rightarrow \quad p - 4 = 0 \text{ or } p + 2 = 0$$
$$\therefore \quad p = 4 \text{ or } p = -2$$

Hence, the value of p is 4 or –2.

2 Try to Solve

1. Find the value of y if the distance between the points (5, –9) and (11, y) is 10 units.
2. For what value of x will the distance between the points (–2, 5) and (x, 19) be $\sqrt{205}$ units.
3. Find the values of y for which the distance between the points $P(2, -3)$ and $Q(10, y)$ is 10 units. **[NCERT]**

Answers

1. –17 or –1   2. –5, 1   3. –9, 3

Example 3 Find the point on the x axis which is equidistant from the points (–2, 5) and (2, 9). **[CBSE 2009]**

Solution: Let $A(-2, 5)$ and $B(2, 9)$ be the given points.
Let $P(x, 0)$ be a point on the x axis which is equidistant from points $A(-2, 5)$ and $B(2, 9)$.

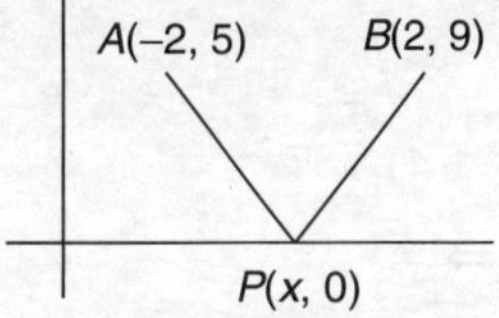

Now $$PA = \sqrt{(x+2)^2 + (0-5)^2}$$
$$= \sqrt{x^2 + 4x + 4 + 25} \quad \text{[Using } (a+b)^2 = a^2 + b^2 + 2ab\text{]}$$
$$= \sqrt{x^2 + 4x + 29}$$

And $$PB = \sqrt{(x-2)^2 + (0-9)^2}$$
$$= \sqrt{x^2 - 4x + 4 + 81} = \sqrt{x^2 - 4x + 85}$$

Since $$PA = PB \quad \text{[Given]}$$
$$\Rightarrow \quad \sqrt{x^2 + 4x + 29} = \sqrt{x^2 - 4x + 85}$$
$$\Rightarrow \quad \cancel{x^2} + 4x + 29 = \cancel{x^2} - 4x + 85 \quad \text{[On squaring both the sides]}$$
$$\Rightarrow \quad 4x + 4x = 85 - 29$$
$$\Rightarrow \quad 8x = 56$$
$$\therefore \quad x = 7$$

Hence, the coordinates of the point on the x axis is (7, 0).

Example 4 Find a point on the y axis which is equidistant from the points $A(6, 5)$ and $B(-4, 3)$.

[CBSE (CCE) 2013]

Solution: Let $P(0, y)$ be the point on the y axis which is equidistant from the points $A(6, 5)$ and $B(-4, 3)$.

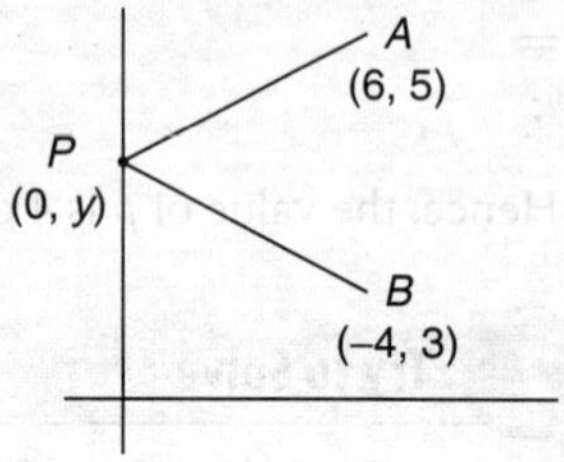

Now $$PA = \sqrt{(6-0)^2 + (5-y)^2}$$

$$= \sqrt{36 + 25 + y^2 - 10y} \quad \left[\text{Using } (a-b)^2 = a^2 - 2ab + b^2\right]$$

$$= \sqrt{61 + y^2 - 10y} = \sqrt{y^2 - 10y + 61}$$

And $$PB = \sqrt{(-4-0)^2 + (3-y)^2}$$

$$= \sqrt{16 + 9 + y^2 - 6y} = \sqrt{y^2 - 6y + 25}$$

$$PA = PB \quad \text{[Given]}$$

$\Rightarrow$ $$\sqrt{y^2 - 10y + 61} = \sqrt{y^2 - 6y + 25}$$

Squaring both sides $$\cancel{y^2} - 10y + 61 = \cancel{y^2} - 6y + 25$$

$\Rightarrow$ $$-10y + 61 = -6y + 25$$

$\Rightarrow$ $$-10y + 6y = 25 - 61$$

$$-4y = -36$$

$$\cancel{-}4y = \cancel{-}36$$

$\Rightarrow$ $$4y = 36$$

$\therefore$ $$y = 9$$

Hence, the coordinates of the point on the y axis is (0, 9).

3 Try to Solve

1. Find the point on the x axis which is equidistant from (2, –5) and (–2, 9). **[NCERT]**
2. If $Q(0, 1)$ is equidistant from $P(5, -3)$ and $R(x, 6)$, find the values of x. Also find the distances QR and PR. **[NCERT]**
3. Find the coordinates of the point on the x axis which is equidistant from the points (–2, 5) and (2, –3). **[CBSE 2017]**

Answers

1. (–7, 0) **2.** ± 4, $QR = \sqrt{41}$, $PR = \sqrt{82}$ when $x = 4$, $PR = 9\sqrt{2}$ when $x = -4$

3. (–2, 0)

Example 5 Prove that the points $A(1, 1)$, $B(-2, 7)$ and $C(3, -3)$ are collinear.

Solution: Let $A(1, 1)$, $B(-2, 7)$ and $C(3, -3)$ be the given points. Then

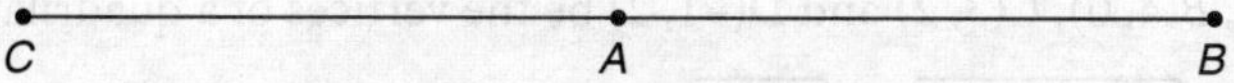

$$AB = \sqrt{(-2-1)^2 + (7-1)^2} = \sqrt{(-3)^2 + 6^2} = \sqrt{45} = 3\sqrt{5} \text{ units}$$

$$BC = \sqrt{(3+2)^2 + (-3-7)^2} = \sqrt{5^2 + (-10)^2} = \sqrt{125} = 5\sqrt{5} \text{ units}$$

$$AC = \sqrt{(3-1)^2 + (-3-1)^2} = \sqrt{2^2 + (-4)^2} = \sqrt{20} = 2\sqrt{5} \text{ units}$$

$$\therefore \quad AB + AC = (3\sqrt{5} + 2\sqrt{5}) \text{ units} = 5\sqrt{5} \text{ units} = BC.$$

Thus, $AB + AC = BC$.

Hence, the given points A, B, C are collinear.

4 Try to Solve

1. Determine whether the points (1, 5), (2, 3) and (–2, –11) are collinear. [NCERT]
2. Using the distance formula, show that the given points are collinear:
   (a) (1, –1), (5, 2) and (9, 5)
   (b) (6, 9), (0, 1) and (–6, –7)
   (c) (–1, –1), (2, 3) and (8, 11)
   (d) (–2, 5), (0, 1) and (2, –3)

Answer

1. No

Example 6 Do the points $A(-5, 6)$, $B(-4, -2)$ and $C(7, 5)$ form a triangle? If so, name the type of the triangle formed.

Solution: The given points are $A(-5, 6)$, $B(-4, -2)$ and $C(7, 5)$.

$$AB = \sqrt{(-4+5)^2 + (-2-6)^2}$$

$$= \sqrt{1^2 + 8^2} = \sqrt{1+64} = \sqrt{65}$$

$$BC = \sqrt{(7+4)^2 + (5+2)^2}$$

$$= \sqrt{11^2 + 7^2} = \sqrt{121+49} = \sqrt{170}$$

$$AC = \sqrt{(7+5)^2 + (5-6)^2}$$

$$= \sqrt{12^2 + (-1)^2} = \sqrt{144+1} = \sqrt{145}$$

Clearly, $AB + BC > AC$; $BC + AC > AB$ and $AB + AC > BC$.

$\therefore$ The three points form a triangle. Neither of the sides are equal, nor do they satisfy Pythagoras' Theorem. Hence, the triangle is scalene.

Example 7 The vertices of a quadrilateral are (–3, 0), (4, 0), (3, 2) and (–1, 2). Which type of quadrilateral is it?

Solution: Let $A(-3, 0)$, $B(4, 0)$, $C(3, 2)$ and $D(-1, 2)$ be the vertices of a quadrilateral.

$$AB = \sqrt{(4+3)^2 + 0^2} = \sqrt{7^2 + 0} = 7$$

$$BC = \sqrt{(3-4)^2 + 2^2} = \sqrt{1+4} = \sqrt{5}$$

$$CD = \sqrt{(-1-3)^2 + (2-2)^2} = \sqrt{16+0} = 4$$

$$DA = \sqrt{(-3+1)^2 + (0-2)^2} = \sqrt{4+4} = \sqrt{8} = 2\sqrt{2}$$

$$AC = \sqrt{(3+3)^2 + 2^2} = \sqrt{36+4} = \sqrt{40} = 2\sqrt{10}$$

$$BD = \sqrt{(-1-4)^2 + 2^2} = \sqrt{25+4} = \sqrt{29}$$

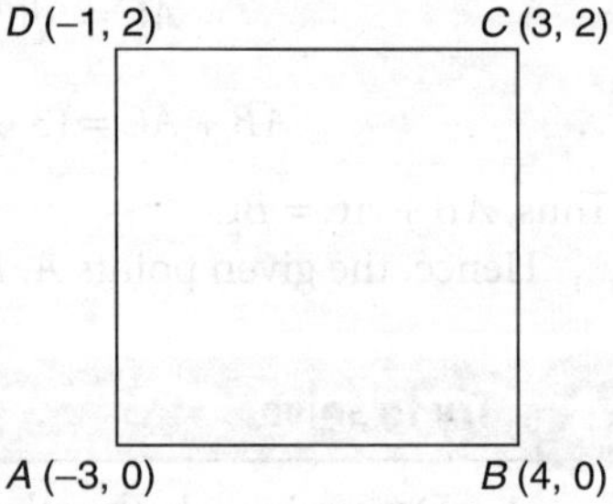

Here, the ordinate of point C and D is the same, that is, 2. A and B are 0.

$\therefore$ $$AB \parallel CD$$

In quadrilateral $ABCD$

(i) Sides are not equal **(ii)** Diagonals are not equal **(iii)** $AB \parallel CD$

$\therefore$ $ABCD$ is a trapezium.

Example 8 Prove that the points (3, 0), (6, 4) and (–1, 3) are the vertices of a right-angled isosceles triangle. **[CBSE 2006 (C)]**

Solution: Let $A(3, 0)$, $B(6, 4)$ and $C(-1, 3)$ be the three given points.

Now

$$AB = \sqrt{(6-3)^2 + (4-0)^2}$$

$$= \sqrt{3^2 + 4^2} = \sqrt{9+16} = \sqrt{25} = 5$$

$$BC = \sqrt{(-1-6)^2 + (3-4)^2} = \sqrt{(-7)^2 + (-1)^2}$$

$$= \sqrt{49+1} = \sqrt{50} = \sqrt{5 \times 5 \times 2} = 5\sqrt{2}$$

$$AC = \sqrt{(-1-3)^2 + (3-0)^2} = \sqrt{(-4)^2 + (3)^2}$$

$$= \sqrt{16+9} = 5$$

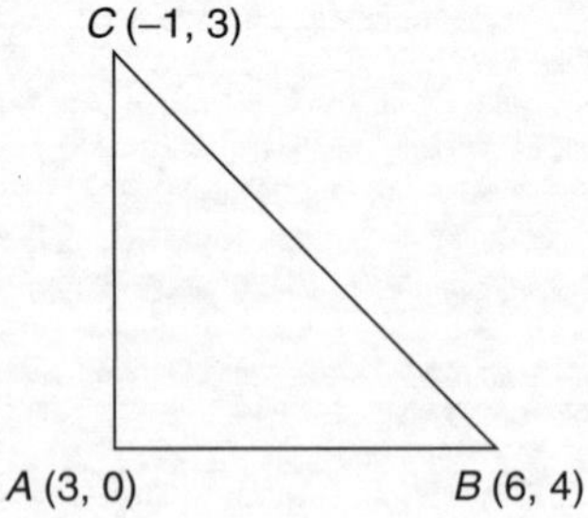

Here, $AB = AC$. $\therefore$ ΔABC is isosceles.

Now $$AB^2 + AC^2 = (5)^2 + (5)^2 = 50$$

$\Rightarrow$ $$= (5\sqrt{2})^2 = BC^2$$

$\therefore$ $$AB^2 + AC^2 = BC^2$$

So, ΔABC is right-angled at A.

Hence, points $A(3, 0)$, $B(6, 4)$ and $C(-1, 3)$ are the vertices of a right-angled isosceles triangle.

5 Try to Solve

1. Do the points (3, 2), (–2, –3) and (2, 3) form a triangle? If so, name the type of triangle formed. [NCERT]
2. Show that the points (1, 7), (4, 2), (–1, –1) and (–4, 4) are the vertices of a square. [NCERT]
3. Check whether (5, –2), (6, 4) and (7, – 2) are the vertices of an isosceles triangle. [NCERT]
4. Name the type of quadrilateral formed, if any, by the following points, and give reasons for your answer:
   (a) (–1, –2), (1, 0), (–1, 2), (–3, 0)
   (b) (–3, 5), (3, 1), (0, 3), (–1, –4)
   (c) (4, 5), (7, 6), (4, 3), (1, 2) [NCERT]
5. If the point $P(2, 2)$ is equidistant from the points $A(-2, k)$ and $B(-2k, -3)$, find k. Also, find the length of AP. [CBSE 2014]

Answers

1. Yes, Right Δ. 3. Yes 4. (a) Square (b) No quadrilateral (c) Parallelogram
5. $k = -3, \Rightarrow AP = \sqrt{41}$, $k = -1 \Rightarrow AP = 5$

Example 9 Find a relation between x and y such that the point (x, y) is equidistant from the points (7, 1) and (3, 5). [CBSE (CCE) 2013]

Solution: Let $P(x, y)$ be a point which is equidistant from $A(7, 1)$ and $B(3, 5)$.

$$\therefore \quad PA = \sqrt{(x-7)^2 + (y-1)^2} = \sqrt{x^2 - 14x + 49 + y^2 - 2y + 1}$$

$$= \sqrt{x^2 - 14x + y^2 - 2y + 50}$$

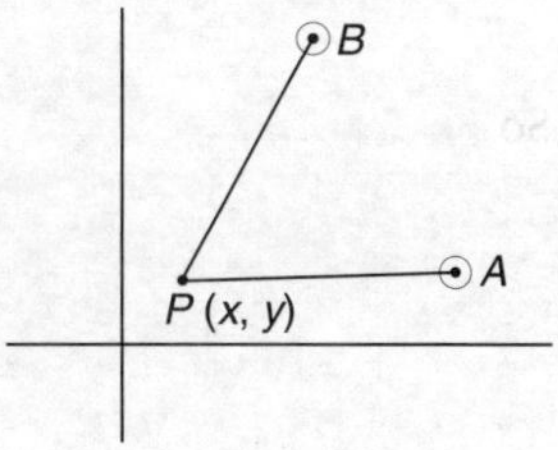

And
$$PB = \sqrt{(x-3)^2 + (y-5)^2}$$

$$= \sqrt{x^2 - 6x + 9 + y^2 - 10y + 25}$$

$$= \sqrt{x^2 - 6x + y^2 - 10y + 34}$$

$$PA = PB \quad \text{[Given]}$$

$$\sqrt{x^2 - 14x + y^2 - 2y + 50} = \sqrt{x^2 - 6x + y^2 - 10y + 34}$$

Squaring both sides

$$\Rightarrow \quad \cancel{x^2} - 14x + \cancel{y^2} - 2y + 50 = \cancel{x^2} - 6x + \cancel{y^2} - 10y + 34$$

$$\Rightarrow \quad -14x - 2y + 50 = -6x - 10y + 34$$

$$\Rightarrow \quad -14x + 6x - 2y + 10y + 50 - 34 = 0$$

$$\Rightarrow \quad -8x + 8y + 16 = 0$$

$$\Rightarrow \quad x - y - 2 = 0 \quad \text{[On dividing both the sides by } -8]$$

Hence, the relation between x and y is $x - y - 2 = 0$.

6 Try to Solve

1. Find a relation between x and y such that the point (x, y) is equidistant from the points (3, 6) and (–3, 4). **[NCERT]**
2. Find the relation between x and y if the points $A(x, y)$, $B(-5, 7)$ and $C(-4, 5)$ are collinear. **[AI 2015]**

Answers

1. $3x + y - 5 = 0$ 2. $2x + y + 3 = 0$

Concept: Section Formula

It is used to find the coordinate of a point which divides a line segment internally.

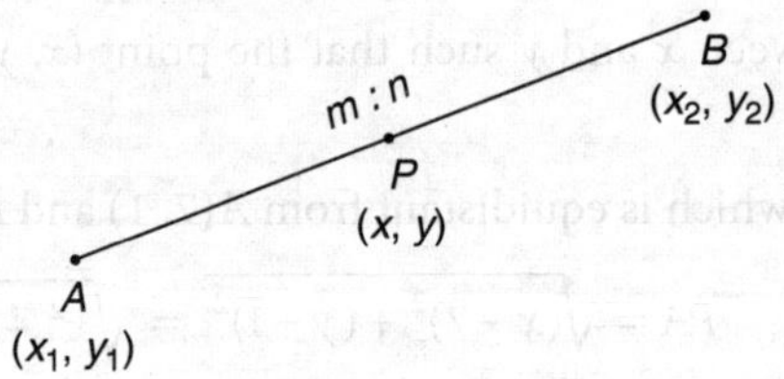

So
$$x = \frac{mx_2 + nx_1}{m+n}, \; y = \frac{my_2 + ny_1}{m+n}$$

OR

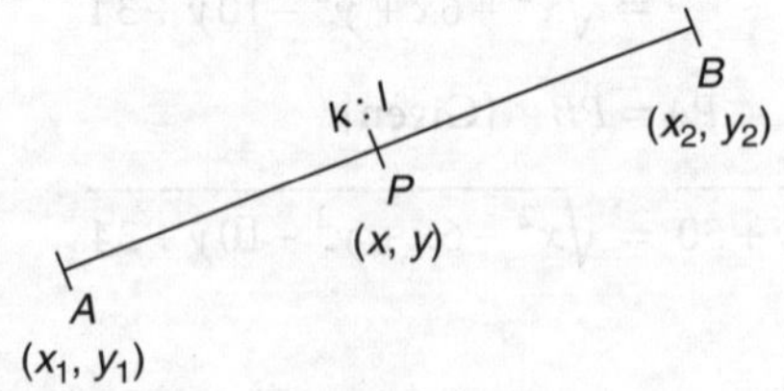

$$x = \frac{kx_2 + x_1.1}{k+1}, \; y = \frac{ky_2 + y_1.1}{k+1}$$

Concept: Mid-point Formula

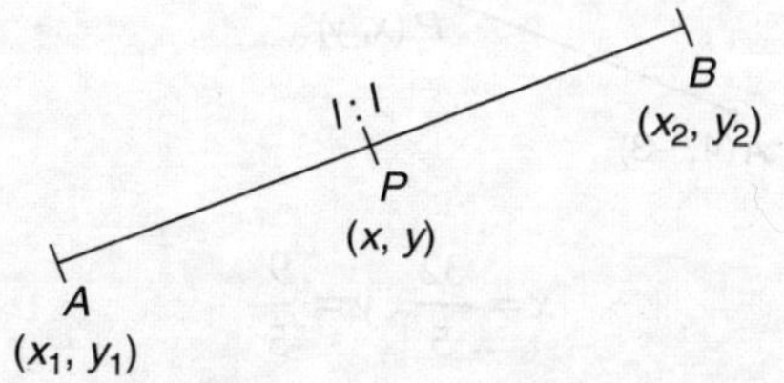

Say, P is the mid-point of AB.

So
$$x = \frac{x_2 + x_1}{2}, y = \frac{y_2 + y_1}{2}$$

Concept: Centroid of a Triangle

It is the point of intersection of the medians of a triangle in the ratio 1:2.

That is $\frac{AG}{GD} = \frac{BG}{GE} = \frac{CG}{GF} = \frac{2}{1}$.

Let $A(x_1, y_1)$, $B(x_2, y_2)$ and $C(x_3, y_3)$ be the coordinates of ΔABC and $G(x, y)$ be the coordinates of the centroid of the Δ. Then

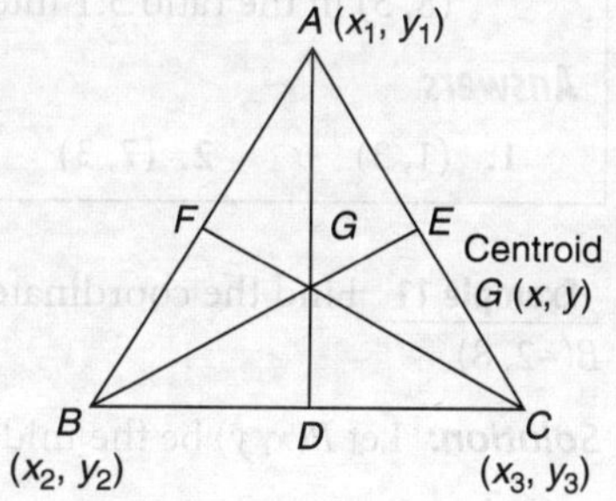

$$x = \frac{x_1 + x_2 + x_3}{3}, y = \frac{y_1 + y_2 + y_3}{3}$$

$\therefore$ Coordinates of centroid of ΔABC is $G\left(\frac{x_1 + x_2 + x_3}{3}, \frac{y_1 + y_2 + y_3}{3}\right)$

Example 10 Find the coordinates of the point which divides the line segment joining points (4, –3) and (8, 5) in the ratio 3:2 internally.

Solution: Let $P(x, y)$ divide the line segment internally in the ratio 3:2.

Here
$$x_1 = 4, y_1 = -3, m = 3$$
$$x_2 = 8, y_2 = 5, n = 2$$

By section formula
$$x = \frac{mx_2 + nx_1}{m+n}, y = \frac{my_2 + ny_1}{m+n}$$

$\Rightarrow$
$$x = \frac{3\times 8 + 2\times 4}{3+2}, y = \frac{3\times 5 + 2\times(-3)}{3+2}$$

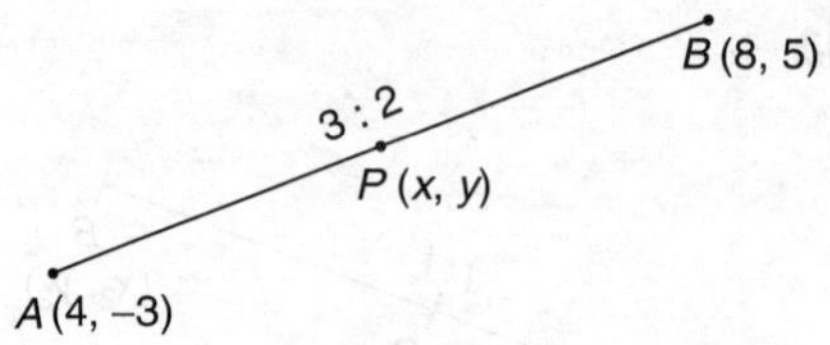

$$\Rightarrow \quad x = \frac{32}{5}, y = \frac{9}{5}$$

$\therefore$ Coordinates of the point P are $\left(\frac{32}{5}, \frac{9}{5}\right)$.

7 Try to Solve

1. Find the coordinates of the point which divides the join of (–1, 7) and (4, –3) in the ratio 2:3. **[NCERT]**
2. Find the coordinates of the point which divides the line segment joining the points (4, –3) and (8, 5) in the ratio 3:1 internally. **[NCERT]**

Answers

1. (1, 3)
2. (7, 3)

Example 11 Find the coordinates of the mid-point of the line segment joining the points $A(6, -2)$ and $B(-2, 8)$.

Solution: Let $P(x, y)$ be the mid-point of line segment AB.

Here $\quad x_1 = 6, y_1 = -2$

$\quad x_2 = -2, y_2 = 8$

We know the coordinates of the mid-point are given by

$$x = \frac{x_1 + x_2}{2} \text{ and } y = \frac{y_1 + y_2}{2}$$

$$\Rightarrow \quad x = \frac{6-2}{2} \text{ and } y = \frac{-2+8}{2}$$

$$x = \frac{4}{2}, y = \frac{6}{2}$$

$$\therefore \quad x = 2 \text{ and } y = 3$$

Therefore, the coordinates of P are (2, 3).

8 Try to Solve

1. Find the coordinates of the mid-point of the line segment joining:
   (a) $A(3, 0)$ and $B(-5, 4)$ (b) $P(-11, -8)$ and $Q(8, -2)$
2. If $(2, p)$ is the mid-point of the line segment joining the points $A(6, -5)$ and $B(-2, 11)$, find the value of p. **[CBSE 2010]**

Answers

1. (a) $(-1, 2)$ (b) $\left(-\frac{3}{2}, -5\right)$ **2.** $p = 3$

Example 12 Find the coordinates of the points of trisection of the line segment joining the points $A(-5, 6)$ and $B(4, -3)$. **[CBSE 2009 (C)]**

Solution: Let P and Q be the points of trisection of AB.

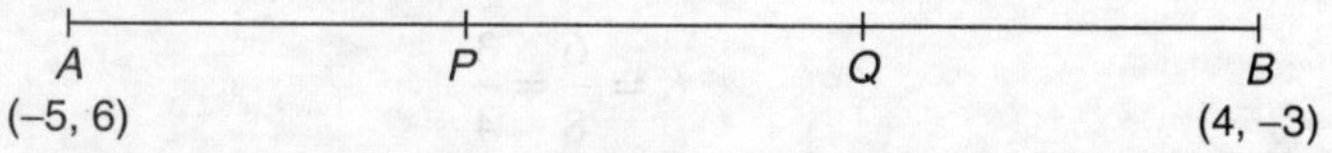

Then, P divides AB in the ratio 1:2. So, the coordinates of P are

$$P\left(\frac{1\times4+2\times(-5)}{1+2}, \frac{1\times(-3)+2\times6}{1+2}\right), \text{ that is, } P(-2, 3)$$

Also, Q divides AB in the ratio 2:1. So, the coordinates of Q are

$$Q\left(\frac{2\times4+1\times(-5)}{2+1}, \frac{2\times(-3)+1\times6}{2+1}\right), \text{ that is, } Q(1, 0)$$

Hence, the points of trisection of AB are $P(-2, 3)$ and $Q(1, 0)$.

9 Try to Solve

1. Find the coordinates of the points of trisection of the line segment joining $(4, -1)$ and $(-2, -3)$. **[NCERT]**
2. Find the coordinates of the points of trisection (points dividing a line in three equal parts) of the line segment joining the points $A(2, -2)$ and $B(-7, 4)$. **[NCERT]**

Answers

1. $\left(2, \frac{-5}{3}\right); \left(0, \frac{-7}{3}\right)$ **2.** $(-1, 0)$ and $(-4, 2)$

Example 13 In what ratio does the point $P(2, 5)$ divide the line joining $A(8, 2)$ and $B(-6, 9)$?

Solution: Let point $P(2, 5)$ divide AB in the ratio K:1.

Coordinates of point $P = \left(\frac{K\cdot(-6)+1.8}{K+1}, \frac{K\cdot9+1.2}{K+1}\right) = (2, 5)$ [Given]

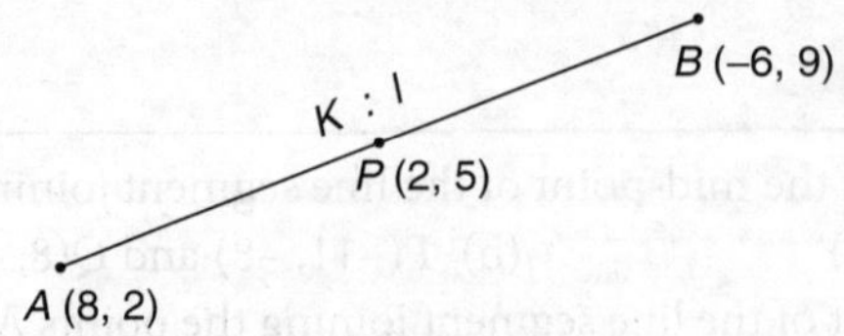

$$\Rightarrow \quad \left(\frac{-6K+8}{K+1}, \frac{9K+2}{K+1}\right) = (2, 5)$$

$$\Rightarrow \quad \frac{-6K+8}{K+1} = 2$$

$$\Rightarrow \quad -6K+8 = 2K+2$$

$$\Rightarrow \quad -6K-2K = 2-8$$

$$\Rightarrow \quad -8K = -6$$

$$\therefore \quad K = \frac{6}{8} = \frac{3}{4}$$

Hence, the required ratio is $\frac{3}{4}$: 1, that is, 3:4.

Example 14 Find the ratio in which the point $P(m, 6)$ divides the line joining $A(-4, 3)$ and $B(2, 8)$. Also find the value of m.

Solution: Let the point $P(m, 6)$ divide AB in the ratio K:1. Coordinate of point P

$$= \left(\frac{K\cdot 2+1(-4)}{K+1}, \frac{K\cdot 8+1.3}{K+1}\right) = (m, 6) \qquad \text{[Given]}$$

$$\Rightarrow \quad \left(\frac{2K-4}{K+1}, \frac{8K+3}{K+1}\right) = (m, 6)$$

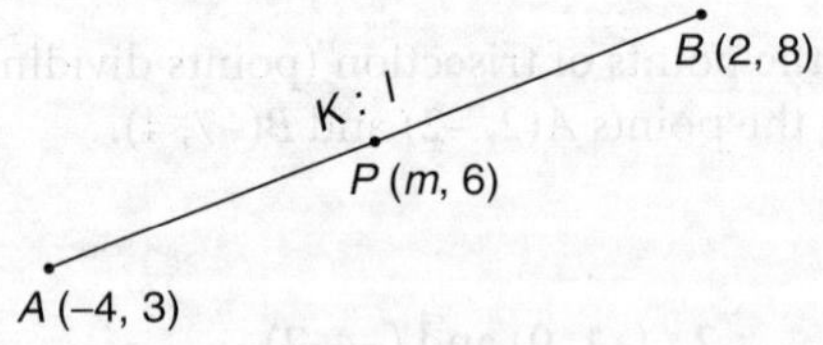

$$\Rightarrow \quad \frac{8K+3}{K+1} = 6$$

$$\Rightarrow \quad 8K+3 = 6K+6$$

$$\Rightarrow \quad 8K - 6K = 6 - 3$$

$$\Rightarrow \quad 2K = 3$$

$$\therefore \quad K = \frac{3}{2}$$

Hence, the required ratio is $\frac{3}{2}$: 1, that is, 3:2.

Again $$\frac{2K-4}{K+1} = m \qquad \text{(i)}$$

Inserting $K = \frac{3}{2}$ in (i), we get $$\frac{2\times\frac{3}{2}-4}{\frac{3}{2}+1} = m$$

$$\Rightarrow \quad \frac{3-4}{\frac{3+2}{2}} = m$$

$$\Rightarrow \quad \frac{-2}{5} = m$$

$$\therefore \quad m = \frac{-2}{5}$$

10 Try to Solve

1. In what ratio does the point (–4, 6) divide the line segment joining the points $A(-6, 10)$ and $B(3, -8)$? **[NCERT]**
2. Find the ratio in which the line segment joining the points (–3, 10) and (6, –8) is divided by (–1, 6). **[NCERT]**
3. If A and B are (–2, –2) and (2, –4), respectively, find the coordinates of P such that $AP = \frac{3}{7}AB$ and P lies on the line segment AB. **[NCERT]**
4. Find the coordinates of the point which divides the line segment joining $A(-2, 2)$ and $B(2, 8)$ into four equal parts. **[NCERT]**

Answers

1. 2:7 **2.** 2:7 **3.** $P\left(-\frac{2}{7}, \frac{-20}{7}\right)$ **4.** $\left(-1, \frac{7}{2}\right)$, (0, 5), $\left(1, \frac{13}{2}\right)$

Example 15 Find the ratio in which the line segment joining $A(1, -5)$ and $B(-4, 5)$ is divided by the x axis. Also find the coordinates of the point of division. **[CBSE (CCE) 2012]**

Solution: Let the line segment AB be divided by the x axis at point P in the ratio K:1.
Therefore, the coordinates of point P

$$= \left[\frac{K\cdot 1 + 1\cdot(-4)}{K+1}, \frac{K\cdot(-5) + 1\times 5}{K+1}\right]$$

$$= \left(\frac{K-4}{K+1}, \frac{-5K+5}{K+1}\right) \quad \ldots\text{(i)}$$

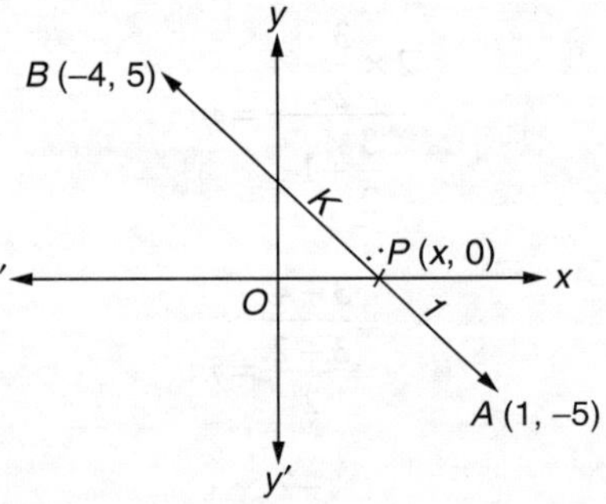

As the point P is on the x axis, coordinates of point $P = (x, 0)$ ...(ii)

From (i) and (ii), we get $\dfrac{-5K+5}{K+1} = 0$

$\Rightarrow$ $-5K + 5 = 0$

$\therefore$ $K = 1$

Inserting $K = 1$ in (i), we get

$$\text{Coordinates of point } P = \left(\frac{1-4}{1+1}, \frac{-5+5}{1+1}\right) = \left(\frac{-3}{2}, 0\right)$$ **[Ans.** $P(-3/2, 0)$**]**

11 Try to Solve

1. Find the ratio in which the y axis divides the line segment joining the points $(5, -6)$ and $(-1, -4)$. Also find the point of intersection.
2. Find the ratio in which the line segment joining the points $(6, 4)$ and $(-1, 7)$ is divided by the x axis.

Answers

1. $5 : 1, \left(0, \dfrac{-13}{3}\right)$ 2. 4:7

Example 16 If points $P(a, 3)$, $Q(6, 1)$, $R(8, 2)$ and $S(9, 4)$ are the vertices of a parallelogram $PQRS$, find the value of a. **[CBSE (CCE) 2011]**

Solution: The vertices of parallelogram $PQRS$ are $P(a, 3)$, $Q(6, 1)$, $R(8, 2)$ and $S(9, 4)$.

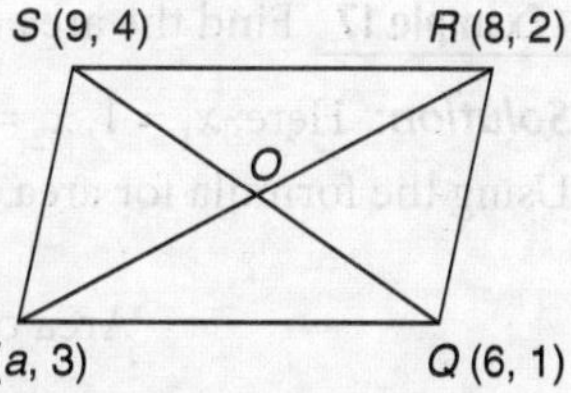

The diagonals of the parallelogram bisect each other at point O.

Therefore, O is the mid-point of SQ and PR.

When O is the mid-point of SQ

$$\text{Coordinates of point } O = \left(\frac{9+6}{2}, \frac{4+1}{2}\right) = \left(\frac{15}{2}, \frac{5}{2}\right) \quad \ldots\text{(i)}$$

When O is the mid-point of PR

$$\text{Coordinates of point } O = \left(\frac{a+8}{2}, \frac{3+2}{2}\right) \quad \ldots\text{(ii)}$$

From (i) and (ii), we get
$$\frac{a+8}{2} = \frac{15}{2}$$

$\Rightarrow$
$$a + 8 = 15$$

Therefore, the value of a is 7.

12 Try to Solve

1. If the points $A(6, 1)$, $B(8, 2)$, $C(9, 4)$ and $D(p, 3)$ are the vertices of a parallelogram taken in order, find the value of p. **[NCERT]**
2. If (1, 2), (4, y), (x, 6) and (3, 5) are the vertices of a parallelogram taken in order, find x and y. **[NCERT]**
3. Find the area of a rhombus if its vertices are (3, 0), (4, 5), (–1, 4) and (–2, –1) taken in order. **[NCERT]**

Answers

1. $p = 7$ **2.** $x = 6, y = 3$ **3.** 24 sq. units

Concept: Area of a Triangle

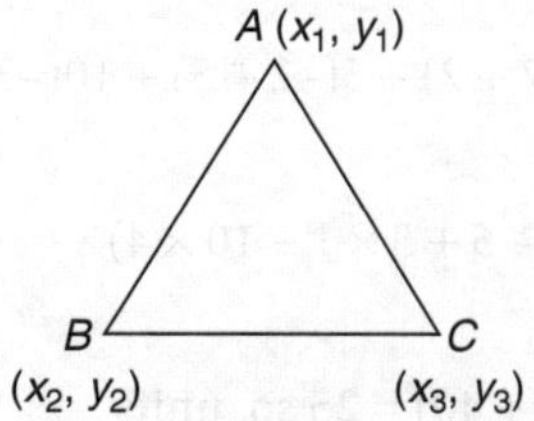

$$\text{Area of } \Delta ABC = \frac{1}{2}\left[x_1(y_2 - y_3) + x_2(y_3 - y_1) + x_3(y_1 - y_2)\right]$$

* Area is a positive quantity.
If all the three points A, B and C are collinear, then area of $\Delta ABC = 0$.

Example 17 Find the area of a triangle whose vertices are (1, –1), (– 4, 6) and (–3, –5).

Solution: Here, $x_1 = 1, x_2 = -4, x_3 = -3, y_1 = -1, y_2 = 6, y_3 = -5$.
Using the formula for area of a triangle

$$\text{Area of } \Delta ABC = \frac{1}{2}[x_1(y_2 - y_3) + x_2(y_3 - y_1) + x_3(y_1 - y_2)]$$

$$= \frac{1}{2}[1(6+5) + (-4)(-5+1) + (-3)(-1-6)]$$

$$= \frac{1}{2}[1 \times 11 + (-4)(-4) + (-3)(-7)]$$

$$= \frac{1}{2}[11 + 16 + 21] = \frac{1}{\not 2} \times \not{48} = 24$$

So, the area of the triangle is 24 square units.

13 Try to Solve

1. Find the area of a triangle formed by the points $A(5, 2)$, $B(4, 7)$ and $C(7, -4)$. [NCERT]
2. Find the area of the triangle whose vertices are: [NCERT]
   (a) (2, 3), (–1, 0), (2, –4) (b) (–5, –1), (3, –5), (5, 2)

Answers

1. 2 sq. units 2. (a) $\frac{21}{2}$ sq. units (b) 32 sq. units

Example 18 Find the area of the quadrilateral $ABCD$ whose vertices are $A(-1, -3)$, $B(5, -7)$, $C(10, -2)$ and $D(5, 17)$.

Solution: The four vertices of quadrilateral $ABCD$ are $A(-1, -3)$, $B(5, -7)$, $C(10, -2)$ and $D(5, 17)$.
Join AC to obtain ΔABC and ΔADC.

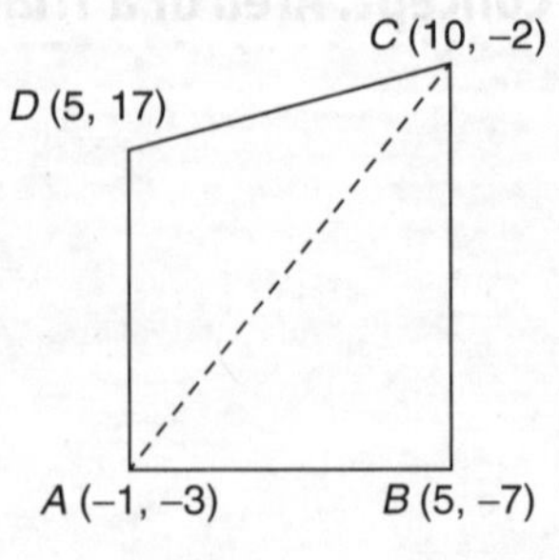

$$\text{Area of } \Delta ABC = \frac{1}{2}[x_1(y_2 - y_3) + x_2(y_3 - y_1) + x_3(y_1 - y_2)]$$

$$\text{Area } (\Delta ABC) = \frac{1}{2}[-1(-7+2) + 5(-2+3) + 10(-3+7)]$$

$$= \frac{1}{2}(-1 \times -5 + 5 \times 1 + 10 \times 4)$$

$$= \frac{1}{2}(5 + 5 + 40) = 25 \text{ sq. units}$$

$$\text{Area } (\Delta ACD) = \frac{1}{2}[-1(-2-17) + 10(17+3) + 5(-3+2)]$$

$$= \frac{1}{2}(-1 \times -19 + 10 \times 20 + 5 \times -1) = \frac{1}{2}(19 + 200 - 5)$$

$$= \frac{1}{2} \times 214 = 107 \text{ sq. units}$$

$$\text{Area } (\Delta ABCD) = \text{ar } (\Delta ABC) + \text{ar}(\Delta ACD)$$

$$= 25 + 107 = 132 \text{ sq. units.}$$

14 Try to Solve

1. Find the area of the quadrilateral whose vertices, taken in order, are (–4, –2), (–3, –5), (3, –2) and (2, 3). **[NCERT]**
2. If $A(-5, 7)$, $B(-4, -5)$, $C(-1, -6)$ and $D(4, 5)$ are the vertices of a quadrilateral, find the area of the quadrilateral $ABCD$. **[NCERT]**

Answers

1. 28 sq. units 2. 72 sq. units

Example 19 Find the value of K so that the points $A(-2, 3)$, $B(3, -1)$ and $C(5, K)$ are collinear.

[CBSE (CCE)2011]

Solution: The points $A(-2, 3)$, $B(3, -1)$ and $C(5, K)$ are collinear. Therefore, the area of ΔABC formed by these points is zero.

$$\text{Area of } \Delta ABC = \frac{1}{2}[x_1(y_2 - y_3) + x_2(y_3 - y_1) + x_3(y_1 - y_2)]$$

$$\therefore \quad \text{Area } (\Delta ABC) = \frac{1}{2}[-2(-1 - K) + 3(K - 3) + 5(3 + 1)] = 0$$

$$\Rightarrow \quad (2 + 2K + 3K - 9 + 20) = 0$$

$$\Rightarrow \quad 5K + 13 = 0$$

$$\therefore \quad K = \frac{-13}{5}$$

Therefore, the value K is $-\frac{13}{5}$.

15 Try to Solve

1. In each of the following, find the value of 'k' for which the points are collinear:
   (a) (7, –2), (5, 1), (3, k) (b) (8, 1), (k, –4), (2, –5) **[NCERT]**
2. Find the value of k if the points $A(2, 3)$, $B(4, k)$ and $C(6, -3)$ are collinear. **[NCERT]**

Answers

1. (a) $k = 4$ (b) $k = 3$ 2. $k = 0$

FOR MORE PRACTICE AND KNOWLEDGE

Distance Formula

The distance between two points $P(x_1, y_1)$ and $Q(x_2, y_2)$ on a plane is given by

$$PQ = \sqrt{(x_2 - x_1)^2 + (y_2 - y_1)^2}$$

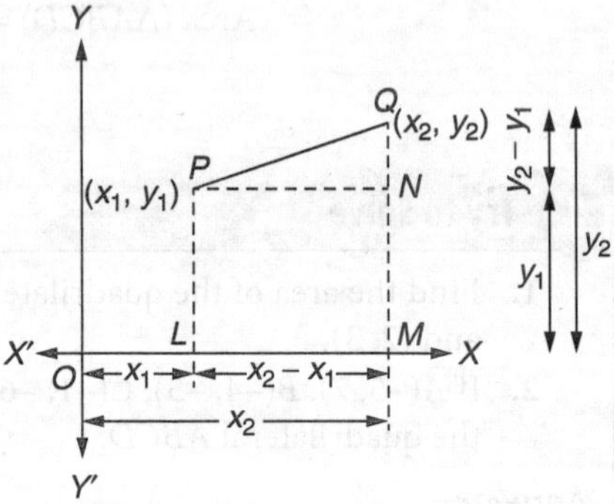

Let $P(x_1, y_1)$ and $Q(x_2, y_2)$ be the two given points. Draw PL and QM perpendicular on OX. Also draw $PN \perp QM$.

Now $\quad OL = x_1,\ OM = x_2,\ PL = y_1,\ QM = y_2$

$\therefore \quad LM = (x_2 - x_1)$ and $QN = (y_2 - y_1)$

Applying Pythagoras' Theorem on ΔPNQ

$$PQ^2 = PN^2 + QN^2$$

$\Rightarrow \quad PQ = (x_2 - x_1)^2 + (y_2 - y_1)^2 \quad [\because PN = LM]$

$\therefore \quad PQ = \sqrt{(x_2 - x_1)^2 + (y_2 - y_1)^2}$

So, the distance between the points $P(x_1, y_1)$ and $Q(x_2, y_2)$ is

$$PQ = \sqrt{(x_2 - x_1)^2 + (y_2 - y_1)^2} \qquad \text{(i)}$$

Equation (i) is called the distance formula.

Derivation of Section Formula

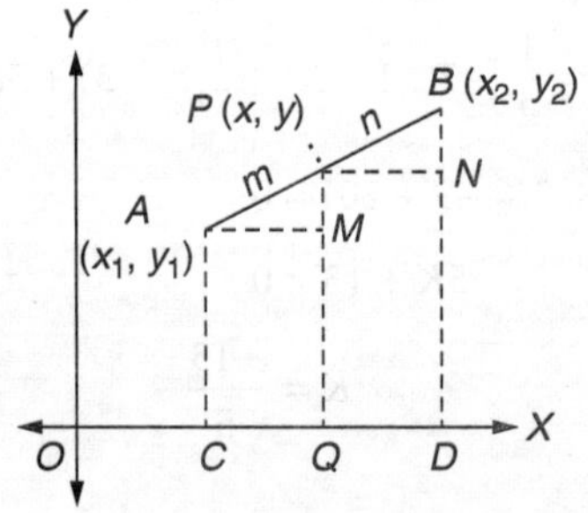

Theorem The coordinates of point $P(x, y)$ which divides the line segment joining $A(x_1, y_1)$ and $B(x_2, y_2)$ internally in the ratio $m{:}n$ are given by

$$\left(x = \frac{mx_2 + nx_1}{m + n}, y = \frac{my_2 + my_1}{m + n}\right)$$

Solution: *Given:* AB is a line segment. The coordinates of A and B are (x_1, y_1) and (x_2, y_2). $P(x, y)$ is a point which divides AB internally in the ratio $m{:}n$.

To prove: $\quad x = \dfrac{mx_2 + nx_1}{m + n}$ and $y = \dfrac{my_2 + ny_1}{m + n}$

Construction: Draw AC, PQ and BD perpendicular on OX from points A, P and B, respectively. Draw $AM \perp PQ$ and $PN \perp BD$.

Proof:

$$OC = x_1 \quad OQ = x \text{ and } OD = x_2$$

$$AC = y_1,\ PQ = y \text{ and } BD = y_2$$

$$AM = CQ = (OQ - OC) = x - x_1$$

$$PN = QD = OD - OQ = x_2 - x$$

$$PM = PQ - MQ = PQ - AC = y - y_1 \qquad [\because MQ = AC]$$

And $$BN = BD - ND = BD - PQ = y_2 - y \qquad [\because ND = PQ]$$

In ΔAMP and ΔPNB

$$\angle M = \angle N \qquad \text{[Each } 90°]$$

$$\angle APM = \angle PBN \qquad \text{[Corresponding angles]}$$

$\therefore$ $$\Delta AMP \sim \angle PNB \qquad \text{[By AA Similarity]}$$

And so $$\frac{AM}{PN} = \frac{PM}{BN} = \frac{AP}{PB}$$

$$\Rightarrow \quad \frac{x - x_1}{x_2 - x} = \frac{y - y_1}{y_2 - y} = \frac{m}{n}$$

$$\Rightarrow \quad \frac{x - x_1}{x_2 - x} = \frac{m}{n} \text{ and } \frac{y - y_1}{y_2 - y} = \frac{m}{n}$$

$$\Rightarrow \quad nx - nx_1 = mx_2 - mx \text{ and } ny - my_1 = my_2 - my.$$

$$\Rightarrow \quad mx + nx = mx_2 + nx_1 \text{ and } my + ny = my_2 + ny_1$$

$$\Rightarrow \quad x(m + n) = mx_2 + nx_1 \text{ and } y(m + n) = my_2 + ny_1$$

$$\therefore \quad x = \frac{mx_2 + nx_1}{m + n} \text{ or } y = \frac{my_2 + ny_1}{m + n}$$

Hence, the coordinates of P are $\left(\dfrac{mx_2 + nx_1}{m + n}, \dfrac{my_2 + ny_1}{m + n}\right)$.

Derivation of Area of Triangle Formula

Theorem The area of a ΔPQR with vertices $P(x_1, y_1)$, $Q(x_2, y_2)$ and $R(x_3, y_3)$ is given by

$$\text{Area } (\Delta PQR) = \frac{1}{2}\ [x_1(y_2 - y_3) + x_2(y_2 - y_1) + x_3(y_1 - y_2)]$$

Solution: *Given*: PQR is a triangle whose vertices are $P(x_1, y_1)$, $Q(x_2, y_2)$ and $R(x_3, y_3)$.

To prove: $$\text{Area } (\Delta PQR) = \frac{1}{2}\ [x_1(y_2 - y_3) + x_2(y_3 - y_1) + x_3(y_1 - y_2)]$$

Construction: Draw QA, PB and RC perpendicular from Q, P and R, respectively, on OX.

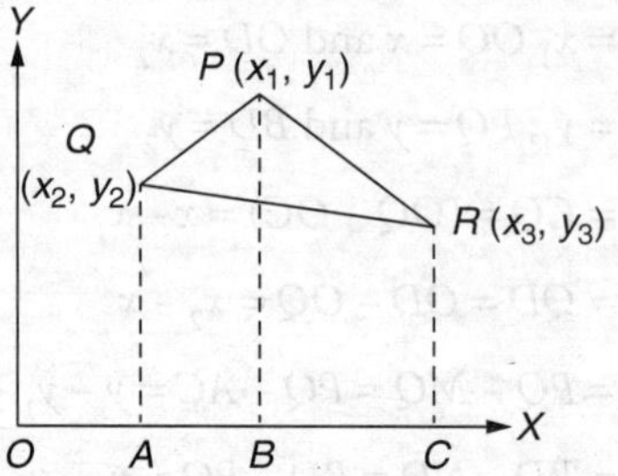

Proof: $AB = OB - OA = (x_1 - x_2)$; $BC = OC - OB = (x_3 - x_1)$

$AC = OC - OA = (x_3 - x_2)$

Area(ΔPQR) = Area of trapezium $ABPQ$ + Area of trapezium $BCRP$ – Area of trapezium $ACRQ$

$$\Rightarrow \quad ar(\Delta PQR) = \left\{\frac{1}{2}(QA + PB) \times AB\right\} + \left\{\frac{1}{2}(PB + RC) \times BC\right\} - \left\{\frac{1}{2}(QA + RC) \times AC\right\}$$

$$= \frac{1}{2}(y_2 + y_1)(x_1 - x_2) + \frac{1}{2}(y_1 + y_3)(x_3 - x_1) - \frac{1}{2}(y_2 + y_3)(x_3 - x_2)$$

$$= \frac{1}{2}[x_1(y_2 + y_1) - x_2(y_2 + y_1) + x_3(y_1 + y_3) - x_1(y_1 + y_3) - x_3(y_2 + y_3) + x_2(y_2 + y_3)]$$

$$= \frac{1}{2}[x_1(y_2 + y_1 - y_1 - y_3) + x_2(y_2 + y_3 - y_2 - y_1) + x_3(y_1 + y_3 - y_2 - y_3)]$$

$$= \frac{1}{2}[(x_1(y_2 - y_3) + x_2(y_3 - y_1) + x_3(y_1 - y_2)]$$

We know that the area of a triangle cannot be negative.

$$\therefore \quad \text{Area } (\Delta PQR) = \frac{1}{2}\,|x_1(y_2 - y_3) + x_2(y_3 - y_1) + x_3(y_1 - y_2)|$$

Example 20 If the point (x, y) is equidistant from the points $(a + b, b - a)$ and $(a - b, a + b)$, prove that $bx = ay$. **[AI 2017 (C)]**

Solution: Let $P(x, y)$, $Q(a + b, b - a)$ and $R(a - b, a + b)$ be the given points. Then

$$PQ = PR \qquad \text{[Given]}$$

$$\Rightarrow \quad \sqrt{\{x-(a+b)\}^2 + \{y-(b-a)\}^2} = \sqrt{\{x-(a-b)\}^2 + \{y-(a+b)\}^2}$$

On squaring both sides, we get

$$\Rightarrow \quad \{x - (a + b)\}^2 + \{y - (b - a)\}^2 = \{x - (a - b)\}^2 + \{y - (a + b)\}^2$$

$$\Rightarrow \quad \cancel{x^2} - 2x(a + b) + (a + \cancel{b})^2 + \cancel{y^2} - 2y(b - a) + (b - \cancel{a})^2 = \cancel{x^2} + (a - \cancel{b})^2 - 2x(a - b) +$$
$$\cancel{y^2} - 2y(a + b) + (a + \cancel{b})^2$$

$\Rightarrow \quad -2x(a+b) - 2y(b-a) = -2x(a-b) - 2y(a+b)$

$\Rightarrow \quad -\cancel{ax} - bx - \cancel{by} + ay = -\cancel{ax} + bx - ay - \cancel{by}$

$\Rightarrow \quad 2bx = 2ay$

$\Rightarrow \quad bx = ay$

Example 21 Find a point on the y axis which is equidistant from the points $A(6, 5)$ and $B(-4, 3)$.

Solution: We know that a point on the y axis is of the form $(0, y)$. So, let the required point be $P(0, y)$. Then

$$PA = PB$$

$$\Rightarrow \quad \sqrt{(0-6)^2 + (y-5)^2} = \sqrt{(0+4)^2 + (y-3)^2}$$

On squaring both sides, we get $\Rightarrow 36 + (y-5)^2 = 16 + (y-3)^2$

$$\Rightarrow \quad 36 + \cancel{y^2} - 10y + 25 = 16 + \cancel{y^2} - 6y + 9$$

$$\Rightarrow \quad 4y = 36$$

$$\Rightarrow \quad y = 9$$

So, the required point is (0, 9).

Example 22 If two vertices of an equilateral triangle are (0, 0) and $(3, \sqrt{3})$, find the third vertex.

Solution: $O(0, 0)$ and $A(3, \sqrt{3})$ are the given points and let $B(x, y)$ be the vertex of equilateral ΔOAB. Then

$$OA = OB = AB$$

$$\Rightarrow \quad OA^2 = OB^2 = AB^2$$

We have $OA^2 = (3-0)^2 + (\sqrt{3}-0)^2 = 12$

$$OB^2 = x^2 + y^2$$

And $AB^2 = (x-3)^2 + (y-\sqrt{3})^2$

$$AB^2 = x^2 + 9 - 6x + y^2 + 3 - 2\sqrt{3}\,y$$

$$\Rightarrow \quad AB^2 = x^2 + y^2 - 6x - 2\sqrt{3}\,y + 12$$

$$\therefore \quad OA^2 = OB^2 = AB^2$$

$$\Rightarrow \quad OA^2 = OB^2 \text{ and } OB^2 = AB^2$$

$$\Rightarrow \quad x^2 + y^2 = 12$$

And $x^2 + y^2 = x^2 + y^2 - 6x - 2\sqrt{3}\,y + 12$

$$\Rightarrow \quad x^2 + y^2 = 12 \text{ and } 6x + 2\sqrt{3}\,y = 12$$

$$\Rightarrow \quad x^2 + y^2 = 12 \text{ and } 3x + \sqrt{3}\,y = 6$$

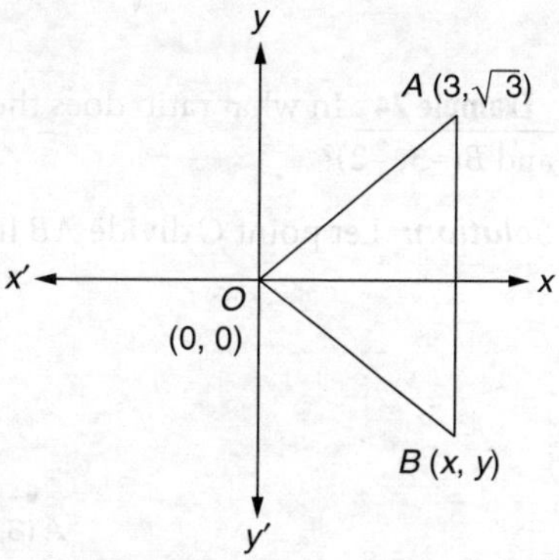

On solving, we get $x = 0, 3$

$\therefore$ On inserting $x = 0$, $3x + \sqrt{3}\,y = 6 \Rightarrow 0 + \sqrt{3}\,y = 6$, $y = \dfrac{6}{\sqrt{3}} = 2\sqrt{3}$ [Inserting $x = 0$ in $3x + \sqrt{3}\,y = 6$]

On inserting $x = 3$, $3x + \sqrt{3}\,y = 6 \Rightarrow 3(3) + \sqrt{3}\,y = 6$, $y = \dfrac{6-9}{\sqrt{3}} = -\sqrt{3}$ [Inserting $x = 3$ in $3x + \sqrt{3}\,y = 6$]

Hence, the coordinates of the third vertex B are $(0, 2\sqrt{3})$ or $(3, -\sqrt{3})$.

Example 23 Find the coordinates of the centre of the circle passing through the points (0, 0), (–2, 1) and (–3, 2). Also find its radius.

Solution: Let $P(x, y)$ be the centre of the circle passing through the points $O(0, 0)$, $A(-2,1)$ and $B(-3, 2)$. Then

$$OP = AP = BP$$

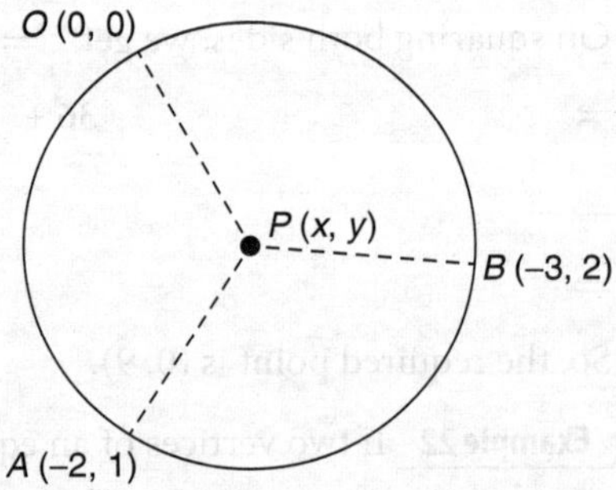

Now $OP = AP$

$\Rightarrow$ $OP^2 = AP^2$

$\Rightarrow$ $x^2 + y^2 = (x + 2)^2 + (y - 1)^2$

$\Rightarrow$ $x^2 + y^2 = x^2 + y^2 + 4x - 2y + 5$

$\Rightarrow$ $4x - 2y + 5 = 0$ ...(i)

And $OP = BP$

$\Rightarrow$ $OP^2 = BP^2$

$\Rightarrow$ $x^2 + y^2 = (x + 3)^2 + (y - 2)^2$

$\Rightarrow$ $x^2 + y^2 = x^2 + y^2 + 6x - 4y + 13$

$\Rightarrow$ $6x - 4y + 13 = 0$ ...(ii)

On solving equations (i) and (ii), we get $x = \dfrac{3}{2}$ and $y = \dfrac{11}{2}$.

$$\text{Centre} = \left(\frac{3}{2}, \frac{11}{2}\right)$$

$$\text{Now, Radius} = OP = \sqrt{x^2 + y^2} = \sqrt{\frac{9}{4} + \frac{121}{4}} = \frac{1}{2}\sqrt{130} \text{ units.}$$

Example 24 In what ratio does the point $C(3/5, 11/5)$ divide the line segment joining the points $A(3, 5)$ and $B(-3, -2)$?

Solution: Let point C divide AB in the ratio $\lambda:1$. Then, the coordinates of C are

$$\left(\frac{-3\lambda + 3}{\lambda + 1}, \frac{-2\lambda + 5}{\lambda + 1}\right)$$

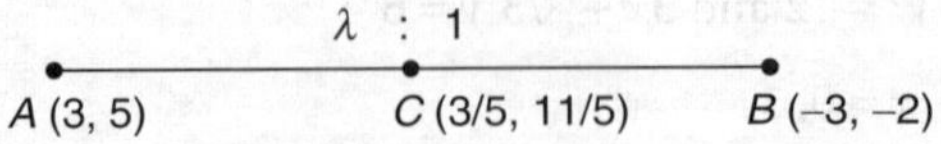

But, the coordinates of C are given as (3/5, 11/5).

$$\therefore \quad \frac{-3\lambda+3}{\lambda+1}=\frac{3}{5} \text{ and } \frac{-2\lambda+5}{\lambda+1}=\frac{11}{5}$$

$$\Rightarrow \quad -15\lambda+15=3\lambda+3 \text{ and } -10\lambda+25=11\lambda+11$$

$$\Rightarrow \quad -18\lambda=-12 \text{ and } -21\lambda=-14$$

$$\Rightarrow \quad \lambda=\frac{12}{18}=\frac{2}{3} \text{ and } \lambda=\frac{14}{21}=\frac{2}{3}$$

$$\therefore \quad \lambda=\frac{2}{3}$$

Hence, the point C divides AB in the ratio 2:3.

Example 25 In what ratio does the y axis divide the line segment joining the points P (–4, 5) and Q(3, –7)? Also find the coordinates of the point of intersection.

Solution: Suppose the y axis divides PQ in the ratio λ:1. Then, the coordinates of the point of division are

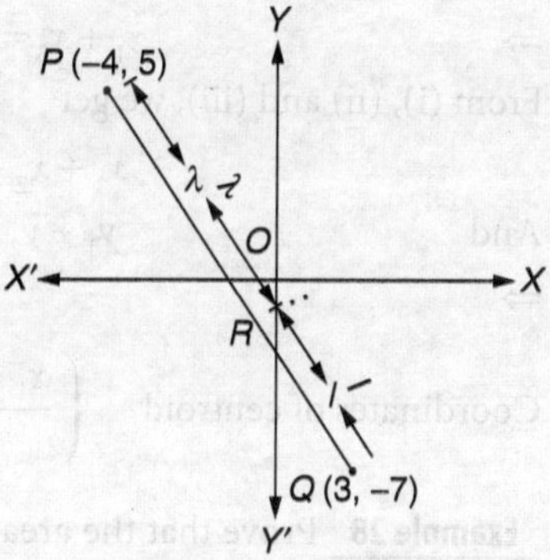

$$R\left(\frac{3\lambda-4}{\lambda+1}, \frac{-7\lambda+5}{\lambda+1}\right)$$

Since R lies on the y axis and the x coordinate of every point on the y axis is zero.

$$\therefore \quad \frac{3\lambda-4}{\lambda+1}=0 \Rightarrow 3\lambda-4=0, \lambda=\frac{4}{3}$$

Hence, the required ratio is $\frac{4}{3}$: 1, that is, 4:3.

Inserting $\lambda = 4/3$ in the coordinates of R, we find that its coordinates are $\left(0, \frac{-13}{7}\right)$.

Example 26 If the points A(6, 1), B(8, 2), C(9, 4) and D(p, 3) are the vertices of a parallelogram, taken in order, find the value of p.

Solution: We know that the diagonals of a parallelogram bisect each other. So, the coordinates of the mid-point of diagonal AC are the same as the coordinates of the mid-point of diagonal BD.

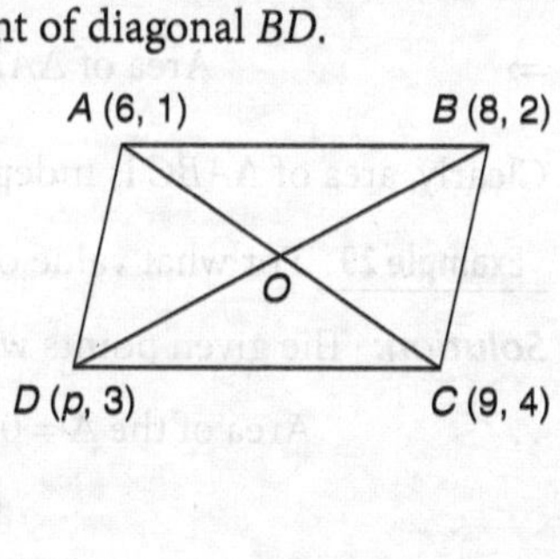

$$\therefore \quad \left(\frac{6+9}{2}, \frac{1+4}{2}\right)=\left(\frac{8+p}{2}, \frac{2+3}{2}\right)$$

$$\Rightarrow \quad \left(\frac{15}{2}, \frac{5}{2}\right)=\left(\frac{8+p}{2}, \frac{5}{2}\right)$$

$$\Rightarrow \quad \frac{15}{2}=\frac{8+p}{2} \Rightarrow 15=8+p \Rightarrow p=7$$

Example 27 If the coordinates of the mid-points of the sides of a triangle are (1, 1), (2, –3) and (3, 4), find its centroid.

Solution: Let $P(1, 1)$, $Q(2, -3)$ and $R(3, 4)$ be the mid-points of sides AB, BC and CA, respectively, of triangle ABC. Let $A(x_1, y_1)$, $B(x_2, y_2)$ and $C(x_3, y_3)$ be the vertices of triangle ABC. Then, P is the mid-point of AB.

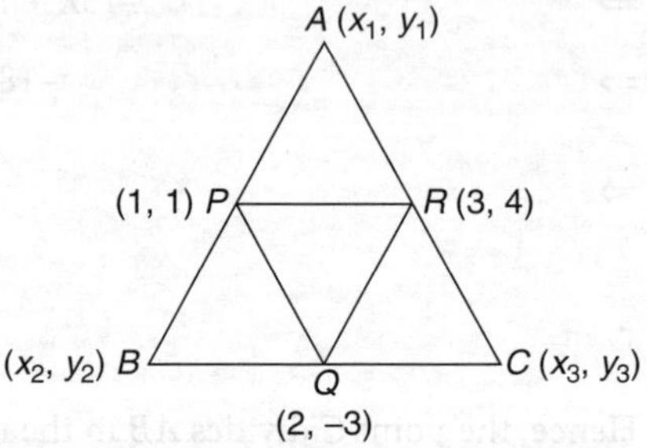

$$\Rightarrow \quad \frac{x_1 + x_2}{2} = 1, \quad \frac{y_1 + y_2}{2} = 1 \quad \ldots\text{(i)}$$

$$\Rightarrow \quad x_1 + x_2 = 2 \text{ and } y_1 + y_2 = 2$$

Q is the mid-point of BC

$$\Rightarrow \quad \frac{x_2 + x_3}{2} = 2, \quad \frac{y_2 + y_3}{2} = -3$$

$$\Rightarrow \quad x_2 + x_3 = 4 \text{ and } y_2 + y_3 = -6 \quad \ldots\text{(ii)}$$

R is the mid-point of AC

$$\Rightarrow \quad \frac{x_1 + x_3}{2} = 3 \text{ and } \frac{y_1 + y_3}{2} = 4$$

$$\Rightarrow \quad x_1 + x_3 = 6 \text{ and } y_1 + y_3 = 8 \quad \ldots\text{(iii)}$$

From (i), (ii) and (iii), we get

$$x_1 + x_2 + x_2 + x_3 + x_1 + x_3 = 2 + 4 + 6$$

And $$y_1 + y_2 + y_2 + y_3 + y_1 + y_3 = 2 - 6 + 8$$

$$\Rightarrow \quad x_1 + x_2 + x_3 = 6 \text{ and } y_1 + y_2 + y_3 = 2 \quad \text{(iv)}$$

Coordinates of centroid $$\left(\frac{x_1 + x_2 + x_3}{3}, \frac{y_1 + y_2 + y_3}{3}\right) = \left(\frac{6}{3}, \frac{2}{3}\right) = \left(2, \frac{2}{3}\right) \quad \text{[From (iv)]}$$

Example 28 Prove that the area of the triangle whose vertices are $(t, t-2)$, $(t + 2, t + 2)$ and $(t + 3, t)$ is independent of t.

Solution: Let $A = (x_1, y_1) = (t, t - 2)$, $B = (x_2, y_2) = (t + 2, t + 2)$ and $C = (x_3, y_3) = (t + 3, t)$ be the vertices of the given triangle. Then

$$\therefore \quad \text{Area of } \Delta ABC = \frac{1}{2}\,|x_1(y_2 - y_3) + x_2(y_3 - y_1) + x_3(y_1 - y_2)|$$

$$\Rightarrow \quad \text{Area of } \Delta ABC = \frac{1}{2}\,\{t(t + 2 - t) + (t + 2)(t - t + 2) + (t + 3)(t - 2 - t - 2)\}|$$

$$\Rightarrow \quad \text{Area of } \Delta ABC = \frac{1}{2}\,|\{(2t + 2t + 4 - 4t - 12)\}| = |-4| = 4 \text{ sq. units}$$

Clearly, area of ΔABC is independent of t.

Example 29 For what value of k are the points $(k, 2 - 2k)(-k + 1, 2k)$ and $(-4 - k, 6 - 2k)$ collinear?

Solution: The given points will be collinear if the area of the triangle formed by them is zero.

$$\therefore \quad \text{Area of the } \Delta = 0$$

$$\frac{1}{2}\,|x_1(y_2 - y_3) + x_2(y_3 - y_1) + x_3(y_1 - y_2)| = 0$$

$$\Rightarrow \quad \frac{1}{2}\,|k(2k-6+2k)+(-k+1)(6-2k-2+2k)+(-4-k)(2-2k-2k)|=0$$

$$\Rightarrow \quad \frac{1}{2}\,|4k^2-6k-4k+4-8+16k-2k+4k^2|=0$$

$$\Rightarrow \quad 8k^2+4k-4=0$$

$$\Rightarrow \quad 4(2k^2+k-1)=0$$

$$\Rightarrow \quad 2k^2+k-1=0$$

$$\Rightarrow \quad k=\frac{1}{2} \text{ or } k=-1$$

Multiple Choice Questions

Choose the correct answer for each of the following questions:

1. A circle drawn with origin as the centre passes through $\left(\frac{13}{2}, 0\right)$. The point which does not lie in the interior of the circle is
   (a) $\left(-\frac{3}{4}, 1\right)$ (b) $\left(2, \frac{7}{3}\right)$ (c) $\left(5, -\frac{1}{2}\right)$ (d) $\left(-6, \frac{5}{2}\right)$ **[NCERT Exemplar]**
2. The points (–4, 0), (4, 0) and (0, 3) are the vertices of a/an
   (a) Right triangle (b) Isosceles triangle (c) Equilateral triangle (d) Scalene triangle **[NCERT Exemplar]**
3. If the distance between the points (4, *p*) and (1, 0) is 5 units, then the value of *p* is
   (a) 4 only (b) ± 4 (c) –4 only (d) 0 **[NCERT Exemplar]**
4. If the segment joining the points (*a*, *b*) and (*c*, *d*) subtends a right angle at the origin, then
   (a) $ac - bd = 0$ (b) $ac + bd = 0$ (c) $ab + cd = 0$ (d) $ab - cd = 0$
5. *AOBC* is a rectangle whose three vertices are *A*(0, 3), *O*(0, 0) and *B*(5, 0). The length of its diagonal is
   (a) 5 (b) 3 (c) $\sqrt{34}$ (d) 4 **[NCERT Exemplar]**
6. The perimeter of a triangle with vertices (0, 4), (0, 0) and (3, 0) is
   (a) 5 (b) 12 (c) 11 (d) $7+\sqrt{5}$ **[NCERT Exemplar]**
7. The triangle whose vertices are (0, 0), (2.7, 0) and (0, 4.9) is a/an
   (a) Equilateral triangle (b) Right-angled triangle
   (c) Isosceles triangle (d) Obtuse-angled triangle
8. The distance of the point (*h*, *k*) from the *x* axis is
   (a) *h* units (b) |*h*| units (c) *k* units (d) |*k*| units
9. The number of points on the *x* axis which are at a distance of 2 units from (2, 4) is
   (a) 1 (b) 2 (c) 0 (d) 3
10. The vertices of a parallelogram taken in order are (–1, –6), (2, –5) and (7, 2). The fourth vertex is
    (a) (1, 4) (b) (1, 1) (c) (4, 4) (d) (4, 1)

11. The points $A\left(0, \frac{8}{3}\right)$, $B(1, 3)$ and $C(82, 30)$
(a) Lie on a circle (b) Are vertices of a right-angled triangle
(c) Are vertices of an obtuse-angled triangle (d) None of these

12. If the distance of the point $(4, a)$ from the x axis is half its distance from the y axis, then $a =$
(a) 4 units (b) 8 units (c) 2 units (d) 6 units

13. If the distance between the points $(8, p)$ and $(4, 3)$ is 5 units, the value of p is
(a) 6 (b) 0 (c) Both (a) and (b) (d) None of these

14. If the four vertices of a parallelogram taken in order are $(-3, -1)$, (a, b), $(3, 3)$ and $(4, 3)$, then $a:b =$
(a) 1:4 (b) 4:1 (c) 1:2 (d) 2:1

15. If $P(1, 2)$, $Q(4, 6)$, $R(5, 7)$ and $S(a, b)$ are the vertices of a parallelogram $PQRS$, then
(a) $a = 2, b = 4$ (b) $a = 3, b = 4$ (c) $a = 2, b = 3$ (d) $a = 3, b = 5$

16. The coordinates of the point which is equidistant from the three vertices of ΔAOB, as shown in the figure, are

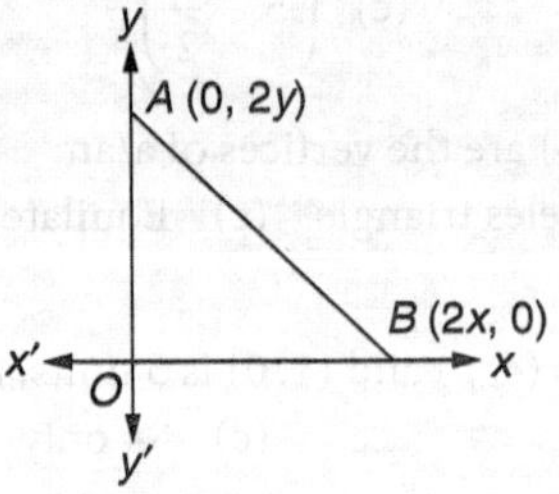

(a) (x, y) (b) (y, x) (c) $\left(\frac{x}{2}, \frac{y}{2}\right)$ (d) $\left(\frac{y}{2}, \frac{x}{2}\right)$ **[NCERT Exemplar]**

17. If $P\left(\frac{a}{3}, 4\right)$ is the mid-point of the line segment joining the points $Q(-6, 5)$ and $R(-2, 3)$, then the value of a is
(a) −4 (b) −12 (c) 12 (d) −6 **[NCERT Exemplar]**

18. If $A(1, 2)$, $B(4, 3)$ and $C(6, 6)$ are the three vertices of a parallelogram $ABCD$, the coordinates of the fourth vertex D are
(a) (1, 2) (b) (3, 4) (c) (3, 5) (d) (3, 3)

19. The vertices of a triangle are (0, 0), (3, 0) and (0, 4). The centroid of the triangle is
(a) $\left(\frac{1}{2}, 2\right)$ (b) $\left(1, \frac{4}{3}\right)$ (c) (1, 1) (d) (0, 3)

20. In the figure, the area of triangle PQR (in sq. units) is

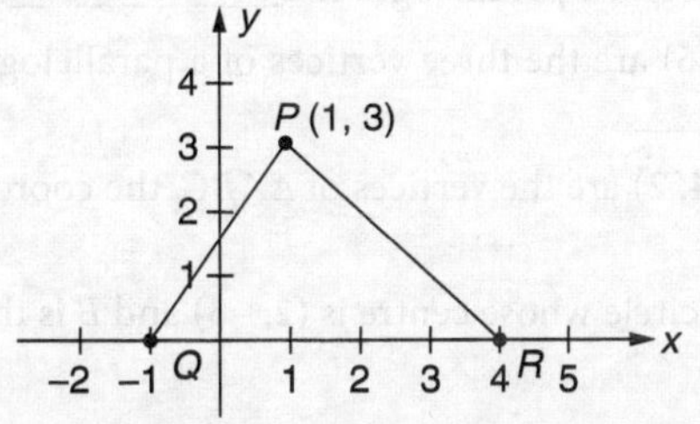

(a) 15 (b) 10 (c) 7.5 (d) 2.5

21. The area of a triangle formed by the points $(p, 2-2p)$, $(1-p, 2p)$ and $(-4-p, 6-2p)$ is 70 square units. How many integral values of p are possible?
(a) 2 (b) 3 (c) 4 (d) None of these

22. A line formed by joining $(-1, 1)$ and $(5, 7)$ is divided by a line $x + y = 4$ in the ratio of
(a) 1:2 (b) 1:3 (c) 3:4 (d) 1:4

23. The area (in square units) of the triangle formed by the points $A(a, 0)$, $O(0, 0)$ and $B(0, b)$ is
(a) ab (b) $\frac{1}{2}ab$ (c) $\frac{1}{2}a^2b^2$ (d) $\frac{1}{2}b^2$

24. The area of the triangle with vertices at the points $(a, b+c)$, $(b, c+a)$ and $(c, a+b)$ is
(a) $(a+b+c)$ sq. units (b) $(a+b-c)$ sq. units
(c) $(a-b+c)$ sq. units (d) 0

25. If the area of the triangle with vertices $(x, 0)$, $(1, 1)$ and $(0, 2)$ is 4 square units, then the value of x is
(a) −2 (b) −4 (c) −6 (d) 8

26. If A and B are the points $(-3, 4)$ and $(2, 1)$, respectively, then the coordinates of the point on AB produced such that $AC = 2BC$ are
(a) (2, 4) (b) (3, 7) (c) (7, −2) (d) None of these

27. The points $(k+1, 1)$, $(2k+1, 3)$ and $(2k+2, 2k)$ are collinear if
(a) $k = -1, 2$ (b) $k = \frac{1}{2}, 2$ (c) $k = 2, 1$ (d) $k = -\frac{1}{2}, 2$

28. The area of ΔABC having vertices $A(1, -1)$, $B(-4, ,6)$ and $C(-3, -5)$ is
(a) 52 square units (b) 24 square units (c) 48 square units (d) 27 square units

Fill in the Blanks

Complete the following statements with appropriate answers(s) in the blank space(s):

29. If $A = (a^2, 2a)$, $B = \left(\frac{1}{a^2}, -\frac{2}{a}\right)$ and $S = (1, 0)$, then $\frac{1}{SA} + \frac{1}{SB} =$ ________.

30. If $P(5, -3)$ and $Q(3, y)$ are the points of trisection of the line segment joining $A(7, -2)$ and $B(1, -5)$, then y equals ________.

31. The three vertices of a parallelogram $ABCD$, taken in order, are $A(1, -2)$, $B(3, 6)$ and $C(5, 10)$. The coordinates of the fourth vertex D are________________.

32. The two vertices of a triangle are $A(6, 3)$ and $B(-1, 7)$ and its centroid is $G(1, 5)$. Then, the third vertex C of ΔABC is ________.

33. Two adjacent sides of a parallelogram are 30 m and 14 m and the diagonal joining the end points of these sides is 40 m. The area of the parallelogram is__________.

34. If $A(1, 2)$, $B(4, 3)$ and $C(6, 6)$ are the three vertices of a parallelogram $ABCD$, the coordinates of the fourth vertex are __________.

35. If $A(-2, 4)$, $B(0, 0)$ and $C(14, 2)$ are the vertices of ΔABC, the coordinates of the centroid of ΔABC are __________.

36. If AB is the diameter of the circle whose centre is $(2, -3)$ and B is the point $(1, 4)$, the coordinates of A are __________.

True/False

State whether the following statements are True/False and also justify:

37. A circle has its centre at the origin and a point $P(5, 0)$ lies on it. The point $Q(6, 8)$ lies outside the circle. State whether true of false. **[NCERT Exemplar]**

38. The point $P(-2, 4)$ lies on a circle of radius 6 units and centre $C(3, 5)$. **[NCERT Exemplar]**

39. The coordinates of the centroid of the triangle whose vertices are $(3, 2)$, $(5, 7)$ and $(8, 8)$ are $\left(\frac{16}{3}, \frac{17}{3}\right)$.

40. If the point $C(-1, 2)$ internally divides the line segment joining $A(2, 5)$ and B in the ratio 3:4, the coordinates of B are $(-5, -2)$.

41. If $A(-1, 3)$, $B(1, -1)$ and $C(5, 1)$ are the vertices of a triangle ABC, the length of the median AD is 4 units.

42. The x coordinate of a point P is twice its y coordinate. If P is equidistant from $Q(2, -5)$ and $R(-3, 6)$, the coordinate of P is $(16, 8)$.

43. The points $A(x_1, y_1)$, $B(x_2, y_2)$ and $C(x_3, y_3)$ are vertices of ΔABC. The centroid of ΔABC is $\left(\frac{x_1+x_2+x_3}{3}, \frac{y_1+y_2+y_3}{3}\right)$.

Assertion–Reason

The following questions consist of two statements – Assertion (A) and Reason (R). Answer these questions by selecting the appropriate option given below.

(A) Both A and R are true and R is the correct explanation for A.
(B) Both A and R are true and R is not the correct explanation for A.
(C) A is true but R is false.
(D) A is false but R is true.

44. **Assertion (A):** Point A is on the y axis at a distance of 4 units from the origin. If the coordinate of the point B is $(-3, 0)$, then the length of AB is 5 units.
Reason (B): Distance between points $A(x_1, y_1)$ and $B(x_2, y_2)$ is $\sqrt{(x_2-x_1)^2+(y_2-y_1)^2}$.

45. **Assertion (A):** The area of the triangle with vertices $(-5, -1)$, $(3, -5)$, $(5, 2)$ is 32 square units.
Reason (B): The point (x, y) divides the line segment of the points (x_1, y_1) and (x_2, y_2) in the ratio k:1 internally; then $x = \frac{kx_2 + x_1}{k+1}$, $y = \frac{ky_2 + y_1}{k+1}$.

46. **Assertion (A):** The distance of the point (3, – 4) from the x axis is 3 units.
Reason (B): The condition that the three points (x_1, y_1), (x_2, y_2) and (x_3, y_3) are collinear is if the area of the triangle = 0.

47. **Assertion (A):** The distance between the points (a, b) and $(-a, -b)$ is $2\sqrt{a^2+b^2}$.
Reason (B): The mid-point of two points (a, b) and $(-a, -b)$ is (0, 0).

48. **Assertion (A):** The area of the triangle with vertices $A(1, 3)$, $B(-1, 0)$ and $C(4, 0)$ is 7.5 square units.
Reason (B): The distance between two given points $A(x_1, y_1)$ and $B(x_2, y_2)$ is $\sqrt{(x_2-x_1)^2+(y_2-y_1)^2}$.

Subjective Questions

49. Check whether (5, –2), (6, 4) and (7, –2) are the vertices of an isosceles triangle. **[NCERT]**

50. Find a point on the y axis which is equidistant from the points $A(6, 5)$ and $B(-4, 3)$. **[NCERT]**

51. Find the distance of a point $P(x, y)$ from the origin. **[CBSE 2018]**

52. Find the coordinates of a point A, where AB is the diameter of a circle whose centre is (2, –3) and B is the point (1, 4). **[NCERT, 2019]**

53. Find the linear relation between x and y such that $P(x, y)$ is equidistant from the points $A(1, 4)$ and $B(-1, 2)$. **[CBSE 2018 (C)]**

54. If the distance of $P(x, y)$ from the points $A(3, 6)$ and $B(-3, 4)$ are equal, prove that $3x + y = 5$.

55. The centre of a circle is $(2\alpha - 1, 7)$ and it passes through the point (–3, –1). If the diameter of the circle is 20 units, then find the value of α.

56. Two points $A(1, 0)$ and $B(-1, 0)$ with a variable point $P(x, y)$ satisfy the relation $AP - BP = 1$. Show that $12x^2 - 4y^2 = 3$.

57. One vertex of an equilateral triangle with side 2 units is at origin. Another vertex lies on the line $x = \sqrt{3}\, y$ and is in the first quadrant. Find the coordinates of the other two vertices.

58. Find the distance between the points $A(2a, 6a)$ and $B(2a + \sqrt{3}\, a, 5a)$.

59. Point A is on the y axis at a distance of 4 units from the origin. If the coordinates of point B are (–3, 0), then find the length of AB. **[Foreign 2013]**

60. Find the value of x such that $PQ = QR$ where the coordinates of P, Q and R are (6, –1), (1, 3) and $(x, 8)$, respectively.

61. The centre of a circle is $(2a, a - 7)$. Find the values of a if the circle passes through the point (11, –9) and has a diameter of $10\sqrt{2}$ units.

62. Show that the points $A(3, 5)$, $B(6, 0)$, $C(1, -3)$ and $D(-2, 2)$ are the vertices of a square $ABCD$. **[Foreign 2011]**

63. Find a relation between x and y if the points (x, y), (1, 2) and (7, 0) are collinear. **[NCERT]**

64. Find the centre of a circle passing through the points (6, –6), (3, –7) and (3, 3). **[NCERT]**

65. The two opposite vertices of a square are (–1, 2) and (3, 2). Find the coordinates of the other two vertices. **[NCERT]**

66. The vertices of a ΔABC are $A(4, 6)$, $B(1, 5)$ and $C(7, 2)$. A line is drawn to intersect sides AB and AC at D and E, respectively, such that $\frac{AD}{AB} = \frac{AE}{AC} = \frac{1}{4}$. Calculate the area of ΔADE and compare it with the area of ΔABC. **[NCERT]**

67. The radius of the circumcircle of a triangle ABC is $5\sqrt{10}$ units. If point P is equidistant from $A(1, 3)$, $B(-3, 5)$ and $C(5, -1)$, then find AP and the coordinates of P.

68. If $A(5, 2)$, $B(2, -2)$ and $C(-2, t)$ are the vertices of a right-angled triangle with $\angle B = 90°$, find the value of t. **[Delhi 2015]**

69. Find the relation between x and y if the points $A(x, y)$, $B(-5, 7)$ and $C(-4, 5)$ are collinear. **[AI 2015]**

70. The x coordinate of a point P is twice its y coordinate. If P is equidistant from $Q(2, -5)$ and $R(-3, 6)$, find the coordinates of P. **[Delhi 2016]**

71. Prove that the points (0, 0), (5, 5) and (–5, 5) are vertices of a right isosceles triangle.

72. If the points $(a, 0)$, $(0, b)$ and $(1, 1)$ are collinear, show that $\frac{1}{a} + \frac{1}{b} = 1$.

73. Find the distance between the points $(\sqrt{5} - 2, \sqrt{3} + 2)$ and $(\sqrt{5} + 1, \sqrt{3} - 1)$.

74. If a points A lies on the x axis and has abscissa $(p + q)$ and another point B lies on the y axis and ordinate $(p - q)$, find the distance between A and B.

75. Find the ratio in which the y axis divides the line segment joining the points (5, –6) and (–1, –4). Also find the point of intersection. **[NCERT]**

76. Find the ratio in which the line segment joining $A(1, -5)$ and $B(-4, 5)$ is divided by the x axis. Also find the coordinates of the point of division. **[NCERT]**

77. Show that the points $A(a, a)$, $B(-a, -a)$ and $C(-a\sqrt{3}, a\sqrt{3})$ form an equilateral triangle. **[Foreign 2015]**

78. Two vertices of a triangle are (3, –5) and (–7, 4). If its centroid is (2, –1), find the third vertex.

79. Point P divides the line segment joining the points $A(2, -5)$ and $B(5, 2)$ in the ratio 2:3. Name the quadrant in which P lies. **[Delhi 2011]**

80. If $P(5, -3)$ and $Q(3, y)$ are the points of trisection of the line segment joining $A(7, -2)$ and $B(1, -5)$, find y. **[Delhi 2013]**

81. Find the ratio in which $P(4, m)$ divides the line segment joining the points $A(2, 3)$ and $B(6, -3)$. Hence find m. **[CBSE 2018]**

82. If $A(4, -8)$, $B(3, 6)$ and $C(5, -4)$ are the vertices of ΔABC, D is the mid-point of BC and P is a point on AD joined such that $\frac{AP}{PD} = 2$, find the coordinates of P.

83. The line segment joining the points $A(2, 1)$ and $B(5, -8)$ is trisected at the points P and Q such that P is nearer to A. If P also lies on the line given by $2x - y + k = 0$, find the value of k.

84. The coordinates of the mid-point of the line joining the points $(3p, 4)$ and $(-2, 2q)$ is $(5, p)$. Find the values of p and q.

85. In the figure, BD bisects $\angle B$. Find the length of BD.

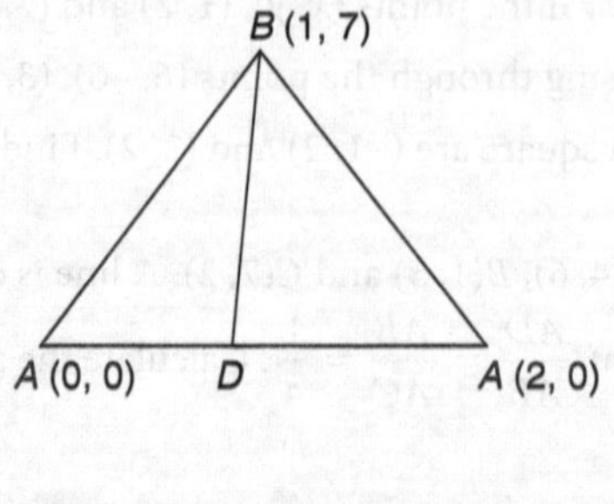

86. If $A(-2, 1)$, $B(a, 0)$, $C(4, b)$ and $D(1, 2)$ are the vertices of a parallelogram $ABCD$, find the values of a and b. Hence find the lengths of its sides. **[CBSE 2018]**

87. If the coordinates of two adjacent vertices of a parallelogram are (3, 2), (1, 0) and the diagonals bisect each other at (2, –5), find the coordinates of the other two vertices. **[CBSE 2018 (C)]**

88. If point $P\left(\frac{1}{2}, y\right)$ lies on the line segment joining the points $A(3, -5)$ and $B(-7, 9)$, then find the ratio in which P divides AB. Also find the value of y.

89. The line segment AB joining the points $A(3, -4)$and $B(1, 2)$ is trisected at the points $P(p, -2)$ and $Q(5/3, q)$. Find the values of p and q. **[Foreign 2013]**

90. Find the value of k if $P(4, -2)$ is the mid-point of the line segment joining the points $A(5k, 3)$ and $B(-k, -7)$.

91. Find the fourth vertex of a rectangle whose three vertices taken in order are (4, 1), (7, 4) and (13, –2).

92. Find the value of m for which the points with coordinates (3, 5), $(m, 6)$ and $\left(\frac{1}{2}, \frac{15}{2}\right)$ are collinear.

93. Point P divides the line segment joining the points $A(-1, 3)$ and $B(9, 8)$ such that $\frac{AP}{PB} = \frac{k}{1}$. If P lies on the line $x - y + 2 = 0$, find the value of k.

94. Point A is on the x axis, point B is on the y axis and the point P lies on line segment AB, such that $P(4, 5)$ and $AP{:}PB = 5{:}3$. Find the coordinates of points A and B.

95. The centroid of the triangle formed by the points (a, b), (b, c) and (c, a) is at origin. If $a^3 + b^3 + c^3 - 3abc = k^2$, then find the value of k.

96. Determine the ratio in which the line $2x + y - 4 = 0$ divides the line segment joining the points $A(2, -2)$ and $B(3, 7)$. **[NCERT]**

97. If the points (10, 5), (8, 4) and (6, 6) are the mid-points of the sides of a triangle, find its vertices.

98. Determine the ratio in which the point $P(a, -2)$ divides the line joining the points $A(-4, 3)$ and $B(2, -4)$. Also find the value of a.

99. The mid-point P of the line segment joining the points $A(-10, 4)$ and $B(-2, 0)$ lies on the line segment joining the points $C(-9, -4)$ and $D(-4, y)$. Find the ratio in which P divides CD. Also find the value of y. **[Foreign 2014]**

100. The base BC of an equilateral triangle ABC lies on the y axis. The coordinates of point C are (0, –3). The origin is the mid-point of the base. Find the coordinates of the points A and B. Also find the coordinates of another point D such that $BACD$ is a rhombus. **[Foreign 2015]**

101. Prove that the straight line joining the points (–1, 3) and (4, –2) passed through the point (a, b) if $a + b = 2$.

102. Find the coordinates of the point which divides internally the line joining the points (p, q) and (q, p) in the ratio $(p - q){:}(p + q)$.

103. Find the area of a triangle whose vertices are (1, –1), (–4, 6) and (–3, –5). **[NCERT]**

104. Find the area of the triangle formed by the points $P(-1.5, 3)$, $Q(6, -2)$ and $R(-3, 4)$. **[NCERT]**

105. Find the area of the triangle formed by joining the mid-points of the sides of the triangle whose vertices are (0, –1), (2, 1) and (0, 3). Find the ratio of this area to the area of the given triangle. **[NCERT]**

106. Let $A(4, 2)$, $B(6, 5)$ and $C(1, 4)$ be the vertices of ΔABC.

(i) The median from A meets BC at D. Find the coordinates of the point D.

(ii) Find the coordinates of the point P on AD such that $AP:PD = 2:1$.

(iii) Find the coordinates of points Q and R on medians BE and CF, respectively, such that $BQ:QE = 2:1$ and $CR:RF = 2:1$.

(iv) What do you observe? **[NCERT]**

107. In the figure, find the area of triangle ABC.

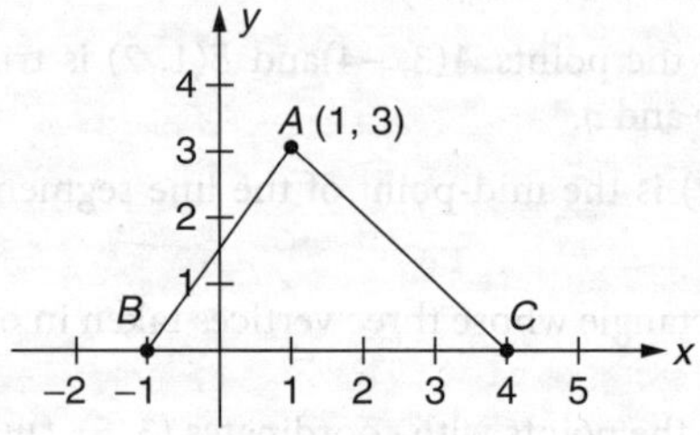

[CBSE 2013]

108. If $P(x, y)$ is any point on the line joining the points $A(a, 0)$ and $B(0, b)$, then show that $\frac{x}{a} + \frac{y}{b} = 1$.

109. Find the area of the triangle formed by joining the mid-points of the sides of the triangle whose vertices are $(0, -1)$, $(2, 1)$ and $(0, 3)$. **[AI 2017(C)]**

110. If the points $(a^2, 2a)$, $(1, -2)$ and $(2, -1)$ never lie on a unique circle, find the value of a.

111. If the area of a triangle with vertices $(x, 3)$, $(4, 4)$ and $(3, 5)$ is 4 square units, find x. **[CBSE 2018(C)]**

112. If $R(x, y)$ is a point on the line segment joining the points $P(a, b)$ and $Q(b, a)$, then prove that $x + y = a + b$.

113. For what value of k, $(k > 0)$ is the area of the triangle with vertices $(-2, 5)$, $(k, -4)$ and $(2k + 1, 10)$ equal to 53 sq. units? **[Foreign 2012]**

114. If the points $A(1, -2)$, $B(2, 3)$, $C(-3, 2)$ and $D(-4, -3)$ are the vertices of parallelogram $ABCD$, then taking AB as the base, find the height of the parallelogram. **[AI 2013]**

115. The area of a triangle is 5 sq. units. Two of its vertices are $(2, 1)$ and $(3, -2)$. The third vertex lies on $y = x + 3$. Find the third vertex.

116. In the figure, find the area of ΔABC.

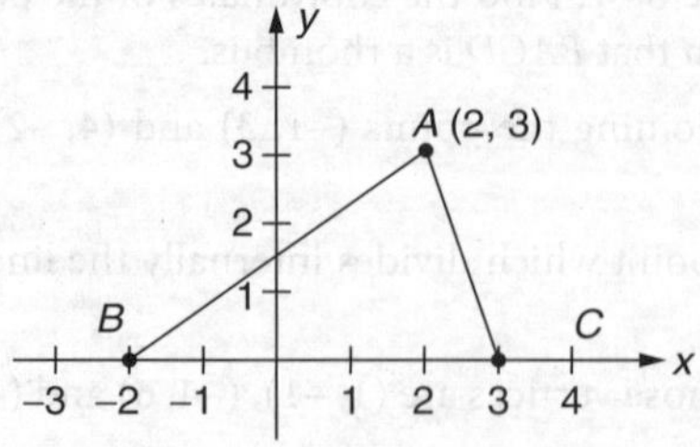

117. The points (p, q), (m, n) and $(p - m, q - n)$ are collinear. Show that $pn = qm$.

118. The area of the triangle formed by the points $(p, 2 - 2p)$, $(1 - p, 2p)$ and $(-4 - p, 6 - 2p)$ is 70 sq. units. How many integral values of p are possible?

119. The three vertices of a parallelogram $ABCD$ taken in order are $A(-1, 0)$, $B(3, 1)$ and $C(2, 2)$. Find the height of the parallelogram with AD as its base.

120. The vertices of a triangle ABC are $A(1, k)$, $B(4, -3)$ and $C(-9, 7)$. Area of the triangle is 15 square units. Find the altitude of the triangle with AB as the base. [k is an integer]

121. The vertices of ΔPQR are $P(4, 6)$, $Q(1, 5)$ and $R(7, 2)$. A line is drawn to intersect sides PQ and PR at M and N, respectively, such that $\frac{PM}{PQ} = \frac{PN}{PR} = \frac{1}{4}$. Calculate the area of triangle PMN.

122. $P(x, y)$, $A(3, 4)$, $B(5, -2)$ are the vertices of triangle PAB such that $PA = PB$ and area of $\Delta PAB = 10$ sq. units. If $PA = k\sqrt{5}$ units, find k.

123. If the centroid of the triangle formed by points $A(a, b)$, $B(b, c)$ and $C(c, a)$ is at the origin, what is the value of $\frac{a^2}{bc} + \frac{b^2}{ca} + \frac{c^2}{ab}$?

124. The coordinates of the points A, B and C are $(6, 3)$, $(-3, 5)$ and $(4, -2)$, respectively. $P(x, y)$ is any point on the plane. Show that $\frac{ar(\Delta PBC)}{ar(\Delta ABC)} = \left|\frac{x + y - 2}{7}\right|$.

125. If $(3, 3)$, $(h, 0)$ and $(0, k)$ are collinear, find the value of $\frac{1}{h} + \frac{1}{k}$.

126. Find the area of the triangle formed by the line $5x - 3y + 15 = 0$ with the coordinate axis.

127. The line $3x + 4y = 24$ cuts the x axis at A and y axis at B. If O is the origin, find the area of ΔAOB.

128. Two consecutive vertices of a rectangle of area 10 sq. units are $(1, 3)$ and $(-2, -1)$. Find the coordinates of the other two vertices.

129. The vertices of a triangle ABC are $A(k, 2 - 2k)$, $B(-k + 1, 2k)$ and $C(-4 - k, 6 - 2k)$. If its area is 70 sq. units, find the integral value of k.

130. If three points (x_1, y_1), (x_2, y_2), (x_3, y_3) lie on the same line, prove that $\frac{y_2 - y_3}{x_2x_3} + \frac{y_3 - y_1}{x_3x_1} + \frac{y_1 - y_2}{x_1x_2} = 0$.

Answers

Multiple Choice Questions

1. (d) **2.** (b) **3.** (b) **4.** (b) **5.** (c) **6.** (b) **7.** (b) **8.** (d) **9.** (c) **10.** (d)
11. (d) **12.** (c) **13.** (c) **14.** (b) **15.** (c) **16.** (a) **17.** (b) **18.** (c) **19.** (b) **20.** (c)
21. (d) **22.** (a) **23.** (b) **24.** (d) **25.** (c) **26.** (c) **27.** (d) **28.** (b)

Fill in the Blanks

29. 1 **30.** –4 **31.** (3, 2) **32.** (–2, 5) **33.** 336 m^3
34. (3, 5) **35.** (4, 2) **36.** (3, –10)

True/False

37. True **38.** False **39.** True **40.** True **41.** False **42.** True **43.** True

Assertion–Reason

44. (A) **45.** (B) **46.** (B) **47.** (B) **48.** (B)

Subjective Questions

49. Yes **50.** (0, 9) **51.** $\sqrt{x^2+y^2}$ units. **52.** (3, –10)

53. $x+y=3$ **55.** – 4, 2 **57.** ($\sqrt{3}$, 1) and ($\sqrt{3}$, –1) **58.** $2a$

59. 5 units **60.** 5 or –3 **61.** 5, 3 **63.** $2x+6y-14=0$ **64.** (3, –2) **65.** (1, 0) and (1, 4)

66. $\frac{15}{32}$ square units; 1 : 16. **67.** AP = 5 $\sqrt{10}$ units, P (–8, –10). **68.** 1. **69.** $2x+y+3=0$

70. (16, 8) **73.** $3\sqrt{2}$ units **74.** $\sqrt{2p^2+2q^2}$ **75.** 5 : 1; $\left(0, -\frac{13}{3}\right)$ **76.** 1 : 1; $\left(\frac{-3}{2}, 0\right)$

78. (10, –2) **79.** P(3.2, –2.2) lies in IV quadrant **80.** –4 **81.** 1 : 1, 0 **82.** (4, –2)

83. –8 **84.** (4, 2) **85.** 7 units. **86.** $a=1$, $b=1$ and $\sqrt{10}$ units **87.** (1, –12) and (3, 10)

88. 1 : 3, $-\frac{3}{2}$ **89.** $p=\frac{7}{3}$, $q=0$ **90.** $k=2$ **91.** (10, –5) **92.** 2 **93.** $\frac{2}{3}$

94. $\left(\frac{32}{3}, 0\right)$ and (0, 8) **95.** $k=0$ **96.** 2 : 9 **97.** (8, 7)(12, 3)(4, 5) **98.** 5:2 and $a=\frac{2}{7}$

99. 3 : 2; $y=6$ **100.** (3 $\sqrt{3}$, 0) (–3 $\sqrt{3}$, 0) **102.** $\left(\frac{p^2-q^2+2pq}{2p}, \frac{p^2+q^2}{2p}\right)$ **103.** 24 square units

104. 0, collinear **105.** 1 square units, 1 : 4

106. (i) $D\left(\frac{7}{2}, \frac{9}{2}\right)$ (ii) $P\left(\frac{11}{3}, \frac{11}{3}\right)$ (iii) $Q\left(\frac{11}{3}, \frac{11}{3}\right)$, $R\left(\frac{11}{3}, \frac{11}{3}\right)$ (iv) Medians are concurrent.

107. 7.5 square units. **109.** 1 square units. **110.** 3, –1 **111.** –3 and 13 **113.** $k=3$ **114.** $\frac{12}{13}\sqrt{26}$ units

115. $\left(\frac{7}{2}, \frac{13}{2}\right)$ or $\left(-\frac{3}{2}, \frac{3}{2}\right)$ **116.** $\frac{15}{2}$ square units **118.** 4, $-\frac{9}{2}$ **119.** $\frac{5}{2}\sqrt{2}$ units

120. $k=-3$, coordinates of A (1, –3), altitude = 10 units **121.** $\frac{15}{32}$ square units **122.** $k=2$ **123.** 3 **125.** $\frac{1}{3}$

126. $\frac{15}{2}$ square units **127.** 24 square units

128. Other vertices are $\left(-\frac{2}{5}, -\frac{11}{5}\right)$ or $\left(-\frac{18}{5}, \frac{1}{5}\right)$ and $\left(\frac{13}{5}, \frac{9}{5}\right)$ or $\left(\frac{-3}{5}, \frac{21}{5}\right)$ **129.** $k=4$, $\frac{-9}{2}$

Coordinate Geometry (Basic)

SELF-ASSESSMENT TEST

Time allowed: 90 min. Max. Marks: 40

General Instructions

1. *All questions are compulsory. There are 20 questions in all.*
2. *This question paper has four sections: Section A, Section B, Section C and Section D.*
3. *Section A contains 10 questions of one mark each, Section B contains three questions of two marks each, Section C contains four questions of three marks each, and Section D contains three questions of four marks each.*

Section A [1 Mark Each]

Direction: (Q. Nos. 1–5) These are Multiple Choice Questions. Select the most appropriate answer for the given questions.

1. In the figure, the area of triangle ABC (in sq. units) is

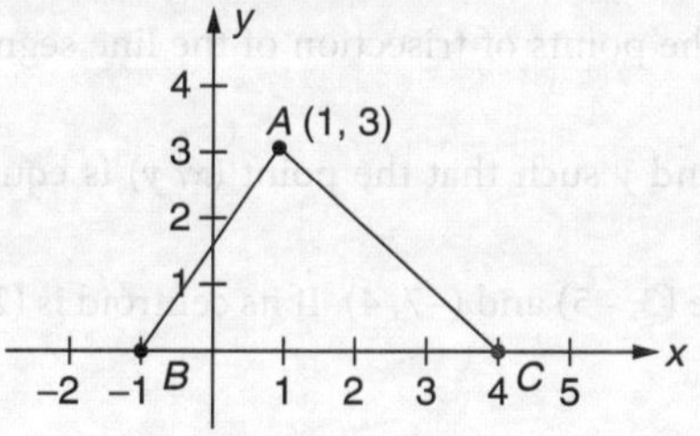

(a) 15 (b) 10 (c) 7.5 (d) 2.5

2. If the distance between the points $(8, p)$ and $(4, 3)$ is 5 units, then the value of p is
(a) 6 (b) 0 (c) both (a) and (b) (d) none of these
3. The vertices of a parallelogram taken in order are $(-1, -6)$, $(2, -5)$ and $(7, 2)$. The fourth vertex is
(a) $(1, 4)$ (b) $(1, 1)$ (c) $(4, 4)$ (d) $(4, 1)$
4. The perimeter of a triangle with vertices $(0, 4)$, $(0, 0)$ and $(3, 0)$ is
(a) 5 (b) 12 (c) 11 (d) $7 + \sqrt{5}$
5. The points $(k + 1, 1)$, $(2k + 1, 3)$ and $(2k + 2, 2k)$ are collinear if
(a) $k = -1, 2$ (b) $k = \frac{1}{2}, 2$ (c) $k = 2, 1$ (d) $k = -\frac{1}{2}, 2$
6. The point $P(-2, 4)$ lies on a circle of radius 6 units and centre $C(3, 5)$. Is this true or false? Justify.
7. Find the point on the x axis that is equidistant from the points $(-2, 5)$ and $(2, 9)$.
8. Do the points $A(-5, 6)$, $B(-4, -2)$ and $C(7, 5)$ form a triangle? If so, name the type of triangle formed.
9. Find the coordinates of the mid-point of the line segment joining the points $A(6, -2)$ and $B(-2, 8)$.
10. In what ratio does the point $P(2, 5)$ divide the line joining $A(8, 2)$ and $B(-6, 9)$?

Section B [2 Marks Each]

11. Find the coordinates of the points of trisection of the line segment joining the points $A(-5, 6)$ and $B(4, -3)$.

12. Find the ratio in which the line segment joining $A(1, -5)$ and $B(-4, 5)$ is divided by the x axis. Also find the coordinates of the point of division.

13. Find the area of the quadrilateral $ABCD$ whose vertices are $A(-1, -3)$, $B(5, -7)$, $C(10, -2)$ and $D(5, 17)$.

Section C [3 Marks Each]

14. Find the ratio in which the point $P(m, 6)$ divides the line joining $A(-4, 3)$ and $B(2, 8)$. Also find the value of m.

15. Find the value of K so that the points $A(-2, 3)$, $B(3, -1)$ and $C(5, K)$ are collinear.

16. If the points $(a, 0)$, $(0, b)$ and $(1, 1)$ are collinear, show that $\frac{1}{a} + \frac{1}{b} = 1$.

17. Prove that the points (3, 0), (6, 4) and (–1, 3) are the vertices of a right-angled isosceles triangle.

Section D [4 Marks Each]

18. If $P(5, -3)$ and $Q(3, y)$ are the points of trisection of the line segment joining $A(7, -2)$ and $B(1, -5)$, find y.

19. Find a relation between x and y such that the point (x, y) is equidistant from the points (7, 1) and (3, 5).

20. Two vertices of a triangle are (3, –5) and (–7, 4). If its centroid is (2, –1), find the third vertex.

Answers

1. (c) **2.** (c) **3.** (d) **4.** (b) **5.** (d) **6.** False **7.** (7, 0) **8.** Yes, Scalene triangle

9. (2, 3) **10.** 3:4 **11.** (–2, 3) and (1, 0) **12.** 1:1, $\left(\frac{-3}{2}, 0\right)$ **13.** 132 sq. units **14.** 3:2 and $\frac{-2}{5}$

15. $\frac{-13}{5}$ **18.** –4 **19.** $x - y - 2 = 0$ **20.** (10, –2)

Coordinate Geometry (Standard)

SELF-ASSESSMENT TEST

Time allowed: 90 min. Max. Marks: 40

General Instructions

1. *All questions are compulsory. There are 20 questions in all.*
2. *This question paper has four sections: Section A, Section B, Section C and Section D.*
3. *Section A contains 10 questions of one mark each, Section B contains three questions of two marks each, Section C contains four questions of three marks each, and Section D contains three questions of four marks each.*

Section A [1 Mark Each]

Direction: (Q. Nos. 1–6) These are Multiple Choice Questions. Select the most appropriate answer for the given questions.

1. If the area of a triangle with vertices $(x, 0)$, $(1, 1)$ and $(0, 2)$ is 4 square units, then the value of x is

(a) -2 (b) -4 (c) -6 (d) 8

2. If A and B are the points $(-3, 4)$ and $(2, 1)$ respectively, then the coordinates of the point on AB produced such that $AC = 2BC$ are

(a) $(2, 4)$ (b) $(3, 7)$ (c) $(7, -2)$ (d) none of these

3. In the figure, the area of triangle ABC (in sq. units) is

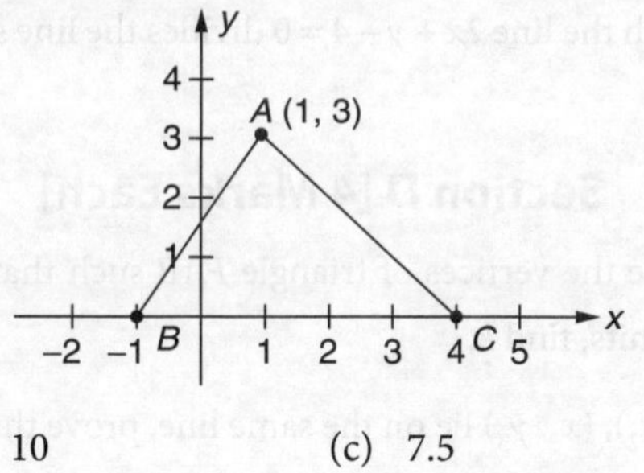

(a) 15 (b) 10 (c) 7.5 (d) 2.5

4. The points $(k + 1, 1)$, $(2k + 1, 3)$ and $(2k + 2, 2k)$ are collinear if

(a) $k = -1, 2$ (b) $k = \frac{1}{2}, 2$ (c) $k = 2, 1$ (d) $k = -\frac{1}{2}, 2$

5. A circle drawn with origin as the centre passes through $\left(\frac{13}{2}, 0\right)$. The point that does not lie in the interior of the circle is

(a) $\left(-\frac{3}{4}, 1\right)$ (b) $\left(2, \frac{7}{3}\right)$ (c) $\left(5, -\frac{1}{2}\right)$ (d) $\left(-6, \frac{5}{2}\right)$

6. The distance of the point (h, k) from the x axis is

(a) h units (b) $|h|$ units (c) k units (d) $|k|$ units

7. If $A = (a^2, 2a)$ and $B = \left(\frac{1}{a^2}, -\frac{2}{a}\right)$ and $S = (1, 0)$, then $\frac{1}{SA} + \frac{1}{SB} =$ __________.

8. The two vertices of a triangle are $A(6, 3)$ and $B(-1, 7)$ and its centroid is $G(1, 5)$. Then, the third vertex C of ΔABC is __________.

9. If the points $(a, 0)$, $(0, b)$ and $(1, 1)$ are collinear, show that $\frac{1}{a}+\frac{1}{b}=1$.

10. Find the value of m for which the points with coordinates $(3, 5)$, $(m, 6)$ and $\left(\frac{1}{2}, \frac{15}{2}\right)$ are collinear.

Section B [2 Marks Each]

11. Find the ratio in which the line segment joining $A(1, -5)$ and $B(-4, 5)$ is divided by the x axis. Also find the coordinates of the point of division.

12. Find the fourth vertex of a rectangle whose three vertices taken in order are $(4, 1)$, $(7, 4)$ and $(13, -2)$.

13. If $(3, 3)$, $(h, 0)$ and $(0, k)$ are collinear, then find the value of $\frac{1}{h} + \frac{1}{k}$.

Section C [3 Marks Each]

14. Find the coordinates of a point A, where AB is the diameter of a circle whose centre is $(2, -3)$ and B is $(1, 4)$.

15. The centroid of the triangle formed by the points (a, b), (b, c) and (c, a) is at origin. If $a^3 + b^3 + c^3 - 3abc = k^2$, then find the value of k.

16. If the points $(10, 5)$, $(8, 4)$ and $(6, 6)$ are the mid-points of the sides of a triangle, find its vertices.

17. Determine the ratio in which the line $2x + y - 4 = 0$ divides the line segment joining the points $A(2, -2)$ and $B(3, 7)$.

Section D [4 Marks Each]

18. $P(x, y)$, $A(3, 4)$, $B(5, -2)$ are the vertices of triangle PAB such that $PA = PB$ and the area of $\Delta PAB =$ 10 sq. units. If $PA = k\sqrt{5}$ units, find k.

19. If three points (x_1, y_1), (x_2, y_2), (x_3, y_3) lie on the same line, prove that $\frac{y_2 - y_3}{x_2x_3} + \frac{y_3 - y_1}{x_3x_1} + \frac{y_1 - y_2}{x_1x_2} = 0$.

20. The area of the triangle formed by the points $(p, 2 - 2p)$, $(1 - p, 2p)$ and $(-4 - p, 6 - 2p)$ is 70 sq. units. How many integral values of p are possible?

Answers

1. (c) **2.** (c) **3.** (c) **4.** (d) **5.** (d) **6.** (d) **7.** 1 **8.** (–2, 5) **10.** 2

11. 1 : 1, $\left(\frac{-3}{2}, 0\right)$ **12.** (10, –5) **13.** $\frac{1}{3}$ **14.** (3, –10) **15.** $k = 0$

16. (8, 7) (12, 3) (4, 5) **17.** 2:9 **18.** $k = 2$ **20.** 4, $-\frac{9}{2}$

CHAPTER

8 Introduction to Trigonometry

Key Concepts

- Relationship between perpendicular, base, hypotenuse and angle of a right triangle
- Problems based on Pythagoras' Theorem
- Application of Pythagoras' Theorem and trigonometric ratios
- Relationship between Pythagoras' Theorem and trigonometry
- Application of trigonometric table
- Problems based on complementary angles
- Problems based on trigonometric identities
- Solving trigonometric equations
- Proving using the given information
- Miscellaneous problems

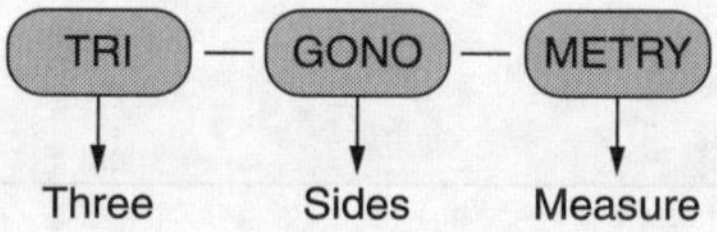

Trigonometry is a branch of mathematics in which we study the relationship between the sides and angles of a triangle.

Concept: Relationship between Perpendicular, Base and Hypotenuse of a Right Triangle

We know that in right ΔABC

$$AC^2 = AB^2 + BC^2$$

$$H^2 = P^2 + B^2 \quad \text{[By Pythagoras' Theorem]}$$

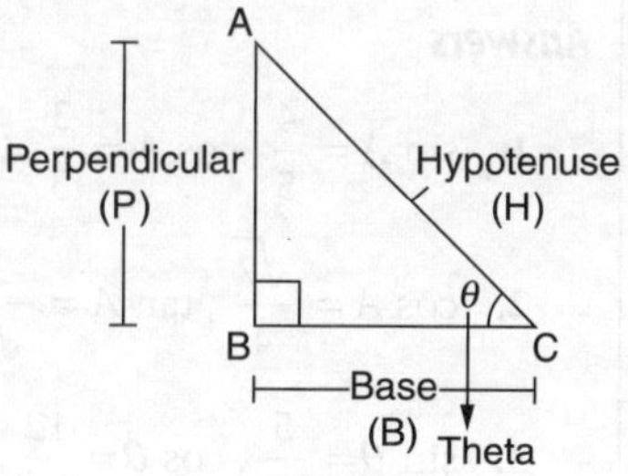

Trigonometric ratios of $\angle C$ of a right ΔABC are defined as:

$$\sin \angle C = \sin \theta = \frac{P}{H}; \ \text{cosec} \angle C = \text{cosec}\, \theta = \frac{H}{P}$$

$$\cos \angle C = \cos \theta = \frac{B}{H}; \ \sec \angle C = \sec \theta = \frac{H}{B}$$

$$\tan \angle C = \tan \theta = \frac{P}{B}; \ \cot \angle C = \cot \theta = \frac{B}{P}$$

Example 1 If $\sin A = \frac{3}{5}$, find the other five trigonometric ratios.

Solution: Let us consider right ΔABC, right-angled at $\angle B$. We have

$$\sin A = \frac{3}{5} = \frac{P}{H} = \frac{CB}{CA}$$

Hence $CB = 3k, CA = 5k$

In right ΔCBA $CA^2 = CB^2 + BA^2$ [$\because$ Pythagoras' Theorem]

$(5k)^2 = (3k)^2 + BA^2$

$25k^2 = 9k^2 + BA^2$

$BA^2 = 25k^2 - 9k^2 = 16k^2 = (4k)^2$

$BA = 4k$

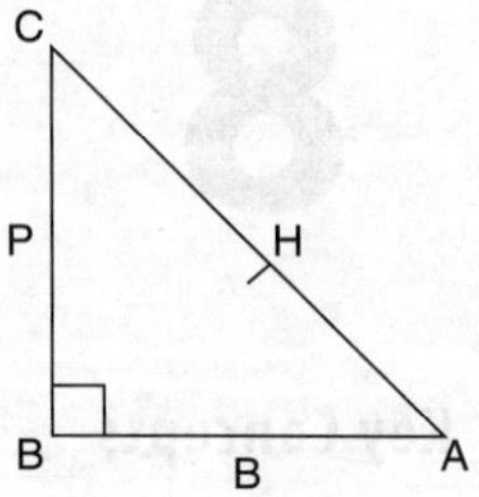

So $P = 3k, B = 4k, H = 5k$

$\sin A = \frac{P}{H} = \frac{3}{5}$ $\quad$ $\text{cosec } A = \frac{H}{P} = \frac{5}{3}$

$\cos A = \frac{B}{H} = \frac{4}{5}$ $\quad$ $\sec A = \frac{H}{B} = \frac{5}{4}$

$\tan A = \frac{P}{B} = \frac{3}{4}$ $\quad$ $\cot A = \frac{B}{P} = \frac{4}{3}$

1 Try to Solve

1. If $\tan A = \frac{4}{3}$, find the other trigonometric ratios of $\angle A$.
2. If $\sin A = \frac{3}{4}$, calculate $\cos A$ and $\tan A$.
3. If $\sec \theta = \frac{13}{12}$, calculate all other trigonometric ratios.

Answers

1. $\sin A = \frac{4}{5}$, $\cos A = \frac{3}{5}$, $\tan A = \frac{4}{3}$, $\cot A = \frac{4}{3}$, $\sec A = \frac{5}{3}$, $\text{cosec } A = \frac{5}{4}$
2. $\cos A = \frac{\sqrt{7}}{4}$, $\tan A = \frac{3}{\sqrt{7}}$
3. $\sin \theta = \frac{5}{13}$, $\cos \theta = \frac{12}{13}$, $\tan \theta = \frac{5}{12}$, $\cot \theta = \frac{12}{5}$, $\text{cosec } \theta = \frac{13}{5}$

Example 2 In a right triangle ABC, right-angled at B, if $\tan A = 1$, then verify that $2 \sin A \cos A = 1$. **[CBSE (CCE) 2009]**

Solution: Since $\tan A = \frac{1}{1} = \frac{P}{B} = \frac{CB}{BA}$

That is $CB = 1k, BA = 1k$

In right ΔCBA $CA^2 = CB^2 + BA^2$ [$\because$ Pythagoras' Theorem]

$$= (1k)^2 + (1k)^2 = 2k^2$$

$$CA = \sqrt{2k^2} = \sqrt{2}k.$$

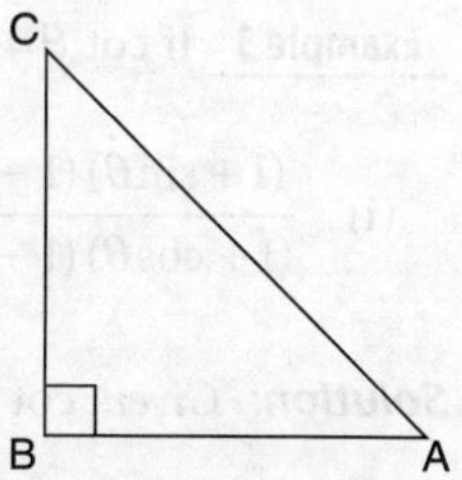

$$\sin A = \frac{P}{H} = \frac{CB}{CA} = \frac{1k}{\sqrt{2}k} = \frac{1}{\sqrt{2}}$$

$$\cos A = \frac{B}{H} = \frac{BA}{CA} = \frac{1k}{\sqrt{2}k} = \frac{1}{\sqrt{2}}$$

We have to prove in LHS.

$$2 \sin A \cdot \cos A = 2 \times \frac{1}{\sqrt{2}} \times \frac{1}{\sqrt{2}} = \cancel{2} \times \frac{1}{\cancel{2}} = 1$$

Hence it is verified.

2 Try to Solve

1. In the figure, find tan P – cot R. **[NCERT, CBSE (CCE) 2009]**

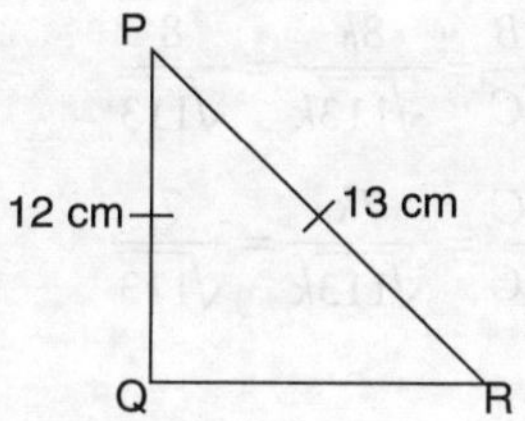

2. In the figure, find
   (a) $\sin^2 \theta + \cos^2 \theta$ (b) $\cos^2\theta - \sin^2\theta$

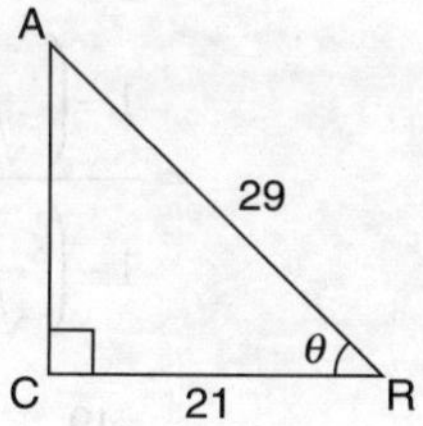

3. If $\sin \theta = \frac{3}{5}$, find the value of $\tan^2 \theta + \sin \theta \cdot \cos \theta + \cot \theta$.

Answers

1. 0 2. (a) 1 (b) $\frac{41}{841}$ 3. $\frac{2851}{1200}$

Example 3 If $\cot\theta = \frac{7}{8}$, evaluate:

(i) $\frac{(1+\sin\theta)(1-\sin\theta)}{(1+\cos\theta)(1-\cos\theta)}$ (ii) $\cot^2\theta$ [NCERT]

Solution: *Given*: $\cot\theta = \frac{7}{8}$

To find: (i) $\frac{(1+\sin\theta)(1-\sin\theta)}{(1+\cos\theta)(1-\cos\theta)}$ (ii) $\cot^2\theta$

Solution: In ΔABC $\cot\theta = \frac{BC}{AB} = \frac{7}{8}$ [Given]

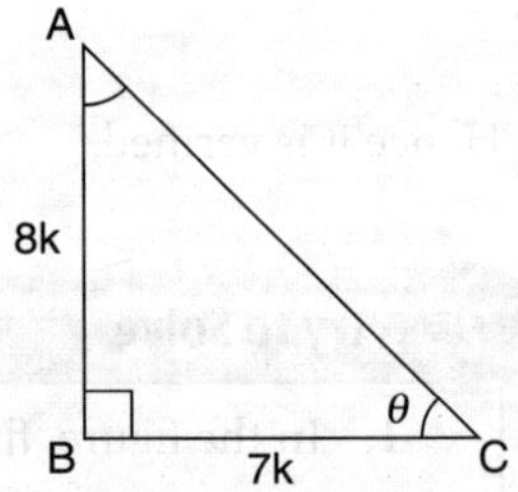

$\therefore$ $BC = 7k$

And $AB = 8k$

$AC^2 = BA^2 + BC^2$ [By Pythagoras' Theorem]

$= 64\,k^2 + 49\,k^2 = k^2(64+49)$

$= k^2 \times 113$

$\therefore$ $AC = \sqrt{113}\,k$

$$\sin\theta = \frac{AB}{AC} = \frac{8k}{\sqrt{113}k} = \frac{8}{\sqrt{113}}$$

$$\cos\theta = \frac{BC}{AC} = \frac{7k}{\sqrt{113}k} = \frac{7}{\sqrt{113}}$$

(i) $\frac{(1+\sin\theta)(1-\sin\theta)}{(1+\cos\theta)(1-\cos\theta)} = \frac{1-\sin^2\theta}{1-\cos^2\theta}$ [$\because (a+b)(a-b) = a^2 - b^2$]

$$= \frac{1-\left(\frac{8}{\sqrt{113}}\right)^2}{1-\left(\frac{7}{\sqrt{113}}\right)^2} = \frac{1-\frac{64}{113}}{1-\frac{49}{113}} = \frac{\frac{113-64}{113}}{\frac{113-49}{113}}$$

$$= \frac{\frac{49}{113}}{\frac{64}{113}} = \frac{49}{\cancel{113}} \times \frac{\cancel{113}}{64} = \frac{49}{64}$$

(ii) $$\cot^2\theta = \left(\frac{\cos\theta}{\sin\theta}\right)^2 = \left(\frac{\frac{7}{\cancel{\sqrt{113}}}}{\frac{8}{\cancel{\sqrt{113}}}}\right)^2 = \left(\frac{7}{8}\right)^2 = \frac{49}{64}$$

Example 4 In ΔABC, right-angled at B, $AB = 5$ cm and $BC = 12$ cm. Find the values of sin A, sec A, sin C and sec C.

Solution: In ΔABC, $\angle B = 90°$, $AB = 5$ cm and $BC = 12$ cm.

By Pythagoras' Theorem, we have

$$AC^2 = (AB^2 + BC^2) = \{(5)^2 + (12)^2\}\text{cm}^2$$

$$= (25 + 144)\text{ cm}^2 = 169\text{ cm}^2$$

$$\therefore \quad AC = \sqrt{169\text{ cm}^2} = 13\text{ cm}$$

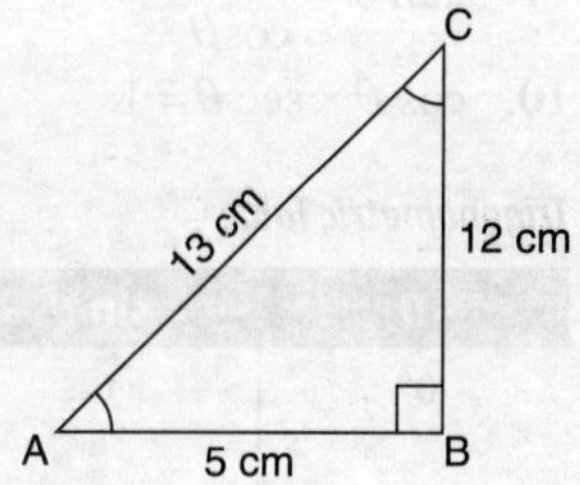

For T-ratios of $\angle A$, we have base $= AB = 5$ cm

perpendicular $= BC = 12$ cm

hypotenuse $= AC = 13$ cm

$$\therefore \quad \sin A = \frac{BC}{AC} = \frac{12}{13} \text{ and } \sec A = \frac{AC}{AB} = \frac{13}{5}$$

For T-ratios of $\angle C$, we have base $= BC = 12$ cm

perpendicular $= AB = 5$ cm

hypotenuse $= AC = 13$ cm

$$\therefore \quad \sin C = \frac{AB}{AC} = \frac{5}{13} \text{ and } \sec C = \frac{AC}{BC} = \frac{13}{12}$$

3 Try to Solve

1. In triangle ABC, right-angled at B, if $\tan A = \frac{1}{\sqrt{3}}$, find the value of:

   (a) $\sin A \cos C + \cos A \sin C$ (b) $\cos A \cos C - \sin A \sin C$ [NCERT]

2. If $3 \cot A = 4$, check whether $\frac{1-\tan^2 A}{1+\tan^2 A} = \cos^2 A - \sin^2 A$ or not. [NCERT]

3. If $\cos\theta = \frac{3}{5}$, find the value of $\frac{\left(\sin\theta - \frac{1}{\tan\theta}\right)}{2\tan\theta}$.

4. If $4\tan\theta = 3$, evaluate $\left(\frac{4\sin\theta - \cos\theta + 1}{4\sin\theta + \cos\theta - 1}\right)$. [CBSE 2018]

Answers

1. (a) 1 (b) 0 **2.** Yes **3.** $\frac{3}{160}$ **4.** $\frac{13}{11}$

Some Formulae to Remember

1. $\sin\theta = \dfrac{1}{\operatorname{cosec}\theta}$
2. $\operatorname{cosec}\theta = \dfrac{1}{\sin\theta}$
3. $\cos\theta = \dfrac{1}{\sec\theta}$
4. $\sec\theta = \dfrac{1}{\cos\theta}$
5. $\tan\theta = \dfrac{1}{\cot\theta}$
6. $\cot\theta = \dfrac{1}{\tan\theta}$
7. $\tan\theta = \dfrac{\sin\theta}{\cos\theta}$
8. $\cot\theta = \dfrac{\cos\theta}{\sin\theta}$
9. $\sin\theta \times \operatorname{cosec}\theta = 1$
10. $\cos\theta \times \sec\theta = 1$
11. $\tan\theta \times \cot\theta = 1$

Trigonometric Table

θ	$\sin\theta$	$\cos\theta$	$\tan\theta$	$\operatorname{cosec}\theta$	$\sec\theta$	$\cot\theta$
0°	0	1	0	n.d	1	n.d
30°	$\frac{1}{2}$	$\frac{\sqrt{3}}{2}$	$\frac{1}{\sqrt{3}}$	2	$\frac{2}{\sqrt{3}}$	$\sqrt{3}$
45°	$\frac{1}{\sqrt{2}}$	$\frac{1}{\sqrt{2}}$	1	$\sqrt{2}$	$\sqrt{2}$	1
60°	$\frac{\sqrt{3}}{2}$	$\frac{1}{2}$	$\sqrt{3}$	$\frac{2}{\sqrt{3}}$	2	$\frac{1}{\sqrt{3}}$
90°	1	0	n.d	1	n.d	0

Concept: Problems Based on the Trigonometric Table

Example 5 Evaluate the following:

(i) $\sin 60° \times \cos 30° + \sin 30° \times \cos 60°$ **(ii)** $\dfrac{5\cos^2 60° + 4\sec^2 30° - \tan^2 45°}{\sin^2 30° + \cos^2 30°}$

(iii) $\dfrac{\sin 30° + \tan 45° - \operatorname{cosec} 60°}{\sec 30° + \cos 60° + \cot 45°}$ **[NCERT]**

Solution: (i) $\sin 60° \cos 30° + \sin 30° \cos 60°$

$$= \frac{\sqrt{3}}{2} \times \frac{\sqrt{3}}{2} + \frac{1}{2} \times \frac{1}{2} = \frac{3}{4} + \frac{1}{4} = \frac{3+1}{4} = 1$$

(ii) $\dfrac{5\cos^2 60° + 4\sec^2 30° - \tan^2 45°}{\sin^2 30° + \cos^2 30°}$

$$= \frac{5\times\left(\frac{1}{2}\right)^2 + 4\left(\frac{2}{\sqrt{3}}\right)^2 - (1)^2}{\left(\frac{1}{2}\right)^2 + \left(\frac{\sqrt{3}}{2}\right)^2} = \frac{\frac{5}{4} + \frac{16}{3} - 1}{\frac{1}{4} + \frac{3}{4}} = \frac{\frac{15+64-12}{12}}{\frac{1+3}{4}} = \frac{67}{12} \times \frac{4}{4} = \frac{67}{12}$$

(iii) $\dfrac{\sin 30° + \tan 45° - \text{cosec } 60°}{\sec 30° + \cos 60° + \cot 45°}$

$$= \frac{\frac{1}{2} + 1 - \frac{2}{\sqrt{3}}}{\frac{2}{\sqrt{3}} + \frac{1}{2} + 1} = \frac{\frac{3}{2} - \frac{2}{\sqrt{3}}}{\frac{2}{\sqrt{3}} + \frac{3}{2}} = \frac{\frac{3\sqrt{3}-4}{2\sqrt{3}}}{\frac{4+3\sqrt{3}}{2\sqrt{3}}} = \frac{3\sqrt{3}-4}{\cancel{2\sqrt{3}}} \times \frac{\cancel{2\sqrt{3}}}{4+3\sqrt{3}} = \frac{3\sqrt{3}-4}{4+3\sqrt{3}}$$

Rationalising the denominator

$$= \frac{(3\sqrt{3}-4)(4-3\sqrt{3})}{(4+3\sqrt{3})(4-3\sqrt{3})}$$ [On multiplying N and D by $(4 - 3\sqrt{3})$]

$$= \frac{12\sqrt{3} - 16 - 9\times 3 + 12\sqrt{3}}{(4)^2 - (3\sqrt{3})^2} = \frac{24\sqrt{3} - 43}{16 - 27} = \frac{-(43 - 24\sqrt{3})}{-11} = \frac{43 - 24\sqrt{3}}{11}$$

4 Try to Solve

Evaluate the following:

1. Find the value of: $\dfrac{\tan 45°}{\sin 30° + \cos 60°}$.
2. Evaluate: $\dfrac{\tan 45°}{\text{cosec} 30°} + \dfrac{\sec 60°}{\cot 45°} - \dfrac{2\sin 90°}{\cos 0°}$.
3. Evaluate: $\cos^2 30° \cos^2 45° + 4 \sec^2 60° + \frac{1}{2} \cos^2 90° - 2 \tan^2 60°$. **[CBSE (CCE) 2011]**
4. Evaluate: $\sin^2 30° \cos^2 45° + 4 \tan^2 30° + \frac{1}{2} \sin^2 90° - 2 \cos^2 90° + \frac{1}{24}$. **[CBSE (CCE) 2011]**
5. Evaluate: $\sin^2 60° + 2 \tan 45° - \cos^2 30°$ **[Allahabad 2019]**

Answers

1. 1 **2.** $\frac{1}{2}$ **3.** $\frac{83}{8}$ **4.** 2 **5.** 2

Example 6 If $\tan (A + B) = \sqrt{3}$ and $\tan (A - B) = \frac{1}{\sqrt{3}}$; $0 < (A + B) < 90°$; $A > B$, find A and B. **[CBSE (CCE) 2009, 12, 15]**

Solution:

$$\tan (A + B) = \sqrt{3}$$

$$\tan (A + B) = \sqrt{3} = \tan 60°$$

$\therefore$ $$A + B = 60° \quad \text{...(i)}$$

$$\tan(A - B) = \frac{1}{\sqrt{3}} = \tan 30°$$

$\therefore$ $$A - B = 30° \quad \text{...(ii)}$$

Adding (i) and (ii), we get

$$A + \cancel{B} = 60°$$
$$A - \cancel{B} = 30°$$

$$2A = 90°$$

$$A = \frac{90}{2} = 45°$$

Inserting the value of A in (i) $\quad 45° + B = 60°$

$$B = 60° - 45° = 15°$$

$$A = 45° \text{ and } B = 15°$$

Example 7 Given that sin $(A + B)$ = sin $A \cdot$ cos B + cos $A \cdot$ sin B, find the value of sin 75°.

[CBSE (CCE) 2012]

Solution: Take $A = 45°$ and $B = 30°$

$$\sin 75° = \sin(45° + 30°) = \sin 45° \cdot \cos 30° + \cos 45° \cdot \sin 30°$$

$$= \frac{1}{\sqrt{2}} \cdot \frac{\sqrt{3}}{2} + \frac{1}{\sqrt{2}} \cdot \frac{1}{2} = \frac{\sqrt{3}}{2\sqrt{2}} + \frac{1}{2\sqrt{2}} = \frac{\sqrt{3}+1}{2\sqrt{2}} = \frac{\sqrt{3}+1}{2\sqrt{2}} \times \frac{\sqrt{2}}{\sqrt{2}} = \frac{\sqrt{6}+\sqrt{2}}{4}$$

5 Try to Solve

1. If $\sin(A - B) = \frac{1}{2}$, $\cos(A + B) = \frac{1}{2}$, $0° < A + B < 90°$, $A > B$, find A and B. **[CBSE (CCE) 2014]**
2. For $\angle A = \angle B = 45°$, verify that sin $(A + B)$ = sin A cos B + cos A sin B.
3. Using the formula sin $(A - B)$ = sin A cos B – cos A sin B, find the value of sin 15°.
4. For $A = 60°$ and $B = 30°$, verify that sin $(A - B)$ = sin A cos B – cos A sin B.
5. If $\sin(A + B) = 1$ and $\tan(A - B) = \frac{1}{\sqrt{3}}$, find the value of:

   (a) tan A + cot B (b) sec A – cosec B **[CBSE 2016]**
6. If $\sin(A + 2B) = \frac{\sqrt{3}}{2}$ and $\cos(A + 4B) = 0$, $A > B$, and $A + 4B < 90°$, then find A and B. **[CBSE 2018 (C)]**

Answers

1. $A = 45°$ and $B = 15°$ **3.** $\frac{\sqrt{6}-\sqrt{2}}{4}$ **5.** (a) $2\sqrt{3}$ (b) 0 **6.** $A = 30°$, $B = 15°$

Concept: Complementary Angles Formulae

$$\sin(90 - \theta) = \cos\theta$$
$$\cos(90 - \theta) = \sin\theta$$
$$\tan(90 - \theta) = \cot\theta$$
$$\text{cosec}(90 - \theta) = \sec\theta$$
$$\sec(90 - \theta) = \text{cosec}\,\theta$$
$$\cot(90 - \theta) = \tan\theta$$

Concept: Trigonometric Identities

1. $$\sin^2\theta + \cos^2\theta = 1$$

$$\sin^2\theta = 1 - \cos^2\theta$$
$$\cos^2\theta = 1 - \sin^2\theta$$

2.
$$\sec^2\theta - \tan^2\theta = 1$$
$$\sec^2\theta = 1 + \tan^2\theta$$
$$\tan^2\theta = \sec^2\theta - 1$$

3.
$$\text{cosec}^2\theta - \cot^2\theta = 1$$
$$\text{cosec}^2\theta = 1 + \cot^2\theta$$
$$\cot^2\theta = \text{cosec}^2\theta - 1$$

Concept: Problems Based on Complementary Angles

Example 8 Express sin 67° + cos 75° in terms of trigonometric ratios of angles between 0° and 45°. [CBSE (CCE) 2014]

Solution:
$$\sin 67° + \cos 75° = \sin(90 - 23) + \cos(90 - 15°)$$
$$= \cos 23° + \sin 15°$$

6 Try to Solve

1. Express cosec 69° + cot 69° in terms of angles between 0° and 45°.
2. Express cos 75° + cot 75° in terms of angles between 0° and 30°. [CBSE (CCE) 2014]

Answers

1. sec 21° + tan 21° **2.** sin 15° + tan 15°

Example 9 Evaluate: [CBSE (CCE) 2014]

(i) $\dfrac{\sin 18°}{\cos 72°}$ (ii) $\dfrac{\tan 26°}{\cot 64°}$ (iii) cos 48° – sin 42° (iv) cosec 31° – sec 59°

Solution:

(i) $\dfrac{\sin 18°}{\cos 72°} = \dfrac{\sin(90° - 72°)}{\cos 72°} = \dfrac{\cos 72°}{\cos 72°} = 1$ (ii) $\dfrac{\tan 26°}{\cot 64°} = \dfrac{\tan(90° - 64°)}{\cot 64°} = \dfrac{\cot 64°}{\cot 64°} = 1$

(iii) $\cos 48° - \sin 42° = \cos(90° - 42°) - \sin 42° = \sin 42° - \sin 42° = 0$

(iv) $\text{cosec } 31° - \sec 59° = \text{cosec }(90° - 59°) - \sec 59° = \sec 59° - \sec 59° = 0$

7 Try to Solve

1. Evaluate: $2\dfrac{\tan 53°}{\cot 37°} - \dfrac{\cot 80°}{\tan 10°}$ **2.** Evaluate: $\dfrac{\tan 35°}{\cot 55°} + \dfrac{\cot 78°}{\tan 12°} - 1$

3. Evaluate: $\dfrac{\cot^2 50°}{\tan^2 40°} - 2\dfrac{\sin^2 75°}{\cos^2 15°}$

Answers

1. 1 **2.** 1 **3.** –1

Example 10 Show that:

(i) $\tan 48° \tan 23° \tan 42° \tan 67° = 1$ (ii) $\cos 38° \cos 52° - \sin 38° \sin 52° = 0$

Solution: (i) $\tan 48° \tan 23° \tan 42° \tan 67° = 1$

$$\text{LHS} = \tan(90° - 42°)\tan(90° - 67°)\tan 42° \tan 67°$$
$$= \cot 42° \cot 67° \tan 42° \tan 67°$$
$$= \frac{1}{\tan 42°} \times \frac{1}{\tan 67°} \times \tan 42° \times \tan 67° = 1 = \text{RHS}$$

(ii) $\cos 38° \cos 52° - \sin 38° \sin 52° = 0$

$$\text{LHS} = \cos(90° - 52°)\cos(90° - 38°) - \sin 38° \sin 52°$$
$$= \sin 52° \times \sin 38° - \sin 38° \times \sin 52° = 0 = \text{RHS.}$$

8 Try to Solve

1. Find the value of: $\sec 50° \sin 40° + \cos 40° \operatorname{cosec} 50°$.
2. Evaluate: $\frac{\cos 75°}{\sin 15°} + \frac{\sin 12°}{\cos 78°} - \cos 18° \operatorname{cosec} 72°$.
3. Show that: $\sin A \cos(90° - A) + \cos A \sin(90° - A) = 1$.
4. Show that: $\frac{\sin\theta}{\sin(90° - \theta)} + \frac{\cos\theta}{\cos(90° - \theta)} = \sec\theta \operatorname{cosec}\theta$.

Answers

1. 2 **2.** 1

Example 11 Evaluate: [CBSE (CCE) 2014]

$$\frac{\sin 32°}{\cos 58°} + \frac{\sin 22°}{\cos 68°} - \frac{\cos 48° \cdot \operatorname{cosec} 42°}{\tan 28° \cdot \tan 25° \tan 30° \tan 62° \cdot \tan 65°}$$

Solution: We have

$$= \frac{\sin 32°}{\cos 58°} + \frac{\sin 22°}{\cos 68°} - \frac{\cos 48° \cdot \operatorname{cosec} 42°}{\tan 28° \tan 25° \tan 30° \tan 62° \tan 65°}$$

$$= \frac{\sin(90° - 58°)}{\cos 58°} + \frac{\sin(90° - 68°)}{\cos 68°} - \frac{\cos(90° - 42°)\operatorname{cosec} 42°}{\tan 28° \tan 25° \times \frac{1}{\sqrt{3}} \times \tan(90° - 28°)\tan(90° - 25°)}$$

$$= \frac{\cos 58°}{\cos 58°} + \frac{\cos 68°}{\cos 68°} - \frac{\sqrt{3}\sin 42° \operatorname{cosec} 42°}{\tan 28° \cdot \tan 25° \cdot \cot 28° \cdot \cot 25°}$$

$$= 1 + 1 - \frac{\sqrt{3} \times 1}{(\tan 28° \cot 28°)(\tan 25° \cot 25°)} \qquad [\because \sin\theta \operatorname{coses}\theta = 1]$$

$$= 2 - \frac{\sqrt{3}}{1 \times 1} \qquad [\because \tan\theta \cdot \cot\theta = 1]$$

$$= 2 - \sqrt{3}$$

9 Try to Solve

1. Evaluate the following:

(a) $\frac{\cos 70^\circ}{\sin 20^\circ} + \frac{\cos 55^\circ \operatorname{cosec} 35^\circ}{\tan 5^\circ \tan 25^\circ \tan 45^\circ \tan 65^\circ \tan 85^\circ}$ [CBSE (CCE) 2015]

(b) $\frac{\cos 58^\circ}{\sin 32^\circ} + \frac{\sin 22^\circ}{\cos 68^\circ} - \frac{\cos 38^\circ \operatorname{cosec} 52^\circ}{\tan 18^\circ \tan 35^\circ \tan 60^\circ \tan 72^\circ \tan 55^\circ}$ [CBSE (CCE) 2015]

2. Without using trigonometric table, prove that:

(a) $\cot 12^\circ \cot 38^\circ \cot 52^\circ \cot 60^\circ \cot 78^\circ = \frac{1}{\sqrt{3}}$

(b) $\tan 7^\circ \tan 23^\circ \tan 60^\circ \tan 67^\circ \tan 83^\circ = \sqrt{3}$

Answers

1. (a) 2 (b) $2 - \frac{1}{\sqrt{3}}$

Concept: Problems Based on Trigonometric Identities

Example 12 Prove the following identities:

(a) $(\operatorname{cosec}^2 A - 1)(\sec A + 1)(\sec A - 1) = 1$ (b) $1 + (1 + \cos A)(1 - \cos A)(1 + \cot^2 A) = 2$

(c) $\tan^2 A + (1 + \tan^2 A)(1 + \sin A)(1 - \sin A) = \sec^2 A$

Solution: (i) LHS $= (\operatorname{cosec}^2 A - 1)(\sec A + 1)(\sec A - 1)$

$= (\operatorname{cosec}^2 A - 1)(\sec^2 A - 1)$ $[\because a^2 - b^2 = (a+b)(a-b)]$

$= \cot^2 A \times \tan^2 A$ $[\because \operatorname{cosec}^2 A - 1 = \cot^2 A \text{ and } \sec^2 A - 1 = \tan^2 A]$

$= \cot^2 A \times \frac{1}{\cot^2 A} = 1 = \text{RHS}$ $\left[\because \tan A = \frac{1}{\cot A}\right]$

(ii) LHS $= 1 + (1 + \cos A)(1 - \cos A)(1 + \cot^2 A)$

$= 1 + (1 - \cos^2 A)\operatorname{cosec}^2 A$ $[\because 1 + \cot^2 A = \operatorname{cosec}^2 A]$

$= 1 + \sin^2 A \times \frac{1}{\sin^2 A}$ $[\because 1 - \cos^2 A = \sin^2 A \text{ and } \operatorname{cosec}^2 A = \frac{1}{\sin^2 A}]$

$= 1 + 1 = 2 = \text{RHS}$

(iii) LHS $= \tan^2 A + (1 + \tan^2 A)(1 + \sin A)(1 - \sin A)$

$= \tan^2 A + \sec^2 A(1 - \sin^2 A)$ $[\because 1 + \tan^2 A = \sec^2 A]$

$= \tan^2 A + \sec^2 A \cdot \cos^2 A$

$= \tan^2 A + 1$ $[\because \sec A \cdot \cos A = 1]$

$= 1 + \tan^2 A = \sec^2 A = \text{RHS}$

10 Try to Solve

1. Prove the following Identities:

(a) $(1 - \sin^2 A)\sec^2 A = 1$ (b) $(1 + \tan^2 A)\cos^2 A = 1$ (c) $(\text{cosec}^2 A - 1)\tan^2 A = 1$

(d) $(1 - \cos^2 A)\text{cosec}^2 A = 1$ (e) $(1 + \cot^2 A)\sin^2 A = 1$ (f) $(\sec^2 A - 1)\cot^2 A = 1$

(g) $\cot^2 \theta - \dfrac{1}{\sin^2 \theta} = -1$ (h) $(1 + \tan^2 \theta)(1 - \sin \theta)(1 + \sin \theta) = 1$

Example 13 Prove the following identities:

(i) $\cot \theta - \tan \theta = \dfrac{2\cos^2\theta - 1}{\sin\theta \times \cos\theta}$ **[CBSE (CCE) 2015]**

(ii) $\dfrac{(1+\sin\theta)^2 + (1-\sin\theta)^2}{2\cos^2\theta} = \dfrac{1+\sin^2\theta}{1-\sin^2\theta}$ **[CBSE (CCE) 2015]**

(iii) $\dfrac{\cos^2\theta}{\sin\theta} - \text{cosec}\,\theta + \sin\theta = 0$ **[CBSE 2006]**

Solution: (i)

$$\text{LHS} = \frac{\cos\theta}{\sin\theta} - \frac{\sin\theta}{\cos\theta} = \frac{(\cos\theta \times \cos\theta) - (\sin\theta \times \sin\theta)}{\sin\theta \cdot \cos\theta}$$

$$= \frac{\cos^2\theta - \sin^2\theta}{\sin\theta \cdot \cos\theta} = \frac{\cos^2\theta - (1 - \cos^2\theta)}{\sin\theta \cdot \cos\theta} \qquad [\because \sin^2\theta = 1 - \cos^2\theta]$$

$$= \frac{\cos^2\theta - 1 + \cos^2\theta}{\sin\theta \cdot \cos\theta} = \frac{2\cos^2\theta - 1}{\sin\theta \cdot \cos\theta} = \textbf{RHS}$$

(ii)

$$\text{LHS} = \frac{(1+\sin\theta)^2 + (1-\sin\theta)^2}{2\cos^2\theta}$$

$$= \frac{1 + \sin^2\theta + \cancel{2\sin\theta} + 1 + \sin^2\theta - \cancel{2\sin\theta}}{2\cos^2\theta} = \frac{2 + 2\sin^2\theta}{2\cos^2\theta}$$

$$= \frac{2(1+\sin^2\theta)}{2(1-\sin^2\theta)} \qquad [\because \cos^2\theta = 1 - \sin^2\theta]$$

$$= \frac{1+\sin^2\theta}{1-\sin^2\theta} = \textbf{RHS}$$

(iii)

$$\text{LHS} = \frac{\cos^2\theta}{\sin\theta} - \text{cosec}\,\theta + \sin\theta$$

$$= \frac{\cos^2\theta}{\sin\theta} - \frac{1}{\sin\theta} + \sin\theta = \frac{\cos^2\theta - 1 + \sin^2\theta}{\sin\theta}$$

$$= \frac{(\sin^2\theta + \cos^2\theta) - 1}{\sin\theta} \qquad \text{[On rearranging the terms]}$$

$$= \frac{1-1}{\sin\theta} \qquad [\because \sin^2\theta + \cos^2\theta = 1]$$

$$= \frac{0}{\sin\theta} = 0 = \textbf{RHS}$$

11 Try to Solve

1. Prove the following identities:

(a) $(\text{cosec}\ \theta - \cot\theta)^2 = \frac{1-\cos\theta}{1+\cos\theta}$ **[CBSE (CCE) 2015]**

(b) $\frac{\cos A}{1+\sin A} + \frac{1+\sin A}{\cos A} = 2\sec A$ **[CBSE (CCE) 2014]**

(c) $\sec A\,(1 - \sin A)(\sec A + \tan A) = 1$ **[CBSE (CCE) 2015]**

Example 14 Evaluate:

(i) $\frac{\sin^2 63° + \sin^2 27°}{\cos^2 17° + \cos^2 73°}$ **(ii)** $\sin 25° \times \cos 65° + \cos 25° \times \sin 65°$.

Solution: **(i)** $\frac{\sin^2 63° + \sin^2 27°}{\cos^2 17° + \cos^2 73°} = \frac{\sin^2 63° + [\sin(90° - 63°)]^2}{\cos^2 17° + [\cos(90° - 17°)]^2}$

$$= \frac{\sin^2 63° + \cos^2 63°}{\cos^2 17° + \sin^2 17°} = \frac{1}{1} = 1 \qquad [\because \sin^2\theta + \cos^2\theta = 1]$$

(ii) $\sin 25° \cdot \cos 65° + \cos 25° \cdot \sin 65°$

$= \sin 25° \times \cos(90° - 25°) + \cos 25° \times \sin(90° - 25°)$

$= \sin 25° \times \sin 25° + \cos 25° \times \cos 25°$

$= \sin^2 25° + \cos^2 25° = 1.$ $[\because \sin^2\theta + \cos^2\theta = 1]$

12 Try to Solve

1. Evaluate: $\sin^2 20° + \sin^2 70° - \tan^2 45°$.
2. Show that $\sin 63° \cos 27° + \cos 63° \sin 27° = 1$.
3. Find the value of $\sin^2 35° + \sin^2 55°$.
4. Find the value of $(\sin^2 33° + \sin^2 57°)$ **[Delhi 2019]**
5. A, B, C are interior angles of ΔABC. Prove that $\text{cosec}\left(\frac{A+B}{2}\right) = \sec\frac{C}{2}$ **[CBSE 2018 (C)]**

Answers

1. 0. **3.** 1 **4.** 1

Example 15 Prove the following identities:

(i) $\dfrac{\operatorname{cosec} A}{\operatorname{cosec} A-1}+\dfrac{\operatorname{cosec} A}{\operatorname{cosec} A+1}=2+2\tan^2 A$ **[CBSE 2007]**

(ii) $\left(1+\dfrac{1}{\tan^2\theta}\right)\left(1+\dfrac{1}{\cot^2\theta}\right)=\dfrac{1}{\sin^2\theta-\sin^4\theta}$ **[CBSE 2006]**

Solution:

(i) $\text{LHS}=\dfrac{\operatorname{cosec} A}{\operatorname{cosec} A-1}+\dfrac{\operatorname{cosec} A}{\operatorname{cosec} A+1}$

$$=\frac{\operatorname{cosec} A(\operatorname{cosec} A+1)+\operatorname{cosec} A(\operatorname{cosec} A-1)}{(\operatorname{cosec} A-1)(\operatorname{cosec} A+1)}$$

$$=\frac{\operatorname{cosec} A(\operatorname{cosec} A+1+\operatorname{cosec} A-1)}{\operatorname{cosec}^2 A-1}$$

$$=\frac{\operatorname{cosec} A\times 2\operatorname{cosec} A}{\cot^2 A}=\frac{2\operatorname{cosec}^2 A}{\cot^2 A} \qquad [\operatorname{cosec}^2 A-1=\cot^2 A]$$

$$=\frac{2(1+\cot^2 A)}{\cot^2 A} \qquad [\because \operatorname{cosec}^2 A=1+\cot^2 A]$$

$$=2\times\left(\frac{1}{\cot^2 A}+\frac{\cot^2 A}{\cot^2 A}\right)=2(\tan^2 A+1)=2+2\tan^2 A=\text{RHS}$$

(ii) $\text{LHS}=\left(1+\dfrac{1}{\tan^2\theta}\right)\left(1+\dfrac{1}{\cot^2\theta}\right)$

$$=(1+\cot^2\theta)(1+\tan^2\theta)$$

$$=\operatorname{cosec}^2\theta\times\sec^2\theta \qquad [\because 1+\cot^2\theta=\operatorname{cosec}^2\theta \text{ and } 1+\tan^2\theta=\sec^2 q]$$

$$=\frac{1}{\sin^2\theta}\times\frac{1}{\cos^2\theta}$$

$$=\frac{1}{\sin^2\theta\times(1-\sin^2\theta)} \qquad [\because \cos^2\theta=1-\sin^2 q]$$

$$=\frac{1}{\sin^2\theta-\sin^4\theta}=\text{RHS}$$

13 Try to Solve

1. Prove the following identities:

(a) $(1-\cos^2 A)\sec^2 A=\tan^2 A$

(b) $\dfrac{1-\tan^2\theta}{\cot^2\theta-1}=\tan^2\theta$ where $\theta\neq 45°$

(c) $\dfrac{\cos\theta}{1+\sin\theta}=\dfrac{1-\sin\theta}{\cos\theta}$

(d) $\sec^2\theta+\operatorname{cosec}^2\theta=\sec^2\theta\operatorname{cosec}^2\theta$

Example 16 Prove: $\dfrac{\tan A+\sec A-1}{\tan A-\sec A+1}=\dfrac{1+\sin A}{\cos A}$. [CBSE (CCE) 2010,13,15]

Solution: LHS $=\dfrac{\tan A+\sec A-1}{\tan A-\sec A+1}$

We know that $\sec^2 A - \tan^2 A = 1$

$$=\frac{(\tan A+\sec A)-(\sec^2 A-\tan^2 A)}{\tan A-\sec A+1}=\frac{(\sec A+\tan A)-(\sec A-\tan A)(\sec A+\tan A)}{\tan A-\sec A+1}$$

$$=\frac{(\sec A+\tan A)[1-(\sec A-\tan A)]}{\tan A-\sec A+1}=\frac{(\sec A+\tan A)[1-\sec A+\tan A]}{\tan A-\sec A+1}$$

$$=\frac{(\sec A+\tan A)[\cancel{1-\sec A+\tan A}]}{[\cancel{\tan A-\sec A+1}]}=\sec A+\tan A.$$

$$=\frac{1}{\cos A}+\frac{\sin A}{\cos A}=\frac{1+\sin A}{\cos A} \qquad \left[\because \sec A=\frac{1}{\cos A},\ \tan A=\frac{\sin A}{\cos A}\right]$$

Hence LHS = RHS.

14 Try to Solve

1. Prove the following identities:

(a) $\dfrac{1+\sec A}{\sec A}=\dfrac{\sin^2 A}{1-\cos A}$ [NCERT]

(b) $(\sin A+\text{cosec } A)^2+(\cos A+\sec A)^2=7+\tan^2 A+\cot^2 A$ [NCERT, CBSE (CCE) 2015]

(c) $\dfrac{\sin\theta-2\sin^3\theta}{2\cos^3\theta-\cos\theta}=\tan\theta$ [NCERT, CBSE 2018]

FOR MORE PRACTICE AND KNOWLEDGE

Trigonometric Ratios of Some Specific Angles

The trigonometric ratio of any particular angle is fixed. Here, we will find the trigonometric ratios for the angles 0°, 30°, 45°, 60° and 90°.

Trigonometric Ratios of 45°

Consider an isosceles right-angled ΔABC in which $AB = BC$ and $\angle B = 90°$.

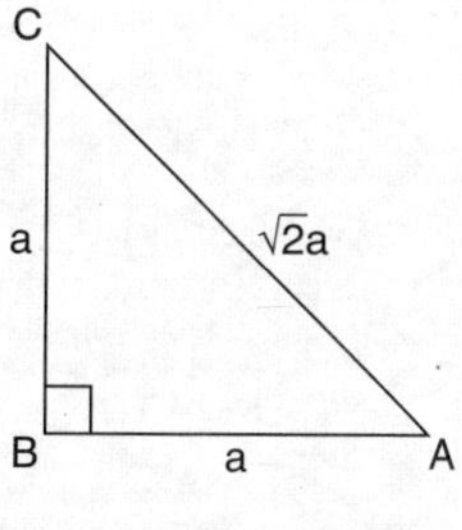

Let $AB = BC = a$

By Pythagoras' Theorem $AC^2 = AB^2 + BC^2$

$\Rightarrow$ $AC^2 = a^2 + a^2$

$\Rightarrow$ $AC^2 = 2a^2$

$\therefore$ $AC=\sqrt{2a^2}=a\sqrt{2}$

In ΔABC $AB = BC$

$\therefore \quad \angle A = \angle C$ [∵ In a Δ, angles opposite to equal sides are equal]

$\angle A + \angle B + \angle C = 180°$ [Angle sum property]

$$\angle A + 90° + \angle A = 180°$$

$$2\angle A = 180° - 90° = 90°$$

$$\therefore \quad \angle A = \frac{90°}{2} = 45°$$

$$\sin A = \frac{BC}{AC}, \sin 45° = \frac{\cancel{a}}{\sqrt{2}\cancel{a}}$$

$$\Rightarrow \quad \sin 45° = \frac{1}{\sqrt{2}}$$

$$\cos A = \frac{AB}{AC}$$

$$\Rightarrow \quad \cos 45° = \frac{a}{a\sqrt{2}}$$

$$\therefore \quad \cos 45° = \frac{1}{\sqrt{2}}$$

$$\tan A = \frac{BC}{AB}$$

$$\therefore \quad \tan 45° = \frac{a}{a} = 1$$

Also, $\operatorname{cosec} 45° = \frac{1}{\sin 45°} = \sqrt{2}$; $\sec 45° = \frac{1}{\cos 45°} = \sqrt{2}$; $\cot 45° = \frac{1}{\tan 45°} = 1.$

Trigonometric Ratios of 30° and 60°

Consider an equilateral ΔABC in which $AB = BC = AC = a$ (say).

$$\angle A = \angle B = \angle C = 60°$$

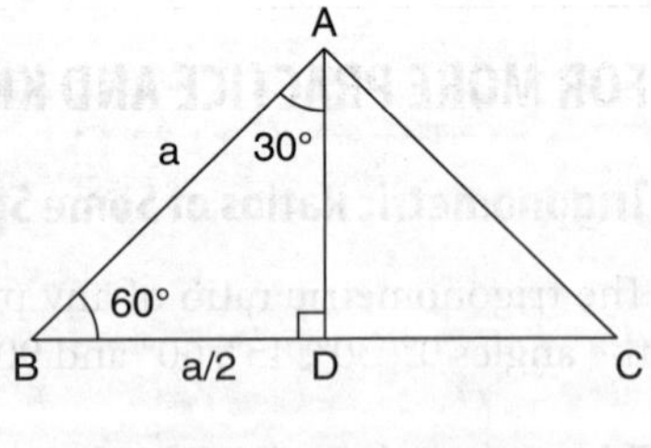

From A, draw $AD \perp BC$.

In ΔADB and ΔADC

$AD = AD$ [Common]

$AB = AC$ [Given]

$\angle ADB = \angle ADC$ [Each 90°]

$\therefore \quad \angle ADB \cong \angle ADC$ [By RHS]

And so $BD = CD$ [By CPCT]

$$\therefore \quad BD = CD = \frac{BC}{2} = \frac{a}{2}$$

$AB^2 = AD^2 + BD^2$ [By Pythagoras' Theorem]

$$a^2 = AD^2 + \left(\frac{a}{2}\right)^2$$

$$a^2 - \frac{a^2}{4} = AD^2$$

$$AD^2 = \frac{3a^2}{4}$$

$$\therefore \quad AD = \frac{\sqrt{3}a}{2}$$

$$\angle B + \angle BAD + \angle ADB = 180° \text{ [By angle sum property]}$$

$$60° + \angle BAD + 90° = 180°$$

$$\therefore \quad \angle BAD = 180° - 150° = 30°$$

For trigonometric ratios of 60°

In ΔADB

$$\sin B = \frac{AD}{AB}$$

$$\Rightarrow \quad \sin 60° = \frac{\frac{\sqrt{3}a}{2}}{a} = \frac{\sqrt{3}}{2}$$

$$\therefore \quad \sin 60° = \frac{\sqrt{3}}{2}$$

$$\cos B = \frac{BD}{AB}$$

$$\cos 60° = \frac{\frac{a}{2}}{a} = \frac{1}{2}$$

$$\Rightarrow \quad \cos 60° = \frac{1}{2}$$

$$\tan B = \frac{AD}{BD}$$

$$\Rightarrow \quad \tan 60° = \frac{\frac{\sqrt{3}a}{2}}{\frac{a}{2}} = \sqrt{3}$$

$$\cos 30° = \frac{AD}{AB} = \frac{\frac{\sqrt{3}a}{2}}{a} = \frac{\sqrt{3}}{2}$$

$$\therefore \quad \tan 60° = \sqrt{3}$$

Also

$$\operatorname{cosec} 60° = \frac{1}{\sin 60°} = \frac{2}{\sqrt{3}}$$

$$\sec 60° = \frac{1}{\cos 60°} = 2$$

$$\cot 60° = \frac{1}{\tan 60°} = \frac{1}{\sqrt{3}}$$

For trigonometric ratios of 30° (refer figure on page 8.16)

In $\angle ADB$ $$\sin \angle BAD = \frac{BD}{AB}$$

$$\Rightarrow \quad \sin 30^\circ = \frac{\frac{a}{2}}{a} = \frac{1}{2}$$

$$\cos 30^\circ = \frac{AD}{AB} = \frac{\frac{\sqrt{3}a}{2}}{a} = \frac{\sqrt{3}}{2}$$

$$\therefore \quad \cos 30^\circ = \frac{\sqrt{3}}{2}$$

$$\tan \angle BAD = \frac{BD}{AD}$$

$$\tan 30^\circ = \frac{\frac{a}{2}}{\frac{\sqrt{3}a}{2}} = \frac{1}{\sqrt{3}}$$

$$\therefore \quad \tan 30^\circ = \frac{1}{\sqrt{3}}.$$

Also, $\operatorname{cosec} 30^\circ = \dfrac{1}{\sin 30^\circ} = 2$; $\sec 30^\circ = \dfrac{1}{\cos 30^\circ} = \dfrac{2}{\sqrt{3}}$; $\cot 30^\circ = \dfrac{1}{\tan 30^\circ} = \sqrt{3}$.

Trigonometric Ratios of 0° and 90°

Consider a ΔABC, right-angled at B.

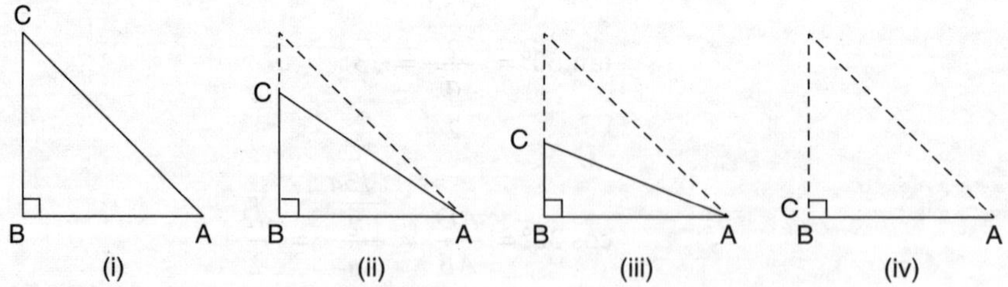

$$\sin A = \frac{BC}{AC}, \cos A = \frac{AB}{AC}, \tan A = \frac{BC}{AB}$$

If the vertex C of $\angle ABC$ comes closer and closer to vertex B, as shown in figures (ii) to (iv), $\angle A$ becomes zero and hypotenuse AC becomes equal to side AB. Side BC becomes zero.

Now $$\sin A = \frac{BC}{AC}$$

$$\Rightarrow \quad \sin 0° = \frac{0}{AB} \qquad [\because BC = 0 \text{ and } AC = AB]$$

$$\therefore \quad \sin 0° = 0$$

$$\cos A = \frac{AB}{AC}$$

$$\Rightarrow \quad \cos 0° = \frac{AB}{AB} = 1$$

$$\therefore \quad \cos 0° = 1$$

$$\tan A = \frac{BC}{AB}$$

$$\Rightarrow \quad \tan 0° = \frac{0}{AB} = 0$$

$$\therefore \quad \tan 0° = 0$$

For trigonometric ratios of 90°

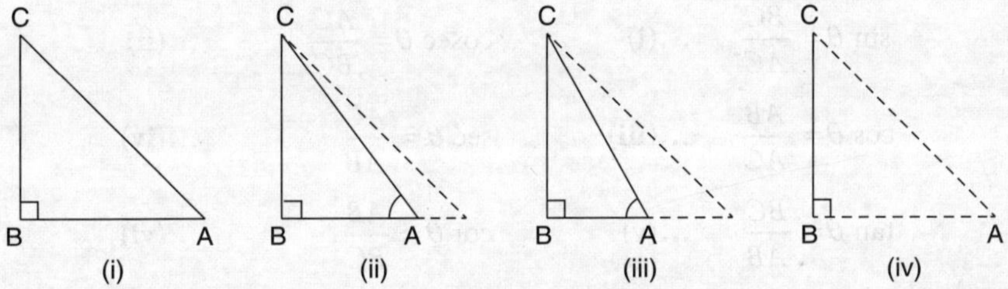

In ΔABC, if vertex A comes closer and closer to vertex B, $\angle A$ increases and approaches 90°. When A coincides with B, $\angle A = 90°$, $AC = BC$ and $AB = 0$.

Now
$$\sin A = \frac{BC}{AC}$$

$$\Rightarrow \quad \sin 90° = \frac{BC}{AB} = 1$$

$$\therefore \quad \sin 0° = 1$$

$$\cos A = \frac{AB}{AC}$$

$$\cos 90° = \frac{AB}{AC} = 0$$

$$\Rightarrow \quad \cos 90° = \frac{0}{AC} = 0$$

$$\therefore \quad \cos 90° = 0$$

$$\tan A = \frac{BC}{AB}$$

$$\tan A = \frac{BC}{AB} = 0$$

$\Rightarrow \quad \tan 90° = \frac{BC}{0} = \text{Not defined}$

Similarly $\quad \operatorname{cosec} 90° = \frac{1}{\sin 90°} = 1; \sec 90° = \frac{1}{\cos 90°} = \frac{1}{0} = \text{Not defined}$

$\cot 90° = \frac{1}{\tan 90°} = \frac{1}{\infty} = 0$

Trigonometric Ratios of Complementary Angles

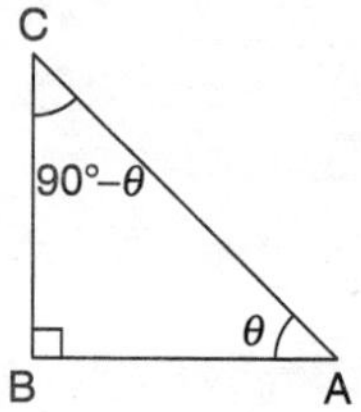

Consider a ΔABC in which $\angle B = 90°$.

$$\angle A + \angle B + \angle C = 180° \text{ [Angle sum property]}$$

$\Rightarrow$ $$\angle A + 90° + \angle C = 180°$$

$$\angle A + \angle C = 90°$$

Let $$\angle A = \theta$$

$$\angle C = 90° - \theta$$

In ΔABC

$\sin\theta = \frac{BC}{AC}$ ...(i) $\quad \operatorname{cosec}\theta = \frac{AC}{BC}$ ...(ii)

$\cos\theta = \frac{AB}{AC}$ ...(iii) $\quad \sec\theta = \frac{AC}{AB}$ ...(iv)

$\tan\theta = \frac{BC}{AB}$ ...(v) $\quad \cot\theta = \frac{AB}{BC}$ ...(vi)

And $\sin(90° - \theta) = \frac{AB}{AC}$ ...(vii) $\quad \cos(90° - \theta) = \frac{BC}{AC}$ ...(viii)

$\operatorname{cosec}(90° - \theta) = \frac{AC}{AB}$ ...(ix) $\quad \sec(90 - \theta) = \frac{AC}{BC}$ ...(x)

$\tan(90° - \theta) = \frac{AB}{BC}$ ...(xi) $\quad \cot(90° - \theta) = \frac{BC}{AB}$ ...(xii)

From (vii) and (iii), we get $\sin(90° - \theta) = \cos\theta$

From (viii) and (i), we get $\cos(90° - \theta) = \sin\theta$

From (xi) and (vi), we get $\tan(90° - \theta) = \cot\theta$

From (ix) and (iv), we get $\operatorname{cosec}(90° - \theta) = \sec\theta$

From (x) and (ii), we get $\sec(90° - \theta) = \operatorname{cosec}\theta$

From (xii) and (v), we get $\cot(90° - \theta) = \tan\theta$

Trigonometric Identities

A trigonometric identity is an equation involving trigonometric ratios of an angle, if it is true for all values of the angle(s) involved.

The basic trigonometric identities are:

(i) $\sin^2\theta + \cos^2\theta = 1$

(ii) $\sec^2\theta - \tan^2\theta = 1$

(iii) $\operatorname{cosec}^2\theta - \cot^2\theta = 1$

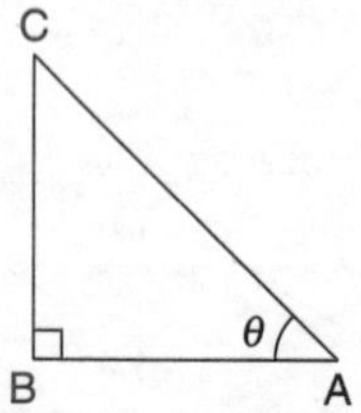

Here, we shall prove these identities.

Consider a ΔABC in which $\angle B = 90°$.

(i) By Pythagoras' Theorem, we have

$$AB^2 + BC^2 = AC^2$$

Dividing both sides by AC^2, we get

$$\left(\frac{AB}{AC}\right)^2 + \left(\frac{BC}{AC}\right)^2 = 1 \qquad \text{...(i)}$$

In ΔABC $\qquad \sin\theta = \dfrac{BC}{AC}$ and $\cos\theta = \dfrac{AB}{AC}$

Inserting the value of $\dfrac{BC}{AC}$ and $\dfrac{AB}{AC}$ in (i), we get

$$(\cos\theta)^2 + (\sin\theta)^2 = 1$$

$$\therefore \quad \sin^2\theta + \cos^2\theta = 1$$

(ii) By Pythagoras' Theorem, we have

$$BC^2 + AB^2 = AC^2$$

Dividing both sides by AB^2, we get

$$\frac{BC^2}{AB^2} + \frac{AB^2}{AB^2} = \frac{AC^2}{AB^2}$$

$$\Rightarrow \quad \frac{AB^2}{AB^2} = \frac{AC^2}{AB^2} - \frac{BC^2}{AB^2} \qquad \text{...(ii)}$$

$$\Rightarrow \quad \frac{AC}{AB} = \sec\theta \text{ and } \frac{BC}{AB} = \tan\theta$$

Inserting the values of $\dfrac{AC}{AB}$ and $\dfrac{BC}{AB}$ in (ii), we get

$$\therefore \quad \sec^2\theta - \tan^2\theta = 1$$

(iii) By Pythagoras' Theorem, we have

$$AB^2 + BC^2 = AC^2$$

Dividing both sides by BC^2, we get

$$\Rightarrow \quad \frac{AB^2}{BC^2} + \frac{BC^2}{BC^2} = \frac{AC^2}{BC^2}$$

$$\Rightarrow \quad 1 = \frac{AC^2}{BC^2} - \frac{AB^2}{BC^2} \qquad \text{...(iii)}$$

In ΔABC $\qquad \dfrac{AC}{BC} = \operatorname{cosec}\theta$ and $\dfrac{AB}{BC} = \cot\theta$

Inserting the values of $\dfrac{AC}{BC}$ and $\dfrac{AB}{BC}$ in (iii), we get

$$1 = \operatorname{cosec}^2\theta - \cot^2\theta$$

$$\therefore \quad \operatorname{cosec}^2\theta - \cot^2\theta = 1$$

Example 17 If $\tan A = \sqrt{2} - 1$, show that $\sin A \cos A = \frac{\sqrt{2}}{4}$.

Solution: We have $\tan A = \frac{\text{Perpendicular}}{\text{Base}} = \frac{\sqrt{2}-1}{1}$

So, we draw a right triangle ABC, right-angled at B such that

Base = $AB = 1$ and Perpendicular = $BC = \sqrt{2} - 1$

In right ΔCBA $\quad AC^2 = AB^2 + BC^2$ [∵ Pythagoras' Theorem]

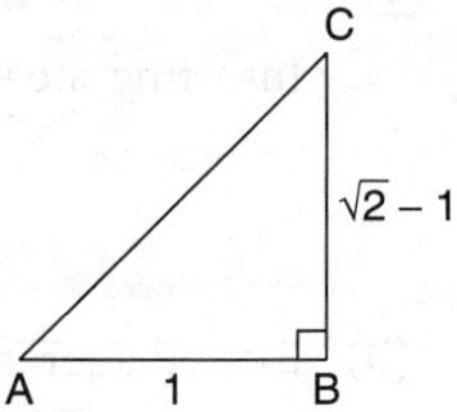

$\Rightarrow \quad AC^2 = 1^2 + (\sqrt{2} - 1)^2$

$\Rightarrow \quad AC^2 = 1 + 2 + 1 - 2\sqrt{2}$

$\Rightarrow \quad AC^2 = 4 - 2\sqrt{2}$

$\Rightarrow \quad AC = \sqrt{4 - 2\sqrt{2}}$

Now $\quad \sin A = \frac{\sqrt{2}-1}{\sqrt{4-2\sqrt{2}}}$

$\cos A = \frac{AB}{AC} = \frac{1}{\sqrt{4-2\sqrt{2}}}$

∴ $\quad \sin A \cos A = \frac{\sqrt{2}-1}{2\sqrt{2}(\sqrt{2}-1)} = \frac{1}{2\sqrt{2}} = \frac{\sqrt{2}}{4}$

Example 18 In a ΔABC right-angled at C, if $\tan A = \frac{1}{\sqrt{3}}$ and $\tan B = \sqrt{3}$, show that $\sin A \cos B + \cos A \sin B = 1$.

Solution: Let us draw a ΔABC, right-angled at C, in which $\tan B = \sqrt{3}$ and $\tan A = \frac{1}{\sqrt{3}}$.

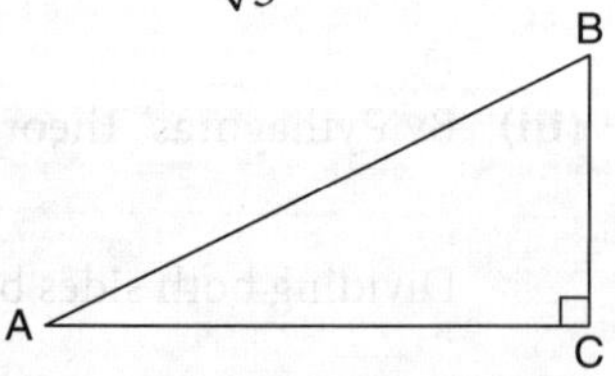

Now $\quad \tan A = \frac{1}{\sqrt{3}}$

$\Rightarrow \quad \frac{BC}{AC} = \frac{1}{\sqrt{3}}$ $\quad \left[\because \tan A = \frac{BC}{AC}\right]$

$\Rightarrow \quad BC = x$ and $AC = \sqrt{3}\,x$ ...(i)

And $\quad \tan B = \sqrt{3}$

$\Rightarrow \quad \frac{AC}{BC} = \frac{\sqrt{3}}{1}$ $\quad \left[\because \tan B = \frac{AC}{BC}\right]$

$\Rightarrow \quad AC = \sqrt{3}x$ and $BC = x$ ...(ii)

From (i) and (ii), we have $\quad BC = x,\ AC = \sqrt{3}\,x$

In right ΔBCA $\quad AB^2 = AC^2 + BC^2$ [∵ Pythagoras' Theorem]

$\Rightarrow \quad AB^2 = (\sqrt{3}\,x)^2 + x^2$

$\Rightarrow \quad AB^2 = 3x^2 + x^2$

$\Rightarrow \quad AB^2 = 4x^2$

$\Rightarrow \quad AB = 2x$

To find the t-ratios of $\angle A$, we have

Base = AC = $\sqrt{3}\,x$, Perpendicular = $BC = x$ and Hypotenuse = $AB = 2x$

$$\therefore \quad \sin A = \frac{BC}{AB} = \frac{x}{2x} = \frac{1}{2} \text{ and } \cos A = \frac{AC}{AB} = \frac{\sqrt{3}x}{2x} = \frac{\sqrt{3}}{2}$$

When we consider the t-ratios of $\angle B$, we have

Base = $BC = x$, Perpendicular = $AC = \sqrt{3}\,x$ and Hypotenuse = $AB = 2x$

$$\therefore \quad \cos B = \frac{BC}{AB} = \frac{x}{2x} = \frac{1}{2} \text{ and } \sin B = \frac{AC}{AB} = \frac{\sqrt{3}x}{2x} = \frac{\sqrt{3}}{2}$$

Now, $\sin A \cos B + \cos A \sin B = \frac{1}{2} \times \frac{1}{2} + \frac{\sqrt{3}}{2} \times \frac{\sqrt{3}}{2} = \frac{1}{4} + \frac{3}{4} = 1.$

Example 19 If $\cot B = \frac{12}{5}$, prove that $\tan^2 B - \sin^2 B = \sin^4 B \sec^2 B$.

Solution: We have $\quad \cot B = \frac{\text{Base}}{\text{Perpendicular}} = \frac{12}{5}$

So, we draw a right triangle ABC, right-angled at C, such that

Base = BC = 12 units and Perpendicular = AC = 5 units.

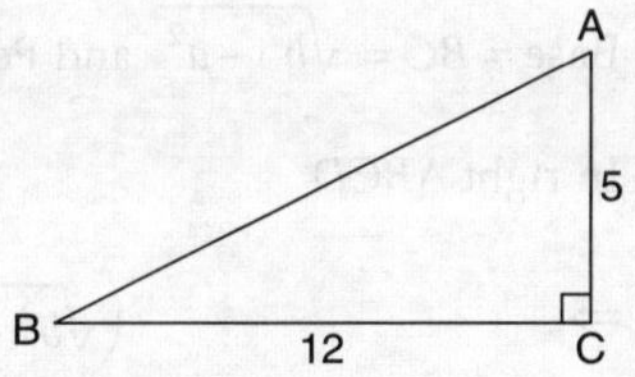

In right ΔACB $\quad AB^2 = BC^2 + AC^2 \quad$ [∵ Pythagoras' Theorem]

Now $\quad AB^2 = 12^2 + 5^2 = 169$

$\Rightarrow \quad AB = \sqrt{169} = 13$

$$\therefore \quad \sin B = \frac{AC}{AB} = \frac{5}{13},\ \tan B = \frac{AC}{BC} = \frac{5}{12} \text{ and, } \sec B = \frac{AB}{BC} = \frac{13}{12}$$

Now $\quad \text{LHS} = \tan^2 B - \sin^2 B$

$$\Rightarrow \quad \text{LHS} = \left(\frac{5}{12}\right)^2 - \left(\frac{5}{13}\right)^2 = \frac{25}{144} - \frac{25}{169}$$

$$\Rightarrow \quad \text{LHS} = 25\left(\frac{1}{144} - \frac{1}{169}\right) = 25\left(\frac{169-144}{144\times169}\right)$$

$$\Rightarrow \quad \text{LHS} = 25 \times \frac{25}{144\times169} = \frac{25\times25}{144\times169}$$

$$\Rightarrow \quad \text{LHS} = \frac{5^2\times5^2}{12^2\times13^2} \qquad \ldots\textbf{(i)}$$

And $\quad \text{RHS} = \sin^4 B \sec^2 B$

$\quad \text{RHS} = (\sin B)^4 (\sec B)^2$

$$\Rightarrow \quad \text{RHS} = \left(\frac{5}{13}\right)^4 \times \left(\frac{13}{12}\right)^2 = \frac{5^4\times13^2}{13^4\times12^2}$$

$$\Rightarrow \quad \text{RHS} = \frac{5^4}{13^2\times12^2} = \frac{5^2\times5^2}{13^2\times12^2} = \sin^2 B \cdot \tan^2 B \qquad \ldots\textbf{(ii)}$$

From (i) and (ii), we have

$$\tan^2 B - \sin^2 B = \sin^4 B \sec^2 B.$$

Example 20 In this figure, $AD = DB$ and $\angle B$ is a right angle. Determine:

(i) $\sin\theta$ (ii) $\cos\theta$ (iii) $\tan\theta$ (iv) $\sin^2\theta + \cos^2\theta$

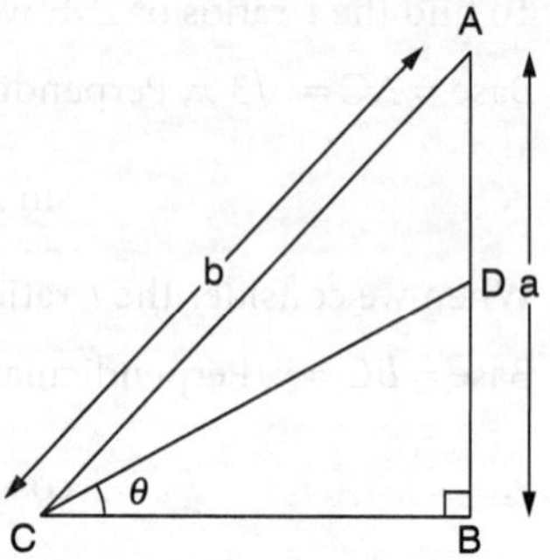

Solution: We have $AB = a$

$\Rightarrow \quad AD + DB = a \quad [\because AD = DB]$

$\Rightarrow \quad 2AD = a$

$\Rightarrow \quad AD = \dfrac{a}{2}$

Thus $AD = DB = \dfrac{a}{2}$

In right ΔABC $\quad AC^2 = AB^2 + BC^2 \quad$ [∵ Pythagoras' Theorem]

$\Rightarrow \quad b^2 = a^2 + BC^2$

$\Rightarrow \quad BC^2 = b^2 - a^2$

$\Rightarrow \quad BC = \sqrt{b^2 - a^2}$

Thus, in ΔBCD, we have

Base $= BC = \sqrt{b^2 - a^2}$ and Perpendicular $= BD = \dfrac{a}{2}$

In right ΔBCD $\quad BC^2 + BD^2 = CD^2 \quad$ [∵ Pythagoras' Theorem]

$\Rightarrow \quad \left(\sqrt{b^2 - a^2}\right)^2 + \left(\dfrac{a}{2}\right)^2 = CD^2$

$\Rightarrow \quad CD^2 = b^2 - a^2 + \dfrac{a^2}{4}$

$\Rightarrow \quad CD^2 = \dfrac{4b^2 - 4a^2 + a^2}{4}$

$\Rightarrow \quad CD^2 = \dfrac{4b^2 - 3a^2}{4}$

$\Rightarrow \quad CD = \dfrac{\sqrt{4b^2 - 3a^2}}{2}$

Now

(i) $\sin\theta = \dfrac{BD}{CD} \Rightarrow \sin\theta = \dfrac{a/2}{\dfrac{\sqrt{4b^2 - 3a^2}}{2}} = \dfrac{a}{\sqrt{4b^2 - 3a^2}}$

(ii) $\cos\theta = \dfrac{BC}{CD} \Rightarrow \cos\theta = \dfrac{\sqrt{b^2 - a^2}}{\dfrac{\sqrt{4b^2 - 3a^2}}{2}} = \dfrac{2\sqrt{b^2 - a^2}}{\sqrt{4b^2 - 3a^2}}$

(iii) $\tan\theta = \dfrac{BD}{BC} \Rightarrow \tan\theta = \dfrac{a/2}{\sqrt{b^2 - a^2}} = \dfrac{a}{2\sqrt{b^2 - a^2}}$

(iv) $\sin^2\theta + \cos^2\theta = \left(\dfrac{a}{\sqrt{4b^2 - 3a^2}}\right)^2 + \left(\dfrac{2\sqrt{b^2 - a^2}}{\sqrt{4b^2 - 3a^2}}\right)^2$

$\Rightarrow \sin^2\theta + \cos^2\theta = \dfrac{a^2}{4b^2 - 3a^2} + \dfrac{4(b^2 - a^2)}{4b^2 - 3a^2} = \dfrac{a^2 + 4b^2 - 4a^2}{4b^2 - 3a^2} = \dfrac{4b^2 - 3a^2}{4b^2 - 3a^2}$

$\Rightarrow \sin^2\theta + \cos^2\theta = 1$

Example 21 Prove that:

(i) $\tan 10° \tan 15° \tan 75° \tan 80° = 1$ **(ii)** $\tan 1° \tan 2° \tan 3° \ldots \tan 89° = 1$

(iii) $\cos 1° \cos 2° \cos 3° \ldots \cos 180° = 0$

Solution:

(i) We have LHS $= \tan 10° \tan 15° \tan 75° \tan 80°$

$= \tan (90° - 80°) \tan (90° - 75°) \tan 75° \tan 80°$

$= \cot 80° \cot 75° \tan 75° \tan 80°$ $[\because \tan (90° - \theta) = \cot \theta]$

$= (\cot 80° \tan 80°) (\cot 75° \tan 75°)$ $[\because \cot\theta . \tan\theta = 1]$

$= 1 \times 1 = 1 =$ RHS

(ii) We have LHS $= \tan 1° \tan 2° \tan 3° \ldots \tan 89°$

$= \tan (90° - 89°) \tan (90° - 88°) \tan (90° - 87°) \ldots \tan 87° \tan 88° \tan 89°$

$= \cot 89° \cot 88° \cot 87° \ldots \tan 87° \tan 88° \tan 89°$

$= (\cot 89° \tan 89°) (\cot 88° \tan 88°) (\cot 87° \tan 87°) \ldots (\cot 44° \tan 44°) . \tan 45°$

$= 1 \times 1 \times 1 \ldots \times 1 = 1 =$ RHS $[\because \cot\theta \tan \theta = 1 \text{ and } \tan 45° = 1]$

(iii) We have LHS $= \cos 1° \cos 2° \cos 3° \ldots \cos 180°$

$= \cos 1° \cos 2° \cos 3° \ldots \cos 89 \cos 90° \cos 91° \ldots \cos 180°$

$= \cos 1° \cos 2° \cos 3° \ldots \cos 89° \cdot 0 \cdot \cos 91° \ldots \cos 180°$

$= 0 =$ RHS $[\because \cos 90° = 0]$

Example 22 If $A + B = 90°$, prove that

$$\sqrt{\frac{\tan A \tan B + \tan A \cot B}{\sin A \sec B} - \frac{\sin^2 B}{\cos^2 A}} = \tan A$$

Solution: We have

$$A + B = 90° \Rightarrow B = 90° - A$$

$\therefore$ $\text{LHS} = \sqrt{\dfrac{\tan A \tan B + \tan A \cot B}{\sin A \sec B} - \dfrac{\sin^2 B}{\cos^2 A}}$

$\Rightarrow$ $\text{LHS} = \sqrt{\dfrac{\tan A \tan (90° - A) + \tan A \cot (90° - A)}{\sin A \sec (90° - A)} - \dfrac{\sin^2 (90° - A)}{\cos^2 A}}$

$\Rightarrow$ $\text{LHS} = \sqrt{\dfrac{\tan A \cot A + \tan A \tan A}{\sin A \operatorname{cosec} A} - \dfrac{\cos^2 A}{\cos^2 A}}$

$\Rightarrow$ $\text{LHS} = \sqrt{1 + \tan^2 A - 1} = \sqrt{\tan^2 A} = \tan A = \text{RHS}$

Example 23 If $\sin 5\theta = \cos 4\theta$, where 5θ and 4θ are acute angles, find the value of θ.

Solution: We have $\sin 5\theta = \cos 4\theta$

$\Rightarrow \sin 5\theta = \sin(90° - 4\theta)$

$\Rightarrow 5\theta = 90° - 4\theta$

$\Rightarrow 9\theta = 90°$

$\Rightarrow \theta = 10°$

Example 24 Evaluate each of the following:

(i) $\cos(40° - \theta) - \sin(50° + \theta) + \dfrac{\cos^2 40° + \cos^2 50°}{\sin^2 40° + \sin^2 50°}$ **[CBSE 2002]**

(ii) $\dfrac{\cos 70°}{\sin 20°} + \dfrac{\cos 55° \operatorname{cosec} 35°}{\tan 5° \tan 25° \tan 45° \tan 65° \tan 85°}$ **[CBSE 2002]**

Solution: (i) We have

$$\cos(40° - \theta) - \sin(50 + \theta) + \frac{\cos^2 40° + \cos^2 50°}{\sin^2 40° + \sin^2 50°}$$

$$= \sin\{90° - (40° - \theta)\} - \sin(50 + \theta) + \frac{\cos^2 40° + \cos^2(90° - 40°)}{\sin^2 40° + \sin^2(90° - 40°)} \quad [\because \cos\alpha = \sin(90 - \theta)]$$

$$= \sin(50° + \theta) - \sin(50 + \theta) + \frac{\cos^2 40° + \sin^2 40°}{\sin^2 40° + \cos^2 40} = 0 + 1 = 1$$

(ii) We have

$$\frac{\cos 70°}{\sin 20°} + \frac{\cos 55° \operatorname{cosec} 35°}{\tan 5° \tan 25° \tan 45° \tan 65° \tan 85°}$$

$$= \frac{\cos(90° - 20°)}{\sin 20°} + \frac{\cos 55° \operatorname{cosec}(90° - 55°)}{\tan 5° \tan 25° \tan 45° \tan(90° - 25°)\tan(90° - 5°)}$$

$$= \frac{\sin 20°}{\sin 20°} + \frac{\cos 55° \sec 55°}{\tan 5° \tan 25° \tan 45° \cot 25° \cot 5°}$$

$$= 1 + \frac{1}{\tan 45°} \quad [\because \tan 5° \cot 5° = 1 \text{ and } \tan 25° \cot 25° = 1]$$

$$= 1 + 1 = 2$$

Example 25 Show that

(i) $2(\cos^2 45° + \tan^2 60°) - 6(\sin^2 45° - \tan^2 30°) = 6$

(ii) $2(\cos^4 60° + \sin^4 30°) - (\tan^2 60° + \cot^2 45°) + 3\sec^2 30° = \dfrac{1}{4}$

Solution:

(i) We have, LHS $2(\cos^2 45° + \tan^2 60°) - 6(\sin^2 45° - \tan^2 30°)$

$$= 2\left\{\left(\frac{1}{\sqrt{2}}\right)^2 + (\sqrt{3})^2\right\} - 6\left\{\left(\frac{1}{\sqrt{2}}\right)^2 - \left(\frac{1}{\sqrt{3}}\right)^2\right\}$$

$$= 2\left(\frac{1}{2} + 3\right) - 6\left(\frac{1}{2} - \frac{1}{3}\right) = 2\left(\frac{1+6}{2}\right) - 6\left(\frac{3-2}{6}\right)$$

$$= 2 \times \frac{7}{2} - 6 \times \frac{1}{6} = 7 - 1 = 6$$

Hence, LHS = RHS.

(ii) We have $\quad 2(\cos^4 60° + \sin^4 30°) - (\tan^2 60° + \cot^2 45°) + 3\sec^2 30°$

$$= 2\left\{\left(\frac{1}{2}\right)^4 + \left(\frac{1}{2}\right)^4\right\} - \left\{(\sqrt{3})^2 + (1)^2\right\} + 3\left(\frac{2}{\sqrt{3}}\right)^2$$

$$= 2\left(\frac{1}{16} - \frac{1}{16}\right) - (3+1) + 3 \times \frac{4}{3} = 2 \times \frac{1}{8} - 4 + 4 = \frac{1}{4}$$

Hence, LHS = RHS.

Example 26 If $x = 30°$, verify that

(i) $\cos 3x = 4\cos^3 x - 3\cos x$ **(ii)** $\sin x = \sqrt{\frac{1-\cos 2x}{2}}$

Solution:

(i) We have $\quad x = 30° \Rightarrow 3x = 90°$

$\therefore$ LHS $\quad \cos 3x = \cos 90° = 0$

$$= 4\cos^3 x - 3\cos x$$

$$\text{RHS} = 4\cos^3 30° - 3\cos 30°$$

$$= 4\left(\frac{\sqrt{3}}{2}\right)^3 - 3\left(\frac{\sqrt{3}}{2}\right) = 4 \times \frac{3\sqrt{3}}{8} - \frac{3\sqrt{3}}{2} = \frac{3\sqrt{3}}{2} - \frac{3\sqrt{3}}{2} = 0$$

$\therefore$ Hence, LHS = RHS.

(ii) We have

RHS $\quad x = 30° \Rightarrow 2x = 60°$

$\therefore \quad \sqrt{\frac{1-\cos 2x}{2}} = \sqrt{\frac{1-\cos 60°}{2}} = \sqrt{\frac{1-\frac{1}{2}}{2}} = \sqrt{\frac{1}{4}} = \frac{1}{2}$

And LHS $\quad \sin x = \sin 30° = \frac{1}{2}$

$\therefore \quad \sin x = \sqrt{\frac{1-\cos 2x}{2}}$

Hence, LHS = RHS.

Example 27 In an acute-angled triangle ABC, if $\tan(A + B - C) = 1$ and $\sec(B + C - A) = 2$, find the value of A, B and C.

Solution: We have $\quad \tan(A + B - C) = 1$ and $\sec(B + C - A) = 2$

$\Rightarrow \quad \tan(A + B - C) = \tan 45°$ and $\sec(B + C - A) = \sec 60°$

$\Rightarrow \quad A + B - C = 45° \quad$ ...(i)

And $\quad B + C - A = 60° \quad$ ...(ii)

Adding (i) and (ii), we get

$$\Rightarrow \quad (A + B - C) + (B + C - A) = 45° + 60°$$

$$\Rightarrow \quad 2B = 105°$$

$$\Rightarrow \quad B = 52\frac{1}{2}°$$

Inserting $B = 52\frac{1}{2}°$ in $B + C - A = 60°$, we get

$$52\frac{1}{2}° + C - A = 60° \Rightarrow C - A = 7\frac{1}{2}° \quad \text{...(iii)}$$

Also, in ΔABC, we have $A + B + C = 180°$ [∵ A.S.P. of Δ's]

$$\Rightarrow \quad A + 52\frac{1}{2}° + C = 180° \quad \left[\because B = 52\frac{1}{2}°\right]$$

$$\Rightarrow \quad C + A = 127\frac{1}{2}° \quad \text{...(iv)}$$

Adding and subtracting (iii) and (iv), we get

$$2C = 135° \text{ and } 2A = 120° \Rightarrow C = 67\frac{1}{2}° \text{ and } A = 60°$$

Hence $A = 60°,\ B = 52\frac{1}{2}°$ and $C = 67\frac{1}{2}°$

Example 28 If α, β and γ are each positive acute angles such that $\sin(\alpha + \beta - \gamma) = \frac{1}{2}$, $\cos(\beta + \gamma - \alpha) = \frac{1}{2}$ and $\tan(\gamma + \alpha - \beta) = 1$, find the values of α, β and γ.

Solution: We have

$$\sin(\alpha + \beta - \gamma) = \frac{1}{2},\ \cos(\beta + \gamma - \alpha) = \frac{1}{2} \text{ and } \tan(\gamma + \alpha - \beta) = 1$$

$$\Rightarrow \quad \sin(\alpha + \beta - \gamma) = \sin 30°, \cos(\beta + \gamma - \alpha) = \cos 60° \text{ and } \tan(\gamma + \alpha - \beta) = \tan 45°$$

$$\Rightarrow \quad \alpha + \beta - \gamma = 30° \quad \text{...(i)}$$

$$\beta + \gamma - \alpha = 60° \quad \text{...(ii)}$$

$$\gamma + \alpha - \beta = 45° \quad \text{...(iii)}$$

Adding (i), (ii) and (iii), we get

$$2\beta = 90° \text{ and } 2\alpha = 75° \Rightarrow \beta = 45° \text{ and } \alpha = 37\frac{1}{2}°$$

Inserting the values of α and β in (i), we get

$$37\frac{1}{2}° + 45 - g = 30 \Rightarrow \gamma = 52\frac{1}{2}°$$

Hence $a = 37\frac{1}{2}°,\ \beta = 45°$ and $\gamma = 52\frac{1}{2}°$.

Example 29 If θ is an acute angle and $\tan\theta + \cot\theta = 2$, find the value of $\tan^7\theta + \cot^7\theta$.

Solution: We have $\tan\theta + \cot\theta = 2$

$$\Rightarrow \quad \tan\theta + \frac{1}{\tan\theta} = 2$$

$$\Rightarrow \quad \frac{\tan^2\theta + 1}{\tan\theta} = 2$$

$$\Rightarrow \quad \tan^2\theta - 2\tan\theta + 1 = 0$$

$$\Rightarrow \quad (\tan\theta - 1)^2 = 0$$

$$\Rightarrow \quad \tan\theta - 1 = 0$$

$$\Rightarrow \quad \tan\theta = 1$$

$$\Rightarrow \quad \tan\theta = \tan 45°$$

$$\Rightarrow \quad \theta = 45°$$

$$\therefore \quad \tan^7\theta + \cot^7\theta = \tan^7 45° + \cot^7 45° = (\tan 45°)^7 + (\cot 45°)^7 = (1)^7 + (1)^7 = 2.$$

Example 30 Prove the following identities:

(i) $(1 + \cot\theta - \operatorname{cosec}\theta)(1 + \tan\theta + \sec\theta) = 2$ **(ii)** $\tan^2\theta + \cot^2\theta + 2 = \sec^2\theta \operatorname{cosec}^2\theta$

(iii) $\sqrt{\sec^2\theta + \operatorname{cosec}^2\theta} = \tan\theta + \cot\theta$

Solution: **(i)** We have

$$\text{LHS} = \left(1 + \frac{\cos\theta}{\sin\theta} - \frac{1}{\sin\theta}\right)\left(1 + \frac{\sin\theta}{\cos\theta} + \frac{1}{\cos\theta}\right)$$

$$\Rightarrow \quad = \left(\frac{\sin\theta + \cos\theta - 1}{\sin\theta}\right)\left(\frac{\cos\theta + \sin\theta + 1}{\cos\theta}\right)$$

$$\Rightarrow \quad = \frac{(\sin\theta + \cos\theta)^2 - 1^2}{\sin\theta\cos\theta}$$

$$\Rightarrow \quad = \frac{\sin^2\theta + \cos^2\theta + 2\sin\theta\cos\theta - 1}{\sin\theta\cos\theta}$$

$$\Rightarrow \quad = \frac{1 + 2\sin\theta\cos\theta - 1}{\sin\theta\cos\theta}$$

$$\Rightarrow \quad = \frac{2\sin\theta\cos\theta}{\sin\theta\cos\theta} = 2 = \text{RHS}$$

Hence, LHS = RHS.

(ii) We have $= \tan^2\theta + \cot^2\theta + 2$

$$\Rightarrow \quad = \tan^2\theta + \cot^2\theta + 2\tan\theta\cot\theta \qquad [\because \tan\theta.\cot\theta = 1]$$

$$\Rightarrow \quad = (\tan\theta + \cot\theta)^2$$

$$\Rightarrow \quad = \left(\frac{\sin\theta}{\cos\theta} + \frac{\cos\theta}{\sin\theta}\right)^2$$

$$\Rightarrow \quad = \left(\frac{\sin^2\theta+\cos^2\theta}{\sin\theta\cos\theta}\right)^2 = \left(\frac{1}{\sin\theta\cos\theta}\right)^2$$

$$= [\text{cosec}\,\theta\cdot\sec\theta]^2 = \text{cosec}^2\theta\cdot\sec^2\theta$$

$$= \text{RHS}$$

Hence LHS = RHS.

(iii) We have $\text{LHS} = \sqrt{\sec^2\theta+\text{cosec}^2\theta}$

$$= \sqrt{(1+\tan^2)+(1+\cot^2\theta)}$$

$$= \sqrt{2+\tan^2\theta+\cot^2\theta}$$

$$= \sqrt{\tan^2\theta+\cot^2\theta+2\times\tan\theta\times\frac{1}{\tan\theta}}$$

$$= \sqrt{\tan^2\theta+\cot^2\theta+2\tan\theta\cdot\cot\theta}$$

$$= \sqrt{(\tan\theta+\cot\theta)^2}$$

$$= \tan\theta+\cot\theta = \text{RHS}$$

Hence, LHS = RHS.

Example 31 Prove the following statements:

(i) $\dfrac{1}{\text{cosec}A-\cot A}-\dfrac{1}{\sin A}=\dfrac{1}{\sin A}-\dfrac{1}{\text{cosec}A+\cot A}$

(ii) $\dfrac{\cot A+\text{cosec}A-1}{\cot A-\text{cosec}A+1}=\dfrac{1+\cos A}{\sin A}$

(iii) $\dfrac{\sin A+\cos A}{\sin A-\cos A}+\dfrac{\sin A-\cos A}{\sin A+\cos A}=\dfrac{2}{\sin^2 A-\cos^2 A}=\dfrac{2}{2\sin^2 A-1}$ **[CBSE 2000]**

(iv) $\dfrac{\cos A}{1-\tan A}+\dfrac{\sin A}{1-\cot A}=\cos A+\sin A$

(v) $\dfrac{\tan A}{1-\cot A}+\dfrac{\cot A}{1-\tan A}=1+\tan A+\cot A=1+\sec A\,\text{cosec}\,A$ **[CBSE 2002C]**

Solution: **(i)** We have $\text{LHS} = \dfrac{1}{\text{cosec}A-\cot A}-\dfrac{1}{\sin A}$

$$\Rightarrow \quad = \frac{1}{(\text{cosec}A-\cot A)}\,\frac{(\text{cosec}A+\cot A)}{\text{cosec}A+\cot A}-\frac{1}{\sin A}$$ [∵ On rationalising the denominator]

$$\Rightarrow \quad = \frac{\text{cosec}A+\cot A}{\text{cosec}^2A-\cot^2A}-\text{cosec}\,A$$

$\Rightarrow \quad = \text{cosec}\,A+\cot A-\text{cosec}\,A$ [∵ $\text{cosec}^2 A-\cot^2 A=1$]

$\Rightarrow \quad = \cot A$ ...(i)

And $\quad \text{RHS} = \dfrac{1}{\sin A} - \dfrac{1}{\operatorname{cosec} A + \cot A}$

$\Rightarrow \quad = \dfrac{1}{\sin A} - \dfrac{\operatorname{cosec} A - \cot A}{(\operatorname{cosec} A + \cot A)(\operatorname{cosec} A - \cot A)}$

$\Rightarrow \quad = \dfrac{1}{\sin A} - \dfrac{\operatorname{cosec} A - \cot A}{\operatorname{cosec}^2 A - \cot^2 A}$

$\Rightarrow \quad = \operatorname{cosec} A - (\operatorname{cosec} A - \cot A)$ $\quad [\because \operatorname{cosec}^2 A - \cot^2 A = 1]$

$\Rightarrow \quad \text{RHS} = \cot A$ ...(ii)

Hence, LHS = RHS.

(ii) We have $\quad \text{LHS} = \dfrac{\cot A + \operatorname{cosec} A - 1}{\cot A - \operatorname{cosec} A + 1}$

$\Rightarrow \quad = \dfrac{(\cot A + \operatorname{cosec} A) - (\operatorname{cosec}^2 A - \cot^2 A)}{(\cot A - \operatorname{cosec} A + 1)}$ $\quad [\because \operatorname{cosec}^2 A - \cot^2 A = 1]$

$\Rightarrow \quad = \dfrac{(\operatorname{cosec} A + \cot A) - (\operatorname{cosec} A + \cot A)(\operatorname{cosec} A - \cot A)}{\cot A - \operatorname{cosec} A + 1}$

$\Rightarrow \quad = \dfrac{(\operatorname{cosec} A + \cot A)[1 - (\operatorname{cosec} A - \cot A)]}{\cot A - \operatorname{cosec} A + 1}$

$\Rightarrow \quad = \dfrac{(\operatorname{cosec} + \cot A)(\cot A - \operatorname{cosec} A + 1)}{(\cot A - \operatorname{cosec} A + 1)}$

$\Rightarrow \quad = \operatorname{cosec} A + \cot A$

$\Rightarrow \quad = \dfrac{1}{\sin A} + \dfrac{\cos A}{\sin A} = \dfrac{1 + \cos A}{\sin A} = \text{RHS}$

(iii) We have $\quad = \dfrac{\sin A + \cos A}{\sin A - \cos A} + \dfrac{\sin A - \cos A}{\sin A + \cos A}$

$\Rightarrow \quad = \dfrac{(\sin A + \cos A)^2 + (\sin A - \cos A)^2}{(\sin A - \cos A)(\sin A + \cos A)}$ $\quad$ [On taking LCM]

$\Rightarrow \quad = \dfrac{(\sin^2 A + \cos^2 A + 2\sin A\cos A) + (\sin^2 A + \cos^2 A - 2\sin A\cos A)}{\sin^2 A - \cos^2 A}$

$\Rightarrow \quad = \dfrac{1 + 2\sin A\cos A + 1 - 2\sin A\cos A}{\sin^2 A - \cos^2 A}$ $\quad [\because \sin^2 A + \cos^2 A = 1]$

$\Rightarrow \quad = \dfrac{2}{\sin^2 A - \cos^2 A} = \dfrac{2}{\sin^2 A - (1 - \sin^2 A)}$

$\Rightarrow \quad = \dfrac{2}{2\sin^2 A - 1} = \text{RHS}$

Hence, LHS = RHS.

(iv) We have $\text{LHS} = \dfrac{\cos A}{1-\tan A} + \dfrac{\sin A}{1-\cot A}$

$\Rightarrow \quad = \dfrac{\cos A}{1-\dfrac{\sin A}{\cos A}} + \dfrac{\sin A}{1-\dfrac{\cos A}{\sin A}}$

$\Rightarrow \quad = \dfrac{\cos^2 A}{\cos A - \sin A} + \dfrac{\sin^2 A}{\sin A - \cos A}$

$\Rightarrow \quad = \dfrac{\cos^2 A}{\cos A - \sin A} - \dfrac{\sin^2 A}{\cos A - \sin A}$

$\Rightarrow \quad = \dfrac{\cos^2 A - \sin^2 A}{\cos A - \sin A}$

$\Rightarrow \quad = \dfrac{(\cos A - \sin A)(\cos A + \sin A)}{(\cos A - \sin A)}$

$\Rightarrow \quad = \cos A + \sin A = \text{RHS}$

Hence, LHS = RHS.

(v) We have $\text{LHS} = \dfrac{\tan A}{1-\cot A} + \dfrac{\cot A}{1-\tan A}$

$\Rightarrow \quad = \dfrac{\tan A}{1-\dfrac{1}{\tan A}} + \dfrac{\dfrac{1}{\tan A}}{1-\tan A}$

$\Rightarrow \quad = \dfrac{\tan A}{\dfrac{\tan A - 1}{\tan A}} + \dfrac{1}{\tan A\,(1-\tan A)}$

$\Rightarrow \quad = \dfrac{\tan^2 A}{\tan A - 1} + \dfrac{1}{\tan A\,(1-\tan A)}$

$\Rightarrow \quad = \dfrac{\tan^3 A - 1}{\tan A\,(\tan A - 1)}$

$\Rightarrow \quad = \dfrac{(\tan A - 1)(\tan^2 + \tan A + 1)}{\tan A(\tan A - 1)}$ $\quad [a^3 - b^3 = (a-b)(a^2+ab+b^2)]$

$\Rightarrow \quad = \dfrac{\tan^2 A + \tan A + 1}{\tan A}$

$\Rightarrow \quad = \dfrac{\tan^2 A}{\tan A} + \dfrac{\tan A}{\tan A} + \dfrac{1}{\tan A}$

$\Rightarrow \quad = \tan A + 1 + \cot A = (1 + \tan A + \cot A)$

$\Rightarrow \quad = 1 + \dfrac{\sin A}{\cos A} + \dfrac{\cos A}{\sin A}$

$\Rightarrow \quad = 1 + \dfrac{\sin^2 A + \cos^2 A}{\sin A \cos A}$

$\Rightarrow \quad = 1 + \dfrac{1}{\sin A \cos A}$ $\quad [\because \sin^2 A + \cos^2 A = 1]$

$\Rightarrow \quad = 1 + \operatorname{cosec} A \sec A = \text{RHS}.$

Hence, LHS = RHS.

Example 32 Prove the following identities:

(i) $\dfrac{\cos A}{1-\sin A} + \dfrac{\sin A}{1-\cos A} + 1 = \dfrac{\sin A \cos A}{(1-\sin A)(1-\cos A)}$

(ii) $\dfrac{(1+\cos A+\tan A)(\sin A-\cos A)}{\sec^3 A-\operatorname{cosec}^3 A} = \sin^2 A \cos^2 A$

Solution:

(i) We have

$\Rightarrow \quad \text{LHS} = \dfrac{\cos A}{1-\sin A} + \dfrac{\sin A}{1-\cos A} + 1$

$\Rightarrow \quad = \dfrac{\cos A(1-\cos A)+\sin A(1-\sin A)+(1-\sin A)(1-\cos A)}{(1-\sin A)(1-\cos A)}$

$\Rightarrow \quad = \dfrac{\cos A-\cos^2 A+\sin A-\sin^2 A+1-\sin A-\cos A+\sin A\cos A}{(1-\sin A)(1-\cos A)}$

$\Rightarrow \quad = \dfrac{(\cos A+\sin A)-(\sin^2 A+\cos^2 A)+1-(\cos A+\sin A)+\sin A\cos A}{(1-\sin A)(1-\cos A)}$

$\Rightarrow \quad = \dfrac{(\cos A+\sin A)-1+1-(\cos A+\sin A)+\sin A\cos A}{(1-\sin A)(1-\cos A)}$

$\Rightarrow \quad = \dfrac{\sin A\cos A}{(1-\sin A)(1-\cos A)} = \text{RHS}$

Hence, LHS = RHS.

(ii) We have

$\Rightarrow \quad \text{LHS} = \dfrac{(1+\cot A+\tan A)(\sin A-\cos A)}{\sec^3 A-\operatorname{cosec}^3 A} = \dfrac{\left(1+\dfrac{\cos A}{\sin A}+\dfrac{\sin A}{\cos A}\right)(\sin A-\cos A)}{\left(\dfrac{1}{\cos^3 A}-\dfrac{1}{\sin^3 A}\right)}$

$$\Rightarrow \quad = \frac{\left(1+\frac{\cos^2 A+\sin^2 A}{\sin A\cos A}\right)(\sin A-\cos A)}{\frac{(\sin^3 A-\cos^3 A)}{\sin^3 A\cos^3 A}}$$

$$\Rightarrow \quad = \frac{\sin A\cos A+\cos^2 A+\sin^2 A}{\sin A\cos A} \times \frac{\sin^3 A\cos^3 A}{\sin^3 A-\cos^3 A}\ (\sin A - \cos A)$$

$$\Rightarrow \quad = \frac{(\sin A\cos A+1)}{\sin^3 A-\cos^3 A} \times (\sin^2 A\cos^2 A)\ (\sin A - \cos A)$$

$$\Rightarrow \quad = \frac{(\sin A\cos A+1)(\sin^2 A\cos^2 A)(\sin A-\cos A)}{(\sin A-\cos A)(\sin^2 A+\cos^2 A+\sin A\cos A)} \quad [\because a^3 - b^3 = (a-b)(a^2+ab+b^2)]$$

$$\Rightarrow \quad = \frac{(\sin A\cos A+1)\sin^2 A\cos^2 A}{(1+\sin A\cos A)}$$

$$\Rightarrow \quad = \sin^2 A\cos^2 A = \text{RHS}.$$

Hence, LHS = RHS.

Example 33 If $\tan\theta + \sin\theta = m$ and $\tan\theta - \sin\theta = n$, show that $m^2 - n^2 = 4\sqrt{mn}$.

Solution: We have LHS $= m^2 - n^2$

$\Rightarrow$ LHS $= (\tan\theta + \sin\theta)^2 - (\tan\theta - \sin\theta)^2$

$= \tan^2\theta + \sin^2\theta + 2\tan\theta . \sin\theta - \tan^2\theta - \sin^2\theta + 2\tan\theta\sin\theta$

$\Rightarrow$ LHS $= 4\tan\theta\sin\theta$

And

$$\Rightarrow \quad \text{RHS} = 4\sqrt{(\tan\theta+\sin\theta)(\tan\theta-\sin\theta)}$$

$$\Rightarrow \quad = 4\sqrt{\tan^2\theta-\sin^2\theta}$$

$$\Rightarrow \quad = 4\sqrt{\frac{\sin^2\theta}{\cos^2}-\sin^2\theta}$$

$$\Rightarrow \quad = 4\sqrt{\frac{\sin^2\theta-\sin^2\theta\cos^2\theta}{\cos^2\theta}}$$

$$\Rightarrow \quad = 4\sqrt{\frac{\sin^2\theta(1-\cos^2\theta)}{\cos^2\theta}}$$

$$\Rightarrow \quad = 4\sqrt{\frac{\sin^4\theta}{\cos^2\theta}} = 4\frac{\sin^2\theta}{\cos\theta} = 4\sin\theta\frac{\sin\theta}{\cos\theta} = 4\sin\theta\tan\theta$$

Thus, LHS = RHS, that is, $m^2 - n^2 = 4\sqrt{mn}$.

Example 34 If $x = r \sin A \cos C$, $y = r \sin A \sin C$ and $z = r \cos A$, prove that $r^2 = x^2 + y^2 + z^2$.

Solution: We have $x^2 + y^2 + z^2 = r^2 \sin^2 A \cos^2 C + r^2 \sin^2 A \sin^2 C + r^2 \cos^2 A$

$\Rightarrow$ $x^2 + y^2 + z^2 = r^2 \sin^2 A (\cos^2 C + \sin^2 C) + r^2 \cos^2 A$

$\Rightarrow$ $x^2 + y^2 + z^2 = r^2 \sin^2 A + r^2 \cos^2 A$ $[\because \cos^2 C + \sin^2 C = 1]$

$\Rightarrow$ $x^2 + y^2 + z^2 = r^2 (\sin^2 A + \cos^2 A)$

$\Rightarrow$ $x^2 + y^2 + z^2 = r^2$

Hence $r^2 = x^2 + y^2 + z^2$

Multiple Choice Questions

Choose the correct answer for each of the following questions:

1. If $\sec x + \tan x = p$, then $\sec x$ is equal to:

(a) $\frac{p^2 - 1}{p}$ (b) $\frac{p^2 - 1}{2p}$ (c) $\frac{p^2 + 1}{p}$ (d) $\frac{p^2 + 1}{2p}$

2. In this figure, if D is the mid-point of BC, the value of $\frac{\cot x^\circ}{\cot y^\circ}$ is

(a) 2 (b) $\frac{1}{4}$

(c) $\frac{1}{3}$ (d) $\frac{1}{2}$

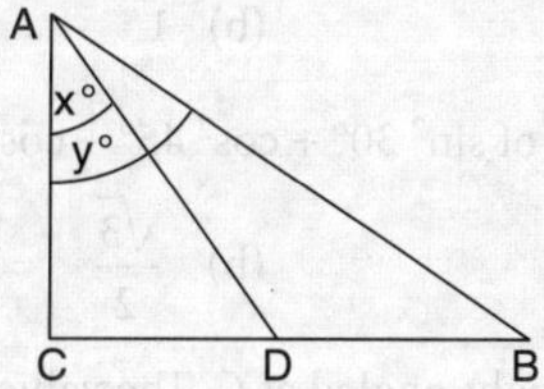

3. If $\sin \theta - \cos \theta = 0$, then the value of $(\sin^4 \theta + \cos^4 \theta)$ is

(a) 1 (b) $\frac{3}{4}$ (c) $\frac{1}{2}$ (d) $\frac{1}{4}$

4. $\cos 1^\circ . \cos 2^\circ . \cos 3^\circ \ldots, \cos 180^\circ$ is equal to

(a) 1 (b) 0 (c) -1 (d) $\frac{1}{2}$

5. The value of $\sin^2 15^\circ + \sin^2 25^\circ + \sin^2 65^\circ + \sin^2 75^\circ$ is

(b) 1 (b) 2 (c) 0 (d) 3

6. If $\sec \theta + \tan \theta = x$, then the value of $\sec \theta - \tan \theta$ in terms of x is

(a) x^2 (b) $\frac{1}{x}$ (c) x^3 (d) $\frac{x}{2}$

7. Express $\sin A$ in terms of $\cot A$

(a) $\frac{\sqrt{1+\cot^2 A}}{\cos^2 \theta}$ (b) $\frac{\sqrt{1-\cot^2 A}}{\cot A}$ (c) $\frac{1}{\sqrt{1+\cot^2 A}}$ (d) $\frac{\sqrt{1-\cot^2 A}}{\cot A}$

8. $\dfrac{(1+\tan^2\theta)}{\cos^2\theta} =$

(a) $\dfrac{1}{\sin^2\theta - \cos^2\theta}$ (b) $\sec^4\theta$ (c) 1 (d) $\dfrac{1}{\sqrt{2}}$

9. If $\tan\theta + \cot\theta = 2$, ($\theta$ is acute), the value of $\tan^2\theta + \cot^2\theta$ is

(a) 0 (b) 1 (c) 4 (d) 2

10. Given that $\sin\alpha = \dfrac{1}{2}$ and $\cos\beta = \dfrac{\sqrt{3}}{2}$, the value of $\alpha + \beta$ is

(a) 30° (b) 45° (c) 60° (d) 90°

11. The value of the expression $\sec^2 50° - \cot^2 40° - \sin^2 45°$ is equal to

(a) 2 (b) $\dfrac{1}{2}$ (c) 1 (d) 0°

12. $\sin(60° + \theta) - \cos(30° - \theta)$ is equal to

(a) $2\cos\theta$ (b) $2\sin\theta$ (c) 0 (d) 1

13. Given that $\tan\theta = \dfrac{1}{\sqrt{3}}$, the value of $\dfrac{\operatorname{cosec}^2\theta - \sec^2\theta}{\operatorname{cosec}^2\theta + \sec^2\theta}$ is

(a) – 1 (b) 1 (c) $\dfrac{1}{2}$ (d) $-\dfrac{1}{2}$

14. The value of $\sin^2 30° + \cos^2 45° + \cos^2 30°$ is

(a) $\dfrac{1}{2}$ (b) $\dfrac{\sqrt{3}}{2}$ (c) $\dfrac{3}{2}$ (d) $\dfrac{2}{3}$

15. ΔABC is right-angled at C. The value of $\cos(A + B)$ is

(a) 0 (b) Not defined (c) – 1 (d) 1

16. If $\tan A = \cot B = \dfrac{15}{7}$, then $A + B$ is equal to

(a) Zero (b) 90° (c) $< 90°$ (d) $> 90°$

17. In the figure, $AD = 3$ cm, $BD = 4$ cm and $CB = 12$ cm, then $\tan\theta$ equals

(a) $\dfrac{3}{4}$ (b) $\dfrac{5}{12}$

(c) $\dfrac{4}{3}$ (d) $\dfrac{12}{5}$

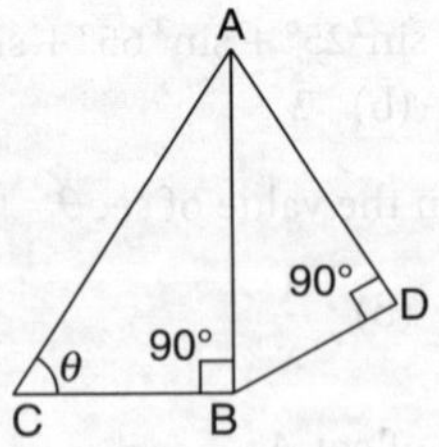

18. The value of $\dfrac{\cot 45°}{\sin 30° + \cos 60°}$ is equal to

(a) 1 (b) $\dfrac{1}{\sqrt{2}}$ (c) $\dfrac{2}{3}$ (d) $\dfrac{1}{32}$

19. Which of the following is not defined?

(a) sin 90° (b) tan 0° (c) cot 90° (d) cosec 0°

20. If $\sin\phi = \frac{1}{3}$, then the value of $2\cot^2\phi + 2$ is equal to

(a) 6 (b) 9 (b) 4 (b) 18

21. If $\sin\theta = \frac{1}{2}$, then the value of $\sin\theta(\sin\theta - \text{cosec}\,\theta)$ is

(a) $\frac{3}{4}$ (b) $\frac{-3}{4}$ (c) $\frac{\sqrt{3}}{2}$ (d) $\frac{-\sqrt{3}}{2}$

22. $(\sec A + \tan A)(1 - \sin A)$ is equal to

(a) sec A (b) tan A (c) sin A (d) cos A

23. If $\sin A + \sin^2 A = 1$, then the value of $\cos^2 A + \cos^4 A$ is

(a) 2 (b) 1 (c) – 2 (d) 0

24. If $\sin(\theta + 36°) = \cos\theta$, where θ and $(\theta + 36°)$ are acute angles, then the value of θ is

(a) 36° (b) 54° (c) 27° (d) 90°

25. If $3\cos\theta = 2\sin\theta$, then the value of $\frac{4\sin\theta - 3\cos\theta}{2\sin\theta + 6\cos\theta}$ is

(a) $\frac{1}{8}$ (b) $\frac{1}{3}$ (c) $\frac{1}{2}$ (d) $\frac{1}{4}$

26. In this figure, ΔABC is right-angled at B and $\tan A = \frac{4}{3}$. If $AC = 15$ cm, the length of BC is

(a) 4 cm (b) 3 cm

(c) 12 cm (d) 9 cm

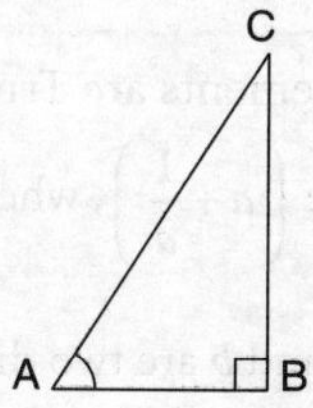

27. In the given figure, $\angle ACB = 90°$, $\angle BDC = 90°$, $CD = 4$ cm, $BD = 3$ cm, $AC = 12$ cm. Then, $\cos A - \sin A$ is equal to

(a) $\frac{5}{12}$ (b) $\frac{5}{13}$

(c) $\frac{7}{12}$ (d) $\frac{7}{13}$

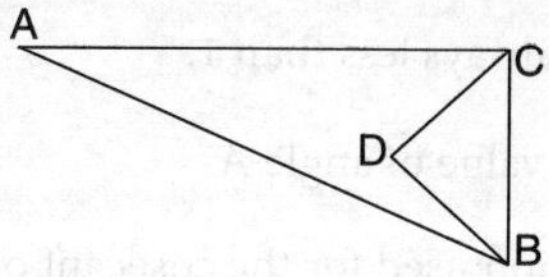

28. $2\cos^2 30° - 1 = ?$

(a) sin 60° (b) cos 60° (c) tan 60° (d) sec 60°

Fill in the Blanks

Complete the following statements with appropriate answer(s) in the blank space(s):

29. When θ increases, the value of sin θ ________, where $0 \le \theta \le 90°$.
30. If x increases from 0° to 90°, then the value of cos x is ________.
31. If θ is a complementary angle, then sec θ is equal to ________.
32. The value of cot θ decreases from ________, where $0 \le \theta \le 90°$.
33. cos (90° – θ) is equal to ________.
34. In a rectangle $ABCD$, $AB = 40$ cm, $\angle BAC = 30°$, then side BC is ________.
35. An equation involving trigonometric ratios of an angle is called ________.
36. The value of $2(\sin^6\theta + \cos^6\theta) - 3(\sin^4\theta + \cos^4\theta) + 1 =$ ________.
37. Suppose $0 < \theta, \phi < 90°$; sin θ and cos ϕ will be equal, when θ and ϕ equal to ________.
38. If $\cos(40° + x) = \sin 30°$, then the value of x is ________.
39. The minimum value of sec A and cosec A is ________.
40. The value of $\frac{\sin^2(90° - \theta)}{\cos^2\theta}$ is equal to ________.
41. If θ is an acute angle and $\sin\theta = \cos\theta$, then the value of $2\tan^2\theta + \cos^2\theta - 2$ is ________.
42. If $\operatorname{cosec}\theta - \cot\theta = \frac{1}{3}$, the value of $(\operatorname{cosec}\theta + \cot\theta)$ is ________.
43. The positive minimum value of sec θ is ________.

True/False

State whether the following statements are True/False and also justify:

44. The value of $2\sin\theta$ can be $\left(a + \frac{1}{a}\right)$, where a is a positive number and $a \ne 1$. **[NCERT Exemplar]**
45. $\cos\theta = \frac{a^2 + b^2}{2ab}$, where a and b are two distinct numbers such that $ab > 0$. **[NCERT Exemplar]**
46. The value of the expression $(\cos^2 23° - \sin^2 67°)$ is positive. **[NCERT Exemplar]**
47. The value of the expression $(\sin 80° - \cos 80°)$ is negative. **[NCERT Exemplar]**
48. (i) The value of tan A is always less than 1.
    (ii) $\sec A = \frac{12}{5}$ for some value of angle A.
    (iii) cos A is the abbreviation used for the cosecant of angle A.
    (iv) cot A is the product of cot and A.
    (v) $\sin\theta = \frac{4}{3}$ for some angle θ. **[NCERT Exemplar]**
49. $(\tan\theta + 2)(2\tan\theta + 1) = 5\tan\theta + 2\sec^2\theta$. **[NCERT Exemplar]**

50. The value of $\sin\theta$ is $\left(a+\frac{1}{a}\right)$, where a is a positive number. **[NCERT Exemplar]**

51. $\tan\theta$ increases faster than $\sin\theta$ as θ increases. **[NCERT Exemplar]**

Assertion–Reason

The following questions consist of two statements – Assertion (A) and Reason (R). Answer these questions by selecting the appropriate option given below.

(A) Both A and R are true and R is the correct explanation for A.
(B) Both A and R are true and R is not the correct explanation for A.
(C) A is true but R is false.
(D) A is false but R is true.

52. **Assertion (A):** The value of the expression $\left(\frac{3\sin 35^\circ}{\cos 55^\circ}\right)^2+\left(\frac{2\cos 55^\circ}{\sin 35^\circ}\right)^2-2\cos 60^\circ$ is 12.
Reason (R): $\sin^2\theta+\cos^2\theta=1$.

53. **Assertion (A):** If $\cos A+\cos^2 A=1$, then $\sin^2 A+\sin^4 A=2$.
Reason (R): $1-\sin^2 A=\cos^2 A$, for any value of A.

54. **Assertion (A):** If $\sec 2A=\operatorname{cosec}(A-27^\circ)$, where $2A$ is an acute angle, then the measure of $\angle A$ is 39°.
Reason (R): $\operatorname{cosec}(90^\circ-\theta)=\sec\theta$.

55. **Assertion (A):** If $\sin\theta=\frac{a}{b}$, then $\cos\theta=\frac{\sqrt{b^2-a^2}}{b}$.
Reason (R): $\sin^2\theta+\cos^2\theta=1$

56. **Assertion (A):** If $\cos A+\cos^2 A=1$, then $\sin^2 A+\sin^4 A=1$.
Reason (R): $1+\sin^2 A=\cos^2 A$, for any value of A.

57. **Assertion (A):** If A, B and C are interior angles of triangle ABC, then $\sin\left(\frac{B+C}{2}\right)=\cos\frac{A}{2}$.
Reason (R): $\sin(90^\circ-\theta)=\cos\theta$.

58. **Assertion (A):** If in a right-angled ΔABC, $\tan B=\frac{12}{5}$, then $\sin B=\frac{12}{13}$.
Reason (R): In a right-angled triangle, $(\text{Hypotenuse})^2=(\text{Perpendicular})^2+(\text{Base})^2$

59. **Assertion (A):** $\operatorname{cosec} 69^\circ+\cot 69^\circ$ can be expressed as $\sec 21^\circ+\tan 21^\circ$ in terms of angles between 0° and 45°.
Reason (R): $\operatorname{cosec}(90^\circ-\theta)=\sec\theta$ and $\cot(90^\circ-\theta)=\tan\theta$.

60. **Assertion (A):** The value of $\sin^2 60^\circ-\sin^2 30^\circ$ is $\frac{1}{2}$.
Reason (R): $\sin^2\theta+\cos^2\theta=1$.

61. **Assertion (A):** In ΔABC, right-angled at A, if $AB=5$, $AC=12$ and $BC=13$, then $\sin B=\frac{5}{13}$.
Reason (R): $\sin\theta=\frac{\text{Perpendicular}}{\text{Hypotenuse}}$

62. **Assertion (*A*):** $(\sin^4\theta - \cos^4\theta + 1)\ \text{cosec}^2\ \theta = 2$.
Reason (*R*): $\sin^4\theta .\text{cosec}^2\theta = 1$.

63. **Assertion (A):** If $\cos\theta = \frac{5}{13}$, then $\frac{\frac{1}{\cos\theta}}{\sin\theta - \tan\theta} = \frac{169}{96}$.

Reason (R): $(\text{Hypotenuse})^2 = (\text{Perpendicular})^2 + (\text{Base})^2$

64. **Assertion (A):** If $\sin\theta - \cos\theta = 0$, then $\sin^4\theta + \cos^4\theta = \frac{1}{2}$.
Reason (R): $\sin\theta = \cos\theta$ at $\theta = 45°$.

65. **Assertion (A):** The value of $\sin\theta = \frac{4}{3}$ is not possible.
Reason (R): Hypotenuse is the largest side in any right-angled triangle.

66. **Assertion (A):** If $\sin x + \cos y = 1$; $x = 30°$ and y is an acute angle, then y is 60°.
Reason (R): $\sin 30° = \frac{1}{2}$ and $\cos 60° = \frac{1}{2}$.

67. **Assertion (A):** If $\cot\theta + \frac{1}{\cot\theta} = 7$, then the value of $\cot^2\theta + \frac{1}{\cot^2\theta}$ is 47.
Reason (R): $\tan\theta\cot\theta = 1$.

Subjective Questions

68. Consider ΔACB, right-angled at C, in which $AB = 29$ units, $BC = 21$ units and $\angle ABC = \theta$ (see figure). Determine the values of

(i) $\cos^2\theta + \sin^2\theta$ (ii) $\cos^2\theta - \sin^2\theta$ **[NCERT]**

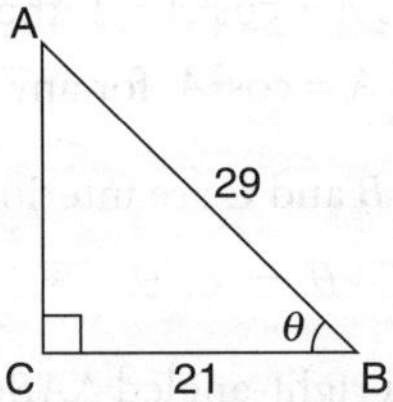

69. In ΔOPQ, right-angled at P, $OP = 7$ cm and $OQ - PQ = 1$ cm (see figure). Determine the values of $\sin Q$ and $\cos Q$. **[NCERT]**

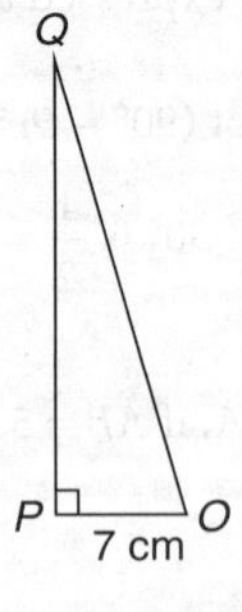

70. If 3 cot $A = 4$, check whether $\dfrac{1-\tan^2 A}{1+\tan^2 A} = \cos^2 A - \sin^2 A$ or not. **[NCERT]**

71. In ΔPQR, right-angled at Q, $PR + QR = 25$ cm and $PQ = 5$ cm. Determine the values of sin P, cos P and tan P. **[NCERT]**

72. In the given figure, $AD = 4$ cm, $BD = 3$ cm and $CB = 12$ cm. Find cot θ. **[Delhi 2008]**

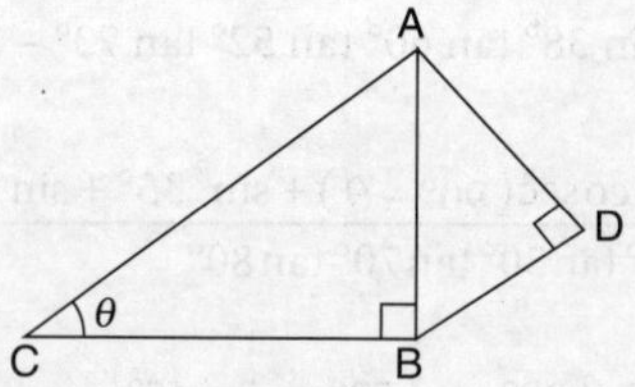

73. (a) If $\sec A = \dfrac{15}{7}$ and $A + B = 90°$, find the value of cosec B. **[Foreign 2009]**

(b) If $3x = \text{cosec}\,\theta$ and $\dfrac{3}{x} = \cot\theta$, find the value of $3\left(x^2 - \dfrac{1}{x^2}\right)$. **[Delhi 2011]**

(c) If $6x = \sec\theta$ and $\dfrac{6}{x} = \tan\theta$, find the value $9\left(x^2 - \dfrac{1}{x^2}\right)$. **[Foreign 2010]**

74. If the ratio of the sine of an angle to its cosine is 3:4, find their actual values.

75. If $\cot\theta = \dfrac{1}{\sqrt{3}}$, show that $\dfrac{\sin^2\theta}{1+\cos^2\theta} = \dfrac{3}{5}$.

76. Find the value of other trigonometric ratios, given that

(i) $\tan\theta = \dfrac{2mn}{m^2-n^2}$ (ii) $\sin\theta = \dfrac{m^2-n^2}{m^2+n^2}$ (iii) $\cos\theta = \dfrac{2m}{m^2+1}$ (iv) $\text{cosec}\,\theta = \sqrt{1+\left(\dfrac{m}{n}\right)^4}$

77. If $\tan\theta = \dfrac{p}{q}$, show that $\dfrac{p\sin\theta - q\cos\theta}{p\sin\theta + q\cos\theta} = \dfrac{p^2-q^2}{p^2+q^2}$.

78. If $\cot\theta = \sqrt{7}$, show that $\dfrac{\text{cosec}^2 - \sec^2\theta}{\text{cosec}^2\theta + \sec^2\theta} = \dfrac{3}{4}$.

79. In the figure, $AD = DB$ and $\angle B$ is a right angle. Determine $\sin^2\theta + \cos^2\theta$.

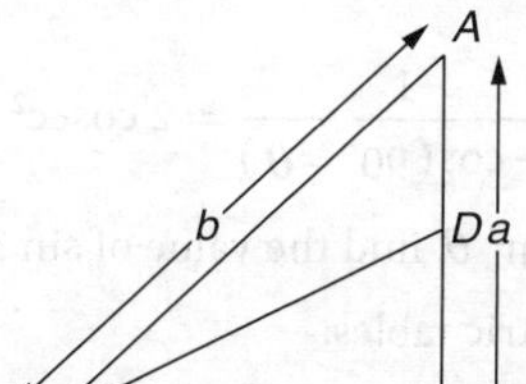

80. If $\tan\theta = \frac{a}{b}$, prove that $\frac{\cos\theta + \sin\theta}{\cos\theta - \sin\theta} = \frac{b+a}{b-a}$.

81. Without using trigonometric tables, evaluate each of the following:

(i) $\frac{\cos^2 20° + \cos^2 70°}{\sec^2 50° - \cot^2 40°} + 2\,\text{cosec}^2\, 58° - 2\cot 58° \tan 32° - 4\tan 13° \tan 37° \tan 45° \tan 53° \tan 77°$ **[CBSE 2006C]**

(ii) $\frac{\sec 39°}{\text{cosec}\, 51°} + \frac{2}{\sqrt{3}} \tan 17° \tan 38° \tan 60° \tan 52° \tan 73° - 3(\sin^2 31° + \sin^2 59°)$ **[CBSE 2006C]**

(iii) $\frac{-\tan\theta \cot(90° - \theta) + \sec\theta\, \text{cosec}(90° - \theta) + \sin^2 35° + \sin^2 55°}{\tan 10° \tan 20° \tan 30° \tan 70° \tan 80°}$ **[CBSE 2005C]**

(iv) $\frac{\sec^2 54° - \cot^2 36°}{\text{cosec}^2 57° - \tan^2 33°} + 2\sin^2 38° \sec^2 52° - \sin^2 45°$ **[CBSE 2005C]**

82. If $\tan A = \cot B$, prove that $A + B = 90°$.

83. If $\sec 4A = \text{cosec}(A - 20°)$, where $4A$ is an acute angle, find the value of A.

84. Without using the trigonometric tables, evaluate

$$\frac{7}{2} \cdot \frac{\cos 70°}{\sin 20°} + \frac{3}{2} \cdot \frac{\cos 55° \,\text{cosec}\, 35°}{\tan 5° \cdot \tan 25° \cdot \tan 45° \cdot \tan 85° \cdot \tan 65°}$$ **[Foreign 2009]**

85. Without using the trigonometric tables, find the given expression:

$$\frac{\sec(90° - \theta) \cdot \text{cosec}\,\theta - \tan(90° - \theta)\cot\theta + \cos^2 25° + \cos^2 65°}{3\tan 27° \cdot \tan 63°}$$

86. Find the value of cosec 30° geometrically. **[Delhi 2010]**

87. If $\sin\theta + \cos\theta = \sqrt{2}\cos(90 - \theta)$, determine $\cot\theta$.

88. Prove that: $\frac{\sin(90° - \theta) \cdot \sin\theta}{\tan\theta} - 1 = -\sin^2\theta$.

89. Prove that: $\tan 10° \cdot \tan 75° \cdot \tan 15° \cdot \tan 80° = 1$.

90. Prove that: $\tan 1° \cdot \tan 2° \cdot \tan 3 \ldots \tan 89° = 1$.

91. Prove that: $\cos 1° \cdot \cos 2° \cdot \cos 3° \ldots \cos 180° = 0$.

92. Prove that: $\sin^2 5° + \sin^2 10° + \ldots + \sin^2 85° + \sin^2 90° = 9\frac{1}{2}$.

93. Prove that: $(\sin 72° + \cos 18°)(\sin 72° - \cos 18°) = 0$.

94. Prove that: $\frac{\cos(90° - A) \cdot \sin(90° - A)}{\tan(90° - A)} = \sin^2 A$.

95. Prove that: $\frac{1}{1 + \cos(90° - \theta)} + \frac{1}{1 - \cos(90° - \theta)} = 2\,\text{cosec}^2(90° - \theta)$.

96. Using the formula $\cos 2\theta = 1 - 2\sin^2\theta$, find the value of sin 30°, if cos 60° = 1/2.

97. Evaluate without using trigonometric tables:

$$\frac{\sec^2(90° - \theta) - \cot^2\theta}{2(\sin^2 25° + \sin^2 65°)} + \frac{2\cos^2 60° \tan^2 28° \tan^2 62°}{3(\sec^2 43° - \cot^2 47°)}$$ **[Delhi 2006]**

98. Evaluate without using trigonometric tables:

$$\frac{\operatorname{cosec}^2(90°-\theta)-\tan^2\theta}{4(\cos^2 48°+\cos^2 42°)}-\frac{2\tan^2 30°\sec^2 52°\sin^2 38°}{\operatorname{cosec}^2 70°-\tan^2 20°}$$ **[AI 2006]**

99. Evaluate without using trigonometric tables:

$$\frac{\sin^2\theta+\sin^2(90°-\theta)}{3(\sec^2 61°-\cot^2 29°)}-\frac{3\cot^2 30°\sin^2 54°\sec^2 36°}{2(\operatorname{cosec}^2 65°-\tan^2 25°)}$$ **[Foreign 2006]**

100. Without using trigonometric tables, evaluate

$$\frac{\cos 60°+\sin 45°-\cot 30°}{\tan 60°+\sec 45°-\operatorname{cosec} 30°}+\frac{\sin 30°+\sin 45°-\tan 60°}{\cot 30°+\operatorname{cosec} 45°-\sec 60°}$$

101. Find the value of x if $4\left(\frac{\sec^2 59°-\cot^2 31°}{3}\right)-\frac{2}{3}\sin 90°+3\tan^2 56°\times\tan^2 34°=\frac{x}{3}$ **[CBSE 2012]**

102. Evaluate without using trigonometric tables:

$$\frac{3\cos 55°}{7\sin 35°}-\frac{4(\cos 70°\cdot\operatorname{cosec} 20°)}{7(\tan 5°\cdot\tan 25°\cdot\tan 45°\cdot\tan 65°\cdot\tan 85°)}$$ **[Delhi 2007]**

103. Evaluate without using trigonometric tables:

$$\tan 7°\,.\,\tan 23°\,.\,\tan 60°\tan 67°\tan 83°+\frac{\cot 54°}{\tan 36°}+\sin 20°\,.\,\sec 70°-2$$ **[AI 2007]**

104. Without using trigonometric table, evaluate $\frac{\cot 58°}{\sin 32°}+\frac{\sin 22°}{\cot 68°}-\frac{\cos 38°\operatorname{cosec} 52°}{\tan 18°\cdot\tan 35°\cdot\tan 60°\tan 72°\cdot\tan 55°}$ **[Delhi 2009]**

105. Evaluate: $\frac{2}{3}\operatorname{cosec}^2 58°-\frac{2}{3}\cot 58°.\tan 32°-\frac{5}{3}\tan 13°.\tan 37°.\tan 45°.\tan 53°.\tan 77°$. **[Delhi 2009]**

106. If A + B = 90°, prove that $\sqrt{\frac{\tan A\tan B+\tan A\cot B}{\sec A\sec B}-\frac{\sin^2 B}{\cos^2 A}}=\tan A=\tan A.$ **[CBSE (CCE) 2012]**

107. If tan $(A+B)=1$ and sin $(2A-B)=1$, find A and B.

108. Show that: $2(\cos^2 45°+\tan^2 60°)-6(\sin^2 45°-\tan^2 30°)=6$.

109. In ΔPQR, right-angled at Q (see figure), $PQ=3$ cm and $PR=6$ cm. Determine $\angle QPR$ and $\angle PRQ$. **[NCERT]**

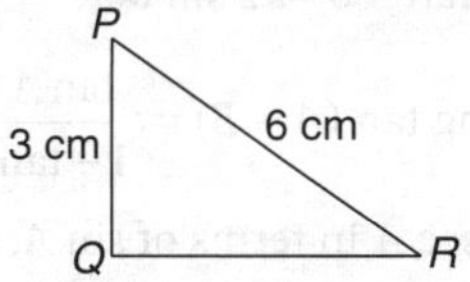

110. Given $A=30°$, verify $\sin 2A=2\sin A\cos A$.

111. If $\cos(40°+x)=\sin 30°$, find the value of x.

112. $\sin 4A=\cos(A-20°)$, where $4A$ is an acute angle. Find the value of A.

113. Given that: $\cos(A-B)=\cos A\cos B+\sin A\sin B$, find the value of cos 15°.

114. If $\tan\theta=\cot(30°+\theta)$, find the value of θ. **[AI 2008]**

115. If $\sin 3\theta = \cos(\theta - 6°)$ and 3θ and $(\theta - 6°)$ are acute angles, find the value of θ. **[AI 2008]**

116. Without using the trigonometric tables, evaluate:

(i) $\dfrac{11\sin 70°}{7\cos 20°} - \dfrac{4}{7}\dfrac{\cos 53° \operatorname{cosec} 37°}{\tan 15° \tan 35° \tan 55° \cdot \tan 75°}$ **[Delhi 2008]**

(ii) $\dfrac{\sin 18°}{\cos 72°} + \sqrt{3} \cdot [\tan 10° \cdot \tan 30° \cdot \tan 40° \cdot \tan 50° \cdot \tan 80°]$ **[Delhi 2008]**

(iii) $(\sin^2 25° + \sin^2 65°) + \sqrt{3}\,(\tan 5° \tan 15° \tan 30° \tan 75° \tan 85°)$ **[AI 2008]**

(iv) $(\cos^2 25° + \cos^2 65°) + \operatorname{cosec}\theta \cdot \sec(90° - \theta) - \cot\theta \cdot \tan(90° - \theta)$ **[AI 2008]**

(v) If $7\sin^2\theta + 3\cos^2\theta = 4$, then show that $\tan\theta = \dfrac{1}{\sqrt{3}}$. **[AI 2008]**

(vi) If $\sec 4A = \operatorname{cosec}(A - 20°)$, where 4A is an acute angle, find the value of A. **[Foreign 2008]**

117. In a ΔABC right-angled at C, if $\tan A = \dfrac{1}{\sqrt{3}}$, find the value of $\sin A \,.\, \cos B + \cos A \sin B$.

118. Evaluate: $\dfrac{\tan^2 60° + 4\sin^2 45° + 3\sec^2 30° + 5\cos^2 90°}{\operatorname{cosec} 30° + \sec 60° - \cot^2 30°}$. **[CBSE (CCE) 2011]**

119. If $A = 60°$ and $B = 30°$, verify that $\sin(A - B) = \sin A \cos B - \cos A \sin B$. **[CBSE (CCE) 2012]**

120. Find A and B, if

(i) $\sin(A + B) = 1$ and $\cos(A - B) = 1$ (ii) $\sin(A + 2B) = \dfrac{\sqrt{3}}{2}$ and $\cos(A + B) = \dfrac{1}{2}$

(iii) $\tan(A - B) = \dfrac{1}{\sqrt{3}}$ and $\tan(A + B) = \sqrt{3}$

121. If $\operatorname{cosec}(A - B) = 2$, $\cot(A + B) = \dfrac{1}{\sqrt{3}}$, $0° < (A + B) < 90°$, $A > B$, then find A and B. **[CBSE (CCE) 2011]**

122. Find the value of x if $4\left(\dfrac{\sec^2 59° - \cot^2 31°}{3}\right)\sin 90° + 3\tan^2 56° \times \tan^2 34° = \dfrac{x}{3}$. **[CBSE (CCE) 2012]**

123. Determine the value of x such that $2\operatorname{cosec}^2 30° + x\sin^2 60° - \dfrac{3}{4}\tan^2 30° = 10$. **[CBSE (CCE) 2011]**

124. In an acute-angled triangle ABC, if $\sin(A + B - C) = \dfrac{1}{2}$ and $\cos(B + C - A) = \dfrac{1}{\sqrt{2}}$ find $\angle A$, $\angle B$ and $\angle C$. **[CBSE (CCE) 2012]**

125. Prove that $(\sqrt{3} + 1)(3 - \cot 30°) = \tan^3 60° - 2\sin 60°$.

126. If $\tan A = \dfrac{1}{2}$ and $\tan B = \dfrac{1}{3}$, by using $\tan(A + B) = \dfrac{\tan A + \tan B}{1 - \tan A \tan B}$, prove that $A + B = 45°$. **[NCERT]**

127. Express the ratios $\cos A$, $\tan A$ and $\sec A$ in terms of $\sin A$. **[NCERT]**

128. Prove that $\dfrac{\sin\theta - \cos\theta + 1}{\sin\theta + \cos\theta - 1} = \dfrac{1}{\sec\theta - \tan\theta}$, using the identity $\sec^2\theta = 1 + \tan^2\theta$. **[NCERT]**

129. Express the trigonometric ratios $\sin A$, $\sec A$ and $\tan A$ in terms of $\cot A$. **[NCERT]**

130. If $a\cos^3\theta + 3a\cos\theta\sin^2\theta = m$, $a\sin^3\theta + 3a\cos^2\theta\sin\theta = n$, prove that $(m + n)^{2/3} + (m - n)^{2/3} = 2a^{2/3}$.

131. Write all the other trigonometric ratios of $\angle A$ in terms of sec A. **[NCERT]**

132. Prove that: $\sec^2\theta - \left[\dfrac{\sin^2\theta - 2\sin^4\theta}{2\cos^4 - \cos^2\theta}\right] = 1.$ **[Foreign 2009]**

133. Prove the following identities, where the angles involved are acute angles for which the expressions are defined:

(i) $\dfrac{\tan\theta}{1-\cot\theta} + \dfrac{\cot\theta}{1-\tan\theta} = 1 + \sec\theta\,\text{cosec}\,\theta$

[**Hint:** With the expression in terms of sin θ and cos θ]

(ii) $\sqrt{\dfrac{1+\sin A}{1-\sin A}} = \sec A + \tan A$

(iii) $(\text{cosec}\,A - \sin A)(\sec A - \cos A) = \dfrac{1}{\tan A + \cot A}$

[**Hint:** Simplify LHS and RHS separately]

(iv) $\left(\dfrac{1+\tan^2 A}{1+\cot^2 A}\right) = \left(\dfrac{1-\tan A}{1-\cot A}\right)^2 = \tan^2 A$

134. Show that $\tan^4\theta + \tan^2\theta = \sec^4\theta - \sec^2\theta$ for $0° < \theta < 90°$.

135. If $\sec^2\theta(1 + \sin\theta)(1 - \sin\theta) = k$, then find the value of k. **[AI 2009]**

136. If $\sec\theta + \tan\theta = m$ and $\sec\theta - \tan\theta = n$, find the value of $\sqrt{mn}$.

137. Show that $(1 + \cot\theta - \text{cosec}\,\theta)(1 + \tan\theta + \sec\theta) = 2$. **[Delhi 2019]**

138. If $\sin\theta + \cos\theta = p$ and $\sec\theta + \text{cosec}\,\theta = q$, show that $q(p^2 - 1) = 2p$.

139. If $\cos\theta + \sin\theta = \sqrt{2}\cos\theta$, show that $\cos\theta - \sin\theta = \sqrt{2}\sin\theta$.

140. Show that: $\dfrac{\sin^2 A}{\cos^2 A} + \dfrac{\cos^2 A}{\sin^2 A} = \dfrac{1}{\sin^2 A.\cos^2 A} - 2.$

141. If $\dfrac{\cos\alpha}{\cos\beta} = m$ and $\dfrac{\cos\alpha}{\sin\beta} = n$, show that $(m^2 + n^2)\cos^2\beta = n^2$. **[CBSE (CCE) 2012]**

142. Prove that $\cos^4 A - \cos^2 A = \sin^4 A - \sin^2 A$.

143. Prove that $\sin^4 A + \cos^4 A = 1 - 2\sin^2 A\cos^2 A$.

144. Prove that $\sin^6 A + \cos^6 A = 1 - 3\sin^2 A\cos^2 A$.

145. Prove that $\sin^8\theta - \cos^8\theta = (\sin^2\theta - \cos^2\theta)(1 - 2\sin^2\theta\cos^2\theta)$

146. Show that $\text{cosec}^2\theta + \sin^2\theta$ can never be less than 2.

147. If $\sqrt{3}\cot^2\theta - 4\cot\theta + \sqrt{3} = 0$, then find the value of $\cot^2\theta + \tan^2\theta$.

148. Prove that $\sin^2 A + \dfrac{1}{1+\tan^2 A} = 1.$

149. Prove that $\sec^6\theta = \tan^6\theta + 3\tan^2\theta\sec^2\theta + 1.$

150. Prove that $\dfrac{1+\cos\theta - \sin^2\theta}{\sin\theta(1+\cos\theta)} = \cot\theta.$

151. Prove that $\left(\tan\theta + \frac{1}{\cos\theta}\right)^2 + \left(\tan\theta - \frac{1}{\cos\theta}\right)^2 = 2\left(\frac{1+\sin^2\theta}{1-\sin^2\theta}\right)$.

152. Prove that (sec A – cosec A) (1 + tan A + cot A) = tan A sec A – cot A cosec A.

153. Prove that $\frac{\tan A + \tan B}{\cot A + \cot B} = \tan A \tan B$.

154. Prove that $\sec^4 A(1 - \sin^4 A) - 2\tan^2 A = 1$.

155. If $x = a \sin\theta$, $y = b \tan\theta$, prove that $\frac{a^2}{x^2} - \frac{b^2}{y^2} = 1$.

156. Prove that:

$$\frac{1}{\operatorname{cosec} A - \cot A} + \frac{1}{\operatorname{cosec} B - \cot B} + \frac{1}{\operatorname{cosec} C - \cot C} = \operatorname{cosec} A + \operatorname{cosec} B + \operatorname{cosec} C + \cot A + \cot B + \cot C$$

157. Show that $\sec^2\theta - \left[\frac{\sin^2\theta - 2\sin^4\theta}{2\cos^4\theta - \cos^2\theta}\right] = 1$. **[Foreign 2009]**

158. Show that: $(\sin\theta + \operatorname{cosec}\theta)^2 + (\cos\theta + \sec\theta)^2 = 7 + \tan^2\theta + \cot^2\theta$. **[NCERT, Delhi 2019]**

159. If $a\cos\theta - b\sin\theta = x$ and a $\sin\theta + b\cos\theta = y$, prove that $a^2 + b^2 = x^2 + y^2$.

160. Prove that: $\frac{\operatorname{cosec} A}{\operatorname{cosec} A - 1} + \frac{\operatorname{cosec} A}{\operatorname{cosec} A + 1} = 2 + 2\tan^2 A = 2\sec^2 A$.

161. If $\cos\alpha = \frac{1}{2}$ and $\tan\beta = \frac{1}{\sqrt{3}}$, find $\sin(\alpha + \beta)$, where α and β are acute angles.

162. Find the value of: $\sin(50° + \theta) - \cos(40° - \theta)$.

163. If $x = \gamma\cos\alpha\sin\beta$, $y = \gamma\cos\alpha\cos\beta$ and $z = \gamma\sin\alpha$, show that $x^2 + y^2 + z^2 = \gamma^2$.

164. In the figure, ABC is a right-angled triangle; D is the mid-point of BC. Show that $\frac{\tan\theta}{\tan\phi} = \frac{1}{2}$.

165. Prove that: $\frac{\sin\theta + \cos\theta}{\sin\theta - \cos\theta} + \frac{\sin\theta - \cos\theta}{\sin\theta + \cos\theta} = \frac{2\sec^2\theta}{\tan^2\theta - 1}$.

166. Prove that: $\frac{1}{\operatorname{cosec}\theta - \cot\theta} - \frac{1}{\sin\theta} = \frac{1}{\sin\theta} - \frac{1}{\operatorname{cosec}\theta + \cot\theta}$.

167. Prove that: $\frac{\cot A - \cos A}{\cot A + \cos A} = \frac{\operatorname{cosec} A - 1}{\operatorname{cosec} A + 1}$. **[Delhi 2008]**

168. Prove that: $\sin\theta(1 + \tan\theta) + \cos\theta(1 + \cot\theta) = \sec\theta + \operatorname{cosec}\theta$. **[AI 2008, 2010]**

169. Prove that: (1 + cot A + tan A) (sin A – cos A) = sin A tan A – cot A cos A. **[Foreign 2008]**

170. Prove that: $(\operatorname{cosec}\theta - \sin\theta).(\sec\theta - \cos\theta) = \frac{1}{\tan\theta + \cot\theta}$. **[AI 2008C, Delhi 2010]**

171. Prove that: $\left(\dfrac{1+\sin\theta-\cos\theta}{1+\sin\theta+\cos\theta}\right)^2 = \dfrac{1-\cos\theta}{1+\cos\theta}$.

172. If 5 sin θ + 3 cos θ = 4, find the value of 3 sin θ – 5 cos θ.

173. Prove that: $(\tan A + \text{cosec } B)^2 - (\cot B - \sec A)^2 = 2 \tan A \cot B (\text{cosec } A + \sec B)$.

174. Prove that $\dfrac{(1+\cot A+\tan A)(\sin A-\cos A)}{\sec^3 A-\text{cosec}^3 A} = \sin^2 A.\cos^2 A$.

175. Prove that $\dfrac{1}{\cot^2\theta}+\dfrac{1}{1+\tan^2\theta} = \dfrac{1}{1-\sin^2\theta}-\dfrac{1}{\text{cosec}^2\theta}$.

176. Prove that $\dfrac{\tan^3\theta}{1+\tan^2\theta}+\dfrac{\cot^3\theta}{1+\cot^2\theta} = \sec\theta . \text{cosec}\,\theta - 2\sin\theta\cos\theta$.

177. Prove that $\dfrac{\sin A}{\sec A+\tan A-1}+\dfrac{\cos A}{\text{cosec} A+\cot A-1} = 1$.

178. Prove that $(1 + \cot A + \tan A)(\sin A - \cos A) = \dfrac{\sec A}{\text{cosec}^2 A}-\dfrac{\text{cosec} A}{\sec^2 A}$.

179. Show that $\dfrac{1}{\sec x-\tan x}-\dfrac{1}{\cos x} = \dfrac{1}{\cos x}-\dfrac{1}{\sec x+\tan x}$.

180. If $\sec\theta = x + \dfrac{1}{4x}$, prove that $\sec\theta + \tan\theta = 2x$ or $\dfrac{1}{2x}$. **[CBSE (CCE 2011)]**

181. If $\sin\theta + \sin^2\theta = 1$, prove that $\cos^2\theta + \cos^4\theta = 1$.

182. Prove that $\dfrac{\cos A}{1-\tan A}-\dfrac{\sin^2 A}{\cos A-\sin A} = \sin A + \cos A$.

183. If $\sec\theta + \tan\theta = p$, prove that $\sin\theta = \dfrac{p^2-1}{p^2+1}$.

184. If $\cot A + \dfrac{1}{\cot A} = 2$, then prove that $\cot^2 A + \dfrac{1}{\cot^2 A} = 2$.

185. If cosec θ – sin θ = l and sec θ – cos θ = m, show $l^2m^2(l^2 + m^2 + 3) = 1$.

186. Given that sin θ + 2 cos θ = 1, then prove that 2 sin θ – cos θ = 2.

187. If $\tan\theta = \dfrac{4}{3}$, find the value of $\dfrac{3\sin\theta+2\cos\theta}{3\sin\theta-2\cos\theta}$.

188. If $1 + \sin^2\theta = 3\sin\theta\cos\theta$, then prove that $\tan\theta = 1$ or $\dfrac{1}{2}$.

189. If $\tan\theta + \sec\theta = l$, then prove that $\sec\theta = \dfrac{l^2+1}{2l}$.

190. If 7 cosec θ – 3 cot θ = 7, prove that 7 cot θ – 3 cosec θ = ± 3.

191. If $\dfrac{x}{a}\cos\theta + \dfrac{y}{b}\sin\theta = 1$ and $\dfrac{x}{a}\sin\theta - \dfrac{y}{b}\cos\theta = 1$, prove that $\dfrac{x^2}{a^2}+\dfrac{y^2}{b^2} = 2 = 2$.

192. If $x = a\cos^3\theta$, $y = b\sin^3\theta$, prove that $\left(\frac{x}{a}\right)^{2/3} + \left(\frac{y}{b}\right)^{2/3} = 1$.

193. If $\sin A + \operatorname{cosec} A = 3$, find the value of $\frac{\sin^4 A + 1}{\sin^2 A}$.

194. If $\cot\theta + \tan\theta = x$ and $\sec\theta - \cos\theta = y$, prove that $(x^2y)^{2/3} - (xy^2)^{2/3} = 1$.

195. 1n ΔABC, right-angled at B, if AB=12 cm and BC = 5 cm, find (i) sin A and tan A, (ii) sin C and cot C. [NCERT]

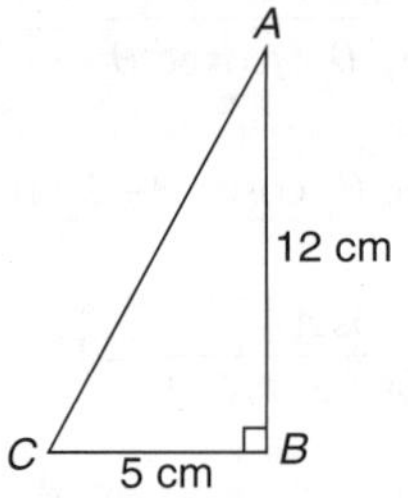

196. If $\angle B$ and $\angle Q$ are acute angles such that sin B = sin Q, then prove that $\angle B = \angle Q$. [NCERT]

197. If tan $2A$ = cot (A – 18°), where $2A$ is an acute angle, find the value of A. [CBSE 2018]

Answers

Multiple Choice Questions

1. (d) **2.** (a) **3.** (c) **4.** (b) **5.** (b) **6.** (b) **7.** (c) **8.** (b) **9.** (d) **10.** (c)
11. (b) **12.** (c) **13.** (c) **14.** (c) **15.** (a) **16.** (b) **17.** (b) **18.** (a) **19.** (d) **20.** (d)
21. (b) **22.** (d) **23.** (b) **24.** (c) **25.** (b) **26.** (c) **27.** (d) **28.** (b)

Fill in the Blanks

29. Increases from 0 to 1. **30.** Decrease **31.** cosec θ **32.** 0° to 90° **33.** sin θ

34. $\frac{40\sqrt{3}}{3}$ cm **35.** Trigonometric identity **36.** 0 **37.** 45° **38.** 20° **39.** 1 **40.** 1

41. $\frac{1}{2}$ **42.** 3 **43.** 1

True/False

44. False **45.** False **46.** False **47.** False
48. (i) False (ii) True (iii) False (iv) False (v) True **49.** True **50.** False **51.** True

Assertion–Reason

52. (B) **53.** (D) **54.** (A) **55.** (A) **56.** (D) **57.** (A) **58.** (A) **59.** (A) **60.** (B) **61.** (D)
62. (C) **63.** (B) **64.** (A) **65.** (A) **66.** (A) **67.** (B)

Subjective Questions

68. (i) 1, (ii) $\frac{41}{841}$ (ii) $\frac{12}{13}, \frac{5}{13}$ **69.** $\frac{7}{25}, \frac{24}{25}$ **70.** Yes

71. $\sin P = \frac{12}{13}$, $\cos P = \frac{5}{13}$, $\tan P = \frac{12}{5}$ **72.** $\frac{12}{5}$ **73.** (a) $\frac{15}{7}$ (b) $\frac{1}{3}$ (c) $\frac{1}{4}$

74. $\sin\theta = \frac{3}{5}$, $\cos\theta = \frac{4}{5}$

76. (i) $\sin\theta = \frac{2mn}{m^2+n^2}$, $\operatorname{cosec}\theta = \frac{m^2+n^2}{2mn}$, $\cot\theta = \frac{m^2-n^2}{2mn}$ (ii) $\cos\theta = \frac{2mn}{m^2+n^2}$, $\tan\theta = \frac{m^2-n^2}{2mn}$

(iii) $\tan\theta = \frac{m^2-1}{2m}$, $\sin\theta = \frac{m^2-1}{m^2+1}$ (iv) $\sin\theta = \frac{n^2}{\sqrt{m^4+n^4}}$, $\cos\theta = \frac{m^2}{\sqrt{m^4+n^4}}$, $\tan\theta = \frac{n^2}{m^2}$

79. 1 **81.** (i) – 1, (ii) 0 (iii) $2\sqrt{3}$ (iv) $\frac{5}{2}$ **83.** 22° **84.** 5. **85.** $\frac{2}{3}$ **86.** 2

87. $\sqrt{2} - 1$ **96.** $\frac{1}{2}$ **97.** $\frac{2}{3}$ **98.** $-\frac{5}{12}$ **99.** $\frac{-25}{6}$ **100.** 1 **101.** 11 **102.** $-\frac{1}{7}$

103. $\sqrt{3}$ **104.** $\frac{2\sqrt{3}-1}{\sqrt{3}}$ **105.** – 1 **107.** $\angle A = 45°, \angle B = 0°$

109. $\angle QPR = 60°$ and $\angle PRQ = 30°$ **111.** $x = 20$

112. 22° **113.** $\frac{\sqrt{3}+1}{2\sqrt{2}}$ **114.** 30° **115.** 24° **116.** (i) 1 (ii) 2 (iii) 2 (iv) 2 (vi) 22°

117. 1 **118.** 9 **120.** (i) $A = 45°, B = 45°$ (ii) $A = 60°, B = 0°$ (iii) $A = 45°, B = 15°$

121. $A = 45°, B = 15°$ **122.** 11 **123.** 3 **124.** 67.5°, 37.5°, 75.5°

127. $\cos A = \sqrt{1-\sin^2 A}$, $\tan A = \frac{\sin A}{\sqrt{1-\sin^2 A}}$, $\sec A = \frac{1}{\sqrt{1-\sin^2 A}}$

129. $\sin A = \frac{1}{\sqrt{1+\cot^2 A}}$, $\tan A = \frac{1}{\cot A}$, $\sec A = \frac{\sqrt{1+\cot^2 A}}{\cot A}$

131. $\sin A = \frac{\sqrt{\sec^2 A - 1}}{\sec A}$, $\cos A = \frac{1}{\sec A}$, $\tan A = \sqrt{\sec^2 A - 1}$, $\cot A = \frac{1}{\sqrt{\sec^2 A - 1}}$, $\operatorname{cosec} A = \frac{\sec A}{\sqrt{\sec^2 A - 1}}$

135. 1 **147.** $\frac{10}{3}$ **136.** + 1 **161.** 1 **162.** 0 **172.** $+3\sqrt{2}$

193. 7 **195.** (i) $\frac{5}{13}, \frac{5}{12}$ (ii) $\frac{12}{13}, \frac{P}{H}$ **197.** 36°

Introduction to Trigonometry (Basic)

SELF-ASSESSMENT TEST

Time allowed: 90 min. Max. Marks: 40

General Instructions

1. *All questions are compulsory. There are 20 questions in all.*
2. *This question paper has four sections: Section A, Section B, Section C and Section D.*
3. *Section A contains 10 questions of one mark each, Section B contains three questions of two marks each, Section C contains four questions of three marks each, and Section D contains three questions of four marks each.*

Section A [1 Mark Each]

Direction: (Q. Nos. 1–5) These are Multiple Choice Questions. Select the most appropriate answer for the given questions.

1. The value of $\sin^2 15° + \sin^2 25° + \sin^2 65° + \sin^2 75°$ is
   (b) 1 (b) 2 (c) 0 (d) 3
2. The value of the expression $\sec^2 50° - \cot^2 40° - \sin^2 45°$ is equal to
   (a) 2 (b) $\frac{1}{2}$ (c) 1 (d) 0°
3. The value of $\frac{\cot 45°}{\sin 30° + \cos 60°}$ is equal to
   (a) 1 (b) $\frac{1}{\sqrt{2}}$ (c) $\frac{2}{3}$ (d) $\frac{1}{32}$
4. Given that $\sin \alpha = \frac{1}{2}$ and $\cos \beta = \frac{\sqrt{3}}{2}$, the value of $\alpha + \beta$ is
   (a) 30° (b) 45° (c) 60° (d) 90°
5. If $\sin \theta = \frac{1}{2}$, then the value of $\sin \theta (\sin \theta - \operatorname{cosec} \theta)$ is
   (a) $\frac{3}{4}$ (b) $\frac{-3}{4}$ (c) $\frac{\sqrt{3}}{2}$ (d) $\frac{-\sqrt{3}}{2}$
6. If θ is a complementary angle, then $\sec \theta$ is equal to __________.
7. If x increases from 0° to 90°, then the value of $\cos x$ is __________.
8. The positive minimum value of $\sec \theta$ is __________.
9. The value of the expression $(\cos^2 23° - \sin^2 67°)$ is positive. Is this true or false? Justify.
10. The value of the expression $(\sin 80° - \cos 80°)$ is negative. Is this true or false? Justify.

Section B [2 Marks Each]

11. If $\cot\theta = \frac{7}{8}$, evaluate: (i) $\frac{(1+\sin\theta)(1-\sin\theta)}{(1+\cos\theta)(1-\cos\theta)}$ (ii) $\cot^2\theta$

12. Evaluate: $\frac{\tan 45^\circ}{\operatorname{cosec} 30^\circ} + \frac{\sec 60^\circ}{\cot 45^\circ} - \frac{2\sin 90^\circ}{\cos 0^\circ}$.

13. Evaluate: $\sin^2 30^\circ \cos^2 45^\circ + 4\tan^2 30^\circ + \frac{1}{2}\sin^2 90^\circ - 2\cos^2 90^\circ + \frac{1}{24}$.

Section C [3 Marks Each]

14. If $\tan(A+B) = \sqrt{3}$ and $\tan(A-B) = \frac{1}{\sqrt{3}}$; $0 < (A+B) < 90^\circ$; $A > B$, find A and B.

15. Evaluate: $\frac{\sin 32^\circ}{\cos 58^\circ} + \frac{\sin 22^\circ}{\cos 68^\circ} - \frac{\cos 48^\circ \cdot \operatorname{cosec} 42^\circ}{\tan 28^\circ \cdot \tan 25^\circ \tan 30^\circ \tan 62^\circ \cdot \tan 65^\circ}$.

16. Prove that $\sin^2 A + \frac{1}{1+\tan^2 A} = 1$.

17. If $\tan A = \frac{1}{2}$ and $\tan B = \frac{1}{3}$, using $\tan(A+B) = \frac{\tan A + \tan B}{1 - \tan A \tan B}$, prove that $A + B = 45^\circ$.

Section D [4 Marks Each]

18. Prove: $\frac{\tan A + \sec A - 1}{\tan A - \sec A + 1} = \frac{1+\sin A}{\cos A}$.

19. If $\cos\theta + \sin\theta = \sqrt{2}\cos\theta$, show that $\cos\theta - \sin\theta = \sqrt{2}\sin\theta$.

20. Prove: $\frac{\sin\theta - 2\sin^3\theta}{2\cos^3\theta - \cos\theta} = \tan\theta$.

Answers

1. (b) **2.** (b) **3.** (a) **4.** (c) **5.** (b) **6.** $\operatorname{cosec}\theta$ **7.** Decrease **8.** 1

9. False **10.** False **11.** (i) $\frac{49}{64}$ (ii) $\frac{49}{64}$ **12.** $\frac{1}{2}$ **13.** 2 **14.** $A = 45^\circ$ and $B = 15^\circ$

15. $2-\sqrt{3}$

Introduction to Trigonometry (Standard)

SELF-ASSESSMENT TEST

Time allowed: 90 min. Max. Marks: 40

General Instructions

1. *All questions are compulsory. There are 20 questions in all.*
2. *This question paper has four sections: Section A, Section B, Section C and Section D.*
3. *Section A contains 10 questions of one mark each, Section B contains three questions of two marks each, Section C contains four questions of three marks each, and Section D contains three questions of four marks each.*

Section A [1 Marks Each]

Direction: (Q. Nos. 1–5) These are Multiple Choice Questions. Select the most appropriate answer for the given questions.

1. If $\sec x + \tan x = p$, then $\sec x$ is equal to

   (a) $\frac{p^2-1}{p}$ (b) $\frac{p^2-1}{2p}$ (c) $\frac{p^2+1}{p}$ (d) $\frac{p^2+1}{2p}$

2. If $\sin\theta - \cos\theta = 0$, then the value of $(\sin^4\theta + \cos^4\theta)$ is

   (a) 1 (b) $\frac{3}{4}$ (c) $\frac{1}{2}$ (d) $\frac{1}{4}$

3. $\cos 1^\circ \cdot \cos 2^\circ \cdot \cos 3^\circ..., \cos 180^\circ$ is equal to

   (a) 1 (b) 0 (c) -1 (d) $\frac{1}{2}$

4. In the figure, $AD = 3$ cm, $BD = 4$ cm and $CB = 12$ cm. Then, $\tan\theta$ equals

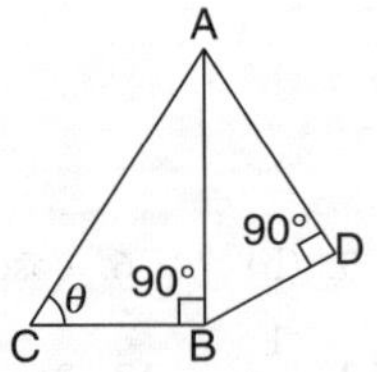

   (a) $\frac{3}{4}$ (b) $\frac{5}{12}$ (c) $\frac{4}{3}$ (d) $\frac{12}{5}$

5. If $\sin A + \sin^2 A = 1$, then the value of $\cos^2 A + \cos^4 A$ is

   (a) 2 (b) 1 (c) -2 (d) 0

6. The values of $\cot\theta$ decreases from __________, where $0 < \theta < 90^\circ$.
7. When θ increases, the value of $\sin\theta$ __________, where $0 \le \theta \le 90^\circ$.
8. The value of $2\sin\theta$ can be $\left(a + \frac{1}{a}\right)$, where a is a positive number, and $a \neq 1$. Is this true or false? Justify.
9. $(\tan\theta + 2)(2\tan\theta + 1) = 5\tan\theta + 2\sec^2\theta$. Is this true or false? Justify.
10. $\tan\theta$ increases faster than $\sin\theta$ as θ increases. Is this true or false? Justify.

Section B [2 Marks Each]

11. If $3x = \text{cosec}\ \theta$ and $\frac{3}{x} = \cot\theta$, find the value of $3\left(x^2 - \frac{1}{x^2}\right)$.

12. In the figure, $AD = DB$ and $\angle B$ is a right angle. Determine $\sin^2\theta + \cos^2\theta$.

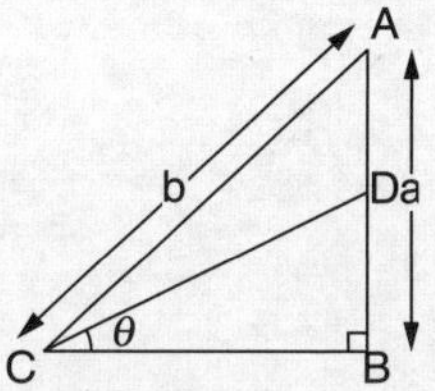

13. Without using trigonometric tables, evaluate:

$$\frac{\cos^2 20° + \cos^2 70°}{\sec^2 50° - \cot^2 40°} + 2\ \text{cosec}^2\ 58° - 2\cot 58° \tan 32° - 4\tan 13° \tan 37° \tan 45° \tan 53° \tan 77°.$$

Section C [3 Marks Each]

14. Find the value of cosec 30° geometrically.

15. Find the value of x if $4\left(\frac{\sec^2 59° - \cot^2 31°}{3}\right) - \frac{2}{3}\sin 90° + 3\tan^2 56° \times \tan^2 34° = \frac{x}{3}$.

16. If $\tan\theta + \sin\theta = m$ and $\tan\theta - \sin\theta = n$, show that $m^2 - n^2 = 4\sqrt{mn}$.

17. Prove that $\frac{\sin\theta - \cos\theta + 1}{\sin\theta + \cos\theta - 1} = \frac{1}{\sec\theta - \tan\theta}$, using the identity $\sec^2\theta = 1 + \tan^2\theta$.

18. Prove the following identities:

(i) $\frac{\cos A}{1 - \sin A} + \frac{\sin A}{1 - \cos A} + 1 = \frac{\sin A \cdot \cos A}{(1 - \sin A)(1 - \cos A)}$

(ii) $\frac{(1 + \cot A + \tan A)(\sin A - \cos A)}{\sec^3 A - \text{cosec}^3 A} = \sin^2 A \cos^2 A$

Section D [4 Marks Each]

19. In an acute-angled triangle ABC, if $\tan(A + B - C) = 1$ and $\sec(B + C - A) = 2$, find the value of A, B and C.

20. (i) If $\sec\theta = x + \frac{1}{4x}$, prove that $\sec\theta + \tan\theta = 2x$ or $\frac{1}{2x}$.

(ii) If $\cos\theta + \sin\theta = \sqrt{2}\cos\theta$, show that $\cos\theta - \sin\theta = \sqrt{2}\sin\theta$.

Answers

1. (d) **2.** (c) **3.** (b) **4.** (b) **5.** (b) **6.** 0° to 90° **7.** Increases from 0 to 1

8. False **9.** True **10.** True **11.** $\frac{1}{3}$ **12.** 1 **13.** −1 **14.** 2 **15.** 11

19. $A = 60°, B = 52\frac{1}{2}°, C = 67\frac{1}{2}°$

CHAPTER

9 Some Applications of Trigonometry

Key Concepts

- Meaning of line of sight
- Meaning of angle of elevation
- Meaning of angle of depression
- To calculate height of building/tower/pole
- To calculate distance of a point from the bottom of a pole/tower/building
- To find angle of elevation or depression
- Distance of the top of a building/tower/pole from the eye of the observer

Meaning of Line of Sight

The figure (i) given below shows a situation in which a person is standing at *A* and is looking at the top of the tower. In figure (ii), the person is at the top of the tower and is looking at an object at point *A*.

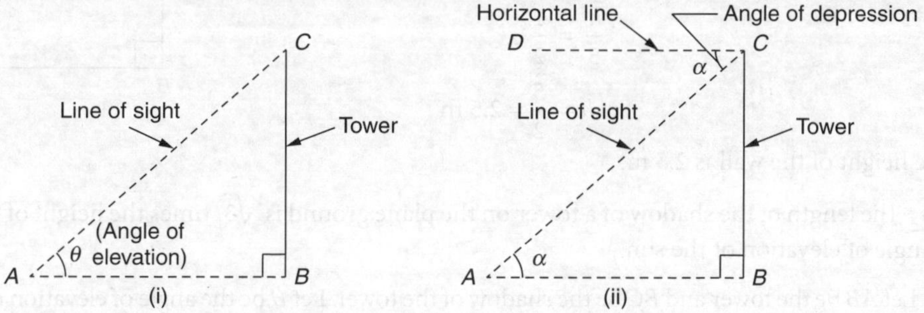

The line drawn from the eye of the person to the point viewed by the person is known as the line of sight. In the given figures, line *AC* is the line of sight.

Meaning of Angle of Elevation

The angle of elevation of a point is the angle made by the line of sight with the horizontal when the point is above the horizontal line. In figure (i), $\angle BAC$ is the angle of elevation as the point viewed, that is, the point *C*, is above the horizontal line *AB*.

Meaning of Angle of Depression

The angle of depression of a point is the angle made by the line of sight with the horizontal when the point is below the horizontal line. In figure (ii), $\angle DCA$ is the angle of depression as the point viewed, that is, the point *A*, is below the horizontal line *CD*. $\angle DCA = \angle BAC$ (alternate interior angles).

Example 1 At a point 25 m away from the foot of the tower, the angle of elevation of the top of the tower is 30°. Find the height of the tower.

Solution: Let *AB* be the tower of height *h* m. *B* is the foot of the tower and point *C* is 25 m away from *B*.

$$\angle ABC = 90°, \angle ACB = 30°$$

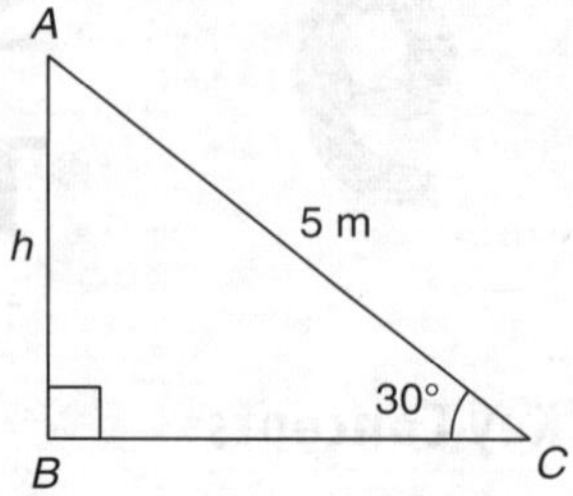

$$BC = 25 \text{ m}$$

In right ΔABC $$\frac{AB}{BC} = \tan 30°$$

$\Rightarrow$ $$\frac{h}{25} = \frac{1}{\sqrt{3}}$$

$\Rightarrow$ $$h = \frac{25}{\sqrt{3}} = 14.43 \text{ m}$$

Hence, the height of the tower is 14.43 m.

Example 2 A ladder is placed against a wall such that it just reaches the top of the wall. The length of the ladder is 5 m and the ladder is inclined at an angle of 30° to the ground. Find the height of the wall.

Solution: Let AC be the ladder and AB be the wall. The top of the ladder is at point A and the foot of the ladder is at C.

Let h be the height of the wall.

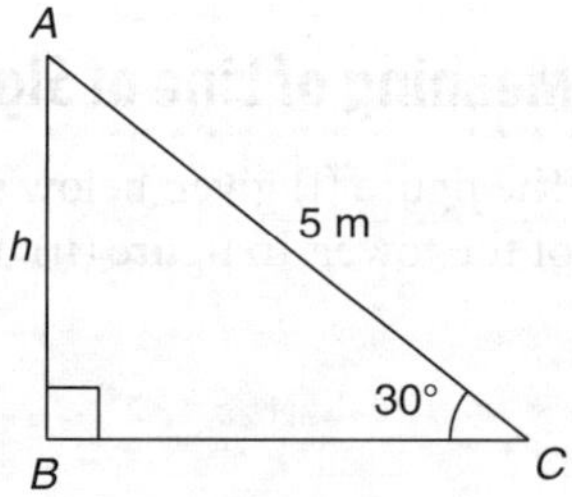

$$\angle ACB = 30°, AC = 5 \text{ m}, AB = h \text{ m}$$

In right ΔABC $$\frac{AB}{AC} = \sin 30°$$

$\Rightarrow$ $$\frac{h}{5} = \frac{1}{2}$$

$\Rightarrow$ $$h = \frac{5}{2} = 2.5 \text{ m}$$

Hence, the height of the wall is 2.5 m.

Example 3 The length of the shadow of a tower on the plane ground is $\sqrt{3}$ times the height of the tower. Find the angle of elevation of the sun.

Solution: Let AB be the tower and BC be the shadow of the tower. Let θ be the angle of elevation of the sun.

Let height of tower = $AB = \sqrt{3}\ h$ units.

$\therefore$ Length of the shadow = $BC = \sqrt{3}\ h$

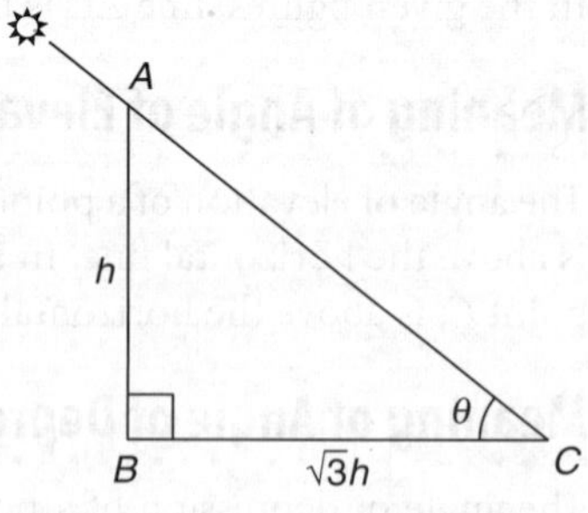

In right ΔABC $$\tan\theta = \frac{AB}{BC}$$

$\Rightarrow$ $$\tan\theta = \frac{h}{\sqrt{3}h} = \frac{1}{\sqrt{3}}$$

$\Rightarrow$ $$\tan\theta = \tan 30°$$

$\therefore$ $$\theta = 30°$$

Hence, the angle of elevation of the sun is 30°.

1 Try to Solve

1. A tower stands vertically on the ground. From a point on the ground, which is 15 m away from the foot of the tower, the angle of elevation of the top of the tower is found to be 60°. Find the height of the tower. **[NCERT]**

2. The length of the shadow of a pillar is $\sqrt{3}$ times its height. Find the angle of elevation of the source of light.
3. A circus artist is climbing a 20-m-long rope, which is tightly stretched and tied from the top of a vertical pole to the ground. Find the height of the pole if the angle made by the rope with the ground is 30° (see figure). **[NCERT]**

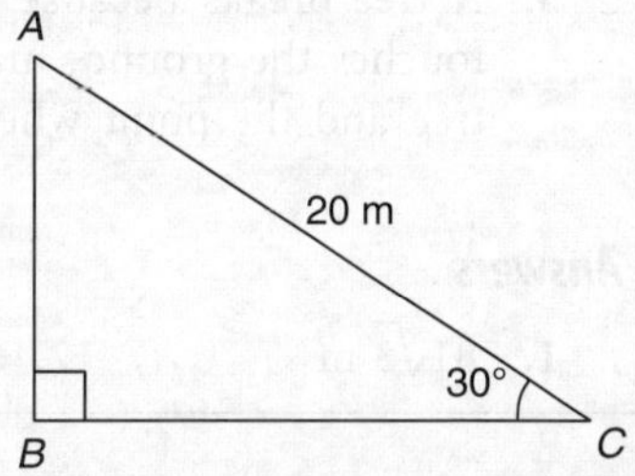

Answers

1. $15\sqrt{3}$ m 2. 30° 3. 10 m

Example 4 A vertical post 9 m high is broken at a certain height, and its upper part, not completely separated, meets the ground at an angle of 30°. Find the height at which the post is broken. **[CBSE (CCE) 2010]**

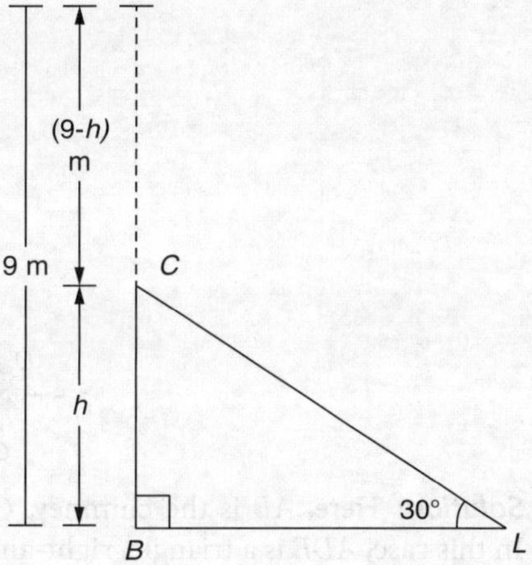

Solution: Let AB be a vertical post of height 9 m which is broken at C at a height of h m from B. The broken part CA meets the ground at D.

$\therefore$ $BC = h$ m

And $AC = DC = (9 - h)$ m

$\angle BDC = 30°$

In right ΔCBD $\sin 30° = \dfrac{CB}{CD}$

$\Rightarrow$ $\dfrac{1}{2} = \dfrac{h}{9-h}$

$\Rightarrow$ $9 - h = 2h$

$\Rightarrow$ $9 = 3h$

$\therefore$ $h = 3$ m

Hence, the post is broken at a height of 3 m from the ground.

2 Try to Solve

1. The shadow of Qutub Minar is 81 m long when the angle of elevation of the sun is 60°. Find the height of Qutub Minar.
2. In the figure, $ABCD$ is a rectangle in which segment AP is drawn as shown. Find the length of AP.

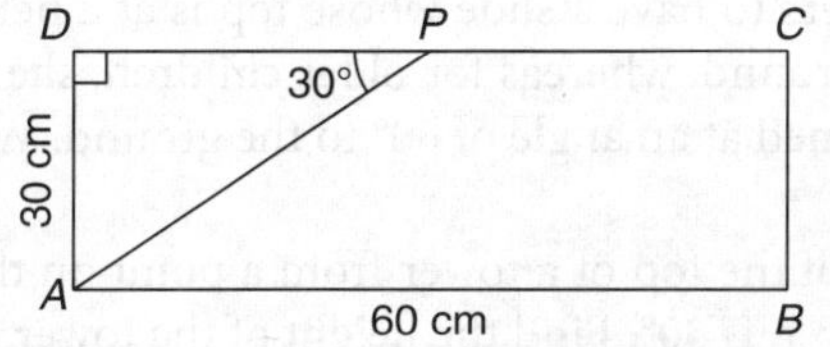

3. A tree breaks because of a storm and the broken part bends so that the top of the tree touches the ground, making an angle of 30° with it. The distance between the foot of the tree and the point where the top touches the ground is 8 m. Find the height of the tree. **[NCERT, CBSE (CCE) 2010]**

Answers

1. $81\sqrt{3}$ m 2. 60 cm 3. $8\sqrt{3}$ m

Example 5 An observer 1.5 m tall is 28.5 m away from a chimney. The angle of elevation of the top of the chimney from her eyes is 45°. What is the height of the chimney?

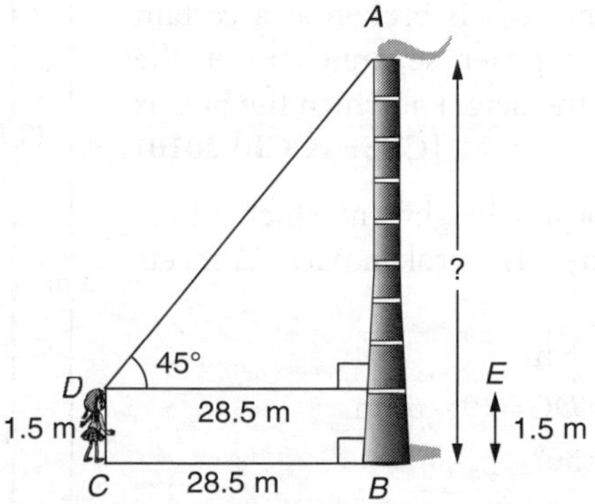

[NCERT]

Solution: Here, AB is the chimney, CD the observer and $\angle ADE$ the angle of elevation (see figure). In this case, ADE is a triangle, right-angled at E, and we are required to find the height of the chimney.

We have $AB = AE + BE = AE + 1.5$. and $DE = CB = 28.5$ m.

To determine AE, we choose a trigonometric ratio that involves both AE and DE. Let us choose the tangent of the angle of elevation.

Now $\tan 45° = \dfrac{AE}{DE}$

That is $1 = \dfrac{AE}{28.5}$

Therefore $AE = 28.5$

So, the height of the chimney (AB) = (28.5 + 1.5) m = 30 m.

3 Try to Solve

1. A contractor plans to install two slides for children to play in a park. For children below the age of 5 years, she prefers to have a slide whose top is at a height of 1.5 m, and is inclined at an angle of 30° to the ground, whereas for older children, she wants to have a steep slide at a height of 3 m, and inclined at an angle of 60° to the ground. What should be the length of the slide in each case? **[NCERT]**
2. The angle of elevation of the top of a tower from a point on the ground, which is 30 m away from the foot of the tower, is 30°. Find the height of the tower. **[NCERT]**
3. An observer 1.6 m tall is $30\sqrt{3}$ m away from a tower. The angle of elevation from his eye to the top of the tower is 30°. Determine the height of the tower. **[CBSE (CCE) 2014]**

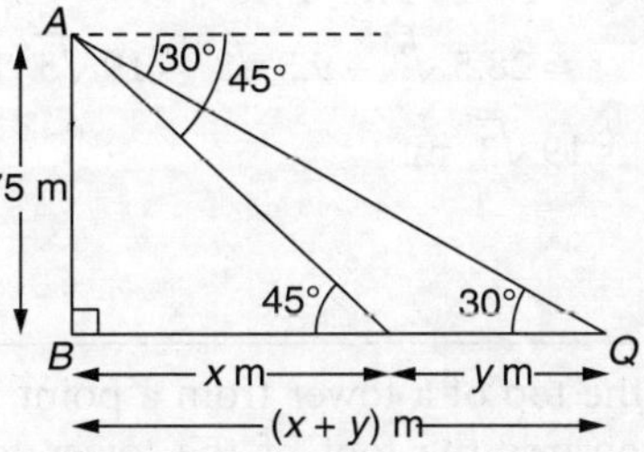

Answers

1. 3 m, $2\sqrt{3}$ m **2.** $10\sqrt{3}$ m **3.** 31.6 m

Example 6 A 1.5-m-tall boy is standing at some distance from a 30-m-high building. The angle of elevation from his eyes to the top of the building increases from 30° to 60° as he walks towards the building. Find the distance he walked towards the building. **[CBSE (CCE) 2015]**

Solution: Let *AB* be the building and *PQ* be the initial position of the boy (see figure) such that

$$\angle APR = 30° \text{ and } AB = 30 \text{ m}$$

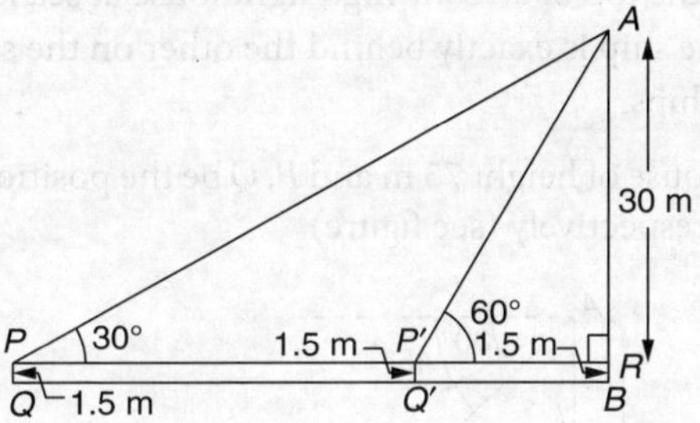

Now, let the new position of the boy be *P′Q′* at a distance *QQ′*.

Here $\angle AP'R = 60°$

In right ΔARP, we have $\tan 30° \frac{AR}{PR} = \frac{AB - RB}{PR}$

$$\Rightarrow \quad \frac{1}{\sqrt{3}} = \frac{30 - 1.5}{PR} = \frac{28.5}{PR}$$

$$PR = 28.5 \times \sqrt{3}$$

Again, in right $\Delta ARP'$, we have $\tan 60° = \frac{AR}{P'R} \Rightarrow \sqrt{3} = \frac{28.5}{P'R}$

$$P'R = \frac{28.5}{\sqrt{3}} \times \frac{\sqrt{3}}{\sqrt{3}} = \frac{28.5\sqrt{3}}{3} = 9.5\sqrt{3}$$

Therefore, the required distance, $QQ' = PP' = PR - P'R$

$$= 28.5\sqrt{3} - 9.5\sqrt{3} = 19\sqrt{3}$$

Hence, distance walked by the boy is $19\sqrt{3}$ m.

4 Try to Solve

1. The angle of elevation of the top of a tower from a point *A* on the ground is 30°. On moving a distance of 20 m towards the foot of the tower to point *B*, the angle of elevation increases to 60°. Find the height of the tower and the distance of the tower from the point *A*. **[CBSE (CCE) 2009]**
2. A person observed that the angle of elevation of the top of a tower is 30°. He walked 50 m towards the foot of the tower along level ground and found that the angle of elevation of the top of the tower was 60°. Find the height of the tower.
3. The shadow of a tower, when the angle of elevation of the sun is 45°, is found to be 10 m longer than when it was 60°. Find the height of the tower.

Answers

1. 17.32 m and 30 m **2.** 43.25 m **3.** 23.66 m

Example 7 As observed from the top of a 75-m-high lighthouse at sea level, the angles of depression of two ships are 30° and 45°. If one ship is exactly behind the other on the same side of the lighthouse, find the distance between the two ships. **[NCERT, CBSE (CCE) 2010]**

Solution: Let *AB* be the lighthouse of height 75 m and *P*, *Q* be the position of the two ships whose angles of depression are 45° and 30°, respectively (see figure).

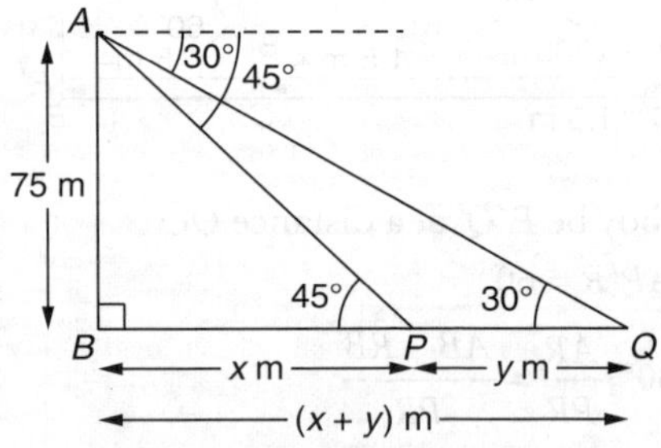

Let $BP = x$ m and $PQ = y$ m

We have $\angle APB = 45°$ and $\angle AQB = 30°$

Now, in right ΔABP, we have $\tan 45° = \dfrac{AB}{BP} \Rightarrow 1 = \dfrac{75}{x}$

$\Rightarrow$ $x = 75$ ...(i)

Again, in right ΔABQ, we have $\tan 30° = \dfrac{AB}{BQ}$

$\Rightarrow$ $\dfrac{1}{\sqrt{3}} = \dfrac{75}{x+y} \Rightarrow x + y = 75\sqrt{3}$ ...(ii)

From (i) and (ii), we get $75 + y = 75\sqrt{3}$

$$y = 75\sqrt{3} - 75$$

⇒ $y = 75(\sqrt{3} - 1)$

Hence, the distance between the two ships is $75(\sqrt{3} - 1)$ metres.

5 Try to Solve

1. From a point on the ground, the angles of elevation of the bottom and the top of a transmission tower fixed at the top of a 20-m-high building are 45° and 60°, respectively. Find the height of the tower. **[NCERT]**
2. A statue, 1.6 m tall, stands on top of a pedestal. From a point on the ground, the angle of elevation of the top of the statue is 60° and from the same point, the angle of elevation of the top of the pedestal is 45°. Find the height of the pedestal. **[NCERT]**
3. The angle of elevation of the top of a building from the foot of the tower is 30° and the angle of elevation of the top of the tower from the foot of the building is 60°. If the tower is 50 m high, find the height of the building. **[NCERT, CBSE (CCE) 2009]**

Answers

1. $20(\sqrt{3} - 1)$ m **2.** $0.8(\sqrt{3} + 1)$ m **3.** $16\frac{2}{3}$ m

Example 8 The angle of elevation of an aeroplane from a point on the ground is 60°. After a flight of 30 seconds, the angle of elevation becomes 30°. If the aeroplane is flying at a constant height of 3000 m, find the speed of the aeroplane.

Solution: Let *A* be the point of observation and *P* and *Q* be positions of the plane. Let *ABC* be the line through *A*, and it is given that the angles of elevation from point *A* to two positions *P* and *Q* are 60° and 30°.

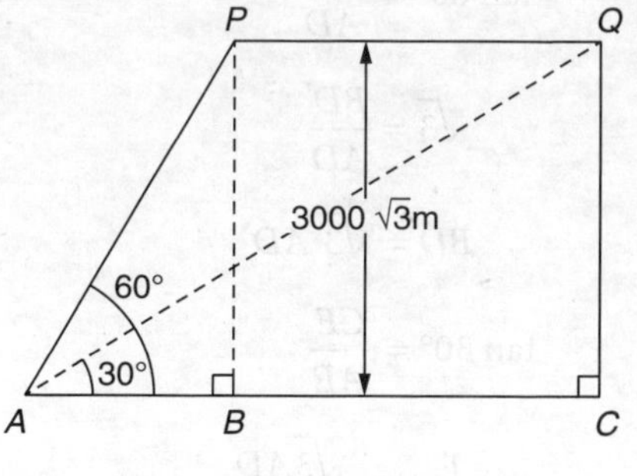

∴ $\angle PAB = 60°, \angle QAB = 30°$

$$\text{Height} = 3000\sqrt{3} \text{ m}$$

So, in right ΔABP, we have $\tan 60° = \frac{BP}{AB}$

$$\sqrt{3} = \frac{3000\sqrt{3}}{AB}$$

⇒ $AB = 3000$ m

In right ΔACQ $\qquad \tan 30° = \dfrac{CQ}{AC}$

$$\frac{1}{\sqrt{3}} = \frac{3000\sqrt{3}}{AC}$$

$$AC = 9000 \text{ m}$$

$\therefore$ $\qquad$ Distance $= BC = AC - AB = 9000 \text{ m} - 3000 \text{ m} = 6000 \text{ m}$

Also, the plane travels for 30 seconds.

Hence, the speed of the plane $= \dfrac{6000}{30} = 200 \text{ m/s} = 200 \times \dfrac{18}{5} \text{ km/h} = 720 \text{ km/h}.$

Example 9 The angle of elevation of a jet fighter from a point A on the ground is 60°. After a flight of 15 seconds, the angle of elevation changes to 30°. If the jet is flying at a speed of 720 km/h, find the constant height at which the jet is flying. [$\sqrt{3} = 1.73$] **[CBSE (CCE) 2011]**

Solution: From a point A on the ground, the jet is first spotted at B. The angle of elevation is 60°. After 15 seconds, the jet is at C and the angle of elevation is 30°.

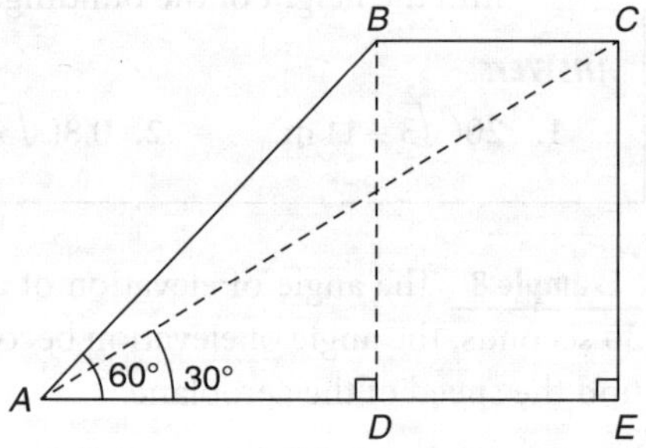

$\therefore$ $\qquad \angle BAD = 60°$ and $\angle CAE = 30°$

Speed of jet $\qquad v = 720 \text{ km/h}$

$$= \frac{720 \times 1000}{60 \times 60} = 200 \text{ m/s}$$

Distance $\qquad BC = 200 \times 15 = 3000 \text{ m}$

$\therefore$ $\qquad DE = BC = 3000 \text{ m}$

Let the constant height be h m from the ground

$\therefore$ $\qquad BD = CE = h \text{ m}$

In right ΔADB $\qquad \tan 60° = \dfrac{BD}{AD}$

$\Rightarrow$ $\qquad \sqrt{3} = \dfrac{BD}{AD}$

$\Rightarrow$ $\qquad BD = \sqrt{3}\, AD$ $\qquad$ ...(i)

In right ΔAEC $\qquad \tan 30° = \dfrac{CE}{AE}$

$\Rightarrow$ $\qquad \dfrac{1}{\sqrt{3}} = \dfrac{\sqrt{3}AD}{AD + DE}$

$\Rightarrow$ $\qquad \dfrac{1}{\sqrt{3}} = \dfrac{\sqrt{3}AD}{AD + 3000}$

$\Rightarrow$ $\qquad AD + 3000 = 3\, AD$

$\Rightarrow$ $\qquad 2\, AD = 3000$

$\therefore$ $\qquad AD = 1500 \text{ m}$

Inserting $AD = 1500$ m in (i), we get $BD = \sqrt{3} \times 1500 = 1500\sqrt{3}$

$= 1500 \times 1.73$

$\therefore$ $BD = 2595$ m

Hence, the jet is flying at a height of 2595 m.

6 Try to Solve

1. The angle of elevation of an aeroplane from a point on the ground is 45°. After a flight of 15 seconds, the elevation changes to 30°. If the aeroplane is flying at a height of 3000 metres, find the speed of the aeroplane. **[CBSE (CCE) 2011]**
2. An aeroplane flying horizontally 1 km above the ground is observed at an elevation of 60°. After 10 seconds, its elevation is observed to be 30°. Find the speed of the aeroplane in km/h.
3. The angle of elevation of a jet fighter from a point A on the ground is 60°. After a flight of 10 seconds, the angle of elevation changes to 30°. If the jet is flying at a speed of 432 km/h, find the constant height at which the jet is flying. **[CBSE (CCE) 2015]**

Answers

1. 527.04 km/h **2.** 415.66 km/h **3.** 1039.2 m

Example 10 The angles of elevation of the top of a tower from two points at a distance of 4 m and 16 m from the base of the tower and in the same straight line with it are complementary. Prove that the height of the tower is 8 m.

Solution: Let AB be a tower of height h m and C and D be two points at a distance of 4 m and 16 m, respectively, from the base of the tower.

$AB = h$ m, $BC = 4$ m and $BD = 16$ m

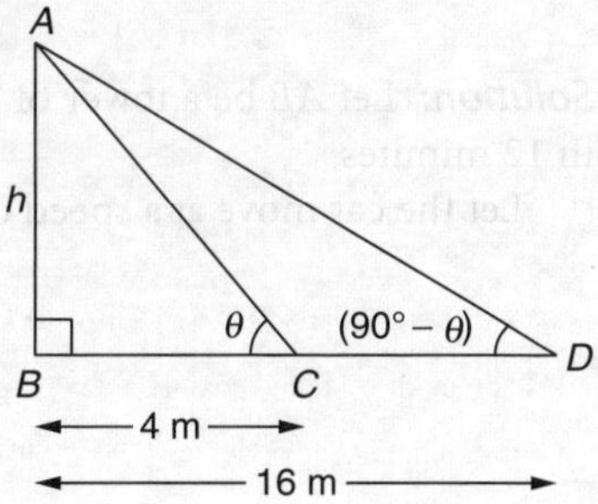

Let $\angle ACB = \theta.$

$\angle ACB + \angle ADB = 90°$ [Given]

$\Rightarrow$ $\theta + \angle ADB = 90°$

$\therefore$ $\angle ADB = (90 - \theta)°$

In right ΔABC $\tan\theta = \dfrac{AB}{BD}$

$\Rightarrow$ $\tan\theta = \dfrac{h}{4}$ (i)

In right ΔABD $\tan(90 - \theta) = \dfrac{AB}{BD}$

$\Rightarrow$ $\cot\theta = \dfrac{h}{16}$ (ii)

Multiplying equations (i) and (ii), we get $\tan\theta \times \cot\theta = \frac{h}{4} \times \frac{h}{16}$

$$1 = \frac{h^2}{64} \quad [\because \tan\theta.\cot\theta = 1]$$

$$h^2 = 64$$

$$h = 8 \text{ m}$$

7 Try to Solve

1. The angle of elevation of the top of a tower from two points at a distance of 4 m and 9 m from the base of the tower and in the same straight line with it are complementary. Prove that the height of the tower is 6 m. **[NCERT]**
2. Two poles of equal height stand opposite each other on either side of the road, which is 80 m wide. From a point between them on the road, the angles of elevation of the top of the poles are 60° and 30°, respectively. Find the height of the pole and the distance of the point from the poles. **[CBSE (CCE) 2015]**
3. From the top of a 7-m-high building, the angle of elevation of the top of a cable tower is 60° and the angle of depression of its foot is 45°. Determine the height of the tower. **[NCERT, CBSE (CCE) 2015]**

Answers

2. $20 \text{ m}\sqrt{3}$, 20 m and 60 m **3.** $7(\sqrt{3}+1)$ m

Example 11 A man at the top of a vertical tower observes a car moving towards the tower. If it takes 12 minutes for the angle of depression to change from 30° to 45°, how soon will the car reach the tower? **[CBSE (CCE) 2011]**

Solution: Let AB be a tower of height h m. The man spotted the car at point C, which reaches point D in 12 minutes.

Let the car move at a speed of x m/min; it reaches the tower in t minutes.

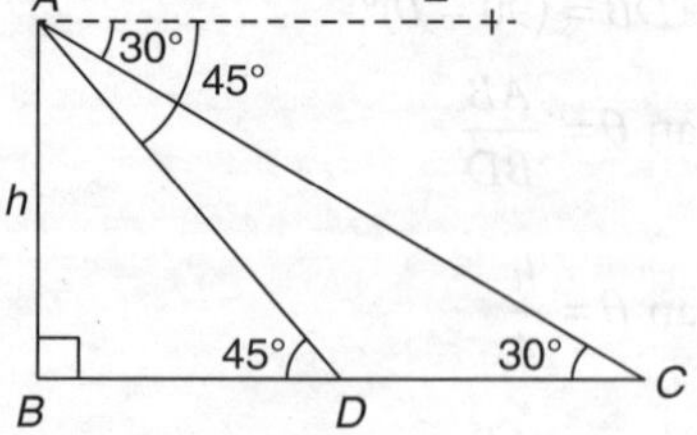

$\therefore \quad CD = 12x$ m

And $\quad BD = xt\ m \quad [\because \text{Distance} = \text{speed} \times \text{time}]$

$\angle ACB = \angle EAC = 30° \quad [\because \text{Alternate interior angles}]$

$\angle ADB = \angle EAD = 45° \quad [\because \text{Alternate interior angles}]$

In right ΔABD $\tan 45° = \frac{AB}{BD}$

$\Rightarrow$ $1 = \frac{AB}{BD}$

$\therefore$ $AB = BD = xt\ m$ $[\because BD = xt\ m]$

In right ΔABC $\tan 30° = \frac{AB}{BC}$

$\Rightarrow$ $\frac{1}{\sqrt{3}} = \frac{AB}{BD + CD}$

$\Rightarrow$ $\frac{1}{\sqrt{3}} = \frac{xt}{xt + 12x}$

$\Rightarrow$ $\frac{1}{\sqrt{3}} = \frac{\cancel{x}t}{\cancel{x}(t+12)}$

$\Rightarrow$ $\frac{1}{\sqrt{3}} = \frac{t}{t+12}$

$\Rightarrow$ $t + 12 = \sqrt{3}\,t$

$\Rightarrow$ $\sqrt{3}\,t - t = 12$

$\Rightarrow$ $t(\sqrt{3} - 1) = 12$

$\Rightarrow$ $t(1.73 - 1) = 12$ $[\sqrt{3} = 1.73]$

$\Rightarrow$ $t \times 0.73 = 12$

$\therefore$ $t = \frac{12}{0.73} = 16.44$ minutes.

8 Try to Solve

1. A straight highway leads to the foot of a tower. A man standing at the top of the tower observes a car at an angle of depression of 30°, which is approaching the foot of the tower at a uniform speed. Six seconds later, the angle of depression of the car is found to be 60°. Find the time taken by the car to reach the foot of the tower. **[NCERT, CBSE (CCE) 2011]**

Answers

1. 3 seconds

FOR MORE PRACTICE AND KNOWLEDGE

Example 12 From the top of a hill, the angles of depression of two consecutive kilometre stones due east are found to be 30° and 45°. Find the height of the hill.

Solution: Let AB be the hill of height h km. Let C and D be two stones due east of the hill at a distance of 1 km from each other such that the angles of depression C and D are 45° and 30°, respectively. Let $AC = x$ km, $CD = 1$ km and $AD = (x + 1)$ km.

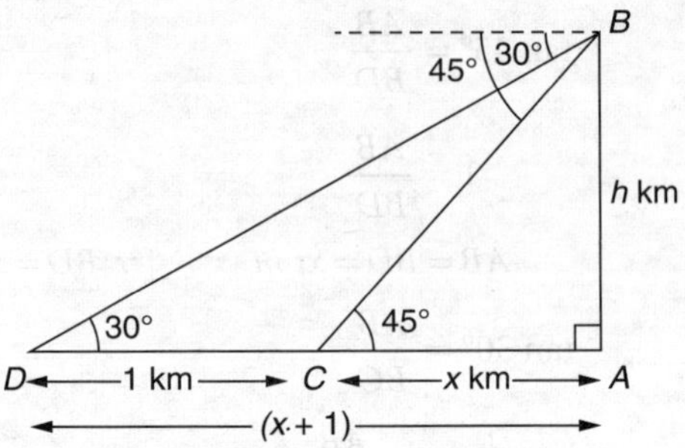

In ΔCAB, we have $\tan 45° = \dfrac{AB}{AC} = \dfrac{h}{x}$

$\Rightarrow$ $1 = \dfrac{h}{x}$ $h = x$ (i)

In a ΔDAB, we have $\tan 30° = \dfrac{AB}{AD}, \dfrac{1}{\sqrt{3}} = \dfrac{h}{x+1}$

$\Rightarrow$ $\sqrt{3}\,h = x + 1$ (ii)

Substituting the value of x from equation (i) in equation (ii), we get

$$\sqrt{3}\,h = h + 1$$

$\Rightarrow$ $h(\sqrt{3} - 1) = 1$

$\Rightarrow$ $h = \dfrac{1}{\sqrt{3}-1} = \dfrac{\sqrt{3}+1}{(\sqrt{3}-1)(\sqrt{3}+1)}$

$\Rightarrow$ $h = \dfrac{\sqrt{3}+1}{2} = \dfrac{2.73}{2} = 1.365$

Hence, the height of the hill is 1.365 km.

Example 13 The angle of elevation of the top Q of a vertical tower PQ from a point X on the ground is 60°. At a point Y, 40 m vertically above X, the angle of elevation is 45°. Find the height of the tower PQ and the distance XQ. **[CBSE 1997 C]**

Solution: Since $XY = 40$ m, $XY = RP = 40$ m

Let $RQ = x\ m$ and $PQ = (40 + x)\ m$.

In ΔYRQ, we have $\tan 45° = \dfrac{QR}{YR}$

$\Rightarrow$ $1 = \dfrac{x}{YR}$

$\Rightarrow$ $YR = x$

$\Rightarrow$ $XP = x$

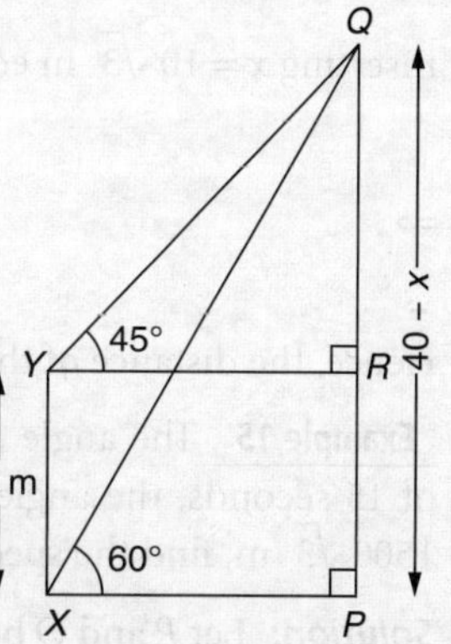

In ΔXPQ, we have $\tan 60° = \dfrac{PQ}{PX}$

$$\Rightarrow \quad \sqrt{3} = \frac{x+40}{x}$$

$$\Rightarrow \quad \sqrt{3}\,x = x + 40$$

$$\Rightarrow \quad x(\sqrt{3} - 1) = 40$$

$$\Rightarrow \quad x = \frac{40}{\sqrt{3}-1}$$

$$\Rightarrow \quad x = \frac{40}{\sqrt{3}-1} \times \frac{\sqrt{3}+1}{\sqrt{3}+1} = 20(\sqrt{3}+1) = 54.64$$

So, the height of the tower $PQ = x + 40 = 54.64 + 40 = 94.64$ metres.

In ΔXPQ, we have $\sin 60° = \dfrac{PQ}{XQ}$

$$\Rightarrow \quad \frac{\sqrt{3}}{2} = \frac{94.64}{XQ}$$

$$\Rightarrow \quad XQ = \frac{94.64 \times 2}{\sqrt{3}}$$

On rationalising the denominator $XQ = \dfrac{94.64 \times 2 \times \sqrt{3}}{3} = 109.3$ metres.

Example 14 A man is standing on the deck of a ship, which is 10 m above water level. He observes the angle of elevation of the top of a hill as 60° and the angle of depression of the base of the hill as 30°. Calculate the distance of the hill from the ship and the height of the hill. **[CBSE 2004, 2005]**

Solution: Suppose the man is standing on the deck of a ship at point A and CD is the hill. It is given that the angle of depression of the base C of the hill CD observed from A is 30° and the angle of elevation of the top D of the hill CD observed from A is 60°. Then, $\angle EAD = 60°$, $\angle BCA = 30°$.

Also $AB = 10$ m

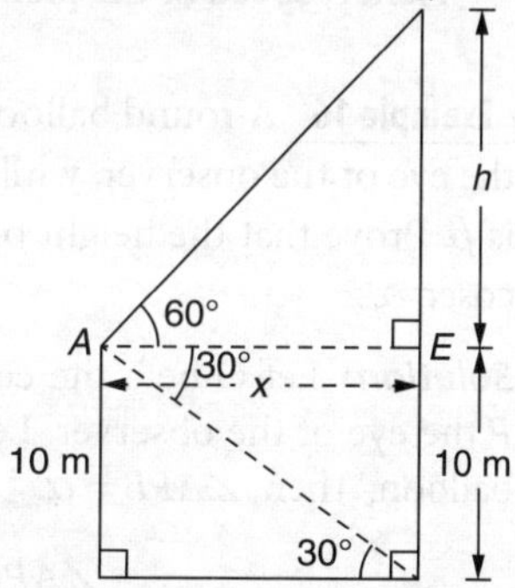

In ΔAED, we have $\tan 60° = \dfrac{DE}{EA}$

$$\Rightarrow \quad \sqrt{3} = \frac{h}{x} \Rightarrow h = \sqrt{3}\,x \quad \ldots\text{(i)}$$

In ΔABC, we have $\tan 30° = \dfrac{AB}{BC}$

$$\Rightarrow \quad \frac{1}{\sqrt{3}} = \frac{10}{x} \quad x = 10\sqrt{3} \quad \ldots\text{(ii)}$$

Inserting $x = 10\sqrt{3}$ in equation (i), we get

$$h = \sqrt{3} \times 10\sqrt{3} = 30$$

$\Rightarrow$ $DE = 30$ m

$\therefore$ $CD = CE + ED = 10 + 30 = 40$ metres

Hence, the distance of the hill from the ship is $10\sqrt{3}$ metres and the height of the hill is 40 metres.

Example 15 The angle of elevation of a jet plane from a point A on the ground is 60°. After a flight of 15 seconds, the angle of elevation changes to 30°. If the jet plane is flying at a constant height of $1500\sqrt{3}$ m, find the speed of the jet plane. **[CBSE 1997]**

Solution: Let P and Q be the two positions of the plane and let A be the point of observation. Let ABC be the horizontal line through A. It is given that the angle of elevation of the plane in two positions P and Q from a point A is 60° and 30°, respectively.

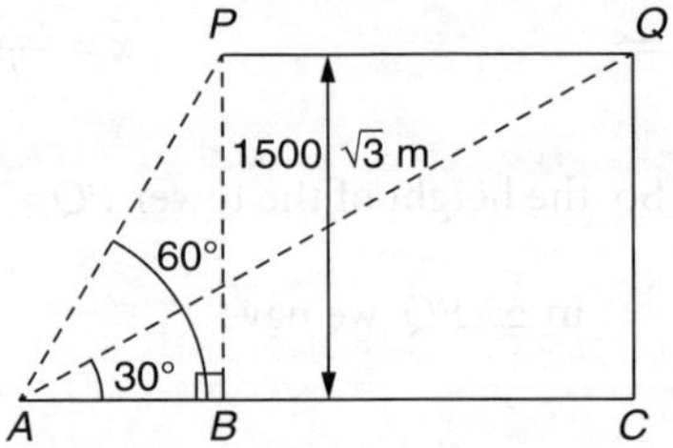

$\therefore$ $PAB = 60°, QAB = 30°$

It is also given that $PB = 1500\sqrt{3}$ metres.

In ΔABP, we have $\tan 60° = \dfrac{BP}{AB}$

$\Rightarrow$ $\sqrt{3} = \dfrac{1500\sqrt{3}}{AB}$

$\Rightarrow$ $AB = 1500$

In ΔACQ, we have $\tan 30° = \dfrac{CQ}{AC}$

$\Rightarrow$ $\dfrac{1}{\sqrt{3}} = \dfrac{1500\sqrt{3}}{AC}$

$\Rightarrow$ $AC = 1500 \times 3 = 4500$ m

$\therefore$ $PQ = BC = AC - AB = 4500 - 1500 = 3000$ m

Thus, the plane travels 3000 m in 15 seconds.

Hence, speed of the plane $= \dfrac{3000}{15} = 200$ m/s $= \dfrac{200}{1000} \times 60 \times 60 = 720$ km/h.

Example 16 A round balloon of radius r subtends an angle α at the eye of the observer, while the angle of elevation of its centre is β. Prove that the height of the centre of the balloon is $r \sin \beta$ cosec $\alpha/2$.

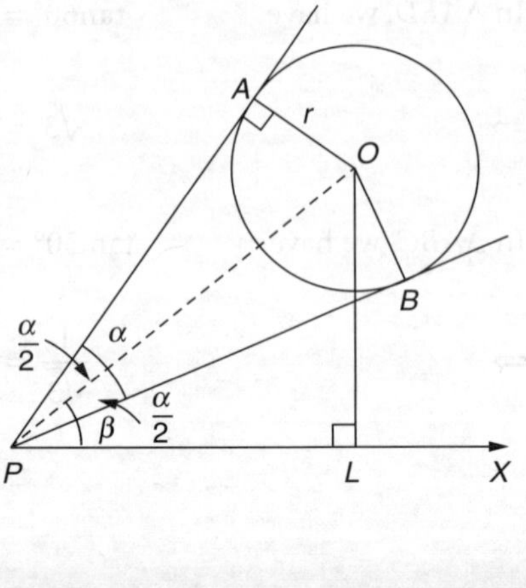

Solution: Let O be a the centre of the balloon of radius r and P the eye of the observer. Let PA, PB be tangents from P to the balloon. Then, $\angle APB = \alpha$.

$$\angle APO = \angle BPO = \frac{\alpha}{2}$$

Let OL be perpendicular from O on the horizontal PX. We are given that the angle of the elevation of the centre of the balloon is β, that is

$$OPL = \beta.$$

In ΔOAP, we have $\sin\dfrac{\alpha}{2} = \dfrac{OA}{OP}$

$\Rightarrow$ $\sin\dfrac{\alpha}{2} = \dfrac{r}{OP}$

$\Rightarrow$ $OP = r \operatorname{cosec}\dfrac{\alpha}{2}$ ...(i)

In ΔOPL, we have$\Rightarrow$ $\sin\beta = \dfrac{OL}{OP}$

$$OL = OP\sin\beta = r\operatorname{cosec}\frac{\alpha}{2}\cdot\sin\beta \quad \text{[From (i)]}$$

Hence, the height of the centre of the balloon is $r\sin\beta\operatorname{cosec}\dfrac{\alpha}{2}$.

Example 17 At the foot of a mountain, the elevation of its summit is 45°; after ascending 1000 m towards the mountain up a slope of 30° inclination, the elevation is found to be 60°. Find the height of the mountain.

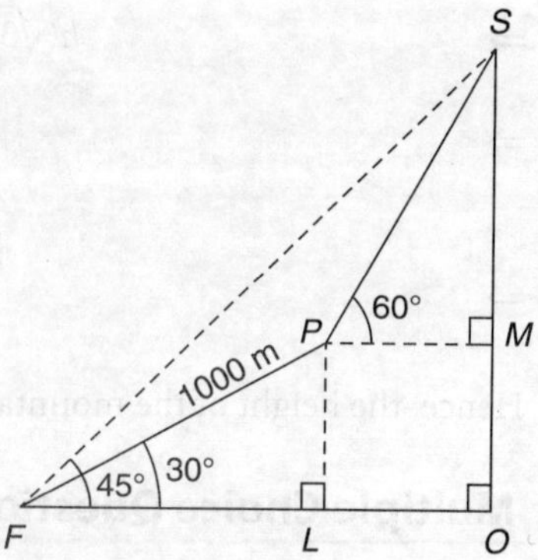

Solution: Let F be the foot and S be the summit of the mountain FOS. Then, $OFS = 45°$, and therefore, $OSF = 45°$.

$$OF = OS = h \text{ km}$$

Let $FP = 1000$ m $= 1$ km be the slope so that $OFP = 30°$.

Draw $PM \perp OS$ and $PL \perp OF$. Join PS. It is given that $\angle MPS = 60°$.

In ΔFPL, we have $\sin 30° = \dfrac{PL}{PF}$

$\Rightarrow$ $PL = PF\sin 30° = \left(1\times\dfrac{1}{2}\right)\text{km} = \dfrac{1}{2}\text{ km}$

$\therefore$ $OM = PL = \dfrac{1}{2}$ km

$\Rightarrow$ $SM = OS - OM = \left(h - \dfrac{1}{2}\right)$ km

Also $\cos 30° = \dfrac{FL}{PF}$

$\Rightarrow$ $FL = PF\cos 30° = \left(1\times\dfrac{\sqrt{3}}{2}\right)\text{km} = \dfrac{\sqrt{3}}{2}\text{ km}$

Now $h = OS = OF = OL + LF$

$\Rightarrow$ $h = OL + \dfrac{\sqrt{3}}{2}$

$$\Rightarrow \qquad OL = \left(h - \frac{\sqrt{3}}{2}\right)\text{km}$$

$$\Rightarrow \qquad PM = \left(h - \frac{\sqrt{3}}{2}\right)\text{km} \qquad \ldots\text{(ii)}$$

In ΔSPM, we have $\tan 60° = \dfrac{SM}{PM}$

$$\Rightarrow \qquad SM = PM \cdot \tan 60°$$

$$\Rightarrow \qquad \left(h - \frac{1}{2}\right) = \left(h - \frac{\sqrt{3}}{2}\right)\sqrt{3} \qquad \text{[From (i) and (ii)]}$$

$$\Rightarrow \qquad h - \frac{1}{2} = h\sqrt{3} - \frac{3}{2}$$

$$\Rightarrow \qquad \sqrt{3}h - h = \frac{3}{2} - \frac{1}{2}$$

$$\Rightarrow \qquad h(\sqrt{3} - 1) = 1$$

$$\Rightarrow \qquad h = \frac{1}{\sqrt{3} - 1}$$

$$\Rightarrow \qquad h = \frac{\sqrt{3} + 1}{(\sqrt{3} - 1)(\sqrt{3} + 1)} = \frac{\sqrt{3} + 1}{2} = \frac{2.732}{2} = 1.366 \text{ km}$$

Hence, the height of the mountain is 1.366 km.

Multiple Choice Questions

Choose the correct answer for each of the following questions:

1. In the figure, if $AD = 7\sqrt{3}$ m, then BC is equal to:

(a) 14 m (b) 27 m
(c) 29 m (d) 28 m

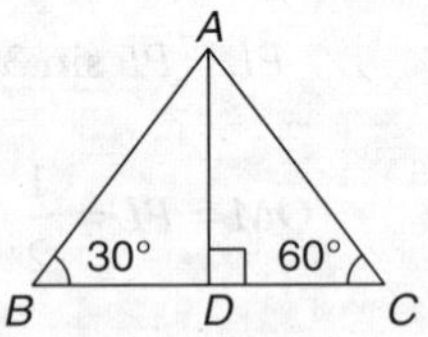

2. In the figure, $ABCD$ is a rectangle. $(AC + AD)$ is equal to

(a) $40\sqrt{3}$ cm (b) $20\sqrt{3}$ cm
(c) $60\sqrt{3}$ cm (d) $80\sqrt{3}$ cm

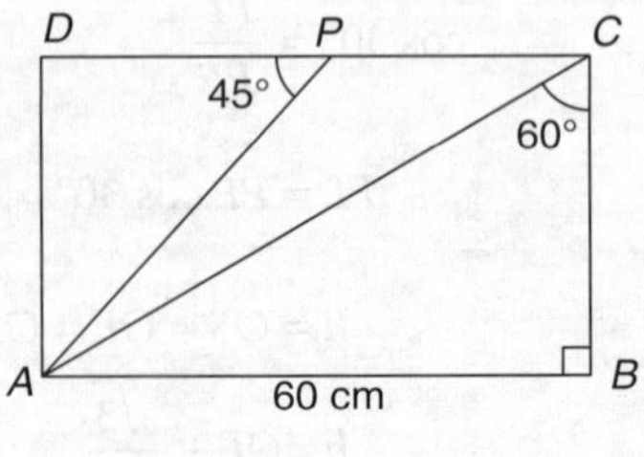

3. A person standing on the bank of a river observes that the angle of elevation of a tree is 60°. When he moves 40 m away, the angle of elevation becomes 30°. At what distance is he now standing from the tree?
(a) 60 m (b) 80 m (c) 69.2 m (d) $40\sqrt{3}$ m

4. The length of the shadow of a tower on plane ground is $\sqrt{3}$ times the height of the tower. The angle of elevation of the sun is
(a) 45° (b) 30° (c) 60° (d) 90°

5. If from a point on the ground, which is 20 m away from the foot of a tower, the angle of elevation of the top of the tower is found to be 60°, then the height of the tower (in metres) is:
(a) $10\sqrt{2}$ (b) $20\sqrt{3}$ (c) $\frac{20}{\sqrt{3}}$ (d) $20\sqrt{2}$

6. If the angles of elevation of a tower from two points at a distance '*a*' and '*b*' from its foot and in the same straight line from it are 30° and 60°, then the height of the tower is
(a) $\sqrt{\frac{a}{b}}$ (b) $\sqrt{a+b}$ (c) $\sqrt{a-b}$ (d) $\sqrt{ab}$

7. In the figure, *PR* is equal to
(a) 32 m (b) 44 m
(c) $32\sqrt{2}$ m (d) $44\sqrt{2}$ m

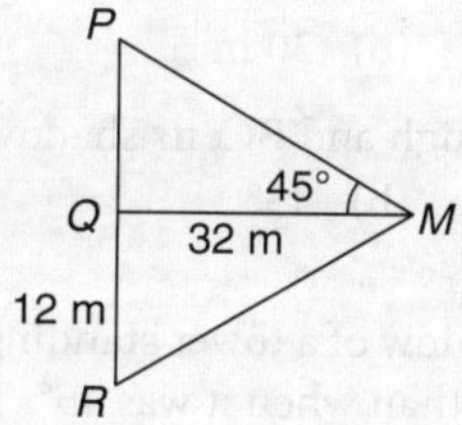

8. The angle of depression of a car standing on the ground from the top of a 75-m-high tower is 30°. The distance of the car from the base of the tower (in m) is:
(a) $25\sqrt{3}$ (b) $50\sqrt{3}$ (c) $75\sqrt{3}$ (d) 150

9. The angle of depression of a car parked on the road from the top of a 150-m-high tower is 30°. The distance of the car from the tower is
(a) $50\sqrt{3}$ m (b) $250\sqrt{3}$ m (c) $150\sqrt{3}$ m (d) 75 m **[CBSE 2014]**

10. From the top of a cliff 20 m high, the angle of elevation of the top of a tower is found to be equal to the angle of depression of the foot of the tower. The height of the tower is
(a) 20 m (b) 40 m (c) 60 m (d) 80 m **[CBSE 2013C]**

11. If a 1.5-m-tall girl stands at a distance of 3 m from a lamp post and casts a shadow of length 4.5 m on the ground, then the height of the lamp post is
(a) 1.5 m (b) 2 m (c) 2.5 m (d) 2.8 m

12. The shadow of a 5-m-long stick is 2 m long. At the same time, the length of the shadow of a 12.5-m-high tree is
(a) 3 m (b) 3.5 m (c) 4.5 m (d) 5 m **[CBSE 2011]**

13. If the height of a vertical pole is equal to the length of its shadow on the ground, the angle of elevation of the sun is
(a) 0° (b) 30° (c) 45° (d) 60° **[CBSE 2014]**

14. A pole casts a shadow of length $2\sqrt{3}$ m on the ground when the sun's elevation is 60°. The height of the pole is

(a) $4\sqrt{3}$ m (b) 6 m (c) 12 m (d) 3 m **[CBSE 2015]**

15. The string of a kite is 100 m long and it makes an angle of 60° with the horizontal. If there is no slack in the string, the height of the kite from the ground is

(a) $50\sqrt{3}$ m (b) $100\sqrt{3}$ m (c) $50\sqrt{2}$ m (d) 100 m

16. On level ground, the angle of elevation of a tower is 30°. On moving 20 m nearer, the angle of elevation is 60°. The height of the tower is

(a) 10 m (b) $10\sqrt{3}$ m (c) 15 m (d) 20 m

17. From the top of a hill, the angles of depression of two consecutive km stones due east are found to be 30° and 45°. The height of the hill is

(a) $\frac{1}{2}(\sqrt{3}-1)$ km (b) $\frac{1}{2}(\sqrt{3}+1)$ km (c) $(\sqrt{3}-1)$ km (d) $(\sqrt{3}+1)$ km

18. An observer 1.5 m tall is 28.5 m away from a tower and the angle of elevation of the top of the tower from the eye of the observer is 45°. The height of the tower is

(a) 27 m (b) 30 m (c) 28.5 m (d) None of these

19. A tower AB is 20 m high and BC, its shadow on the ground, is $20\sqrt{3}$ m long. The sun's altitude is

(a) 30° (b) 45° (c) 60° (d) None of these

[CBSE 2015]

20. The length of the shadow of a tower standing on level ground is found to be $2x$ metres longer when the sun's elevation is 30° than when it was 45°. The height of the tower is

(a) $(2\sqrt{3}x)$ m (b) $(2\sqrt{2}x)$ m (c) $(\sqrt{3}-1)\,x$ m (d) $(\sqrt{3}+1)x$ m

21. The angle of elevation of the top of a tower from a point 20 metres away from its base is 45°. The height of the tower is

(a) 10 m (b) 20 m (c) 30 m (d) $20\sqrt{3}$ m

22. The length of the shadow of a person is x when the angle of elevation of the sun is 45°. If the length of the shadow increases by $(\sqrt{3}-1)\,x$, then the angle of elevation of the sun should become

(a) 60° (b) 45° (c) 30° (d) 20°

23. Two poles are 25 m and 15 m high and the line joining their tops makes an angle of 45° with the horizontal. The distance between these poles is

(a) 5 m (b) 8 m (c) 9 m (d) 10 m

Fill in the Blanks

Complete the following statements with appropriate answers(s) in the blank space(s):

24. If a flagstaff 6 m high placed at the top of a tower throws a shadow $2\sqrt{3}$ m along the ground, then the angle of elevation of the sun is ________.

25. If the height of a tower is doubled and the distance between the observer and the foot of the tower is also doubled, then the angle of elevation remains ________.

26. The angle of elevation and depression are ________.

Assertion–Reason

The following questions consist of two statements – Assertion (A) and Reason (R). Answer these questions by selecting the appropriate option given below.

(A) Both A and R are true and R is the correct explanation for A.
(B) Both A and R are true and R is not the correct explanation for A.
(C) A is true but R is false.
(D) A is false but R is true.

27. **Assertion:** The angle of depression of a car parked at the top of a 150-m-high tower is 30°. Then the distance of the car from the tower is 150 m.

Reason: $\tan\theta = \dfrac{\text{Perpendicular}}{\text{Base}}$.

28. **Assertion:** A vertical pole 6 m long casts a shadow of length 3.6 m on the ground. The height of a tower which casts a shadow of length 18 m at the same time is 30 m.

Reason: The sun makes equal angular elevations with different objects at the same time.

29. **Assertion:** TF is a tower with F on the ground. The angle of elevation T from A is x such that $\tan x = \dfrac{2}{5}$ and $AF = 200$ m. The angle of elevation of T from a nearer point B is y with $BF = 80$ m. The value of y is 45°.

Reason: $\tan^2 A = \cot (A - 18)°$, where $2A$ is a more acute angle than $A = 36°$.

Subjective Questions

30. From a point P on the ground, the angle of elevation of the top of a 10-m-tall building is 30°. A flag is hoisted at the top of the building and the angle of elevation of the top of the flagstaff from P is 45°. Find the length of the flagstaff and the distance of the building from the point P. [You may take $\sqrt{3} = 1.732$] [NCERT]

31. The shadow of a tower standing on level ground is found to be 40 m longer when the sun's altitude is 30° than when it is 60°. Find the height of the tower. [NCERT]

32. The angles of depression of the top and bottom of an 8-m-high building from the top of a multi-storeyed building are 30° and 45°, respectively. Find the height of the multi-storeyed building and the distance between the two buildings. [NCERT]

33. From a point on a bridge across a river, the angles of depression of the banks on opposite sides of the river are 30° and 45°, respectively. If the bridge is at a height of 3 m from the banks, find the width of the river. [NCERT]

34. A kite is flying at a height of 60 m above the ground. The string attached to the kite is temporarily tied to a point on the ground. The inclination of the string to the ground is 60°. Find the length of the string, assuming that there is no slack in the string. [NCERT]

35. A 1.5-m-tall boy is standing at some distance from a 30-m-high building. The angle of elevation from his eyes to the top of the building increases from 30° to 60° as he walks towards the building. Find the distance he walked towards the building. [NCERT]

36. Two poles of equal height are standing opposite each other on either side of the road, which is 80 m wide. From a point between them on the road, the angles of elevation of the top of the poles are 60° and 30°, respectively. Find the height of the poles and the distances of the point from the poles. **[NCERT]**

37. A TV tower stands vertically on the bank of a canal. From a point on the other bank directly opposite the tower, the angle of elevation of the top of the tower is 60°. From another point 20 m away from this point on the line joining this point to the foot of the tower, the angle of elevation of the top of the tower is 30° (see figure). Find the height of the tower and the width of the canal. **[NCERT]**

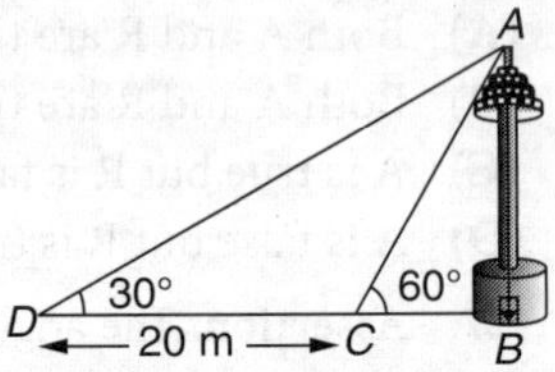

38. A 1.2-m-tall girl spots a balloon moving with the wind in a horizontal line at a height of 88.2 m from the ground. The angle of elevation of the balloon from the eyes of the girl at any instant is 60°. After some time, the angle of elevation reduces to 30° (see figure). Find the distance travelled by the balloon during the interval.

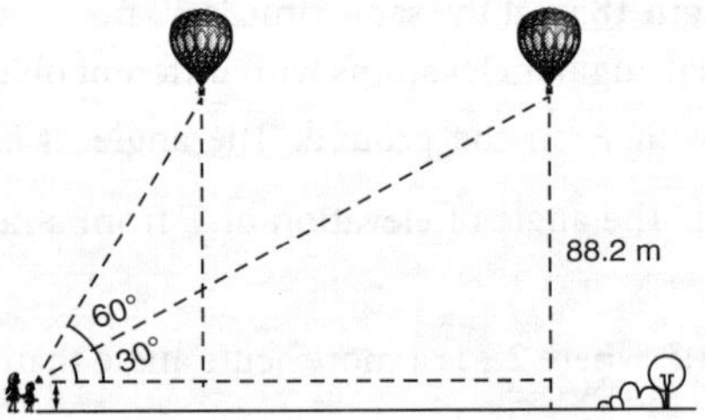

[NCERT]

39. A man on the deck of a ship 12 m above water level observes that the angle of elevation of the top of a cliff is 60° and the angle of depression of the base of the cliff is 30°. Find the distance of the cliff from the ship and the height of the cliff. [Use $\sqrt{3} = 1.732$]

40. From the top of a tower 50 m high, the angles of depression of the top and bottom of a pole are observed to be 45° and 60°, respectively. Find the height of the pole. **[Foreign 2016]**

41. As observed from the top of a lighthouse, 100 m above sea level, the angle of depression of a ship, sailing directly towards it, changes from 30° to 60°. Determine the distance travelled by the ship during the period of observation. [Use $\sqrt{3} = 1.732$] **[All India, Foreign 2016]**

42. A bridge across a river makes an angle of 45° with the river bank (see figure). If the length of the bridge across the river is 150 m, what is the width of the river?

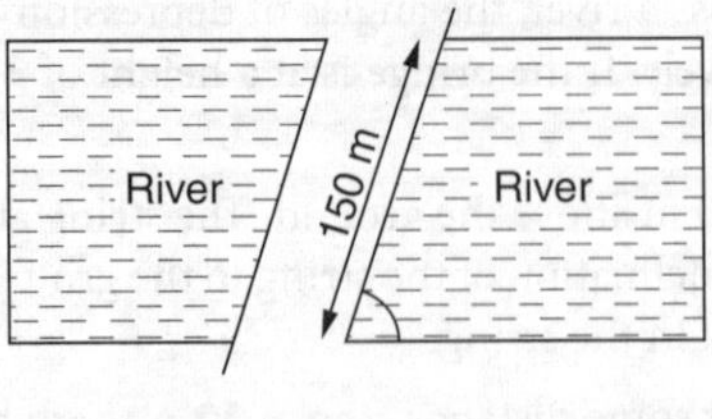

43. Two ships are in the sea on either side of a lighthouse in such a way that the ships and the lighthouse are in the same straight line. The angles of depression of the two ships as observed from the top of the lighthouse are 60° and 45°. If the height of the lighthouse is 200 m, find the distance between the two ships. [Use $\sqrt{3} = 1.73$] **[Delhi 2014]**

44. The angle of elevation of the top of a vertical tower from a point on the ground is 60°. From another point 10 m vertically above the first, its angle of elevation is 30°. Find the height of the tower. **[AI 2011]**

45. The angles of elevation and depression of the top and bottom of a lighthouse from the top of a 60-m-high building are 30° and 60°, respectively. Find
(i) The difference between the heights of the lighthouse and the building
(ii) The distance between the lighthouse and the building **[AI 2012]**

46. A parachutist is descending vertically and makes angles of depression of 45° and 60° at two observation points 100 m apart on the left side of himself. Find, in metres, the approximate height from which he falls and also find, in metres, the approximate distance of the point where he falls on the ground from the first observation point.

47. The angle of elevation of the top of a tower at a distance of 120 m from a point *A* on the ground is 45°. If the angle of elevation of the top of a flagstaff fixed at the top of the tower from *A* is 60°, find the height of the flag staff. [Use $\sqrt{3} = 1.73$] **[AI 2014]**

48. A highway leads to the foot of a 300-m-high tower. An observatory is set at the top of the tower. It sees a car moving towards it at an angle of depression of 30°. After 15 seconds, the angle of depression becomes 60°. Find the distance travelled by the car during this time.

49. A fire in a building *B* is reported by telephone to two fire stations *P* and *Q*, 20 km apart on a straight road. *P* observes that the fire is at an angle of 60° to the road and *Q* observes that it is at an angle of 45° to the road. Which station should send its team and how much will this team have to travel?

50. The angle of elevation of the top *Q* of a vertical tower *PQ* from a point *X* on the ground is 60°. At a point *Y*, 40 m vertically above *X*, the angle of elevation is 45°. Find the height of the tower *PQ* and the distance *XQ*.

51. A ladder of length 6 m makes an angle of 45° with the floor while leaning against one wall of a room. If the foot of the ladder is kept fixed on the floor and it is made to lean against the opposite wall of the room, it makes an angle of 60° with the floor. Find the distance between these two walls of the room. **[Foreign 2011]**

52. The horizontal distance between two poles is 15 m. The angle of depression of the top of the first pole, as seen from the top of second pole, is 30°. If the height of the second pole is 24 m, find the height of the first pole. [Use $\sqrt{3} = 1.732$] **[Delhi 2013]**

53. From a window (60 metres above the ground) of a house on a street, the angles of elevation and depression of the top and the foot of another house on the opposite side of the street are 60° and 45°, respectively. Show that the height of the opposite house is $60(1 + \sqrt{3})$ metres.

54. The length of the shadow of a tower on level ground is found to be $2x$ metres longer when the sun's altitude is 30° than when it was 45°. Prove that the height of tower is $x(\sqrt{3} + 1)$ metres.

55. The angle of elevation θ of the top of a lighthouse, as seen by a person on the ground, is such that tan $\theta = \frac{5}{12}$. When the person moves a distance of 240 m towards the lighthouse, the angle of elevation becomes ϕ, such that tan $\phi = \frac{3}{4}$. Find the height of the lighthouse.

56. A path separates two walls. A ladder leaning against one wall rests at a point on the path. It reaches a height of 90 m on the wall and makes an angle of 60° with the ground. If, while resting at the same point on the path, it were made to lean against the other wall, it would have made an angle of 45° with the ground. Find the height it would have reached on the second wall.

57. The pilot of an aircraft flying horizontally at a speed of 1200 km/h observes that the angle of depression of a point on the ground changes from 30° to 45° in 15 seconds. Find the height at which the aircraft is flying.

58. The lower window of a house is at a height of 2 m above the ground and its upper window is 4 m vertically above the lower window. At a certain instant, the angles of elevation of a balloon from these windows are observed to be 60° and 30°, respectively. Find the height of the balloon above the ground.

59. The angles of depression of two ships from the top of a lighthouse and on the same side of it are found to be 45° and 30°. If the ships are 200 m apart, find the height of the lighthouse. **[Delhi 2012, NCERT Exemplar]**

60. An observer 1.5 m tall is 20.5 m away from a tower 22 m high. Determine the angle of elevation of the top of the tower from the eye of the observer. **[NCERT]**

61. The length of a string between a kite and a point on the staff of the roof of a building 10 m high is 180 m. If the string makes an angle θ with the level ground such that $\tan\theta = \frac{4}{3}$, how high is the kite from the ground? **[NCERT]**

62. Two lamp posts are of equal height. A boy measured the elevation of the top of each lamp post from the mid-point of the line segment joining the feet of the lamp posts as 30°. After walking 15 m towards one of them, he measured the elevation of the top of nearest lamp post at the point where he stands as 60°. Determine the height of each lamp post and the distance between them.

63. An aeroplane is observed at the same time by two anti-aircraft guns 6000 m apart to be at elevations of 30° and 45°, respectively. Find the height of the plane and its horizontal distance from the nearer gun.

64. A vertical tower stands on a horizontal plane and is surmounted by a vertical flagstaff of height h. At a point on the plane, the angles of elevation of the bottom and top of the flagstaff are α and β, respectively. Prove that the height of the tower is $\frac{h\tan\alpha}{\tan\beta - \tan\alpha}$.

65. Two stations are due south of a leaning tower which leans towards the north at distances a and b from its foot. If α, β are the elevations of the top of the tower from the stations, prove that its inclination θ to the horizontal is given by

$$\cot\theta = \frac{b\cot\alpha - a\cot\beta}{b - a}$$

66. The angle of elevation of a tower from a point on the same level as the foot of the tower is 30°. On advancing 150 metres towards the foot of the tower, the angle of elevation of the tower becomes 60°. Show that the height of the tower is 129.9 m. [Use $\sqrt{3} = 1.732$]

67. If the angle of the elevation of a cloud from a point h metres above a lake is α and the angle of depression of its reflection in the lake is β, prove that the distance of the cloud from the point of observation is $\frac{2h\sec\alpha}{\tan\beta - \tan\alpha}$. **[CBSE 2004]**

68. From an aeroplane vertically above a straight horizontal road, the angles of depression of two consecutive mile stones on opposite sides of the aeroplane are observed to be α and β. Show that the height in miles of the aeroplane above the road is given by $\frac{\tan\alpha\tan\beta}{\tan\alpha + \tan\beta}$. **[CBSE 2004]**

69. From the top of a lighthouse, the angles of depression of two ships on opposite sides of it are observed to be α and β. If the height of the lighthouse is h metres and the line joining the ships passes through the foot of the lighthouse, show that the distance between the ships is $\frac{h(\tan\alpha + \tan\beta)}{\tan\alpha\tan\beta}$ metres.

70. The angle of elevation of the top of a hill at the foot of a tower is 60° and the angle of elevation of the top of the tower from the foot of the hill is 30°. If the tower is 50 m high, what is the height of the hill? **[CBSE 2006C]**

71. From the top of a 50-m-high tower, the angles of depression of the top and bottom of a pole are observed to be 45° and 60°, respectively. Find the height of the pole.

72. A balloon is connected to a meteorological ground station by a cable of length 215 m inclined at 60° to the horizontal. Determine the height of the balloon from the ground. Assume that there is no slack in the cable.

73. A man sitting at a height of 20 m on a tall tree in a small island in the middle of a river observes two poles directly opposite each other on the two banks of the river and in line with the foot of the tree. If the angles of depression of the feet of the poles from a point at which the man is sitting on the tree on either side of the river are 60° and 30°, respectively. Find the width of the river.

74. Two men standing on either side of a tower 60 m high observe the angles of elevation of the top of the tower to be 45° and 60°, respectively. Find the distance between the two men.

75. On a horizontal plane, there is a vertical tower with a flag pole at the top of tower. At a point 9 metres away from the foot of the tower, the angle of elevation of the top and bottom of the flag pole are 60° and 30°, respectively. Find the height of the tower and the flag pole mounted on it. **[CBSE 2005]**

76. A bird is sitting at the top of a tree, which is 80 m high. The angle of elevation of the bird, from a point on the ground is 45°. The bird flies away from the point of observation horizontally and remains at a constant height. After 2 seconds, the angle of elevation of the bird from the point of observation becomes 30°. Find the speed of flying of the bird. **[Delhi 2016]**

77. A moving boat is observed from the top of a 150-m-high cliff moving away from the cliff. The angle of depression of the boat changes from 60° to 45° in 2 minutes. Find the speed of the boat in m/min. **[Delhi 2017]**

78. As observed from the top of a 100-m-high house from sea level, the angles of depression of two ships are 30° and 45°. If one ship is exactly behind the other on the same side of the lighthouse, find the distance between the two ships. [Use $\sqrt{3} = 1.732$] **[Delhi 2018]**

79. The angle of elevation of the top of a hill at the foot of a tower is 60° and the angle of depression from the top of a tower to the foot of hill is 30°. If the tower is 50 m high, find the height of the hill. **[CBSE 2006 (C), 2018 (C), Delhi 2018]**

80. An aeroplane is flying at a height of 300 m above the ground. Flying at this height, the angles of depression from the aeroplane of two points on both banks of a river in opposite directions are 45° and 60°, respectively. Find the width of the river. [Use $\sqrt{3} = 1.732$] **[AI 2017]**

81. A man at the top of a vertical tower observes a car moving at a uniform speed coming directly towards it. If it takes 12 minutes for the angle of depression to change from 30° to 45°, how soon after this will the car reach the tower?

82. Amit standing on a horizontal plane finds a bird flying at a distance of 200 m from him at a deviation of 30°. Deepak standing on the roof of a 50-m-high building finds the angle elevation of the same bird to be 45°. Amit and Deepak are on opposite sides of the bird. Find the distance of the bird from Deepak. **[CBSE, Allahabad 2019]**

Answers

Multiple Choice Questions

1. (d) **2.** (c) **3.** (a) **4.** (b) **5.** (b) **6.** (d) **7.** (b) **8.** (c) **9.** (c) **10.** (a)
11. (c) **12.** (d) **13.** (c) **14.** (b) **15.** (a) **16.** (b) **17.** (b) **18.** (b) **19.** (a) **20.** (d)
21. (b) **22.** (c) **23.** (d)

Fill in the Blanks

24. 60° **25.** same **26.** equal

Assertion–Reason

27. (D) **28.** (A) **29.** (B)

Subjective Questions

30. 7.32 m, 17.32 m **31.** $(20\sqrt{3})$ m **32.** $4(3+\sqrt{3})$ m, $4(3+\sqrt{3})$ m **33.** $3(\sqrt{3}+1)$ m
34. $40\sqrt{3}$ m **35.** $19\sqrt{3}$ m **36.** $20\sqrt{3}$ m, 20 m, 60 m **37.** $10\sqrt{3}$ m, 10 m
38. $58\sqrt{3}$ m **39.** 20.78 m, 48 m **40.** $\frac{50(\sqrt{3}-1)}{\sqrt{3}}$ m **41.** 115.466 m **42.** $75\sqrt{2}$ m
43. 315.33 m **44.** 15m **45.** 20 m, $20\sqrt{3}$ m **46.** 236.6 m **47.** 87.6 m **48.** $200\sqrt{3}$ m
49. *P*, 14.66 km **50.** 94.64 m, 109.3 m **51.** 7.23 m **52.** 15.34 m **55.** 225 m
56. 73.47 m **57.** 6.83 m **58.** 8 m **59.** $100(\sqrt{3}+1)$ m. **60.** 45° **61.** 154 m
62. $\frac{15\sqrt{3}}{2}$ m, 45 m **63.** $3000(\sqrt{3}+1)$ m **70.** 150 m **71.** 21.13 m **72.** 186 m **73.** $\frac{80}{\sqrt{3}}$ m
74. 94.64 m **75.** $3\sqrt{3}$ m, $6\sqrt{3}$ m **76.** 29.28 m/s **77.** 31.7 m/min
78. 73.2 m **79.** 150 m **80.** 473.2 m **81.** 16 min. 39 sec. **82.** 70.7 m

Some Applications of Trigonometry (Basic)

SELF-ASSESSMENT TEST

Time allowed: 90 min. Max. Marks: 40

General Instructions

1. *All questions are compulsory. There are 17 questions in all.*
2. *This question paper has three sections: Section A, Section B and Section C.*
3. *Section A contains 8 questions of one mark each, Section B contains four questions of three marks each, Section C contains five questions of four marks each.*

Section A [1 Mark Each]

Direction: (Q. Nos. 1–5) These are Multiple Choice Questions. Select the most appropriate answer for the given questions.

1. The length of the shadow of a tower on the plane ground is $\sqrt{3}$ times the height of the tower. The angle of elevation of the sun is
   (a) 45° (b) 30° (c) 60° (d) 90°
2. The angle of elevation of the top of a tower from a point 20 metres from its base is 45°. The height of the tower is
   (a) 10 m (b) 20 m (c) 30 m (d) $20\sqrt{3}$ m
3. The string of a kite is 100 m long and it makes an angle of 60° with the horizontal. If there is no slack in the string, the height of the kite from the ground is
   (a) $50\sqrt{3}$ m (b) $100\sqrt{3}$ m (c) $50\sqrt{2}$ m (d) 100 m
4. From the top of a cliff 20 m high, the angle of elevation of the top of a tower is found to be equal to the angle of depression of the foot of the tower. The height of the tower is
   (a) 20 m (b) 40 m (c) 60 m (d) 80 m
5. If the height of a vertical pole is equal to the length of its shadow on the ground, the angle of elevation of the sun is
   (a) 0° (b) 30° (c) 45° (d) 60°
6. The angles of elevation and depression are ________.
7. At a point 25 m from the foot of a tower, the angle of elevation of the top of the tower is 30°. Find the height of the tower.
8. The length of the shadow of a tower on the plane ground is $\sqrt{3}$ times the height of the tower. Find the angle of elevation of the sun.

Section B [3 Marks Each]

9. The angle of elevation of the top of a tower from a point on the ground, which is 30 m from the foot of the tower, is 30°. Find the height of the tower.
10. The shadow of a tower, when the angle of elevation of the sun is 45°, is found to be 10 m longer than when it was 60°. Find the height of the tower.

11. The angle of elevation of the top of a building from the foot of the tower is 30° and the angle of elevation of the top of the tower from the foot of the building is 60°. If the tower is 50 m high, find the height of the building.

12. The angles of depression of the top and bottom of an 8-m-high building from the top of a multi-storeyed building are 30° and 45°, respectively. Find the height of the multi-storeyed building and the distance between the two buildings.

Section C [4 Marks Each]

13. The angle of elevation of an aeroplane from a point on the ground is 60°. After a flight of 30 seconds, the angle of elevation becomes 30°. If the aeroplane is flying at a constant height of $3000\sqrt{3}$ m, find the speed of the aeroplane.

14. The angle of elevation of the top of a tower from two points at a distance of 4 m and 9 m from the base of the tower and in the same straight line with it are complementary. Prove that the height of the tower is 6 m.

15. As observed from the top of a 100-m-high house from sea level, the angles of depression of two ships are 30° and 45°. If one ship is exactly behind the other on the same side of the lighthouse, find the distance between the two ships. [Use $\sqrt{3} = 1.732$]

16. Two ships are in the sea on either side of a lighthouse in such a way that the ships and the lighthouse are in the same straight line. The angles of depression of the two ships as observed from the top of the lighthouse are 60° and 45°. If the height of the lighthouse is 200 m, find the distance between the two ships. [Use $\sqrt{3} = 1.73$]

17. The angle of elevation of the top of a building from the foot of the tower is 30° and the angle of elevation of the top of tower from the foot of the building is 60°. If the tower is 50 m high, find the height of the building.

Answers

1. (b) **2.** (b) **3.** (a) **4.** (a) **5.** (c) **6.** Equal **7.** 14.43 m **8.** 30° **9.** $10\sqrt{3}$ m

10. 23.66 m **11.** $16\frac{2}{3}$ m **12.** $4(3+\sqrt{3})$ m, $4(3+\sqrt{3})$ m **13.** 720 km/h

15. 73.2 m **16.** 315.33 m **17.** $0.8(\sqrt{3}+1)$ m

Some Applications of Trigonometry (Standard)

SELF-ASSESSMENT TEST

Time allowed: 90 min. Max. Marks: 40

General Instructions

1. *All questions are compulsory. There are 17 questions in all.*
2. *This question paper has three sections: Section A, Section B and Section C.*
3. *Section A contains 8 questions of one mark each, Section B contains four questions of three marks each, Section C contains five questions of four marks each.*

Section A [1 Mark Each]

Direction: (Q. Nos. 1–5) These are Multiple Choice Questions. Select the most appropriate answer for the given questions.

1. The length of the shadow of a tower on the plane ground is $\sqrt{3}$ times the height of the tower. The angle of elevation of the sun is

   (a) 45° (b) 30° (c) 60° (d) 90°

2. In the figure, *PR* is equal to

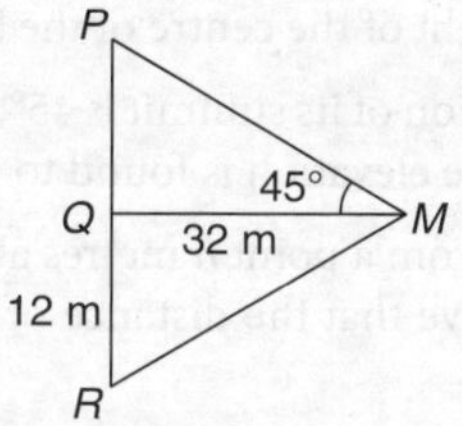

   (a) 32 m (b) 44 m (c) $32\sqrt{2}$ m (d) $44\sqrt{2}$ m

3. The length of the shadow of a tower standing on level ground is found to be 2*x* metres longer when the sun's elevation is 30° than when it is 45°. The height of the tower is

   (a) $(2\sqrt{3}x)$ m (b) $(2\sqrt{2}x)$ m (c) $(\sqrt{3}-1)\,x$ m (d) $(\sqrt{3}+1)x$ m

4. A tower AB is 20 m high and BC, its shadow on the ground, is $20\sqrt{3}$ m long. The sun's altitude is

   (a) 30° (b) 45° (c) 60° (d) none of these

5. The length of the shadow of a person is *x* when the angle of elevation of the sun is 45°. If the length of the shadow increases by $(\sqrt{3}-1)\,x$, then the angle of elevation of the sun should become

   (a) 60° (b) 45° (c) 30° (d) 20°

6. If the height of a tower is doubled and the distance between the observer and the foot of the tower is also doubled, then the angle of elevation remains ____________.

7. The length of the shadow of a tower on the plane ground is $\sqrt{3}$ times the height of the tower. Find the angle of elevation of the sun.

8. A vertical post 9 m high is broken at a certain height, and its upper part, not completely separated, meets the ground at an angle of 30°. Find the height at which the post is broken.

Section B [3 Mark Each]

9. A 1.5-m-tall boy is standing at some distance from a 30-m-high building. The angle of elevation from his eyes to the top of the building increases from 30° to 60° as he walks towards the building. Find the distance he walked towards the building.

10. The angles of elevation of the top of a tower from two points at a distance of 4 m and 16 m from the base of the tower and in the same straight line with it are complementary. Prove that the height of the tower is 8 m.

11. A straight highway leads to the foot of a tower. A man standing at the top of the tower observes a car at an angle of depression of 30°, which is approaching the foot of the tower at a uniform speed. Six seconds later, the angle of depression of the car is found to be 60°. Find the time further taken by the car to reach the foot of the tower.

12. The angle of elevation of the top Q of a vertical tower PQ from a point X on the ground is 60°. At a point Y, 40 m vertically above X, the angle of elevation is 45°. Find the height of the tower PQ and the distance XQ.

Section C [4 Mark Each]

13. The angle of elevation of a jet fighter from a point A on the ground is 60°. After a flight of 15 seconds, the angle of elevation changes to 30°. If the jet is flying at a speed of 720 km/h, find the constant height at which the jet is flying. [$\sqrt{3} = 1.73$]

14. A round balloon of radius r subtends an angle α at the eye of the observer, while the angle of elevation of its centre is β. Prove that the height of the centre of the balloon is $r \sin \beta \operatorname{cosec} \alpha/2$.

15. At the foot of a mountain, the elevation of its summit is 45°; after ascending 1000 m towards the mountain up a slope of 30° inclination, the elevation is found to be 60°. Find the height of the mountain.

16. If the angle of elevation of a cloud from a point h metres above a lake is α and the angle of depression of its reflection in the lake is β, prove that the distance of the cloud from the point of observation is $\dfrac{2h \sec\alpha}{\tan\beta - \tan\alpha}$.

17. From the top of a lighthouse, the angles of depression of two ships on opposite sides of it are observed to be α and β. If the height of the lighthouse is h metres and the line joining the ships passes through the foot of the lighthouse, show that the distance between the ships is $\dfrac{h(\tan\alpha + \tan\beta)}{\tan\alpha \tan\beta}$ metres.

Answers

1. (b) 2. (b) 3. (d) 4. (a) 5. (c) 6. Same 7. 30° 8. 3 m 9. $19\sqrt{3}$ m
11. 3 sec 12. 94.64 m, 109.3 m 13. 2595 m 15. 1.366 km

CHAPTER

10 Circles

Key Concepts

- Definition of circle, chord, secant and tangent
- The tangent at any point of a circle is perpendicular to the radius through the point of contact
- The tangents drawn from an external point to a circle are equal in length and equally inclined

Definitions

(i) Circle: A collection of all the points on a plane that are at a constant distance from a fixed point on a plane. The fixed point is known as the centre and the constant distance is called the radius. It is represented as $C(O, r)$.

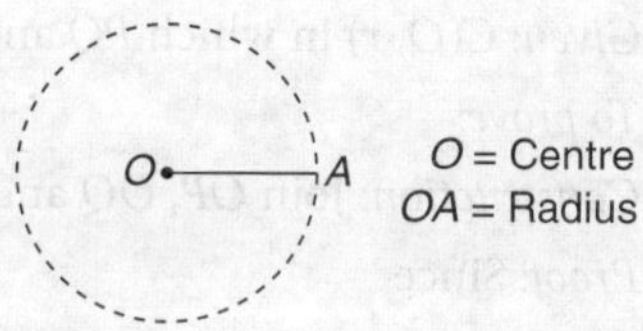

(ii) Chord: A line segment joining any two points on a circle.

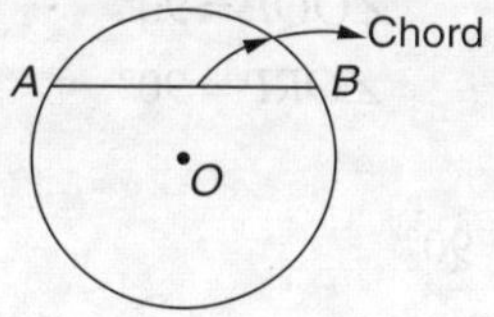

(iii) Secant: A line that intersects a circle at two distinct points.

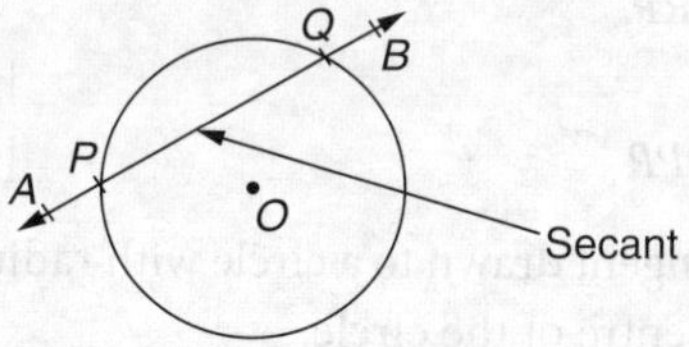

(iv) Tangent: A line that meet the circle at exactly one point. Such a point is called the point of contact.

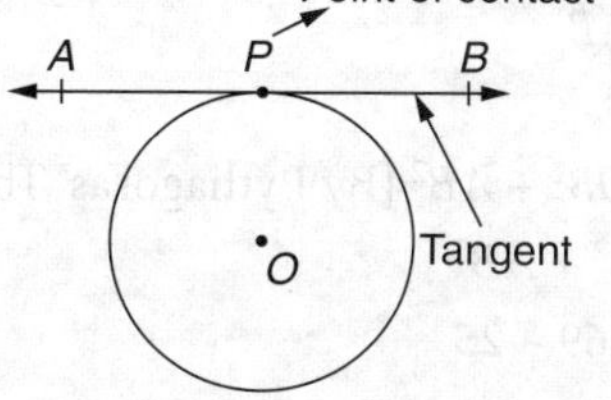

Theorem 1 The tangent at any point of a circle is perpendicular to the radius through the point of contact. **[CBSE Frequently Asked]**

Given: $C(O, r)$ in which XY is a tangent to the circle and 'P' is the point of contact.

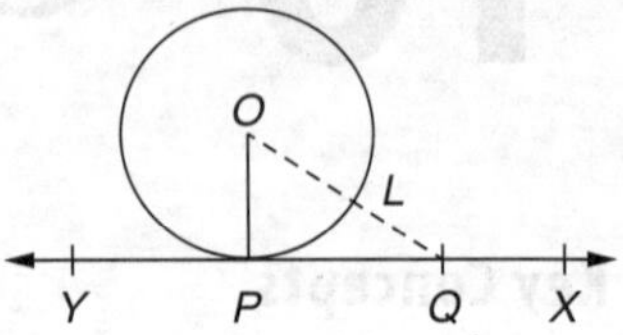

To prove: $OP \perp XY$

Proof: Let us suppose OP is not $\perp$ on XY, say $OQ \perp XY$.

$OP = OL$ [$\because$ Equal radii of the circle] ...(i)

$OQ > OL$ [From the figure]

$OQ > OP$ [From (i)]

We know that $\perp$ distance is the shortest distance. So, our assumption is wrong. Hence, $OP \perp XY$.

Theorem 2 The tangents drawn from an external point to a circle are equal in length and equally inclined. **[CBSE Frequently Asked]**

Given: $C(O, r)$ in which PQ and PR are tangents drawn from an external point P.

To prove: $PQ = PR$

Construction: Join OP, OQ and OR.

Proof: Since $OQ \perp PQ$ [$\because$ Radius is $\perp$ to the point of contact]

$OR \perp PR$ [$\because$ Radius is $\perp$ to the point of contact]

That is $\angle OQP = 90°$ ...(i)

And $\angle ORP = 90°$ ...(ii)

Consider ΔOQP and ΔORP

$\angle OQP = \angle ORP = 90°$ [From (i) and (ii)]

$OP = OP$ [Common]

$OQ = OR$ [$\because$ Equal radii]

$\Delta OQP \cong \Delta ORP$ [RHS]

$PQ = PR$ [CPCT]

$\angle OPQ = \angle OPR$ [CPCT]

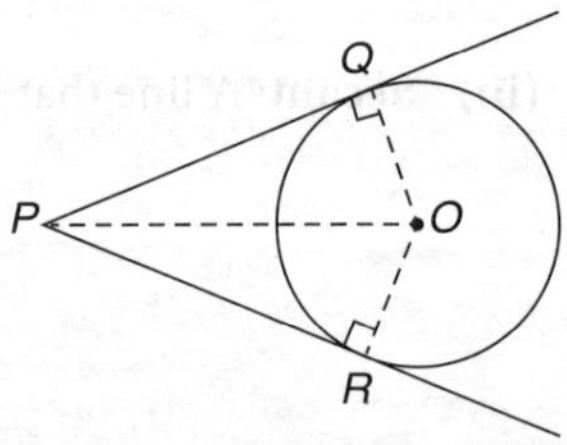

Example 1 Find the length of the tangent drawn to a circle with radius 5 cm from a point 13 cm away from the centre of the circle.

Solution: Given: $OA = 13$ cm, $OB = 5$ cm. We know that the tangent at any point of a circle is perpendicular to the radius through the point of contact.

That is $OB \perp AB$

So, OBA is a Δ right-angled at B.

$\therefore$ $OA^2 = OB^2 + AB^2$ [By Pythagoras' Theorem]

$\Rightarrow$ $13^2 = 5^2 + AB^2$

$\Rightarrow$ $AB^2 = 169 - 25$

$\Rightarrow$ $AB^2 = 144$

$\therefore$ $AB = 12$ cm

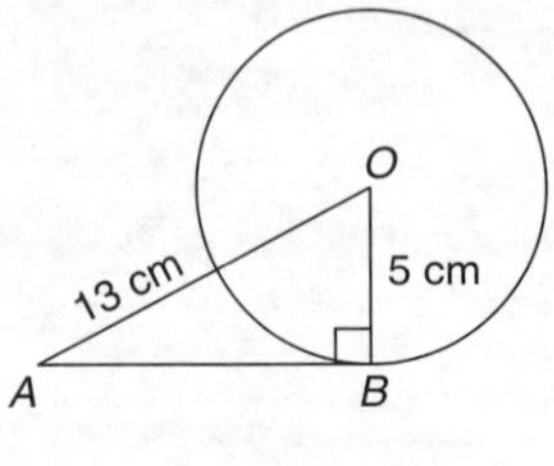

Therefore, the length of the tangent AB is 12 cm.

Example 2 A point M is 26 cm away from the centre of a circle and the length of the tangent drawn from M to the circle is 24 cm. Find the radius of the circle.

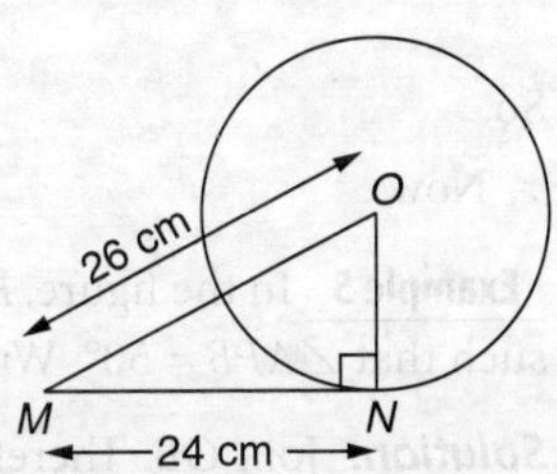

Solution: Given: $OM = 26$ cm, $MN = 24$ cm. We know that the tangent at any point to a circle is perpendicular to the radius through the point of contact.

$\therefore$ ONM is a Δ right-angled at N.

$OM^2 = ON^2 + MN^2$ [By Pythagoras' Theorem]

$\Rightarrow \quad (26)^2 = ON^2 + (24)^2$

$\Rightarrow \quad ON^2 = (26)^2 - (24)^2$

$\Rightarrow \quad ON^2 = 676 - 576 = 100$

$\therefore \quad ON = 10$ cm

Hence, the radius of the circle is 10 cm.

Example 3 A line through the centre O of a circle of radius 7 cm cuts the tangent at a point P on the circle and at Q such that $PQ = 24$ cm. Find OQ.

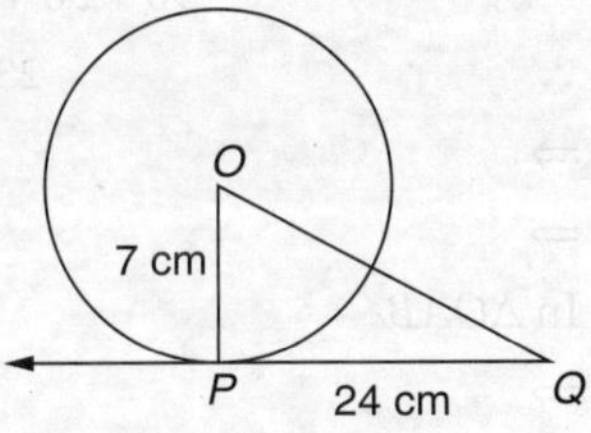

Solution: The tangent at a point on a circle is perpendicular to the radius through the point of contact.

Therefore, OP is perpendicular at PQ. In right ΔOPQ

$OQ^2 = OP^2 + PQ^2$ [$\because$ Pythagoras' Theorem]

$\Rightarrow \quad OQ^2 = 7^2 + 24^2$

$\Rightarrow \quad OQ^2 = 49 + 576$

$\Rightarrow \quad OQ^2 = 625$

$\Rightarrow \quad OQ = 25$ cm.

1 Try to Solve

1. A tangent PQ at a point P of circle of radius 5 cm meets a line through the centre O at a point Q so that $OQ = 12$ cm. Length PQ is:
   (a) 12 cm (b) 13 cm (c) 8.5 cm (d) $\sqrt{119}$ cm **[NCERT]**
2. O is the centre of a circle of radius 8 cm. The tangent at a point A on the circle cuts a line through O at B such that $AB = 15$ cm. Find OB.
3. If the tangent at a point P to a circle with centre O cuts a line through O at Q such that $PQ = 24$ cm and $OQ = 25$ cm, find the radius of the circle. **[CBSE (CCE) 2015]**

Answers

1. (d) 2. 17 cm 3. 7 cm

Example 4 In the figure, CP and CQ are tangents from an external point C to a circle with centre O. AB is another tangent that touches the circle at R. If $CP = 11$ cm and $BR = 4$ cm, find the length of BC. **[CBSE (CCE) 2010]**

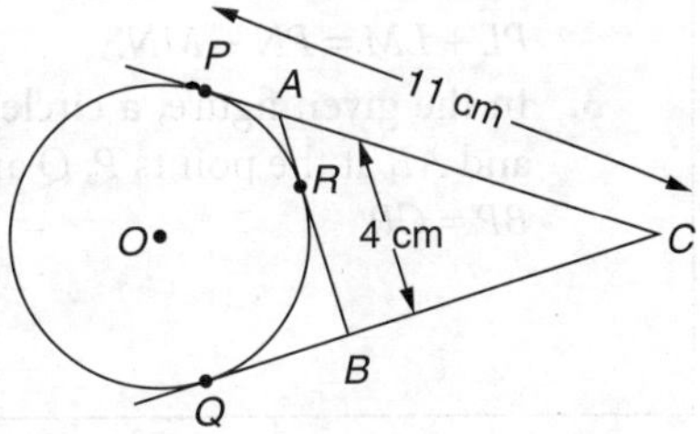

Solution: In the given figure, $CP = CQ$ [$\because$ Tangents drawn from an external point to a circle are equal in length]

So $CP = CQ = 11$ cm

Also $BR = BQ$ [∵ Tangents drawn from an external point to a circle are equal in length]

So $BR = BQ = 4$ cm

∴ Now $BC = CQ - BQ = (11 - 4)$ cm $= 7$ cm

Example 5 In the figure, *PA* and *PB* are tangents to the circle with centre *O* such that $\angle APB = 50°$. Write the measure of $\angle OAB$. **[CBSE (CCE) 2015]**

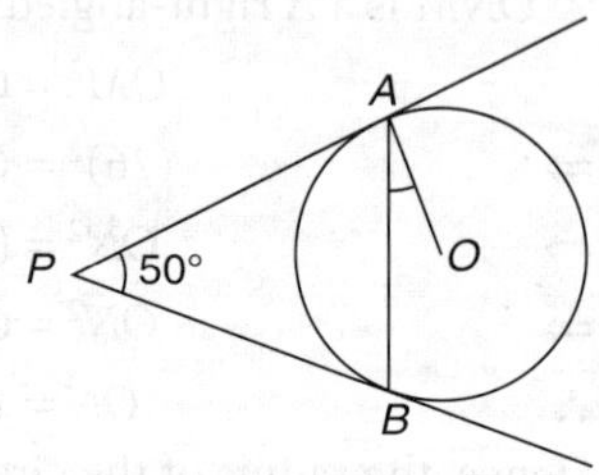

Solution: Join *OB*. Therefore, *PA* and *PB* are tangents to the circle drawn from an external point *P*. We know that the tangent is perpendicular to the radius (*r*).

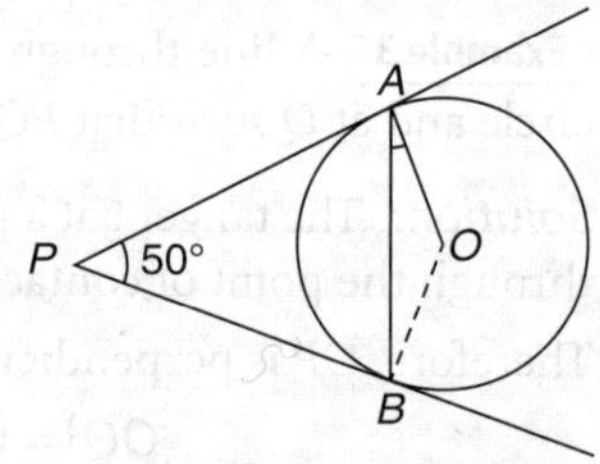

$$\angle OAP = \angle OBP = 90°$$

Consider quadrilateral *OAPB*. Then

$\angle OAP + \angle APB + \angle OBP + \angle AOB = 360°$ [∵ Angle sum property (ASP) of quadrilaterals]

$$90 + 50 + 90 + \angle AOB = 360°$$

∴ $230° + \angle AOB = 360°$

⇒ $\angle AOB = 360° - 230°.$

⇒ $\angle AOB = 130°$

In ΔOAB $OA = OB$ [∵ Equal radii]

⇒ $\angle A = \angle B = x$ (say) [∵ Angles opposite to equal sides are equal]

$\angle A + \angle B + \angle AOB = 180°$ [∵ ASP of triangles]

⇒ $x + x + 130° = 180°$

⇒ $2x = 50°$

⇒ $x = 25°$

∴ $\angle OAB = 25°$

2 Try to Solve

1. Two tangents *TP* and *TQ* are drawn to a circle with centre *O* from an external point *T*. Prove that $\angle PTQ = 2\angle OPQ$. **[NCERT, CBSE (CCE) 2016]**
2. Prove that the lengths of the tangents drawn from an external point to a circle are equal.
   Using the above, prove the following:
   In the figure, *PA* and *PB* are tangents from an external point *P* to a circle with centre *O*. *LN* touches the circle at *M*. Prove that $PL + LM = PN + MN$. **[CBSE (CCE) 2010]**

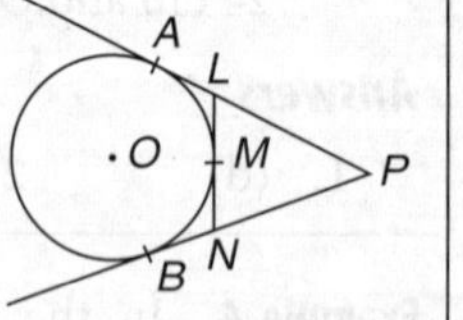

3. In the given figure, a circle inscribed in ΔABC, touches its sides *BC*, *CA* and *AB* at the points *P*, *Q* and *R*, respectively. If $AB = AC$, then prove that $BP = CP$. **[Foreign 2013]**

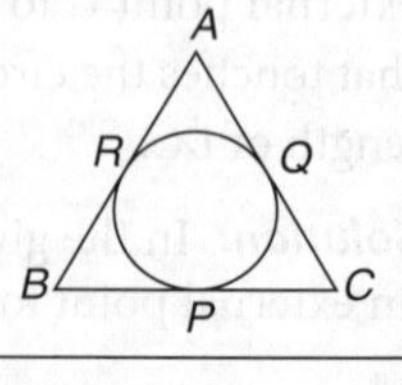

Example 6 If the radii of two concentric circles are 15 cm and 17 cm, show that the length of the chord of one circle, which is tangent to the other circle, is 16 cm. **[CBSE (CCE) 2012]**

Solution: *Given:* $OA = 17$ cm, $OC = 15$ cm. AB is a chord of the larger circle, which is the tangent to the smaller circle at point C. Therefore, $OC \perp AB$ ($\because$ A tangent is perpendicular to the radius at the point of contact).

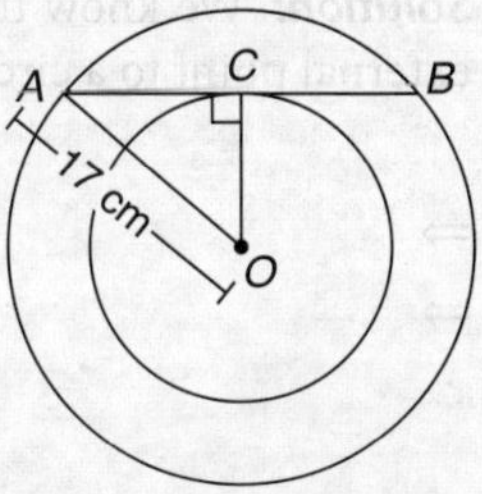

In right ΔAOC $\quad OA^2 = OC^2 + AC^2$ $\quad$ [$\because$ Pythagoras' Theorem]

$\Rightarrow \quad 17^2 = 15^2 + AC^2$

$\Rightarrow \quad AC^2 = 17^2 - 15^2$

$\Rightarrow \quad AC^2 = 289 - 225 = 64$

$\therefore \quad AC = 8$ cm

We know that the perpendicular drawn from the centre to a chord bisects the chord.

$\therefore$ C is the mid-point of AB

So $\quad AB = 2AC = 2 \times 8 = 16$ cm.

Example 7 In the figure, PA and PB are tangents to a circle with centre O drawn from external point P. QR is a tangent at point S. Prove that $PQ + QS = PR + RS$. **[CBSE 2008]**

Solution: *Given*: PA and PB are tangents to a circle with centre O. QR is a tangent at S.

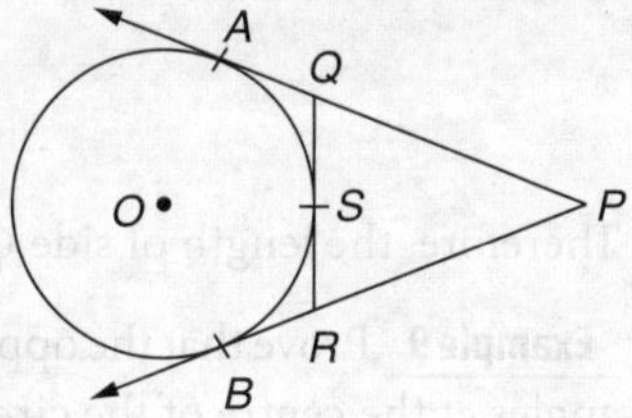

To prove: $\quad PQ + QS = PR + RS$

Proof: We know that the lengths of tangents drawn from the same external point to a circle are equal.

$QA = QS \quad$ ...(i)

$RB = RS \quad$ ...(ii)

And $\quad PA = PB$

$\Rightarrow \quad PQ + QA = PR + RB$

$\therefore \quad PQ + QS = PR + RS \quad$ [$\because$ $QA = QS$ and $RB = RS$ from (i) and (ii)]

3 Try to Solve

1. Two concentric circles are of radii 5 cm and 3 cm. Find the length of the chord of the larger circle which touches the smaller circle. **[NCERT]**
2. A quadrilateral $ABCD$ is drawn to circumscribe a circle (see figure). Prove that $AB + CD = AD + BC$. **[NCERT, CBSE 2015]**

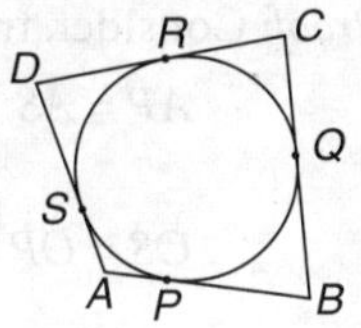

3. In the figure, XP and XQ are two tangents to the circle with centre O, drawn from an external point X. ARB is another tangent touching the circle at R. Prove that $XA + AR = XB + BR$. **[CBSE (CCE) 2014]**

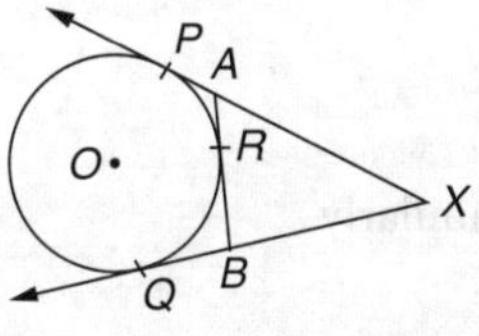

Answer

1. 8 cm

Example 8 In the figure, quadrilateral *ABCD* circumscribes the circle. Find the length of the side *CD*. **[CBSE (CCE) 2011]**

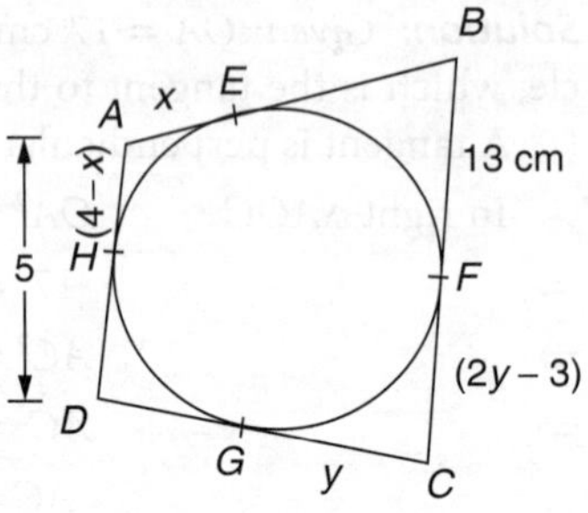

Solution: We know that the length of tangents drawn from the same external point to a circle are equal in length.

$$AE = AH$$

$\Rightarrow \quad x = 4 - x$

$\Rightarrow \quad 2x = 4$

$\therefore \quad x = 2$

$$AH = 4 - x = 4 - 2 = 2 \text{ cm}$$

$$HD = AD - AH$$

$$= 5 - 2 = 3 \text{ cm}$$

Now $\quad DG = DH = 3 \text{ cm}$

And $\quad CG = CF$

$\Rightarrow \quad y = 2y - 3$

$\Rightarrow \quad -2y + y = -3$

$\therefore \quad y = 3$

$$CG = y = 3 \text{ cm}$$

$$CD = CG + DG = 3 + 3 = 6 \text{ cm}$$

Therefore, the length of side $CD = 6$ cm.

Example 9 Prove that the opposite sides of a quadrilateral circumscribing a circle subtend supplementary angles at the centre of the circle. **[NCERT, CBSE (CCE) 2014]**

Solution: *Given*: *ABCD* is a quadrilateral, circumscribing a circle with centre *O* and touching the quadrilateral at *P*, *Q*, *R* and *S*, respectively.

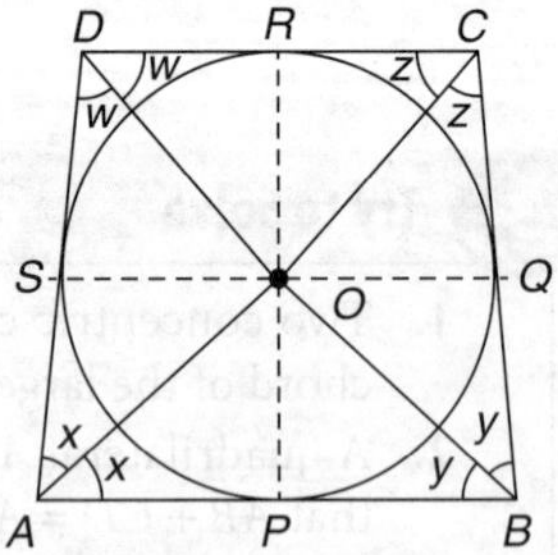

To prove:

(i) $\angle AOB + \angle COD = 180°$

(ii) $\angle BOC + \angle AOD = 180°$

Construction: Join *OP*, *OQ*, *OR* and *OS*.

Proof: Consider, triangles *APO* and *ASO*

$AP = AS$ [Lengths of the tangents drawn from the same external point to the circle are equal]

$OS = OP$ [Radii]

$OA = OA$ [Common]

$\therefore \quad \Delta APO \cong \angle ASO$ [SSS congruency rule]

$\therefore \quad \angle OAP = \angle OAS = x$ (say) [CPCT]

Similarly $\quad \angle OBP = \angle OBQ = y$ (say)

$\angle OCQ = \angle OCR = z$ (say)

And $\quad \angle ODR = \angle ODS = w$ (say)

We have $\angle DAB + \angle ABC + \angle BCD + \angle CDA = 360°$ [∵ Angle sum property of quadrilateral]

$\Rightarrow \quad 2x + 2y + 2z + 2w = 360°$

$\Rightarrow \quad x + y + z + w = 180°$ ...(i)

Consider $\angle AOB + \angle COD = [180° - x - y] + [180° - w - z]$ [Angle sum property of Δ]

$\Rightarrow \quad = 360° - (x + y + z + w)$

$\Rightarrow \quad = 360° - 180°$ [Using (i)]

$\therefore \quad \angle AOB + \angle COD = 180°$

Again consider $\angle BOC + \angle AOD = [180 - y - z] + [180° - x - w]$ [Angle sum property of Δ]

$= 360° - (x + y + z + w)$

$\therefore \quad = 360 - 180° = 180°$ [Using (i)]

Hence proved.

4 Try to Solve

1. Prove that the angle between the two tangents drawn from an external point to a circle is supplementary to the angle subtended by the line segment joining the points of contact at the centre. **[NCERT]**
2. Prove that the parallelogram circumscribing a circle is a rhombus. **[NCERT]**

Example 10 In the figure, a triangle *ABC* is drawn to circumscribe a circle of radius 2 cm such that the segments *BD* and *DC* into which *BC* is divided by the point of contact *D* are the lengths 4 cm and 3 cm, respectively. If area of $\Delta ABC = 21\text{ cm}^2$, find the lengths of sides *AB* and *AC*. **[CBSE (CCE) 2011]**

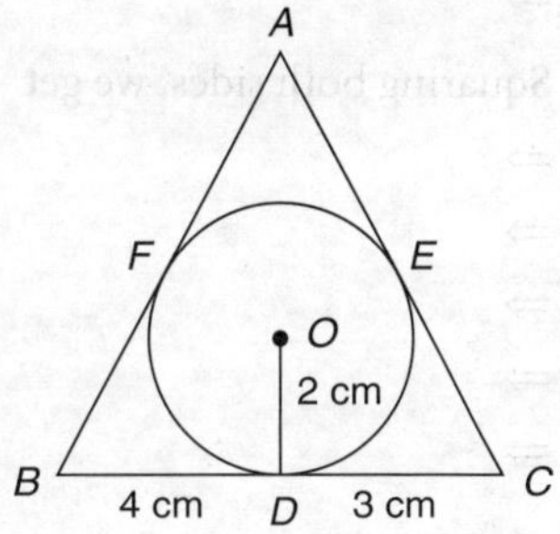

Solution: Let $AE = AF = y$ (say) [∵ Tangents drawn from the same external point are equal in length]

$BD = BF = 4$ cm

$CD = CE = 3$ cm

$$ar\,\Delta BOC = \frac{1}{2} \times 7 \times 2 = 7\text{ cm}^2 = b\text{ (say)}$$

$$ar\,\Delta AOB = \frac{1}{2} \times (4 + y) \times 2 = (4 + y)\text{cm}^2 = a\text{ (say)}$$

$$ar\,\Delta AOC = \frac{1}{2} \times (3 + y) \times 2 = (3 + y)\text{cm}^2 = c\text{ (say)}$$

Now $\quad ar\,\Delta ABC = ar\,\Delta AOB + ar\,\Delta BOC + ar\,\Delta AOC$

$\Rightarrow \quad = 4 + y + 7 + 3 + y$

$ar\,\Delta ABC = 14 + 2y$ ...(i)

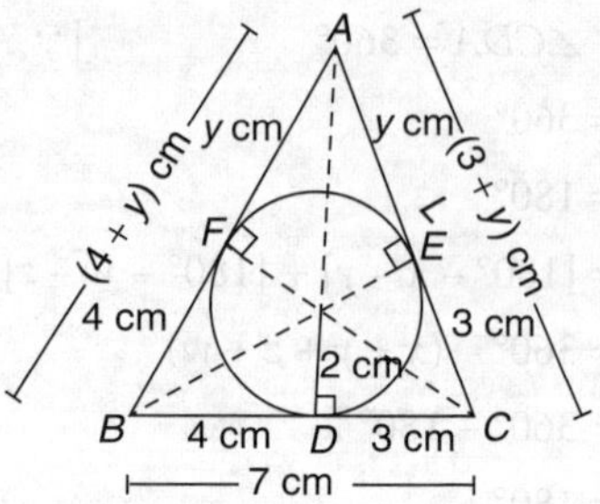

For ΔABC

$$\text{Semi-perimeter } s = \frac{a+b+c}{2} = \frac{4+y+7+3+y}{2} = \frac{14+2y}{2} = 7+y$$

$$ar\,\Delta ABC = \sqrt{s(s-a)(s-b)(s-c)} \quad [\because \text{ By Heron's formula}]$$

$$= \sqrt{(7+y)(7+y-4-y)(7+y-7)(7+y-3-y)}$$

$$= \sqrt{(7+y)\times 3\times y\times 4}$$

$$ar\,\Delta ABC = 2\sqrt{3y(7+y)} \quad \text{...(ii)}$$

From (i) and (ii) $\Rightarrow$ $2\sqrt{3y(7+y)} = 14+2y$

$\Rightarrow$ $\sqrt{3y(7+y)} = 7+y$

Squaring both sides, we get $\Rightarrow$ $3y\,(7+y) = (7+y)^2$

$\Rightarrow$ $21y + 3y^2 = 49 + y^2 + 14y$

$\Rightarrow$ $2y^2 + 7y - 49 = 0$

$\Rightarrow$ $2y^2 + 14y - 7y - 49 = 0$

$\Rightarrow$ $2y(y+7) - 7(y+7) = 0$

$\Rightarrow$ $(2y-7)\,(y+7) = 0$

$\Rightarrow$ $y = \frac{7}{2},\ y = -7$

Since, $y = -7$ is not possible, $y = 3.5$ cm.

Hence, the length of side $AB = 4 + 3.5 = 7.5$ cm and $AC = 3 + 3.5 = 6.5$ cm.

Example 11 In the figure, a right triangle *ABC* circumscribes a circle of radius *r*. If *AB* and *BC* are of lengths 8 cm and 6 cm, respectively, find the value of *r*. **[CBSE (CCE) 2012]**

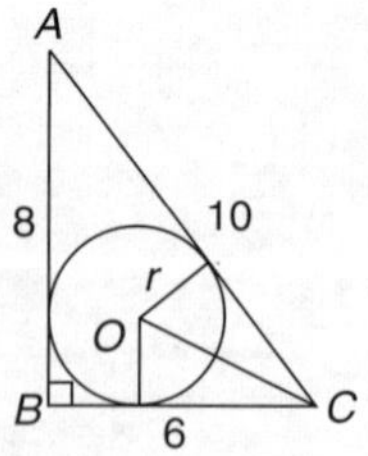

Solution: $\because$ *ABC* is Δ right-angled at *B*. So, by Pythagoras' Theorem

$$AC^2 = AB^2 + BC^2 = 8^2 + 6^2 = 100$$

$$AC = 10 \text{ cm}$$

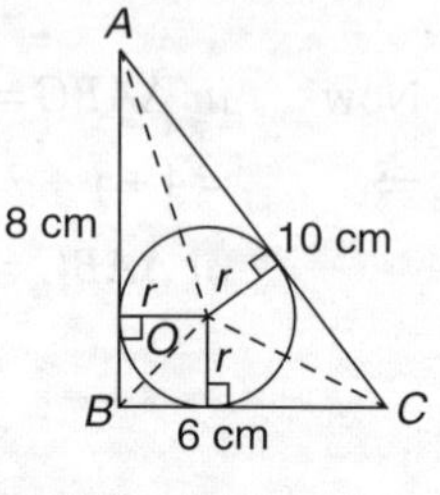

So $\quad ar\,(\Delta ABC) = \frac{1}{2}\times 6\times 8 = 24 \text{ cm}^2$

Also $\quad ar\,(\Delta ABC) = ar(\Delta OBC) + ar(\Delta OAC) + ar(\Delta OAB)$

$$\Rightarrow \quad 24 = \frac{1}{2} \times 6 \times r + \frac{1}{2} \times 10 \times r + \frac{1}{2} \times 8 \times r \text{ [Using ar } \Delta = \frac{1}{2} \times \text{Base} \times \text{Height]}$$

$$\Rightarrow \quad 24 = 3r + 5r + 4r$$

$$\Rightarrow \quad 12r = 24$$

$$\Rightarrow \quad r = 2 \text{ cm}$$

5 Try to Solve

1. In the figure, a circle inscribed in ΔABC touches its sides AB, BC and AC at points D, E and F, respectively. If $AB = 12$ cm, $BC = 8$ cm and $AC = 10$ cm, find the length of AD, BE and CF. **[CBSE (CCE) 2013]**

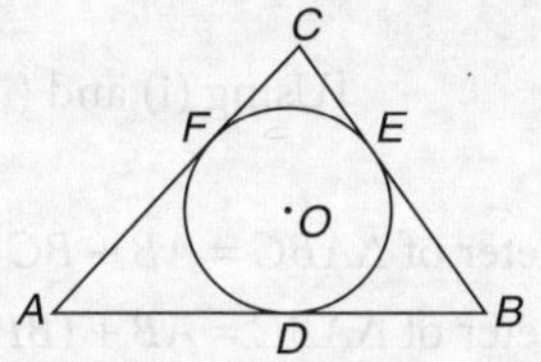

2. A triangle ABC is drawn to circumscribe a circle of radius 4 cm, such that the segments BD and DC into which BC is divided by the point of contact D are of lengths 8 cm and 6 cm, respectively (see figure). Find the sides AB and AC. **[NCERT]**

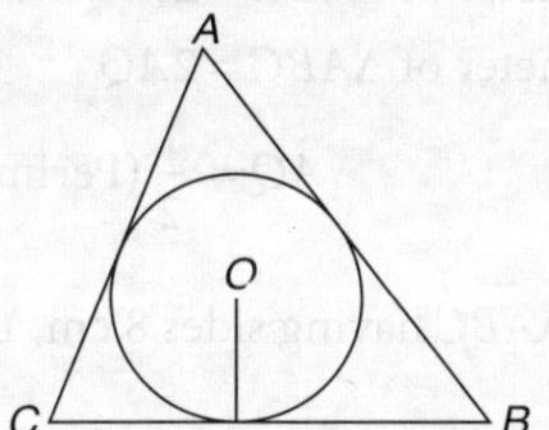

3. In the figure, a triangle ABC is drawn to circumscribe a circle of radius 3 cm, such that the segments BD and DC are of length 6 cm and 9 cm, respectively. If the area of ΔABC is 54 cm^2, find the length of sides AB and AC. **[CBSE (CCE) 2015]**

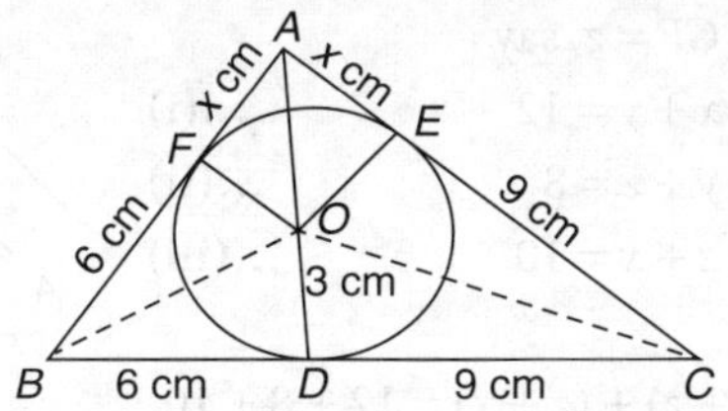

Answers

1. 7 cm, 5 cm, 3 cm
2. $AB = 15$ cm, $AC = 13$ cm
3. $AB = 9$ cm, $AC = 12$ cm

FOR MORE PRACTICE AND KNOWLEDGE

Example 12 A circle is touching the side BC of ΔABC at P and touching AB and AC produced at Q and R, respectively. Prove that $AQ = \frac{1}{2}$ (Perimeter of ΔABC). **[CBSE 1993C, 1995, 2001, 2002]**

Solution: As tangents drawn from the same external point to a circle are equal in length

$\therefore$ $BP = BQ$ [From B] ...(i)

$CP = CR$ [From C] ...(ii)

$AQ = AR$ [From A] ...(iii)

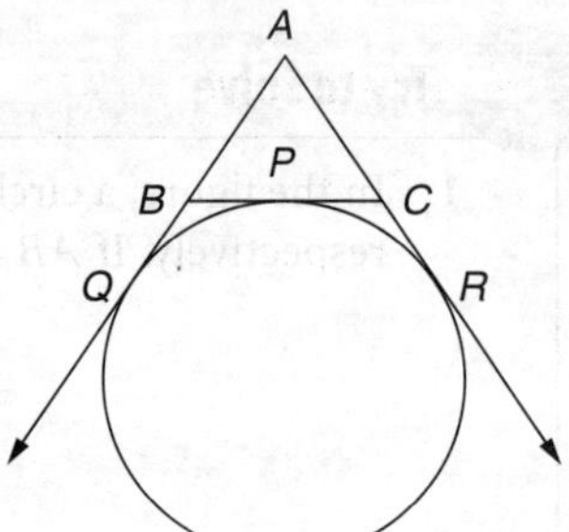

From (iii), we have

$AQ = AR$

$\Rightarrow$ $AB + BQ = AC + CR$

$\Rightarrow$ $AB + BP = AC + CP$ [Using (i) and (ii)] ...(iv)

Now

$\Rightarrow$ Perimeter of $\Delta ABC = AB + BC + AC$

$\Rightarrow$ Perimeter of $\Delta ABC = AB + (BP + PC) + AC$

$\Rightarrow$ Perimeter of $\Delta ABC = (AB + BP) + (AC + PC)$

$\Rightarrow$ Perimeter of $\Delta ABC = 2(AB + BP)$ [From (iv)]

$\Rightarrow$ Perimeter of $\Delta ABC = 2(AB + BQ)$ [From (i)]

$\Rightarrow$ Perimeter of $\Delta ABC = 2AQ$

$\therefore$ $AQ = \frac{1}{2}$ (Perimeter of ΔABC)

Example 13 A circle is inscribed in a ΔABC having sides 8 cm, 10 cm and 12 cm, as shown in the figure. Find AD, BE and CF. **[CBSE 2001]**

Solution: We know that tangents drawn from the same external point to a circle are equal in length. Therefore

$AD = AF = x$, say

$BD = BE = y$, say

$CE = CF = z$, say

$AB = x + y = 12$ ...(i)

$BC = y + z = 8$ ...(ii)

$CA = z + x = 10$ ...(iii)

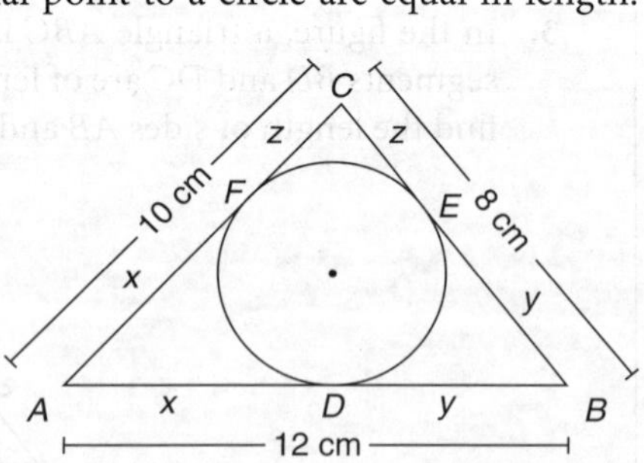

Adding (i), (ii) and (iii), we get

$\Rightarrow$ $(x + y) + (y + z) + (z + x) = 12 + 8 + 10$

$\Rightarrow$ $x + x + y + y + z + z = 30$

$\Rightarrow$ $2x + 2y + 2z = 30$

$\Rightarrow$ $2(x + y + z) = 30$

$\Rightarrow$ $x + y + z = 15$ ...(iv)

Now $x + y = 12$ and $x + y + z = 15$ [From (i) and (iv)]

$\Rightarrow$ $12 + z = 15$

$\Rightarrow$ $z = 3$

$y + z = 8$ and $x + y + z = 15$ [From (ii) and (iv)]

$\Rightarrow$ $x + 8 = 15$

$\Rightarrow$ $x = 7$

$z + x = 10$ and $x + y + z = 15$ [From (iii) and (iv)]

$\Rightarrow$ $y + 10 = 15$

$\Rightarrow$ $y = 5.$

Hence, $AD = x = 7$ cm, $BE = y = 5$ cm and $CF = z = 3$ cm.

Example 14 Two tangents *TP* and *TQ* are drawn to a circle with centre *O* from an external point *T*. Prove that $\angle PTQ = 2\angle OPQ$.

Solution: We know that the lengths of tangents drawn from the same external point to a circle are equal in length.

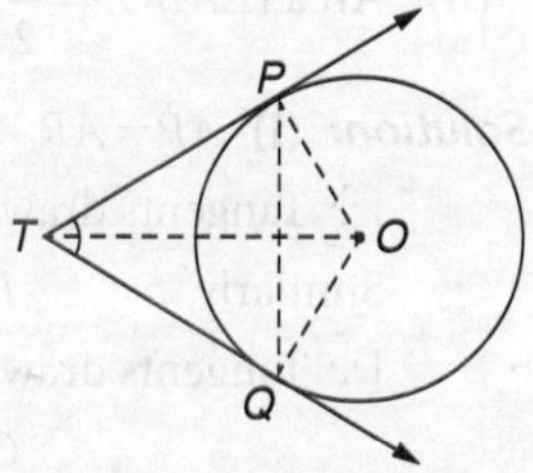

$\therefore$ $TP = TQ$

$\Rightarrow \Delta TPQ$ is an isosceles triangle.

$\Rightarrow$ $\angle TPQ = \angle TQP$ ...(i)

In ΔTPQ, we have $\angle TPQ + \angle TQP + \angle PTQ = 180°$ [$\because$ ASP of Δs]

$\Rightarrow$ $2\angle TPQ = 180° - \angle PTQ$ [From (i)]

$\Rightarrow$ $\angle TPQ = 90° - \frac{1}{2}\angle PTQ$

$\Rightarrow$ $\frac{1}{2}\angle PTQ = 90° - \angle TPQ$ ...(ii)

Since $OP \perp TP$.

$\therefore$ $\angle OPT = 90°$ [$\because$ Radius is $\perp$ to point of contact]

$\Rightarrow$ $\angle OPQ + \angle TPQ = 90°$

$\Rightarrow$ $\angle OPQ = 90° - \angle TPQ$ ...(iii)

From (ii) and (iii), we get $\frac{1}{2}\angle PTQ = \angle OPQ$

$\Rightarrow$ $\angle PTQ = 2\angle OPQ$

Example 15 *PQ* is a chord of length 8 cm of a circle of radius 5 cm. The tangents at *P* and *Q* intersect at a point *T* (see figure). Find the length *TP*. [NCERT]

Solution: Join *OT*. Let it intersect *PQ* at the point *R*. Then, ΔTPQ is isosceles and *TO* is the angle bisector of *PTQ*. So, $OT \perp PQ$ and *OT* bisects *PQ* ,which gives $PR = RQ = 4$ cm.

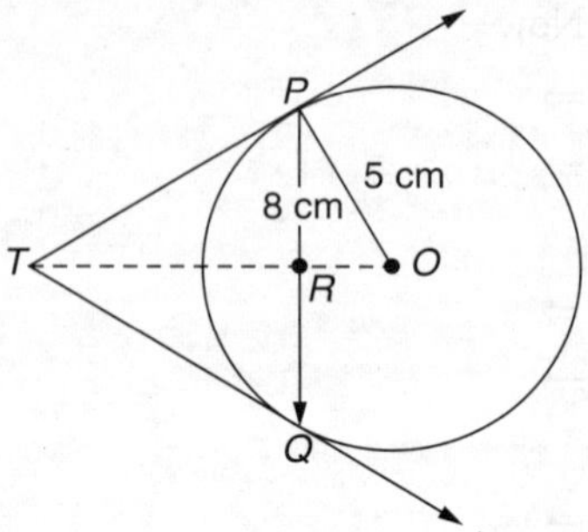

Also $OR = \sqrt{OP^2 - PR^2} = \sqrt{5^2 - 4^2}$ cm = 3 cm

Now $\angle TPR + \angle RPO = 90° = \angle TPR + \angle PTR$

So $\angle RPO = \angle PTR$

Therefore, right triangle *TRP* is similar to right triangle *PRO* by AA similarity.

This gives $\frac{TP}{PO} = \frac{RP}{RO}$, that is, $\frac{TP}{5} = \frac{4}{3}$ or $TP = \frac{20}{3}$ cm.

Example 16 In the figure, the sides *AB*, *BC* and *CA* of triangle *ABC* touch a circle with centre *O* and radius *r* at *P*, *Q* and *R*, respectively. **[AI 2013]**

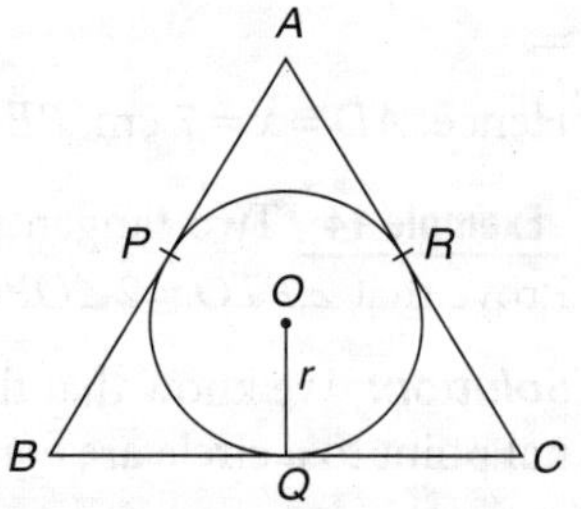

Prove that

(i) $AB + CQ = AC + BQ$

(ii) Area $(\Delta ABC) = \frac{1}{2}$ (perimeter of ΔABC) $\times r$

Solution: (i) $AP = AR$

[∵ Tangents drawn from the same external point to a circle are equal in length] ...(i)

Similarly, $BP = BQ$

[∵ Tangents drawn from the same external point to a circle are equal in length] ...(ii)

$CR = CQ$

[∵ Tangents drawn from the same external point to a circle are equal in length] ...(iii)

Now $AP = AR$

$\Rightarrow (AB - BP) = (AC - CR)$

$\Rightarrow AB + CR = AC + BP$

$\Rightarrow AB + CQ = AC + BQ$ [Using equations (ii) and (iii)]

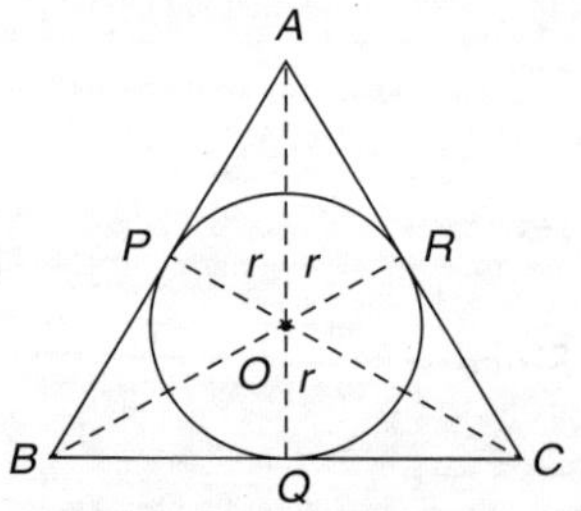

(ii) Let $AB = x, BC = y, AC = z$

∴ Perimeter of $\Delta ABC = x + y + z$ ...(iv)

$$\text{Area of } \Delta ABC = \frac{1}{2} (\text{Area of } \Delta AOB + \text{Area of } \Delta BOC + \text{Area of } \Delta AOC)$$

$$\Rightarrow \text{Area of } \Delta ABC = \frac{1}{2} AB \times OP + \frac{1}{2} \times BC \times OQ + \frac{1}{2} \times AC \times OR$$

$$\Rightarrow \text{Area of } \Delta ABC = \frac{1}{2} x \times r + \frac{1}{2} y \times r + \frac{1}{2} z \times r$$

$$\Rightarrow \text{Area of } \Delta ABC = \frac{1}{2} (x + y + z) \times r$$

$$\Rightarrow \text{Area of } \Delta ABC = \frac{1}{2} (\text{Perimeter of } \Delta ABC) \times r$$ [From iv]

Example 17 In the figure, *TAS* is a tangent to the circle with centre *O*, at the point *A*. If $\angle OBA = 32°$, find the value of *x*.

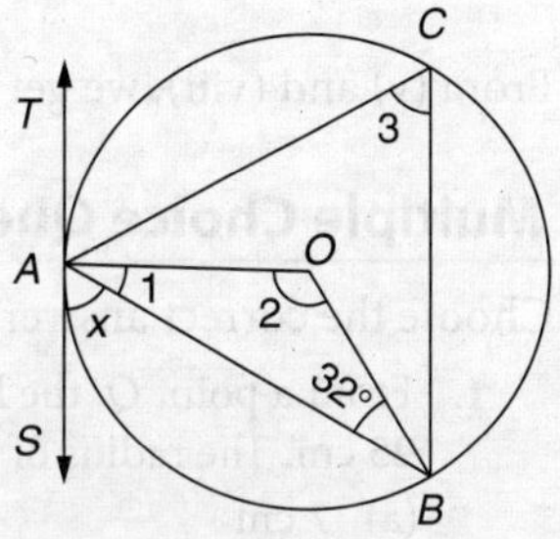

Solution: *Given*: *TAS* is tangent to the circle with centre *O* at point *A*.

$$\angle OBA = 32°$$

To find: x

In ΔOAB $\quad OA = OB \quad$ [Radii of the same circle]

$\Rightarrow \quad \angle 1 = 32° \quad$ [Angle opposite to equal sides of a triangle are equal]

In $\Delta OAB \quad \angle 1 + \angle 2 + 32° = 180° \quad$ [ASP of a triangle]

$\Rightarrow \quad \angle 2 + 32° + 32° = 180°$

$\Rightarrow \quad \angle 2 = 180° - 64° = 116°$

$$\angle 3 = \frac{1}{2} \; 2 = \frac{1}{2} \times 116° = 58° \quad \text{[Degree Measure Theorem]}$$

TAS is tangent to the circle $\quad \angle x = \angle ACB = 58° \quad$ [Angles in alternate segments are equal]

Example 18 In right ΔABC, a circle with side *AB* as diameter is drawn to intersect the hypotenuse *AC* at *P*. Prove that the tangent to the circle at *P* bisects *BC*.

Solution: *To prove:* $\quad BQ = QC$

$\because$ *AB* is a diameter

$\therefore \quad \angle APB = 90° \quad$ [Angle in semicircle]

$\therefore \quad \angle BPC = 90°, \angle 5 + \angle 6 = 90° \quad$ **...(i)**

Also $\quad \angle 1 + \angle 2 = 90° \quad$ **...(ii)** $\quad$ [Sum of acute angles of a right Δ]

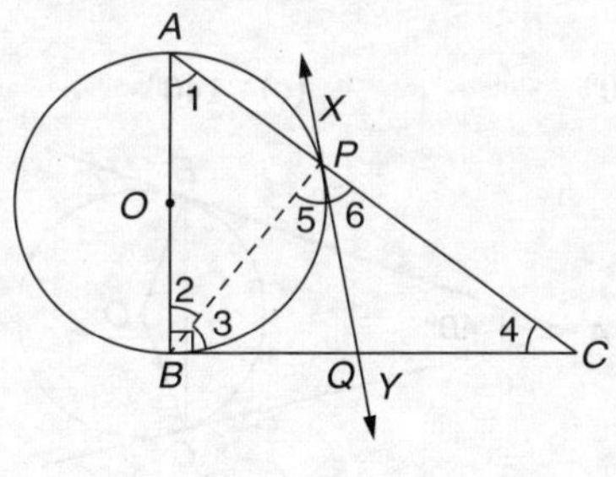

Also $\quad \angle 2 + \angle 3 = 90° \quad$ [Given] **...(iii)**

From (ii) and (iii) $\quad \angle 1 + \angle 2 = \angle 2 + \angle 3 = 90°$

$\Rightarrow \quad \angle 1 = \angle 3 \quad$ **...(iv)**

Also $\quad BQ = PQ \quad$ [Tangents to the circle from Q] **...(v)**

$\therefore \quad \angle 3 = \angle 5 \quad$ [Angles opposite equal sides are equal]

Substituting this value in equation (i), we get

$$\angle 3 + \angle 6 = 90°$$

$\Rightarrow \quad \angle 1 + \angle 6 = 90° \quad$ **...(vi)** [Using (iv)]

Also $\quad \angle 1 + \angle 4 = 90° \quad$ [Sum of acute angles of a right ΔABC] **...(vii)**

From (vi) and (vii), we get $\angle 6 = \angle 4$

$\Rightarrow$ $PQ = CQ$ [Opposite sides of equal angles in a triangle are equal] ...(viii)

From (v) and (viii), we get $BQ = CQ$

Multiple Choice Questions

Choose the correct answer for each of the following questions:

1. From a point Q, the length of the tangent to a circle is 24 cm and the distance of Q from the centre is 25 cm. The radius of the circle is
   (a) 7 cm (b) 12 cm (c) 15 cm (d) 24.5 cm **[NCERT]**
2. In the figure, if TP and TQ are the two tangents to a circle with centre O so that $\angle POQ = 110°$, then $\angle PTQ$ is equal to
   (a) 60° (b) 70° (c) 80° (d) 90° **[NCERT]**

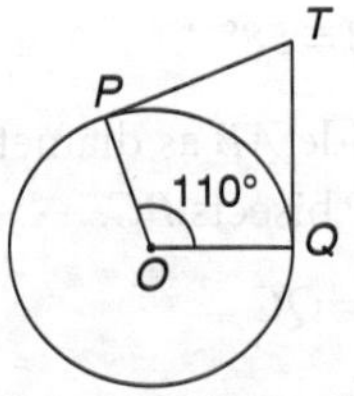

3. If tangents PA and PB from a point P to a circle with centre O are inclined to each other at angle of 80°, then $\angle POA$ is equal to
   (a) 50° (b) 60° (c) 70° (d) 80° **[NCERT]**
4. In the figure, AB and AC are tangents to the circle with centre O such that $\angle BAC = 40°$. Then, $\angle BOC$ is equal to
   (a) 40° (b) 50° (c) 140° (d) 150° **[AI 2011]**

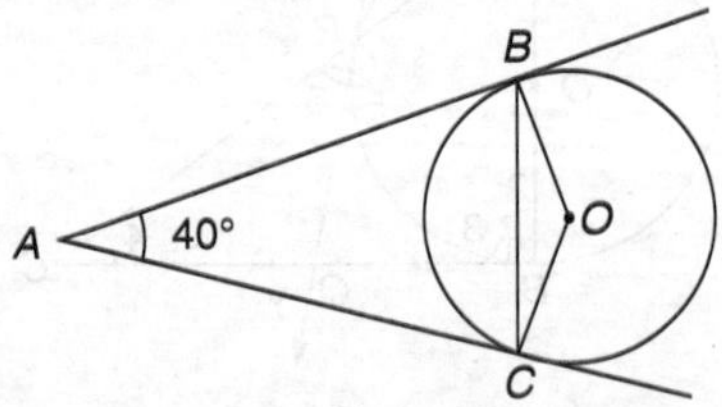

5. A circle touches the x axis at A and the y axis at B. If O is origin and $OA = 5$ units, the diameter of the circle is
   (a) 8 units (b) 10 units (c) 10 units (d) $8\sqrt{2}$ units
6. From a point P which is at a distance of 13 cm from the centre O of a circle of radius 5 cm, the pair of tangents PQ and PR to the circle are drawn. The area of the quadrilateral $PQOR$ is
   (a) 60 cm^2 (b) 65 cm^2 (c) 30 cm^2 (d) 32.5 cm^2 **[NCERT Exemplar]**
7. In the figure, AT is a tangent to the circle with centre O such that $OT = 4$ cm and $\angle OTA = 30°$. AT is equal to
   (a) 4 cm (b) 2 cm (c) $2\sqrt{3}$ cm (d) $4\sqrt{3}$ cm
   [NCERT Exemplar]

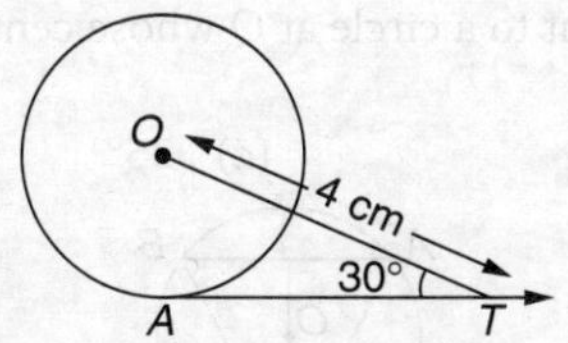

8. In the figure, if O is the centre of a circle, PQ is a chord and the tangent PR at P makes an angle of 50° with PQ, $\angle POQ$ is equal to
(a) 100° (b) 80° (c) 90° (d) 75° **[NCERT Exemplar]**

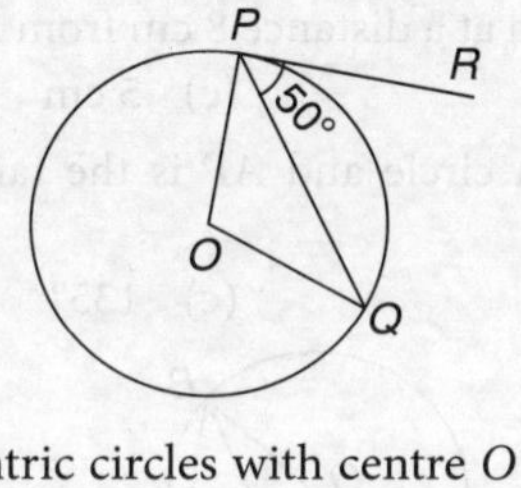

9. In the figure, there are two concentric circles with centre O and radii 5 cm and 3 cm. PA and PB are tangents to these circles from an external point P. If $PA = 12$ cm, the length of PB (in cm) is

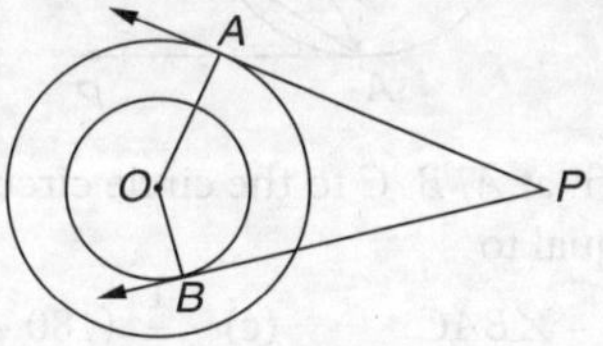

(a) 10 (b) $4\sqrt{10}$ (c) 12 (d) $\sqrt{178}$ **[Foreign 2013]**

10. In the figure, AP, AQ and BC are tangents to the circle. If $AB = 5$ cm, $AC = 6$ cm and $BC = 4$ cm, the length of AP (in cm) is
(a) 7.5 (b) 15 (c) 10 (d) 9 **[AI 2012]**

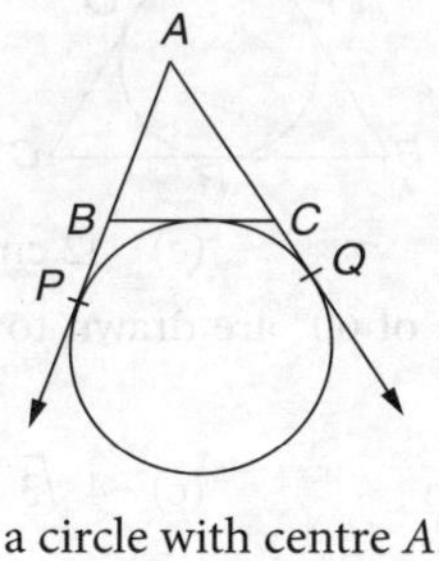

11. In figure, PQ and PR are tangents to a circle with centre A. If $\angle QPA = 27°$, then $\angle QAR$ equals
(a) 63° (b) 153° (c) 126° (d) 117° **[Foreign 2012]**

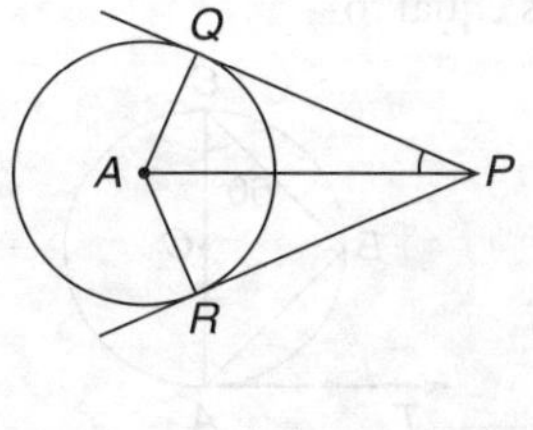

12. In the figure, if *PQR* is the tangent to a circle at *Q* whose centre is *O*, *AB* is a chord parallel to *PR* and $\angle BQR = 70°$, $\angle AQB$ is equal to

(a) 20° (b) 40° (c) 35° (d) 45° **[NCERT Exemplar]**

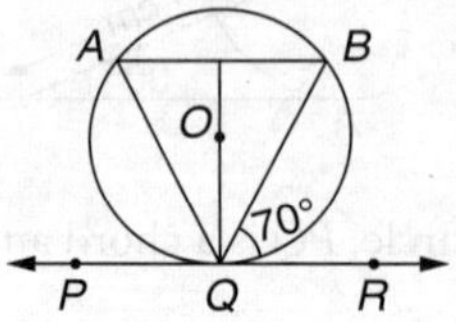

13. At one end *A* of a diameter *AB* of a circle of radius 5 cm, tangent *XAY* is drawn to the circle. The length of the chord *CD* parallel to *XY* and at a distance 8 cm from *A* is

(a 8 cm (b) 6 cm (c) 5 cm (d) 4 cm

14. In the figure, *AB* is a chord of a circle and *AP* is the tangent at point *A* such that $\angle BAP = 60°$. Then, $\angle ACB$ is

(a) 150° (b) 120° (c) 135° (d) 110°

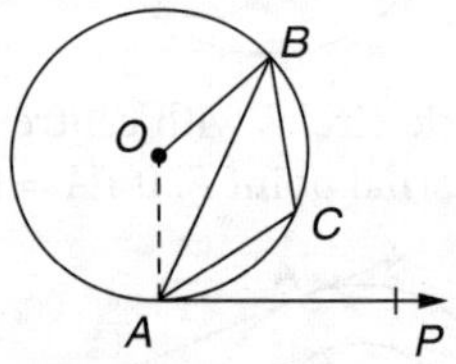

15. If tangents *QR*, *PR*, *PQ* are drawn at *A*, *B*, *C* to the circle circumscribing an acute-angled ΔABC, so as to form ΔPQR, then $\angle RPQ$ is equal to

(a) $\angle BAC$ (b) $180 - \angle BAC$ (c) $\frac{1}{2}(180 - \angle BAC)$ (d) $180 - 2\angle BAC$

16. A circle is drawn touching the sides of ΔABC at *P*, *Q*, *R*. If $AP + BC = 13$ cm, the perimeter of ΔABC is

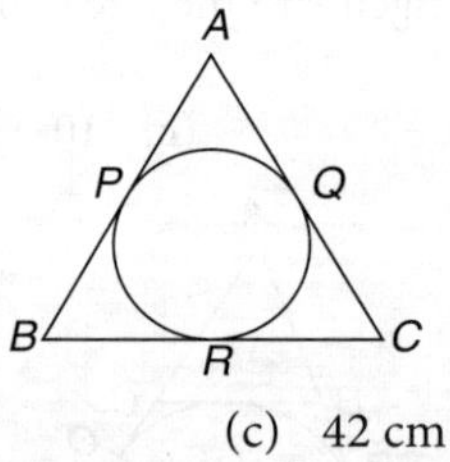

(a) 39 cm (b) 26 cm (c) 42 cm (d) 20 cm

17. If two tangents inclined at an angle of 60° are drawn to a circle of radius 4 cm, the length of each tangent is equal to

(a) $\frac{4}{\sqrt{3}}$ cm (b) $3\sqrt{3}$ cm (c) $4\sqrt{3}$ cm (d) $2\sqrt{3}$ cm

18. *AB* is a chord of the circle and *AOC* is its diameter such that $\angle ACB = 50°$. If *AT* is the tangent to the circle at the point *A*, then $\angle BAT$ is equal to

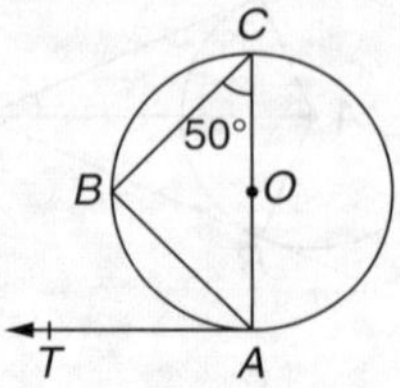

(a) 40° (b) 50° (c) 60° (d) 65°

19. To draw a pair of tangents to a circle which are inclined to each other at an angle of 100°, it is required to draw tangents at end points of those two radii of the circle, the angle between which should be

(a) 100° (b) 50° (c) 80° (d) 260°

20. The distance between two parallel tangents in a circle of radius 3.5 cm is

(a) 7 cm (b) 14 cm (c) 3.5 cm (d) 1.75 cm

21. In the figure, if *PA* and *PB* are tangents to the circle with centre *O*, such that $\angle APB = 50°$, then $\angle OAB$ is equal to

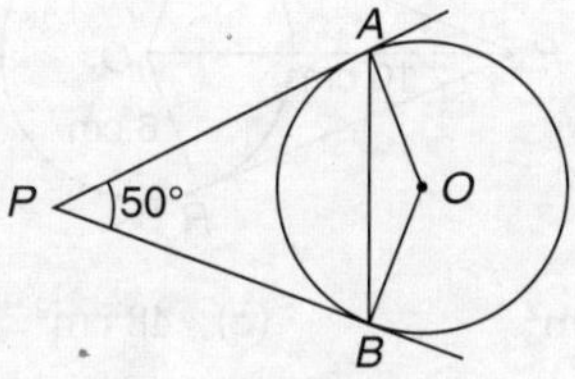

(a) 25° (b) 30° (c) 60° (d) 90°

22. In the figure, $\angle AOB = 110°$. Then, $\angle PAB$ is

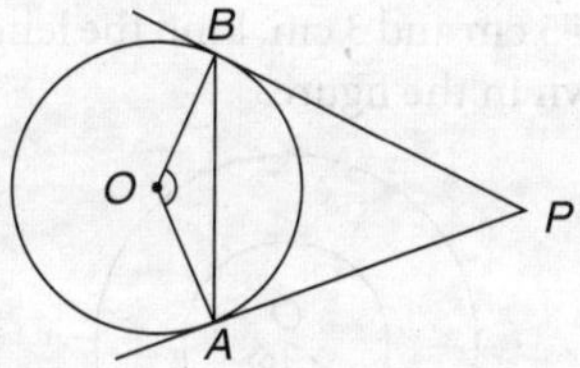

(a) 70° (b) 35° (c) 45° (d) 55°

23. In the figure, if $PL = 5$ cm, the perimeter of ΔPQR is

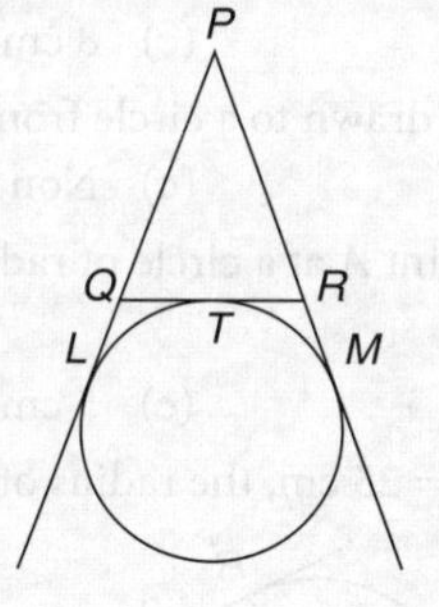

(a) 8 cm (b) 10 cm (c) 25 cm (d) 15 cm

24. In the figure, a circle with centre *O* is inscribed in a quadrilateral *ABCD* such that it touches the sides *BC*, *AB*, *AD* and *CD* at points *P*, *Q*, *R* and *S*, respectively. If $AB = 29$ cm, $AD = 23$ cm, $\angle B = 90°$ and $DS = 5$ cm, the radius of the circle (in cm) is

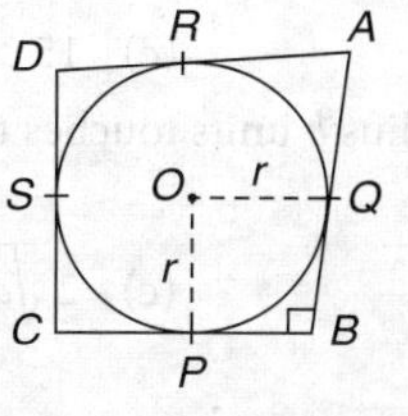

(a) 11 (b) 18 (c) 6 (d) 15

25. The distance between two parallel lines is 14 cm. The radius of the circle which will touch both the lines is

(a) 6 cm (b) 7 cm (c) 12 cm (d) 14 cm

26. In the figure, O is the centre of a circle of radius 6 cm. At a distance of 10 cm from O, a point P is taken. Two tangents PQ and PR are drawn to the circle from this point. The area of quadrilateral $PQOR$ is

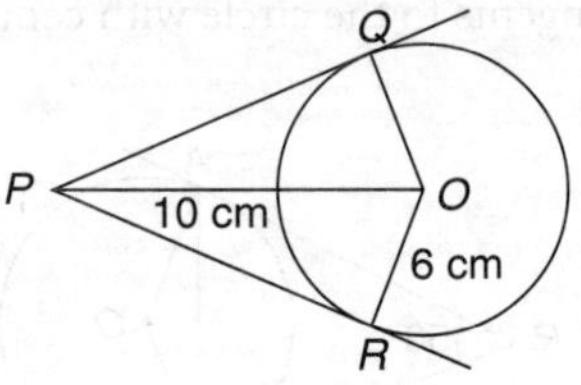

(a) 60 cm^2 (b) 28 cm^2 (c) 48 cm^2 (d) 30 cm^2

27. AT is a tangent to a circle at A with centre O from an external point T such that $OT = 8$ cm and $\angle OTA = 30°$. The length of AT (in cm) is

(a) $\sqrt{2}$ (b) $3\sqrt{2}$ (c) $4\sqrt{3}$ (d) 4

28. Two concentric circles are of radii 5 cm and 3 cm. Find the length of the chord of the larger circle which touches the smaller circle as shown in the figure.

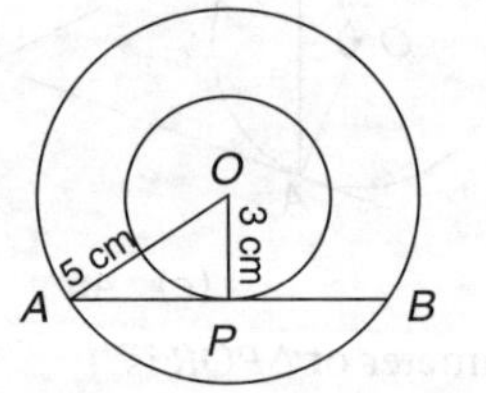

(a) 6 cm (b) 9 cm (c) 8 cm (d) 2 cm

29. The number of tangents that can be drawn to a circle from a point inside the circle are

(a) 1 (b) Infinite (c) None (d) 2

30. The length of the tangent from a point A at a circle of radius 3 cm is 4 cm. The distance of A from the centre of the circle is

(a) $\sqrt{7}$ cm (c) 7 cm (c) 5 cm (d) 25 cm

31. In the figure, if $PR = 24$ cm and $OP = 25$ cm, the radius of circle is

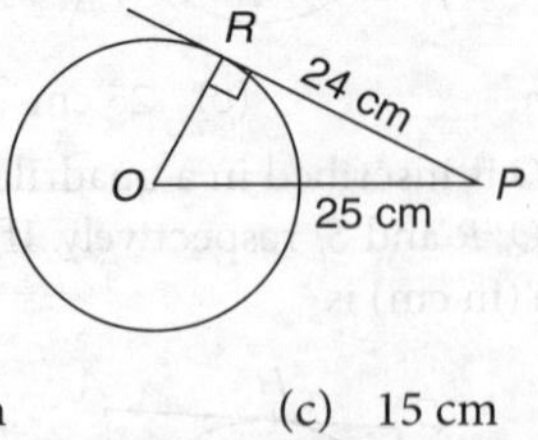

(a) 7 cm (b) 12 cm (c) 15 cm (d) 49 cm

32. If the chord of the larger circle of radius b units touches the smaller concentric circle of radius a units, its length is

(a) $\sqrt{a^2 - b^2}$ (b) $\sqrt{b^2 - a^2}$ (c) $2\sqrt{a^2 - b^2}$ (d) $2\sqrt{b^2 - a^2}$

33. In the figure, PA and PB are two tangents drawn from an external point P to a circle with centre C and radius 4 cm. If $PA \perp PB$, the length of each tangent is

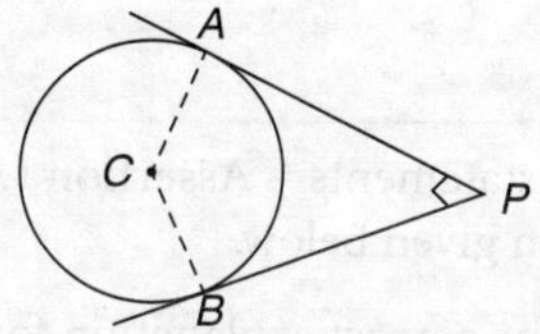

(a) 3 cm (b) 4 cm (c) 5 cm (d) 6 cm

34. If PQ and PR are tangents to a circle with centre O, inclined at an angle of 70°, then $\angle QAR$ is

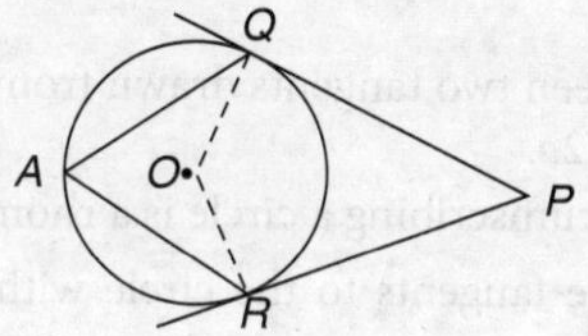

(a) 110° (b) 35° (c) 45° (d) 55°

35. In the figure, O is the centre of a circle, AB is a chord and AT is the tangent at A. If $\angle AOB = 100°$, then $\angle BAT$ is equal to

(a) 100° (b) 40° (c) 50° (d) 90° **[Delhi 2011]**

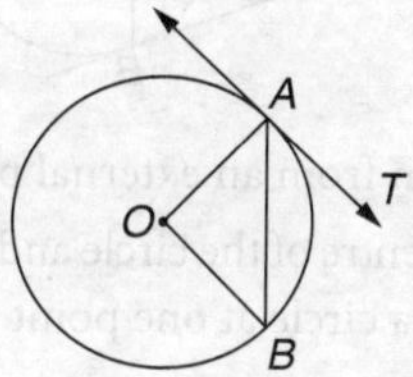

Fill in the Blanks

Complete the following statements with appropriate answers(s) in the blank space(s):

36. Two tangents drawn at the end points of the diameter of a given circle are always __________.

37. When we draw any line from the centre to a tangent, it will always be __________.

38. If point P lies inside the circle, the number of tangents(s) drawn from point P is __________.

39. If two tangents are drawn to a circle from an external point, they subtend __________ angles at the centre.

40. The __________ to a circle is a special case of __________, when the two end points of its corresponding chord coincide.

41. In the figure, $AP = AQ$ and $PB = QC$. Then, AB is equal to __________.

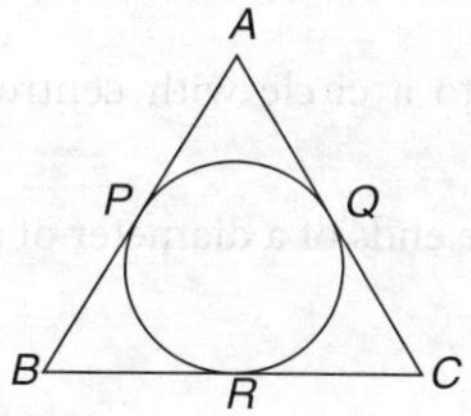

42. In two concentric circles, if the chords are drawn in the outer circle which touches the inner circle, then all chords are of __________ length.

Assertion–Reason

The following questions consist of two statements – Assertion (A) and Reason (R). Answer these questions by selecting the appropriate option given below.

(A) Both A and R are true and R is the correct explanation for A.
(B) Both A and R are true and R is not the correct explanation for A.
(C) A is true but R is false.
(D) A is false but R is true.

43. **Assertion (A):** If the angle between two tangents drawn from an external point P to a circle of radius a and centre O is 60°, then $OP = 2a$.
Reason (R): A parallelogram circumscribing a circle is a rhombus.

44. **Assertion (A):** If PA and PB are tangents to the circle with centre O such that $\angle APB = 50°$, then $\angle OAB = 25°$.

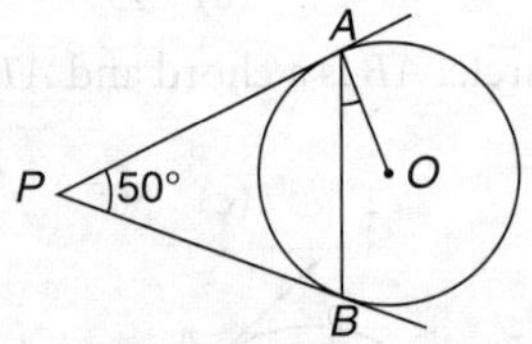

Reason (R): Length of tangents drawn from an external point to a circle are equal.

45. **Assertion (A):** In the figure, O is the centre of the circle and AT is a tangent at point A. Then, $\angle BAT = 60°$.
Reason (R): A straight line can meet a circle at one point only.

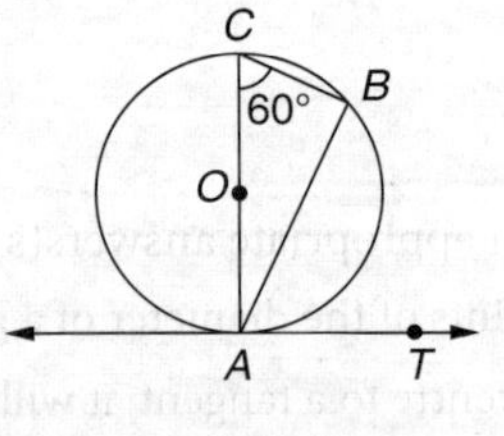

46. **Assertion (A):** In two concentric circles, the chord of the larger circle which touches the smaller circle is bisected at the point of contact.
Reason (R): The distance between two parallel tangents of a circle of radius 4 cm is 8 cm.

Subjective Questions

47. Prove that in two concentric circles, the chord of the larger circle, which touches the smaller circle, is bisected at the point of contact. **[NCERT]**

48. Two tangents TP and TQ drawn to a circle with centre O form an external point T. Prove that $\angle PTO = 2\angle OPQ$. **[NCERT]**

49. Prove that the tangents drawn at the ends of a diameter of a circle are parallel. **[NCERT]**

50. Prove that the perpendicular at the point of contact to the tangent to a circle passes through the centre. **[NCERT]**

51. The length of a tangent from a point *A* at distance 5 cm from the centre of the circle is 4 cm. Find the radius of the circle. **[NCERT]**

52. Two concentric circles are of radii 5 cm and 3 cm. Find the length of the chord of the larger circle which touches the smaller circle. **[NCERT]**

53. A quadrilateral *ABCD* is drawn to circumscribe a circle (see figure). Prove that $AB + CD = AD + BC$.

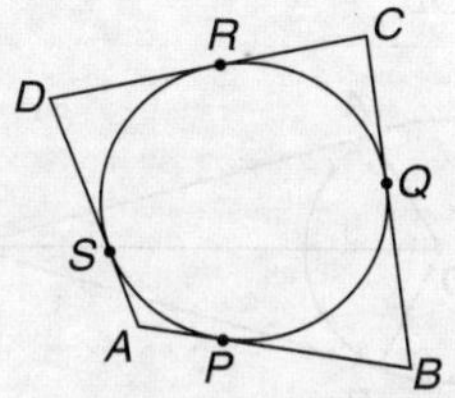

[NCERT]

54. In the figure. *XY* and *X′Y′* are two parallel tangents to a circle with centre *O* and another tangent *AB* with point of contact *C* intersects *XY* at *A* and *X′Y′* at *B*. Prove that $\angle AOB = 90°$.

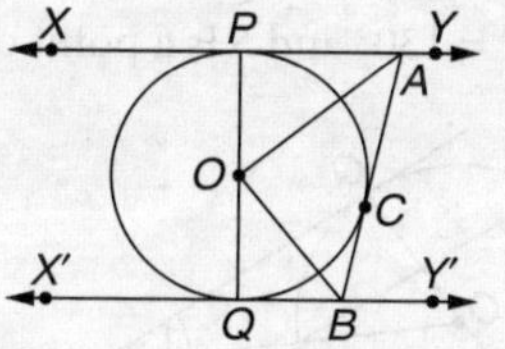

[NCERT]

55. Two concentric circles are of radii 13 cm and 5 cm. The length of the chord of the larger circle which touches the smaller circle is ___________.

56. In the figure, *O* is the centre of the circle, *PQ* is a chord and *PT* is tangent to the circle at *P*. If $\angle POQ = 70°$, find $\angle TPQ$.

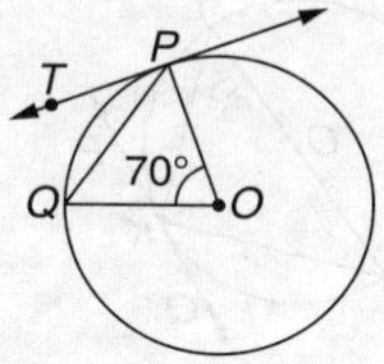

[AI 2017 (C)]

57. What is the distance between two parallel tangents of a circle of radius 7 cm?

58. In the figure, *AB*, *AC* and *AD* are tangents. If $AB = 5$ cm, find *AD*.

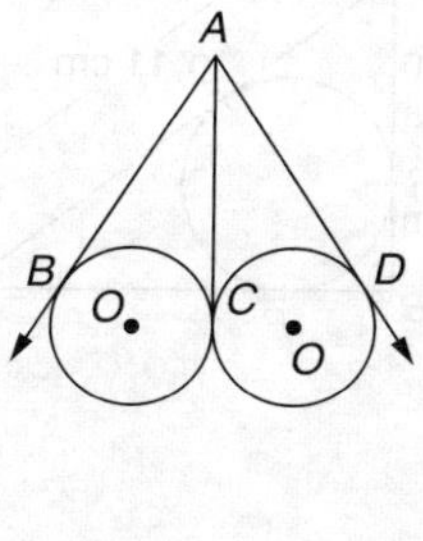

59. In the figure, AB and CD are common tangents to two circles of unequal radii. Prove that $AB = CD$.

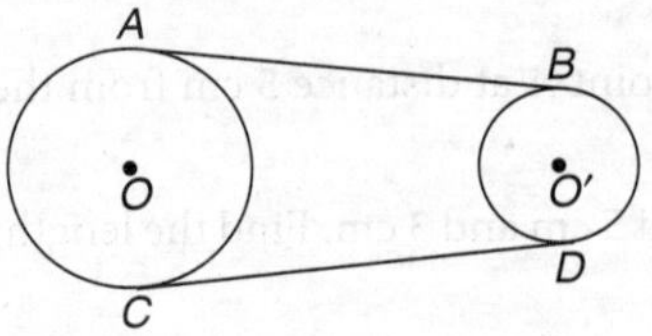

[NCERT Exemplar, Delhi 2017]

60. In the figure, if $\angle ATO = 40°$, find $\angle AOB$.

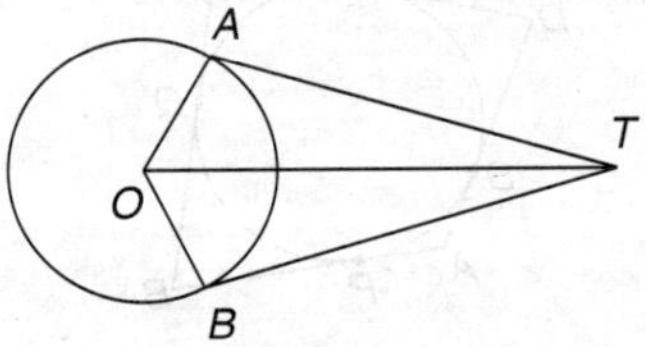

61. Prove that the tangents drawn at the end points of a chord of a circle make equal angles with the chord.

62. In the figure, PQ is a tangent from an external point P to a circle with centre O and OP cuts the circle at T and QOR is a diameter. If $\angle POR = 130°$ and S is a point on the circle, find $\angle 1 + \angle 2$.

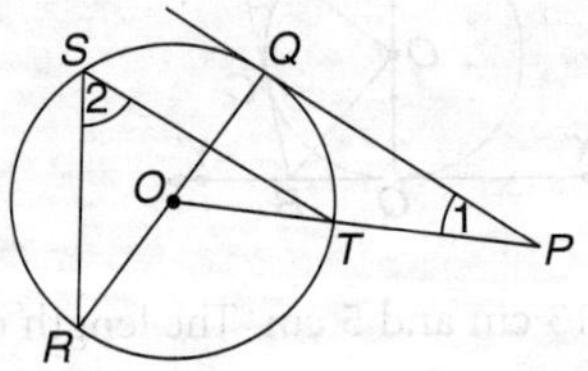

[AI 2017 (C)]

63. In the figure, O is the centre of the circle and LN is a diameter. If PQ is a tangent to the circle at K and $\angle KLN = 30°$, find $\angle PKL$.

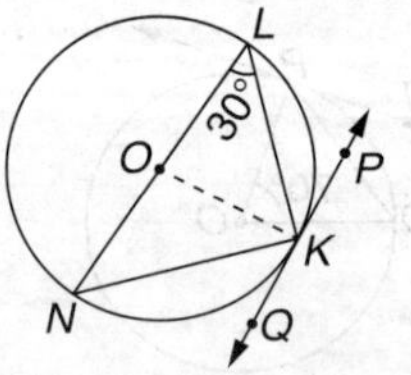

[AI 2017 (C)]

64. In the figure, ΔABC is circumscribing a circle. Find the length of BC.

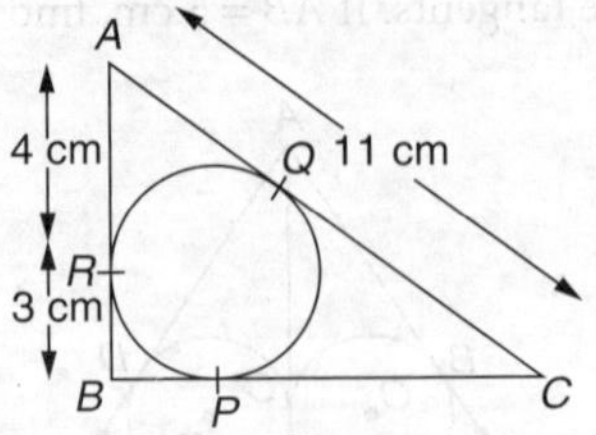

65. In the figure, there are two concentric circles, with centre O and of radii 5 cm and 3 cm. From an external point P, tangents PA and PB are drawn to these circles. If $AP = 12$ cm, find the length of BP.

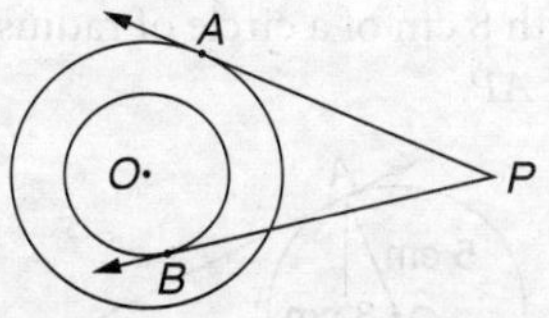

66. In the figure, a circle touches all the four sides of a quadrilateral $ABCD$ whose sides are $AB = 6$ cm, $BC = 9$ cm and $CD = 8$ cm. Find the length of side AD. **[AI 2011]**

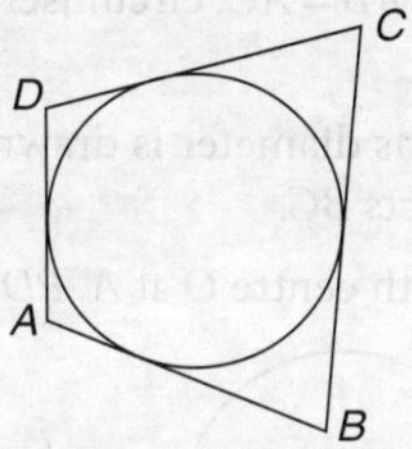

67. In the figure, O is the centre of the circle, PQ is a tangent to the circle at A. If $\angle PAB = 58°$, find $\angle ABQ$ and $\angle AQB$.

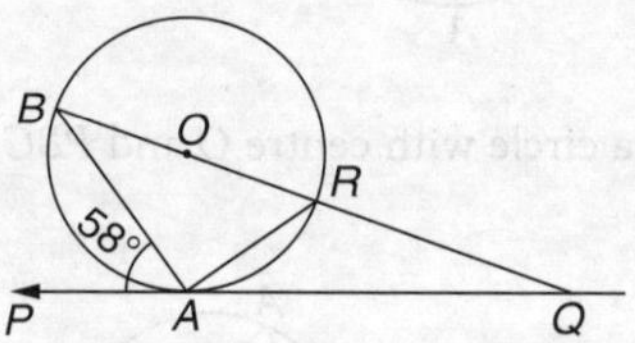

68. In the figure, a circle touches the side BC of ΔABC at P and touches AB and AC produced at Q and R, respectively. If $AQ = 5$ cm, find the perimeter of ΔABC.

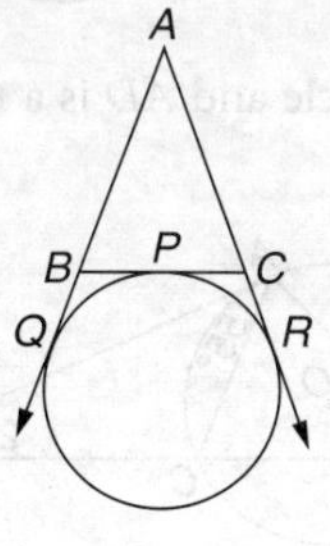

69. In the figure, TAS is a tangent to the circle, with centre O, at the point A. If $\angle OBA = 32°$, find the value of x.

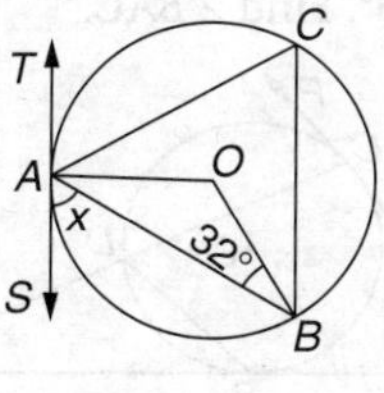

70. Two tangents PA and PB are drawn to the circle with centre O, such that $\angle APB = 120°$. Prove that $OP = 2AP$. **[Foreign 2016]**

71. In the figure, AB is a chord of length 8 cm of a circle of radius 5 cm. The tangents to the circle at A and B intersect at P. Find the length of AP. **[CBSE 2019, 2018 (C)]**

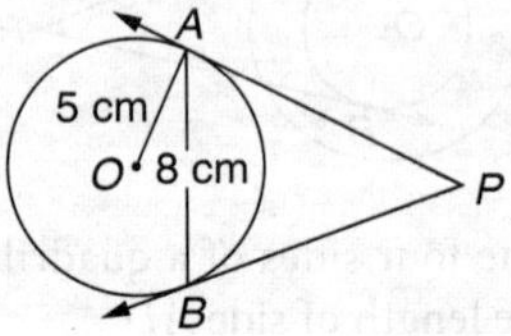

72. ABC is an isosceles triangle, in which $AB = AC$, circumscribed about a circle. Show that BC is bisected at the point of contact. **[Foreign 2012]**

73. In right ΔABC, a circle with side AB as diameter is drawn to intersect the hypotenuse AC at P. Prove that the tangent to the circle at P bisects BC.

74. In the figure, PA touches the circle with centre O at A. PD bisects $\angle APB$. Prove that $AD = AE$.

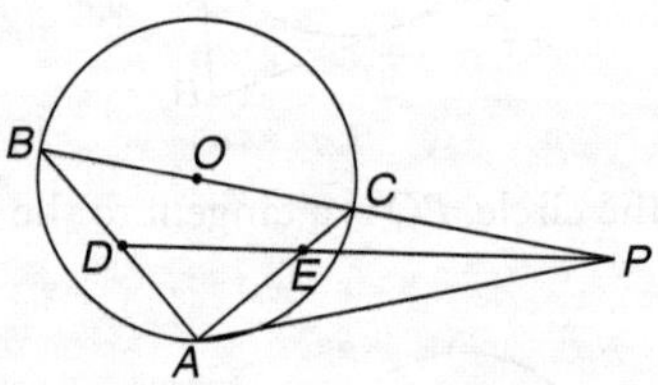

75. In the figure, PA is a tangent to a circle with centre O and PBC is a secant so that $PB = AB$. Prove that $PB.PC = AC^2$.

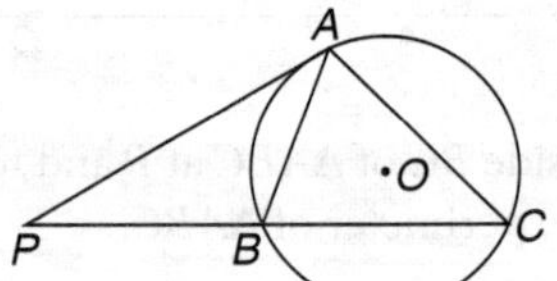

76. In the figure, O is the centre of the circle and AD is a tangent to the circle at A. If $\angle CAD = 55°$ and $\angle ADC = 25°$, then find $\angle ABO$.

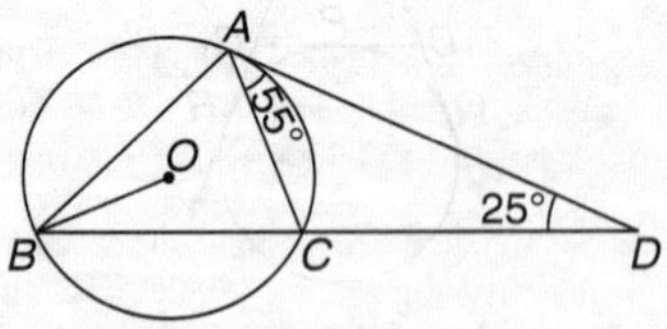

77. In the figure, ABE and ACP are tangents to the circle with centre O at B and C, respectively. CDE is a straight line. $DE = BD$ and $\angle DEB = 25°$. Find $\angle BAC$.

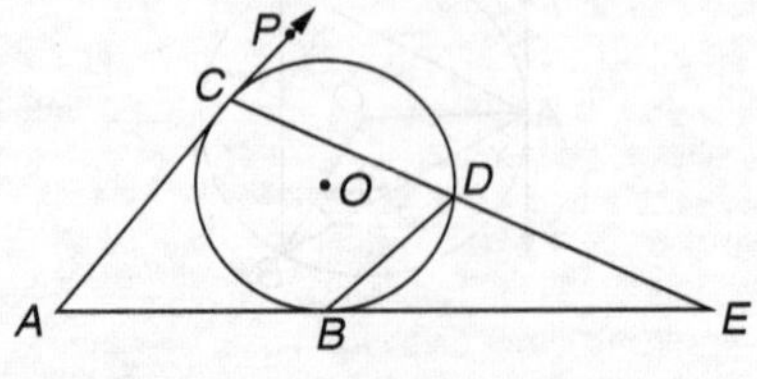

78. In the figure, AB is a diameter of a circle with centre O and AC is a tangent. If $\angle AOD = 60°$, find $\angle ACD$.

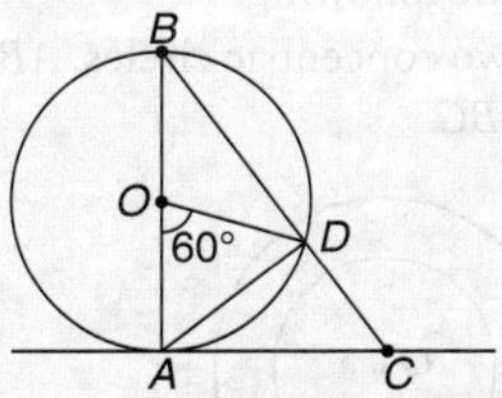

79. In the figure, PQ is a diameter of a circle with centre O and PT is a tangent at P. QT intersects the circle at R. If $\angle POR = 72°$, find $\angle PTR$.

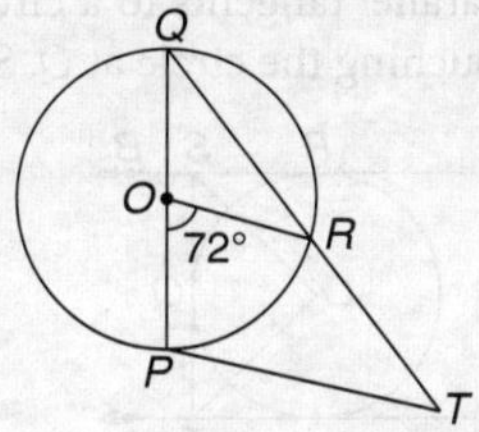

80. In the figure, $ABCD$ is a quadrilateral circumscribing a circle. DCY, ABY, ADX and BCX are straight lines. Prove that $XB + BY = XD + DY$.

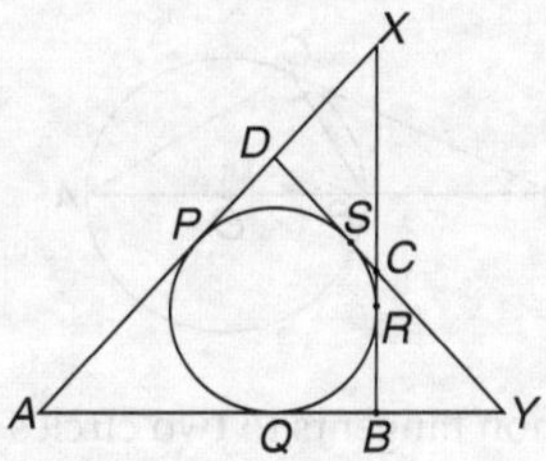

81. In the figure, two tangents PA and PB are drawn to a circle with centre O from an external point P. Prove that $\angle APB = 2\ \angle OAB$.

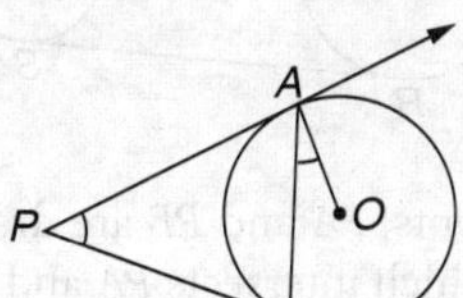

82. Prove that the lengths of tangents drawn from an external point to a circle are equal. Using the above, prove the following:

(i) A quadrilateral $ABCD$ is drawn to circumscribe a circle **[Delhi 2012, AI 2012, Foreign 2012]**

(ii) Prove that $AB + CD = AD + BC$.

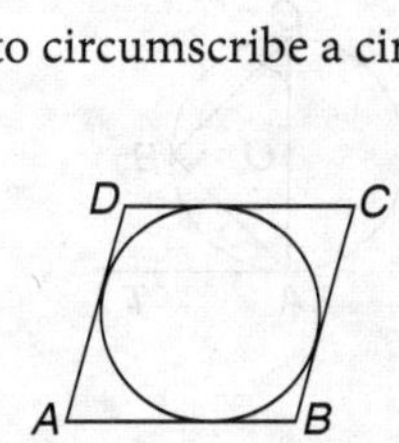

83. Prove that the tangent at any point of a circle is perpendicular to the radius through the point of contact. Using the above, perform the following:

In the figure, O is the centre of the two concentric circles. AB is a chord of the larger circle touching the smaller circle at C. Prove that $AC = BC$.

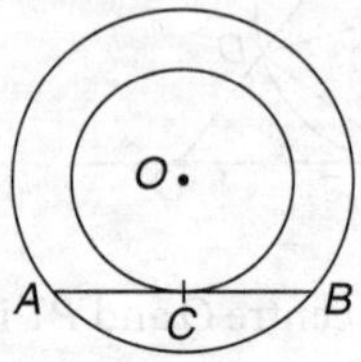

[Delhi 2016, 2013; Foreign 2012; AI 2011]

84. In the figure, AB and CD are two parallel tangents to a circle with centre O. ST is a tangent segment between the two parallel tangents touching the circle at Q. Show that $\angle SOT = 90°$.

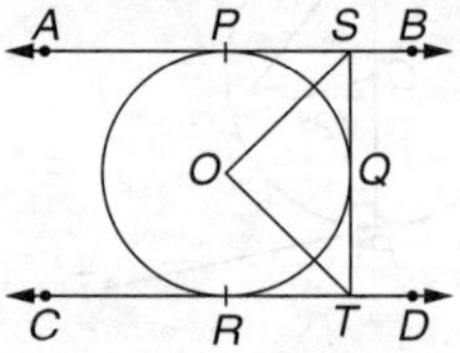

[Delhi 2013]

85. The tangent at a point C of a circle and a diameter AB when extended intersect at P. If $\angle PCA = 110°$, find $\angle CBA$ (see figure).

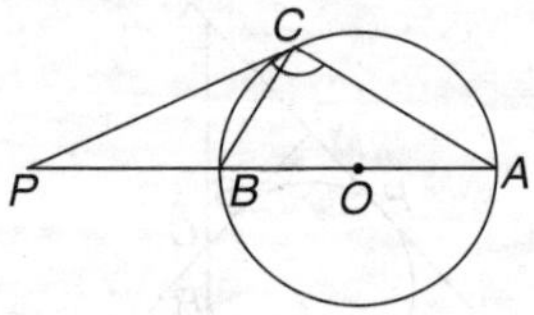

[NCERT Exemplar]

86. In the figure, PQ and RS are common tangents to two circles of unequal radii. Prove that $PQ = RS$.

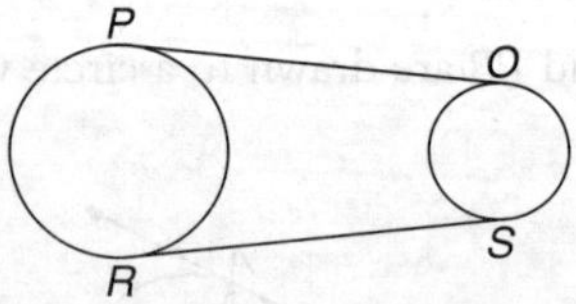

87. From an external point P, two tangents, PA and PB are drawn to a circle with centre O. At one point E on the circle, a tangent is drawn which intersects PA and PB at C and D, respectively. If $PA = 10$ cm, find the perimeter of the triangle PCD. **[NCERT Exemplar]**

88. If AB is a chord of a circle with centre O, AOC is a diameter and AT is the tangent at A, as shown in figure, prove that $\angle BAT = \angle ACB$.

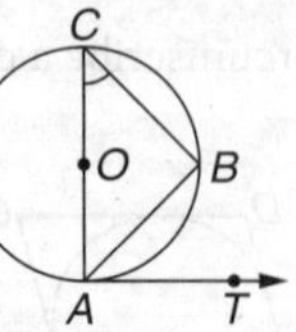

[NCERT Exemplar]

89. In the figure, tangents *PQ* and *PR* are drawn to a circle such that ∠*RPQ* = 30°. A chord *RS* is drawn parallel to the tangent *PQ*. Find the ∠*RQS*.

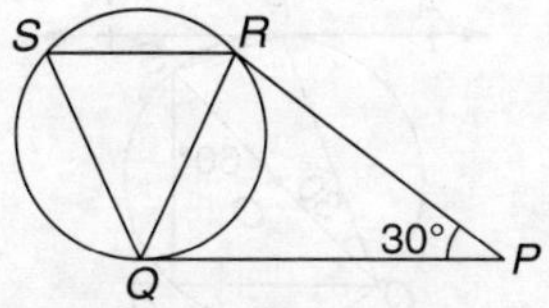

90. In the figure, the common tangent, *AB* and *CD* are tangents to two circles with centres *O* and *O'* intersect at *E*. Prove that the points *O, E, O'* are collinear.

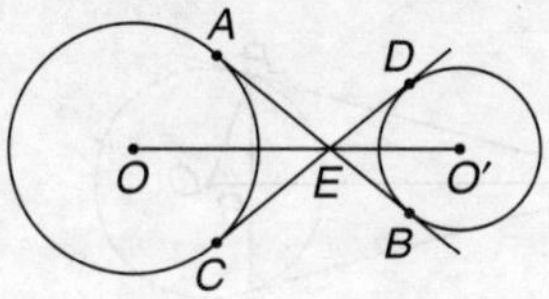

91. In the figure, *O* is the centre of the circle, *AB* is a chord and *AT* is the tangent at *A*. If ∠*AOB* = 100°, then find ∠*BAT*.

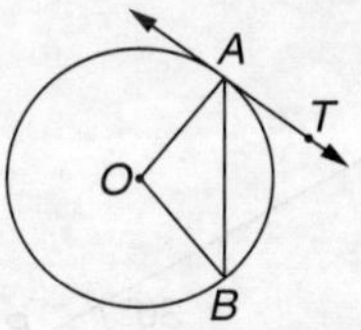

[Delhi 2011]

92. In the figure, *PA* and *PB* are tangents to the circle with centre *O*. If ∠*APB* = 60°, then find ∠*OAB*.

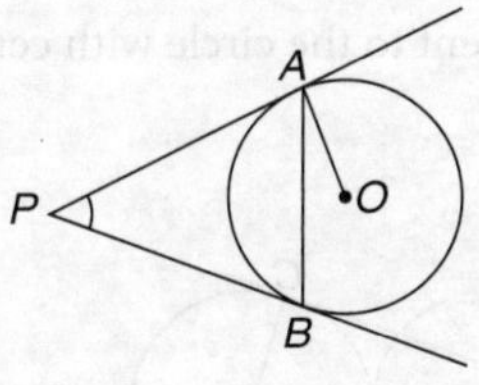

[Delhi 2011]

93. In the figure, *PQ* and *PR* are tangents to circle with centre *A*. If ∠*QPA* = 27°, then find ∠*QAR*.

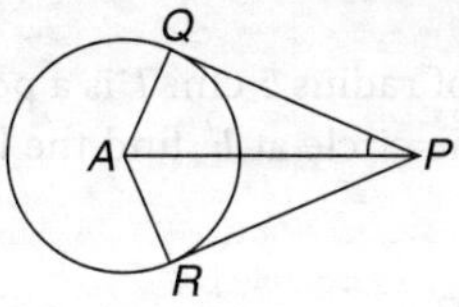

[Foreign 2012]

94. In the figure, find *x* if ∠*EBD* = 146°.

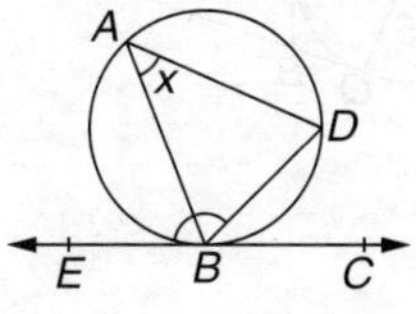

95. In the figure, QS is the diameter and O is the centre of circle. APT is the tangent at P. Find $\angle APQ$.

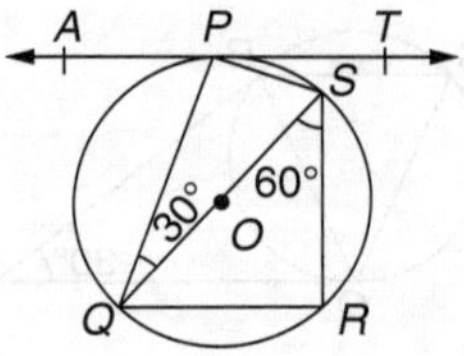

96. In the figure, PQ is a chord of length 16 cm, of a circle of radius 10 cm. The tangents at P and Q intersect at a point T. Find the length of TP.

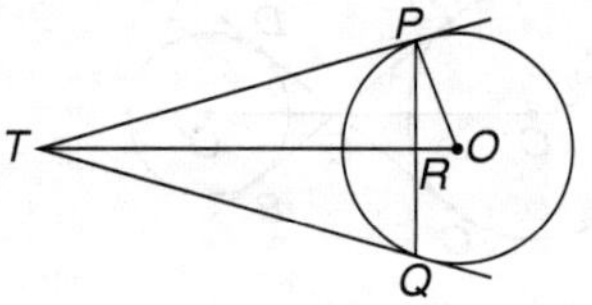

[AI 2012]

97. In the figure, tangents PQ and PR are drawn from an external point P to a circle with centre O, such that $\angle RPQ = 30°$. A chord RS is drawn parallel to the tangent PQ. Find $\angle RQS$.

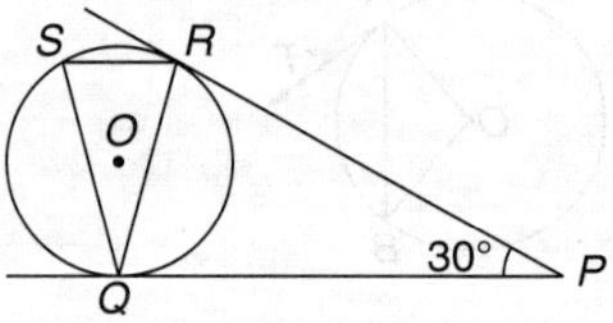

[Delhi 2015]

98. In the figure, two equal circles, with centres O and O' touch each other at X. OO' produced meets the circle with centre O' at A. AC is a tangent to the circle with centre O, at the point C. $O'D$ is perpendicular to AC. Find the value of $\dfrac{DO'}{CO}$.

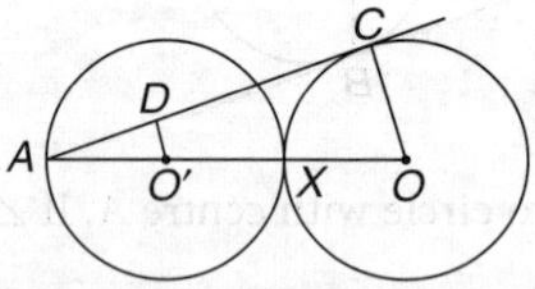

[AI 2016]

99. In the figure, O is the centre of a circle of radius 5 cm. T is a point such that $OT = 13$ cm and OT intersects circle at E. If AB is a tangent to the circle at E, find the length of AB, where TP and TQ are two tangents to the circle.

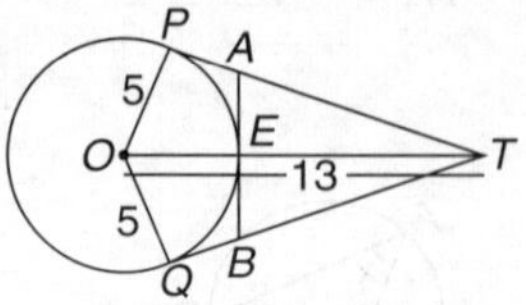

[Delhi 2016]

100. In the figure, AB is the diameter of a circle with centre O and AT is a tangent. If $\angle AOQ = 58°$, find $\angle ATQ$.

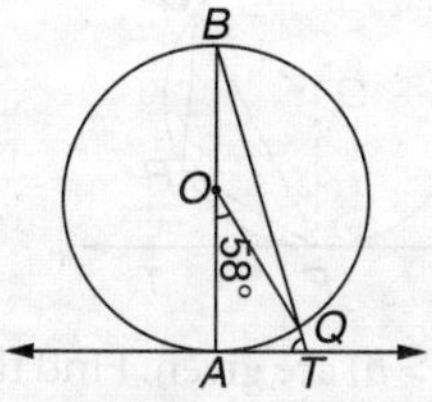

[Delhi 2015]

101. In the figure, two tangents RQ and RP are drawn from an external point R to the circle with centre O. If $\angle PRQ = 120°$, prove that $OR = PR + RQ$.

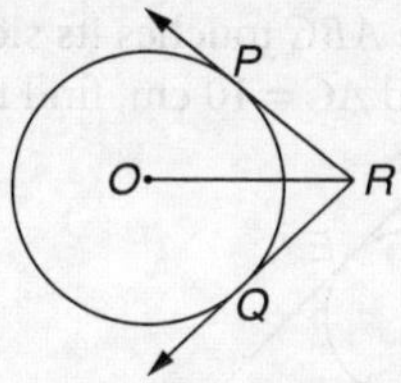

[AI 2015]

102. In the figure, a triangle ABC is drawn to circumscribe a circle of radius 3 cm, such that the segments BD and DC are, respectively, of length 6 cm and 9 cm. If the area of ΔABC is 54 cm^2, find the length of sides AB and AC.

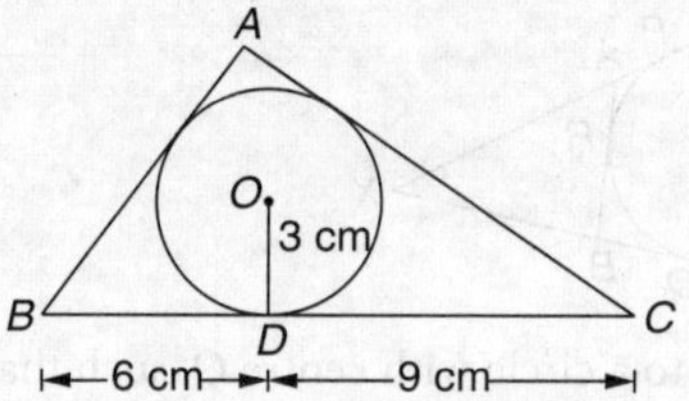

[AI 2015]

103. In the figure, AOB is a diameter of a circle with centre O and AC is a tangent to the circle at A. If $\angle BOC = 130°$, then find $\angle ACO$.

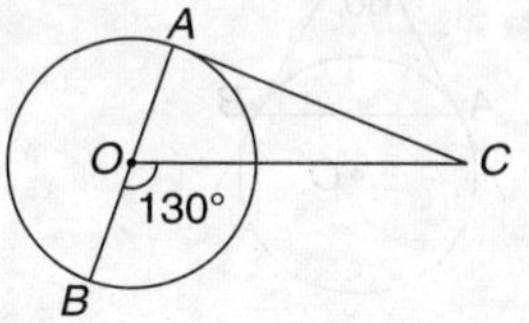

[Foreign 2016]

104. In the figure, PA and PB are tangents to the circle with centre O such that $\angle APB = 50°$. Write the measure of $\angle OAB$.

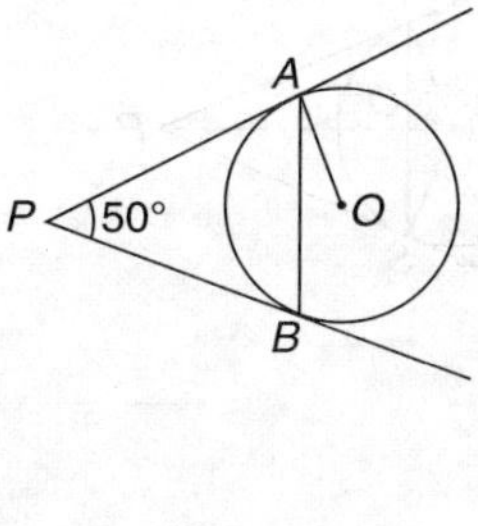

[Delhi 2015]

105. In the figure, PQ is a chord of a circle with centre O and PT is a tangent. If $\angle QPT = 60°$, find $\angle PRQ$.

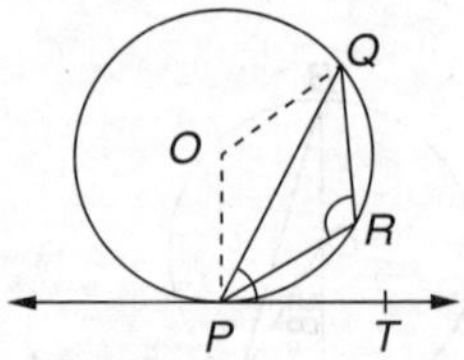

[AI 2015]

106. Two concentric circles of radii a and b $(a > b)$ are given. Find the length of the chord of the larger circle which touches the smaller circle. **[Foreign 2015]**

107. If d_1, d_2 $(d_2 > d_1)$ are the diameters of two concentric circles and c is the length of a chord of a circle which is tangent to the other circle, prove that $d_2^2 = c^2 + d_1^2$. **[Foreign 2011]**

108. In the figure, a circle inscribed in triangle ABC touches its sides AB, BC and AC at points D, E and F, respectively. If $AB = 12$ cm, $BC = 8$ cm and $AC = 10$ cm, find the lengths of AD, BE and CF.

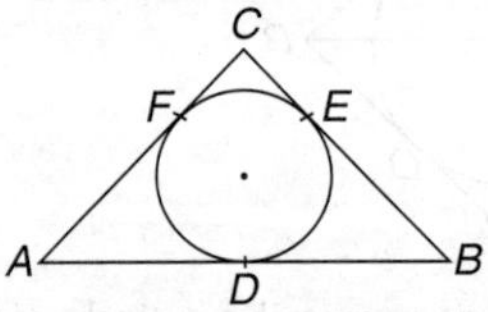

[Delhi 2013]

109. In the figure, XP and XQ are two tangents to a circle with centre O from a point X outside the circle. ARB is tangent to the circle at R. Prove that $XA + AR = XB + BR$.

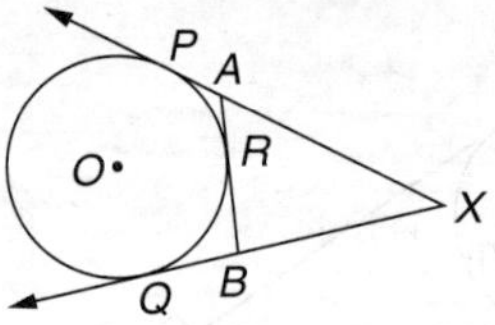

[Foreign 2014]

110. In the figure, AP and BP are tangents to a circle with centre O, such that $AP = 5$ cm and $\angle APB = 60°$. Find the length of chord AB.

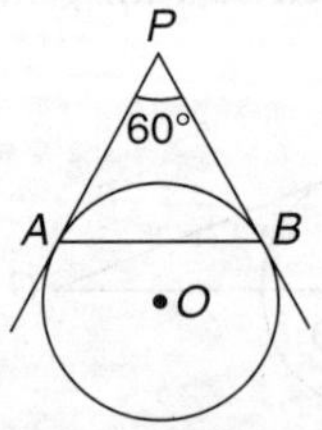

[Delhi 2016]

111. In the figure, from an external point P, two tangents PT and PS are drawn to a circle with centre O and radius r. If $OP = 2r$, show that $\angle OTS = \angle OST = 30°$.

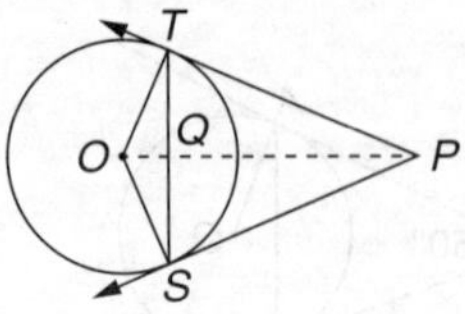

[AI 2016]

112. In the figure, PQ is a tangent at a point C to a circle with centre O. If AB is a diameter and $\angle CAB = 30°$, find $\angle PCA$.

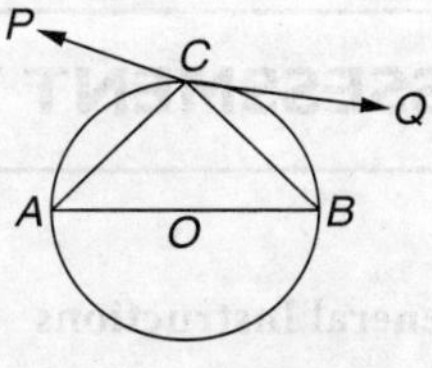

[AI 2016]

113. Two tangents TP and TQ are drawn to a circle with centre O from an external point T. Prove that $\angle PTQ = 2\angle OPQ$. **[Delhi 2017]**

Answers

Multiple Choice Questions

1. (a) **2.** (b) **3.** (a) **4.** (c) **5.** (b) **6.** (a) **7.** (c) **8.** (a) **9.** (b) **10.** (a)
11. (c) **12.** (b) **13.** (a) **14.** (b) **15.** (d) **16.** (b) **17.** (c) **18.** (b) **19.** (c) **20.** (a)
21. (a) **22.** (d) **23.** (b) **24.** (a) **25.** (b) **26.** (c) **27.** (c) **28.** (c) **29.** (c) **30.** (c)
31. (a) **32.** (d) **33.** (b) **34.** (d) **35.** (c)

Fill in the Blanks

36. Parallel **37.** Perpendicular **38.** No tangent **39.** Equal **40.** Tangent, Secant **41.** AC
42. Same

Assertion–Reason

43. (B) **44.** (B) **45.** (C) **46.** (B)

Subjective Questions

51. 3 cm **52.** 8 cm **55.** 24 cm **56.** 35° **57.** 14 cm **58.** 5 cm **60.** 100° **62.** 105° **63.** 60°
64. 10 cm **65.** 12.65 cm **66.** 5 cm **67.** 32°, 26° **68.** 10 cm **69.** 58° **71.** $\frac{20}{3}$ cm
76. 10° **77.** 80° **78.** 60° **79.** 54° **85.** 70° **87.** 20 cm **89.** 30° **91.** 50° **92.** 30°
93. 126° **94.** 34° **95.** 60° **96.** $\frac{40}{3}$ cm **97.** 30° **98.** $\frac{1}{3}$ **99.** $\frac{20}{3}$ cm, $\frac{10}{3}$ cm **100.** 61°
102. 15 cm, 12 cm and 9 cm **103.** 40° **104.** 25° **105.** 120° **106.** $2\sqrt{a^2 - b^2}$
108. 7 cm, 5 cm, 3 cm **110.** 5 cm **112.** 60°

Circles (Basic)

SELF-ASSESSMENT TEST

Time allowed: 90 min. Max. Marks: 40

General Instructions

1. *All questions are compulsory. There are 20 questions in all.*
2. *This question paper has four sections: Section A, Section B, Section C and Section D.*
3. *Section A contains 10 questions of one mark each, Section B contains three questions of two marks each, Section C contains four questions of three marks each, and Section D contains three questions of four marks each.*

Section A [1 Mark Each]

Direction: (Q. Nos. 1–5) These are Multiple Choice Questions. Select the most appropriate answer for the given questions.

1. In the figure, if *TP* and *TQ* are the two tangents to a circle with centre *O* so that $\angle POQ = 110°$, then $\angle PTQ$ is equal to

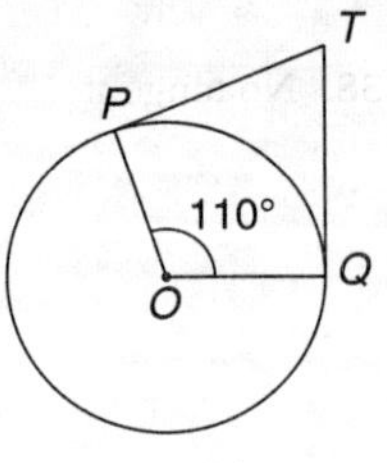

(a) 60° (b) 70° (c) 80° (d) 90°

2. In the figure, *AB* and *AC* are tangents to the circle with center *O* such that $\angle BAC = 40°$. Then, $\angle BOC$ is equal to

(a) 40° (b) 50° (c) 140° (d) 150°

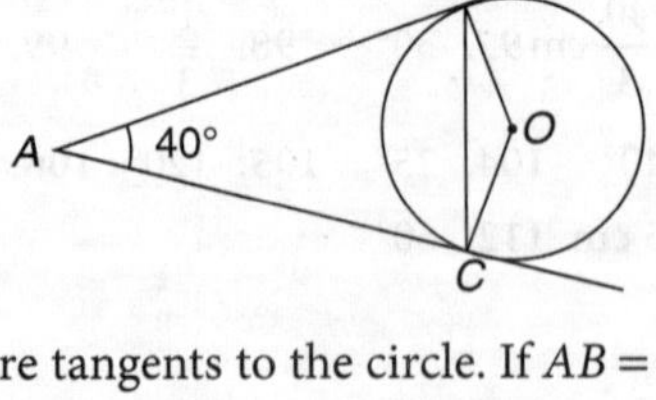

3. In the figure, *AP*, *AQ* and *BC* are tangents to the circle. If $AB = 5$ cm, $AC = 6$ cm and $BC = 4$ cm, then the length of *AP* (in cm) is

(a) 7.5 (b) 15 (c) 10 (d) 9

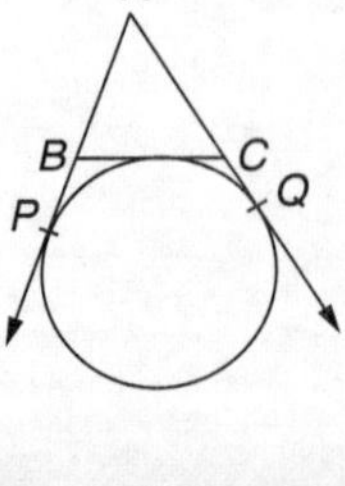

4. If two tangents inclined at an angle of 60° are drawn to a circle of radius 4 cm, the length of each tangent is equal to

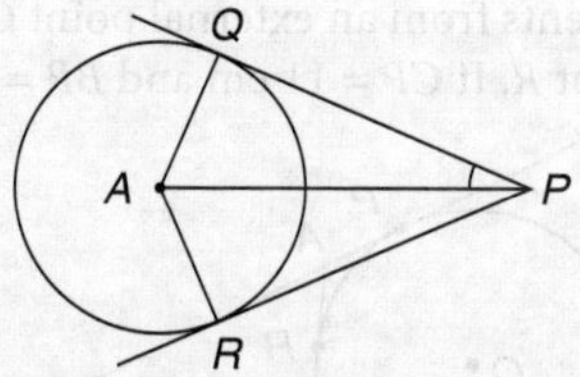

(a) $\frac{4}{\sqrt{3}}$ cm (b) $3\sqrt{3}$ cm (c) $4\sqrt{3}$ cm (d) $2\sqrt{3}$ cm

5. In the figure, *PQ* and *PR* are tangents to a circle with centre *A*. If $\angle QPA = 27°$, then $\angle QAR$ equals

(a) 63° (b) 153° (c) 126° (d) 117°

6. Two tangents, drawn at the end points of diameter of a given circle, are always __________.

7. If two tangents are drawn to a circle from an external point, they subtend __________ angles at the centre.

8. A quadrilateral *ABCD* is drawn to circumscribe a circle (see figure). Prove that $AB + CD = AD + BC$.

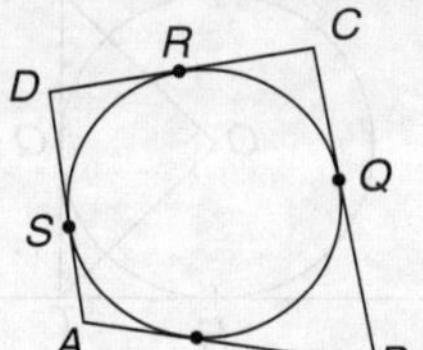

9. If the radii of two concentric circles are 15 cm and 17 cm, show that the length of the chord of one circle that is tangent to the other circle is 16 cm.

10. In the figure, find *x* if $\angle EBD = 146°$.

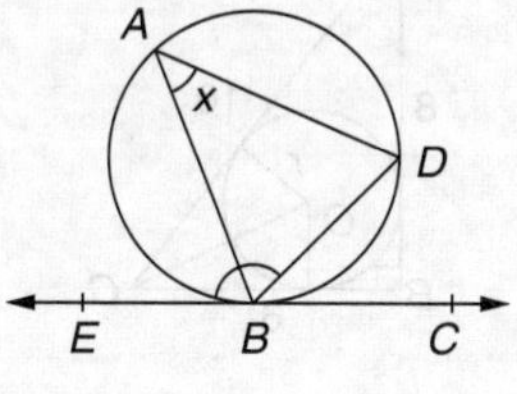

Section B [2 Marks Each]

11. In the figure, *TAS* is a tangent to the circle with centre *O* at the point *A*. If $\angle OBA = 32°$, find the value of *x*.

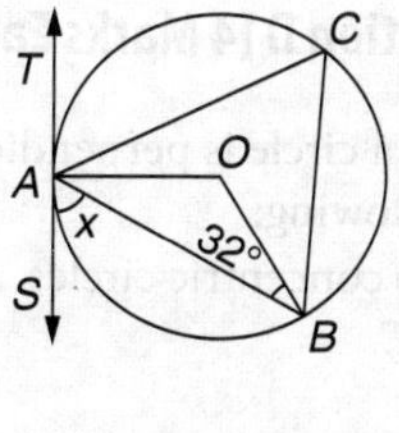

12. A point M is 26 cm away from the centre of a circle and the length of the tangent drawn from M to the circle is 24 cm. Find the radius of the circle.

13. In the figure, CP and CQ are tangents from an external point C to a circle with centre O. AB is another tangent which touches the circle at R. If $CP = 11$ cm and $BR = 4$, find the length of BC.

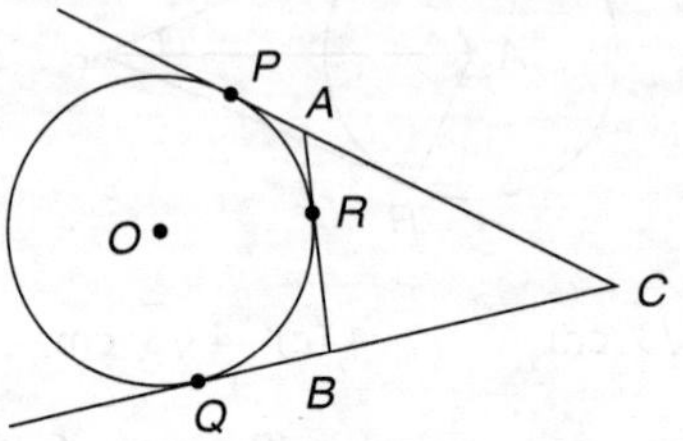

Section C [3 Marks Each]

14. In the figure, AB and CD are two parallel tangents to a circle with centre O. ST is a tangent segment between the two parallel tangents, touching the circle at Q. Show that $\angle SOT = 90°$.

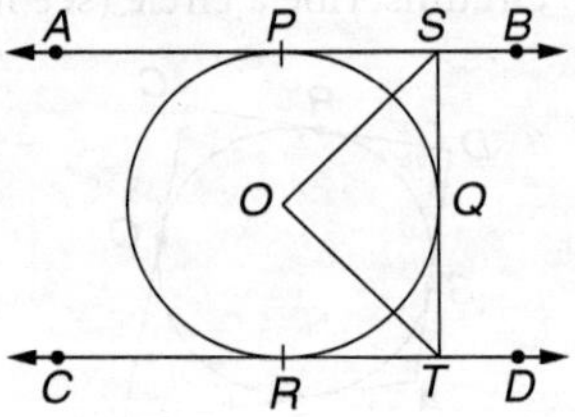

15. In the figure, a right triangle ABC circumscribes a circle of radius r. If AB and BC are of length 8 cm and 6 cm, respectively, find the value of r.

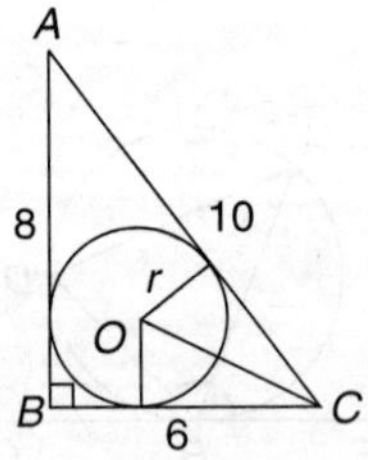

16. Two tangents TP and TQ are drawn to a circle with centre O from an external point T. Prove that $\angle PTQ = 2\angle OPQ$.

17. Prove that the angle between the two tangents drawn from an external point to a circle is supplementary to the angle subtended by the line segment joining the points of contact at the centre.

Section D [4 Marks Each]

18. Prove that the tangent at any point of a circle is perpendicular to the radius through the point of contact. Using the above, perform the following:
In the figure, O is the centre of the two concentric circles. AB is a chord of the larger circle touching the smaller circle at C. Prove that $AC = BC$.

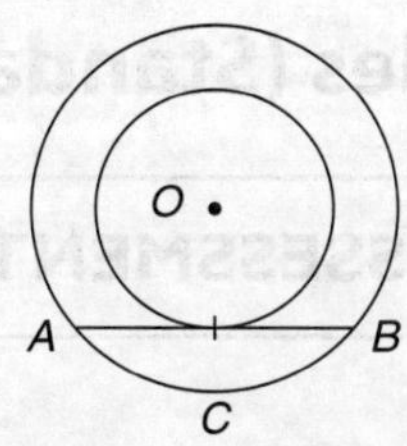

19. Prove that the lengths of tangents drawn from an external point to a circle are equal. Using the above, prove the following:
A quadrilateral *ABCD* is drawn to circumscribe a circle. Prove that $AB + CD = AD + BC$.

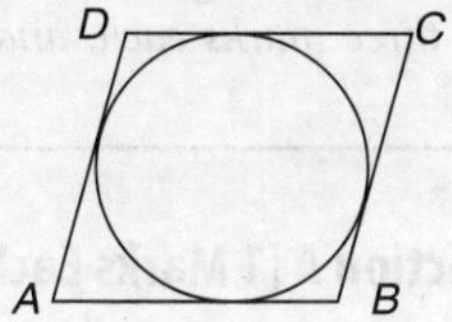

20. In the figure, a triangle *ABC* is drawn to circumscribe a circle of radius 2 cm such that the segments *BD* and *DC* into which *BC* is divided by the point of contact *D* are the lengths 4 cm and 3 cm, respectively. If the area of $\Delta ABC = 21\ \text{cm}^2$, then find the length of the sides *AB* and *AC*.

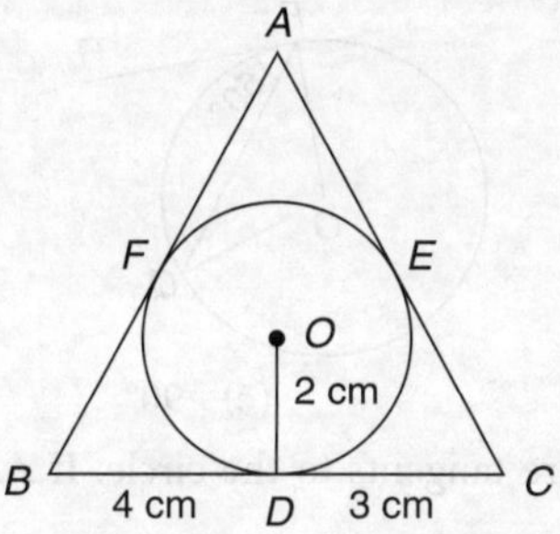

Answers

1. (b) **2.** (c) **3.** (a) **4.** (c) **5.** (c) **6.** Parallel **7.** Equal **10.** 34° **11.** 58°
12. 10 cm **13.** 7 cm **15.** 2 cm **20.** 7.5 cm, 7 cm and 6.5 cm

Circles (Standard)

SELF-ASSESSMENT TEST

Time allowed: 90 min. Max. Marks: 40

General Instructions

1. *All questions are compulsory. There are 20 questions in all.*
2. *This question paper has four sections: Section A, Section B, Section C and Section D.*
3. *Section A contains 10 questions of one mark each, Section B contains three questions of two marks each, Section C contains four questions of three marks each, and Section D contains three questions of four marks each.*

Section A [1 Marks Each]

Direction: (Q. Nos. 1–5) These are Multiple Choice Questions. Select the most appropriate answer for the given questions.

1. In the figure, if O is the centre of a circle, PQ is a chord and the tangent PR at P makes an angle of 50° with PQ, then $\angle POQ$ is equal to

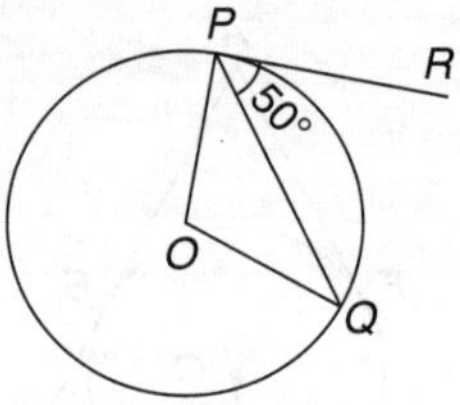

(a) 100° (b) 80° (c) 90° (d) 75°

2. In the figure, AP, AQ and BC are tangents to the circle. If $AB = 5$ cm, $AC = 6$ cm and $BC = 4$ cm, then the length of AP (in cm) is

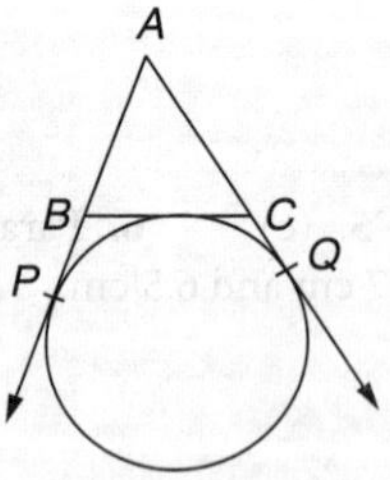

(a) 7.5 (b) 15 (c) 10 (d) 9

3. In the figure, if PQR is the tangent to a circle at Q whose centre is O, AB is a chord parallel to PR and $\angle BQR = 70°$, then $\angle AQB$ is equal to

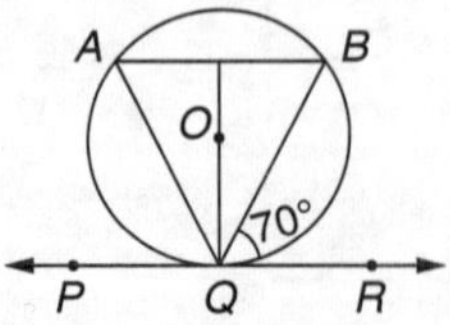

(a) 20° (b) 40° (c) 35° (d) 45°

4. In the figure, O is the centre of a circle, AB is a chord and AT is the tangent at A. If $\angle AOB = 100°$, then $\angle BAT$ is equal to

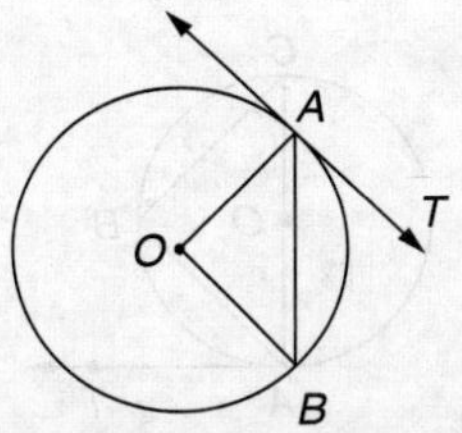

(a) 100° (b) 40° (c) 50° (d) 90°

5. A circle is drawn touching the sides of ΔABC at P, Q, R. If $AP + BC = 13$ cm, then the perimeter of ΔABC is

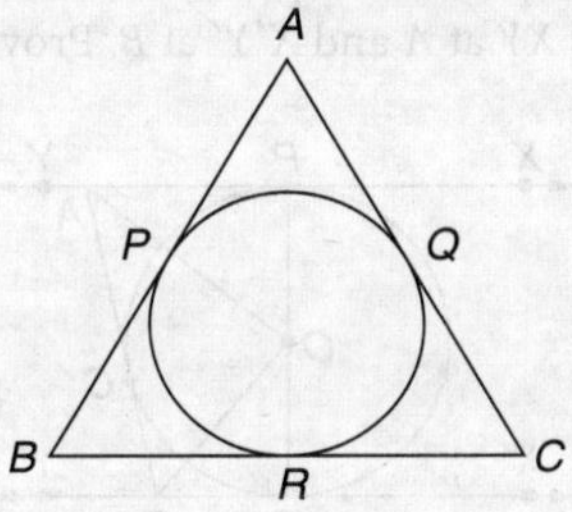

(a) 39 cm (b) 26 cm (c) 42 cm (d) 20 cm

6. In two concentric circles, if chords are drawn in the outer circle which touch the inner circle, then all the chords are of __________ length.

7. If two tangents are drawn to a circle from an external point, then they subtend __________ angles at the centre.

8. In the figure, AB, AC and AD are tangents. If $AB = 5$ cm, find AD.

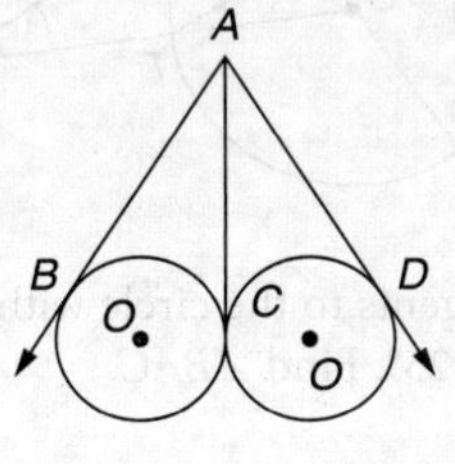

9. In the figure, AB is a diameter of a circle with centre O and AC is a tangent. If $\angle AOD = 60°$, find $\angle ACD$.

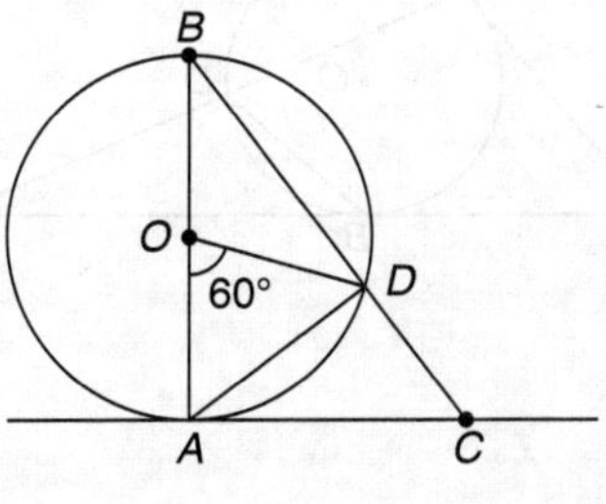

10. If AB is the chord of a circle with centre O, AOC is the diameter and AT is the tangent at A, as shown in the figure, prove that $\angle BAT = \angle ACB$.

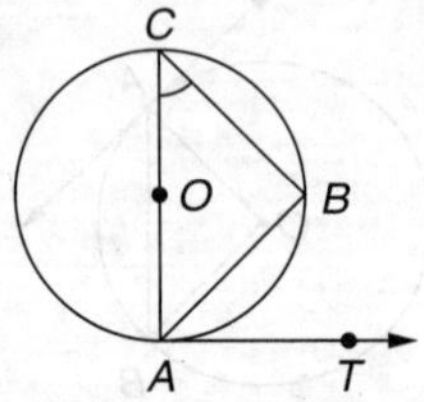

Section B [2 Marks Each]

11. In the figure, XY and $X'Y'$ are two parallel tangents to a circle with centre O and another tangent AB with point of contact C intersects XY at A and $X'Y'$ at B. Prove that $\angle AOB = 90°$.

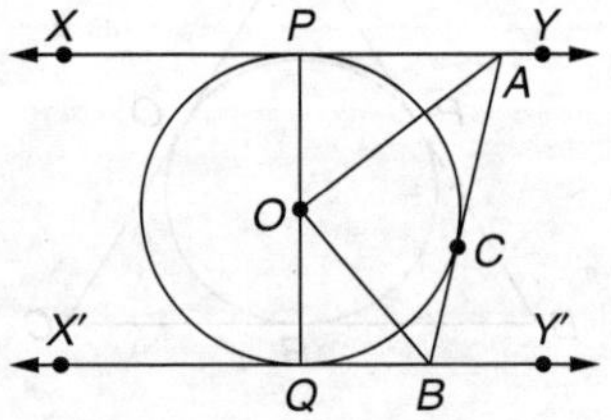

12. In the figure, PQ is a tangent from an external point P to a circle with centre O and OP cuts the circle at T and QOR is a diameter. If $\angle POR = 130°$ and S is a point on the circle, find $\angle 1 + \angle 2$.

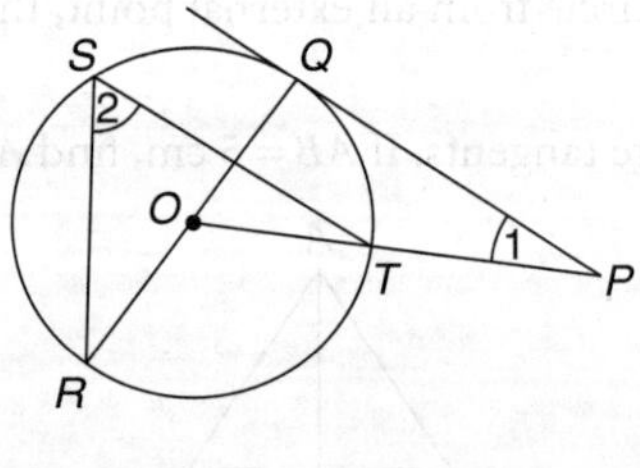

13. In the figure, ABE and ACP are tangents to the circle with centre O at B and C, respectively. CDE is a straight line. $DE = BD$ and $\angle DEB = 25°$. Find $\angle BAC$.

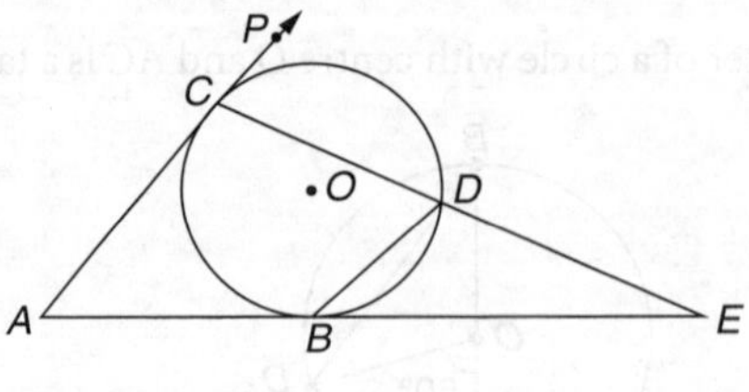

Section C [3 Marks Each]

14. In the figure, two tangents *PA* and *PB* are drawn to a circle with centre *O* from an external point *P*. Prove that $\angle APB = 2\angle OAB$.

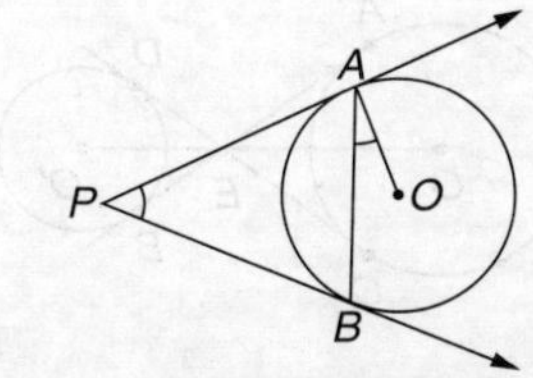

15. Prove that the tangent at any point of a circle is perpendicular to the radius through the point of contact. Using the above, perform the following:
In the figure, *O* is the centre of the two concentric circles. *AB* is a chord of the larger circle touching the smaller circle at *C*. Prove that $AC = BC$.

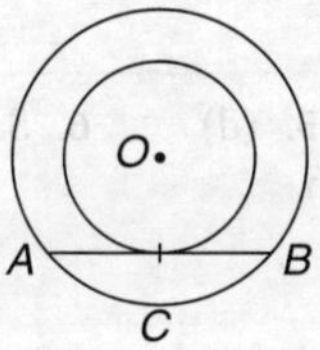

16. Prove that the lengths of the tangents drawn from an external point to a circle are equal. Using the above, prove the following:
A quadrilateral *ABCD* is drawn to circumscribe a circle. Prove that $AB + CD = AD + BC$.

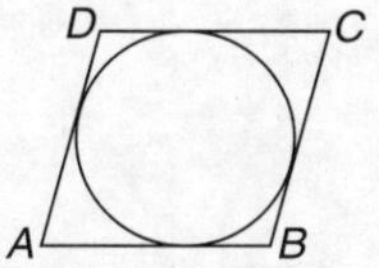

17. In the figure, a triangle *ABC* is drawn to circumscribe a circle of radius 2 cm such that the segments *BD* and *DC* into which *BC* is divided by the point of contact *D* are of length 4 cm and 3 cm, respectively. If the area of $\Delta ABC = 21\text{ cm}^2$, then find the length of the sides *AB* and *AC*.

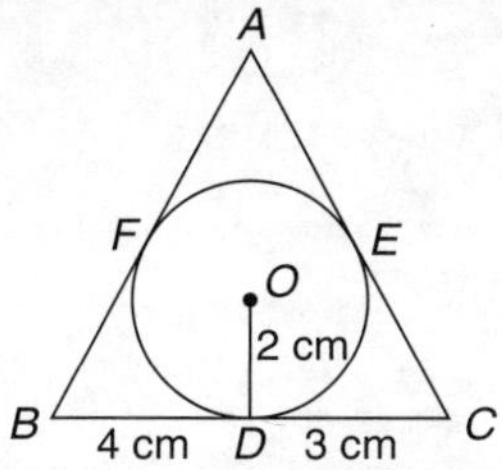

Section D [4 Marks Each]

18. In the figure, AB and CD are tangents to two circles with centres O and O' and intersect at E. Prove that the points O, E, O' are collinear.

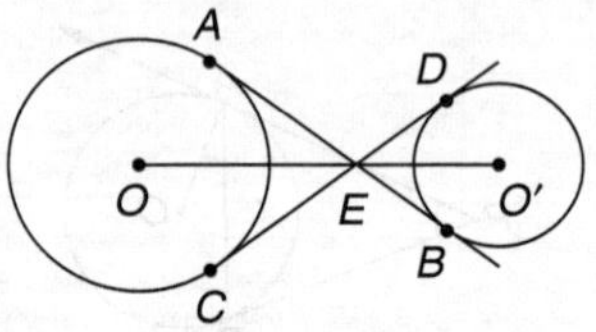

19. Two tangents TP and TQ are drawn to a circle with centre O from an external point T. Prove that $\angle PTQ = 2\angle OPQ$.

20. In right ΔABC, a circle with side AB as diameter is drawn to intersect the hypotenuse AC at P. Prove that tangent to the circle at P bisects BC.

Answers

1. (a) **2.** (a) **3.** (b) **4.** (c) **5.** (d) **6.** Same **7.** Equal **8.** 5 cm **9.** 60°
12. 105° **13.** 80° **17.** 7. 5 cm, 7 cm, 6.5 cm

CHAPTER

11 Constructions

Key Concepts

- Bisector of a line segment
- To draw a triangle similar to a given triangle as per the given scale factor, say $\frac{m}{n}$ (i) $m > n$ (ii) $m < n$
- To draw a pair of tangents to a given circle

Concept: Division of a Line Segment in a Given Ratio

Example 1 Draw a line segment of length 6.5 cm and divide it in the ratio 3:4. Measure the two parts.

Solution: *Steps of Construction*

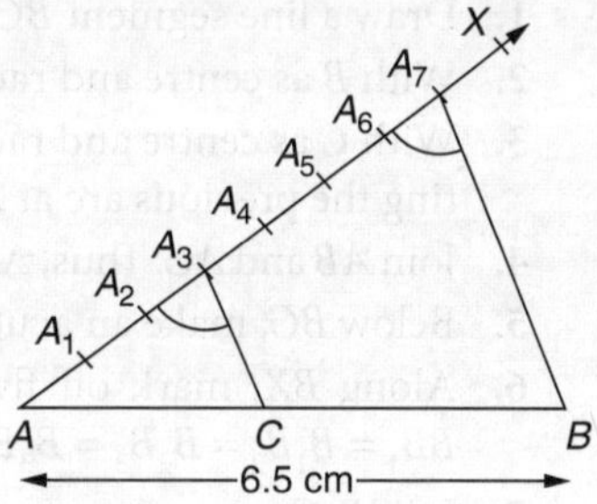

1. Draw the line segment $AB = 6.5$ cm.
2. Draw a ray AX, making an acute angle with AB.
3. Locate 7 points on AX such that $AA_1 = A_1A_2 = A_2A_3 = A_3A_4 = A_4A_5 = A_5A_6 = A_6A_7$.
4. Join A_7B.
5. From A_3, draw $A_3C \parallel A_7B$ meeting AB at C.

$$\therefore \quad \frac{AC}{CB} = \frac{3}{4}$$

*See Justification part of construction in *For More Practice and Knowledge*

Example 2 Divide a line segment 6 cm long in the ratio 4:3. Prove your assertion.

Solution: *Steps of Construction*

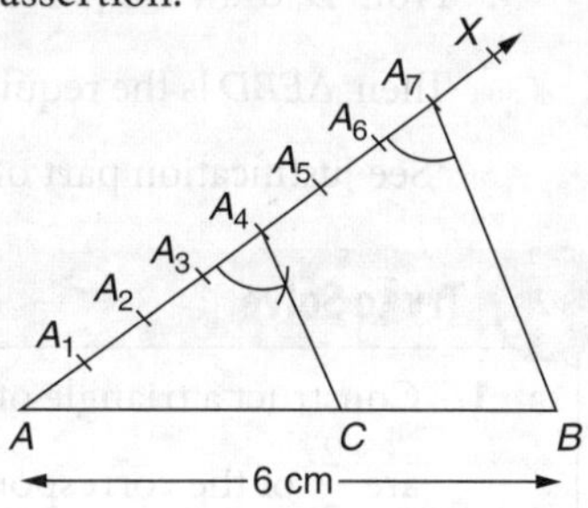

1. Draw a line segment $AB = 6$ cm.
2. Draw a ray AX, making an acute angle $\angle BAX$.
3. Along AX, mark $(4 + 3) = 7$ points A_1, A_2, A_3, A_4, A_5, A_6 and A_7 such that $AA_1 = A_1A_2 = A_2A_3 = A_3A_4 = A_4A_5 = A_6A_7$.
4. Join A_7B.
5. From A_4, draw A_4C parallel to A_7B (by making an angle equal to $\angle AA_7B$), meeting AB at C.
   Then, C is the point on AB that divides it in the ratio 4:3.
   Thus, $AC:CB = 4:3$. [BPT]

Alternative Method

Steps of Construction

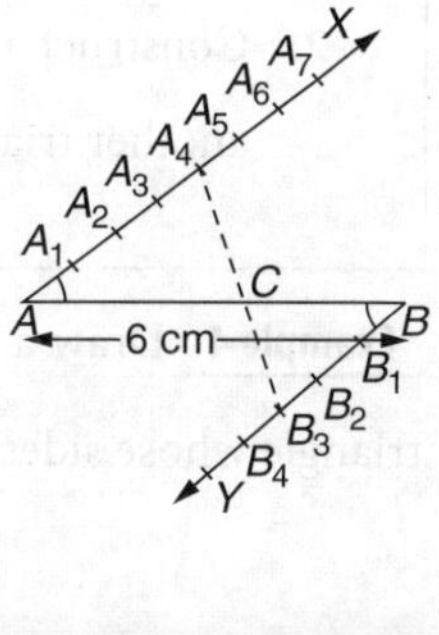

1. Draw a line segment $AB = 6$ cm.
2. Draw a ray AX, making an acute $\angle BAX$.
3. Draw a ray BY parallel to AX by making $\angle ABY = \angle BAX$.

4. Locate the points A_1, A_2, A_3, A_4 on AX and B_1, B_2, B_3 on BY such that $AA_1 = A_1A_2 = A_2A_3 = A_3A_4 = BB_1 = B_1B_2 = B_2B_3$.
5. Join A_4B_3, intersecting AB at a point C.
   Then, $AC:CB = 4:3$.
   *See Justification part of construction in *For More Practice and Knowledge*

1 Try to Solve

1. Draw a line segment of length 7.6 cm and divide it in the ratio 5:8. Measure the two parts. **[NCERT]**
2. Draw a line segment 8 cm long and divide it in the ratio 5:3. Measure the two parts.

Concept: To Construct a Triangle Similar to a Given Triangle as Per the Given Scale Factor

Example 3 Construct a triangle with sides 5 cm, 5.5 cm and 6.5 cm. Then, construct another triangle whose sides are $\frac{3}{5}$ times the corresponding sides of the first triangle.

Solution: *Steps of Construction*

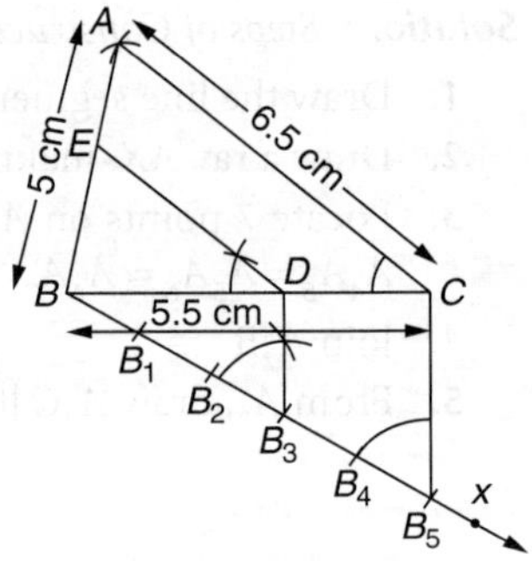

1. Draw a line segment $BC = 5.5$ cm.
2. With B as centre and radius equal to 5 cm, draw an arc.
3. With C as centre and radius equal to 6.5 cm, draw another arc cutting the previous arc at A.
4. Join AB and AC. Thus, ΔABC with given lengths of sides is obtained.
5. Below BC, make an acute angle $\angle CBX$.
6. Along BX, mark off five points B_1, B_2, B_3, B_4 and B_5 such that $BB_1 = B_1B_2 = B_2B_3 = B_3B_4 = B_4B_5$.
7. Join B_5C.
8. From B_3, draw $B_3D \parallel B_5C$ (by making an angle $\angle BB_3D = \angle BB_5C$) meeting BC at D.
9. From D, draw $DE \parallel CA$ (by making an angle $\angle BDE = \angle BCA$) meeting AB at E.

Then, ΔEBD is the required triangle, each of whose sides is $\frac{3}{5}$ of the corresponding side of ΔABC.

*See Justification part of construction in *For More Practice and Knowledge*

2 Try to Solve

1. Construct a triangle of sides 4 cm, 5 cm and 6 cm, and then a triangle similar to it whose sides are $\frac{2}{3}$ of the corresponding sides of the first triangle. **[NCERT]**
2. Construct a triangle ABC in which $BC = 6$ cm, $AB = 5$ cm and $\angle ABC = 60°$. Then, construct another triangle whose sides are $\frac{3}{4}$ times the corresponding sides of ΔABC. **[CBSE (CCE) 2009]**

Example 4 Draw a triangle ABC with side $BC = 6$ cm, $AB = 5$ cm and $\angle ABC = 60°$. Then, construct a triangle whose sides are $\frac{3}{4}$ of the corresponding sides of the triangle ABC.

Solution: *Steps of Construction*

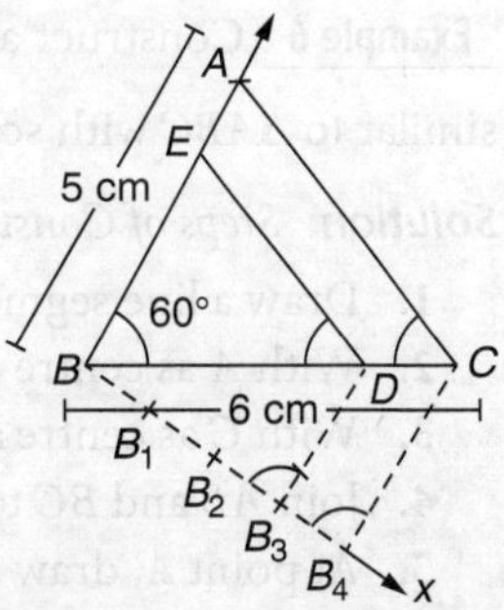

1. Construct a ΔABC in which $BC = 6$ cm and, $AB = 5$ cm and $\angle ABC = 60°$.
2. Below BC, make an acute $\angle CBX$.
3. Along BX, mark off four arcs B_1, B_2, B_3, B_4 such that $BB_1 = B_1B_2 = B_2B_3 = B_3B_4$.
4. Join B_4C.
5. From B_3, draw $B_3D \parallel B_4$ meeting BC at D.
6. From D, draw $ED \parallel AC$ meeting BA at E.
   Now, we have ΔEBD, which is the required triangle whose sides are $\frac{3}{4}$ of the corresponding sides of ΔABC.
   *See Justification part of construction in *For More Practice and Knowledge*

3 Try to Solve

1. Construct ΔABC in which $BC = 6.5$ cm, $AB = 4.5$ cm and $\angle ABC = 60°$. Construct a triangle similar to this triangle whose sides are $\frac{4}{5}$ of the corresponding sides of ΔABC. **[CBSE (CCE) 2012]**
2. Construct a ΔDEF with side $EF = 6.5$ cm, $\angle D = 105°$, $\angle E = 45°$. Then, construct a triangle whose sides are 3/5 times the corresponding sides of ΔDEF. **[NCERT, CBSE (CCE) 2012]**
3. Draw a ΔABC in which $AB = 5$ cm, $BC = 6$ cm and $\angle ABC = 60°$. Construct a triangle similar to ΔABC using scale factor 5/7. **[CBSE (CCE) 2011]**

Example 5 Construct a ΔABC in which $AC = 5$ cm, $\angle C = 60°$ and altitude $BM = 4$ cm. Construct a triangle similar to ΔABC with scale factor $\frac{3}{2}$.

Solution: *Steps of Construction*

1. Draw a line segment $AC = 5$ cm.
2. Construct $\angle ACP = 60°$.
3. At point A, draw a perpendicular on AC.
4. With A as centre, cut an arc equal to 4 cm, that is, $AN = 4$ cm.
5. At point N, draw another perpendicular which intersects CP at B.
6. Join AB.
7. At point A, draw a ray AX, making an acute $\angle CAX$.
8. Locate three points on PX such that $AA_1 = A_1A_2 = A_2A_3 = x$ (say).
9. Join A_2C.
10. Draw $A_3C_1 \parallel A_2C$ which meet reproduced at C_1.
11. Draw $C_1B_1 \parallel CB$ which meets AB produced at B_1.
    Then, AB_1C_1 is the required triangle.
    *See Justification part of construction in *For More Practice and Knowledge*

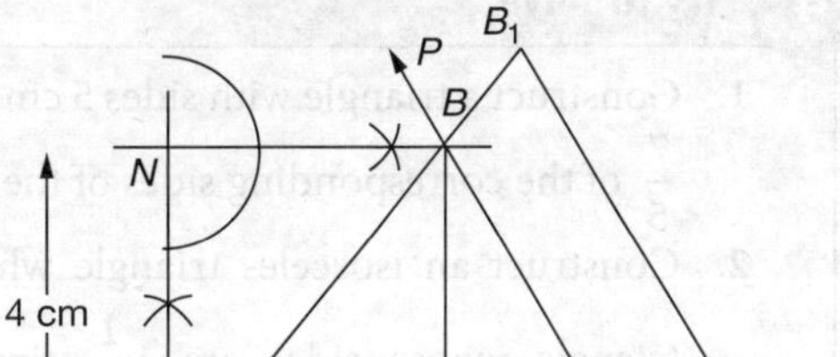

Example 6 Construct a ΔABC in which $AB = 4$ cm, $BC = 6$ cm and $AC = 9$ cm. Construct a triangle similar to ΔABC with scale factor $\frac{3}{2}$.

Solution: *Steps of Construction*

1. Draw a line segment $AC = 9$ cm.
2. With A as centre and radius = 4 cm, draw an arc.
3. With C as centre and radius = 6 cm, draw another arc. The two arcs intersect at point B.
4. Join AB and BC to obtain ΔABC.
5. At point A, draw a ray AX, making an acute $\angle CAX$.
6. Locate three points on AX such that $AA_1 = A_1A_2 = A_2A_3 = x$ (say).
7. Join A_2C.
8. Draw $A_3C_1 \parallel A_2C$ which meets AC produced at C_1.
9. Draw $C_1B_1 \parallel CB$ which meets AB produced at B_1.

Then, AB_1C_1 is the required triangle.

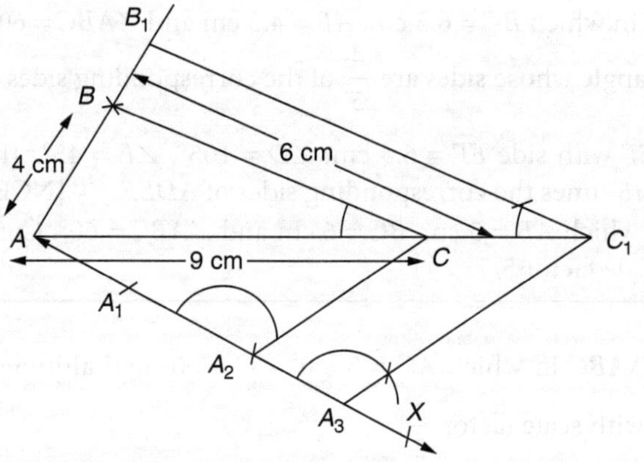

*See Justification part of construction in *For More Practice and Knowledge*

4 Try to Solve

1. Construct a triangle with sides 5 cm, 6 cm and 7 cm and then another triangle whose sides are $\frac{7}{5}$ of the corresponding sides of the first triangle. **[NCERT]**
2. Construct an isosceles triangle whose base is 8 cm and altitude 4 cm and then another triangle whose sides are $1\frac{1}{2}$ times the corresponding sides of the isosceles triangle. **[NCERT, CBSE (CCE) 2010]**
3. Draw a triangle ABC with side $BC = 7$ cm, $\angle B = 45°$, $\angle A = 105°$. Then, construct a triangle whose sides are $\frac{4}{3}$ times the corresponding sides of ΔABC. **[NCERT, CBSE (CCE) 2010]**

Example 7 Draw a right triangle in which the sides (other than hypotenuse) are of length 4 cm and 3 cm. Then, construct another triangle whose sides are $\frac{5}{3}$ times the corresponding sides of the given triangle. **[NCERT, CBSE (CCE)]**

Solution: *Steps of Construction*

1. Draw a right-angled ΔABC with $BC = 4$ cm, $AB = 3$ cm.
2. Draw an acute $\angle CBX$.
3. Draw five equal intercepts on BX.
4. Join b_3C and draw $b_5C' \parallel b_3C$, which intersects BC produced at C'.
5. Draw $C'A' \parallel CA$, which intersects BA produced at A'.

$\therefore A'BC''$ is the required Δ.

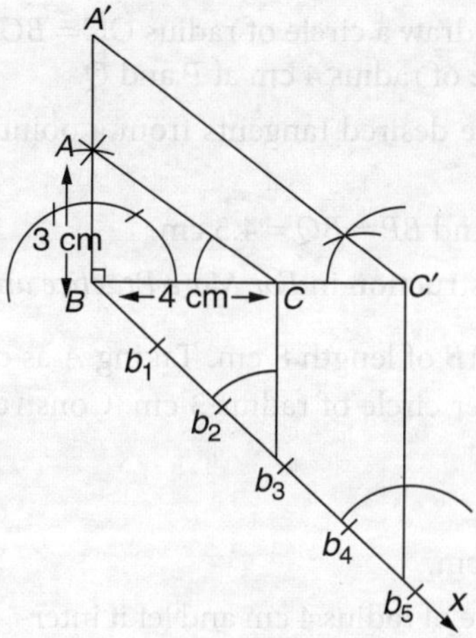

*See Justification part of construction in *For More Practice and Knowledge*

5 Try to Solve

1. Draw a triangle ABC in which $AB = 6$ cm, $BC = 5.8$ cm and $\angle ABC = 75°$. Construct a triangle similar to ΔABC with scale factor $\frac{7}{5}$. **[NCERT]**
2. Draw an isosceles triangle ABC in which $AB = AC = 6$ cm and $BC = 5$ cm. Construct a triangle PQR similar to ΔABC in which $PQ = 8$ cm. **[NCERT]**
3. Construct an equilateral triangle of side 4.8 cm and then another triangle whose sides are $\frac{7}{4}$ of the corresponding sides of the first triangle. **[NCERT]**

Concept: Construction of Tangents to a Circle

Example 8 Draw a circle of radius 6 cm. From a point 10 cm away from its centre, construct a pair of tangents to the circle and measure their length. **[CBSE, (CCE) 2010]**

Solution: *Steps of Construction*

1. Take a point O and draw a circle of radius 6 cm.
2. Take a point P at a distance of 10 cm from the centre O.
3. Join OP and bisect it. Let M be the mid-point.
4. With M as centre and MP as radius, draw a circle to intersect the circle at Q and R.

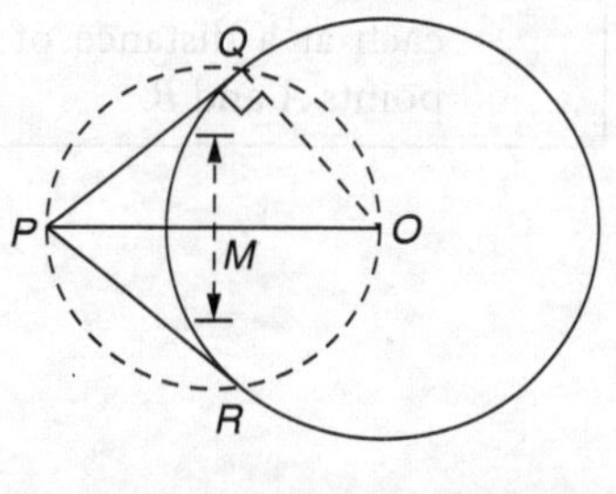

5. Join PQ and PR. Then, PQ and PR are the required tangents. On measuring, we find $PQ = PR = 8$ cm.
*See Justification part of construction in *For More Practice and Knowledge*

Example 9 Construct a tangent to a circle of radius 4 cm from a point on the concentric circle of radius 6 cm and measure its length. Also verify the measurement by actual calculation. **[CBSE, (CCE) 2010]**

Solution: *Steps of Construction*

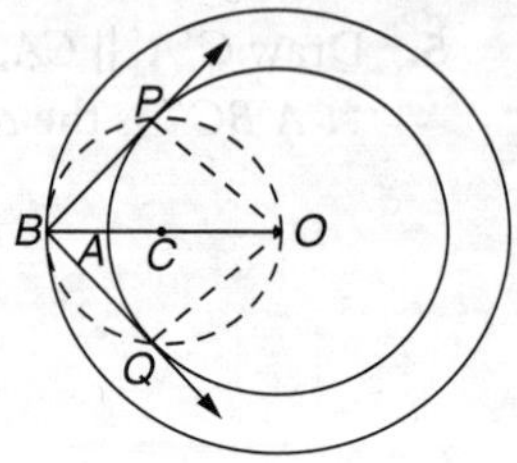

1. Take a point O and draw a circle of radius $OA = 4$ cm. Also draw a concentric circle of radius $OB = 6$ cm.
2. Find the mid-point C of OB, draw a circle of radius $OC = BC$. Suppose this circle intersects the circle of radius 4 cm at P and Q.
3. Join BP and BQ to obtain the desired tangents from a point B on the circle of radius 6 cm.

By actual measurement, we find $BP = BQ = 4.5$ cm.

*See Justification part of construction in *For More Practice and Knowledge*

Example 10 Draw a line segment AB of length 8 cm. Taking A as centre, draw a circle of radius 4 cm, and taking B as centre, draw another circle of radius 3 cm. Construct tangents to each circle from the centre of the other circle.

Solution: *Steps of Construction*

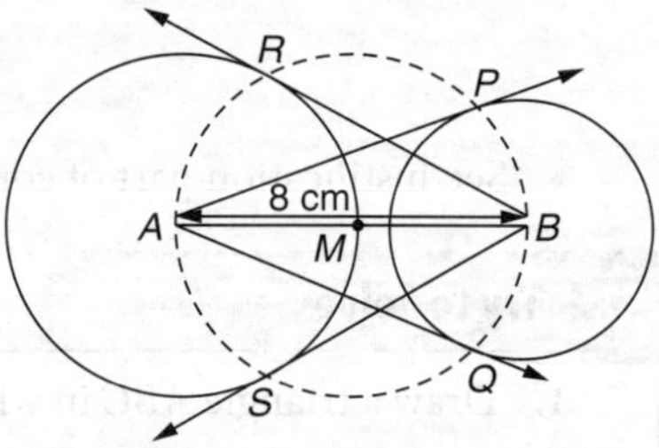

1. Draw a line segment $AB = 8$ cm.
2. With A as centre, draw a circle of radius 4 cm and let it intersect the line segment AB at M.
3. With B as centre, draw a circle of radius 3 cm.
4. With M as centre, draw a circle of radius AM and let it intersect the given two circles at P, Q and R, S.
5. Join AP, AQ, BR and BS.

These are the required tangents.

*See Justification part of construction in *For More Practice and Knowledge*

6 Try to Solve

1. Draw a circle of radius 5 cm. Form a point P 8 cm away from its centre and construct a pair of tangents to the circle.
2. Draw a line segment AB of length 7 cm. Taking A as centre, draw a circle of radius 3 cm, and taking B as centre, draw another circle of radius 2 cm. Construct tangents to each circle from the centre of the other circle.
3. Draw a circle of radius 3 cm. Take two points A and B on one of its extended diameters, each at a distance of 6 cm from its centre. Draw tangents to the circle from these two points A and B.

Example 11 Draw a pair of tangents to a circle of radius 5 cm that are inclined to each other at an angle of 60°. **[NCERT, CBSE, (CCE) 2012]**

Solution: *Steps of Construction*

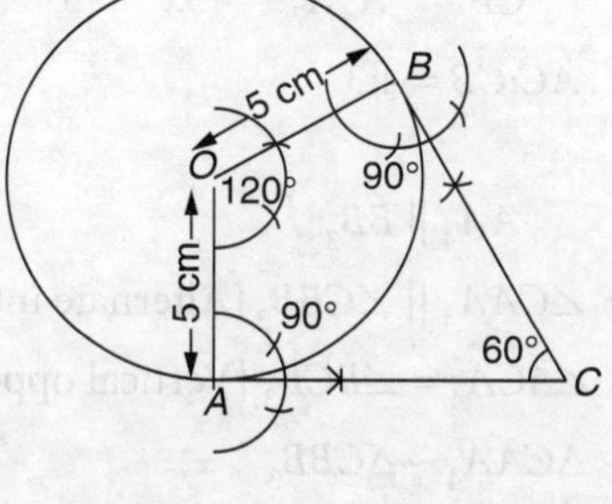

1. Draw a circle with radius $OA = 5$ cm.
2. Construct $\angle AOB = 120°$.
3. Draw perpendiculars at points A and B, which intersect at point C.
4. By actual measurement, we find that $\angle ACB = 60°$.

*See Justification part of construction in *For More Practice and Knowledge*

7 Try to Solve

1. Draw a pair of tangents to a circle of radius 3 cm, which are inclined to each other at an angle of 75°. **[CBSE, (CCE) 2011]**
2. Draw a pair of tangents to a circle of radius 4 cm, which are inclined to each other at an angle 60°. **[CBSE, (CCE) 2012]**
3. Construct two tangents AB and AC from an external point A to a circle of radius 3.5 cm and centre O such that $\angle BOC = 105°$. **[CBSE, (CCE) 2011]**

FOR MORE PRACTICE AND KNOWLEDGE

To improve our knowledge of the topic, we need to be able to understand the justification of each question and its proof. Here, we present the justification or proof for Examples 1–11.

Example 1 Justification $AA_1 = A_1A_2 = \ldots = A_6A_7 = x$

$\therefore$ $AA_3 = 3x$ and $A_3A_7 = 4x$

In ΔAA_7B, $A_3C \parallel A_7B$. $\dfrac{AC}{CB} = \dfrac{AA_3}{A_3A_7}$

$\Rightarrow$ $\dfrac{AC}{CB} = \dfrac{3x}{4x} = \dfrac{3}{4}$

$\therefore$ $AC{:}CB = 3{:}4$

Measure of $AC = 2.79$ cm and $BC = 3.71$ cm.

Example 2 **Justification** Let $AA_1 = A_1A_2 = \ldots = A_6A_7 = x.$

In ΔABA_7, we have $A_4C \parallel A_7B$.

$$\therefore \quad \frac{AC}{CB} = \frac{AA_4}{A_4A_7} = \frac{4x}{3x} = \frac{4}{3} \quad \text{[By BPT]}$$

Hence $\quad AC:CB = 4:3$

Alternative Justification:

Here $\quad AA_4 \parallel BB_3$.

$\therefore \quad \angle CAA_4 \parallel \angle CBB_3$ [Alternate interior angle]

Also $\quad \angle ACA_4 = \angle BCB_3$ [Vertical opposite angle]

$\therefore \quad \Delta CAA_4 \sim \Delta CBB_3$ [By *AA* similarity]

And so $$\frac{AC}{CB} = \frac{AA_4}{BB_3} = \frac{4}{3}$$

Hence $\quad AC:CB = 4:3$

Example 3 **Justification** Since $DE \parallel CA$, we have $\Delta ABC \sim \Delta EBD$.

$$\therefore \quad \frac{EB}{AB} = \frac{DE}{CA} = \frac{BD}{BC} = \frac{3}{5}$$

Example 4 **Justification**

Here $\quad DE \parallel CA$.

$$\therefore \quad \Delta ABC \sim \Delta EBD \text{ and } \frac{EB}{AB} = \frac{BD}{BC} = \frac{DE}{CA} = \frac{3}{4}$$

Hence, we obtain the new triangle similar to the given triangle. The sides are equal to $\frac{3}{4}$ of the corresponding sides of ΔABC.

Example 5 **Justification** $\Delta AB_1C_1 \sim \Delta ABC$ [By AA similarity]

$$\frac{AB_1}{AB} = \frac{B_1C_1}{BC} = \frac{AC_1}{AC}$$

But $$\frac{AC_1}{AC} = \frac{AA_3}{AA_2}$$

$$\frac{AC_1}{AC} = \frac{3x}{2x} = \frac{3}{2}$$

From (i) and (ii), we get $$\frac{AB_1}{AB} = \frac{B_1C_1}{BC} = \frac{AC_1}{AC} = \frac{3}{2}$$

Example 6 **Justification** $\quad \Delta AB_1C_1 \sim \Delta ABC$ [By *AA* similarity]

$$\frac{AB_1}{AB} = \frac{B_1C_1}{BC} = \frac{AC_1}{AC}$$

But $$\frac{AC_1}{AC} = \frac{AA_3}{AA_2}$$

$$\frac{AC_1}{AC} = \frac{3x}{2x} = \frac{3}{2}$$

From (i) and (ii), we get $\frac{AB_1}{AB} = \frac{B_1C_1}{BC} = \frac{AC_1}{AC} = \frac{3}{2}$

Example 7 **Justification** Since $CA \parallel DE$, we have $\Delta ABC \sim \Delta EBD$.

$\therefore$ $\frac{EB}{AB} = \frac{DE}{CA} = \frac{BD}{BC} = \frac{4}{3}$

Example 8 **Justification** On joining OQ, we find that $\angle PQO = 90°$, as $\angle PQO$ is the angle in the semicircle.

$\therefore$ $PQ \perp OQ$

Since, OQ is the radius of the given circle, PQ must be a tangent to the circle. Similarly, PR is also a tangent to the circle.

Example 9 **Justification** In ΔBPO, we have

$$\angle BPO = 90°, OB = 6 \text{ cm and } OP = 4 \text{ cm}$$

$\therefore$ $OB^2 = BP^2 + OP^2$ [Using Pythagoras' Theorem]

$\Rightarrow$ $BP = \sqrt{OB^2 - OP^2}$

$\Rightarrow$ $P = \sqrt{36-16} = \sqrt{20} \text{ cm} = 4.47 \text{ cm}$

Similarly $BP = 4.47$ cm.

Example 10 **Justification** On joining BP, we have $\angle BPA = 90°$, as $\angle BPA$ is the angle in the semicircle.

$\therefore$ $AP \perp BP$

Since BP is the radius of the given circle, AP must be a tangent to the circle.

Similarly, AQ, BR and BS are the tangents.

Example 11 **Justification**

$$\angle AOB = (180 - 60°) = 120°$$

Consider quadrilateral $OACB$.

$$\angle OAC = 90°, \angle OBC = 90°$$

$$\angle AOB = 120°$$

According to the property of quadrilaterals

$$\angle OAC + \angle OBC + \angle AOB + \angle ACB = 360°$$

$$90° + 90° + 120° + \angle ACB = 360°$$

$$\angle ACB = 60°$$

Example 12 Let ABC be a right triangle in which $AB = 6$ cm, $BC = 8$ cm and $\angle B = 90°$. BD is the perpendicular from B on AC. The circle through B, C, D is drawn. Construct the tangents from A to this circle. [NCERT]

Solution: *Steps of Construction*

1. Draw a right triangle ABC with $AB = 6$ cm, $BC = 8$ cm and $\angle B = 90°$.

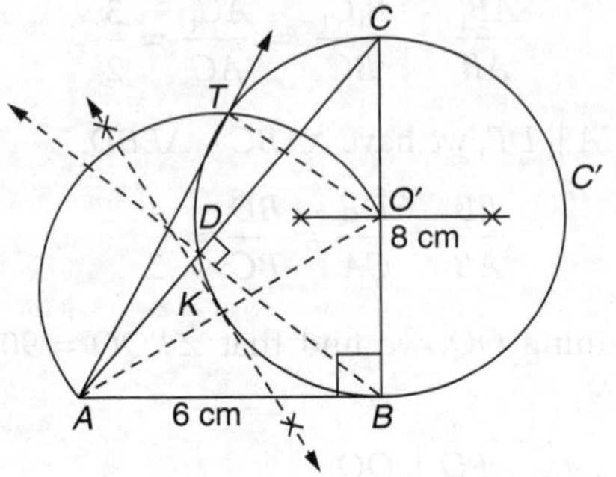

2. From B, draw BD perpendicular to AC.
3. Draw the perpendicular bisector of BC, which intersects BC at point O'.
4. Take O' as centre and $O'B$ as radius and draw a circle C' that passes through points B, C and D.
5. Join $O'A$ and draw a perpendicular bisector of $O'A$, which intersects $O'A$ at point K.
6. Take K as centre and draw an arc of radius KO' that intersects the previous circle C' at T.
7. Join AT. AT is the required tangent.

Justification:

$$\angle BDC = 90°$$

$\Rightarrow$ BC acts as diameter.

AB is a tangent to the circle having centre O'. Join $O'T$.

In $\Delta AO'T$ $\qquad AO'^2 = AT^2 + O'T^2$

BC is diameter $\qquad O'B = \frac{1}{2}BC = \frac{1}{2}(8) = 4$ cm

In $\Delta AO'B$ $\qquad AO'^2 = AB^2 + O'B^2 = (6)^2 + (4)^2$

$$= 36 + 16 = 52 \text{ cm}^2$$

$\because$ As $\qquad AT^2 + O'T^2 = O'A^2$

$\therefore \Rightarrow$ $\qquad AT^2 + O'T^2 = 52$

$\Rightarrow$ $\qquad AT^2 + (4)^2 = 52$

$\Rightarrow$ $\qquad AT^2 = 52 - 16 = 36$

$\Rightarrow$ $\qquad AT = 6$ cm

$\Rightarrow$ $\qquad AT = AB = 6$ cm

A pair of tangents can be drawn to a circle from an external point outside the circle. These two tangents are equal in length.

$\therefore$ $\qquad AB = AT.$

Example 13 Draw a circle with the help of a bangle. Take a point outside the circle. Construct a pair of tangents from this point to the circle. **[NCERT]**

Solution: *Steps of Construction*

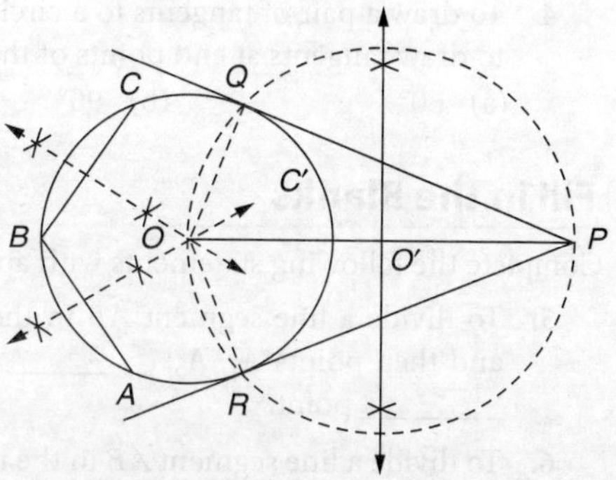

1. Draw a circle C using a bangle. T find the centre, take three non-collinear points A, B and C, lying on the circle. Join AB and BC and draw a perpendicular bisector of AB and BC, intersecting at a point O; 'O' is the centre of the circle.
2. Take a point P outside the circle. Join OP.
3. Draw a perpendicular bisector of OP, which intersects OP at point O'.
4. Take O' as the centre with OO' as radius and draw a circle which passes through O and P, intersecting the previous circle at points R and Q.
5. Join PQ and PR.
6. PQ and PR are the required pair of tangents.

Justification:

Join OQ and OR. In ΔOQP and ΔOPR

$OQ = OR$ [Radii of the circle]

$OP = OP$ [Common]

$\angle Q = \angle R = 90°$ [Radius is $\perp$ to tangent]

$\Delta OQP \cong \Delta ORP$ [By RHS]

$\therefore \quad PQ = PR$

A pair of tangents can be drawn to a circle from an external point lying outside the circle. These two tangents are equal in length.

$\therefore \quad PQ = PR$

Multiple Choice Questions

Choose the correct answer for each of the following questions:

1. In the figure, P divides AB internally in the ratio

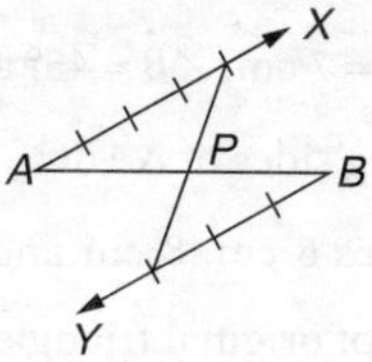

(a) 3:4 (b) 4:3
(c) 3:7 (d) 4:7

2. To divide a line segment AB in the ratio 4:7, a ray AX is drawn such that $\angle BAX$ is an acute angle and then points $A_1, A_2, A_3 \ldots$ are located at equal distances on ray A and the point B is joined to

(a) A_{12} (b) A_{11} (c) A_{10} (d) A_9

3. To construct a triangle similar to a given ΔABC, with its sides 8/5 of the corresponding sides of ΔABC, draw a ray BX such that $\angle CBX$ is an acute angle and X is on the opposite side of A with respect to BC. The minimum number of points to be located at equal distances on ray BX is

(a) 3 (b) 5 (c) 8 (d) 13

4. To draw a pair of tangents to a circle which are inclined to each other at an angle of 60°, it is required to draw tangents at end points of those two radii of the circle. The angle between them should be

(a) 60° (b) 90° (c) 120° (d) 135°

Fill in the Blanks

Complete the following statements with appropriate answers(s) in the blank space(s):

5. To divide a line segment *AB* in the ratio 5:8, a ray *AX* is drawn such that $\angle BAX$ is an acute angle and then points A_1, A_2, ________ are located at equal distances on the ray *AX*. Point *B* is joined to ________ point.

6. To divide a line segment *AB* in the ratio 3:7, ray *AX* is drawn, so that $\angle BAX$ is an acute angle and then points A_1, A_2, ________ are located at equal distances. The number of points located on the ray *AX* is

Subjective Questions

7. Construct a triangle similar to the given triangle *ABC* with its sides equal to $\frac{3}{4}$ of the corresponding sides of the triangle *ABC* (scale factor $\frac{3}{4}$). **[NCERT]**

8. Construct a triangle similar to a given triangle *ABC* with its sides equal to $\frac{5}{3}$ of the corresponding sides of the triangle *ABC* (scale factor $\frac{5}{3}$). **[NCERT]**

9. Draw a triangle *ABC* with sides $BC = 6$ cm, $AB = 5$ cm and $\angle ABC = 60°$. Then, construct a triangle whose sides are $\frac{3}{4}$ of the corresponding sides of triangle *ABC*.

10. Draw a line segment of length 8 cm and divide it internally in the ratio 4:5. **[CBSE 2017]**

11. Construct a triangle *ABC* in which $AB = 5$ cm, $BC = 6$ cm and $AC = 7$ cm. Construct another triangle similar to ΔABC such that its sides are $\frac{3}{5}$ of the corresponding sides of ΔABC.

12. Draw a line segment of length 8 cm and divide it in the ratio 3:2. Measure the two parts.

13. Construct an isosceles triangle whose base is 8 cm and altitude 4 cm and then construct another triangle whose sides are $\frac{3}{4}$ times the corresponding sides of the isosceles triangle.

14. Draw a triangle *ABC* with $BC = 7$ cm, $\angle B = 45°$ and $\angle A = 105°$. Then, construct a triangle whose sides are $\frac{4}{5}$ times the corresponding sides of ΔABC.

15. Construct a triangle with sides 6 cm, 8 cm and 10 cm. Construct another triangle whose sides are $\frac{3}{5}$ of the corresponding sides of original triangle. **[CBSE 2018 C]**

16. Construct an isosceles triangle with base 8 cm and altitude 4 cm. Construct another triangle whose sides are $\frac{2}{3}$ times the corresponding sides of the isosceles triangle. **[CBSE 2017]**

17. Construct a triangle *ABC* with side $BC = 7$ cm, $\angle B = 45°$, $\angle A = 105°$. Then, construct another triangle whose sides are $\frac{3}{4}$ times the corresponding sides of the ΔABC. **[AI 2017]**

18. Draw an isosceles triangle *ABC* in which the base *BC* is 8 cm long and its altitude AD through *A* is 4 cm long. Then, draw another triangle whose sides are $\frac{2}{3}$ of the corresponding sides of the ΔABC. **[AI 2017 C]**

19. Construct a ΔABC in which $CA = 6$ cm, $AB = 5$ cm and $\angle BAC = 45°$. Then, construct a triangle whose sides are $\frac{3}{5}$ of the corresponding sides of ΔABC. **[Delhi 2019]**

20. Draw a right triangle in which sides (other than hypotenuse) are of length 8 cm and 6 cm. Then, construct another triangle whose sides are $\frac{3}{4}$ times the corresponding sides of the first triangle. **[Delhi 2017 (C), AI 2012]**

21. Draw a right triangle in which the sides (other than the hypotenuse) are of length 4 cm and 3 cm. Now construct another triangle whose sides are $\frac{3}{5}$ times the corresponding sides of the given triangle. **[Delhi 2017]**

22. Draw an isosceles ΔABC in which $BC = 5.5$ cm and altitude $AL = 3$ cm. Then, construct another triangle whose sides are $\frac{3}{4}$ of the corresponding sides of ΔABC. **[AI 2016]**

23. Two line segment AB and AC include an angle of 60° where $AB = 5$ cm and $AC = 7$ cm. Locate points P and Q on AB and AC, respectively, such that $AP = \frac{3}{4} AB$ and $AQ = \frac{1}{4} AC$. Join P and Q and measure the length PQ. **[NCERT Exemplar]**

24. Draw a parallelogram $ABCD$ in which $BC = 5$ cm, $AB = 3$ cm and $\angle ABC = 60°$. Divide it into triangles BCD and ABD using the diagonal BD. Construct triangle $BD'C'$ similar to ΔBDC with scale factor $\frac{4}{3}$. Draw a line segment $D'A'$ parallel to DA, where A' lies on the extended side BA. Is $A'BC'D$ a parallelogram? **[NCERT Exemplar]**

25. Draw an isosceles triangle ABC in which $AB = AC = 6$ and $BC = 5$ cm. Construct a triangle PQR similar to ΔABC in which $PQ = 8$ cm. Also justify the construction. **[NCERT Exemplar]**

26. By geometrical construction, it is possible to divide a line segment in the ratio $\sqrt{3} : \frac{1}{\sqrt{3}}$. Justify.

27. Construct a triangle of sides 4 cm, 5 cm and 6 cm and then a triangle similar to it whose sides are $\frac{2}{3}$ of the corresponding sides of the first triangle. **[NCERT]**

28. Construct a triangle with sides 5 cm, 6 cm and 7 cm and then another triangle whose sides are $\frac{7}{5}$ of the corresponding sides of the first triangle. **[NCERT]**

29. Draw a triangle ABC with side $BC = 6$ cm, $AB = 5$ cm and $\angle ABC = 60°$. Then, construct a triangle whose sides are $\frac{3}{4}$ of the corresponding sides of the triangle ABC. **[NCERT]**

30. Draw a pair of tangents to a circle of radius 4 cm which are inclined to each other at an angle of 45°.

31. Let PQR be a right triangle in which $PQ = 3$ cm, $QR = 4$ cm and $\angle Q = 90°$. QS is the perpendicular from Q on PR. The circle through Q, R, S, is drawn. Construct the tangents from P to this circle.

32. Draw two concentric circles of radii 2 cm and 5 cm. Take a point P on the outer circle and construct a pair of tangents PA and PB to the smaller circle. Measure PA.

33. Construct tangents to a circle of radius 3 cm from a point on a concentric circle of radius 5 cm and measure its length. **[AI 2016]**

34. Draw a line segment AB of length 7 cm. Taking A as centre, draw a circle of radius 3 cm, and taking B as centre, draw another circle of radius 2 cm. Construct tangents to each circle from the centre of the other circle. **[Delhi 2015, Foreign 2011]**

35. Draw a triangle ABC with $BC = 7$ cm, $\angle B = 45°$ and $\angle C = 60°$. Then, construct another triangle whose sides are $\frac{3}{5}$ times the corresponding sides of ΔABC. **[AI 2012]**

36. A pair of tangents can be constructed to a circle inclined at angle of 170°. Justify. **[NCERT Exemplar]**

37. Draw a line segment of length 7.6 cm and divide it in the ratio 5:8. Measure the two parts. **[NCERT]**

Answers

Multiple Choice Questions

1. (b) 2. (b) 3. (c) 4. (c)

Fill in the Blanks

5. (A_{13}) 6. 10

Constructions (Basic)

SELF-ASSESSMENT TEST

Time allowed: 90 min. Max. Marks: 40

General Instructions

1. *All questions are compulsory. There are 12 questions in all.*
2. *This question paper has two sections: Section A and Section B.*
3. *Section A contains 8 questions of three marks each, Section B contains four questions of four marks each.*

Section A [3 Marks Each]

1. Draw a line segment of length 7.6 cm and divide it in the ratio 5:8. Measure the two parts.
2. Construct a triangle with sides 5 cm, 5.5 cm and 6.5 cm. Then, construct another triangle whose sides are $\frac{3}{5}$ times the corresponding sides of the first triangle.
3. Construct a ΔABC in which $AB = 4$ cm, $BC = 6$ cm and $AC = 9$ cm. Construct a triangle similar to ΔABC with scale factor $\frac{3}{2}$.
4. Draw a circle of radius 6 cm. From a point 10 cm from its centre, construct the pair of tangents to the circle and measure their length.
5. Construct an isosceles triangle whose base is 8 cm and altitude is 4 cm and then construct another triangle whose sides are $\frac{3}{4}$ times the corresponding sides of the isosceles triangle.
6. Construct a triangle with sides 5 cm, 6 cm and 7 cm and then another triangle whose sides are $\frac{7}{5}$ of the corresponding sides of the first triangle.
7. Draw two concentric circles of radii 2 cm and 5 cm. Take a point P on the outer circle and construct a pair of tangents PA and PB to the smaller circle. Measure PA.
8. Construct tangents to a circle of radius 3 cm from a point on a concentric circle of radius 5 cm and measure its length.

Section B [4 Marks Each]

9. Construct a ΔABC in which $AC = 5$ cm, $\angle C = 60°$ and altitude $BM = 4$ cm. Construct a triangle similar to ΔABC with scale factor $\frac{3}{2}$.
10. Construct a tangent to a circle of radius 4 cm from a point on the concentric circle of radius 6 cm and measure its length. Also verify the measurement by actual calculation.
11. Draw a pair of tangents to a circle of radius 5 cm that are inclined to each other at an angle of 60°.
12. Draw a triangle ABC with $BC = 7$ cm, $\angle B = 45°$ and $\angle A = 105°$. Then, construct a triangle whose sides are $\frac{4}{5}$ times the corresponding sides of ΔABC.

Constructions (Standard)

SELF-ASSESSMENT TEST

Time allowed: 90 min. Max. Marks: 40

General Instructions

1. *All questions are compulsory. There are 12 questions in all.*
2. *This question paper has two sections: Section A and Section.*
3. *Section A contains 8 questions of three marks each, Section B contains four questions of four marks each.*

Section A [3 Marks Each]

1. Draw a line segment of length 6.5 cm and divide it in the ratio 3:4. Measure the two parts.
2. Draw a line segment of length 7.6 cm and divide it in the ratio 5:8. Measure the two parts.
3. Two line segments AB and AC include an angle of 60°, where $AB = 5$ cm and $AC = 7$ cm. Locate points P and Q on AB and AC, respectively, such that $AP = \frac{3}{4}AB$ and $AQ = \frac{1}{4}AC$. Join P and Q and measure the length PQ.
4. Draw a line segment AB of length 7 cm. Taking A as centre, draw a circle of radius 3 cm, and taking B as centre, draw another circle of radius 2 cm. Construct tangents to each circle from the centre of the other circle.
5. Draw a triangle ABC with side $BC = 6$ cm, $AB = 5$ cm and $\angle ABC = 60°$. Then, construct a triangle whose sides are $\frac{3}{4}$ th of the corresponding sides of the triangle ABC.
6. Construct a ΔABC in which $AC = 5$ cm, $\angle C = 60°$ and altitude $BM = 4$ cm. Construct a triangle similar to ΔABC with scale factor $\frac{3}{2}$.
7. Draw a right triangle in which the sides (other than hypotenuse) are of length 4 cm and 3 cm. Then, construct another triangle whose sides are $\frac{5}{3}$ times the corresponding sides of the given triangle.
8. Draw a line segment AB of length 8 cm. Taking A as centre, draw a circle of radius 4 cm, and taking B as centre, draw another circle of radius 3 cm. Construct tangents to each circle from the centre of the other circle.

Section B [4 Marks Each]

9. Let ABC be a right triangle in which $AB = 6$ cm, $BC = 8$ cm and $\angle B = 90°$. BD is the perpendicular from B on AC. The circle through B, C, D is drawn. Construct the tangents from A to this circle.
10. Draw a triangle ABC with $BC = 7$ cm, $\angle B = 45°$ and $\angle A = 105°$. Then, construct a triangle whose sides are $\frac{4}{5}$ times the corresponding sides of ΔABC.
11. A pair of tangents can be constructed to a circle inclined at an angle of 170°. Justify.
12. Let PQR be a right triangle in which $PQ = 3$ cm, $QR = 4$ cm and $\angle Q = 90°$. QS is the perpendicular from Q on PR. The circle through Q, R, S is drawn. Construct the tangents from P to this circle.

CHAPTER 12 Areas Related to Circles

Key Concepts

- Circumference of circle
- Area of circle
- Length of arc
- Area of minor and major sectors
- Area of minor and major arcs
- Area of shaded regions
- Area enclosed by combination of plane figures

Important Formulae with Concepts

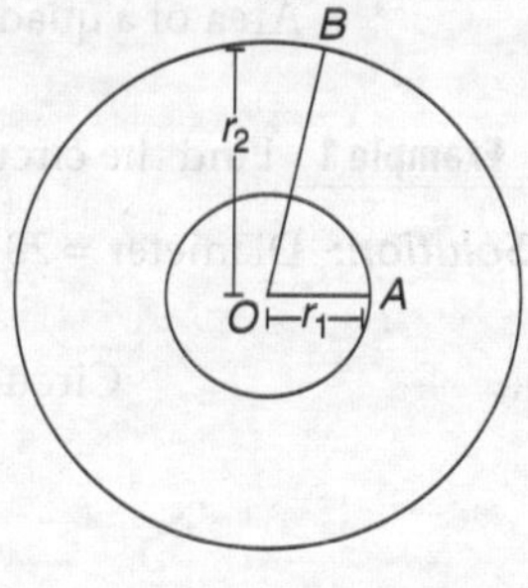

1. Perimeter or circumference = $2\pi r = \pi d$ unit of a circle, where d is the diameter of a circle.
2. Area of circle = πr^2 unit^2.
3. Perimeter of semicircle = $(\pi r + 2r)$ unit.
4. Area of semicircle = $\frac{1}{2}\pi r^2$ unit^2.
5. Area enclosed by two concentric circles = $\pi(r_2^2 - r_1^2)$ unit^2.
6. Rotation of the hands of a clock
   (i) Angle described by minute hand in 60 minutes = 360°.
   ∴ Angle described by minute hand in 1 minute = 6°.
   (ii) Angle described by hour hand in 12 hours = 360°.
   ∴ Angle described by hour hand in 1 hour = 30°.

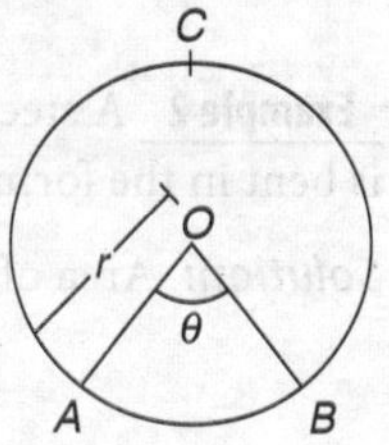

7. Length of arc $\overparen{AB}$ is called minor arc. $\overparen{BCA}$ is called major arc.

   Length of arc $(l) = \frac{\theta}{360°} \times 2\pi r$ unit.

8. **Area of sector:** This can be explained using a slice of a of pizza. The slice is the minor sector and the rest of the pizza is the major sector.

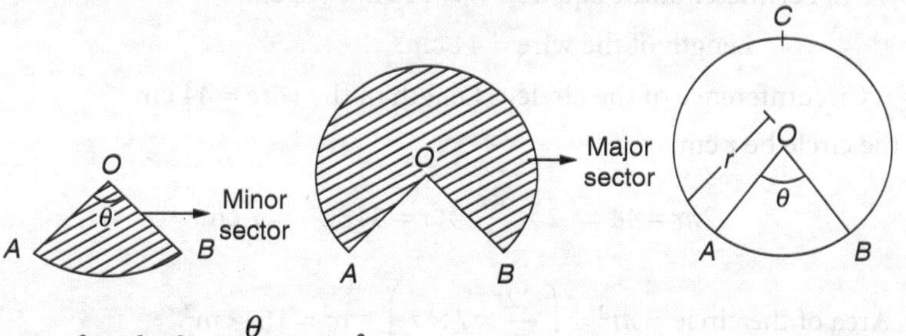

Area of sector of angle $\theta = \frac{\theta}{360°} \times \pi r^2$

Or $= \frac{1}{2}(lr)$

9. Area of segment

Area of minor ADB segment = Area of sector $OADB$ – Area of ΔOAB.

$$= \frac{\theta}{360°} \times \pi r^2 - \text{Area of } \Delta OAB$$

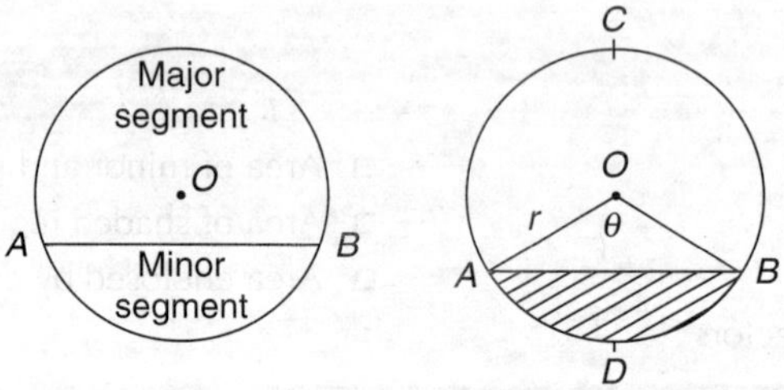

Area of major segment BCA = Area of circle – Area of minor segment ADB.

* $\sin 120° = \frac{\sqrt{3}}{2}$

* Area of a quadrant = $\frac{1}{4}\pi r^2$

Example 1 Find the circumference and area of a circle of diameter 28 cm.

Solution: Diameter = 28 cm ⇒ radius r = 14 cm.

∴ Circumference of the circle = $2\pi r = \left(2 \times \frac{22}{7} \times 14\right)$ cm = 88 cm

Area of the circle = $\pi r^2 = \left(\frac{22}{7} \times 14 \times 14\right)$ cm^2 = 616 cm^2.

Example 2 A steel wire, when bent in the form of a square, encloses an area of 121 cm^2. The same wire is bent in the form of a circle. Find the area of the circle. **[CBSE (CCE) 2015]**

Solution: Area of the square = 121 cm^2.

Side of the square = $\sqrt{121}$ cm = 11 cm.

Perimeter of the square = 4 × 11 cm = 44 cm.

∴ Length of the wire = 44 cm.

∴ Circumference of the circle = Length of the wire = 44 cm.

Let the radius of the circle be r cm.

Then $2\pi r = 44 \Rightarrow 2 \times \frac{22}{7} \times r = 44 \Rightarrow r = 7$ cm.

∴ Area of the circle = $\pi r^2 = \left(\frac{22}{7} \times 7 \times 7\right)$ cm^2 = 154 cm^2.

Example 3 If the perimeter of a semicircular protractor is 36 cm, find its diameter.

Solution: Let the radius of the protractor be r cm.

Then, perimeter = $(\pi r + 2r)$ cm = $(\pi + 2)r$ cm.

$$= \left(\frac{22}{7}+2\right) r \text{ cm} = \frac{36}{7} r \text{ cm}.$$

$$\therefore \quad \frac{36}{7} r = 36 \Rightarrow r = \left(36 \times \frac{7}{36}\right) = 7.$$

Hence, diameter = $2r$ cm = 14 cm.

1 Try to Solve

1. Find the area of a circle whose circumference is 22 cm.
2. The circumference of a circle exceeds its diameter by 16.8 cm. Find the radius of the circle.
3. If the diameter of a semicircular protractor is 14 cm, find its perimeter.
4. Find the area of a quadrant of a circle whose circumference is 22 m.
5. A wire, when bent in the form of an equilateral triangle, encloses an area of 121 cm^2. If the wire is bent in the form of a circle, find the area enclosed by the circle. [Use $\pi = \frac{22}{7}$][**AI 2017 (C)**]

Answers

1. 38.5 cm^2 **2.** 3.92 cm **3.** 36 cm **4.** 9.625 m^2 **5.** 346.5 cm^2

Example 4 A road which is 7 m wide surrounds a circular park whose circumference is 352 m. Find the area of the road.

Solution: Let the radius of the park be r m.

Then, its circumference = $2\pi r$ m.

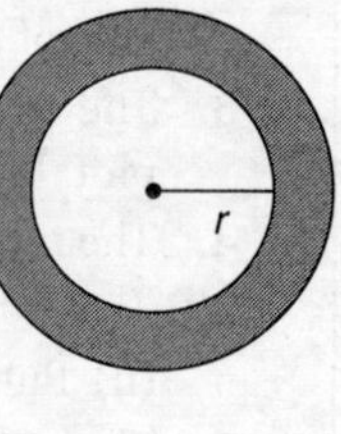

$$\therefore \quad 2\pi r = 352 \Rightarrow 2 \times \frac{22}{7} \times r = 352$$

$$\Rightarrow \quad r = 352 \times \frac{7}{44} = 56.$$

Thus, inner radius = 56 m, outer radius = $(56 + 7)$m = 63 m.

$$\text{Area of the road} = \pi[(63)^2 - (56)^2]\text{m}^2$$

$$= \frac{22}{7} \times (63 + 56)(63 - 56)\text{m}^2$$

$$= \left(\frac{22}{7} \times 119 \times 7\right) \text{m}^2 = 2618 \text{ m}^2.$$

Example 5 A racetrack is in the form of a ring whose inner and outer circumferences are 437 m and 503 m, respectively. Find the width of the track and also its area. **[CBSE (CCE) 2015]**

Solution: Let r m and R m be the radii of the inner and outer boundaries, respectively.

Then, $2\pi r = 437$ and $2\pi R = 503$

$$\Rightarrow \quad r = \frac{437}{2\pi} \text{ and } R = \frac{503}{2\pi}.$$

Width of the track = $(R - r)$ m

$$= \left(\frac{503}{2\pi} - \frac{437}{2\pi}\right) \text{m}$$

$$= \frac{1}{2\pi} \times (503 - 437) \text{ m}$$

$$= \left(\frac{1}{2} \times \frac{7}{22} \times 66\right) \text{m} = 10.5 \text{ m}$$

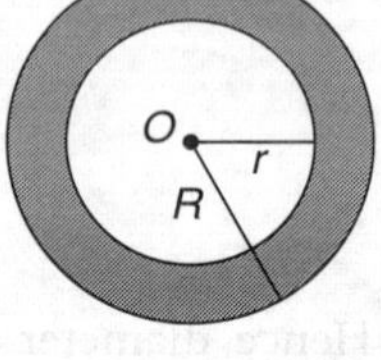

Area of the track = $\pi(R^2 - r^2)m^2 = \pi(R + r)(R - r)m^2$

$$= \left[\pi\left(\frac{503}{2\pi} + \frac{437}{2\pi}\right) \times 10.5\right] \text{m}^2 \qquad [\because (R - r) = 10.5]$$

$$= \left[\left(\frac{940}{2}\right) \times 10.5\right] \text{m}^2 = \left(470 \times \frac{21}{2}\right) \text{m}^2$$

$$= (235 \times 21)\text{m}^2 = 4935 \text{ m}^2.$$

2 Try to Solve

1. A path 7 metres wide lies outside a circular park whose radius is 18 metres. Find the area of the path.
2. A road which is 7 m wide surrounds a circular park whose circumference is 152 m. Find the area of the road.
3. The circumference of a circular plot is 220 m. A 15-m-wide concrete track runs outside the plot. Find the area of the track.
4. The inner circumference of a circular track is 440 m. The track is 14 m wide. Find the cost of
   (i) Levelling it at 20 paise/sq. metre. **[CBSE (CCE) 2011]**
   (ii) Putting up a fence along the outer circle at ₹ 2/metre.
5. The radii of two circles are 8 cm and 6 cm, respectively. Find the radius of the circle having an area equal to the sum of the areas of the two circles. **[NCERT]**

Answers

1. 946 m^2 **2.** 1218 m^2 **3.** 4007.14 m^2 **4.** ₹ 1355.20, ₹ 1056 **5.** 10 cm

Example 6 The diameter of each wheel of a bus is 140 cm. How many revolutions per minute must a wheel make in order to move at a speed of 66 km per hour? **[CBSE (CCE) 2012]**

Solution: Distance covered by a wheel in 1 minute

$$= \left(\frac{66 \times 1000 \times 100}{60}\right) \text{cm} = 110000 \text{ cm}.$$

$$\text{Circumference of wheel} = \left(2 \times \frac{22}{7} \times 70\right) \text{cm} = 440 \text{ cm}.$$

$$\text{Number of revolutions} = \frac{\text{Distance covered by a wheel in 1 min.}}{\text{Circumference of the wheel}}$$

$$\text{Number of revolutions in 1 minute} = \left(\frac{110000}{440}\right) = 250 \text{ revolutions}.$$

Example 7 The minute hand of a clock is 12 cm long. Find the area of the face of the clock described by the minute hand in 35 minutes.

Solution: Angle described by the minute hand in 60 minutes = 360°.

Angle described by the minute hand in 35 minutes

$$= \left(\frac{360}{60} \times 35\right)^{\circ} = 210^{\circ}.$$

$\therefore$ $\theta = 210^{\circ}$ and $r = 12$ cm.

Area swept by the minute hand in 35 minutes

$$= \left(\frac{\pi r^2 \theta}{360}\right) = \left(\frac{22}{7} \times 12 \times 12 \times \frac{210}{360}\right) \text{cm}^2 = 264 \text{ cm}^2.$$

3 Try to Solve

1. The wheels of a car are of diameter 80 cm each. How many complete revolutions does each wheel make in 10 minutes when the car is travelling at a speed of 66 km per hour? **[NCERT]**
2. The short and long hands of a clock are 4 cm and 6 cm long, respectively. Find the sum of the distances travelled by their tips in two days. **[CBSE 2018 (C)]**
3. A bicycle wheel makes 5,000 revolutions in moving 11 km. Find the diameter of the wheel. **[CBSE (CCE) 2013]**
4. The length of the minute hand of a clock is 14 cm. Find the area swept by the minute hand in one minute.
5. A pendulum swings through an angle of 30° and describes an arc 8.8 cm in length. Find the length of the pendulum. **[CBSE (CCE) 2012]**

Answers

1. 4375 **2.** 1910.86 cm **3.** 70 cm **4.** 10.27 cm^2 **5.** 16.8 cm

Example 8 The perimeter of a sector of a circle of radius 5.3 cm is 16.6 cm. Find the area of the sector.

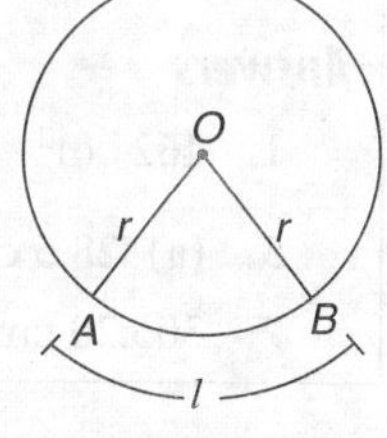

Solution: Radius of circle, $r = 5.3$ cm

$$\text{Perimeter of sector} = r + r + \text{arc } AB$$

$$16.6 = 5.3 + 5.3 + l$$

$\Rightarrow$ $$l = 16.6 - 10.6 = 6 \text{ cm}$$

$$\text{Area of sector} = \frac{1}{2}lr$$

$$= \frac{1}{2} \times 6 \times 5.3 = 15.9 \text{ cm}^2.$$

Example 9 In the figure, sectors of two concentric circles of radii 7 cm and 3.5 cm are given. Find the area of the shaded region.

Solution: Area of shaded $ABDC$ = Area of sector OCD – Area of sector OAB

$$= \frac{30°}{360°} \times \pi \times 7^2 - \frac{30°}{360°} \times \pi \times (3.5)^2$$

$$= \frac{\pi}{12} [7^2 - (3.5)^2]$$

$$= \frac{\pi}{12} (7 + 3.5)(7 - 3.5)$$

$$= \frac{22}{7} \times \frac{1}{12} 10.5 \times 3.5 = 9.625 \text{ cm}^2.$$

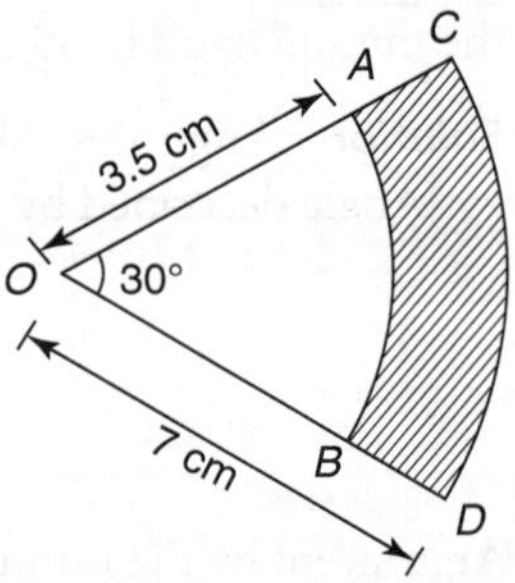

4 Try to Solve

1. Find the area of the sector of a circle whose central angle is 30°, and the radius of the circle is 42 cm.
2. A horse is tethered to one corner of a rectangular grass field 40 m by 24 m by a rope 14 m long. Over how much area of the field can it graze? **[CBSE (CCE) 2015]**
3. A horse is placed for grazing inside a rectangular field 70 m by 52 m and is tethered to one corner by a rope 21 m long. On how much area can it graze? **[CBSE (CCE) 2015]**
4. Find the area of a sector of a circle with radius 6 cm if the angle of the sector is 60°. **[NCERT]**
5. A chord of a circle of radius 10 cm subtends a right angle at the centre. Find the area of the corresponding:
   (a) Minor segment **[NCERT]**
   (b) Major sector [Use $\pi = 3.14$]
6. In a circle of radius 21 cm, an arc subtends an angle of 60° at the centre. Find: **[NCERT]**
   (a) The length of the arc
   (b) Area of the sector formed by the arc
   (c) Area of the segment formed by the corresponding chord
7. A chord PQ of a circle of radius 10 cm subtends an angle of 60° at the centre of the circle. Find the area of the major and minor segments of the circle. **[Delhi 2017]**

Answers

1. 462 cm² **2.** 154 cm² **3.** 346.5 m² **4.** 132/7 cm²

5. (a) 28.5 cm² (b) 285.5 cm² **6.** (a) 22 cm (b) 231 cm² (c) $231 - \frac{441\sqrt{3}}{4}$ cm²

7. 305.20 cm², 9.08 cm²

Example 10 In the figure, AB is the diameter of the circle, $AC = 6$ cm and $BC = 8$ cm. Find the area of the shaded region. [Use $\pi = 3.14$] **[CBSE (CCE) 2012]**

Solution: Given: $AC = 6$ cm, $BC = 8$ cm. AB is the diameter of the circle.

$\therefore$ $\angle ACB = 90°$ [$\because$ Angle in a semicircle is a right angle]

In right ΔACB $AB^2 = AC^2 + BC^2$ [$\because$ Pythagoras' Theorem]

$= (6)^2 + (8)^2$

$AB^2 = 36 + 64 = 100$

$\therefore$ $AB = 10$ cm

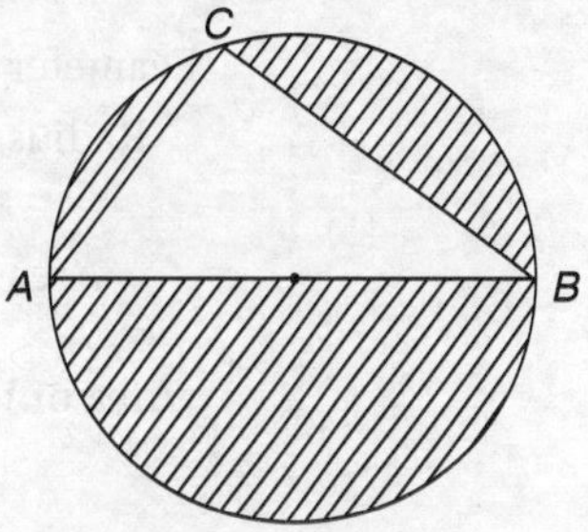

Radius of circle $r = \frac{AB}{2} = \frac{10}{2} = 5$ cm

Areas of right $\Delta ABC = \frac{1}{2} \times AC \times BC$

$= \frac{1}{2} \times 6 \times 8 = 24 \text{ cm}^2$

Area of circle $= \pi r^2 = 3.14 \times 5 \times 5 = 78.5 \text{ cm}^2$

Area of shaded portion = Area of circle – Area of triangle

$= 78.5 - 24 = 54.5 \text{ cm}^2$.

Example 11 In the figure, $ABCD$ is a square with side 14 cm. Find the area of the shaded region.

Solution: Side of a square = 14 cm. **[NCERT, CBSE (CCE) 2008]**

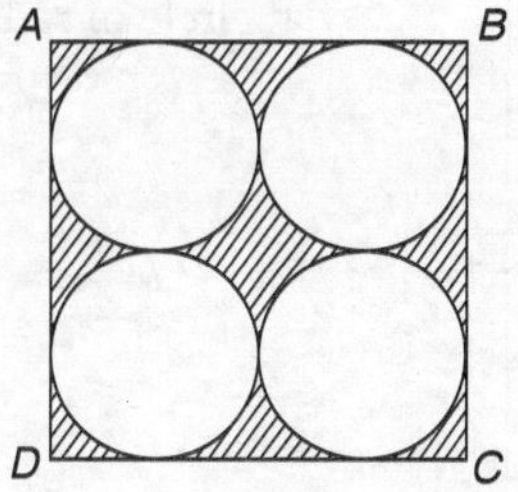

Radius of a circle $= \frac{14}{4} = 3.5$ cm

Area of a square $= 14 \times 14 = 196 \text{ cm}^2$

Area of four circles $= 4 \times \pi r^2$

$= 4 \times \frac{22}{7} \times 3.5 \times 3.5 = 154 \text{ cm}^2$

Area of shaded region $= 196 - 154 = 42 \text{ cm}^2$.

Example 12 Find the area of the shaded design, as shown in figure (i), where $ABCD$ is a square of side 12 cm and semicircles are drawn with each side of the square as diameter. [Use $\pi = 3.14$]

[NCERT, CBSE (CCE) 2012]

Solution: Mark the unshaded region as I, II, III and IV, as shown in figure (ii).

Area of square $= 12 \times 12 = 144 \text{ cm}^2$

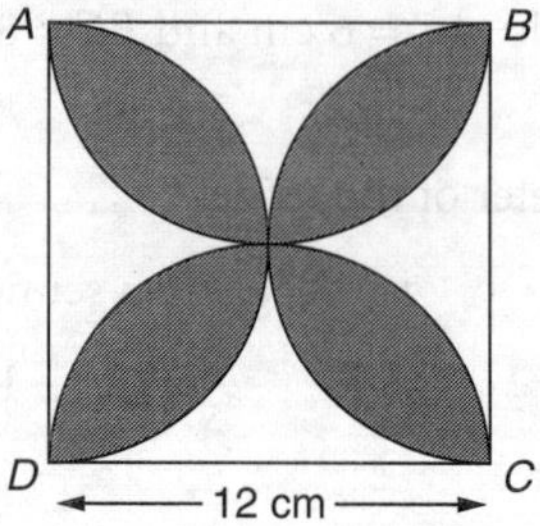

(i)

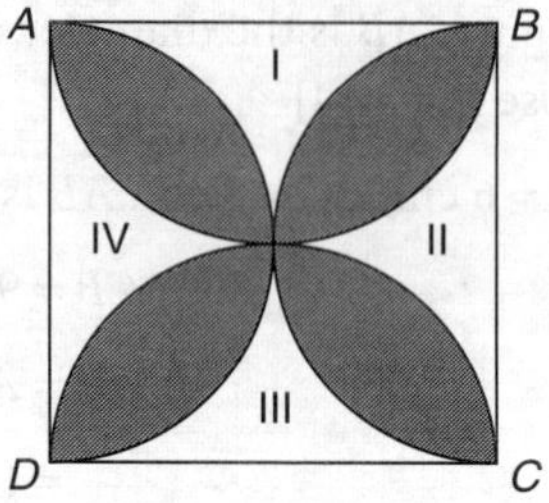

(ii)

Diameter of semicircle = Side of square = 12 cm

$\therefore$ Radius of semicircle = 6 cm

$$\text{Area of semicircle} = \frac{\pi r^2}{2} = \frac{3.14 \times 6 \times 6}{2} = 56.52 \text{ cm}^2$$

$$\text{Area of I} + \text{Area of III} = \text{Area of square} - 2 \times \text{Area semicircle}$$
$$= 144 - (2 \times 56.52) = 30.96 \text{ cm}^2$$

Similarly Area of II + Area of IV = 30.96 cm²

$$\text{Area of shaded region} = \text{Area of square} - \text{Area of } (I + II + III + IV)$$
$$= 144 - 2 \times 30.96 = 82.08 \text{ cm}^2.$$

5 Try to Solve

1. An umbrella has 8 ribs which are equally spaced (see figure). Assuming the umbrella to be a flat circle of radius 45 cm, find the area between two consecutive ribs. **[NCERT]**

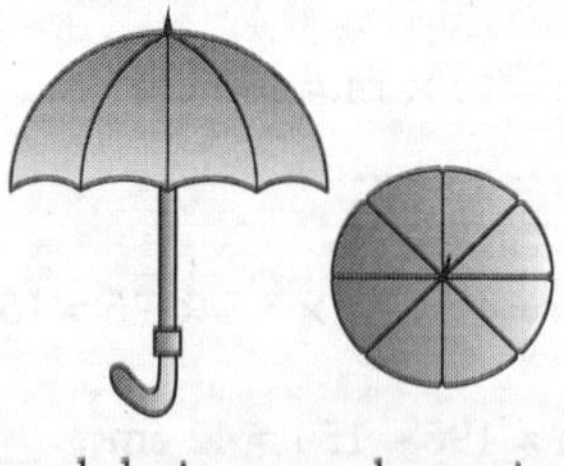

2. A round table cover has six equal designs, as shown in the figure. If the radius of the cover is 28 cm, find the cost of making the designs at the rate of ₹0.35 per cm². [Use $\sqrt{3} = 1.7$] **[NCERT, CBSE (CCE) 2013]**

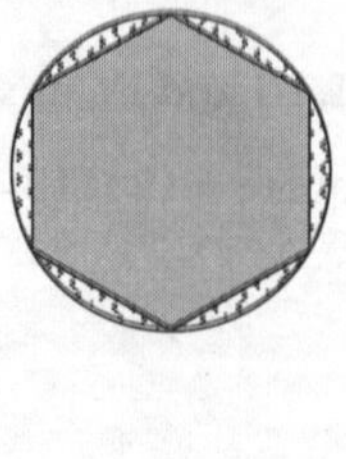

3. In the figure, a square $OABC$ is inscribed in a quadrant $OPBQ$ of a circle. If $OA = 20$ cm, find the area of the shaded region. [Use $\pi = 3.14$] **[NCERT, CBSE (CCE) 2014]**

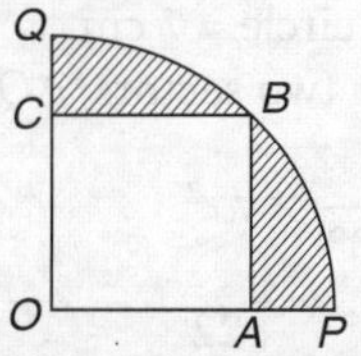

4. In the figure, find the area of the shaded region enclosed between two concentric circles of radii 7 cm and 14 cm, where $\angle AOC = 40°$. $\left[\text{Use } \pi = \frac{22}{7}\right]$ **[NCERT, CBSE (CCE) 2015]**

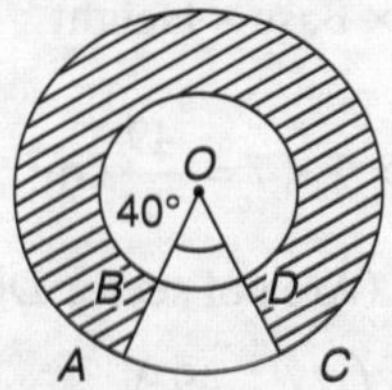

5. In the figure, find the area of the shaded region. [Use $\pi = 3.14$] **[CBSE (CCE) 2014]** **[Delhi 2015]**

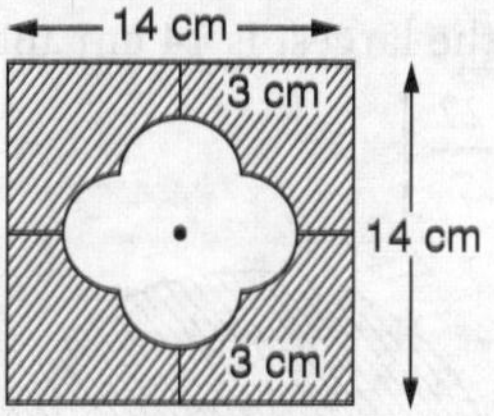

6. In the figure, two concentric circles with centre O have radii 21 cm and 42 cm. If $\angle AOB = 60°$, find the area of the shaded region. [Use $\pi = \frac{22}{7}$] **[CBSE (CCE) 2014]**

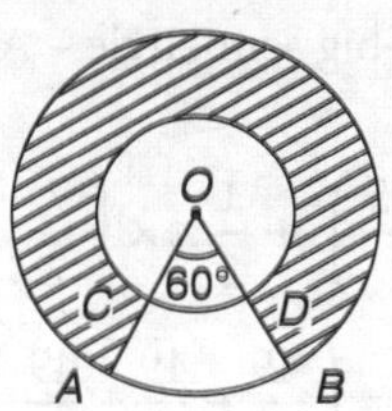

Answers

1. $\frac{22275}{28}$ cm² = 795.53 cm²
2. ₹162.68
3. 228 cm²
4. 410.67 cm²
5. 154.88 cm²
6. 3465 cm²

Example 13 In the figure, $ABCD$ is a square of side 7 cm. $DPBA$ and $DQBC$ are quadrants of circles, each of radius 7 cm. Find the area of the shaded region. [Use = $\pi = \frac{22}{7}$] **[CBSE (CCE) 2012]**

Solution: Side of square = 7 cm. Radius of circle = 7 cm.
Shaded area is the common region between two sectors, ΔQBC and ΔPBA.

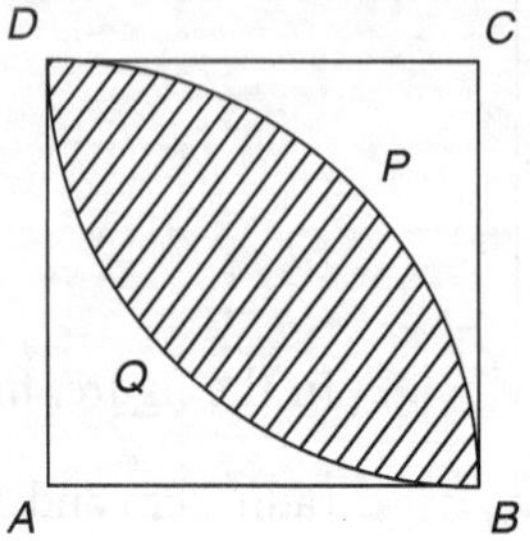

$$\text{Area of sector} = \frac{\theta}{360°} \times \pi r^2$$

$$= \frac{90°}{360°} \times \frac{22}{7} \times (7)^2$$

$$= \frac{1}{2} \times 11 \times 7 = \frac{77}{2}\ \text{cm}^2$$

$$\text{Area of a triangle} = \frac{1}{2} \times \text{Base} \times \text{Height}$$

$$\text{Area of } \Delta DCB = \frac{1}{2} \times 7 \times 7 = \frac{49}{2}\ \text{cm}^2$$

$$\text{Area of the shaded region} = 2 \times (\text{Area of sector } DQBC - \text{Area of } \Delta DBC)$$

$$= 2 \times \left(\frac{77}{2} - \frac{49}{2}\right) = 2 \times \left(\frac{77-49}{2}\right) = 28\ \text{cm}^2$$

Example 14 In the figure, the boundary of the shaded region consists of four semicircular arcs, the wo smallest being equal. If the diameter of the largest is 14 cm and that of the smallest is 3.5 cm, calculate the area of the shaded region. $\left[\text{Use } \pi = \frac{22}{7}\right]$ **[CBSE (CCE) 2016]**

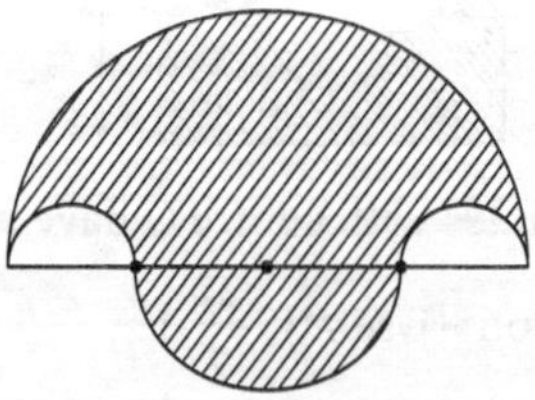

Solution: Area of shaded region = Area of big semicircle – Area of 2 small semicircles + Area of middle semicircles.

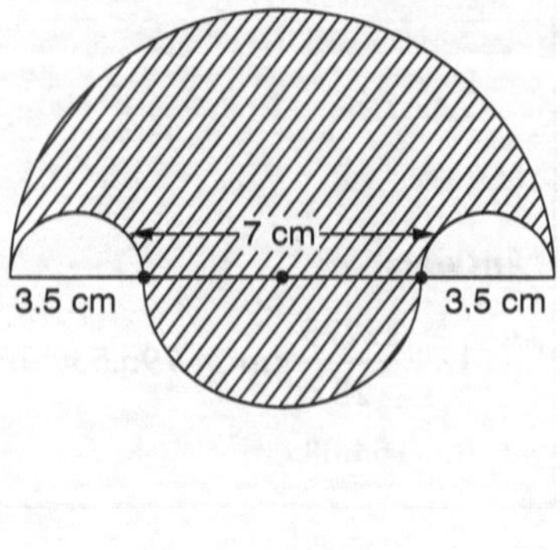

$$= \frac{1}{2}\pi \times (7)^2 - 2 \times \frac{1}{2} \times \pi \times \left(\frac{3.5}{2}\right)^2 + \frac{1}{2}\pi \times \left(\frac{7}{2}\right)^2$$

$$= \frac{49}{2}\pi - \left(\frac{7}{4}\right)^2 \pi + \frac{1}{2}\left(\frac{49}{4}\right)\pi = \left(\frac{49}{2} - \frac{49}{16} + \frac{49}{8}\right)\pi$$

$$= \left(\frac{1}{2} - \frac{1}{16} + \frac{1}{8}\right) \times 49 \times \frac{22}{7} = \left(\frac{8-1+2}{16}\right) \times 7 \times 22$$

$$= \frac{9}{16} \times 7 \times 22 = 86.625\ \text{cm}^2.$$

Example 15 The figure depicts a racing track whose left and right ends are semicircular.

[NCERT, CBSE (CCE) 2014]

The distance between the two inner parallel line segments is 60 m and they are each 106 m long. If the track is 10 m wide, find:

(i) The distance around the track along its inner edge

(ii) The area of the track

Solution: (i) Distance around of track $= 2 \times 106 + 2\pi r$

(Length of two parallel lines + Circumference of two semicircles)

$$= 212 + 2 \times 3.14 \times 30$$

$$= 212 + 188.4 = 400.4 \text{ m}$$

(ii) Area of bigger semicircle $= \dfrac{\pi r^2}{2} = \dfrac{22}{7} \times \dfrac{1}{2} \times 40 \times 40$

$$= \frac{22 \times 1600}{14} \text{m}^2$$

Area of smaller semicircle $= \dfrac{\pi r^2}{2} = \dfrac{22}{7} \times \dfrac{1}{2} \times 30 \times 30$

$$= \frac{22 \times 900}{14} \text{ m}^2$$

A_1 = Total area of semicircular track

$$= 2\left(\frac{22}{14} \times 1600 - \frac{22}{14} \times 900\right)$$

$$= 2 \times \frac{22}{14} (1600 - 900)$$

$$= \frac{22}{7} \times 700 = 2200 \text{ m}^2$$

A_2 = Area of rectangular track

$$= 2 \times 10 \times 106 = 2120 \text{ m}^2$$

Total area of track $= A_1 + A_2 = 2200 + 2120 = 4320 \text{ m}^2$

6 Try to Solve

1. In the figure, $OACB$ is a circle with centre O and radius 3.5 cm. If $OD = 2$ cm, find the area of the
   (a) Quadrant $OACB$ (b) Shaded region **[NCERT]**

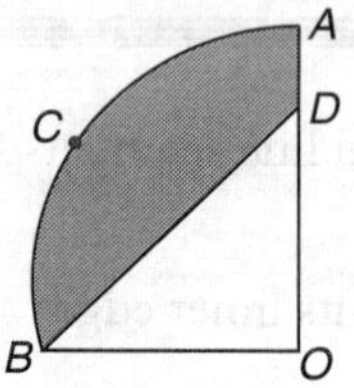

2. The area of an equilateral triangle ABC is 17320.5 cm^2. With each vertex of the triangle as centre, a circle is drawn with radius equal to half the length of the side of the triangle (see figure). Find the area of the shaded region. [Use $\pi = 3.14$ and $\sqrt{3} = 1.73205$]
   [NCERT, CBSE (CCE) 2014]

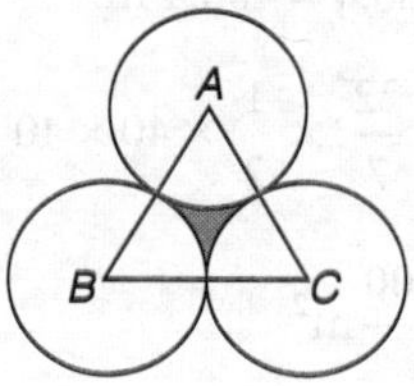

3. In the figure, ABC is a quadrant of a circle of radius 14 cm and a semicircle is drawn with BC as diameter. Find the area of the shaded region. **[NCERT, CBSE (CCE) 2015]**

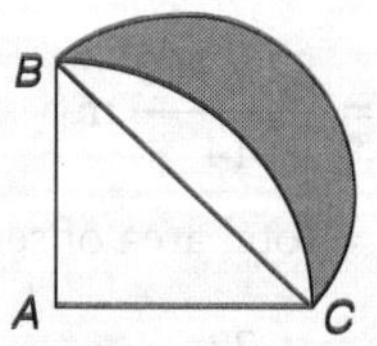

4. Find the area of the shaded region in the figure, if $ABCD$ is a square of side 14 cm and APC and BPC are semicircles. **[NCERT]**

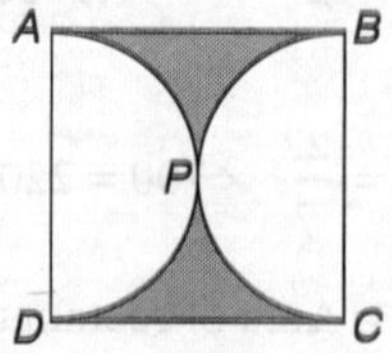

5. From each corner of a square of side 4 cm, a quadrant of a circle of radius 1 cm is cut, as is a circle of diameter 2 cm, as shown in the figure. Find the area of the remaining portion of the square. **[NCERT, CBSE (CCE) 2015]**

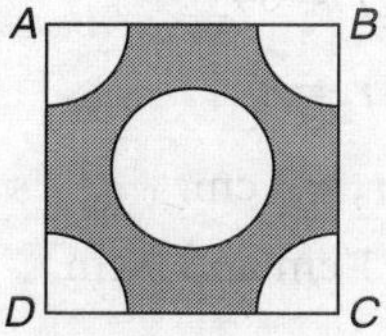

Answers

1. (a) $\frac{77}{8}$ cm^2 (b) $\frac{49}{8}$ cm^2 2. 1620.5 cm^2 3. 98 cm^2 4. 42 cm^2

5. $\frac{68}{7}$ cm^2

FOR MORE PRACTICE AND KNOWLEDGE

Example 16 Two circles touch externally. The sum of their areas is 130 π sq. cm and the distance between their centres is 14 cm. Find the radii of the circles.

Solution: If two circles touch externally, the distance between their centres is equal to the sum of their radii.

Suppose the radii of the two circles is r_1 cm and r_2 cm, respectively. Suppose, C_1 and C_2 are the centres of the given circles. Then

$$C_1C_2 = r_1 + r_2$$

$\Rightarrow$ $$14 = r_1 + r_2 \quad [\because C_1C_2 = 14 \text{ cm (given)}]$$

$\Rightarrow$ $$r_1 + r_2 = 14 \quad ...(i)$$

It is given that the sum of the areas of two circles is equal to 130 π cm^2.

$\therefore$ $$\pi r_1^2 + \pi r_2^2 = 130\pi$$

$\Rightarrow$ $$r_1^2 + r_2^2 = 130 \quad ...(ii)$$

Now $$(r_1 + r_2)^2 = r_1^2 + r_2^2 + 2r_1r_2$$

$\Rightarrow$ $$14^2 = 130 + 2r_1r_2 \quad \text{[From (i) and (ii)]}$$

$\Rightarrow$ $$196 - 130 = 2r_1r_2$$

$\Rightarrow$ $$r_1r_2 = 33 \quad ...(iii)$$

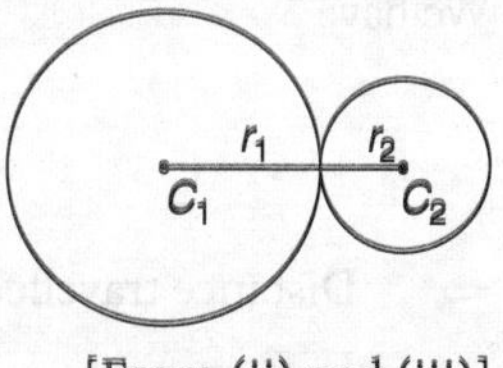

Now $$(r_1 - r_2)^2 = r_1^2 + r_2^2 - 2r_1r_2$$

$\Rightarrow$ $$(r_1 - r_2)^2 = 130 - 2 \times 33 \quad \text{[From (ii) and (iii)]}$$

$\Rightarrow \qquad (r_1 - r_2)^2 = 64$

$\Rightarrow \qquad r_1 - r_2 = 8 \qquad$ ...(iv)

Solving (i) and (iv), we get $r_1 = 11$ cm and $r_2 = 3$ cm.

Hence, the radii of the two circles are 11 cm and 3 cm.

Example 17 A boy is cycling such that the wheels of the cycle make 140 revolutions per minute. If the diameter of the wheel is 60 cm, calculate the speed per hour at which the boy is cycling.

Solution: We have Radius of the wheel $= r = \frac{60}{2}$ cm = 30 cm.

$$\text{Circumference of the wheel} = 2\pi r = 2 \times \frac{22}{7} \times 30 \text{ cm} = \frac{1320}{7} \text{ cm}$$

$$\text{Distance covered in one revolution} = \text{Circumference} = \frac{1320}{7} \text{ cm}$$

$$\therefore \quad \text{Distance covered in 140 revolutions} = \frac{1320}{7} \times 140 \text{ cm} = (1320 \times 20) \text{ cm} = 26400 \text{ cm} = \frac{26400}{100} \text{ m}$$

$$= 264 \text{ m} = \frac{264}{1000} \text{km}$$

As the wheels make 140 revolutions per minute

Distance covered in one minute = Distance covered in 140 revolutions

$$\Rightarrow \quad \text{Distance covered in one minute} = \frac{264}{1000} \text{ km}$$

$$\Rightarrow \quad \text{Distance covered in one hour} = \frac{264}{1000} \times 60 \text{ km} = 15.84 \text{ km}$$

Hence, the speed at which the boy is cycling is 15.84 km/h.

Example 18 A car has wheels which are 80 cm in diameter. How many complete revolutions does each wheel make in 10 minutes when the car is travelling at a speed of 66 km per hour?

Solution: We have speed of the car = 66 km/h.

$\therefore \quad$ Distance travelled by the car in 1 hour = 66 km

$$\Rightarrow \quad \text{Distance travelled by the car in 10 minutes} = \left(\frac{66}{60} \times 10\right) \text{km} = 11 \text{ km} = 11 \times 1000 \times 100 \text{ cm}$$

We have Radius of car wheels = 40 cm

$$\therefore \quad \text{Circumference of the wheels} = 2 \times \frac{22}{7} \times 40 \text{ cm}$$

$$\Rightarrow \quad \text{Distance travelled by the car when its wheels make one complete revolution} = 2 \times \frac{22}{7} \times 40 \text{ cm}$$

$\therefore$ Number of revolutions made by the wheels in 10 minutes.

$$= \frac{\text{Distance covered by the car in 10 minutes}}{\text{Distance covered by the car when its wheels make one complete revolution}}$$

$$= \frac{11 \times 1000 \times 100}{2 \times \frac{22}{7} \times 40} = \frac{11 \times 1000 \times 100 \times 7}{2 \times 22 \times 40} = 4375$$

Hence, each wheel makes 4375 revolutions in 10 minutes.

Example 19 An elastic belt is placed around the rim of a pulley of radius 5 cm. One point on the belt is pulled directly away from the centre O of the pulley until it is at P, 10 cm from O. Find the length of the belt that is in contact with the rim of the pulley. Also find the shaded area.

Solution: In the figure, let $\angle AOP = \angle BOP = \theta$. Clearly, portion AB of the belt is not in contact with the rim of the pulley in right triangle OAP.

$$\cos\theta = \frac{OA}{OP} = \frac{5}{10} = \frac{1}{2} \quad \theta = 60° \quad \angle AOB = 2\theta = 120°$$

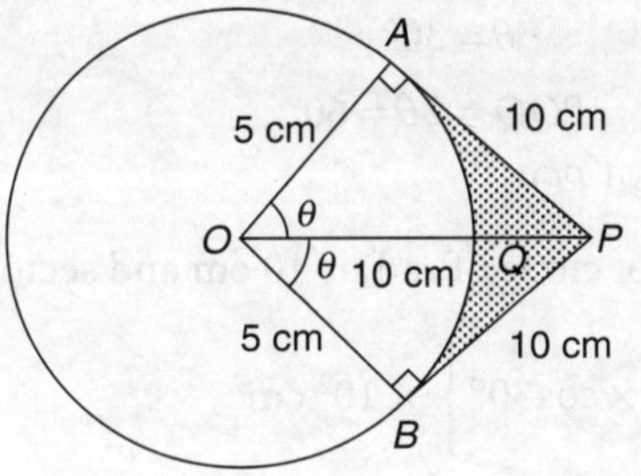

$\therefore$ $$\text{Arc } AB = \frac{120° \times 2 \times \pi \times 5}{360} \text{ cm} = \frac{10\pi}{3} \text{ cm} \qquad \left[l = \frac{\theta}{360} \times 2\pi r \right]$$

Hence, length of the belt that is in contact with the rim of the pulley

$$= \text{Circumference of the rim} - \text{Length of arc } AB$$

$$= 2\pi \times 5 \text{ cm} - \frac{10\pi}{3} \text{ cm} = \frac{20\pi}{3} \text{ cm}$$

Now $$\text{Area of sector } OAQB = \frac{120}{360} \times \pi \times 5^2 \text{ cm}^2 = \frac{25\pi}{3} \text{ cm}^2 \qquad \left[\text{Area} = \frac{\theta}{360} \times \pi r^2 \right]$$

Area of quadrilateral $OAPB = 2(\text{Area of } \Delta OAP)$

$$= 2 \times \left(\frac{1}{2} \times OA \times AP \right)$$

$$= 5 \times 5\sqrt{3} \text{ cm}^2 \qquad \left[\because OP^2 = OA^2 + AP^2 \Rightarrow AP = \sqrt{100 - 25} = 5\sqrt{3} \right]$$

$$= 25\sqrt{3} \text{ cm}^2$$

Hence Shaded area = Area of quadrilateral *OAPB* – Area of sector *OAQB*.

$$= \left(25\sqrt{3} - \frac{25\pi}{3}\right) \text{cm}^2 = \frac{25}{3}(3\sqrt{3} - \pi)\text{cm}^2$$

Example 20 The figure shows two arcs, *A* and *B*. Arc *A* is part of the circle with centre *O* and radius *OP*. Arc *B* is part of the circle with centre *M* and radius *PM*, where *M* is the mid-point of *PQ*. Show that the area enclosed by the two arcs is equal to $25\left(\sqrt{3} - \frac{\pi}{6}\right)$ cm^2.

Solution: Area enclosed by arc *B* and chord *PQ* = Area of semicircle of radius 5 cm

$$= \frac{1}{2} \times \pi \times 5^2 \text{ cm}^2 = \frac{25\pi}{2} \text{ cm}^2$$

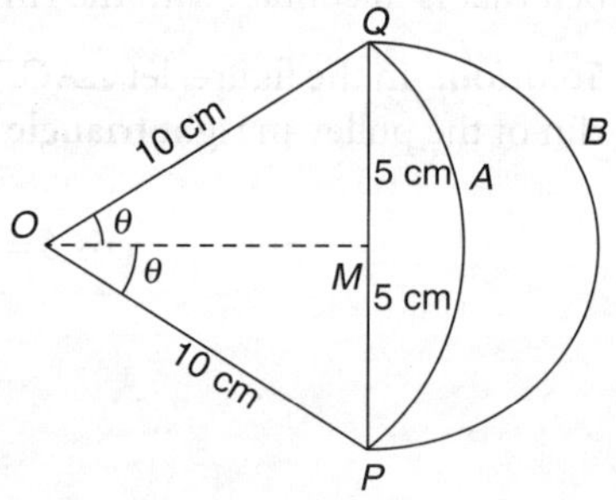

Let $\angle MOQ = \angle MOP = \theta$

In ΔOMP, we have $\sin\theta = \frac{PM}{OP} = \frac{5}{10} = \frac{1}{2}$

$\Rightarrow$ $\theta = 30°$

$\Rightarrow$ $POQ = 2\theta = 60°$

$\therefore$ Area enclosed by arc *A* and chord *PQ*

= Area of segment of circle of radius 10 cm and sector containing angle 60°

$$= \left\{\frac{\pi \times 60}{360} - \sin 30° \times \cos 30°\right\} \times 10^2 \text{ cm}^2 \qquad \left[\because A = \left\{\frac{\pi\theta}{360} - \sin\frac{\theta}{2}\cos\frac{\theta}{2}\right\} r^2\right]$$

$$= \left\{\frac{50\pi}{3} - 25\sqrt{3}\right\} \text{cm}^2$$

Hence $\text{Required area} = \left\{\frac{25\pi}{2} - \left(\frac{50\pi}{3} - 25\sqrt{3}\right)\right\} \text{cm}^2$

$\Rightarrow$ $\text{Required area} = \left\{25\sqrt{3} - \frac{25\pi}{6}\right\} \text{cm}^2 = 25\left\{\sqrt{3} - \frac{\pi}{6}\right\} \text{cm}^2$

Example 21 *PQRS* is the diameter of a circle of radius 6 cm. The lengths *PQ*, *QR* and *RS* are equal. Semicircles are drawn with *PQ* and *QS* as diameter, as shown in the figure. Find the perimeter and area of the shaded region.

Solution: We have

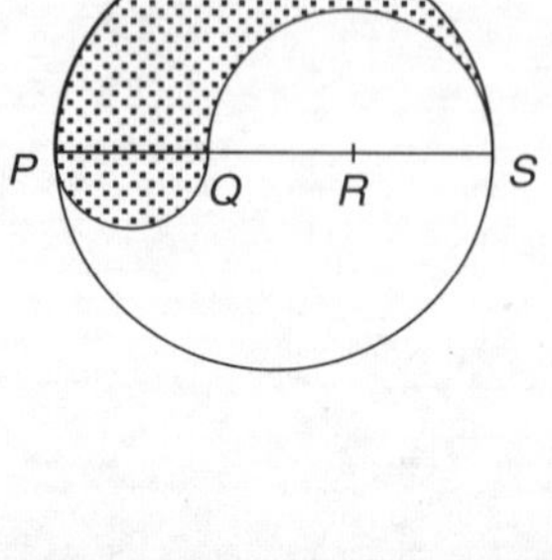

PS = Diameter of a circle of radius 6 cm = 12 cm

$\therefore$ $PQ = QR = RS = \frac{12}{3} = 4\text{cm}$

$QS = QR + RS = (4 + 4)$ cm = 8 cm

Hence, required perimeter

$$= \text{Arc of semicircle of radius 6 cm} + \text{Arc of semicircle of radius 4 cm}$$

$$+ \text{Arc of semicircle of radius 2 cm}$$

$$= (\pi \times 6 + \pi \times 4 + \pi \times 2)\text{cm} = 12\pi \text{ cm}$$

Required area = Area of semicircle with *PS* as diameter + Area of semicircle with *PQ* as diameter – Area of semicircle with *QS* as diameter.

$$= \frac{1}{2} \times \frac{22}{7} \times (6)^2 + \frac{1}{2} \times \frac{22}{7} \times 2^2 - \frac{1}{2} \times \frac{22}{7} \times (4)^2$$

$$= \frac{1}{2} \times \frac{22}{7}(6^2 + 2^2 - 4^2)$$

$$= \frac{1}{2} \times \frac{22}{7} \times 24 = \frac{264}{7} \text{ cm}^2 = 37.71 \text{ cm}^2$$

Example 22 In the figure, *AOBCA* represents a quadrant of a circle of radius 3.5 cm with centre *O*. Calculate the area of the shaded portion. [Take $\pi = 22/7$] **[Delhi 2017, NCERT]**

Solution: We have

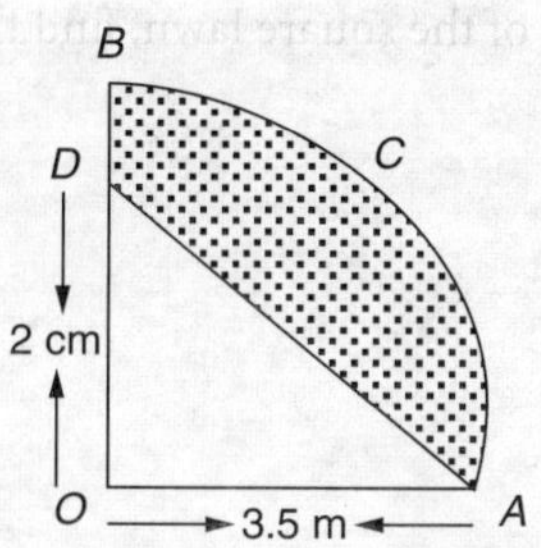

Area of quadrant *AOBCA* $= \frac{1}{4}\pi r^2$

$$= \frac{1}{4} \times \frac{22}{7} \times (3.5)^2$$

$$= \frac{1}{4} \times \frac{22}{7} \times \frac{7}{2} \times \frac{7}{2} = \frac{77}{8} \text{ cm}^2 = 9.625 \text{ cm}^2$$

Area of $\Delta AOD = \frac{1}{2} \times \text{Base} \times \text{Height}$

$$= \frac{1}{2}(OA \times OD) = \frac{1}{2}[3.5 \times 2] \text{ cm}^2 = 3.5\text{cm}^2$$

Hence Area of the shaded portion = Area of quadrant – Area of ΔAOD

$\Rightarrow$ Area of the shaded portion $= (9.625 - 3.5) \text{ cm}^2 = 6.125 \text{ cm}^2$

Example 23 The area of an equilateral triangle is 1732.05 cm². With each angular point as centre, a circle is described with radius equal to half the length of the side of the triangle. Find the area of the triangle not included in the circles. [Use $\pi = 3.14$]

Solution: Let each side of the equilateral triangle be *a* cm.

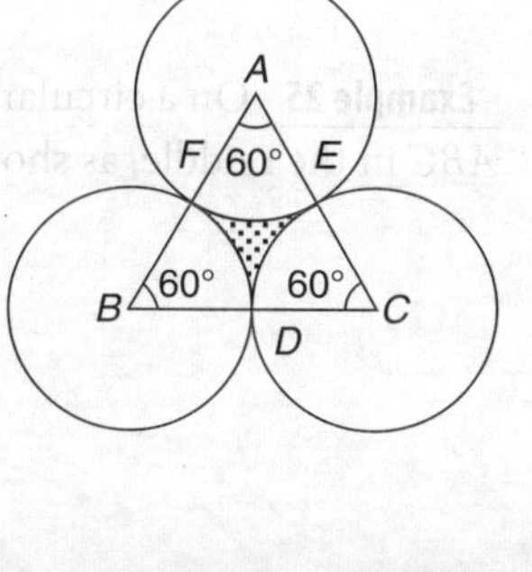

Then Area = 1732.05 cm²

$$\Rightarrow \quad \frac{\sqrt{3}}{4}a^2 = 1732.05$$

$$\Rightarrow \quad \frac{a^2}{4} = \frac{1732.05}{\sqrt{3}} \Rightarrow \left(\frac{a}{2}\right)^2 = \frac{1732.05}{\sqrt{3}} \qquad \text{...(i)}$$

Clearly, radius of each circle is $\frac{a}{2}$ cm.

Let A be the area of three sectors, each of angle 60°, in a circle of radius $\frac{a}{2}$ cm. Then

$$A = 3\left\{\frac{60}{360} \times 3.14 \times \left(\frac{a}{2}\right)^2\right\} \text{cm}^2$$

$$\Rightarrow \quad A = \frac{1}{2} \times 3.14 \times \left(\frac{a}{2}\right)^2 = \frac{1}{2} \times 3.14 \times \frac{1730.05}{\sqrt{3}} = 1570.05 \text{ cm}^2 \qquad \text{[From (i)]}$$

Now

Required area = Area of ΔABC – Area of three sectors each of angle 60° in a circle of radius $\frac{a}{2}$

$\Rightarrow$ Required area = (1732.05 – 1570.04) cm^2 = 162.01 cm^2

Example 24 In the figure, two circular flower beds have been shown on two sides of a square lawn $ABCD$ of side 56 m. If the centre of each circular flower bed is the point of intersection of the diagonals of the square lawn, find the sum of the areas of the lawns and the flower beds. **[NCERT]**

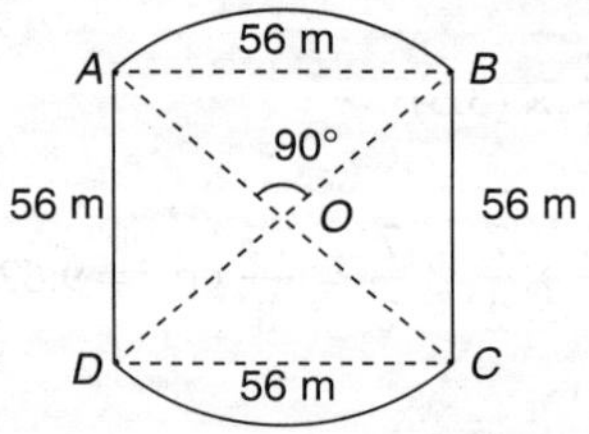

Solution: Total area = Area of sector OAB + Area of sector ODC + Area of ΔOAD + Area of ΔOBC

$$= \left(\frac{90}{360} \times \frac{22}{7} \times 28\sqrt{2} \times 28\sqrt{2} + \frac{90}{360} \times \frac{22}{7} \times 28\sqrt{2} \times 28\sqrt{2} + \frac{1}{4} \times 56 \times 56 + \frac{1}{4} \times 56 \times 56\right) \text{m}^2$$

$$= \frac{1}{4} \times 28 \times 56\left(\frac{22}{7} + \frac{22}{7} + 2 + 2\right) \text{m}^2$$

$$= \frac{7 \times 56}{7}(22 + 22 + 14 + 14)\text{m}^2 = 56 \times 72 \text{ m}^2 = 4032 \text{ m}^2$$

Example 25 On a circular table cover of radius 32 cm, a design is formed leaving an equilateral triangle ABC in the middle, as shown in the figure. Find the area of the design (shaded region). **[NCERT]**

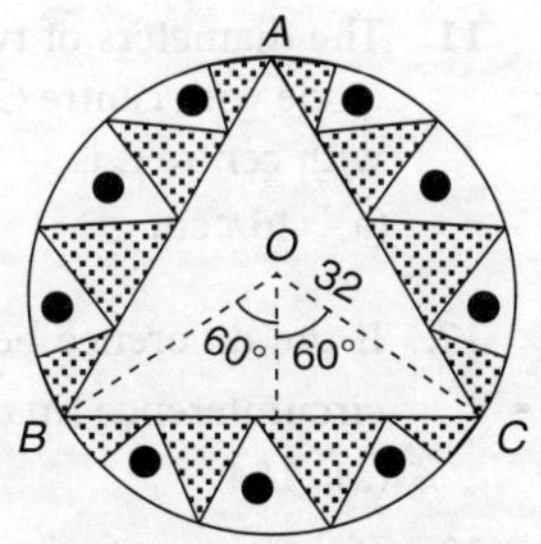

Solution: In ΔOBD, we have

$$\cos 60° = \frac{OD}{OB} \text{ and } \sin 60° = \frac{BD}{OB} \Rightarrow \frac{1}{2} = \frac{OD}{32} \text{ and } \frac{\sqrt{3}}{2} = \frac{BD}{32}$$

$$\Rightarrow \quad OD = 16 \text{ and } BD = 16\sqrt{3}$$

$$\Rightarrow \quad BC = 2BD = 32\sqrt{3}$$

Area of the shaded region

$$= \text{Area of the circle} - \text{Area of } \Delta ABC = \left\{\pi \times 32^2 - \frac{\sqrt{3}}{4} \times (32\sqrt{3})^2\right\} \text{cm}^2$$

$$= \left\{\frac{22}{7} \times 32 \times 32 - 768\sqrt{3}\right\} \text{cm}^2 = \left\{\frac{22528}{7} - 768\sqrt{3}\right\} \text{cm}^2$$

Multiple Choice Questions

Choose the correct answer for each of the following questions:

1. The perimeter of a sector of a circle of radius 5.2 cm is 16.4 cm. The area of the sector is
(a) 31.2 cm^2 (b) 15.01 cm^2 (c) 15.6 cm^2 (d) 16.6 cm^2

2. The area of a square that can be inscribed in a circle of radius 6 cm is
(a) 72 cm^2 (6) 144 cm^2 (c) 216 cm^2 (d) 288 cm^2

3. If the diameter of a circle is 7 cm, then the perimeter of the semicircle is
(a) 36 cm (b) 7π (c) 18 cm (d) 14π

4. If the circumference and area of a circle are numerically equal, the radius of the circle is equal to
(a) 2 (b) 1 (c) 7 (d) π

5. The area of the largest triangle that can be inscribed in a semicircle of radius 'r' units is
(a) $\sqrt{2}\, r^2$ sq. units (b) $\frac{1}{2} r^2$ sq. units (c) r^2 sq. units (d) $2r^2$sq. units

6. The circumference of two circles is in the ratio 4:5. What is the ratio of the areas of these circles?
(a) 64:125 (6) 16:25 (c) 4:10 (d) 4:5

7. The area of a circle whose diameter is d is
(a) $\frac{\pi d^2}{4}$ (b) πd^2 (c) $\frac{\pi d^2}{2}$ (d) πd

8. The area of the quadrant of a circle of radius 'r' is
(a) $\frac{1}{2}\pi r^2$ (b) πr (c) $\frac{1}{4}\pi r^2$ (d) $\left(\frac{\pi}{2} + 2r\right)$

9. The area of the largest circle that can be drawn inside a rectangle with sides 7 m × 3.5 m is
(a) $\frac{77}{2}$ m^2 (b) 77 cm^2 (c) $\frac{77}{8}$ m^2 (d) $\frac{77}{4}$ m^2

10. The area of a quadrant of a circle of circumference 22 cm is [Take $\pi = 22/7$]
(a) 3.05 cm^2 (b) 3.5 cm^2 (c) 9.625 cm^2 (d) 35.5 cm^2

11. The diameters of two circles with centres A and B are 16 cm and 30 cm, respectively. If the area of a circle with centre C is equal to the sum of areas of the other two circles, the circumference of the circle with centre C is

(a) 46π cm (b) 28π cm (c) 34π cm (d) 32π cm

12. If the difference between the circumference and the radius of a circle is 37 cm, using $\pi = \frac{22}{7}$, the circumference (in cm) of the circle is

(a) 154 (b) 44 (c) 14 (d) 1

13. If the circumference of a circle is 44 cm, the area of a circle is

(a) 276 cm^2 (b) 44 cm^2 (c) 176 cm^2 (d) 154 cm^2

14. If the circumference of a circle is equal to the perimeter of a square, the ratio of their areas is

(a) 22:7 (b) 14:11 (c) 7:22 (d) 7:11

15. Find the perimeter of a half-circle

(a) $\frac{1}{2}(2\pi r)$ (b) $r(2+\pi)$ (c) $\frac{\pi r}{2}$ (d) $\pi r^2 \times \frac{1}{2}$

16. If the sum of the areas of two circles with radii R_1 and R_2 is equal to the area of a circle of radius R, then

(a) $R_1 + R_2 = R$ (b) $R_1^2 + R_2^2 = R^2$ (c) $R_1 + R_2 < R$ (d) $R_1^2 + R_2^2 < R^2$

17. If the sum of the circumferences of two circles with radii R_1 and R_2 is equal to the circumference of a circle of radius R, then

(a) $R_1 + R_2 = R$ (b) $R_1 + R_2 > R$

(c) $R_1 + R_2 < R$ (d) Nothing definite can be said about the relation among R_1, R_2 and R

18. The area of the largest triangle that can be inscribed in a semicircle of radius r units is

(a) r^2 sq. units (b) $\frac{1}{2} r^2$ sq. units (c) $2r^2$ sq. units (d) $\sqrt{2}\, r^2$ sq. units

19. An arc of a circle is of length 5π cm and the sector it bounds has an area of 20π cm^2. Then, the radius of the circle is

(a) 4 cm (b) 8 cm (c) 12 cm (d) 16 cm **[NCERT Exemplar]**

20. The radius of a circle whose circumference is equal to the sum of the circumferences of the two circles of diameters 36 cm and 20 cm is

(a) 56 cm (b) 42 cm (c) 28 cm (d) 16 cm **[NCERT Exemplar]**

21. The area of a sector of angle p (in degrees) of a circle with radius R is

(a) $\frac{p}{180} \times 2\pi R$ (b) $\frac{p}{180} \times \pi R^2$ (c) $\frac{p}{360} \times 2\pi R$ (d) $\frac{p}{720} \times 2\pi R^2$ **[NCERT]**

22. In the figure, ABC is an equilateral triangle inscribed in a circle of radius 4 cm with centre O. The area of the shaded region is

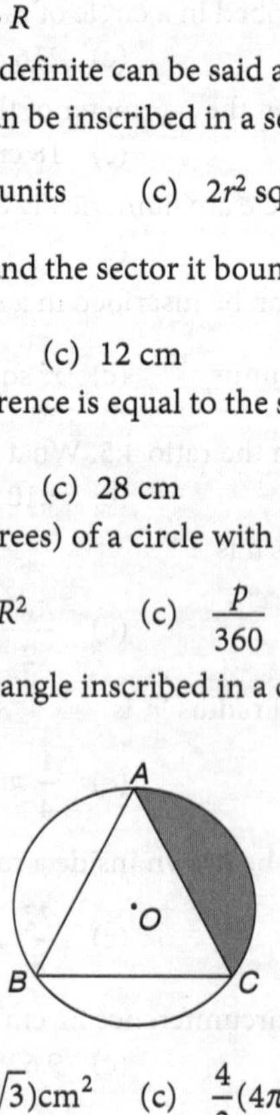

(a) $\frac{5}{3}(5\pi - 3\sqrt{3})\text{cm}^2$ (b) $\frac{2}{3}(2\pi - \sqrt{3})\text{cm}^2$ (c) $\frac{4}{3}(4\pi - 3\sqrt{3})\text{cm}^2$ (d) $\frac{7}{3}(7\pi - 3\sqrt{3})\text{cm}^2$

Fill in the Blanks

Complete the following statements with appropriate answer (s) in the blank space (s):

23. The circumference of a circle is equal to the sum of the circumferences of two circles having diameter 34 cm and 28 cm. The radius of the new circle is ________.

24. If the area of a sector of a circle is $\frac{7}{20}$ of the area of the circle, then the sector angle is ________.

25. As the radius of circle increases, the area of circle ________.

26. The area of sector of a circle with radius r and angle measuring θ is ________.

27. The area of the quadrant of a circle is ________.

28. The perimeter of a square whose diagonal is 20 cm is ________.

True/False

State whether the following statements are True/False and also justify:

29. If the length of an arc of a circle of radius r is equal to that of an arc of a circle of radius $2r$, then the angle of the corresponding sector of the first circle is double the angle of the corresponding sector of the other circle. Is this statement false? Why? **[NCERT Exemplar]**

Assertion–Reason

The following questions consist of two statements – Assertion (A) and Reason (R). Answer these questions by selecting the appropriate option given below.

(A) Both A and R are true and R is the correct explanation for A.
(B) Both A and R are true and R is not the correct explanation for A.
(C) A is true but R is false.
(D) A is false but R is true.

30. **Assertion (*A*):** The area of the largest triangle that is inscribed in a semicircle of diameter $2r$ unit is $2r^2$ sq. units.
Reason (*R*): Angle in a semicircle is a right angle.

31. **Assertion (*A*):** A chord AB of a circle of radius 20 cm subtends a right angle at the centre. Area of the minor segment formed is 114 sq. cm.
Reason (*R*): Area of minor segment = Area of circle – Area of sector

32. **Assertion (*A*):** In a rhombus of side 10 cm, one of the diagonals is 12 cm long. The length of the second diagonal is 16 cm.
Reason (*R*): The diagonals of a rhombus bisect each other at right angles.

33. **Assertion (*A*):** In a circle of radius 6 cm, the angle of the sector is 60°. The area of the sector is $18\frac{6}{7}$ cm^2.
Reason (*R*): Area of a circle with radius r is πr^2.

34. **Assertion (*A*):** The area covered by a 21-cm-long minute hand in 20 minutes is 462 cm^2.
Reason (*R*): Area of circle is πr^2.

35. **Assertion (*A*):** If the length of the diagonals of a rhombus are 24 cm and 10 cm, then the length of each side of the rhombus is 13 cm.
Reason (*R*): The diagonals of a rhombus bisect each other perpendicularly.

36. **Assertion (A):** The radius of the outer circle is 21 cm and the radius of the inner circle is 14 cm. Then, the area enclosed between two concentric circles is 770 cm^2.
Reason (R): If R and r are the outer and inner radii of a ring, then the area of the ring is $\pi(R^2 - r^2)$.

37. **Assertion (A):** The circumference of the base of a cone is 44 cm and the slant height is 25 cm. Then, its curved surface area is 550 cm^2.
Reason (R): Circumference of the base of a cone = $2\pi r$ and its curved surface area πrl.

Subjective Questions

38. In the figure, PQR is a right-angled triangle at P. Find the area of the shaded region if PR = 4 cm, RQ = 5 cm and I is centre of the circle in ΔPQR. **[CBSE 2009]**

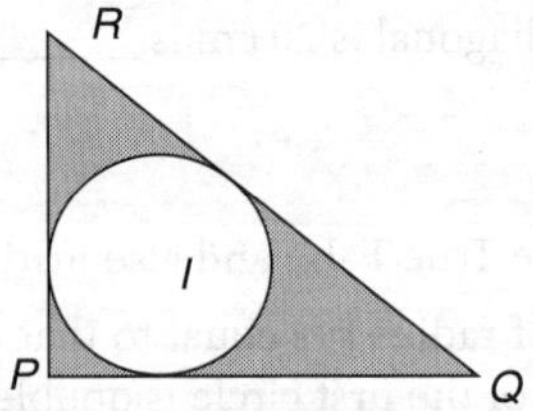

39. The cost of fencing a circular field at the rate of ₹24 per metre is ₹5280. The field is to be ploughed at the rate of ₹0.50 per m^2. Find the cost of ploughing the field. [Take $\pi = \frac{22}{7}$] **[NCERT]**

40. The radii of two circles are 19 cm and 9 cm, respectively. Find the radius of the circle which has a circumference equal to the sum of the circumferences of the two circles. **[NCERT]**

41. The figure depicts an archery target marked with five scoring regions from the centre outwards, as Gold, Red, Blue, Black and White. The diameter of the region representing Gold is 21 cm and each of the other bands is 10.5 cm wide. Find the area of each of the five scoring regions. **[NCERT]**

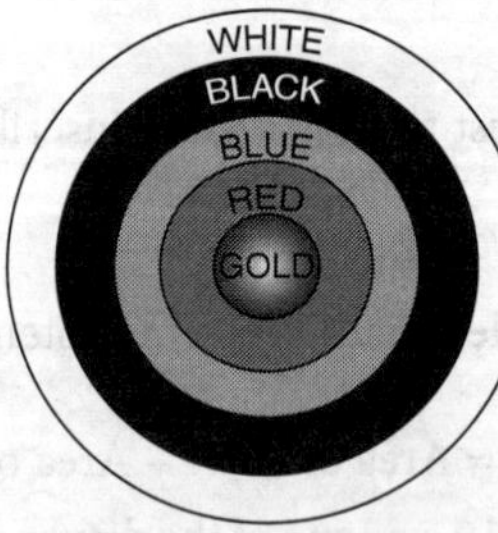

42. The cost of fencing a circular field at the rate of ₹36 per m is ₹11,880. The field is to be ploughed at the rate of ₹0.60 per m^2. Find the cost of ploughing the field. [Take π = 22/7]

43. A circular footpath of width 2 m is constructed at the rate of ₹20 per m^2 around a circular park of radius 1500 m. Find the total cost of construction of the footpath. [Take π = 3.14] **[CBSE 2012]**

44. If the area of a semicircular field is 30,800 sq m, find the perimeter of the field.

45. The outer and inner diameters of a circular ring are 34 cm and 32 cm, respectively. Then, find the area of the ring. **[CBSE 2011]**

46. The shaded area in the figure between the circumference of two concentric circles is 346.5 cm^2. The circumference of the inner circle is 88 cm. Calculate the radius of the outer circle. **[CBSE 2012]**

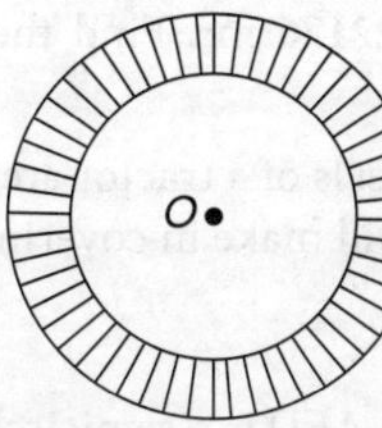

47. The ratio of the outer and inner circumference of a circular path is 23:22. If the path is 5 m wide, find the diameter of the inner circle. **[CBSE 2013]**

48. The diameter of a cycle wheel is 21 cm. How many revolutions will it make to travel 1.98 km? **[CBSE 2011]**

49. Two parallel lines touch the circle at points *A* and *B*, respectively. If the area of the circle is 25π cm^2, find the length of *AB*. **[CBSE 2011]**

50. A wire, when bent in the form of a square, encloses an area of 121 sq. cm. If the wire was bent in the form of a circle, find the area enclosed by the circle. $\left[\text{Take } \pi = \frac{22}{7}\right]$ **[CBSE 2011]**

51. With the vertices *A*, *B* and *C* of a triangle *ABC* as centres, arcs are drawn with radii 5 cm each, as shown in the figure. If *AB* = 14 cm, *BC* = 48 cm and *CA* = 50 cm, find the area of the shaded region. [Use $\pi = 3.14$] **[NCERT Exemplar]**

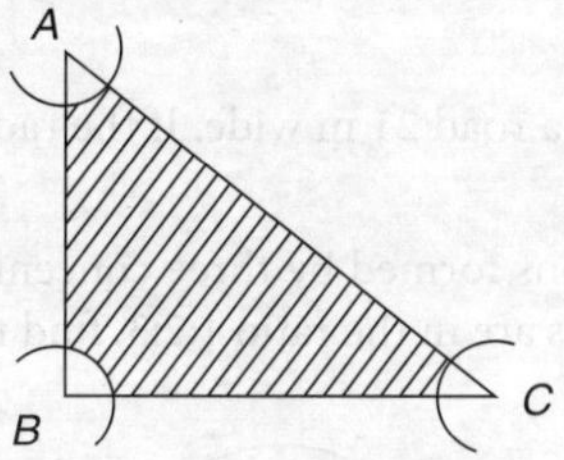

52. If the circumference of a circle and the perimeter of a square are equal, check whether it is correct that area of the circle > area of the square. **[NCERT Exemplar]**

53. The wheel of a motor cycle is of radius 35 cm. How many revolutions per minute must the wheel make so as to keep a speed of 66 km/h? **[NCERT Exemplar]**

54. What will be the perimeter of a quadrant of a circle of radius *r*? **[CBSE 2013]**

55. If the perimeter of a semicircular protractor is 36 cm, find its diameter. **[CBSE 2008]**

56. Find the radius of a circle whose circumference is equal to the sum of the circumference of two circles of radii 15 cm and 18 cm. **[NCERT Exemplar]**

57. If the circumference of two concentric circles forming a ring are 88 cm and 66 cm, find the width of the ring.

58. If the area of a circle is numerically equal to twice its circumference, find the diameter of the circle. **[CBSE 2013]**

59. Find the area of a quadrant of a circle whose circumference is 22 cm. **[NCERT, CBSE 2013]**

60. A garden roller has a circumference of 4 m. Find the number of revolutions it makes to move 40 m. **[CHSE 2011]**

61. The area of a circular playground is 22176 m^2. Find the cost of fencing this ground at the rate of ₹50 per m. **[NCERT Exemplar]**

62. The diameters of the front and rear wheels of a tractor are 80 cm and 2 m, respectively. Find the number of revolutions that the rear wheel will make in covering a distance in which the front wheel makes 1400 revolutions. **[NCERT Exemplar]**

63. Find the perimeter of the figure, where *AED* is a semicircle and *ABCD* is a rectangle. $\left[\text{Take } \pi = \frac{22}{7}\right]$ **[CBSE 2015]**

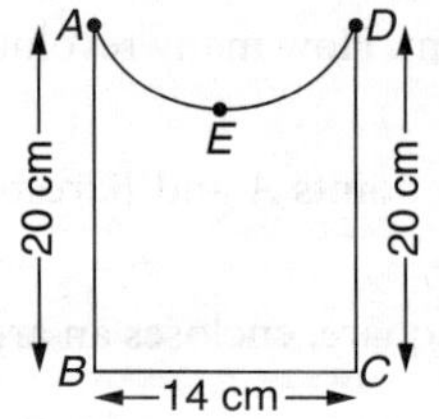

64. An athlete runs on a circular track of radius 49 m and covers a distance of 3080 m along its boundary. How many rounds has he taken to cover this distance? $\left[\text{Take } \pi = \frac{22}{7}\right]$ **[CBSE 2015]**

65. A road that is 7 m wide surrounds a circular park whose circumference is 352 m. Find the area of the road. **[CBSE 2012]**

66. A circular park is surrounded by a road 21 m wide. If the radius of the park is 105 m, find the area of the road. **[NCERT Exemplar]**

67. An archery target has three regions formed by three concentric circles, as shown in the figure. If the diameters of the concentric circles are in the ratio 1:2:3, find the ratio of the areas of the three regions. **[NCERT Exemplar]**

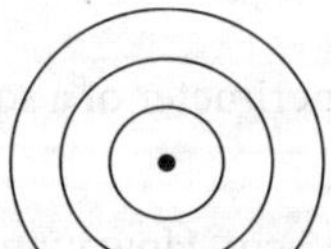

68. If three circles of radius '*a*' each are drawn such that each touches the other two, prove that the area included between them is equal to $\frac{4}{25}a^2$. [Take $\sqrt{3} = 1.73$ and $\pi = 3.14$]

69. Find the area of the sector of a circle with radius 4 cm and of angle 30°. Also find the area of the corresponding major sector. [Use $\pi = 3.14$] **[NCERT]**

70. Find the area of the segment *AYB*, shown in the figure, if the radius of the circle is 21 cm and $\angle AOB = 120°$. [Use $\pi = \frac{22}{7}$] **[NCERT]**

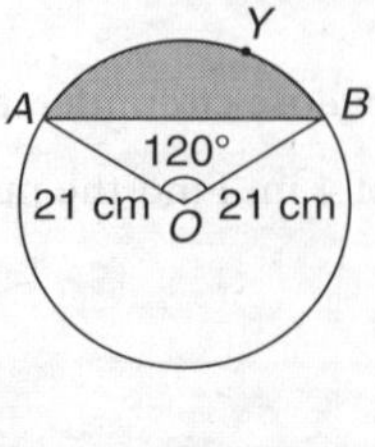

71. Find the area of a quadrant of a circle whose circumference is 44 cm. **[NCERT]**

72. The length of the minute hand of a clock is 14 cm. Find the area swept by the minute hand in 5 minutes. **[NCERT]**

73. A chord of a circle of radius 15 cm subtends an angle of 60° at the centre. Find the areas of the corresponding minor and major segments of the circle. [Use $\pi = 3.14$ and $\sqrt{3} = 1.73$] **[NCERT]**

74. A chord of a circle of radius 12 cm subtends an angle of 120° at the centre. Find the area of the corresponding segment of the circle. [Use $\pi = 3.14$ and $\sqrt{3} = 1.73$] **[NCERT]**

75. A horse is tied to a peg at one corner of a square-shaped grass field of side 15 m by means of a 5-m-long rope (see figure). Find **[NCERT]**

(i) The area of that part of the field in which the horse can graze.

(ii) The increase in the grazing area if the rope were 10 m long instead of 5 m. [Use $\pi = 3.14$]

76. A brooch is made with silver wire in the form of a circle with diameter 35 mm. The wire is also used in making 5 diameters which divide the circle into 10 equal sectors, as shown in the figure. Find:

(i) The total length of the silver wire required.

(ii) The area of each sector of the brooch. **[NCERT]**

77. A car has two wipers which do not overlap. Each wiper has a blade of length 25 cm sweeping through an angle of 115°. Find the total area cleaned at each sweep of the blades. **[NCERT]**

78. To warn ships for underwater rocks, a lighthouse spreads a red light over a sector of angle 80° to a distance of 16.5 km. Find the area of the sea over which the ships are warned. [Use $\pi = 3.14$] **[NCERT]**

79. The minute hand of a clock is 12 cm long. Find the area of the face of the clock described by the minute hand between 9 am and 9:35 am. **[CBSE 2012]**

80. The area of a sector of a circle of radius 36 cm is 54π cm^2. Find the length of the corresponding arc of the sector. **[CBSE 2011]**

81. A cow is tied with a rope of length 14 m at the corner of a rectangular field of dimensions 20 m × 16 m. Find the area of the field in which the cow can graze. **[NCERT Exemplar]**

82. Find the area of the minor segment of a circle of radius 14 cm, when the angle of the corresponding sector is 60°. **[NCERT Exemplar]**

83. A chord of a circle of radius 20 cm subtends an angle of 90° at the centre. Find the area of the corresponding major segment of the circle. [Take $\pi = 3.14$] **[NCERT Exemplar]**

84. A chord 10 cm long is drawn in a circle whose radius is $\sqrt{50}$ cm. Find the area of the segment. **[CBSE 2013]**

85. Find the area of the sector of central angle $x°$ of a circle with radius $4r$. **[CBSE 2011]**

86. If the area of a sector of a circle is $\frac{5}{18}$ of the area of that circle, find the central angle of the sector. **[CBSE 2015]**

87. Find the area of the sector of a circle of radius 5 cm, if the corresponding arc length is 3.5 cm. **[NCERT Exemplar]**

88. The area of a sector of a circle of radius 18 cm is 54π cm². Find the length of the corresponding arc of the sector.

89. The length of the minute hand of a clock is 5 cm. Find the area swept by the minute hand during the time period 6:05 am and 6:40 am. **[NCERT Exemplar]**

90. In the figure, a square of diagonal 8 cm is inscribed in a circle. Find the area of the shaded region. **[NCERT Exemplar]**

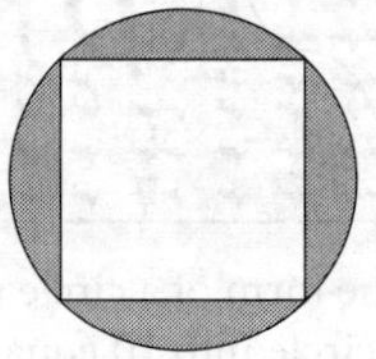

91. All the vertices of a rhombus lie on a circle. Find the area of the rhombus if the area of the circle is 1256 cm². **[NCERT Exemplar]**

92. Find the area of the largest circle that can be drawn inside the given rectangle of length 'a' cm and breadth 'a' cm ($a > b$). **[(NCERT Exemplar), CBSE 2015]**

93. In the figure, a circle is inscribed in a square of side 4 cm and another circle circumscribes the square. Prove that the area of the circumscribed circle is two times the area of the inscribed circle. **[NCERT Exemplar]**

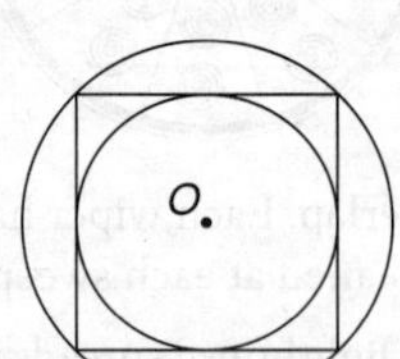

94. In the figure, arcs are drawn by taking vertices A, B and C of an equilateral triangle of side 10 cm, to intersect the sides BC, CA and AB at their respective mid-points D, E and F. Find the area of the shaded region. **[NCERT Exemplar]**

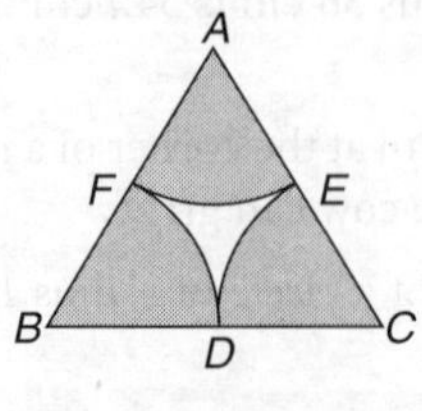

95. Find the area of the shaded region in the figure, where arcs drawn with centres *A*, *B*, *C* and *D* intersect in pairs at mid-points *P*, *Q*, *R* and *S* of the sides *AB*, *BC*, *CD* and *DA*, respectively of a square *ABCD* of side 12 cm. [Use $\pi = 3.14$] **[CBSE 2018]**

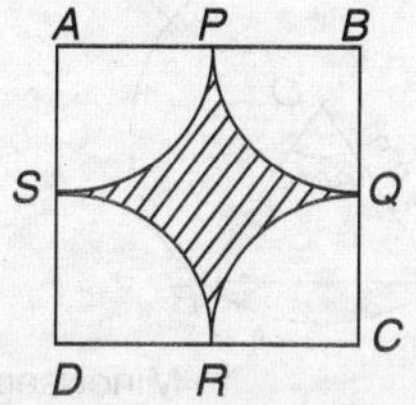

96. The area of a sector of central angle 200° of a circle is 770 cm^2. Find the length of the corresponding arc of this sector. **[NCERT Exemplar]**

97. A flower bed is laid in a square park of side 10 m, as shown (shaded). Find the area of the flower bed if the portions left out are the quadrants of a circle of the same radius. The diameter of the circle is equal to the side of the square. **[CBSE 2015]**

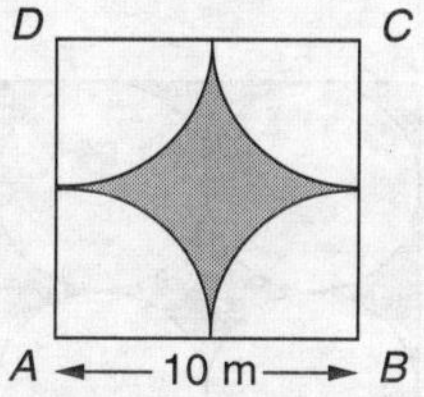

98. In the figure, *AC* = 24 cm, *BC* = 10 cm and *O* is the centre of the circle. Find the area of the shaded region. [Take $\pi = 3.14$] **[CBSE 2010]**

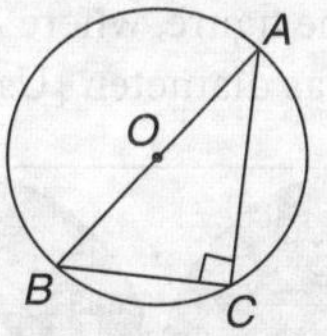

99. A ceiling fan has three wings, as shown in the figure. Find the length of arc described between two consecutive wings, where the length of each wing is 0.98 m. **[CBSE 2014]**

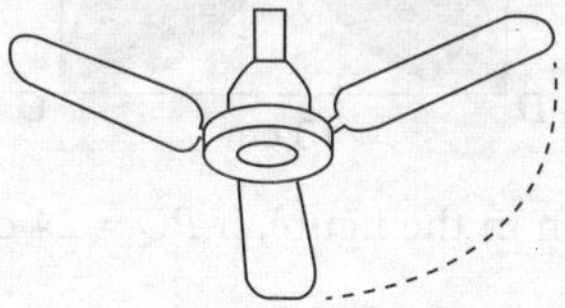

100. The radius of the circumcircle of a right-angled triangle is 3 cm and the altitude drawn to the hypotenuse is 2 cm long. Find the sum of the areas of the three disjoint segments of the circumcircle formed by the sides of this triangle. **[CBSE 2015]**

101. Find the difference of the areas of a sector of angle 120° and its corresponding major sector of a circle of radius 21 cm. **[NCERT Exemplar]**

102. The central angles of two sectors of circles of radii 7 cm and 21 cm are, respectively, 120° and 40°. Find the area of the two sectors as well as the lengths of the corresponding arcs. What do you observe? **[NCERT Exemplar]**

103. Find the area of the minor segment of a circle of radius 14 cm when the angle of the corresponding sector is 60°. **[NCERT Exemplar]**

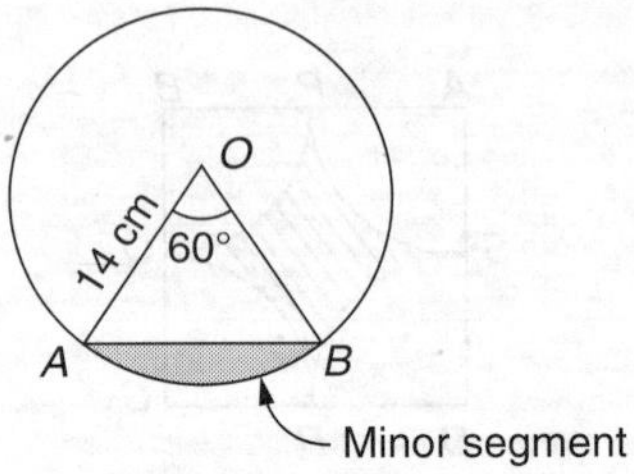

104. Find the area of the segment of a circle of radius 12 cm whose corresponding sector has a central angle of 60°. [Take $\pi = 3.14$]

105. Three circles each of radius 3.5 cm are drawn in such a way that each of them touches the other two. Find the area enclosed between these circles. **[NCERT Exemplar]**

106. Find the area of the shaded region in the figure, where *PQRS* is a square of side 14 cm. **[NCERT]**

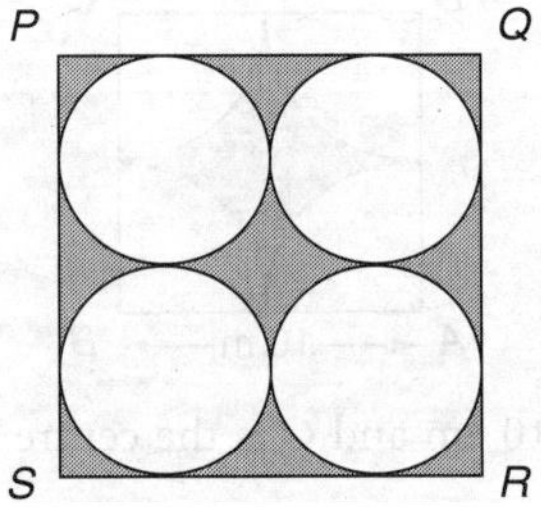

107. Find the area of the shaded design in the figure, where *ABCD* is a square of side 10 cm and semicircles are drawn with each side of the square as diameter. [Use $\pi = 3.14$] **[NCERT]**

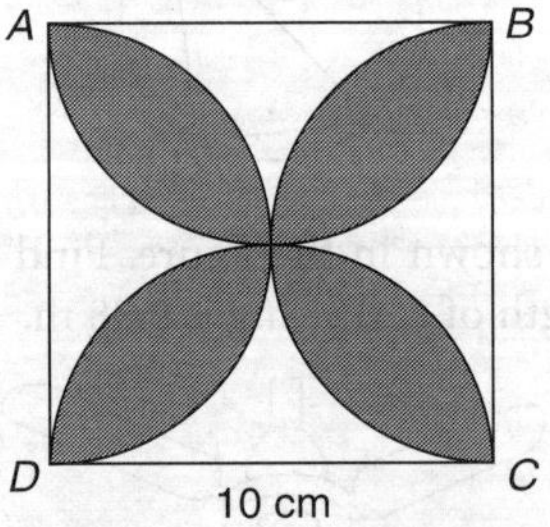

108. Find the area of the shaded region in the figure, if $PQ = 24$ cm, $PR = 7$ cm and O is the centre of the circle. **[NCERT]**

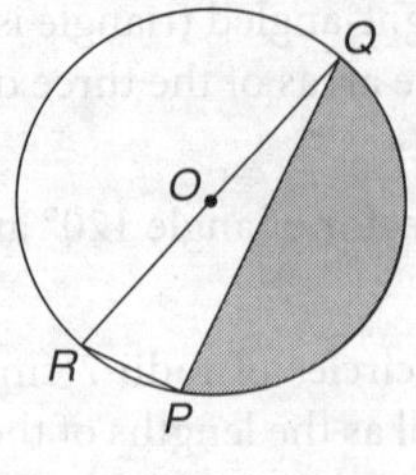

109. Find the area of the shaded region in the figure if the radii of the two concentric circles with centre O are 7 cm and 14 cm, respectively, and $\angle AOC = 40°$. **[AI 2016, NCERT]**

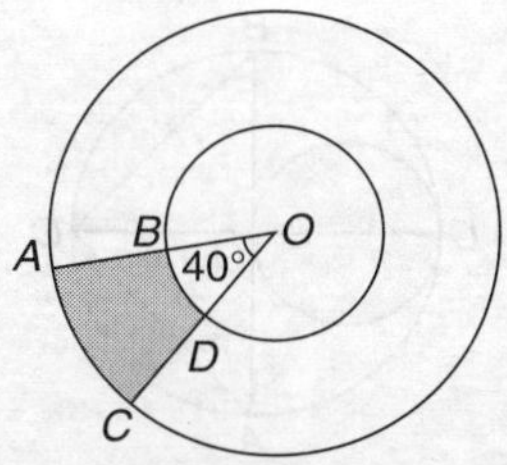

110. Find the area of the shaded region in the figure, where a circular arc of radius 6 cm has been drawn with vertex O of an equilateral triangle OAB of side 12 cm as centre. **[NCERT]**

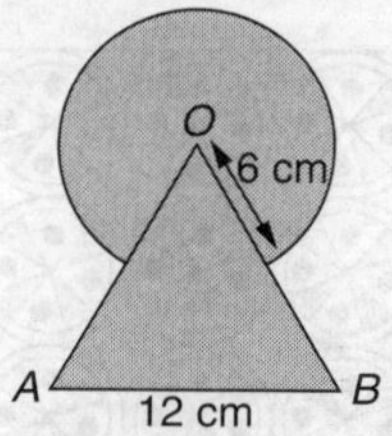

111. A memento is made, as shown in the figure. Its base $PBCR$ is silver plated from the front at the rate of ₹20 per cm^2. Find the total cost of silver plating. $\left[\text{Take } \pi = \frac{22}{7}\right]$

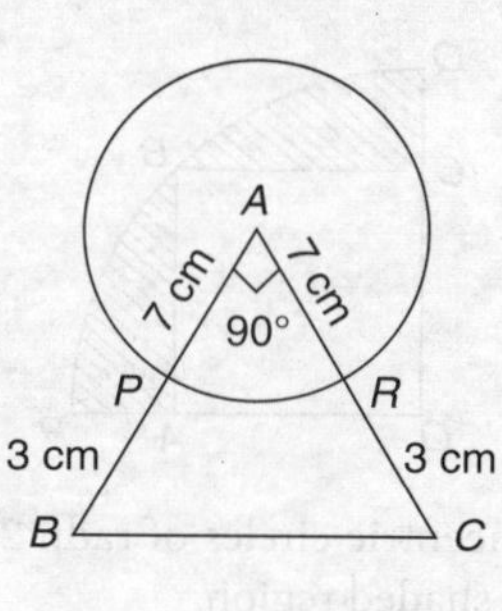

112. In the figure, $ABCD$ is a square of side 14 cm. With centres A, B, C and D, four circles are drawn such that each circle touches externally two of the remaining three circles. Find the area of the shaded region. **[NCERT]**

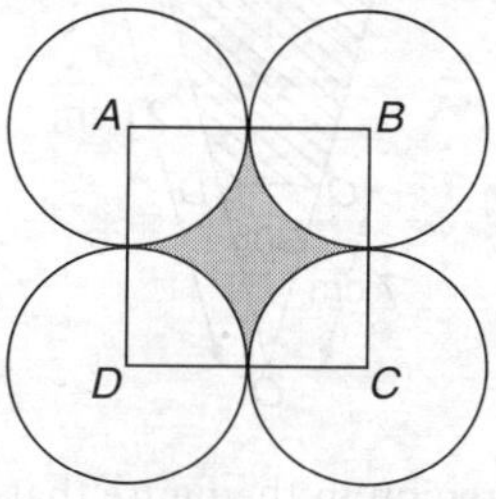

113. In the figure, AB and CD are two diameters of a circle (with centre O) perpendicular to each other and OD is the diameter of the smaller circle. If $OA = 7$ cm, find the area of the shaded region. **[NCERT]**

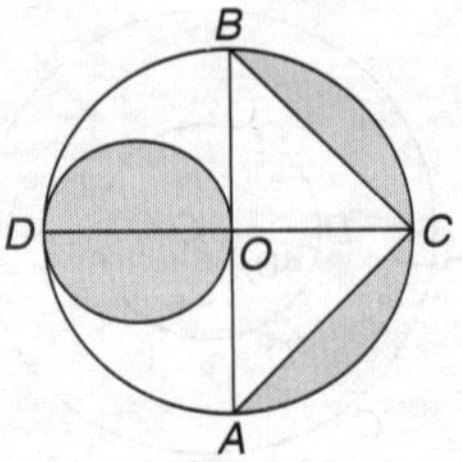

114. On a square handkerchief, nine circular designs each of radius 7 cm are made (see figure). Find the area of the remaining portion of the handkerchief. **[NCERT]**

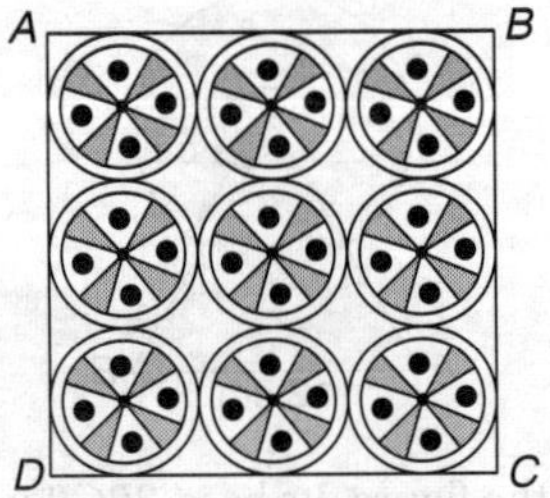

115. In the figure, a square $OABC$ is inscribed in a quadrant $OPBQ$. If $OA = 30$ cm, find the area of the shaded region. [Use $\pi = 3.14$] **[Delhi 2014]**

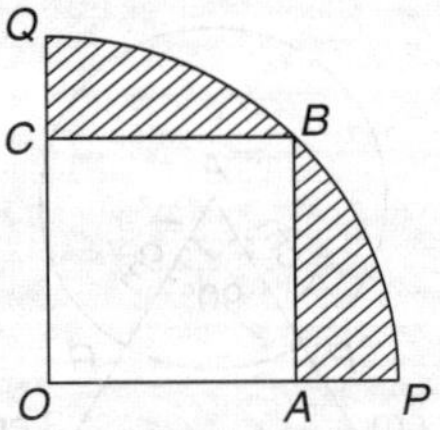

116. AB and CD are the arcs of two concentric circles of radii 21 cm and 7 cm and centre O (see figure). If $\angle AOB = 30°$, find the area of the shaded region. **[NCERT]**

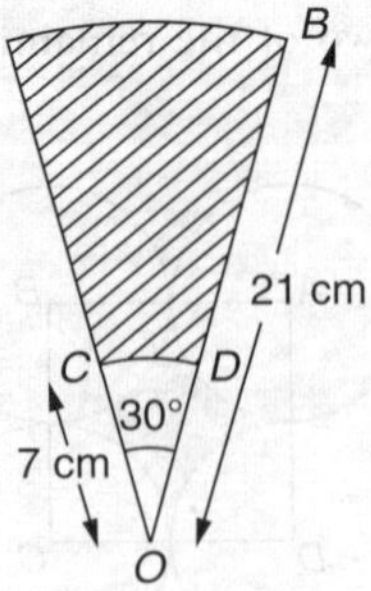

117. Calculate the area of the designed region in the figure that is common between the two quadrants of circles of radius 8 cm each. **[NCERT]**

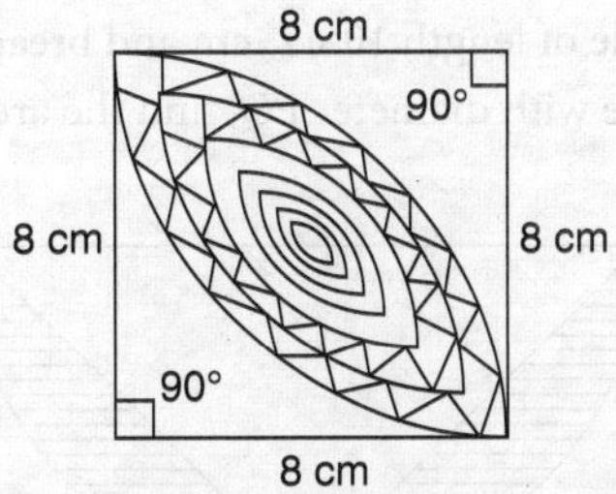

118. In the figure, *ABCD* is a square of side 7 cm and *A, B, C* and *D* are centres of equal circles touching externally in pairs. Find the area of the shaded region.

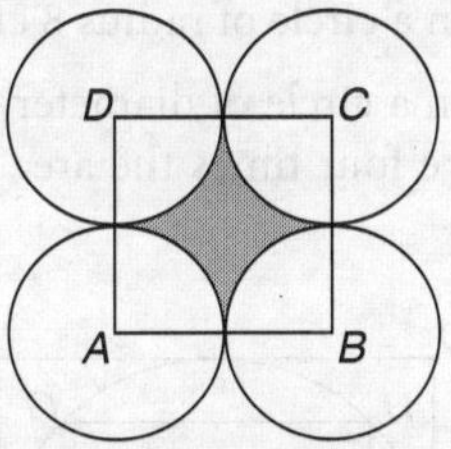

119. In the figure, diameter *AB* is 12 cm long. *AB* is trisected at points *P* and *Q*. Find the area of the shaded region. [Take $\pi = 3.14$]

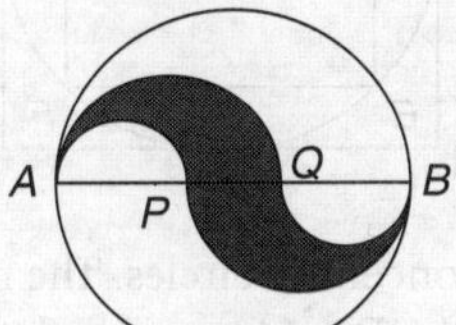

120. Find the area of the shaded region in the figure.

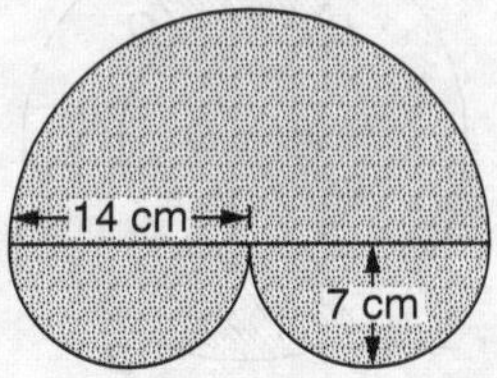

121. There are three semicircles, *A, B* and *C* having diameter 3 cm each and another semi-circle *E* having a circle *D* with diameter 45 cm, as shown in the figure. Find the area of the shaded region.

[AI 2017, CBSE 2013]

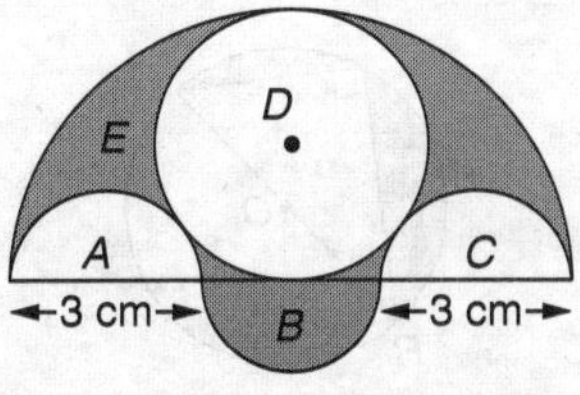

122. In the figure, $PQRS$ is a rectangle of length $10\sqrt{2}$ cm and breadth $5\sqrt{2}$ cm. If PEO is an isosceles triangle inscribed in the semicircle with diameter PQ, find the area of the shaded region. **[CBSE 2015]**

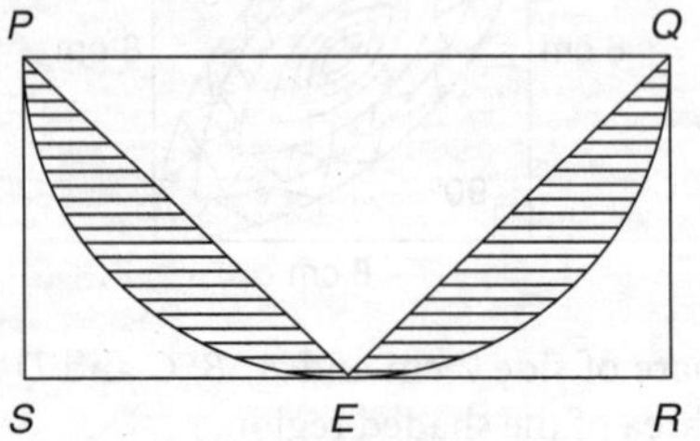

123. Find the area of a square inscribed in a circle of radius 8 cm. **[NCERT Exemplar]**

124. In the figure, a square is inscribed in a circle of diameter d and another square is circumscribing the circle. Is the area of the outer square four times the area of the inner square? Give reasons for your answer. **[NCERT Exemplar]**

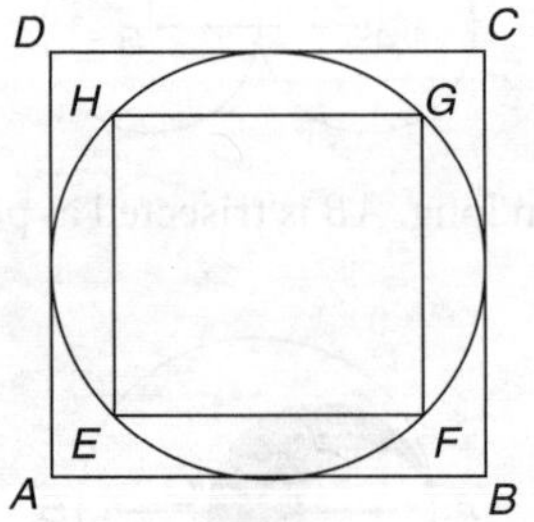

125. In the figure, O is the centre of the concentric circles. The radius of the inner circle is half the radius of the outer circle. If $\angle DOE = 135°$ and $OD = 14$ cm, calculate the area of the shaded region.

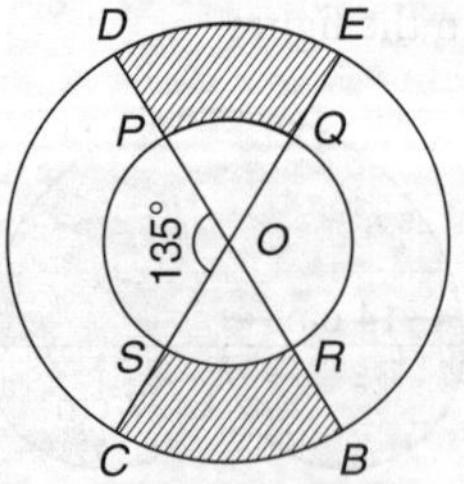

126. Find the area of the shaded region in the figure, if $PR = 12$ cm, $PQ = 5$ cm and O is the centre of the circle. [Take $\pi = \frac{22}{7}$] **[CBSE 2008, 2009]**

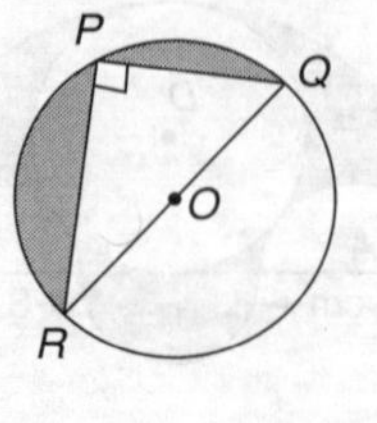

127. Calculate the area of the shaded region shown in the figure that is common between two quadrants of a circle of radius 6 cm each.

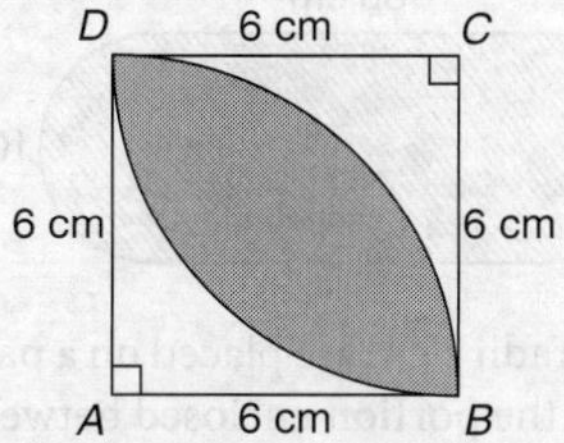

128. If the figure, $OACBO$ represents a quadrant of a circle of radius 7 cm with centre O. If $OD = 5$ cm, find the area of the shaded region.

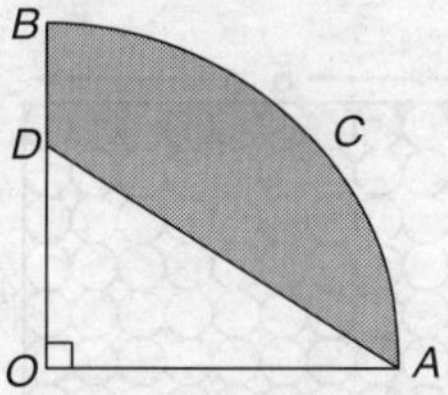

129. In the figure, arcs have been drawn with radius 14 cm each and with centres P, O and R. Find the area of the shaded region. **[NCERT Exemplar]**

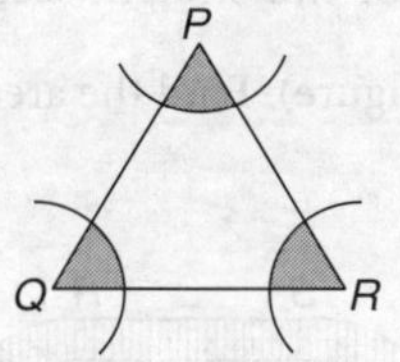

130. Find the area of the shaded region in the figure, if $BC = BD = 8$ cm, $AC = AD = 15$ cm and O is the centre of the circle. [Take $\pi = 3.14$]

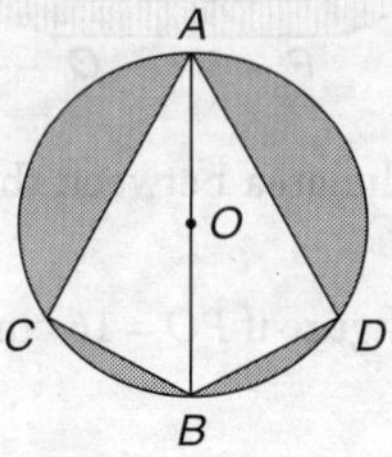

131. In the figure, ABC is a quadrant of a circle of radius 10 cm and a semicircle is drawn with BC as diameter. Find the area of the shaded region.

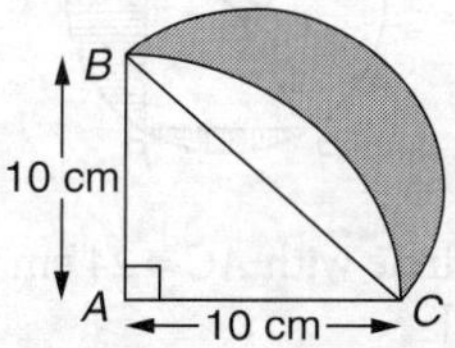

132. Find the area of the flower bed (with semicircular ends) shown in the figure.

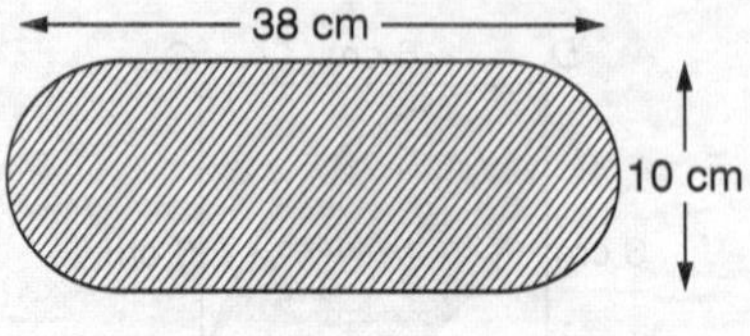

133. Four circular cardboard pieces of radii 7 cm are placed on a paper in such a way that each piece touches other two pieces. Find the area of the portion enclosed between these pieces. **[NCERT Exemplar]**

134. The floor of a room is of dimensions 5 m × 4 m and it is covered with circular tiles of diameter 50 cm each, as shown in the figure. Find the area of the floor that remains uncovered with tiles. [Take $\pi = 3.14$] **[NCERT Exemplar]**

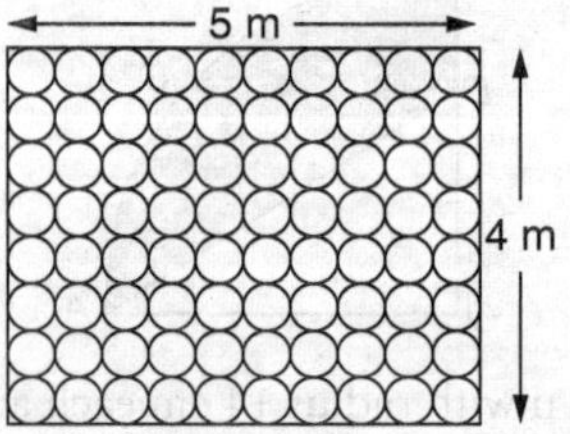

135. A 21-m-wide athletic track consists of two straight sections 150 m long joining semi-circular ends whose diameters are 84 m each (see figure). Find the area of the track. $\left[\text{Take } \pi = \frac{22}{7} \text{ and } \sqrt{3} = 1.73\right]$

[CBSE 2015]

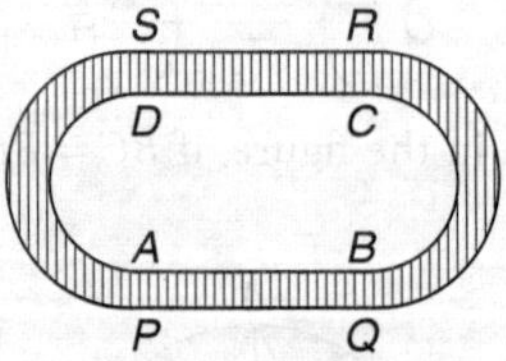

136. The side of a square is 10 cm. Find the area between the inscribed and circumscribed circles of the square. **[CBSE 2018 (C)]**

137. Find the area of shaded region in the figure if $PQ = 16$ cm, $PR = 12$ cm and O is the centre of the circle. [Take $\pi = 3.14$] **[CBSE 2011]**

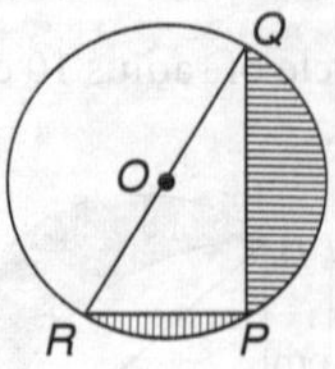

138. In the figure, O is the centre of the circle with $AC = 24$ cm, $AB = 7$ cm and $\angle BOD = 90°$. Find the area of the shaded region. [Take $\pi = 3.14$] **[AI 2012]**

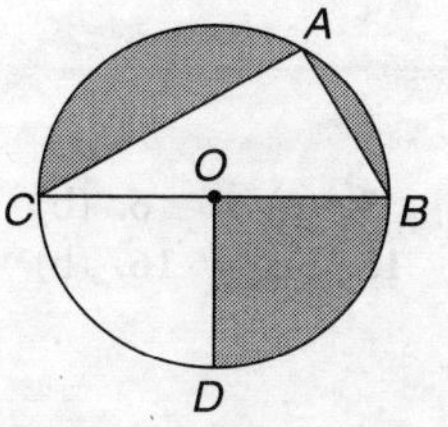

139. The sides of a triangular field are 15 m, 16 m and 17 m. With the three corners of field, a cow, a buffalo and a horse are tied separately with ropes of length 7 m each to graze in the field. Find the area of field which cannot be grazed on by the three animals. **[(NCERT Exemplar), CBSE 2009]**

140. Find the area of the shaded region in the figure, if *ABCD* is a rectangle with sides 8 cm and 6 cm and *O* is the centre of the circle. [Take $\pi = 3.14$] **[Delhi 2019]**

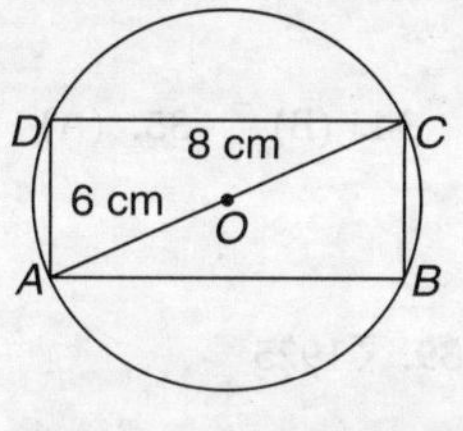

141. In the figure, a square *OABC* is inscribed in a quadrant *OPBQ*. If *OA* = 15 cm, find the area of the shaded region. [Use $\pi = 3.14$] **[Allahabad 2019]**

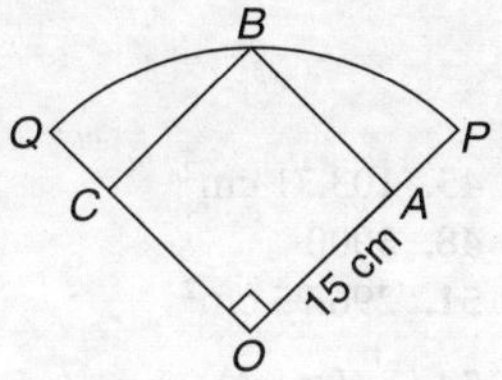

142. In the figure, *ABCD* is a trapezium with $AB \parallel DC$ and $\angle BCD = 30°$. If BGEC is a sector of a circle with centre *C* and *AB* = *BC* = 7 cm, *DE* = 4 cm and *BF* = 3.5 cm, find the area of the shaded region. $\left[\text{Use } = \pi \dfrac{22}{7}\right]$ **[AI 2017 C]**

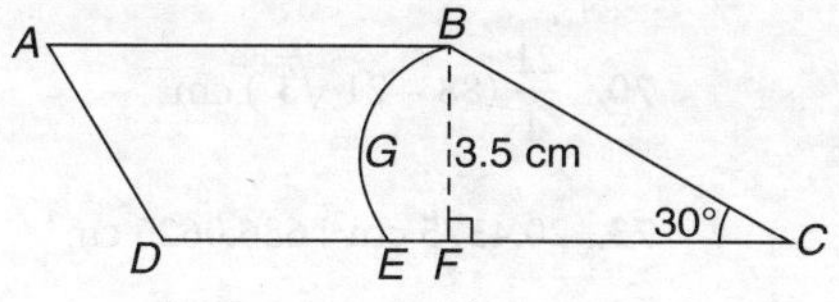

143. In the figure, *DABC* is a right-angled triangle in which $\angle A$ is 90°. Semicircles are drawn on *AB, AC* and *BC* as diameters. Find the area of the shaded region. **[AI 2017]**

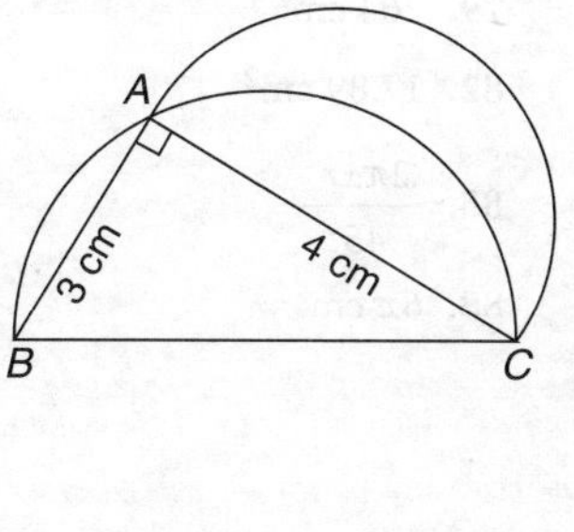

Answers

Multiple Choice Questions

1. (c)	**2.** (a)	**3.** (c)	**4.** (a)	**5.** (c)	**6.** (b)	**7.** (a)	**8.** (c)	**9.** (c)	**10.** (c)
11. (c)	**12.** (b)	**13.** (d)	**14.** (b)	**15.** (a)	**16.** (b)	**17.** (a)	**18.** (a)	**19.** (b)	**20.** (c)
21. (d)	**22.** (c)								

Fill in the Blanks

23. 31 cm **24.** 126° **25.** Increases **26.** $\frac{\pi r^2\theta}{360}$ **27.** $\frac{1}{4}\pi r^2$ **28.** $40\sqrt{2}$ cm

True/False

29. True

Assertion–Reason

30. (D) **31.** (C) **32.** (A) **33.** (B) **34.** (B) **35.** (A) **36.** (A) **37.** (A)

Subjective Questions

38. $\frac{20}{7}$ cm^2

39. ₹1925

40. 28 cm

41. Gold = 346.5 cm^2
Red = 1039.5 cm^2
Blue = 1732.5 cm^2
Black = 2425.5 cm^2
White = 3118.5 cm^2

42. ₹51975

43. ₹377051.2

44. 720 cm

45. 103.71 cm^2

46. 175 cm

47. 220 cm

48. 3000

49. 10 cm

50. 154 cm^2

51. 296.75 cm^2

52. Yes

53. 500 m

54. $\frac{r}{2}(\pi+4)$

55. 14 cm

56. 33 cm

57. 3.5 cm

58. 8 units

59. 9.625 cm^2

60. 10

61. ₹26400

62. 560

63. 76 cm.

64. 10

65. 2618 m^2

66. 15246 cm^2

67. 1 : 3 : 5

69. 46.1 cm^2

70. $\frac{21}{4}(88-21\sqrt{3})$ cm^2

71. $\frac{77}{2}$ cm^2

72. $\frac{154}{3}$ cm^2

73. 20.4375 cm^2; 686.0625 cm^2

74. 88.44 cm^2

75. (i) 19.625 m^2 (ii) 58.875 m^2

76. (i) 285 mm (ii) $\frac{385}{4}$ mm^2

77. $\frac{158125}{126}$ cm^2

78. 189.97 cm^2

79. 264 cm^2

80. 3π cm.

81. 154 m^2

82. 17.89 cm^2

83. 1142 cm^2

84. 14.28 cm^2

85. $\frac{2\pi x r^2}{45°}$

86. 100°

87. 8.75 cm^2

88. 6π cm

89. $45\frac{5}{6}$ cm^2

90. $(16\pi - 32)\text{cm}^2$ **91.** 800 cm^2 **92.** $\frac{b^2\pi}{4}\text{cm}^2$

94. 39.29 cm^2 **95.** 30.96 cm^2 **96.** $73\frac{1}{3}$ cm **97.** 21.4 cm^2

98. 145.33 cm^2 **99.** 2.05 m **100.** 22.29 cm^2

101. 462 cm^2 **102.** $\frac{44}{3}$ cm **103.** $(102.6 + 49\sqrt{3})\text{cm}^2$

104. $(24\pi - 36\sqrt{3})$ cm^2 **105.** 1.97 cm^2 **106.** 42 cm^2

107. 57 cm^2 **108.** $\frac{4523}{28}$ cm^2

109. $\frac{154}{3}$ cm^2 **110.** $\left(\frac{660}{7} + 36\sqrt{3}\right)$ cm^2 **111.** ₹230

112. 42 cm^2 **113.** 66.5 cm^2 **114.** 378 cm^2

115. 513 cm^2 **116.** $\frac{308}{3}$ cm^2 **117.** $\frac{256}{7}$ cm^2

118. 10.5 cm^2 **119.** 37.68 $\pi\,\text{cm}^2$ **120.** 462 cm^2

121. $\frac{63}{16}\pi$ cm^2 **122.** $\frac{200}{7}$ cm^2 **123.** 128 cm^2

124. No **125.** $\frac{147}{4}\pi$ cm^2 **126.** 36.39 cm^2

127. 20.57 cm^2 **128.** 21 cm^2 **129.** 308 cm^2

130. 106.87 cm^2 **131.** 50 cm^2 **132.** $(280 + 25\pi)$ cm^2

133. 42 cm^2 **134.** 4.3 m^2 **135.** 10458 cm^2

136. $25\pi\,\text{cm}^2$ **137.** 61 cm^2 **138.** 283.97 cm^2

139. $(24\sqrt{21} - 77)$ m^2 **140.** 30.5 cm^2 **141.** 128.25 cm^2

142. 18.67 cm^2 **143.** 6 cm^2

Areas Related to Circles (Basic)

SELF-ASSESSMENT TEST

Time allowed: 90 min. Max. Marks: 40

General Instructions

1. *All questions are compulsory. There are 20 questions in all.*
2. *This question paper has four sections: Section A, Section B, Section C and Section D.*
3. *Section A contains 10 questions of one mark each, Section B contains three questions of two marks each, Section C contains four questions of three marks each, and Section D contains three questions of four marks each.*

Section A [1 Mark Each]

Direction (Q. Nos. 1–5): These are Multiple Choice Questions. Select the most appropriate answer for the given questions.

1. If the diameter of a circle is 7 cm, then the perimeter of the semicircle is
   (a) 36 cm (b) 7π (c) 18 cm (d) 14π
2. The area of a quadrant of a circle of circumference 22 cm is (Take $\pi = 22/7$)
   (a) $3.05\ cm^2$ (b) $3.5\ cm^2$ (c) $9.625\ cm^2$ (d) $35.5\ cm^2$
3. If the circumference of a circle is 44 cm, then the area of the circle is
   (a) $276\ cm^2$ (b) $44\ cm^2$ (c) $176\ cm^2$ (d) $154\ cm^2$
4. The area of the quadrant of a circle of radius '*r*' is
   (a) $\frac{1}{2}\pi r^2$ (b) πr (c) $\frac{1}{4}\pi r^2$ (d) $\left(\frac{\pi}{2} + 2r\right)$
5. The area of a sector of angle *p* (in degrees) of a circle with radius *R* is
   (a) $\frac{p}{180} \times 2\pi R$ (b) $\frac{p}{180} \times \pi R^2$ (c) $\frac{p}{360} \times 2\pi R$ (d) $\frac{p}{720} \times 2\pi R^2$
6. As the radius of the circle increases, the area of the circle ____________.
7. The area of the quadrant of a circle is ____________.
8. If the perimeter of a semicircular protractor is 36 cm, find its diameter.
9. A road that is 7 m wide surrounds a circular park whose circumference is 352 m. Find the area of the road.
10. The perimeter of a sector of a circle of radius 5.3 cm is 16.6 cm. Find the area of the sector.

Section B [2 Marks Each]

11. The diameter of each wheel of a bus is 140 cm. How many revolutions per minute must a wheel make in order to move at a speed of 66 km per hour?
12. In the figure, *ABCD* is a square of side 14 cm. Find the area of the shaded region.

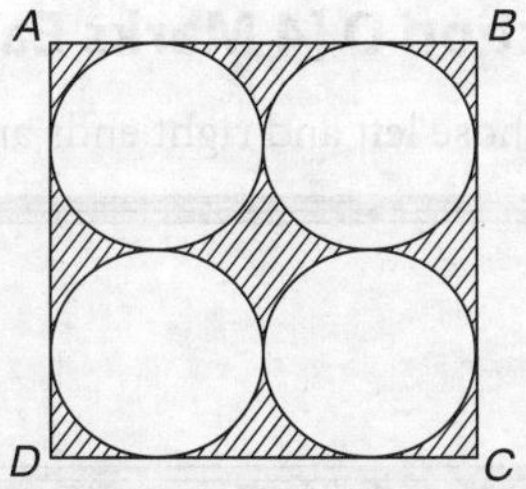

13. If the area of a circle is numerically equal to twice its circumference, find the diameter of the circle.

Section C [3 Marks Each]

14. In the figure, *ABCD* is a square of side 7 cm. *DPBA* and *DQBC* are quadrants of circles, each of radius 7 cm. Find the area of the shaded region. [Use = $\pi = \frac{22}{7}$]

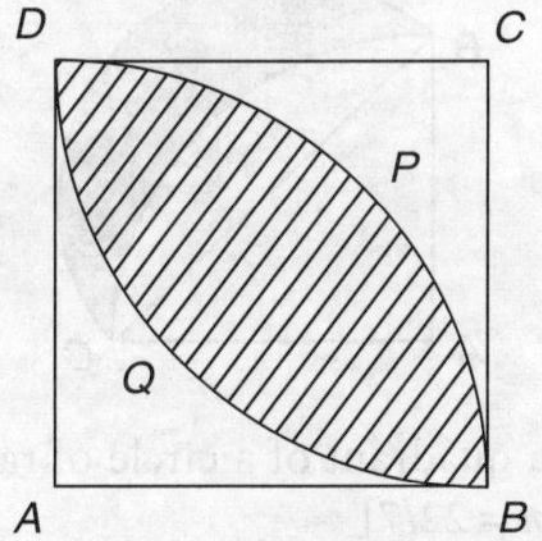

15. A wire when bent in the form of a square encloses an area of 121 sq. cm. If the wire was bent in the form of a circle, find the area enclosed by the circle. $\left[\text{Take } \pi = \frac{22}{7}\right]$

16. The cost of fencing a circular field at the rate of ₹ 36 per m is ₹ 11,880. The field is to be ploughed at the rate of ₹ 0.60 per m^2. Find the cost of ploughing the field. [Take $\pi = 22/7$]

17. A horse is tied to a peg at one corner of a square-shaped grass field of side 15 m by means of a 5-m-long rope (see figure). Find

(i) The area of that part of the field in which the horse can graze

(ii) The increase in the grazing area if the rope were 10 m long instead of 5 m [Use $\pi = 3.14$]

Section D [4 Marks Each]

18. The figure depicts a racing track whose left and right ends are semicircular.

The distance between the two inner parallel line segments is 60 m and they are each 106 m long. If the track is 10 m wide, find:
(i) The distance around the track along its inner edge
(ii) The area of the track

19. In the figure, ABC is the quadrant of a circle of radius 14 cm and a semicircle is drawn with BC as diameter. Find the area of the shaded region.

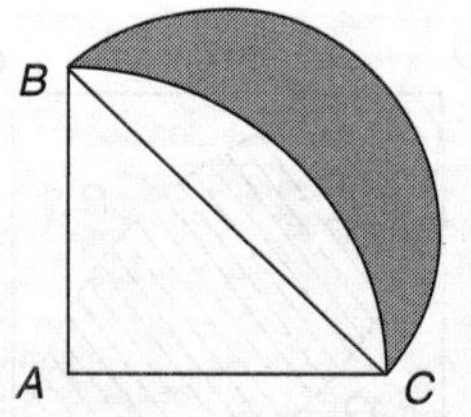

20. In the figure, $AOBCA$ represents a quadrant of a circle of radius 3.5 cm with centre O. Calculate the area of the shaded portion. [Take $\pi = 22/7$]

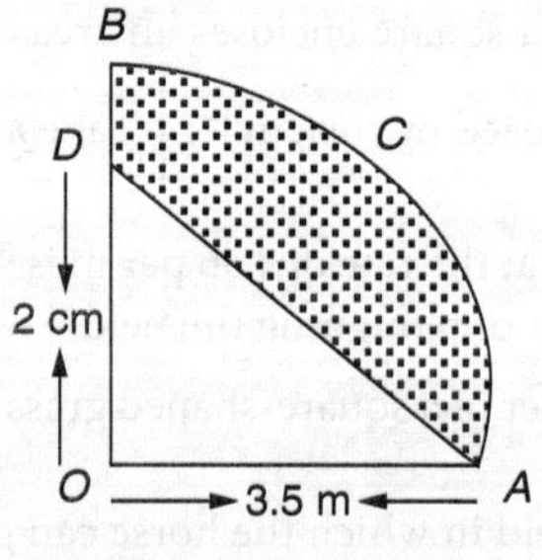

Answers

1. (c)	2. (c)	3. (d)	4. (c)	5. (d)
6. Increase	7. $\frac{1}{4}\pi r^2$	8. 14 cm	9. 2618 cm^2	10. 15.9 cm
11. 250 revolutions	12. 42 cm^2	13. 8 units	14. 28 cm^2	15. 154 cm^2
16. ₹51,975	17. (i) 19.625 m^2, (ii) 58.875 m^2		18. (i) 400.4 m, (ii) 4320 m^2	
19. 98 m^2	20. 6.125 cm^2			

Areas Related to Circles (Standard)

SELF-ASSESSMENT TEST

Time allowed: 90 min. Max. Marks: 40

General Instructions

1. *All questions are compulsory. There are 20 questions in all.*
2. *This question paper has four sections: Section A, Section B, Section C and Section D.*
3. *Section A contains 10 questions of one mark each, Section B contains three questions of two marks each, Section C contains four questions of three marks each, and Section D contains three questions of four marks each.*

Section A [1 Mark Each]

Direction (Q. Nos. 1–5): These are Multiple Choice Questions. Select the most appropriate answer for the given questions.

1. The area of a circle whose diameter is d is

(a) $\frac{\pi d^2}{4}$ (b) πd^2 (c) $\frac{\pi d^2}{2}$ (d) πd

2. If the difference between the circumference and the radius of a circle is 37 cm, then using $\pi = \frac{22}{7}$, the circumference (in cm) of the circle is

(a) 154 (b) 44 (c) 14 (d) 1

3. If the sum of the circumferences of two circles with radii R_1 and R_2 is equal to the circumference of a circle of radius R, then

(a) $R_1 + R_2 = R$
(b) $R_1 + R_2 > R$
(c) $R_1 + R_2 < R$
(d) nothing definite can be said about the relation among R_1, R_2 and R.

4. The area of the largest triangle that can be inscribed in a semicircle of radius r units is

(a) r^2 sq. units (b) $\frac{1}{2} r^2$ sq. units (c) $2r^2$ sq. units (d) $\sqrt{2}\, r^2$ sq. units

5. The area of the largest circle that can be drawn inside a rectangle with sides 7×3.5 m is

(a) $\frac{77}{2}$ m^2 (b) 77 cm^2 (c) $\frac{77}{8}$ m^2 (d) $\frac{77}{4}$ m^2

6. The perimeter of a square whose diagonal is 20 cm is ________________ .

7. If the area of a sector of a circle is $\frac{7}{20}$ of the area of the circle, then the sector angle is ________________ .

8. AB and CD are, respectively, the arcs of two concentric circles of radii 21 cm and 7 cm and centre O (see figure). If $\angle AOB = 30°$, find the area of the shaded region.

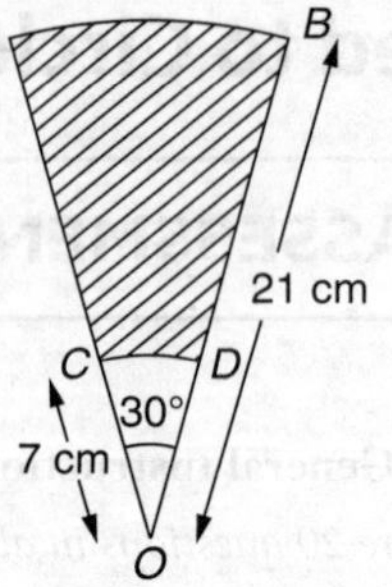

9. The length of the minute hand of a clock is 5 cm. Find the area swept by the minute hand during the time period 6 : 05 am to 6 : 40 am.

10. The wheel of a motor cycle is of radius 35 cm. How many revolutions per minute must the wheel make so as to maintain a speed of 66 km/h?

Section B [2 Marks Each]

11. With the vertices *A, B* and *C* of a triangle *ABC* as centres, arcs are drawn with radii 5 cm each, as shown in the figure. If $AB = 14$ cm, $BC = 48$ cm and $CA = 50$ cm, find the area of the shaded region. [Use $\pi = 3.14$] **[NCERT Exemplar]**

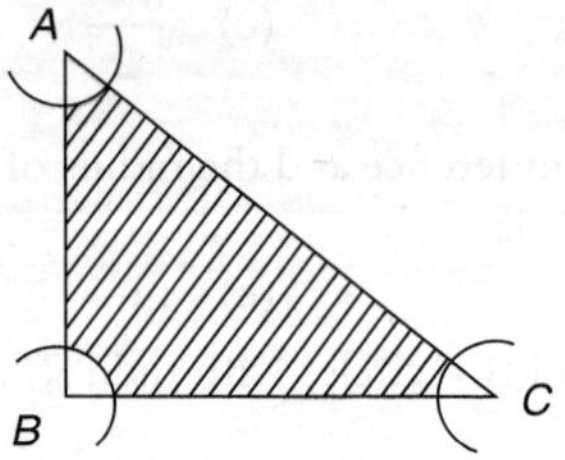

12. An archery target has three regions formed by three concentric circles, as shown in the figure. If the diameters of the concentric circles are in the ratio 1:2:3, find the ratio of the areas of the three regions. **[NCERT Exemplar]**

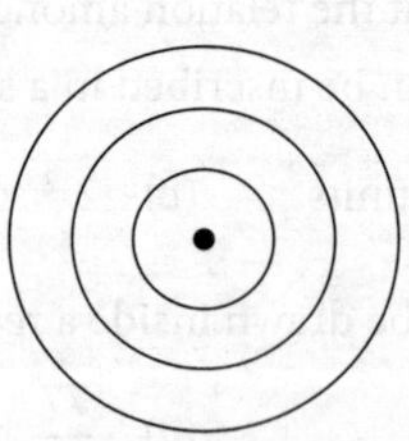

13. Find the area of the segment *AYB*, shown in the figure, if the radius of the circle is 21 cm and $\angle AOB = 120°$. [Use $\pi = \frac{22}{7}$]

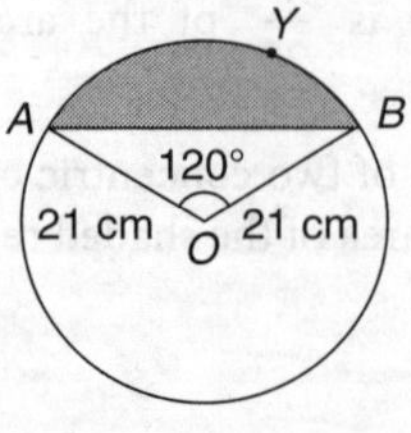

Section C [3 Marks Each]

14. Find the area of the shaded design in the figure, where $ABCD$ is a square of side 10 cm and semicircles are drawn with each side of the square as diameter. [Use $\pi = 3.14$]

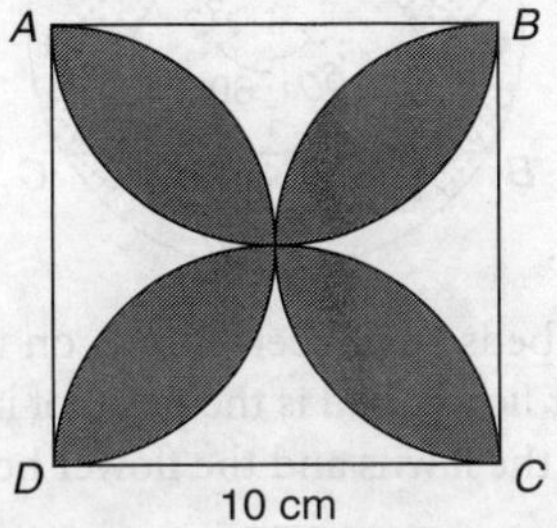

15. There are three semicircles, A, B and C having a diameter of 3 cm each and another semicircle E having a circle D with diameter 45 cm, as shown in the figure. Find the area of the shaded region.

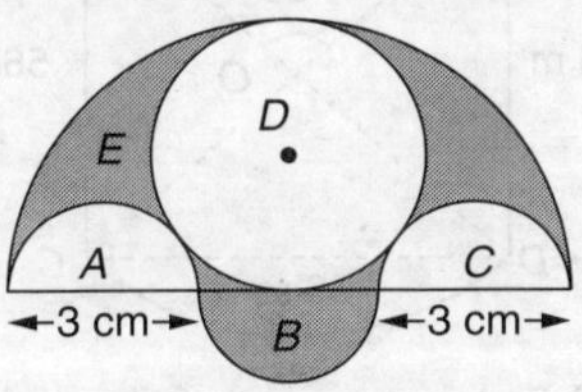

16. Calculate the area of the designed region in the figure that is common to the two quadrants of the circles of radius 8 cm each.

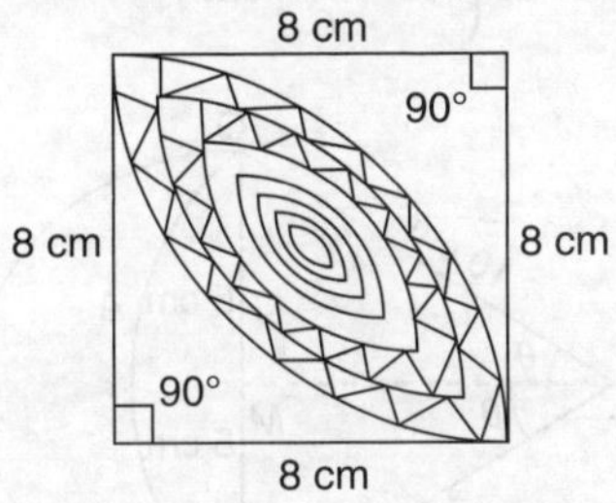

17. In the figure, O is the centre of the circle with $AC = 24$ cm, $AB = 7$ cm and $\angle BOD = 90°$. Find the area of the shaded region. [Take $\pi = 3.14$]

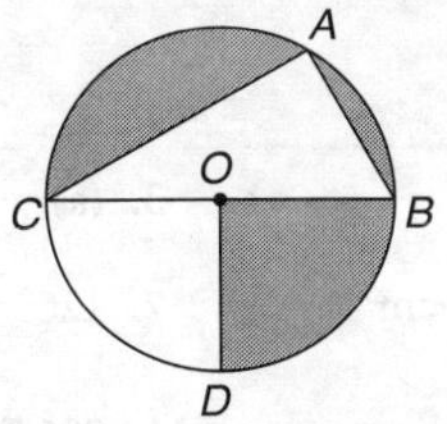

Section D [4 Marks Each]

18. On a circular table cover of radius 32 cm, a design is formed, leaving an equilateral triangle ABC in the middle, as shown in the figure. Find the area of the design (shaded region).

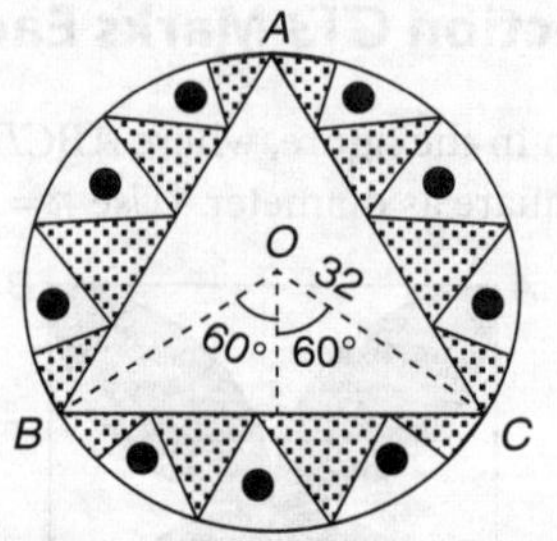

19. In the figure, two circular flower beds have been shown on two sides of a square lawn *ABCD* of side 56 m. If the centre of each circular flower bed is the point of intersection of the diagonals of the square lawn, find the sum of the areas of the lawns and the flower beds.

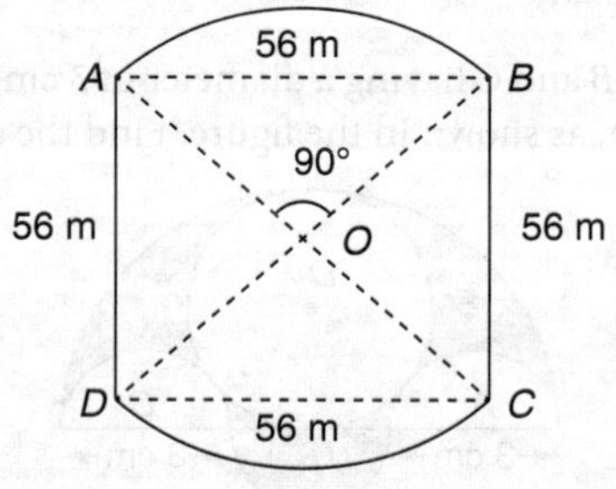

20. The figure shows two arcs, *A* and *B*. Arc *A* is part of the circle with centre *O* and radius *OP*. Arc *B* is part of the circle with centre *M* and radius *PM*, where *M* is the mid-point of *PQ*. Show that the area enclosed by the two arcs is equal to $25\left(\sqrt{3}-\frac{\pi}{6}\right)$ cm².

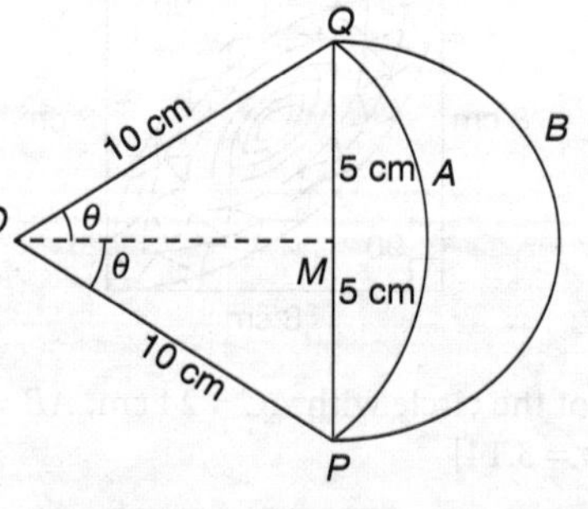

Answers

1. (a)	**2.** (b)	**3.** (a)	**4.** (a)
5. (c)	**6.** $40\sqrt{2}$ cm	**7.** 126°	**8.** $\frac{308}{3}$ cm²
9. $45\frac{5}{6}$ cm²	**10.** 500 m	**11.** 296.75 cm²	**12.** 1:3:5
13. $\frac{21}{4}(88-21\sqrt{3})$ cm²	**14.** 57 cm²	**15.** $\frac{63}{16}\pi$ cm²	**16.** $\frac{256}{7}$ cm²
17. 283.97 cm²	**18.** $\left\{\frac{22528}{7}-768\sqrt{3}\right\}$ cm²	**19.** 4032 m²	

CHAPTER

13 Surface Areas and Volumes

Key Concepts

- Problems based on surface area of figures such as cube, cuboid, right circular cylinder, right circular cone, sphere and hemisphere
- Problems based on surface area of combined figures
- Problems based on volume of figures such as cube, cuboid, right circular cylinder, right circular cone, sphere and hemisphere
- Volume of combined figures
- Problems based on melting, change in shape, embankment, etc.
- Problems based on frustum – CSA, TSA, volume, change in shape, etc.

Useful Formulae

S. No.	Figure	Lateral Surface Area	Total Surface Area	Volume	Diagonal	Shape
1.	**Cube**	$4l^2$	$6l^2$	l^3	$\sqrt{3}l$	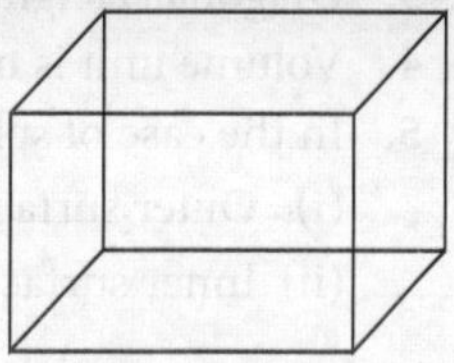
2.	**Cuboid**	$2h(l+b)$	$2(lh+bh+lb)$	lbh	$\sqrt{l^2+b^2+h^2}$	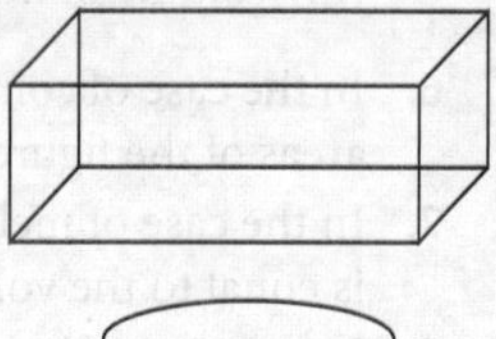
3.	**Right Circular Cylinder**	$2\pi rh$	$2\pi r(h+r) = 2\pi r^2 + 2\pi rh$	$\pi r^2 h$	_____	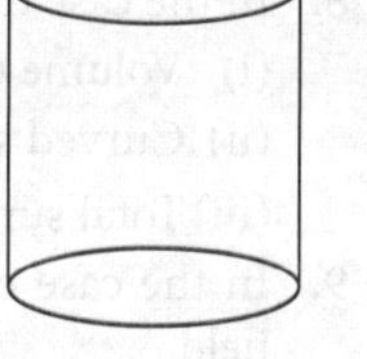
4.	**Right Circular Cone**	πrl	$\pi r(l+r)$	$\frac{1}{3}\pi r^2 h$	_____	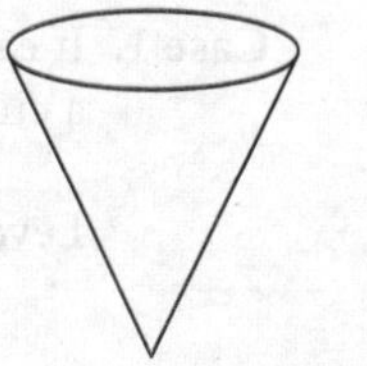

(Continued)

S. No.	Figure	Lateral Surface Area	Total Surface Area	Volume	Diagonal	Shape
5.	**Sphere**	$4\pi r^2$	$4\pi r^2$	$\frac{4}{3}\pi r^3$	——	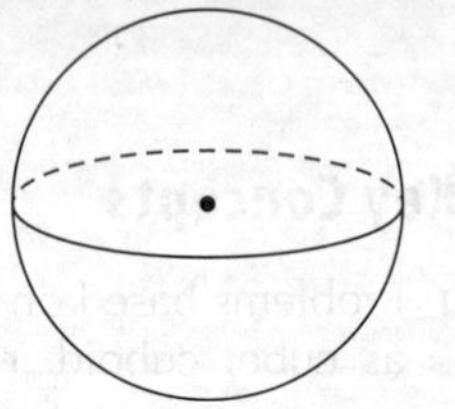
6.	**Hemisphere**	$2\pi r^2$	$3\pi r^2$	$\frac{2}{3}\pi r^3$	——	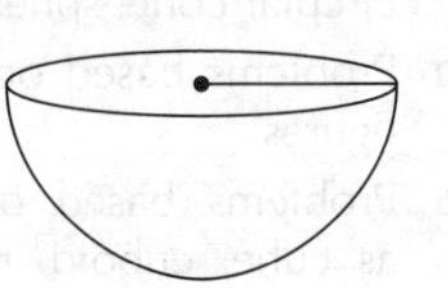
7.	**Frustum**	$\pi l(r_1 + r_2)$	$\pi l(r_1 + r_2) + \pi(r_1^2 + r_2^2)$	$\frac{1}{3}\pi h (r_1^2 + r_2^2 + r_1 r_2)$	——	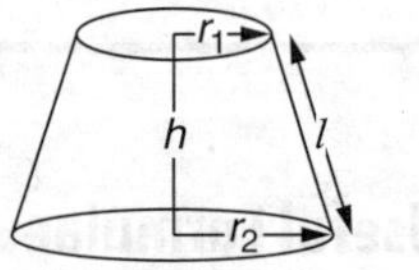

1. Slant height of frustum $(l) = \sqrt{h^2 + (r_1 - r_2)^2}$.
2. Surface area unit is always unit^2.
3. Diagonal/Height/Length unit is always – unit.
4. Volume unit is unit^3.
5. In the case of spherical shell:
   (i) Outer surface area $= 4\pi R^2$ square units.
   (ii) Inner surface area $= 4\pi r^2$ square units.
   (iii) Volume of hollow sphere $= \frac{4}{3}\pi(R^3 - r^3)$ unit^3.
6. In the case of combined figure total surface area, this denotes the curved surface areas of the figure combined.
7. In the case of melting total volume of solids, the volume of solids before melting is equal to the volume of solids after melting.

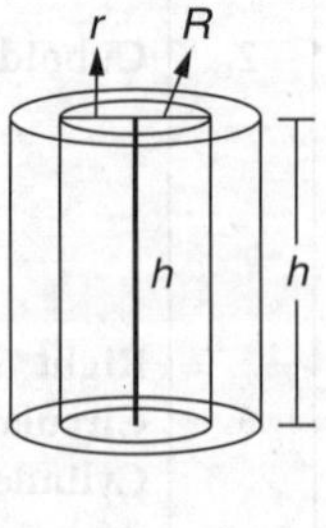

8. In the case of hollow cylinders:
   (i) Volume of $\pi h(R^2 - r^2)$ unit^3.
   (ii) Curved surface area $= 2\pi Rh + 2\pi rh = 2\pi h(R + r)$ unit^2.
   (iii) Total surface area $= 2\pi h(R + r) + 2\pi(R^2 - r^2)$ unit^2.
9. In the case of earth dug out, we must find the level raised in the field.

   Case 1. If earth is dug out from a rectangular field in the shape of a cuboid.

   $$\text{Level raised } (H) = \frac{l \times b \times h}{L \times B - l \times b} \text{ unit.}$$

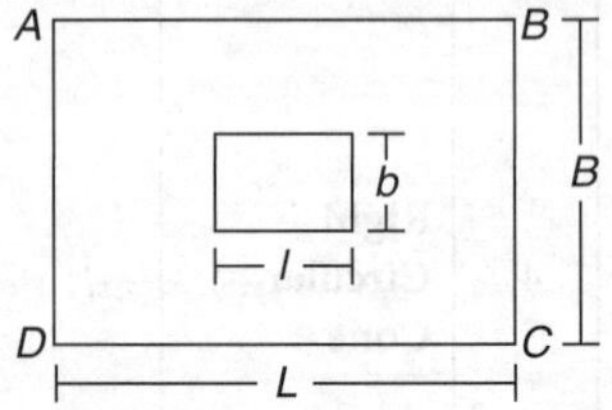

Case 2. If earth is dug out from a rectangular field in the shape of a cylinder inside the field.

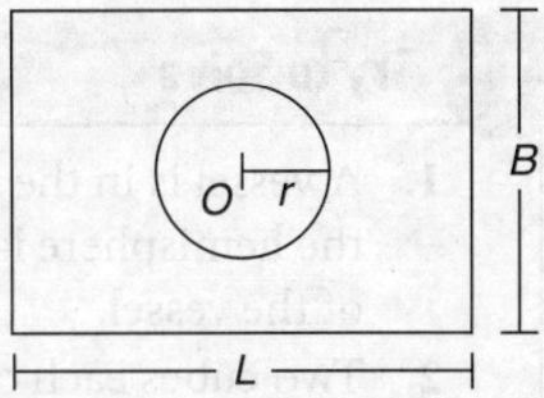

$$\text{Level raised } (H) = \frac{\pi r^2 h}{L \times B - \pi r^2} \text{ unit.}$$

Case 3. If earth is dug out from outside the field in the shape of a cuboid, then level raised of the rectangular field is given by

$$\text{Level raised } (H) = \frac{l \times b \times h}{L \times B} \text{ unit.}$$

Case 4. If earth is dug out from outside the field in the shape of a cylinder, then level raised of the rectangular field is given by

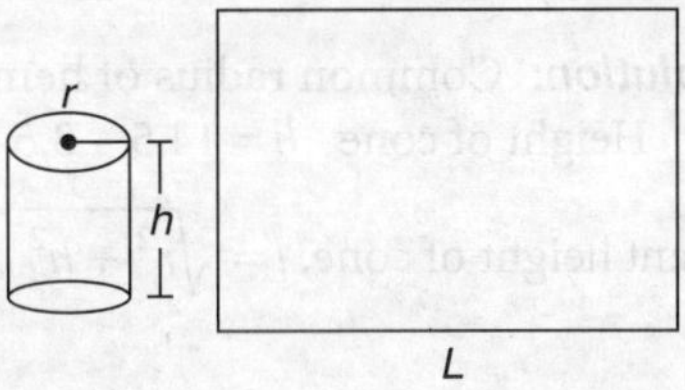

$$\text{Level raised } (H) = \frac{\pi r^2 h}{L \times B}.$$

Case 5. If earth is dug out cylindrically to form an embankment (Volume of embankment = Volume of the wall), then the volume of earth taken out is given by volume of earth taken out.

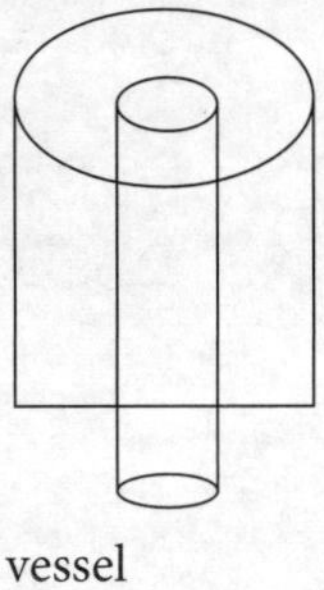

$$\pi(R^2 - r^2)h = \pi r^2 H$$

$$\text{Height of embankment} = \frac{\pi r^2 H}{\pi(R^2 - r^2)}$$

where R = radius of the embankment; r = radius of the earth dug out; H = height of the well, that is, height of the earth dug out; and h = height of the embankment.

10. In the case of some balls dropped in a vessel,
Number of balls × Volume of each ball = Volume of water that flows out from the vessel

$$\therefore \quad \text{Number of balls} = \frac{\text{Volume of water flows out}}{\text{Volume of one ball}}.$$

Example 1 Atish made a bird bath for his garden in the shape of a cylinder with a hemispherical depression at one end, as shown in the figure. The height of the cylinder is 2 m and its diameter is 84 cm. Find the total surface area of the bird bath. $\left[\text{Use } \pi = \frac{22}{7}\right]$. **[CBSE (CCE) 2011]**

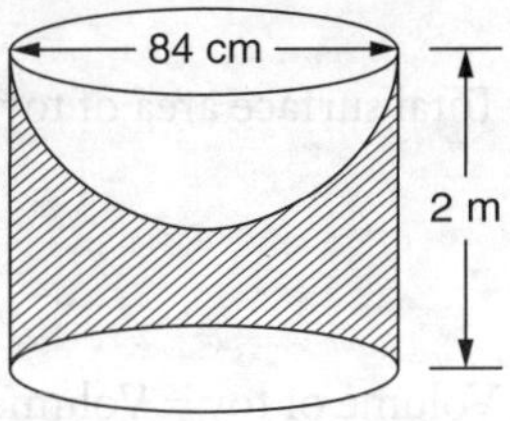

Solution: Height of cylinder $h = 2 \text{ m} = 200 \text{ cm}$
Radius of cylinder and hemisphere $r = 42$ cm
Total surface area of bird bath = *CSA* of cylinder + *CSA* of hemisphere

$$= 2\pi rh + 2\pi r^2 = 2\pi r\,(h + r) = 2 \times \frac{22}{7} \times 42 \times (200 + 42) = 2 \times 22 \times 6 \times 242$$

$$= 63888 \text{ cm}^2 = \frac{63888}{10000} \text{ m}^2 = 6.39 \text{ m}^2. \qquad \left[\because \frac{1}{10000}\text{ cm}^2 = 1\text{ m}^2\right]$$

1 Try to Solve

1. A vessel is in the form of a hollow hemisphere mounted by a hollow cylinder. The diameter of the hemisphere is 14 cm and the total height of the vessel is 13 cm. Find the inner surface area of the vessel. [NCERT]
2. Two cubes each of volume 64 cm^3 are joined end to end. Find the surface area of the resulting cuboid. [NCERT]

Answers

1. 572 m^2 2. 160 cm^2

Example 2 A toy is in the form of a cone mounted on a hemisphere of diameter 7 cm. The total height of the toy is 14.5 cm. Find the volume and total surface area of the toy. [CBSE 2007]

Solution: Common radius of hemisphere and cone $r = 3.5$ cm

Height of cone $h = 14.5 - 3.5 = 11$ cm

Slant height of cone, $l = \sqrt{r^2 + h^2} = \sqrt{3.5^2 + 11^2} = \sqrt{12.25 + 121} = \sqrt{133.25} = 11.54$ cm

```
            11.54
   1)      133.25
            1
           ---
  21)      033
            21
           ---
 225)      1225
           1125
           ----
2304)      10000
            9216
           -----
             784
```

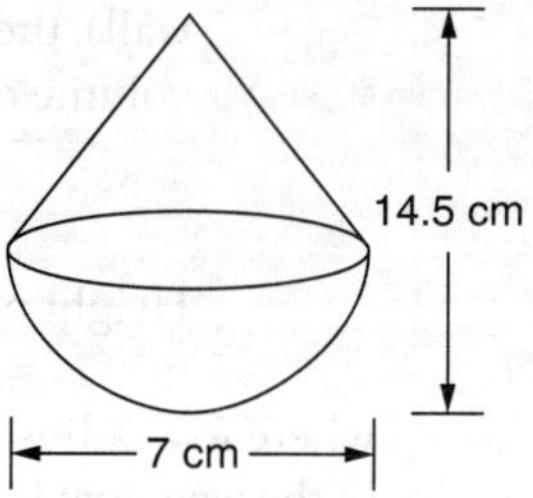

Total surface area of toy = *CSA* of cone + *CSA* of hemisphere

$= \pi rl + 2\pi r^2 = \pi r\,(l + 2r)$

$= \frac{22}{7} \times 3.5 \times (11.54 + 2 \times 3.5) = 22 \times 0.5 \times 18.54 = 203.94\ cm^2$

Volume of toy = Volume of cone + Volume of hemisphere

$= \frac{1}{3}\pi r^2 h + \frac{2}{3}\pi r^3 = \frac{1}{3}\pi r^2 (h + 2r)$

$= \frac{1}{3} \times \frac{22}{7} \times 3.5 \times 3.5\ (11 + 2 \times 3.5)$

$= \frac{1}{3} \times 22 \times 0.5 \times 3.5 \times 18 = 231\ cm^3.$

2 Try to Solve

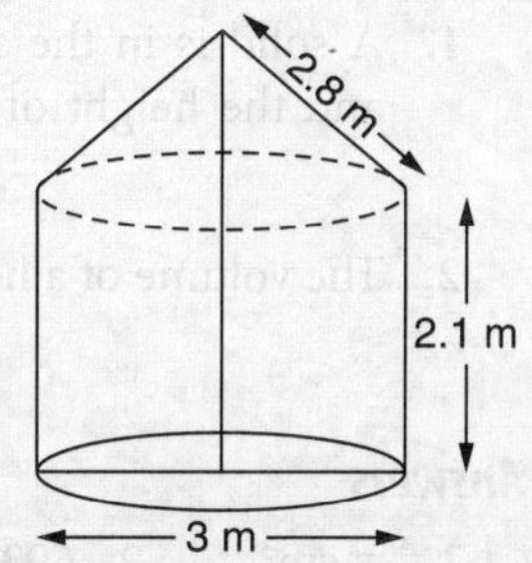

1. A toy is in the form of a cone of radius 3.5 cm mounted on a hemisphere of the same radius. The total height of the toy is 15.5 cm. Find the total surface area of the toy. **[NCERT, CBSE (CCE) 2010, 12, 13, 15]**
2. In the figure, a tent is in the shape of a cylinder surmounted by a conical top of the same diameter. If the height and diameter of the cylindrical part is 2.1 m and 3 m, respectively and the slant height of the conical part is 2.8 m, find the cost of the canvas needed to make the tent if the canvas is available at a rate of ₹ 500/sq. metre. $\left[\text{Use } \pi = \dfrac{22}{7}\right]$ **[CBSE (CCE) 2016]**
3. A toy is in the form of a cone mounted on a hemisphere of the same radius of 7 cm. If the total height of the toy is 31 cm, find its total surface area. $\left[\text{Use } \pi = \dfrac{22}{7}\right]$ **[CBSE (CCE) 2013]**

Answers

1. 214.5 cm^2 2. ₹ 16,500 3. 858 cm^2

Example 3 A vessel is in the form of a hemispherical bowl surmounted by a hollow cylinder. The diameter of the hemisphere is 21 cm and the total height of the vessel is 14.5 cm. Find the capacity. **[CBSE (CCE) 2014]**

Solution: Total height of vessel = 14.5 cm

Common radius of hemisphere and cylinder = 10.5 cm

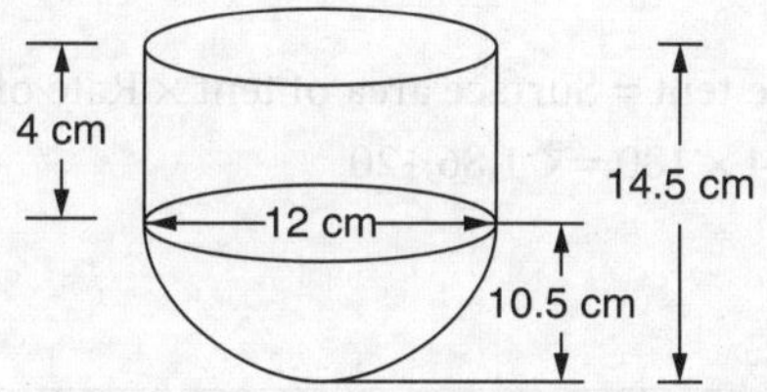

Height of cylinder $h = 14.5 - 10.5 = 4$ cm

Capacity of the vessel = Volume of hemisphere + Volume of cylinder

$$= \frac{2}{3}\pi r^3 + \pi r^2 h = \pi r^2\left(\frac{2}{3}r + h\right)$$

$$= \frac{22}{7} \times 10.5 \times 10.5\left(\frac{2}{3}\times 10.5 + 4\right)$$

$$= 22 \times 1.5 \times 10.5\ (2 \times 3.5 + 4)$$

$$= 3811.5 \text{ cm}^3.$$

3 Try to Solve

1. A solid is in the shape of a cone standing on a hemisphere, with both radii equal to 1 cm and the height of the cone equal to its radius. Find the volume of the solid in terms of π. **[NCERT]**
2. The volume of a hemisphere is $2425\ \frac{1}{2}$ cm³. Find its curved surface area. $\left[\text{Use } \pi = \frac{22}{7}\right]$ **[CBSE (CCE) 2012]**

Answers

1. π cm³ 2. 693 cm²

Example 4 A tent is in the shape of a right circular cylinder to a height of 3 m and it is conical above it. The total height of the tent is 13.5 m and the radius of the base is 14 m. Find the cost of the cloth required to make the tent at the rate of ₹ 180 per m². **[CBSE (CCE) 2012]**

Solution: Height of cylinder, $h = 3$ m. Common radius of cone and cylinder, $r = 14$ m.

Height of cone $h_1 = 13.5 - 3 = 10.5$ m

Slant height of cone $l = \sqrt{r^2 + h_1^2} = \sqrt{14^2 + 10.5^2}$

$= \sqrt{306.25} = 17.5$ cm

Surface area of tent = *CSA* of cylinder + *CSA* of cone

$= 2\pi rh + \pi rl = \pi r\,(2h + l)$

$= \frac{22}{7} \times 14\,(2 \times 3 + 17.5) = 22 \times 2 \times 23.5 = 1034 \text{ m}^2$

Cost of cloth required to make the tent = Surface area of tent × Rate of cloth

$= 1034 \times 180 =$ ₹ 1,86,120

4 Try to Solve

1. A solid is in the shape of a cone surmounted by a hemisphere, the radius of each being 3.5 cm and the total height of the solid being 9.5 cm. Find the volume of the solid. $\left[\text{Use } \pi = \frac{22}{7}\right]$ **[NCERT, CBSE (CCE) 2012]**
2. A military tent of height 8.25 m is in the form of a right circular cylinder of base diameter 30 m and height 5.5 m surmounted by a right circular cone of the same base radius. Find the length of the canvas used in making the tent, if the breadth of the canvas is 1.5 m. **[CBSE (CCE) 2012]**

Answers

1. 166.83 cm³ 2. 825 m

Example 5 A decorative block, as shown in the figure, is made up of two solids: a cube and a hemisphere. The base of the block is the cube with an edge of 5 cm and the hemisphere attached on the top has a diameter of 4.2 cm. If the block is to be painted, find the total area to be painted. $\left[\text{Take } \pi = \dfrac{22}{7}\right]$ **[CBSE (CCE) 2012]**

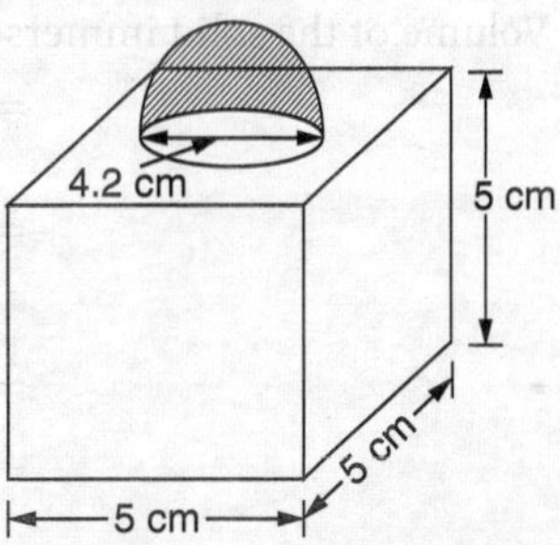

Solution: Side of cube (a) = 5 cm

Diameter of hemisphere = 4.2 cm

Radius of hemisphere (r) = 2.1 cm

Total area to be painted

$$= TSA \text{ of cube} + CSA \text{ of hemisphere} - \text{Area of base of hemisphere}$$

$$= 6a^2 + 2\pi r^2 - \pi r^2 = 6 \times 5 \times 5 + \pi r^2$$

$$= 150 + \frac{22}{7} \times 2.1 \times 2.1 = 163.86 \text{ cm}^2$$

5 Try to Solve

1. A cubical block of side 7 cm is surmounted by a hemisphere. What is the greatest diameter the hemisphere can have? Find the surface area of the solid. **[NCERT]**
2. From a solid cylinder whose height is 2.4 cm and diameter 1.4 cm, a conical cavity of the same height and same diameter is hollowed out. Find the total surface area of the remaining solid to the nearest cm^2. **[NCERT]**

Answers

1. 7 cm, 332.5 cm^2 2. 18 cm^2

Example 6 A cylindrical tub of radius 5 cm and length 9.8 cm is full of water. A solid in the form of a right circular cone mounted on a hemisphere is immersed in the tub. If the radius of the hemisphere is 3.5 cm and the total height of the solid is 8.5 cm, find the volume of water left in the tub. **[CBSE (CCE) 2014]**

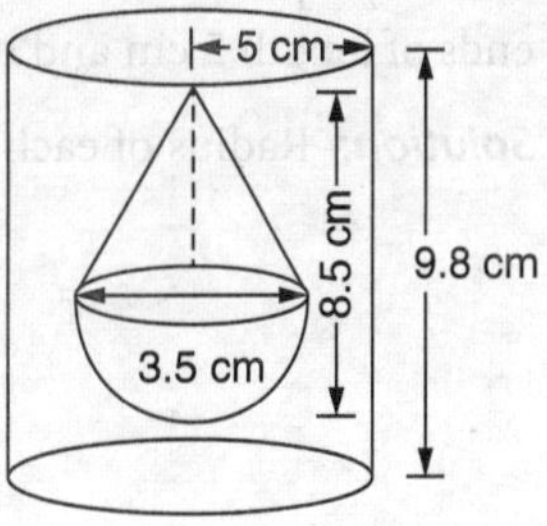

Solution: Radius of the cylinder, $r = 5$ cm.

Height of the cylinder, $h = 9.8$ cm.

Radius of the hemisphere, $R = 3.5$ cm.

Radius of the cone = Radius of the hemisphere = 3.5 cm = $\dfrac{7}{2}$ cm.

Height of the cone, $H = (8.5 - 3.5)$ cm = 5 cm.

Volume of the water in the cylindrical tub

$$= \pi r^2 h = \left(\frac{22}{7} \times 5 \times 5 \times 9.8\right) \text{cm}^3 = 770 \text{ cm}^3.$$

Volume of the solid immersed in the tub

= Volume of the hemisphere + Volume of the cone

$$= \frac{2}{3}\pi R^3 + \frac{1}{3}\pi R^2 H$$

$$= \left[\left(\frac{2}{3}\times\frac{22}{7}\times\frac{7}{2}\times\frac{7}{2}\times\frac{7}{2}\right)+\left(\frac{1}{3}\times\frac{22}{7}\times\frac{7}{2}\times\frac{7}{2}\times 5\right)\right]\text{cm}^3$$

$$= \left(\frac{539}{6}+\frac{385}{6}\right)\text{cm}^3 = \left(\frac{924}{6}\right)\text{cm}^3 = 154\ \text{cm}^3$$

$\therefore$ Volume of water left in the tub = (770 − 154) cm^3 = 616 cm^3.

6 Try to Solve

1. A solid toy is in the form of a hemisphere by a right circular cone. The height of the cone is 2 cm and the diameter of the base is 4 cm. Determine the volume of the toy. If a right circular cylinder circumscribes the toy, find the difference of the volumes of the cylinder and the toy. [Take $\pi = 3.14$] **[NCERT]**
2. A solid consisting of a right circular cone of height 120 cm and radius 60 cm standing on a hemisphere of radius 60 cm is placed upright in a right circular cylinder full of water such that it touches the bottom. Find the volume of water left in the cylinder, if the radius of the cylinder is 60 cm and its height is 180 cm. **[NCERT, CBSE (CCE) 2009]**

Answers

1. 25.12 cm^3, 25.12 cm^3 2. 1.131 m^3

Example 7 A gulab jamun contains sugar syrup up to about 30% of its volume. Find approximately how much syrup would be found in 45 gulab jamuns, each shaped like a cylinder with two hemispherical ends of length 5 cm and diameter 2.8 cm. **[NCERT, CBSE (CCE) 2015]**

Solution: Radius of each hemispherical part, $r = \left(\frac{2.8}{2}\right)$ cm = 1.4 cm.

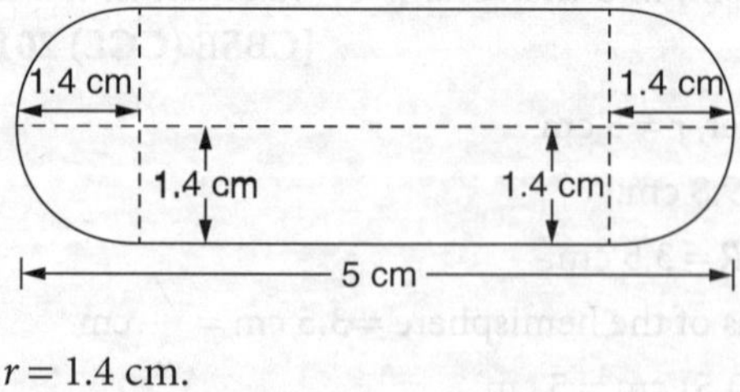

Radius of the cylindrical part, r = 1.4 cm.

Length of the cylindrical part, h = (5 − 2 × 1.4) cm = 2.2 cm.

Volume of one gulab jamun

= Volume of two hemispherical parts + Volume of the cylindrical part

$$= \left(2 \times \frac{2}{3}\pi r^3 + \pi r^2 h\right) = \left(\frac{4}{3}\pi r^3 + \pi r^2 h\right)$$

$$= \pi r^2\left(\frac{4}{3}r + h\right) = \left[\frac{22}{7} \times 1.4 \times 1.4 \times \left(\frac{4}{3} \times 1.4 + 2.2\right)\right] \text{cm}^3$$

$$= \left(6.16 \times \frac{12.2}{3}\right) \text{cm}^3 = \left(\frac{75.152}{3}\right) \text{cm}^3$$

Volume of 45 gulab jamuns

$$= \left(\frac{75.152}{3} \times 45\right) \text{cm}^3 = 1127.28 \text{ cm}^3$$

Volume of the syrup = (30% of 1127.28) cm^3

$$= \left(\frac{30}{100} \times 1127.28\right) \text{cm}^3 = 338.184 \text{ cm}^3.$$

Hence, 45 gulab jamuns contain approximately 338 cm^3 of syrup.

7 Try to Solve

1. A toy is in the form of a cylinder with hemispherical ends. If the whole length of the toy is 90 cm and its diameter is 42 cm, find the cost of painting the toy at the rate of 70 paise per sq. cm.
2. A juice seller was serving his customers using glasses, as shown in the figure. The inner diameter of the cylindrical glass was 5 cm, but the bottom of the glass had a hemispherical raised portion which reduced the capacity of the glass. If the height of the glass was 10 cm, find the apparent capacity of the glass and its actual capacity. [Use $\pi = 3.14$] **[NCERT, CBSE (CCE) 2013]**

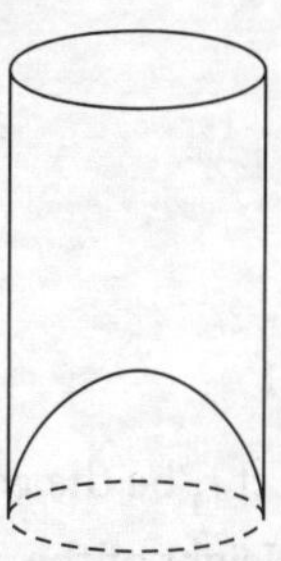

Answers

1. ₹ 8,316 2. 163.54 cm^3

Example 8 The dimensions of a metallic cuboid are 100 × 80 × 64 cm. It is melted and recast into a cube. Find the surface area of the cube.

Solution: Let each edge of the cube be a. Then, volume of the cube = a^3.

Volume of the cuboid = (100 × 80 × 64) cm^3 = 512000 cm^3.

Volume of the cube = Volume of the cuboid

$$\Rightarrow \quad a^3 = 512000 \text{ cm}^3$$

$$\Rightarrow \quad \sqrt[3]{512 \times 1000} \text{ cm} = \sqrt[3]{8 \times 8 \times 8 \times 10 \times 10 \times 10}$$

$$= (8 \times 10) \text{ cm} = 80 \text{ cm}.$$

Surface area of the cube = $6a^2$ = (6 × 80 × 80) cm^2 = 38400 cm^2.

8 Try to Solve

1. A metallic sphere of radius 4.2 cm is melted and recast into the shape of a cylinder of radius 6 cm. Find the height of the cylinder. [NCERT]
2. Metallic spheres of radii 6 cm, 8 cm and 10 cm, respectively, are melted to form a single solid sphere. Find the radius of the resulting sphere. [NCERT]

Answers

1. 2.74 cm 2. 12 cm

Example 9 A cylindrical container, whose radius is 6 m and height 15 m, is filled with ice cream. The ice cream is distributed to 10 children in equal cones having hemispherical tops. If the height of the conical portion is 4 times the radius of its base, find the radius of the cone. [CBSE (CCE) 2012]

Solution: Radius of cylindrical container, $r = 6$ m

Height of cylindrical container, $h = 15$ m

Volume of cylindrical container, $V = \pi r^2 h$

$$= \pi \times 6 \times 6 \times 15 = 540\,\pi\,\text{m}^3.$$

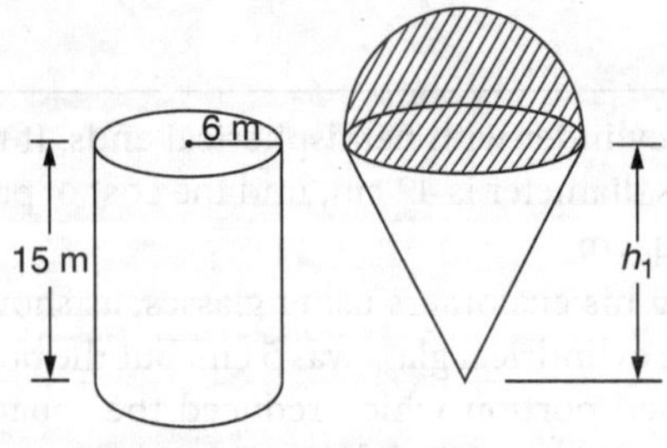

Let r_1 be the common radius of the cone and hemisphere

Height of the cone, $h_1 = 4r_1$ [Given]

Volume of ice cream cone with hemispherical top

$$V_1 = \text{Volume of cone} + \text{Volume of hemisphere}$$

$$= \frac{1}{3}\pi r_1^2 h_1 + \frac{2}{3}\pi r_1^3 = \frac{1}{3}\pi r_1^2 (h_1 + 2r_1)$$

$$= \frac{1}{3}\pi r_1^2 (4r_1 + 2r_1) = \frac{1}{3}\pi r_1^2 \times 6r_1 = 2\pi r_1^3\ \text{m}^3.$$

$$10 \times \text{Volume of ice cream cones} = \text{Volume of cylindrical container}$$

$$10 \times 2\pi r_1^3 = 540\,\pi$$

$\Rightarrow$ $$r_1^3 = \frac{540\pi}{20\pi}$$

$\Rightarrow$ $$r_1^3 = 27$$

$\therefore$ $$r_1 = 3\ \text{m}$$

Hence, the radius of the cone = 3 m.

9 Try to Solve

1. A vessel is in the form of a hollow hemisphere mounted by a hollow cylinder of the same radius. The diameter of the hemisphere is 14 cm and the total height of the vessel is 13 cm. Find the inner surface area of the vessel.
2. A solid toy is in the form of a right circular cylinder with a hemispherical shape at one end and a cone at the other end. Their common diameter is 4.2 cm, and the height of the cylindrical and conical portions is 12 cm and 7 cm, respectively. Find the volume of the given toy. [Take $\pi = 22/7$]
3. A container shaped like a right circular cylinder having diameter 12 cm and height 15 cm is full of ice cream. The ice cream is to be filled into cones of height 12 cm and diameter 6 cm, having a hemisphere shape on the top. Find the number of such cones that can be filled with the ice cream. **[NCERT]**

Answers

1. 572 cm^2 2. 218.064 cm^3 3. 10

Example 10 A hemispherical bowl of internal radius 9 cm is full of liquid. The liquid is to be filled into cylindrical-shaped small bottles, each of diameter 3 cm and height 4 cm. How many bottles are needed to empty the bowl?

Solution: Radius of the hemispherical bowl, $R = 9$ cm

Volume of liquid in the bowl $= \frac{2}{3}\pi R^3 = \left(\frac{2}{3}\pi \times 9 \times 9 \times 9\right)$ cm^3

$= (486\,\pi)$ cm^3

Radius of each cylindrical bottle $r = \frac{3}{2}$ cm

Height of each cylindrical bottle, $h = 4$ cm

Volume of each cylindrical bottle

$$= \pi r^2 h = \left(\pi \times \frac{3}{2} \times \frac{3}{2} \times 4\right) \text{ cm}^3 = (9\pi) \text{ cm}^3.$$

$$\text{Required number of bottles} = \frac{\text{Volume of liquid in the bowl}}{\text{Volume of each bottle}} = \frac{486\pi}{9\pi} = 54.$$

10 Try to Solve

1. How many silver coins, 1.75 cm in diameter and 2 mm in thickness, must be melted to form a cuboid of dimension $5.5 \times 10 \times 3.5$ cm? **[NCERT]**
2. The dimensions of a metallic cuboid are $100 \times 90 \times 81$ cm. It is melted and recast into a cube. Find the surface area of the cube. **[CBSE (CCE) 2011]**

Answers

1. 400 2. 48600 cm^2

Example 11 A well of diameter 4 m is dug 14 m deep. The earth taken out is spread evenly all around the well to form a 40-cm-high embankment. Find the width of the embankment. **[CBSE (CCE) 2015]**

Solution: Given, the diameter of the well = 4 m

Radius of the well = 2 m

Depth of the well = 14 m = Height

Volume of earth taken out from the well = $\pi r^2 h$

$$= \pi \times 2 \times 2 \times 14 = 56\pi \text{ m}^3$$

$\because$ Earth taken out from the well is evenly spread to form an embankment having height 40 cm = 0.4 m.

Let the external radius of the embankment be R.

Internal radius = 2 m = radius of well

Volume of embankment (cylindrical) = $\pi(R^2 - r^2)h$

$$= \pi(R^2 - r^2) \times 0.4 = \pi(R^2 - 4) \times \frac{4}{10} \text{ m}^3$$

Volume of embankment = Volume of earth taken out from well

According to the question $\quad \pi (R^2 - 4) \times \frac{4}{10} = 56\pi$

$$R^2 - 4 = \frac{56 \times 10}{4} = 140$$

$\Rightarrow \quad R^2 = 140 + 4 = 144 \Rightarrow R = 12$ m

$\therefore$ Width of the embankment = $R - r = 12 - 2 = 10$ m.

Example 12 A 21-m-deep well with diameter 6 m is dug and the earth dug out is spread evenly to form a platform of 27 × 11 m. Find the height of the platform. $\left[\text{Use} = \frac{22}{7}\right]$ **[CBSE (CCE) 2015]**

Solution: Diameter of cylinder = 6 m

Radius of cylinder, $r = \frac{6}{2} = 3$ m

Height of cylinder, $h = 21$ m

Cuboidal platform: Length = 27 m, breadth = 11 m and let height = h'.

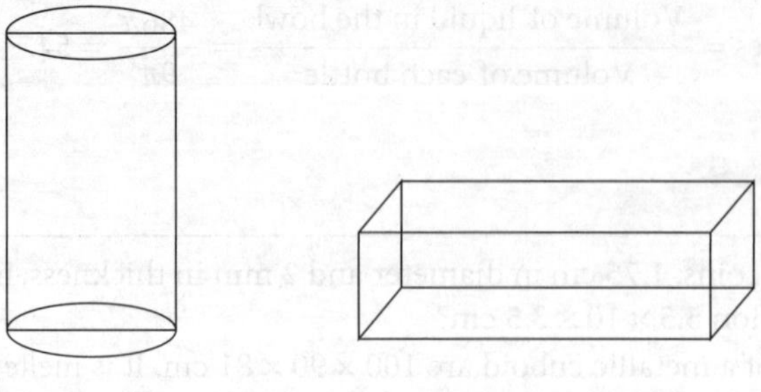

According to the question

$$\text{Volume of cuboidal platform} = \text{Volume of cylindrical well}$$

$$l \times b \times h' = \pi r^2 h$$

$$27 \times 11 \times h' = \frac{22}{7} \times 3 \times 3 \times 21$$

$$h' = \frac{22 \times 3 \times 3 \times 21}{7 \times 27 \times 11} = 2 \text{ m}$$

$\therefore$ Height of platform = 2 m

11 Try to Solve

1. A well of diameter 3 m is dug 14 m deep. The earth taken out of it has been spread evenly all around it in the shape of a circular ring of width 4 m to form an embankment. Find the height of the embankment. **[NCERT]**
2. A 20-m-deep well with diameter 14 m is dug up and the earth dug out is spread out evenly to form a platform of 22 × 14 m. Find the height of the platform.

Answers

1. 1.125 m 2. 10 m

Example 13 If the radii of the circular ends of a conical bucket which is 45 cm high, are 28 cm and 7 cm, find the capacity of the bucket. [Use $\pi = 22/7$]

Solution: Clearly, the bucket forms a frustum of a cone such that the radii of its circular ends are $r_1 =$ 28 cm, $r_2 = 7$ cm and height $h = 45$ cm. Therefore

Capacity of the bucket = Volume of the frustum

$\Rightarrow$ $$\text{Capacity of the bucket} = \frac{1}{3}\pi h\,(r_1^2 + r_2^2 + r_1 r_2)$$

$\Rightarrow$ $$= \frac{1}{3} \times \frac{22}{7} \times 45(28^2 + 7^2 + 28 \times 7)$$

$\Rightarrow$ $$= 22 \times 15 \times (28 \times 4 + 7 + 28) = 330 \times 147 \text{ cm}^3$$

$\Rightarrow$ $$\text{Capacity of the bucket} = 48510 \text{ cm}^3$$

Example 14 The perimeters of the ends of a frustum are 48 cm and 36 cm. If the height of the frustum is 11 cm, find its volume.

Solution: Let r_1 and r_2 be the radii of the circular ends of the frustum and h be its height. Then

$$2\pi r_1 = 48,\ 2\pi r_2 = 36 \text{ and } h = 11 \text{ cm} \Rightarrow r_1 = \frac{24}{\pi},\ r_2 = \frac{18}{\pi} \text{ and } h = 11 \text{ cm}$$

Let V be the volume of the frustum. Then

$$V = \frac{1}{3}\ \pi\,(r_1^2 + r_2^2 + r_1 r_2)h$$

$\Rightarrow$ $$V = \frac{1}{3} \times \pi \times 11 \times \left\{\left(\frac{24}{\pi}\right)^2 + \left(\frac{18}{\pi}\right)^2 + \frac{24}{\pi} \times \frac{18}{\pi}\right\}$$

$$\Rightarrow \quad V = \frac{1}{3} \times \pi \times 11 \times \frac{(576+324+432)}{\pi^2} \text{ cm}^3$$

$$\Rightarrow \quad V = \frac{11}{3} \times \frac{1332}{\pi} \text{ cm}^3 = \frac{11}{3} \times \frac{1332}{22} \times 7 \text{ cm}^3 = 1554 \text{ cm}^3$$

12 Try to Solve

1. A drinking glass is the shape of a frustum of a cone of height 14 cm. The diameters of its two circular ends are 4 cm and 2 cm. Find the capacity of the glass. **[NCERT]**
2. The slant height of the frustum of a cone is 4 cm and the perimeters (circumference) of its circular ends are 18 cm and 6 cm. Find the curved surface area of the frustum. **[AI 2017, Delhi 2010]**

Answers

1. $102\frac{2}{3}$ cm^3 2. 48 cm^2

FOR MORE PRACTICE AND KNOWLEDGE

Frustum of a Right Circular Cone

Frustum: If a cone is cut by a plane parallel to its base and the cone that is formed on one side of the plane is removed, the part that is left on the other side of the plane is called the frustum of the cone (see figure). Here, *PAB* is a cone and *ABDC* is the frustum of the cone.

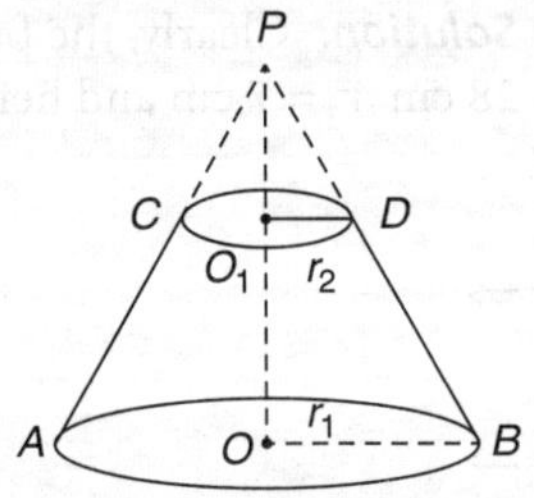

Height of the frustum of a cone: It is the perpendicular distance between its two circular ends.

$$OO_1 = \text{height} = OP - O_1P$$

Slant height: It is the length of the line joining the extremities of two parallel radii, that is, *D* and *B* or *C* and *A*.

$\therefore$ Slant height = BD or $AC = \sqrt{h^2 + (r_1 - r_2)^2}$ $(r_1 > r_2)$.

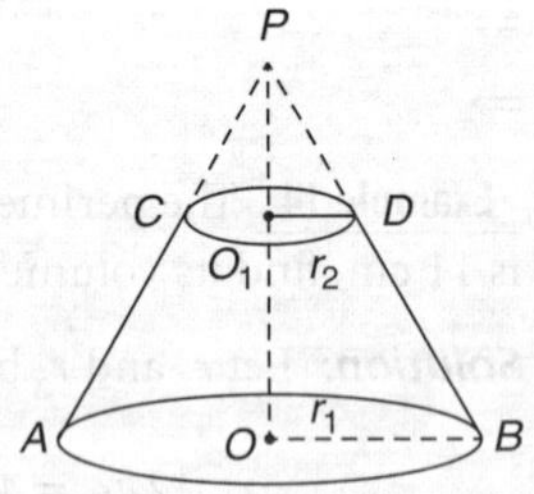

Derivation of the Volume and Surface Area of a Frustum of a Right Circular Cone

Let r_1, h_1, l_1 and r_2, h_2, l_2 be the radius, height and slant height of two cones *APB* and *CPD*, respectively. Let *h* and *l* be the height and slant height of the frustum, *ABDC*.

$$r_1 = OB,\ r_2 = O_1D,\ h_1 = PO,\ h_2 = PO_1,\ l_1 = PB \text{ and } l_2 = PD$$

$$h = OO_1 = h_1 - h_2 \text{ and } l = BD = l_1 - l_2$$

In ΔPO_1D and ΔPOB $\quad \angle P = \angle P$ [Common]

$\angle PO_1D = \angle POB$ [Each 90°]

$\therefore \quad \Delta PO_1D \sim \Delta POB$ [By AA similarity]

$\frac{PO_1}{PO} = \frac{PD}{PB} = \frac{O_1D}{OB}$ [∵ If two Ds are similar, then their corresponding sides are in proportion]

$\Rightarrow \quad \frac{h_2}{h_1} = \frac{l_2}{l_1} = \frac{r_2}{r_1}$

$\Rightarrow \quad \frac{h_1 - h}{h_1} = \frac{l_1 - l}{l_1} = \frac{r_2}{r_1}$ [∵ $h_2 = h_1 - h$ and $l_2 = l_1 - l$]

$\Rightarrow \quad 1 - \frac{h}{h_1} = 1 - \frac{l}{l_1} = \frac{r_2}{r_1}$

$\Rightarrow \quad 1 - \frac{h}{h_1} = \frac{r_2}{r_1}$ and $1 - \frac{l}{l_1} = \frac{r_2}{r_1}$

$\Rightarrow \quad \frac{h}{h_1} = 1 - \frac{r_2}{r_1}$ and $\frac{l}{l_1} = 1 - \frac{r_2}{r_1}$

$\Rightarrow \quad \frac{h}{h_1} = \frac{r_1 - r_2}{r_1}$ and $\frac{l}{l_1} = \frac{r_1 - r_2}{r_1}$

$\Rightarrow \quad \frac{h_1}{h} = \frac{r_1}{r_1 - r_2}$ and $\frac{l_1}{l} = \frac{r_1}{r_1 - r_2}$

$\therefore \quad h_1 = \frac{hr_1}{r_1 - r_2}$ and $l_1 = \frac{lr_1}{r_1 - r_2}$

$$h_2 = h_1 - h = \frac{hr_1}{r_1 - r_2} - h = h\,\frac{(r_1 - r_1 + r_2)}{r_1 - r_2} = \frac{hr_2}{r_1 - r_2}$$

And $$l_2 = l_1 - l = \frac{lr_1}{r_1 - r_2} - l = l\,\frac{(r_1 - r_1 + r_2)}{r_1 - r_2} = \frac{lr_2}{r_1 - r_2}$$

Volume of the frustum of a cone = Volume of cone PAB – Volume of cone PCD

$$= \frac{1}{3}\pi r_1^2 h_1 - \frac{1}{3}\pi r_2^2 h_2 = \frac{1}{3}\pi r_1^2 h_1 - \frac{1}{3}\pi r_2^2 h_2$$

$$= \frac{\pi}{3}[r_1^2 h_1 - r_2^2 h_2]$$

$$= \frac{\pi}{3}\left(\frac{hr_1^3}{r_1 - r_2} - \frac{hr_2^3}{r_1 - r_2}\right) \qquad \left[\because \; h_2 = \frac{hr_2}{r_1 - r_2} \text{ and } h_1 = \frac{hr_1}{r_1 - r_2}\right]$$

$$= \frac{\pi h}{3(r_1 - r_2)}(r_1^3 - r_2^3) = \frac{\pi h}{3(r_1 - r_2)}(r_1 - r_2)(r_1^2 + r_1 r_2 + r_2^2)$$

$$= \frac{\pi h}{3}(r_1^2 + r_1 r_2 + r_2^2)$$

$$\boxed{\text{Volume of frustum of a cone} = \frac{\pi h}{3}(r_1^2 + r_1 r_2 + r_2^2)}$$

Curved surface area (*CSA*) of the frustum

$$= CSA \text{ of cone } PAB - CSA \text{ of cone } PCD$$

$$= \pi r_1 l_1 - \pi r_2 l_2 = \pi\left[\frac{r_1 \times r_1 l}{r_1 - r_2} - \frac{r_2 \times r_2 l}{r_1 - r_2}\right]$$

$$= \pi l\left(\frac{r_1^2}{r_1 - r_2} - \frac{r_2^2}{r_1 - r_2}\right) = \pi l \frac{r_1^2 - r_2^2}{r_1 - r_2}$$

$$= \pi l \times \frac{(r_1 - r_2)(r_1 + r_2)}{(r_1 - r_2)}$$

$$\boxed{CSA = \pi l(r_1 + r_2)}$$

Total surface area (*TSA*) of the frustum

$$= CSA \text{ of frustum} + \text{Surface area of circular ends}$$

$$TSA = \pi l(r_1 + r_2) + \pi r_1^2 + \pi r_2^2 = \pi l r_1 + \pi l r_2 + \pi r_1^2 + \pi r_2^2$$

$$= \pi l r_1 + \pi r_1^2 + \pi l r_2 + \pi r_2^2$$

$$\therefore \quad \boxed{TSA = \pi r_1(l + r_1) + \pi r_2(l + r_2)}$$

Example 15 A solid sphere of radius 3 cm is melted and then cast into small spherical balls, each of diameter 0.6 cm. Find the number of balls thus obtained. **[CBSE (CCE) 2012]**

Solution: Let the total number of balls be *n*.

$$\text{Volume of the solid sphere} = \frac{4}{3}\pi r^3 = \frac{4}{3}\pi \times 3^3 \text{ cm}^3 = 36\,\pi \text{ cm}^3$$

$$\text{Radius of a spherical ball} = \frac{0.6}{2} \text{ cm} = 0.3 \text{ cm}$$

$$\text{Volume of a spherical ball} = \frac{4}{3}\pi \times (0.3)^3 \text{ cm}^3 = \frac{4}{\cancel{3}}\pi \times \frac{\cancel{3}}{10} \times \frac{3}{10} \times \frac{3}{10} \text{ cm}^3 = \frac{36\pi}{1000} \text{ cm}^3$$

$$\therefore \quad \text{Volume of } n \text{ spherical balls} = \frac{36\pi}{1000} n \text{ cm}^3$$

Volume of the solid sphere = Volume of *n* spherical balls

$$\therefore \quad 36\text{p} = \frac{36\pi}{1000} n \Rightarrow n = 1000$$

Hence, 1000 spherical balls are obtained by melting the given solid sphere.

Example 16 A solid iron rectangular block of dimensions 4.4 m, 2.6 m and 1 m is cast into a hollow cylindrical pipe of internal radius 30 cm and thickness 5 cm. Find the length of the pipe.

Solution: Suppose the length of the pipe is *l* cm. The volume of iron in the pipe is equal to the volume of iron in the block.

$$\text{Volume of the block} = (4.4 \times 2.6 \times 1) \text{ m}^3 = (440 \times 260 \times 100) \text{ cm}^3$$

$$r = \text{Internal radius of the pipe} = 30 \text{ cm}$$

R = External radius of the pipe = (30 + 5) cm = 35 cm

$\therefore$ Volume of iron in the pipe = (External volume of pipe) – (Internal volume of pipe)

$$\Rightarrow \quad = \pi R^2 h - \pi r^2 h = \pi(R^2 - r^2)h = \pi(R + r)(R - r)h$$

$$\Rightarrow \quad = \pi \times (35 + 30) \times (35 - 30) \times h \text{ cm}^3$$

Volume of iron in the pipe $= \pi \times 65 \times 5 \times h$ cm^3

Now Volume of iron in the pipe = Volume of iron in the block

$$\Rightarrow \quad \pi \times 65 \times 5 \times h = 440 \times 260 \times 100$$

$$\Rightarrow \quad \frac{22}{7} \times 65 \times 5 \times h = 440 \times 260 \times 100$$

$$\Rightarrow \quad h = \left(440 \times 260 \times 100 \times \frac{7}{22} \times \frac{1}{65} \times \frac{1}{5}\right) \text{ cm}$$

$$\Rightarrow \quad h = 11200 \text{ cm} = 112 \text{ m.}$$

Hence, the length of the pipe is 112 m.

Example 17 A well, whose diameter is 7 m, has been dug 22.5 m deep and the earth dug out is used to form an embankment around it. If the height of the embankment is 1.5 m, find the width of the embankment.

Solution: Radius of the well $= \frac{7}{2}$ m = 3.5 m

Depth of the well = 22.5 m

$\therefore$ Volume of the earth dug out $= \pi \times (3.5)^2 \times 22.5$ m^3

$$= \pi \times \frac{7}{2} \times \frac{7}{2} \times \frac{45}{2} \text{ m}^3$$

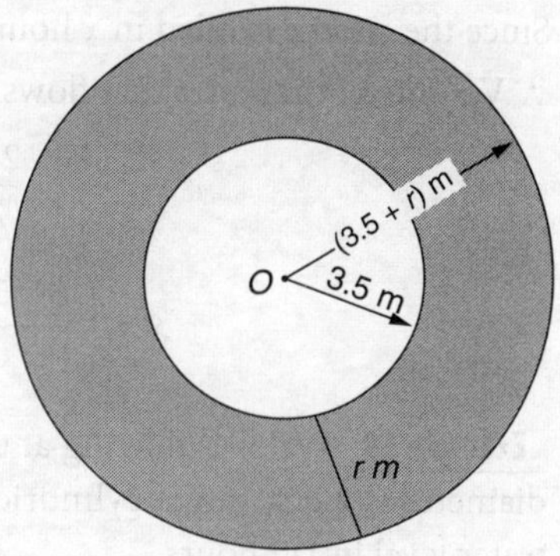

Suppose the width of the embankment is *r* metres.

The embankment forms a cylindrical shell whose inner and outer radii are 3.5 m and (r + 3.5) m, respectively, and height 1.5 m.

$\therefore$ Volume of the embankment $= \pi\{(r + 3.5)^2 - (3.5)^2\} \times 1.5$ m^3

$$\pi\{(r + 3.5 - 3.5)(r + 3.5 + 3.5)\} \times \frac{3}{2} = \pi(r + 7)r \times \frac{3}{2} \text{ m}^3$$

But Volume of the embankment = Volume of the earth dug out

$$\Rightarrow \quad \cancel{\pi} r(r + 7) \times \frac{\cancel{3}}{\cancel{2}} = \cancel{\pi} \times \frac{7}{\cancel{2}} \times \frac{7}{2} \times \frac{\cancel{45}}{2}$$

$$\Rightarrow \quad r(r + 7) = \frac{49}{4} \times 15$$

$$\Rightarrow \quad 4r^2 + 28r = 735$$

$$\Rightarrow \quad 4r^2 + 28r = 735$$

$$4r^2 + 28r - 735 = 0$$

On solving, we get
$$r = \frac{-28 \pm \sqrt{784 + 11760}}{8}$$

$$\Rightarrow \quad r = \frac{-28 \pm \sqrt{12544}}{8} = \frac{-28 \pm 112}{8} = \frac{84}{8} = 10.5 \quad [\because r > 0]$$

Hence, the width of the embankment is 10.5 m.

Example 18 Water is flowing at the rate of 3 km/h through a circular pipe of 20 cm internal diameter into a circular cistern of diameter 10 m and depth 2 m. In how much time will the cistern be filled?

Solution: Suppose the cistern is filled in x hours. Since water is flowing at the rate of 3 km/h

Length of the water column in x hours = $3x$ km = 3000 x metres.

The water column forms a cylinder of radius

$$r = \frac{20}{2} \text{ cm} = 10 \text{ cm} = \frac{1}{10} \text{ m and } h \text{ m height (length)} = 3000x \text{ metres}$$

∴ Volume of the water that flows in the cistern in x hours

$$= \pi r^2 h = \left(\frac{22}{7} \times \frac{1}{10} \times \frac{1}{10} \times 3000x\right) \text{m}^3$$

Also, volume of the cistern = $\left(\frac{22}{7} \times 5 \times 5 \times 2\right) \text{m}^3$ $\quad [\because r = 5 \text{ m}, h = 2 \text{ m}]$

Since the cistern is filled in x hours.

∴ Volume of the water that flows in the cistern in x hours = Volume of the cistern

$$\Rightarrow \quad \frac{22}{7} \times \frac{1}{10} \times \frac{1}{10} \times 3000x = \frac{22}{7} \times 5 \times 5 \times 2$$

$$\Rightarrow \quad x = \left(\frac{5 \times 5 \times 2 \times 10 \times 10}{3000}\right) \text{hours} = \frac{5}{3} \text{ hours} = 1 \text{ hour } 40 \text{ minutes.}$$

Example 19 Water is flowing at the rate of 7 metres per second through a circular pipe whose internal diameter is 2 cm, into a cylindrical tank the radius of whose base is 40 cm. Determine the increase in water level in 1/2 hours. **[CBSE 2006C]**

Solution: Rate of flow of water = 7 m/sec = 700 cm/sec.

Length of the water column in $\frac{1}{2}$ hours = $(700 \times 30 \times 60)$ cm

Internal radius of circular pipe = 1 cm.

Clearly, the water column forms a cylinder of radius 1 cm and length = $(700 \times 30 \times 60)$.

∴ Volume of the water that flows in the tank in $\frac{1}{2}$ h.

$$= \left(\frac{22}{7} \times 1 \times 1 \times 700 \times 30 \times 60\right) \text{cm}^3 \quad [\text{Using: } V = \pi r^2 h]$$

Let h cm be the rise in the level of water in the tank. Then

$$\text{Volume of the water in the tank} = \frac{22}{7} \times 40 \times 40 \times h \text{ cm}^3$$

From (i) and (ii), we have $\frac{22}{7} \times 40 \times 40 \times h = \frac{22}{7} \times 1 \times 1 \times 700 \times 30 \times 60$

$\Rightarrow$ $h = \frac{700 \times 30 \times 60}{40 \times 40}$ cm = 787.5 cm

Hence, the rise in the level of water in the tank in $\frac{1}{2}$ h is 787.5 cm

Example 20 The barrel of a fountain pen, cylindrical in shape, is 7 cm long and 5 mm in diameter. A full barrel of ink in the pen will be used up in writing 330 words on an average. How many words would use a bottle of ink containing one fifth of a litre?

Solution: Volume of a barrel = $\left(\frac{22}{7} \times 0.25 \times 0.25 \times 7\right)$ cm^3 = 1.375 cm^3

Volume of an ink bottle = $\frac{1}{5}$ litre = $\frac{1000}{5}$ cm^3 = 200 cm^3

$\therefore$ Total number of barrels that can be filled from the given volume of ink = $\frac{200}{1.375}$

So, required number of words = $\frac{200}{1.375} \times 330$ = 48000 words.

Example 21 A conical vessel of radius 6 cm and height 8 cm is completely filled with water. A sphere is lowered into the water and its size is such that when it touches the sides, it is just immersed, as shown in the figure. What fraction of water overflows?

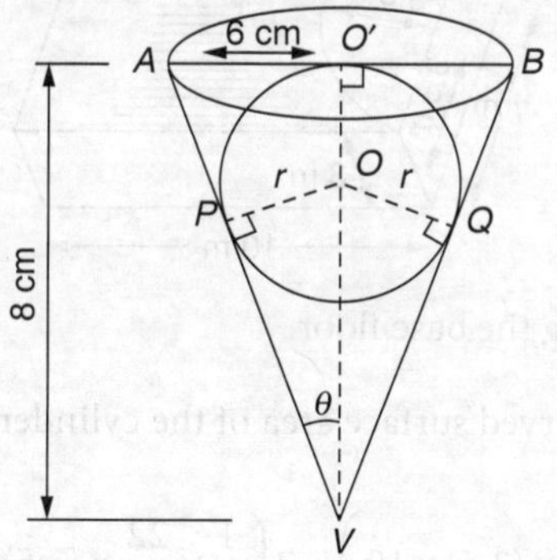

Solution: Let the radius of the sphere be r cm.

In $\Delta VO'A$, we have $\tan\theta = \frac{6}{8} = \frac{3}{4} \Rightarrow \sin\theta = \frac{3}{5}$

In ΔVPO, we have $\sin\theta = \frac{r}{VO}$

$\Rightarrow$ $\frac{3}{5} = \frac{r}{8 - r}$

$\Rightarrow$ $24 - 3r = 5r$

$\Rightarrow \quad 8r = 24$

$\Rightarrow \quad r = 3$ cm

$\therefore \quad V_1$ = Volume of the sphere $= \frac{4}{3}\pi \times 3^3 \text{ cm}^3 = 36\pi \text{ cm}^3$

V_2 = Volume of the water = Volume of the cone $= \frac{1}{3}\pi \times 6^2 \times 8 \text{ cm}^3 = 96\pi \text{ cm}^3$

Volume of the water that flows out of the cone is the same as the volume of the sphere, that is, V_1.

$\therefore$ Fraction of the water that flows out $= V_1 : V_2 = 36\pi : 96\pi = 3:8$.

Example 22 A godown is of the form shown in the figure. The vertical cross-section parallel to the width of the building is a rectangle of 7 × 3 m, mounted by a semicircle of radius 3.5 m. The inner measurements of the cuboidal portion of the building are 10 × 7 × 3 m. Find the volume of the godown and the total interior surface area excluding the floor (base). [Take $\pi = 22/7$]

Solution: As the top of the building is in the form of half of a cylinder of radius 3.5 m and length 10 m, split along the diameter.

$\therefore \quad V$ = Volume of the godown

$\Rightarrow \quad V$ = Volume of the cuboid $+ \frac{1}{2}$ (Volume of the cylinder of radius 3.5 m and length 10 m)

$$= \left\{10 \times 7 \times 3 + \frac{1}{2}\left(\frac{22}{7} \times 3.5 \times 3.5 \times 10\right)\right\} \text{ m}^3 = (210 + 192.5)\text{m}^3 = 402.5 \text{ m}^3$$

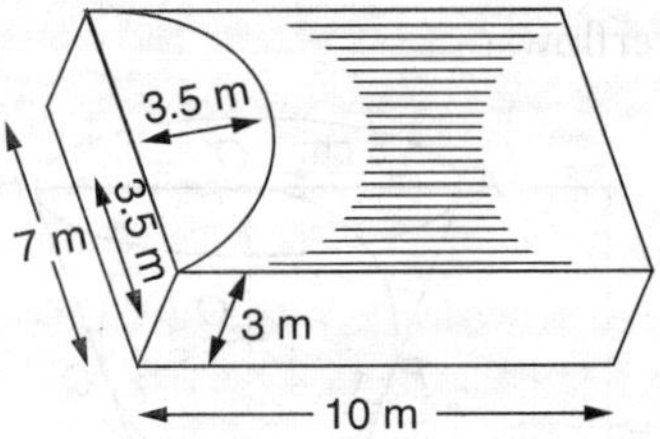

Total interior surface area excluding the base floor

$= $ Area of four walls $+ \frac{1}{2}$ (Curved surface area of the cylinder) + 2 (Area of the semicircles)

$$= \left[2(10+7) \times 3 + \frac{1}{2}\left(2 \times \frac{22}{7} \times (3.5) \times 10\right) + 2\left(\frac{1}{2} \times \frac{22}{7} \times (3.5)^2\right)\right] \text{ m}^2$$

$= (102 + 110 + 38.5) \text{ m}^2 = 250.5 \text{ m}^2$

Example 23 The interior of a building is in the form of a cylinder of diameter 4.3 m and height 3.8 m, surmounted by a cone whose vertical angle is a right angle. Find the area of the surface and the volume of the building.

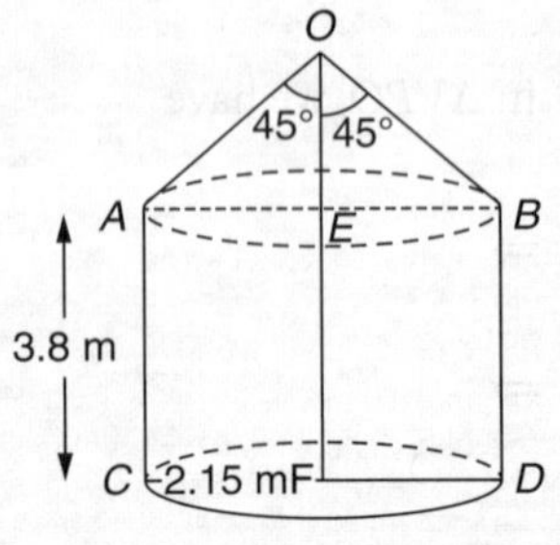

Solution: Let *OACDBO* be the building such that $AE = EB = 2.15$ m.

$AC = 3.8$ m and $\angle AOB = 90°$

$\therefore \quad \angle AOE = \angle BOE = 45°$

$\Rightarrow \quad OE = AE = 2.15$ m

$\therefore \quad OA = \sqrt{AE^2 + OE^2} = \sqrt{(2.15)^2 + (2.15)^2} = 2.15\sqrt{2} = 3.04$ m.

Volume of the building

$$= \text{Volume of the cylindrical portion} + \text{Volume of the conical portion}$$

$$= \pi \times (2.15)^2 \times 3.8 + \frac{1}{3}\pi \times (2.15)^2 \times 2.15$$

$$= \frac{1}{3}\pi \times (2.15)^2 \{3 \times 3.8 + 2.15\} = \frac{1}{3} \times 3.14 \times (2.15)^2 \times 13.55 \text{ m}^3$$

$$= \frac{1}{3} \times 3.14 \times 4.6225 \times 13.55 \text{ m}^3 = 65.55 \text{ m}^3$$

Area of the surface of the building

$$= \text{Curved surface area of conical portion} + \text{Curved surface area of cylindrical portion}$$

$$= \pi \times 2.15 \times 3.04 + 2\pi \times 2.15 \times 3.8 \text{ m}^2$$

$$= (3.14 \times 2.15 \times 3.04 + 2 \times 3.14 \times 2.15 \times 3.8) \text{ m}^2$$

$$= (20.52 + 51.30) \text{ m}^2 = 71.82 \text{ m}^2$$

Example 24 A wooden toy rocket is in the shape of a cone mounted on a cylinder, as shown in the figure. The height of the entire rocket is 26 cm, while the height of the conical part is 6 cm. The base of the conical portion has a diameter of 5 cm, while the base diameter of the cylindrical portion is 3 cm. If the conical portion is to be painted orange and the cylindrical portion yellow, find the area of the rocket painted with each of these colours. [Take $\pi = 3.14$]

Solution: Let r be the radius of the base of the cone and its slant height be l. Further, let r_1 be the radius of the cylinder and h_1 be its height.

We have $\quad r = 2.5$ cm, $h = 6$ cm,

$r_1 = 1.5$ cm and $h_1 = 20$ cm

Now $\quad l = \sqrt{r^2 + h^2}$

$\Rightarrow \quad l = \sqrt{(2.5)^2 + 6^2}$

$\Rightarrow \quad l = \sqrt{6.25 + 36} = \sqrt{42.25} = 6.5$ cm

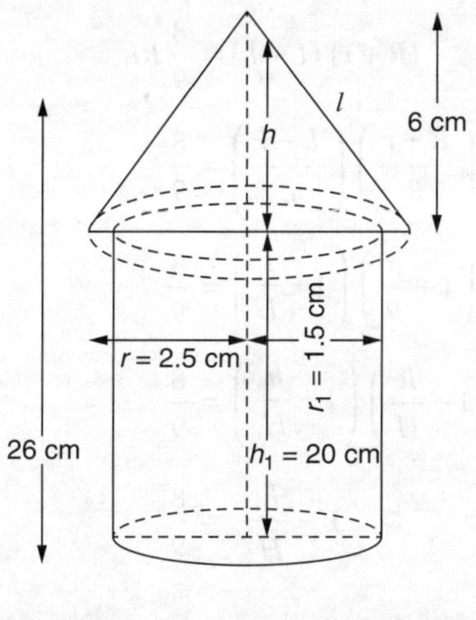

Area to be painted orange

= Curved surface area of the cone + Base area of the cone – Base area of the cylinder

$$= \pi rl + \pi r^2 - \pi r_1^2 = \pi\{rl + r^2 - r_1^2\}$$
$$= \pi\{2.5 \times 6.5 + (2.5)^2 - (1.5)^2\} \text{ cm}^2$$
$$= 3.14\,(16.25 + 6.25 - 2.25) \text{ cm}^2$$
$$= 3.14 \times 20.25 \text{ cm}^2 = 63.585 \text{ cm}^2$$

Area to be painted yellow = Curved surface area of the cylinder + Area of the base of the cylinder

$$= 2\pi r_1 h_1 + \pi r_1^2 = \pi r_1(2h_1 + r_1)$$
$$= 3.14 \times 1.5\,(2 \times 20 + 1.5) \text{ cm}^2 = 3.14 \times 1.5 \times 41.5 \text{ cm}^2$$
$$= 4.71 \times 41.5 \text{ cm}^2 = 195.465 \text{ cm}^2$$

Example 25 A hollow cone is cut by a plane parallel to the base and the upper portion is removed. If the curved surface of the remainder is $\frac{8}{9}$ of the curved surface of the whole cone, find the ratio of the line segment into which the cone's altitude is divided by the plane. **[CBSE 2004]**

Solution: Let VAB be a hollow cone of height H, slant height L and base radius R. Suppose this cone is cut by a plane parallel to the base such that O' is the centre of the circular section of the cone. Let h be the height, l the slant height and r the base radius of the smaller cone $VA'B'$.

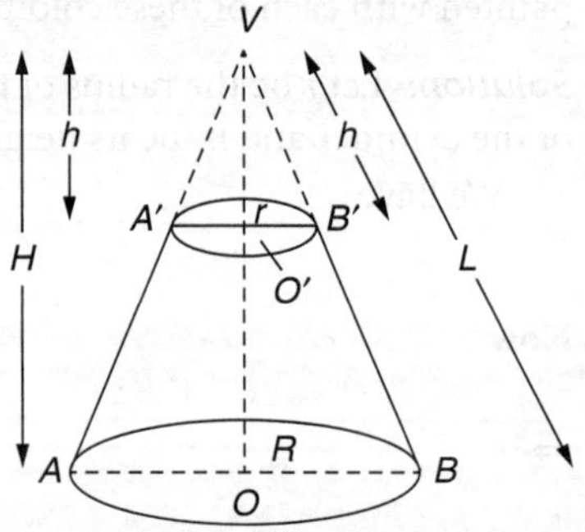

Clearly $\quad \Delta VO'A' \sim \Delta VOA$

$$\therefore \quad \frac{VO'}{VO} = \frac{O'A'}{OA} = \frac{VA'}{VA}$$

$$\Rightarrow \quad \frac{h}{H} = \frac{r}{H} = \frac{l}{L} \quad \ldots\text{(i)}$$

It is given that

Curved surface area of the frustum $ABB'A' = \frac{8}{9} \times$ Curved surface area of the cone

$$\Rightarrow \quad \not{\pi}\,(R + r)\,(L - l) = \frac{8}{9}\,\not{\pi}\,RL$$

$$\Rightarrow \quad (R + r)\,(L - l) = \frac{8}{9}\,RL$$

$$\Rightarrow \quad \left(\frac{R + r}{R}\right)\left(\frac{L - l}{L}\right) = \frac{8}{9}$$

$$\Rightarrow \quad \left(1 + \frac{r}{R}\right)\left(1 - \frac{l}{L}\right) = \frac{8}{9}$$

$$\Rightarrow \quad \left(1 + \frac{h}{H}\right)\left(1 - \frac{h}{H}\right) = \frac{8}{9} \quad \text{[Form (i)]}$$

$$\Rightarrow \quad 1 - \frac{h^2}{H^2} = \frac{8}{9}$$

$$\Rightarrow \quad \frac{h^2}{H^2} = 1 - \frac{8}{9}$$

$$\Rightarrow \quad \frac{h^2}{H^2} = \frac{1}{9}$$

$$\Rightarrow \quad \frac{h}{H} = \frac{1}{3} \Rightarrow h = \frac{H}{3}$$

Hence, the required ratio $= \dfrac{h}{H-h} = \dfrac{\frac{H}{3}}{H - \frac{H}{3}} = \dfrac{1}{2}$

Example 26 The height of a cone is 30 cm. A small cone is cut off at the top by a plane parallel to the base. If its volume is $\frac{1}{27}$ of the volume of the given cone, at what height above the base is the section made?

Solution: Let VAB be a cone of height 30 cm and base radius r cm. Suppose it is cut off by a plane parallel to the base at a height h from the base of the cone.

Clearly, $\Delta VOA \sim \Delta VOA'$

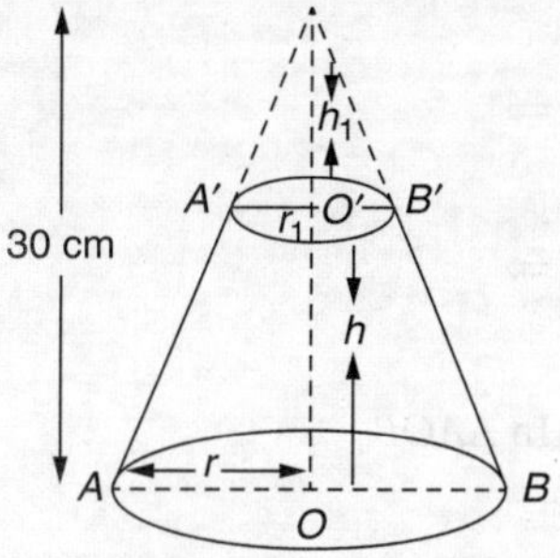

$$\therefore \quad \frac{VO}{VO'} = \frac{AO}{O'A'} \quad \ldots\text{(i)}$$

$$\Rightarrow \quad \frac{30}{h_1} = \frac{r}{r_1}$$

Volume of cone $VA'B' = \dfrac{1}{27}$ Volume of cone VAB

$$\Rightarrow \quad \frac{1}{3}\pi r_1^2 h_1 = \frac{1}{27} \times \frac{1}{3}\pi r^2 \times 30$$

$$\Rightarrow \quad \left(\frac{r_1}{r}\right)^2 h_1 = \frac{10}{9}$$

$$\Rightarrow \quad \left(\frac{h_1}{30}\right)^2 h_1 = \frac{10}{9} \quad \text{[From (i)]}$$

$$\Rightarrow \quad h_1^3 = 1000$$

$$\Rightarrow \quad h_1 = 10 \text{ cm}$$

$$\therefore \quad h = 30 - h_1 = (30 - 10) \text{ cm} = 20 \text{ cm}$$

Hence, the section is made 20 cm from the base of the cone.

Example 27 An oil funnel of tin sheet consists of a cylindrical portion 10 cm long attached to the frustum of a cone. If the total height is 22 cm, diameter of the cylindrical portion is 8 cm and the diameter of the top of the funnel is 18 cm, find the area of tin required to make the funnel.

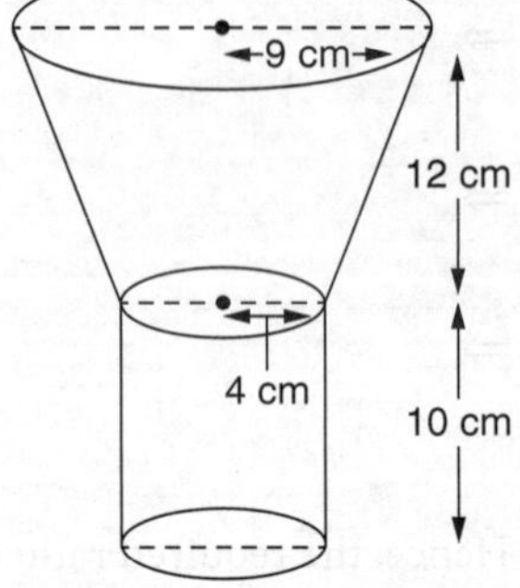

Solution: Let l be the slant height of the frustum part of the funnel. Then

$$l = \sqrt{(9-4)^2 + 12^2} = \sqrt{25+144} \text{ cm} = 13 \text{ cm}$$

Tin required = Curved surface area of cylindrical portion + Curved surface area of frustum portion

$$\Rightarrow \quad = 2\pi r_1 h + \pi(r_1 + r_2)l$$

$$= \{2\pi \times 4 \times 10 + \pi(4+9) \times 13\}\text{cm}^2$$

$$\Rightarrow \quad \text{Tin required} = (80\,\pi + 169\,\pi)\text{ cm}^2 = 249\,\pi\text{ cm}^2$$

Example 28 A metallic right circular cone 20 cm high and whose vertical angle is 60° is cut into two parts at the middle of its height by a plane parallel to its base. If the frustum so obtained is drawn into a wire of diameter $\frac{1}{16}$ cm, find the length of the wire.

Solution: Let ADC be a cone with vertical angle 60°. The cone is cut into two parts parallel to its base at height 10 cm.

The radius of the larger end of the frustum = R_1

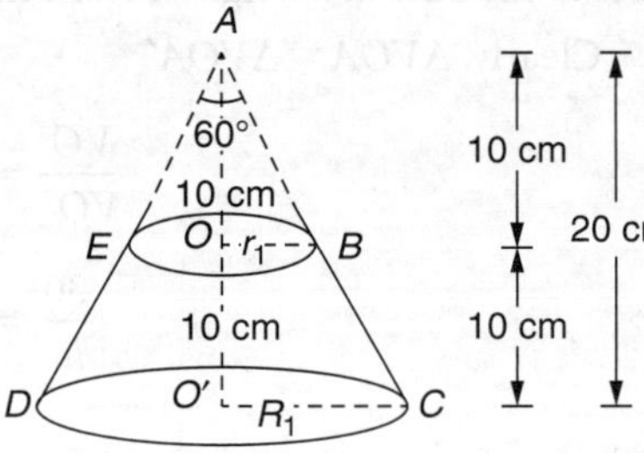

In $\Delta AO'C$ $\quad \tan 30° = \frac{O'C}{O'A}$

$$\Rightarrow \quad \frac{1}{\sqrt{3}} = \frac{O'C}{20} = \frac{R_1}{20}$$

$$\Rightarrow \quad R_1 = \frac{20}{\sqrt{3}} \text{ cm}$$

In ΔAOB $\quad \tan 30° = \frac{r_1}{OA}$

$$\Rightarrow \quad \frac{1}{\sqrt{3}} = \frac{r_1}{10} \quad r_1 = \frac{10}{\sqrt{3}} \text{ cm}$$

(where $h = 10$ cm is the height of frustum and the radius of the smaller end = r_1).

$$\text{Volume of frustum} = \frac{\pi h}{3}(R_1^2 + r_1^2 + R_1 r_1)$$

$$= \frac{22}{7} \times \frac{10}{3}\left[\left(\frac{20}{\sqrt{3}}\right)^2 + \left(\frac{10}{\sqrt{3}}\right)^2 + \frac{20 \times 10}{\sqrt{3} \times \sqrt{3}}\right]$$

$$= \frac{22}{7} \times \frac{10}{3}\left[\frac{400}{3} + \frac{100}{3} + \frac{200}{3}\right] = \frac{22}{7} \times \frac{10}{3} \times \frac{700}{3}$$

$$= \frac{22 \times 10 \times 100}{9}$$

A wire is formed, having a diameter of $\frac{1}{16}$ cm and length h cm.

Volume of the wire so obtained $= \pi r^2 h = \frac{22}{7} \times \left(\frac{1}{16 \times 2}\right)^2 h = \frac{22}{7} \times \frac{1}{32} \times \frac{1}{32} \times h$

According to the question, Volume of the frustum = Volume of the wire

$$= \frac{22 \times 10 \times 100}{9} = \frac{22}{7} \times \frac{1}{32} \times \frac{1}{32} \times h$$

$$\Rightarrow \quad = \frac{22 \times 10 \times 100 \times 32 \times 32 \times 7}{22 \times 9} = h$$

$$\Rightarrow \quad h = 796444.44 \text{ cm or } 7964.4 \text{ m.}$$

Multiple Choice Questions

Choose the correct answer for each of the following questions:

1. The radii of the circular ends of a bucket of height 40 cm are 24 cm and 15 cm. The slant height (in cm) of the bucket is
   (a) 51 (b) 49 (c) 43 (d) 41 **[Foreign 2012]**
2. A sphere of diameter 18 cm is dropped into a cylindrical vessel of diameter 36 cm, partly filled with water. If the sphere is completely submerged, the water level rises (in cm) by
   (a) 3 (b) 4 (c) 5 (d) 6 **[Delhi 2011]**
3. The shape of a gilli, in the gilli-danda game (see figure), is a combination of

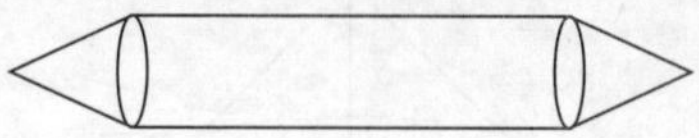

   (a) Two cylinders (b) A cone and a cylinder
   (c) Two cones and a cylinder (d) Two cylinders and a cone **[NCERT Exemplar]**
4. A right circular cylinder of radius r cm and height h cm ($h > 2r$) just encloses a sphere of diameter
   (a) r cm (b) $2r$ cm (c) h cm (d) $2h$ cm **[NCERT Exemplar]**
5. During conversion of a solid from one shape to another, the volume of the new shape will
   (a) Increase (b) Decrease (c) Remain unaltered (d) Be doubled **[NCERT Exemplar]**
6. A rectangular block $6 \times 12 \times 15$ cm is cut into an exact number of equal cubes. The least possible number of cubes will be
   (a) 6 (b) 11 (c) 3 (d) 40
7. The number of coins, 1.5 cm in diameter and 0.2 cm thick, to be melted to form a right circular cylinder of height 10 cm and diameter 4.5 cm is
   (a) 350 (b) 400 (c) 450 (d) 500
8. A cone is cut through a plane parallel to its base and the cone that is formed on one side of that plane is removed. The new part that is left over on the other side of the plane is called
   (a) The frustum of a cone (b) Cone (c) Cylinder (d) Sphere **[NCERT Exemplar]**

9. In a right circular cone, the cross-section made by a plane parallel to the base is a
(a) Circle (b) Frustum of a cone (c) Sphere (d) Hemisphere **[NCERT Exemplar]**

10. A metallic spherical shell of internal and external diameters 4 cm and 8 cm, respectively, is melted and recast into a cone of base diameter 8 cm. The height of the cone is
(a) 18 cm (b) 15 cm (c) 14 cm (d) 12 cm

11. The volume of two cubes is in the ratio 8:125. The ratio of their surface area is
(a) 2:5 (b) 4:25 (c) 16:25 (d) 8:125

12. The volume of two spheres is in the ratio 64:27. The ratio of their surface area is
(a) 3:4 (b) 4:3 (c) 9:16 (d) 16:9 **[NCERT Exemplar]**

13. Three cubes, each of volume 216 m^3, are joined end to end. The surface area of the resulting solid is
(a) 216 m^2 (b) 480 m^2 (c) 432 m^2 (d) 504 m^2

14. If the surface area of two spheres is in the ratio 16:9, their volumes will be in the ratio
(a) 27:64 (b) 64:27 (c) 4:3 (d) 3:4

15. A solid is hemispherical at the bottom and conical (of the same radius) above it. If the surface area of the two parts is equal, then the ratio of its radius and slant height of the conical part is
(a) 2:1 (b) 1:2 (c) 1:4 (d) 4:1

16. The number of revolutions taken by a wheel of diameter 84 cm to cover a distance of 792 m is
(a) 260 (b) 42p (c) 300 (d) 84p

17. A toy is in the form of a right circular cylinder surmounted by a right circular cone, as shown in the figure. If $AD = DE = DC = 1$ cm, the volume of the toy in cm^3 is

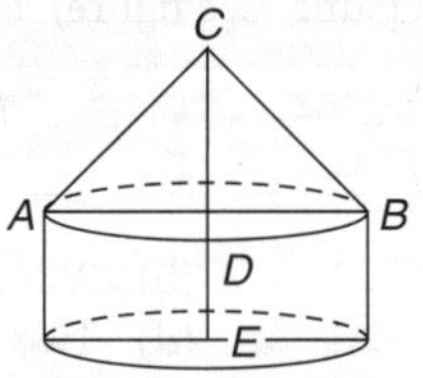

(a) $\frac{4}{3}\pi$ (b) $\frac{3}{4}\pi$ (c) 3π (d) 4π

18. The volume of the largest right circular cone that can be cut out from a cube of edge 4.2 cm is
(a) 77.6 cm^3 (b) 58.6 cm^3 (c) 19.7 cm^3 (d) 19.4 cm^3

19. How many spherical balls, each of radius 1 cm, can be made from a solid sphere of lead of radius 8 cm?
(a) 256 (b) 512 (c) 1024 (d) 576

20. If the radius of the base of a cylinder is doubled and the height remains unchanged, its curved surface area becomes
(a) Double (b) Three times (c) Half (b) No change

21. If two solid hemispheres of the same base radius 'r' are joined along their bases, the curved surface area of this new solid is
(a) $6\pi r^2$ (b) $4\pi r^2$ (c) $2\pi r^2$ (d) $8\pi r^2$

22. The radius of a wire is decreased to one-third. If the volume remains the same, the length will become
(a) 3 times (b) 6 times (c) 9 times (d) 27 times

23. A solid sphere of radius 10 cm is moulded into 8 spherical solid balls of equal radius. The radius of each ball is
(a) 2 cm (b) 3 cm (c) 4 cm (d) 5 cm

24. The diameter of a wheel is 1.26 m. How far will it travel in 500 revolutions?
(a) 2670 m (b) 2880 m (c) 1980 m (d) 1596 m

25. Twelve solid spheres of the same size are made by melting a solid metallic cylinder of base diameter 2 cm and height 16 cm. The diameter of each sphere is
(a) 2 cm (b) 3 cm (c) 4 cm (d) 6 cm

Fill in the Blanks

Complete the following statements with appropriate answer (s) in the blank space(s):

26. If h is the height, l is the slant height and r_1 and r_2 are the radii of the circular ends ($r_1 > r_2$) of the frustum of a cone, the volume of the frustum of the cone is ________.

27. The total surface area of the frustum of a cone is ________.

28. A cone-shaped container is filled with water. A spherical ball is dropped in the container. The volume of water flowing out of the container is equal to the volume of ________.

29. The longest diameter of a cone that can be fully fitted in a cube of edge 8 cm is ________.

30. The volume of water standing in a field = Area of field × ________.

31. If a semicircle is rolled along a diameter, then the figure formed is ________.

32. The radii of the bases of two right circular solid cones of the same height are r_1 and r_2, respectively. If both cones are melted and recast into a solid sphere of radius R, the height of each cone is ________.

33. If two solids hemispheres of the same base of radius r are joined along their bases, then the curved surface area of this new solid is ________.

34. If a cylinder is filled with water and a spherical ball of radius r is dropped into the cylinder, the quantity of water spread out is ________.

35. When we rotate a right triangle about the perpendicular line, the three-dimensional figure formed is ________.

Subjective Questions

36. If a right circular cylinder just encloses a sphere of radius r, find the curved surface area of the cylinder.

37. A solid is in the shape of a cone mounted on a hemisphere of the same base radius. If the curved surface areas of the hemispherical part and the conical part are equal, find the ratio of the radius and the height of the conical part. **[Foreign 2012]**

38. The cost of preparing the walls of a room 12 m long at the rate of ₹1.35 per m^2 is ₹ 340.20 and the cost of matting the floor at 85 paise per m^2 is ₹ 91.80. Find the height of the room.

39. There are two cones. The curved surface area of one is twice that of the other. The slant height of the latter is twice that of the former. Find the ratio of their radii.

40. A tent of height 77 dm is in the form of a right circular cylinder of diameter 36 m and height 44 dm surmounted by a right circular cone. Find the cost of the canvas at ₹ 3.50 per m^2.

41. A toy is in the shape of a right circular cylinder with a hemisphere at one end and a cone at the other. The radius and height of the cylindrical part are 5 cm and 13 cm, respectively. The radii of the hemispherical and conical parts are the same as that of the cylindrical part. Find the surface area of the toy if the total height of the toy is 30 cm.

42. A class room is 7 m long, 6.5 m wide and 4 m high. It has one door of 3×1.4 m and three windows, each measuring 2×1 m. The interior walls are to be colour washed. The contractor charges ₹ 5.25 per m^2. Find the cost of colour washing.

43. Rasheed received a playing top (*lattu*) as his birthday present, which surprisingly had no colour on it. He wanted to colour it with his crayons. The top is shaped like a cone surmounted by a hemisphere (see figure). The entire top is 5 cm in height and the diameter of the top is 3.5 cm. Find the area he has to colour. [Take $\pi = \frac{22}{7}$] **[NCERT]**

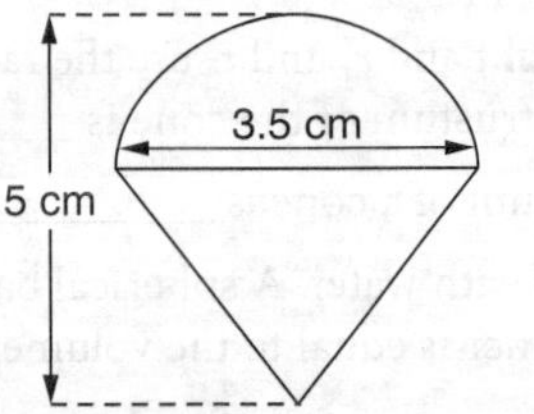

44. The decorative block shown in the figure is made of two solids: a cube and a hemisphere. The base of the block is a cube with edge 5 cm, and the hemisphere fixed on the top has a diameter of 4.2 cm. Find the total surface area of the block. [Take $\pi = \frac{22}{7}$] **[NCERT]**

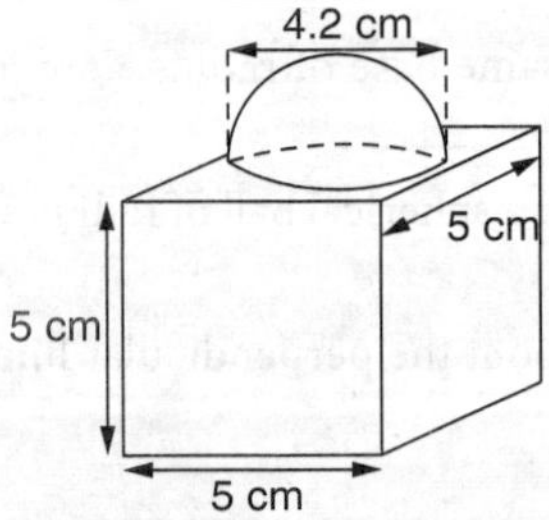

45. A wooden toy rocket is in the shape of a cone mounted on a cylinder, as shown in the figure. The height of the entire rocket is 26 cm, while the height of the conical part is 6 cm. The base of the conical portion has a diameter of 5 cm, while the base diameter of the cylindrical portion is 3 cm. If the conical portion is to be painted orange and the cylindrical portion yellow, find the area of the rocket painted with each of these colours. [Take $\pi = 3.14$] **[NCERT]**

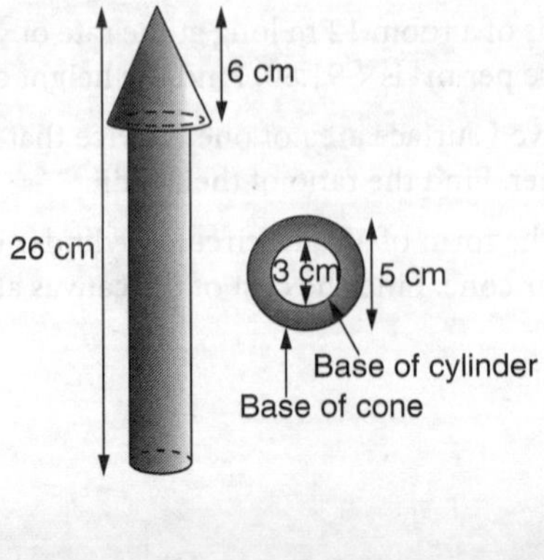

46. A hemispherical depression is cut out from one face of a cubical wooden block such that the diameter l of the hemisphere is equal to the edge of the cube. Determine the surface area of the remaining solid. **[NCERT]**

47. A medicine capsule is in the shape of a cylinder with two hemispheres stuck to each of its ends (see figure). The length of the entire capsule is 14 mm and the diameter of the capsule is 5 mm. Find its surface area. **[NCERT]**

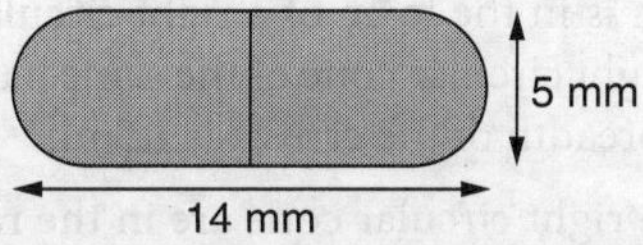

48. A tent is in the shape of a cylinder surmounted by a conical top. If the height and diameter of the cylindrical part are 2.1 m and 4 m, respectively, and the slant height of the top is 2.8 m, find the area of the canvas used for making the tent. Also find the cost of the canvas of the tent at the rate of ₹ 500 per m^2. **[NCERT]**

49. A wooden article was made by scooping out a hemisphere from each end of a solid cylinder, as shown in the figure. If the height of the cylinder is 10 cm and its base is of radius 3.5 cm, find the total surface area of the article. **[NCERT]**

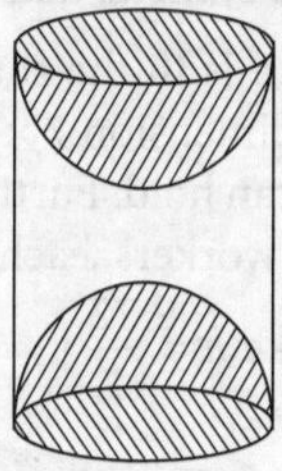

50. Two cubes have their volume in the ratio 1:64. What is the ratio of their surface area?

51. If the total surface area of a solid hemisphere is 462 cm^2, find its volume. [Take $\pi = \frac{22}{7}$] **[AI 2014]**

52. If h, c and V, respectively, are the height, curved surface area and volume of a cone, prove that $3\pi Vh^3 - c^2h^2 + 9V^2 = 0$.

53. Water flows in a tank of 150 × 100 m at the base through a pipe whose cross-section is 2 × 1.5 dm at a speed of 15 km/h. In what time will the water be 3 m deep?

54. A solid wooden toy is in the form of a hemisphere surmounted by a cone of the same radius. The radius of the hemisphere is 3.5 cm and the total wood used in the making of the toy is $166\frac{5}{6}$ cm^3. Find the height of the toy. Also find the cost of painting the hemispherical part of the toy at the rate of ₹ 10 per cm^2. [Take $\pi = \frac{22}{7}$] **[Delhi 2015]**

55. In the figure, from a cuboidal solid metallic block of dimensions 15 × 10 × 5 cm, a cylindrical hole of diameter 7 cm is drilled out. Find the surface area of the remaining block. [Take $\pi = \frac{22}{7}$] **[Delhi 2015]**

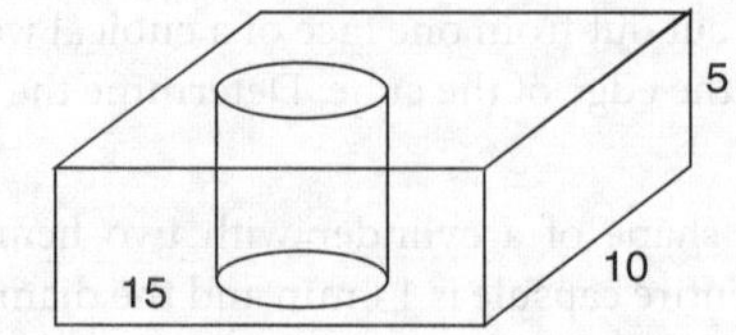

56. A military tent of height 8.25 m is in the form of a right circular cylinder of base diameter 30 m and height 5.5 m surmounted by a right circular cone of the same base radius. Find the length of the canvas used in making the tent, if the breadth of the canvas is 1.5 m. **[AI 2012]**

57. The radius and slant height of a right circular cone are in the ratio 7:13 and its curved surface area is 286 cm^2. Find the radius of the cone. [Take $\pi = \frac{22}{7}$]

58. A solid rectangular piece of iron measures $5 \times 6 \times 2$ m. Find the weight of this piece in kg if 1 m^3 of iron weighs 50 g. Also find the cost of the piece of iron at the rate of ₹ 750/ kg.

59. A metallic cylinder has radius 3 cm and height 5 cm. To reduce its weight, a conical hole is drilled in the cylinder. The conical hole has a radius of $\frac{3}{2}$ cm and its depth is $\frac{8}{9}$ cm. Calculate the ratio of the volume of metal left in the cylinder to the volume of metal taken out in the conical shape. **[Foreign 2015]**

60. Shanta runs an industry in a shed, which is in the shape of a cuboid surmounted by a half cylinder (see figure). If the base of the shed is of dimension 7×15 m, and the height of the cuboidal portion is 8 m, find the volume of air that the shed can hold. Further, suppose the machinery in the shed occupies a total space of 300 m^3, and there are 20 workers, each of whom occupy about 0.08 m^3 of space on an averag, how much air is in the shed? [Take $\pi = \frac{22}{7}$] **[NCERT]**

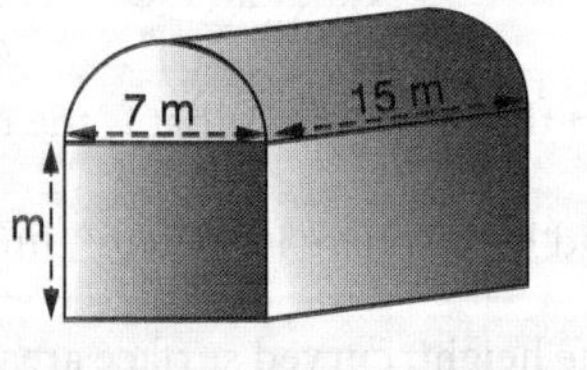

61. Rachel, an engineering student, was asked to make a model shaped like a cylinder, with two cones attached at its two ends using a thin aluminum sheet. The diameter of the model is 3 cm and its length is 12 cm. If each cone has a height of 2 cm, find the volume of air contained in the model that Rachel made. (Assume the outer and inner dimensions of the model to be nearly the same.) **[NCERT]**

62. A pen stand made of wood is in the shape of a cuboid with four conical depressions to hold pens. The dimensions of the cuboid are $15 \times 10 \times 3.5$ cm. The radius of each of the depressions is 0.5 cm and the depth is 1.4 cm. Find the volume of wood in the entire stand (see figure). **[NCERT]**

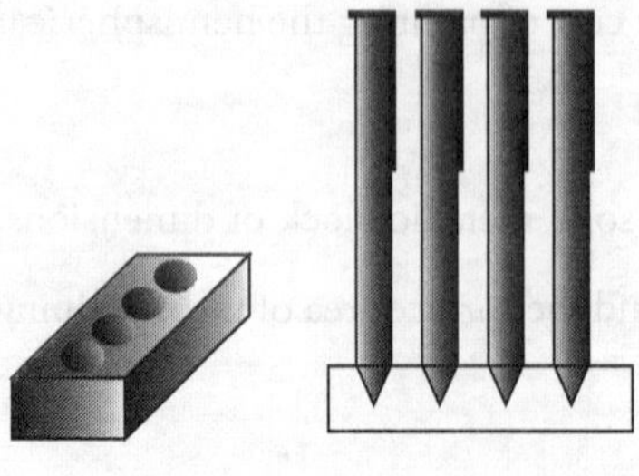

63. A vessel is in the form of an inverted cone. Its height is 8 cm and the radius of its top, which is open, is 5 cm. It is filled with water up to the brim. When lead shots, each of which is a sphere of radius 0.5 cm, are dropped into the vessel, one-fourth of the water flows out. Find the number of lead shots dropped in the vessel. **[NCERT]**

64. A solid iron pole consists of a cylinder of height 220 cm and base diameter 24 cm, which is surmounted by another cylinder of height 60 cm and radius 8 cm. Find the mass of the pole, given that 1 cm^3 of iron has approximately 8 g mass. [Use $\pi = 3.14$] **[NCERT]**

65. A spherical glass vessel has a cylindrical neck 8 cm long, 2 cm in diameter; the diameter of the spherical part is 8.5 cm. By measuring the amount of water it holds, a child finds its volume to be 345 cm^3. Check whether she is correct, taking the above as the inside measurements, and $\pi = 3.14$. **[NCERT]**

66. A copper wire, 3 mm in diameter, is wound about a cylinder whose length is 12 cm and diameter 10 cm, so as to cover the curved surface of the cylinder. Find the length and mass of the wire, assuming the density of copper to be 8.88 g per cm^3. **[NCERT]**

67. Two cylindrical cans have equal base areas. If one of the cans is 15 cm high and other is 20 cm high, find the ratio of their volumes.

68. A right circular cylinder and a cone have equal bases and equal heights. If their curved surface areas are in the ratio 8:5, show that the ratio between the radius of their bases to their height is 3:4. **[CBSE 2018 (C)]**

69. The sum of the radius of the base and height of a solid cylinder is 37 cm. If the total surface area of the solid cylinder is 1628 cm^2, find the volume of the cylinder. [$\pi = 22/7$] **[Delhi 2016]**

70. The length of a hall is 20 m and width 16 m. The sum of the areas of the floor and the flat roof is equal to the sum of the areas of the four walls. Find the height and volume of the hall.

71. The cost of painting the total outside surface of a closed cylindrical oil tank at 60 paise per sq. dm is ₹ 237.60. The height of the tank is six times the radius of the base of the tank. Find its volume.

72. A circus tent is in the shape of a cylinder surmounted by a conical top of the same diameter. If their common diameter is 56 cm, height of cylindrical part is 6 m and the total height of the tent above the ground is 27 m, find the area of canvas used in making the tent. **[Delhi 2017 (C)]**

73. A sector of a circle of radius 12 cm has an angle 120°. It is rolled up so that two bounding radii are joined together to form a cone. Find the volume of the cone.

74. The dimensions of a solid iron cuboid are $4.4 \times 2.6 \times 1.0$ m. It is melted and recast into a hollow cylindrical pipe of 30 cm inner radius and thickness 5 cm. Find the length of the pipe. **[AI 2017]**

75. A solid metallic cylinder of diameter 12 cm and height 15 cm is melted and recast into toys, each in the shape of a cone of radius 3 cm and height 9 cm. Find the number of toys so formed. **[AI 2017 (C)]**

76. A juice seller serves his customers using a glass, as shown in the figure. The inner diameter of the cylindrical glass is 5 cm, but the bottom of the glass has a raised hemispherical portion which reduces the capacity of the glass. If the height of the glass is 10 cm, find the apparent capacity of the glass and its actual capacity. [$\pi = 3.14$]

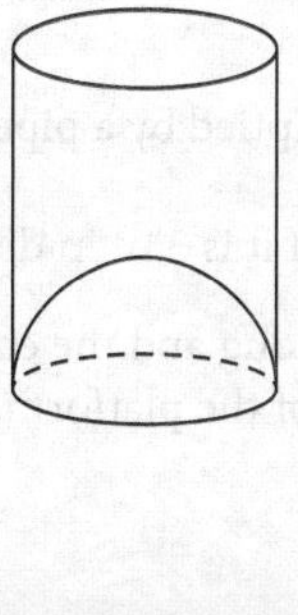

77. The interior of a building is in the form of a cylinder of diameter 4.3 m and height 3.8 m, surmounted by a cone whose vertical angle is a right angle. Find the area of the surface and volume of the building.

78. A solid is in the form of a cylinder with hemispherical ends. The total height of the solid is 20 cm and the diameter of the cylinder is 7 cm. Find the total volume of the solid. [Take $\pi = \frac{22}{7}$] **[Allahabad 2017]**

79. The areas of three adjacent faces of a rectangular block are in the ratio 2:3:4 and its volume is 9000 cu. cm. Find the length of the shortest side.

80. The volumes of two spheres are in the ratio 64:27. Find their radii if the sum of their radii is 21 cm.

81. A heap of wheat is in the form of a cone of diameter 9 m and height 3.5 m. Find its volume. How much canvas cloth is required to cover the heap? [Use $\pi = 3.14$]

82. From a solid cylinder whose height is 8 cm and radius 6 cm, a conical cavity of height 8 cm and of base radius 6 cm is hollowed out. Find the volume of the remaining solid correct to two decimal places. Also find the total surface area of the remaining solid. [Use $\pi = 3.1416$]

83. A cylindrical road roller made of iron is 1 m long. Its inner diameter is 54 cm and the thickness of the iron sheet rolled into the road roller is 9 cm. Find the weight of the roller if 1 cubic cm of iron weighs 8 g. [Use $\pi = 3.14$]

84. The height of a cone is 30 cm. A small cone is cut off at the top by a plane parallel to the base. If its volume is 1/27 of the volume of the given cone, at what height above the base is the section made? **[Foreign 2016]**

85. A rectangular block $6 \times 12 \times 15$ cm is cut into an exact number of equal cubes. Find the least number of such cubes.

86. A lead pencil consists of a cylinder of wood with a solid cylinder of graphite filled into it. The diameter of the pencil is 7 mm; the diameter of the graphite is 1 mm and the length of the pencil is 10 cm. Calculate the weight of the whole pencil if the specific gravity of the wood is 0.7 g/cm^3 and that of the graphite is 2.1 g/cm^3.

87. An iron pillar has some part in the form of a right circular cylinder and the remaining in the form of a right circular cone. The radius of the base of the cone and cylinder is 8 cm. The cylindrical part is 240 cm high and the conical part is 36 cm high. Find the weight of the pillar if one cubic cm of iron weighs 10 g. [Take $\pi = \frac{22}{7}$]

88. A cone of height 24 cm and radius of base 6 cm is made up of modelling clay. A child reshapes it in the form of a sphere. Find the radius of the sphere. **[NCERT]**

89. Selvi's house has an overhead tank in the shape of a cylinder. This is filled by pumping water from a sump (an underground tank), which is in the shape of a cuboid. The sump has dimensions 1.57 m $\times$ 1.44 m $\times$ 95 cm. The overhead tank has a radius 60 cm and height 95 cm. Find the height of the water left in the sump after the overhead tank has been completely filled with water from the sump which had been full. Compare the capacity of the tank with that of the sump. [Use $\pi = 3.14$] **[NCERT]**

90. A copper rod of diameter 1 cm and length 8 cm is drawn into a wire of length 18 m of uniform thickness. Find the thickness of the wire. **[NCERT]**

91. A hemispherical tank full of water is emptied by a pipe at the rate of $3\frac{4}{7}$ litres per second. How much time will it take to empty half the tank if it is 3 m in diameter? [Take $\pi = \frac{22}{7}$] **[NCERT]**

92. A 20-m-deep well with diameter 7 m is dug and the earth from digging is evenly spread out to form a platform of 22×14 m. Find the height of the platform. **[NCERT]**

93. A cylindrical bucket, 32 cm high and with radius of base 18 cm, is filled with sand. This bucket is emptied on the ground and a conical heap of sand is formed. If the height of the conical heap is 24 cm, find the radius and slant height of the heap. **[NCERT]**

94. The water in a canal, 6 m wide and 1.5 m deep, flows at a speed of 10 km/h. How much area will it irrigate in 30 minutes if 8 cm of standing water is needed? **[Delhi 2019, AI 2017 (C), NCERT]**

95. A farmer connects a pipe of internal diameter 20 cm from a canal into a cylindrical tank in her field, which is 10 m in diameter and 2 m deep. If water flows through the pipe at the rate of 3 km/h, in how much time will the tank be filled? **[NCERT]**

96. The radii of the ends of the frustum of a cone 45 cm high are 28 cm and 7 cm (see figure). Find its volume, the curved surface area and the total surface area. [Take $\pi = \frac{22}{7}$] **[NCERT]**

97. A cistern, internally measuring 150 × 120 × 110 cm, has 129,600 cm^3 of water in it. Porous bricks are placed in the water until the cistern is full to the brim. Each brick absorbs 1/17 of its own volume of water. How many bricks can be put in without the water overflowing, each brick being 22.5 × 7.5 × 6.5 cm? **[NCERT]**

98. A right triangle, whose sides are 3 cm and 4 cm (other than hypotenuse) is made to revolve about its hypotenuse. Find the volume and surface area of the double cone so formed. [Choose value of π as found appropriate] **[NCERT]**

99. In one fortnight of a given month, there was rainfall of 10 cm in a river valley. If the area of the valley is 7280 km^2, show that the total rainfall was approximately equivalent to the addition to the normal water of three rivers, each 1072 km long, 75 m wide and 3 m deep. **[NCERT]**

100. A cone of height 20 cm and radius of base 5 cm is made up of modelling clay. A child reshapes it in the form of a sphere. Find the diameter of the sphere. **[AI 2011]**

101. A solid sphere of radius 10.5 cm is melted and recast into smaller solid cones, each of radius 3.5 cm and height 3 cm. Find the number of cones so formed. [Take $\pi = \frac{22}{7}$] **[AI 2012]**

102. A hemispherical bowl of internal radius 9 cm is full of water. Its contents are emptied in a cylindrical vessel of internal radius 6 cm. Find the height of water in the cylindrical vessel. **[AI 2012]**

103. Solid spheres of diameter 6 cm are dropped into a cylindrical beaker containing some water and are fully submerged. If the diameter of the beaker is 18 cm and the water rises by 40 cm, find the number of solid spheres dropped in the water.

104. A teak wood log is cut first in the form of a cuboid of length 2.3 m, width 0.75 m and of a certain thickness. Its volume is 1.104 m^3. How many rectangular planks of size 2.3 × 0.75 × 0.04 m can be cut from the cuboid?

105. To construct a wall 24 m long, 0.4 m thick and 6 m high, bricks of dimensions 25 × 16 × 10 cm each are used. If the mortar occupies $1/10^{th}$ of the volume of the wall, find the number of bricks used.

106. A sphere of diameter 12 cm is dropped in a right circular cylindrical vessel, partly filled with water. If the sphere is completely submerged in water, the water level in the cylindrical vessel rises by $3\frac{5}{9}$ cm. Find the diameter of the cylindrical vessel.

107. About $\frac{3}{4}$th part of a conical vessel of internal radius 5 cm and height 24 cm is full of water. The water is emptied into a cylindrical vessel with internal radius 10 cm. Find the height of water in the cylindrical vessel. **[Delhi 2017]**

108. Water flows out through a circular pipe whose internal radius is 1 cm, at the rate of 80 cm/second into an empty cylindrical tank, the radius of whose base is 40 cm. By how much will the level of water rise in the tank in half an hour?

109. In a rainwater harvesting system, the rainwater from a roof of 22 × 20 m drains into a cylindrical tank having diameter of base 2 m and height 3.5 m. If the tank is full, find the rainfall in cm. **[AI 2017]**

110. Water is flowing at the rate of 15 km/hour through a pipe of diameter 14 cm into a cuboidal pond, which is 50 m long and 44 m wide. In what time will the level of water in the pond rise by 21 cm? **[AI 2011]**

111. Find the cost of sinking a tube well 280 m deep having diameter 3 m at the rate of ₹ 3.60 per m^3. Find the cost of cementing its inner curved surface at ₹ 2.50 per m^2.

112. An iron spherical ball has been melted and recast into smaller balls of equal size. If the radius of each of the smaller balls is 1/4 of the radius of the original ball, how many such balls are made? Compare the surface area of all the smaller balls combined with that of the original ball.

113. A 10 g of wood in the shape of a cylinder of base radius 1 ft and height 8 ft is cut all around to form a cuboid. What volume of the cylinder is removed?

114. A wax cylinder of diameter 21 cm and height 441 cm is chipped off and shaped to form a cone of maximum volume. The chipped-off wax is recast into a solid sphere. Find the diameter of the sphere.

115. The radii of the bases of two right circular solid cones of the same height are r_1 and r_2, respectively. The cones are melted and recast into a solid sphere of radius R. Show that the height of each cone is given by $h = \dfrac{4R^3}{r_1^2 + r_2^2}$.

116. An open box is made from a square lamina of side 18 cm, by cutting equal squares of side 3 cm at the corners and folding up the remaining flaps. Find the volume of the box.

117. A granary is in the shape of a cuboid of size 8 × 6 × 3 m. If a bag of grain occupies a space of 0.65 m^3, how many bags can be stored in the granary?

118. A rectangular water reservoir is 10.8 × 3.75 m at the base. Water flows into it at the rate of 18 m/s through a pipe having a cross-section 7.5 × 4.5 cm. Find the height to which the water will rise in the reservoir in 30 minutes.

119. A cylindrical tub of radius 16 cm contains water to a depth of 30 cm. A spherical iron ball is dropped into the tub and the level of water is raised by 9 cm. What is the radius of the ball?

120. A solid cylinder of diameter 12 cm and height 15 cm is melted and recast into 12 toys in the shape of a right circular cone mounted on a hemisphere. Find the radius of the hemisphere and the total height of the toy if the height of the cone is three times the radius.

121. A plot of land in the form of a rectangle has dimensions 240 × 180 m. A drainlet 10 m wide is dug all around it (on the outside) and the earth dug out is evenly spread over the plot, increasing its surface level by 25 cm. Find the depth of the drainlet.

122. A rocket is in the form of a right circular cylinder closed at the lower end and surmounted by a cone with the same radius as that of the cylinder. The diameter and height of the cylinder are 6 cm and 12 cm, respectively. If the slant height of the conical portion is 5 cm, find the total surface area and volume of the rocket. [Use $\pi = 3.14$] **[NCERT Exemplar]**

123. A copper sphere of radius 3 cm is beaten and drawn into a wire of diameter 0.2 cm. Find the length of the wire.

124. Two spheres of the same metal weigh 1 kg and 7 kg. The radius of the smaller sphere is 3 cm. The two spheres are melted to form a single big sphere. Find the diameter of the new sphere. **[Foreign 2015]**

125. Water is flowing at the rate of 2.52 km/h through a cylindrical pipe into a cylindrical tank, the radius of whose base is 40 cm. If the increase in the level of water in the tank in half an hour is 3.15 m, find the internal diameter of the pipe. **[Delhi 2015]**

126. From each end of a solid metal cylinder, metal was scooped out in hemispherical form of the same diameter. The height of the cylinder is 10 cm and its base is of radius 4.2 cm. The rest of the cylinder is melted and converted into a cylindrical wire of 1.4 cm thickness. Find the length of the wire. [Use $\pi = \frac{22}{7}$] **[Delhi 2015]**

127. Hanumappa and his wife Gangamma are busy making jaggery out of sugarcane juice. They have processed the sugarcane juice to make the molasses, which is poured into moulds in the shape of a frustum of a cone having the diameters of its two circular faces as 30 cm and 35 cm; the vertical height of the mould is 14 cm (see figure). If each cm^3 of molasses has a mass of about 1.2 g, find the mass of the molasses that can be poured into each mould. [Take $\pi = \frac{22}{7}$] **[NCERT]**

128. An open metal bucket is in the shape of a frustum of a cone, mounted on a hollow cylindrical base made of the same metallic sheet (see figure). The diameters of the two circular ends of the bucket are 45 cm and 25 cm, the total vertical height of the bucket is 40 cm and that of the cylindrical base is 6 cm. Find the area of the metallic sheet used to make the bucket, where we do not take into account the handle of the bucket. Also, find the volume of water the bucket can hold. [Take $\pi = \frac{22}{7}$] **[NCERT]**

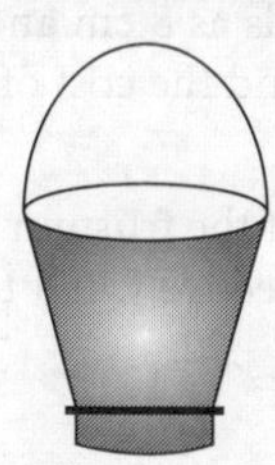

129. A *fez*, the cap used by Turks, is shaped like the frustum of a cone (see figure). If its radius on the open side is 10 cm, radius at the upper base is 4 cm and its slant height is 15 cm, find the area of material used for making it. **[NCERT]**

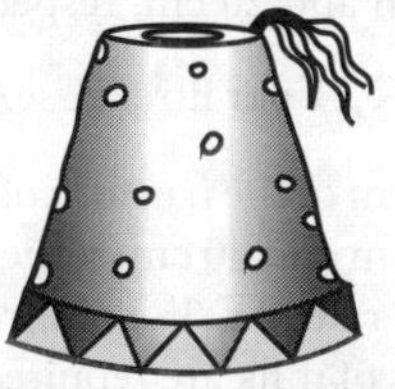

130. A container, opened from the top and made up of a metal sheet, is in the form of the frustum of a cone of height 16 cm with the radii of its lower and upper ends as 8 cm and 20 cm, respectively. Find the cost of the milk which can completely fill the container, at the rate of ₹ 20 per litre. Also find the cost of the metal sheet used to make the container, if it costs ₹ 8 per 100 cm^2. [Take $\pi = 3.14$] **[NCERT]**

131. A metallic right circular cone 20 cm high and whose vertical angle is 60° is cut into two parts at the middle of its height by a plane parallel to its base. If the frustum so obtained is drawn into a wire of diameter $\frac{1}{16}$ cm, find the length of the wire. **[NCERT]**

132. An oil funnel made of tin sheet consists of a 10-cm-long cylindrical portion attached to the frustum of a cone. If the total height is 22 cm, diameter of the cylindrical portion is 8 cm and the diameter of the top of the funnel is 18 cm, find the area of the tin sheet required to make the funnel (see figure). **[NCERT]**

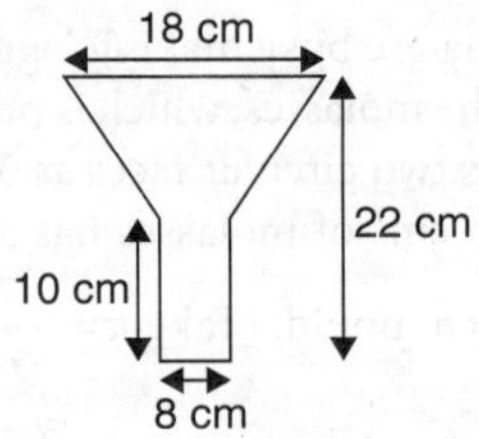

133. The diameters of the lower and upper ends of a bucket in the form of the frustum of a cone are 10 cm and 30 cm, respectively. If its height is 24 cm, find the area of the metal sheet used to make the bucket. [Use $\pi = 3.14$] **[CBSE 2018]**

134. A tent consists of the frustum of a cone, surmounted by a cone. If the diameter of the upper and lower circular ends of the frustum are 14 m and 26 m, respectively, the height of the frustum is 8 m and the slant height of the surmounted conical portion is 12 m, find the area of canvas required to make the tent. (Assume that the radii of the upper circular end of the frustum and the base of the surmounted conical portion are equal).

135. A bucket of height 16 cm is made up of metal sheets in the form of the frustum of a right circular cone with the radii of its lower and upper ends as 8 cm and 20 cm, respectively. Find the volume of milk which can be filled in the bucket. Also find the cost of making the bucket when the metal sheet costs ₹ 15 per 100 cm^2.

136. A bucket open at the top is of the form of the frustum of a cone. The diameters of its upper and lower circular ends are 40 cm and 20 cm, respectively. If a total of 17,600 cm^3 of water can be filled in the bucket, find its total surface area. [Use $\pi = \frac{22}{7}$] **[Foreign 2013]**

137. A bucket has a top and bottom diameter of 40 cm and 20 cm, respectively. Find the volume of the bucket if its depth is 12 cm. Also find the cost of the tin sheet used for making the bucket at the rate of ₹ 1.20 per cm^2.

138. A container open at the top is in the form of the frustum of a cone of height 24 cm with the radii of its lower and upper circular ends as 8 cm and 20 cm, respectively. Find the cost of milk which can completely fill the container at the rate of ₹ 21 per litre. [Use $\pi = \frac{22}{7}$] **[AI 2014]**

139. Milk in a container, which is in the form of the frustum of a cone of height 30 cm and the radii of whose lower and upper circular ends are 20 cm and 40 cm, respectively, is to be distributed in a camp for flood victims. If this milk is available at the rate of ₹ 35 per litre and 880 litres of milk are needed daily for a camp, find how many such containers of milk are required for a camp and what cost it will place on the donor agency for this. **[Foreign 2015]**

140. In the given figure, from the top of a solid cone of a height 12 cm and base radius 6 cm, a cone of height 4 cm is removed by a plane parallel to the base. Find the total surface area of the remaining solid.
$\left[\text{Use}\ \pi = \frac{22}{7}\ \text{and}\ \sqrt{5} = 2.236\right]$ **[Delhi 2015]**

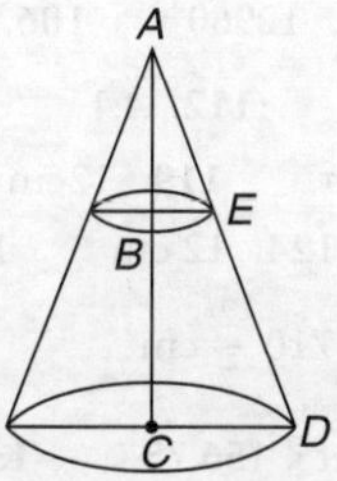

141. A solid metallic right circular cone 20 cm high and whose vertical angle is 60° is cut into two parts at the middle of its height by a plane parallel to its base. If the frustum so obtained is drawn into a wire of diameter $\frac{1}{12}$ cm, find the length of the wire. **[Delhi 2014]**

Answers

Multiple Choice Questions

1. (d) **2.** (a) **3.** (c) **4.** (b) **5.** (c) **6.** (d) **7.** (c) **8.** (a) **9.** (a) **10.** (c)
11. (b) **12.** (d) **13.** (d) **14.** (b) **15.** (b) **16.** (c) **17.** (a) **18.** (d) **19.** (b) **20.** (a)
21. (b) **22.** (c) **23.** (d) **24.** (c) **25.** (a)

Fill in the Blanks

26. $\frac{1}{3}\pi h(r_1^2 + r_2^2 + r_1 r_2)$ **27.** $\pi l(r_1 + r_2) + \pi r_1^2 + \pi r_2^2$ **28.** Spherical ball **29.** 8 cm
30. Height of standing water **31.** Cone **32.** $h = \frac{4R^3}{r_1^2 + r_2^2}$ **33.** $4\pi r^2$ **34.** $\frac{4}{3}\pi r^2$ **35.** Cone

Subjective Questions

36. 4 πr^2 sq units. **37.** 1: $\sqrt{3}$ **38.** 6 m **39.** 4:1 **40.** ₹ 5,365.80 **41.** 770 cm^2.
42. ₹ 513.45 **43.** 39.6 cm^2 **44.** 163.86 cm^2.
45. Area painted orange = 63.585 cm^2. Area painted yellow = 195.464 cm^2 **46.** $\frac{1}{4}\, l^2(\pi + 24)$, unit^2
47. 220 m^2 **48.** 44 m^2, ₹ 22000 **49.** 374 cm^2. **50.** 1:16 **51.** 718.67 cm^3
53. 100 hours **54.** 9.5 cm, ₹ 770. **55.** 583 cm^2. **56.** 825 m **57.** 7 m
58. Weight = 3 kg, Cost = ₹ 2250. **59.** 133:2 **60.** 1128. 75 m^3, 827.15 m^2. **61.** 61 cm^3
62. 523.53 cm^3 **63.** 100 **64.** 892.86 kg **65.** Not correct, 346.51 cm^3
66. = 1256 cm; 788 gm (approx) **67.** 3 : 4 **69.** 4620 cm^3 **70.** 8.88 m, 2844.4 m^3 **71.** 509.14 dm^3
72. 4136 m^2 **73.** 189.61 cm^3 **74.** 112 m **75.** 20 **76.** 196.25 cm^3, 163.55 cm^3
77. 71.896 cm^2, 65.61 m^3 **78.** Volume of the solid = 680.167 cm^3 **79.** 15 cm. **80.** 12 cm and 9 cm.

81. 80.541 cm^2. **82.** 603.19 cm^3, 603.1872 cm^2 **83.** 1424.304 kg **84.** 10 cm **85.** 40
86. 2.805gm **87.** 506.88 kg **88.** 6 cm **89.** 1:2 **90.** 0.67 mm **91.** 16.5 min **92.** 2.5 m
93. 36 cm; $12\sqrt{13}$ cm **94.** 562500 m^2, or 56.25 hectares **95.** 100 minutes.
96. 48510 cm^3, 5461.5 cm^2, 8079.5 cm^2 **97.** 1792 **98.** 30.14 cm^2, 52.78 cm^2 **100.** 10 cm **101.** 126
102. 13.5 cm **103.** 90 **104.** 16 **105.** 12960 **106.** 18 cm **107.** 1.5 m **108.** 90 cm
109. 2.5 cm **110.** 2 hours **111.** ₹ 6600 **112.** 4:1 **113.** $8(\pi - 2)$ ft^3 **114.** $21\sqrt[3]{21}$ cm
116. 432 cm^3 **117.** 221 bags **118.** 2.7 m **119.** 12cm **120.** 3 cm, 12 cm **121.** 1.23 m
122. 301.44 cm^2, 377.1 cm^3 **123.** 36 m **124.** 12 cm **125.** 4 cm **126.** 158.4 cm **127.** 14 kg
128. 4860.9 cm^2, 33.62 litres (approx) **129.** $710\frac{2}{7}$ cm^2
130. Cost of milk is ₹ 209 and cost of metal sheet ₹ 156.75. **131.** 7964.4 m **132.** $782\frac{4}{7}$ cm^2
133. 1711.3 cm^2 **134.** 284π m^2 **135.** 10,459.42 cm^3, ₹ 294.17 **136.** 2765.714 cm^2 **137.** ₹ 21.44
138. ₹ 329.47 **139.** ₹ 30,800 **140.** 350.592 cm^2 **141.** 4.48 km

Surface Areas and Volumes (Basic)

SELF-ASSESSMENT TEST

Time allowed: 90 min. Max. Marks: 40

General Instructions

1. *All questions are compulsory. There are 20 questions in all.*
2. *This question paper has four sections: Section A, Section B, Section C and Section D.*
3. *Section A contains 10 questions of one mark each, Section B contains three questions of two marks each, Section C contains four questions of three marks each, and Section D contains three questions of four marks each.*

Section A [1 Mark Each]

Direction (Q. Nos. 1–5): These are Multiple Choice Questions. Select the most appropriate answer for the given questions.

1. The volume of two cubes is in the ratio 8:125. The ratio of their surface area is
(a) 2:5 (b) 4:25 (c) 16:25 (d) 8:125

2. The number of revolutions taken by a wheel of diameter 84 cm to cover a distance of 792 m is
(a) 260 (b) 42p (c) 300 (d) 84p

3. The number of coins, 1.5 cm in diameter and 0.2 cm thick, to be melted to form a right circular cylinder of height 10 cm and diameter 4.5 cm is
(a) 350 (b) 400 (c) 450 (d) 500

4. The diameter of a wheel is 1.26 m. How far will it travel in 500 revolutions?
(a) 2670 m (b) 2880 m (c) 1980 m (d) 1596 m

5. How many spherical balls, each of radius 1 cm, can be made from a solid sphere of lead of radius 8 cm?
(a) 256 (b) 512 (c) 1024 (d) 576

6. The total surface area of the frustum of a cone is ___________.

7. The volume of water standing in a field = Area of field × ___________.

8. The dimensions of a metallic cuboid are $100 \times 80 \times 64$ cm. It is melted and recast into a cube. Find the surface area of the cube.

9. The dimensions of a metallic cuboid are $100 \times 90 \times 81$ cm. It is melted and recast into a cube. Find the surface area of the cube.

10. Two cubes have their volume in the ratio 1:64. What is the ratio of their surface area?

Section B [2 Marks Each]

11. There are two cones. The curved surface area of one is twice that of the other. The slant height of the latter is twice that of the former. Find the ratio of their radii.

12. Atish made a bird bath for his garden in the shape of a cylinder with a hemispherical depression at one end, as shown in the figure. The height of the cylinder is 2 m and its diameter is 84 cm. Find the total surface area of the bird bath. $\left[\text{Use } \pi = \dfrac{22}{7}\right]$

13. The volume of a hemisphere is $2425\frac{1}{2}$ cm^3. Find its curved surface area. $\left[\text{Use } \pi = \dfrac{22}{7}\right]$

Section C [3 Marks Each]

14. A medicine capsule is in the shape of a cylinder with two hemispheres stuck to each of its ends (see figure). The length of the entire capsule is 14 mm and the diameter of the capsule is 5 mm. Find its surface area.

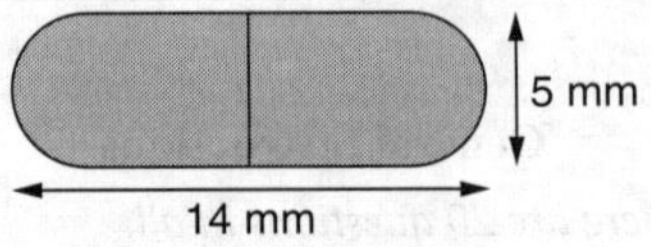

15. A wooden article was made by scooping out a hemisphere from each end of a solid cylinder, as shown in the figure. If the height of the cylinder is 10 cm and its base is of radius 3.5 cm, find the total surface area of the article.

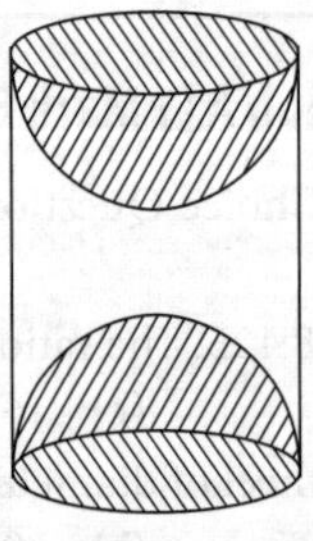

16. The sum of the radius of the base and the height of a solid cylinder is 37 cm. If the total surface area of the solid cylinder is 1628 cm^2, find the volume of the cylinder. [$\pi = 22/7$]

17. In a rainwater harvesting system, the rainwater from a roof of 22 × 20 m drains into a cylindrical tank having diameter of base 2 m and height 3.5 m. If the tank is full, find the rainfall in cm.

Section D [4 Marks Each]

18. The perimeters of the ends of a frustum are 48 cm and 36 cm. If the height of the frustum is 11 cm, find its volume.

19. A gulab jamun contains sugar syrup up to about 30% of its volume. Find approximately how much syrup would be found in 45 gulab jamuns, each shaped like a cylinder with two hemispherical ends of length 5 cm and diameter 2.8 cm.

20. A toy is in the form of a cone of radius 3.5 cm mounted on a hemisphere of the same radius. The total height of the toy is 15.5 cm. Find the total surface area of the toy.

Answers

1. (b) **2.** (c) **3.** (c) **4.** (c) **5.** (b) **6.** $\pi l(r_1 + r_2) + \pi r_1^2 + \pi r_2^2$

7. height of standing water **8.** 38400 cm^2 **9.** 48600 cm^2 **10.** 1:16 **11.** 4:1

12. 6.39 cm^2 **13.** 693 cm^2 **14.** 220 m^2 **15.** 374 cm^2 **16.** 4620 cm^3

17. 2.5 cm **18.** 1554 cm^3 **19.** 338 cm^3 **20.** 214.5 cm^2

Surface Areas and Volumes (Standard)

SELF-ASSESSMENT TEST

Time allowed: 90 min. Max. Marks: 40

General Instructions

1. *All questions are compulsory. There are 20 questions in all.*
2. *This question paper has four sections: Section A, Section B, Section C and Section D.*
3. *Section A contains 10 questions of one mark each, Section B contains three questions of two marks each, Section C contains four questions of three marks each, and Section D contains three questions of four marks each.*

Section A [1 Mark Each]

Direction (Q. Nos. 1–5): These are Multiple Choice Questions. Select the most appropriate answer for the given questions.

1. The shape of a gilli, in the gilli-danda game (see figure), is a combination of

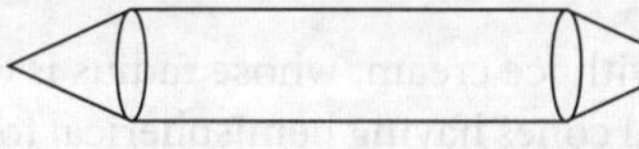

(a) two cylinders (b) a cone and a cylinder
(c) two cones and a cylinder (d) two cylinders and a cone

2. A rectangular block $6 \times 12 \times 15$ cm is cut into an exact number of equal cubes. The least possible number of cubes will be
(a) 6 (b) 11 (c) 3 (d) 40

3. A solid is hemispherical at the bottom and conical (of the same radius) above it. If the surface areas of the two parts are equal, then the ratio of its radius and slant height of the conical part is
(a) 2:1 (b) 1:2 (c) 1:4 (d) 4:1

4. The radius of a wire is decreased to one-third. If the volume remains the same, the length will become
(a) 3 times (b) 6 times (c) 9 times (d) 27 times

5. Twelve solid spheres of the same size are made by melting a solid metallic cylinder of base diameter 2 cm and height 16 cm. The diameter of each sphere is.
(a) 2 cm (b) 3 cm (c) 4 cm (d) 6 cm

6. The volume of water standing in a field = Area of field × ____________.

7. When we rotate a right triangle about the perpendicular line, then the three-dimensional figure formed is ____________.

8. The dimensions of a metallic cuboid are $100 \times 80 \times 64$ cm. It is melted and recast into a cube. Find the surface area of the cube.

9. If h, c and V, respectively, are the height, curved surface area and volume of a cone, prove that $3\pi Vh^3 - c^2h^2 + 9V^2 = 0$.

10. If the total surface area of a solid hemisphere is 462 cm², find its volume. [Take $\pi = \frac{22}{7}$]

Section B [2 Marks Each]

11. A decorative block, as shown in the figure, is made of two solids: a cube and a hemisphere. The base of the block is the cube with an edge of 5 cm and the hemisphere attached on the top has a diameter of 4.2 cm. If the block is to be painted, find the total area to be painted $\left[\text{Take } \pi = \frac{22}{7}\right]$.

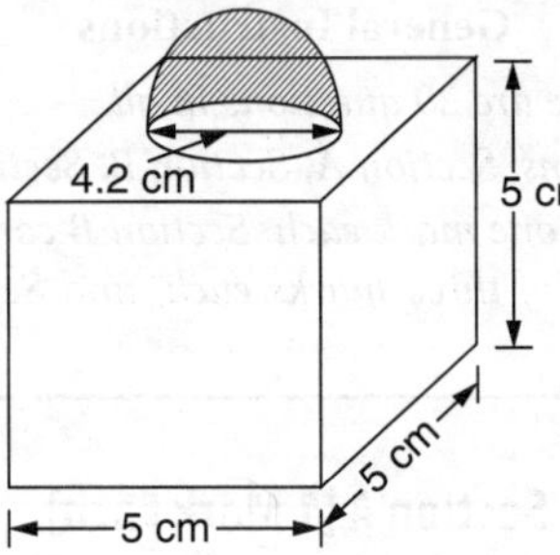

12. A gulab jamun contains sugar syrup up to about 30% of its volume. Find approximately how much syrup would be found in 45 gulab jamuns, each shaped like a cylinder with two hemispherical ends of length 5 cm and diameter 2.8 cm.

13. A cylindrical container is filled with ice cream, whose radius is 6 m and height 15 m. The ice cream is distributed to 10 children in equal cones having hemispherical tops. If the height of the conical portion is 4 times the radius of its base, find the radius of the cone.

Section C [3 Marks Each]

14. A toy is in the form of a cone of radius 3.5 cm mounted on a hemisphere of the same radius. The total height of the toy is 15.5 cm. Find the total surface area of the toy.

15. A well, whose diameter is 7m, has been dug 22.5 m deep and the earth dug out is used to form an embankment around it. If the height of the embankment is 1.5 m, find the width of the embankment.

16. Water is flowing at the rate of 7 metres per second through a circular pipe whose internal diameter is 2 cm into a cylindrical tank the radius of whose base is 40 cm. Determine the increase in the water level in 1/2 hour.

17. The interior of a building is in the form of a cylinder of diameter 4.3 m and height 3.8 m, surmounted by a cone whose vertical angle is a right angle. Find the area of the surface and the volume of the building.

Section D [4 Marks Each]

18. A hollow cone is cut by a plane parallel to the base and the upper portion is removed. If the curved surface of the remainder is $\frac{8}{9}$ of the curved surface of the whole cone, find the ratio of the line segment into which the cone's altitude is divided by the plane.

19. A metallic right circular cone 20 cm high and whose vertical angle is 60° is cut into two parts at the middle of its height by a plane parallel to its base. If the frustum so obtained is drawn into a wire of diameter $\frac{1}{16}$ cm, find the length of the wire.

20. A right circular cylinder and a cone have equal bases and equal heights. If their curved surface areas are in the ratio 8:5, show that the ratio between the radius of their bases to their height is 3:4.

Answers

1. (c) **2.** (d) **3.** (b) **4.** (c) **5.** (a) **6.** Height of standing water **7.** Cone

8. 38400 cm^2 **10.** 718.67 cm^3 **11.** 163.86 cm^2 **12.** 338 cm^3 **13.** 3 m

14. 214.5 cm^2 **15.** 10.5 m **16.** 787.5 cm **17.** 71.82 cm^2, 65.55 m^3

18. $\frac{1}{2}$ **19.** 7964.4 m

CHAPTER 14 Statistics

Key Concepts

- ❑ Mean of grouped data
- ❑ Median of grouped data
- ❑ Mode of grouped data
- ❑ Ogive of grouped data

Definitions

Statistics is the branch of mathematics that deals with CPAI, where

C ⇒ Represents collection of data
P ⇒ Represents presentation of data
A ⇒ Represents analysis of data
I ⇒ Represents interpretation of data

If all the steps of CPAI are fulfilled, then decision-making becomes very strong.

Types of Series

1. **Individual:** Only data is given in ascending or descending order.
2. **Discrete:** Frequency is given, along with data.
3. **Continuous:** Frequency is given, along with continuous data.

Concept: Mean

Mean of Grouped Data

The following methods are used to calculate the mean of grouped data:

(i) Direct method
(ii) Assumed mean method
(iii) Step deviation method

Direct Method

Steps

1. Find the class marks (mid-value) x_i of each class.

   Class mark of a class $= \dfrac{\text{Lower limit} + \text{Upper limit}}{2}$.
2. Calculate $f_i x_i$ for each x_i.
3. Find $\Sigma f_i x_i$ and Σf_i.
4. Calculate mean using the formula, $\bar{X} = \dfrac{\sum f_i x_i}{\sum f_i}$.

Example 1 Find the mean of the following frequency distribution: [CBSE (CCE) 2011]

Class	0–10	10–20	20–30	30–40	40–50
Frequency (*f*)	7	8	15	9	10

Solution: The frequency distribution table is as follows:

Class	Frequency	Mid-value	$f_i x_i$
0–10	7	5	35
10–20	8	15	120
20–30	15	25	375
30–40	9	35	315
40–50	10	45	450
	$\Sigma f_i = 49$		$\Sigma f_i x_i = 1295$

$$\text{Mean, } \overline{X} = \frac{\sum f_i x_i}{\sum f_i} = \frac{1295}{49} = 26.43.$$

Example 2 The arithmetic mean of the following distribution is 25. Determine the value of p. [CBSE 2006]

Class	0–10	10–20	20–30	30–40	40–50
Frequency	5	18	15	p	6

Solution: The frequency distribution table is as follows:

Class	Frequency f_i	Mid-value x_i	$f_i x_i$
0–10	5	5	25
10–20	18	15	270
20–30	15	25	375
30–40	p	35	$35p$
40–50	6	45	270
	$\Sigma f_i = 44 + p$		$\Sigma f_i x_i = 940 + 35p$

$$\text{Mean, } \overline{X} = \frac{\sum f_i x_i}{\sum f_i}$$

$$25 = \frac{940 + 35p}{44 + p}$$

$$\Rightarrow \quad 25(44 + p) = 940 + 35p$$
$$\Rightarrow \quad 1100 + 25p = 940 + 35p$$
$$\Rightarrow \quad 1100 - 940 = 35p - 25p$$
$$\Rightarrow \quad 160 = 10p$$
$$\Rightarrow \quad p = 16$$

Example 3 Find the missing frequencies f_1 and f_2 if the mean of 50 observations given below is 38.2.

Class	0–10	10–20	20–30	30–40	40–50	50–60	60–70
Frequency	4	4	f_1	10	f_2	8	5

Solution: The frequency distribution table is as follows:

Class	Frequency f_i	Mid-value x_i	f_ix_i
0–10	4	5	20
10–20	4	15	60
20–30	f_1	25	$25f_1$
30–40	10	35	350
40–50	f_2	45	$45f_2$
50–60	8	55	440
60–70	5	65	325
	$\Sigma f_i = 31 + f_1 + f_2$		$\Sigma f_ix_i = 1195 + 25f_1 + 45f_2$

$$\Sigma f_i = 31 + f_1 + f_2 = 50 \text{ [Given]}$$

$$\therefore \quad f_1 + f_2 = 19 \qquad \ldots\text{(i)}$$

Mean $$\bar{X} = \frac{\sum f_ix_i}{\sum f_i}$$

$$38.2 = \frac{1195 + 25f_1 + 45f_2}{50}$$

$$\Rightarrow \quad 38.2 \times 50 = 1195 + 25f_1 + 45f_2$$
$$\Rightarrow \quad 1910 - 1195 = 25f_1 + 45f_2$$
$$\Rightarrow \quad 25f_1 + 45f_2 = 715$$
$$\Rightarrow \quad 5f_1 + 9f_2 = 143 \qquad \ldots\text{(ii)} \quad \text{[On dividing both sides by 5]}$$

Multiplying (i) by 5, on both sides

$$5f_1 + 5f_2 = 95 \qquad \ldots\text{(iii)}$$

Subtracting (ii) and (iii), we get $5f_1 + 9f_2 = 143$

$$5f_1 + 5f_2 = 95$$

$$- \quad - \quad -$$

$$4f_2 = 48$$

∴ $f_2 = 12$

Inserting $f_2 = 12$ in (i), we get

$$x + 12 = 19$$

∴ $f_1 = 7$

Hence $f_1 = 7$ and $f_2 = 12$.

1 Try to Solve

1. Find the mean of the frequency distribution:

Class interval	0–10	10–20	20–30	30–40	40–50	50–60
Frequency	12	18	27	20	17	6

2. Find the mean of the following frequency distribution: **[CBSE 2006]**

Class	0–20	20–40	40–60	60–80	80–100
Frequency	15	18	21	29	2

3. The mean of the following distribution is 18. Find the missing frequency f. **[NCERT]**

Class interval	11–13	13–15	15–17	17–19	19–21	21–23	23–25
Frequency	3	6	9	13	f	5	4

4. Find the missing frequencies x and y, if the mean of the given distribution is 14.4. **[NCERT]**

Class interval	0–6	6–12	12–18	18–24	24–30	Total
Frequency	4	x	5	y	1	20

5. The mark obtained by 120 students in a mathematics test is given in the following distribution:

Marks obtained	0–20	20–40	40–60	60–80	80–100	Total
No. of students	17	f_1	32	f_2	19	120

The number of students in the marks groups 20–40 and 60–80 are f_1 and f_2, respectively. Find f_1 and f_2 if the mean marks is 50.

Answers

1. 28 2. 53 3. 8 4. $x = 3, y = 7$ 5. $f_1 = 28, f_2 = 24$

Assumed Mean Method (Short-cut Method)

Steps

1. Find the class mark (or mid-value) x_i of each class.

   $$\text{Class mark} = \frac{\text{Lower limit} + \text{Upper limit}}{2}.$$

2. Choose a suitable value of x_i in the middle of the x_i column, known as assumed mean, and denote it as 'a'.
3. Calculate the deviation of each x_i from the assumed mean, that is, $d_i = (x_i - a)$.
4. Calculate $f_i d_i$ for each d_i.
5. Find $\sum f_i d_i$ and $\sum f_i$.
6. Calculate mean using the formula

   $$\bar{X} = a + \frac{\sum f_i d_i}{\sum f_i}$$

Example 4 Find the mean of the following frequency distribution: **[CBSE 2007 C]**

Class	0–100	100–200	200–300	300–400	400–500
Frequency	6	9	15	12	8

Solution:

Class	Frequency (f_i)	Mid-value (x_i)	$d_i = x_i - a$	$f_i d_i$
0–100	6	50	–200	–1200
100–200	9	150	–100	–900
200–300	15	(250)	0	0
300–400	12	350	100	1200
400–500	8	450	200	1600
	$\Sigma f_i = 50$			$\Sigma f_i d_i = 700$

Let the assumed mean (a) be 250.

Mean $$\bar{X} = a + \frac{\sum f_i d_i}{\sum f_i} = 250 + \frac{700}{50}$$

$$= 250 + 14 = 264$$

2 Try to Solve

1. Find the mean of the frequency distribution:

Class interval	0–6	6–12	12–18	18–24	24–30
Frequency	6	8	10	9	7

2. Thirty women were examined in a hospital by a doctor and the number of heart beats per minute were recorded and summarised as follows. Find the mean heart beats per minute for these women. [NCERT]

No. of beats per minute	65–68	68–71	71–74	74–77	77–78	80–83	83–86
No. of women	2	4	3	8	7	4	2

3. The table below shows the daily expenditure on food of 25 households in a locality. Find the mean expenditure on food by a suitable method.

Daily expenditure (in ₹)	100–150	150–200	200–250	250–300	300–350
Number of households	4	5	12	2	2

4. To find out the concentration of SO_2 in the air (in parts per million or ppm), data was collected for 30 localities in a certain city and is presented below: [NCERT]

Concentration of SO_2 (in ppm)	Frequency
0.00–0.04	4
0.04–0.08	9
0.08–0.12	9
0.12–0.16	2
0.16–0.20	4
0.20–0.24	2

Find the mean concentration of SO_2 in the air.

5. A class teacher has the following absentee record of 40 students of a class for the whole term. Find the mean number of days a student was absent. [NCERT]

No. of days	0–6	6–10	10–14	14–20	20–28	28–38	38–40
No. of students	11	10	7	4	4	3	1

Answers

1. 15.45 2. 75.9 3. ₹ 211 4. 0.099 ppm 5. 12.38 days

Step Deviation Method

This method is used when the values of f_i and x_i are large.

Steps

1. Find out the mid-values for all classes.

$$\text{Mid-value} = \frac{\text{Lower limit} + \text{Upper limit}}{2}.$$

2. Choose a suitable value of x_i in the middle of the x_i column, known as assumed mean, and denote it as 'a'.
3. Calculate $d_i = x_i - a$ for all x_i.

h is the difference between the upper and lower limits.

4. Calculate $u_i = \dfrac{x_i - a}{h}$ for each $d_i = x_i - a$, where h is the class size or common factor for all d_i.
5. Calculate $f_i u_i$ for each u_i.
6. Find $\Sigma f_i u_i$ and Σf_i.
7. Calculate mean using the formula

$$\bar{X} = a + \frac{\sum f_i u_i}{\sum f_i} \times h$$

Example 5 Calculate mean for the following distribution by step deviation method:

Class	0–20	20–40	40–60	60–80	80–100	100–120
Frequency	20	35	52	44	38	31

Solution: The frequency distribution table is

Class	Frequency (f_i)	Mid-value (x_i)	$x_i - a$	$u_i = \dfrac{x_i - a}{h}$	$f_i u_i$
0–20	20	10	– 40	– 2	– 40
20–40	35	30	–20	–1	–35
40–60	52	50	0	0	0
60–80	44	70	20	1	44
80–100	38	90	40	2	76
100–120	31	110	60	3	93
	$\Sigma f_i = 220$				$\Sigma f_i u_i = 138$

Let the assumed mean (a) be 50.

Class size $h = 20$

Mean $\bar{X} = a + \dfrac{\sum f_i u_i}{\sum f_i} \times h$

$$= 50 + \frac{138}{220} \times 20 = 50 + 12.54 = 62.54$$

Example 6 Find the mean age from the following frequency distribution:

Age (in years)	25–29	30–34	35–39	40–44	45–49	50–54	55–59
Numbers of persons	4	14	22	16	6	5	3

Solution: The given frequency distribution is in inclusive form. We first convert it to exclusive form.

Age (in years)	No. of persons (f_i)	Mid-value (x_i)	$x_i - a$	$u_i = \frac{x_i - a}{h}$	$f_i u_i$
24.5–29.5	4	27	– 15	– 3	– 12
29.5–34.5	14	32	–10	–2	–28
34.5–39.5	22	37	–5	–1	–22
39.5–44.5	16	(42)	0	0	0
44.5–49.5	6	47	5	1	6
49.5–54.5	5	52	10	2	10
54.5–59.5	3	57	15	3	9
	$\Sigma f_i = 70$				$\Sigma f_i u_i = -37$

Let the assumed mean (a) be 42.

Class size $h = 5$

Mean $\overline{X} = a + \dfrac{\sum f_i u_i}{\sum f_i} \times h$

$= 42 + \dfrac{-37}{70} \times 5 = 42 - 2.64 = 39.36$

∴ Mean age = 39.36 years.

3 Try to Solve

1. Find the mean of the frequency distribution: **[NCERT]**

Class interval	0–6	6–12	12–18	18–24	24–30
Frequency	6	8	10	9	7

2. Thirty women were examined in a hospital by a doctor and the number of heart beats per minute were recorded and summarised as follows. Find the mean heart beats per minute for these women. **[NCERT]**

No. of beats per minute	65–68	68–71	71–74	74–77	77–80	80–83	83–86
No. of women	2	4	3	8	7	4	2

Find the mean for each of the following frequency distributions: (5–14)

3. The table below shows the daily expenditure on food of 25 households in a locality. **[NCERT]**

Daily expenditure (in ₹)	100–150	150–200	200–250	250–300	300–350
Number of households	4	5	12	2	2

Find the mean expenditure on food by a suitable method.

4. To find out the concentration of SO_2 in the air (in parts per million or ppm), data was collected for 30 localities in a certain city and is presented below: [NCERT]

Concentration of SO_2 (in ppm)	Frequency
0.00–0.04	4
0.04–0.08	9
0.08–0.12	9
0.12–0.16	2
0.16–0.20	4
0.20–0.24	2

Find the mean concentration of SO_2 in the air.

5. A class teacher has the following absentee record of 40 students of a class for the whole term. Find the mean number of days a student was absent. [NCERT]

No. of days	0–6	6–10	10–14	14–20	20–28	28–38	38–40
No. of students	11	10	7	4	4	3	1

Answers

1. 15.45 2. 75.9 3. ₹ 211 4. 0.009 ppm 5. 12.38 days

Concept: Median

For median of individual series, where the number of terms determines whether it is odd or even, we must first arrange the terms in ascending or descending order.

For odd number of terms: $\text{Median} = \left(\frac{n+1}{2}\right)^{\text{th}}$ term

For even number of terms: $\text{Median} = \dfrac{\left(\frac{n}{2}\right)^{\text{th}} \text{term} + \left(\frac{n}{2}+1\right)^{\text{th}} \text{term}}{2}$

Finding Median for Grouped Data

Steps

1. For the given frequency distribution, obtain $N = \Sigma f_i$.
2. Find $\frac{N}{2}$.
3. Look at the cumulative frequency just greater than $\frac{N}{2}$ and find the corresponding class, known as median class (as the middle-most observation lies in this class).

4. Compute the median using the formula:

$$M_e = l + \left[\frac{h \times \left(\frac{N}{2} - cf\right)}{f}\right],$$

where l = lower limit of median class; h = width of median class; f = frequency of median class; cf = cumulative frequency of the class preceding the median class; and $N = \Sigma f_i$.

Example 7 Find the median of the following data:

Marks	20–30	30–40	40–50	50–60	60–70	70–80	80–90
No. of students	5	15	25	20	7	8	10

Solution: We prepare the cumulative frequency table, as given below:

Class	Frequency (f_i)	Cumulative frequency
20–30	5	5
30–40	15	20
40–50	25	45
50–60	20	65
60–70	7	72
70–80	8	80
80–90	10	90
	$N = \Sigma f_i = 90$	

Now $N = 90 \Rightarrow \frac{N}{2} = 45$

The cumulative frequency just greater than 45 is 65 and the corresponding class is 50–60.

Thus, the median class is 50–60.

∴ $l = 50$, $h = 10$, $f = 20$, $cf = cf$ of the preceding class $= 45$.

Median $$M_e = l + \left\{\frac{h \times \left(\frac{N}{2} - cf\right)}{f}\right\} = 50 + \left\{10 \times \frac{(45 - 45)}{20}\right\} = 50 + 0 = 50.$$

Hence, the median mark is 50.

4 Try to Solve

1. In a hospital, the ages of diabetic patients were recorded as follows. Find the median age.

Age (in years)	0–15	15–30	30–45	45–60	60–75
Number of patients	5	20	40	50	25

2. The following table shows the daily wages of workers in a factory:

Daily wages (in ₹)	0–100	100–200	200–300	300–400	400–500
Number of workers	40	32	48	22	8

Find the median daily wage income of the workers.

3. The following frequency distribution gives the monthly consumption of electricity of 68 consumers of a locality. Find the median. **[NCERT]**

Monthly consumption (in units)	Number of consumers
65–85	4
85–105	5
100–125	13
125–145	20
145–165	14
165–185	8
185–205	4

Answers

1. 46.5 years 2. ₹ 206.25 3. 137 units

Example 8 If the median of the following frequency distribution is 24, find the missing frequency:

Class interval	0–10	10–20	20–30	30–40	40–50
Frequency	5	25	?	18	7

Solution: Let the missing frequency be x. The cumulative frequency distribution table is as follows:

Class interval	Frequency	Cumulative frequency
0–10	5	5
10–20	25	30
20–30	x	$30 + x$ ←Median class
30–40	18	$48 + x$
40–50	7	$55 + x$
Total frequency = $55 + x$		

As median = 24, it lies in class 20–30. ∴ The median class is 20–30.

l (lower limit of median class) = 20;

cf (cumulative frequency of class preceding the median class) = 30; f (frequency of the median class) = x; h (class size) = 10.

$$N = 55 + x \Rightarrow \frac{N}{2} = \frac{55 + x}{2}$$

$$\text{Median} = l + \left(\frac{\frac{N}{2} - cf}{f} \right) \times h$$

$$24 = 20 + \frac{\frac{55 + x}{2} - 30}{x} \times 10$$

$$24 - 20 = \frac{55 + x - 60}{2x} \times 10$$

$$4 = \frac{x - 5}{\cancel{2}x} \times \cancel{10}$$

$$4 = \frac{(x - 5) \times 5}{x}$$

$$4x = 5x - 25$$

$$\therefore \quad x = 25$$

Hence, the missing frequency = 25.

5 Try to Solve

1. Find the missing frequency of the given distribution:

Class interval	0–10	10–20	20–30	30–40	40–50
Frequency	5	25	f	18	7

The median given is 24.

Answer

1. 25

Example 9 The median of the following data is 525. Find the values of x and y, if the total frequency is 100. [NCERT]

Class interval	Frequency
0–100	2
100–200	5
200–300	x

(Continued)

Class interval	Frequency
300–400	12
400–500	17
500–600	20
600–700	y
700–800	9
800–900	7
900–1000	4

Solution:

Class interval	Frequency	Cumulative frequency
0–100	2	2
100–200	5	7
200–300	x	$7 + x$
300–400	12	$19 + x$
400–500	17	$36 + x$
500–600	20	$56 + x$
600–700	y	$56 + x + y$
700–800	9	$65 + x + y$
800–900	7	$72 + x + y$
900–1000	4	$76 + x + y$

It is given that $N = 100$. So, $76 + x + y = 100$, that is, $x + y = 24$.

The median is 525, which lies in the class 500–600.

So $$l = 500, f = 20, cf = 36 + x, h = 100$$

Using the formula: $$\text{Median} = l + \left(\frac{\frac{N}{2} - cf}{f}\right) h, \text{ we get}$$

$$525 = 500 + \left[\frac{50 - (36 + x)}{20}\right] \times 100$$

$$525 = 500 + \left(\frac{50 - 36 - x}{\cancel{20}}\right) \times \cancel{100}$$

That is $525 - 500 = (14 - x) \times 5$

$$25 = 70 - 5x$$

$$5x = 70 - 25 = 45$$

So $x = 9$

Therefore $9 + y = 24$

That is $y = 15$

6 Try to Solve

1. If the median of the distribution given below is 28.5, find the values of x and y. [NCERT]

Class interval	Frequency
0–10	5
10–20	x
20–30	20
30–40	15
40–50	y
50–60	5
Total	60

2. The median of the following data is 32.5. Find the values of x and y.

Class interval	0–10	10–20	20–30	30–40	40–50	50–60	60–70	Total
Frequency	x	5	9	12	y	3	2	40

3. If the median of the following distribution is 46, find the missing frequencies, f_1 and f_2.

Class interval	10–20	20–30	30–40	40–50	50–60	60–70	70–80	Total
Frequency	12	30	f_1	65	f_2	25	18	229

Answers

1. $x = 8, y = 7$ 2. $x = 3, y = 6$ 3. $f_1 = 34, f_2 = 45$

Concept: Mode

The mode of any frequency distribution is the value of the observation for which the frequency is maximum. A given frequency distribution may have the same maximum frequency for more than one observation. Then the given data is multimodal.

Mode of Grouped Data

Steps

1. From the given frequency distribution, determine the class of maximum frequency distribution. This is known as modal class. Mode is a value inside the modal class.

2. Calculate mode using the formula

$$\text{Mode} = l + \frac{f_1 - f_0}{2f_1 - f_0 - f_2} \times h$$

where l = lower limit of the modal class; f_1 = frequency of the modal class; f_0 = frequency of the class preceding the modal class; f_2 = frequency of the class succeeding the modal class; and h = class size.

Empirical Relationship between the Three Measures of Central Tendency

The empirical relationship between mean, median and mode is

Mode = 3 Median – 2 Mean

This relationship can be used to find a central tendency when the other two central tendencies are given.

Example 10 Find the mode of the following data: **[CBSE (CCE) 2012]**

Marks	50–60	60–70	70–80	80–90	90–100
No. of students	3	12	32	20	6

Solution: As the class interval 70–80 has maximum frequency, that is 32, the modal class is 70–80.

l (lower limit of modal class) = 70; f_1 (frequency of the modal class) = 32; f_0 (frequency of the class preceding the modal class) = 12; f_2 (frequency of the class succeeding the modal class) = 20; h (class size) = 10.

$$\text{Mode} = l + \left(\frac{f_1 - f_0}{2f_1 - f_0 - f_2}\right) h$$

$$= 70 + \left(\frac{32 - 12}{2 \times 32 - 12 - 20}\right) \times 10 = 70 + \frac{20}{32} \times 10 = 70 + 6.25 = 76.25$$

Example 11 What is the mode of the data if median = 3 and mean = 2?

Solution: We know that empirical relationship

$$\text{Mode} = 3 \text{ Median} - 2 \text{ Mean}$$
$$\text{Mode} = 3 \times 3 - 2 \times 2 = 9 - 4$$
$$\text{Mode} = 5$$

Example 12 What is the median of the data if mode = 5 and mean = 2?

Solution: We know that empirical relationship

$$\text{Mode} = 3 \text{ Median} - 2 \text{ Mean}$$
$$5 = 3 \text{ Median} - 2\,(2)$$
$$5 = 3 \text{ Median} - 4$$
$$5 + 4 = 3 \text{ Median}$$
$$9 = 3 \text{ Median}$$
$$\text{Median} = \frac{9}{3} = 3$$

7 Try to Solve

1. The following data gives information on the observed lifetimes (in hours) of 225 electrical components: [NCERT]

Lifetimes (in hours)	0–20	20–40	40–60	60–80	80–100	100–120
Frequency	10	35	52	61	38	29

Determine the modal lifetimes and the components.

2. The following distribution gives the state-wise teacher–student ratio in the higher secondary schools of India. Find the mode and mean of this data. Interpret the two measures. [NCERT]

Number of students per teacher	Number of states/UT
15–20	3
20–25	8
25–30	9
30–35	10
35–40	3
40–45	0
45–50	0
50–55	2

3. The given distribution shows the number of runs scored by some top batsman of the world in one-day international cricket matches. Find the mode of the data. [NCERT]

Runs scored	Number of batsmen
3000–4000	4
4000–5000	18
5000–6000	9
6000–7000	7
7000–8000	6
8000–9000	3
9000–10000	1
10000–11000	1

4. What is the mean of the data if median = 3 and mode = 5?
5. What is the mode of the data if median = 2 and mean = 1?
6. What is the median of the data if mode = 4 and mean = 1?

Answers

1. 65.625 hours
2. Mode = 30.6, Mean = 29.2, Most states/UT have a student–teacher ratio of 30.6 and on an average, this ratio is 29.2
3. 4608.7 runs. 4. 2 5. 4 6. 2

Concept: Cumulative Frequency Curve (Or Ogive)

An ogive is a freehand graph showing the curve of a cumulative frequency distribution. The term 'ogive' is derived from the word 'ogee', meaning a shape consisting of a concave arc flowing into a convex arc, thus forming an S-shaped curve with vertical ends.

Let a grouped frequency distribution be given.

For a 'less than' series:
On a graph paper, we mark the lower class limits along the x axis and the corresponding cumulative frequencies along the y axis.

(i) On joining these points successively by line segments, we get a polygon, called the cumulative frequency polygon.
(ii) On joining these points successively by smooth curves, we get a curve, known as the cumulative frequency curve or an ogive.

For a 'greater than' series:
On a graph paper, we mark the lower class limits along the x axis and the corresponding cumulative frequencies along the y axis.

(i) On joining these points successively by line segments, we get a polygon, called the cumulative frequency polygon.
(ii) On joining these points successively by smooth curves, we get a curve, known as the cumulative frequency curve or an ogive.

Obtaining Median from Cumulative Frequency Curve

Steps – Method 1

1. Draw either the less than or the greater than cumulative frequency curve for the given frequency distribution.
2. Locate $\frac{N}{2}$ on the y axis. Let it be A.
3. From point A, draw a line AB parallel to the x axis, cutting the curve at a point B.
4. From B, draw BC perpendicular to the x axis, cutting the x axis at C.

The point of intersection of this perpendicular with the x axis, that is, the abscissa or x coordinate of C gives the median of the data.

Steps – Method 2

1. Draw both ogives (less than and greater than) for the given frequency distribution on the same axis.
2. The two ogives intersect each other at a point, say A.

3. From A, draw AB perpendicular to the x axis, meeting the x axis at B.
   Clearly, the abscissa or x coordinate of point B gives the median of the data.

Example 13 Draw a cumulative frequency curve (less than) for the following data and find median from it:

Class intervals	200–220	220–240	240–260	260–280	280–300	300–320
Frequency	7	3	6	8	2	4

Solution: The cumulative frequency table is given below:

Class	Frequency	Cumulative frequency
Less than 220	7	7
Less than 240	3	10
Less than 260	6	16
Less than 280	8	24
Less than 300	2	26
Less than 320	4	30

The curve is shown in the given figure.

$$N = 30$$

$$\therefore \quad \frac{N}{2} = 15$$

Median from the given figure = 258

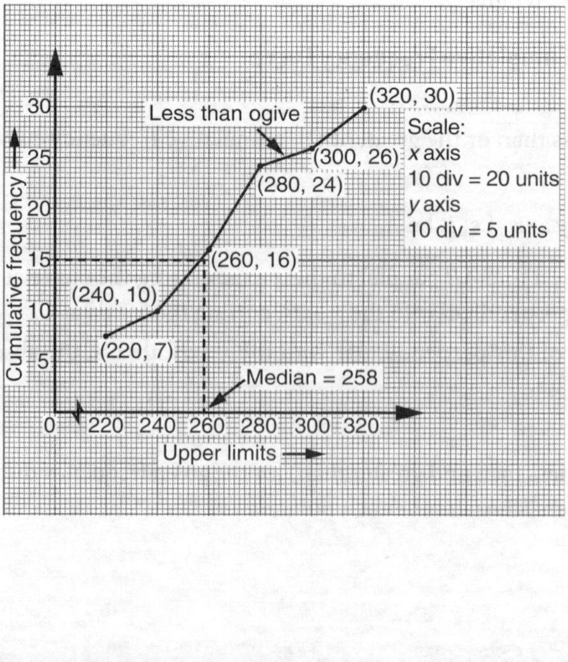

8 Try to Solve

1. During the medical check-up of 35 students of a class, their weights were recorded as follows: **[NCERT]**

Weight (in kg)	Number of students
Less than 38	0
Less than 40	3
Less than 42	5
Less than 44	9
Less than 46	14
Less than 48	28
Less than 50	32
Less than 52	35

Draw a less than ogive for the given data. Then obtain the median weight from the graph and verify the result using the formula.

2. The following distribution gives the daily wages of 58 workers of a factory: **[CBSE (CCE) 2011]**

Wages (in ₹)	20–40	40–60	60–80	80–100	100–120	120–140	140–160
No. of workers	4	6	10	16	12	7	3

Convert the above distribution to a less than cumulative frequency distribution. Draw its ogive.

Answers

1. Draw the ogive by plotting the points: (38, 0), (40, 3), (42, 5), (44, 9), (46, 14), (48, 28), (50, 32) and (52, 35). Here $\frac{N}{2} = 17.5$. Locate the point on the ogive whose ordinate is 17.5.

Example 14 The heights of 50 plants in a garden were recorded and the data is presented below. Draw a more than ogive for this data. **[CBSE (CCE) 2012]**

Height (in cm)	135–140	140–145	145–150	150–155	155–160	160–165
Number of plants	4	7	18	11	6	4

Solution: The more than cumulative frequency table is shown below:

Height (in cm)	Frequency	Cumulative frequency
More than 135	4	50
More than 140	7	46
More than 145	18	39

(*Continued*)

Height (in cm)	Frequency	Cumulative frequency
More than 150	11	21
More than 155	6	10
More than 160	4	4
	Total = 50	

The more than ogive is shown in the given figure.

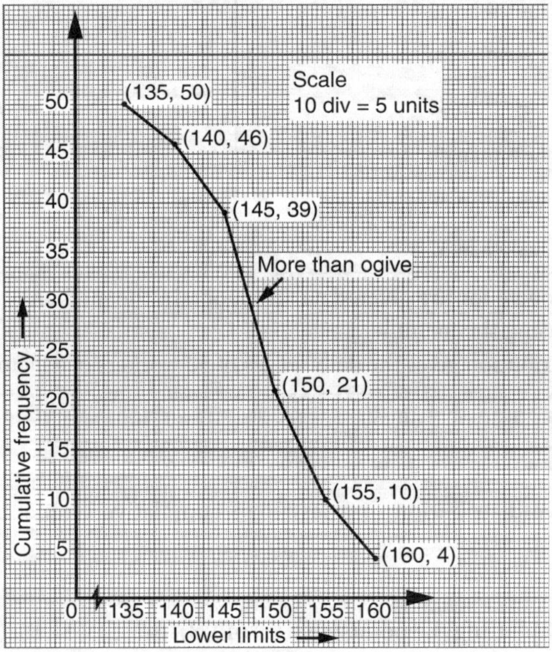

9 Try to Solve

1. Change the following distribution to a more than distribution and draw its ogive. From the graph, find the median. **[CBSE (CCE) 2011]**

Class	0–5	5–10	10–15	15–20	20–25
Frequency	6	8	10	6	4

2. The annual profits earned by 30 shops of a shopping complex in a locality are represented by the following distribution: **[CBSE (CCE) 2012]**

Profit (in Lakhs ₹)	No. of shops
More than or equal to 5	30
More than or equal to 10	28

Profit (in Lakhs ₹)	No. of shops
More than or equal to 15	16
More than or equal to 20	14
More than or equal to 25	10
More than or equal to 30	7
More than or equal to 35	3

Draw a more than ogive and find the median.

Answers

1. 11.5 2. 17.5

FOR MORE PRACTICE AND KNOWLEDGE

Example 15 The following table gives the enrollment in Higher Secondary Schools in 2012. Find the average enrollment per school.

Enrollment	No. of schools
20–39	526
40–59	620
60–79	674
80–99	717
100–119	681
120–139	612
140–159	540
160–179	517
180–199	552
Total	**5439**

Solution: Here, $a = 109.5$, $h = 20$. Then, we have the following table for computation of mean.

Class interval	Class frequency (f_i)	Class mid-point (x_i)	$u_i = \frac{x_i - 109.5}{20}$	$f_i u_i$
20–39	526	29.5	–4	–2104
40–59	620	49.5	–3	–1860
60–79	674	69.5	–2	–1348
80–99	717	89.5	–1	–717

(Continued)

Class interval	Class frequency (f_i)	Class mid-point (x_i)	$u_i = \frac{x_i - 109.5}{20}$	$f_i u_i$
100–119	681	109.5	0	0
120–139	612	129.5	+1	+612
140–159	540	149.5	+2	+1080
160–179	517	169.5	+3	+1551
180–199	552	189.5	+4	+2208
Total	5439			$\sum_{i=1}^{9} f_i u_i = -578$

$$\therefore \quad \bar{X} = a + \frac{\sum f_i u_i}{\sum f_i} \times h = 109.5 + \frac{(-578)}{5439} \times 20 = 109.5 - 2.12 = 107.38$$

$\therefore$ Mean enrollment = 107.38.

Example 16 The average score of boys in an examination of a school is 71 and that of girls is 73. The average score of the school in the examination is 71.8. Find the ratio of the number of boys to the number of girls who appeared in the examination.

Solution: Let the number of boys be x and the number of girls be y.

Average score of x boys = 71

$\therefore$ Total score of x boys = $71x$

Average score of y girls = 73

$\therefore$ Total score of y girls = $73y$

Average score of the school = 71.8

$$\therefore \quad \frac{71x + 73y}{x + y} = 71.8$$

$$\Rightarrow \quad 71x + 73y = 71.8x + 71.8y$$

$$\Rightarrow \quad 73y - 71.8y = 71.8x - 71x$$

$$\Rightarrow \quad 1.2y = 0.8x$$

$$\Rightarrow \quad \frac{x}{y} = \frac{1.2}{0.8} = \frac{3}{2}$$

$$\therefore \quad x : y = 3 : 2$$

$\Rightarrow$ Number of boys:Number of girls = 3:2.

Example 17 Find the median for the following data:

Class interval	Frequency
110–119	5
120–129	25
130–139	40
140–149	60
150–159	40
160–169	25
170–179	5

Solution: To compute the median, we need to convert the given distribution to a continuous one.

Class interval	Frequency	Cumulative frequency
109.5–119.5	5	5
119.5–129.5	25	30
129.5–139.5	40	70
139.5–149.5	60	130 ← Median class
149.5–159.5	40	170
159.5–169.5	25	195
169.5–179.5	5	200
Total	***N* = 200**	

$$\frac{N}{2} = \frac{200}{2} = 100$$

∴ Median class is 139.5 – 149.5.

$$L = 139.5, f = 60, cf = 70, h = 10$$

$$\therefore \qquad \text{Median} = L + \left(\frac{\frac{N}{2} - cf}{f}\right) \times h$$

$$= 139.5 + \left(\frac{100 - 70}{60}\right) \times 10 = 144.5.$$

Example 18 The frequency distribution of marks obtained by 53 students out of 100 in a certain examination is given below:

Marks	No. of students
0–10	5
10–20	3

(Continued)

Marks	No. of students
20–30	4
30–40	3
40–50	3
50–60	4
60–70	7
70–80	9
80–90	7
90–100	8

(i) Draw the cumulative frequency curve or an ogive (less than).

(ii) Draw the cumulative frequency curve or an ogive (greater than).

Solution:

(i)

Marks obtained	Number of students (Cumulative frequency)
Less than 10	5
Less than 20	$5 + 3 = 8$
Less than 30	$8 + 4 = 12$
Less than 40	$12 + 3 = 15$
Less than 50	$15 + 3 = 18$
Less than 60	$18 + 4 = 22$
Less than 70	$22 + 7 = 29$
Less than 80	$29 + 9 = 38$
Less than 90	$38 + 7 = 45$
Less than 100	$45 + 8 = 53$

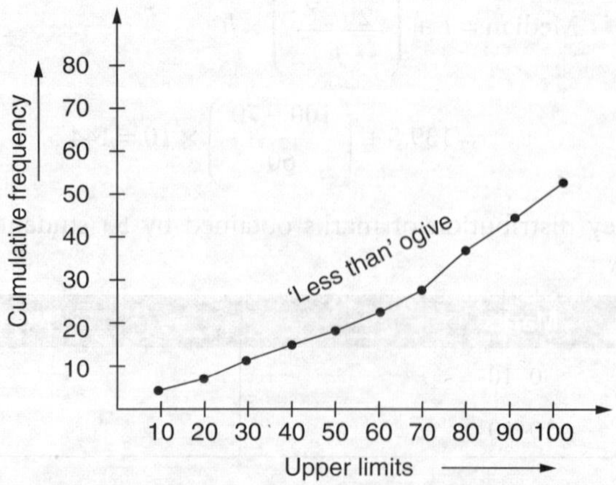

(ii)

Marks obtained	Number of students (Cumulative frequency)
More than or equal to 0	53
More than or equal to 10	53 – 5 = 48
More than or equal to 20	48 – 3 = 45
More than or equal to 30	45 – 4 = 41
More than or equal to 40	41 – 3 = 38
More than or equal to 50	38 – 3 = 35
More than or equal to 60	35 – 4 = 31
More than or equal to 70	31 – 7 = 24
More than or equal to 80	24 – 9 = 15
More than or equal to 90	15 – 7 = 8

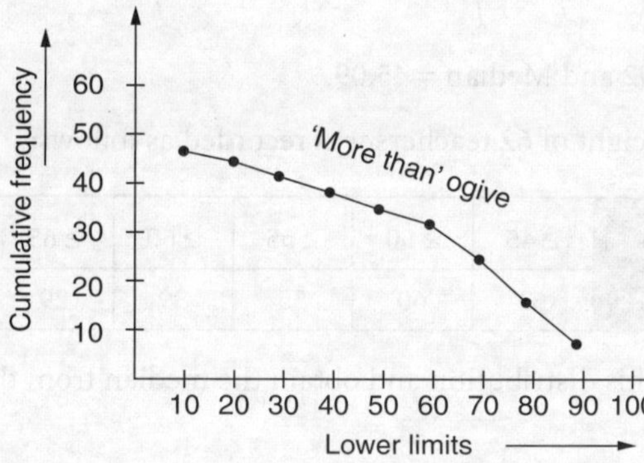

Example 19 In a hospital during October 2013, the number of patients admitted for dengue and their ages are as follows:

Age (in years)	0–8	8–16	16–24	24–32	32–40	40–48	48–56	56–64	64–72
No. of patients	10	12	8	25	15	11	21	30	22

Find the mean and median age of the patients.

Solution:

Class interval	x_i	f_i	c.f.	$u_i = \dfrac{x_i - 36}{8}$	f_iu_i	
0–8	4	10	10	–4	–40	
8–16	12	12	22	–3	–36	
16–24	20	8	30	–2	–16	
24–32	28	25	55	–1	–25	
32–40	36	15	70	0	0	
40–48	44	11	81	1	11	→ Median (C.I)

(*Continued*)

Class interval	x_i	f_i	c.f.	$u_i = \frac{x_i - 36}{8}$	f_iu_i
48–56	52	21	102	2	42
56–64	60	30	132	3	90
64–72	68	22	154	4	88
Total		$\Sigma f_i = 154$			$\Sigma f_iu_i = 114$

Here, assumed mean $(a) = 36$; class size $(h) = 8$, Total frequency $(\Sigma f_i) = 154$.

Mean $$(\bar{x}) = a + \frac{\Sigma f_i u_i}{\Sigma f_i} \times h = 36 + 8 \times \frac{114}{154} = 36 + 5.92 = 41.92$$

$$\text{Median} = L + \left(\frac{\frac{N}{2} - c.f.}{f}\right) \times h = 40 + \left(\frac{77 - 70}{11}\right) \times 8 = 40 + \frac{56}{11} = 45.09$$

Hence Mean = 41.92 and Median = 45.09.

Example 20 In a school, the weight of 62 teachers was recorded as follows:

Weight (in kg)	≥ 45	≥ 50	≥ 55	≥ 60	≥ 65	≥ 70	≥ 75	≥ 80
Number of teachers	62	60	55	39	29	9	4	1

Draw a greater than ogive for this distribution and obtain the median from the curve.

Solution:

Weight (in kg)	No. of teachers
More than or equal to 45	62
More than or equal to 50	60
More than or equal to 55	55
More than or equal to 60	39
More than or equal to 65	29
More than or equal to 70	9
More than or equal to 75	4
More than or equal to 80	1

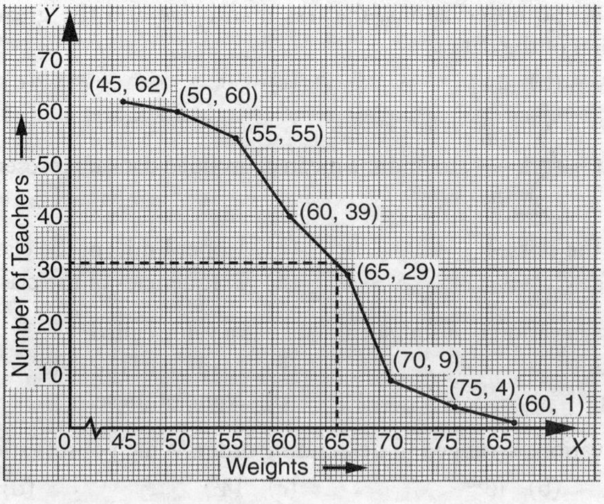

From the graph, median = 64.

Multiple Choice Questions

Choose the correct answer for each of the following questions:

1. If the mean of the following frequency distribution is 5, then the value of P is

x_i	2	4	6	10	$P + 5$
f_i	3	2	2	1	1

(a) 7 (b) 8 (c) 9 (d) 4

2. The mode of a frequency distribution can be determined graphically from

(a) Ogive (b) Histogram (c) Frequency polygon (d) Bar diagram

3. If $\bar{x}$ is the arithmetic mean of n observations $x_1, x_2, \ldots, x_n$, then the arithmetic mean of $ax_1, ax_2, \ldots, ax_n$ is

(a) $a\bar{x}$ (b) $\frac{\bar{x}}{a}$ (c) $\frac{a}{\bar{x}}$ (d) None of these

4. The mean and median of a set of data is 14 and 16, respectively. The value of mode is

(a) 13 (b) 16 (c) 18 (d) 20

5. The relation connecting the measures of central tendencies is

(a) Mode = 2 Median – 3 Mean (b) Mode = 3 Median – 2 Mean
(c) Mode = 2 Median + 3 Mean (d) Mode = 3 Median + 2 Mean

6. The measure of central tendency which cannot be found graphically is

(a) Mean (b) Median (c) Mode (d) None of these

7. The mean of 20 numbers is 13. The new mean, if each observation is increased by 5, is

(a) 13 (b) 18 (c) 65 (d) 8

8. The class mark of a class is equal to

(a) Upper limit – Lower limit (b) Lower limit + Upper limit
(c) $\frac{1}{3}$ (Lower limit + Upper limit) (d) $\frac{1}{2}$ (Lower limit + Upper limit)

9. For a given data with 50 observations, the less than ogive and the greater than ogive intersect at (38.5, 34). The median of the data is
(a) 38.5 (b) 34 (c) 50 (d) 4.5

10. In the formula $\overline{x} = a + \frac{\sum fidi}{\sum fi}$, to find the mean of grouped data, d represents the deviations from a of:
(a) Lower limits of the classes (b) Upper limits of the classes
(c) Mid-points of the classes (d) Frequencies of the class marks

11. The middle-most observation of a statistical data has a value called
(a) Mean of the data (b) Median of the data
(c) Mode of the data (d) None of these

12. A data has 25 observations arranged in descending order. Which observation represents the median?
(a) 12th (b) 13th (c) 14th (d) 15th

13. Which of the following is not a measure of central tendency?
(a) Mode (b) Frequency (c) Median (d) Mean

14. If a is mode, b is mean and c is median of a given data, then
(a) $3c = a + 2b$ (b) $3b = 2a + c$ (c) $3b = a + 2c$ (d) $a = b + c$

15. The mean of 5 observations x, $x + 2$, $x + 4$, $x + 6$ and $x + 8$ is 11, then the value of x is
(a) 4 (b) 7 (c) 11 (d) 6

16. The lower limit of modal class of the following data is

CI	0–10	10–20	20–30	30–40	40–50
Frequency	5	8	13	7	6

(a) 10 (b) 30 (c) 20 (d) 50

Fill in the Blanks

Complete the following statements with appropriate answers(s) in the blank space(s):

17. Median class is the class whose cumulative frequency is greater than (and nearest to) ________.

18. Suppose the mean of 10 observations is 20, if we add 5 in each observation, then the new mean will be ________.

19. The class in which mode lies is called the ________.

20. The median of a frequency distribution is found graphically with the help of ________.

21. While drawing a less than ogive, take a ________ limit.

22. The mid-point of a class interval is called its ________.

Subjective Questions

23. A survey was conducted by a group of students as a part of their environment awareness programme, in which they collected the following data regarding the number of plants in 20 houses in a locality. Find the mean number of plants per house.

Number of plants	0–2	2–4	4–6	6–8	8–10	10–12	12–14
Numbers of houses	1	2	1	5	6	2	3

Which method did you use for finding the mean, and why? **[NCERT]**

24. Consider the following distribution of daily wages of 50 workers of a factory: **[NCERT]**

Daily wages (in ₹)	100–120	120–140	140–160	160–180	180–200
Number of workers	12	14	8	6	10

Find the mean daily wages of the workers of the factory by using an appropriate method.

25. In a retail market, fruit vendors were selling mangoes kept in packing boxes. These boxes contained varying numbers of mangoes. The following was the distribution of mangoes according to the number of boxes:

Number of mangoes	50–52	53–55	56–58	59–61	62–64
Numbers of boxes	15	110	135	115	25

Find the mean number of mangoes kept in a packing box. Which method of finding mean did you choose? **[NCERT]**

26. The table below shows the daily expenditure on food of 25 households in a locality:

Daily expenditure (in ₹)	100–150	150–200	200–250	250–300	300–350
Numbers of households	4	5	12	2	2

Find the mean daily expenditure on food by a suitable method. **[NCERT]**

27. The following table gives the literacy rate (in percentage) of 35 cities. Find the mean literacy rate. **[NCERT]**

Literacy rate (in %)	45–55	55–65	65–75	75–85	85–95
Number of cities	3	10	11	8	3

28. The distribution below shows the number of wickets taken by bowlers in one-day cricket matches. Find the mean number of wickets by choosing a suitable method. What does the mean signify? **[NCERT]**

Numbers of wickets	20–60	60–100	100–150	150–250	250–350	350–450
Numbers of bowlers	7	5	16	12	2	3

29. The table below gives the percentage distribution of female teachers in the primary schools of rural areas of various states and union territories of India. Find the mean percentage of female teachers by all the three methods discussed in this section. **[NCERT]**

Percentage of female teachers	15–25	25–35	35–45	45–55	55–65	65–75	75–85
Numbers of States/UT	6	11	7	4	4	2	1

Source: Seventh All India School Education Survey conducted by NCERT

30. The marks obtained by 30 students of Class X of a certain school in a Mathematics paper consisting of 100 marks are presented in table below. Find the mean of the marks obtained by the students. [NCERT]

Marks obtained (x_i)	10	20	36	40	50	56	60	70	72	80	88	92	95
Numbers of students (f_i)	1	1	3	4	3	2	4	4	1	1	2	3	1

31. Consider the following distribution:

Marks obtained	Less than 10	Less than 20	Less than 30	Less than 40	Less than 50
Numbers of students	02	10	15	30	40

Find the number of students having marks in the range 30–40.

32. Consider the following distribution:

Monthly income range (in ₹)	Number of families
More than or equal to 5,000	150
More than or equal to 10,000	132
More than or equal to 15,000	115
More than or equal to 20,000	85
More than or equal to 25,000	68
More than or equal to 30,000	42

Find the number of families having an income range (in ₹) of 15,000–20,000.

33. Find the mean age of 100 residents of a town from the following data:

Age equal and above (in years)	0	10	20	30	40	50	60	70
No. of persons	100	90	75	50	25	15	5	0

34. The table below shows the daily expenditure on food of 25 households in a locality:

Daily expenditure (in ₹)	100–150	150–200	200–250	250–300	300–350
No. of households	4	5	12	2	2

Find the mean daily expenditure on food by a suitable method.

35. An aircraft has 120 passenger seats. The number of seats occupied during 100 flights is given in the following table:

Number of seats	100–104	104–108	108–112	112–116	116–120
Frequency	15	20	32	18	15

Determine the mean number of seats occupied over the flights.

36. The mileage (km/l) of 50 cars of the same model was tested by a manufacturer and details are tabulated as given below:

Mileage (km/l)	10–12	12–14	14–16	16–18
Number of cars	7	12	18	13

Find the mean mileage. The manufacturer claimed that the mileage of the model was 16 km/l. Do you agree with this claim?

37. If the median height of 50 students of class in the following frequency distribution is 144 cm, find the missing frequencies *x* and *y*. **[CBSE 2014]**

Height (in cm)	125–130	130–135	135–140	140–145	145–150	150–155	155–160
No. of students	2	4	x	y	8	9	5

38. Find the missing frequencies (f_1, f_2 and f_3) in the following frequency distribution when it is given that $f_2 : f_3 = 4:3$ and mean = 50. **[CBSE 2014]**

Class interval	0–20	20–40	40–60	60–80	80–100	Total
Frequency	17	f_1	f_2	f_3	19	**120**

39. The literacy rates of 40 cities is given in the following table. If it is given that mean literacy rate is 63.5, then find the missing frequencies *x* and *y*. **[CBSE 2014]**

Literacy rate	35–40	40–45	45–50	50–55	55–60	60–65	65–70	70–75	75–80	80–85	85–90
No. of cities	1	2	3	x	y	6	8	4	2	3	2

40. Consider the following frequency distribution:

Class	0–5	6–11	12–17	18–23	24–29
Frequency	13	10	15	8	11

Find the upper limit of the median class. **[NCERT Exemplar]**

41. Consider the following frequency distribution of the height of 60 students of a class:

Height (in cm)	150–155	155–160	165–170	170–175	170–175	175–180
Number of students	15	13	10	8	9	5

What is the sum of the lower limit of the modal class and upper limit of the median class? **[NCERT Exemplar]**

42. The following table shows the cumulative frequency distribution of marks of 800 students in an examination:

Marks	Number of students	Marks	Number of students
Below 10	10	Below 60	570
Below 20	50	Below 70	670
Below 30	130	Below 80	740
Below 40	270	Below 90	780
Below 50	440	Below 100	800

Construct a frequency distribution table for the above data.

43. Find the unknown entries a, b, c, d, e, f in the following distribution of height of students in a class:

Height (in cm)	Frequency	Cumulative frequency
150–155	20	a
155–160	18	b
160–165	c	59
165–170	24	d
170–175	e	94
175–180	6	f
Total	**100**	

44. The following are the ages of 300 patients getting medical treatment in a hospital on a particular day:

Age (in years)	10–20	20–30	30–40	40–50	50–60	60–70
No. of patients	60	42	55	70	53	20

Form:

(i) Less than cumulative frequency distribution. (ii) Greater than cumulative frequency distribution.

45. Calculate the median from the following data:

Rent (in ₹)	Number of Tenants
1500–2500	8
2500–3500	10
3500–4500	15
4500–5500	25
5500–6500	40
6500–7500	20
7500–8500	15
8500–9500	7

46. If the median of the following distribution is 28.5, find the missing frequencies:

Class interval	0–10	10–20	20–30	30–40	40–50	50–60	Total
Frequency	5	f_1	20	15	f_2	5	**60**

47. The annual rainfall record of a city of 66 days is given in the following table:

Rainfall (in cm)	0–10	10–20	20–30	30–40	40–50	50–60
Number of days	22	10	8	15	5	6

Calculate the median rainfall using ogive (greater than and less than).

48. Find the mean, mode and median of the following frequency distribution:

Class	0–50	50–100	100–150	150–200	200–250	250–300	300–350
Frequency	2	3	5	6	5	3	1

49. The time, in seconds, taken by 150 athletes to run a 110 m hurdle race are tabulated below:

Class	13.8–14	14–14.2	14.2–14.4	14.4–14.6	14.6–14.8	14.8–15
Frequency	2	4	5	71	48	20

Find the number of athletes who completed the race in less than 14.6 seconds.

50. The following table shows the cumulative frequency distribution of marks of 800 students in an examination:

Marks	Number of students	Marks	Number of student
Below 10	10	Below 60	570
Below 20	50	Below 70	670
Below 30	130	Below 80	740
Below 40	270	Below 90	780
Below 50	440	Below 100	800

Construct a frequency distribution table for the data.

51. Find the frequencies and the median for the following distribution if the mean is 1.46.

No. of Accidents	0	1	2	3	4	5	**Total**
Frequency (No. of days)	46	x	y	25	10	5	**200**

52. Find the mean, mode and median of the following frequency distribution:

Class	0–10	10–20	20–30	30–40	40–50	50–60	60–70
Frequency	4	4	7	10	12	8	5

53. A survey regarding the height (in cm) of 51 girls of Class X of a school was conducted and the following data was obtained: **[NCERT]**

Height (in cm)	Number of girls
Less than 140	4
Less than 145	11
Less than 150	29
Less than 155	40
Less than 160	46
Less than 165	51

Find the median height.

54. A life insurance agent sent the following data about the distribution of ages of 100 policy holders. Calculate the median age, if policies are given only to persons of age 18 years and upwards but less than 60 years. **[NCERT]**

Age (in years)	Numbers of policy holders
Below 20	2
Below 25	6
Below 30	24
Below 35	45
Below 40	78
Below 45	89
Below 50	92
Below 55	98
Below 60	100

55. The lengths of 40 leaves of a plant are measured correct to the nearest millimetre, and the data obtained is represented in the following table: **[NCERT]**

Length (in mm)	Number of leaves
118–126	3
127–135	5
136–144	9
145–153	12
154–162	5
163–171	4
172–180	3

Find the median length of the leaves.

56. The following table gives the distribution of the lifetime of 400 neon lamps:

Life time (in hours)	Numbers of lamps
1500–2000	14
2000–2500	56
2500–3000	60
3000–3500	86
3500–4000	74
4000–4500	62
4500–5000	48

Find the median life time of a lamp. **[NCERT]**

57. 100 surnames were randomly picked up from a local telephone directory and the frequency distribution of the number of letters in the English alphabets in the surnames was obtained as follows: **[NCERT]**

Number of letters	1-4	4–7	7–10	10–13	13–16	16–19
Number of surnames	6	30	40	16	4	4

Determine the median number of letters in the surnames. Find the mean number of letters in the surnames. Also find the modal size of the surnames.

58. The distribution below gives the weight of 30 students of a class. Find the median weight of the students. **[NCERT]**

Weight (in kg)	40–45	45–50	50–55	55–60	60–65	65–70	70–75
Number of students	2	3	8	6	6	3	2

59. What is the value of the median of the data represented by the following graph of less than ogive and greater than ogive?

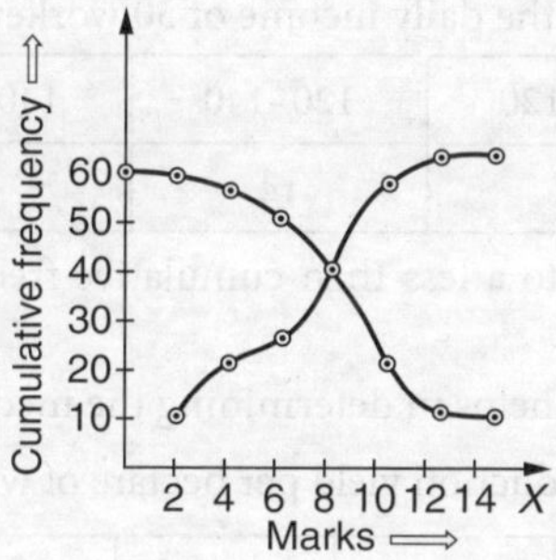

60. Draw a 'less than' ogive and 'greater than' ogive for the following distribution and hence find its median:

Classes	20–30	30–40	40–50	50–60	60–70	70–80	80–90
Frequency	10	8	12	24	6	25	15

61. The following is the frequency distribution of the duration of 100 calls made on a mobile phone:

Duration (in seconds)	95–125	125–155	155–185	185–215	215–245
Number of calls	14	22	28	21	15

Calculate the average duration (in sec.) of a call and also find the median from a cumulative frequency curve.

62. Fifty students enter for a school javelin throw competition. The distances thrown (in metres) are recorded below:

Distance (in m)	0–20	20–40	40–60	60–80	80–100
Number of students	6	11	17	12	4

(i) Construct a cumulative frequency table.
(ii) Draw a cumulative frequency (less than) curve and calculate the median distance thrown using this curve.

(iii) Calculate the median distance using the formula for median.
(iv) Are the median distances calculated in (ii) and (iii) the same?

63. The annual profits earned by 30 shops of a shopping complex in a locality give rise to the following distribution: [NCERT]

Profit (in lakhs in ₹)	Number of shops (frequency)
More than or equal to 5	30
More than or equal to 10	28
More than or equal to 15	16
More than or equal to 20	14
More than or equal to 25	10
More than or equal to 30	7
More than or equal to 35	3

Draw both ogives for the above data and hence obtain the median.

64. The following distribution gives the daily income of 50 workers of a factory:

Daily income (in ₹)	100–120	120–140	140–160	160–180	180–200
Number of workers	12	14	8	6	10

Convert the distribution above to a less than cumulative frequency distribution and draw its ogive. Find the median from this ogive. [NCERT]

65. Which graphical representation helps in determining the median of a given frequency distribution?

66. The following table gives the production yield per hectare of wheat of 100 farms of a village: [NCERT]

Production yield (in kg/ha)	50–55	55–60	60–65	65–70	70–75	75–80
Number of farms	2	8	12	24	38	16

Change the distribution to a greater than distribution, and draw its ogive.

67. For the following distribution, draw a less than ogive and from the curve, find the median. [CBSE 2014]

Marks obtained	Less than 20	Less than 30	Less than 40	Less than 50	Less than 60	Less than 70	Less than 80	Less than 90	Less than 100
No. of Students	2	7	17	40	60	82	85	90	100

68. The lifetimes of electric bulbs are given in the following frequency distribution table:

Life time (in hours)	More than or equal to 240	More than or equal to 280	More than or equal to 320	More than or equal to 360	More than or equal to 400	More than or equal to 440	More than or equal to 480
Numbers of bulbs	100	98	87	77	47	22	10

Draw a greater than ogive and from it, find the median. Verify it by actual calculations. [CBSE 2016]

69. Find the value of f_1 from the following data, if its mode is 65.

Class	Frequency
0–20	6
20–40	8
40–60	f_1
60–80	12
80–100	6
100–120	5

The frequencies 6, 8, f_1 and 12 are in ascending order. **[CBSE (CCE) 2012]**

70. For the following distribution, find the modal class: **[NCERT Exemplar]**

Marks	Below 10	Below 20	Below 30	Below 40	Below 50	Below 60
Number of students	3	12	27	57	75	80

71. A survey conducted on 20 households in a locality by a group of students resulted in the following frequency table for the number of family members in a household:

Family size	1-3	3-5	5-7	7-9	9-11
Number of families	7	8	2	2	1

Find the mode of this data. **[NCERT]**

72. The following table shows the ages of the patients admitted in a hospital during a year: **[NCERT]**

Age (in years)	5–15	15–25	25–35	35–45	45–55	55–65
Number of patients	6	11	21	23	14	5

Find the mode and the mean of the data given above. Compare and interpret the two measures of central tendency.

73. The following data gives the distribution of the total monthly household expenditure of 200 families of a village. Find the modal monthly expenditure of the families. Also find the mean monthly expenditure. **[NCERT]**

Expenditure (in ₹)	Number of families
1000–1500	24
1500–2000	40
2000–2500	33
2500–3000	28
3000–3500	30
3500–4000	22
4000–4500	16
4500–5000	7

74. A student noted the number of cars passing through a spot on a road for 100 periods, each of 3 minutes, and summarised it in the table given below. Find the of mode the data. **[NCERT]**

Numbers of cars	0–10	10–20	20–30	30–40	40–50	50–60	60–70	70–80
Frequency	7	14	13	12	20	11	15	8

75. Find the mode of the following frequency distribution: **[CBSE 2014]**

Class	0–6	6–12	12–18	18–24	24–30
Frequency	7	5	10	12	6

76. Determine the missing frequency x, from the following data, when mode is 67: **[CBSE 2016]**

Class	40–50	50–60	60–70	70–80	80–90
Frequency	5	x	15	12	7

77. Find mode using an empirical relation, when it is given that mean and median are 10.5 and 9.6, respectively. **[CBSE 2016]**

78. If the mode of the following distribution is 57.5, find the value of x. **[CBSE 2013]**

Class	30–40	40–50	50–60	60–70	70–80	80–90	90–100
Frequency	6	10	16	x	10	5	2

79. There were 35 students in a hostel. Due to the admission of 7 new students, the expenses of the mess were increased by ₹ 42 per day, while the average expenditure per head diminished by ₹1. What was the original expenditure of the mess?

80. In a school, 85 boys and 35 girls appeared in a public examination. The mean marks of boys was found to be 40%, whereas the mean marks of girls was 60%. Determine the average marks of the school.

81. The table below shows the salaries of 280 persons: **[CBSE 2018]**

Salary (in thousands ₹)	No. of Persons
5–10	49
10–15	133
15–20	63
20–25	15
25–30	6
30–35	7
35–40	4
40–45	2
45–50	1

Calculate the median salary of data.

82. By changing the following frequency distribution to 'less than', draw its ogive. **[CBSE 2018 (C)]**

Classes	0 –15	15–30	30–45	45–60	60–75
Frequency	6	8	10	6	4

83. The following distribution gives the daily income of 50 workers of a factory: **[CBSE 2018]**

Daily Income (in ₹)	100–120	120 -140	140–160	160–180	180–200
Number of workers	12	14	8	6	10

Convert the distribution above to a less than cumulative frequency distribution and draw its ogive.

84. If the median of the following frequency distribution is 32.5, find the value of f_1 and f_2. **[Delhi 2019]**

Classes	0–10	10–20	20–30	30–40	40–50	50–60	60–70	Total
Frequency	f_1	5	9	12	f_2	3	2	40

Find the mean, median and mode of the above data.

85. Change the following distribution to a greater than distribution. Hence, draw the ogive for this distribution. **[Chennai 2019]**

Intervals	20–30	30–40	40–50	50–60	60–70	70–80	80–90
Frequency	10	8	12	24	6	25	15

86. Which measure of central tendency is given by the x coordinate of the point of intersection of the greater than ogive and the less than ogive?

Answers

Multiple Choice Questions

1. (d) **2.** (c) **3.** (a) **4.** (d) **5.** (b) **6.** (a) **7.** (b) **8.** (d) **9.** (a) **10.** (c)
11. (b) **12.** (b) **13.** (b) **14.** (a) **15.** (b) **16.** (c)

Fill in the Blanks

17. $\frac{N}{2}$ **18.** 25 **19.** Modal class **20.** Ogive **21.** Upper limit **22.** Class mark

Subjective Questions

23. 8.1 plants **24.** ₹ 145.20 **25.** 57.19 **26.** ₹ 211 **27.** 69.43% **28.** 152.89 **29.** 39.71
30. 59.3 **31.** 15 **32.** 30 **33.** 31 years **34.** ₹ 211 **35.** 110 (approx).
36. 14.48 km/l, No. **37.** $x = 7, y = 15$ **38.** $f_1 = 28, f_2 = 32, f_3 = 24$. **39.** $x = 5$, and $y = 4$. **40.** 17.5
41. 315
42.

Marks	0–10	10–20	20–30	30–40	40–50	50–60	60–70	70–80	80–90	90–100
No. of students	10	40	80	140	170	130	100	70	40	20

43. $a = 20, b = 38, c = 21, d = 83, e = 11, f = 100$.

44.

(i) Less than Type		(ii) More than Type	
Age (in years)	**Number of students**	**Age (in years)**	**Number of students**
Less than 20	60	More than or equal to 10	300
Less than 30	102	More than or equal to 20	240
Less than 40	157	More than or equal to 30	198
Less than 50	227	More than or equal to 40	143
Less than 60	280	More than or equal to 50	73
Less than 70	300	More than or equal to 60	20

45. ₹5800 46. $f_1 = 8, f_2 = 7$

47. 21.25 cm **48.** Mean = 169, median = 171, Mode = 175. **49.** 82

50.

Marks	0–10	10–20	20–30	30–40	40–50	50–60	60–70	70–80	80–90	90–100
No. of students	10	40	80	140	170	130	100	70	40	20

51. $x = 76$, $y = 38$, Median = 1 **52.** Mean = 38.2, Median = 40, Mode = 43.33. **53.** 149.03 cm

54. Median age = 35.76 years **55.** 146.75 mm **56.** 3406.98 hours

57. Median = 8.05, Mean = 8.32, Mode = 7.88 **58.** 56.67 kg **59.** 8 **60.** 58.33

61. 170.3 sec, 170 sec.

62. (i)

Distance (in m)	Number of students	Cumulative frequency
Less than 20	6	6
Less than 40	11	17
Less than 60	17	34
Less than 80	12	46
Less than 100	4	50

(ii) 49.41 m (iii) 49.41 (iv) yes **63.** 17.5 **64.** ₹ 138 **65.** Ogive **67.** 54 **68.** 396

69. $f_1 = 10$ **70.** 30 – 40 **71.** 3.286

72. Mode = 36.8 years, Mean = 35.37 years. Maximum number of patients admitted in the hospital are of the age 36.8 years (approx.), while on an average the age of a patient admitted to the hospital is 35.37 years.

73. Mode = ₹ 1,847.83, Mean = ₹ 2,662.5 **74.** 44.7 cars. **75.** 19.5 **76.** $x = 8$.

77. 7.8 **78. 14.** **79.** ₹ 420 **80.** 45.83

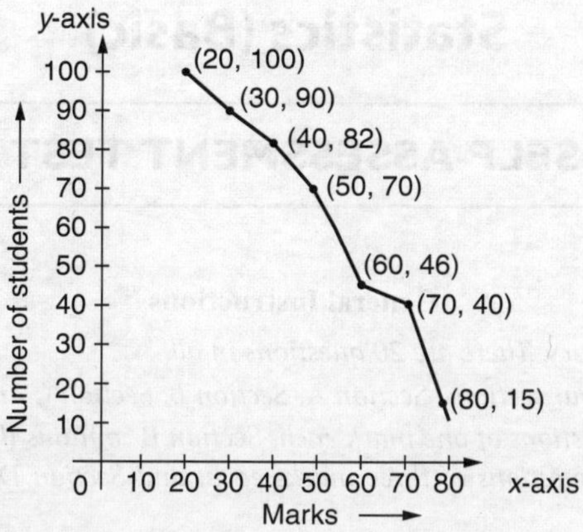

81. 13.42 (in thousands ₹)

84. $f_1 = 3, f_2 = 6$

85.

Marks obtained more than or equal to	No. of students
20	100
30	90
40	82
50	70
60	46
70	40
80	15

86. Median

Statistics (Basic)

SELF-ASSESSMENT TEST

Time allowed: 90 min. Max. Marks: 40

General Instructions

1. *All questions are compulsory. There are 20 questions in all.*
2. *This question paper has four sections: Section A, Section B, Section C and Section D.*
3. *Section A contains 10 questions of one mark each, Section B contains three questions of two marks each, Section C contains four questions of three marks each, and Section D contains three questions of four marks each.*

Section A [1 Mark Each]

Direction (Q. Nos. 1–5): These are Multiple Choice Questions. Select the most appropriate answer for the given questions.

1. The class mark of a class is equal to
   (a) upper limit – lower limit (b) lower limit + upper limit
   (c) $\frac{1}{3}$ (lower limit + upper limit) (d) $\frac{1}{2}$ (lower limit + upper limit)
2. The relation connecting the measures of central tendencies is
   (a) Mode = 2 Median – 3 Mean (b) Mode = 3 Median – 2 Mean
   (c) Mode = 2 Median + 3 Mean (d) Mode = 3 Median + 2 Mean
3. The mean of 5 observations x, $x + 2$, $x + 4$, $x + 6$ and $x + 8$ is 11. Then, the value of x is
   (a) 4 (b) 7 (c) 11 (d) 6
4. The lower limit of the modal class of the following data is

CI	0–10	10–20	20–30	30–40	40–50
Frequency	5	8	13	7	6

   (a) 10 (b) 30 (c) 20 (d) 50
5. For some data with 50 observations 'the less than ogive' and the 'greater than ogive' intersect at (38.5, 34). The median of the data is
   (a) 38.5 (b) 34 (c) 50 (d) 4.5
6. While drawing the less than ogive, take a __________ limit.
7. The mid-point of a class interval is called its __________ .
8. Consider the following distribution:

Marks obtained	Less than 10	Less than 20	Less than 30	Less than 40	Less than 50
Numbers of students	02	10	15	30	40

   Find the number of students having marks in the range 30–40.

9. Consider the following frequency distribution:

Class	0–5	6–11	12–17	18–23	24–29
Frequency	13	10	15	8	11

Find the upper limit of the median class.

10. What is the mode of the data if median = 3 and mean = 2.

Section B [2 Marks Each]

11. The arithmetic mean of the following distribution is 25. Determine the value of p.

Classes	0–10	10–20	20–30	30–40	40–50
Frequency (f)	7	8	15	9	10

12. Find the median of the following data:

Marks	20–30	30–40	40–50	50–60	60–70	70–80	80–90
No. of students	5	15	25	20	7	8	10

13. Find the value of f_1 from the following data, if its mode is 65.

Class	Frequency
0–20	6
20–40	8
40–60	f_1
60–80	12
80–100	6
100–120	5

where frequencies 6, 8, f_1 and 12 are in ascending order.

Section C [3 Marks Each]

14. The mark obtained by 120 students in a mathematics test is given in the following distribution:

Marks obtained	0–20	20–40	40–60	60–80	80–100	Total
No. of students	17	f_1	32	f_2	19	120

The number of students in the marks group 20–40 and 60–80 are f_1 and f_2, respectively. Find f_1 and f_2 if the mean marks is 50.

15. Draw a cumulative frequency curve (less than) for the following data and find median from it.

Class intervals	200–220	220–240	240–260	260–280	280–300	300–320
Frequency	7	3	6	8	2	4

16. If the mode of the following distribution is 57.5, find the value of x.

Class	30–40	40–50	50–60	60–70	70–80	80–90	90–100
Frequency	6	10	16	x	10	5	2

17. Find the mean age from the following frequency distribution:

Age (in years)	25–29	30–34	35–39	40–44	45–49	50–54	55–59
Number of persons	4	14	22	16	6	5	3

Section D [4 Marks Each]

18. If the median of the following frequency distribution is 32.5, find the value of f_1 and f_2.

Classes	0–10	10–20	20–30	30–40	40–50	50–60	60–70	Total
Frequency	f_1	5	9	12	f_2	3	2	40

Find the mean, median and mode of the above data.

19. The median of the following data is 525. Find the values of x and y, if the total frequency is 100.

Class interval	Frequency
0–100	2
100–200	5
200–300	x
300–400	12
400–500	17
500–600	20
600–700	y
700–800	9
800–900	7
900–1000	4

20. Find the mean, mode and median of the following frequency distribution:

Classes	0–10	10–20	20–30	30–40	40–50	50–60	60–70
Frequency	4	4	7	10	12	8	5

Answers

1. (d) **2.** (b) **3.** (b) **4.** (c) **5.** (a) **6.** Upper limit **7.** Class-Mark **8.** 15
9. 17.5 **10.** 5 **11.** $p = 16$ **12.** 50 **13.** $f_1 = 10$ **14.** $f_1 = 28, f_2 = 24$ **15.** 258 **16.** 14
17. 39.36 years **18.** $f_1 = 3, f_2 = 6$ **19.** $x = 9$ and $y = 15$
20. Mean = 38.2, Median = 40, Mode = 43.33

Statistics (Standard)

SELF-ASSESSMENT TEST

Time allowed: 90 min. Max. Marks: 40

General Instructions

1. *All questions are compulsory. There are 20 questions in all.*
2. *This question paper has four sections: Section A, Section B, Section C and Section D.*
3. *Section A contains 10 questions of one mark each, Section B contains three questions of two marks each, Section C contains four questions of three marks each, and Section D contains three questions of four marks each.*

Section A [1 Mark Each]

Direction (Q. Nos. 1–5): These are Multiple Choice Questions. Select the most appropriate answer for the given questions.

1. The mode of a frequency distribution can be determined graphically from
(a) ogive (b) histogram (c) frequency polygon (d) bar diagram
2. The mean of 20 numbers is 13. The new mean if each observation is increased by 5 is
(a) 13 (b) 18 (c) 65 (d) 8
3. For some data with 50 observations, 'the less than ogive' and 'greater than ogive' intersect at (38.5, 34). The median of the data is
(a) 38.5 (b) 34 (c) 50 (d) 4.5
4. The middle-most observation of a statistical data has a value which is called
(a) mean of the data (b) median of the data
(c) mode of the data (d) none of the these
5. The lower limit of the modal class of the following data is

CI	0–10	10–20	20–30	30–40	40–50
Frequency	5	8	13	7	6

(a) 10 (b) 30 (c) 20 (d) 50
6. The mid-point of a class interval is called its ________.
7. Suppose the mean of 10 observations is 20. If we add 5 in each observation, then the new mean will be ________.
8. The average score of boys in an examination of a school is 71 and that of girls is 73. The average score of the school in the examination is 71.8. Find the ratio of the number of boys to the number of girls who appeared in the examination.
9. What is the median of the data if mode = 5 and mean = 2.
10. Find the mode of the following data:

Marks	50–60	60–70	70–80	80–90	90–100
No. of students	3	12	32	20	6

Section B [2 Marks Each]

11. Find the mean of the following frequency distribution:

Class	0–100	100–200	200–300	300–400	400–500
Frequency	6	9	15	12	8

12. If the median height of 50 students of a class in the following frequency distribution is 144 cm, find missing frequencies x and y.

Height (in cm)	125–130	130–135	135–140	140–145	145–150	150–155	155–1
No. of students	2	4	x	y	8	9	5

13. Find the frequencies and the median for the following distribution, if the mean is 1.46:

No. of accidents	0	1	2	3	4	5	Tot
Frequency (No. of days)	46	x	y	25	10	5	20

Section C [3 Marks Each]

14. Find the missing frequencies (f_1, f_2 and f_3) in the following frequency distribution when it is given f_2:f_3 = 4:3 and mean = 50.

Class interval	0–20	20–40	40–60	60–80	80–100	Tota
Frequency	17	f_1	f_2	f_3	19	120

15. If the median of the following distribution is 28.5, find the missing frequencies:

Class interval	0–10	10–20	20–30	30–40	40–50	50–60	Total
Frequency	5	f_1	20	15	f_2	5	60

16. Find the mean age from the following frequency distribution:

Age (in years)	25–29	30–34	35–39	40–44	45–49	50–54	55–
Number of persons	4	14	22	16	6	5	3

17. During the medical check-up of 35 students of a class, their weights were recorded as follows:

Weight (in kg)	Number of students
Less than 38	0
Less than 40	3
Less than 42	5
Less than 44	9
Less than 46	14
Less than 48	28
Less than 50	32
Less than 52	35

Draw a less than ogive for the given data. Hence obtain the median weight from the graph and the result by using the formula.

Section D [4 Marks Each]

8. The frequency distribution of marks obtained by 53 students out of 100 in a certain examination is given below:

Marks	No. of students
0–10	5
1–20	3
20–30	4
30–40	3
40–50	3
50–60	4
60–70	7
70–80	9
80–90	7
90–100	8

(i) Draw the cumulative frequency curve or an ogive (less than).
(ii) Draw the cumulative frequency curve or an ogive (greater than).

9. In a hospital during the month of October 2013, the number of patients admitted for dengue and their ages are as follows:

Age (in years)	0–8	8–16	16–24	24–32	32–40	40–48	48–56	56–64	64–72
No. of patients	10	12	8	25	15	11	21	30	22

Find the mean and median age of patients.

0. The median of the following data is 525. Find the values of x and y, if the total frequency is 100.

Class interval	Frequency
0–100	2
100–200	5
200–300	x
300–400	12
400–500	17
500–600	20
600–700	y
700–800	9
800–900	7
900–1000	4

Answers

1. (c) **2.** (b) **3.** (a) **4.** (b) **5.** (c) **6.** Class - mark **7.** 25 **8.** 3:2 **9.** 3

10. 76.25 **11.** 264 **12.** $x = 7$ and $y = 15$ **13.** $x = 76, y = 38$, Median = 1 **14.** $f_1 = 28, f_2 = 32, f_3 = 24$

15. $f_1 = 8, f_2 = 7$ **16.** 39.36 years

17. Draw the ogive by plotting the points: (38, 0), (40, 3), (42, 5), (44, 9), (46, 14), (48, 28), (50, 32) and (52, 35).

Here $\frac{N}{2} = 17.5$. Locate the point on the ogive whose ordinate is 17.5.

18. (i)

Marks obtained	Number of students (Cumulative frequency)
Less than 10	5
Less than 20	$5 + 3 = 8$
Less than 30	$8 + 4 = 12$
Less than 40	$12 + 3 = 15$
Less than 50	$15 + 3 = 18$
Less than 60	$18 + 4 = 22$
Less than 70	$22 + 7 = 29$
Less than 80	$29 + 9 = 38$
Less than 90	$38 + 7 = 45$
Less than 100	$45 + 8 = 53$

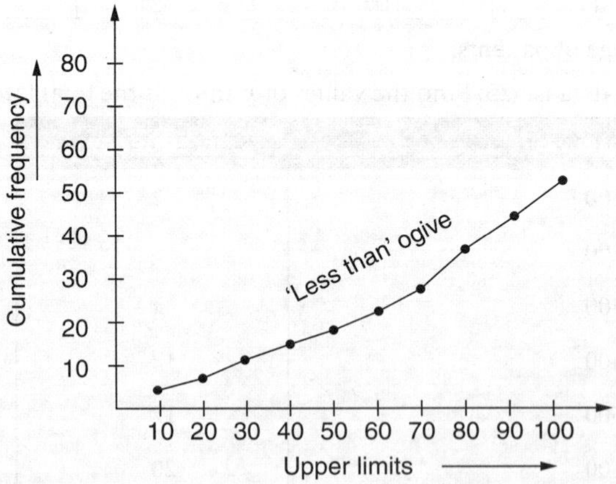

(ii)

Marks obtained	Number of students (Cumulative frequency)
More than or equal to 0	53
More than or equal to 10	$53 - 5 = 48$
More than or equal to 20	$48 - 3 = 45$

(Continued)

Marks obtained	Number of students (Cumulative frequency)
More than or equal to 30	45 – 4 = 41
More than or equal to 40	41 – 3 = 38
More than or equal to 50	38 – 3 = 35
More than or equal to 60	35 – 4 = 31
More than or equal to 70	31 – 7 = 24
More than or equal to 80	24 – 9 = 15
More than or equal to 90	15 – 7 = 8

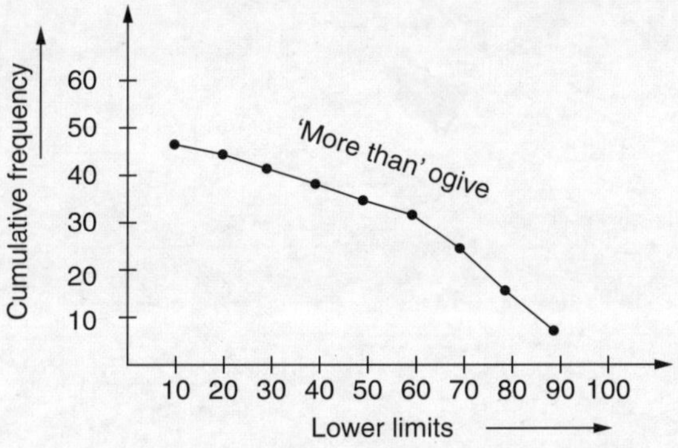

19. Mean = 41.92 and Median = 45.09

20. $x = 9$ and $y = 15$

CHAPTER

15 Probability

Key Concepts

- To find the probability of throwing a coin, two coins and three coins simultaneously
- To find the probability of throwing a die, a paired dice, etc.
- To find the probability of drawing a playing card from a well-shuffled deck
- Miscellaneous problems

Definitions

1. **Event:** The collection of outcomes of an experiment.
2. **Trial:** An action that results in one or several outcomes.
3. **Favourable outcomes of an event:** Those possible outcomes of an experiment that favour the occurrence of a particular event.
4. **Equally likely outcomes:** If in a random experiment, each outcome is as likely to occur as the other, then the outcomes are called equally likely.
5. $P(E) \rightarrow$ probability of the occurrence an event.

   $P(\bar{E}) \rightarrow$ probability of the non-occurrence an event.

   $P(E) + P(\bar{E}) = 1.$

   $P(\bar{E}) = 1 - P(E).$
6. $P(E) = 0$ means impossible event.
7. $P(E) = 1$ means sure event.
8. $0 \leq P(E) \leq 1$, probability of an event never less then 0 and nor more than 1.
9. $P(E) = \dfrac{\text{Number of favourable outcomes of } E}{\text{Total number of possible outcomes}}$

Concept: Coins

If a single coin is thrown, then the total number of outcomes is $\{H, T\}$.

If a pair of coins is thrown simultaneously, then the outcomes are

$$\{HH, HT, TH, TT\}.$$

If three coins are thrown simultaneously, then the total number of outcomes are

$$\begin{Bmatrix} H & H & H \\ H & H & T \\ H & T & H \\ H & T & T \\ T & H & H \\ T & H & T \\ T & T & H \\ T & T & T \end{Bmatrix}$$

At least means 'greater than or equal to'.
At most means 'less than or equal to'.
Unbiased means 'having equal probability of each outcome'.

Example 1 If a coin is tossed, what is the probability of getting: [NCERT]

(i) Head (ii) Tail

Solution: Total number of outcomes = 2

(i) Favourable outcomes = 1 $\{H\}$

$$P(E) = \frac{\text{Number of favourable outcomes}}{\text{Total number of outcomes}} = \frac{1}{2}$$

(ii) Favourable outcomes = 1$\{T\}$

$$P(E) = \frac{\text{Number of favourable outcomes}}{\text{Total number of outcomes}} = \frac{1}{2}$$

Example 2 A coin is tossed two times. Find the probability of getting at most one head.

Solution: When a coin is tossed two times, the outcomes are (*HH, HT, TH, TT*).

∴ Number of possible outcomes = 4

Number of favourable outcomes = 3 (*HT, TH, TT*)

$$P\text{ (of getting at most one head)} = \frac{3}{4}$$

Example 3 Two coins are tossed simultaneously. Find the probability of getting exactly one head. [CBSE (CCE) 2011]

Solution: When two coins are tossed simultaneously, the possible outcomes are (*HH, HT, TH, TT*).

∴ Number of possible outcomes = 4

Number of favourable outcomes = 2

$$P\text{ (of getting exactly one head)} = \frac{2}{4} = \frac{1}{2}$$

Example 4 A game consists of tossing a one-rupee coin three times and noting its outcome each time. Anand wins if all the tosses give the same result, that is, three heads or three tails, and loses otherwise. Calculate the probability that Anand will lose the game.

Solution: Number of outcomes = 8

Number of outcomes in which the same result does not occur = 8 – 2 = 6

Number of favourable outcomes = 6

$$\therefore P\text{ (that Anand will lose the game)} = \frac{6}{8} = \frac{3}{4}$$

Example 5 A piggy bank contains hundred 50 p coins, fifty ₹ 1 coins, twenty ₹ 2 coins and ten ₹ 5 coins. If it is equally likely that one of the coins will fall out when the bank is turned upside down, what is the probability that the coin (i) Will be a 50 p coin? (ii) Will not be a ₹ 5 coin? [NCERT]

Solution: Total no. of coins = 100 + 50 + 20 + 10 = 180

(i) Number of favourable outcomes = 100

$$\therefore P\text{ (that the coin is a 50 p coin)} = \frac{100}{180} = \frac{5}{9}$$

(ii) Number of favourable outcomes = 170 (100 + 50 + 20)

$\therefore P$ (that the coin is not a ₹ 5 coin) $= \frac{170}{180} = \frac{17}{18}$.

Example 6 A coin is tossed three times. Then find the probability of:

(i) 2 heads **(ii)** 2 tails **(iii)** 3 heads

Solution: When a coin is tossed three times, the outcomes are (*HHH*), (*HHT*), (*HTH*), (*HTT*), (*THH*), (*THT*), (*TTH*), (*TTT*).

Number of outcomes = 8

(i) The outcomes having 2 heads are (*HHT*), (*HTH*), (*THH*)

Number of favourable outcomes = 3

$\therefore P$ (of getting 2 heads) $= \frac{3}{8}$

(ii) The outcomes having 2 tails are (*HTT*),(*THT*),(*TTH*)

Number of favourable outcomes = 3

$\therefore P$ (of getting 2 tails) $= \frac{3}{8}$

(iii) The outcome having 3 heads is (*HHH*)

Number of favourable outcomes = 1

$\therefore P$ (of getting 3 heads) $= \frac{1}{8}$

Example 7 Three coins are tossed together. Find the probability of getting:

(i) Exactly two heads **(ii)** At least two heads **(iii)** No tails

Solution: When three coins are tossed, the possible outcomes are (*HHH*), (*HHT*), (*HTH*), (*HTT*), (*THH*), (*THT*), (*TTH*), (*TTT*).

Number of possible outcomes = 8

(i) Number of favourable outcomes = 3{*HHT, HTH, THH*}

P (of getting exactly two heads) $= \frac{3}{8}$

(ii) Number of favourable outcomes = 4 {*HHT, HTH, THH, HHH*}

P (of getting at least two heads) $= \frac{4}{8} = \frac{1}{2}$

(iii) Number of favourable outcomes = 1 {*HHH*}

P (of getting no tails) $= \frac{1}{8}$.

1 Try to Solve

1. Find the probability of getting a head when a coin is tossed once. Also find the probability of getting a tail.
2. Two unbiased coins are tossed simultaneously. Find the probability of getting two heads. [NCERT]

3. Harpreet tosses two different coins simultaneously (say, one is of ₹ 1 and other of ₹ 2). What is the probability that she gets at least one head?
4. Three unbiased coins are tossed together. Find the probability of getting:
   (i) All heads (ii) Exactly two heads (iii) Exactly one head
   (iv) At least two heads (v) At least two tails
5. A coin is tossed three times. Find the probability of getting:
   (i) All tails (ii) At least two tails (iii) At most two tails

Answers

1. $\frac{1}{2}, \frac{1}{2}$ **2.** $\frac{1}{4}$ **3.** $\frac{3}{4}$ **4.** (i) $\frac{1}{8}$ (ii) $\frac{3}{8}$ (iii) $\frac{3}{8}$ (iv) $\frac{1}{2}$ (v) $\frac{1}{2}$

5. (i) $\frac{1}{8}$ (ii) $\frac{1}{2}$ (iii) $\frac{7}{8}$

Concept: Dice

In the first case, if a die is thrown, the total number of outcomes is {1, 2, 3, 4, 5, 6}.

In the second case, if a pair of dice is thrown, then the total number of outcomes is

$$\begin{bmatrix} (1,1) & (1,2) & (1,3) & (1,4) & (1,5) & (1,6) \\ (2,1) & (2,2) & (2,3) & (2,4) & (2,5) & (2,6) \\ (3,1) & (3,2) & (3,3) & (3,4) & (3,5) & (3,6) \\ (4,1) & (4,2) & (4,3) & (4,4) & (4,5) & (4,6) \\ (5,1) & (5,2) & (5,3) & (5,4) & (5,5) & (5,6) \\ (6,1) & (6,2) & (6,3) & (6,4) & (6,5) & (6,6) \end{bmatrix}$$

So, there are 36 cases in total.

Example 8 A die is thrown. Find the probability of getting:

(i) A prime number (ii) 2 or 4 (iii) A multiple of 2 or 3

Solution:

(i) Number of possible outcomes = 6 {1, 2, 3, 4, 5, 6}
Number of favourable outcomes = 3 {2, 3, 5}

P (of getting a prime number) = $\frac{3}{6} = \frac{1}{2}$

(ii) Number of favourable outcomes = 2

P (of getting 2 or 4) = $\frac{2}{6} = \frac{1}{3}$

(iii) Multiples of 2 or 3 are 2, 3, 4, 6
Number of favourable outcomes = 4

P (of getting a multiple of 2 or 3) = $\frac{4}{6} = \frac{2}{3}$

Example 9 A student says that if you throw a die, it will show up 1 or not 1. Therefore, the probability of getting 1 and the probability of getting not 1 is equal to 1/2. Is this correct? Give reasons. **[CBSE (CCE) 2012]**

Solution: For getting '1'

Number of possible outcomes = 6

Number of favourable outcome = 1

P (of getting 1) = $\frac{1}{6}$

For getting 'not 1'

Number of favourable outcome = 5 (Number 2, 3, 4, 5 and 6)

P (of getting not 1) = $\frac{5}{6}$

No, the outcomes are not equally likely. The probability of getting 1 is $\frac{1}{6}$ and getting not 1 is $\frac{5}{6}$.

Example 10 A die is thrown once. Find the probability of getting: **[NCERT]**

(i) A prime number (ii) A number lying between 2 and 6 (iii) An odd number

Solution: No. of possible outcomes = 6 {1, 2, 3, 4, 5, 6}

(i) Favourable outcomes = 3{2, 3, 5}

$\therefore P$ (of getting a prime number) = $\frac{3}{6} = \frac{1}{2}$

(ii) P (of getting a number lying between 2 and 6) = $\frac{3}{6} = \frac{1}{2}$ {3, 4, 5}

(iii) P (of getting an odd number) = $\frac{3}{6} = \frac{1}{2}$ {1, 3, 5}

Example 11 A die is thrown twice. What is the probability that 4 will come up at least once. **[CBSE (CCE) 2012]**

Solution: Number of possible outcomes = 36

Number of favourable outcomes = 11 {(1, 4), (2,4), (3, 4), (4,4), (5, 4), (6, 4), (4, 1), (4, 2), (4, 3), (4, 5), (4, 6)}

P (4 will come up at least once) = $\frac{11}{36}$

Example 12 In a simultaneous throw of a pair of dice, find the probability of getting

(i) 8 as sum (ii) A doublet (iii) A sum greater than 9 (iv) A doublet of prime numbers
(v) A sum less than 6 (vi) A sum more than 7

Solution: The possible outcomes in a simultaneous throw of a pair of dice is shown.

Number of possible outcomes = 36

	1	2	3	4	5	6
1	(1, 1)	(1, 2)	(1, 3)	(1, 4)	(1, 5)	(1, 6)
2	(2, 1)	(2, 2)	(2, 3)	(2, 4)	(2, 5)	(2, 6)
3	(3, 1)	(3, 2)	(3, 3)	(3, 4)	(3, 5)	(3, 6)
4	(4, 1)	(4, 2)	(4, 3)	(4, 4)	(4, 5)	(4, 6)
5	(5, 1)	(5, 2)	(5, 3)	(5, 4)	(5, 5)	(5, 6)
6	(6, 1)	(6, 2)	(6, 3)	(6, 4)	(6, 5)	(6, 6)

(i) Outcomes having 8 as sum are (2, 6), (3, 5), (4, 4), (5, 3), (6, 2).
Number of favourable outcomes = 5

P (of getting 8 as sum) = $\frac{5}{36}$

(ii) The doublets are (1, 1), (2, 2), (3, 3), (4, 4), (5, 5), (6, 6).
Number of favourable outcomes = 6

P (of getting a doublet) = $\frac{6}{36} = \frac{1}{6}$

(iii) Outcomes having sum greater than 9 are (4, 6), (5, 5), (5, 6), (6, 4), (6, 5) and (6, 6).
Number of favourable outcomes = 6

P (of getting a sum greater than 9) = $\frac{6}{36} = \frac{1}{6}$

(iv) Doublets of prime numbers are (2, 2), (3, 3), (5, 5).
Number of favourable outcomes = 3

P (of getting a doublet of a prime number) = $\frac{3}{36} = \frac{1}{12}$

(v) Outcomes having sum less than 6 are (1, 1), (1, 2), (1, 3), (1, 4), (2, 1), (2, 2), (2, 3), (3, 1), (3, 2), (4, 1).
Number of favourable outcomes =10

P (of getting a sum less than 6) = $\frac{10}{36} = \frac{5}{18}$

(vi) Outcomes having sum more than 7 are (2, 6), (3, 5), (3, 6), (4, 4), (4, 5), (4, 6), (5, 3), (5, 4), (5, 5), (5, 6), (6, 2), (6, 3), (6, 4), (6, 5), (6, 6).
Number of favourable outcomes = 15

P (of getting a sum more than 7) = $\frac{15}{36} = \frac{5}{12}$

2 Try to Solve

1. A die is thrown once. What is the probability of getting a prime number?
2. A die is thrown once. What is the probability that it shows: (i) an odd number (ii) a number greater than 4?
3. A pair of dice is thrown once. Find the probability of getting the same number on each die.
4. Two dice are thrown simultaneously. Find the probability that the sum of the two numbers appearing on the top is less than or equal to 10.
5. Two dice are thrown together. Find the probability that product of the sum of the numbers on the top on the dice is: (i) 6, (ii) 12, (iii) 7.

Answers

1. $\frac{1}{2}$ **2.** (i) $\frac{1}{2}$, (ii) $\frac{1}{3}$ **3.** $\frac{1}{6}$ **4.** $\frac{11}{12}$ **5.** (i) $\frac{1}{9}$, (ii) $\frac{1}{9}$, (iii) 0

Example 13 In a single throw of a pair of different dice, what is the probability of getting: [CBSE (CCE) 2016]

(i) A prime number on each dice (ii) A total of 9 or 11

Solution:

(i) Total possible cases when two dice are thrown together = $6 \times 6 = 36$

Favourable cases when both numbers are prime are (2, 2), (2, 3), (2, 5), (3, 2), (3, 3), (3, 5), (5, 2), (5, 3), (5, 5), that is, 9 outcomes

$$P\text{ (a prime number on each dice)} = \frac{\text{Favourable cases}}{\text{Total cases}} = \frac{9}{36} = \frac{1}{4}$$

(ii) Favourable cases when the sum of numbers is 9 or 11 are (3, 6), (4, 5), (5, 4), (5, 6), (6, 3), (6,5), that is, 6 outcomes

$$P\text{ (a total of 9 or 11)} = \frac{\text{Favourable outcomes}}{\text{Total possible outcomes}} = \frac{6}{36} = \frac{1}{6}.$$

Example 14 Two different dice are thrown together. Find the probability of: [CBSE (CCE) 2016]

(i) Getting a number greater than 3 on each die

(ii) Getting a total of 6 or 7 of the numbers on two dice

Solution:

(i) When two dice are thrown together, the total possible outcomes = $6 \times 6 = 36$.

Favourable outcomes when both dice have number more than 3 are (4, 4), (4, 5), (4, 6), (5, 4), (5, 5), (5, 6), (6, 4), (6, 5), (6, 6), that is, 9 outcomes.

P (a number greater than 3 on each dice)

$$= \frac{\text{Number of favourable outcomes}}{\text{Number of total possible outcomes}}$$

$$= \frac{9}{36} = \frac{1}{4}.$$

(ii) Favourable outcomes when the sum of the numbers appearing on the dice is 6 or 7 are (1, 5), (1, 6), (2, 4), (2, 5), (3, 3), (3, 4), (4, 2), (4, 3), (5, 1), (5, 2), (6, 1), that is, 11 outcomes.

$$P\text{ (a total of 6 or 7)} = \frac{11}{36}.$$

3 Try to Solve

1. Two different dice are rolled together. Find the probability of getting: [CBSE (CCE) 2015]
   (i) The sum of numbers on two dice to be 5
   (ii) Even numbers on both dice
2. Two different dice are tossed together. Find the probability that: [CBSE (CCE) 2014]
   (i) The number on each die is odd
   (ii) The sum of numbers appearing on the two dice is 5
3. Two different dice are rolled simultaneously. Find the probability that the sum of numbers appearing on the two dice is 10. [NCERT]

4. A dice is rolled twice. Find the probability that: [NCERT, CBSE (CCE) 201
   (i) 5 will not come up either time
   (ii) 5 will come up exactly one time
5. Two dice are thrown simultaneously. Determine the probability that the difference of t numbers on the two dice is 2. [NCERT, CBSE (CCE) 201

Answers

1. (i) $\frac{1}{9}$ (ii) $\frac{1}{4}$ **2.** (i) $\frac{1}{4}$ (ii) $\frac{1}{9}$ **3.** $\frac{1}{12}$ **4.** (i) $\frac{25}{36}$ (ii) $\frac{5}{18}$ **5.** $\frac{2}{9}$

Concept: Playing Cards

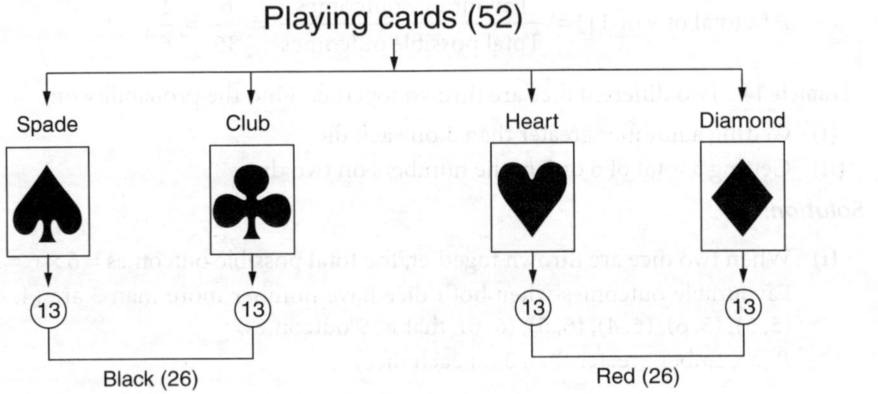

Probability of Playing Cards

A deck of paying cards consists of 52 cards. The deck of cards is divided into four suits (groups) c cards each: spades, hearts, diamonds and clubs. Clubs and spades are black in colour, while hearts diamonds are red. The cards in each suite (group) are Ace, King, Queen, Jack, 10, 9, 8, 7, 6, 5, 4, 3, ar King, Queen and Jack are called face cards. A deck of playing cards has 12 face cards (four in each gro in which 6 are black and 6 are red.

Example 15 A card is drawn at random from a well-shuffled deck of playing cards. Find the probab that the card drawn is:

(i) A red king **(ii)** Either a king or a queen **(iii)** Neither a king nor a queen

Solution: Number of outcomes = 52

(i) Number of favourable outcomes = 2

$\therefore P$ (of getting a red king) $\frac{2}{52} = \frac{1}{26}$.

(ii) Number of favourable outcomes = 4 (king) + 4 (queen) = 8

$\therefore P$ (of getting either king or queen) = $\frac{8}{52} = \frac{2}{13}$

(iii) Number of favourable outcomes = 52 – 4 (king) – 4 (queen) = 44

$\therefore P$ (of getting neither a king nor a queen) = $\frac{44}{52} = \frac{11}{13}$.

Example 16 One card is drawn from a pack of 52 cards, each of the 52 cards being equally likely to be drawn. Find the probability that the card drawn is:

(i) An ace **(ii)** A face card **(iii)** Red and a king **(iv)** A spade or an ace **(v)** A red face card

Solution: Number of outcome = 52

(i) There are four ace cards in a pack of 52 cards

$\therefore$ Number of favourable outcomes = 4

$$P\text{ (of getting an ace)} = \frac{4}{52} = \frac{1}{13}$$

(ii) There are 12 face cards in a pack of 52 cards

$\therefore$ Number of favourable outcomes = 12

$$P\text{ (of getting a face card)} = \frac{12}{52} = \frac{3}{13}$$

(iii) There are 2 red kings

$\therefore$ Number of favourable outcomes = 2

$$P\text{ (of getting red and a king)} = \frac{2}{52} = \frac{1}{26}$$

(iv) There are 13 spades, including ace, in a pack of 52 cards. There are 4 ace cards in total. To get a spade or an ace, we should have all the spades and all the ace cards (13 + 3 = 16 cards).

$\therefore$ Number of favourable outcomes = 16

$$P\text{ (of getting a spade or an ace)} = \frac{16}{52} = \frac{4}{13}$$

(v) There are two groups of red cards. Each group has three face cards.

Number of favourable outcomes = 6

$$P\text{ (of getting a red face card)} = \frac{6}{52} = \frac{3}{26}$$

Example 17 The queen, jack and 10 of spades are removed from a pack of 52 cards. A card is drawn from the remaining well-shuffled pack. Find the probability of getting: **[CBSE (CCE) 2011]**

(i) A red card **(ii)** A king **(iii)** A black card

Solution: Number of possible outcomes = 49

(a) Number of favourable outcomes = 26

(number of red cards)

$$\therefore P\text{ (of getting a red card)} = \frac{26}{49}$$

(b) Number of favourable outcomes = 4

(number of kings)

$$P\text{ (of getting a king)} = \frac{4}{49}$$

(c) Total number of black in a pack of 52 cards = 26

Out of which, three cards of spades are removed.

$\therefore$ Number of favourable outcomes = 26 – 3 = 23

$$P\text{ (of getting a black card)} = \frac{23}{49}$$

4 Try to Solve

1. A card is drawn from a well-shuffled deck of 52 cards. Find the probability of getting:
   (a) A black face card (b) A jack of hearts **[CBSE (CCE) 2010]**
2. All the jacks, queens and kings are removed from a deck of 52 playing cards. The remaining cards are well shuffled and then one card is drawn at random. Giving ace a value 1 and similar values for other cards, find the probability that the card has a value:
   (a) 7 (b) Greater than 7 (c) Less than 7 **[NCERT Exemplar]**
3. King, Queen and Jack of Clubs are removed from a deck of 52 playing cards and the remaining cards are shuffled. A card is drawn from the remaining cards. Find the probability of getting a card of:
   (a) Hearts (b) Queen (c) Clubs

Answers

1. (a) $\frac{3}{26}$, (b) $\frac{1}{52}$ **2.** (a) $\frac{1}{10}$, (b) $\frac{3}{10}$, (c) $\frac{3}{5}$ **3.** (a) 13/49, (b) 3/49, (c) 10/49

Example 18 All the red face cards are removed from a pack of 52 playing cards. A card is drawn at random from the remaining cards, after reshuffling them. Find the probability that the drawn card is

(i) Of red colour **(ii)** A queen **(iii)** An ace **(iv)** A face card **[CBSE (CCE) 2014]**

Solution: Face cards are taken as (4 jacks, 4 queens and 4 kings).

As all the red face cards are removed, 6 cards (2 jacks, 2 queens and 2 kings) are removed from the 52 cards.

∴ Remaining cards = 52 – 6 = 46 cards

∴ Total outcomes of drawing a card from 46 cards = 46

(i) Favourable outcomes of drawing a red card
= 20 (as 6 red cards from 26 have been removed).

∴ Probability of drawing a red card = $\frac{20}{46} = \frac{10}{23}$.

(ii) Favourable outcomes of drawing a queen
= 2 (as 2 red queens have been removed out of 4)

∴ Probability of drawing a queen = $\frac{2}{46} = \frac{1}{23}$

(iii) Favourable outcomes of drawing an ace = 4

∴ Probability of drawing an ace = $\frac{4}{46} = \frac{2}{23}$

(iv) Favourable outcomes of drawing a face card
= 6 (as 2 jacks, 2 queens and 2 kings have been removed)

∴ Probability of drawing a face cards = $\frac{6}{46} = \frac{3}{23}$.

Example 19 Five cards, ten, jack, queen, king and ace of diamonds, are well shuffled with their faces downwards. One card is then picked at random.

(a) What is the probability that the drawn card is a queen?

(b) If a queen is drawn and set aside, and a second card is drawn, find the probability that the second card is (i) an ace (ii) a queen. **[NCERT, CBSE (CCE) 2014]**

Solution: Five cards – ten, jack, queen, king and ace of diamonds – are shuffled with faces down.

Total number of possibilities of selecting one card = 5

(a) Favourable outcomes for drawing a queen = 1

$\therefore$ Probability of drawing a queen = $\frac{1}{5}$

(b) When the queen is kept aside, the remaining cards = 4

So, total number of outcomes of drawing a card = 4

(i) Favourable outcome of drawing an ace is 1.

$\therefore$ Probability of drawing an ace = $\frac{1}{4}$

(ii) Favourable outcome of drawing a queen = 0 (as there is no card of queen).

$\therefore$ Probability of drawing a queen = $\frac{0}{4}$, that is, 0

5 Try to Solve

1. One card is drawn from a well-shuffled deck of 52 cards. Find the probability of getting:

(a) A king of red colour (b) A face card (c) A red face card

(d) Jack of hearts (e) A spade (f) Queen of diamond

2. Three cards of spades are lost from a pack of 52 playing cards. The remaining cards were well shuffled and then a card was drawn at random from them. Find the probability that the drawn card is of black colour. **[CBSE (CCE) 2013]**

3. The ace, jack, queen and king of clubs are removed from a pack of 52 playing cards. The remaining cards are well shuffled and a card is drawn from them at random. Find the probability that the drawn card is of black colour. **[CBSE (CCE) 2013]**

Answers

1. (a) 1/26, (b) 3/13, (c) 3/26, (d) 1/52, (e) 1/4, (f) 1/52 **2.** 23/49 **3.** 11/24

Miscellaneous Problems

Example 20 A bag contains 12 balls, out of which x are white. **[NCERT, CBSE 2013]**

(i) If one ball is drawn at random, find the probability that it is a white ball.

(ii) If 6 more white balls are put in the bag, the probability of drawing a white ball is double that in (i). Find x.

Solution: Total balls = 12. White balls are x.

(i) Probability of getting a white ball = $\frac{x}{12}$

(ii) If 6 more balls are put into the bag, total balls are $12 + 6 = 18$

White balls are $6 + x$

Probability of getting white balls = $\frac{x+6}{18}$

According to the question $$\frac{x+6}{18} = 2\left(\frac{x}{12}\right)$$

$$\Rightarrow \quad \frac{x+6}{18} = \frac{\cancel{2}\,x}{\cancel{12}}$$

$$\frac{x+6}{\cancel{18}\,3} = \frac{x}{\cancel{6}}$$

$$x + 6 = 3x$$

$$6 = 2x$$

$$\Rightarrow \quad x = 3$$

Example 21 A box contains 70 cards numbered from 1 to 70. If one card is drawn at random from the box, find the probability that it bears **[CBSE, (CCE) 2012]**

(i) A perfect square number **(ii)** A number divisible by 2 and 3

Solution: Total cards = 70

(i) Cards with a perfect square number = 1, 4, 9, 16, 25, 36, 49, 64

$$P\text{ (a perfect square number)} = \frac{\text{Number of favourable outcomes}}{\text{Number of total outcomes}} = \frac{8}{70} = \frac{4}{35}$$

(ii) Cards with number divisible by 2 and 3 = 6, 12, 18, 24, 30, 36, 42, 48, 54, 60, 66.

$$P\text{ (a number divisible by 2 and 3)} = \frac{11}{70}$$

Example 22 Cards marked with numbers 1, 3, 5, ..., 101 are placed in a bag and mixed thoroughly. A card is then drawn at random from the bag. Find the probability that the number on the drawn card is:

(i) Less than 19

(ii) A prime number less than 20 **[CBSE (CCE) 2012]**

Solution: Total number of cards in the bag = 51

(i) Number of cards bearing a number less than 19 are 9 (1, 3, 5, 7, 9, 11, 13, 15, 17).

$\therefore$ Probability that the number on the drawn card is less than 19 = $\frac{9}{51} = \frac{3}{17}$

(ii) Number of cards bearing a prime number less than 20 are 7 (3, 5, 7, 11, 13, 17, 19).

Probability that the number on the drawn card is a prime number less than 20 = $\frac{7}{51}$.

Example 23 For any event E, if $P(E) = 0.98$, find the value of $P(\bar{E})$.

Solution: Since we know that

$$P(E) + P(\bar{E}) = 1$$
$$0.98 + P(\bar{E}) = 1$$
$$P(\bar{E}) = 1 - 0.98 = 0.02$$

Probability of $P(\bar{E}) = 0.02$.

Example 24 Find the probability of getting 53 Mondays in a non-leap year.

Solution: Total number of days in a non-leap year = 365.

Total number of complete weeks in a non-leap year = 52.

Total number of days left in a non-leap year after completing 52 weeks = 1.

So, this one day could be any day of the 7 days of the week.

Favourable outcomes = 1

Total number outcomes = 7

$$P(E) = \frac{\text{Number of favourable outcomes}}{\text{Total number of outcomes}} = \frac{1}{7}.$$

6 Try to Solve

1. Cards marked with numbers 5, 6, 7, …, 74 are placed in a bag and mixed thoroughly. One card is drawn at random from the bag. Find the probability that the number on the card is a perfect square. **[CBSE (CCE) 2011]**
2. Cards bearing numbers 1, 3, 5, …, 35 are kept in a bag. A card is drawn at random from the bag. Find the probability of getting a card bearing
   (a) A prime number less than 15
   (b) A number divisible by 3 and 5 **[CBSE (CCE) 2010]**
3. A bag contains 20 balls, out of which x balls are red. **[CBSE (CCE) 2014]**
   (i) If one ball is drawn at random from the bag, find the probability that it is not red.
   (ii) If 4 more red balls are put into the bag, the probability of drawing a red ball will be $\frac{5}{4}$ times the probability of drawing a red ball in the first case. Find the value of x.
4. A bag contains 18 balls, out of which x balls are red. **[CBSE (CCE) 2013]**
   (i) If one ball is drawn at random from the bag, what is the probability that it is not red?
   (ii) If 2 more red balls are put in the bag, the probability of drawing a red ball will be $\frac{9}{8}$ times the probability of drawing a red ball in the first case. Find the value of x.
5. A game of chance consists of spinning an arrow on a circular board, divided into 8 equal parts, which comes to rest pointing at one of the numbers 1, 2, 3, …, 8 (see figure), which are equally likely outcomes. What is the probability that the arrow will point at (i) an odd number, (ii) a number greater than 3, (iii) a number less than 9. **[NCERT, CBSE (CCE) 2016]**

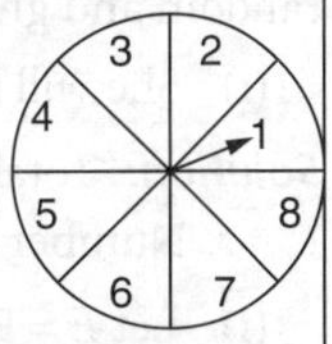

6. For any event E, if $P(E) = 0.99$, find the value of P (not E).
7. Find the probability of getting 53 Sundays in a non-leap year.

Answers

1. $\frac{3}{35}$ 2. (a) $\frac{5}{18}$, (b) $\frac{1}{18}$ 3. (i) $\frac{20-x}{20}$, (ii) 8 4. (i) $\frac{18-x}{18}$, (ii) 8

5. (i) $\frac{1}{2}$, (ii) $\frac{5}{8}$, (iii) 1 6. 0.01 7. $\frac{1}{7}$

FOR MORE PRACTICE AND KNOWLEDGE

Example 25 A game of chance consists of spinning an arrow which comes to rest pointing at one of the numbers 1, 2, 3, 4, 5, 6, 7 and 8 (see figure) and these are equally likely outcomes. What is the probability that it will point at

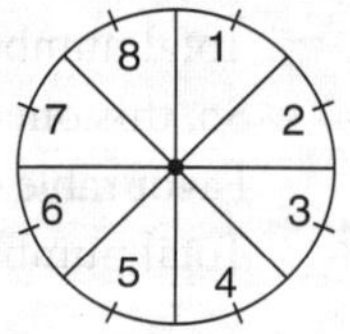

(i) 8 **(ii)** An odd number **(iii)** A number greater than 2
(iv) A number less than 9 [NCERT]

Solution: Total number of points on the circle = 8

(i) Let E_1 = Event of getting an arrow at number 8
∴ Number of outcomes favourable to E_1 =1
Probability that arrow points to number 8 $P(E_1) = \frac{1}{8}$

(ii) Let E_2 = Event of getting an arrow at an odd number. Here, odd numbers are 1, 3, 5 and 7.
∴ Number of outcomes favourable to E_2 = 4. Probability that the arrow points to an odd number

$$P(E_2) = \frac{4}{8} = \frac{1}{2}$$

(iii) Let E_2 = Event of getting an arrow at a number greater than 2, that is, at 3, 4, 5, 6, 7 or 8.
∴ Number of outcomes favourable to E_3 = 6. Probability that the arrow points to a number greater than 2

$$P(E_2) = \frac{6}{8} = \frac{3}{4}$$

(iv) $P(E_4) = \frac{\not{8}}{\not{8}} = 1$

Example 26 A lot consists of 144 ball pens of which 20 are defective and the others are of good quality. Nuri will buy a pen if it is good, but will not buy if it is defective. The shopkeeper draws one pen at random and gives it to her. What is the probability that [NCERT]
(i) She will buy it **(ii)** She will not buy it

Solution: Total number of ball pens = 144
∴ Number of all possible outcomes = 144

(i) Let E = Event of getting a good ball pen. Then, the number of outcomes favourable to E = Total number of ball pens – Defective ball pens = 144–20 = 124
She will buy a pen, if it is a good pen.
∴ Required probability = $P(E) = \frac{124}{144} = \frac{31}{36}$

(ii) She will not buy a pen, if it is not a good pen.

$\therefore$ Required probability = 1 – Probability of getting a good pen $[\because P(E) + P(\bar{E}) = 1]$

$$= 1 - P(E) = 1 - \frac{31}{36} = \frac{5}{36}$$

Example 27 Two dice, one blue and one grey, are thrown at the same time. Then **[NCERT]**

(i) Complete the following table:

Event: Sum on 2 dice	2	3	4	5	6	7	8	9	10	11	12
Probability	$\frac{1}{36}$						$\frac{5}{36}$				$\frac{1}{36}$

(ii) A student argues that there are 11 possible outcomes 2, 3, 4, 5, 6, 7, 8, 9, 10, 11 and 12. Therefore, each of them has a probability $\frac{1}{11}$. Do you agree with this argument? Justify your answer.

Solution:

(i) Total possible outcomes on throwing two dice are

(1, 1), (1, 2), (1, 3), (1, 4), (1, 5), (1, 6) (2, 1), (2, 2), (2, 3), (2, 4), (2, 5), (2, 6) (3, 1), (3, 2), (3, 3), (3, 4), (3, 5), (3, 6) (4, 1), (4, 2), (4, 3), (4, 4), (4, 5), (4, 6) (5, 1), (5, 2), (5, 3), (5, 4), (5, 5), (5, 6) (6, 1), (6, 2), (6, 3), (6, 4), (6, 5) and (6, 6)

$\therefore$ Number of all possible outcomes = 36

(a) Let E_1 = Sum on two dice is 3. Then, E_1 would consist of two outcomes, namely, (1, 2) and (2, 1).

$\therefore$ Number of outcomes favourable to $E = 2$

Hence $P(E_1) = \frac{2}{36} = \frac{1}{18}$.

(b) Let E_2 = Sum on two dice is 4. Then, E_2 would consist of three outcomes, namely, (1, 3), (2, 2) and (3, 1).

$\therefore$ Number of outcomes favourable to $E_2 = 3$

Hence $P(E_2) = \frac{3}{36} = \frac{1}{12}$.

(c) Let E_3 = Sum on two dice is 5. Then, E_1 would consist of four outcomes, namely, (1, 4), (2, 3), (3, 2) and (4, 1).

$\therefore$ Number of outcomes favourable to $E_3 = 3$

Hence $P(E_3) = \frac{4}{36} = \frac{1}{9}$.

(d) Let E_4 = Sum on two dice is 6. Then, E_4 would consist of five outcomes, namely, (1, 5), (2, 4), (3, 3), (4, 2) and (5, 1).

$\therefore$ Number of outcomes favourable to $E_4 = 5$

Hence $P(E_4) = \frac{5}{36}$.

(e) Let E_5 = Sum on two dice is 7. Then, E_5 would consist of six outcomes, namely, (1, 6), (2, 5), (3, 4), (4, 3), (5, 2) and (6, 1).

$\therefore$ Number of outcomes favourable to $E_5 = 6$

Hence $P(E_5) = \frac{6}{36} = \frac{1}{6}$.

(f) Let E_6 = Sum on two dice is 9. Then, E_6 would consist of four outcomes, namely, (3, 6), (4, 5), (5, 4) and (6,3).

∴ Number of outcomes favourable to $E_6 = 4$.

Hence $$P(E_6) = \frac{4}{36} = \frac{1}{9}.$$

(g) Let E_7 = Sum on two dice is 10. Then, E_7 would consist of three outcomes, namely, (4, 6), (5, 5) and (6, 4).

∴ Number of outcomes favourable to $E_7 = 3$

Hence $$P(E_7) = \frac{3}{36} = \frac{1}{12}.$$

(h) Let E_8 = Sum on two dice is 11. Then, E_8 would consist of two outcomes, namely, (6, 5) and (5,6).

∴ Number of outcomes favourable to $E_8 = 2$

Hence $$P(E_8) = \frac{2}{36} = \frac{1}{18}.$$

(ii) No, we do not agree with the given argument because the events of eleven sums are not equally likely.

Example 28 Two customers, Shyam and Ekta, are visiting a particular shop in the same week (Tuesday to Saturday). Each is equally likely to visit the shop on any day as on another day. What is the probability that both will visit the shop on

(i) The same day **(ii)** Consecutive days **(iii)** Different days **[NCERT]**

Solution: Number of days from Tuesday to Saturday = 5. Total number of possible outcomes for both = 5 × 5 = 25

(i) If both visit the shop on the same day, then there are 5 days. So, the number of favourable outcomes = 5

Now, P (both will visit on the same day) $= \frac{5}{25} = \frac{1}{5}$.

(ii) If they visit the shop on consecutive days, then favourable outcomes are (*T*, *W*), (*W*, *Th*), (*Th*, *F*), (*F*,*S*), (*S*, *F*), (*F*, *Th*), (*Th*, *W*) and (*W*, *T*)

∴ Number of favourable outcomes = 8

Hence, P (both will visit on consecutive days) $= \frac{8}{25}$.

(iii) P (both will visit on different days) = 1−P (both will visit on the same day) $= 1 - \frac{1}{5} = \frac{4}{5}$.

Example 29 A bag contains 5 red balls and some blue balls. If the probability of drawing a blue ball is double that of a red ball, determine the number of blue balls in the bag. **[NCERT]**

Solution: Let the number of blue balls be n. Total number of red balls = 5.

∴ Total number of balls = 5 + n

$$P(\text{red ball}) = \frac{\text{Number of red balls}}{\text{Total Number of balls}} = \frac{5}{5+n} \text{ and } P(\text{blue ball}) = \frac{\text{Number of blue balls}}{\text{Total number of balls}} = \frac{n}{5+n}$$

According to the question, P (drawing a blue ball) = $2 \times P$ (drawing a red ball)

$$\Rightarrow \qquad \frac{n}{\cancel{5+n}} = 2 \times \frac{5}{\cancel{5+n}} \Rightarrow n = 10$$

Hence, the number of blue balls in the bag is 10.

Example 30 A number x is chosen at random from the numbers – 4, – 3, – 2, – 1 0, 1 2, 3, 4. What is the probability that $|x| < 2$? **[CBSE 2016]**

Solution: Number of all possible outcomes = 9; Numbers favourable to $|x| < 2$ are – 1, 0 and 1.

$\therefore$ Number of favourable outcomes = 3, So, required probability $= \frac{3}{9} = \frac{1}{3}$.

Example 31 A letter is chosen at random from the English alphabet. What is the probability that it is a letter of the word 'RAMANUJAN'? **[CBSE 2015]**

Solution: Number of all possible outcomes = 26

Letters in the word 'RAMANUJAN' are R, A, M, N, U, J.

$\therefore$ Number of favourable outcomes = 6. Hence, required probability $= \frac{6}{26} = \frac{3}{13}$.

Example 32 A single letter is selected at random from the word 'PROBABILITY'. Find the probability that it is a vowel. **[CBSE 2012]**

Solution: Total number of outcomes = 11. There are 4 vowels namely O, A, I, I.

So, number of favourable outcomes = 4. Hence $P(\text{vowels}) = \frac{4}{11}$

Example 33 A die has its six faces marked 0, 1, 1, 1, 6, 6. Two such dice are thrown together and the total score is recorded. **[NCERT Exemplar]**

(i) How many different scores are possible? **(ii)** What is the probability of getting a total of 7?

Solution: Two such dice are thrown together, therefore number of all possible outcomes = 36

(i) The different scores which are possible = 6 scores, namely 0, 1, 2, 6, 7 and 12.

(ii) Let E = Event of getting a sum 7, that is

{(1,6),(1,6), (1,6), (1,6),(1,6), (1,6), (6,1), (6,1), (6,1), (6,1), (6,1), (6,1)}.

$\therefore$ Number of outcomes favourable to E = 12

Hence $$P(E) = \frac{12}{36} = \frac{1}{3}$$

Example 34 A child's game has 8 triangles of which 3 are blue and the rest are red, and 10 squares of which 6 are blue and rest are red. One piece is lost at random. Find the probability that it is a

(i) Triangle (ii) Square (iii) Blue square (iv) Red triangle **[NCERT Exemplar]**

Solution: Total number of figures = 8 triangles + 10 squares = 18

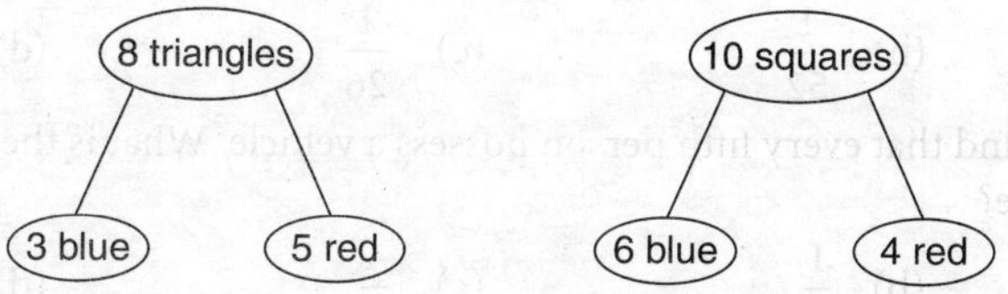

(i) P (lost piece is a triangle) $= \dfrac{8}{18} = \dfrac{4}{9}$ **(ii)** P (lost piece is a square) $= \dfrac{10}{18} = \dfrac{5}{9}$

(iii) P (lost piece is blue square) $= \dfrac{6}{18} = \dfrac{1}{3}$ **(iv)** P (lost piece is red triangle) $= \dfrac{5}{18}$.

Multiple Choice Questions

Choose the correct answer for each of the following questions:

1. Which of the following cannot be the probability of an event? **[Delhi 2011]**

(a) 1.5 (b) $\dfrac{3}{5}$ (c) 25% (d) 0.3

2. The probability of throwing a number greater than 2 with a fair die is **[Foreign 2011]**

(a) $\dfrac{2}{3}$ (b) $\dfrac{5}{6}$ (c) $\dfrac{1}{3}$ (d) $\dfrac{2}{5}$

3. The set of all possible outcomes of a random experiment is called

(a) Sample space (b) Elementary events

(c) Complementary events (d) Favorable events

4. The probability of getting a bad egg in a lot of 400 is 0.035. The number of bad eggs in the lot is **[NCERT Exemplar]**

(a) 7 (b) 14 (c) 21 (d) 28

5. A girl calculates that the probability of her winning the first prize in a lottery is 0.08. If 6000 tickets are sold, how many tickets has she bought? **[NCERT Exemplar]**

(a) 40 (b) 240 (c) 480 (d) 750

6. Cards bearing numbers 2, 3, 4, …, 11 are kept in a bag. A card is drawn at random from the bag. The probability of getting a card with a prime number is **[AI 2012]**

(a) $\dfrac{1}{2}$ (b) $\dfrac{2}{5}$ (c) $\dfrac{3}{10}$ (d) $\dfrac{5}{9}$

7. There are 25 tickets bearing numbers from 1 to 25. One ticket is drawn at random. The probability that the number on it is a multiple of 5 or 6 is

(a) $\dfrac{7}{25}$ (b) $\dfrac{9}{25}$ (c) $\dfrac{11}{12}$ (d) $\dfrac{13}{25}$

8. The probability of getting a number between 1 and 100 which is divisible by 1 and itself only is

(a) $\dfrac{29}{98}$ (b) $\dfrac{1}{2}$ (c) $\dfrac{25}{98}$ (d) $\dfrac{23}{98}$

9. The probability expressed as a percentage of a particular occurrence can never be **[NCERT Exemplar]**

(a) Less than 100 (b) Less than 0

(c) Greater than 1 (d) Anything but a whole number

10. If $P(A)$ denotes the probability of an event A, then **[NCERT Exemplar]**

(a) $P(A) < 0$ (b) $P(A) > 1$ (c) $0 \le P(A) \le 1$ (d) $-1 \le P(A) \le 1$

11. What is the probability of getting a king in a well-shuffled pack of 52 cards?

(a) $\dfrac{1}{13}$ (b) $\dfrac{1}{52}$ (c) $\dfrac{1}{26}$ (d) 0

12. In a survey, it is found that every fifth person possess a vehicle. What is the probability of a person not possessing a vehicle?

(a) 0 (b) $\dfrac{1}{5}$ (c) $\dfrac{4}{5}$ (d) 1

13. If $P(E)$ is 42% of an event, what is the probability of failure of the event?
(a) 8 % (b) 1 (c) 0 (d) 58%

14. The probability of a sure event is
(a) 2 (b) 1 (c) 0 (d) ½

15. The probability of getting an even number, when a die is thrown once, is
(a) $\frac{1}{2}$ (b) $\frac{1}{3}$ (c) $\frac{1}{6}$ (d) $\frac{5}{6}$

16. Which of the following can be the probability of an event?
(a) – 0.04 (b) 1.004 (c) $\frac{18}{23}$ (d) $\frac{8}{7}$

17. If two coins are tossed simultaneously, then the probability of getting at least one head is
(a) $\frac{1}{4}$ (b) $\frac{3}{4}$ (c) $\frac{1}{2}$ (d) 1

18. If a die is rolled once, then the probability of obtaining a number greater than 3 is
(a) $\frac{1}{2}$ (b) $\frac{1}{3}$ (c) 0 (d) $\frac{2}{3}$

19. In a throw of a pair of dice, the probability of getting a doublet is
(a) $\frac{1}{3}$ (b) $\frac{1}{6}$ (c) $\frac{5}{12}$ (d) $\frac{2}{3}$

20. If three coins are tossed simultaneously, then the probability of getting no head is
(a) $\frac{3}{4}$ (b) $\frac{3}{8}$ (c) $\frac{1}{8}$ (d) $\frac{1}{4}$

21. In a single throw of a pair of dice, the probability of getting the sum 9 is
(a) 1/18 (b) 1/9 (c) 1/6 (d) 2/9

22. A box contains 90 discs, numbered from 1 to 90. If one disc is drawn at random from the box, the probability that it bears a prime number less than 23 is
(a) $\frac{7}{90}$ (b) $\frac{10}{90}$ (c) $\frac{4}{45}$ (d) $\frac{9}{89}$

23. If $P(E)$ is 0.75, what is $P(\text{not } E)$?
(a) 0.35 (b) 0.25 (c) 0 (d) 1

24. A bag contains 8 red and 9 blue marbles. A marble is taken out at random. What is P (red marble)?
(a) $\frac{8}{9}$ (b) $\frac{8}{17}$ (c) $\frac{9}{17}$ (d) $\frac{16}{17}$

25. Ashmita and Shreya are sisters. What is the probability that both have a birthday on 14th September (ignoring leap years)?
(a) $\frac{1}{30}$ (b) $\frac{2}{365}$ (c) $\frac{1}{366}$ (d) $\frac{1}{365}$

26. If the probability of an event is p, then the probability of its complementary event will be
(a) $1 - p$ (b) $p - 1$ (c) $\frac{1}{p}$ (d) $1 - \frac{1}{p}$

27. Which of the following cannot be the probability of an event?
(a) $\frac{1}{5}$ (b) 0.3 (c) 4% (d) $\frac{5}{4}$

Fill in the Blanks

Complete the following statements with appropriate answers(s) in the blank space(s):

28. The probability of getting 101 marks out of 100 marks is ________.
29. The probability of getting a face card from a pack of cards is ________.
30. The sum of probabilities of all the elementary events is always ________.
31. The sum of the probability of an event happening and not happening is equal to ________.
32. A single letter is selected at random from the word 'PROBABILITY'. The probability that it is a vowel is ________.
33. The probability of getting the same number 2 on tossing two dice is ________.
34. In the throwing of a die, the probability of getting a multiple of 7 is ________.
35. The probability of an event that cannot happen is ________.
36. If the occurrence of one event cannot be expected in preference of the other event, the events are called ________.
37. A die is thrown once. The probability of getting a prime number is ________.
38. A box contains cards numbered 6 to 50. A card is drawn at random from the box. The probability that the drawn card has a number which is a perfect square is ________. **[AI 2013]**

Subjective Questions

39. Two different dice are tossed together. Find the probability:
    (i) Of getting a doublet
    (ii) Of getting a sum 10, of the numbers on the two dice. **[CBSE 2018]**
40. An integer is chosen at random between 1 and 100. Find the probability that it is: **[CBSE 2018]**
    (i) Divisible by 8 (ii) Not divisible by 8
41. The probability of selecting a blue marble at random from a jar that contains only blue, black and green marbles is $\frac{1}{5}$. The probability of selecting a black marble at random from the same jar is $\frac{1}{4}$. If the jar contains 11 green marbles, find the total number of marbles in the jar. **[CBSE, Allahabad 2019]**
42. A bag contains 5 white balls, 7 red balls, 4 black balls and 2 blue balls. One ball is drawn at random from the bag. What is the probability that the ball drawn is
    (i) White or blue (ii) Red or black
    (iii) Not white (iv) Neither white nor black
43. A card is drawn at random from a well-shuffled deck of playing cards. Find the probability that the card drawn is
    (i) A king or a jack (ii) A non-ace
    (iii) A red card (iv) Neither a king nor a queen
44. Two dice are thrown simultaneously. What is the probability that 5 will come up on both dice?
45. Peter throws two different dice together and finds the product of the two numbers obtained. Rina throws a die and squares the number obtained. Who has the better chance to get the number 25? **[Delhi 2017]**
46. A number is selected at random from the numbers 3, 5, 5, 7, 7, 7, 9, 9, 9, 9. Find the probability that the selected number is their average.

47. If a number x is chosen from the number 1, 2, 3 and a number y is selected from the numbers 1, 4, 9, find the probability that $xy = 10$.

48. A two-digit number is selected at random. Find the probability that it is multiple of 4 but not divisible by 5.

49. A box contains cards numbered from 1 to 20. A card is drawn at random from the box. Find the probability that number on the drawn card is
(i) A prime number (ii) A composite number
(iii) A number divisible by 3 **[CBSE 2018(C)]**

50. Two dice are numbered 1, 2, 3, 4, 5, 6 and 1, 2, 2, 3, 3, 4, respectively. They are thrown and the sum of the numbers on them is noted. Find the probability of getting: (i) Sum 7 (ii) Sum is a perfect square.

51. A child has a die whose six faces show the letters as given below:

A	B	C	A	D	A

The die is thrown once. Find the probability of getting
(i) A (ii) D

52. What is the probability that a leap year will have 53 Mondays and 53 Tuesdays?

53. What is the probability that the month of December has exactly 5 Sundays, 5 Mondays and 5 Tuesdays?

54. A three-digit number is chosen. Find the probability that all three digits are not the same.

55. A number x is chosen from the numbers – 4, – 3,– 2, 1, 0, 1, 2, 3, 4. Find the probability that $|x| < 3$.

56. Write the name of the experiment whose outcome has to be among the set of events that are completely known but whose exact outcome is unknown.

57. A man is known to speak truth 5 out of 7 times. He throws a die and a number other than six comes up. Find the probability that he reports it is a six.

58. Archana calculates that probability of her winning the first prize in a lottery is 0.04. If 12,000 tickets are sold, how many tickets has she bought?

59. A letter is chosen at random from the letters of the word 'ASSASSINATION'. Find the probability that the letter chosen is a vowel?

60. A letter of the English alphabet is chosen at random. Determine the probability that the chosen letter is a consonant. **[Delhi 2015]**

61. A game of chance consists of spinning an arrow which comes to rest pointing at one of the numbers 1,2,3, 4, 5, 6, 7, 8 and these are equally likely outcomes. Find the probability that the arrow will point at any factor of 8. **[Foreign 2015]**

62. A ticket is drawn at random from a bag containing tickets numbered from 1 to 40. Find the probability that the selected ticket has a number which is a multiple of 5. **[AI 2011]**

63. Three cards of spades are lost from a pack of 52 playing cards. The remaining cards were well shuffled and then a card was drawn at random from them. Find the probability that the drawn card is of black colour. **[Foreign 2013]**

64. Two different dice are tossed together. Find the probability **[AI 2014]**
(i) That the number on each dice is even (ii) That the sum of numbers appearing on two dice is 5

65. Find the probability that a leap year should have exactly 52 Tuesdays.

66. A card is drawn at random from a well-shuffled deck of playing cards. Find the probability that the card drawn is
(i) A card of spade or an ace (ii) A red king **[AI 2015]**
(iii) Neither a king nor a queen (iv) Either a king or queen

67. Cards marked with numbers 3,4,5, ..., 50 are placed in a box and mixed thoroughly. One card is drawn at random from the box. Find the probability that number on the drawn card is **[Delhi 2016]**
(i) Divisible by 7 (ii) A number which is a perfect square

68. All kings, queens and aces are removed from a pack of 52 cards. The remaining cards are well shuffled and then a card is drawn from it. Find the probability that the drawn card is **[AI 2012]**
(i) A black face card (ii) A red card

69. The probability of selecting a red ball at random from a jar that contains only red, blue and orange balls is $\frac{1}{4}$. The probability of selecting a blue ball at random from the same jar is $\frac{1}{3}$. If the jar contains 10 orange balls, find the total number of balls in the jar. **[AI 2015]**

70. If odds in favour of an event is 2:3. Find the probability of the non-occurrence of this event.

71. A bag contains 24 balls, out of which x are white. If one ball is drawn at random, the probability of drawing a white ball is y. 12 more white balls are added to the bag. Now if a ball is drawn from the bag, the probability of drawing the white ball is $\frac{5}{3}y$. Find the value of x. **[Delhi 2013]**

72. At a fete, cards bearing numbers 1 to 1000, one number on one card, are put in a box. Each player selects one card at random and that card is not replaced. If the selected card has a perfect square greater than 500, the player wins a prize. What is the probability that **[NCERT Exemplar]**
(i) The first player wins a prize (ii) The second player wins a prize, if the first has won

73. A card is drawn at random from a well-shuffled deck of playing cards. Find the probability that the card drawn is
(i) A card of spade or an ace (ii) A black king **[AI 2015]**
(iii) Neither a jack nor a king (iv) Either a king or a queen

74. A group consists of 12 persons, of which 3 are extremely patient, 6 are extremely honest and the rest are extremely kind. A person from the group is selected at random. Assuming that each person is equally likely to be selected, find the probability of selecting a person who is: (i) Extremely patient (ii) Extremely kind or honest. **[Delhi 2013]**

75. Honey goes to school either by a car driven by his driver or on his bicycle. The probability that he will use the car is $\frac{3}{7}$. What is the probability that he will use his bicycle for going to school? **[Foreign 2011]**

76. A card is drawn at random from a well-shuffled pack of 52 playing cards. Find the probability of getting neither a red card nor a queen. **[AI 2016]**

77. A card is drawn at random from a well-shuffled pack of 52 cards. Find the probability of getting: (i) A red king (ii) A queen or a jack. **[AI 2012]**

78. The probability of guessing the correct answer to a certain test is $\frac{p}{12}$. If the probability of not guessing the correct answer to this question is $\frac{1}{3}$, then find the value of p.

79. There are 1000 sealed envelopes in a box, 10 of them contain a cash prize of ₹ 100 each, 100 of them contain a cash prize of ₹ 50 each, 200 of them contain a cash prize of ₹ 10 each and the rest do not contain any cash prize. If they are well shuffled and an envelope is picked up, what is the probability that it contains no cash prize? **[NCERT Exemplar]**

80. A bag contains slips numbered from 1 to 100. If Fatima chooses a slip at random from the bag, it will either bear odd number or an even number. Since this situation has only two possible outcomes, the probability of each is $\frac{1}{2}$. Justify. **[NCERT Exemplar]**

81. In a family having three children, there may be no girl, one girl, two girls or three girls. So, the probability of each is $\frac{1}{4}$. Is this correct? Justify your answer. **[NCERT Exemplar]**

82. Apoorv throws two dice once and computes the product of the numbers appearing on the dice. Peehu throws one die and squares the number that appears on it. Who has the better chance of getting the number 36? Why? **[NCERT Exemplar]**

83. A die has its six faces marked 0, 1, 1, 1, 6, 6. Two such dice are thrown together and the total score is recorded. **[NCERT Exemplar]**

(i) How many different scores are possible?

(ii) What is the probability of getting a total of 7?

84. A lot consists of 48 mobiles phones of which 42 are good, 3 have only minor defects and 3 have major defects. Varnika will buy a phone if it is of good quality, but the trader will only buy a mobile if it has no major defect. One phone is selected at random from the lot. What is the probability that it is **[NCERT Exemplar]**

(i) Acceptable to Varnika? (ii) Acceptable to the trader?

85. From a pack of 52 playing cards. Jacks, Queens and Kings of red colours are removed. From the remaining, a card is drawn at random. Find probability that the drawn card is: **[AI 2016]**

(i) A black King (ii) A card of red colour (iii) A card of black colour.

86. In a single throw of a pair of different dice, what is probability of getting: (i) A prime number on each dice (ii) A total of 9 or 11. **[Delhi 2016]**

87. Two different dice are thrown together. Find the probability of: **[Delhi 2016]**

(i) Getting a number greater than 3 on each die

(ii) Getting a total of 6 or 7 of the numbers on two dice

88. A dice is thrown twice. Find the probability that **[Foreign 2013]**

(i) 5 may not appear either time

(ii) The same number may not appear on the dice thrown two times

89. The figure shows a disc on which a player spins an arrow twice. The fraction $\frac{a}{b}$ is formed, where 'a' is the number of sectors on which the arrow stops on the first spin and 'b' is the number of the sectors on which the arrow stops on the second spin. On each spin, each sector has an equal chance of selection by the arrow. Find the probability that the fraction $\frac{a}{b} > 1$. **[Foreign 2016]**

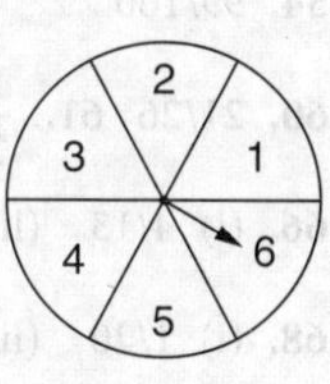

90. A number x is selected at random from the numbers 1, 4, 9, 16 and another number is selected at random from the numbers 1, 2, 3, 4. Find the probability that the value of xy is more than 16. **[AI 2016]**

91. A number x is selected at random from the numbers 1, 2, 3 and 4. Another number y is selected at random from the numbers 1, 4, 9 and 16. Find the probability that product of x and y is less than 16. **[AI 2016]**

92. A lot consists of 144 ball pens, of which 20 are defective and the others are goods sold by the shopkeeper. Nuri will buy a pen if it is good, but will not buy if it is defective. The shopkeeper draws one pen at random and gives it to her. What is probability that:

(i) She will buy it (ii) She will not buy it

93. A box contains cards numbered 3, 5, 7, 9, …., 35, 37. A card is drawn at random from the box. Find the probability that the number on the drawn card is a prime number. **[AI 2013]**

94. Find the probability that the month of June has exactly 5 Tuesdays.

95. A box contains 40 balls out of which x are white. If one ball is drawn at random from the bag, what is the probability it is white. If 20 more white balls are put in the box, the probability of drawing a white ball is the same as what it was before. Find x.

96. A box contains 50 marbles, some are red and others are green. If a marble is drawn from the box, the probability that it is red is $\frac{2}{5}$. Find the number of green marbles in the box.

Answers

Multiple Choice Questions

1. (a) **2.** (a) **3.** (a) **4.** (b) **5.** (c) **6.** (a) **7.** (b) **8.** (c) **9.** (b) **10.** (c)
11. (a) **12.** (c) **13.** (d) **14.** (b) **15.** (a) **16.** (c) **17.** (b) **18.** (a) **19.** (b) **20.** (c)
21. (b) **22.** (c) **23.** (b) **24.** (b) **25.** (d) **26.** (a) **27.** (d)

Fill in the Blanks

28. Impossible event **29.** 3/13 **30.** = 1 **31.** One **32.** $\frac{4}{11}$ **33.** 1/36 **34.** 0 **35.** = 0
36. Equally likely events **37.** $\frac{1}{2}$ **38.** $\frac{1}{9}$

Subjective Questions

39. (i) $\frac{1}{6}$, (ii) $\frac{1}{12}$ **40.** (i) $\frac{6}{49}$ (ii) $\frac{43}{49}$ **41.** 20 **42.** (i) 7/18 (ii) 11/18 (iii) 13/18 (iv) $\frac{1}{2}$
43. (i) 2/13 (ii) 12/13 (iii) 1/2 (iv) 11/13 **44.** $\frac{1}{36}$ **45.** Rina **46.** 3/10 **47.** 0 **48.** 1/5
49. (i) 2/5 (ii) 11/20 (iii) 3/10 **50.** (i) $\frac{1}{6}$ (ii) 7/36 **51.** (i) $\frac{1}{2}$ (ii) $\frac{1}{6}$ **52.** 1/7 **53.** 1/7
54. 99/100 **55.** 5/9 **56.** Random Experiment **57.** 2/7 **58.** 480 **59.** 6/13
60. 21/26 **61.** $\frac{1}{2}$ **62.** $\frac{1}{5}$ **63.** $\frac{23}{49}$ **64.** (i) 1/4 (ii) 1/9 **65.** 5/7
66. (i) 4/13 (ii) 1/26 (iii) 11/13 (iv) 2/13 **67.** (i) 7/48 (ii) $\frac{1}{8}$
68. (i) 1/20 (ii) $\frac{1}{2}$ **69.** = 24 **70.** $\frac{3}{5}$ **71.** 8 **72.** (i) 9/1000 (ii) 8/999
73. (i) 4/13 (ii) 1/26 (iii) 11/13 (iv) 2/13 **74.** (i) 1/4 (ii) 3/4 **75.** 4/7 **76.** 6/13
77. (i) 1/26 (ii) 2/13 **78.** 8 **79.** 0.69
80. Yes, the outcomes 'odd number' and 'even number' are equally likely in the situation considered.
81. No, the outcomes are not equally likely. For example, outcome 'one girl' means *gbb*, *bgb*, *bbg*, 'three girls' means *ggg* and so on.
82. Peehu **83.** (i) 6 scores (0, 1, 2, 6, 7, 12) (ii) 1/3 **84.** (i) 7/8 (ii) 15/16
85. (i) 1/23 (ii) 10/23 (iii) 13/23 **86.** (i) ¼ (ii) 1/6 **87.** (i) $\frac{1}{4}$ (ii) $\frac{11}{36}$
88. (i) 25/36 (ii) $\frac{5}{6}$ **89.** $\frac{5}{12}$ **90.** 3/8 **91.** 1/2 **92.** (i) $\frac{31}{36}$ (ii) $\frac{5}{36}$ **93.** 11/18
94. 2/7 **95.** 40 **96.** 30

Probability (Basic)

SELF-ASSESSMENT TEST

Time allowed: 90 min. Max. Marks: 40

General Instructions

1. *All questions are compulsory. There are 20 questions in all.*
2. *This question paper has four sections: Section A, Section B, Section C and Section D.*
3. *Section A contains 10 questions of one mark each, Section B contains three questions of two marks each, Section C contains four questions of three marks each, and Section D contains three questions of four marks each.*

Section A [1 Mark Each]

Direction (Q. Nos. 1–5): These are Multiple Choice Questions. Select the most appropriate answer for the given questions.

1. Which of the following cannot be the probability of an event?

(a) 1.5 (b) $\frac{3}{5}$ (c) 25% (d) 0.3

2. The probability of getting a bad egg in a lot of 400 is 0.035. The number of bad eggs in the lot is

(a) 7 (b) 14 (c) 21 (d) 28

3. What is the probability of getting a king in a well-shuffled pack of 52 cards?

(a) $\frac{1}{13}$ (b) $\frac{1}{52}$ (c) $\frac{1}{26}$ (d) 0

4. Which of the following cannot be the probability of an event?

(a) $\frac{1}{5}$ (b) 0.3 (c) 4% (d) $\frac{5}{4}$

5. If $P(E)$ is 0.75, what is $P(\text{not } E)$?

(a) 0.35 (b) 0.25 (c) 0 (d) 1

6. The probability of getting a face card from a pack of cards is ________.

7. The sum of probability of an event happening and not happening is equal to ________.

8. A child has a die whose six faces show letters, as given below:

A	B	C	A	D	A

The die is thrown once. Find the probability of getting

(i) A (ii) D

9. What is the probability that the month of December had exactly 5 Sundays, 5 Mondays and 5 Tuesdays?

10. A letter of the English alphabet is chosen at random. Determine the probability that the chosen letter is a consonant.

Section B [2 Marks Each]

11. A piggy bank contains hundred 50 p coins, fifty ₹ 1 coins, twenty ₹ 2 coins and ten ₹ 5 coins. If it is equally likely that one of the coins will fall out when the bank is turned upside down, what is the probability that the coin

(i) Will be a 50 p coin (ii) Will not be a ₹5 coin **[NCERT]**

12. A coin is tossed three times. Find the probability of getting:

(i) All tails (ii) At least two tails (iii) At most two tails

13. Three coins are tossed together. Find the probability of getting:

(i) Exactly two heads (ii) At least two heads (iii) No tails

Section C [3 Marks Each]

14. In a simultaneous throw of a pair of dice, find the probability of getting:

(i) 8 as sum (ii) A doublet (iii) A sum greater than 9 (iv) A doublet of a prime number

(v) A sum less than 6 (vi) A sum more than 7

15. One card is drawn from a pack of 52 cards, each of the 52 cards being equally likely to be drawn. Find the probability that the card drawn is:

(i) An ace (ii) A face card (iii) Red and a king (iv) A spade or an ace (v) A red face card

16. Five cards - ten, jack, queen, king and ace of diamonds - are well shuffled with their faces down. One card is then picked at random.

(a) What is the probability that the drawn card is the queen?

(b) If the queen is drawn and set aside, and a second card is drawn, find the probability that the second card is (i) An ace (ii) A queen

17. A bag contains 12 balls, out of which x are white.

(i) If one ball is drawn at random, find the probability that it is a white ball.

(ii) If 6 more white balls are put in the bag, the probability of drawing a white ball is double that in (i). Find x.

Section D [4 Marks Each]

18. A bag contains 18 balls, of which x balls are red.

(i) If one ball is drawn at random from the bag, what is the probability that it is not red?

(ii) If 2 more red balls are put in the bag, the probability of drawing a red ball will be $\frac{9}{8}$ times the probability of drawing a red ball in the first case. Find the value of x.

19. A lot consists of 144 ball pens of which 20 are defective and the others are of good quality. Nuri will buy a pen if it is good, but will not buy if it is defective. The shopkeeper draws one pen at random and gives it to her. What is the probability that:

(i) She will buy it (ii) She will not buy it

20. Two customers, Shyam and Ekta, are visiting a particular shop in the same week (Tuesday to Saturday). Each is equally likely to visit the shop on one day as on another day. What is the probability that they visit on:

(i) The same day (ii) Consecutive days (iii) Different days

Answers

1. (a) **2.** (b) **3.** (a) **4.** (d) **5.** (b) **6.** 3/13 **7.** One **8.** (i) $\frac{1}{2}$, (ii) $\frac{1}{6}$

9. $\frac{1}{7}$ **10.** 21/26 **11.** (i) $\frac{5}{9}$, (ii) $\frac{17}{18}$ **12.** (i) $\frac{1}{8}$, (ii) $\frac{1}{2}$, (iii) $\frac{7}{8}$

13. (i) $\frac{3}{8}$, (ii) $\frac{1}{2}$, (iii) $\frac{7}{8}$ **14.** (i) $\frac{5}{36}$, (ii) $\frac{1}{6}$, (iii) $\frac{1}{6}$, (iv) $\frac{1}{12}$, (v) $\frac{5}{18}$, (vi) $\frac{5}{12}$

15. (i) $\frac{1}{13}$, (ii) $\frac{3}{13}$, (iii) $\frac{1}{26}$, (iv) $\frac{4}{13}$, (v) $\frac{3}{26}$ **16.** (a) $\frac{1}{5}$, (b) (i) $\frac{1}{4}$, (ii) 0

17. (i) $\frac{x}{12}$, (ii) $x = 3$ **18.** (i) $\frac{18-x}{18}$, (ii) 8 **19.** (i) $\frac{31}{36}$, (ii) $\frac{5}{36}$ **20.** (i) $\frac{1}{5}$, (ii) $\frac{8}{25}$, (iii) $\frac{4}{5}$

Probability (Standard)

SELF-ASSESSMENT TEST

Time allowed: 90 min. Max. Marks: 40

General Instructions

1. *All questions are compulsory. There are 20 questions in all.*
2. *This question paper has four sections: Section A, Section B, Section C and Section D.*
3. *Section A contains 10 questions of one mark each, Section B contains three questions of two marks each, Section C contains four questions of three marks each, and Section D contains three questions of four marks each.*

Section A [1 Mark Each]

Direction (Q. Nos. 1–5): These are Multiple Choice Questions. Select the most appropriate answer for the given questions.

1. The set of all possible outcomes of a random experiment is called
   (a) sample space (b) elementary events
   (c) complementary events (d) favorable events
2. A girl calculates that the probability of her winning the first prize in a lottery is 0.08. If 6000 tickets are sold, how many tickets has she bought?
   (a) 40 (b) 240 (c) 480 (d) 750
3. If two coins are tossed simultaneously, then the probability of getting at least one head is
   (a) $\frac{1}{4}$ (b) $\frac{3}{4}$ (c) $\frac{1}{2}$ (d) 1
4. In a throw of a pair of dice, the probability of getting a doublet is
   (a) $\frac{1}{3}$ (b) $\frac{1}{6}$ (c) $\frac{5}{12}$ (d) $\frac{2}{3}$
5. If $P(A)$ denotes the probability of an event A, then
   (a) $P(A) < 0$ (b) $P(A) > 1$ (c) $0 \le P(A) \le 1$ (d) $-1 \le P(A) \le 1$
6. The probability of getting 101 marks out of 100 marks is ____________.
7. A single letter is selected at random from the word 'PROBABILITY'. The probability that it is a vowel is____________.
8. Two different dice are tossed together. Find the probability:
   (i) Of getting a doublet
   (ii) Of getting a sum 10, of the numbers on the two dice
9. If a number x is chosen from the numbers 1, 2, 3 and a number y is selected from the numbers 1, 4, 9. Find the probability that $xy = 10$.
10. What is the probability that a leap year will have 53 Mondays and 53 Tuesdays?

Section B [2 Marks Each]

11. A lot consists of 144 ball pens of which 20 are defective and the others are of good quality. Nuri will buy a pen if it is good, but will not buy if it is defective. The shopkeeper draws one pen at random and gives it to her. What is the probability that:

(i) She will buy it (ii) She will not buy it

12. A letter is chosen at random from the English alphabet. What is the probability that it is a letter of the word 'RAMANUJAN'?

13. A bag contains 5 red balls and some blue balls. If the probability of drawing a blue ball is double that of a red ball, determine the number of blue balls in the bag.

Section C [3 Marks Each]

14. A bag contains 12 balls, of which x are white.

(i) If one ball is drawn at random, find the probability that it is a white ball.

(ii) If 6 more white balls are put in the bag, the probability of drawing a white ball is double that in (i). Find x.

15. Five cards - ten, jack, queen, king and ace of diamonds - are well shuffled with their faces down. One card is then picked at random.

(a) What is the probability that the drawn card is the queen?

(b) If the queen is drawn and set aside, and a second card is drawn, find the probability that the second card is (i) An ace (ii) A queen

16. All the red face cards are removed from a pack of 52 playing cards. A card is drawn at random from the remaining cards, after reshuffling them. Find the probability that the drawn card is

(i) Of red colour (ii) A queen (iii) An ace (iv) A face card

17. Two customers, Shyam and Ekta, are visiting a particular shop in the same week (Tuesday to Saturday). Each is equally likely to visit the shop on one day as on another. What is the probability that both will visit the shop on

(i) The same day (ii) Consecutive days (iii) Different days

Section D [4 Marks Each]

18. A card is drawn at random from a well-shuffled deck of playing cards. Find the probability that the card drawn is

(i) A spade or an ace (ii) A black king

(iii) Neither a jack nor a king (iv) Either a king or a queen

19. A group consists of 12 persons, of which 3 are extremely patient, 6 are extremely honest and the rest are extremely kind. A person from the group is selected at random. Assuming that each person is equally likely to be selected, find the probability of selecting a person who is (i) Extremely patient (ii) Extremely kind or honest

20. Two dice, one blue and one grey, are thrown at the same time. Then

(i) Complete the following table:

Event: Sum on 2 dice	2	3	4	5	6	7	8	9	10	11	12
Probability	$\frac{1}{36}$						$\frac{5}{36}$				$\frac{1}{36}$

(ii) A student argues that there are 11 possible outcomes: 2, 3, 4, 5, 6, 7, 8, 9, 10, 11 and 12. Therefore, each of them has a probability $\frac{1}{11}$. Do you agree with this argument? Justify your answer.

Answers

1. (a) **2.** (c) **3.** (b) **4.** (b) **5.** (c) **6.** Impossible event **7.** $\frac{4}{11}$ **8.** (i) $\frac{1}{6}$, (ii) $\frac{1}{12}$

9. 0 **10.** $\frac{1}{7}$ **11.** (i) $\frac{31}{36}$, (ii) $\frac{5}{36}$ **12.** $\frac{3}{13}$ **13.** 10 **14.** (i) $\frac{x}{12}$, (ii) 3

15. (a) $\frac{1}{5}$, (b) (i) $\frac{1}{4}$, (ii) 0 **16.** (i) $\frac{10}{23}$, (ii) $\frac{1}{23}$, (iii) $\frac{2}{23}$, (iv) $\frac{3}{23}$ **17.** (i) $\frac{1}{5}$, (ii) $\frac{8}{25}$, (iii) $\frac{4}{5}$

18. (i) $\frac{4}{13}$, (ii) $\frac{1}{26}$, (iii) $\frac{11}{13}$, (iv) $\frac{2}{13}$ **19.** (i) $\frac{1}{4}$, (ii) $\frac{3}{4}$ **20.** $\frac{1}{18}, \frac{1}{12}, \frac{1}{9}, \frac{5}{36}, \frac{1}{6}, \frac{1}{9}, \frac{1}{12}, \frac{1}{18}$, (ii) No

HINTS AND SOLUTIONS

Chapter 1: Real Numbers

5 Try to Solve

4. Smallest prime number = 2. Smallest composite number = 4. HCF of 2 and 4 is 2.

Multiple Choice Questions

1. (b) Because 18 and 25 are both coprime numbers (having no common factor).

4. (c) Find the HCF.

5. (a) Taking 70 and 125, subtract 5 and 8, respectively, to obtain 65 and 117. Now, find the HCF of 65 and 117 to obtain 13.

6. (b) 245 – 5 = 240 and 1029 – 5 = 1024. Now, find the HCF of 240 and 1024, which is 16.

7. (d) By Euclid's division lemma

$$a = bq + r, 0 \le r < b$$

$$a = 143 \times q + 31 = 13 \times 11 \times q + 13 \times 2 + 5 = 13[11 \times q + 2] + 5$$

Hence, the remainder is 5.

10. (a) Since 2 is the smallest prime number and the least prime factor of a is 3 and the least prime factor of b is 5. Both are odd, so $(a + b)$ is even. ∴ Least prime factor of $(a + b)$ is 2.

11. (c)

$$1 = 1 \times 1$$
$$2 = 2 \times 1$$
$$3 = 3 \times 1$$
$$4 = 2^2$$
$$5 = 5 \times 1$$
$$6 = 2 \times 3$$
$$7 = 7 \times 1$$
$$8 = 2^3$$
$$9 = 3^2$$
$$10 = 2 \times 5$$

So, LCM of 1 to 10 (both inclusive) = $2^3 \times 3^2 \times 5 \times 7 = 2520$

12. (b)

$$0.\overline{68} + 0.\overline{73} = x + y \quad \text{(i)}$$

$$x = 0.\overline{68} \quad \text{(ii)}$$

$$x = 0.68\overline{68}$$

Multiplying both sides by 100, we get $100x = 68.\overline{68}$ (iii)

Subtracting (ii) from (iii), we get $99x = 68$

$$x = \frac{68}{99} \quad \text{(iv)}$$

Similarly $y = \frac{73}{99}$ (v)

Inserting (iv) and (v) in (i), we get $0.\overline{68} + 0.\overline{73} = \frac{68}{99} + \frac{73}{99} = \frac{68 + 73}{99} = \frac{141}{99} = 1.\overline{42}$

13. (b)

$$65 = 13 \times 5$$
$$117 = 13 \times 3 \times 3$$
$$\text{HCF} = 13$$
$$65m - 117 = 13$$
$$65m = 13 + 117$$
$$65m = 130$$
$$m = \frac{130}{65} = 2$$

14. (d)

$$a = x^5y^2,\ b = x^2y^3$$
$$\text{HCF}\ (a, b) = x^2y^2$$

16. (a) Say, $a = 7$ and $b = 8$ are coprime. $a^2 = 49$ and $b^2 = 64$ are also coprime.

17. (a) $\left(\sqrt{2}+\sqrt{3}\right)^{10}\left(\sqrt{2}-\sqrt{3}\right)^{10} = \left[\left(\sqrt{2}\right)^2-\left(\sqrt{3}\right)^2\right]^{10}$

$= [2 - 3]^{10} = (-1)^{10} = 1$, which is a rational number.

19. (b) $\frac{17}{18}, \frac{43}{45}, \frac{59}{60}, \frac{31}{36}$

LCM of 18, 45, 60 and 36 is 180.

$$\frac{17}{18} = \frac{17}{18}\times\frac{10}{10} = \frac{170}{180}$$
$$\frac{43}{45} = \frac{43}{45}\times\frac{4}{4} = \frac{172}{180}$$
$$\frac{59}{60} = \frac{59}{60}\times\frac{3}{3} = \frac{177}{180}$$
$$\frac{31}{36} = \frac{31}{36}\times\frac{5}{5} = \frac{155}{180}$$
$$\frac{155}{180} < \frac{170}{180} < \frac{172}{180} < \frac{177}{180}$$
$$\frac{31}{36} < \frac{17}{18} < \frac{43}{45} < \frac{59}{60}$$

20. (c) Let the numbers be $3x$, $4x$ and $5x$. LCM = $60x$. So, $60x = 1200$ or $x = 20$.
Hence, the numbers are 60, 80 and 100. HCF = 20.

Subjective Questions

37.

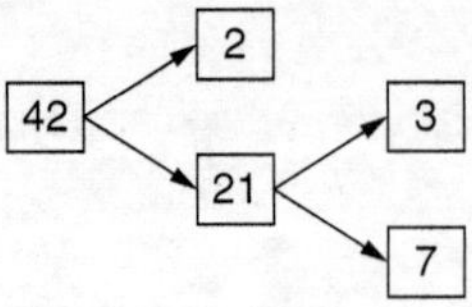

39. HCF of 66, 88 and 110 is 22. Number of rows = $\frac{66}{22}+\frac{88}{22}+\frac{110}{22} = 12$.

40. HCF of 336, 240 and 96 is 48. Total no. of rooms = $\frac{336}{48}+\frac{240}{48}+\frac{96}{48} = 14$.

41. $2472 = 2^3 \times 3 \times 103$

$1284 = 2^2 \times 3 \times 107$

$\text{LCM} = 2^3 \times 3^2 \times 5 \times 103 \times 107$

$N = 2^2 \times 3^2 \times 5 = 180$

42. Find the LCM.

43. $2520 = 2^3 \times 3^2 \times 5 \times 7$ (i)

Given $2520 = 2^3 \times 3^p \times q \times 7$ (ii)

From (i) and (ii), we get $p = 2$ and $q = 5$

45. Find the LCM.

46. (a) HCF of 28, 16 and 12 is 4. No. of books each student received = 4.

(b) Total no. of students = $\frac{28}{4} + \frac{16}{4} + \frac{12}{4} = 14$.

47. Find the HCF.

48. $A = 2n + 13$ and $B = n + 7$

For finding HCF, say $n = 2$.

$A = 2 \times 1 + 13 = 15$, $B = 1 + 7 = 8$ and $n = 2$

$A = 2 \times 2 + 13 = 17$, $B = 2 + 7 = 9$

In both cases, A and B are coprime. So, HCF is 1.

49. $445 - 4 = 441$

$572 - 5 = 567$

$699 - 6 = 693$

Now, find the HCF of 441, 567 and 693, which is 63.

50. LCM of 36, 42 and 72 seconds is 504 seconds or 8 min 24 s.

Hence, at 9 : 08 : 24, they will change simultaneously.

51. Find the HCF.

53. Find the HCF.

63. $\sqrt{180} = \sqrt{2 \times 2 \times 3 \times 3 \times 5} = 6\sqrt{5}$. In order to make a number rational, we need to multiply it with $\sqrt{5}$ and 1. We obtain 30, which is rational, in which $\sqrt{5}$ is irrational and 1 is rational.

68. Let $n = 3k$; $3k + 1$ or $3k + 2$.

(i) When $n = 3k \Rightarrow n$ is divisible by 3

If we add 2 on both sides, $n + 2 = 3k + 2$, where $n + 2$ is not divisible by 3.

If we add 4 on both sides, $n + 4 = 3k + 4 = 3k + 3 + 1 = 3(k + 1) + 1$, where $n + 4$ is not divisible by 3.

(ii) When $n = 3k + 1 \Rightarrow n$ is not divisible by 3

If we add 2 on both sides, $n + 2 = 3k + 1 + 2 = 3k + 3 = 3(k + 1)$, where $(n + 2)$ is not divisible by 3.

If we add 4 on both sides, $n + 4 = 3k + 1 + 4 = 3k + 3 + 2 = 3k + 3 + 2 = 3(k + 1) + 2$, where $(n + 4)$ is not divisible by 3.

(iii) When $n = 3k + 2 \Rightarrow n$ is not divisible by 3

Adding 2 on both sides, $n + 2 = 3k + 2 = 3k + 2 + 2 = 3k + 4 = 3k + 3 + 1$

$n + 2 = 3(k + 1) + 1$, where $n + 2$ is not divisible by 3.

Adding 4 on both sides, $n + 4 = 3k + 2 + 4 = 3k + 6 = 3(k + 2)$, where $(n + 4)$ is divisible by 3.

So, we conclude that only one n, $(n + 2)$ and $(n + 4)$ is divisible by 3.

69. Since p is an odd integer, $p = 4q + 1$ or $4q + 3$ for some integer q.

$$p = 4q + 1$$
$$p^2 = 16q^2 + 8q + 1 = 8(2q^2 + q) + 1$$
$$p^2 = 8a + 1, \text{ where } a = 2q^2 + q$$

$\Rightarrow p^2$ leaves remainder 1 when divided by 8.

When $p = 4q + 3$

$$p^2 = 16q^2 + 24q + 9 = 16q^2 + 24q + 8 + 1$$
$$= 8(2q^2 + 3q + 1) + 1 = 8a + 1, \text{ where } a = 2q^2 + 3q + 1$$

$\therefore p^2$ leaves a remainder 1 when divided by 8.

70. Let a, $a + 1$ and $a + 2$ be three consecutive positive integers. Let $b = 6$, $0 \le r < 6$.
Using Euclid's division lemma, 'a' will be of the form $6q$, $6q + 1$, $6q + 2$, $6q + 3$, $6q + 4$, $6q + 5$, where q is an integer.

Case I: When $a = 6q$, so the other positive integers are $a + 1 = 6q + 1$ and $a + 2 = 6q + 2$.
Hence, product of three consecutive integers

$$= a(a + 1)(a + 2) = 6q(6q + 1)(6q + 2)$$

This shows that the product is a multiple of 6.

Case II: When $a = 6q + 1$, so the other positive integers are $a + 1 = 6q + 2$ and $a + 2 = 6q + 3$.
Hence, product of three consecutive integers

$$= (6q + 1)(6q + 2)(6q + 3) = (6q + 1)2(3q + 1)3(2q + 1)$$
$$= 6(6q + 1)(3q + 1)(2q + 1)$$

This shows that the product is a multiple of 6.

Case III: When $a = 6q + 2$, so the other positive integers are $a + 1 = 6q + 3$ and $a + 2 = 6q + 4$.

$$= (6q + 2)(6q + 3)(6q + 4) = 2(3q + 1)3(2q + 1)(6q + 4)$$
$$= 6(3q + 1)(2q + 1)(6q + 4)$$

This shows that the product is a multiple of 6.
Similarly, it can be proved for all the other values of a.
Hence, the product of three consecutive positive integers is divisible by 6.

80. No, by Euclid's division lemma, $0 \le r < b$

$a = bq + r$

Here $b = 4$, so $a = 4q + r$ for $0 \le r < 4$

That is $r = 0, 1, 2, 3$

So $a = 4q + 0$, $a = 4q + 1$, $a = 4q + 2$ and $a = 4q + 3$.

81. Let a be any positive integer and $b = 5$. Using Euclid's division lemma, $a = bq + r$, $0 \le r < 5$.

That is $r = 0, 1, 2, 3, 4$

Case I: Where $r = 0$

$$a = 5q + 0 = 5q$$

On squaring both sides, we get $a^2 = (5q)^2 = 5(5q^2) = 5m$, where $5q^2 = m$

Case II: Where $r = 1$

$$a = 5q + 1$$

On squaring both sides, we get

$$a^2 = (5q + 1)^2 = 25q^2 + 10q + 1 = 5(5q^2 + 2q) + 1$$
$$a^2 = 5m + 1, \text{ where } m = 5q^2 + 2q$$

Case III: Where $r = 2$

$a = 5q + 2$

On squaring both sides, we get $a^2 = (5q + 2)^2 = 25q^2 + 20q + 4 = 5(5q^2 + 4q) + 4$

$a^2 = 5m + 4$, where $m = 5q^2 + 4q$

Case IV: Where $r = 3$

$a = 5q + 3$

On squaring both sides, we get $a^2 = (5q + 3)^2 = 25q^2 + 30q + 9 = 25q^2 + 30q + 5 + 4$

$= 5(5q^2 + 6q + 1) + 4 = 5m + 4$, where $m = 5q^2 + 6q + 1$

Case V: Where $r = 4$

$a = 5q + 4$

On squaring both sides, we get $a^2 = (5q + 4)^2 = 25q^2 + 40q + 16$

$= 25q^2 + 40q + 15 + 1 = 5(5q^2 + 8q + 3) + 1$

$a^2 = 5m + 1$, where $m = 5q^2 + 8q + 3$

Therefore, the square of any positive integer cannot be of the form $5m + 2$ or $5m + 3$.

83. Since we know that every odd integer is of the form $4q + 1$ or $4q + 3$, where q is some integer.

Case I: When $n = 4q + 1$

$n^2 - 1 = (4q + 1)^2 - 1 = 16q^2 + 1 + 8q - 1$

$= 16q^2 + 8q = 8(2q^2 + 1)$, which is divisible by 8.

Case II: When $n = 4q + 3$

$n^2 - 1 = (4q + 3)^2 - 1 = 16q^2 + 9 + 24q - 1$

$= 16q^2 + 24q + 8 = 8(2q^2 + 3q + 1)$, which is divisible by 8.

Hence, if n is an odd integer of $n^2 - 1$, then it is divisible by 8.

84. It is given that we need two odd positive integers, say $x = 2q + 1$ and $y = 2q + 3$, and q may be any positive integer.

So $x^2 + y^2 = (2q + 1)^2 + (2q + 3)^2 = 4q^2 + 1 + 4q + 4q^2 + 9 + 12q$

$= 8q^2 + 16q + 10$

$= 2(4q^2 + 8q + 5)$

Shows that it is always even but not divisible by 4.

87. In the given question, $n^3 - n$

$= n(n^2 - 1) = n(n - 1)(n + 1) = (n - 1)(n)(n + 1)$

Now, this question becomes the same as Q70 above.

Chapter 2: Polynomials

Multiple Choice Questions

6. (a) $f(x) = x^3 + ax^2 + bx + c$

$\alpha = x = -1$ is the root of $f(x)$ means $f(x = -1) = (-1)^3 + a(-1)^2 + b(-1) + c = 0$

$= -1 + a - b + c = 0$

$c = b - a + 1$ (i)

We know that for cubic polynomial $\alpha\beta\gamma = -\frac{d}{a}$

$$\alpha\beta\gamma = -c$$

$$(-1)\beta\gamma = -c \qquad [\because \alpha = -1]$$

$$\beta\gamma = c$$

$$\beta\gamma = b - a + 1 \qquad \text{[From (i)]}$$

11. (d) $f(x) = kx^2 + 2x + 3k$

We have given $\alpha + \beta = \alpha\beta$

$$-\frac{2}{k} = \frac{3k}{k}$$

$$k = -\frac{2}{3}$$

15. (d) $f(x) = 2x^2 + 5x + k$

Here $\alpha + \beta = -\frac{5}{2}$ (i)

$$\alpha\beta = \frac{k}{2} \qquad \text{(ii)}$$

We know that $(\alpha+\beta)^2 = \alpha^2 + \beta^2 + 2\alpha\beta$

$$\alpha^2 + \beta^2 = (\alpha+\beta)^2 - 2\alpha\beta$$

$$= \left(-\frac{5}{2}\right)^2 - 2\frac{k}{2} \qquad \text{[From (i) and (ii)]}$$

$$\alpha^2 + \beta^2 = \frac{25}{4} - k \qquad \text{(iii)}$$

We have given $\alpha^2 + \beta^2 + \alpha\beta = \frac{21}{4}$

$$\frac{25}{4} - k + \frac{k}{2} = \frac{21}{4} \qquad \text{[From (ii) and (iii)]}$$

$$-\frac{k}{2} = \frac{21}{4} - \frac{25}{4} = -1$$

$$k = 2$$

24. (b) $f(x) = 4\sqrt{3}x^2 + 5x - 2\sqrt{3} = 4\sqrt{3}x^2 + 8x - 3x - 2\sqrt{3}$

$$= 4x\left(\sqrt{3}x + 2\right) - \sqrt{3}\left(\sqrt{3} + 2\right) = \left(4x - \sqrt{3}\right)\left(\sqrt{3}x + 2\right)$$

Zeros are $4x - \sqrt{3} = 0$

$$x = \frac{\sqrt{3}}{4}$$

$$\sqrt{3}x + 2 = 0$$

$$x = -\frac{2}{\sqrt{3}}$$

Subjective Questions

37. $f(x) = x^2 + (a + 1)x + b$

$x = 2$ and $x = -3$ are the roots of $f(x)$. This means

$f(x = 2) = (2)^2 + (a + 1) \times 2 + b = 0$

$\Rightarrow \quad 4 + 2a + 2 + b = 0$

$\Rightarrow \quad 2a + b = -6$ (i)

$f(x = -3) = (-3)^2 + (a + 1)(-3) + b = 0$

$\Rightarrow \quad 9 - 3a - 3 + b = 0$

$\Rightarrow \quad 3a - b = 6$ (ii)

Adding (i) and (ii), we get $5a = 0$

$a = 0$

Inserting the value a in (i), we get $b = -6$ [**Ans.** $a = 0$ and $b = -6$]

41. $f(x) = ax^2 - 3(a - 1)x - 1$

Since 1 is zero of $f(x)$ means $x = 1$ is the factor of $f(x)$

$f(x = 1) = a(1)^2 - 3(a - 1)(1) - 1 = a - 3a + 3 - 1 = 0$

$= -2a + 2 = 0$

$a = 1$

43. In cubic polynomial $ax^3 + bx^2 + cx + d$ and cubic polynomial $3x^3 - 5x^2 - 11x - 3$, on comparing $a = 3$, $b = -5$, $c = -11$ and $d = -3$, we need to check if 3, –1 and $-\frac{1}{3}$ are zeros. This means

$f(x = 3) = 3(3)^3 - 5(3)^2 - 11(3) - 3 = 81 - 45 - 33 - 3 = 0$

$f(x = -1) = 3(-1)^3 - 5(-1)^2 - 11(-1) - 3 = -3 - 5 + 11 - 3 = 0$

$$f\left(x = -\frac{1}{3}\right) = 3\left(-\frac{1}{3}\right)^3 - 5\left(-\frac{1}{3}\right)^2 - 11\left(-\frac{1}{3}\right) - 3 = -\frac{1}{9} - \frac{5}{9} + \frac{11}{3} - 3 = 0$$

Therefore, 3, –1 and $-\frac{1}{3}$ are zeros of $f(x)$. So, $\alpha = 3$, $\beta = -1$ and $\gamma = -\frac{1}{3}$.

Now $$\alpha + \beta + \gamma = 3 + (-1) + \left(-\frac{1}{3}\right) = 2 - \frac{1}{3} = \frac{5}{3} = \frac{-(-5)}{3} = \frac{-b}{a}$$

$$\alpha\beta + \beta\gamma + \gamma\alpha = 3(-1) + (-1) \times \left(-\frac{1}{3}\right) + \left(-\frac{1}{3}\right) \times 3 = -3 + \frac{1}{3} - 1 = \frac{-11}{3} = \frac{c}{a}$$

$$\alpha\beta\gamma = 3 \times (-1) \times \left(-\frac{1}{3}\right) = 1 = \frac{-(-3)}{3} = \frac{-d}{a}$$

Hence it is verified.

44. Let α, β, γ be the three zeros. And we have given

$$\alpha + \beta + \gamma = 2, \alpha\beta + \beta\gamma + \gamma\alpha = -7, \alpha\beta\gamma = -14$$

We know that for cubic polynomial

$$f(x) = [x^3 - (\alpha + \beta + \gamma)x^2 + (\alpha\beta + \beta\gamma + \gamma\alpha)x - \alpha\beta\gamma]$$

$$= k[x^3 - 2x^2 - 7x - (-14)] = k[x^3 - 2x^2 - 7x + 14)]$$

46. Let honest people (α) = 12 and dishonest people (β) = 8.

So $\quad S = \alpha + \beta = 12 + 8 = 20,\ P = \alpha\beta = 12 \times 8 = 96$

We know that for quadratic polynomial

$$k(x^2 - Sx + P) = k(x^2 - 20x + 96)$$

58. In polynomial $5x^2 - (3 + k)x + 7$, it is given that $\alpha + \beta = 0$.

$$-\frac{b}{a} = -\left[-\frac{(3+k)}{5}\right] = 0$$

$$3 + k = 0$$

$$k = -3$$

The second polynomial $2x^2 - 2(k + 11)x + 30$ now becomes $2x^2 - 2(-3 + 11)x + 30$.

$$2x^2 - 16x + 30$$

$$2x^2 - 10x - 6x + 30$$

$$2x(x - 5) - 6(x - 5)$$

$$(2x - 6)(x - 5)$$

Zeros are $x = 3$ and $x = 5$.

62. $\quad p(x) = x^2 - p(x + 1) - c = x^2 - px - p - c = x^2 - px - (p + c)$

Here $\quad \alpha + \beta = \dfrac{-(-p)}{1} = p \qquad$ (i)

$\quad \alpha\beta = \dfrac{-(p+c)}{1} = -(p + c) \qquad$ (ii)

We have given $(\alpha+1)(\beta+1) = 0$

$\quad \alpha\beta + \alpha + \beta + 1 = 0 \qquad$ (iii)

Inserting (i) and (ii) in (iii), we get $-p - c + p + 1 = 0$

$$c = 1$$

64. $f(x) = ax^2 - 5x + c$ and we are given that m and n are zeros.

So $\quad m + n = \dfrac{-(-5)}{a} = \dfrac{5}{a} \qquad$ (i)

$\quad mn = \dfrac{c}{a} \qquad$ (ii)

It is given that $\quad m + n = 10 \qquad$ (iii)

From (i) and (iii), we get $\dfrac{5}{a} = 10$

$\quad a = \dfrac{1}{2} \qquad$ (iv)

It is also given that $mn = 10 \qquad$ (v)

From (ii), (iv) and (v), we get $\dfrac{c}{\frac{1}{2}} = 10$

$\quad c = 5 \qquad$ **[Ans.** $a = \dfrac{1}{2}$ and $c = 5$**]**

65. $$f(x) = x^2 + 7x + 7$$

$$\alpha + \beta = \frac{-(+7)}{1} = -7 \qquad \text{(i)}$$

$$\alpha\beta = \frac{7}{1} = 7 \qquad \text{(ii)}$$

To find: $\frac{1}{\alpha} + \frac{1}{\beta} - 2\alpha\beta$

$$= \frac{\beta + \alpha - 2\alpha^2\beta^2}{\alpha\beta} = \frac{(\alpha+\beta) - 2(\alpha\beta)^2}{\alpha\beta}$$

$$= \frac{-7 - 2(7)^2}{7} = \frac{-7-98}{7} = \frac{-105}{7} = -15$$

67. $$f(x) = x^2 - (k + 6)x + 2(2k - 1)$$

Here $$\alpha + \beta = \frac{-(-(k+6))}{1} = k + 6$$

$$\alpha\beta = \frac{2(2k-1)}{1} = 4k - 2$$

We have given $\alpha + \beta = \frac{1}{2}\alpha\beta$

$$k + 6 = \frac{1}{2}(4k - 2)$$

$$2k + 12 = 4k - 2$$

$$14 = 2k$$

$$k = 7$$

68. $$f(x) = x^3 - 12x^2 + 44x + c$$

$$\alpha + \beta + \gamma = \frac{-(-12)}{1} = 12 \qquad \text{(i)}$$

$$\alpha\beta + \beta\gamma + \gamma\alpha = \frac{44}{1} = 44 \qquad \text{(ii)}$$

$$\alpha\beta\gamma = -c \qquad \text{(iii)}$$

$$2\beta = \alpha + \gamma \quad \text{[Given]} \qquad \text{(iv)}$$

Inserting (iv) in (i), we get $3\beta = 12$

$$\beta = 4$$

Inserting the value of β in (i), we get

$$\alpha + \gamma = 8 \qquad \text{(v)}$$

Inserting the value of β in (ii), we get $4\alpha + 4\gamma + \gamma\alpha = 44$

$$4(\alpha + \gamma) + \gamma\alpha = 44$$

$$4 \times 8 + \gamma\alpha = 44 \qquad \text{[From (v)]}$$

$$\gamma\alpha = 12 \qquad \text{(vi)}$$

Inserting the value of β and (vi) in (iii), we get $4 \times 12 = -c$

$$c = -48$$

69.

$$f(x) = x^3 - 2x^2 + qx - r$$

$$\alpha + \beta = 0 \quad \text{[Given]} \qquad \text{(i)}$$

$$\alpha + \beta + \gamma = 0 + \gamma = \frac{-(-2)}{1} \qquad \text{[From (i)]}$$

$$\gamma = 2 \qquad \text{(ii)}$$

$$\alpha\beta + \beta\gamma + \gamma\alpha = q$$

$$\alpha\beta + 2\beta + 2\alpha = q \qquad \text{[From (ii)]}$$

$$\alpha\beta + 2(\beta + \alpha) = q$$

$$\alpha\beta + 2 \times 0 = q \qquad \text{[From (i)]}$$

$$\alpha\beta = q \qquad \text{(iii)}$$

$$\alpha\beta\gamma = -(-r) = r$$

$$2q = r$$

70.

$$f(x) = x^3 + px^2 + qx + 2$$

We have $\alpha\beta + 1 = 0$. Find $2p + q + 5 = ?$

$$\alpha\beta + 1 = 0 \qquad \text{[Given]}$$

$$\alpha\beta = -1 \qquad \text{(i)}$$

$$\alpha\beta\gamma = -2$$

$$-1\gamma = -2 \qquad \text{[From (i)]}$$

$$\gamma = 2 \qquad \text{(ii)}$$

$$\alpha\beta + \beta\gamma + \gamma\alpha = q$$

$$-1 + \gamma(\beta + \alpha) = q$$

$$\gamma(\alpha + \beta) = q + 1$$

$$2(\alpha + \beta) = q + 1 \qquad \text{[From (ii)]}$$

$$\alpha + \beta = \frac{q+1}{2} \qquad \text{(iii)}$$

$$\alpha + \beta + \gamma = -p$$

$$\frac{q+1}{2} + 2 = -p \qquad \text{[From (ii) and (iii)]}$$

$$2p + q + 5 = 0$$

72.

$$f(x) = 4x^2 - 8kx + 8x - 9$$

$$f(x) = 4x^2 - (8k - 8)x - 9$$

We have given $\beta = -\alpha$.

So

$$\alpha + \beta = -\left[\frac{-(8k-8)}{4}\right].$$

$$\alpha + (-\alpha) = 2k - 2$$

$$0 = 2k - 2$$

$$k = 1 \qquad \text{(i)}$$

$$p(x) = kx^2 + 3kx + 2$$

Now $p(x)$ becomes $p(x) = x^2 + 3x + 2$ [From (i)]

$$= x^2 + 2x + x + 2 = x(x + 2) + 1(x + 2) = (x + 1)(x + 2)$$

$$x = -1 \text{ or } x = -2$$

73. $f(x) = 2x^2 - 5x - 3$ (i)

Say α and β are the roots of $f(x)$.

$$p(x) = x^2 + px + q \quad \text{(ii)}$$

Here, roots are 2α and 2β.

From (i)

$$\alpha + \beta = \frac{-(-5)}{2} = \frac{5}{2} \quad \text{(iii)}$$

$$\alpha\beta = \frac{-3}{2} \quad \text{(iv)}$$

From (ii)

$$2\alpha + 2\beta = -p$$

$$2(\alpha + \beta) = -p$$

$$2 \times \frac{5}{2} = -p \quad \text{[From (iii)]}$$

$$p = -5$$

$$(2\alpha)(2\beta) = q$$

$$4\alpha\beta = q$$

$$4 \times \frac{-3}{2} = q$$

$$-6 = q$$

Hence $p = -5$ and $q = -6$.

76. $f(x) = x^4 - 6x^3 - 26x^2 + 138x - 35$

$2 \pm \sqrt{3}$ are the zeros of $f(x)$. Here, factors are $\left\{x - \left(2 + \sqrt{3}\right)\right\}$ and $\left\{x - \left(2 - \sqrt{3}\right)\right\}$.

That is, $\left(x - 2 - \sqrt{3}\right)\left(x - 2 + \sqrt{3}\right) = \left(x - 2\right)^2 - \left(\sqrt{3}\right)^2 = x^2 - 4x + 1$

$\therefore$ $(x^2 - 4x + 1)$ is a factor of the polynomial.

$$
\begin{array}{r|l}
 & x^2 - 2x - 35 \\
\hline
x^2 - 4x + 1 & x^4 - 6x^3 - 26x^2 + 138x - 35 \\
 & x^4 - 4x^3 + x^2 \\
 & - \quad + \quad - \\
\hline
 & -2x^3 - 27x^2 + 138x - 35 \\
 & -2x^3 + 8x^2 - 2x \\
 & + \quad - \quad + \\
\hline
 & -35x^2 + 140x - 35 \\
 & -35x^2 + 140x - 35 \\
 & + \quad - \quad + \\
\hline
 & 0
\end{array}
$$

$$p(x) = (x^2 - 4x + 1)(x^2 - 2x - 35) = (x^2 - 4x + 1)(x^2 - 7x + 5x - 35)$$
$$= (x^2 - 4x + 1)\{x(x - 7) + 5(x - 7)\} = (x^2 - 4x + 1)(x + 5)(x - 7)$$

The zeros are $2+\sqrt{3}, 2-\sqrt{3}$, 7 and –5.

78. $f(x) = x^4 - 6x^3 + 16x^2 - 25x + 10$

$g(x) = x^2 - 2x - k,\ \gamma(x) = x + a$

Dividing $f(x)$ by $g(x)$, we get the remainder $(x + a)$.

$$\begin{array}{r|l} & x^2 - 4x + (8 - k) \\ x^2 - 2x + k & x^4 - 6x^3 + 16x^2 - 25x + 10 \\ & x^4 - 2x^3 + kx^2 \\ & - \quad + \quad - \\ \hline & -4x^3 + (16 - k)x^2 - 25x + 10 \\ & -4x^3 + \quad 8x^2 - 4kx \\ & + \quad - \quad + \\ \hline & (8 - k)x^2 + (4k - 25)x + 10 \\ & (8 - k)x^2 + (2k - 16)x + 8k - k^2 \\ & - \quad - \quad - \quad + \\ \hline & (2k - 9)x + (k^2 - 8k + 10) \end{array}$$

Now, the degree of $r(x)$ is less than $g(x)$. So

$$x + a = (2k - 9)x + k^2 - 8k + 10$$

Comparing LHS and RHS, we get

$2k - 9 = 1$

$k = 5$ (i)

$k^2 - 8k + 10 = a$

$(5)2 - 8(5) + 10 = a$ [From (i)]

$-5 = a$ [**Ans.** $a = -5$ and $k = 5$]

79. $f(x) = x^3 - 3x^2 + x + 1$

$\alpha = a - b, \beta = a, \gamma = a + b$

We have $\alpha + \beta + \gamma = -\dfrac{(-3)}{1}$

$a - b + a + a + b = 3$

$3a = 3$

$a = 1$ (i)

$\alpha\beta\gamma = -1$

$(a - b)(a)(a + b) = -1$

$(1 - b)(1)(1 + b) = -1$

$1 - b^2 = -1$

$b^2 = 2$

$b = \pm\sqrt{2}$ [**Ans.** $a = 1$ and $b = \pm\sqrt{2}$]

80. $f(x) = x^3 + 2x^2 + kx + 3$

Here, $x - 3$ on division gives remainder 21.

$$f(x = +3) = (3)^3 + 2(3)^2 + k(3) + 3 = 21$$
$$= 27 + 18 + 3k + 3 = 21$$
$$3k = -27$$
$$k = -9$$

$$
\begin{array}{l}
\qquad\qquad\qquad 2x^2 - 3x - (8 + 2k) \\
x^2 + 2x + k \,\big)\, 2x^4 + x^3 - 14x^2 + 5x + 6 \,\big(\\
\qquad\qquad\quad 2x^4 + 4x^3 + 2x^2 \\
\qquad\qquad\quad - \quad - \quad - \\
\hline
\qquad\qquad\quad -3x^3 - (14 + 2k)x^2 + 2x + 6 \\
\qquad\qquad\quad -3x^3 - 6x^2 - 3kx \\
\qquad\qquad\quad + \quad + \quad + \\
\hline
\qquad\qquad\qquad -(8 + 2k)x^2 + (5 + 3k)x + 6 \\
\qquad\qquad\qquad -(8 + 2k)x^2 + (-16 - 4k)x + (-8k - 2k^2) \\
\qquad\qquad\qquad + \quad - \quad - \\
\hline
\qquad\qquad\qquad\qquad (21 + 7k)x + (2k^2 + 8k + 6)
\end{array}
$$

Here, remainder must be equal to zero. Therefore
$(21 + 7k)x + (2k^2 + 8k + 6) = 0.x + 0$
Comparing both sides, we get $21 + 7k = 0$ or $2k^2 + 8k + 6 = 0$

$$k = -3 \text{ or } k = -1 \text{ or } k = -3$$

Here, $k = -3$ satisfies the condition that the remainder should be zero. So $f(x)$ becomes $x^3 + 2x^2 - 9x + 3$

$$
\begin{array}{l}
x - 3 \,\big)\, x^3 + 2x^2 - 9x + 3 \,\big(\, x^2 + 5x + 6 \\
\qquad\quad x^3 - 3x^2 \\
\qquad\quad - \quad + \\
\hline
\qquad\qquad 5x^2 - 9x + 3 \\
\qquad\qquad 5x^2 - 15x \\
\qquad\qquad - \quad + \\
\hline
\qquad\qquad\qquad 6x + 3 \\
\qquad\qquad\qquad 6x - 18 \\
\qquad\qquad\qquad - \quad + \\
\hline
\qquad\qquad\qquad\qquad 21
\end{array}
$$

Quotient is $x^2 + 5x + 6$.

That is $x^3 + 2x^2 - 9x + 3 = (x^2 + 5x + 6)(x - 3) + 21$

$$x^3 + 2x^2 - 9x - 18 = (x + 2)(x + 3)(x - 3)$$

So, zeros of $x^3 + 2x^2 + kx - 18$ are -2, 3, -3.

81. Since $x^2 + 2x + k$ is the factor of $2x^4 + x^3 - 14x^2 + 5x + 6$, the remainder = 0.

$$2x^4 + x^3 - 14x^2 + 5x + 6 = (x^2 + 2x - 3)(2x^2 - 3x + (-8 - 2(-3)))$$
$$= (x^2 + 2x - 3)(2x^2 - 3x - 2) = (x - 1)(x + 3)(x - 2)(2x + 1)$$

Hence, the zeros are 1, -3, 2 and $-1/2$.

83. Since $x^3 + 2x^2 + a$ is the factor of $p(x) = x^5 - x^4 - 4x^3 + 3x^2 + 3x + b$, the remainder = 0.

$$
\begin{array}{l}
\qquad\qquad\qquad\quad x^2 - 3x + 2 \\
x^3 + 2x^2 + a \overline{\big)\, x^5 - x^4 - 4x^3 + 3x^2 + 3x + b \,\big(} \\
\qquad\qquad\quad\ \ x^5 + \ 2x^4 \quad + ax^2 \\
\qquad\qquad\quad\ \underline{-\quad -\qquad\quad -\qquad\qquad} \\
\qquad\qquad\qquad - 3x^4 - 4x^3 + (3 - a)x^2 + 3x + b \\
\qquad\qquad\qquad - 3x^4 - \ 6x^3 \qquad\qquad - 3ax \\
\qquad\qquad\qquad \underline{+\qquad +\qquad\qquad\qquad +\qquad} \\
\qquad\qquad\qquad\qquad 2x^3 + (3 - a)x^2 + (3 + 3a)x + b \\
\qquad\qquad\qquad\qquad 2x^3 + 4x^2 \qquad\qquad\quad + 2a \\
\qquad\qquad\qquad\qquad \underline{-\qquad -\qquad\qquad\qquad -\quad} \\
\qquad\qquad\qquad\qquad - (1 + a)x^2 + (3 + 3a)x + (b - 2a)
\end{array}
$$

Since remainder is zero, $-(1 + a)x^2 + (3 + 3a)x + (b - 2a) = 0.x^2 + 0.x + 0$

On comparing both sides $\quad 1 + a = 0$

$$a = -1 \qquad \text{(i)}$$

$$b - 2a = 0$$

$$b - 2(-1) = 0 \qquad \text{[From (i)]}$$

$$b = -2$$

Now $\quad q(x) = x^3 + 2x^2 - 1$

$$p(x) = x^5 - x^4 - 4x^3 + 3x^2 + 3x + b$$

$$r(x) = 0,\ g(x) = x^3 - 3x + 2$$

We know that Dividend = Divisor × Quotient + Remainder

$$p(x) = (x^3 + 2x^2 - 1)(x^2 - 3x + 2) + 0$$

$$= (x^3 + 2x^2 - 1)(x - 2)(x - 1)$$

Hence, the zeros of $p(x)$ are 1 and 2, which are not the zeros of $q(x)$.

84.

$$
\begin{array}{l}
\qquad\quad\ \ ax + b \\
x + c \overline{\big)\, ax^2 + (b + ac)x + bc \,\big(} \\
\qquad\ \ ax^2 + acx \\
\qquad \underline{-\qquad -\qquad\qquad} \\
\qquad\qquad bx + bc \\
\qquad\qquad bx + bc \\
\qquad\qquad \underline{-\quad\ -\quad} \\
\qquad\qquad\quad\ \ \underline{x\qquad}
\end{array}
$$

Quotient = $ax + b$ and remainder = 0.

87.

$$
\begin{array}{l}
x^2 - x + 1 \overline{\big)\, x^4 + 2x^3 - 4x^2 + 6x - 3 \,\big(} x^2 + 3x - 2 \\
\qquad\qquad\ x^4 - x^3 + x^2 \\
\qquad\qquad \underline{-\quad\ +\quad -\qquad\qquad} \\
\qquad\qquad\quad 3x^3 - 5x^2 + 6x \\
\qquad\qquad\quad 3x^3 - 3x^2 + 3x \\
\qquad\qquad\quad \underline{-\qquad +\qquad -\quad} \\
\qquad\qquad\qquad -2x^2 + 3x - 3 \\
\qquad\qquad\qquad -2x^2 + 2x - 2 \\
\qquad\qquad\qquad \underline{+\qquad -\qquad +} \\
\qquad\qquad\qquad\qquad\quad \underline{x - 1}
\end{array}
$$

Ans. $(x - 1)$

88. $f(x) = x^3 - 6x^2 + 3x + 10$

Given $\alpha = a, \beta = a + b$ and $\gamma = a + 2b$

$$\alpha + \beta + \gamma = \frac{-(-6)}{1}$$

$$a + a + b + a + 2b = 6$$

$$3a + 3b = 6$$

$$a + b = 2$$

$$b = 2 - a$$

So $\alpha = a, \beta = a + 2 - a, \gamma = a + 2(2 - a)$

$$\alpha\beta + \beta\gamma + \gamma\alpha = 3$$

$$\Rightarrow a \times 2 + 2 \times (4 - a) + (4 - a) \times a = 3$$

$$2a + 8 - 2a + 4a - a^2 = 3$$

$$a^2 - 4a - 5 = 0$$

$$(a - 5)(a + 1) = 0$$

$$a = -1, 5$$

If $a = -1$, then $b = 3$. If $a = 5$, then $b = -3$.
Inserting the values of a and b, we obtain the roots –1, 2 and 5 or 5, 2 and –1.

Chapter 3: Linear Equations in Two Variables

3 Try to Solve

1. (c) $\frac{x+1}{2} + \frac{y-1}{3} = 8, \frac{3x+3+2y-2}{6} = 8$

$3x + 2y = 47$ (i)

$$\frac{x-1}{3} + \frac{y+1}{2} = 9$$

$$\frac{2x-2+3y+3}{6} = 9$$

$2x + 3y = 53$ (ii)

Solve (i) and (ii) using the elimination method.

(d) $\frac{3}{2}x - \frac{5}{3}y = -2, \frac{9x-10y}{6} = -2$

$9x - 10y = -12$ (i)

$$\frac{x}{3} + \frac{y}{2} = \frac{13}{6}, \frac{2x+3y}{6} = 13$$

$2x + 3y = 78$ (ii)

Solve (i) and (ii) by the elimination method.

4 Try to Solve

1. (d) $\frac{2xy}{x+y} = \frac{3}{2}$, $4xy = 3x + 3y$

Dividing both sides by xy, we get

$$4 = \frac{3}{y} + \frac{3}{x} \quad \text{(i)}$$

Similarly, $-\frac{6}{y} + \frac{3}{x} = 10$ (ii)

Now, insert $\frac{1}{x} = u$ and $\frac{1}{y} = v$ and then simplify using the elimination method.

17 Try to Solve

1. Let the numerator of the fraction be 'x' and the denominator of the fraction be 'y'. Hence, the fraction is $\frac{x}{y}$. Equations are $\frac{x+2}{y+2} = \frac{9}{11}$ and $\frac{x+3}{y+3} = \frac{5}{6}$. Simplify and solve both equations.

2. Equations are $\frac{3x}{y-3} = \frac{18}{11}$ and $\frac{x+8}{2y} = \frac{2}{5}$. Simplify and solve both equations.

3. Equations are $x + y = 12$ and $\frac{x}{y+3} = \frac{1}{2}$. Simplify and solve both equations.

4. Equations are $\frac{x+1}{y-1} = 1$ and $\frac{x}{y+1} = \frac{1}{2}$. Simplify and solve both equations.

18 Try to Solve

1. Let the digit at the ten's place = x and the digit at the unit's place = y.

Two-digit number = $10x + y$

Reversing the two-digit number = $10y + x$

According to the question, $x + y = 10$ (i)

$10x + y - 72 = 10y + x$

$9x - 9y = 72$

$x - y = 8$ (ii)

Solve (i) and (ii).

2. $10x + y + 10y + x = 99$

$x + y = 9$ (i)

$5 + 10x + y = 6(x + y) - 4$

$4x - 5y = -9$ (ii)

Solve (i) and (ii).

3. $10x + y + 10y + x = 66$

$x + y = 6$ (i)

If $x > y$ $\quad x - y = 2$ (ii)

Or if $y > x$ $\quad y - x = 2$ (iii)

Solve (i) and (ii) and then solve (i) and (iii). We obtain a double answer.

19 Try to Solve

1. Let the present age of the father = x years and the present age of the son = y years.
   According to the question, $x = 3y$

$$x - 3y = 0 \quad \text{(i)}$$
$$x + 12 = 2(y + 12)$$
$$x - 2y = 12 \quad \text{(ii)}$$

Solve (i) and (ii).

2. Let the present age of $A = x$ years and the present age of $B = y$ years.
   According to the question, $x + 10 = 2(y + 10)$

$$x - 2y = 10 \quad \text{(i)}$$
$$x - 5 = 3(y - 5)$$
$$x - 3y = -10 \quad \text{(ii)}$$

Solve (i) and (ii).

3. Let the present age of Nuri = x years and the present age of Sonu = y years.
   According to the question, $x - 5 = 3(y - 5)$

$$x - 3y = -10 \quad \text{(i)}$$
$$x + 10 = 2(y + 10)$$
$$x - 2y = 10 \quad \text{(ii)}$$

Solve (i) and (ii).

21 Try to Solve

1. Let the speed of the train = x km/h. The speed of the bus = y km/h.

According to the question, $\dfrac{60}{x} + \dfrac{240}{y} = 4$ (i)

$$\frac{100}{x} + \frac{200}{y} = 4 + \frac{10}{60} = \frac{25}{6} \quad \text{(ii)}$$

Solve (i) and (ii).

2. Let the original speed of Ajeet = x km/h. The original speed of Amit = y km/h.

According to the question, $\dfrac{30}{x} - \dfrac{30}{y} = 2$ (i)

$$\frac{30}{y} - \frac{30}{2x} = 1 \quad \text{(ii)}$$

Solve (i) and (ii).

3. Let the speed of the train = x km/h. The speed of the car = y km/h.

According to the question, $\dfrac{160}{x} + \dfrac{600}{y} = 8$ (i)

$$\frac{240}{x} + \frac{520}{y} = 8 + \frac{12}{60} = \frac{41}{5} \quad \text{(ii)}$$

Solve (i) and (ii).

22 Try to Solve

1. Let the speed of the steamer in still water = x km/h. Speed of stream = y km/h.
Speed in case of upstream = $(x - y)$ km/h. Speed in case of downstream = $(x + y)$ km/h.

According to the question, $\frac{45}{x-y}+\frac{65}{x+y}=14$ (i)

$$\frac{35}{x-y}+\frac{52}{x+y}=11 \quad \text{(ii)}$$

Solve (i) and (ii).

2. $\frac{8}{x+y}=\frac{\not{40}}{\not{60}}=\frac{2}{3}$

$$24 = 2x + 2y \quad \text{(i)}$$

$$\frac{8}{x-y}=1,\ x-y=8 \quad \text{(ii)}$$

Solve (i) and (ii).

3. $\frac{30}{x-y}+\frac{44}{x+y}=10$ (i)

$$\frac{40}{x-y}+\frac{55}{x+y}=13 \quad \text{(ii)}$$

Solve (i) and (ii).

23 Try to Solve

1. Let the usual speed of the train = x km/h. Usual time of the train = y h. Distance covered = xy km.
According to the question, $(x + 10)(y - 2) = xy$

$$-2x + 10y = 20 \quad \text{(i)}$$

$$(x - 10)(y + 3) = xy$$

$$3x - 10y = 30 \quad \text{(ii)}$$

Solve (i) and (ii).

2. Let the original uniform speed = x h. Original time = y h.
Original distance covered = xy km.
According to the question, $(x + 6)(y - 4) = xy$

$$-4x + 6y = 24 \quad \text{(i)}$$

$$(x - 6)(y + 6) = xy$$

$$6x - 6y = 36 \quad \text{(ii)}$$

Solve (i) and (ii).

3. Let the original speed = x km/h. Original time = y h. Original distance = xy km.

According to the question, $\left(x+\frac{1}{2}\right)(y-1)=xy$

$$-x+\frac{1}{2}y=\frac{1}{2}$$

$$-2x + y = 1 \quad \text{(i)}$$
$$(x - 1)(y + 3) = xy$$
$$3x - y = 3 \quad \text{(ii)}$$

Solve (i) and (ii).

Multiple Choice Questions

1. (c) $2^{x+y} = 2^{x-y} = \sqrt{8} = \left(2^3\right)^{\frac{1}{2}} = (2)^{\frac{3}{2}}$

On comparing powers, we get

$$x + y = \frac{3}{2} \quad \text{(i)}$$

And $$x - y = \frac{3}{2} \quad \text{(ii)}$$

Solving (i) and (ii), we get $y = 0$.

5. (b) $\angle C = 3\angle B = 2\angle A + 2\angle B$

Since $3\angle B = 2\angle A + 2\angle B$

$$2\angle A - \angle B = 0 \quad \text{(i)}$$

In ΔABC, $\angle A + \angle B + \angle C = 180°$

$$\angle A + \angle B + 3\angle B = 180° \quad [\because \angle C = 3\angle B \text{ given}]$$
$$\angle A + 4\angle B = 180° \quad \text{(ii)}$$

Solving (i) and (ii), we get $\angle B = 40°$.

9. (d) Let the present age of a man = x years. Present age of the son = y years.

According to the question, $x + 5 = 3(y + 5)$

$$x - 3y = 10 \quad \text{(i)}$$
$$x - 5 = 7(y - 5)$$
$$x - 7y = -30 \quad \text{(ii)}$$

Solving (i) and (ii), we get age of man = 40 years.

11. (d) Because $x = -y$ means $-\frac{1}{y} - \frac{1}{y} = -\frac{2}{y} \neq 0$ if $y > 0$.

14. (a) $$\frac{2x + y + 2}{5} = \frac{3x - y + 1}{3}$$

$$6x + 3y + 6 = 15x - 5y + 5$$
$$9x - 8y = 1 \quad \text{(i)}$$
$$\frac{3x - y + 1}{3} = \frac{3x + 2y + 1}{6}$$
$$18x - 6y + 6 = 9x + 6y + 3$$
$$9x - 12y = -3 \quad \text{(ii)}$$

Solving (i) and (ii), we get $x = 1$ and $y = 1$.

17. (a) Intersect y axis means $x = 0$

$x - y = 8$ means $0 - y = 8$, $y = -8$

27. (c) Let the present age of father = x years. Present age of son = y years.

According to the question, $x = 6y$

$$x - 6y = 0 \quad \text{(i)}$$

$$x + 4 = 4(y + 4),\ x - 4y = 12 \quad \text{(ii)}$$

Solving (i) and (ii), we get $x = 36$ and $y = 6$.

Subjective Questions

57. $$\sqrt{2}x + \sqrt{3}y = 0 \quad \text{(i)}$$

$$\sqrt{3}x - \sqrt{8}y = 0 \quad \text{(ii)}$$

From (i), we get $x = -\dfrac{\sqrt{3}}{\sqrt{2}}y$ (iii)

Substituting (iii) in (ii), we get $x = 0$ and $y = 0$.

64. $$\frac{2}{\sqrt{x}} + \frac{3}{\sqrt{y}} = 2 \quad \text{(i)}$$

$$\frac{4}{\sqrt{x}} - \frac{9}{\sqrt{y}} = -1 \quad \text{(ii)}$$

Insert $\dfrac{1}{\sqrt{x}} = a$ and $\dfrac{1}{\sqrt{y}} = b$.

On solving we get, $x = 4$ and $y = 9$.

65. Insert $\dfrac{1}{x-1} = a$ and $\dfrac{1}{x-2} = b$, then solve to get $x = 4$ and $y = 5$.

66. Divide both sides by xy and then take $\dfrac{1}{x} = a$ and $\dfrac{1}{y} = b$. Solve to get $x = 1$ and $y = 2$.

$\dfrac{1}{x} = a$ and $\dfrac{1}{y} = b$, then solve and get $x = 1$ and $y = 2$.

67. Insert $\dfrac{1}{3x+y} = a$ and $\dfrac{1}{3x-y} = b$, then solve and get $x = 1$ and $y = 1$.

68. Insert $\dfrac{1}{x} = a$ and $\dfrac{1}{y} = b$, then solve and get $x = a^2$ and $y = b^2$.

74. $$\frac{b^2}{a}x - \frac{a^2}{b}y = ab(a+b)$$

$$\frac{b^3 - a^3y}{ab} = ab\,(a + b)$$

$$b^3x - a^3y = a^3b^2 + a^2b^3 \quad \text{(i)}$$

$$b^3x - a^3y = 2a^2b^2 \quad \text{(ii)}$$

Solving (i) and (ii), we get $x = a^2$ and $y = -b^2$.

76. $$a\,(x + y) + b\,(x - y) = a^2 - ab + b^2$$

$$ax + ay + bx - by = a^2 - ab + b^2$$

$$(a + b)\,x + (a - b)\,y = a^2 - ab + b^2 \quad \text{(i)}$$

$$a(x + y) - b(x - y) = a^2 + ab + b^2$$

$$ax + ay - bx + by = a^2 + ab + b^2$$

$$(a - b)x + (a + b)y = a^2 + ab + b^2 \qquad \text{(ii)}$$

Solving (i) and (ii) carefully, we get $x = \dfrac{a^2 + b^2}{b(a+b)}$ and $y = \dfrac{2b}{a+b}$.

77.

$$bx + ay = a + b \qquad \text{(i)}$$

$$ax\left(\frac{1}{a-b} - \frac{1}{a+b}\right) + by\left(\frac{1}{b-a} - \frac{1}{b+a}\right) = \frac{2a}{a+b}$$

$$ax\left(\frac{\cancel{a} + b - \cancel{a} + b}{(a-b)(a+b)}\right) + by\left[\frac{\cancel{b} + a - \cancel{b} + a}{(b-a)(b+a)}\right] = \frac{2a}{a+b}$$

$$\frac{2ab}{a^2 - b^2}x + \frac{2ab}{b^2 - a^2}y = \frac{2a}{a+b}$$

$$\frac{2ab}{a^2 - b^2}x - \frac{2ab}{a^2 - b^2}y = \frac{2ab}{a+b}$$

$$\frac{2ab}{a^2 - b^2}(x - y) = \frac{2a}{a+b}$$

$$\frac{2ab}{(a-b)(a+b)}(x - y) = \frac{2a}{a+b}$$

$$bx - by = a - b \qquad \text{(ii)}$$

Solving (i) and (ii), we get $x = \dfrac{a^2 - b^2}{b(a+b)}$ and $y = \dfrac{2b}{a+b}$.

83.

$$6x + 3y = 8x + 9y - 5$$

$$2x + 6y = 5 \qquad \text{(i)}$$

$$8x + 9y - 5 = 10x + 12y - 8$$

$$2x + 3y = 3 \qquad \text{(ii)}$$

Solving (i) and (ii), we get $x = \dfrac{1}{2}$ and $y = \dfrac{2}{3}$.

131. Let the speed of the train = x km/h. Speed of the car = y km/h.

According to the question, $\dfrac{320}{x} + \dfrac{280}{y} = 8 + \dfrac{40}{60} = \dfrac{26}{3}$ (i)

$$\frac{200}{x} + \frac{400}{y} = 8 + \frac{40}{60} + \frac{30}{60} = \frac{55}{6} \qquad \text{(ii)}$$

Insert $\dfrac{1}{x} = a$ and $\dfrac{1}{y} = b$ and then solve (i) and (ii).

132. Let the distance of the journey covered = x km. Original speed of the train = y km/h.

Original time taken = $\dfrac{x}{y}$ h.

According to the question, $\frac{30}{y}+\frac{x-30}{\frac{4}{5}y}=\frac{x}{y}+\frac{45}{60}$

$$\frac{30}{y}+\frac{5x-150}{4y}=\frac{x}{y}+\frac{3}{4}$$

$$\frac{120+5x-150}{4y}=\frac{4x+3y}{4y}$$

$$x-3y=30 \quad \text{(i)}$$

$$\frac{48}{y}+\frac{x-48}{\frac{4}{5}y}=\frac{x}{y}+\frac{36}{60}$$

$$\frac{48}{y}+\frac{5x-240}{4y}=\frac{x}{y}+\frac{3}{5}$$

$$\frac{192+5x-240}{4y}=\frac{5x+3y}{5y}$$

$$5x-12y=240 \quad \text{(ii)}$$

Solve (i) and (ii).

133. Let the present age of the father = x years. Present age of the son = y years.

According to the question, $x+y=55$ (i)

$$[x+(x-y)]+x=93$$

$$3x-y=93 \quad \text{(ii)}$$

Solve (i) and (ii).

134. Let the first solution = x litres

Second solution = y litres

According to the question, $x+y=10$ (i)

$$\frac{50}{100}x+\frac{25}{100}y=10\times\frac{40}{100}$$

$$50x+25y=400$$

$$2x+5y=16 \quad \text{(ii)}$$

Solve (i) and (ii).

135. Let lot $A=x$ bananas and lot $B=y$ bananas.

According to the question, $x\times\frac{2}{3}+y\times 1=400$

$$2x+3y=1200 \quad \text{(i)}$$

$$x\times 1+\frac{4}{5}y=460$$

$$5x+4y=2300 \quad \text{(ii)}$$

Solve (i) and (ii).

136. Let the fixed charge of the taxi = ₹x

Variable charge of the taxi = ₹y

According to the question, $x + 10y = 75$ (i)

$$x + 15y = 110 \quad \text{(ii)}$$

Solve (i) and (ii) and insert the value of x and y in $x + 25y =$ ________

137. Let the number of candidates in room $A = x$

Number of candidates in room $B = y$

According to the question, $x - 10 = y + 10$

$$x - y = 20 \quad \text{(i)}$$

$$x + 20 = 2(y - 20)$$

$$x - 2y = -60 \quad \text{(ii)}$$

Solve (i) and (ii).

138. Let the length of the rectangle = x units. Breadth of the rectangle = y units.

Area of the rectangle = (xy) sq. units

According to the question, $(x - 5)(y + 2) = xy - 80$

$$2x - 5y = -70 \quad \text{(i)}$$

$$(x + 10)(y - 5) = xy + 50$$

$$-5x + 10y = 100 \quad \text{(ii)}$$

Solve (i) and (ii).

139. Let the present age of the father = x years. Sum of present age of the sons = y years.

According to the question, $\frac{1}{3}(x - 10) = y - 20$

$$x - 3y = -50 \quad \text{(i)}$$

$$y = x - 14$$

$$x - y = 14 \quad \text{(ii)}$$

Solving (i) and (ii), we get $x = 46$ and $y = 32$ where 'y' is the sum of ages of two sons, that is, elder son's age = 17 years and younger son's age = 15 years.

140. Let the distance covered uphill = x km

Distance covered downhill = y km

According to the question, $x + y = 650$ (i)

$$\frac{x}{30} + \frac{y}{50} = 15 \quad \text{(ii)}$$

Solve (i) and (ii).

141. Let the CP of 1 horse = ₹x and the CP of 1 cow = ₹y

According to the question, $4x + 9y = 1340$ (i)

$$\frac{10}{100}x + \frac{20}{100}y = 188$$

$$x + 2y = 1880 \quad \text{(ii)}$$

Solve (i) and (ii).

142. We know that Dividend = Divisor × Quotient + Remainder

Let the larger number = x and the smaller number = y.

According to the question, $3x = y \times 4 + 3$

$$3x - 4y = 3 \quad \text{(i)}$$

$$7y = x \times 5 + 1$$

$$5x - 7y = -1 \quad \text{(ii)}$$

Solve (i) and (ii).

143. Let the CP of a tea set = ₹x and the CP of a lemon set = ₹y.

According to the question, $-\frac{5}{100}x + \frac{15}{100}y = 7$

$$-x + 3y = 140 \quad \text{(i)}$$

$$\frac{5}{100}x + \frac{10}{100}y = 13$$

$$x + 2y = 260 \quad \text{(ii)}$$

Solve 'x' and 'y'.

144. Since $AB + BC + CD + DE + EA = 21$

$$3 + (x - y) + 5 + (x - y) + 3 = 21$$

$$2(x - y) = 10$$

$$x - y = 5 \quad \text{(i)}$$

$$x + y = 5 \quad [\because \text{Opposite sides of rectangle are equal}] \quad \text{(ii)}$$

Solve (i) and (ii).

146. Let the number of questions attempted correctly = x

Number of questions attempted wrongly = y

According to the question, $x + y = 120$ (i)

$$1 \times x - \frac{1}{4} \times y = 90 \quad \text{(ii)}$$

Solve (i) and (ii).

147. Let the total number of poor people = x

Total money a man has = ₹y

According to the question, $120 \times x - 60 = y$

$$120x - y = 60 \quad \text{(i)}$$

$$90 \times x + 90 = y$$

$$90x - y = -90 \quad \text{(ii)}$$

Solve (i) and (ii).

148. Let the money invested in the first bond = ₹x

Money invested in the second bond = ₹y

According to the question, $x + y = 30{,}000$ (i)

$$\frac{5}{100}x + \frac{7}{100}y = 1800$$

$$5x + 7y = 180000 \quad \text{(ii)}$$

Solve (i) and (ii).

149. Let the speed of rowing in still water = x km/h. Speed of stream in still water = y km/h. Speed in case of upstream = $(x - y)$ km/h. Speed in case of downstream = $(x + y)$ km/h.

According to the question, $\frac{20}{x+y} = 2$

$$2x + 2y = 20$$

$$x + y = 10 \quad \text{(i)}$$

$$\frac{4}{x-y} = 2$$

$$x - y = 2 \quad \text{(ii)}$$

Solve (i) and (ii).

150. Let a man finish the work in x days, and a woman finish the same work in y days.

So 1 man's one-day work = $\frac{1}{x}$

$\therefore$ 5 men's one-day work = $\frac{5}{x}$

1 woman's one-day work = $\frac{1}{y}$

2 women's one-day work = $\frac{1}{x}$

According to the question, $\frac{5}{x} + \frac{2}{y} = \frac{1}{4}$ (i)

$$\frac{6}{x} + \frac{3}{y} = \frac{1}{3} \quad \text{(ii)}$$

Insert $\frac{1}{x}$ = u and $\frac{1}{y}$ = v, then solve (i) and (ii).

154. Let the total number of rows = x

Total number of students in each row = y

Total number of students = xy

According to the question, $(x - 2)(y + 4) = xy$

$$4x - 2y = 8 \quad \text{(i)}$$

$$(x + 4)(y - 4) = xy$$

$$-4x + 4y = 16$$

$$x - y = -4 \quad \text{(ii)}$$

Solve (i) and (ii).

156. Let the number of ₹50 notes $= x$. Number of ₹100 notes $= y$.

According to the question, $x + y = 25$

$$50x + 100y = 2000$$

157. Let the fixed charge of the library for first 3 days = ₹x

Additional charge of library per day = ₹y

According to the question, $x + 4y = 27$ ...(i)

$$x + 2y = 21 \quad \text{...(ii)}$$

Solve (i) and (ii).

158. Let the larger angle $= x$. Smaller angle $= y$.

According to the question, $x - y = 18°$ (i)

$$x + y = 180° \quad \text{(ii)}$$

Solve (i) and (ii).

161. Let the present age of Jacob $= x$ years. Present age of Jacob's son $= y$ years.

According to the question, $x + 5 = 3(y + 5)$

$$x - 3y = 10 \quad \text{(i)}$$

$$x - 5 = 7(y - 5)$$

$$x - 7y = -30 \quad \text{(ii)}$$

Solve (i) and (ii).

163. $x + y + 40° = 180°$ [∵ ASP of Δs]

$$x + y = 140° \quad \text{(i)}$$

$$x - y = 30° \quad \text{(ii)}$$

Solve (i) and (ii).

164. Let the present age of father $= x$ years. Sum of present age of two sons $= y$ years.

According to the question, $x = 2y$

$$x - 2y = 0 \quad \text{(i)}$$

$$x + 20 = y + 40$$

$$x - y = 20 \quad \text{(ii)}$$

169. Let the CP of each table = ₹x and the CP of each chair = ₹y.

According to the question, $x + \frac{10}{100}x + y + \frac{25}{100}y = 1050$

$$22x + 25y = 21000 \quad \text{(i)}$$

$$x + \frac{25}{100}x + y + \frac{10}{100}y = 1065$$

$$25x + 22y = 21300 \quad \text{(ii)}$$

Solve (i) and (ii).

170. Let the time taken by the pipe of large diameter to fill the pool be x hours and the time taken by the pipe of smaller diameter be y hours.

So, in 1 hour, the pipe of larger diameter fills $\frac{1}{x}$ part of the pool and, in 4 hours, the pipe of larger diameter fills $\frac{4}{x}$ parts of the pool.

Similarly, in 9 hours, the pipe of smaller diameter fills $\frac{9}{y}$ parts of the pool.

According to the question, $\frac{4}{x}+\frac{9}{y}=\frac{1}{2}$ (i)

$$\frac{12}{x}+\frac{12}{y}=1 \quad \text{(ii)}$$

Insert $\frac{1}{x}=a$ and $\frac{1}{y}=b$ and then solve (i) and (ii).

171. Let the CP of one pen = ₹x and CP of one pencil box = ₹y.
According to the question, $4x + 4y = 100$ (i)

$$3x - y = 15 \quad \text{(ii)}$$

175. Let the digit at the ten's place = x and the digit at the unit's place = y.

Two-digit number = $10x + y$

Reversing the two-digit number = $10y + x$

According to the question, $10x + y = 8(x + y) - 5$

$$2x - 7y = -5 \quad \text{(i)}$$

If $x > y$ $\quad 10x + y = 16(x - y) + 3$

$$6x - 17y = -3 \quad \text{(ii)}$$

If $y > x$ $\quad 10x + y = 16(y - x) + 3$

$$26x - 15y = 3 \quad \text{(iii)}$$

First solve (i) and (ii) and then solve (i) and (iii). You may get a double answer.

176. Let the cost of the first-class full fare = ₹$2x$

Cost of half-ticket = ₹x; Reservation charge = ₹y

According to the question, $2x + y = 2530$ (i)

$$2x + y + x + y = 3810$$

$$3x + 2y = 3810 \quad \text{(ii)}$$

Solve (i) and (ii).

178. Let the CP of a saree = ₹x. List price of a sweater = ₹y.

According to the question, $x+\frac{8}{100}x+y-\frac{10}{100}y=1008$

$$108x + 90y = 100800 \quad \text{(i)}$$

$$x+\frac{10}{100}x+y-\frac{8}{100}y=1028$$

$$110x + 92y = 102800 \quad \text{(ii)}$$

Solve (i) and (ii).

Chapter 4: Quadratic Equations

16 Try to Solve

1. Let the first positive integer = x. Second consecutive positive integer = $x + 1$.
   According to the question, $(x)(x + 1) = 182$
   $$x^2 + x = 182$$
   $$x^2 + x - 182 = 0$$
   Solve the equation by yourself.
2. $(x)(x + 1) = 420$. Solve the equation by yourself.
3. $(x)(x + 1) = 360$. Solve the equation by yourself.
4. Let Johan have x marbles. Jivanti has $(45 - x)$ marbles.
   According to the question, $(x - 5)(45 - x - 5) = 124$
   $$(x - 5)(40 - x) = 124$$
   $$40x - x^2 - 200 + 5x = 124$$
   $$x^2 - 45x + 324 = 0$$
   Solve this equation by yourself.

17 Try to Solve

1. Let the first part be = x. Second part = $15 - x$.
   According to the question, $x^2 + (15 - x)^2 = 765$. Solve this equation by yourself.
2. $x^2 + (39 - x)^2 = 765$. Solve the equation by yourself.
3. $(x)(27 - x) = 182$. Solve the equation by yourself.

18 Try to Solve

1. Let the first odd positive integer be $(2x + 1)$. Second consecutive odd integer = $(2x + 3)$.
   According to the question, $(2x + 1)^2 + (2x + 3)^2 = 34$. Solve the equation by yourself.
2. Let the first even positive integer be $2x$. Second consecutive even positive integer = $(2x + 2)$.
   According to the question, $(2x)^2 + (2x + 2)^2 = 52$. Solve the equation by yourself.
3. Let the first positive integer be (x). Consecutive second positive integer = $(x + 1)$.
   According to the question, $(x)^2 + (x + 1)^2 = 365$. Solve the equation by yourself.

19 Try to Solve

1. Let the digit at the ten's place be x. The digit at the unit's place = $\frac{12}{x}$.

   Two-digit number = $10 \times x + \frac{12}{x}$

   Reversing, we get $10 \times \frac{12}{x} + x$

According to the question, $9+10x+\frac{12}{x}=10\times\frac{12}{x}+x$

$$\frac{9x+10x^2+12}{x}=\frac{120+x^2}{x}$$

$$9x^2+9x-108=0$$

$$x^2-x-12=0$$

Solve the equation by yourself.

2. $$45+10x+\frac{14}{x}=10\times\frac{14}{x}+x$$

$$\frac{45x+10x^2+14}{x}=\frac{140+x^2}{x}$$

$$9x^2+45x-126=0$$

$$x^2+5x-14=0$$

Solve the equation by yourself.

20 Try to Solve

1. Let the first number = x. Second number = $\frac{16}{x}$.

According to the question, $x+\frac{16}{x}=\frac{1}{3}$.

$$\frac{x^2+16}{x}=\frac{1}{3}$$

$$3x^2-x+48=0.$$

Solve the equation by yourself.

2. $x+\frac{1}{x}=\frac{17}{4}$. Solve the equation by yourself.

21 Try to Solve

1. Let the usual speed of a passenger train = x km/h. Increased speed of passenger train = $(x + 15)$ km/h.

Time taken originally = $\frac{150}{x}$ h. Time taken now = $\frac{150}{x+15}$ h.

Since Original time taken > Time taken now

According to the question, $\frac{150}{x}-\frac{150}{x+15}=1$. Solve the equation by yourself.

2. $\frac{90}{x}-\frac{90}{x+15}=\frac{30}{60}$. Solve the equation by yourself.

22 Try to Solve

1. Let the speed of the fast train be x km/h. Speed of the slow train = $(x - 10)$ km/h.

Time taken by fast train for journey = $\frac{200}{x}$ h. Time taken by slow train for journey = $\frac{200}{x-10}$ h.

Since the slow train takes more time than the fast train, according to the question

$$\frac{200}{x-10} - \frac{200}{x} = 1$$

Solve the equation by yourself.

2. $\frac{360}{x} - \frac{360}{x+10} = 3$. Solve the equation by yourself.

3. $\frac{132}{x} - \frac{132}{x+11} = 1$. Solve the equation by yourself.

23 Try to Solve

1. Let the original speed of aeroplane be x km/h. Reduced speed of aeroplane = $(x - 400)$ km/h.

   Original time of aeroplane = $\frac{6000}{x}$ h. Time of aeroplane now = $\frac{6000}{x-400}$ h.

   Since, the time taken now is more than the original time taken

$$\frac{6000}{x-400} - \frac{6000}{x} = \frac{30}{60}$$

$$\frac{6000[x-(x-400)]}{(x-400)(x)} = \frac{1}{2}$$

$$6000 \times 400 \times 2 = x^2 - 400x$$

$$x^2 - 400x - 4800000 = 0$$

$$x^2 - 2400x + 2000x - 4800000 = 0$$

   Solve the equation by yourself.

2. $\frac{360}{x} - \frac{360}{x+5} = 1$. Solve the equation by yourself.

24 Try to Solve

1. Let the speed of the stream be x km/h. Speed of boat upstream = $(15 - x)$ km/h.
   Speed of boat downstream = $(15 + x)$ km/h.

   According to the question, $\frac{30}{15-x} + \frac{30}{15+x} = 4 + \frac{30}{60}$

$$\frac{30}{15-x} + \frac{30}{15+x} = \frac{9}{2}$$

$$\frac{30[15+x+15-x]}{(15-x)(15+x)} = \frac{9}{2}$$

$$1800 = 9[225 - x^2]$$

$$9x^2 = 2025 - 1800$$

$$9x^2 = 225$$

   Solve the equation by yourself.

2. $\frac{12}{11-x}+\frac{12}{11+x}=2+\frac{45}{60}$. Solve the equation by yourself.

3. $\frac{24}{18-x}-\frac{24}{18+x}=1$. Solve the equation by yourself.

25 Try to Solve

1. Let the original duration of tour be x days. Original daily expense = ₹ $\frac{360}{x}$.

   Daily expense when tour extends 4 days = ₹ $\frac{360}{x+4}$.

   According to the question, $\frac{360}{x}-\frac{360}{x+4}=3$. Solve the equation by yourself.

2. $\frac{80}{x}-\frac{80}{x+1}=1$. Solve the equation by yourself.

26 Try to Solve

1. Let the present age of the father be x years. Present age of Puneet = $(45 - x)$ years.
   According to the question, $(x)(45 - x) = 126$. Solve the equation by yourself.
2. $(x - 5)(x + 8) = 30$. Solve the equation by yourself.
3. Let the present age of Rohan be x years. Present age of Rohan's mother = $(x + 26)$ years.
   According to the question, $(x + 3)(x + 26 + 3) = 360$

   $$(x + 3)(x + 29) = 360$$

   Solve the equation by yourself.

27 Try to Solve

1. Let perpendicular (AB) be x m. Hypotenuse (AC) = $(2x + 6)$ m.
   Base (BC) = $(2x + 4)$ m.

   In right ΔABC $\quad AC^2 = AB^2 + BC^2$ [∵ Pythagoras' Theorem]

   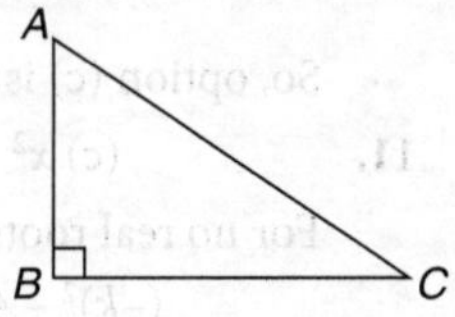

   $$(2x + 6)^2 = x^2 + (2x + 4)^2$$

   $$(2x)^2 + (6)^2 + 2 \times 2x \times 6 = x^2 + (2x)^2 + (4)^2 + 2 \times 2x \times 4$$

   $$4x^2 + 36 + 34x = x^2 + 4x^2 + 16 + 16x$$

   $$x^2 - 8x - 20 = 0$$

   Solve the equation by yourself.
2. Let base (BC) be x m. Perpendicular = $(x - 7)$ m. Hypotenuse (AC) = 13 m.

   In right ΔABC $\quad AC^2 = AB^2 + BC^2$ [∵ Pythagoras' Theorem]

   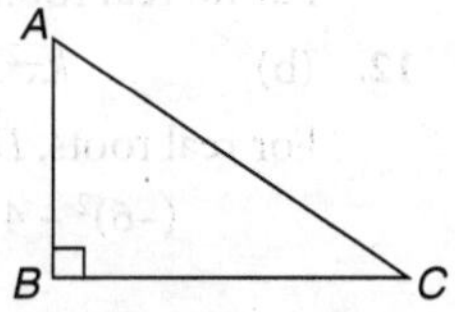

   $$(13)^2 = (x - 7)^2 + x^2$$

   Solve the equation by yourself.

3. 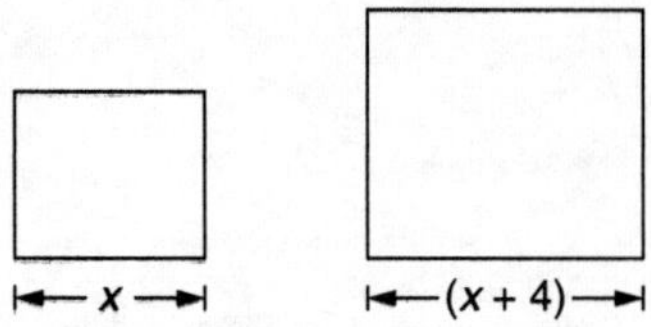

Let the side of the big square be $(x + 4)$ m. Side of the small square $= x$ m.
Area of big square $= (x + 4)^2$ m^2. Area of small square $= x^2$ m^2.
According to the question, $(x + 4)^2 + (x)^2 = 400$. Solve the equation by yourself.

Multiple Choices Question

3. (b)

$$9x^2 + \frac{3}{4}x - \sqrt{2} = 0$$

$$(3x)^2 + \frac{1}{4}(3x) - \sqrt{2} = 0$$

$$(3x)^2 + 2\times(3x)\times\frac{1}{8} - \sqrt{2} = 0$$

$$(3x)^2 + 2\times(3x)\times\frac{1}{8} + \left(\frac{1}{8}\right)^2 - \left(\frac{1}{8}\right)^2 - \sqrt{2} = 0$$

So, option (b) is correct. $\frac{1}{64}$ is the constant.

7. (c) $ax^2 + bx + c = 0$

Say, one root $= \alpha$ and the second root $= \frac{1}{\alpha}$.

So, product of root $= \not{\alpha} \times \frac{1}{\not{\alpha}} = \frac{c}{a}$

$$1 = \frac{c}{a}$$

$$c = a$$

So, option (c) is correct.

11. (c) $x^2 - kx + 1 = 0$

For no real roots, $D < 0$.

$$(-k)^2 - 4 \times 1 \times 1 < 0$$

$$k^2 - 4 < 0$$

$$k^2 < 4$$

For no real roots, $-2 < k < 2$.

12. (b) $kx^2 - 6x - 2 = 0$

For real roots, $D \geq 0$.

$$(-6)^2 - 4 \times k(-2) \geq 0$$

$$36 + 8k \geq 0$$

$$k \geq -\frac{9}{2}$$

14. (b) Let one root of the equation be α. Second root of equation = 6α.

$$\alpha + 6\alpha = \frac{14}{p}$$

$$7\alpha = \frac{14}{p}$$

$$\Rightarrow \quad \alpha = \frac{2}{p} \quad \text{(i)}$$

$$\alpha \times 6\alpha = \frac{8}{p}$$

$$6\alpha^2 = \frac{8}{p}$$

$$6 \times \left(\frac{2}{p}\right)^2 = \frac{8}{p} \quad \text{[From (i)]}$$

$$6 \times \frac{4}{p^2} = \frac{8}{p}$$

$$\therefore \quad p = 3$$

15. (b) $x^2 - 2x - r + 1 = 0$

$$\Rightarrow \quad x^2 - 2x + 1 - r = 0$$

$$\Rightarrow \quad x^2 - x(1-r) - x(1+r) + (1-r) = 0$$

$$\Rightarrow \quad x[x-(1-r)] - (1+r)[x-(1-r)] = 0$$

$$\Rightarrow \quad [x-(1-r)][x-(1+r)] = 0$$

$$x = 1 - r \text{ or } x = 1 + r$$

16. (b) $ax^2 - 5x + c = 0$

$$S = \alpha + \beta = \frac{5}{a}$$

$$\frac{5}{a} = 10$$

$$a = \frac{1}{2}$$

$$P = \alpha\beta = \frac{c}{a} = 10$$

$$c = 20$$

$$\frac{c}{\frac{1}{2}} = 10$$

$$c = 20$$

Option (b) is correct.

20. (b)

$$\alpha = 2 + \sqrt{3}, \beta = 2 - \sqrt{3}$$

$$S = \alpha + \beta = 4, P = \alpha\beta = 1$$

$$x^2 - Sx + P = 0$$

$$x^2 - 4x + 1 = 0$$

Option (b) is correct.

22. (a)

$$f(x) = x^2 + px - 4 = 0$$

$$f(x = -4) = (-4)^2 + p(-4) - 4 = 0$$

$$= 16 - 4p - 4 = 0$$

$$p = 3$$

$$x^2 + px + k = 0$$

$$x^2 + 3x + k = 0$$

For equal roots, $D = 0$.

$$D = 9 - 4k = 0$$

$$k = \frac{9}{4}$$

24. (b)

$$x^2 - x - p(p + 1) = 0$$

$$x^2 - (p + 1)x + px - p(p + 1) = 0$$

$$x[x - (p + 1)] + p[x - (p + 1)] = 0$$

$$(x + p)[x - (p + 1)] = 0$$

either $x + p = 0$, $x = -p$ or $x - (p + 1) = 0$, $x = (p + 1)$

Option (b) is correct.

25. (b)

$$x^2 - 2x + 1 = 0$$

$$(x - 1)^2 = 0$$

$$x = 1$$

$$x + \frac{1}{x} = 1 + \frac{1}{1} = 2$$

Option (b) is correct.

26. (b) $ax^2 + bx + c = 0$. Say, one root is α. Second root is $\beta = -\alpha$.

$$S = \alpha + \beta = \alpha - \alpha = 0 = -\frac{b}{a}$$

$$b = 0$$

Hence, option (b) is correct.

30. (d)

$$(x^2 + 1)^2 - x^2 = 0$$

$$x^4 + 1 + 2x^2 - x^2 = 0$$

$$x^4 + x^2 + 1 = 0$$

Insert $x^2 = t$, so $x^4 = t^2$

$$t^2 + t + 1 = 0$$

$$D = (1)^2 - 4 \times 1 \times 1 = 1 - 4 = -3 < 0$$

No real roots are present.

Subjective Question

50.

$$\sqrt{3}x^2 - 2\sqrt{2}x - 2\sqrt{3} = 0$$

$$\sqrt{3}x^2 - 3\sqrt{2}x + \sqrt{2}x - 2\sqrt{3} = 0$$

$$\sqrt{3}x\left(x - \sqrt{6}\right) + \sqrt{2}\left(x - \sqrt{6}\right) = 0$$

$$\left(\sqrt{3}x + \sqrt{2}\right)\left(x - \sqrt{6}\right) = 0$$

$$x = -\frac{\sqrt{2}}{\sqrt{3}} \text{ and } x = \sqrt{6}$$

$$x = -\frac{\sqrt{2}}{\sqrt{3}} \times \frac{\sqrt{3}}{\sqrt{3}} = -\frac{\sqrt{6}}{3}$$

51. $\dfrac{2x}{x-3} + \dfrac{1}{2x+3} + \dfrac{3x+9}{(x-3)(2x+3)} = 0$

$$\frac{4x^2 + 6x + x - 3 + 3x + 9}{(x-3)(2x+3)} = 0$$

$$4x^2 + 10x + 6 = 0$$

$$4x^2 + 4x + 6x + 6 = 0$$

$$4x\,(x + 1) + 6\,(x + 1) = 0$$

$$(4x + 6)\,(x + 1) = 0$$

$$x = -\frac{3}{2} \text{ and } x = -1$$

53.

$$x^2 + \left(\frac{a}{a+b} + \frac{a+b}{a}\right)x + 1 = 0$$

$$x^2 + \frac{a}{a+b}x + \frac{a+b}{a}x + 1 = 0$$

$$x\left[x + \frac{a}{a+b}\right] + \frac{a+b}{a}\left[x + \frac{a}{a+b}\right] = 0$$

$$\left(x + \frac{a+b}{a}\right)\left(x + \frac{a}{a+b}\right) = 0$$

$$x = -\frac{a+b}{a} \text{ or } x = -\frac{a}{a+b}$$

54. $9x^2 - 9(a + b)\,x + (2a^2 + 5ab + 2b^2) = 0$

Using the quadratic formula $x = \dfrac{-b \pm \sqrt{b^2 - 4ac}}{2a}$

$$x = \frac{-[-9(a+b)] \pm \sqrt{[-9(a+b)]^2 - 4\times 9 \times (2a^2+5ab+2b^2)}}{2\times 9} = \frac{9(a+b) \pm 3\sqrt{9a^2+9b^2+18ab-8a^2-20ab-8b^2}}{18}$$

$$x = \frac{9(a+b) \pm 3\sqrt{a^2+b^2-2ab}}{18} = \frac{9(a+b) \pm 3\sqrt{(a-b)^2}}{18} = \frac{9(a+b) \pm 3(a-b)}{18}$$

$$x = \frac{9a+9b+3a-3b}{18} \quad \text{or} \quad x = \frac{9a+9b-3a+3b}{18}$$

$$x = \frac{12a+6b}{18} \quad \text{or} \quad x = \frac{6a+12b}{18}$$

$$x = \frac{2a+b}{3} \quad \text{or} \quad x = \frac{a+2b}{3}$$

56. $$\frac{a}{x-a} + \frac{b}{x-b} = \frac{2c}{x-c}$$

$$\Rightarrow \frac{a(x-b)+b(x-a)}{(x-a)(x-b)} = \frac{2c}{x-c}$$

$$\Rightarrow \frac{ax+bx-2ab}{x^2-ax-bx+ab} = \frac{2c}{x-c}$$

$\Rightarrow ax^2 + bx^2 - 2abx - acx - bcx + 2abc = 2cx^2 - 2acx - 2bcx + 2abc$

$\Rightarrow (a + b - 2c)x^2 + (ac + bc - 2ab)x = 0$

$\Rightarrow x\{(a + b - 2c)x + (ac + bc - 2ab)\} = 0$

$$x = 0 \text{ or } x = \frac{2ab-ac-bc}{a+b-2c}$$

57. $4x^2 - 4a^2x + (a^4 - b^4) = 0$

$4x^2 - 2(a^2 - b^2)x - 2(a^2 + b^2)x + (a^2 - b^2)(a^2 + b^2) = 0$

$2x[2x - (a^2 - b^2)] - (a^2 + b^2)[2x - (a^2 - b^2)] = 0$

$[2x - (a^2 - b^2)][2x - (a^2 + b^2)] = 0$

$$x = \frac{a^2-b^2}{2} \text{ or } x = \frac{a^2+b^2}{2}$$

58. $x^2 - (2b - 1)x + (b^2 - b - 20) = 0$

Using the quadratic formula, $D = [-(2b - 1)]^2 - 4 \times 1 \times (b^2 - b - 20)$

$D = 4b^2 + 1 - 4b - 4b^2 + 4b + 80$

$D = 81$

$$x = \frac{-[-(2b-1)] \pm \sqrt{81}}{2\times 1} = \frac{2b-1\pm 9}{2}$$

$$x = \frac{2b-10}{2} \text{ or } x = \frac{2b+8}{2}$$

$x = b - 5$ or $x = b + 4$

60. $\left(\frac{2x-1}{x+3}\right)-3\left(\frac{x+3}{2x-1}\right)=5$

Insert $\frac{2x-1}{x+3}=t$ and $\frac{x+3}{2x-1}=\frac{1}{t}$. Solve the equation by yourself.

61. $$\frac{x+3}{x+2}=\frac{3x-7}{2x-3}$$

$$(x+3)(2x-3)=(3x-7)(x+2)$$

$$2x^2-3x+6x-9=3x^2+6x-7x-14$$

$$x^2-4x-5=0$$

Solve the equation by yourself.

64. $$\frac{1}{2a+b+2x}=\frac{1}{2a}+\frac{1}{b}+\frac{1}{2x}$$

$$\Rightarrow \frac{1}{2a+b+2x}-\frac{1}{2x}=\frac{1}{2a}+\frac{1}{b}$$

$$\Rightarrow \frac{2x-(2a+b+2x)}{(2a+b+2x)(2x)}=\frac{b+2a}{2ab}$$

$$\frac{-\cancel{(2a+b)}}{(2a+b+2x)(2x)}=\frac{\cancel{(b+2a)}}{2ab}$$

$$-2ab=4ax+2bx+4x^2$$

$$4x^2+2bx+4ax+2ab=0$$

$$2x(2x+b)+2a(2x+b)=0$$

$$(2x+2a)(2x+b)=0$$

$$x=-a\ x=-\frac{b}{2}$$

69. Yes, we may take $2x^2+x-4=0$ with rational coefficients, but its roots are $\frac{-1+\sqrt{33}}{4}$ and $\frac{-1-\sqrt{33}}{4}$.

70. Yes, we may take $\sqrt{3}x^2-7\sqrt{3}x+12\sqrt{3}=0$. Its roots are 3 and 4 which are rational.

73. $(p+1)x^2-6(p+1)x+3(p+9)=0$

Here $$D=[-6(p+1)]^2-4\times(p+1)3(p+9)$$

$$=36(p+1)^2-12(p+1)(p+9)=12(p+1)[3(p+1)-(p+9)]$$

$$=12(p+1)[2p-6]=24(p+1)(p-3)$$

For equal roots, $D=0$.

$$24(p+1)(p-3)=0$$

$$p=-1 \text{ or } p=3$$

75. Since $x=-2$ is the root of $3x^2+7x+p=0$

$$3(-2)^2+7(-2)+p=0$$

$$12-14+p=0$$

$$\Rightarrow p=2$$

$$\therefore x^2+k(4x+k-1)+2=0$$

$$x^2 + 4kx + k^2 - k + 2 = 0$$

$$D = (4k)^2 - 4 \times 1 \times (k^2 - k + 2) = 0$$
$$= 16k^2 - 4k^2 + 4k - 8 = 0 = 12k^2 + 4k - 8 = 0$$
$$= 3k^2 + k - 2 = 0$$

Solve the equation by yourself.

85. $(1 + m^2)x^2 + 2mcx + c^2 - a^2 = 0$

Here $$D = (2mc)^2 - 4(1 + m^2)(c^2 - a^2) = 0$$
$$= \cancel{4m^2c^2} - 4c^2 + 4a^2 - \cancel{4m^2c^2} + 4m^2a^2 = 4a^2 - 4c^2 + 4m^2a^2 = 0$$
$$4a^2 + 4m^2a^2 = 4c^2$$
$$a^2(1 + m^2) = c^2$$

86. $(a^2 + b^2)x^2 + 2(ac + bd)x + (c^2 + d^2) = 0$

$$D = [2(ac+bd)]^2 - 4\times(a^2+b^2)(c^2+d^2) = 4(ac+bd)^2 - 4(a^2+b^2)(c^2+d^2)$$
$$D = 4a^2c^2 + 4b^2d^2 + 8abcd - 4(a^2c^2 + a^2d^2 + b^2c^2 + b^2d^2)$$
$$D = \cancel{4a^2c^2} + \cancel{4b^2d^2} + 8abcd - \cancel{4a^2c^2} - 4a^2d^2 - 4b^2c^2 - \cancel{4b^2d^2}$$
$$D = -4a^2d^2 - 4b^2c^2 + 8abcd = -[2ad - 2bc]^2 < 0$$

The overall value of 'D' always remains negative because $ad \neq bc$.
So, $D < 0$ has no real roots.

87. $(a^2 + b^2)x^2 + 2(ac + bd)x + (c^2 + d^2) = 0$

For equal roots, $D = 0$.

$$D = [2(ac + bd)]^2 - 4(a^2 + b^2)(c^2 + d^2) = 0$$
$$= 4(a^2c^2 + b^2d^2 + 2abcd) - 4(a^2c^2 + a^2d^2 + b^2c^2 + b^2d^2) = 0$$
$$= -a^2d^2 - b^2c^2 + 2abcd = 0$$
$$= -1(ad - bc)^2 = 0$$
$$ad = bc$$

94. Let the number of toys produced on that day be x.
Cost of production of each toy that day = ₹$(55 - x)$. Total cost of production that day = ₹$x(55 - x)$.
According to the question, $x(55 - x) = 750$

$$55x - x^2 = 750$$
$$x^2 - 55x + 750 = 0$$

95. Let the number of articles produced on a particular day be x.
Cost of each article produced on a particular day = ₹$(3 + 2x)$.
According to the question, $x(3 + 2x) = 90$

$$2x^2 + 3x - 90 = 0$$

Solve the equation by yourself.

96. Let Shefali's marks in mathematics be x. Shefali's marks in English = $30 - x$.
According to the question, $(x + 2)(30 - x - 3) = 210$

$$(x + 2)(27 - x) = 210$$

Solve the equation by yourself.

97. Let the larger pipe fill the tank in x h. Smaller pipe fills the tank = $(x + 10)$ h.

The tank filled by the larger pipe in 1 hour = $\frac{1}{x}$ parts.

The tank filled by the smaller pipe in 1 hour = $\frac{1}{x+10}$ parts.

Both pipes together can fill the tank in $9\frac{3}{8}$ hours = $\frac{75}{8}$ hours.

According to the question, $\frac{1}{x} + \frac{1}{x+10} = \frac{1}{75/8}$

$$\frac{1}{x} + \frac{1}{x+10} = \frac{8}{75}$$

Solve the equation by yourself.

98. Let the side of the larger square be x m. Let the side of the smaller square be y m.

Perimeter of larger square = $4x$ m. Perimeter of smaller square = $4y$ m.

Area of larger square = x^2 m^2. Area of smaller square = y^2 m^2.

According to the question, $4x - 4y = 24$

$$4(x - y) = 24$$

$$x - y = 6$$

$$x = 6 + y \quad \text{(i)}$$

$$x^2 + y^2 = 468$$

$$(6 + y)^2 + y^2 = 468 \quad \text{[From (i)]}$$

$$36 + y^2 + 12y + y^2 = 468$$

$$2y^2 + 12y + 432 = 0$$

$$y^2 + 6y + 216 = 0$$

Solve the equation by yourself.

99. Let her actual marks be x. According to the question

$$\therefore \quad 9(x + 10) = x^2$$

$$x^2 - 9x - 90 = 0$$

Solve the equation by yourself.

100. Let the original average speed be x km/h. Increased speed = $(x + 6)$ km/h.

According to the question, $\frac{63}{x} + \frac{72}{x+6} = 3$. Solve the equation by yourself.

101. Let the natural number be x. According to the question

$$x^2 - 84 = 3(8 + x)$$

$$x^2 - 3x - 108 = 0$$

Solve the equation by yourself.

102. Let the actual age of Zeba be x years. When she was 5 years younger = $(x - 5)$ years.

According to the question, $(x - 5)^2 = 11 + 5x$

$$x^2 - 15x + 14 = 0$$

Solve the equation by yourself.

103. Let the present age of Nisha be x years. Present age of Asha = $(x^2 + 2)$ years.
Age of Nisha when she becomes her mother's age, that is, $\{(x^2 + 2) - x\}$ years.
Then, Asha's age also increases = $\{(x^2 + 2) - x\}$ years.

$$(x^2 + 2) + \{(x^2 + 2) - x\} = 10x - 1$$
$$2x^2 - 11x + 5 = 0$$

Solve the equation by yourself.

104.

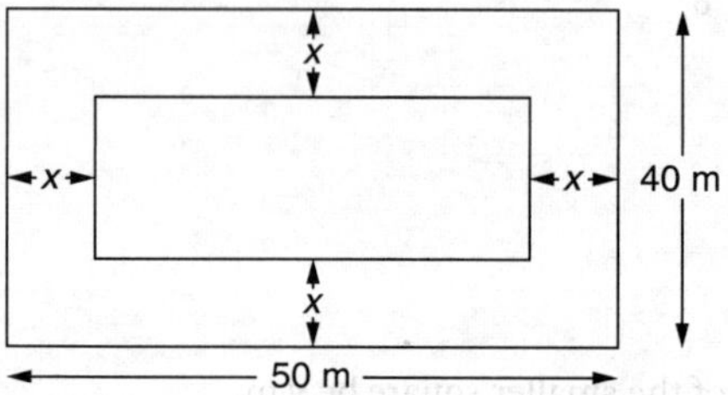

Length of rectangular lawn = 50 m. Breadth of rectangular lawn = 40 m.
Length of rectangular pond = $(50 - 2x)$ m. Breadth of rectangular pond = $(40 - 2x)$ m.
According to the question, $50 \times 40 - (50 - 2x)(40 - 2x) = 1184$

$$2000 - [2000 - 100x - 80x + 4x^2] = 1184$$
$$4x^2 - 180x + 1184 = 0$$

Solve the equation by yourself.

105. Since, time between 2 pm and 3 pm = 1 h = 60 min.
We have given 't' minutes past 2 pm and the time needed by the minute hand of the clock to show 3 pm was 3 minutes less than $t^2/4$ minutes.

According to the question, $t + \left(\dfrac{t^2}{4} - 3\right) = 60$

$$4t + t^2 - 12 = 240$$
$$t^2 + 4t - 252 = 0$$

Solve the equation by yourself.

106. Same as Q97.

107. Let the speed of covering 150 km be x km/h. Speed of returning 150 km = $(x + 10)$ km/h.

Time taken for travelling 150 km = $\dfrac{150}{x}$ h. Time taken for returning 150 km = $\dfrac{150}{x+10}$ h.

Time of going > Time of returning

According to the question, $\dfrac{150}{x} - \dfrac{150}{x+10} = \dfrac{5}{2}$. Solve the equation by yourself.

108. Let the length of the rectangular park be x m. Breadth of rectangular park = $(x - 3)$ m.
Base of isosceles triangle = $(x - 3)$ m.
Height of isosceles triangle = 12 m.

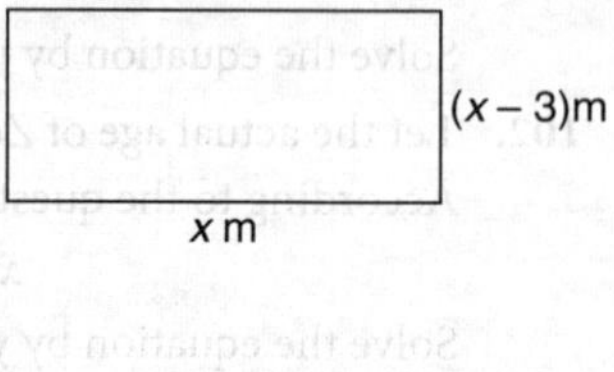

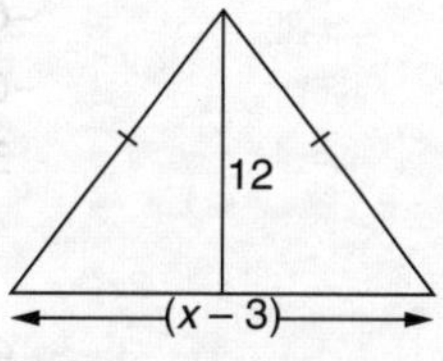

According to the question, $(x)(x-3) = \frac{1}{2}(x-3)\times 12+4$

$$x^2 - 9x + 14 = 0$$

Solve the equation by yourself.

113. Let the number of students attending the picnic be x. Total budget for food = ₹2000.
According to the question, $\frac{2000}{x-5} - \frac{2000}{x} = 20$. Solve the equation by yourself.

119. Cost price = ₹x. Profit = $\frac{(x+10)}{100}\times x$.

We know that SP = CP + Profit

$$2x - 20 = x + \left(\frac{x+10}{100}\right)\times x$$

We have $2x - 20 = \frac{100x + x^2 + 10x}{100}$

$$200x - 2000 = x^2 + 110x$$

$$x^2 - 90x + 2000 = 0$$

Solve the equation by yourself.

Chapter 5: Arithmetic Progression

12 Try to Solve

1. 6, 13, 20, …, 2, 16

$$a = 6,\ d = 13 - 6 = 7,\ a_n = 216$$

$$a_n = a + (n-1)d$$

$$216 = 6 + (n-1)7$$

$$216 - 6 = (n-1)7$$

$$210 = (n-1)7$$

$$\frac{210}{7} = n - 1$$

$$30 = n - 1$$

$31 = n$, which is odd. So, middle term = $\frac{1}{2}(31+1)$ = 16th term.

$$a_{16} = a + (16-1)d = 6 + 15\times 7 = 111$$

2. 10, 7, 4, …, (–62)

$$a = 10,\ d = 7 - 10 = -3,\ a_n = -62$$

$$a_n = a + (n-1)d$$

$$-62 = 10 + (n - 1)(-3)$$
$$-62 = 10 - 3n + 3$$
$$3n = 75$$
$n = 25$, which is odd. So, middle term $= \frac{1}{2}(25+1) = 13^{th}$ term.

$$a_{13} = a + (13 - 1)d = 10 + 12 \times (-3) = 10 - 36 = -26$$

13 Try to Solve

3. 108, 117, … , 999

$$a = 108, d = 9, a_n = 999, n = ?$$
$$a_n = a + (n - 1)d$$
$$999 = 108 + (n - 1)9$$
$$999 - 108 = (n - 1)9$$
$$891 = (n - 1)9$$
$$\frac{891}{9} = n - 1$$
$$99 = n - 1$$
$$n = 100$$

16 Try to Solve

3. 63, 60, 57, 54, …

$$a = 63, d = 60 - 63 = -3, S_n = 693, n = ?$$
$$S_n = \frac{n}{2}[2a + (n-1)d]$$
$$693 = \frac{n}{2}[2 \times 63 + (n-1)(-3)]$$
$$693 = \frac{n}{2}[126 - 3n + 3]$$
$$1386 = n[129 - 3n]$$
$$1386 = 129n - 3n^2$$
$$1386 = 129n - 3n^2$$
$$3n^2 - 129n + 1386 = 0$$
$$3[n^2 - 43n + 462] = 0$$
$$n^2 - 21n - 22n + 462 = 0$$
$$n(n - 21) - 22(n - 21) = 0$$
$$(n - 22)(n - 21) = 0$$

Either $\quad$ Or

$n - 22 = 0 \quad n - 21 = 0$

$n = 22 \quad n = 21$

17 Try to Solve

2. $a_4 + a_8 = 24$ $\quad$ $a_6 + a_{10} = 44$

$a + 3d + a + 7d = 24$ $\quad$ $a + 5d + a + 9d = 44$

$2a + 10d = 24$ $\quad$ $2a + 14d = 44$

$a + 5d = 12$ (i) $\quad$ $a + 7d = 22$ (ii)

Solving (i) and (ii), we get

$$a + 5d = 12$$
$$a + 7d = 22$$
$$-2d = -10$$
$$d = 5$$

Inserting the value of d in (i), we get $a + 5 \times 5 = 12$

$$a = -13$$
$$a_{10} = a + 9d = -13 + 9 \times 5 = 32$$
$$S_n = \frac{n}{2}[a + a_n] = \frac{10}{2}[-13 + 32] = 5 \times 19 = 95$$

3. $S_{51} = ?$ $a_2 = 14$ and $a_3 = 18$

$$a_2 = 14$$
$$a + d = 14 \quad \text{(i)}$$
$$a_3 = 18$$
$$a + 2d = 18 \quad \text{(ii)}$$

Solving (i) and (ii), we get $a = 10$ and $d = 4$

$$S_{51} = \frac{51}{2}[2 \times 10 + (51 - 1) \times 4] = \frac{51}{2}[20 + 200] = 5610$$

20 Try to Solve

1. Two-digit numbers divisible by 2 are 10, 12, …, 98.

Here $a = 10, d = 2, a_n = 98.$

$$a_n = a + (n - 1)d$$
$$98 = 10 + (n - 1) \times 2$$
$$88 = (n - 1)$$
$$n = 45$$
$$S_1 = \frac{45}{2}[10 + 98] \qquad \left[\because S_n = \frac{n}{2}[a + a_n]\right]$$
$$= 2430$$

Two-digit numbers divisible by 3 and not by 2 are 15, 21, …, 99.

$$b = 15, d_1 = 6, b_n = 99$$
$$b_n = b + (n_1 - 1)d_1$$
$$99 = 15 + (n_1 - 1) \times 6$$

$$\frac{84}{6} = (n_1 - 1)$$

$$n_1 = 15$$

$$S_2 = \frac{15}{2}[15+99] = 855$$

We need $= S_1 + S_2 = 2430 + 855 = 3285$

Multiple Choice Questions

8. (c) $\frac{1}{p}, \frac{1-p}{p}, \frac{1-2p}{p}, \ldots$

$$d = \frac{1-p}{p} - \frac{1}{p} = \frac{1-p-1}{p} = \frac{-p}{p} = -1$$

(c) is the correct option.

9. (c) $\frac{1}{3}, \frac{1-3b}{3}, \frac{1-6b}{3}$

$$d = \frac{1-3b}{3} - \frac{1}{3} = \frac{1}{3} - \frac{3b}{3} - \frac{1}{3} = -b$$

(c) is the correct option.

10. (d) $\sqrt{7}, \sqrt{28}, \sqrt{63} \ldots$ or $\sqrt{7}, 2\sqrt{7}, 3\sqrt{7}, \ldots$

$$d = 2\sqrt{7} - \sqrt{7} = \sqrt{7}$$

$$a_4 = a_3 + d = 3\sqrt{7} + \sqrt{7} = 4\sqrt{7} = \sqrt{112}$$

(d) is the correct option.

11. (b) $a_n = 2n + 1$

$$a_1 = 2(1) + 1 = 3$$

$$a_2 = 2(2) + 1 = 5$$

$$a_3 = 2(3) + 1 = 7$$

$$S_3 = \frac{3}{2}[3+7] = 15$$

(b) is the correct option.

12. (a) $S_n = 3n^2 + 6n$

$$S_1 = a_1 = 3(1)^2 + 6(1) = 9$$

$$S_2 = a_1 + a_2 = a_1 + a_1 + d = 2a_1 + d = 3(2)^2 + 6(2)$$

$$\Rightarrow 2 \times 9 + d = 24$$

$$d = 24 - 18 = 6$$

(a) is the correct option.

14. (b) $a_{13} = 4a_3$

$$a + 12d = 4(a + 2d)$$

$$a + 12d = 4a + 8d$$

$$4d = 3a$$

$$d = \frac{3}{4}a \qquad \text{(i)}$$

$$a_5 = 16$$

$$a + 4d = 16$$

$$a + 4 \times \frac{3}{4}a = 16$$

$$4a = 16$$

$$a = 4$$

Inserting the value of a in (i), we get

$$d = \frac{3}{4} \times 4 = 3$$

$$S_{10} = \frac{10}{2}\left[2 \times 4 + (10-1) \times 3\right] = 5[8 + 27] = 5 \times 35 = 175$$

(b) is the correct option.

20. (a)

$$a_{17} - a_{10} = 21$$

$$(a + 16d) - (a + 9d) = 21$$

$$7d = 21$$

$$d = 3$$

(a) is the correct option.

24. (c) $\sqrt{2} + \sqrt{8} + \sqrt{18} + \ldots$

That is $\sqrt{2} + 2\sqrt{2} + 3\sqrt{2} + \ldots$

$$a = \sqrt{2},\ d = 2\sqrt{2} - \sqrt{2} = \sqrt{2}$$

$$S_n = \frac{n}{2}\left[2 \times \sqrt{2} + (n-1)\sqrt{2}\right] = \frac{n}{2}\left[2\sqrt{2} + \sqrt{2}n - \sqrt{2}\right]$$

$$= \frac{n}{2}\left[\sqrt{2} + \sqrt{2}n\right] = \frac{1}{\sqrt{2}}(n)(1+n)$$

(c) is the correct option

27. (c) $(8k + 4), (6k - 2), (2k - 7)$

We know that in an AP, the common difference is equal. Therefore

$$(6k - 2) - (8k + 4) = (2k - 7) - (6k - 2)$$

$$-2k - 6 = -4k - 5$$

$$2k = 1$$

$$k = \frac{1}{2}$$

(c) is the correct option.

31. (c)

$$S_{2n} = 3S_n$$

$$\frac{2n}{2}\left[2a + (2n-1)d\right] = 3 \times \frac{n}{2}\left[2a + (n-1)d\right]$$

$$4an + 2n(2n-1)d = 6an + 3n(n-1)d$$
$$2an + 3n(n-1)d - 2n(2n-1)d = 0$$
$$n\{2a + d(3n - 3 - 4n + 2)\} = 0$$
$$2a - (n+1)d = 0$$
$$2a = (n+1)d \quad \text{(i)}$$
$$\frac{S_{3n}}{S_n} = \frac{\frac{3n}{2}[2a+(3n-1)d]}{\frac{n}{2}[2a+(n-1)d]} = \frac{3[(n+1)d+(3n-1)d]}{[(n+1)d+(n-1)d]}$$
$$= \frac{3[n+1+3n-1]d}{[n+1+n-1]d} = \frac{3\times 4n}{2n} = \frac{6}{1}$$

(c) is the correct option.

32. (b) $\sqrt{3}+\sqrt{12}+\sqrt{27}+\sqrt{48}+\ldots$ or $\sqrt{3}+2\sqrt{3}+3\sqrt{3}+4\sqrt{3}+\ldots$

$$a = \sqrt{3},\ d = \sqrt{3}$$
$$S_n = \frac{n}{2}\left[2\sqrt{3}+(n-1)\sqrt{3}\right] = \frac{n}{2}\left[2\sqrt{3}+\sqrt{3}n-\sqrt{3}\right]$$
$$= \frac{n}{2}\left[\sqrt{3}+\sqrt{3}n\right] = \frac{\sqrt{3}n}{2}(1+n)$$

(b) is the correct option.

33. (a)
$$(S_n)_{\text{even}} = \frac{n}{2}[2\times 2+(n-1)\times 2] = \frac{n}{2}[4+2n-2]$$
$$(S_n)_{\text{even}} = \frac{n}{2}[2n+2] \quad \text{(i)}$$
$$(S_n)_{\text{odd}} = \frac{n}{2}[2\times 1+(n-1)\times 2] = \frac{n}{2}[2+2n-2] = \frac{n}{2}[2n]$$
$$(S_n)_{\text{odd}} = n^2 \quad \text{(ii)}$$
$$(S_n)_{\text{even}} = k(S_n)_{\text{odd}} \quad \text{[Given]}$$
$$\frac{n}{2}[2n+2] = k(n^2)$$
$$n(n+1) = k(n^2)$$
$$\frac{n+1}{n} = k$$

(a) is the correct option.

34. (c) 20, 17, 14, …

$$a = 20,\ d = 17 - 20$$

First negative term, $a_n < 0$
$$a + (n-1)d < 0$$
$$20 + (n-1)(-3) < 0$$
$$20 - 3n + 3 < 0$$

$$23 < 3n$$

$$\frac{23}{3} < n$$

$n > \frac{23}{3}$ means the first negative term is the 8th term.

(c) is the correct option.

Subjective Questions

54. $n - 2, 4n - 1$ and $5n + 2$

Since all the three terms are in an AP, their common difference is equal. Therefore

$$(4n - 1) - (n - 2) = (5n + 2) - (4n - 1)$$

$$3n + 1 = n + 3$$

$$n = 1$$

56. Let the first term be $= a - d$. Second term $= a$. Third term $= a + d$.

According to the question, $a - d + a + a + d = 33$

$$3a = 33$$

$$a = 11 \quad \text{(i)}$$

$$(a - d)(a + d) - a = 29$$

$$a^2 - d^2 - a = 29$$

$$(11)^2 - d^2 - 11 = 29 \quad \text{[From (i)]}$$

$$110 - 29 = d^2$$

$$81 = d^2$$

$$d = \pm 9$$

If $d = 9$ and $a = 11$

2, 11, 20, …

If $d = -9$ and $a = 11$

20, 11, 2, …

65. 5, 6.75, …, 20.75

$a = 5$, $d = 1.75$, $a_n = 20.75$. Find $n = ?$

Use $a_n = a + (n - 1)d$.

67. $a_{26} = 0$, $a_{11} = 3$, $a_n = -\frac{1}{5}$, $d = ?$ and $n = ?$

$$a + 25d = 0 \quad \text{(i)}$$

$$a + 10d = 3 \quad \text{(ii)}$$

Use (i) and (ii) to find a and d and then insert in a_n to find n.

70. Since the common difference is equal

$$(2k^2 + 3k + 6) - (k^2 + 4k + 8) = (3k^2 + 4k + 4) - (2k^2 + 3k + 6)$$

Solve and find k.

71. Let the first angle (smallest) be $a - d$. Second angle (smallest) $= a$. Third angle (largest) $= a + d$

According to the question, $a + d = 2(a - d)$

$$a + d = 2a - 2d$$

$$3d = a \quad \text{(i)}$$

From the angle sum property of triangles

$$a - d + a + a + d = 180°$$
$$3a = 180$$
$$a = 60°$$

Inserting the value of 'a' in (i), we get $d = 20°$

So, the first angle = 40°, second angle = 60° and the third angle = 80°.

76. $$a = 5 \text{ and } S_4 = \frac{1}{2}[S_8 - S_4]$$

$$2S_4 = S_8 - S_4$$
$$3S_4 = S_8$$
$$3 \times \frac{4}{2}[2 \times 5 + (4-1)d] = \frac{8}{2}[2 \times 5 + (8-1)d]$$
$$6[10 + 3d] = 4[10 + 7d]$$
$$60 + 1.8d = 40 + 28d$$
$$20 = 10d$$
$$d = 2$$

80. 200, 250, 300, …

$$a = 200, d = 50 \text{ and } n = 30. S_{30} = ?$$

Use the formula to solve the problem.

81. $x, x - 20, x - 40, \ldots$

$$a = x, n = 7, d = -20, S_n = 700$$

Find x using the formula for S_n.

82. The circumference of a semicircle = πr

0.5, 1π, 1.5π, 2.0π, … are in AP.

∴ Sum of circumference of 13 semicircles = π (0.5 + 1 + 1.5 + 2 + … 13 terms)

Use the formula for S_n to solve the problem.

83. 20, 19, 18, … are in AP.

$$S_n = 200, a = 20, d = -1$$

First find n and then find a_n.

84. The distance covered to pick up the first, second and third potato, and so on is as follows:

$$2 \times 5, 2 \times 8, 2 \times 11, \ldots$$

⇒ 10, 16, 22, … are in AP. Find $S_{10} = ?$

85. $$a_3 + a_7 = 6 \quad \text{(i)}$$

$$a_3 \cdot a_7 = 8 \quad \text{(ii)}$$

Use the first equation to express a in terms of d. Then, inserting the value of a in (ii), we get

$$d = \frac{1}{2} \text{ and } a = 1$$

Therefore $$S_{16} = 76.$$

However, if $d = -\frac{1}{2}$ and $a = 5$

$$S_{16} = 20$$

86. Height of 1st step = $\frac{1}{4}$ m. Height of 2nd step = $\frac{1}{4}+\frac{1}{4} = \frac{1}{2}$ m. Height of 3rd step = $\frac{1}{2}+\frac{1}{4} = \frac{3}{4}$ m, and so on.

Volume of concrete required for 1st step

$$(a_1) = l \times b \times h = \frac{1}{2} \times 50 \times \frac{1}{4} = \frac{50}{8} \text{ m}^3$$

Volume of concrete required for 2nd step

$$(a_2) = \frac{1}{2} \times 50 \times \frac{1}{2} = \frac{50}{4} \text{ m}^3$$

Volume of concrete required for 3rd step

$$(a_3) = \frac{1}{2} \times 50 \times \frac{3}{4} = \frac{150}{8} \text{ m}^3$$

The volume of concrete required for each step is $\frac{50}{8}, \frac{50}{4}, \frac{150}{8}, \ldots$ are in AP.

$$a = \frac{50}{8}, d = \frac{50}{8}$$

Find S_{15} using the formula.

88. (ii) $\left(4-\frac{1}{n}\right)+\left(4-\frac{2}{n}\right)+\left(4-\frac{3}{n}\right)+\ldots n$ terms

$$a = 4-\frac{1}{n}, d = 4-\frac{2}{n}-4+\frac{1}{n} = -\frac{1}{n}$$

$$S_n = \frac{n}{2}\left[2\times\left(4-\frac{1}{n}\right)+(n-1)\times\left(-\frac{1}{n}\right)\right]$$

Solve using the formula.

(iii) $\frac{a-b}{a+b}+\frac{3a-2b}{a+b}+\frac{5a-3b}{a+b}+\ldots$ up to 11 terms

$$a = \frac{a-b}{a+b}, d = \frac{3a-2b}{a+b}-\frac{a-b}{a+b} = \frac{2a-b}{a+b}, n = 11$$

Find S_{11} using the formula.

92. $S_{12} = 3(S_8 - S_4)$

LHS $S_{12} = \frac{12}{2}[2a+(12-1)d] = 6[2a+11d]$

RHS $3(S_8 - S_4) = 3\left[\frac{8}{2}(2a+7d)-\frac{4}{2}(2a+3d)\right]$

Solving, we get LHS = RHS.

93. $n = 11$, which is odd. So, the middle term $= \frac{1}{2}(n+1) = \frac{1}{2}(11+1) = 6^n$ term

So $a_6 = 30$

$a + 5d = 30$ (i)

We need to find out S_{11}.

$$S_{11} = \frac{11}{2}[2a + (11-1)d] = \frac{11}{2}[2a + 10d] = \frac{11}{2} \times 2[a + 5d]$$

$= 11\ [30]$ [From (i)]

$S_{11} = 330$

99. Let the first term be $a - 3d$. Second term $= a - d$. Third term $= a + d$. Fourth term $= a - 3d$.

According to the question

$a - 3d + a - d + a + d + a + 3d = 32$

$4a = 32$

$a = 8$ (i)

$$\frac{(a-3d)(a+3d)}{(a-d)(a+d)} = \frac{7}{15}$$

$$\frac{a^2 - 9d^2}{a^2 - d^2} = \frac{7}{15}$$

$15\ [(8)^2 - 9d^2] = 7[(8)^2 - d^2]$

$15\ (8)^2 - 135\ d^2 = 7(8^2) - 7d^2$

$15\ (8)^2 - 7(8^2) = 135d^2 - 7d^2$

$8^2(15 - 7) = 128d^2$

$8 \times 8 \times 8 = 128d^2$

$$\frac{\cancel{8} \times \cancel{8} \times \cancel{8}}{\cancel{128}_{\ 2}} = d^2$$

$4 = d^2$

$d = \pm 2.$

If $a = 8$ and $d = 2$, the terms are 2, 6, 10, 14, …

If $a = 8$ and $d = -2$, the terms are 14, 10, 6, 2, …

102. $n = 37$

Middle term $= \frac{1}{2}[37 + 1] = 19^{th}$ term

According to the question

$a_{18} + a_{19} + a_{20} = 225$ (i)

$a_{35} + a_{36} + a_{37} = 429$ (ii)

Solving (i) and (ii), we get $a = 3$ and $d = 4$. Find the AP.

103. $$\frac{a_{11}}{a_{18}} = \frac{2}{3}$$

$$\frac{a + 10d}{a + 17d} = \frac{2}{3}$$

$$a = 4d \quad \text{(i)}$$

$$\frac{a_5}{a_{21}} = \frac{a+4d}{a+20d}$$

$$= \frac{4d+4d}{4d+20d} = \frac{8d}{24d} = \frac{1}{3} \quad \text{[From (i)]}$$

Now, we need to find $\dfrac{S_5}{S_{21}} = \dfrac{\frac{5}{2}[2a+(5-1)d]}{\frac{21}{2}[2a+(21-1)d]}$

$$= \frac{5}{21}\left[\frac{8d+4d}{8d+20d}\right] \quad \text{[From (ii)]}$$

$$= \frac{5}{21} \times \frac{12d}{28d} = \frac{5}{49}$$

104. $a_1 = a, a_2 = b$ and $a_n = c$

$$d = b - a$$

$$c = a + (n-1)(b-a)$$

$$c - d = (n-1)(b-a)$$

$$n - 1 = \frac{c-a}{b-a}$$

$$n = \frac{c+b-2a}{b-a}$$

$$S_n = S_{\frac{c+b-2a}{b-a}} = \frac{\frac{c+b-2a}{b-a}}{2}\left[2a + \left(\frac{c+b-2a}{b-a} - 1\right)(b-a)\right]$$

$$= \frac{b+c-2a}{2(b-a)}\left[2a+c+b-2a-b+a\right] = \frac{b+c+2a}{2(b-a)}(c+a)$$

106. 1000, 1100, 1200, …

$$a = 1000, n = 30, d = 100$$

Using $a_n = a + (n-1)d$

$$a_{30} = 1000 + (30-1) \times 100$$

$$= 1000 + 2900 = 3900$$

$$S_n = \frac{n}{2}\left[a + a_n\right]$$

$$S_{30} = \frac{30}{2}\left[1000 + 3900\right] = 15 \times 4900 = 73{,}500$$

Total loan amount = ₹ 1,18,000.

Balance to be paid = 1,18,00 – 73,500 = ₹ 44,500.

107. Here, $n = 27$, which is odd. So, middle term = $\frac{1}{2}(27+1) = 14$

$$a_{13} + a_{15} = 2 + 2 = 4$$

$$a_{12} + a_{16} = 4 + 4 = 8$$

So, AP becomes 4, 8, 12, …, 52

Here $a = 4, d = 4, a_n = 52, n = 13$

$$S_{13} = \frac{13}{2}[4+52] = 364 \text{ m}$$

Distance covered to the books = 364 m. ∴ Total distance covered = 364 + 364 = 728 m.

Maximum distance she travelled carrying a flag

$$= 2 + 2 + \ldots + 13 \text{ times.}$$

$$= 2 \times 13 = 26 \text{ m}$$

108. There are two APs, say: $a_1, a_2, \ldots, a_n$ and $b_1, b_2, \ldots, b_n$

In the first AP, common difference = d

In the second AP, common difference = d'

In the first AP, sum of AP = S_n

In the second AP, sum of AP = S_n'

So $$\frac{S_n}{S'_n} = \frac{\frac{n}{2}[2a_1 + (n-1)d]}{\frac{n}{2}[2b_1 + (n-1)d']}$$

$$\frac{S_n}{S'_n} = \frac{a_1 + \frac{(n-1)}{2}d}{b_1 + \frac{(n-1)}{2}d'} \quad \text{(i)}$$

And $$\frac{S_n}{S'_n} = \frac{7n+1}{4n+27} \quad \text{[Given]} \quad \text{(ii)}$$

From (i) and (ii), we get $$\frac{7n+1}{4n+27} = \frac{a_1 + \left(\frac{n-1}{2}\right)d}{b_1 + \left(\frac{n-1}{2}\right)d'} \quad \text{(iii)}$$

We need to find $\frac{a_m}{b_m} = ?$

$$\frac{a_m}{b_m} = \frac{a_1 + (m-1)d}{b_1 + (n-1)d'} \quad \text{(iv)}$$

Now, we need to replace $\frac{n-1}{2}$ with $m - 1$, that is, $n = 2m - 1$ in (iii).

$$\frac{a_m}{a_n} = \frac{a_1 + (m-1)d}{b_1 + (m-1)d} = \frac{7(2m-1)+1}{4(2m-1)+27} = \frac{14m-6}{8m+23}$$

Hence, the ratio of the mth term is $14m - 6 : 8m + 23$.

109. $S_1 = \frac{n}{2}[2+(n-1)\cdot 1] = \frac{n(n+1)}{2}$ (i)

$S_2 = \frac{n}{2}[2+(n-1)\cdot 2] = n^2$ (ii)

$S_3 = \frac{n}{2}[2+(n-1)\cdot 3] = \frac{n}{2}[3n+1]$ (iii)

Now, $S_1 + S_3 = \frac{n(n+1)}{2} + \frac{n}{2}(3n-1) = 2n^2 = 2S_2$ [From (i), (ii) and (iii)]

112. Let us suppose that the policeman catches the thief after t minutes. Uniform speed of thief = 50 m/min.
Since thief began running 2 minutes before the policeman started running
Distance covered by thief in $(t + 2)$ min.

$= 50 \times (t + 2)$ m (i)

So, the AP is 60, 65, 70, … (in minutes)
Here, the total distance covered by the policeman

$= \frac{t}{2}[2\times 60+(t-1)\times 5]$

$= 60t + \frac{5t}{2}(t-1)$ (i)

When the policeman catches the thief, equation (i) = equation (ii)

$60 + \frac{5}{2}t^2 - \frac{5t}{2} = 50t + 100$

$t^2 + 3t - 40 = 0$

Solve this to find the solution.

117. $\text{RHS} = 3[S_{20} - S_{10}] = 3\left[\frac{20}{2}(2a+19d) - \frac{10}{2}(2a+9d)\right]$

$= 3[5(2a + 29d)] = \frac{30}{2}[2a+(30-1)d] = S_{30}$

Hence, LHS = RHS.

120. Three-digit numbers leaving remainder 3 when divided by 4 are 103, 107, 111, …, 999
Here $a = 103, d = 4, n = ?, a_n = 999$

$a_n = a + (n - 1)d$

$999 = 103 + (n - 1) \times 4$

$n = 225$

The middle term $= \frac{1}{2}(n+1) = \frac{1}{2}(225+1) = 113$

$a_{113} = a + (113 - 1)d = 103 + 112 \times 4 = 551$

There are 112 terms before the 113th term.

$S_{112} = \frac{112}{2}[2\times 103\times 111\times 4] = 36400$

Similarly, $S_{225} = 123975$

Sum of terms after middle term = 123975 – 36400 = 87024

132. $\frac{1}{1+\sqrt{x}}, \frac{1}{1-x}, \frac{1}{1-\sqrt{x}}, \ldots$

$$d = \frac{1}{1-x} - \frac{1}{1+\sqrt{x}} = \frac{1+\sqrt{x}-(1-x)}{(1-x)(1+\sqrt{x})}$$

$$\frac{1+\sqrt{x}-1+x}{(1-x)\left(1+\sqrt{x}\right)} = \frac{x+\sqrt{x}}{(1-x)\left(1+\sqrt{x}\right)} = \frac{\sqrt{x}\left(\sqrt{x}+1\right)}{(1-x)\left(1+\sqrt{x}\right)}$$

$$d = \frac{\sqrt{x}}{1-x}$$

$$a_4 = a_3 + d = \frac{1}{1-\sqrt{x}} + \frac{\sqrt{x}}{1-x}$$

$$= \frac{1}{1-\sqrt{x}} + \frac{\sqrt{x}}{\left(1-\sqrt{x}\right)\left(1+\sqrt{x}\right)} = \frac{1+\sqrt{x}+\sqrt{x}}{\left(1-\sqrt{x}\right)\left(1+\sqrt{x}\right)} = \frac{1+2\sqrt{x}}{1-x}$$

135. $$\frac{a_{11}}{a_{18}} = \frac{2}{3}$$

$$\frac{a+10d}{a+17d} = \frac{2}{3}$$

$$3a + 30d = 2a + 34d$$

$$a = 4d$$

$$\frac{S_5}{S_{10}} = \frac{\frac{5}{2}[2\times 4d + (5-1)\times d]}{\frac{10}{2}[2\times 4d + (10-1)d]} = \frac{12d}{34d} = \frac{6}{17}$$

136. $$S_n = 3n^2 - 4n$$

$$S_{n-1} = 3(n-1)^2 - 4(n-1)$$

$$a_n = S_n - S_{n-1} = 3n^2 - 4n - 3(n-1)^2 + 4(n-1)$$

$$= 3[n^2 - (n-1)^2] - 4[n - (n-1)]$$

$$= 3[n^2 - n^2 - 1 + 2n] - 4(1)$$

$$= 6n - 3 - 4 = 6n - 7$$

Chapter 6: Triangles

4 Try to Solve

3. Consider ΔPQR and ΔQST

$\angle PRQ = \angle STQ$ [∵ Each 90°]

$\angle Q = \angle Q$ [Common]

$\Delta PQR \sim \Delta SQT$ [∵ AA]

$\frac{PQ}{SQ} = \frac{QR}{QT} = \frac{PR}{ST}$ [If two Δs are similar, then their corresponding sides are in proportion]

$PQ \times QT = QS \times QR$

5 Try to Solve

1. Given: Height of lamp post = 4.5 m; Height of boy = 150 cm = 1.5 m
   Let the length of the shadow of the boy = $DE = x$ m
   Length of the shadow of the lamp post = $BE = (x + y)$ m
   where $BD = y = 1.6 \times 5 = 8$ m.

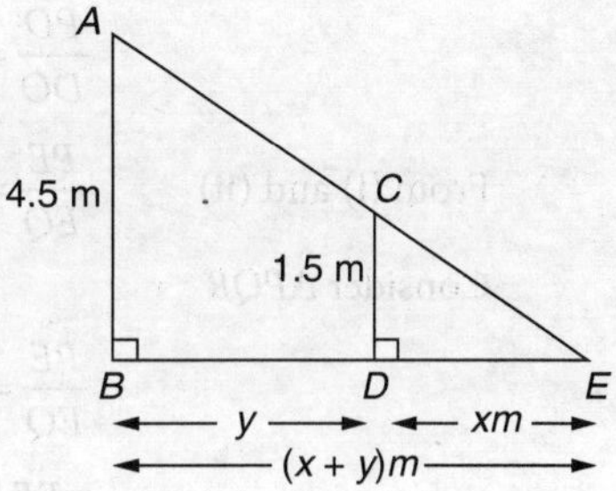

$\Delta CDE \sim \Delta ABE$ [∵ AA]

$\frac{1.5}{4.5} = \frac{x}{x+8}$

$\frac{1}{3} = \frac{x}{x+8}$

$x + 8 = 3x$

$8 = 3x - x$

$8 = 2x$

$4 = x$

Hence, shadow = 4 m.

7 Try to Solve

1. Consider ΔBAC, $DE \parallel AC$

$\frac{BD}{DA} = \frac{BE}{EC}$ [∵ BPT] (i)

Consider ΔBAE, $DF \parallel AE$

$\frac{BD}{DA} = \frac{BF}{FE}$ [∵ BPT] (ii)

From (i) and (ii) $\frac{BE}{EC} = \frac{BF}{FE}$

8 Try to Solve

1. Consider ΔCAB, $DE \parallel AB$

$\frac{CD}{AC} = \frac{CE}{EB}$ [∵ BPT] (i)

Consider ΔCAB, $BD \parallel EF$

$\frac{CF}{CD} = \frac{CE}{EB}$ [∵ BPT] (ii)

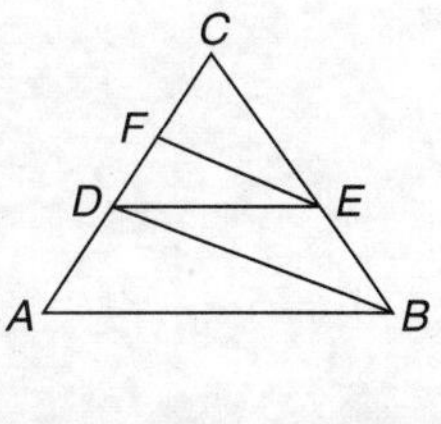

From (i) and (ii) we get. $\frac{CD}{AC} = \frac{CF}{CD}$

$CD^2 = AC \times CF$

9 Try to Solve

2. Consider ΔPQO, $DE \parallel OQ$

$$\frac{PE}{EQ} = \frac{PD}{DO} \qquad [\because BPT] \qquad \text{(i)}$$

Consider ΔPOR, $DF \parallel OR$

$$\frac{PD}{DO} = \frac{PF}{FR} \qquad [\because BPT] \qquad \text{(ii)}$$

From (i) and (ii) $\quad \dfrac{PE}{EQ} = \dfrac{PF}{FR} \qquad$ (ii)

Consider ΔPQR

$$\frac{PE}{EQ} = \frac{PF}{FR} \qquad \text{[From (ii)]}$$

$$EF \parallel QR \qquad \text{[Converse of } BPT\text{]}$$

12 Try to Solve

1. Given: ΔABC, in which $XY \parallel AC$ and Area of trapezium $ACYX$ = Area ΔBXY

To find: $\dfrac{AX}{AB} = ?$

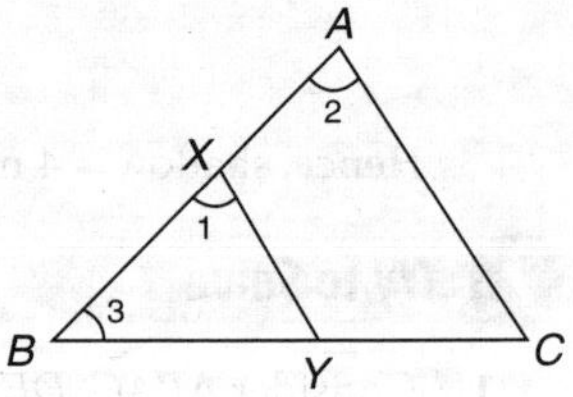

Solution: Since $XY \parallel AC \quad \angle 1 = \angle 2 \quad [\because$ Corresponding angles$]$ (i)

Consider ΔBXY and ΔBAC $\quad \angle 1 = \angle 2 \quad$ [From (i)]

$$\angle 3 = \angle 3 \qquad \text{[Common]}$$

$$\Delta BXY \sim \Delta BAC \qquad [\because AA]$$

$$\frac{\text{Area } \Delta BXY}{\text{Area } \Delta BAC} = \frac{BX^2}{BA^2} \qquad [\because \text{Area Theorem}]$$

$$\frac{\text{Area } \Delta BXY}{\text{Area } \Delta BXY + \text{Area trapezium } XACY} = \frac{BX^2}{BA^2} \qquad [\because \text{Area } \Delta BXY = \text{area of trapezium } ACYX]$$

$$\frac{\text{Area } \Delta BXY}{2 \text{ Area } \Delta BXY} = \frac{BX^2}{BA^2}$$

$$\frac{1}{2} = \frac{BX^2}{BA^2}$$

$$\frac{1}{\sqrt{2}} = \frac{BX}{BA}$$

$$\frac{1}{\sqrt{2}} = \frac{BA - AX}{BA}$$

$$\frac{1}{\sqrt{2}} = \frac{BA}{BA} - \frac{AX}{AB}$$

$$\frac{1}{\sqrt{2}} = 1 - \frac{AX}{AB}$$

$$\frac{AX}{AB} = 1 - \frac{1}{\sqrt{2}} = \frac{\sqrt{2}-1}{\sqrt{2}}$$

2. $\Delta AQB \sim \Delta APC$ [∵ Both are equilateral Δ]

$$\frac{\text{Area } \Delta AQB}{\text{Area } \Delta APC} = \frac{AB^2}{AC^2} \quad [\because \text{Area theorem}] \quad \text{(i)}$$

Similarly $$\frac{\text{Area } \Delta BRC}{\text{Area } \Delta APC} = \frac{BC^2}{AC^2} \quad \text{(ii)}$$

Adding (i) and (ii), we get

$$\frac{\text{Area } \Delta AQB}{\text{Area } \Delta APC} + \frac{\text{Area } \Delta BRC}{\text{Area } \Delta APC} = \frac{AB^2}{AC^2} + \frac{BC^2}{AC^2} = \frac{AB^2 + BC^2}{AC^2}$$

$$\frac{\text{Area } \Delta AQB + \text{Area } \Delta BRC}{\text{Area } \Delta APC} = \frac{AC^2}{AC^2}$$

[∵ ΔABC is a right Δ using Pythagoras' Theorem $AC^2 = AB^2 + BC^2$]

$$\frac{\text{Area } \Delta AQB + \text{Area } \Delta BRC}{\text{Area } \Delta APC} = 1$$

$$\text{Area } \Delta AQB + \text{Area } \Delta BRC = \text{Area } \Delta APC$$

13 Try to Solve

6. $t = 1\frac{1}{2}\text{ h} = \frac{3}{2}\text{ h}$

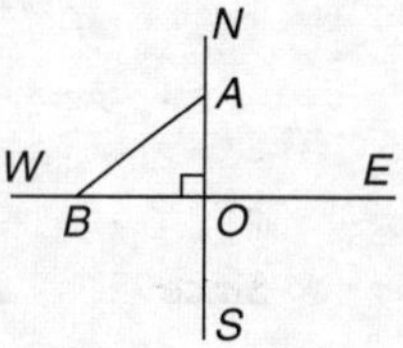

$$OA = 1000 \times \frac{3}{2} = 1500 \text{ km}$$

$$OB = 1200 \times \frac{3}{2} = 1800 \text{ km}$$

$$AB^2 = (1500)^2 + (1800)^2 = 2250000 + 3240000$$

$$AB^2 = 5490000, \ AB = \sqrt{5490000}$$

$$AB = 300\sqrt{61} \text{ km.}$$

Multiple Choice Questions

3. (c) Since we know that the diagonals of a rhombus bisect each other at right angles.

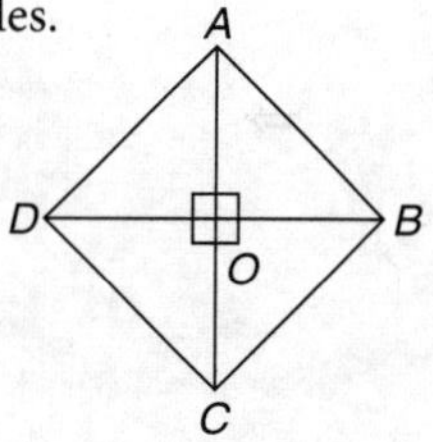

$AC = 24$ cm

So $OA = 12$ cm, and $OC = 12$ cm

$BD = 32$ cm

$OB = OD = 16$ cm

In right ΔAOB, $AB^2 = OA^2 + OB^2$ [Pythagoras' Theorem]

$= (12)^2 + (16)^2 = 144 + 256$

$$AB^2 = 400$$
$$AB = 20 \text{ cm}$$

Perimeter of $ABCD = 4 \times 20 = 80$ cm

Hence, (c) is the correct option.

Subjective Questions

34. No, as their corresponding angles are not equal.

41. (ii) Since quad. $ABCD \sim$ quad. $PQRS$

$$\frac{AB}{PQ} = \frac{BC}{QR} = \frac{CD}{RS} = \frac{AD}{PS}$$

$$\Rightarrow \quad \frac{16}{x} = \frac{5}{y} = \frac{\frac{50}{3}}{z} = \frac{20}{7}$$

On comparing, we get $x = \frac{28}{5}, y = \frac{7}{4}, z = \frac{35}{6}$

42. Proof:

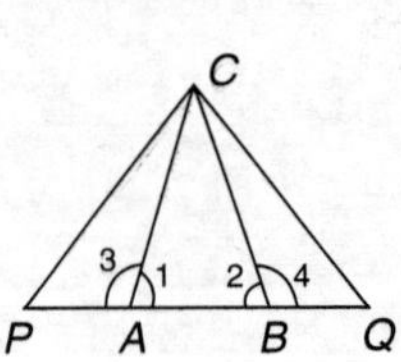

$\angle 1 + \angle 3 = \angle 2 + \angle 4$ [∵ Linear pair axiom]

$AC = BC$ [Given]

$\angle 1 = \angle 2$ [∵ Angle opposite to equal sides are equal] (ii)

∴ $\angle 3 = \angle 4$ [From (i) and (ii)]

Since $AP \times BC = (AC)^2$

$\Rightarrow \quad AP \times BQ = AC \times BC$

$$\Rightarrow \quad \frac{AP}{AC} = \frac{BC}{BQ}$$

$\angle 3 = \angle 4$

∴ $\Delta ACP \sim \Delta BCQ$ (SAS)

43. $\Delta AMB \sim \Delta CMD$ [Given]

$$\Rightarrow \quad \frac{BM}{MD} = \frac{AM}{CM}$$ [∵ If two Δs are similar, then their corresponding sides are in proportion]

$$\Rightarrow \quad \frac{x}{y} = \frac{MD}{z}$$

$$MD = \frac{xz}{y}$$

45. Since $\angle 1 = \angle 2$ [Given]

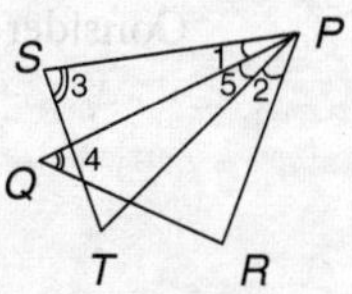

So $\angle 1 + \angle 5 = \angle 2 + \angle 5$

$\angle SPT = \angle RPQ$ (i)

Consider ΔPST and ΔPQR

$\angle SPT = \angle RPQ$ [From (i)]

$\angle PST = \angle PQR$ [Given]

$\Delta PST \sim \Delta PQR$ (AA)

$\frac{ST}{QR} = \frac{PT}{PR}$ [∵ If two Δs are similar, then their corresponding sides are in proportion]

$ST \times PR = PT \times QR$

49. $\angle 1 = \angle 2$ [∵ Opposite sides of || are equal]

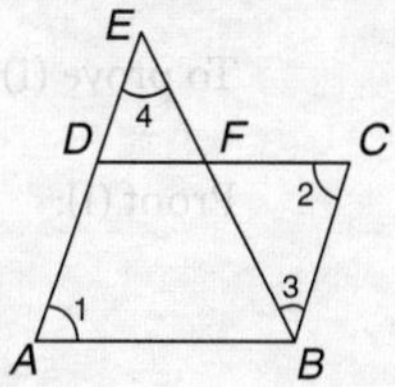

$\angle 3 = \angle 4$ [∵ Alternate interior angles]

$\Delta ABE \sim \Delta CFB$ (AA)

60. Firstly, make $\Delta LMN \sim \Delta QNP$ (AA)

$\frac{\text{Area } \Delta LMN}{\text{Area } \Delta QNP} = \frac{MN^2}{NP^2}$ $\left[\because NP = \frac{2}{3}MN\right]$

$= \frac{MN^2}{\left(\frac{2}{3}MN\right)^2} = \frac{9}{4}$

62. Firstly, prove $\Delta MDE \cong \Delta MCB$ (ASA)

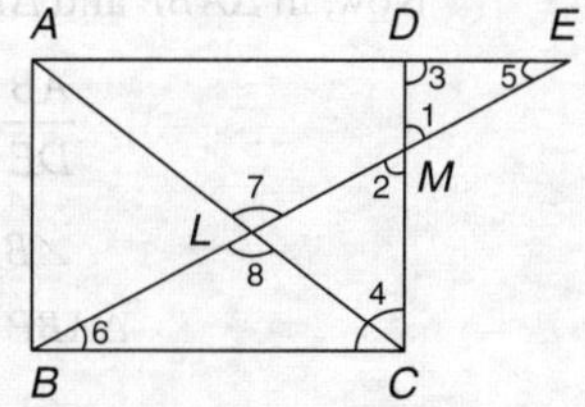

$DE = BC$ [CPCT]

But $AD = BC$ [∵ Opposite sides of || are equal]

∴ $AD = DE$ and $AE = 2AD$

$AE = 2BC$ [∵ $AD = BC$] (i)

Now $\Delta BLC \sim \Delta ELA$ (AA)

$\frac{BL}{EL} = \frac{BC}{EA}$ [∵ If two Δs are similar, then their corresponding sides are in proportion]

$\frac{BL}{EL} = \frac{BC}{2BC}$ [From (i)]

$EL = 2BL$

63. In ΔCBA, $PQ \parallel BA$

$\frac{CP}{PB} = \frac{CQ}{OA}$ [∵ BPT] (i)

Consider ΔCBD, $PR \parallel BD$

$$\frac{CP}{PB} = \frac{CR}{RD} \qquad [\because \text{BPT}] \qquad \text{(ii)}$$

From (i) and (ii) $\dfrac{CQ}{QA} = \dfrac{CR}{RD}$ (iii)

Consider ΔCAD, $\dfrac{CQ}{QA} = \dfrac{CR}{RD}$ [From (iii)]

$QR \parallel AD$ [Converse of BPT]

65. (i) Given ΔABC and ΔDEF in which $\angle A = \angle D$ and $\angle B = \angle E$ $\angle C = \angle F$ in which AP and DQ are medians, that is, $BP = PC$ and $EQ = QF$.

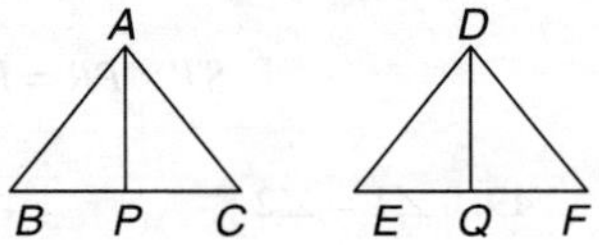

To prove (i): $\dfrac{BC}{EF} = \dfrac{AP}{DQ}$

Proof (i): $\Delta ABC \sim \Delta DEF$ (AAA)

$\dfrac{AB}{DE} = \dfrac{BC}{EF}$ [∵ If two Δs are similar, then their corresponding sides are in proportion]

So $\dfrac{AB}{DE} = \dfrac{2BP}{2EQ}$ [∵ P and Q are the mid-points of BC and EF, respectively]

$\Rightarrow$ $\dfrac{AB}{DE} = \dfrac{BP}{EQ}$ (ii)

Now, in ΔABP and ΔDEQ

$\dfrac{AB}{DE} = \dfrac{BP}{EQ}$ [From (i)]

$\angle B = \angle E$ [Given]

$\Delta ABP \sim \Delta DEQ$ (SAS)

$\dfrac{AE}{AD} = \dfrac{EM}{DN}$ [∵ If two Δs are similar, then their corresponding sides are in proportion]

But $\dfrac{AB}{DE} = \dfrac{BC}{EF}$ [From (i)]

$\Rightarrow$ $\dfrac{BC}{EF} = \dfrac{AP}{DQ}$

Similarly, you can prove other parts: (ii) AP and DQ are bisectors of $\angle BAC$ and $\angle EDF$, (iii) AP and DQ are altitudes, that is, $AP \perp BC$ and $DQ \perp EF$.

67. Given: Trapezium $ABCD$, $AB \parallel CD$. Diagonals AC and BD intersect each other at 'O' such that $\frac{DQ}{QB} = \frac{1}{2}$

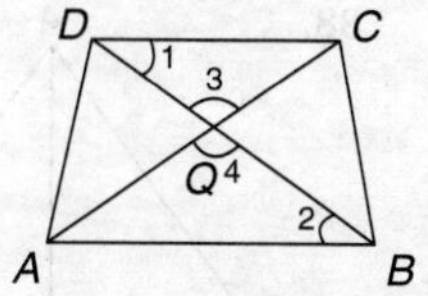

To prove: $AB = 2CD$

Proof: Consider ΔDQC and ΔBQA

$\angle 1 = \angle 2$ [∵ $DC \parallel AB$ and AB are transversal, alternate interior angles]

$\angle 3 = \angle 4$ [V.O. angles]

$\Delta DQC \sim \Delta BQA$ (AA)

$$\frac{DQ}{BQ} = \frac{DC}{BA}$$

$$\frac{1}{2} = \frac{DC}{BA} \qquad \left[\because \frac{DQ}{QB} = \frac{1}{2} \text{ is given}\right]$$

$$BA = 2\,DC$$

68. Given: ΔABC in which D is the mid-point of BC, that is, $BD = DC$, and E is the mid-point of AD, that is, $AE = ED$.

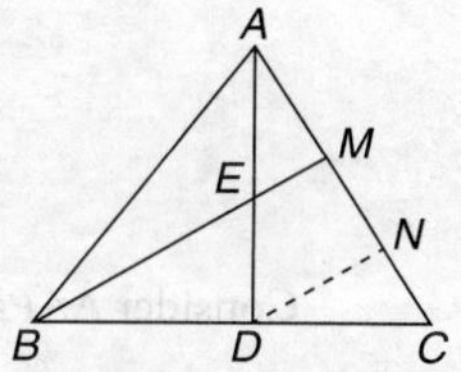

To prove: $BE = 3EM$

Construction: Draw $DN \parallel BM$

Proof: In ΔADN, $EM \parallel DN$

$\Delta AEM \sim \Delta ADN$ (AA)

$$\frac{AE}{AD} = \frac{EM}{DN}$$ [∵ If two Δs are similar, then their corresponding sides are in proportion]

$$\frac{AE}{2AE} = \frac{EM}{DN}$$ [∵ E is the mid-point of AD]

$$DN = 2EM \qquad \text{(i)}$$

Similarly, $\Delta CDN \sim \Delta CBM$

$$\frac{CD}{CB} = \frac{DN}{BM}$$

$$\frac{CD}{2CD} = \frac{DN}{BM}$$

$$BM = 2DN \qquad \text{(ii)}$$

From (i) and (ii)

$$BM = 2\,(2\,EM)$$

$$BM = 4\,EM$$

$$BE + EM = 4EM$$

$$BE = 3\,EM$$

88.

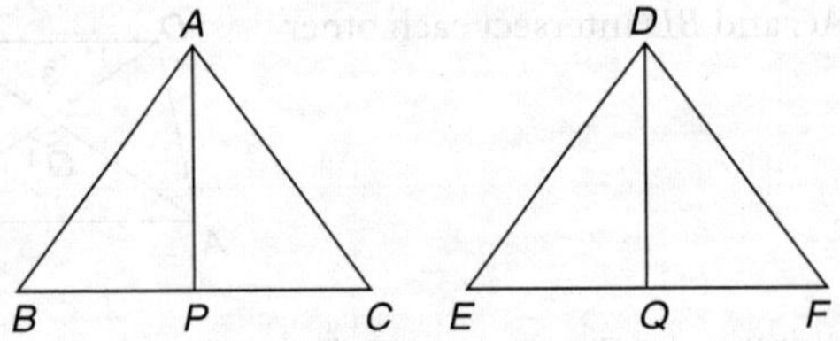

Given: $\Delta ABC \sim \Delta DEF$. AP and DQ are medians of sides BC and EF, respectively. $BP = PC$ and $EQ = QF$.

To prove: $\dfrac{\text{Area } \Delta ABC}{\text{Area } \Delta DEF} = \dfrac{AP^2}{DQ^2}$

Proof: $\Delta ABC \sim \Delta DEF$ [Given]

$\dfrac{AB}{DE} = \dfrac{BC}{EF}$ [∵ If two Δs are similar, then their corresponding sides are in proportion]

$\dfrac{AB}{DE} = \dfrac{2BP}{2EQ}$ [∵ P and Q are the mid-points]

$\dfrac{AB}{DE} = \dfrac{BP}{EQ}$ (i)

Consider ΔAPB and ΔDQE $\dfrac{AB}{DE} = \dfrac{BP}{EQ}$ [From (i)]

$\angle B = \angle E$ [∵ If two Δs are similar, then their corresponding angles are equal]

$\Delta APB \sim \Delta DQE$ [SAS]

$\dfrac{AP}{DQ} = \dfrac{AB}{DE}$ (ii)

$\dfrac{\text{Area } \Delta ABC}{\text{Area } \Delta DEF} = \dfrac{AB^2}{DE^2}$

$= \dfrac{AP^2}{DQ^2}$ [From (ii)]

99. Given: ΔABC in which D, E and F are the mid-points of sides BC, CA and AB, respectively.

To prove: $3(AB^2 + BC^2 + CA^2) = 4(AD^2 + BE^2 + CF^2)$

Construction: $AG \perp BC$

Proof: Using Pythagoras' Theorem, we get

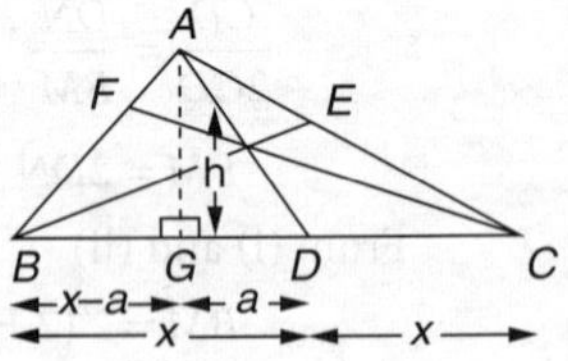

$AB^2 + AC^2 = 2AD^2 + \dfrac{1}{2} BC^2$ (i)

$BC^2 + CA^2 = 2CF^2 + \dfrac{1}{2} AB^2$ (ii)

$AB^2 + BC^2 = 2BE^2 + \dfrac{1}{2} AC^2$ (iii)

Adding (i), (ii) and (iii), we get

$$2\,(AB^2 + BC^2 + CA^2) = 2\left(AD^2 + CF^2 + BE^2 \right) + \frac{1}{2}\left(AB^2 + BC^2 + CA^2 \right)$$

On simplification, we get

$$3\,(AB^2 + BC^2 + CA^2) = 4(AD^2 + BE^2 + CF^2)$$

102. $\dfrac{P_1}{P_2} = \dfrac{AB}{LM}$

$\dfrac{60}{48} = \dfrac{AB}{8}$

On solving, we get $AB = 10$ cm.

103. Since $BD = DC$

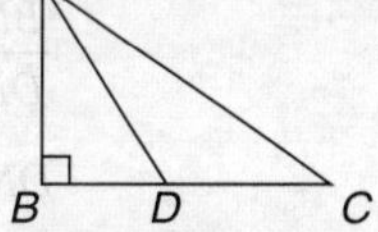

$BD = \dfrac{1}{2}BC$ (i)

In ΔABC $AC^2 = AB^2 + BC^2$ [∵ Pythagoras' theorem]

$BC^2 = AC^2 + AB^2$ (ii)

In ΔABD $AD^2 = AB^2 + BD^2$ [Pythagoras' theorem]

$$AD^2 = AB^2 + \left(\frac{BC}{2}\right)^2$$

$$4AD^2 = 4AB^2 + BC^2$$

$$4AD^2 = 4AB^2 + AC^2 - AB^2$$

$$AC^2 = 4AD^2 - 3AB^2$$

104. $q - p = 2,\ p = q - 2$

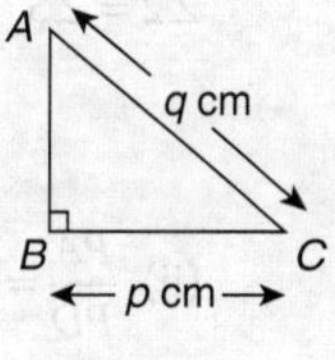

$$AC^2 = AB^2 + BC^2$$

$$AB^2 = AC^2 - BC^2$$

$$= q^2 - p^2 = q^2 - (q-2)^2 = q^2 - (q^2 + 4 - 4q)$$

$$AB^2 = 4q - 4 = 4(q-1)$$

$$AB = 2\sqrt{q-1}\ \text{cm}$$

117. In ΔBRD, $BR \parallel PQ$

$\therefore \dfrac{BD}{PD} = \dfrac{RD}{QD}$ [∵ BPT] (i)

Consider ΔRDP, $PR \parallel QC$

$\therefore \dfrac{RD}{QD} = \dfrac{PD}{CD}$ [∵ BPT] (ii)

From (i) and (ii), we get $\frac{PD}{CD} = \frac{BD}{PD}$

$\Rightarrow \quad BD \times CD = PD \times PD = 12 \times 12 = 144 \text{ cm}^2$

121. (i) $\Delta DMC \sim \Delta BMD$ [Perpendicular drawn from the vertex of a right Δ divides into two Δs which are similar to each other]

$\frac{DM}{BM} = \frac{MC}{MD}$ [∵ If two Δs are similar then their corresponding sides are in proportion]

$DM^2 = CM \times BM$

$DM^2 = CM \times DN$ [∵ $BM = DN$]

(ii) Similarly, $\Delta BND \sim \Delta DNA$

$\frac{BN}{DN} = \frac{ND}{NA}$

$DN^2 = AN \times BN$

$DN^2 = AN \times DM$ [∵ $BN = DM$]

122. To prove $AC^2 = AB^2 + BC^2 + 2BC.BD$

Proof: In right ΔADC, $AC^2 = AD^2 + DC^2$ [∵ Pythagoras' theorem]

$= AB^2 - DB^2 + (DB + BC)^2$

$= AB^2 + BC^2 + 2DB \times BC$

127. (i) Consider ΔPAC and ΔPDB

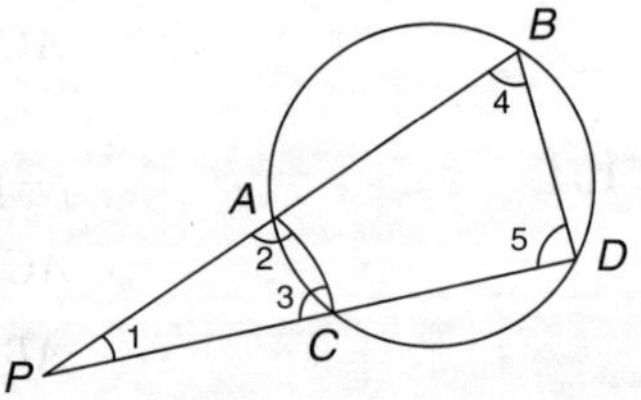

$\angle 1 = \angle 1$ [Common]

$\angle 2 = \angle 5$ [∵ Exterior angle of a cyclic quadrilateral is equal to interior opposite angle]

$\Delta PAC \sim \Delta PDB$ [∵ AA]

(ii) $\frac{PA}{PD} = \frac{PC}{PB}$ [∵ If two Δs are similar, then their corresponding sides are in proportion]

$PA \cdot PB = PC \cdot PD$

129. In right ΔADC $b^2 = h^2 + (a - x)^2$

$b^2 = h^2 + a^2 + x^2 - 2ax$

133. In right ΔABC, $AC^2 = AB^2 + BC^2$ [∵ Pythagoras' Theorem]

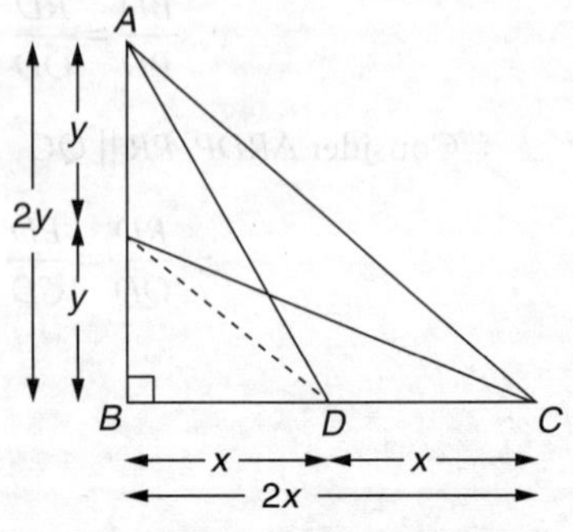

$(5)^2 = (2y)^2 + (2x)^2$

$25 = 4y^2 + 4x^2$

$\frac{25}{4} = x^2 + y^2 \quad$ (i)

In right ΔABD, $\left(\frac{3\sqrt{5}}{2}\right)^2 = x^2 + (2y)^2$

$$\frac{45}{4} = x^2 + 4y^2 \qquad \text{(ii)}$$

Subtract (i) from (ii) $\frac{20}{4} = 3y^2$

$$y^2 = \frac{5}{3} \qquad \text{(iii)}$$

Insert (iii) in (i) $\frac{25}{4} = x^2 + \frac{5}{3}$

$$x^2 = \frac{25}{4} - \frac{5}{3} = \frac{75-20}{12} = \frac{55}{12}$$

In right ΔEBC $\quad EC^2 = EB^2 + BC^2$

$$= y^2 + (2x)^2 = y^2 + 4x^2 = \frac{5}{3} + \cancel{4} \times \frac{55}{\cancel{12}_{3}} = \frac{5+55}{3}$$

$$EC^2 = \frac{60}{3} = 20$$

$$EC = \sqrt{20} = 2\sqrt{5} \text{ cm}$$

134. $QS = \frac{1}{3}QR$

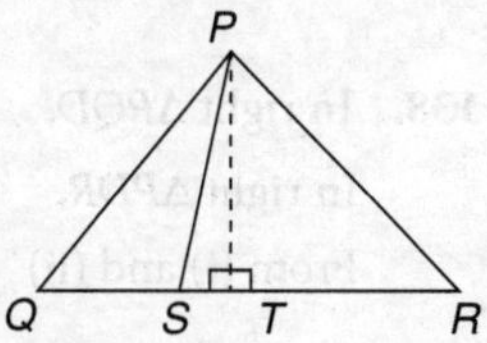

$SR = \frac{2}{3}QR$

$QT = TR$

$QT = \frac{1}{2}QR$ and $TR = \frac{1}{2}QR$; $QP = QR = PR$

In right ΔPQT, $PQ^2 = PT^2 + QT^2$

$$= PT^2 + \left(QS + ST\right)^2 = PT^2 + QS^2 + ST^2 + 2QS \cdot ST$$

$$= PS^2 + \left(\frac{1}{3}QR\right)^2 + 2\times\frac{1}{3}QR\times\left[SR - TR\right] = PS^2 + \frac{QR^2}{9} + \frac{2}{3}QR\left(\frac{2}{3}QR - \frac{1}{2}QR\right)$$

$$= PS^2 + \frac{QR^2}{9} + \frac{2}{3}QR\left(\frac{QR}{6}\right)$$

$$PQ^2 = PS^2 + \frac{QR^2}{9} + \frac{QR^2}{9}$$

$$PQ^2 = \frac{9PS^2 + PQ^2 + PQ^2}{9} \quad [\because PQ = QR \text{ given}]$$

$$7\,PQ^2 = 9\,PS^2$$

135. We have to use the theorem: If a perpendicular is drawn from the vertex of the right angle of a right triangle, then it divides into two triangles which are similar to each other and similar to the big triangles.

So $\Delta ADB \sim \Delta ABC$

$$\frac{AD}{AB} = \frac{AB}{AC}$$

$$\frac{4}{x} = \frac{x}{9}$$

$$x^2 = 36$$

$$x = 6$$

$$\Delta BDC \sim \Delta ADB$$

$$\frac{BD}{AD} = \frac{DC}{DB}$$

$$\frac{y}{4} = \frac{5}{y}$$

$$y^2 = 20$$

$$y = 2\sqrt{5}$$

138. In right ΔPQD, $PD^2 = a^2 - c^2$ (i)

In right ΔPDR, $PD^2 = b^2 - d^2$ (ii)

From (i) and (ii) $a^2 - c^2 = b^2 - d^2$

$$a^2 - b^2 = c^2 - d^2$$

$$(a - b)(a + b) = (c - d)(c + d)$$

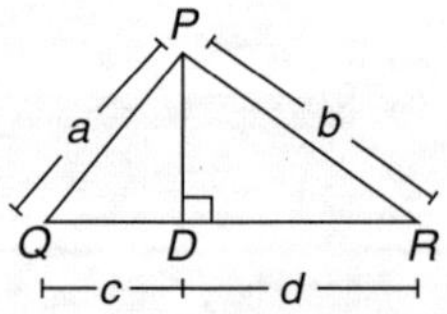

Chapter 7: Coordinate Geometry

Multiple Choice Questions

1. (d) Given: Centre of circle is O (0, 0) and circle passes through the point $\left(\frac{13}{2}, 0\right)$.

Radius of circle = Distance between (0, 0) and $\left(\frac{13}{2}, 0\right)$

$$= \sqrt{\left(\frac{13}{2} - 0\right)^2 + (0-0)^2} = \frac{13}{2} = 6.5$$

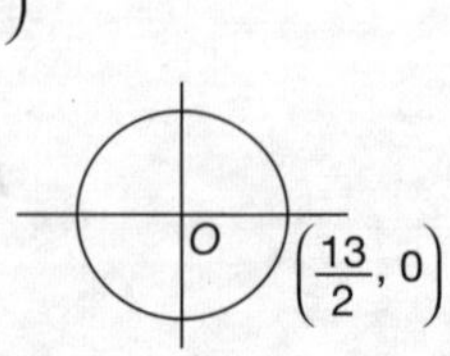

Any point in the interior of the circle will have radius less than 6.5.

(a) Distance between (0, 0) and $\left(\frac{-3}{4}, 1\right)$

$$= \sqrt{\left(\frac{-3}{4} - 0\right)^2 + (1-0)^2} = \sqrt{\frac{9}{16} + 1} = \sqrt{\frac{25}{16}} = \frac{5}{4} = 1.25, \text{ which is less than 6.5.}$$

(b) Distance between (0, 0) and $\left(2, \frac{7}{3}\right)$

$$= \sqrt{(2-0)^2 + \left(\frac{7}{3} - 0\right)^2} = \sqrt{4 + \frac{49}{9}} = \sqrt{\frac{36+49}{9}} = \sqrt{\frac{85}{9}} = \frac{9.22}{3}$$

= 3.1 (approx.), which is less than 6.5.

(c) Distance between (0, 0) and $\left(5, \frac{-1}{2}\right) = \sqrt{(5-0)^2 + \left(\frac{-1}{2} - 0\right)^2}$

$$= \sqrt{25 + \frac{1}{4}} = \sqrt{\frac{101}{2}} = \frac{10.04}{2} = 5.02, \text{ which is less than 6.5.}$$

(d) Distance between (0, 0) and $\left(-6, \frac{5}{2}\right)$

$$= \sqrt{(-6-0)^2 + \left(\frac{5}{2} - 0\right)^2} = \sqrt{36 + \frac{25}{4}} = \frac{13}{4} = 6.5, \text{ which is equal to 6.5.}$$

Hence, the point $\left(-6, \frac{5}{2}\right)$ lies on the circle.

2. (b) Let A (–4, 0), B (4, 0) and C (0, 3) be the given vertices.

$$AB = \sqrt{(4+4)^2 + (0-0)^2} = \sqrt{(8)^2} = 8$$

$$BC = \sqrt{(0-4)^2 + (3-0)^2} = \sqrt{16+9} = \sqrt{25} = 5$$

$$AC = \sqrt{(0+4)^2 + (3-0)^2} = \sqrt{16+9} = \sqrt{25} = 5$$

As $BC = AC$, the given points are the vertices of an isosceles triangle.

3. (b) Let A (4, p) and B (1, 0) be the given points.

$$\text{AB} = \sqrt{(1-4)^2 + (0-p)^2} = 5 \text{ [Given]}$$

$$\Rightarrow \sqrt{(-3)^2 + (-p)^2} = 5$$

$\Rightarrow \quad \sqrt{9+p^2} = 5$

$\Rightarrow \quad 9 + p^2 = 25$ [On squaring both sides]

$\Rightarrow \quad p^2 = 16$

$\Rightarrow \quad p = \pm 4$

4. (b) Let O (0, 0), P (a, b), Q (c, d) be the given points and $\angle POQ$ be the right angle (given). OP is perpendicular to OQ. As the two lines are perpendicular, the product of their slopes is –1.

That is, Slope of $OP \times$ Slope of $OQ = -1$

$\Rightarrow \left(\dfrac{b-0}{a-0}\right)\times\left(\dfrac{d-0}{c-0}\right) = -1$ [Slope of a line joining points (x_1, y_1) and $(x_2, y_2) = \dfrac{y_2 - y_1}{x_2 - x_1}$]

$\Rightarrow \quad \dfrac{b}{a}\times\dfrac{d}{c} = -1$

$\Rightarrow \quad \dfrac{bd}{ac} = -1 \Rightarrow bd = -ac$

$\Rightarrow \quad ac + bd = 0.$

5. (c) A (0, 3), O (0, 0) and B (5, 0) are the given points.

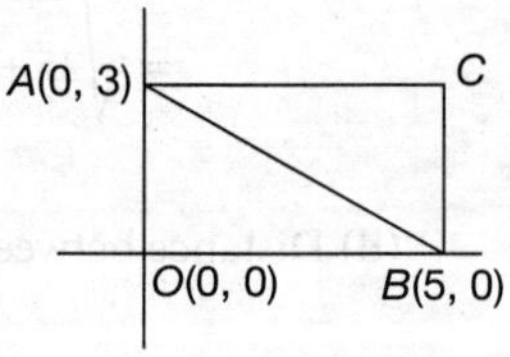

As the lengths of both the diagonals of a rectangle are the same, $AB = OC$.

So $\quad AB = \sqrt{(5-0)^2 + (0-3)^2} = \sqrt{25+9} = \sqrt{34}$ units.

6. (b) Let A (0, 4), B (0, 0) and C (3, 0) be the given points.

$$AB = \sqrt{(0-0)^2 + (0-4)^2} = \sqrt{0+16} = \sqrt{16} = 4$$

$$BC = \sqrt{(3-0)^2 + (0-0)^2} = \sqrt{9+0} = 3$$

$$AC = \sqrt{(3-0)^2 + (0-4)^2} = \sqrt{9+16} = \sqrt{25} = 5$$

Perimeter of triangle $ABC = AB + BC + AC = 4 + 3 + 5 = 12$ units

7. (b) Let A (0, 0), B (2.7, 0) and C (0, 4.9) be the given vertices of the triangle.

$$AB = \sqrt{(2.7-0)^2 + (0-0)^2} = \sqrt{(2.7)^2 + 0} = 2.7$$

$$BC = \sqrt{(0-2.7)^2 + (4.9-0)^2} = \sqrt{7.29 + 24.01} = \sqrt{31.3} = 5.59 \text{ (approx.)}$$

$$AC = \sqrt{(0-0)^2 + (4.9-0)^2} = \sqrt{(4.9)^2} = 4.9$$

As $AB^2 + AC^2 = BC^2$, it is a right-angled triangle.

8. (d) The point (h, k) will be at a height h from the x axis and height h' from the y axis.
So, distance of (h, k) from the x axis = height of the point from the x axis = $|k|$ units.

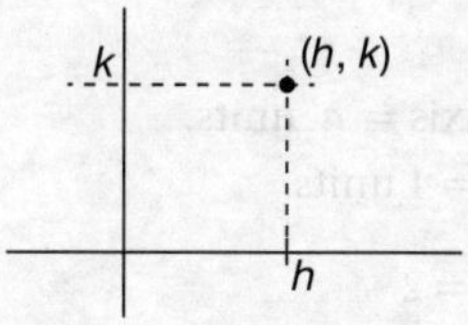

9. (c) There is no such point. The x axis is 4 units from (2, 4), so there are no points on the x axis that are 2 units from (2, 4).

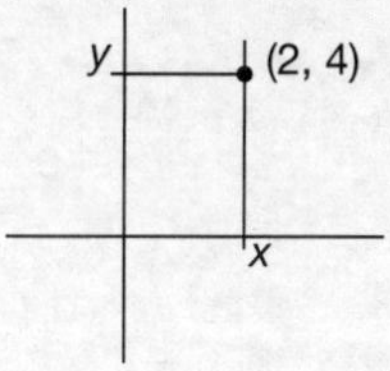

10. (d) Let $A(-1, -6)$, $B(2, -5)$ and $C(7, 2)$ be the given vertices of a parallelogram.

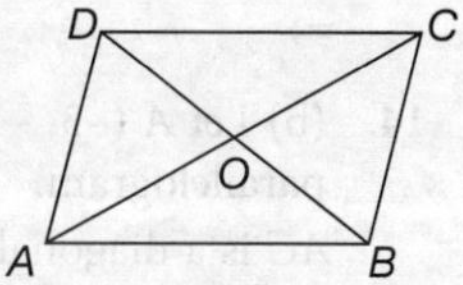

We know that the diagonals of a parallelogram bisect each other. So, the coordinates of the mid-point of diagonal AC are the same as the coordinates of the mid-point of the diagonal BD.

$$O(x, y) = \left(\frac{-1+7}{2}, \frac{-6+2}{2}\right)$$

$$\Rightarrow \quad (x, y) = \left(\frac{6}{2}, \frac{-4}{2}\right) = (3, -2)$$

Let the coordinates of point D be (x_1, y_1).

$$\Rightarrow \quad (x, y) = \left(\frac{2+x_1}{2}, \frac{-5+y_1}{2}\right)$$

$$\Rightarrow \quad (3, -2) = \left(\frac{2+x_1}{2}, \frac{-5+y_1}{2}\right)$$

$$\Rightarrow \quad \frac{2+x_1}{2} = 3; \quad \frac{-5+y_1}{2} = -2$$

$$\Rightarrow \quad 2 + x_1 = 6; \ -5 + y_1 = -4$$

$$\Rightarrow \quad x_1 = 4; \ y_1 = 1$$

$\therefore$ $D(4, 1)$ is the fourth vertex of the parallelogram.

11. (d) Let $A\left(0, \frac{8}{3}\right)$, $B(1, 3)$ $C(82, 30)$ be the given points.

$$\text{Area of } \Delta ABC = \frac{1}{2}\left[x_1(y_2 - y_3) + x_2(y_3 - y_1) + x_3(y_1 - y_2)\right]$$

$$= \frac{1}{2}\left[0(3-30) + 1\left(30 - \frac{8}{3}\right) + 82\left(\frac{8}{3} - 3\right)\right] = \frac{1}{2}\left[0 + 30 - \frac{8}{3} + \frac{656}{3} - 246\right]$$

$$= \frac{1}{6}\times(746-746) = 0$$

∴ Points A, B, C are collinear.

12. (c) Distance of the point $(4, a)$ from the x axis = 'a' units.
Distance of the point $(4, a)$ from the y axis = 4 units.

According to the question, $a = \frac{1}{2}(4) \Rightarrow a = 2$

13. (c) Let $A\ (8, p)$ and $B\ (4, 3)$ be the given points.

$$AB = 5 \qquad \text{[Given]}$$

$$\Rightarrow \sqrt{(4-8)^2+(3-p)^2} = 5$$

$$\Rightarrow (-4)^2 + 9 + p^2 - 6p = 25 \qquad \text{[On squaring both sides]}$$

$$\Rightarrow 16 + 9 + p^2 - 6p = 25$$

$$\Rightarrow 25 + p^2 - 6p = 25$$

$$\Rightarrow p^2 - 6p = 0$$

$$\Rightarrow p\ (p-6) = 0 \Rightarrow p = 0, 6$$

14. (b) Let $A\ (-3, -1)$, $B\ (9, b)$, $C\ (3, 3)$ and $D\ (4, 3)$ be the given vertices of a parallelogram.

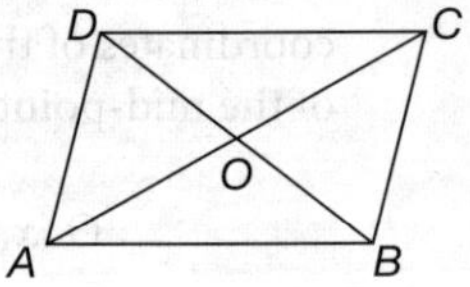

AC is a diagonal and BD is a diagonal (taken in order). We know that the diagonals bisect each other in a parallelogram.

Let $O\ (x, y)$ be the intersection point of diagonal $O\ (x, y) = \left(\frac{-3+3}{2}, \frac{-1+3}{2}\right)$ (with diagonal AC).

$$\Rightarrow (x, y) = (0, 1)$$

Now, with diagonal BD $\left(\frac{a+4}{2}, \frac{b+3}{2}\right) = (0, 1)$

$$\Rightarrow \frac{a+4}{2} = 0; \frac{b+3}{2} = 1$$

$$\Rightarrow a = -4; b = 2 - 3 = -1$$

∴ $D\ (-4, -1)$ and $a : b = 4 : 1$.

15. (c) We know that the vertices of a parallelogram bisect each other.
∴ Let $O\ (x, y)$ be the intersection point of both diagonals.

∴ $O\ (x, y)$ = Mid-point of PR = Mid-point of QS

$$(x, y) = \left(\frac{1+5}{2}, \frac{2+7}{2}\right) = \left(\frac{6}{2}, \frac{9}{2}\right) = \left(3, \frac{9}{2}\right)$$ and (x, y) = Mid-point of QS.

$$\Rightarrow \left(3, \frac{9}{2}\right) = \left(\frac{4+a}{2}, \frac{6+b}{2}\right)$$

$$\Rightarrow \frac{4+a}{2} = 3; \frac{6+b}{2} = \frac{9}{2}$$

$\Rightarrow \quad 4 + a = 6; 6 + b = 9$

$\Rightarrow \quad a = 2; b = 3$

16. (a) The mid-point of the hypotenuse is equidistant from the vertices of a right-angled triangle.

$\therefore$ Required point $= \left(\frac{0+2x}{2}, \frac{0+2y}{2}\right) = \left(\frac{2x}{2}, \frac{2y}{2}\right) = (x, y)$

17. (b) $P\left(\frac{a}{3}, 4\right)$ = Mid-point of O (–6, 5) and R (–2, 3).

$$\Rightarrow \left(\frac{a}{3}, 4\right) = \left(\frac{-6-2}{2}, \frac{5+3}{2}\right)$$

$$\Rightarrow \left(\frac{a}{3}, 4\right) = \left(\frac{-8}{2}, \frac{8}{2}\right)$$

$$\Rightarrow \left(\frac{a}{3}, 4\right) = (-4, 4)$$

$$\Rightarrow \frac{a}{3} = -4 = -12$$

18. (c) Let A (1, 2), B (4, 3), C (6, 6) be the three vertices of a parallelogram $ABCD$ and let D (x, y) be the fourth vertex.

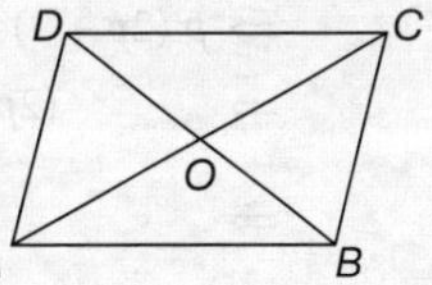

AC is the diagonal and BD is the diagonal. We know that the diagonals bisect each other in a parallelogram.

Let O (x, y) be the intersection point of diagonal O (x, y) = $\left(\frac{6+1}{2}, \frac{6+2}{2}\right)$ (with diagonal AC)

$$\Rightarrow (x, y) = \left(\frac{7}{2}, 4\right)$$

Now, with diagonal BD $\left(\frac{4+x_1}{2}, \frac{3+y_1}{2}\right) = \left(\frac{7}{2}, 4\right)$

$$\frac{4+x_1}{2} = \frac{7}{2}; \frac{3+y_1}{2} = 4$$

$$\Rightarrow x_1 = 3; y_1 = 5$$

D (3, 5) is the required point.

19. (b) Let A (0, 0), B (3, 0) and C (0, 4) be the vertices of a triangle.

Centroid of the triangle $= \left(\frac{x_1+x_2+x_3}{3}, \frac{y_1+y_2+y_3}{3}\right) = \left(\frac{0+3+0}{3}, \frac{0+0+4}{3}\right) = \left(1, \frac{4}{3}\right)$

20. (c) ar (ΔABC) $= \frac{1}{2}\left[x_1(y_2 - y_3) + x_2(y_3 - y_1) + x_3(y_1 - y_2)\right]$

A (1, 3), B (–1, 0), C (4, 0)

$$= \frac{1}{2}\left[1(0-0) + (-1)(0-3) + (4)(3-0)\right] = \frac{1}{2}[0+3+12] = \frac{15}{2} = 7.5 \text{ sq. units.}$$

21. (a) Let $A\ (p, 2-2p)$, $B\ (1-p, 2p)$ and $C\ (-4-p, 6-2p)$ be the vertices of a triangle ABC.

Here $x_1 = p, x_2 = 1 - p, x_3 = -4 - p$

$y_1 = 2 - 2p, y_2 = 2p, y_3 = 6 - 2p$

$$\text{ar}\ (\Delta ABC) = \frac{1}{2}\left[x_1(y_2 - y_3) + x_2(y_3 - y_1) + x_3(y_1 - y_2)\right]$$

$$70 = \frac{1}{2}\left[p(2p-6+2p) + (1-p)(6-2p-2+2p) + (-4-p)(2-2p-2p)\right]$$

$$70 = \frac{1}{2}\left[p(4p-6) + (1-p)(4) + (-4-p)(2-4p)\right]$$

$$70 = \frac{1}{2}\left[4p^2 - 6p + 4 - 4p - 8 + 16p - 2p + 4p^2\right]$$

$$70 = \frac{1}{2}\left[8p^2 + 4p - 4\right] = \frac{4}{2}\left[2p^2 + p - 1\right]$$

$$35 = 2p^2 + p - 1$$

$\Rightarrow \quad 2p^2 + p - 36 = 0$

$\Rightarrow \quad 2p^2 + 9p - 8p - 36 = 0$

$\Rightarrow \quad p\ (2p + 9) - 4\ (2p + 9) = 0$

$\Rightarrow \quad (2p + 9)(p - 4) = 0$

$\Rightarrow \quad p = \frac{-9}{3}, 4$

23. (b) $\text{Area}\ (\Delta ABC) = \frac{1}{2}\left[a(0-b) + 0(b-0) + 0(0-0)\right] = \frac{1}{2}|-ab| = \frac{ab}{2}$

24. (d) Let $A\ (a, b + c)$, $B\ (b, c + a)$ and $C\ (c, a + b)$ be the vertices of the ΔABC.

$$\text{ar}\ (\Delta ABC) = \frac{1}{2}\left[a(c+a-a-b) + b(a+b-b-c) + c(b+c-c-a)\right]$$

$$= \frac{1}{2}\left[a(c-b) + b(a-c) + c(b-a)\right] = \frac{1}{2}\left[ac - ab + ab - bc + bc - ca\right] = \frac{1}{2}(0) = 0$$

25. (c) Let $A\ (x, 0)$, $B\ (1, 1)$ and $C\ (0, 2)$ be the vertices of a triangle with area = 4 sq. units.

$$\text{ar}\ (\Delta ABC) = \frac{1}{2}\left[x_1(y_2 - y_3) + x_2(y_3 - y_1) + x_3(y_1 - y_2)\right]$$

$$4 = \frac{1}{2}\left[x(1-2) + 1(2-0) + 0(0-1)\right] = \frac{1}{2}\left[-x + 2 + 0\right]$$

$\Rightarrow \quad 8 = -x + 2$

$\Rightarrow \quad x = -6$

26. (c) $AC = 2\ BC$

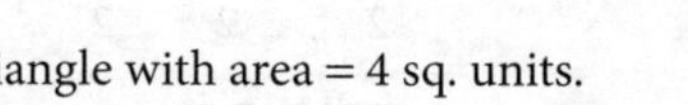

$\Rightarrow \quad AB + BC = 2\ BC$

$\Rightarrow \quad AB = BC$

$\Rightarrow$ Point B is the mid-point of AC.

$$\Rightarrow \left(\frac{x+(-3)}{2}, \frac{y+y}{2}\right) = (2, 1)$$

$$\Rightarrow \quad \frac{-3+x}{2} = 2; \ \frac{y+y}{2} = 1$$

$$= \frac{1}{\sqrt{\left(1+a^2\right)^2}} + \frac{1}{\sqrt{\left(1+\frac{1}{a^2}\right)^2}} = \frac{1}{1+a^2} + \frac{1}{1+\frac{1}{a^2}}$$

$$= \frac{1}{1+a^2} + \frac{a^2}{1+a^2} = \frac{1+a^2}{1+a^2} = 1$$

$$\Rightarrow \quad -3 + x = 4; \ y + 4 = 2$$

$$\Rightarrow \quad x = 7; \ y = -2$$

$\therefore$ Required point is C (7, –2)

27. (d) Let A $(k + 1, 1)$, B $(2k + 1, 3)$ and C $(2k + 2, 2k)$ be the three points.

These three points are collinear if ar (ΔABC) = 0.

$$\therefore \quad \text{ar}(\Delta ABC) = \frac{1}{2}\left[x_1\left(y_2 - y_3\right) + x_2\left(y_3 - y_1\right) + x_3\left(y_1 - y_2\right)\right]$$

$$0 = \frac{1}{2}\left[(k+1)[3-2k] + (2k+1)(2k-1) + (2k+2)(1-3)\right]$$

$$\Rightarrow \quad 0 = [3k - 2k^2 + 3 - 2k + 4k^2 - 1 - 4k - 4]$$

$$\Rightarrow \quad 0 = 2k^2 - 3k - 2$$

$$\Rightarrow \quad 2k^2 - 3k - 2 = 0$$

$$2k^2 - 4k + k - 2 = 0$$

$$\Rightarrow \quad 2k\,(k - 2) + (k - 2) = 0$$

$$(2k + 1)\,(k - 2) = 0$$

$$\Rightarrow \quad k = \frac{-1}{2}, 2$$

28. (b) $\text{ar}(\Delta ABC) = \frac{1}{2}\left[1(6+5) + (-4)(-5+1) + (-3)(-1-6)\right]$

$$= \frac{1}{2}\left[11 + (-4)(-4) + (-3)(-7)\right] = \frac{1}{2}\left[11 + 16 + 21\right]$$

$$= \frac{1}{2} \times 48 = 24 \text{ sq. units}$$

Fill in the Blanks

29. (1). Explanation: $A = (a^2, 2a)$; $B = \left(\frac{1}{a^2}, \frac{-2}{a}\right)$; $S = (1, 0)$

$$\frac{1}{SA}+\frac{1}{SB} = \frac{1}{\sqrt{(1-a^2)^2+(0-2a)^2}}+\frac{1}{\sqrt{\left(1-\frac{1}{a^2}\right)^2+\left(0+\frac{2}{a}\right)^2}}$$

$$= \frac{1}{\sqrt{1+a^4-2a^2+4a^2}}+\frac{1}{\sqrt{1+\frac{1}{a^4}-\frac{2}{a^2}+\frac{4}{a^2}}} = \frac{1}{\sqrt{1+a^4+2a^2}}+\frac{1}{\sqrt{1+\frac{1}{a^4}+\frac{2}{a^2}}}$$

$$= \frac{1}{\sqrt{(1+a^2)^2}}+\frac{1}{\sqrt{\left(1+\frac{1}{a^2}\right)^2}} = \frac{1}{1+a^2}+\frac{1}{1+\frac{1}{a^2}} = \frac{1}{1+a^2}+\frac{a^2}{1+a^2} = \frac{1+a^2}{1+a^2} = 1$$

30. (–4). Explanation: A(7, –2) P(5, –3) Q(3, y) B(1, –5)

Given: $P(5, -3)$ and $Q(3, y)$ are the points of trisection of the line segment $A(7, -2)$ and $B(1, -5)$.

$$AP = PQ = QB$$

$$\Rightarrow \quad PQ = QB$$

∴ Q is the mid-point of PB.

$$\Rightarrow \quad \text{Coordinates of } Q = \left(\frac{5+1}{2}, \frac{-3+(-5)}{2}\right)$$

$$\text{Coordinates of } (3, y) = (3, -4)$$

$$\Rightarrow \quad y = -4$$

31. (3, 1). Explanation: Let $A(1, 2)$, $B(4, 3)$ and $C(6, 6)$ be the given points. Let $D(x, y)$ be the fourth point.

$$\text{Mid-point of } AC = \left(\frac{1+6}{2}, \frac{2+6}{2}\right) = \left(\frac{7}{2}, 4\right)$$

$$\text{Mid point of } BD = \left(\frac{x+4}{2}, \frac{y+3}{2}\right)$$

$$\Rightarrow \quad \left(\frac{7}{2}, 4\right) = \left(\frac{x+y}{2}, \frac{y+3}{2}\right)$$ [∵ Diagonals of a parallelogram bisect each other]

$$\Rightarrow \quad \frac{x+y}{2} = \frac{7}{2}; \quad \frac{y+3}{2} = 4$$

$\Rightarrow \quad x + y = 7; y + 3 = 4$

$\Rightarrow \quad x = 3; y = 1$

$\therefore$ (3, 1) is the fourth point.

32. (–2,5). Explanation: Centroid G (1, 5) = $\left(\frac{x_1+x_2+x_3}{3}, \frac{y_1+y_2+y_3}{3}\right)$

$\Rightarrow \quad (1, 5) = \left(\frac{6-1+x_3}{3}, \frac{3+7+y_3}{3}\right)$

$\Rightarrow \quad \frac{5+x_3}{3} = 1; \frac{10+y_3}{3} = 5$

$\Rightarrow \quad 5 + x_3 = 3; 10 + y_3 = 15$

$\Rightarrow \quad x_3 = -2; y_3 = 5$

$\therefore$ Required point is C (–2, 5)

33. D(3,0). Explanation: $AB = 30$; $AD = 14$.

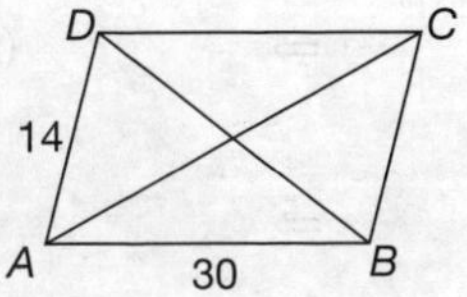

As the opposite sides of a parallelogram are equal, $AB = CD = 30$ m; $AD = BC = 14$ m.

Area of $ABCD$ = Area of (ΔABD) + Area of (ΔBCD)

= Area of (ΔABD) + Area of (ΔABD) [As the length of the sides of both the triangles are the same, their area will be equal]

= 2 Area of (ΔABD)

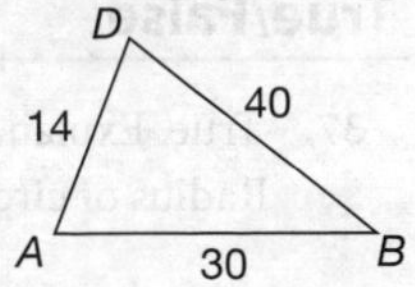

Now, for area of ΔABD

$$s = \frac{a+b+c}{2} = \frac{1}{2}(30+14+40) = 42$$

$$\text{Area } (\Delta ABD) = \sqrt{s(s-a)(s-b)(s-c)} = \sqrt{42(12)(28)(2)} = 168 \text{ sq. m}$$

Area of parallelogram = 2 × 168 = 336 sq. m

34. (3, 0). Explanation: Let $D(a, b)$ be the fourth vertex.

$$\text{Midpoint of } AC = \left(\frac{1+6}{2}, \frac{2+3}{2}\right) = \left(\frac{7}{2}, \frac{5}{2}\right)$$

$$\text{Midpoint of } BD = \left(\frac{a+4}{2}, \frac{b+3}{2}\right)$$

$$\Rightarrow \quad \left(\frac{7}{2}, \frac{5}{2}\right) = \left(\frac{a+4}{2}, \frac{b+5}{2}\right)$$

$$\Rightarrow \quad \frac{a+4}{2} = \frac{7}{2};\ \frac{b+5}{2} = \frac{5}{2}$$

$$\Rightarrow \quad a = 3;\ b = 0$$

$\therefore$ D (3, 0) is the required point.

35. (4, 2). Explanation: Centroid of $\Delta ABC = \left(\frac{x_1+x_2+x_3}{3}, \frac{y_1+y_2+y_3}{3}\right) = \left(\frac{-2+0+14}{3}, \frac{4+0+2}{3}\right)$

$$= \left(\frac{12}{3}, \frac{6}{3}\right) = (4, 2)$$

$\therefore$ Centroid of ΔABC is (4, 2).

36. (3, –10). Explanation: As O (2, –3) is the centre of the circle and B (1, 4) is a point on the circle.

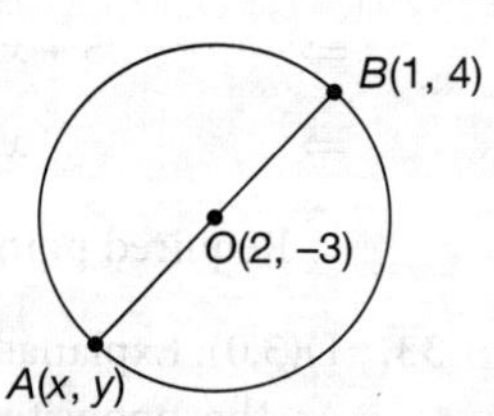

O is the mid-point of AB.

$$\Rightarrow \quad (2, -3) = \left(\frac{x+1}{2}, \frac{y+4}{2}\right)$$

$$\Rightarrow \quad \frac{x+1}{2} = 2;\ \frac{y+4}{2} = -3$$

$$\Rightarrow \quad x = 3;\ y = -10$$

$\therefore$ The coordinates of A are (3, 10).

True/False

37. True. Explanation: Given: Centre of circle O (0, 0), point that lies on the circle P (5, 0).

Radius of circle = Distance between points P (5, 0) and O (0, 0)

$$= \sqrt{(0-5)^2+(0-0)^2} = \sqrt{(5)^2} = 5$$

Now, any point with distance from the centre of the circle more than the radius (5) lies outside that circle.

$\therefore$ Distance between Q (6, 8) and O (0, 0) = $\sqrt{(0-6)^2+(0-8)^2}$

$= \sqrt{36+64} = \sqrt{100} = 10$ units, which is more than 5. $\therefore$ Q (6, 8) lies outside the circle.

38. False. Explanation: Distance between P (–2, 4) and C (3, 5) = $\sqrt{(3+2)^2+(5-4)^2} = \sqrt{25+1}$ $= \sqrt{26} = 5.09$ (approx.), which is less than the radius of the circle, 6 units.

$\therefore$ Point P (–2, 4) lies inside the circle.

39. True. Explanation: Coordinates of centroid = $\left(\frac{x_1+x_2+x_3}{3}, \frac{y_1+y_2+y_3}{3}\right) = \left(\frac{3+5+8}{3}, \frac{2+7+8}{3}\right)$

$$= \left(\frac{16}{3}, \frac{17}{3}\right)$$

$\therefore$ The given statement is true.

40. True. Explanation: Let the coordinates of point $B\,(x_1, y_1)$

As C divides AB in 3 : 4

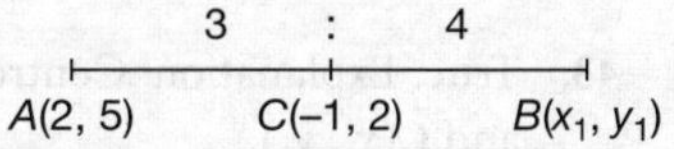

$\therefore$ By section formula, $\left(\dfrac{3x_1 + 4(2)}{3+4}, \dfrac{3y_1 + 4(5)}{3+4}\right) = (-1, 2)$

$$\Rightarrow \quad \left(\frac{3x_1 + 8}{7}, \frac{3y_1 + 20}{7}\right) = (-1, 2)$$

$$\Rightarrow \quad \frac{3x_1 + 8}{7} = -1 \quad \frac{3y_1 + 20}{7} = 2$$

$$\Rightarrow \quad 3x_1 + 8 = -7;\ 3y_1 + 20 = 14$$

$$\Rightarrow \quad 3x_1 = -15;\ 3y_1 = -6$$

$$\Rightarrow \quad x_1 = -5;\ y_1 = -2$$

$\therefore$ $B\,(-5, -2)$ is the required point.

41. False. Explanation: Let AD be the median on BC. As the median divides a line segment in two equal parts

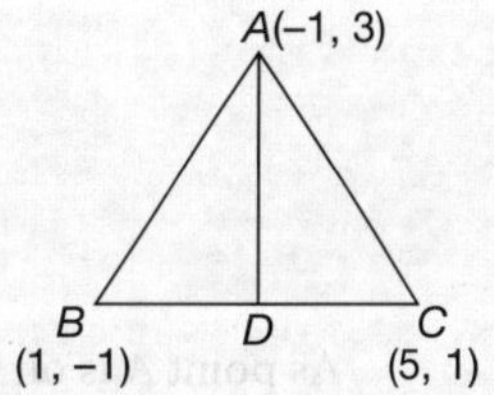

$\Rightarrow \quad BD = DC$

$\Rightarrow$ D is the mid-point of BC.

$$\therefore\ D\,(x, y) = \left(\frac{1+5}{2}, \frac{-1+1}{2}\right) = \left(\frac{6}{2}, 0\right) = (3, 0)$$

$\therefore$ Coordinates of D are (3, 0).

We now need to find the length of median AD.

$$AD = \sqrt{(3+1)^2 + (0-3)^2} = \sqrt{(4)^2 + (3)^2} = \sqrt{16+9} = \sqrt{25} = 5 \text{ units}$$

42. True. Explanation: x coordinate of the point P is twice the y coordinate.

$$\Rightarrow \quad x = 2y$$

$\therefore$ $P\,(2y, y)$ is the point.

Now, P is equidistant from $Q\,(2, -5)$ and $R\,(-3, 6)$.

$$\Rightarrow \quad PQ = PR$$

$$\Rightarrow \quad \sqrt{(2-2y)^2 + (-5-y)^2} = \sqrt{(-3-2y)^2 + (6-y)^2}$$

On squaring both sides, we get $(2 - 2y)^2 + (-5 - y)^2 = (-3 - 2y)^2 + (6 - y)^2$

$$\Rightarrow 4 + \cancel{4y^2} - 8y + 25 + \cancel{y^2} + 10y = 9 + \cancel{4y^2} + \cancel{12y} + 36 + \cancel{y^2} - \cancel{12y}$$

$$\Rightarrow \quad 2y + 29 = 45$$

$$\Rightarrow \quad 2y = 16$$

$$y = 8$$

$\therefore$ $P\,(16, 8)$ is the required point.

43. True. Explanation: Centroid of ΔABC is $\left(\frac{x_1+x_2+x_3}{3}, \frac{y_1+y_2+y_3}{3}\right)$ with points $A(x_1, y_1)$, $B(x_2, y_2)$ and $C(x_3, y_3)$.

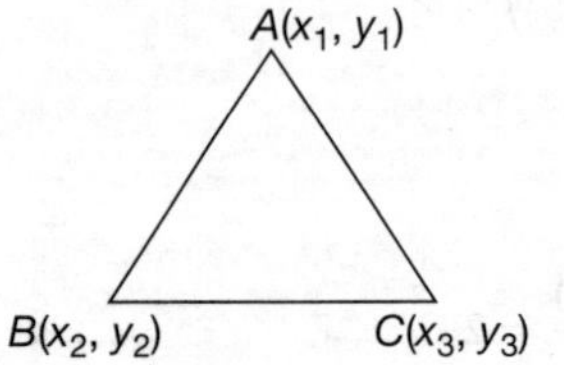

Assertion–Reason

44. A. Explanation: We will understand this with the help of a diagram.

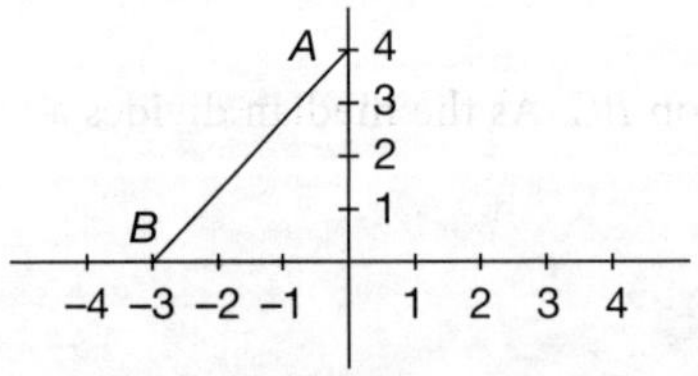

As point A is on the y axis at a distance of 4 units from the origin.

$\Rightarrow$ A (0, 4) is the point B (–3, 0).

To find length of AB, we need to use the distance formula.

$\therefore$ A and R both are correct and R is the correct explanation of A.

45. B. Explanation: To find the area of a triangle with three points, we use the formula

$$\text{ar }(\Delta ABC) = \frac{1}{2}\left[x_1(y_2 - y_3) + x_2(y_3 - y_1) + x_3(y_1 - y_2)\right]$$

$\therefore$ A is true.

In reason B, the section formula is given.

If the point (x, y) divides the line segment of the point (x_1, y_1) and (x_2, y_2) in the ratio $k : 1$ externally, then $x = \frac{kx_2 + 1}{k+1}$ and $y = \frac{ky_2 + y_1}{k+1}$.

$\therefore$ R is true. But R is not the correct explanation of A.

46. B. Explanation: Distance of a point (3, –4) from the x axis is 4 units (not 3 units). $\therefore$ A is false.

The formula for three points to be collinear is

$$\text{ar }(\Delta ABC) = \frac{1}{2}\left[x_1(y_2 - y_3) + x_2(y_3 - y_1) + x_3(y_1 - y_2)\right] = 0$$

$\therefore$ Reason B is also false.

47. B. Explanation: The distance between (a, b) and $(-a, -b)$ is $\sqrt{(-a-a)^2 + (-b-b)^2} = \sqrt{a^2 + b^2}$

$\therefore$ A is true.

Mid-point of (a, b) and $(-a, -b)$ is $\left(\frac{a+(-a)}{2}, \frac{b+(-b)}{2}\right) = \left(\frac{a-a}{2}, \frac{b-b}{2}\right) = (0, 0)$

∴ R is true. But R is not the explanation for *A*.

48. B. Explanation: ar $(\Delta ABC) = \frac{1}{2}[x_1(y_2 - y_3) + x_2(y_3 - y_1) + x_3(y_1 - y_2)]$

$$= \frac{1}{2}[1(0-0)+(-1)(0-3)+4(3-0)] = \frac{1}{2}[0+3+12] = \frac{15}{2} = 7.5 \text{ sq. units.}$$

∴ A is true.

In the reason, the distance formula is given. ∴ R is also true. But A is not the explanation for R.

Subjective Questions

51. Distance of point *P* from origin *O* where *P* (x, y) is the point $= \sqrt{(0-x)^2 + (0-y)^2} = \sqrt{x^2 + y^2}$

52. Let *O* $(2, -3)$ be the centre of the circle and *B* $(1, 4)$ be the point on the circle.

Let *A* (x, y) be the other point on the circle such that *AB* is the diameter.

⇒ *O* is the mid-point of the line *AB*.

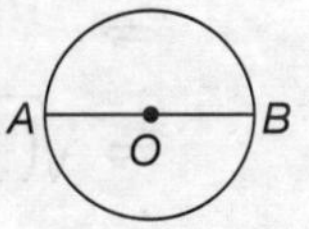

$$\Rightarrow \quad (2, -3) = \left(\frac{1+x}{2}, \frac{4+y}{2}\right)$$

$$\Rightarrow \quad \frac{x+1}{2} = 2; \quad \frac{4+y}{2} = -3 \text{ [On compairing both sides]}$$

$$\Rightarrow \quad x + 1 = 4; \; 4 + y = -6$$

$$\Rightarrow \quad x = 3; \; y = -10$$

∴ The coordinates of *A* are $(3, -10)$

53. As *P* (x, y) is equivalent from the points *A* $(1, 4)$ and *B* $(-1, 2)$.

$$\Rightarrow \quad PA = PB$$

$$\Rightarrow \quad \sqrt{(1-x)^2 + (4-y)^2} = \sqrt{(-1-x)^2 + (2-y)^2}$$

On squaring both sides, we get

$$(1 - x)^2 + (4 - y)^2 = [-(1 + x)]^2 + (2 - y)^2$$

$$\Rightarrow \cancel{1} + \cancel{x^2} - 2x + 16 + \cancel{y^2} - 8y = \cancel{1} + \cancel{x^2} + 2x + 4 + \cancel{y^2} - 4y$$

$$\Rightarrow \quad 16 - 4 = 2x - 4y + 2x + 8y$$

$$\Rightarrow \quad 12 = 4x + 4y$$

$$\Rightarrow \quad 4(x + y) = 12$$

$$\Rightarrow \quad x + y = \frac{12}{4}$$

$$\Rightarrow \quad x + y = 3$$

∴ The linear relation between *x* and *y* is given by $x + y = 3$

54. $P(x, y)$ is equidistant from the points $A(3, 6)$ and $B(-3, 4)$.

$$\Rightarrow \quad PB = PA$$

$$\Rightarrow \quad \sqrt{(-3-x)^2+(4-y)^2} = \sqrt{(3-x)^2+(6-y)^2}$$

On squaring both sides, we get

$$[-(3+x)]^2 + (4-y)^2 = (3+x)^2 + (6-y)^2$$

$$\Rightarrow \cancel{9} + \cancel{x^2} + 6x + 16 + \cancel{y^2} - 8y = \cancel{9} + \cancel{x^2} - 6x + 36 + \cancel{y^2} - 12y$$

$$\Rightarrow \quad 6x + 6x - 8y + 12y = 36 - 16$$

$$\Rightarrow \quad 12x + 4y = 20$$

$$\Rightarrow \quad 4(3x + y) = 20$$

$$\Rightarrow \quad 3x + y = 5$$

55. Centre of circle is $O(2\alpha, -1, 7)$ and circle passes through point $A(-3, -1)$.

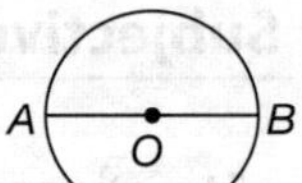

OA is the radius of the circle. Diameter = 20 units (given).

$$\Rightarrow \quad \text{Radius} = 10 \text{ units} \left(\frac{20}{2}\right)$$

$$\Rightarrow \quad OA = 10 \text{ units}$$

$$\Rightarrow \quad \sqrt{(-3-2\alpha+1)^2+(-1-7)^2} = 10$$

$$\Rightarrow \quad (-2-2\alpha)^2 + (-8)^2 = 100 \qquad \text{[On squaring both sides]}$$

$$\Rightarrow \quad 4 + 4\alpha^2 + 8\alpha + 64 = 100$$

$$\Rightarrow \quad 4\alpha^2 + 8\alpha = 32$$

$$\Rightarrow \quad 4(\alpha^2 + 2\alpha) = 32$$

$$\Rightarrow \quad \alpha^2 + 2\alpha = 8$$

$$\Rightarrow \quad \alpha^2 + 2\alpha - 8 = 0$$

$$\Rightarrow \quad \alpha^2 + 4\alpha - 2\alpha - 8 = 0$$

$$\Rightarrow \quad \alpha(\alpha + 4) - 2(\alpha + 4) = 0$$

$$\Rightarrow \quad (\alpha - 2)(\alpha + 4) = 0$$

$$\Rightarrow \quad \alpha = 2, -4$$

56. As $A(1, 0)$, $B(-1, 0)$ $P(x, y)$ are the given points and $AP - BP = 1$ (given)

$$\Rightarrow \quad \sqrt{(x-1)^2+(y-0)^2} - \sqrt{(x+1)^2+(y-0)^2} = 1$$

$$\Rightarrow \quad \sqrt{(x-1)^2+(y)^2} = 1 + \sqrt{(x+1)^2+y^2}$$

On squaring both sides $\quad (x-1)^2 + y^2 = \left(1+\sqrt{(x+1)^2+y^2}\right)^2$

$$\Rightarrow \quad x^2 + 1 - 2x + y^2 = 1 + (x+1)^2 + y^2 + 2\sqrt{(x+1)^2+y^2}$$

$$\Rightarrow \quad x^2 + \cancel{1} - 2x + \cancel{y^2} - \cancel{1} - \cancel{y^2} - (x+1)^2 = 2\sqrt{(x+1)^2+y^2}$$

$$\Rightarrow \quad x^2 - 2x - \left(x^2 + 1 + 2x\right) = 2\sqrt{(x+1)^2+y^2}$$

$\Rightarrow \quad -(4x+1) = 2\sqrt{(x+1)^2 + y^2}$

On squaring both sides $\quad (4x + 1)^2 = 4(x + 1)^2 + 4y^2$

$\Rightarrow \quad 16x^2 + 1 + 8x = 4(x^2 + 1 + 2x) + 4y^2$

$\Rightarrow \quad 16x^2 + 8x + 1 = 4x^2 + 4 + 8x + 4y^2$

$12x^2 - 4y^2 = 3$

57. One vertex is at origin O (0, 0).

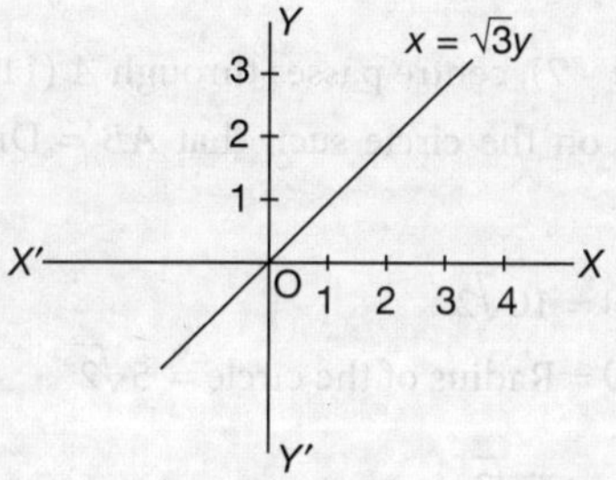

Let the other vertex be on the limit $P(x, y)$. Then $PO = 2$ units (given).

$\sqrt{(x-0)^2 + (y-0)^2} = 2$

$\Rightarrow \quad x^2 + y^2 = 4$ [On squaring on both sides]

$\Rightarrow \quad (\sqrt{3}y)^2 + y^2 = 4$ [$\because$ The point $P(x, y)$ lies on the line $x = \sqrt{3}\,y$. This point must satisfy the equation of the line]

$\Rightarrow \quad 3y^2 + y^2 = 4$

$\Rightarrow \quad 4y^2 = 4$

$\Rightarrow \quad y^2 = 1 \Rightarrow y = \pm 1$

If $\quad y = 1 \Rightarrow x = \sqrt{3}(1) \Rightarrow x = \sqrt{3}$

If $\quad y = -1 \Rightarrow x = \sqrt{3}(-1) \Rightarrow x = -\sqrt{3}$

As the second vertex lies in the first quadrant. The second vertex is $(\sqrt{3}, 1)$ and the other vertex is $(\sqrt{3}, -1)$.

$\therefore (\sqrt{3}, 1), (\sqrt{3}, -1)$ are the other two vertices.

58. Distance between $AB = \sqrt{\left((2a + \sqrt{3}a) - 2a\right)^2 + (5a - 6a)^2} = \sqrt{(\sqrt{3}a)^2 + (-a)^2}$

$= \sqrt{3a^2 + a^2} = \sqrt{4a^2} = 2a$ units

59. Point A is on the y axis at a distance of 4 units from the origin.

$\Rightarrow$ Point A is (0, 4)

And B (–3, 0) is the given point

$\therefore AB = \sqrt{(-3-0)^2 + (0-4)^2} = \sqrt{(-3)^2 + (-4)^2} = \sqrt{9+16} = \sqrt{25} = 5$ units

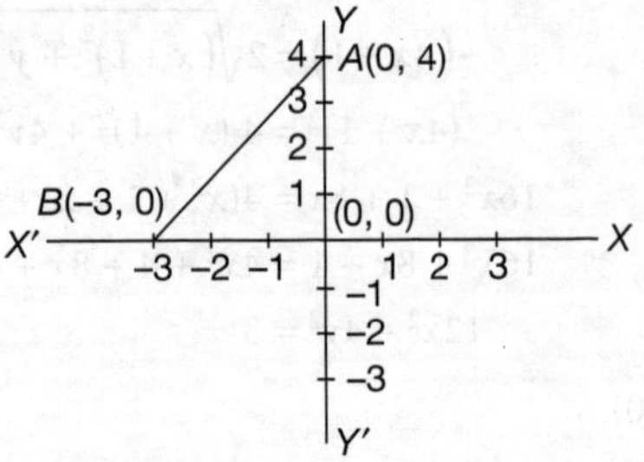

61. Centre with the circle O $(2a, a - 7)$, centre passes through A $(11, -9)$.

Let B (x, y) be another point on the circle such that AB = Diameter of the circle $\left(10\sqrt{2} \text{ units}\right)$.

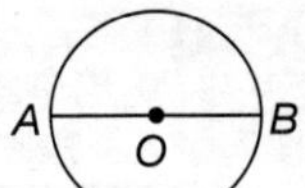

$$\Rightarrow \quad AB = 10\sqrt{2}$$

$$\Rightarrow \quad AO = \text{Radius of the circle} = 5\sqrt{2}$$

$$\sqrt{(11-2a)^2 + (-9-(a-7))^2} = 5\sqrt{2}$$

On squaring on both sides, $(11 - 2a)^2 + (-9 - (a - 7))^2 = \left(5\sqrt{2}\right)^2$

$$\Rightarrow \quad 121 + 4a^2 - 44a + 81 + (a - 7)^2 + 18(a - 7) = 50$$

$$\Rightarrow \quad 121 + 4a^2 - 44a + 81 + a^2 + 49 - 14a + 8a - 126 = 50$$

$$\Rightarrow \quad 5a^2 - 40a + 125 = 50$$

$$\Rightarrow \quad 5a^2 - 40a + 75 = 50$$

$$\Rightarrow \quad 5(a^2 - 8a + 15) = 0$$

$$\Rightarrow \quad a^2 - 8a + 15 = 0$$

$$\Rightarrow \quad a^2 - 5a - 3a + 15 = 0$$

$$\Rightarrow \quad a(a - 5) - 3(a - 5) = 0$$

$$\Rightarrow \quad (a - 5)(a - 3) = 0$$

$$a = 5, 3$$

62. The points A, B, C, D are vertices of a square. If $AB = BC = CD = DA$ and the diagonal $AC = BD$

$$AB = \sqrt{(6-3)^2 + (0-5)^2} = \sqrt{(3)^2 + (-5)^2} = \sqrt{9+25} = \sqrt{34}$$

$$BC = \sqrt{(1-6)^2 + (-3-0)^2} = \sqrt{(-5)^2 + (-3)^2} = \sqrt{25+9} = \sqrt{34}$$

$$CD = \sqrt{(-2-1)^2 + (2+3)^2} = \sqrt{(-3)^2 + (5)^2} = \sqrt{9+25} = \sqrt{34}$$

$$DA = \sqrt{(3+2)^2 + (5-2)^2} = \sqrt{(5)^2 + (3)^2} = \sqrt{25+9} = \sqrt{34}$$

$$AC = \sqrt{(1-3)^2 + (-3-5)^2} = \sqrt{(-2)^2 + (-8)^2} = \sqrt{4+64} = \sqrt{68}$$

$$BD = \sqrt{(-2-6)^2 + (2-0)^2} = \sqrt{(-8)^2 + (2)^2} = \sqrt{64+4} = \sqrt{68}$$

As the sides $AB = BC = CD = AD$ and diagonals $AC = BD$, A, B, C, D are the vertices of a square $ABCD$.

63. If the given points are collinear, the area of the triangle formed by the three points will be O.

$$\text{Area of triangle} = \frac{1}{2}\left[x_1(y_2-y_3)+x_2(y_3-y_1)+x_3(y_1-y_2)\right]$$

$$0 = \frac{1}{2}\left[x(2-0)+1(0-y)+7(y-2)\right]$$

$$\Rightarrow \quad 0 = \frac{1}{2}\left[2x+6y-14\right]$$

$$\Rightarrow \quad 2x + 6y - 14 = 0$$

$$\Rightarrow \quad x + 3y - 7 = 0$$

64. Let $O(x, y)$ be the centre of the circle and let the points $A(6, -6)$, $B(3, -7)$ and $C(3, 3)$ represent the points A, B and C on the circumference of the circle.

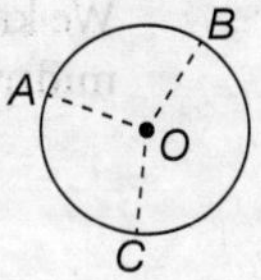

$$\therefore \quad OA = OB = OC$$

$$OA = OB$$

$$\Rightarrow \quad OA^2 = OB^2$$

$$\Rightarrow \quad (x-6)^2 + (y+6)^2 = (x-3)^2 + (y+7)^2$$

$$\Rightarrow \cancel{x^2} + 36 - 12x + \cancel{y^2} + 36 + 12y = \cancel{x^2} + 9 - 6x + \cancel{y^2} + 49 + 14y$$

$$\Rightarrow \quad -6x - 2y + 14 = 0$$

$$\Rightarrow \quad 3x + y = 7 \qquad \text{(i)}$$

Similarly, $OA = OC$

$$\Rightarrow \quad OA^2 = OC^2$$

$$(x-6)^2 + (y+6)^2 = (x-3)^2 + (y-3)^2$$

$$\Rightarrow \cancel{x^2} + 36 - 12x + \cancel{y^2} + 36 + 12y = \cancel{x^2} + 9 - 6x + \cancel{y^2} + 9 - 6y$$

$$\Rightarrow \quad -6x + 18y + 54 = 0$$

$$\Rightarrow \quad -3x + 9y = -27 \qquad \text{(ii)}$$

On adding (i) and (ii) $\quad 10y = -20$

$$\Rightarrow \quad y = -2$$

From (i), we obtain $\quad 3x - 2 = 7$

$$\Rightarrow \quad 3x = 9$$

$$\Rightarrow \quad x = 3$$

$\therefore$ Center of the circle is $(3, -2)$

65. Let $ABCD$ be a square having $(-1, 2)$ and $(3, 2)$ as vertices A and C, respectively. Let (x, y), (x_1, y_1) be the coordinates of vertex B and D, respectively.

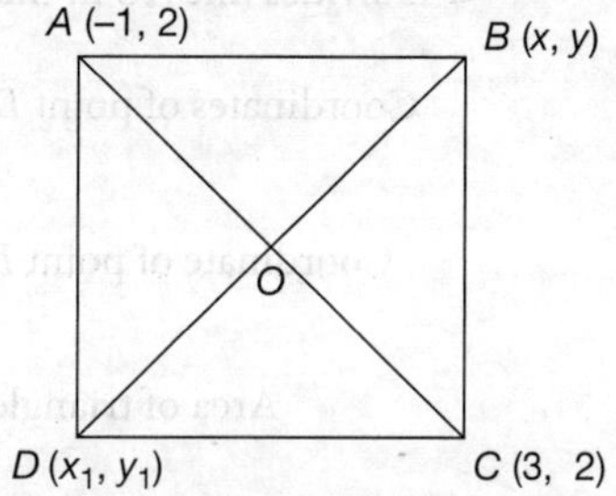

We know that the sides of a square are equal to each other:

$\therefore AB = BC$

$$\Rightarrow \quad \sqrt{(x+1)^2+(y-2)^2} = \sqrt{(x-3)^2+(y-2)^2}$$

$$\Rightarrow \quad x^2 + 2x + 1 + y^2 - 4y + 4 = x^2 + 9 - 6x + y^2 + 4 - 4y$$

$$\Rightarrow \quad 8x = 8$$

$$\Rightarrow \quad x = 1$$

We know that in a square, all interior angles are 90°.

$$\Rightarrow \quad AB^2 + BC^2 = AC^2$$

$$\Rightarrow \left(\sqrt{(1+1)^2+(y-2)^2}\right)^2+\left(\sqrt{(1-3)^2+(y-2)^2}\right)^2=\left(\sqrt{(3+1)^2+(2-2)^2}\right)^2$$

$$\Rightarrow 4 + y^2 + 4 - 4y + 4 + y^2 - 4y + 4 = 16$$

$$\Rightarrow \quad 2y^2 + 6 - 8y = 16$$

$$\Rightarrow \quad 2y^2 - 8y = 0$$

$$\Rightarrow \quad y(y-4) = 0$$

$$\Rightarrow \quad y = 0 \text{ or } 4$$

We know that in a square, the diagonals are of equal length and bisect each other at 90°. Let O be the mid-point of AC. It will also be the mid-point of BD.

$$\text{Coordinates of } O = \left(\frac{-1+3}{2}, \frac{2+2}{2}\right)$$

$$\Rightarrow \quad \left(\frac{1+x_1}{2}, \frac{y+y_1}{2}\right) = (1, 2)$$

$$\Rightarrow \quad \frac{1+x_1}{2} = 1; \ \frac{y+y_1}{2} = 2$$

$$\Rightarrow \quad x_1 + 1 = 2; \ y + y_1 = 4$$

$$\Rightarrow \quad x_1 = 1; \text{ If } y = 0 \Rightarrow y_1 = 4$$

If $\quad y = 4 \Rightarrow y_1 = 0$

∴ The required coordinates are (1, 0) and (1, 4).

66. As

$$\frac{AD}{AB} = \frac{AE}{AC} = \frac{1}{4}$$

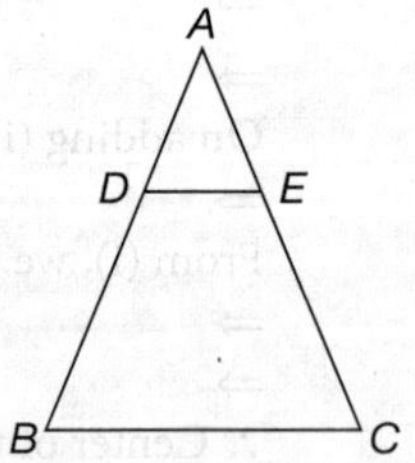

$$\Rightarrow \quad \frac{AD}{AD+DB} = \frac{AE}{AE+EC} = \frac{1}{4}$$

$$\Rightarrow \quad \frac{AD}{BD} = \frac{AE}{EC} = \frac{1}{3}$$

⇒ D divides line AB in the ratio 1 : 3

⇒ E divides line AC in the ratio 1 : 3

$$\text{Coordinates of point } D = \left(\frac{1\times1+3\times4}{1+3}, \frac{1\times5+3\times6}{1+3}\right) = \left(\frac{13}{4}, \frac{23}{4}\right)$$

$$\text{Coordinate of point } E = \left(\frac{1\times7+3\times4}{1+3}, \frac{1\times2+3\times6}{1+3}\right) = \left(\frac{19}{4}, \frac{20}{4}\right)$$

$$\text{Area of triangle} = \frac{1}{2}\left[x_1(y_2-y_3)+x_2(y_3-y_1)+x_3(y_1-y_2)\right]$$

$$\therefore \quad \text{Area of } (\Delta ADE) = \frac{1}{2}\left[4\left(\frac{23}{4}-\frac{20}{4}\right)+\frac{13}{4}\left(\frac{20}{4}-6\right)+\frac{19}{4}\left(6-\frac{23}{4}\right)\right]$$

$$= \frac{1}{2}\left[3-\frac{13}{4}+\frac{19}{16}\right] = \frac{1}{2}\left[\frac{48-52+19}{16}\right] = \frac{15}{32} \text{ sq. units.}$$

$$\text{Area } (\Delta ABC) = \frac{1}{2}\left[4(5-2)+1(2-6)+7(6-5)\right] = \frac{1}{2}[12-4+7] = \frac{15}{2} \text{ sq. units}$$

Clearly, the ratio between area (ΔADE) and ar (ΔABC) is 1 : 16.

67. Radius of circumcentre = $5\sqrt{10}$ units

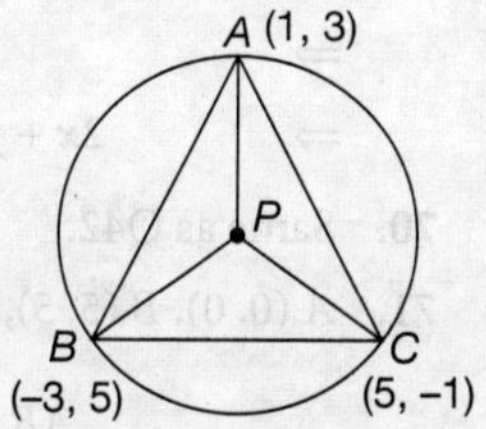

$$AP = PC = PB = 5\sqrt{10}$$

We first need to find the centre of the circle, $P(x, y)$.

As $AP = BP$

$$\Rightarrow \sqrt{(x-1)^2+(y-3)^2} = \sqrt{(-3-x)^2+(5-y)^2}$$

On squaring both sides $(x-1)^2 + (y-3)^2 = (3+x)^2 + (5-y)^2$

$$\Rightarrow \cancel{x^2} + 1 - 2x + \cancel{y^2} + \cancel{9} - 6y = \cancel{9} + \cancel{x^2} + 6x + 25 + \cancel{y^2} - 10y$$

$$8x - 4y = -24$$

$$\cancel{4}(2x-y) = \cancel{-24}$$

$$2x - y = -6 \quad \text{(i)}$$

$$AP = CP$$

$$\Rightarrow AP^2 = CP^2$$

$$\Rightarrow (x-1)^2 + (y-3)^2 = (x-5)^2 + (y+1)^2$$

$$\Rightarrow \cancel{x^2} + \cancel{1} - 2x + \cancel{y^2} + 9 - 6y = \cancel{x^2} + 25 - 10x + \cancel{y^2} + \cancel{1} + 2y$$

$$\Rightarrow 8x - 8y = 16$$

$$\cancel{8}(x-y) = \cancel{16}$$

$$x - y = 2 \quad \text{(ii)}$$

Solving (i) and (ii), we get $x = -8$ and $y = -10$

On calculating, $AP = 5\sqrt{10}$ units

68. $A(5, 2)$; $B(2, -2)$; $C(-2, t)$ are vertices of a right-angled triangle.

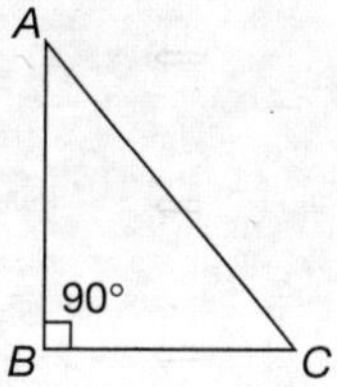

$$\Rightarrow AB^2 + BC^2 = AC^2$$

$$\Rightarrow \left(\sqrt{(2-5)^2+(-2-2)^2}\right)^2 + \left(\sqrt{(-2-2)^2+(t+2)^2}\right)^2 = \left(\sqrt{(-2-5)^2+(t-2)^2}\right)^2$$

$$\Rightarrow 9 + 16 + 16 + (t+2)^2 = 49 + (t-2)^2$$

$$\Rightarrow 41 + t^2 + 4 + 4t = 49 + t^2 + 4 - 4t$$

$$\Rightarrow 8t = 8$$

$$\Rightarrow t = 1$$

69. Three points are collinear, if ar (ΔABC) = 0.

$$\text{Ar }(\Delta ABC) = \frac{1}{2}\left[x_1(y_2-y_3)+x_2(y_3-y_1)+x_3(y_1-y_2)\right]$$

$$0 = \frac{1}{2}\left[x(7-5)+(-5)(5-y)+(-4)(y-7)\right]$$

$$\Rightarrow \quad 0 = 2x - 25 + 5y - 4y + 28$$

$$\Rightarrow \quad 2x + y + 3 = 0$$

70. Same as Q42.

71. A (0, 0), B (5, 5), C (–5, 5) are the given vertices of a triangle.

$$AB = \sqrt{(5-0)^2+(5-0)^2} = \sqrt{25+25} = \sqrt{50} = 5\sqrt{2}$$

$$BC = \sqrt{(-5-5)^2+(5-0)^2} = \sqrt{100+0} = \sqrt{100} = 10$$

$$CA = \sqrt{(0+5)^2+(0-5)^2} = \sqrt{25+25} = \sqrt{50} = 5\sqrt{2}$$

As $AB = CA$, A, B, C are vertices of an isosceles triangle.

Now, A, B, C will be vertices of a right isosceles triangle if $AB^2 + CA^2 = BC^2$

Taking LHS $\left(5\sqrt{2}\right)^2+\left(5\sqrt{2}\right)^2 = 50 + 50 = 100 = (10)^2 = BC^2$

$$\therefore \quad AB^2 + CA^2 = BC^2$$

$\therefore$ A, B, C are vertices of a right isosceles triangle.

72. Points A (a, 0), B (0, b), C (1, 1) are collinear if ar (ΔABC) = 0.

$$\text{Ar }(\Delta ABC) = \frac{1}{2}\left[x_1(y_2-y_3)+x_2(y_3-y_1)+x_3(y_1-y_2)\right]$$

$$\Rightarrow \quad \frac{1}{2}\left[a(b-1)+0(1-0)+1(0-b)\right] = 0$$

$$\Rightarrow \quad ab - a - b = 0$$

$$\Rightarrow \quad ab = a + b$$

$$\Rightarrow \quad ab \times 1 = a + b$$

$$\Rightarrow \quad \frac{a+b}{ab} = 1$$

$$\Rightarrow \quad \frac{a}{ab} + \frac{b}{ab} = 1$$

$$\Rightarrow \quad \frac{1}{b} + \frac{1}{a} = 1$$

$$\Rightarrow \quad \frac{1}{a} + \frac{1}{b} = 1$$

Hence proved.

73. $AB=\sqrt{\left[\left(\sqrt{5}+1\right)-\left(\sqrt{5}-2\right)\right]^2+\left[\left(\sqrt{3}-1\right)-\left(\sqrt{3}+2\right)\right]^2}=\sqrt{\left(\sqrt{5}+1-\sqrt{5}+2\right)^2+\left(\sqrt{3}-1-\sqrt{3}-2\right)^2}$

$=\sqrt{(3)^2+(-3)^2}=\sqrt{9+9}=\sqrt{18}=3\sqrt{2}$ units

74. Point A lies on the x axis with abscissa '$p + q$'.

$\Rightarrow$ A is given by $(p + q, 0)$ and point B lies on the y axis with ordinate $p - q$.

$\Rightarrow$ B is given by $(0, p - q)$.

$\therefore AB=\sqrt{(0-(p+q))^2+(p-q-0)^2}=\sqrt{p^2+q^2+2pq+p^2+q^2-2pq}=\sqrt{2p^2+2q^2}$ units.

75. Suppose the y axis divides PQ in the ratio λ: 1. Then, the coordinates of the point of division are $R\left(\frac{5\lambda-1}{\lambda+1},\frac{-6\lambda-4}{\lambda+1}\right)$.

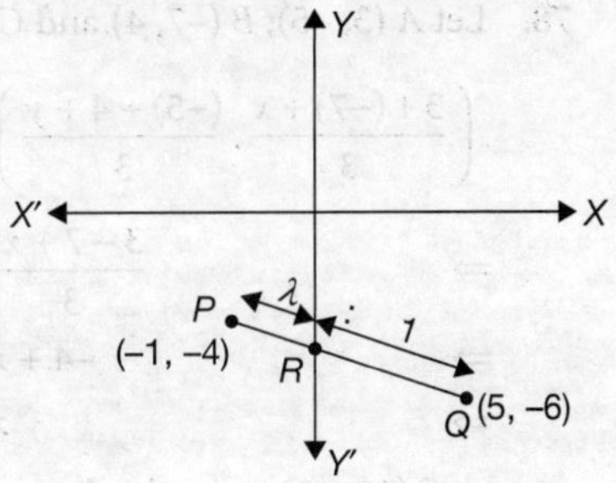

Since R lies on the y axis and the x coordinate of every point on the y axis is zero

$\Rightarrow \quad \frac{5\lambda-1}{\lambda+1}=0$

$\Rightarrow \quad 5\lambda-1=0$

$\Rightarrow \quad 5\lambda=1$

$\Rightarrow \quad \lambda=\frac{1}{5}$

Hence, the required ratio is $\frac{1}{5}:1$

Inserting $\lambda=\frac{1}{5}$ in the coordinates of R, we find that its coordinates are $\left(0,\frac{-13}{3}\right)$.

76. Suppose the x axis divides the line segment joining BA in the ratio $\lambda:1$. Then, the coordinates of the point of division are R $\left(\frac{\lambda-4}{\lambda+1},\frac{-5\lambda+5}{\lambda+1}\right)$.

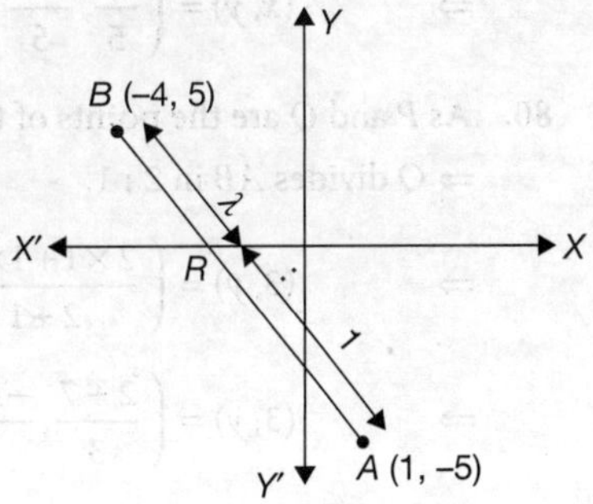

Since R lies on the x axis and the y coordinate of every point on the x axis is O

$\Rightarrow \quad \frac{-5\lambda+5}{\lambda+1}=0$

$\Rightarrow \quad -5\lambda+5=0$

$\Rightarrow \quad 5\lambda=5$

$\Rightarrow \quad \lambda=1$

Hence, the required ratio is 1 : 1.

Inserting $\lambda=1$ in the coordinates of R, we find that its coordinates are $\left(\frac{-3}{2},0\right)$.

77. The centre of the circle is O (2, –3) and A (x, y) and B (1, 4) are any points on the circle such that AB is the diameter.

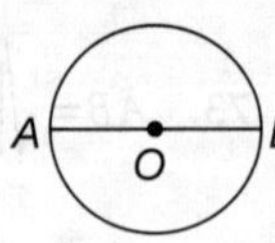

$\Rightarrow$ $AO = OB$ = Radius of the circle. O is the mid-point of AB.

$$(2, -3) = \left(\frac{x+1}{2}, \frac{y+4}{2}\right)$$

$$\Rightarrow \frac{x+1}{2} = 2; \frac{y+4}{2} = -3$$

$$\Rightarrow x + 1 = 4; y + 4 = -6$$

$$\Rightarrow x = 3; y = -10$$

$\therefore$ A (3, –10) are the coordinates of A.

78. Let A (3, –5); B (–7, 4) and C (x, y) be the vertices of the triangle. Centroid of triangle ABC = (2, –1).

$$\left(\frac{3+(-7)+x}{3}, \frac{(-5)+4+y}{3}\right) = (2, -1)$$

$$\Rightarrow \frac{3-7+x}{3} = 2; \frac{-5+4+y}{3} = -1$$

$$\Rightarrow -4 + x = 6; -1 + y = -3$$

$$\Rightarrow x = 10; y = -2$$

$\therefore$ C (10, –2) is the third vertex.

79. P divides AB in 2 : 3. Let the coordinates of P be (x, y).

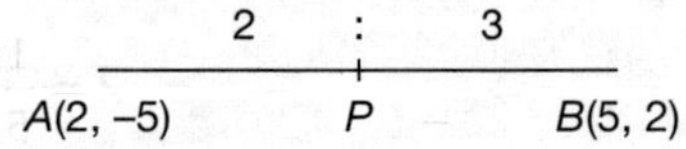

$$\Rightarrow (x, y) = \left(\frac{2\times5+3\times2}{2+3}, \frac{2\times2+3\times(-5)}{3+2}\right)$$

$$\Rightarrow (x, y) = \left(\frac{10+6}{5}, \frac{4-15}{5}\right)$$

$$\Rightarrow (x, y) = \left(\frac{16}{5}, \frac{-11}{5}\right)$$ which lies in the fourth quadrant.

80. As P and Q are the points of trisection of the line AB

$\Rightarrow$ Q divides AB in 2 : 1.

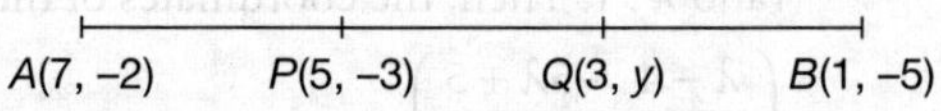

$$\Rightarrow (3, y) = \left(\frac{2\times1+1\times7}{2+1}, \frac{2\times(-5)+1(-2)}{2+1}\right)$$

$$\Rightarrow (3, y) = \left(\frac{2+7}{3}, \frac{-10-2}{3}\right)$$

$$\Rightarrow (3, y) = \left(\frac{9}{3}, \frac{-12}{3}\right)$$

$$\Rightarrow (3, y) = (3, -4)$$

$$\Rightarrow y = -4$$

81. Let P divide AB in λ: 1.

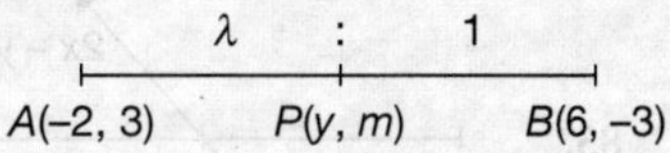

$$\Rightarrow \quad (4, m) = \left(\frac{\lambda \times (6) + 1 \times 2}{\lambda + 1}, \frac{-3\lambda + 3}{\lambda + 1}\right)$$

$$\Rightarrow \quad (4, m) = \left(\frac{6\lambda + 2}{\lambda + 1}, \frac{-3\lambda + 3}{\lambda + 1}\right)$$

$$\Rightarrow \quad \frac{6\lambda + 2}{\lambda + 1} = 4; \ \frac{-3\lambda + 3}{\lambda + 1} = m \qquad \text{(i)}$$

$$\Rightarrow \quad 6\lambda + 2 = 4\lambda + 4$$

$$\Rightarrow \quad 2\lambda = 2$$

$$\Rightarrow \quad \lambda = 1$$

Inserting the value of $\lambda = 1$ in (i) $\dfrac{-3(1) + 3}{1 + 1} = m$

$$\Rightarrow \quad m = 0$$

82. As D is the mid-point of BC.

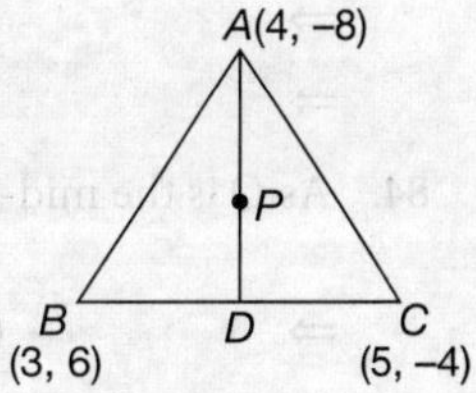

$$\Rightarrow \quad D\,(x, y) = \left(\frac{3 + 5}{2}, \frac{6 - 4}{2}\right)$$

$$\Rightarrow \quad D\,(x, y) = \left(\frac{8}{2}, \frac{2}{2}\right)$$

$$\Rightarrow \quad D\,(x, y) = (4, 1)$$

As P is any point on AD such that $\dfrac{AP}{PD} = 2$

$$\Rightarrow \quad \frac{AP}{PD} = \frac{2}{1}$$

$$\Rightarrow \quad AP : PD = 2 : 1$$

By section formula,

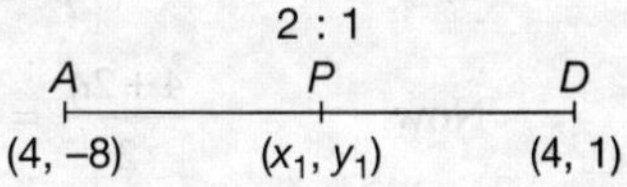

$$P\,(x_1, y_1) = \left(\frac{2 \times 4 + 1 \times 4}{2 + 1}, \frac{2 \times 1 + 1 \times (-8)}{2 + 1}\right)$$

$$\Rightarrow \quad P\,(x_1, y_1) = \left(\frac{8 + 4}{3}, \frac{2 - 8}{3}\right)$$

$$\Rightarrow \quad P\,(x_1, y_1) = \left(\frac{12}{3}, \frac{-6}{3}\right)$$

$$\Rightarrow \quad P\,(x_1, y_1) = (4, -2)$$

83.

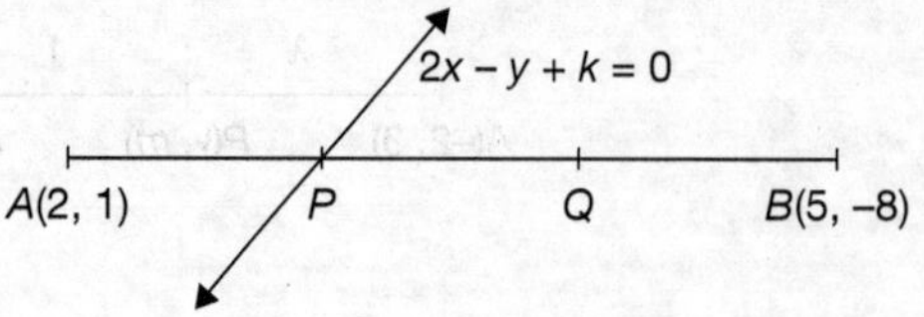

The given points are A (2, 1) and B (5, –8). P and Q trisect the line segment AB.

$$AP = PQ = QB$$

$$PB = PQ + QB = AP + AP = 2AP$$

$$AP : PB = AP : 2AP = 1 : 2$$

$\Rightarrow$ P divides line segment AB in the ratio 1 : 2.

$$\text{Coordinates of } P \text{ are} = \left(\frac{1\times5+2\times2}{1+2}, \frac{1\times(-8)+1\times2}{1+2}\right) = \left(\frac{5+4}{3}, \frac{-8+2}{3}\right) = \left(\frac{9}{3}, \frac{-6}{3}\right) = (3, -2)$$

As P (3, –2) lies on $2x - y + k = 0$

$\Rightarrow$ $2(3) - (-2) + k = 0$

$\Rightarrow$ $6 + 2 + k = 0$

$\Rightarrow$ $k = -8$

84. As Q is the mid-point of AB.

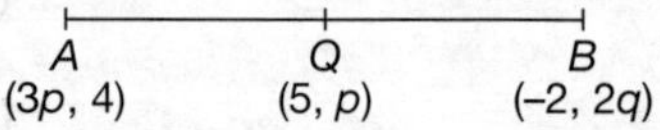

$\Rightarrow$ $(5, p) = \left(\frac{3p-2}{2}, \frac{4+2q}{2}\right)$

$\Rightarrow$ $\frac{3p-2}{2} = 5;\ \frac{4+2q}{2} = p$

$\Rightarrow$ $3p - 2 = 10$

$\Rightarrow$ $3p = 12$

$\Rightarrow$ $p = 4$

Now $\frac{4+2q}{2} = 4(= p)$

$\Rightarrow$ $4 + 2q = 8$

$\Rightarrow$ $2q = 4$

$\Rightarrow$ $q = 2$

85. As BD bisects $\angle B$, D is the mid-point of AC.

$\Rightarrow$ Coordinates of $D = \left(\frac{2+0}{2}, \frac{0+0}{2}\right) = \left(\frac{2}{2}, 0\right) = (1, 0)$

Length of $BD = \sqrt{(1-1)^2+(0-7)^2} = \sqrt{(0)^2+(-7)^2} = \sqrt{0+49} = \sqrt{49} = 7$ units.

86. In a parallelogram, the diagonals bisect each other.

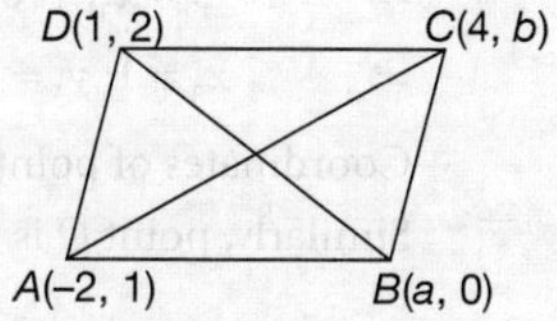

$\Rightarrow$ Mid-point of AC = Mid-point of BD

$$\Rightarrow \left(\frac{-2+4}{2}, \frac{1+b}{2}\right) = \left(\frac{1+a}{2}, \frac{2+0}{2}\right)$$

$$\Rightarrow \left(\frac{2}{2}, \frac{1+b}{2}\right) = \left(\frac{1+a}{2}, \frac{2}{2}\right)$$

$$\Rightarrow \left(1, \frac{1+b}{2}\right) = \left(\frac{1+a}{2}, 1\right)$$

$$\Rightarrow 1 = \frac{1+a}{2}; \frac{1+b}{2} = 1$$

$$\Rightarrow a + 1 = 2; b + 1 = 2$$

$$\Rightarrow a = 1; b = 1$$

Now, using the distance formula, we can find out the length of the sides of the parallelogram.

$$AB = \sqrt{(1+2)^2 + (0-1)^2} = \sqrt{(3)^2 + (-1)^2} = \sqrt{9+1} = \sqrt{10} \text{ units.}$$

$$BC = \sqrt{(4-1)^2 + (1-0)^2} = \sqrt{(3)^2 + (1)^2} = \sqrt{9+1} = \sqrt{10} \text{ units}$$

$$CD = \sqrt{(1-4)^2 + (2-1)^2} = \sqrt{(-3)^2 + (1)^2} = \sqrt{9+1} = \sqrt{10} \text{ units}$$

$$DA = \sqrt{(-2-1)^2 + (1-2)^2} = \sqrt{(-3)^2 + (-1)^2} = \sqrt{9+1} = \sqrt{10} \text{ units.}$$

$$AD = AB = BC = CD = DA$$

$\therefore$ The given parallelogram is a rhombus.

87. We know that the diagonals of a parallelogram bisect each other. Hence, point O is the mid-point of line AC and BD.

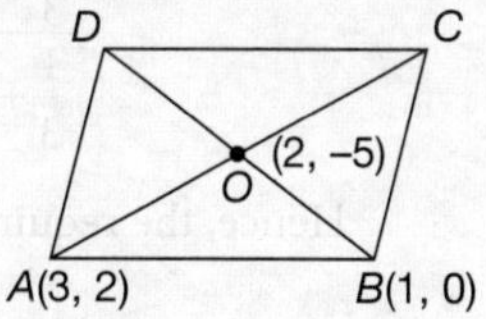

$$\underset{(x_1, y_1)}{A(3, 2)} \qquad \underset{(x, y)}{O(2, -5)} \qquad \underset{(x_2, y_2)}{B(1, 0)}$$

Let C (x_3, y_3), D (x_4, y_4) be the other two vertices. Since O is the mid-point of line AC, according to the mid-point formula

$$x = \frac{x_1 + x_3}{2}; y = \frac{y_1 + y_3}{2}$$

$$2 = \frac{3 + x_3}{2}; -5 = \frac{2 + y_3}{2}$$

$\Rightarrow \quad 4 - 3 = x_3; -10 - 2 = y_3$

$\Rightarrow \quad x_3 = 1; y_3 = -12$

Coordinates of point C are (1, –12).

Similarly, point O is the mid-point of line segment BD, according to the mid-point formula

$$x = \frac{x_2 + x_4}{2}; y = \frac{y_2 + y_4}{2}$$

$$\Rightarrow \quad 2 = \frac{1 + x_4}{2}; -5 = \frac{0 + y_4}{2}$$

$\Rightarrow \quad x_4 = 3; y_4 = -10$

∴ Coordinates of point D are (3, –10).

88. Let the point $P\left(\frac{1}{2}, y\right)$ divide the line segment joining the points A (3, –5) and B (–7, 9) in the ratio $k : 1$.

$$\left(\frac{1}{2}, y\right) = \left(\frac{k(-7)+3}{k+1}, \frac{k(9)-3}{k+1}\right)$$

$$\Rightarrow \quad \frac{-7k+3}{k+1} = \frac{1}{2}; \frac{9k-5}{k+1} = y$$

$\Rightarrow \quad k + 1 = -14k + 6$

$$\Rightarrow \quad k = \frac{1}{3}$$

Now, substituting $k = \frac{1}{3}$ in $\frac{9k-5}{k+1} = y$, we get

$$\frac{\frac{9}{3} - 5}{\frac{1}{3} + 1} = y \Rightarrow y = \frac{9 - 15}{1 + 3} \Rightarrow y = \frac{-3}{2}$$

Hence, the required ratio is 1 : 3 and $y = \frac{-3}{2}$.

89. We are given the line segment joining the points A (1, –2) and B (–3, 4). Let P and Q be the points of trisection of AB.

$$AP = PQ = QB$$

∴ P divides AB internally in the ratio 1 : 2. The coordinates of P, by applying the section formula, are

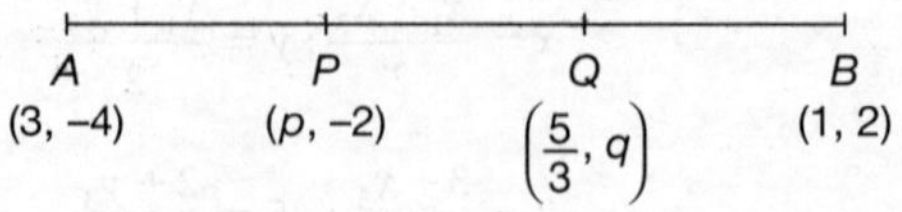

$$\left(\frac{1\times1+2\times3}{1+2}, \frac{1\times2+(-4)\times2}{1+2}\right) = (p, -2)$$

$$\Rightarrow \quad \left(\frac{7}{3}, \frac{-6}{3}\right) = (p, -2)$$

$$\Rightarrow \quad \left(\frac{7}{3}, -2\right) = (p, -2)$$

$$\Rightarrow \quad p = \frac{7}{3}$$

Now, Q divides AB in 2 : 1. By applying the section formula, we get

$$\left(\frac{2\times1+1\times3}{2+1}, \frac{2\times2+1\times(-4)}{2+1}\right) = \left(\frac{5}{3}, q\right)$$

$$\Rightarrow \quad \left(\frac{5}{3}, 0\right) = \left(\frac{5}{3}, q\right)$$

$$\Rightarrow \quad q = 0$$

90. $P(4, -2)$ is the mid-point of line segment joining the point $A(5k, 3)$ and $B(-k, -7)$.
∴ By the mid-point formula

$$(4, -2) = \left(\frac{5k+(-k)}{2}, \frac{3+(-7)}{2}\right)$$

$$\Rightarrow \quad (4, -2) = \left(\frac{5k-k}{2}, \frac{3-7}{2}\right)$$

$$\Rightarrow \quad (4, -2) = \left(\frac{4k}{2}, \frac{-4}{2}\right)$$

$$\Rightarrow \quad (4, -2) = (2k, -2)$$

$$\Rightarrow \quad 2k = 4 \Rightarrow k = 2$$

91. Let $D(x_1, y_1)$ be the fourth vertex and let $O(x, y)$ be the intersection point of diagonals AC and BD.
As O is the mid-point of AC.

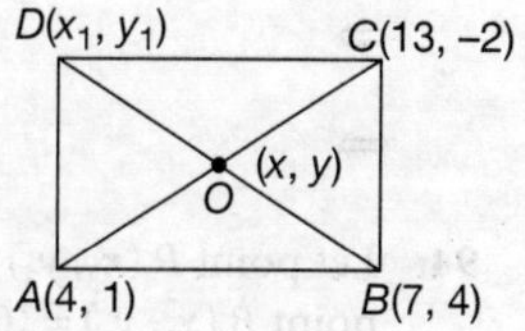

$$\Rightarrow \quad (x, y) = \left(\frac{4+13}{2}, \frac{1+(-2)}{2}\right)$$

$$\Rightarrow \quad (x, y) = \left(\frac{17}{2}, \frac{-1}{2}\right)$$

Now, O is also the mid-point of BD.

$$\Rightarrow \quad \left(\frac{17}{2}, \frac{-1}{2}\right) = \left(\frac{x_1+7}{2}, \frac{y_1+4}{2}\right)$$

$$\Rightarrow \quad \frac{x_1+7}{2} = \frac{17}{2}; \frac{y_1+4}{2} = \frac{-1}{2}$$

$$\Rightarrow \quad x_1 + 7 = 17; y_1 + 4 = -1$$

$$\Rightarrow \quad x_1 = 10; y_1 = -5$$

∴ $D(10, -5)$ is the fourth vertex.

92. The given points A (3, 5); B (m, 6); $C\left(\frac{1}{2}, \frac{15}{2}\right)$ are collinear.

$$\Rightarrow \quad \text{ar}\,(\Delta ABC) = 0$$

$$\Rightarrow \quad \frac{1}{2}\left[x_1(y_2 - y_3) + x_2(y_3 - y_1) + x_3(y_1 - y_2)\right] = 0$$

$$\Rightarrow \quad \frac{1}{2}\left[3\left(6 - \frac{15}{2}\right) + m\left(\frac{15}{2} - 5\right) + \frac{1}{2}(5 - 6)\right] = 0$$

$$\Rightarrow \quad 3\left(\frac{-3}{2}\right) + m\left(\frac{5}{2}\right) + \frac{1}{2}(-1) = 0$$

$$\Rightarrow \quad \frac{-9}{2} + \frac{5m}{2} - \frac{1}{2} = 0$$

$$\Rightarrow \quad \frac{5m}{2} = \frac{10}{2}$$

$$\Rightarrow \quad 5m = 10$$

$$\Rightarrow \quad m = 2$$

93. P (x, y) divides AB in k : 1.

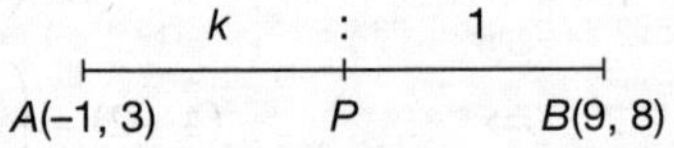

$$\Rightarrow \quad (x, y) = \left(\frac{9k - 1}{k + 1}, \frac{8k + 3}{k + 1}\right)$$

Now, P lies on $x - y + 2 = 0$.

$$\Rightarrow \quad \left(\frac{9k - 1}{k + 1}\right) - \left(\frac{8k + 3}{k + 1}\right) + 2 = 0$$

$$\Rightarrow \quad 9k - 1 - 8k - 3 + 2(k + 1) = 0$$

$$\Rightarrow \quad k - 4 + 2k + 2 = 0$$

$$\Rightarrow \quad 3k - 2 = 0$$

$$\Rightarrow \quad k = \frac{2}{3}$$

94. Let point $P\,(x_3, y_3) = (4, 5)$ which divides the line segment joining the point $A\,(x_1, y_1) = (a, 0)$ and the point $B\,(x_2, y_2) = (0, b)$ in the ratio $AP : BP = 5 : 3$. Then, using the section formula

$$(x_3, y_3) = \left(\frac{x_1 n + x_2 m}{m + n}, \frac{y_1 n + y_2 m}{m + n}\right)$$

$$\Rightarrow \quad (4, 5) = \left(\frac{3a + 0 \times 5}{3 + 5}, \frac{0(3) + b(5)}{5 + 3}\right)$$

$$\Rightarrow \quad (4, 5) = \left(\frac{3a}{8}, \frac{5b}{8}\right)$$

$$\Rightarrow \quad \frac{3a}{8} = 4; \quad \frac{5b}{8} = 5$$

$$\Rightarrow \quad 3a = 32; \ b = 8$$

$$\Rightarrow \quad a = \frac{32}{3}$$

$\therefore \quad A = (a, 0) = \left(\frac{32}{8}, 0\right)$

$B = (0, b) = (0, 8)$

95. Let the vertices of the triangle ABC be $A\ (a, b)$, $B\ (b, c)$ and $C\ (c, a)$.

Centroid of the triangle ABC is $\left(\frac{a+b+c}{3}, \frac{b+c+a}{3}\right)$.

Given that the centroid of the triangle ABC is at the origin.

$$\left(\frac{a+b+c}{3}, \frac{a+b+c}{3}\right) = (0, 0)$$

$$\Rightarrow \quad a + b + c = 0$$

Now, consider the identity $a^3 + b^3 + c^3 - 3abc = (a + b + c)\ (a^2 + b^2 + c^2 - ab - bc - ca)$.

Since, $a + b + c = 0$, the above equation becomes

$$a^3 + b^3 + c^3 - 3abc = (0) \times (a^2 + b^2 + c^2 - ab - bc - ca)$$

$$a^3 + b^3 + c^3 - 3abc = 0$$

$$\Rightarrow \quad k = 0 \qquad [\text{As it is given that } a^3 + b^3 + c^3 - 3abc = k^2]$$

96. Let the line $C\ (x, y)$ divide the line segment AB in the ratio $k : 1$. Then, by section formula

$$(x, y) = \left(\frac{2(1)+3k}{k+1}, \frac{-2(1)+7k}{k+1}\right)$$

$$\Rightarrow \quad x = \frac{2+3k}{k+1};\ y = \frac{-2+7k}{k+1}$$

As C is a point on line $2x + y - 4 = 0$

$$\Rightarrow 2\left(\frac{2+3k}{k+1}\right) + \left(\frac{-2+7k}{k+1}\right) - 4 = 0$$

$$\Rightarrow \quad \frac{4+6k}{k+1} + \frac{(-2+7k)}{k+1} - 4 = 0$$

$$\Rightarrow \quad \frac{4+6k-2+7k-4(k+1)}{k+1} = 0$$

$$\Rightarrow \quad \frac{2+13k-4k-4}{k+1} = 0$$

$$\Rightarrow \quad 9k - 2 = 0$$

$$\Rightarrow \quad k = \frac{2}{9}$$

$\therefore \quad$ Required ratio = 2 : 9

97. D, E, F are the mid-points of the line segments AB, BC and AC, respectively.

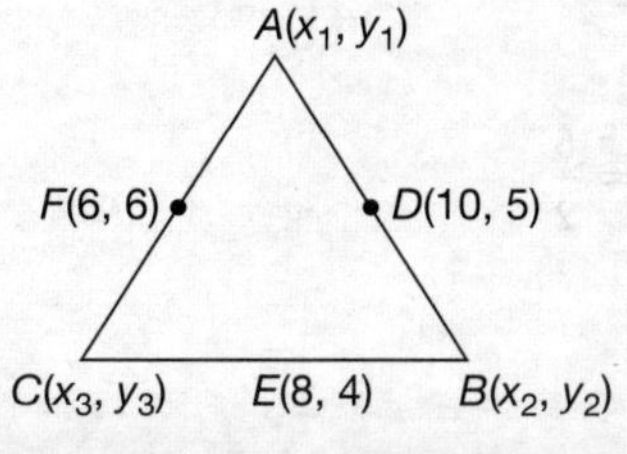

$$\Rightarrow \quad \left(\frac{x_1+x_2}{2}, \frac{y_1+y_2}{2}\right) = (10, 5)$$

$$\Rightarrow \quad (x_1+x_2, y_1+y_2) = (20, 10) \qquad \text{(i)}$$

Similarly $$\left(\frac{x_1+x_3}{2}, \frac{y_1+y_3}{2}\right) = (6, 6)$$

$$\Rightarrow \quad (x_1+x_3, y_1+y_3) = (12, 12) \qquad \text{(ii)}$$

Similarly

$$\left(\frac{x_2+x_3}{2}, \frac{y_2+y_3}{2}\right) = (8, 4)$$

$$\Rightarrow \quad (x_2+x_3, y_2+y_3) = (16, 8) \qquad \text{(iii)}$$

Adding (i), (ii) and (iii), we get

$$(x_1+x_2+x_1+x_3+x_2+x_3, y_1+y_2+y_1+y_3+y_2+y_3) = (20+12+16, 10+12+8)$$

$$\Rightarrow \quad (2(x_1+x_2+x_3), 2(y_1+y_2+y_3)) = (48, 30)$$

$$\Rightarrow \quad (x_1+x_2+x_3, y_1+y_2+y_3) = (24, 15)$$

$$\Rightarrow \quad x_1+x_2+x_3 = 24;\ y_1+y_2+y_3 = 15$$

$$\Rightarrow \quad 20 + x_3 = 24 \quad \text{[From (i)]};\ 10 + y_3 = 5 \text{ [from (i)]}$$

$$\Rightarrow \quad x_3 = 4;\ y_3 = 5 \qquad \therefore (x_3, y_3) = (4, 5)$$

Inserting the value of $x_3 = 4$ and $y_3 = 5$ in (ii)

$$(x_1+4, y_1+5) = (12, 12)$$

$$x_1 + 4 = 12;\ y_1 + 5 = 12$$

$$\Rightarrow \quad x_1 = 8;\ y_1 = 7$$

$$\therefore \quad (x_1, y_1) = (8, 7)$$

Inserting the value of $x_3 = 4$ and $y_3 = 5$ in (iii)

$$(x_2+4, y_2+5) = (16, 8)$$

$$\Rightarrow \quad x_2 + 4 = 16;\ y_2 + 5 = 8$$

$$\Rightarrow \quad x_2 = 12;\ y_2 = 3$$

$$\Rightarrow \quad (x_2, y_2) = (12, 3)$$

98. Let P divide AB in the ratio $k : 1$.

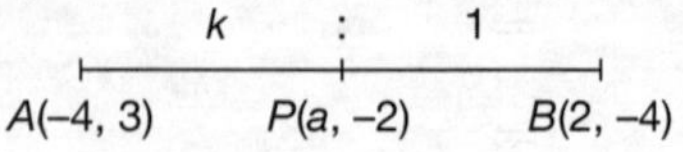

$$\Rightarrow \quad (a, -2) = \left(\frac{2k-4}{k+1}, \frac{-4k+3}{k+1}\right)$$

$$\Rightarrow \quad \frac{-4k+3}{k+1} = -2$$

$$\Rightarrow \quad -4k + 3 = -2k - 2$$

$$\Rightarrow \quad 3 + 2 = 2k$$

$$\Rightarrow \quad 2k = 5 \Rightarrow k = \frac{5}{2}$$

$\therefore$ Required ratio $= \frac{5}{2} : 1$ or $5 : 2$

Inserting the value of $k = \frac{5}{2}$ in $\frac{2k-4}{k+1} = a$, we get

$$\frac{2\left(\frac{5}{2}\right)-4}{\left(\frac{5}{2}\right)+1} = a \Rightarrow \frac{5-4}{\frac{7}{2}} = a$$

$$\Rightarrow \quad a = \frac{1}{\frac{7}{2}} \quad \Rightarrow a = \frac{2}{7}$$

99. As P is the mid-point of AB

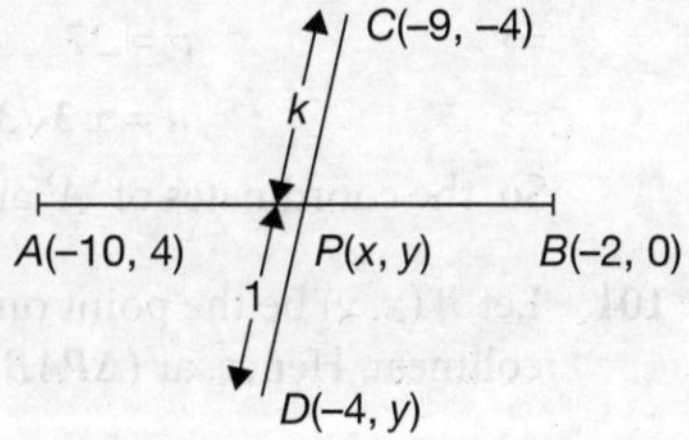

$$\Rightarrow \quad (x, y) = \left(\frac{-10-2}{2}, \frac{4+0}{2}\right)$$

$$\Rightarrow \quad (x, y) = \left(\frac{-12}{2}, \frac{4}{2}\right)$$

$$\Rightarrow \quad (x, y) = (-6, 2)$$

Now, let P divide CD in $k : 1$.

$$\Rightarrow \quad (-6, 2) = \left(\frac{-4k-9}{k+1}, \frac{yk-4}{k+1}\right)$$

$$\Rightarrow \quad -6 = \frac{-4k-9}{k+1}$$

$$\Rightarrow \quad -6k - 6 = -4k - 9$$

$$\Rightarrow \quad 2k = 3$$

$$\Rightarrow \quad k = \frac{3}{2}$$

$\therefore$ Required ratio $= \frac{3}{2} : 1$, that is, $= 3 : 2$

Inserting the value of $k = \frac{3}{2}$ in $\frac{yk-4}{k+1} = 2$, we get $\frac{y\left(\frac{3}{2}\right)-4}{\left(\frac{3}{2}\right)+1} = 2$

$$\Rightarrow \quad \frac{3y-8}{5} = 2 \Rightarrow 3y - 8 = 10$$

$$\Rightarrow \quad 3y = 18$$

$$\Rightarrow \quad y = 6$$

100. Since (0, 0) is the mid-point of the base BC, let the coordinates of point B (x_1, y_1) and the coordinates of C be (0, –3).

$$\Rightarrow \quad (0, 0) = \left(\frac{x_1+0}{2}, \frac{y_1-3}{2}\right)$$

$$\Rightarrow \quad \frac{4+0}{2} = 0; \quad \frac{y_1-3}{2} = 0$$

∴ Coordinates of point B are (0, 3). Clearly, the vertex A lies on the x axis. Let the point be A $(a, 0)$. Since ABC is an equilateral triangle, we have

$$AC^2 = AB^2 = BC^2$$

$$\Rightarrow \quad AC^2 = BC^2$$

$$\Rightarrow (a - 0)^2 + (0 - 3)^2 = (0 - 0)^2 + (3 + 3)^2$$

$$\Rightarrow \quad a^2 + 9 = 36$$

$$\Rightarrow \quad a^2 = 27$$

$$\Rightarrow \quad a = \pm 3\sqrt{3}$$

So, the coordinates of 'A' are $\left(3\sqrt{3}, 0\right)$ or $\left(-3\sqrt{3}, 0\right)$.

101. Let P (x, y) be the point on the line joining the given points A (1, –3) and B (–4, 2). ∴ P, A and B are collinear. Hence, ar $(\Delta PAB) = 0$. We know that

$$\text{Area of triangle} = \frac{1}{2}\left[x_1\left(y_2 - y_3\right) + x_2\left(y_3 - y_1\right) + x_3\left(y_1 - y_2\right)\right]$$

As $\quad$ ar $(\Delta PAB) = 0$

$$\Rightarrow \frac{1}{2}\left[x(-3-2) + 1(2-y) + (-4)\left(y - (-3)\right)\right] = 0$$

$$\Rightarrow \quad x(-5) + 1(2 - y) + (-4)(y + 3) = 0$$

$$\Rightarrow \quad -5x + 2 - y - 4y - 12 = 0$$

$$\Rightarrow \quad -5x - 5y - 10 = 0$$

$$\Rightarrow \quad -5(x + y + 2) = 0$$

$$\Rightarrow \quad x + y + 2 = 0$$

Hence proved.

102. Let point P (x, y) divide the line segment AB in $(p - q)$, $(p + q)$.

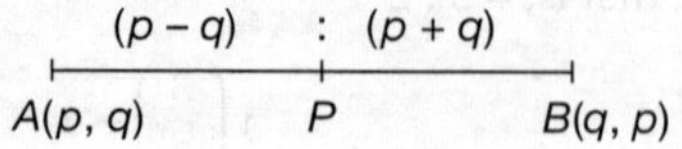

$$\Rightarrow \quad (x, y) = \left(\frac{(p-q)q + (p+q)p}{p-q+p+q}, \frac{(p-q)p + (p+q)q}{p-q+p+q}\right)$$

$$\Rightarrow \quad (x, y) = \left(\frac{pq - q^2 + p^2 + pq}{2p}, \frac{p^2 - pq + pq + q^2}{2p}\right)$$

$$\Rightarrow \quad (x, y) = \left(\frac{p^2 - q^2 + 2pq}{2p}, \frac{p^2 + q^2}{2p}\right)$$

105. Let the vertices of the triangle be A (0, –1), B (2, 1), C (0, 3).

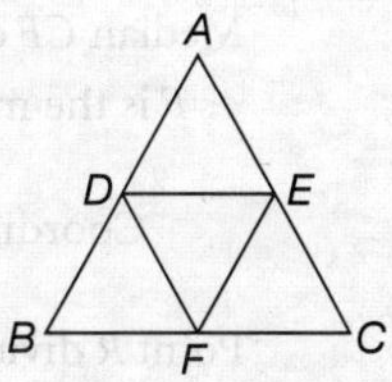

Let D, E, F be the mid-points of the sides of this triangle. Coordinates of D, E and F are given by

$$D = \left(\frac{0+2}{2}, \frac{-1+1}{2}\right) = (1, 0)$$

$$E = \left(\frac{0+0}{2}, \frac{3-1}{2}\right) = (0, 1)$$

$$F = \left(\frac{2+0}{2}, \frac{1+3}{2}\right) = (1, 2)$$

$$\text{Area }(\Delta DEF) = \frac{1}{2}\left[1(2-1)+1(1-0)+0(0-2)\right]$$

$$= \frac{1}{2}[1+1] = 1 \text{ sq units.}$$

$$\text{Area }(\Delta ABC) = \frac{1}{2}\left[0(1-3)+2(3-(-1))+0(-1-1)\right]$$

$$= \frac{1}{2}(8) = 4 \text{ sq. units}$$

∴ Required ratio = 1 : 4

106. (i) Median AD of the triangle will divide the side BC in two equal parts.

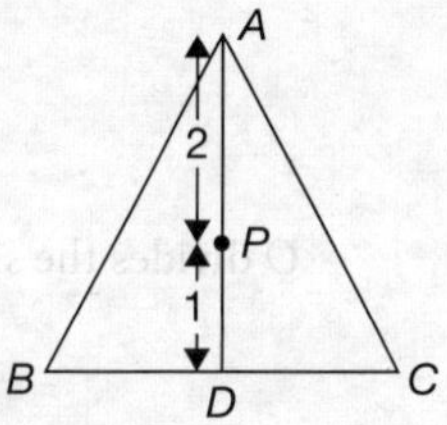

D is the mid point of side BC coordinates of $D = \left(\frac{6+1}{2}, \frac{5+4}{2}\right) = \left(\frac{7}{2}, \frac{9}{2}\right)$

(ii) Point P divides the side AD in the ratio 2 : 1

$$\text{Coordinates of } P = \left(\frac{2\times\frac{7}{2}+1\times4}{2+1}, \frac{2\times\frac{9}{2}+1\times2}{2+1}\right) = \left(\frac{11}{3}, \frac{11}{3}\right)$$

(iii) Median BE of the triangle will divide the side AC in 2 equal parts

∴ E is the mid point of side AC

$$\text{Coordinates of } E = \left(\frac{4+1}{2}, \frac{2+4}{2}\right)$$

$$= \left(\frac{5}{2}, 3\right)$$

Point Q divides side BE in the ratio 2 : 1

$$\text{Coordinates of } Q = \left(\frac{2\times\frac{5}{2}+1\times6}{2+1}, \frac{2\times3+1\times5}{2+1}\right) = \left(\frac{11}{3}, \frac{11}{3}\right)$$

Median CF of the triangle will divide the side AB in two equal parts.

$\therefore$ F is the mid-point of side AB.

$$\text{Coordinates of } F = \left(\frac{4+6}{2}, \frac{2+5}{2}\right) = \left(5, \frac{7}{2}\right)$$

Point R divides the side CF in the ratio 2 : 1.

$$\text{Coordinates of } R = \left(\frac{2\times5+1\times1}{2+1}, \frac{2\times\frac{7}{2}+1\times4}{2+1}\right) = \left(\frac{11}{3}, \frac{11}{3}\right)$$

(iv) It can be observed that the coordinates of P, Q, R are the same. $\therefore$ All of them represent the same point on the plane, the centroid of the triangle.

(v) Consider a triangle, ΔABC, having its vertices as $A(x_1, y_1)$, $B(x_2, y_2)$ and $C(x_3, y_3)$. Median AD of the triangle will divide the side BC in two equal parts.

$\therefore$ D is the mid-point of BC.

$$\text{Coordinates of } D = \left(\frac{x_2+x_3}{2}, \frac{y_2+y_3}{2}\right)$$

Let the centroid of this triangle be O.

107. $A(1, 3)$; $B(-1, 0)$ and $C(4, 0)$ are the three vertices of the ΔABC.

$$\text{Ar}(\Delta ABC) = \frac{1}{2}\left[1(0-0)+(-1)(0-3)+4(3-0)\right] = \frac{1}{2}\left[0+(-1)(-3)+4(3)\right]$$

$$= \frac{1}{2}[3+12] = \frac{15}{2} \text{ sq. units} = 7.5 \text{ sq. units.}$$

O divides the sides AD in the ratio 2 : 1.

$$\text{Coordinates of } O = \left(\frac{2\times\left(\frac{x_2+x_3}{2}\right)+1\times x_1}{2+1}, \frac{2\times\left(\frac{y_2+y_3}{2}\right)+1\times y_1}{2+1}\right) = \left(\frac{x_1+x_2+x_3}{3}, \frac{y_1+y_2+y_3}{3}\right)$$

108. Let $P(x, y)$ divide AB in the ratio $k : 1$.

$$\Rightarrow \quad (x, y) = \left(\frac{0\times k+a}{k+1}, \frac{kb+0}{k+1}\right)$$

$$\Rightarrow \quad (x, y) = \left(\frac{a}{k+1}, \frac{kb}{k+1}\right)$$

$$\text{Consider } \frac{x}{a}+\frac{x}{b} = \frac{\frac{a}{k+1}}{\frac{a}{1}}+\frac{\frac{kb}{k+1}}{\frac{b}{1}} = \frac{a}{k+1}\times\frac{1}{a}+\frac{kb}{k+1}\times\frac{1}{b} = \frac{1}{k+1}+\frac{k}{k+1} = \frac{k+1}{k+1} = 1$$

$$\therefore \quad \frac{x}{a}+\frac{y}{b} = 1$$

Hence proved.

110. Three points never lie on a unique circle if they are collinear. The three points $A(a^2, 2a)$, $B(1, -2)$, $C(2, -1)$ are collinear if ar $(\Delta ABC) = 0$.

$$\text{Ar}(\Delta ABC) = \frac{1}{2}\left[a^2(-2+1)+1(-1-2a)+2(2a+2)\right]$$

$$\Rightarrow \quad 0 = \frac{1}{2}\left[-a^2-1-2a+4a+4\right]$$

$$\Rightarrow \quad 0 = -a^2 + 2a + 3$$

$$\Rightarrow \quad a^2 - 2a - 3 = 0$$

$$\Rightarrow \quad a^2 - 3a + a - 3 = 0$$

$$\Rightarrow \quad a(a-3) + 1(a-3) = 0$$

$$\Rightarrow \quad (a+1)(a-3) = 0$$

$$\Rightarrow \quad a = -1, 3$$

111. $\text{ar}(\Delta ABC) = \frac{1}{2}\left[x_1(y_2-y_3)+x_2(y_3-y_1)+x_3(y_1-y_2)\right]$

$$4 = \frac{1}{2}\left[x(4-5)+4(5-3)+3(3-4)\right]$$

$$4 = \frac{1}{2}\left[-x+8-3\right]$$

$$8 = -x + 5$$

$$\Rightarrow \quad -x = 3$$

$$\Rightarrow \quad x = -3$$

113. ar $(\Delta ABC) = 53$ sq units

$$\Rightarrow \quad \frac{1}{2}\left[-2(-4-10)+k(10-5)+(2k+1)(5+4)\right] = 53$$

$$\Rightarrow \quad 28 + 5k + (2k+1)9 = 106$$

$$\Rightarrow \quad 28 + 5k + 18k + 9 = 106$$

$$\Rightarrow \quad 23k = 106 - 37$$

$$\Rightarrow \quad k = 3$$

112. Since $R(x, y)$ is a point on the line segment joining the points $P(a, b)$ and $Q(b, a)$, $P(a, b)$, $Q(b, a)$ and $R(x, y)$ are collinear.

$$\Rightarrow \quad \text{ar}(\Delta PQR) = 0$$

$$\Rightarrow \quad \frac{1}{2}\left[a(a-y)+b(y-b)+x(b-a)\right] = 0$$

$$\Rightarrow \quad a^2 - ay + by - b^2 + bx - ax = 0$$

$$\Rightarrow \quad -ay - ax + by + bx = b^2 - a^2$$

$$\Rightarrow \quad -a(y+x) + b(y+x) = (b-a)(b+a) \qquad \left[\because a^2-b^2=(a+b)(a-b)\right]$$

$$\Rightarrow \quad (b-a)(x+y) = (b-a)(b+a)$$

$$\Rightarrow \quad x + y = b + a$$

Hence proved.

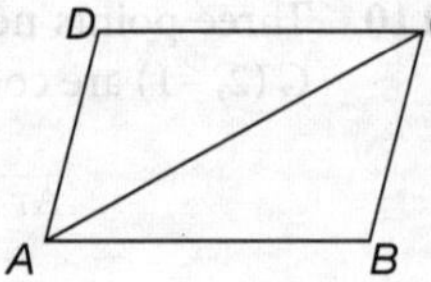

114. Since parallelogram $ABCD$ and ΔABC lie on the same base and between the same parallels

$\therefore$ ar (parallelogram $ABCD$) = 2 ar (ΔABC)

$\therefore$ Using the area formula of Δ

$$\text{ar of } \Delta ABC = \frac{1}{2}\left[x_1(y_2 - y_3) + x_2(y_3 - y_1) + x_3(y_1 - y_2)\right] = \frac{1}{2}\left[1(3-2)+2(2+2)+(-3)(-2-3)\right]$$

$$= \frac{1}{2}[1+8+15] = \frac{1}{2} \times 24 = 12 \text{ sq. units}$$

$\therefore$ ar (parallelogram $ABCD$) = 2 × 12 = 24 sq. units

$$AB = \sqrt{(x_2 - x_1)^2 + (y_2 - y_1)^2} = \sqrt{(2-1)^2 + (3+2)^2} = \sqrt{1+25} = \sqrt{26}$$

$\therefore$ Area of parallelogram = Height × Base

$$24 = \text{Height} \times \sqrt{26}$$

$$\text{Height} = \frac{24}{\sqrt{26}} \times \frac{\sqrt{26}}{\sqrt{26}} = \frac{24\sqrt{26}}{26} = \frac{12\sqrt{26}}{13} \text{ sq. units.}$$

115. According to question, let the third vertex be $(x, x + 3)$.

As ar (ΔABC) = 5 sq. units

$$\Rightarrow \frac{1}{2}\left[x_1(y_2 - y_3) + x_2(y_3 - y_1) + x_3(y_1 - y_2)\right] = 5$$

$$\Rightarrow \quad 2(-2 - x - 3) + 3(x + 3 - 1) + x(1 + 2) = 10$$

$$\Rightarrow \quad 2(-x - 5) + 3(x + 2) + 3x = 10$$

$$\Rightarrow \quad -2x - 10 + 3x + 6 + 3x = 10$$

$$\Rightarrow \quad 4x - 4 = 10$$

$$\Rightarrow \quad 4x = 14$$

$$\Rightarrow \quad x = \frac{7}{2}$$

$\therefore$ The third vertex is $\left(\frac{7}{2}, \frac{13}{2}\right)$.

117. The points $A(p, q)$; $B(m, n)$; $C(p - m, q - n)$ are collinear.

$$\Rightarrow \quad \text{ar}(\Delta ABC) = 0$$

$$\Rightarrow \quad \frac{1}{2}\left[p(n-(q-n)) + m(q-n-q) + (p-m)(q-n)\right] = 0$$

$$\Rightarrow \cancel{pn} - \cancel{pq} + np + \cancel{mq} - \cancel{mn} - \cancel{mq} + \cancel{pq} - \cancel{pn} - mq + \cancel{mn} = 0$$

$$pn - mq = 0$$

$$pn = mq$$

Hence proved.

118. A $(p, 2 - 2p)$; B $(1 - p, 2p)$ and $C(-4 - p, 6 - 2p)$ are the given vertices of triangle ABC and ar (ΔABC) = 70 sq. units (given)

$$\Rightarrow \frac{1}{2}\left[p(2p-6+2p) + (1-p)(6-2p-2+2p) + (-4-p)(2-2p-2p)\right] = 70$$

$$\Rightarrow \quad p(4p-6)+(1-p)4+(-4-p)(2-4p) = 140$$

$$\Rightarrow \quad 4p^2-6p+4-4p-8+16p-2p+4p^2 = 140$$

$$\Rightarrow \quad 8p^2+4p-4 = 140$$

$$\Rightarrow \quad 8p^2+4p-144 = 0$$

$$\Rightarrow \quad 4(2p^2+p-36) = 0$$

$$\Rightarrow \quad 2p^2+p-36 = 0$$

$$\Rightarrow \quad 2p^2+9p-8p-36 = 0$$

$$\Rightarrow \quad p(2p+9)-4(2p+9) = 0$$

$$\Rightarrow \quad (2p+9)(p-4) = 0$$

$$\Rightarrow \quad p = \frac{-9}{2}, 4$$

119. We know that the diagonals of parallelogram bisect each other.
$\Rightarrow$ O is the mid-point of AC and BD. Let $D(x_1, y_1)$ be the fourth vertex and let $O(x, y)$ be the intersection point of both the diagonals. As O is the mid-point of AC, by the mid-point formula

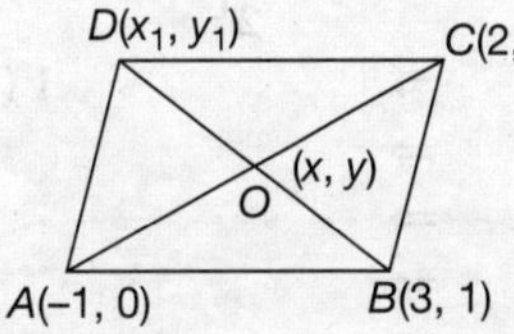

$$\Rightarrow \quad (x, y) = \left(\frac{-1+2}{2}, \frac{0+2}{2}\right)$$

$$\Rightarrow \quad (x, y) = \left(\frac{1}{2}, 1\right)$$

Now, as O is the mid-point of BD

$$\Rightarrow \quad \left(\frac{1}{2}, 1\right) = \left(\frac{x_1+3}{2}, \frac{y_1+1}{2}\right)$$

$$\Rightarrow \quad \frac{x_1+3}{2} = \frac{1}{2};\ \frac{y_1+1}{2} = 1$$

$$\Rightarrow \quad x_1+3 = 1;\ y_1+1 = 2$$

$$\Rightarrow \quad x_1 = -2;\ y_1 = 1$$

$\therefore$ $D(-2, 1)$ is the fourth vertex of parallelogram $ABCD$.

Since parallelogram $ABCD$ and ΔABC lie on the same base and between the same parallels

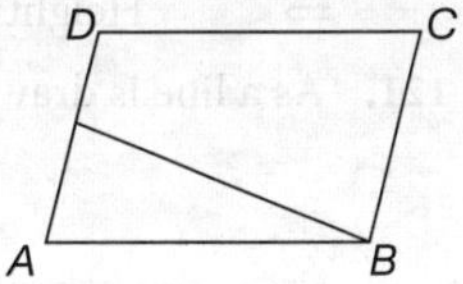

$\therefore$ ar (parallelogram $ABCD$) = 2ar (ΔABC)

$\therefore$ Using area formula of Δ

$$\text{ar}(\Delta ABC) = \frac{1}{2}\left[x_1(y_2-y_3)+x_2(y_3-y_1)+x_3(y_1-y_2)\right]$$

$$= \frac{1}{2}\left[(-1)(1-2)+3(2-0)+2(0-1)\right] = \frac{1}{2}[1+6-2] = \frac{5}{2}$$

Area of parallelogram $ABCD = 2\times\frac{5}{2} = 5$ sq. units.

Now $$AD = \sqrt{(x_2 - x_1)^2 + (y_2 - y_1)^2} = \sqrt{(-1+2)^2 + (0-1)^2}$$

$$= \sqrt{(1)^2 + (-1)^2} = \sqrt{1+1} = \sqrt{2}.$$

$\therefore$ Area of parallelogram = Height × base

$$5 = \text{Height} \times \sqrt{2}$$

$$\Rightarrow \quad \text{Height} = \frac{5}{\sqrt{2}} \times \frac{\sqrt{2}}{\sqrt{2}}$$

$$\text{Height} = \frac{5\sqrt{2}}{2}$$

120. ar (ΔABC) = 15 sq. units (given)

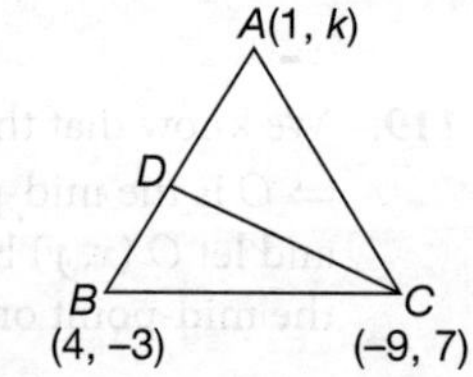

$$\Rightarrow \quad \frac{1}{2}\left[1(-3-7) + 4(7-k) + (-9)(k+3)\right] = 15$$

$$\Rightarrow \quad 1(-10) + 4(7-k) - 9(k+3) = 30$$

$$\Rightarrow \quad -10 + 28 - 4k - 9k - 27 = 30$$

$$\Rightarrow \quad -13k - 9 = 30$$

$$\Rightarrow \quad -13k = 39$$

$$\Rightarrow \quad k = -3$$

$\therefore$ Coordinates of A are (1, –3).

Now $$AB = \sqrt{(x_2 - x_1)^2 + (y_2 - y_1)^2} = \sqrt{(4-1)^2 + [(-3)-(-3)]^2} = \sqrt{(3)^2 + (-3+3)^2}$$

$$= \sqrt{(3)^2 + 0} = 3 \text{ units.}$$

The altitude drawn from point C to the base AB is the height of triangle ABC with corresponding base as AB. And we know

$$\text{Area of triangle} = \frac{1}{2} \times \text{Base} \times \text{Height}$$

$$\Rightarrow \quad 15 = \frac{1}{2} \times AB \times \text{Height}$$

$$\Rightarrow \quad 30 = 3 \times \text{Height}$$

$$\Rightarrow \quad \text{Height} = 10 \text{ units}$$

121. As a line is drawn to intersect sides PQ and PR at M and N, respectively

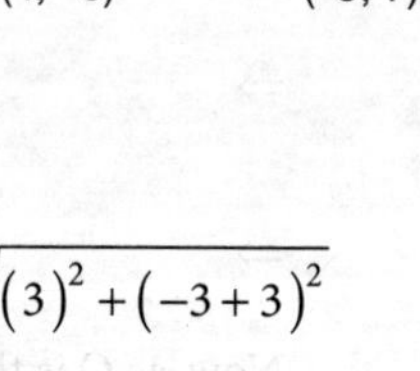

$$\frac{PM}{PQ} = \frac{PN}{PR} = \frac{1}{4}$$

$\Rightarrow$ Line MN is parallel to QR.

$\Rightarrow \Delta PMN \sim \Delta PQR$ and side ratio $= \frac{1}{4}$ (given)

$$\text{Area of } \Delta POR = \frac{1}{2}\left[4(5-2) + 1(2-6) + 7(6-5)\right] = \frac{1}{12}[12 - 4 + 7] = \frac{15}{2}$$

$$\text{Area of } \Delta PMN = \frac{1}{16} \times \frac{15}{2} = \frac{15}{32} \text{ sq. units.}$$

Another Method

Find the coordinates of *M* and *N*.

$$\frac{PM}{PQ} = \frac{1}{4} \Rightarrow PM : MQ = 1 : 3$$

$$\text{Coordinates of } M = \left(\frac{1\times1+3\times4}{4}, \frac{1\times5+3\times6}{4}\right) = \left(\frac{13}{4}, \frac{23}{4}\right)$$

$$\text{Similarly, coordinates of } N = \left(\frac{1\times7+3\times4}{4}, \frac{1\times2+3\times6}{4}\right) = \left(\frac{19}{4}, 5\right)$$

$$\text{Area of } (\Delta PMN) = \frac{1}{2}\left[x_1(y_2-y_3)+x_2(y_3-y_1)+x_3(y_1-y_2)\right]$$

$$= \frac{1}{2}\left[4\left(\frac{23}{4}-5\right)+\frac{13}{4}(5-6)+\frac{19}{4}\left(6-\frac{23}{4}\right)\right] = \frac{1}{2}\left[3-\frac{13}{4}+\frac{19}{16}\right] = \frac{15}{32}$$

122. $$\text{ar}(\Delta PAB) = \frac{1}{2}\left[x_1(y_2-y_3)+x_2(y_3-y_1)+x_3(y_1-y_2)\right]$$

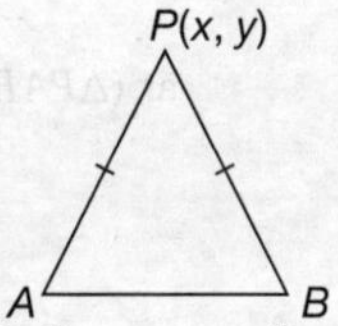

$\Rightarrow$ $10 = \frac{1}{2}\left[x(4+2)+3(-2-y)+5(y-4)\right]$

$\Rightarrow$ $20 = 6x + 2y - 26$

$\Rightarrow$ $3x + y - 13 = 10$

$\Rightarrow$ $3x + y - 23 = 0$

$\Rightarrow$ $3x + y = 23$ (i)

Now, as $PA = PB$

$\Rightarrow$ $PA^2 = PB^2$

$\Rightarrow$ $(3 - x)^2 + (4 - y)^2 = (5 - x)^2 + (-2 - y)^2$

$\Rightarrow$ $9 + x^2 - 6x + 16 + y^2 - 8y = 25 + x^2 - 10x + 4 + y^2 + 4y$

$\Rightarrow$ $25 - 6x - 8y = 29 - 10x + 4y$

$\Rightarrow$ $4x - 12y = 4$

$\Rightarrow$ $x - 3y = 1$ (ii)

From (i) and (ii)

$$3x + y = 23$$
$$(x - 3y = 1) \times 3$$

$$\cancel{3x} + y = 23$$
$$\cancel{3x} - 9y = 3$$
$$- \quad + \quad -$$
$$10y = 20$$

$\Rightarrow$ $y = 2$

Inserting the value of $y = 2$ in (ii)

$$x - 3(2) = 1$$

$\Rightarrow \quad x - 6 - 1$

$\Rightarrow \quad x = 7$

$\therefore$ Coordinates of P are (7, 2)

Now, $PA = k\sqrt{5}$ units (given)

$$\Rightarrow \quad \sqrt{(3-7)^2+(4-2)^2} = k\sqrt{5}$$

$$\Rightarrow \quad \sqrt{(-4)^2+(2)^2} = k\sqrt{5}$$

$$\Rightarrow \quad \sqrt{16+4} = k\sqrt{5}$$

$$\Rightarrow \quad \sqrt{20} = k\sqrt{5}$$

$$\Rightarrow \quad 2\sqrt{5} = k\sqrt{5}$$

$$\Rightarrow \quad k = 2$$

123. $A(t, t-2)$; $B(t+2, t+2)$; $C(t+3, t)$ are the given vertices of ΔABC.

$$\text{ar}(\Delta PAB) = \frac{1}{2}\left|\left[x_1(y_2-y_3)+x_2(y_3-y_1)+x_3(y_1-y_2)\right]\right|$$

$$= \frac{1}{2}\left|\left[t(t+2-t)+(t+2)(t-t+2)+(t+3)(t-2-t-2)\right]\right|$$

$$= \frac{1}{2}\left|\left[2t+2(t+2)+(t+3)(-4)\right]\right|$$

$$= \frac{1}{2}\left|\left[2t+2t+4-4t-12\right]\right| = \frac{1}{2}|-8| = \frac{1}{2}\times 8 = 4 \text{ sq. units, which is independent of 't'.}$$

124. $$\text{ar}(\Delta ABC) = \frac{1}{2}\left[x_1(y_2-y_3)+x_2(y_3-y_1)+x_3(y_1-y_2)\right]$$

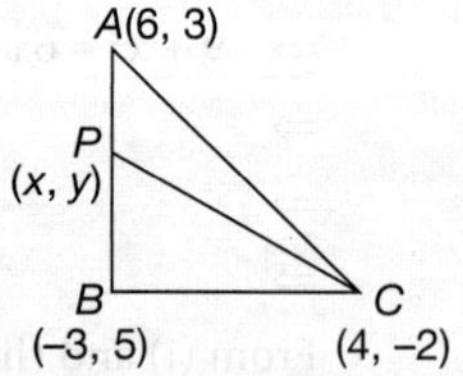

$$= \frac{1}{2}\left[6(5+2)+(-3)(-2-3)+4(3-5)\right]$$

$$= \frac{1}{2}\left[42+15-8\right] = \frac{49}{2}$$

Now $$\text{ar}(\Delta PBC) = \frac{1}{2}\left[x(5+2)+(-3)(-2-y)+4(y-5)\right] = \frac{1}{2}\left[5x+2x+6+3y+4y-20\right]$$

$$= \frac{1}{2}\left[7x+7y-14\right] = \frac{7}{2}(x+y-2)$$

Now $$\frac{\text{ar}(\Delta PBC)}{\text{ar}(\Delta ABC)} = \frac{\frac{7}{2}(x+y-2)}{\frac{49}{2}}$$

$$\Rightarrow \quad \frac{\text{ar}(\Delta PBC)}{\text{ar}(\Delta ABC)} = \frac{x+y-2}{7}$$

Hence proved

125. As points A (3, 3); B (h, 0); C (0, k) are collinear.

$\Rightarrow \quad \text{ar}(\Delta ABC) = 0$

$\Rightarrow \quad \frac{1}{2}[3(0-k)+h(k-3)+0(3-0)] = 0$

$\Rightarrow \quad -3k + hk - 3h + 0 = 0$

$\Rightarrow \quad kh = 3k + 3h$

$\Rightarrow \quad kh = 3(k + h)$

$\Rightarrow \quad k + h = \frac{kh}{3}$

$\Rightarrow \quad \frac{k+h}{kh} = \frac{1}{3}$

$\Rightarrow \quad \frac{1}{h} + \frac{1}{k} = \frac{1}{3}$

126. The intersection point of $5x - 3y + 15 = 0$ with the x axis is given by (x, 0).

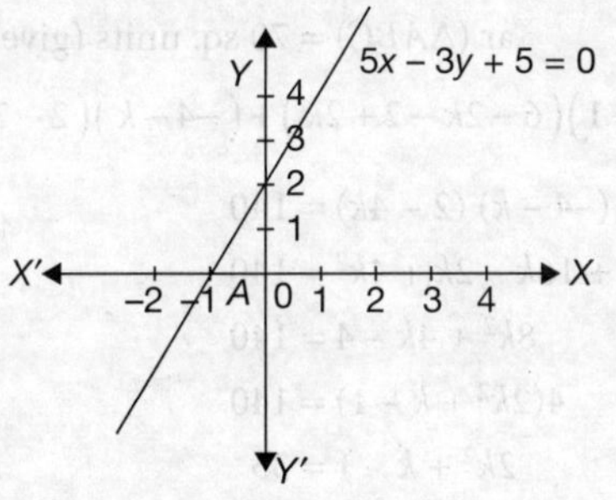

$\Rightarrow \quad 5x - 3(0) + 15 = 0$

$\Rightarrow \quad 5x + 15 = 0$

$\Rightarrow \quad x = -3$

$\Rightarrow$ The line $5x - 3y + 15 = 0$ intersects the x axis at $(-3, 0)$.

Similarly, the intersection point of $5x - 3y + 15 = 0$ with the y axis is given by (0, y).

$\Rightarrow \quad 5(0) - 3y + 15 = 0$

$\Rightarrow \quad -3y + 15 = 0$

$\Rightarrow \quad 3y = 15$

$\Rightarrow \quad y = \frac{15}{3}$

$\Rightarrow \quad y = 5$

$\therefore$ The line $5x - 3y + 15 = 0$ intersects the y axis at (0, 5).

$\therefore$ A (–3, 0); B (0, 5); O (0, 0) are the vertices of the triangle ABC.

$$\therefore \quad \text{ar}(\Delta ABC) = \frac{1}{2}\left|\left[(-3)(5-0)+0(0-0)+0(0-5)\right]\right|$$

$$= \frac{1}{2}\left|(-15)\right| = \frac{1}{2}\times 15 = \frac{15}{2} \text{ sq. units}$$

127. Coordinates of the point A at which the x axis cuts the line $3x + 4y = 24$ is given by $(x, 0)$

$\Rightarrow \quad 3x + 4(0) = 24$

$\Rightarrow \quad x = 8$

$\therefore$ Coordinates of A are (8, 0).

Coordinates of point B at which the line $3x + 4y = 24$ cuts the y axis is given by $(0, y)$.

$\Rightarrow \quad 3(0) + 4y = 24$

$\Rightarrow \quad y = 6$

$\therefore$ Coordinates of B are (0, 6).

$$\therefore \quad \text{ar}(\Delta AOB) = \frac{1}{2}\left[8(6-0)+0(0-0)+0(0-6)\right] = \frac{1}{2}(48) = 24 \text{ sq. units}$$

129. A $(k, 2 - 2k)$; B $(-k + 1, 2k)$ and C $(-4 - k, 6 - 2k)$ are the given vertices of the triangle ABC.

$$\text{ar}(\Delta ABC) = 70 \text{ sq. units (given)}$$

$$\Rightarrow \frac{1}{2}\left[k(2k-6+2k)+(-k+1)(6-2k-2+2k)+(-4-k)(2-2k-2k)\right] = 70$$

$\Rightarrow \quad k(4k - 6) + 4(-k + 1) + (-4 - k)(2 - 4k) = 140$

$\Rightarrow \quad 4k^2 - 6k - 4k + 4 - 8 + 16k - 2k + 4k^2 = 140$

$\Rightarrow \quad 8k^2 + 4k - 4 = 140$

$\Rightarrow \quad 4(2k^2 + k - 1) = 140$

$\Rightarrow \quad 2k^2 + k - 1 = 35$

$\Rightarrow \quad 2k^2 + k - 36 = 0$

$\Rightarrow \quad 2k^2 + 9k - 8k - 36 = 0$

$\Rightarrow \quad k(2k + 9) - 4(2k + 9) = 0$

$\Rightarrow \quad (2k + 9)(k - 4) = 0$

$\Rightarrow \quad k = \frac{-9}{2}, 4$

130. We know that $\frac{1}{2}\left[x_1(y_2 - y_3) + x_2(y_3 - y_1) + x_3(y_1 - y_2)\right] = 0$

$x_1(y_2 - y_3) + x_2(y_3 - y_1) + x_3(y_1 - y_2) = 0$ (i)

We have to prove $\frac{y_2 - y_3}{x_2 x_3} + \frac{y_3 - y_1}{x_3 x_1} + \frac{y_1 - y_2}{x_1 x_2} = 0$

$$\text{LHS} = \frac{x_1(y_2 - y_3) + x_2(y_3 - y_1) + x_3(y_1 - y_2)}{x_1 x_2 x_3}$$

$$= \frac{0}{x_1 x_2 x_3} \qquad \text{[From (i)]}$$

$$= 0$$

Chapter 8: Introduction to Trigonometry

3 Try to Solve

4. $4 \tan \theta = 3$

$$\tan \theta = \frac{3}{4} \qquad \text{(i)}$$

Using Pythagoras' Theorem, we get $\sec \theta = \frac{5}{4}$ (ii)

To find $= \frac{4\sin\theta - \cos\theta + 1}{4\sin\theta + \cos\theta - 1}$

Dividing the numerator and denominator by cos θ, we get

$$= \frac{4\frac{\sin\theta}{\cos\theta} - \frac{\cos\theta}{\cos\theta} + \frac{1}{\cos\theta}}{4\frac{\sin\theta}{\cos\theta} + \frac{\cos\theta}{\cos\theta} - \frac{1}{\cos\theta}} = \frac{4\tan\theta - 1 + \sec\theta}{4\tan\theta + 1 - \sec\theta}$$

$$= \frac{4 \times \frac{3}{4} - 1 + \frac{5}{4}}{4 \times \frac{3}{4} + 1 - \frac{5}{4}} \qquad \text{[From (i) and (ii)]}$$

$$= \frac{\frac{12 - 4 + 5}{4}}{\frac{12 + 4 - 5}{4}} = \frac{13}{11}$$

4 Try to Solve

4. $\left(\frac{1}{2}\right)^2 \times \left(\frac{1}{\sqrt{2}}\right)^2 + 4 \times \left(\frac{1}{\sqrt{3}}\right)^2 + \frac{1}{2} \times (1)^2 - 2 \times (0)^2 + \frac{1}{24}$

$$= \frac{1}{8} + \frac{4}{3} + \frac{1}{2} + \frac{1}{24} = \frac{3 + 32 + 12 + 1}{24} = \frac{48}{24} = 2$$

5 Try to Solve

6. If $\sin(A + 2B) = \frac{\sqrt{3}}{2} = \sin 60°$

$$A + 2B = 60 \quad \text{(i)}$$

$$\cos(A + 4B) = 0 = \cos 0°$$

$$A + 4B = 90 \quad \text{(ii)}$$

On solving (i) and (ii), we get $A = 30°$ and $B = 15°$

13 Try to Solve

1. (c) $\frac{\cos\theta}{1+\sin\theta} = \frac{1-\sin\theta}{\cos\theta}$

LHS, rationalise the denominator

$$= \frac{\cos\theta}{1+\sin\theta} \times \frac{1-\sin\theta}{1-\sin\theta} = \frac{\cos\theta(1-\sin\theta)}{1-\sin^2\theta} = \frac{\cos\theta(1-\sin\theta)}{\cos^2\theta}$$

$$= \frac{1-\sin\theta}{\cos\theta} = \text{RHS}$$

(d) $\sec^2\theta + \text{cosec}^2\,\theta = \sec^2\,\theta\,\text{cosec}^2\,\theta$

LHS

$$= \frac{1}{\cos^2\theta} + \frac{1}{\sin^2\theta} = \frac{\sin^2\theta + \cos^2\theta}{\sin^2\theta \cdot \cos^2\theta} = \frac{1}{\sin^2\theta \cdot \cos^2\theta}$$

$$= \frac{1}{\sin^2\theta} \times \frac{1}{\cos^2\theta} = \text{cosec}^2\theta \times \sec^2\theta$$

14 Try to Solve

1. (ii) $(\sin A + \text{cosec}\, A)^2 + (\cos A + \sec A)^2 = 7 + \tan^2 A + \cot^2 A$

LHS

$$= \sin^2 A + \text{cosec}^2 A + 2\sin A \cdot \text{cosec}\, A + \cos^2 A + \sec^2 A + 2\cos A \cdot \sec A$$

$$= (\sin^2 A + \cos^2 A) + 1 + \cot^2 A + 2\sin A \times \frac{1}{\sin A} + 1 + \tan^2 A + 2\cos A \times \frac{1}{\cos A}$$

$$= 1 + 1 + \cot^2 A + 2 + 1 + \tan^2 A + 2$$

$$= 7 + \tan^2 A + \cot^2 A$$

RHS

(iii) $\frac{\sin\theta - 2\sin^3\theta}{2\cos^3\theta - \cos\theta} = \tan\theta$

LHS

$$= \frac{\sin\theta\left[1-2\sin^2\theta\right]}{\cos\theta\left[2\cos^2\theta - 1\right]} = \tan\theta\left[\frac{1-2\left(1-\cos^2\theta\right)}{2\cos^2\theta - 1}\right]$$

$$= \tan\theta\left[\frac{1-2+2\cos^2\theta}{2\cos^2\theta-1}\right] = \tan\theta\left[\frac{2\cos^2\theta-1}{2\cos^2\theta-1}\right] = \tan\theta.$$

RHS

Multiple Choice Questions

1. (d) $\sec x + \tan x = p$

$\tan x = p - \sec x$

S.B.S

$\tan^2 x = p^2 + \sec^2 x - 2p\sec x$

$\sec^2 x - 1 = p^2 + \sec^2 x - 2p\sec x$

$-1 = p^2 - 2p\sec x$

$2p\sec x = p^2 + 1$

$\sec x = \dfrac{p^2+1}{2p}$

Hence, option (d) is correct.

2. (a) Since $CD = DB$, $CD = \frac{1}{2}BC$ and $DB = \frac{1}{2}BC$

Here $\cot x^\circ = \dfrac{AC}{CD}$ (i)

$\cot y^\circ = \dfrac{AC}{CB}$ (ii)

Dividing (i) by (ii), we get $\dfrac{\cot x^\circ}{\cot y^\circ} = \dfrac{\frac{AC}{CD}}{\frac{AC}{CB}} = \dfrac{CB}{CD} = \dfrac{CB}{\frac{1}{2}BC} = \dfrac{2}{1}$

So, option (a) is correct.

3. (c) $\sin\theta - \cos\theta = 0$

$\sin\theta = \cos\theta$

$\dfrac{\sin\theta}{\cos\theta} = 1$

$\tan\theta = 1$

$\tan\theta = \tan 45^\circ$

So $\theta = 45^\circ$

So $\sin^4\theta + \cos^4\theta = (\sin 45^\circ)^4 + (\cos 45^\circ)^4 = \left(\dfrac{1}{\sqrt{2}}\right)^4 + \left(\dfrac{1}{\sqrt{2}}\right)^4 = \dfrac{1}{4} + \dfrac{1}{4} = \dfrac{1}{2}$

Option (c) is correct.

4. (b) $\cos 1^\circ, \cos 2^\circ \ldots \cos 90^\circ \ldots \cos 180^\circ$

$\cos 1^\circ, \cos 2^\circ \ldots \cos 90^\circ \ldots \cos 180^\circ = 0$ [$\because \cos 90^\circ = 0$]

So, option (b) is correct.

6. (b) $\sec\theta + \tan\theta = x$

So $\dfrac{1}{\sec\theta + \tan\theta} = \dfrac{1}{x}$

$$\frac{1}{\sec\theta + \tan\theta} \times \frac{\sec\theta - \tan\theta}{\sec\theta - \tan\theta} = \frac{1}{x}$$

$$\frac{\sec\theta - \tan\theta}{\sec^2\theta - \tan^2\theta} = \frac{1}{x}$$

$$\sec\theta - \tan\theta = \frac{1}{x} \quad [\because \sec^2\theta - \tan^2\theta = 1]$$

Option (b) is correct.

7. (c) $\sin A = \dfrac{1}{\text{cosec} A} = \dfrac{1}{\sqrt{1+\cot^2 A}}$. So, option (c) is correct.

9. (c) $\tan\theta + \cot\theta = 2$

S.B.S $\tan^2\theta + \cot^2\theta + 2\tan\theta \cdot \cot\theta = 4$

$\tan^2\theta + \cot^2\theta + 2 = 4$ $\quad [\because \tan\theta . \cot\theta = 1]$

$\tan^2\theta + \cot^2\theta = 2$

Hence, (c) is the correct option.

12. (c) $\sin(60 + \theta) - \cos(30 - \theta)$

$= \sin[90 - (30 - \theta)] - \cos(30 - \theta) = \cos(30 - \theta) - \cos(30 - \theta) = 0$

Option (c) is correct.

16. (b) $\tan A = \cot B = \dfrac{15}{7}$

$\tan A = \tan(90 - B)$

On compairing $A = 90 - B$

$A + B = 90°$

Option (b) is correct.

17. (b) $AB = \sqrt{3^2 + 4^2} = 5.$

$\tan\theta = \dfrac{AB}{BC} = \dfrac{5}{12}$

Option (b) is correct.

23. (b) $\sin A + \sin^2 A = 1$ (i)

$\sin A = 1 - \sin^2 A$

$\sin A = \cos^2 A$ (ii)

$\sin^2 A = \cos^4 A$ (iii)

Inserting (ii) and (iii) in (i), we get $\cos^2 A + \cos^4 A = 1$

Option (b) is correct.

24. (c) $\sin(\theta + 36^\circ) = \cos\theta$

$\sin(\theta + 36^\circ) = \sin(90 - \theta)$

On comparing both sides, $\theta + 36^\circ = 90 - \theta$

$\theta = 27^\circ$

Option (c) is correct.

Fill in the Blanks

36. $2(\sin^6\theta + \cos^6\theta) - 3(\sin^4\theta + \cos^4\theta) + 1$

$$= 2\left[(\sin^2\theta)^3 + (\cos^2\theta)^3\right] - 3\left[(\sin^2\theta)^2 + (\cos^2\theta)^2\right] + 1$$

$$= 2\left[(\sin^2\theta + \cos^2\theta)^3 - 3\sin^2\theta\cdot\cos^2\theta(\sin^2\theta + \cos^2\theta)\right] - 3\left[(\sin^2\theta + \cos^2\theta)^2 - 2\sin^2\theta\cdot\cos^2\theta\right] + 1$$

$$= 2\left[(1)^3 - 3\sin^2\theta\cdot\cos^2\theta \times 1\right] - 3\left[(1)^2 - 2\sin^2\theta\cdot\cos^2\theta\right] + 1$$

$$= 2 - 6\sin^2\theta\cdot\cos^2\theta - 3 + 6\sin^2\theta\cdot\cos^2\theta + 1 = 0$$

41. $\sin\theta = \cos\theta$

$$\frac{\sin\theta}{\cos\theta} = 1$$

$\tan\theta = 1 = \tan 45^\circ$

$\theta = 45^\circ$

$$= 2\tan^2\theta + \cos^2\theta - 2 = 2(\tan 45^\circ)^2 + (\cos 45^\circ)^2 - 2 = 2(1)^2 + \left(\frac{1}{\sqrt{2}}\right)^2 - 2 = 2 + \frac{1}{2} - 2 = \frac{1}{2}$$

42. $$\operatorname{cosec}\theta - \cot\theta = \frac{1}{3}$$

$$\frac{1}{\operatorname{cosec}\theta - \cot\theta} = 3$$

$$\frac{1}{\operatorname{cosec}\theta - \cot\theta} \times \frac{\operatorname{cosec}\theta + \cot\theta}{\operatorname{cosec}\theta + \cot\theta} = 3$$

$$\frac{\operatorname{cosec}\theta + \cot\theta}{\operatorname{cosec}^2\theta - \cot^2\theta} = 3$$

$$\operatorname{cosec}\theta + \cot\theta = 3$$

True/False

47. In the case of sine function, as 'θ' increases between 0° and 90°, the value of sine also increases, but in the case of the cosine function, as θ increases between 0° and 90°, the value of the cosine function decreases. The result is (+)ve for expression (sin 80° – cos 80°). Hence, the statement is false.

49. $(\tan\theta+2)(2\tan\theta+1) = 2\tan^2\theta+\tan\theta+4\tan\theta+2$

$$= 2\tan^2\theta+5\tan\theta+2 = 2(\tan^2\theta+1)+5\tan\theta = 2\sec^2\theta+5\tan\theta$$

Hence, the statement is true.

Subjective Questions

73. (a) $\sec A = \dfrac{15}{7}$ (i)

$A + B = 90°$

$A = 90 - B$ (ii)

Inserting (ii) in (i) $\sec(90-B) = \dfrac{15}{7}$ and $\operatorname{cosec} B = \dfrac{15}{7}$

(b) $3x = \operatorname{cosec}\theta,\ \dfrac{3}{x} = \cot\theta$

S.B.S

$(3x)^2 = \operatorname{cosec}^2\theta,\ \left(\dfrac{3}{x}\right)^2 = \cot^2\theta$

$9x^2 = \operatorname{cosec}^2\theta$ (i)

$\dfrac{9}{x^2} = \cot^2\theta$ (ii)

Subtracting (ii) from (i), we get $9x^2 - \dfrac{9}{x^2} = \operatorname{cosec}^2\theta - \cot^2\theta$

$$9\left(x^2 - \frac{1}{x^2}\right) = 1$$

$$3\times 3\left(x^2 - \frac{1}{x^2}\right) = 1$$

$$3\left(x^2 - \frac{1}{x^2}\right) = \frac{1}{3}$$

76. (i) $\tan\theta = \dfrac{2mn}{m^2-n^2}$

We know that $\sec^2\theta - \tan^2\theta = 1$

$$\sec^2\theta = 1+\tan^2\theta = 1+\left(\frac{2mn}{m^2-n^2}\right)^2 = 1+\frac{4m^2n^2}{(m^2-n^2)^2} = \frac{(m^2-n^2)^2+4m^2n^2}{(m^2-n^2)^2}$$

$$= \frac{m^4+n^4-2m^2n^2+4m^2n^2}{(m^2-n^2)^2} = \frac{m^4+n^4+2m^2n^2}{(m^2-n^2)^2} = \frac{(m^2+n^2)^2}{(m^2-n^2)^2}$$

$$\sec^2\theta = \left(\frac{m^2+n^2}{m^2-n^2}\right)^2$$

$$\sec\theta = \frac{m^2+n^2}{m^2-n^2}$$

$$\frac{1}{\cos\theta} = \frac{m^2+n^2}{m^2-n^2}$$

$$\cos\theta = \frac{m^2-n^2}{m^2+n^2}$$

$$\sin^2\theta = 1-\cos^2\theta = 1-\left(\frac{m^2-n^2}{m^2+n^2}\right)^2$$

$$= \frac{\left(m^2+n^2\right)^2-\left(m^2-n^2\right)^2}{\left(m^2+n^2\right)^2} = \frac{m^4+n^4+2m^2n^2-m^4-n^4+2m^2n^2}{\left(m^2+n^2\right)^2}$$

$$\sin^2\theta = \frac{4m^2n^2}{\left(m^2+n^2\right)^2} = \left(\frac{2mn}{m^2+n^2}\right)^2$$

$$\sin\theta = \frac{2mn}{m^2+n^2},\ \operatorname{cosec}\theta = \frac{m^2+n^2}{2mn},\ \cot\theta = \frac{1}{\tan\theta} = \frac{m^2-n^2}{2mn}$$

80. $\tan\theta = \dfrac{a}{b}$

$$\frac{\cos\theta+\sin\theta}{\cos\theta-\sin\theta}$$

Dividing the numerator and denominator by cos θ, we get

$$= \frac{1+\tan\theta}{1-\tan\theta} = \frac{1+\frac{a}{b}}{1-\frac{a}{b}} = \frac{b+a}{b-a}$$

86. In order to prove cosec 30°, we have to consider an equivalent ΔABC.

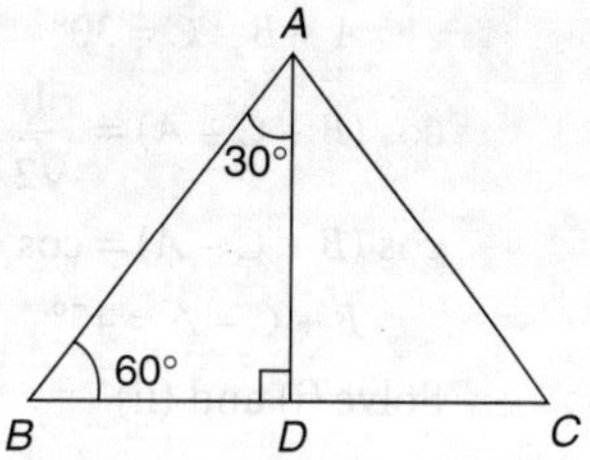

$\Delta ABD \cong \Delta ACD$

So $BD = DC$ [CPCT]

That is $BD = \frac{1}{2}BC$ and $\angle BAD = \angle CAD = \frac{1}{2}\times 60° = 30°$

So $\operatorname{cosec} 30° = \dfrac{AB}{BD} = \dfrac{BC}{\frac{1}{2}BC}$ [∵ ΔABC is equilateral]

$\operatorname{cosec} 30° = \dfrac{2}{1} = 2.$

92. LHS

$$= (\sin^2 5° + \sin^2 85°) + (\sin^2 10° + \sin^2 80°) + \ldots + (\sin^2 45° + \sin^2 90°)$$

$$= [\sin^2 5° + \sin^2 (90 - 5°)] + [\sin^2 10° + \sin^2 (90 - 10°)] + \ldots + \left[\left(\frac{1}{\sqrt{2}}\right)^2 + (1)^2\right]$$

$$= (\sin^2 5° + \cos^2 5°) + (\sin^2 10° + \cos^2 10°) + \ldots + \left[\frac{1}{2}+1\right]$$

$$= 1 + 1 + \ldots + 8 \text{ times} + \frac{1}{2} + 1 = 9\frac{1}{2}$$

101. $4\left[\frac{\sec^2 59° - \cot^2 31°}{3}\right] - \frac{2}{3}\sin 90° + 3\tan^2 56° \times \tan^2 34° = \frac{x}{3}$

LHS

$$= 4\left[\frac{\sec^2(90-31°) - \cot^2 31°}{3}\right] - \frac{2}{3}\times 1 + 3\tan^2(90-34°)\times\tan^2 34°$$

$$= 4\left[\frac{\operatorname{cosec}^2 31 - \cot^2 31°}{3}\right] - \frac{2}{3} + 3\cot^2 34°\cdot\tan^2 34°$$

$$= 4\times\frac{1}{3} - \frac{2}{3} + 3\times\frac{1}{\tan^2 34}\times\tan^2 34 \qquad \left[\because \operatorname{cosec}^2\theta - \cot^2\theta = 1, \cot\theta = \frac{1}{\tan\theta}\right]$$

$$= \frac{4}{3} - \frac{2}{3} + 3\times 1 = \frac{4-2+9}{3} = \frac{11}{3}$$

RHS

$$= \frac{x}{3} \text{ means } \frac{11}{3} = \frac{x}{3}$$

$$x = 11$$

124. $\sin(A + B - C) = \frac{1}{2}$

$\sin(A + B - C) = \sin 30°$

$A + B - C = 30°$ (i)

$\cos(B + C - A) = \frac{1}{\sqrt{2}}$

$\cos(B + C - A) = \cos 45°$

$B + C - A = 45°$ (ii)

Solve (i) and (ii).

128. $\frac{\sin\theta - \cos\theta + 1}{\sin\theta + \cos\theta - 1} = \frac{1}{\sec\theta - \tan\theta}$

LHS

Divide the numerator and denominator by $\cos\theta$.

$$= \frac{\tan\theta - 1 + \sec\theta}{\tan\theta + 1 - \sec\theta} = \frac{\tan\theta + \sec\theta - 1}{\tan\theta - \sec\theta + 1}$$

We have given:

$$\sec^2\theta = 1 + \tan^2\theta$$

$$\sec^2\theta - \tan^2\theta = 1$$

$$= \frac{\tan\theta + \sec\theta - \left(\sec^2\theta - \tan^2\theta\right)}{\tan\theta - \sec\theta + 1} = \frac{(\tan\theta + \sec\theta) - (\sec\theta + \tan\theta)(\sec\theta - \tan\theta)}{\tan\theta - \sec\theta + 1}$$

$$= \frac{(\tan\theta + \sec\theta)(1 - \sec\theta + \tan\theta)}{\tan\theta - \sec\theta + 1} = \tan\theta + \sec\theta$$

$$= (\tan\theta + \sec\theta) \times \frac{\sec\theta - \tan\theta}{\sec\theta - \tan\theta}$$ [On multiplying and dividing by sec θ – tan θ]

$$= \frac{\sec^2\theta - \tan^2\theta}{\sec\theta - \tan\theta} = \frac{1}{\sec\theta - \tan\theta} = \text{RHS}$$

132. $\sec^2\theta - \left[\dfrac{\sin^2\theta - 2\sin^4\theta}{2\cos^2\theta - \cos^2\theta}\right] = 1$

$$\text{LHS} = \sec^2\theta - \left[\frac{\sin^2\theta\left(1 - 2\sin^2\theta\right)}{\cos^2\theta\left(2\cos^2\theta - 1\right)}\right] = \sec^2\theta - \left[\tan^2\theta\left(\frac{1 - 2\left(1 - \cos^2\theta\right)}{2\cos^2\theta - 1}\right)\right]$$

$$= \sec^2\theta - \left[\tan^2\theta\left(\frac{2\cos^2\theta - 1}{2\cos^2\theta - 1}\right)\right] = \sec^2\theta - \tan^2\theta = 1$$ [$\because \sec^2\theta - \tan^2\theta = 1$]

137. Convert $\cot\theta = \dfrac{\cos\theta}{\sin\theta}$, $\text{cosec}\,\theta = \dfrac{1}{\sin\theta}$

$\sec\theta = \dfrac{1}{\cos\theta}$, $\tan\theta = \dfrac{\sin\theta}{\cos\theta}$

Solve by yourself.

139. $\cos\theta + \sin\theta = \sqrt{2}\cos\theta$

To prove: $\cos\theta - \sin\theta = \sqrt{2}\sin\theta$

$$\cos\theta + \sin\theta = \sqrt{2}\cos\theta \quad \text{(i)}$$

S.B.S

$$\cos^2\theta + \sin^2\theta + 2\sin\theta\cos\theta = 2\cos^2\theta$$

$$2\sin\theta\cos\theta = \cos^2\theta - \sin^2\theta$$

$$2\sin\theta\cos\theta = (\cos\theta - \sin\theta)(\cos\theta + \sin\theta)$$

$$2\sin\theta\cos\theta = (\cos\theta - \sin\theta) \times \sqrt{2}\cos\theta$$ [From (i)]

$$\frac{2\sin\theta \cdot \cos\theta}{\sqrt{2}\cos\theta} = (\cos\theta - \sin\theta)$$

$$\sqrt{2}\sin\theta = (\cos\theta - \sin\theta)$$

145. $\sin^8\theta-\cos^8\theta = \left(\sin^2\theta-\cos^2\theta\right)\left(1-2\sin^2\theta\cdot\cos^2\theta\right)$

LHS $= (\sin^4\theta)^2 - (\cos^4\theta)^2 = (\sin^4\theta - \cos^4\theta)(\sin^4\theta+\cos^4\theta)$

$= (\sin^2\theta-\cos^2\theta)(\sin^2\theta+\cos^2\theta)[(\sin^2\theta+\cos^2\theta)^2 - 2\sin^2\theta\cos^2\theta]$

$= (\sin^2\theta-\cos^2\theta)\times 1[(1)^2 - 2\sin^2\theta\cos^2\theta]$

$= (\sin^2\theta-\cos^2\theta)\times(1-2\sin^2\theta\cos^2\theta)$

149. $\sec^6\theta=\tan^6\theta+3\tan^2\theta\sec^2\theta+1$

LHS $= (\sec^2\theta)^3 = (1+\tan^2\theta)^3 = (1)^3+(\tan^2\theta)^3+3\times1\times\tan^2\theta\left[1+\tan^2\theta\right]$

$= 1+\tan^6\theta+3\tan^2\theta\times\sec^2\theta$

152. $(\sec A - \operatorname{cosec} A)(1+\tan A+\cot A) = \tan A\sec A - \cot A\operatorname{cosec} A$

$$\text{LHS} = \left(\frac{1}{\cos A}-\frac{1}{\sin A}\right)\left(1+\frac{\sin A}{\cos A}+\frac{\cos A}{\sin A}\right) = \frac{(\sin A-\cos A)}{\sin A\cdot\cos\theta}\left[\frac{\sin A\cdot\cos A+\sin^2 A+\cos^2 A}{\sin A\cdot\cos A}\right]$$

$$=\frac{\sin^3 A-\cos^3 A}{\sin^2 A\cdot\cos^2 A} = \frac{\sin^3 A}{\sin^2 A\cdot\cos^2 A}-\frac{\cos^3 A}{\sin^2 A\cdot\cos^2 A} = \frac{\sin A\times 1}{\cos A\times\cos A}-\frac{\cos A}{\sin A}\times\frac{1}{\sin A}$$

$= \tan A\cdot\sec A - \cot A\cdot\operatorname{cosec} A$

156. LHS

$$\frac{1}{\operatorname{cosec}A-\cot A}+\frac{1}{\operatorname{cosec}B-\cot B}+\frac{1}{\operatorname{cosec}C-\cot C}$$

$$=\frac{1}{\operatorname{cosec}A-\cot A}\times\frac{\operatorname{cosec}A+\cot A}{\operatorname{cosec}A+\cot A}+\frac{1}{\operatorname{cosec}B-\cot B}\times\frac{\operatorname{cosec}B+\cot B}{\operatorname{cosec}B+\cot B}+$$

$$\frac{1}{\operatorname{cosec}C-\cot C}\times\frac{\operatorname{cosec}C+\cot C}{\operatorname{cosec}C+\cot C}$$

$$=\frac{\operatorname{cosec}A+\cot A}{\operatorname{cosec}^2 A-\cot^2 A}+\frac{\operatorname{cosec}B+\cot B}{\operatorname{cosec}^2 B-\cot^2 B}+\frac{\operatorname{cosec}C+\cot C}{\operatorname{cosec}^2 C-\cot^2 C}$$

$= \operatorname{cosec} A+\cot A+\operatorname{cosec} B+\cot B+\operatorname{cosec} C+\cot C$

163. $x=\gamma\cos\alpha\cdot\sin\beta$

S.B.S

$x^2=\gamma^2\cos^2\alpha\cdot\sin^2\beta$ (i)

$y=\gamma\cos\alpha\cdot\cos\beta$

S.B.S

$y^2=\gamma^2\cos^2\alpha\cdot\cos^2\beta$ (ii)

$z=\gamma\sin\alpha$

S.B.S

$z^2=\gamma^2\sin^2\alpha$ (iii)

Adding (i), (ii) and (iii), we get

$$x^2+y^2+z^2=\gamma^2\cos^2\alpha\cdot\sin^2\beta+\gamma^2\cos^2\alpha\cdot\cos^2\beta+\gamma^2\sin^2\alpha$$

$$= \gamma^2\left[\cos^2\alpha\cdot\sin^2\beta+\cos^2\alpha\cdot\cos^2\beta+\sin^2\alpha\right]$$

$$= \gamma^2\left[\cos^2\alpha\left(\sin^2\beta+\cos^2\beta\right)+\sin^2\alpha\right] = \gamma^2\left[\cos^2\alpha+\sin^2\alpha\right] \quad \left[\because \sin^2\theta+\cos^2\theta=1\right]$$

$$= \gamma^2\cdot 1 = \gamma^2$$

164. $BD = DC$

That is $DC = \frac{1}{2}BC$ (i)

$$\tan\theta = \frac{AC}{DC} \quad \text{(ii)}$$

$$\tan\phi = \frac{AC}{BC} \quad \text{(iii)}$$

Dividing (ii) by (iii), we get $\frac{\tan\theta}{\tan\phi} = \frac{\frac{AC}{DC}}{\frac{AC}{BC}} = \frac{BC}{DC} = \frac{BC}{\frac{1}{2}BC}$ [From (i)]

$$\frac{\tan\theta}{\tan\phi} = \frac{2}{1}$$

168. LHS

$$\sin\theta(1+\tan\theta)+\cos\theta(1+\cot\theta) = \sin\theta\left[1+\frac{\sin\theta}{\cos\theta}\right]+\cos\theta\left[1+\frac{\cos\theta}{\sin\theta}\right]$$

$$= \sin\theta\left[\frac{\cos\theta+\sin\theta}{\cos\theta}\right]+\cos\theta\left[\frac{\sin\theta+\cos\theta}{\sin\theta}\right] = \left(\sin\theta+\cos\theta\right)\left[\frac{\sin\theta}{\cos\theta}+\frac{\cos\theta}{\sin\theta}\right]$$

$$= \left(\sin\theta+\cos\theta\right)\left[\frac{\sin^2\theta+\cos^2\theta}{\sin\theta\cdot\cos\theta}\right] = \frac{\sin\theta+\cos\theta}{\sin\theta\cdot\cos\theta}$$

$$= \frac{\sin\theta}{\sin\theta\cdot\cos\theta}+\frac{\cos\theta}{\sin\theta\cdot\cos\theta} = \frac{1}{\cos\theta}+\frac{1}{\sin\theta} = \sec\theta+\text{cosec}\,\theta$$

171. LHS

$$\left(\frac{(1+\sin\theta)-\cos\theta}{(1+\sin\theta)+\cos\theta}\right)^2 = \frac{(1+\sin\theta)^2+\cos^2\theta-2\cos\theta(1+\sin\theta)}{(1+\sin\theta)^2+\cos^2\theta+2\cos\theta(1+\sin\theta)}$$

$$= \frac{1+\sin^2\theta+2\sin\theta+\cos^2\theta-2\cos\theta\,(1+\sin\theta)}{1+\sin^2\theta+2\sin\theta+\cos^2\theta+2\cos\theta\,(1+\sin\theta)}$$

$$= \frac{1+1+2\sin\theta-2\cos\theta(1+\sin\theta)}{1+1+2\sin\theta+2\cos\theta(1+\sin\theta)} \quad \left[\because \sin^2\theta+\cos^2\theta=1\right]$$

$$= \frac{2+2\sin\theta-2\cos\theta(1+\sin\theta)}{2+2\sin\theta+2\cos\theta(1+\sin\theta)} = \frac{2(1+\sin\theta)-2\cos\theta(1+\sin\theta)}{2(1+\sin\theta)+2\cos\theta(1+\sin\theta)}$$

$$= \frac{2(1+\sin\theta)(1-\cos\theta)}{2(1+\sin\theta)(1+\cos\theta)} = \frac{1-\cos\theta}{1+\cos\theta}$$

173. LHS

$$(\tan A+\text{cosec}B)^2-(\cot B-\sec A)^2$$

$$=\left[\tan^2 A+\text{cosec}^2B+2\tan A\cdot\text{cosec}\,B\right]-\left[\cot^2 B+\sec^2 A-2\cot B\cdot\sec A\right]$$

$$=\tan^2 A+\text{cosec}^2B+2\tan A\cdot\text{cosec}\,B-\cot^2 B-\sec^2 A+2\cot B\cdot\sec A$$

$$=\left(\text{cosec}^2B-\cot^2 B\right)-\left(\sec^2 A-\tan^2 A\right)+2\tan A\cdot\text{cosec}B+2\cot B\cdot\sec A$$

$$=1-1+2\tan A\cdot\text{cosec}\,B+2\cot B\cdot\sec A=2\tan A\cdot\text{cosec}B+2\cot B\cdot\sec A$$

$$=2\left[\frac{\sin A}{\cos A}\times\frac{1}{\sin B}+\frac{\cos B}{\sin B}\times\frac{1}{\cos A}\right]=2\left[\frac{\sin A+\cos B}{\cos A\cdot\sin B}\right]$$

RHS

$$=2\tan A\cdot\cot B(\text{cosec}A+\sec B)$$

$$=2\tan A\cdot\cot B\cdot\text{cosec}A+2\tan A\cdot\cot B\cdot\sec B$$

$$=2\times\frac{\sin A}{\cos A}\times\frac{\cos B}{\sin B}\times\frac{1}{\sin A}+2\times\frac{\sin A}{\cos A}\times\frac{\cos B}{\sin B}\times\frac{1}{\cos B}=2\left[\frac{\cos B}{\cos A\cdot\sin B}+\frac{\sin A}{\cos A\cdot\sin B}\right]$$

$$=2\left[\frac{\cos B+\sin A}{\cos A\cdot\sin B}\right]$$

180. We know that

$$\tan\theta=\pm\sqrt{\sec^2\theta-1}=\pm\sqrt{\left(x-\frac{1}{4x}\right)^2-1}=\pm\left(x-\frac{1}{4x}\right)$$

$$\sec\theta+\tan\theta=x+\frac{1}{4x}+x-\frac{1}{4x}=2x$$

Or

$$\sec\theta+\tan\theta=x+\frac{1}{4x}-\left(x-\frac{1}{4x}\right)=\frac{1}{2x}$$

183. $\sec\theta+\tan\theta=p$

$$\frac{1}{\cos\theta}+\frac{\sin\theta}{\cos\theta}=p$$

$$\frac{1+\sin\theta}{\cos\theta}=p$$

S.B.S

$$\frac{(1+\sin\theta)^2}{\cos^2\theta}=p^2$$

$$\frac{(1+\sin\theta)^2}{1-\sin^2\theta}=p^2$$

$$\frac{(1+\sin\theta)^2}{(1-\sin\theta)(1+\sin\theta)}=p^2$$

$$\frac{1+\sin\theta}{1-\sin\theta}=p^2 \qquad \text{(i)}$$

$$\frac{p^2-1}{p^2+1}=\frac{\dfrac{1+\sin\theta}{1-\sin\theta}-1}{\dfrac{1+\sin\theta}{1-\sin\theta}+1}=\frac{\dfrac{1+\sin\theta-1+\sin\theta}{1-\sin\theta}}{\dfrac{1+\sin\theta+1-\sin\theta}{1-\sin\theta}}$$

$$=\frac{2\sin\theta}{2}=\sin\theta$$

185. $\operatorname{cosec}\theta-\sin\theta=l$

$$\frac{1}{\sin\theta}-\sin\theta=l$$

$$\frac{1-\sin^2\theta}{\sin\theta}=l$$

$$\frac{\cos^2\theta}{\sin\theta}=l \quad \text{(i)}$$

And $\sec\theta-\cos\theta=m$

$$\frac{1}{\cos\theta}-\cos\theta=m$$

$$\frac{\sin^2\theta}{\cos\theta}=m \quad \text{(ii)}$$

LHS

$$l^2m^2(l^2+m^2+3)=\left(\frac{\cos^2\theta}{\sin\theta}\right)^2\left(\frac{\sin^2\theta}{\cos\theta}\right)^2\left[\left(\frac{\cos^2\theta}{\sin\theta}\right)^2+\left(\frac{\sin^2\theta}{\cos\theta}\right)^2+3\right]$$

$$\frac{\cos^4\theta}{\sin^2\theta}\times\frac{\sin^4\theta}{\cos^2\theta}\left[\frac{\cos^4\theta}{\sin^2\theta}+\frac{\sin^4\theta}{\cos^2\theta}+3\right]=\cancel{\sin 2\theta\cdot\cos 2\theta}\left[\frac{\cos^6\theta+\sin^6\theta+3\sin^2\theta\cos^2\theta}{\cancel{\sin^2\theta}\cdot\cancel{\cos^2\theta}}\right]$$

$$=\cos^6\theta+\sin^6\theta+3\sin^2\theta\cdot\cos^2\theta=\left(\cos^2\theta\right)^3+\left(\sin^2\theta\right)^3+3\sin^2\theta\cdot\cos^2\theta$$

$$=\left(\cos^2\theta+\sin^2\theta\right)^3-3\cos^2\theta\cdot\sin^2\theta+3\sin^2\theta\cdot\cos^2\theta$$

$[\because (a+b)^3=a^3+b^3+3ab(a+b)]$

$=(1)^3=1$ RHS

193. $\sin A+\cos A=3$

$$\sin A+\frac{1}{\sin A}=3$$

$$\frac{\sin^2A+1}{\sin A}=3$$

S.B.S

$$\frac{\sin^4A+1+2\sin^2A}{\sin^2A}=9$$

$$\frac{\sin^4A+1}{\sin^2A}+\frac{2\sin^2A}{\sin^2A}=9$$

$$\frac{\sin^4 A + 1}{\sin^2 A} + 2 = 9$$

$$\frac{\sin^4 A + 1}{\sin^2 A} = 7$$

Chapter 9: Some Applications of Trigonometry

5 Try to Solve

1. In right ΔBCD, $\frac{x}{20} = \cot 45° = 1$

$x = 20$ (i)

In right ΔACD, $\frac{h+20}{x} = \tan 60°$

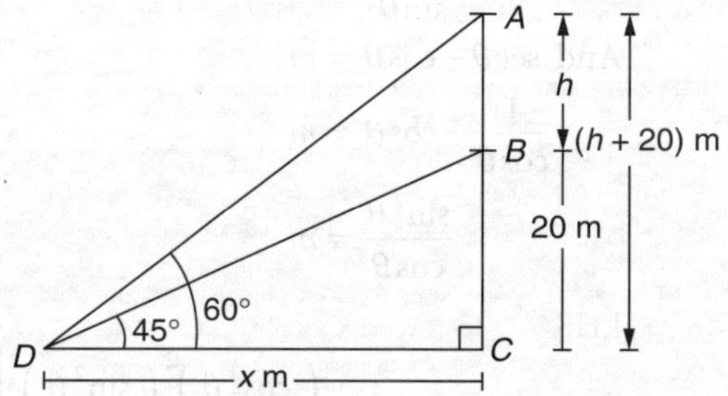

$$\frac{h+20}{x} = \sqrt{3}$$

$$h + 20 = x\sqrt{3}$$

$$h + 20 = 20\sqrt{3} \qquad \text{[From (i)]}$$

$$h = 20\sqrt{3} - 20 = 20\left(\sqrt{3} - 1\right) \text{ m}$$

2. We need to find h.

In right ΔABC, $\frac{h}{x} = \tan 45° = 1$

$h = x$ (i)

In right ΔDBC, $\frac{h+1.6}{x} = \tan 60° = \sqrt{3}$

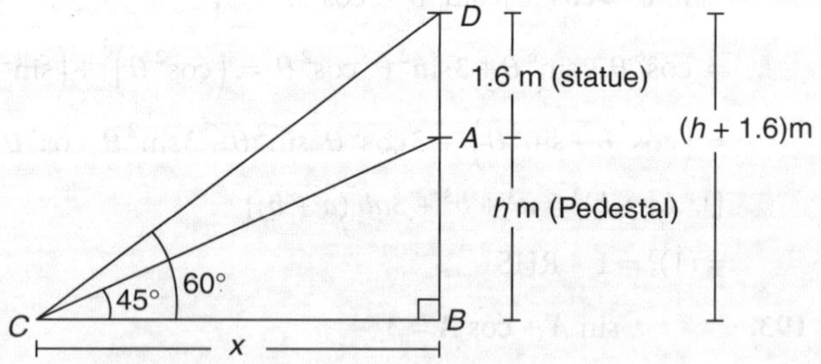

$$h + 1.6 = \sqrt{3}\,x$$

$$h + 1.6 = \sqrt{3}\,h \qquad \text{[From (i)]}$$

$$1.6 = h\left(\sqrt{3} - 1\right)$$

$$h = \frac{1.6}{\sqrt{3}-1} \times \frac{\sqrt{3}+1}{\sqrt{3}+1} = \frac{1.6\left(\sqrt{3}+1\right)}{2} = 0.8\left(\sqrt{3}+1\right) \text{ m}$$

3. In right ΔABD, $\frac{x}{50} = \cot 60° = \frac{1}{\sqrt{3}}$

$$x = \frac{50}{\sqrt{3}} \text{ m} \qquad \text{(i)}$$

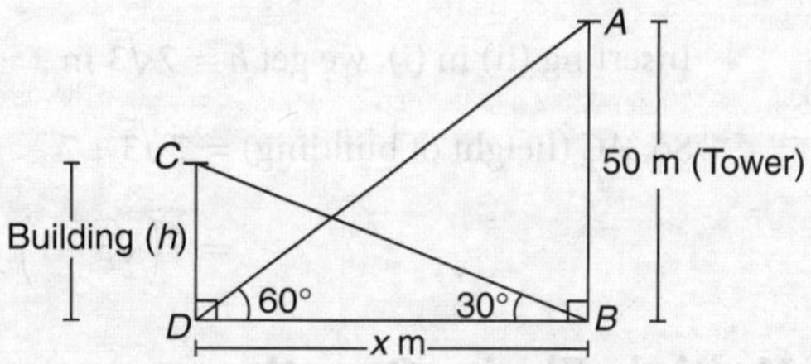

In right ΔCDB, $\frac{h}{x} = \tan 30° = \frac{1}{\sqrt{3}}$

$$h = \frac{x}{\sqrt{3}} = \frac{\frac{50}{\sqrt{3}}}{\sqrt{3}} \qquad \text{[From (i)]}$$

$$h = \frac{50}{3} = 16\frac{1}{3} \text{ m}$$

7 Try to Solve

2. We have to find out $AB = CD = h = ?$ $BE = ?$ $ED = ?$

In right ΔABE, $\frac{h}{x} = \tan 60° = \sqrt{3}$

$$h = \sqrt{3}\,x \text{ m} \qquad \text{(ii)}$$

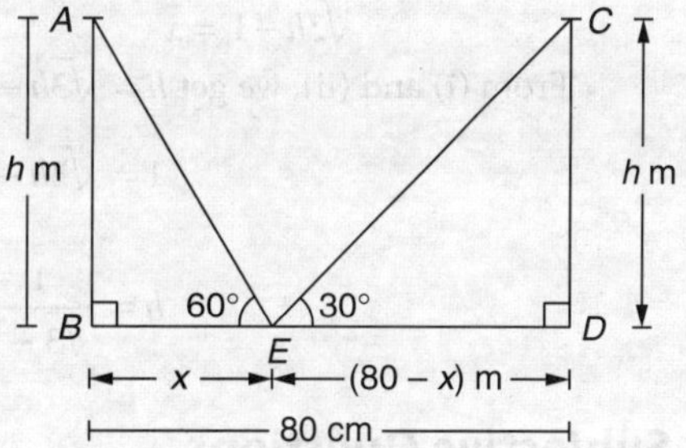

In right ΔCDE, $\frac{h}{80 - x} = \tan 30° = \frac{1}{\sqrt{3}}$

$$h = \frac{80 - x}{\sqrt{3}} \qquad \text{(ii)}$$

From (i) and (ii), we get $\sqrt{3}x = \frac{80 - x}{\sqrt{3}}$

$$3x = 80 - x$$
$$4x = 80$$
$$x = 20 \text{ m}$$

Inserting the value value of x in (i), we get $h = 20\sqrt{3}$ m

Hence, the height of the pole = $20\sqrt{3}$ m and the distance of the point $BE = 20$ m and $ED = 60$ m.

3. In right ΔABD, $\frac{h}{x} = \tan 60° = \sqrt{3}$

$$h = \sqrt{3}x \qquad \text{(i)}$$

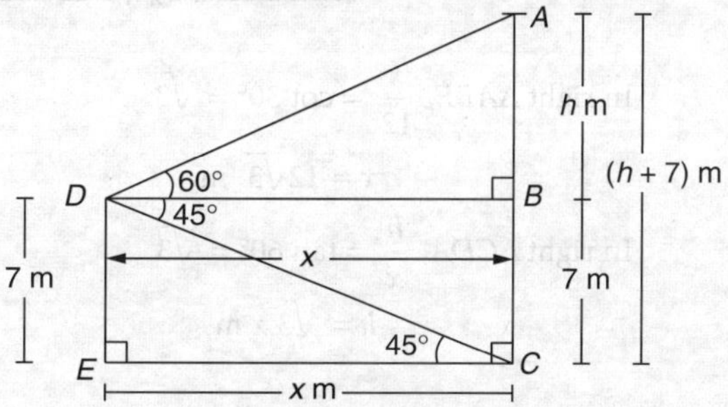

In right ΔDEC, $\frac{x}{7} = \cot 45° = 1$

$$x = 7 \qquad \text{(ii)}$$

Inserting (ii) in (i), we get $h = 7\sqrt{3}$ m

So, AC (height of building) $= 7\sqrt{3} + 7$

$$= 7\left(\sqrt{3}+1\right) \text{ m}$$

Multiple Choice Questions

17. (b) In right ΔABC, $\frac{h}{x} = \tan 45° = 1$

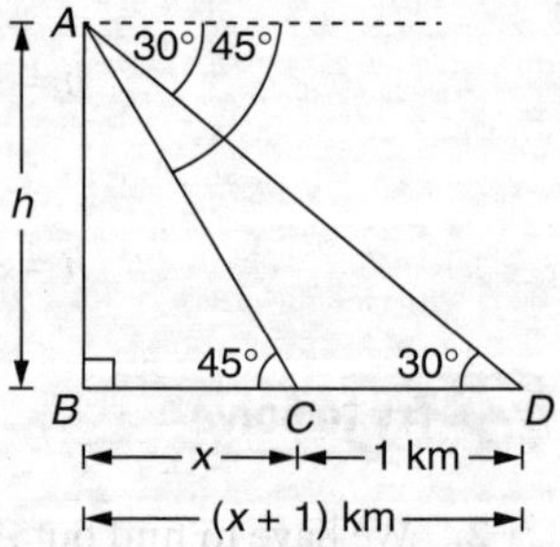

$$h = x \quad \text{(i)}$$

In right ΔABD, $\frac{h}{x+1} = \tan 30° = \frac{1}{\sqrt{3}}$

$$h = \frac{x+1}{\sqrt{3}}$$

$$\sqrt{3}h - 1 = x \quad \text{(ii)}$$

From (i) and (ii), we get $h = \sqrt{3}h - 1$

$$1 = \sqrt{3}h - h = h\left(\sqrt{3}-1\right)$$

$$h = \frac{1}{\sqrt{3}-1} \times \frac{\sqrt{3}+1}{\sqrt{3}+1} = \frac{\sqrt{3}+1}{2} \text{ km}$$

Subjective Questions

39.

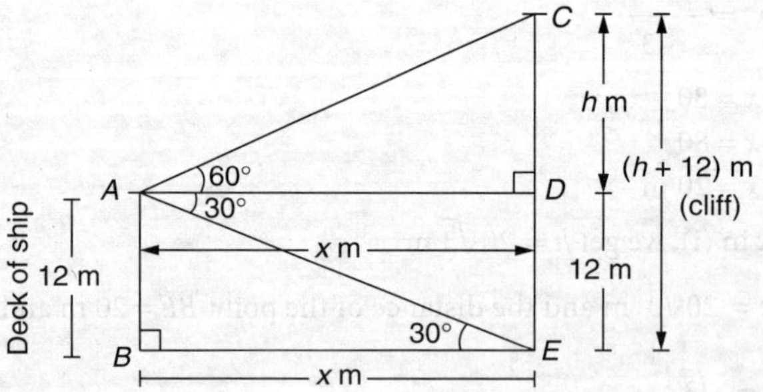

In right ΔABE, $\frac{x}{12} = \cot 30° = \sqrt{3}$

$$x = 12\sqrt{3} \text{ m} \quad \text{(i)}$$

In right ΔCDA, $\frac{h}{x} = \tan 60° = \sqrt{3}$

$$h = \sqrt{3}\,x \text{ m} \quad \text{(ii)}$$

Inserting (i) in (ii), we get $h = \sqrt{3} \times 12\sqrt{3} = 36$ m

Height of cliff = 36 + 12 = 48 m

Distance of the cliff = $12\sqrt{3} = 12 \times 1.73 = 20.78$ m

41. We need to find out $CD = x = ?$

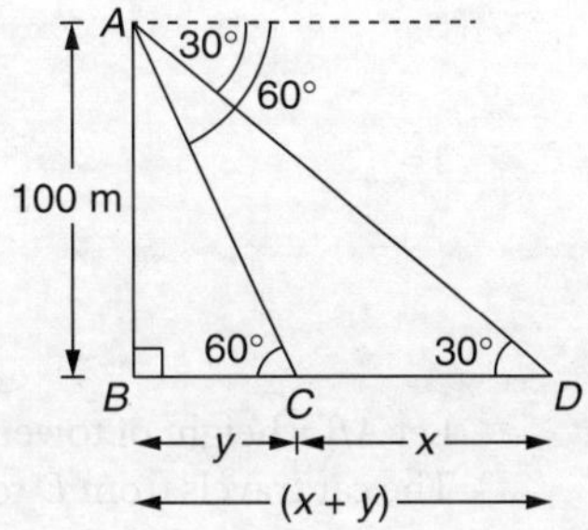

In right ΔABC, $\frac{y}{100} = \cot 60° = \frac{1}{\sqrt{3}}$

$$y = \frac{100}{\sqrt{3}} \text{ m} \quad \text{(i)}$$

In right ΔABD, $\frac{x+y}{100} = \cot 30° = \sqrt{3}$

$$x + y = 100\sqrt{3}$$

$$x + \frac{100}{\sqrt{3}} = 100\sqrt{3}$$

$$x = \frac{100\sqrt{3}}{1} - \frac{100}{\sqrt{3}} = \frac{200}{\sqrt{3}} \times \frac{\sqrt{3}}{\sqrt{3}} = \frac{200\sqrt{3}}{3} \text{ m}$$

$$x = 115.466 \text{ m}$$

46. In right ΔABC, $\frac{h}{x} = \tan 60° = \sqrt{3}$

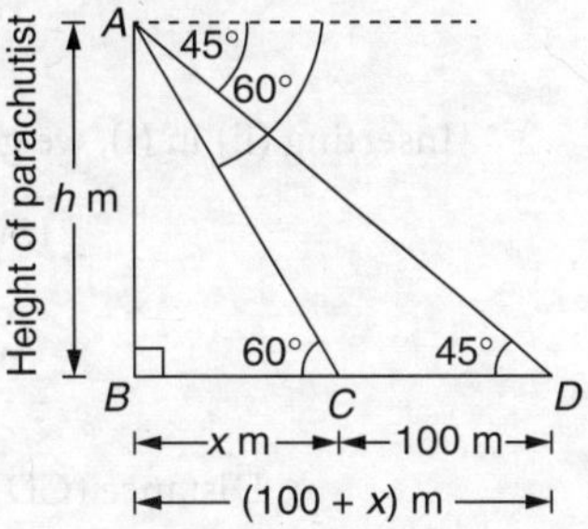

$$h = \sqrt{3}x \quad \text{(i)}$$

In right ΔABD, $\frac{h}{100+x} = \tan 45° = 1$

$$h = 100 + x$$

$$h - 100 = x \quad \text{(ii)}$$

Inserting (ii) in (i), we get

$$h = \sqrt{3}(h - 100) = \sqrt{3}h - 100\sqrt{3}$$

$$100\sqrt{3} = h(\sqrt{3} - 1)$$

$$\frac{100\sqrt{3}}{\sqrt{3}-1} = h$$

$$\frac{100\sqrt{3}}{\sqrt{3}-1} \times \frac{\sqrt{3}+1}{\sqrt{3}+1} = h$$

$$\frac{\cancel{100}\sqrt{3}(\sqrt{3}+1)}{\cancel{2}} = h$$

$$50\left(3+\sqrt{3}\right)=h$$

$$236.6 \text{ m}=h$$

48.

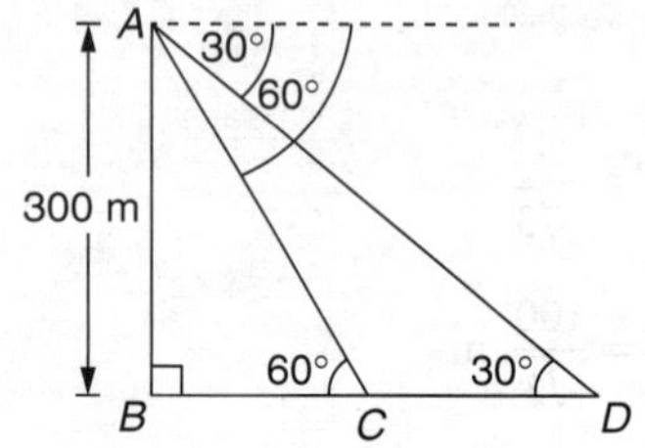

Let AB = height of tower = 300 m. Originally, the car is at position D with an angle of depression of 30°. The car travels from D to C at a velocity of v m/sec. Time taken from D to C is 15 seconds; so, distance = 15 v m. From C to B, time taken is t seconds; hence, the C to B distance becomes $v\,t$ m.

In right ΔABD, $\dfrac{vt+15v}{300}=\cot 30° = \sqrt{3}$

$$vt + 15v = 300\sqrt{3} \qquad \text{(i)}$$

In right ΔABC, $\dfrac{vt}{300}=\cot 60° = \dfrac{1}{\sqrt{3}}$

$$vt = \frac{300}{\sqrt{3}} = 100\sqrt{3} \qquad \text{(i)}$$

Inserting (ii) in (i), we get $100\sqrt{3}+15v=300\sqrt{3}$

$$15v = 300\sqrt{3}-100\sqrt{3} = 200\sqrt{3}$$

$$v = \frac{200\sqrt{3}}{15} \text{ m/sec}$$

$$\text{Distance (CD)} = vt = \frac{200\sqrt{3}}{15}\times 15 = 200\sqrt{3} \text{ m}$$

49. Calculate PB and BQ. Solve this problem by yourself.

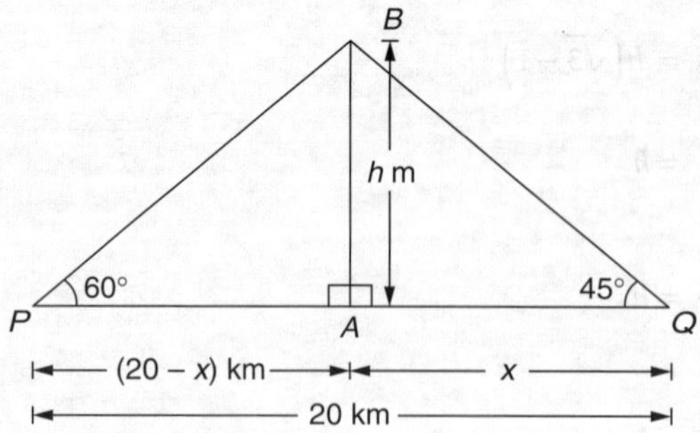

50. Find PQ = ? XQ = ?

In right ΔQLY, $\dfrac{h}{x}=\tan 45° = 1$

$$h=x \qquad \text{(i)}$$

In right ΔQPX, $\dfrac{h+40}{x} = \tan 60° = \sqrt{3}$

$$h + 40 = \sqrt{3}x \qquad \text{(ii)}$$

Inserting (i) in (ii), we get $h + 40 = \sqrt{3}h$

$$40 = h\left(\sqrt{3}-1\right)$$

$$h = \frac{40}{\sqrt{3}-1} \times \frac{\sqrt{3}+1}{\sqrt{3}+1} = \frac{40\left(\sqrt{3}+1\right)}{2} = 20\left(\sqrt{3}+1\right)\text{ m}$$

$$x = 20\left(\sqrt{3}+1\right)\text{ m} \qquad \text{[From (i)]}$$

Height of tower $= h + 40 = 20\sqrt{3} + 20 + 40$

$$= 20\left(\sqrt{3}+3\right)\text{ m} = 20(1.73 + 3) = 94.64 \text{ m}$$

In right ΔQPX, $\dfrac{XQ}{PX} = \sec 60°$

$$\frac{XQ}{20\left(\sqrt{3}+1\right)} = 2$$

$$XQ = 40\left(\sqrt{3}+1\right)\text{ m} = 40(1.73 + 1) = 109.3 \text{ m}$$

54. In right ΔABC, $\dfrac{h}{y} = \tan 45° = 1$

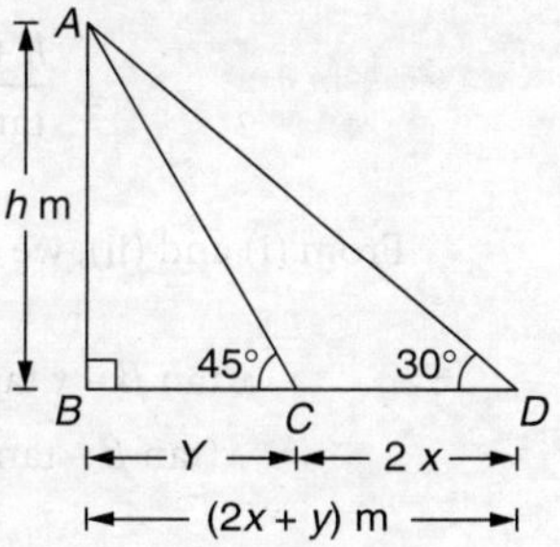

$$h = y \qquad \text{(i)}$$

In right ΔABD, $\dfrac{h}{2x+y} = \tan 30° = \dfrac{1}{\sqrt{3}}$

$$\sqrt{3}h = 2x + y$$

$$\sqrt{3}h = 2x + h \qquad \text{[From (i)]}$$

$$h = \frac{2x}{\sqrt{3}-1} \times \frac{\sqrt{3}+1}{\sqrt{3}+1} = \frac{2x\left(\sqrt{3}+1\right)}{2} = x\left(\sqrt{3}+1\right)\text{ m}$$

55. $\tan \theta = \dfrac{5}{12}$

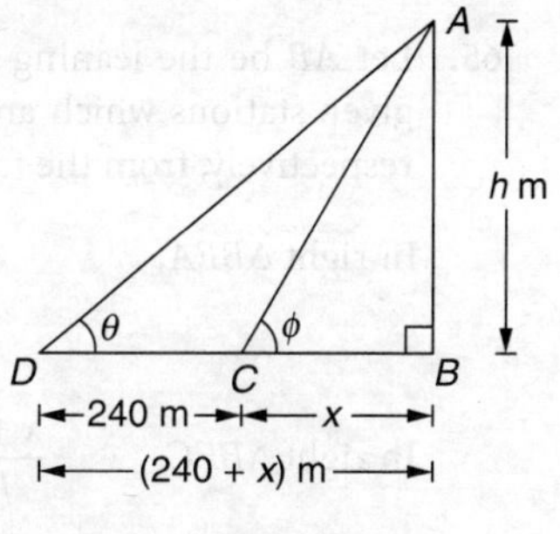

$$\tan \phi = \frac{3}{4}$$

In right ΔABC, $\dfrac{h}{x} = \tan \phi = \dfrac{3}{4}$

$$h = \frac{3}{4}x$$

$$x = \frac{4}{3}h \qquad \text{(i)}$$

In right ΔABD, $\dfrac{h}{240+x} = \tan\theta = \dfrac{5}{12}$

$$h = \frac{5}{12}[240+x]$$

$$h = \frac{5}{12}\left[240+\frac{4}{3}h\right] \qquad \text{[From (i)]}$$

$$h = 100 + \frac{5}{9}h$$

$$h - \frac{5}{9}h = 100$$

$$\frac{4}{9}h = 100$$

$$h = 225 \text{ m}$$

64. In right ΔBCD, $\dfrac{x}{y} = \tan\alpha$

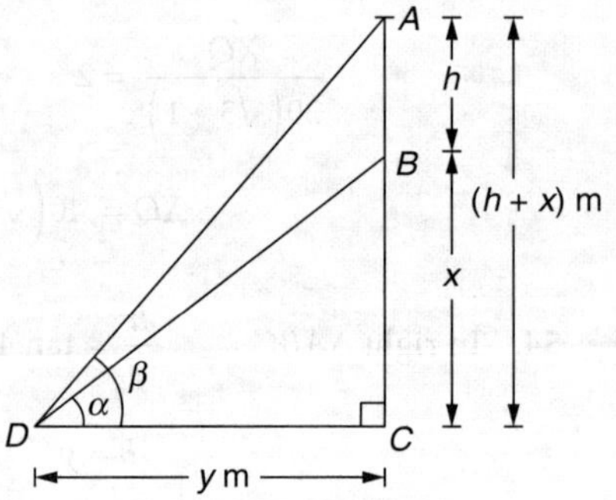

$$\frac{x}{\tan\alpha} = y \qquad \text{(i)}$$

In right ΔACD, $\dfrac{h+x}{y} = \tan\beta$

$$\frac{h+x}{\tan\beta} = y \qquad \text{(ii)}$$

From (i) and (ii), we get $\dfrac{x}{\tan\alpha} = \dfrac{h+x}{\tan\beta}$

$$x\tan\beta - x\tan\alpha = h\tan\alpha$$

$$x(\tan\beta - \tan\alpha) = h\tan\alpha$$

$$x = \frac{h\tan\alpha}{\tan\beta - \tan\alpha}$$

$$\text{Height of tower} = \frac{h\tan\alpha}{\tan\beta - \tan\alpha}$$

65. Let AB be the leaning tower and let C and D be two given stations which are at a distance of a and b units, respectively, from the foot A of the tower.

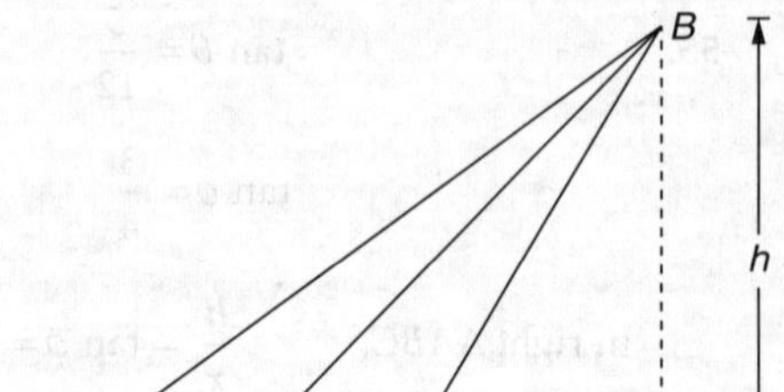

In right ΔBEA, $\dfrac{x}{h} = \cot\theta$

$$x = h\cot\theta \qquad \text{(i)}$$

In right ΔBEC, $\dfrac{x+a}{h} = \cot\alpha$

$$x + a = h\cot\alpha$$

$$x = h\cot\alpha - a \qquad \text{(ii)}$$

In right ΔBED, $\dfrac{x+b}{h} = \cot \beta$

$$x + b = h \cot \beta$$

$$x = h \cot \beta - b \qquad \text{(iii)}$$

From (i) and (ii), we get $h \cot \theta = h \cot \alpha - a$

$$\frac{a}{\cot\alpha - \cot\theta} = h \qquad \text{(iv)}$$

From (i) and (iii), we get $h \cot \beta - b = h \cot \theta$

$$h = \frac{b}{\cot\beta - \cot\theta} \qquad \text{(v)}$$

From (iv) and (v), we get $\dfrac{a}{\cot\alpha - \cot\theta} = \dfrac{b}{\cot\beta - \cot\theta}$

$$a\cot\beta - a\cot\theta = b\cot\alpha - b\cot\theta$$

$$\cot\theta(b-a) = b\cot\alpha - a\cot\beta$$

$$\cot\theta = \frac{b\cot\alpha - a\cot\beta}{b-a}$$

67. We need to find AC.

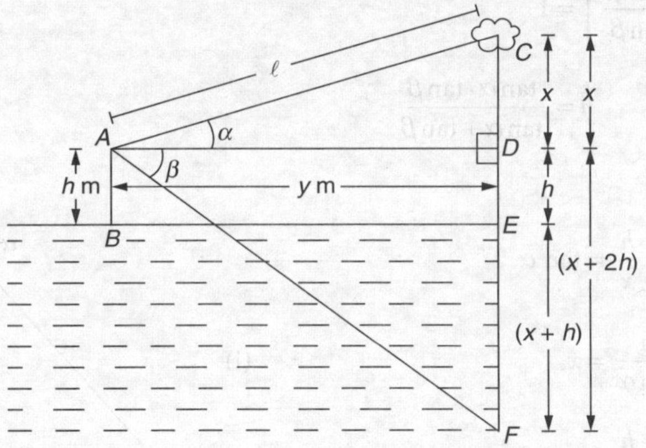

In right ΔCDA, $\dfrac{x}{y} = \tan \alpha$

$$x = y \tan \alpha \qquad \text{(i)}$$

In right ΔADF, $\dfrac{x+2h}{y} = \tan \beta$

$$x = y \tan \beta - 2h \qquad \text{(ii)}$$

From (i) and (ii), we get $y \tan \alpha = y \tan \beta - 2h$

$$\frac{2h}{(\tan\beta - \tan\alpha)} = y \qquad \text{(iii)}$$

In right ΔCDA, $\frac{l}{y} = \sec\alpha$

$$l = y\sec\alpha \quad \text{(iv)}$$

Inserting (iii) in (iv), we get $AC = l = \frac{2h\sec\alpha}{\tan\beta - \tan\alpha}$

68. In right ΔADB, $\frac{h}{x} = \tan\alpha$

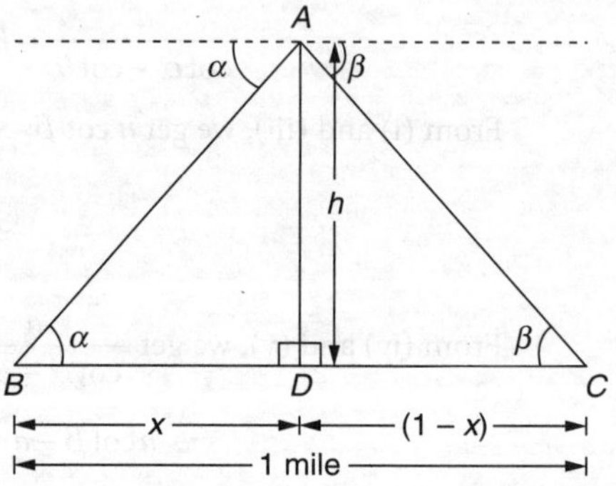

$$\frac{h}{\tan\alpha} = x \quad \text{(i)}$$

In right ΔADC, $\frac{h}{1-x} = \tan\beta$

$$\frac{h}{\tan\beta} = 1 - x \quad \text{(iii)}$$

Inserting (i) in (ii), we get $\frac{h}{\tan\beta} = 1 - \frac{h}{\tan\alpha}$

$$\frac{h}{\tan\beta} + \frac{h}{\tan\alpha} = 1$$

$$h\left[\frac{\tan\alpha + \tan\beta}{\tan\alpha \cdot \tan\beta}\right] = 1$$

$$h = \frac{\tan\alpha \cdot \tan\beta}{\tan\alpha + \tan\beta}$$

69. Find $CD = x + y$.

In right ΔABC, $\frac{h}{x} = \tan\alpha$

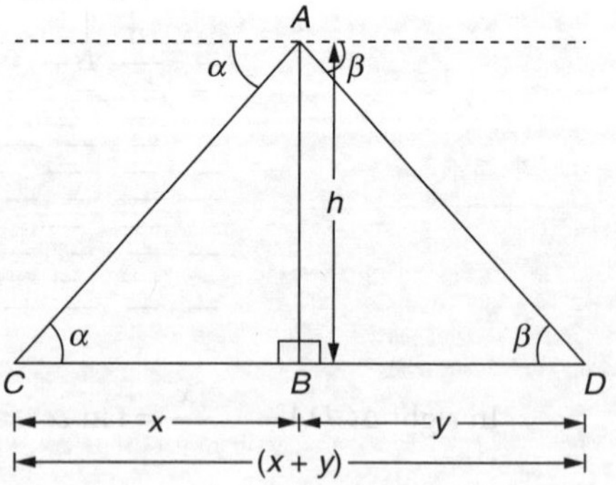

$$\frac{h}{\tan\alpha} = x \quad \text{(i)}$$

In right ΔABD, $\frac{h}{y} = \tan\beta$

$$\frac{h}{\tan\beta} = y \quad \text{(ii)}$$

Adding (i) and (ii), we get

$$x + y = \frac{h}{\tan\alpha} + \frac{h}{\tan\beta} = h\left[\frac{\tan\beta + \tan\alpha}{\tan\alpha \cdot \tan\beta}\right]$$

82. Find $BE = l = ?$

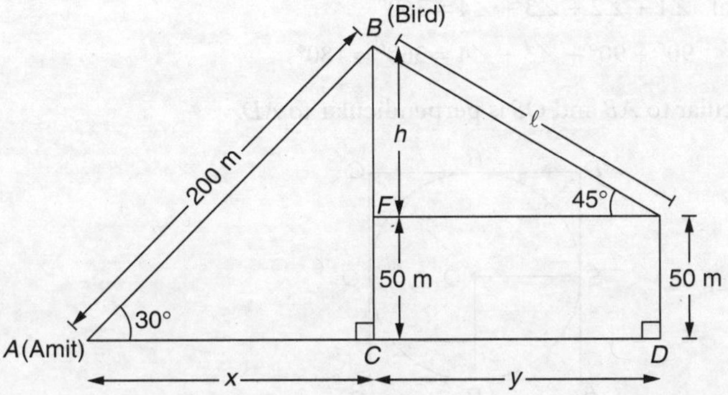

In right ΔBCA, $\frac{h+50}{200} = \sin 30° = \frac{1}{2}$

$h + 50 = 100$

$h = 50$ m (i)

In right ΔBFE, $\frac{l}{h} = \cos 45° = \sqrt{2}$

$l = \sqrt{2}h$

$l = \sqrt{2} \times 50 = 50\sqrt{2}$ [From (i)]

$l = 70.7$ m

Hence, $BE = 70.7$ m.

Chapter 10: Circles

4 Try to Solve

1. To prove: $\angle 3 + \angle 4 = 180°$

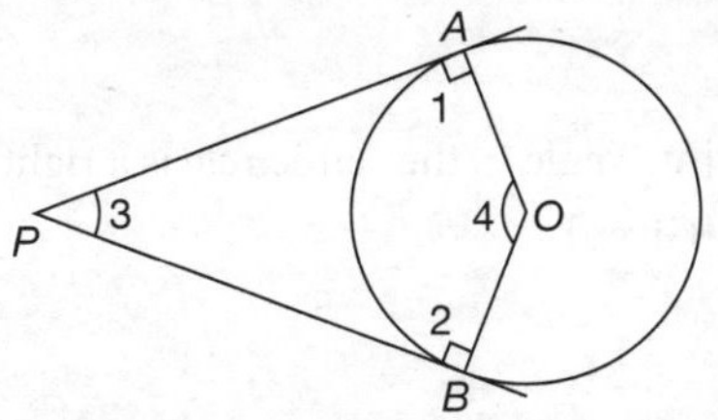

Solution: $\angle 1 = \angle 2 = 90°$ [∵ Radius is perpendicular to the point of contact] (i)

In a quadrilateral, $\angle 1 + \angle 2 + \angle 3 + \angle 4 = 360°$ [ASP of quadrilateral]

$$90° + 90° + \angle 3 + \angle 4 = 360° = 180°$$

2. *OP* is perpendicular to *AB* and *OS* is perpendicular to *AD*.

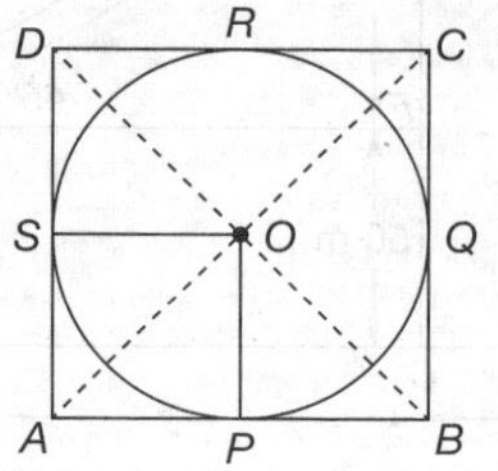

$\Delta OPB \cong OSD$

$PB = SD$ [CPCT] (i)

$AP = AS$ [Lengths of tangents] (ii)

Adding (i) and (ii) $AP + PB = AS + DS \Rightarrow AB = AD$

Similarly $AB = BC = CD = DA$

Hence parallelogram *ABCD* is a rhombus.

Multiple Choice Questions

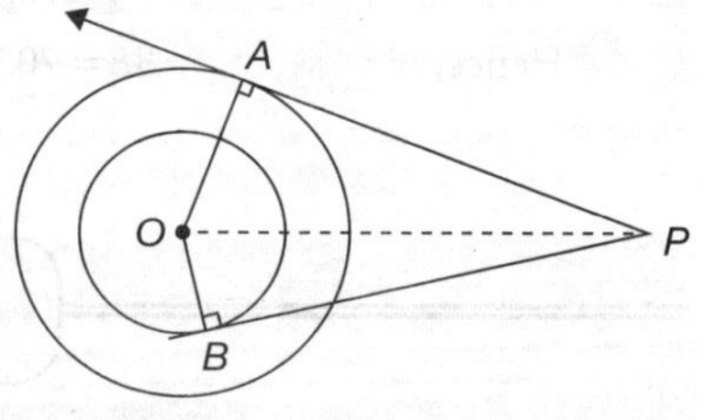

9. (b) $PA = 12$ cm, $OA = 5$ cm, $OB = 3$ cm

$$OP^2 = OA^2 + AP^2 = OB^2 + BP^2$$

$$25 + 144 = 9 + BP^2$$

$$BP^2 = 160$$

$$BP = \sqrt{160} = 4\sqrt{10} \text{ cm}$$

12. (b) $AB \parallel PR$, *BQ* is a transversal

$\angle ABQ = \angle BQR = 70°$ [∵ Alternate interior angles] (i)

$\angle BQR = \angle BAQ = 70°$ [∵ Angle in the alternate segment] (ii)

In ΔABQ, $\angle BAQ + \angle ABQ + \angle AQB = 180°$ [∵ ASP of Δs]

$70° + 70° + \angle AQB = 180°$ [From (i) and (ii)]

$\angle AQB = 40°$

18. (b) Since *AC* is the diameter

$\angle ABC = 90°$ [∵ Angle in the semicircle is a right angle] (i)

$\angle ACB = 50°$ [Given] (ii)

In ΔABC

$\angle ABC + \angle ACB + \angle BAC = 180°$ [∵ ASP of Δs]

$90° + 50° + \angle BAC = 180°$

$\angle BAC = 40°$ (iii)

OA is perpendicular to *TA*.

$\angle BAC + \angle BAT = 90°$ [Radius is perpendicular to point to contact]

$40° + \angle BAT = 90°$

$\angle BAT = 50°$

22. (d) In a quadrilateral, $\angle AOB + \angle OBP + \angle BPA + \angle PAO = 360°$ [∵ ASP of quadrilateral]

$110° + 90° + \angle BPA + 90° = 360°$

$\angle BPA = 70°$ (i)

$AP = BP$ [∵ Tangents drawn from the same external point are equal in length]

$\angle PBA = \angle PAB = x$ [Angles opposite to equal sides are equal] (ii)

In ΔPBA, $x + x + \angle BPA = 180°$ [∵ ASP of Δs]

$x = 55°$

35. (c) OA is perpendicular to AT.

$\angle OAB + \angle BAT = 90°$

In ΔOAB $OA = OB$ [Equal radii]

$\angle OAB = \angle OBA = x$ [∵ Angles opposite to equal sides are equal]

$\angle OAB + \angle OBA + \angle AOB = 180°$ [∵ ASP of Δs]

$x + x + 100° = 180°$

$x = 40°$ (ii)

From (i) and (ii), we get $\angle BAT = 50°$

Subjective Questions

47. OC is perpendicular to AB.

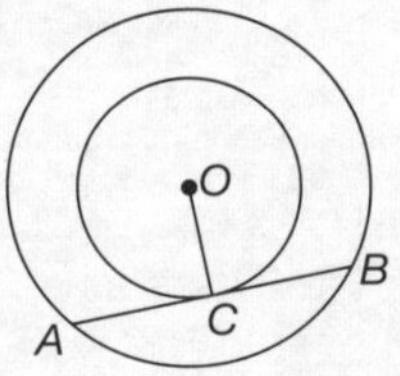

$AC = CB$ [∵ Perpendicular draws from the centre of a circle to the chord bisects the chord]

54. $\Delta POA \cong \Delta COA$

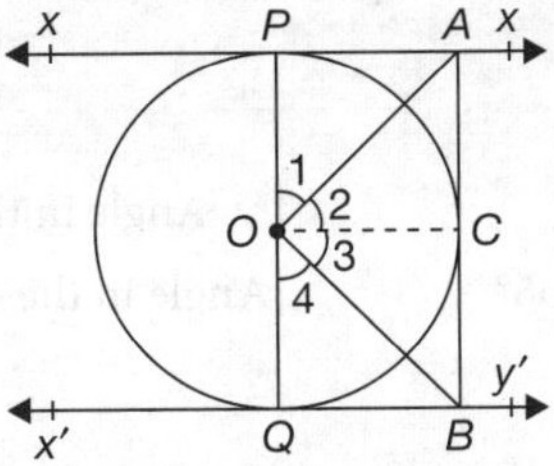

$\angle 1 = \angle 2$ [CPCT] (i)

Similarly $\angle 3 = \angle 4$ [CPCT] (ii)

Since POQ is a line, $\angle 1 + \angle 2 + \angle 3 + \angle 4 = 180°$ [LPA] (iii)

From (i), (ii) and (iii) $\angle AOB = 90°$

58. $AB = AC = 5$ [∵ Tangents drawn from the same external point are equal in length] (i)

$AC = AD$ [∵ Tangents drawn from the same external point are equal in length] (ii)

From (i) and (ii), $AD = 5$ cm

59. $PA = PC$ (i)

$PB = PD$ (ii)

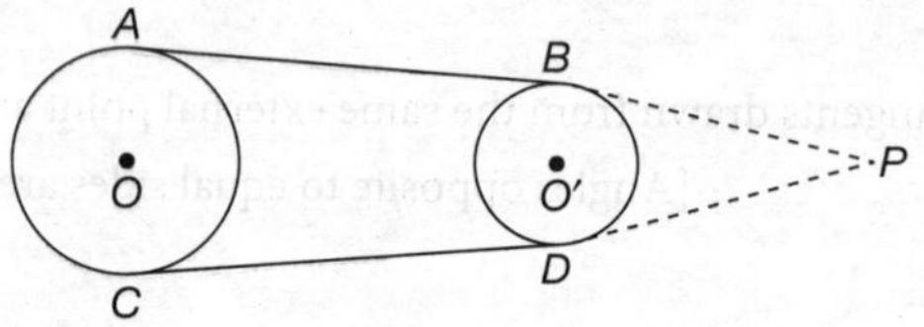

Subtracting (ii) from (i), $PA - PB = PC - PD$

$$AB = CD$$

62. $\angle POR = \angle ROT = 130°$

$$\angle ROT = 2\angle 2$$

$$\angle 2 = 65° \quad \text{(i)}$$

$$\angle ROT + \angle TOQ = 180°$$

$$\angle TOQ = 50° \quad [\angle ROT = 180°]$$

$$\angle TOQ + \angle OQP + \angle 1 = 180°$$

$$\angle 1 = 40° \quad \text{(ii)}$$

Adding (i) and (ii), we get $\angle 1 + \angle 2 = 40° + 65° = 105°$

65.

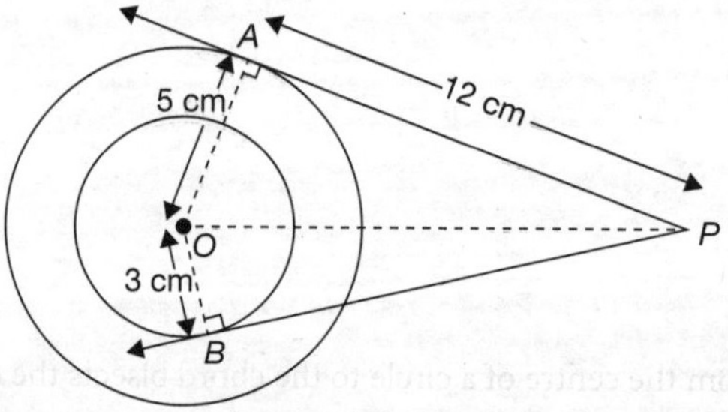

$$OP^2 = OA^2 + AP^2$$

$$OP = 13 \text{ cm}, OB = 3 \text{ cm}$$

$$BP = \sqrt{160} = 12.65 \text{ cm}$$

67. $\angle BAR = 90°$ [∵ Angle in the semicircle] (i)

$\angle PAB = \angle ARB = 58°$ [Angle in the alternate segment] (ii)

In ΔABR, $\angle BAR + \angle ARB + \angle ABR = 180°$ [ASP of Δs]

$90° + 58° + \angle ABR = 180°$ [From (i) and (ii)]

$\angle ABR = 32°$ (iii)

AP and AQ are opposite rays.

$\angle BAP + \angle BAQ = 180°$ [∵ L pair]

$58° + \angle BAQ = 180°$

$\angle BAQ = 122°$ (iv)

In ΔBAQ, $\angle ABQ + \angle BAQ + \angle AQB = 180°$ [∵ ASP of Δs]

$$32° + 122° + \angle AQB = 180°$$

$$\angle AQB = 26°$$

71.

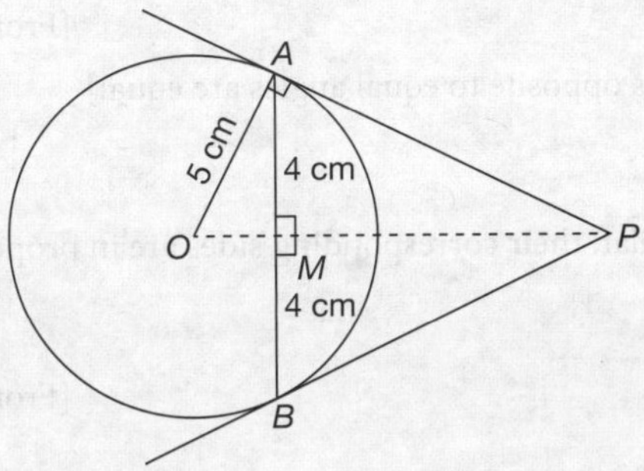

In right ΔOMA, $OM = 3$ cm

$$\Delta OMA \sim \Delta OAP$$

$$\frac{3}{5} = \frac{4}{AP}$$

So $$AP = \frac{20}{3} \text{ cm}$$

74.

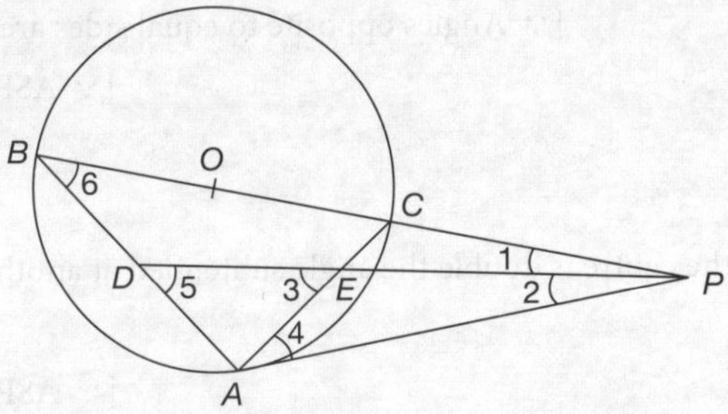

$\angle 1 = \angle 2$ [Given] (i)

$\angle 4 = \angle 6$ [Angle in tangents and chord is equal to angle in alternate segment] (ii)

$\angle 3 = \angle 4 + \angle 2$ [Exterior angle property of a triangle]

$\angle 3 = \angle 6 + \angle 1$ [From (i) and (ii)]

$\angle 3 = \angle 5$ [∵ $\angle 5 = \angle 1 + \angle 6$ [∵ of exterior angle property of a triangle] (iii)

In ΔADE, $\angle 3 = \angle 5$ [From (iii)]

$AD = AE$ [∵ Sides opposite to equal angles are equal]

75. $\angle 1 = \angle 2$ [∵ Angle between tangent and chord is equal to angle in the alternate segment] (i)

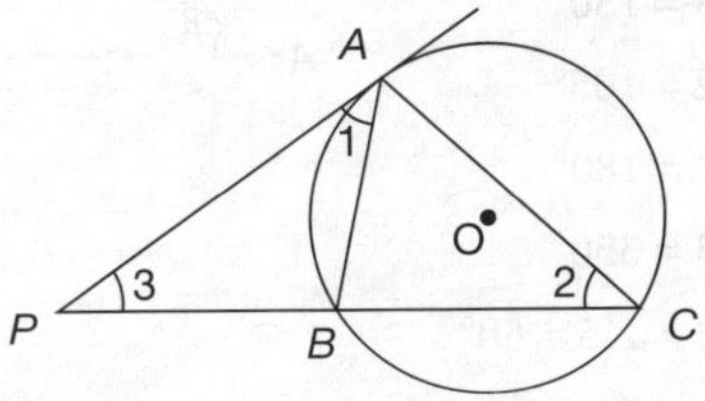

In ΔAPB, $PB = AB$ [Given]

$\angle 1 = \angle 3$ [$\because$ Angles opposite to equal sides are equal] (ii)

From (i) and (ii), $\angle 2 = \angle 3$ (iii)

In ΔPAC, $\angle 2 = \angle 3$ [From (iii)]

$AP = AC$ [$\because$ Sides opposite to equal angles are equal] (iv)

Therefore $\Delta PAC \sim \Delta PBA$ [$\because$ AA]

$\frac{PA}{PB} = \frac{PC}{PA}$ [$\because$ If two Δs are similar, their corresponding sides are in proportion]

$PA^2 = PB \times PC$

$AC^2 = PB \times PC$ [From (iv)]

76.

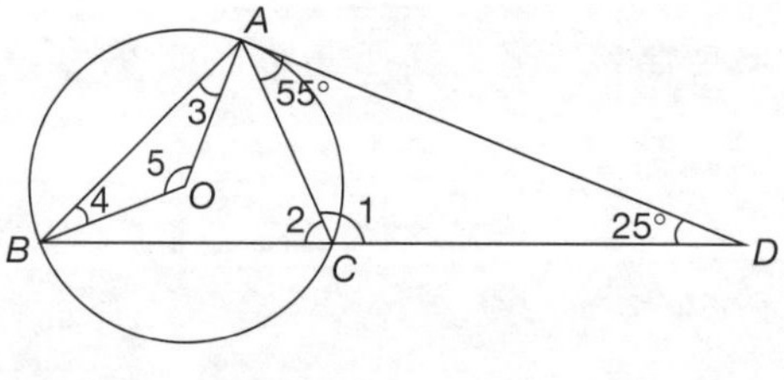

In ΔAOB, $OA = OB$ [$\because$ Equal radii]

$\angle 3 = \angle 4$ [$\because$ Angles opposite to equal sides are equal]

In ΔACD, $55° + \angle 1 + 25° = 180°$ [$\because$ ASP of Δs]

$\angle 1 = 100°$

$\angle 2 = 80°$ (ii)

$\angle 5 = 2\angle 2$ [$\because$ Angle subtended at the centre is double the angle subtended at another arc]

$\angle 5 = 2 \times 80 = 160°$ (iii)

In ΔAOB, $\angle 3 + \angle 4 + \angle 5 = 180°$ [$\because$ ASP of Δs]

$\angle 4 + \angle 4 + 160° = 180°$ [From (i) and (ii)]

$\angle 4 = 10°$

$\angle ABO = 10°$

77. $DE = BD$ [Given]

$\angle 1 = \angle 8 = 25°$ (i)

$\angle 7 = 130°$ [$\because$ ASP of Δs]

$\angle 9 = 50°$

$\angle 1 = \angle 4 = 25°$

In ΔCBE, $\angle 4 + \angle 2 + \angle 1 + \angle 8 = 180°$

$\angle 2 = 105°$

$\angle 1 + \angle 2 + \angle 3 = 180°$

$\angle 3 = 350°$

$\angle 3 = \angle 5 = 50°$

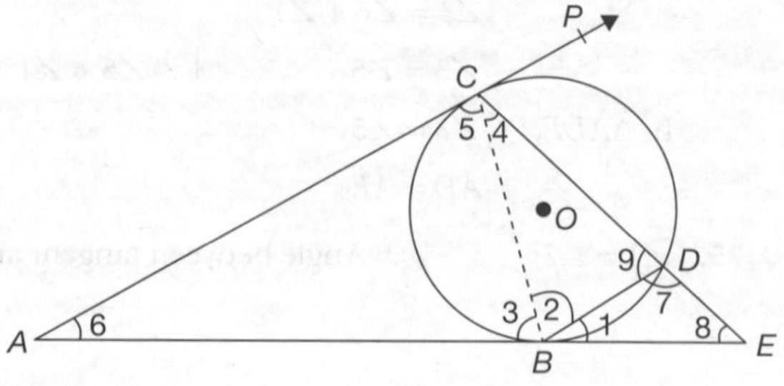

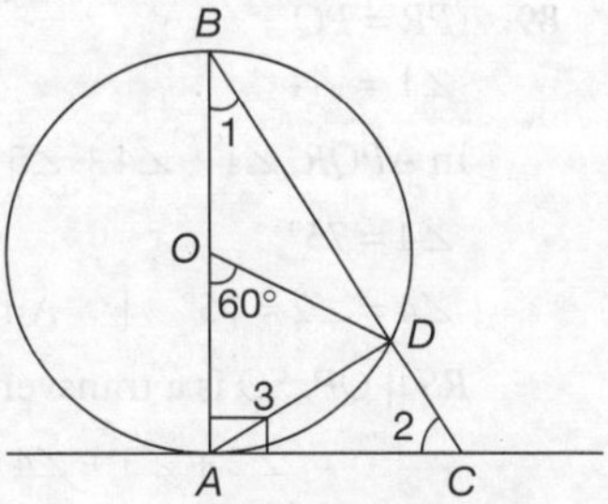

In ΔACB, $\angle 3 + \angle 5 + \angle 6 = 180°$

$\angle 6 = 80°$

78. $\angle 1 = \frac{1}{2}\ \angle AOD = \frac{1}{2} \times 60$

$\angle 1 = 30°$ (i)

$\angle 3 = 90°$ (ii)

In ΔBAC, $\angle 1 + \angle 2 + \angle 3 = 180°$

$\angle 2 = 60°$ [From (i) and (ii)]

79. Since PQ is the diameter

$\angle 2 = 90°$ [∵ Angle in the semicircle is a right angle] (i)

RQ and RT are opposite rays

$\angle 2 + \angle 4 = 180°$ [∵ LPA]

$\angle 4 = 90°$ [From (i)] (ii)

$\angle POR = 2\angle PQR$ [∵ Angle subtended at the centre is double the angle subtended at another arc]

$72° = 2\angle PQR$

$\angle 3 = \angle PQR = 36°$ (iii)

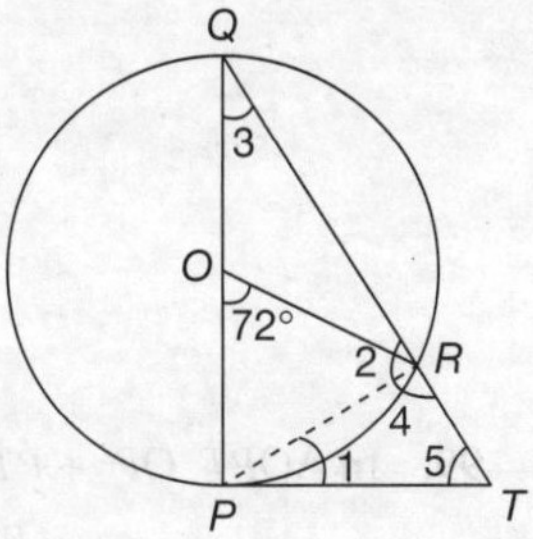

Since TP is a tangent, $\angle 1 = \angle 3 = 36°$ [∵ Angle in the alternate segment] (iv)

In ΔPRT, $\angle 1 + \angle 4 + \angle 5 = 180°$ [∵ ASP of Δs]

$36° + 90° + \angle 5 = 180°$ [From (ii) and (iv)]

$\angle 5 = 54°$

Hence $\angle PTR = 54°$

80. To prove: $XB + BY = XD + DY$

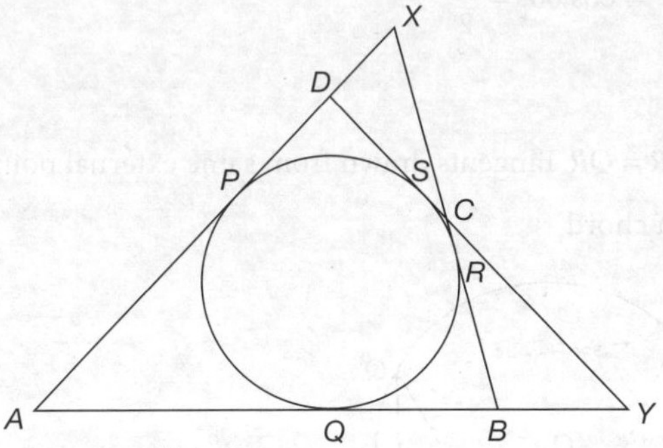

Solution: LHS: $XB + BY = XR + RB + BY = XP + QB + BY$

$= XD + DP + QY = XD + DS + YS = XD + DY$

89. $PR = PQ$

$\angle 1 = \angle 4$ (i)

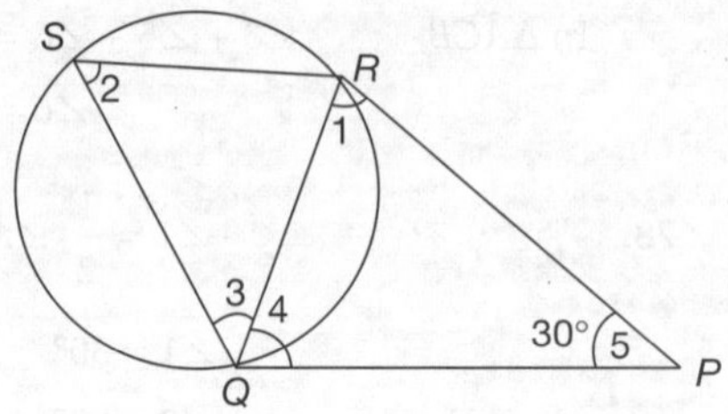

In ΔPQR, $\angle 1 + \angle 4 + \angle 5 = 180°$ [∵ ASP of Δs]

$\angle 4 = 75°$

$\angle 4 = \angle 2 = 75°$ [∵ Angle in the alternate segment] (ii)

$RS \parallel QP$, SQ is a transversal

$\angle 2 + \angle 3 + \angle 4 = 180°$ [∵ Co-interior angles]

$\angle 3 = 30°$ [From (ii)]

98. $\Delta AO'D \sim \Delta AOC$ [∵ AA]

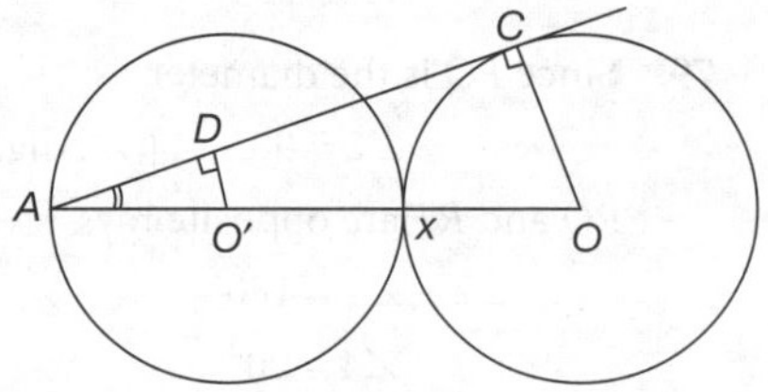

$$\frac{AO'}{AO} = \frac{O'D}{OC}$$

$$\frac{r}{3r} = \frac{O'D}{OC}$$

$$\frac{1}{3} = \frac{O'D}{OC}$$

99. In ΔOPT, $OP^2 + PT^2 = OT^2$ [∵ Pythagoras' Theorem]

$PT = 12$ cm

$TE = OT - OE = 13 - 5 = 8$ cm

Let $PA = AE = x$.

In ΔTEA, $(8)^2 + x^2 = (12 - x)^2$

On solving we get, $x = 3.3$ cm. $AB = 2x = 6.6$ cm.

101. $\angle PRO = \angle QRO = \frac{1}{2} \angle PRQ = 60°$ [∵ Equally inclined tangents are equal]

In right ΔOPR, $\cos \angle ORP = \frac{PR}{OR} = \cos 60° = \frac{1}{2}$

$OR = 2PR$

$OR = PR + PR$

$OR = PR + OR$ [∵ $PR = QR$ Tangents drawn from same external point are equal in length]

105. Since PT is a tangent and PQ is a chord

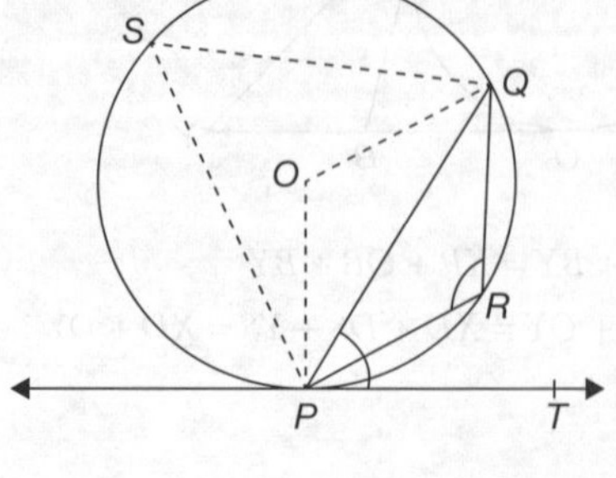

$$\angle QPT = \angle PSQ = 60° \quad [\because \text{ Angles in the alternate segment are equal}] \quad \text{(i)}$$

Since *PRQS* is a cyclic quadrilateral

$$\angle PSQ + \angle PRQ = 180° \quad [\text{Sum of opposite angles of cyclic quadrilateral are supplementary}]$$

$$60° + \angle PRQ = 180° \quad [\text{From (i)}]$$

$$\angle PRQ = 120°$$

Chapter 12: Areas Related to Circles

3 Try to Solve

3. Distance covered in 1 revolution = $2\pi a = \pi d$ unit.

Distance covered in 5000 revolutions = 5000 πd unit.

Total distance covered by bicycle = 11 km = 11,000,00 cm

According to the question 5000 πd = 11, 000, 00

$$d = \frac{11{,}000{,}00}{5{,}000 \times 22} \times 7 = 70 \text{ cm}$$

5. $l = 8.8$ cm, $\theta = 30°$, $r = ?$

$$l = 2\pi r \frac{\theta}{360}$$

$$8.8 = 2 \times \frac{22}{7} \times r \times \frac{30}{360}$$

12

$$\frac{8.8 \times 7 \times 12}{2 \times 22} = r$$

$$r = 16.8 \text{ cm}$$

5 Try to Solve

1. $r = 45$ cm, $\theta = \dfrac{360°}{8} = 45°$

Area between two consecutive ribs = $\pi r^2 \dfrac{\theta}{360}$

$$= \frac{22}{7} \times 45 \times 45 \times \frac{45}{360} = 795.53 \text{ cm}^2$$

2. $r = 28$ cm, $\theta = \dfrac{360}{6} = 60°$

Area of one design = $\pi r^2 \dfrac{\theta}{360} - \dfrac{1}{2} r^2 \sin\theta$

$$= \frac{22}{7} \times 28 \times 28 \times \frac{60}{360} - \frac{1}{2} \times 28 \times 28 \times \frac{\sqrt{3}}{2} = 77.47 \text{ cm}^2$$

Area of design $= 6 \times 77.47$

Cost of making the design $= 0.35 \times 6 \times 77.47 = ₹\,162.69$

3. *OABC* is a square, as shown in the figure.

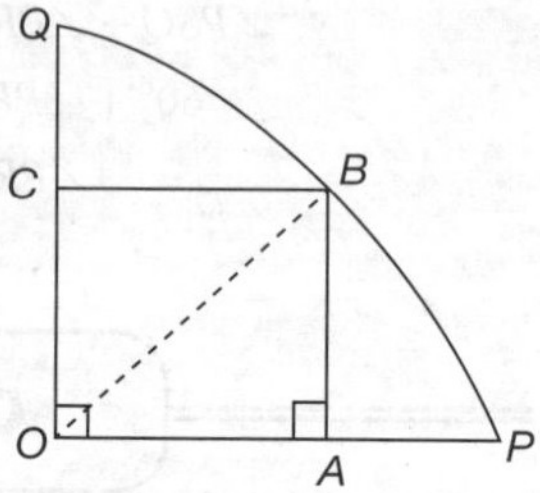

$\therefore \quad \angle OAB = 90°$

In right $\Delta OAB \quad OB^2 = OA^2 + AB^2$ [By Pythagoras' Theorem]

$= (20)^2 + (20)^2$

$OB = 20\sqrt{2}$ cm

$r = 20\sqrt{2}$ cm

Area of shaded portion $= \frac{1}{4}\pi r^2 - (20)^2 = \frac{1}{4} \times 3.14 \times 20 \times 20 \times 2 - 400 = 228\ \text{cm}^2$

5. Area of square $= (14)^2 = 196\ \text{cm}^2$.

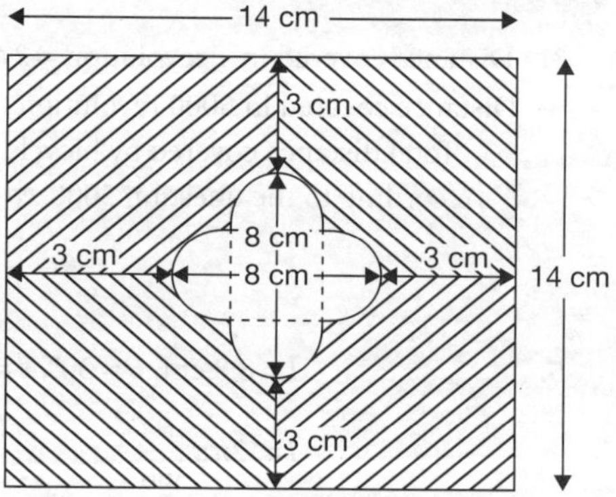

Let the radius of a semicircle $= x$.

Radius of two semicircle $= 2x$.

Side of inner square $= 2x$.

According to the figure $2x + 2x = 8$

$x = 2$ cm

Side of inner square $= 4$ cm

Area of unshaded region = Area of inner square + 4 (Area of semicircle)

$= (4)^2 + 4\left[\frac{\pi}{4}(2)^2\right] = (16 + 8\pi)\ \text{cm}^2$

$\therefore$ Area of shaded region = Area of square – Area of unshaded region

$= 196 - [16 + 8\pi] = 154.88\ \text{cm}^2$

6 Try to Solve

1. $r = 3.5\ \text{cm} = \frac{7}{2}$ cm

(i) Area of quadrant $= \frac{1}{4} \times \frac{22}{7} \times \frac{7}{2} \times \frac{7}{2} = \frac{77}{8}\ \text{cm}^2$

(ii) Area of shaded region = Area of quadrant – Area of ΔBOD

$= \frac{77}{8} - \frac{1}{2} \times 2 \times \frac{7}{2} = \frac{49}{8}\ \text{cm}^2$

2. $\frac{\sqrt{3}}{4}a^2 = 17320.5$, $a = 200$ cm, $r = 100$ cm

Area of shaded region = Area of Δ – 3 × Area of sector

$= 17320.5 - 3 \times 3.14 \times 100 \times 100 \times \frac{60}{360} = 1620.5\ \text{cm}^2$

3. Area of shaded region = Area of ΔBAC + Area of semicircle with BC as diameter – Area of quadrant *BACDB*.

$BC^2 = (14)^2 + (14)^2$

$$BC = 14\sqrt{2} \text{ cm}$$
$$r = 7\sqrt{2} \text{ cm}$$

Compute the rest by yourself.

Subjective Questions

38. Using Pythagoras' Theorem $PQ = 3$ cm

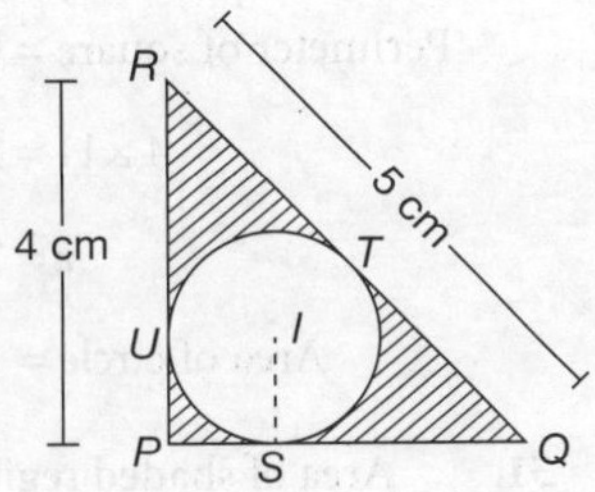

$$\text{Area of } \Delta RPQ = \frac{1}{2} \times 3 \times 4 = 6 \text{ cm}^2$$

$$\text{Area of } \Delta RPQ = \text{Area } \Delta RIP + \text{Area } \Delta PIQ + \text{Area } \Delta QIR$$

$$6 = \frac{1}{2} \times r \times 4 + \frac{1}{2} \times r \times 3 + \frac{1}{2} \times r \times 5$$

$$6 = \frac{1}{2} \times r \times (4 + 3 + 5)$$

$$r = 1 \text{ cm}$$

$$\text{Area of shaded region} = \text{Area of } \Delta RPQ - \text{Area of circle} = 6 - \frac{22}{7} \times (1)^2 = \frac{20}{7} \text{ cm}^2$$

41. r_1 (Radius of gold region) $= \frac{21}{2}$ cm

r_2 (Radius of red region) $= 21$ cm

r_3 (Radius of blue region) $= \frac{63}{2}$ cm

r_4 (Radius of black region) $= 42$ cm

r_5 (Radius of white region) $= \frac{105}{2}$ cm

$$\text{Area bounded by gold region} = \pi r_1^2 = \frac{22}{7} \times \frac{21}{2} \times \frac{21}{2} = 346.5 \text{ cm}^2$$

$$\text{Area bounded by red region} = \pi r_2^2 - \pi r_1^2 = \pi\left(r_2^2 - r_1^2\right) = \frac{22}{7}\left[(21)^2 - \left(\frac{21}{2}\right)^2\right] = 1039.5 \text{ cm}^2$$

$$\text{Area bounded by blue region} = \pi r_3^2 - \pi r_2^2 = \pi\left[r_3^2 - r_2^2\right] = \frac{22}{7}\left[\left(\frac{63}{2}\right)^2 - (21)^2\right] = 1732.5 \text{ cm}^2$$

$$\text{Area bounded by black region} = \pi r_4^2 - \pi r_3^2 = \pi\left[r_4^2 - r_3^2\right] = \frac{22}{7}\left[(42)^2 - \left(\frac{63}{2}\right)^2\right] = 2424.5 \text{ cm}^2$$

$$\text{Area bounded by white region} = \pi\left[r_5^2 - r_4^2\right] = \frac{22}{7}\left[\left(\frac{105}{2}\right)^2 - (42)^2\right] = 3118.5 \text{ cm}^2$$

47. $\dfrac{2\pi R}{2\pi r} = \dfrac{23}{22}, \ \dfrac{R}{r} = \dfrac{23}{22}$

$$22R - 23r = 0 \qquad \text{(i)}$$

$$R - r = 5 \qquad \text{(ii) [Given]}$$

Solve (i) and (ii) and find $2r = ?$

50. Area of square $(a^2) = 121$. Side of square $(a) = 11$ cm.

Perimeter of square = Circumference of circle

$$4 \times 11 = 2 \times \frac{22}{7} \times r$$

$$r = 7 \text{ cm}$$

$$\text{Area of circle} = \pi r^2 = \frac{22}{7} \times 7 \times 7 = 154 \text{ cm}^2$$

51. Area of shaded region = Area of ΔABC – Area of three sectors

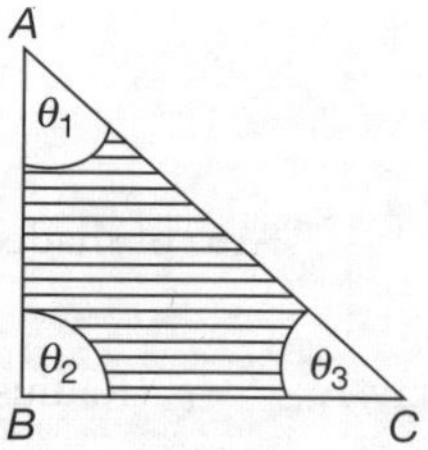

Δ is right Δ because. $AC^2 = AB^2 + BC^2$

So $$\text{ar } \Delta ABC = \frac{1}{2} \times 14 \times 48$$ (with 2 and 14 cancelled to 7)

$$\text{Area of three sectors} = \pi r^2 \left(\frac{\theta_1 + \theta_2 + \theta_3}{360} \right)$$

$$= \frac{22}{7} \times 5 \times 5 \times \frac{180}{360} \qquad [\because \text{ ASP of } \Delta s]$$

$$= \frac{275}{7} \text{ cm}^2$$

$$\text{Area of shaded region} = 336 - \frac{275}{7} = 296.75 \text{ cm}^2$$

63. Perimeter $= 20 + 20 + 14 + \dfrac{22}{7} \times 7 = 76$ cm

76. Diameter = 35 mm

Total length of 5 diameters = 5 × 35 = 175 mm

$$\text{Circumference of circle} = 2 \times \frac{22}{7} \times \frac{35}{2} = 110 \text{ mm}$$

(i) Total length of wire used = Length of 5 diameters + Circumference of brooch

= 175 + 110 = 285 mm

(ii) $\theta = \dfrac{360}{10} = 36°$

$$\text{Area of each sector} = \frac{22}{7} \times \frac{35}{2} \times \frac{35}{2} \times \frac{36}{360} = \frac{385}{4} \text{ mm}^2.$$

79. Angle swept by minute hand in 60 min = 360°

$$\text{Angle swept by minute hand in 1 min} = \frac{360}{60} = 6°$$

Angle swept by minute hand in 35 min = 6 × 35 = 210°

Area swept by minute hand in 35 min $= \frac{22}{7} \times 12 \times 12 \times \frac{210}{360} = 264 \text{ cm}^2$

99. $\theta = \frac{360}{3} = 120°, r = 0.98$ cm

Length of arc. $= 2 \times \frac{22}{7} \times 0.98 \times \frac{120}{360} = 2.05$ m

105. Area of shaded region = Area of equilateral Δ – Area of three sectors.

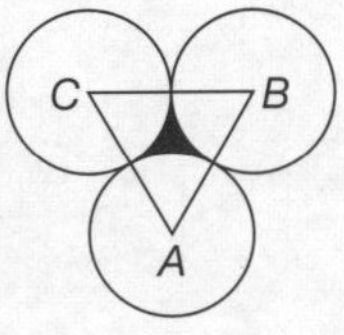

$$= \frac{\sqrt{3}}{7} \times 7 \times 7 - \frac{22}{7} \times \frac{7}{2} \times \frac{7}{2} \times \frac{180}{360} = 1.97 \text{ cm}^2$$

110. Area of shaded region = Area of equilateral Δ + Area of major sector

$$= \frac{\sqrt{3}}{4} \times 12 \times 12 + \frac{22}{7} \times 6 \times 6 \times \frac{300}{360} = \left(36\sqrt{3} + \frac{660}{7}\right) \text{cm}^2$$

117. Area of square = 8 × 8 = 64 cm²

Area of quadrant $= \frac{22}{7} \times 8 \times 8 \times \frac{90}{360} = \frac{352}{7}$ cm²

Area of square left on subtracting area of one quadrant $= 64 - \frac{352}{7} = \frac{96}{7}$ cm²

Area of quadrant $= \frac{352}{7}$ cm²

Total area of the shaded portion = Area of quadrant – Area of square left

$$= \frac{352}{7} - \frac{96}{7} = \frac{256}{7} \text{ cm}^2$$

119. Area of shaded region = 2 × [Area of semicircle with diameter *AQ* – Area of semicircle with diameter *AP*]

121. Area of shaded region = [Area of semi-circle *E* + Area of semicircle *B*] – [Area circle *D* + 2 × Area of semicircle *A*]

124. No; because

Diagonal of inner square = Diameter of a circle = *d*

Let the side of inner square *EFGH* be *x*.

$$d^2 = x^2 + x^2 = 2x^2$$

So $\quad x^2 = \frac{d^2}{2}$ = Area of inner square

Since, side of outer square = *d*, area of outer square = d^2.

Since, area of outer square is not equal to four times the area of the inner square.

134. Area of floor remains uncovered with tiles = 5 × 4 – 80 × Area of circular tiles

$$= 5 \times 4 - 80 \times \frac{22}{7} \times \frac{25}{100} \times \frac{25}{100} = 4.3 \text{ m}^2$$

138. $BC^2 = (24)^2 + (7)^2 = 625, BC = 25 \text{ cm}, r = \frac{25}{2}$ cm

Area of shaded region = Area of circle – Area of quadrant – Area of Δ*BAC*

142. Area of shaded region = Area of Trapezium – Area of sector $BGEC$

$$= \frac{1}{2} \times (7 \times 11) \times 3.5 - \frac{22}{7} \times 7 \times 7 \times \frac{30°}{360°} = 18.67 \text{ cm}^2$$

143. $BC^2 = (3)^2 + (4)^2 = 25, BC = 5$ cm

Area of shaded region = Area of semicircle on side AB + Area of semicircle on side AC – Area of semicircle on side BC + Area of ΔABC.

$$= \frac{1}{2} \times \frac{22}{7} \times \left(\frac{3}{2}\right)^2 + \frac{1}{2} \times \frac{22}{7} \times (2)^2 - \frac{1}{2} \times \frac{22}{7} \times \left(\frac{5}{2}\right)^2 + \frac{1}{2} \times 3 \times 4 = 6 \text{ cm}^2$$

Chapter 13: Surface Areas and Volumes

2 Try to Solve

1. This is a combined figure.

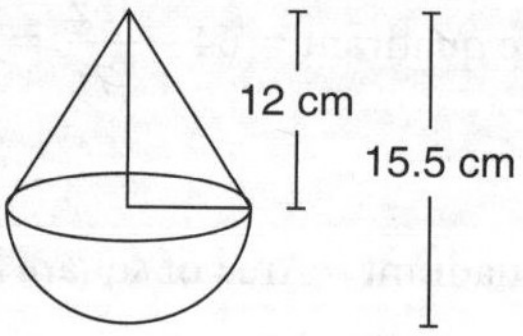

Hemisphere + Cone

Dimension of hemisphere $r = 3.5$ cm. Dimension of cone $r = 3.5 = \frac{7}{2}$ cm.

$h = 12$ cm

$$l = \sqrt{(12)^2 + \left(\frac{7}{2}\right)^2} = \frac{25}{2} = 12.5 \text{ cm}$$

Total surface area of the figure = CSA of cone + CSA of hemisphere

$$= \pi rl + 2\pi r^2$$

Insert the values to obtain the answer.

2. This is same as Q1. First, calculate

Total surface area = CSA of cone + CSA of cylinder

Then, in order to calculate total cost, multiply total surface area by ₹500.

3 Try to Solve

1. Dimension of cone $r = 1$ cm, $h = 1$ cm

Dimension of hemisphere $r = 1$ cm

Volume of solid = Volume of cone + Volume of hemisphere

Perform the computation by yourself.

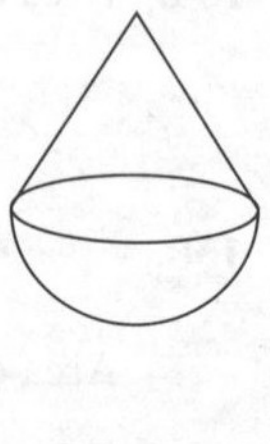

4 Try to Solve

2. This is a combined figure.

$$\text{TSA of figure} = \text{CSA of cone} + \text{CSA of cylinder}$$

After calculating the TSA, make it equal to area of the canvas. This means $l \times b$, where $b = 1.5$ m.

Insert the value to obtain the length of the canvas.

5 Try to Solve

1. Diameter of hemisphere = Side of cubical block = 7 cm.

$$\text{TSA of solid} = \text{TSA of cube} + \text{CSA of hemisphere} - \text{Area of base of hemisphere}$$
$$= 6a^2 + 2\pi r^2 - \pi r^2$$

Insert the values and perform the calculation by yourself.

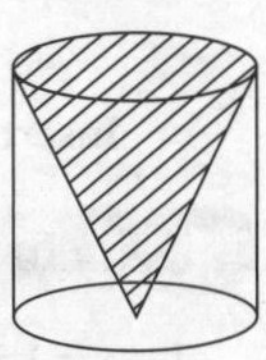

2. $h = 2.4$ cm, $r = 0.7$ cm

$$l = \sqrt{(0.7)^2 + (2.4)^2} = 2.5 \text{ cm}$$

$$\text{TSA of solid} = \text{CSA of cylinder} + \text{CSA of cone} + \text{Area of base of cylinder}$$
$$= 2\pi rh + \pi rl + \pi r^2$$

Insert the values and perform the calculation by yourself.

6 Try to Solve

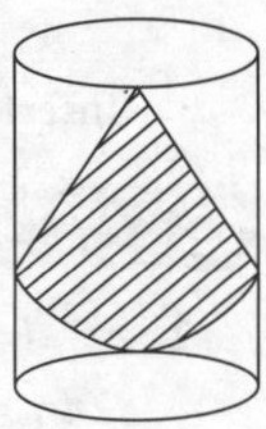

2. $r = 60$ cm, height of cone = 120 cm, height of cylinder = 180 cm

Volume of water left in the cylinder = Volume of cylinder – Volume of solid

$$= \pi r^2 h - \left[\frac{1}{3}\pi r^2 h + \frac{2}{3}\pi r^3\right]$$

Insert the values and perform the calculation by yourself.

7 Try to Solve

2. $2r = 5$, $h = 10$ cm, $r = \frac{5}{2}$ cm

Actual capacity of the glass = Apparent capacity of glass – Volume of hemisphere

$$= \pi r^2 h - \frac{2}{3}\pi r^3$$

Insert the values and perform the calculation by yourself.

9 Try to Solve

2. $2r = 4.2$, $r = 2.1$ cm, $h_1 = 12$ cm, $h_2 = 7$ cm

Volume of toy = Volume of hemisphere + Volume of cylinder + Volume of cone

$$= \frac{2}{3}\pi r^3 + \pi r^2 h_1 + \frac{1}{3}\pi r^2 h_2$$

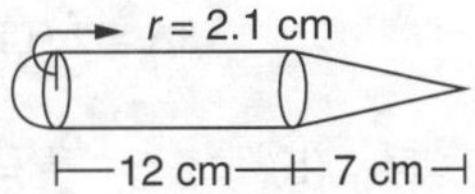

Insert the values and perform the calculation by yourself.

3. Radius of base of cylinder $r = 6$ cm, $h = 15$ cm
For cone, $h_1 = 12$ cm, $r_1 = 3$ cm

$$\text{No. of cones required} = \frac{\text{Volume of ice cream in cylinder}}{\text{Volume of ice cream in cone}} = \frac{\pi r^2 h}{\frac{1}{3}\pi r_1^2 h_1 + \frac{2}{3}\pi r_1^3}$$

Insert the values and perform the calculation by yourself.

10 Try to Solve

1. Silver coin: $r = \frac{1.75}{2} = \frac{7}{8}$ cm, $h_1 = \frac{2}{10}$ cm

Cuboidal box: $l = 5.5$ cm, $b = 10$ cm, $h = 3.5$ cm

$n \times$ Volume of silver coins = Volume of cuboidal box

$$n = \frac{lbh}{\pi r^2 h_1}$$

Insert the value and find 'n'.

11 Try to Solve

1. $r = 1.5$ m, $h = 14$ m
$R = 1.5 + 4 = 5$ m, H = height of embankment = ?
Volume of embankment = Volume of well

$$\pi\left(R^2 - r^2\right)H = \pi \times r^2 \times H$$

Insert the value and find H.

12 Try to Solve

2. $l = 4$ cm

$2\pi r_1 = 18$ and $2\pi r_2 = 6$

$r_1 = \frac{63}{22}$ cm and $r_2 = \frac{21}{22}$ cm

CSA of the frustum of the cone $= \pi\left(r_1 + r_2\right)l$

Insert the values and perform the calculation by yourself.

Multiple Choice Questions

13. (d) $l^3 = 216 = (6)^3$

$l = 6$ m

Cuboid $l = 6 + 6 + 6 = 18$ cm

$B = 6$ cm, $H = 6$ cm

SA = 2 (LB + BH + LH) = 2 [18 × 6 + 6 × 6 + 18 × 6] = 504 m^2

14. (b) $\frac{S_1}{S_2} = \frac{4\pi r_1^2}{4\pi r_2^2} = \frac{16}{9}$

$\frac{r_1}{r_2} = \frac{4}{3}$

$$\frac{V_1}{V_2} = \frac{\frac{4}{3}\pi r_1^3}{\frac{4}{3}\pi r_2^3} = \frac{64}{27}$$

20. (a)

	Old cylinder	New cylinder
Radius	r	$2r$
CSA	$2\pi rh$	$2\pi(2r)h$ $2\times(2\pi rh)$

So it becomes double.

Subjective Questions

39.

Cones	I	II
Slant height	l	$2l\ (l_1)$
Radius	r	r_1

$$\frac{\text{C.S.A.I}}{\text{C.S.A.II}} = \frac{\pi rl}{\pi r_1 l_1} = \frac{r}{r_1}\times\frac{l}{2l}$$

$$\frac{2}{1} = \frac{r}{2r_1}$$

$$\frac{r}{r_1} = \frac{4}{1}$$

41.

	Cone	Cylinder	Hemisphere
Radius (r)	5 cm	5 cm	5 cm
Height (h)	12 cm(h_1)	13 cm(h_2)	–
Slant height (l)	13 cm	–	–

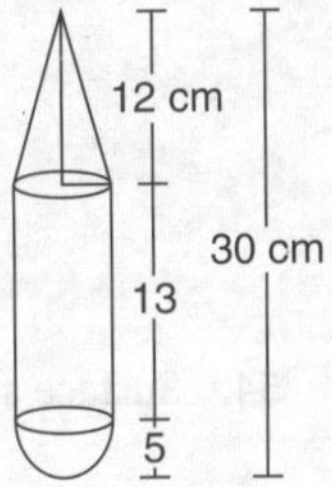

Surface area of toy = CSA of cone + CSA of cylinder + CSA of hemisphere

$$= \pi rl + 2\pi rh_2 + 2\pi r^2$$

Insert the values and perform the calculation by yourself.

43. Height of cone $= \left(5 - \frac{3.5}{2}\right) = 3.25$ cm, $r = \frac{3.5}{2}$ cm

$$l = \sqrt{r^2 + h^2} = \sqrt{\left(\frac{3.5}{2}\right)^2 + (3.25)^2}\ \text{cm} = 3.7\ \text{cm}$$

TSA of the toy = CSA of hemisphere + CSA of cone

$$= 2\pi r^2 + \pi rl$$

Insert the values and perform the calculation by yourself.

44. Surface area of the block = TSA of cube – Base area of hemisphere + CSA of hemisphere

45. $r = 2.5$ cm, $h = 6$ cm, $r' = 1.5$ cm and $h' = 26 - 6 = 20$ cm, $l = \sqrt{(2.5)^2 + (6)^2} = 6.5$ cm.

Area to be painted orange = CSA of cone + Base area of the cone – Base area of the cylinder

$$= \pi rl + \pi r^2 - \pi(r')^2$$

Area to be painted yellow = CSA of cylinder + Area of one base of the cylinder = $2\pi r'h' + \pi(r')^2$

Insert the values and perform the calculation by yourself.

49. $h = 10$ cm and $r = 3.5$ cm

TSA of article = CSA of cylinder + 2 × CSA of hemisphere

52. To prove: $3\pi Vh^3 - C^2h^3 + 9V^2 = 0$

$$\text{LHS} = 3\pi\left(\frac{1}{3}\pi r^2 h\right)h^3 - (\pi rl)^2 h^2 + 9\left(\frac{1}{3}\pi r^2 h\right)^2$$

$$= 3\pi \times \frac{1}{3}\pi r^2 h \times h^3 - \pi^2 r^2 l^2 h^2 + 9 \times \frac{1}{9}\pi^2 r^4 h^2$$

$$= \pi^2 r^2 h^4 - \pi^2 r^2 l^2 h^2 + \pi^2 r^4 h^2 = \pi^2 r^2 h^2\left[h^2 - l^2 + r^2\right]$$

$$= \pi^2 r^2 h^2\left[l - l^2\right] \qquad [\because r^2 + h^2 = l^2]$$

$$= \pi^2 r^2 h^2 (0) = 0$$

53. Volume of tank = 150 × 100 × 3

Volume of water passing through the pipe = $\frac{2}{10} \times \frac{1.5}{10} \times h$

Volume of tank = Volume of water passing through the pipe

$$150 \times 100 \times 3 = \frac{2}{10} \times \frac{15}{100} \times h$$

$$h = \frac{\overset{10}{\cancel{150}} \times 100 \times 3 \times 10 \times 100}{2 \times \cancel{15}} = 1500000 \text{ m} = 1500 \text{ km}$$

Speed = 15 km/h.

$$\text{Time} = \frac{\text{Distance}}{\text{speed}} = \frac{1500}{15} = 100 \text{ h.}$$

54. Surface area of remaining block = SA of cuboid + CSA of cylinder – SA of two circular ends

$$= 2\,[lb + bh + lh] + 2\pi rh - 2\pi r^2$$

where $l = 15$ cm, $b = 10$ cm, $h = 5$ cm and $r = \frac{5}{2}$ cm.

Insert the values and perform the calculation by yourself.

60. Total volume of air a shed can hold = Volume of cuboid + $\frac{1}{2}$ Volume of cylinder

$$= 15 \times 7 \times 8 + \frac{1}{2} \times \frac{22}{7} \times \frac{7}{2} \times \frac{7}{2} \times 15 = 1128.75 \text{ m}^3$$

Total space occupied by the machinery = 300 m^3

Total space occupied by the worker = 2 × 0.08 = 1.6 m^3

Volume of air when there are machinery and workers = 1128.75 – (300 + 1.6) = 827.15 m^3

63. Volume of four conical depression $= 4 \times \frac{1}{3} \times \frac{22}{7} \times \frac{1}{2} \times \frac{1}{2} \times 1.4 \text{ cm}^3 = 1.465 \text{ cm}^3$

$$\text{Volume of cuboidal box} = 15 \times 10 \times 3.5 = 525 \text{ cm}^3$$
$$\text{Remaining volume of box} = 525 - 1.465 = 523.5 \text{ cm}^3$$

64. Total volume of solid = Volume of 1st cylinder + Volume of 2nd cylinder

Mass of pole = Volume of solid × Mass of 1 cm^3 of iron

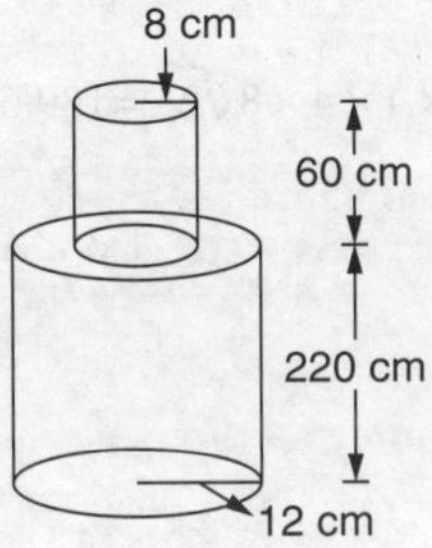

65. Total volume of the glass vessel = Volume of cylindrical part + Volume of spherical part

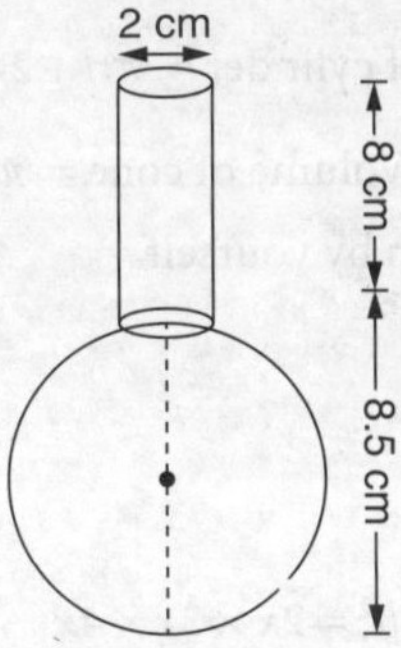

68.
$$\frac{2\pi rh}{\pi rl} = \frac{8}{5}$$
$$\frac{h}{\sqrt{r^2 + h^2}} = \frac{4}{5}$$

S.B.S $\quad 25\,h^2 = 16\,(r^2 + h^2)$

$$\frac{r}{h} = \frac{3}{4}$$

69. $\quad r + h = 37 \quad$ (i)

$$2\pi r\,(h + r) = 1628$$

Inserting (i) in (ii), we get $\quad 2\pi r \times 37 = 1628$

$$r = 7 \text{ cm}, h = 37 - 7 = 30 \text{ cm}$$

73. Length of arc $= \dfrac{2\pi r\theta}{360}$ = Circumference of the base of the cone

$$2\pi r = 2\pi \times 12 \times \frac{120}{360}$$

$$r = 4 \text{ cm and } l = 12$$

$$h = \sqrt{l^2 - r^2} = 8\sqrt{2} \text{ cm}$$

Volume of cone $= \dfrac{1}{3}\pi r^2 h = \dfrac{1}{3} \times \dfrac{22}{7} \times 4 \times 4 \times 8\sqrt{2} = 189.61 \text{ cm}^3$

77. Since $OA = OB$ $[\because \angle A = \angle B = 45°]$

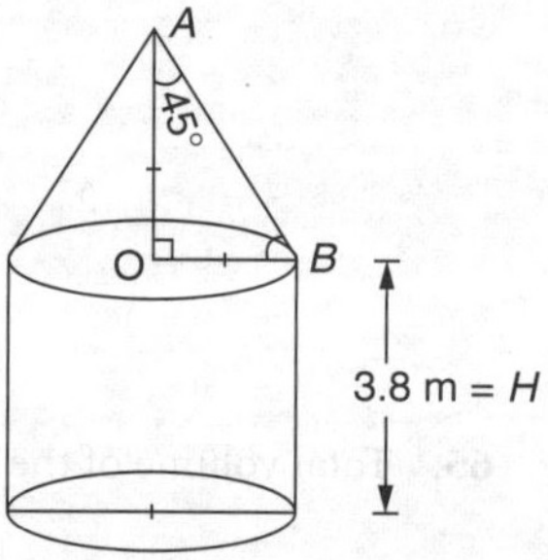

Cone: $r = \dfrac{4.3}{2}$ m, $h = \dfrac{4.3}{2}$ m

$$l = \sqrt{r^2 + h^2} = \frac{4.3}{2}\sqrt{2} \text{ m}$$

$$r = \frac{4.3}{2}$$

Surface area of solid = CSA of cone + CSA of cylinder $= \pi rl + 2\pi rH$

Volume of building = Volume of cylinder + Volume of cone $= \pi r^2 H + \dfrac{1}{3}\pi r^2 h$

Insert the values and perform the calculation by yourself.

79.

$$lh = 2x \quad \text{(i)}$$

$$bh = 3x \quad \text{(ii)}$$

$$lb = 4x \quad \text{(iii)}$$

$$lbh = 9000 \text{ cm}^3 \quad \text{(iv)}$$

Multiplying (i), (ii) and (iii), we get $l^2b^2h^2 = 2x \times 3x \times 4x$

$$(lbh)^2 = 24\,x^3$$

$$9000 \times 9000 = 24\,x^3$$

$$x^3 = \frac{9000 \times 9000}{24} = 337500 \text{ cm}, x = 150 \text{ cm}$$

$$\frac{l \times b \times b \times h}{l \times h} = \frac{2x \times 3x}{4x}$$

$\Rightarrow$ $$b^2 = \frac{3}{2} \times 150 = 225$$

$$b = 15 \text{ cm}, h = 30 \text{ cm}, l = 40 \text{ cm}$$

83. $h = 100$ cm, $r = 27$ cm

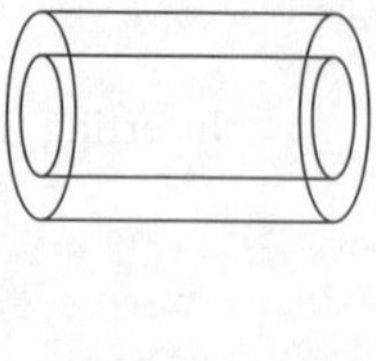

$R = 27 + 9 = 36$ cm

Volume of cylinder $= \pi\,(R^2 - r^2)\,h$

Total weight of roller = Volume of cylinder × 8 g

Insert the values and perform the calculation by yourself.

84. Volume of small cone = Volume of big cone

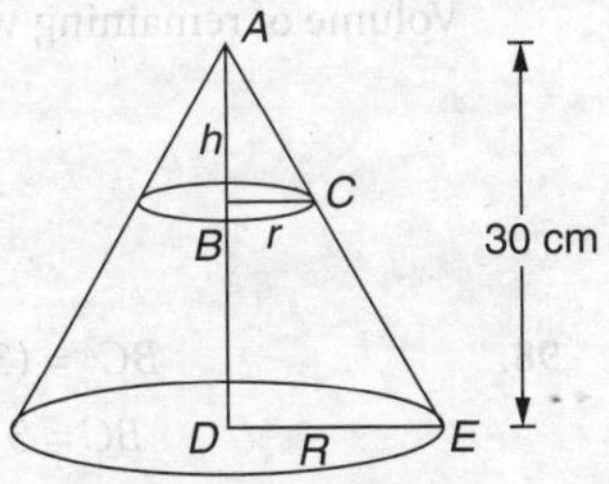

$$\frac{1}{3}\pi r^2 h = \frac{1}{27} \times \frac{1}{3}\pi R^2 \times 30 \qquad \Delta ABC \sim \Delta ADE$$

$$h = \frac{1}{27} \times \frac{R^2}{r^2} \times 30 \qquad \frac{r}{R} = \frac{h}{30} \qquad \text{(i)}$$

$$h = \frac{1}{27} \times \frac{h^2}{(30)^2} \times 30 \qquad \text{[From (i)]}$$

$$h = 10 \text{ cm}$$

Plane is cut at = 30 − 10 = 20 cm

94. $b = 6$ m, $h = 1.5$ m

$$\text{Rate of flowing water} = 10 \text{ km/h} = \frac{10 \times 1000}{60} \text{ m/min}$$

$$\text{Length (distance) covered in 30 min} = \frac{10 \times 1000}{60} \times 30$$

$$\text{Volume of water flowing in 30 min} = \frac{6 \times 15 \times 10 \times 1000 \times 30}{10 \times 60} = 45000 \text{ m}^3$$

$$h = 8 \text{ cm} = \frac{8}{100} \text{ m}$$

$$\text{Volume of water required} = A\left(\frac{8}{100}\right)$$

$$\text{According to the question, } 45000 = \frac{A \times 8}{100}$$

$$A = 562500 \text{ m}^2$$

95. $r = 10$ cm, speed = 3 km/h. Let it fill the tank in t hours.

$$\text{Volume of water flowing in } t \text{ hours} = \pi \times \left(\frac{10}{100}\right)^2 \times \frac{3 \times 1000 \times t}{1} = 30\pi t \text{ m}^3$$

$$\text{Volume of water in cylinder tank} = \pi r^2 h = \pi(5)^2 \times 2 \text{ m}^3$$

$$\text{According to the question, } \pi(30)t = \pi(25)(2)$$

$$t = \frac{5}{3} \text{ h} = \frac{5}{3} \times 60 = 100 \text{ min}$$

97. Volume of cistern = $150 \times 120 \times 110 = 1980000$ cm^3

Volume of water in cistern = 129600 cm^3

Remaining volume of cistern = 1980000 − 129600 = 1850400 cm^3

Volume of each brick = $22.5 \times 7.5 \times 6.5 = 1096.875$ cm^3

$$\text{Water absorbed by one brick} = \frac{1096.875}{17} \text{ cm}^3$$

$$\text{Water absorbed by } n \text{ bricks} = \left(\frac{1096.875}{17}\right) n \text{ cm}^3$$

Volume of remaining water + Water absorbed by n bricks = $n \times$ Volume of one brick

$$1850400 + n\left(\frac{1096.875}{17}\right) = n(1096.875)$$

$$1792 = n \text{ (approxminately)}$$

98.

$$BC^2 = (3)^2 + (4)^2 = 25$$

$$BC = 5 \text{ cm}$$

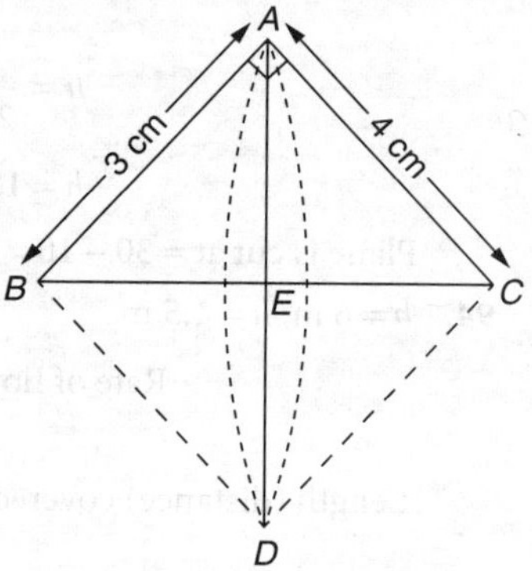

ΔABC revolves about the hypotenuse BC. It forms two cones ABD and ACD.

$$\Delta AEB \sim \Delta CAB$$

$$\frac{AE}{4} = \frac{3}{5}$$

$$AE = 2.4 \text{ cm}$$

$$BE = \sqrt{(3)^2 - (2.4)^2} = 1.8 \text{ cm}$$

$$CE = BC - BE = 5 - 1.8 = 3.2 \text{ cm}$$

$$\text{Volume of cone } ABD = \frac{1}{3} \times \frac{22}{7} \times (2.4)^2 \times 1.8 = 10.86 \text{ cm}^3$$

$$\text{Volume of cone ACD} = \frac{1}{3} \times \frac{22}{7} \times (2.4)^2 \times 3.2 = 19.31 \text{ cm}^3$$

$$\text{Volume of double cone} = 10.86 + 19.31 = 30.17 \text{ cm}^3$$

$$\text{CSA of } ABD = \pi rl = \frac{22}{7} \times 2.4 \times 3 = 22.63 \text{ cm}^2$$

$$\text{CSA of } ACD = \frac{22}{7} \times 2.4 \times 4 = 30.17 \text{ cm}^2$$

$$\text{TSA of double cone} = 22.63 + 30.17 = 52.8 \text{ cm}^2$$

99.

$$\text{Area of valley} = 7280 \text{ km}^2$$

$$\text{Height of rainfall in the valley} = 10 \text{ cm} = \frac{10}{100 \times 1000} \text{ km}$$

$$\text{Amount of rainfall} = \text{Area of valley} \times \text{Height of rainfall}$$

$$= \frac{7280 \times 10}{100 \times 1000} = 0.728 \text{ km}^3 \qquad \text{(i)}$$

Dimensions of river $\quad l = 1072 \text{ km}, b = 75 \text{ m} = \frac{75}{1000} \text{ km}, h = \frac{3}{1000} \text{ km}$

$$\text{Volume cuboidal river} = 1072 \times \frac{75}{1000} \times \frac{3}{1000} = 0.2412 \text{ km}^3$$

$$\text{Amount of water in 3 rivers} = 3 \times 0.2412 = 0.7236 \text{ km}^3 \qquad \text{(ii)}$$

This shows that total rainfall is equivalent to the addition to the normal water of three rivers.

115. Volume of first cone + Volume of second cone = Volume of sphere

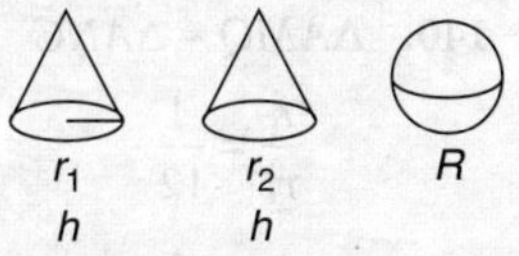

$$\frac{1}{3}\pi r_1^2 h + \frac{1}{3}\pi r_2^2 h = \frac{4}{3}\pi R^3$$

$$\frac{1}{3}\pi h\left[r_1^2 + r_2^2\right] = \frac{4}{3}\pi R^3$$

$$h = \frac{4R^3}{r_1^2 + r_2^2}$$

125. Speed of water = 2.52 km/h = 2520 m/h

In 30 minutes, water passes $= 2520 \times \frac{30}{60} = 1260 \text{ m} = 126000 \text{ cm}$

Volume of water passes in 30 min. $= \frac{22}{7} \times r^2 \times 126000 \text{ cm}^3$

Volume of water in cylinder tank $= \frac{22}{7} \times 40 \times 40 \times 315$

Since Volume of water passes in 30 min = Volume of water in cylinder tank

$$\frac{22}{7} \times r^2 \times 126000 = \frac{22}{7} \times 40 \times 40 \times 315$$

$r = 2$ cm, Internal diameter = 4 cm.

129. Radius of open side $(r_1) = 10$ cm

Radius of upper base $(r_2) = 4$ cm

$l = 15$ cm

CSA of frustum $= \pi(r_1 + r_2)l$

Surface area of upper base cap $= \pi r_2^2$

TSA of cap = CSA of frustum + Area of upper base

Insert the values and perform the calculation by yourself.

132. Height of cylinder, $h = 10$ cm

Radius of cylinder, $r = 4$ cm

Height of frustum of cone, $h_1 = 12$ cm

Radius of top of frustum, $(r_1) = 9$ cm

Radius of bottom of frustum, $(r_2) = 4$ cm

Area of tin required = CSA of cylinder + CSA of frustum of cone

$$= 2\pi rh + \frac{1}{3}\pi h\left(r_1^2 + r_2^2 + r_1 r_2\right)$$

Insert the values and perform the calculation by yourself.

140. $\Delta AMQ \sim \Delta ANC$

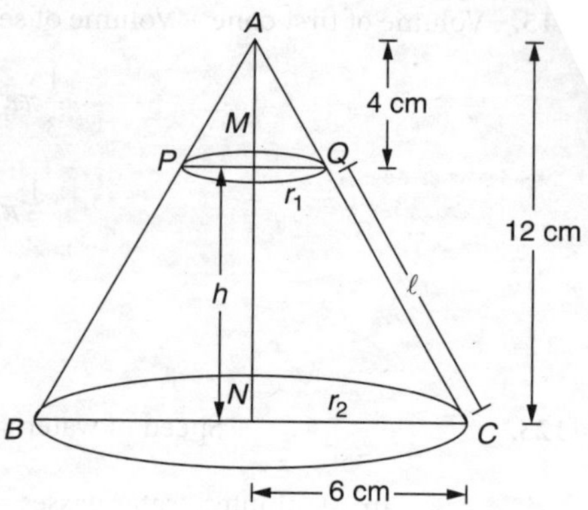

$$\frac{r_1}{r_2} = \frac{4}{12}$$

$$\frac{r_1}{6} = \frac{4}{12}$$

$$r_1 = 2 \text{ cm}$$

$$l = \sqrt{(r_2 - r_1)^2 + h^2} = \sqrt{(6-2)^2 + 8^2} = 4\sqrt{5} \text{ cm}$$

$$\text{Total SA of frustum} = \pi(r_1 + r_2)l + \pi r_1^2 + \pi r_2^2$$

Insert the values and perform the calculation by yourself.

Chapter 14: Statistics

Subjective Questions

38.

Class interval	x_i	f_i	$f_i x_i$
0–20	10	17	170
20–40	30	f_1	$30f_1$
40–60	50	f_2	$50f_2$
60–80	70	f_3	$70f_3$
80–100	90	19	1710
Total		120	$1880 + 30f_1 + 50f_2 + 70f_3$

According to the question

$$17 + f_1 + f_2 + f_3 + 19 = 120$$

$$\Rightarrow \quad f_1 + f_2 + f_3 = 84 \qquad \text{(i)}$$

Now $$\text{Mean} = \frac{\sum f_i x_i}{\sum f_i}$$

$$50 = \frac{1880 + 30f_1 + 50f_2 + 70f_3}{120}$$

$$\Rightarrow 30f_1 + 50f_2 + 70f_3 + 1880 = 6000$$

$$\Rightarrow \quad 30f_1 + 50f_2 + 70f_3 = 4120 \qquad \text{(ii)}$$

Also $$\frac{f_2}{f_3} = \frac{4}{3}$$

$$\Rightarrow \quad 3f_2 - 4f_3 = 0 \qquad \text{(iii)}$$

On solving (i), (ii) and (iii), we get

$$f_1 = 28; f_2 = 32 \text{ and } f_3 = 24.$$

79. Let us suppose that total expenditure of the mess is ₹ x.

$\Rightarrow$ Average expenditure per student = ₹ $\frac{x}{35}$

Average expenditure per student = $\frac{x+42}{42} = \frac{x}{42} + 1$

According to the question, $\frac{x}{42} + 1 = \frac{x}{35} - 1$

$$\Rightarrow \quad 1 + 1 = \frac{x}{35} - \frac{x}{42}$$

$$\Rightarrow \quad 2 = \frac{6x - 5x}{210}$$

$$\Rightarrow \quad x = 420$$

Original expenditure of mess was ₹420.

80. Here, $n_1 = 85$ and $n_2 = 35$, $\bar{x}_1 = 40$ and $\bar{x}_2 = 60$

Combined mean (average marks percentage of the school) = $\frac{85 \times 40 + 35 \times 60}{85 + 35}$

$$= \frac{3400 + 2100}{120} = \frac{5500}{120} = 45.83.$$

Chapter 15: Probability

6 Try to Solve

1. Here, number on card having perfect square = {9, 16, 25, 36, 49, 64}

Total number of outcomes = 70

Number of favourable outcomes (perfect square) = 6

$$P(E) = \frac{6}{70} = \frac{3}{35}.$$

2. Total number of outcomes = 18

(i) Prime number = {3, 5, 7, 11, 13}

Favourable outcomes $P(E) = \frac{5}{18}$

(ii) Number divisible by 3 and 5 = {15}

Favourable outcomes = 1

$$P(E) = \frac{1}{18}$$

Subjective Questions

41. Probability of blue marble $P(E_1) = \frac{1}{5}$

Probability of black marble $P(E_2) = \frac{1}{4}$.

Let us suppose probability of green marble $P(E_3) = x$. We know that

$$P(E_1) + P(E_2) + P(E_3) = 1$$

$$\frac{1}{5} + \frac{1}{4} + x = 1$$

$$x = 1 - \frac{1}{5} - \frac{1}{4} = \frac{20-4-5}{20} = \frac{11}{20}$$

$$P(E_3) = \frac{11}{20}$$

$$P(E_3) = \frac{\text{No. of green marbles}}{\text{Total no. of marbles}}$$

$$\frac{11}{20} = \frac{11}{\text{Total no. of marbles}}$$

Total no. of marbles = 20.

47. x can be chosen from 1, 2, 3. y can be chosen from 1, 4, 9.

Number of cases of $xy = \{1 \times 1, 1 \times 4, 1 \times 9, 2 \times 1, 2 \times 4, 2 \times 9, 3 \times 1, 3 \times 4, 3 \times 9$

$= \{1, 4, 9, 2, 8, 18, 3, 12, 27\}$

Number of favourable cases = 0, $xy = 10$

Total number of cases $(xy) = 9$

$$P(E) = \frac{0}{9} = 0$$

49. Cards are $\{1, 2, 3, \ldots, 20\}$

(i) Prime numbers are {2, 3, 5, 7, 11, 13, 17, 19}

$$P(\text{prime numbers}) = \frac{8}{20} = \frac{2}{5}$$

(ii) Composite numbers are {4, 6, 8, 9, 10, 12, 14, 15, 16, 18, 20}

$$P(\text{composite numbers}) = \frac{11}{20}$$

(iii) Numbers divisible by 3 are {3, 6, 9, 12, 15, 18}

$$P(\text{a number divisible by 3}) = \frac{6}{20} = \frac{3}{10}$$

52. Number of days in a leap year = 366

366 days = 52 weeks + 2 days

The remaining two days may be

(i) Sunday and Monday (ii) Monday and Tuesday (iii) Tuesday and Wednesday

(iv) Wednesday and Thursday (v) Thursday and Friday (vi) Friday and Saturday

(vii) Saturday and Sunday

Total outcomes = 7

Favourable outcomes = 1

∴ Required probability = $\frac{1}{7}$

53. Total number of days in December = 31

31 days = 4 week + 3 days

These remaining days may be

(i) Sunday, Monday and Tuesday (ii) Monday, Tuesday and Wednesday

(iii) Tuesday, Wednesday and Thursday (iv) Wednesday, Thursday and Friday

(v) Thursday, Friday and Saturday (vi) Friday, Saturday and Sunday

(vii) Saturday, Sunday and Monday

Total outcomes = 7

Favourable outcomes = 1

∴ Required probability = $\frac{1}{7}$

59. ASSASSINATION

In which $A = 3, S = 4, I = 2, N = 2, T = 1, O = 1.$

Here, the vowels are A, I and O.

Total favourable outcomes = 3 + 2 + 1 = 6

Total letter as outcome = 13

Probability of getting a vowel = $\frac{6}{13}$

60. Total number of alphabets = 26.

Consonants = Total alphabets – Vowels = 26 – 5 = 21

Probability of consonant = $\frac{21}{26}$

71. Total number of balls in the bag = 24

White balls = x

According to the question, P (White balls) = $\frac{x}{24} = y$ (i)

Now, 12 white balls are added; so

White balls = $(x + 12)$

Total balls = 24 + 12 = 36

P (White balls) = $\frac{x+12}{36} = \frac{5}{3}y$ (ii)

Inserting (i) in (ii), we get $\frac{x+12}{36} = \frac{5}{3} \times \frac{x}{24}$

$2x + 24 = 5x$

$24 = 3x$

$x = 8$

78. We know that $p(E) + p(E') = 1$.

$$\frac{p}{12} + \frac{1}{3} = 1$$

$$\frac{p}{12} = 1 - \frac{1}{3} = \frac{2}{3}$$

$$p = \frac{2}{3} \times 12 = 8$$

79. Total number of envelopes = 1000

Envelopes containing cash price = 10 + 100 + 200 = 310

Envelopes containing no cash prize (E) = 1000 − 310 = 690

$$P(E) = \frac{690}{1000} = 0.69$$

90. Number '*x*' is selected at random from numbers 1, 4, 9, 16. Another number '*y*' is selected at random from numbers 1, 2, 3 and 4.

Possible outcomes of *xy* are

{(1, 1), (1, 2), (1, 3), (1, 4), (4, 1), (4, 2), (4, 3), (4, 4), (9, 1), (9, 2), (9, 3), (9, 4), (16, 1), (16, 2), (16, 3), (16, 4)}

Total number of positive outcomes = 16.

Favourable outcomes (E) = {(9, 2), (9, 3), (9, 4), (16, 2), (16, 3) or (16, 4)}

$$\therefore \quad P(A) = \frac{6}{16} = \frac{3}{8}$$